HIGHER LEVEL

# Mathematics
## Applications and Interpretation
### for the IB Diploma

IBRAHIM WAZIR
TIM GARRY
JIM NAKAMOTO
KEVIN FREDERICK
STEPHEN LUMB

Published by Pearson Education Limited, 80 Strand, London, WC2R 0RL

www.pearsonglobalschools.com

Text © Pearson Education Limited 2019
Theory of Knowledge chapter authored by Ric Sims
Development edited by Jim Newall
Copy edited by Linnet Bruce
Proofread by Eric Pradel and Linnet Bruce
Indexed by Georgie Bowden
Designed by © Pearson Education Limited 2019
Typeset by © Tech-Set Ltd, Gateshead, UK
Original illustrations © Pearson Education Limited 2019
Illustrated by © Tech-Set Ltd, Gateshead, UK
Cover design by © Pearson Education Limited 2019
**Cover images**: Front: © **Getty Images**: Busà Photography
Inside front cover: **Shutterstock.com**: Dmitry Lobanov

The rights of Ibrahim Wazir, Tim Garry, Jim Nakamoto, Kevin Frederick and Stephen Lumb to be identified as the authors of this work have been asserted by them in accordance with the Copyright, Designs and Patents Act 1988**.**

First published 2019

24  23  22  21  20  19
IMP  10  9  8  7  6  5  4  3  2

**British Library Cataloguing in Publication Data**
A catalogue record for this book is available from the British Library

ISBN 978 0 435 19344 7

Printed in Slovakia by Neografia

**Acknowledgements**
The authors and publisher would like to thank the following individuals and organisations for their kind permission to reproduce copyright material.

**Photographs**
(Key: b-bottom; c-centre; l-left; r-right; t-top)

**Getty Images:** Chris eberhardt/Getty Images1, J Broughton Photography/Moment/Getty Images 15, Howard Pugh (Marais)/Moment/Getty Images 57, aaaaimages/Moment/Getty Images 99, Ryan Trussler/Getty Images 123, Johner Images/Getty Images 173, Gabriel Perez/Moment/Getty Images 201, Witthaya Prasongsin/Moment/Getty Images 273, David Crunelle/EyeEm/Getty Images 297, Kuninobu Sato/EyeEm/Getty Images 361, Wonsup Im.EyeEm.Getty Images 407, Daniela White Images/Getty Images 449, JPL/Moment/Getty Images 525, baxsyl/Moment/Getty Images 577, d3sign/Moment/Getty Images 633, Image Source/Ditto/Getty Images 665, Paul Biris/Moment/Getty Images 773, Alberto Manuel Urosa Toledano/Moment/Getty Images 751, Jamil Caram/EyeEm/Getty Images 781, Westend61/Getty Images 833.

All other images © Pearson Education

We are grateful to the following for permission to reproduce copyright material:

**Text**

**pages 904–905, Edge Foundation Inc.:** What Kind of Thing Is a Number? A Talk with Reuben Hersh, Wed, Oct 24, 2018. Used with permission of Edge Foundation Inc.

Text extracts relating to the IB syllabus and assessment have been reproduced from IBO documents. Our thanks go to the International Baccalaureate for permission to reproduce its copyright.

This work has been developed independently from and is not endorsed by the International Baccalaureate (IB). International Baccalaureate® is a registered trademark of the International Baccalaureate Organization.

This work is produced by Pearson Education and is not endorsed by any trademark owner referenced in this publication.

**Dedications**

*I dedicate this work to the memory of my parents and my brother, Saeed, who passed away during the early stages of work on this edition.*

*My special thanks go to my wife, Lody, for standing beside me throughout writing this book. She has been my inspiration and motivation for continuing to improve my knowledge and move my career forward. She is my rock, and I dedicate this book to her.*

*My thanks go to all the students and teachers who used the earlier editions and sent us their comments.*

Ibrahim Wazir

*In loving memory of my parents.*

*I wish to express my deepest thanks and love to my wife, Val, for her unflappable good nature and support — and for smiling and laughing with me each day. I am infinitely thankful for our wonderful and kind-hearted children — Bethany, Neil and Rhona. My love for you all is immeasurable.*

Tim Garry

*To Penny*

Jim Nakamoto

*I have inexpressible gratitude and appreciation for my wife Julie who has supported me and our family as I took on this second job. The only life raft here is love.*

*Lena and William, my heart ached every time I had to turn you away to get work done. Thank you for your acceptance and I hope you forgive me.*

*Ibrahim, thank you for your confidence in me and thinking I am worthy of this task.*

*Thank you to my grade 12 students who worked through some early chapter drafts and corrected plenty of typos. Slackerz unite!*

*Finally, to everyone else who picked up loose ends while I was otherwise occupied: thank you for easing the burden of this task.*

Kevin Frederick

*I would like to thank my family for their love and invaluable support. I would also like to thank all of my students and colleagues who make teaching Mathematics such a joy.*

Stephen Lumb

# Contents

# Introduction

This textbook comprehensively covers all of the material in the syllabus for the two-year *Mathematics: Applications and Interpretation Higher Level* course of the International Baccalaureate (IB) Diploma Programme (DP). First teaching of this course starts in the autumn of 2019 with first exams occurring in May 2021. We, the authors, have strived to thoroughly explain and demonstrate the mathematical concepts and methods listed in the course syllabus.

## Content

As you will see when you look at the table of contents, the five syllabus topics (see margin note) are fully covered, though some are split over different chapters in order to group the information as logically as possible. This textbook has been designed so that the chapters proceed in a manner that supports effective learning of the course content. Thus – although not essential – it is recommended that you read and study the chapters in numerical order. It is particularly important that you thoroughly review and understand all of the content in the first chapter, *Algebra and function basics*, before studying any of the other chapters.

Other than the final two chapters (**Theory of knowledge** and **Internal assessment**), each chapter has a set of **exercises** at the end of every section. Also, at the end of each chapter there is a set of **practice questions,** which are designed to expose you to questions that are more 'exam-like'. Many of the end-of-chapter practice questions are taken from past IB exam papers. Near the end of the book, you will find answers to all of the exercises and practice questions. There are also numerous **worked examples** throughout the book, showing you how to apply the concepts and skills you are studying.

The Internal assessment chapter provides thorough information and advice on the required **mathematical exploration component**. Your teacher will advise you on the timeline for completing your exploration and will provide critical support during the process of choosing your topic and writing the draft and final versions of your exploration.

The final chapter in the book will support your involvement in the **Theory of knowledge** course. It is a thought-provoking chapter that will stimulate you to think more deeply and critically about the nature of knowledge in mathematics and the relationship between mathematics and other areas of knowledge.

### eBook

Included with this textbook is an eBook that contains a digital copy of the textbook and additional high-quality enrichment materials to promote your understanding of a wide range of concepts and skills encountered throughout the course. These materials include:

- Interactive *GeoGebra* **applets** demonstrating key concepts
- **Worked solutions** for all exercises and practice questions
- Graphical display calculator (**GDC**) support

To access the eBook, please follow the instructions located on the inside cover.

# Information boxes

As you read this textbook, you will encounter numerous boxes of different colours containing a wide range of helpful information.

## Learning objectives

You will find learning objectives at the start of each chapter. They set out the content and aspects of learning covered in the chapter.

### Learning objectives

By the end of this chapter, you should be familiar with...

- different forms of equations of lines and their gradients and intercepts
- parallel and perpendicular lines
- different methods to solve a system of linear equations (maximum of three equations in three unknowns)

## Key facts

Key facts are drawn from the main text and highlighted for quick reference to help you identify clear learning points.

 A function is **one-to-one** if each element $y$ in the range is the image of exactly one element $x$ in the domain.

## Hints

Specific hints can be found alongside explanations, questions, exercises, and worked examples, providing insight into how to analyse/answer a question. They also identify common errors and pitfalls.

 If you use a graph to answer a question on an IB mathematics exam, you must include a clear and well-labelled sketch in your working.

## Notes

Notes include general information or advice.

 Quadratic equations will be covered in detail in Chapter 2.

## Examples

Worked examples show you how to tackle questions and apply the concepts and skills you are studying.

### Example 1.5

Find $x$ such that the distance between points $(1, 2)$ and $(x, -10)$ is 13 units.

#### Solution

$d = 13 = \sqrt{(x - 1)^2 + (-10 - 2)^2} \Rightarrow 13^2 = (x - 1)^2 + (-12)^2$

$\Rightarrow 169 = x^2 - 2x + 1 + 144 \Rightarrow x^2 - 2x - 24 = 0$

$\Rightarrow (x - 6)(x + 4) = 0 \Rightarrow x - 6 = 0 \text{ or } x + 4 = 0$

$\Rightarrow x = 6 \text{ or } x = -4$

# How to use this book

This book is designed to be read by you – the student. It is very important that you read this book carefully. We have strived to write a readable book – and we hope that your teacher will routinely give you reading assignments from this textbook, thus giving you valuable time for productive explanations and discussions in the classroom. Developing your ability to read and understand mathematical explanations will prove to be valuable to your long-term intellectual development, while also helping you to comprehend mathematical ideas and acquire vital skills to be successful in the *Applications and Interpretation* HL course. Your goal should be understanding, not just remembering. You should always read a chapter section thoroughly before attempting any of the exercises at the end of the section.

Our aim is to support genuine inquiry into mathematical concepts while maintaining a coherent and engaging approach. We have included material to help you gain insight into appropriate and wise use of your GDC and an appreciation of the importance of proof as an essential skill in mathematics. We endeavoured to write clear and thorough explanations supported by suitable worked examples, with the overall goal of presenting sound mathematics with sufficient rigour and detail at a level appropriate for a student of HL mathematics.

Our thanks go to Jim Nakamoto, Kevin Frederick and Stephen Lumb who joined our team for this edition, helping us to add richness and variety to the series.

For over 10 years, we have been writing successful textbooks for IB mathematics courses. During that time, we have received many useful comments from both teachers and students. If you have suggestions for improving this textbook, please feel free to write to us at globalschools@pearson.com. We wish you all the best in your mathematical endeavours.

Ibrahim Wazir and Tim Garry

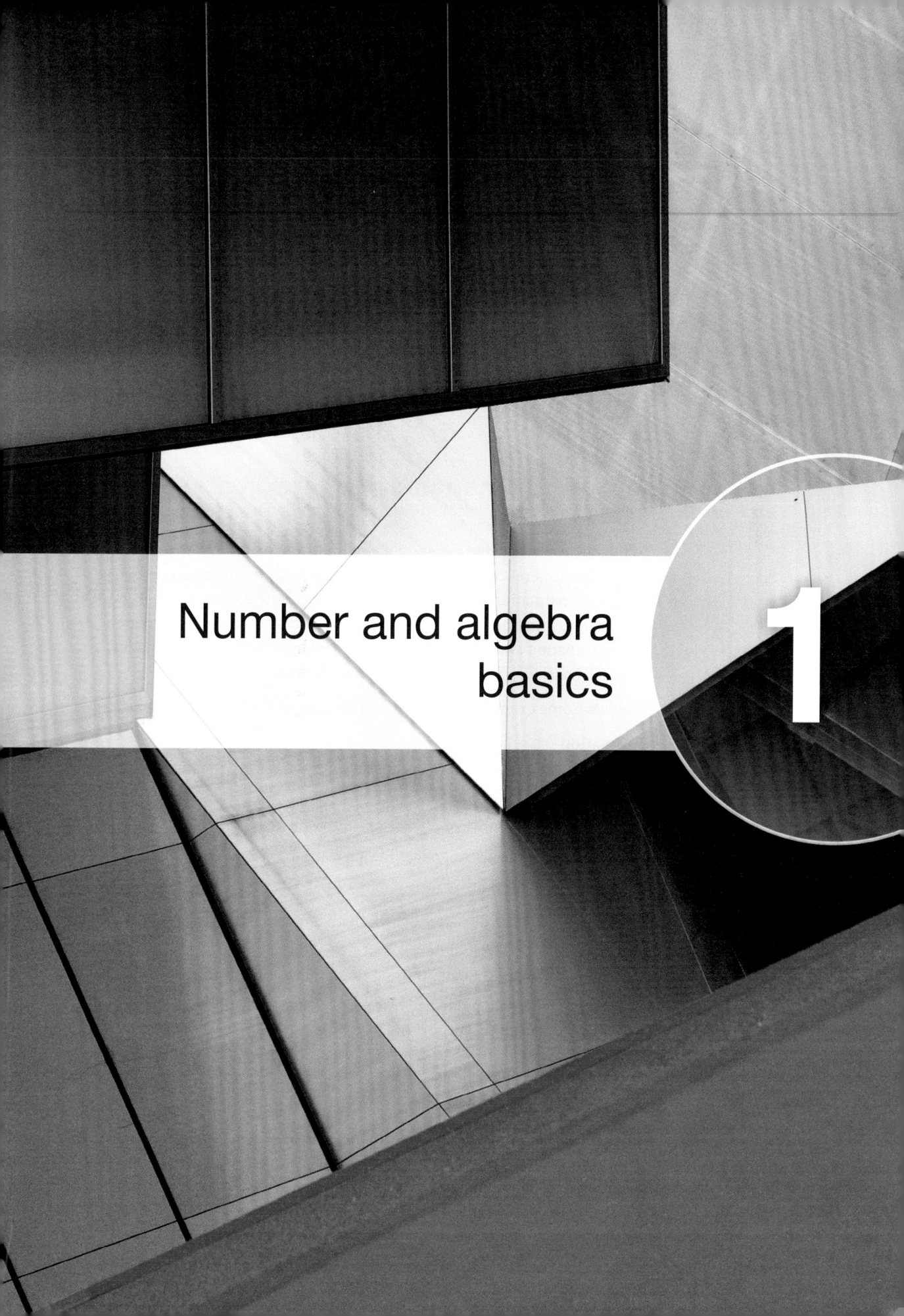

# Number and algebra basics

**1**

# Number and algebra basics

## Learning objectives

By the end of this chapter, you should be familiar with…

- making reasonable estimations and better approximations
- demonstrating an understanding of the rules of exponents
- using correct scientific notation
- demonstrating an understanding of the rules of logarithms.

This chapter revises and consolidates previous knowledge of scientific notation, exponential expressions, logarithms and estimation skills.

## 1.1 Estimation and approximation

While the terms **estimation** and **approximation** are often used to mean a guess, their inferences are different. Although both terms suggest a lack of precision, estimation infers a lack of precision in the process of measurement, while approximation lacks precision in the statement of the measurement. Both estimation and approximation skills are important in mathematics, but they are skills that are practiced every day in many contexts.

Here are some examples of estimation and approximation.

### Estimation

1. You are cycling to a campsite that is 100 km away. Estimate your arrival time if you depart at 08:00.
2. Estimate the number of olives that would fill a litre jar.
3. Estimate the number of pages in this textbook.

### Approximation

1. Approximate the diameter of a circle that has a circumference of 109.2 cm.
2. Your bank has a digital device that scans the waiting line of customers and suggests the approximate waiting time in line is 6 minutes.
3. According to data published by our local airport, "Approximately 2 million passengers used the airport in December".

## Rounding answers

In giving an estimation or approximation, measurements are often rounded to some level of accuracy, with the rule simply being that digits less than 5 are rounded to 0; and digits that are 5 or greater increase the preceding digit by 1.

## Example 1.1

(a) Round each value to the nearest unit.

   (i) 256.4    (ii) 1.49    (iii) 63.5    (v) 700.9

(b) Round each value to the nearest one-hundredth.

   (i) 1.006    (ii) 7.295    (iii) 67.085    (iv) 34.113

### Solution

(a) (i)  256.4   **4** is less than 5, so round down to 0
                 256.4 rounded to the nearest unit is 256

   (ii)  1.49   **4** is less than 5, so round down to 0
                 1.49 rounded to the nearest unit is 1

   (iii) 63.5   **5** is rounded up, adding 1 to the units digit
                 63.5 rounded to the nearest unit is 64

   (iv)  700.9   **9** is rounded up, adding 1 to the units digit
                 700.9 rounded to the nearest unit is 701

(b) (i)  1.006   **6** is rounded up, adding 1 to the hundredths digit
                 1.006 rounded to the nearest one-hundredth is 1.01

   (ii)  7.295   **5** is rounded up, adding 1 to the hundredths digit,
                 which in turn adds 1 to the tenths digits in this case
                 7.295 rounded to the nearest one-hundredth is 7.30

   (iii) 67.085  **5** is rounded up, adding 1 to the hundredths digit
                 67.085 rounded to the nearest one-hundredth is 67.09

   (iv)  34.113  **3** is less than 5, so round down to 0
                 34.113 rounded to the nearest one-hundredth is 34.11

## Percentage error

The approximate answer produced as a result of rounding depends on the digit to which it is rounded, and may or may not be appropriate.

## Example 1.2

The rounded values in part (a) of Example 1.1 produced differences of

(i)   $256.4 - 256 = 0.4$          (ii)  $1.49 - 1 = 0.49$

(iii) $63.5 - 64 = -0.5$          (iv)  $700.9 - 701 = -0.1$

What are the percentage errors in rounding if the original values are assumed to be precise measurements?

**Solution**

Dividing the differences by the original values we obtain:

(i) $\dfrac{0.4}{256.4} \approx 0.156\%$

(ii) $\dfrac{0.49}{1.49} \approx 32.9\%$

(iii) $\dfrac{-0.5}{63.5} \approx -0.787\%$

(iv) $\dfrac{-0.1}{700.9} \approx -0.0143\%$

The errors are all quite small except for the second one. (Note that this percentage is rounded, too!) Choosing an arbitrary decimal place to which a measurement is rounded produces inaccuracies that may not be acceptable.

In IB mathematics, where an exact final answer is not required, an approximate answer, to the required accuracy, is important. To achieve this, a thorough understanding of the notion of **significant figures (s.f.)** is critical. We will revisit percentage errors after studying significant figures.

## Significant figures (s.f.)

| Rule | Example |
|------|---------|
| All non-zero digits are significant | 74 818 226 has 8 s.f.<br>123.45 has 5 s.f. |
| All zeros between non-zero digits are significant | 103.05 has 5 s.f.<br>780 002 has 6 s.f. |
| Zeros to the left of an implied decimal point are **not** significant, whereas zeros to the right of an explicit decimal **are** significant. | 23 000 has 2 s.f., while 23 000.0 has 6 s.f. |
| To the right of a decimal point, all leading zeros are **not** significant, whereas all zeros that follow non-zero digits **are** significant. | 0.0043 has 2 s.f., while 0.0043000 has 5 s.f. |

**Table 1.1** Significant figures rules and examples

### Example 1.3

Indicate the number of significant figures in each value.

(a) 30 020   (b) 30 020.0   (c) 0.008   (d) 1000.0

(e) 1.09   (f) 7.00101   (g) 0.02   (h) 0.020

**Solution**

(a) 4: The non-zero digits and the zeros in between are significant.

(b) 6: All digits between the leading non-zero digit and the decimal point are significant. The zero after the decimal point is also significant.

(c) 1: Only the '8' is significant.

(d) 5: All digits between the leading non-zero digit and the decimal point are significant. The zero after the decimal point is also significant.

(e) 3: The non-zero digits and the zero in between are significant.

(f) 6: The non-zero digits and the zeros in between are significant.

(g) 1: Only the '2' is significant.

(h) 2: The '2' and the trailing zero are significant.

## Percentage error revisited

In Example 1.1, we looked at the following values rounded to the nearest unit:

| 256.4 | 1.49 | 63.5 | 700.9 |

Now, instead of rounding to the nearest unit, consider these values given to 3 significant figures:

| 256 | 1.49 | 63.5 | 701 |

The differences of these values from the original values would be:

$$256.4 - 256 = 0.4 \qquad 1.49 - 1.49 = 0$$
$$63.5 - 63.5 = 0 \qquad 700.9 - 701 = -0.1$$

Hence, their percentage errors would be:

$$\frac{0.4}{256.4} \approx 0.156\% \qquad \frac{0}{1.49} = 0\%$$
$$\frac{0}{63.5} = 0\% \qquad \frac{-0.1}{700.9} \approx -0.0143\%$$

## Exercise 1.1

1. Round each value to the nearest unit.
   (a) 25.8     (b) 0.61     (c) 1200.7     (d) 83.47

2. Round each value to the nearest one-hundredth.
   (a) 27.047     (b) 800.008     (c) 3.14159     (d) 0.0009

3. Give an example that would justify the use of measurements given to the following levels of precision.
   (a) 1.72 m     (b) 0.014 s     (c) 250 km     (d) 2.43 mB
   (e) 1.27 cm     (f) 1200 g     (g) 23°C     (h) 0.2 A

4. If the original values in questions **1** and **2** were precise measurements, find the percentage errors in their rounded values, to two decimal places.

5. Determine the number of significant figures in each value.
   (a) 3910     (b) 3901     (c) 8200     (d) 8200.0
   (e) 100.3     (f) 100.03     (g) 0.002     (h) 0.0020

6. Give each value correct to 3 significant figures.
   (a) 5627     (b) 3098     (c) 4762 311     (d) 3.14159
   (e) 0.0002070     (f) 100.03     (g) 0.02013     (h) 0.020003

# 1 Number and algebra basics

## 1.2 Rules of exponents

Exponents and the rules for their use will be required throughout this course.

The rules of exponents for real values $a$, $m$ and $n$ are:

1. $a^m \cdot a^n = a^{m+n}$

2. $\dfrac{a^m}{a^n} = a^{m-n}$, provided $a \neq 0$

3. $(a^m)^n = a^{m \cdot n}$

4. $a^0 = 1$, provided $a \neq 0$

5. $a^{-m} = \dfrac{1}{a^m}$

6. $a^{\frac{m}{n}} = \sqrt[n]{a^m} = \left(\sqrt[n]{a}\right)^m$, provided $a > 0$

### Example 1.4

Use the rules of exponents to write each of the following in the form $2^n$, where $n \in \mathbb{Z}$.

(a) $2^4 \cdot 2^2$
(b) $2^8 \cdot 2^{-2}$
(c) $4^2 \cdot 2^3$
(d) $2^{-4} \cdot 4^{\frac{1}{2}}$

(e) $\dfrac{2^8}{2^2}$
(f) $\dfrac{2^{-2}}{2^{-6}}$
(g) $(2^4)^2$
(h) $(2^4)^0$

### Solution

(a) $2^6$    The bases are the same. Add the exponents.

(b) $2^6$    The bases are the same. Add the exponents.

(c) $2^7$    The bases are different. First convert $4^2$ to $2^4$, then add the exponents.

(d) $2^{-3}$    First convert $4^{\frac{1}{2}}$ to $2^1$, then add the exponents.

(e) $2^6$    The bases are the same. Subtract the exponents.

(f) $2^4$    The bases are the same. Subtract the exponents.

(g) $2^8$    Multiply the exponents.

(h) $2^0$    Multiply the exponents.

### Exercise 1.2

Use the rules of exponents to write each of the following in the form $3^n$, where $n \in \mathbb{Z}$.

1. $3^4 \cdot 3^2$
2. $3^4 \cdot 3^{-2}$
3. $\dfrac{3^6}{3^2}$
4. $\dfrac{3^{-6}}{3^2}$

5. $(3^4)^2$
6. $(3^{-1})^{-3}$
7. $9^4 \cdot 3^2$
8. $9^4 \cdot 81^2$

9. $9^4 \cdot 81^{-2}$
10. $\sqrt{27}$
11. $\sqrt{27} \cdot 9\sqrt{3}$
12. $\sqrt[3]{9} \cdot 3\sqrt[3]{9}$

**13.** $\sqrt{3} \cdot \sqrt[3]{3}$  **14.** $\sqrt{3} \cdot \sqrt[4]{9}$  **15.** $9^{-2} \cdot (3\sqrt{27})^2$  **16.** 1

**17.** $9^2 \cdot 3^{\frac{1}{2}}$  **18.** $\sqrt{3} \cdot \sqrt[3]{81}$  **19.** $3^{\frac{1}{2}} \cdot \sqrt[3]{9}$  **20.** $\sqrt{27} \cdot 3^{\frac{3}{2}}$

**21.** $\dfrac{3^6}{\sqrt{3}}$  **22.** $\dfrac{3^2}{\sqrt[3]{9}}$  **23.** $\dfrac{3^{\frac{2}{3}}}{\sqrt[3]{9}}$  **24.** $\dfrac{9^{-\frac{1}{2}}}{\sqrt[3]{3}}$

# 1.3 Scientific notation

Scientific notation is used to represent very large and very small measurements without having to count decimal places. For example, the approximate distance from the Earth to the sun is 149 600 000 km. Using scientific notation this would be written as $1.496 \times 10^8$ km.

The ångström (Å) is a unit of length equal to one ten-billionth of a metre. In scientific notation, it is written as $1 \times 10^{-10}$ m. It is very useful to have a notation that immediately shows the magnitude of this number that would otherwise be written as 0.0000000001 m.

Provided that measurements with comparable units are used, addition and subtraction is straightforward. If the units are not comparable, they need to be converted to a common unit.

The least number of significant figures in any measurement determines the number of significant figures in the answer.

## Example 1.5

Find the perimeter of a rectangle with length $l = 2.3 \times 10^{-1}$ m and width $w = 9.5 \times 10^{-2}$ m.

### Solution

There is a difference in the order of magnitude between the length and width, so a conversion is required.

$l = 2.3 \times 10^{-1}$ m $= 23 \times 10^{-2}$ m

$w = 9.5 \times 10^{-2}$ m

The perimeter, $p$, is given by:

$$p = 2(l + w)$$
$$= 2(23 \times 10^{-2} + 9.5 \times 10^{-2})$$
$$= 2(32.5 \times 10^{-2})$$
$$= 65 \times 10^{-2} \text{ m}$$

or $\quad = 6.5 \times 10^{-1}$ m

### Example 1.6

Find the area of the rectangle in Example 1.5

**Solution**

Area $A = l \times w = (2.3 \times 10^{-1}) \times (9.5 \times 10^{-2})$
$$= (2.3 \times 9.5) \times (10^{-1} \times 10^{-2})$$
$$= 21.85 \times 10^{-3}$$
$$= 2.185 \times 10^{-2}$$

But the given measurements were given to 2 s.f. so the answer should use the same degree of accuracy.

Area $A = 2.2 \times 10^{-2}\, \text{m}^2$ (2 s.f.)

Don't forget to include units.

### Exercise 1.3

1. Express the following in scientific notation.

   (a) 1203     (b) 7 billion     (c) 0.000301     (d) 20.01

   (e) 2000     (f) 0.00070     (g) 12.03     (h) 10006

   (i) 10.001     (j) 1 googol

2. Work out each calculation and give your answer in scientific notation.

   (a) $210 \times 8000$        (b) $200 \times 0.00018$

   (c) $(2.3 \times 10^9) \times (8 \times 10^2)$        (d) $(2.3 \times 10^{-3}) \times (8 \times 10^3)$

3. Find each value to 3 significant figures and express the answer in scientific notation.

   (a) $2^{30}$     (b) $2^{31} - 1$     (c) $e^{\pi}$     (d) $\pi^e$

Consider the number of significant figures in each value.

## 1.4   Exponents and logarithms

Exponents and logarithms are inverses of each other. For instance, should a relation be expressed as $y = a^x$, then $x$ is the exponent of the base $a$ which yields the quantity $y$, written simply as $x = \log_a y$

$y = a^x \Rightarrow x = \log_a y$

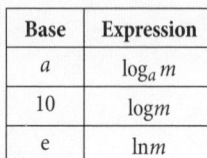

When the base is 10, we typically do not write it. We can simply write **log y**. A logarithm with base 10 is called a **common logarithm**.

When the base is **e**, we write **ln y**. A logarithm with base **e** is called a **natural logarithm**.

| Base | Expression |
|------|------------|
| $a$ | $\log_a m$ |
| 10 | $\log m$ |
| e | $\ln m$ |

## Example 1.7

The squares of a chessboard are numbered consecutively from 1 to 64. If one grain of rice is placed on the first square, two grains on the second square, four grains on the third square, doubling the number of grains with each square, on which square would there first be more than a million ($10^6$) grains of rice? Theoretically, of course!

### Solution

The number of grains of rice follows an exponential pattern.

| Square ($x$) | Grains of rice ($y$) | In exponential form |
|:---:|:---:|:---:|
| 1 | 1 | $2^0$ |
| 2 | 2 | $2^1$ |
| 3 | 4 | $2^2$ |
| 4 | 8 | $2^3$ |
| Noting the number of the square and the exponent... | | |
| $n$ | $\geqslant 10^6$ | $2^{n-1}$ |

So, we need to find a **solution** to $y = 2^{x-1}$ when $y = 10^6$. Stating this as a logarithm problem, we must solve the equation $x - 1 = \log_2 10^6$ or $x = \log_2 (10^6) + 1$

We can use our GDC to solve the equation.

Since the number of the square must be a positive integer, the 21st square will be the first to hold in excess of one million grains of rice.

We can do the calculation using the Solver feature of a GDC (see Figure 1.1).

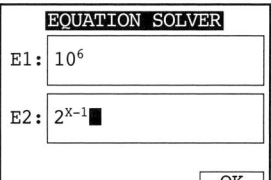

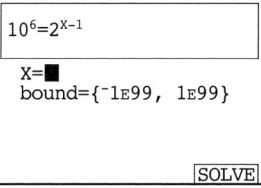

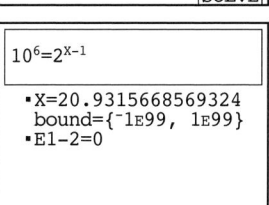

**Figure 1.1** GDC screens for the solution to Example 1.7

## Example 1.8

Earthquake magnitudes ($R$) are measured on the Richter scale which is a base-10 logarithmic scale, and relative comparisons are useful. For example, an earthquake of magnitude $R = 4$ is ten times as strong as an earthquake of magnitude $R = 3$. What is the magnitude of an earthquake $R_1$ that is twice as strong as another of magnitude $R = 3$?

### Solution

Comparing the relative magnitudes, the equation to be solved for $R_1$ is:

$$2 = \frac{10^{R_1}}{10^3}$$

$$\Rightarrow 2 \cdot 10^3 = 10^{R_1}$$

$$\Rightarrow R_1 = \log(2 \cdot 10^3) \text{ and using a GDC gives}$$

$$\Rightarrow R_1 \approx 3.301$$

## Exercise 1.4

1. Write each equation in logarithmic form.

   (a) $1000 = 10^3$

   (b) $64 = 4^3$

   (c) $100^{\frac{3}{2}} = 1000$

   (d) $9^{\frac{1}{2}} = 3$

   (e) $2\sqrt{2} = 8^{\frac{1}{2}}$

   (f) $10^0 = 1$

   (g) $e^0 = 1$

   (h) $6^{-2} = \dfrac{1}{36}$

   (i) $\left(\sqrt{2}\right)^{-2} = \dfrac{1}{2}$

   (j) $3^{-\frac{1}{2}} = \dfrac{1}{\sqrt{3}}$

   (k) $\left(\dfrac{1}{2}\right)^{-3} = 8$

   (l) $8^{-\frac{1}{2}} = \dfrac{\sqrt{2}}{4}$

   (m) $(-2)^3 = -8$

   (n) $(0.01)^{-1} = 100$

   (o) $\left(\dfrac{\sqrt{2}}{2}\right)^3 = \dfrac{\sqrt{2}}{4}$

2. Express each equation in the form $x = \ldots$

   (a) $y = 2^x$

   (b) $y = 10^x$

   (c) $y = e^x$

   (d) $y = 2^{3x}$

   (e) $y = 3 \cdot 2^x$

   (f) $y = 5 - 2^x$

   (g) $y = 3^{2x}$

   (h) $y = 3^{\frac{x}{2}}$

   (i) $y = e^{2x}$

   (j) $y = 2^{x-3}$

   (k) $y = e^{\frac{x}{2}}$

   (l) $y = \dfrac{1}{2}e^{2x}$

3. Consider Example 1.7 about the grains of rice on a chess board. Would any one of the 64 squares ever hold more than exactly 1 billion grains of rice?

4. Using the earthquake context of Example 1.8, find the magnitude of an earthquake that is:

   (a) ten times as powerful as one of magnitude $R = 5.2$

   (b) twice as powerful as one of magnitude $R = 5.2$

## 1.5 Rules of logarithms

Logarithms may seem to be just mirror images of exponents, and rules regarding logarithms may appear to be similar to those for exponents. However, logarithms are defined only when the base $b$ and its arguments are positive.

1. $\log_b(m \cdot n) = \log_b m + \log_b n$

2. $\log_b\left(\dfrac{m}{n}\right) = \log_b m - \log_b n$

3. $\log_b(m^n) = n \cdot \log_b m$

4. $\log_b 1 = 0$

5. $\log_b m = \dfrac{1}{\log_m b}$

6. $\log_b m = \dfrac{\log_a m}{\log_a b}$, for any $a > 0$

## Example 1.9

Let's revisit Example 1.8, but this time we'll use the first rule of logarithms. What is the magnitude of an earthquake $R_1$ that is twice as strong as another of magnitude $R = 3$?

### Solution

Comparing the relative magnitudes, the equation to be solved for $R_1$ is:

$$2 = \frac{10^{R_1}}{10^3}$$

$$\Rightarrow 2 \cdot 10^3 = 10^{R_1}$$

$$\Rightarrow R_1 = \log(2 \cdot 10^3) \text{ and this time, by using the first rule of logarithms:}$$

$$\Rightarrow R_1 = \log 2 + \log 10^3$$

$$\Rightarrow R_1 = \log 2 + 3 \text{ and since } \log 2 \approx 0.301$$

$$\Rightarrow R_1 \approx 3.301$$

It is worth noting that regardless of any other magnitude to which one compares an earthquake, an earthquake that is twice as strong has a magnitude that is greater by $R \approx 0.301$.

## Exercise 1.5

1. Determine the value of each of the following.

   (a) $\log_2 16$      (b) $\log_{16} 2$      (c) $\log_{\sqrt{2}} 16$

   (d) $\log_2 \sqrt{2}$      (e) $\log_2(-16)$      (f) $\log_2 2\sqrt{2}$

   (g) $\log_{\sqrt{2}} 2\sqrt{2}$      (h) $\log_{2\sqrt{2}} 2$      (i) $\log 4 + \log 25$

   (j) $\log 30 - \log 300$      (k) $\ln\left(\frac{1}{e^2}\right)$      (l) $\ln \sqrt{e}$

2. Simplify each expression.

   (a) $\log_a a^3$      (b) $\log_a \sqrt{a}$      (c) $\log_{\sqrt{a}} a\sqrt{a}$

   (d) $\log_{\sqrt{a}} \sqrt[3]{a}$      (e) $\log_{a^2} a^3$      (f) $\log_{a^2} \sqrt{a}$

   (g) $\log_{a^2} \sqrt[3]{a}$      (h) $\log_{a^2} a\sqrt{a}$      (i) $\log_a a^{-3} + \log_a a^4$

   (j) $\log_{a^2} a^{-3} + \log_{a^2} a^4$      (k) $\log_a a^3 - \log_a a^2$      (l) $\log_a a^3 - \log_a \sqrt{a}$

3. Solve each equation for $x$.

   (a) $\log_3 x = -2$          (b) $\log_2(x - 3) = 5$

   (c) $\log_x 3 = -2$          (d) $\log_3(x^2 + 2x + 1) = 0$

**Optional**

Which is larger, $e^{\pi}$ or $\pi^{e}$?

The numerical values of $e^{\pi}$ and $\pi^{e}$ were calculated in Exercise 1.3. Now, consider an analysis using the graph of $f(x) = \dfrac{\ln x}{x}$

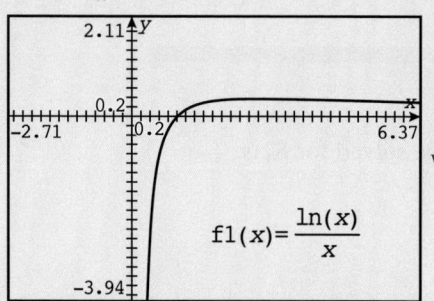

with its maximum at

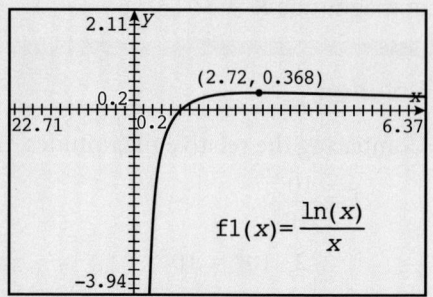

Note that the maximum is found at $x = e$ (which can also be found using calculus later), and $\pi$ is further to its right, below the maximum.

Hence, $f(e) > f(\pi) \Rightarrow \dfrac{\ln e}{e} > \dfrac{\ln \pi}{\pi}$

Now, $\ln e = 1 \Rightarrow \qquad \dfrac{1}{e} > \dfrac{\ln \pi}{\pi}$

$$\pi > e \ln \pi$$

$$\pi > \ln \pi^{e}$$

Then $\qquad\qquad e^{\pi} > e^{\ln \pi^{e}}$

But $e^{\ln \pi^{e}} = \pi^{e}$

And so $\qquad\qquad e^{\pi} > \pi^{e}$

## Chapter 1 practice questions

1. Determine the number of significant figures in each of the following.

   (a) 2308      (b) 2300      (c) 1000      (d) 4000.9

   (e) 570.03      (f) 0.0003      (g) 0.00320

2. Give each value correct to 3 significant figures.

   (a) 58 261      (b) 6107      (c) 123 807

   (d) 1.618      (e) 0.003051      (f) 400.01

3. Use the rules of exponents to write each of the following in the form $2^{n}$, where $n \in \mathbb{Z}$.

   (a) $2^{6} \cdot 2^{2}$      (b) $2^{8} \cdot 2^{-2}$      (c) $\dfrac{2^{9}}{2^{3}}$      (d) $\dfrac{2^{-6}}{2^{3}}$

   (e) $(2^{3})^{2}$      (f) $(2^{-3})^{-2}$      (g) $4^{4} \cdot 2^{2}$      (h) $4^{4} \cdot 16^{4}$

   (i) $8^{6} \cdot 16^{-3}$      (j) $\sqrt{8}$      (k) $\sqrt{8} \cdot 4\sqrt{2}$      (l) $\sqrt[3]{4} \cdot 2\sqrt[3]{16}$

   (m) $\sqrt{2} \cdot \sqrt[3]{2}$      (n) $\sqrt{2} \cdot \sqrt[4]{8}$      (o) $16^{-2} \cdot (4\sqrt{8})^{2}$      (p) $\left(\dfrac{\sqrt{2}}{2}\right)^{-3}$

   (q) $4^{2} \cdot 2^{\frac{1}{2}}$      (r) $\sqrt[3]{16} \cdot \sqrt[4]{4}$      (s) $\sqrt{8} \cdot 2^{\frac{3}{2}}$      (t) $2^{\frac{1}{3}} \cdot \sqrt[3]{4}$

   (u) $\dfrac{2}{\sqrt{2}}$      (v) $\dfrac{2^{\frac{1}{3}}}{\sqrt[3]{64}}$      (w) $\dfrac{2^{\frac{2}{3}}}{\sqrt[3]{4}}$      (x) $\dfrac{4^{-\frac{1}{3}}}{\sqrt[3]{2}}$

4. Express each number in scientific notation.

   (a) 52 270

   (b) 13.1401

   (c) 0.0000604

   (d) 0.0009

   (e) 0.0090

   (f) 32.001

   (g) 500 003

   (h) 100.00

   (i) 1 $\mu$m

5. State each value in scientific notation, correct to 3 significant figures.

   (a) $10^{18}$

   (b) $e^e$

   (c) one nanosecond

   (d) $\dfrac{1 + \sqrt{5}}{2}$

6. Write each equation in logarithmic form.

   (a) $243 = 3^5$

   (b) $256 = 2^8$

   (c) $100^{\frac{1}{2}} = 10$

   (d) $64^{\frac{1}{6}} = 2$

   (e) $9\sqrt{3} = 3^{\frac{5}{2}}$

   (f) $10^{-3} = 0.001$

   (g) $e^0 = 1$

   (h) $5^{-3} = \dfrac{1}{125}$

   (i) $(3\sqrt{3})^{-2} = \dfrac{1}{27}$

   (j) $8^{-\frac{1}{2}} = \dfrac{1}{2\sqrt{2}}$

   (k) $\left(\dfrac{1}{4}\right)^{-3} = 64$

   (l) $27^{-\frac{1}{2}} = \dfrac{\sqrt{3}}{9}$

   (m) $(-2)^{-3} = -\dfrac{1}{8}$

   (n) $(0.1)^{-2} = 100$

   (o) $\left(\dfrac{\sqrt{3}}{3}\right)^3 = \dfrac{\sqrt{3}}{9}$

   (p) $\left(\dfrac{1}{\sqrt{2}}\right)^{-3} = 2\sqrt{2}$

7. Express each equation with $x$ as the subject.

   (a) $y = 5^x$

   (b) $y = 10^x$

   (c) $y = e^x$

   (d) $y = 2^{2x}$

   (e) $y = 3 \cdot 3^x$

   (f) $y = 7 + 3^x$

   (g) $y = 2^{-2x}$

   (h) $y = 2^{\frac{x}{3}}$

   (i) $y = e^{\frac{x}{2}}$

   (j) $y = 5^{x+3}$

   (k) $y = e^{x-1}$

   (l) $y = \dfrac{1}{e^{2x}}$

8. Determine the value of each of the following.

   (a) $\log_3 243$

   (b) $\log_{243} 3$

   (c) $\log_{\frac{1}{2}} 16$

   (d) $\log_3 3\sqrt{3}$

   (e) $\log(-16)$

   (f) $\log_4 2\sqrt{2}$

   (g) $\log 50 + \log 20$

   (h) $\log 4000 - \log 4$

   (i) $\ln e^{-2}$

   (j) $\ln\left(\dfrac{1}{\sqrt{e}}\right)$

9. Write down the value of:

   (a) $2\log_{64}8$

   (b) $\log_8\sqrt{8}$

   (c) $\log_{\sqrt{3}}3\sqrt{3}$

   (d) $\log_{\sqrt{16}}\sqrt[3]{16}$

   (e) $\log_{\sqrt{8}}8^3$

   (f) $\log_{5^2}\sqrt{5}$

   (g) $\log_9 9^{-3} + \log_3 9^4$

   (h) $\log_{\sqrt{8}}2^{-3} + \log_8 4^4$

   (i) $\log_{2\sqrt{2}}4^3 - \log_{\sqrt{2}}2$

   (j) $\log_{\frac{1}{3}}3 - \log_3\sqrt{3}$

10. Solve each equation for $x$.

    (a) $\log_5 x = -3$

    (b) $\log_x\dfrac{1}{4} = -2$

    (c) $\log_3(x^2 - 2x - 5) = 1$

11. Given that $2^m = 8$ and $2^n = 16$,

    (a) write down the value of $m$ and of $n$

    (b) hence or otherwise solve $8^{2x+1} = 16^{2x-3}$

12. Consider $a = \log_2 3 \times \log_3 4 \times \log_4 5 \times \ldots \times \log_{31} 32$
    Given that $a \in \mathbb{Z}$, find the value of $a$.

13. Given that $\log_x y = 4\log_y x$, find all the possible expressions of $y$ as a function of $x$.

Functions

2

## Learning objectives

By the end of this chapter, you should be familiar with…

- Understanding the concepts of a function, domain, range and graph
- Using function notation, e.g., $f(x)$, $v(t)$, $C(n)$
- Finding composite functions in context; the notation $(f \circ g)(x) = f(g(x))$
- Finding an inverse function $f^{-1}$, including domain restriction
- Using piecewise functions
- Understanding the graph of a function; its equation $y = f(x)$
- Creating a sketch from information given or a context, including transferring a graph from screen to paper
- Determining key features of graphs; using technology to graph functions and find points of intersection of curves
- Understanding these transformations of graphs:
  - Translations: $y = f(x) + b$ and $y = f(x - a)$
  - Reflections: in the $x$-axis $y = -f(x)$, and in the $y$-axis $y = f(-x)$
  - Vertical stretch with scale factor $p$: $y = pf(x)$
  - Horizontal stretch with scale factor $\frac{1}{q}$: $y = f(qx)$
  - Composite transformations.

The relationship between two quantities – how the value of one quantity depends on the value of another quantity – is the key behind the concept of a function. Functions and how we use them are at the very foundation of many topics in mathematics, and are essential to our understanding of much of what will be covered later in this book. This chapter will look at some general characteristics and properties of functions. We will consider composite and inverse functions, and investigate how the graphs of functions can be transformed by means of translations, stretches and reflections.

## 2.1 Concept of a function, domain and range

Consider finding a mathematical model for the population growth of a country. Suppose a country has population $P$ at time $t$ (in years) and is growing at a rate of 2% per year. Then the two variables $P$ and $t$ are related by the formula $P = P_0(1.02)^t$, where $P_0$ is the population at time $t = 0$. Therefore, assuming the population growth remains constant, the formula can be used to calculate $P$ for any value of $t$.

In general, suppose that the values of a particular independent variable, for example, $x$, determine the values of a dependent variable $y$ in such a way that for a specific value of $x$, a single value of $y$ is determined.

Then we say that $y$ is a function of $x$ and we write $y = f(x)$ (read '$y$ equals $f$ of $x$') or $y = g(x)$, etc. where the letters $f$, $g$, etc. represent the name of the function. For the relationship at the start of this section, we can write:

Population $P$ is a function of time $t$: $P(t) = P_0(1.02)^t$

As with the population $P$ and the time $t$, many mathematical relationships concern how the value of one variable determines the value of a second variable. Other examples include:

The future value ($V$) of an investment with an annual contribution of $1000 earning 2.5% interest per year, compounded annually, determined by the number of years $n$ is given by $V(n) = 1000\left(\dfrac{1.025^n - 1}{0.025}\right)$

The amount ($A$) remaining of 8 micrograms of radioactive isotope Nitrogen-16 with a half-life of 10 minutes determined by the number of minutes $m$ is given by $A(m) = 8\left(\dfrac{1}{2}\right)^{\frac{m}{10}}$

The distance, $d$, that a number, $x$, is from the origin, is determined by its absolute value is given by $d(x) = |x|$

Along with equations, other useful ways of representing a function include a graph of the equation on a Cartesian coordinate system (also called a rectangular coordinate system), a table, a set of ordered pairs, or a mapping.

The largest possible set of values for the independent variable (the input set) is called the domain – and the set of resulting values for the dependent variable (the output set) is called the range. In the context of a mapping, each value in the domain is mapped to its image in the range.

All the various ways of representing a mathematical function illustrate that its defining characteristic is that it is a rule by which each number in the domain determines a *unique* number in the range.

Not all equations represent a function. The solution set for the equation $x^2 + y^2 = 1$ is the set of ordered pairs $(x, y)$ on the circle of radius equal to 1 and centre at the origin (see Figure 2.1). If we solve the equation for $y$, we get $y = \pm\sqrt{1 - x^2}$

It is clear that any value of $x$ between $-1$ and $1$ will produce two different values of $y$. Since at least one value in the domain ($x$) determines more than one value in the range ($y$), then the equation does not represent a function. A correspondence between two sets that does not satisfy the definition of a function is called a relationship or a relation.

For many physical phenomena, we observe that one quantity depends on another. The word function is used to describe this dependence of one quantity on another, i.e. how the value of an independent variable determines the value of a dependent variable. A common mathematical task is to find how to express one variable as a function of another variable.

The coordinate system for the graph of an equation has the independent variable on the horizontal axis and the dependent variable on the vertical axis.

A function is a correspondence (mapping) between two sets $X$ and $Y$ in which each element of set $X$ corresponds to (maps to) exactly one element of set $Y$. The domain is set $X$ (independent variable) and the range is set $Y$ (dependent variable).

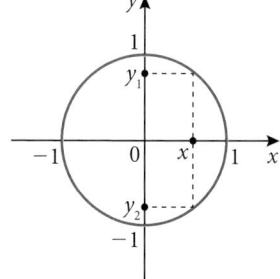

**Figure 2.1** Graph of $x^2 + y^2 = 1$

A vertical line intersects the graph of a function at no more than one point (this is known as the vertical line test).

### Example 2.1

A company wants to find the minimum cost of building an oil pipeline from point $A$ to point $B$ across two types of soil. The first type of soil is made of clay and the cost of building the pipeline through this type of soil is \$7 million per kilometre. The second type of soil is sandy and the cost of building the pipeline is \$4 million per km. The boundary $XY$ between the two types of soil is a straight line running from east to west and point $A$ lies 70 km directly north of it. Point $B$ is 100 km to the east of point $A$ and 30 km directly south of the boundary. This is shown in the diagram along with a possible route $APB$.

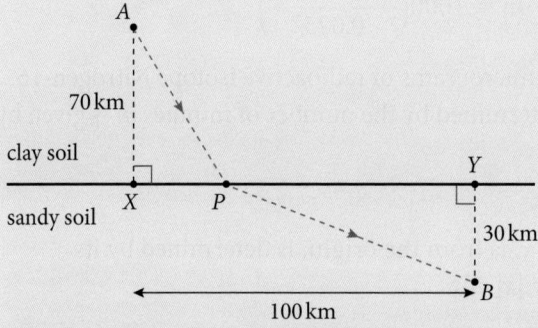

Find a function to model the cost of the general route from $A$ to $B$ and use it to find the minimum cost of building the pipeline.

### Solution

We need to introduce a variable to represent the possible position of point $P$ on the boundary, so let distance $XP = d$

We can now find expressions for other lengths in the diagram.

Using Pythagoras' theorem, $AP = \sqrt{AX^2 + XP^2} = \sqrt{70^2 + d^2} = \sqrt{4900 + d^2}$

Since $XY = 100$, then $PY = 100 - d$

Using this in triangle $PYB$ gives

$$PB = \sqrt{PY^2 + YB^2} = \sqrt{(100 - d)^2 + 30^2} = \sqrt{10\,900 - 200d + d^2}$$

Multiplying the cost per kilometre by the distance will give the total cost of the pipeline, so the total cost $C(d) = 7AP + 4PB$ (in millions of dollars.)

Substituting the expressions above:

$$C(d) = 7\sqrt{4900 + d^2} + 4\sqrt{10\,900 - 200d + d^2}$$

Note that this function shows how the cost $C$ of the pipeline, depends on the distance $d$ of $P$ from $X$. To find the minimum cost of building the pipeline, we can graph the function and use a graphing calculator to find the minimum point.

From a GDC we get the minimum cost as \$833 million.

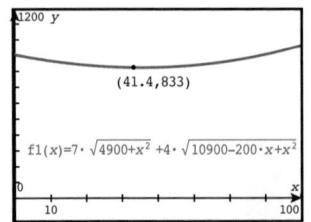

**Figure 2.2** Using a GDC to find the minimum cost

## Domain and range of a function

In Example 2.1 we found the function

$$C(d) = 7\sqrt{4900 + d^2} + 4\sqrt{10\,900 - 200d + d^2}$$

Since point $P$ lies on the boundary between $X$ and $Y$, we can see that
$$0 \leqslant d \leqslant 100$$

This is called the domain of the function.

 The range of a function can be found graphically or by analysing the function algebraically.

We already found the minimum value using a GDC and we can find the maximum value to determine the range.

So the range is $833 \leqslant C \leqslant 975$

In general, when writing a function, you should also state its domain.

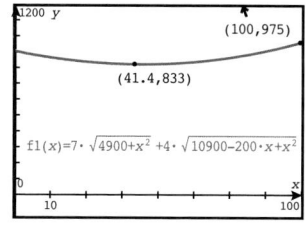

**Figure 2.3** Using a GDC to find the range

### Exercise 2.1

1. For each equation:

   **(i)** match it with its graph

   **(ii)** state, with justification, whether or not the equation represents a function. Assume that $x$ is the independent variable and $y$ is the dependent variable.

   **(a)** $y = 2x$

   **(b)** $y = -3$

   **(c)** $x - y = 2$

   **(d)** $x^2 + y^2 = 4$

   **(e)** $y = 2 - x$

   **(f)** $y = x^2 + 2$

   **(g)** $y^3 = x$

   **(h)** $y = \dfrac{2}{x}, x \neq 0$

   **(i)** $x^2 + y = 2$

A.
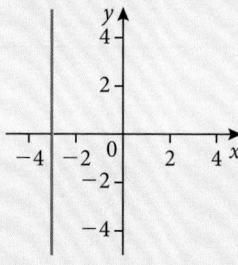

B.
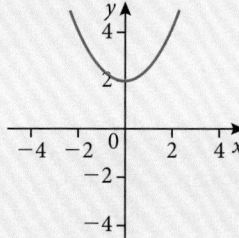

C.
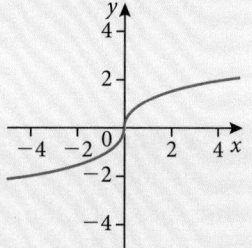

D.
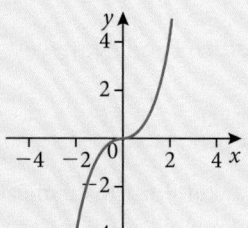

E.
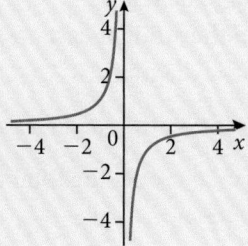

F.
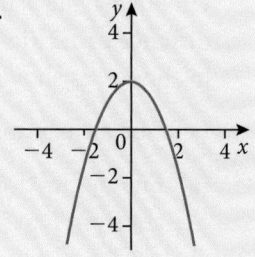

G. H. I.

J. K. L.

2. Consider the relationship $h(x) = \sqrt{x - 4}$, $x \geq a$

   (a) Find the least value of $a$ which makes $h$ a function.

   (b) Find

       (i) $h(29)$       (ii) $h(53)$

   (c) State the range of $h$.

3. A rectangular swimming pool with dimensions 12 m by 18 m is surrounded by a pavement of uniform width $x$ m. Find the area of the pavement, $A$, as a function of $x$.

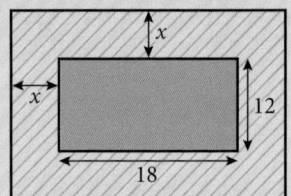

4. A farmer, $F$, uses a tractor to travel to his house, $H$. The tractor can travel at a speed of 25 km per hour on roads but only at 15 km per hour over fields. On one day he is in a large field at a distance of 1.2 km from the straight east-west road $GH$ and 4 km west of his house, as shown in the diagram.

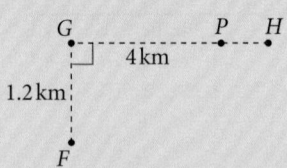

   Let $P$ be an arbitrary point on the road $GH$ and let $GP = x$

   (a) Find a function $T(x)$ to model the total time taken for the farmer to get from $F$ to $H$.

   (b) Find the minimum time for his journey.

   (c) List any assumptions you have made about the farmer's journey.

5. The thickness ($t$ mm) of a brake pad on a car, which has been driven $k$ kilometres, can be modelled by the function

$$t(k) = a\sqrt{120\,000 - k}, \text{ where } 0 \leqslant k \leqslant 120\,000 \text{ and } a \in \mathbb{R}$$

(a) Find the value of $a$, given that the initial thickness is 12 mm.

(b) It is recommended to replace the brake pads after the car has driven 90 000 km. Find the thickness when the brake pads should be replaced.

6. A hospital patient is given an IV drip to replace lost fluids. The IV bag initially contains 0.5 litres and the drip rate is set at 2 m$l$ per minute.

(a) Find a function for the amount, $a$ m$l$, remaining in the bag at time $t$ minutes.

(b) An alarm sounds when the amount of fluid remaining is 50 m$l$. Find the time taken until the alarm sounds.

7. The volume of liquid, $v$ cm$^3$ in a conical flask is a function of the height, $h$ cm of the liquid, given by

$$v(h) = 3\pi\left(10 - \frac{(10-h)^3}{100}\right) \text{ where } 0 \leqslant h \leqslant 8$$

(a) Find the greatest volume of liquid that the flask can contain.

(b) Find the height of liquid in the flask when it is half full.

8. Sarah is spending her holiday camping. She looks out of her tent and notices that another tent on the other side of the large field is on fire. She picks up her largest cooking pot and runs to the river to fill it up, before running

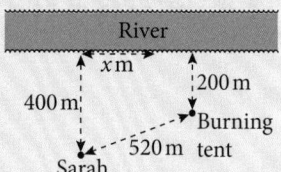

to the burning tent to pour the water onto the fire. On the way to the river she runs at 22 km h$^{-1}$, but on the way to the tent, she only runs at 12 km h$^{-1}$, otherwise she will spill too much water. Her tent is 400 m from the river, the burning tent is 200 m from the river and the distance between the tents is 520 m, as shown in the diagram.

Let the point where Sarah reaches the river be a distance $x$ m along the river, as shown.

(a) Find a function to model the time taken (in minutes) for Sarah to reach the burning tent.

(b) State the domain of the function.

(c) Find the least time it takes Sarah to reach the burning tent.

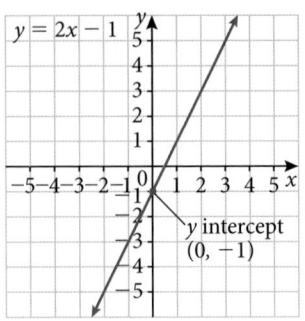

$y = 2x - 1$

**Figure 2.4** Graph of the line $y = 2x - 1$

## 2.2 Linear and piecewise functions

### Linear functions

You will probably be familiar with linear functions from your previous studies in Mathematics. In this section we review several key concepts and consider some applications. Recall that the linear function $f(x) = mx + c$ has a gradient (slope) given by $m$, and a $y$-intercept at $(0, c)$. The equation $y = mx + c$ is called the gradient-intercept form. For example, the graph of the function $f(x) = 2x - 1$ is shown in Figure 2.4.

Recall that the gradient is given by $\dfrac{\text{change in } y}{\text{change in } x}$ or $\dfrac{\Delta y}{\Delta x}$ or $\dfrac{y_2 - y_1}{x_2 - x_1}$

In the graph of $f(x) = 2x - 1$ we can see that each time the $x$ value increases by 1 unit, the $y$ value increases by 2 units.

Hence the gradient $= 2$

There are two other useful forms of the equation of a line. The equation $ax + by + d = 0$ is called the general form of a line. In this form $a$, $b$ and $d$ are usually written as integers. For example the equation $y = \dfrac{1}{2}x + \dfrac{2}{3}$ can be rearranged to give $3x - 6y + 4 = 0$

In this case we have $a = 3$, $b = -6$ and $d = 4$

The equation $y - y_1 = m(x - x_1)$ is called the point-gradient form of a line, and is particularly useful when you are not given the $y$-intercept. In the equation $m$ represents the gradient and $(x_1, y_1)$ represents a point on the line.

The point-gradient form of a line can be derived algebraically starting with the gradient-intercept form. Given that a straight line has gradient, $m$, and passes through the point $(x_1, y_1)$, we can first write the equation using gradient–intercept form as

$$y = mx + c$$

substituting the given point for $(x, y)$ gives

$$y_1 = mx_1 + c$$

rearranging for $c$ gives

$$c = y_1 - mx_1$$

which can be substituted into the original equation to give

$$y = mx + y_1 - mx_1$$

rearranging and factorising gives

$$y - y_1 = m(x - x_1)$$

### Example 2.2

Find the equation of the straight line passing through the points $(1, 4)$ and $(-3, 10)$. Give your answer in the form $ax + by + d = 0$, where $a, b, d \in \mathbb{Z}$.

## Solution

We can find the gradient, $m$, using $m = \dfrac{y_2 - y_1}{x_2 - x_1}$

$$m = \frac{4 - 10}{1 - (-3)} = -\frac{6}{4} = -\frac{3}{2}$$

Then find the equation using the point–gradient form.

Substituting $m = -\dfrac{3}{2}$ and $(1, 4)$ gives:

$$y - 4 = \frac{3}{2}(x - 1)$$

The equation can now be rearranged into general form

$$3x - 2y + 5 = 0$$

## Piecewise functions

In this section we explore functions that are made out of several sub-functions. Consider a taxi company that has the following structure for charging customers.

A piecewise function is a function that is defined on a sequence of intervals.

  Initial cost of starting a journey = \$3

  Cost per kilometre for the first 8 kilometres = 50 cents

  Cost per kilometre for subsequent kilometres = 30 cents

We want to find a function to model the cost, $C$ (\$), of a journey for a distance $d$ (km).

Since the cost per kilometre is a constant, we can model the journey using linear functions. For the first 8 km, the cost is \$0.50 per km and the starting cost is \$3.

This gives $C = 3 + 0.5d, \ 0 \leqslant d \leqslant 8$

A journey of exactly 8 km would cost $C = 3 + 0.5(8) = \$7$. We can use this value to help model the cost for subsequent kilometres. For $d > 8$, we can use \$7 as the starting cost and each kilometre costs \$0.30. So we get $C = 7 + 0.3(d - 8), \ d > 8$

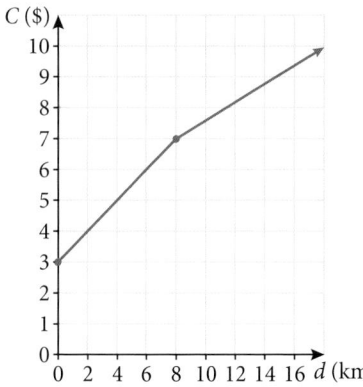

Figure 2.5 Graph of the piecewise function to model the cost of a taxi journey

We can write these two results as one piecewise function:

$$C = \begin{cases} 3 + 0.5d, & 0 \leqslant d \leqslant 8 \\ 7 + 0.3d - 8, & d > 8 \end{cases}$$

Note that the two functions meet at the point $(8, 7)$, so the function is continuous.

A continuous curve is one for which its graph can be sketched without lifting your pen.

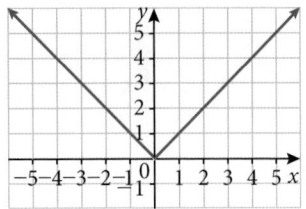

**Figure 2.6** Graph of $y = |x|$

A common piecewise function that you may have encountered before is the absolute value function. The absolute value function $f(x) = |x|$ is defined by

$$|x| = \begin{cases} x, & x \geqslant 0 \\ -x, & x < 0 \end{cases}$$

So, for example, $|7| = 7$ and $|-5| = 5$. The graph of $y = |x|$ is shown in Figure 2.6

## Example 2.3

Let $f$ be the function defined by $f(x) = \begin{cases} x^2, & 0 \leqslant x \leqslant 4 \\ 2x + c, & x > 4 \end{cases}$

(a) Find the value of $c$ that makes $f$ continuous at $x = 4$

(b) Sketch the graph of $f$.

### Solution

(a) First calculate the value of $f$ at $x = 4$

$$f(4) = 4^2 = 16$$

To make $f$ continuous, we need the value of $2x + c$ to equal 16 when $x = 4$, so we can solve

$$2(4) + c = 16$$

$$\Rightarrow c = 8$$

(b)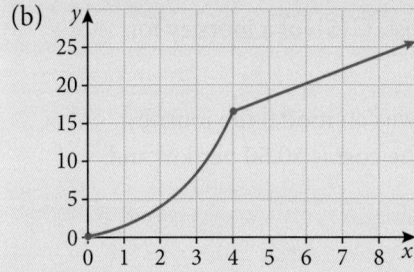

## Graphs of functions

| $x$ | $-2$ | $-1$ | $0$ | $1$ | $2$ |
|-----|------|------|-----|-----|-----|
| $f(x)$ | $\frac{1}{4}$ | $\frac{1}{4}$ | $1$ | $2$ | $4$ |

**Table 2.1** Table of values for $f(x) = 2^x$

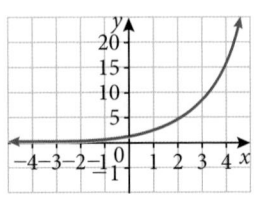

**Figure 2.7** Graph of $f(x) = 2^x$

In this section we explore the graphs of several functions and consider their domain and range. We also consider GDC techniques to find important features of graphs.

In Chapter 1 we reviewed the rules of exponents. We now want to consider the graphs of exponential functions. Consider $f(x) = 2^x$

We can make a table of values or use a GDC to sketch the function.

Either from the table or the graph we can see that the $y$-intercept is $(0, 1)$. The domain of $f$ is $x \in \mathbb{R}$ and its range is $y > 0$. (Note $y$ is not equal to 0, because there is no value of $x$ such that $2^x = 0$) We can also see that as the value of $x$

decreases, the value of $y$ also decreases and approaches zero. We say that the graph has a horizontal asymptote at $y = 0$.

For the general exponential function, $f(x) = a^x$, $a > 0$, we can make the following observations.

If a curve approaches a fixed line, but does not meet it at any finite distance, then the line is said to be an asymptote to the curve.

- $f(0) = a^0 = 1$, so the graph always has a $y$-intercept at $(0, 1)$
- If $a > 1$, then as $x$ increases, so does $f$. We say that $f$ is an increasing function. An increasing exponential function is said to grow exponentially.
- If $0 < a < 1$, then as $x$ increases, $f$ decreases. We say that $f$ is a decreasing function. A decreasing exponential function is said to decay exponentially.
- If $a = 1$, then $f(x) = 1$, which is a constant, so this is not an exponential function.
- $y = 0$ is a horizontal asymptote for the graph of $y = f(x)$
- The domain of $f$ is $x \in \mathbb{R}$ and its range is $y > 0$

A function $f$ is an increasing function if $x_1 < x_2$ implies $f(x_1) < f(x_2)$. It is a decreasing function if $x_1 < x_2$ implies $f(x_1) > f(x_2)$. If a function is either increasing or decreasing, it is said to be monotonic.

## Example 2.4

The activity, $a$ becquerels (Bq), of a radioactive substance can be modelled by the function $a(t) = \dfrac{50}{2^t}$, $t \geq 0$, where $t$ is the time in days.

(a) Sketch the graph of $a$ against $t$, for $0 \leq t \leq 10$

(b) Find the initial activity of the substance.

(c) Find the time it takes for the activity of the substance to reach 20 Bq.

(d) Write down the equation of the horizontal asymptote of the graph of $a$.

(e) Describe what happens to the activity of the substance after a long period of time.

   The activity of a second radioactive substance can be modelled by the function $b(t) = \dfrac{80}{3^t}$, $t \geq 0$, where $t$ is the time in days.

(f) Find the time when the two substances have equal activity.

### Solution

(a)

a (Bq) graph showing exponential decay from 50 at t=0, decreasing toward 0 as t increases to 10, with t (days) on horizontal axis.

(b) $a(0) = \dfrac{50}{1} = 50\,\text{Bq}$

(c) We need to solve the equation

$$\frac{50}{2^t} = 20$$

Using a GDC we can find the intersection of the curves $a(t) = \frac{50}{2^t}$ and $a(t) = 20$

The curves intersect at $(1.32, 20)$, so it takes 1.32 days for the activity to reach 20 Bq.

(d) From the graph we can see that the horizontal asymptote is $a = 0$

(e) From part (d) we can deduce that after a long period of time, the activity of the substance tends to zero.

(f) We need to solve the equation $\frac{50}{2^t} = \frac{80}{3^t}$

Using a GDC we can find the intersection of the curves $a(t) = \frac{50}{2^t}$ and $b(t) = \frac{80}{3^t}$

The curves intersect at $(1.16, 22.4)$, so it takes 1.16 days for the activity to be equal.

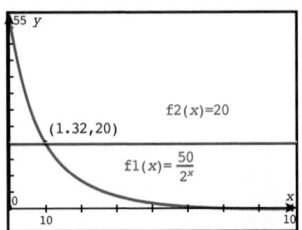

**Figure 2.8** GDC screen for the solution to Example 2.4 (c)

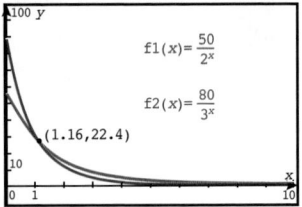

**Figure 2.9** GDC screen for the solution to Example 2.4 (f)

> The equation $f(x) = g(x)$ can be solved graphically on a GDC by sketching the graphs $y = f(x)$ and $y = g(x)$, then finding the points of intersection of the two graphs.
>
> Alternatively, the equation can be written as $f(x) - g(x) = 0$
>
> This equation can be solved graphically on a GDC by sketching the graph $y = f(x) - g(x)$, then finding the zeros.

In Chapter 1 you learned the rules of logarithms. We now want to consider the graphs of logarithmic functions. Consider $f(x) = \log_3 x$, $x > 0$. We can sketch the graph to explore its features.

Note that the domain of $f$ is $x > 0$ and its range is $y \in \mathbb{R}$

It is tempting to think that the graph has a horizontal asymptote, but this is not the case. As the value of $x$ increases, the value of $y$ also increases and there is no limit on how large $y$ can become. That is why the range is $y \in \mathbb{R}$.

The graph does have a vertical asymptote at $x = 0$

The graph has an $x$-intercept at $(1, 0)$.

For the general logarithmic function, $f(x) = \log_a x$, $a > 0$, we can make the following observations.

**Figure 2.10** The graph of $f(x) = \log_3 x$, $x > 0$

○ $f(1) = \log_a 1 = 0$, so the graph always has an $x$-intercept at $(1, 0)$

○ The graph $y = f(x)$ has a vertical asymptote at $x = 0$

○ The domain of $f$ is $x > 0$ and its range is $y \in \mathbb{R}$

**Example 2.5**

The decibel (dB) is a unit for measuring the relative loudness of sounds. If $s$ is the intensity of a sound (measured in watts per $m^2$), then we can calculate the relative loudness in decibels, $n$, using the formula

$$n(s) = 10 \log_{10}\left(\frac{s}{10^{-12}}\right)$$

The number $10^{-12}$ in the formula represents the threshold of human hearing.

(a) The sound intensity of a flushing toilet is approximately $3.2 \times 10^{-5}\,\text{W}\,\text{m}^{-2}$.
Calculate the relative loudness in decibels.

(b) The relative loudness of a firecracker is usually around 145 dB.
Calculate the intensity of the sound.

(c) Sound $a$ has an intensity 1 million times greater than sound $b$.
Calculate the difference in decibels between the two sounds.

**Solution**

(a) Substituting into the formula gives:

$$n(3.2 \times 10^{-5}) = 10 \log_{10}\left(\frac{3.2 \times 10^{-5}}{10^{-12}}\right) = 10 \log_{10}(3.2 \times 10^{7}) = 75.1\,\text{dB}$$

(b) We need to solve the equation $10 \log_{10}\left(\dfrac{s}{10^{-12}}\right) = 145$

The intersection is at (316, 145), so the sound intensity is 316 W m$^{-2}$.

(c) Suppose sound $b$ has an intensity $x$ W m$^{-2}$. Then sound $a$ has an intensity given by $1\,000\,000x$ W m$^{-2}$. The difference in decimals is:

$$10 \log_{10}\left(\frac{1\,000\,000x}{10^{-12}}\right) - 10 \log_{10}\left(\frac{x}{10^{-12}}\right)$$

$$= 10\left(\log_{10}\left(\frac{1\,000\,000x}{10^{-12}}\right) - \log_{10}\left(\frac{x}{10^{-12}}\right)\right)$$

$$= 10 \log_{10}\left(\frac{1\,000\,000x}{10^{-12}} \cdot \frac{10^{-12}}{x}\right)$$

$$= 10 \log_{10} 1\,000\,000 = 60\,\text{dB}$$

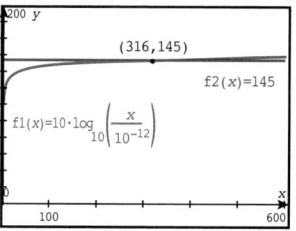

**Figure 2.11** GDC screen for the solution to Example 2.5

# 2 Functions

## Exercise 2.2

1. Find the equation of each line. Give your answer in the form
   $ax + by + d = 0$, where $a, b, d \in \mathbb{Z}$

   (a) A line with gradient 3 and $y$-intercept $(0, -4)$

   (b) A line with gradient 2, which passes through the point $(3, -1)$

   (c) A line which passes through the points $(-3, 5)$ and $(2, -10)$

2. The freezing point of water is 0°C (Celsius) and 32°F (Fahrenheit).

   The boiling point of water is 100°C (Celsius) and 212°F (Fahrenheit).

   (a) Use these values to find a linear function that converts Celsius to Fahrenheit.

   (b) Normal body temperature is 37°C. Use your answer to part (a) to convert this into Fahrenheit.

   (c) The temperature of the sun is 9941°F. Convert this into Celsius.

   In the Kelvin temperature scale, the freezing point of water is 273 K and the boiling point of water is 373 K.

   (d) Find a linear function that converts Kelvin to Fahrenheit.

   (e) Absolute zero is 0 K. Convert this to:

       (i) Fahrenheit     (ii) Celsius.

3. A moneychanger exchanges 250 THB (Thai Baht) into 400 PHP (Philippine Pesos).

   (a) Find a linear function to convert Thai Baht into Philippine Pesos.

   (b) Gen want to exchange 520 THB. Find the amount of Philippine Pesos she receives.

   (c) Glenn wants to exchange 1280 PHP. Find the amount of Thai Baht he receives.

4. A company supplies drinking water dispensers to office buildings. It has two models that customers can choose from.

   Model A costs $400 to install and has a monthly fee of $50.

   Model B costs $150 to install and has a monthly fee of $80.

   (a) Find a linear function to model the cost of model A.

   (b) Find a linear function to model the cost of model B.

   (c) A customer needs a water dispenser for one year. Determine which model he should choose to minimise his cost.

5. A taxi driver charges $4 to pick up a customer, plus $0.80 per kilometre.

   (a) Find an equation to model the total cost $C$ in terms of the distance travelled $d$.

**(b)** Find the total cost for a journey of distance 12 km.

The total cost charged by a different taxi driver can be modelled by $C = d + 2$

**(c)** Explain what the gradient of the line $C = d + 2$ represents.

**(d)** Explain what the $y$-intercept of the line $C = d + 2$ represents.

**6.** Function $f$ is defined by $f(x) = \begin{cases} 2x + 1, & x \leqslant 3 \\ ax + b, & x > 3 \end{cases}$

Given that $f$ is continuous at $x = 3$:

**(a)** find the value of $a$ when $b = 8$

**(b)** find the value of $b$ when $a = 4$

**7.** Function $g$ is defined by $g(x) = \begin{cases} 1 - x^2, & 0 \leqslant x \leqslant 2 \\ ax + b, & x > 2 \end{cases}$

Given that $g$ is continuous at $x = 2$:

**(a)** find the value of $a$ when $b = 2$

**(b)** find the value of $b$ when $a = 3$

**8.** The cost of a journey can be modelled by

$$C(d) = \begin{cases} 60 + 12d, & 1 \leqslant d \leqslant 5 \\ x + 9d, & d > 5 \end{cases}$$

where $d$ is the distance travelled in kilometres.

**(a)** Given that $C$ is a continuous function, find the value of $x$.

**(b)** Find the cost of a journey of distance 9 km.

**9.** Function $f$ is defined by $f(x) = \begin{cases} x^3 - 6x^2 + 3x + 10, & x \leqslant 5 \\ -5x^2 + 70x - 225, & x > 5 \end{cases}$

**(a)** Show that $f$ is continuous at $x = 5$

**(b)** Sketch the graph of $y = f(x)$ for $0 \leqslant x \leqslant 9$

**(c)** Find the minimum value of $f$ for $0 \leqslant x \leqslant 9$

**(d)** Find the maximum value of $f$ for $0 \leqslant x \leqslant 9$

**(e)** Solve $f(x) = 4$ for $0 \leqslant x \leqslant 9$

**10.** The activity, $m$ becquerels, of a radioisotope can be modelled by the function $m(t) = \dfrac{30}{3^t}$, where $t$ is the time in days.

**(a)** Sketch the graph of $m$ against $t$, for $0 \leqslant t \leqslant 6$

**(b)** Find the initial activity of the substance.

**(c)** Find the time it takes for the activity of the substance to halve.

**(d)** Write down the equation of the horizontal asymptote of the graph of $m$.

**(e)** Describe what happens to the activity of the substance after a long period of time.

11. A species of rare bird is introduced into a wilderness area in the hope that the bird population will grow. The size of the population, $P$, can be modelled by

$$P(t) = \frac{100}{1 + e^{3-t}}$$

where $t$ is the time in years.

(a) Sketch the graph of $P$ against $t$, for $0 \leqslant t \leqslant 12$

(b) Find the initial number of birds that are introduced into the wilderness area.

(c) Find the time it takes for the number of birds to reach 20.

(d) Write down the equation of the horizontal asymptote of the graph of $P$.

(e) Describe what happens to the population of birds after a long period of time.

12. An earthquake creates waves. The magnitude, $M$, of an earthquake can be modelled by measuring the amplitude, $A$ of these waves. One model is:

$$M = \log_{10}(A) + 4$$

(a) Sketch the graph of $M$ against $A$, for $0 \leqslant A \leqslant 200$

(b) Find the amplitude of a wave, when the magnitude measures 5.3

(c) Find the magnitude of an earthquake, when the amplitude is 10 times greater than your answer in part (b).

(d) Find the ratio of the amplitudes of the waves created by earthquakes that have magnitudes 4.3 and 4.9

## 2.3 Composite functions

### Composition of functions

Consider the function $f(x) = \sqrt{x + 4}, x \geqslant -4$. When you evaluate $f(x)$ for a certain value of $x$ in the domain, for example $x = 5$, it is necessary for you to perform computations in two separate steps in a certain order.

$f(5) = \sqrt{5 + 4} \Rightarrow f(5) = \sqrt{9}$     Step 1: compute the sum of $5 + 4$

$f(5) = 3$     Step 2: compute the square root of 9

Given that the function has two separate evaluation steps, $f(x)$ can be seen as a combination of two simpler functions that are performed in a specified order. According to how $f(x)$ is evaluated (as shown above), the simpler function to be performed first is the rule of 'adding 4' and the second is the rule of 'taking the square root'. If $h(x) = x + 4$ and $g(x) = \sqrt{x}$, then we can create (compose) the function $f(x)$ from a combination of $h(x)$ and $g(x)$ as follows:

$$f(x) = g(h(x))$$
$$= g(x + 4) \text{ Step 1: substitute } x + 4 \text{ for } h(x)$$
$$= \sqrt{x + 4} \text{ Step 2: apply the function } g(x) \text{ on the input } x + 4$$

We obtain the rule $\sqrt{x + 4}$ by first applying the rule $x + 4$ and then applying the rule $\sqrt{x}$. A function that is obtained from simpler functions by applying one after another in this way is called a composite function. In the example above, $f(x) = \sqrt{x + 4}$ is the composition of $h(x) = x + 4$ followed by $g(x) = \sqrt{x}$. In other words, $f$ is obtained by substituting $h$ into $g$, and can be denoted in function notation by $g(h(x))$, which is read '$g$ of $h$ of $x$'.

We start with a number $x$ in the domain of $h$ and find its image $h(x)$. If this number $h(x)$ is in the domain of $g$, we then compute the value of $g(h(x))$. The resulting composite function is denoted as $(g \circ h)(x)$. See mapping illustration in Figure 2.12.

> From the explanation on how $f$ is the composition of $g$ and $h$, you can see why a composite function is sometimes referred to as a function of a function. Also note that in the notation $g(h(x))$, the function $h$ that is applied first is written inside, and the function $g$ that is applied second is written outside.

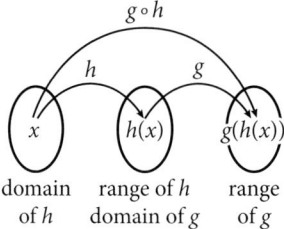

Figure 2.12  Mapping for composite function $g(h(x))$

The composition of two functions, $g$ and $h$, such that $h$ is applied first and $g$ second is given by

$$(g \circ h)(x) = g(h(x))$$

The domain of the composite function $g \circ h$ is the set of all $x$ in the domain of $h$ such that $h(x)$ is in the domain of $g$.

The notations $(g \circ h)(x)$ and $g(h(x))$ are both commonly used to denote a composite function where $h$ is applied first then followed by applying $g$. Since we are reading this from left to right, it is easy to apply the functions in the incorrect order. It may be helpful to read $g \circ h$ as $g$ following $h$, or as $g$ composed with $h$ to emphasise the order in which the functions are applied. Also, in both notations, $(g \circ h)(x)$ or $g(h(x))$ the function applied first is closest to the variable $x$.

## Example 2.6

Let $f(x) = x^2$ and $g(x) = 2x - 5$

(a)  Calculate $(f \circ g)(7)$

(b)  Find an expression for $(g \circ f)(x)$

(c)  Calculate $(f \circ f)(5)$

(d)  Find an expression for $(g \circ g)(x)$ in its simplest form.

(e)  Solve $(f \circ g)(x) = (g \circ f)(x)$

## Solution

(a)  $(f \circ g)(7) = f(g(7)) = f(2(7) - 5) = f(9) = 9^2 = 81$

(b)  $(g \circ f)(x) = g(f(x)) = g(x^2) = 2x^2 - 5$

(c)  $(f \circ f)(5) = f(f(5)) = f(5^2) = f(25) = 25^2 = 625$

(d)  $(g \circ g)(x) = g(g(x)) = g(2x - 5) = 2(2x - 5) - 5 = 4x - 15$

Use the polynomial root finder on a GDC to help solve polynomial equations.

`polyRoots(x²-10·x+15,x)`
$\{1.83772, 8.16228\}$

It is important to note in part (e) of Example 2.6, that $f \circ g$ is *not* equal to $g \circ f$. This is because the functions are applied in reverse order. In general
$$f \circ g \neq g \circ f$$

(e)  First find an expression for $(f \circ g)(x)$
$$(f \circ g)(x) = f(g(x)) = f(2x - 5) = (2x - 5)^2 = 4x^2 - 20x + 25$$
using part (b): $4x^2 - 20x + 25 = 2x^2 - 5$

so $2x^2 - 20x + 30 = 0$
$$x^2 - 10x + 15 = 0$$
Solve using the quadratic formula or a GDC:
$$x = 1.83 \text{ or } x = 8.16$$

## Finding the domain of a composite function

Referring back to Figure 2.12, it is important to note that in order for a value of $x$ to be in the domain of the composite function $g \circ h$, two conditions must be met:

- $x$ must be in the domain of $h$
- $h(x)$ must be in the domain of $g$.

Likewise, it is also worth noting that $g(h(x))$ is in the range of $g \circ h$ only if $x$ is in the domain of $g \circ h$. Example 2.7 illustrates these points – and also that, in general, the domains of $g \circ h$ and $h \circ g$ are not the same.

### Example 2.7

Let $g(x) = x^2 - 4$ and $h(x) = \sqrt{x}, x \geq 0$

Find:

(a)  $(g \circ h)(x)$ and its domain and range

(b)  $(h \circ g)(x)$ and its domain and range.

**Solution**

First, establish the domain and range for both $g$ and $h$. For $g(x) = x^2 - 4$, the domain is $x \in \mathbb{R}$ and the range is $y \geq -4$. For $h(x) = \sqrt{x}$, the domain is $x \geq 0$ and the range is $y \geq 0$

(a)  $(g \circ h)(x) = g(h(x))$
$= g(\sqrt{x})$      To be in the domain of $g \circ h$, $\sqrt{x}$ must be
$= (\sqrt{x})^2 - 4$    defined for $x \Rightarrow x \geq 0$
$= x - 4$        Therefore, the domain of $g \circ h$ is $x \geq 0$

Since $x \geq 0$, then the range for $y = x - 4$ is $y \geq -4$

Therefore, $(g \circ h)(x) = x - 4$, and its domain is $x \geq 0$, and its range is $y \geq -4$

(b) $(h \circ g)(x) = h(g(x))$     $g(x) = x^2 - 4$ must be in the domain of $h$

$\Rightarrow x^2 - 4 \geqslant 0 \Rightarrow x^2 \geqslant 4$

$\phantom{(b)} = h(x^2 - 4)$     Therefore, the domain of $h \circ g$ is

$x \leqslant -2$ or $x \geqslant 2$

$\phantom{(b)} = \sqrt{x^2 - 4}$     and with $x \leqslant -2$ or $x \geqslant 2$, the range for

$y = \sqrt{x^2 - 4}$ is $y \geqslant 0$

Therefore, $(h \circ g)(x) = \sqrt{x^2 - 4}$, and its domain is $x \leqslant -2$ or $x \geqslant 2$, and its range is $y \geqslant 0$

## Exercise 2.3

1. Let $f(x) = 2x$ and $g(x) = \dfrac{1}{x - 3}, x \neq 0$

   (a) Find the value of:

      (i) $(f \circ g)(5)$     (ii) $(g \circ f)(5)$

   (b) Find an expression for:

      (i) $(f \circ g)(x)$     (ii) $(g \circ f)(x)$

2. Let $f(x) = 2x - 3$ and $g(x) = 2 - x^2$

   (a) Calculate:

      (i) $(f \circ g)(0)$     (ii) $(g \circ f)(0)$     (iii) $(f \circ f)(4)$

      (iv) $(g \circ g)(-3)$     (v) $(f \circ g)(-1)$     (vi) $(g \circ f)(-3)$

   (b) Find an expression for:

      (i) $(f \circ g)(x)$     (ii) $(g \circ f)(x)$     (iii) $(f \circ f)(x)$     (iv) $(g \circ g)(x)$

3. For each pair of functions, find $(f \circ g)(x)$ and state its domain.

   (a) $f(x) = 4x - 1, g(x) = 2 + 3x$

   (b) $f(x) = x^2 + 1, g(x) = -2x$

   (c) $f(x) = 1 + x^2, g(x) = \sqrt{x + 1}, x \geqslant -1$

   (d) $f(x) = \dfrac{2}{x + 4}, x \neq -4, g(x) = x - 1$

   (e) $f(x) = 3x + 5, g(x) = \dfrac{x - 5}{3}$

   (f) $f(x) = 2 - x^3, g(x) = \sqrt[3]{1 - x^2}$

   (g) $f(x) = \dfrac{2x}{4 - x}, x \neq 4, g(x) = \dfrac{1}{x^2}, x \neq 0$

   (h) $f(x) = \dfrac{2}{x + 3}, x \neq -3, g(x) = \dfrac{5}{x - 4}, x \neq 4$

4. Let $f(x) = \dfrac{x}{x-1}$, $x \neq 1$ and $g(x) = x^2 - 1$

   Solve $(f \circ g)(x) = (g \circ f)(x)$

5. The graphs of two functions $f$ and $g$ are shown.

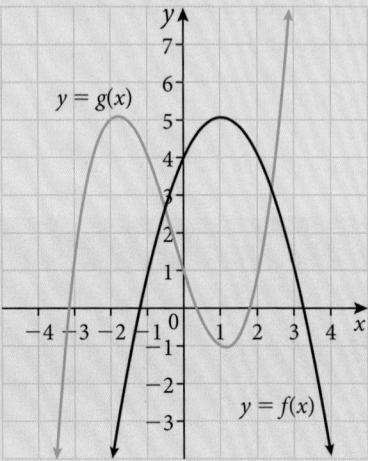

   Use the graphs to find the value of:

   **(a)** $(f \circ g)(-3)$        **(b)** $(g \circ g)(1)$        **(c)** $(g \circ f)(-1)$

6. Let $f(x) = x + 3$

   **(a)** Find $(f \circ f)(x)$

   **(b)** Find $(f \circ f \circ f)(x)$

   **(c)** Find $f^{(4)}(x)$, where $f^{(4)} = f \circ f \circ f \circ f$

   **(d)** Use your answers to write down an expression for $f^{(n)}(x)$

   **(e)** Hence solve $f^{(50)}(x) = 20$

7. Let $f(x) = 2x$

   **(a)** Find $(f \circ f)(x)$

   **(b)** Find $(f \circ f \circ f)(x)$

   **(c)** Find $f^{(4)}(x)$, where $f^{(4)} = f \circ f \circ f \circ f$

   **(d)** Use your answers to write down an expression for $f^{(n)}(x)$

   **(e)** Hence find the least value of $n$ such that $f^{(n)}(5) > 1\,000\,000$

8. Let $f(x) = 3x - 2$

   **(a)** Find $(f \circ f)(x)$

   **(b)** Find $(f \circ f \circ f)(x)$

   **(c)** Find $f^{(4)}(x)$, where $f^{(4)} = f \circ f \circ f \circ f$

   **(d)** Use your answers to write down an expression for $f^{(n)}(x)$

   **(e)** Hence solve $f^{(5)}(x) = f^{(7)}(2)$

9. A company produces fabric to sell to clothing manufacturers. One of their knitting machines produces 2 metres of fabric every 5 minutes. After 2 hours of continuous use, the machine requires stopping for 10 minutes of cleaning. Company staff work in shifts to operate the machine 24 hours a day, 7 days a week.

   (a) Show that a function to model the length, $L$ m, of fabric produced by the machine is given by $L(t) = \dfrac{288}{13} t$, where $t$ is the time in hours.

   The company sells the fabric at \$12 per metre. Each sale incurs an administration fee. The company has found the income from sales, \$S, in term of $L$, can be modelled by the function $S(L) = 12L + 10\sqrt{L}$. The company sells all the fabric produced by the machine.

   (b) Find a function to model the income from sales, \$S, in term of $t$.

   The company believes there is demand for greater sales and considers investing in a faster machine that can produce 3 metres of fabric every 5 minutes. This machine also requires stopping for 10 minutes of cleaning after 2 hours of continuous use.

   (c) Assuming all the fabric is sold, show that a function to model the income from sales from this new machine is given by
   $$S_2(t) = \dfrac{5184}{13} t + 10\sqrt{\dfrac{432}{13} t}$$

   (d) Find a function to model the difference, $D(t)$, in sales between the two machines.

   The company decides it will only invest in the new machine if it can recover the cost of the machine through the difference in sales over a one-year period.

   (e) Find the greatest amount the company is willing to invest in the new machine.

# 2.4 Inverse functions

## Pairs of inverse functions

If we choose a number and cube it (raise it to the power of 3), and then take the cube root of the result, the answer is the original number. The same result would occur if we applied the two rules in the reverse order. That is, first take the cube root of a number and then cube the result – and again the answer is the original number. Let's write each of these rules as a function with function notation. Write the cubing function as $f(x) = x^3$, and the cube root function as $g(x) = \sqrt[3]{x}$

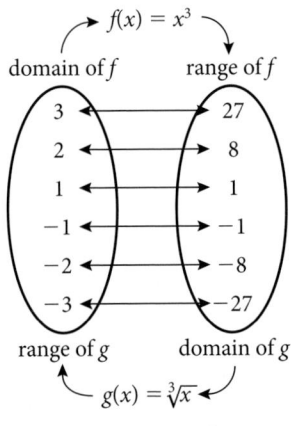

**Figure 2.13** Mapping diagram for the cubing and cube root functions

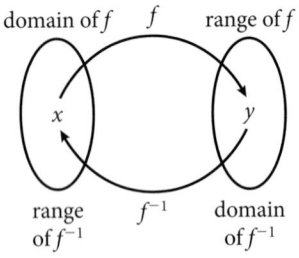

**Figure 2.14** Mapping diagram for a pair of inverse functions $f(x) = y$ and $f^{-1}(y) = x$

Now using what we know about composite functions and operations with radicals and exponents we can write what was described above in symbolic form.

Cube a number and then take the cube root of the result:

$$g(f(x)) = \sqrt[3]{x^3} = (x^3)^{\frac{1}{3}} = x^1 = x$$

For example, $g(f(-2)) = \sqrt[3]{(-2)^3} = \sqrt[3]{-8} = -2$

Take the cube root of a number and then cube the result:

$$f(g(x)) = (\sqrt[3]{x})^3 = (x^{\frac{1}{3}})^3 = x^1 = x$$

For example, $f(g(27)) = (\sqrt[3]{27})^3 = 3^3 = 27$

Because function $g$ has this reverse (inverse) effect on function $f$, we call function $g$ the inverse of function $f$. Function $f$ has the same inverse effect on function $g$ $[g(27) = 3$ and then $f(3) = 27]$, making $f$ the inverse function of $g$. The functions $f$ and $g$ are inverses of each other. The cubing and cube root functions are an example of a pair of inverse functions. The mapping diagram for functions $f$ and $g$ in Figure 2.13 illustrates the relationship for a pair of inverse functions where the domain of one is the range for the other.

The composite of two inverse functions is the function that always produces the same number that was first substituted into the function. This function is called the identity function because it assigns each number in its domain to itself, and is denoted by $I(x) = x$

If $f$ and $g$ are two functions such that $(f \circ g)(x) = x$ for every $x$ in the domain of $g$ and $(g \circ f)(x) = x$ for every $x$ in the domain of $f$, then the function $g$ is the inverse of the function $f$. The notation to indicate the function that is the inverse of function $f$ is $f^{-1}$. Therefore,

$$(f \circ f^{-1})(x) = x \text{ and } (f^{-1} \circ f)(x) = x$$

The domain of $f$ must be equal to the range of $f^{-1}$, and the range of $f$ must be equal to the domain of $f^{-1}$.

Remember that the notation $(f \circ g)(x)$ is equivalent to $f(g(x))$

It follows from the definition that if $g$ is the inverse of $f$, then it must also be true that $f$ is the inverse of $g$.

Do not mistake the $-1$ in the notation $f^{-1}$ for an exponent. It is not an exponent. If a superscript of $-1$ is applied to the name of a function, as in $f^{-1}$ or $\sin^{-1}$, then it denotes the function that is the inverse of the named function (e.g. $f$ or $\sin$). If a superscript of $-1$ is applied to an expression, as in $7^{-1}$ or $(2x + 5)^{-1}$, then it is an exponent and denotes the reciprocal of the expression.

For a pair of inverse functions, $f$ and $g$, the composite functions $f(g(x))$ and $g(f(x))$ are equal, a relationship that we learned in Section 2.3 is not generally true for an arbitrary pair of functions.

In general, the functions $f(x)$ and $g(x)$ are a pair of inverse functions if the following two statements are true:

$$g(f(x)) = x \text{ for all } x \text{ in the domain of } f$$

$$f(g(x)) = x \text{ for all } x \text{ in the domain of } g$$

## Example 2.8

Given $h(x) = \dfrac{x-3}{2}$ and $p(x) = 2x + 3$, show that $h$ and $p$ are inverse functions.

### Solution

Since the domain and range of both $h(x)$ and $p(x)$ is the set of all real numbers, then:

for any real number $x$, $p(h(x)) = p\left(\dfrac{x-3}{2}\right) = 2\left(\dfrac{x-3}{2}\right) + 3 = x - 3 + 3 = x$

for any real number $x$, $h(p(x)) = h(2x + 3) = \dfrac{(2x + 3) - 3}{2} = \dfrac{2x}{2} = x$

since $p(h(x)) = h(p(x)) = x$ then $h$ and $p$ are a pair of inverse functions.

Returning to our initial example, it is clear that both $f(x) = x^3$ and $g(x) = \sqrt[3]{x}$ satisfy the definition of a function because for both $f$ and $g$ every number in its domain determines exactly one number in its range. Since they are a pair of inverse functions then the reverse is also true for both, that every number in its range is determined by exactly one number in its domain. Such a function is called a one-to-one function. The phrase 'one-to-one' is appropriate because each value in the domain corresponds to exactly one value in the range, and each value in the range corresponds to exactly one value in the domain.

The mapping diagram for $f$ and $g$ in Figure 2.13 illustrates this one-to-one correspondence between the domain and range for each function.

A function is one-to-one if each element $y$ in the range is the image of exactly one element $x$ in the domain.

## The existence of an inverse function

Determining whether a function is one-to-one is very useful because the inverse of a one-to-one function will also be a function. Analysing the graph of a function is the most effective way to determine whether a function is one-to-one. Look at the graph of the one-to-one function $f(x) = x^3$ (Figure 2.15). It is clear that as the value of $x$ increases over the domain (i.e. from $-\infty$ to $\infty$), the function values are always increasing. A function that is always increasing, or always decreasing, throughout its domain is one-to-one and has an inverse function.

The function $f(x) = x^2$ is not one-to-one for all real numbers as can be seen in Figure 2.16. However, the function $g(x) = x^2$ with domain $x \leqslant 0$ is always decreasing (one-to-one) and the function $h(x) = x^2$ with domain $x \geqslant 0$ is always increasing (one-to-one), as can be seen in figures 2.17 and 2.18.

This demonstrates that a function that is not one-to-one (always increasing or always decreasing) can be made so by restricting its domain.

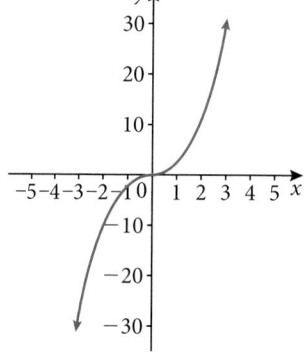

**Figure 2.15** Graph of $f(x) = x^3$ which increases as $x$ goes from $-\infty$ to $\infty$

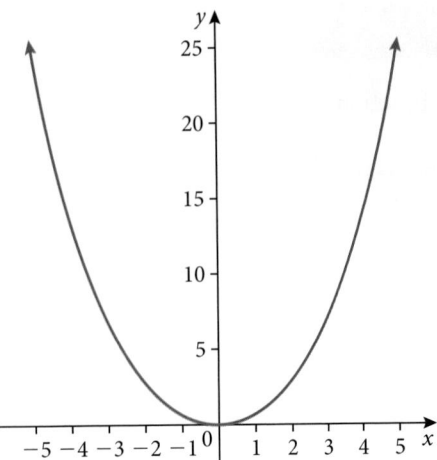

**Figure 2.16** The graph of $f(x) = x^2$

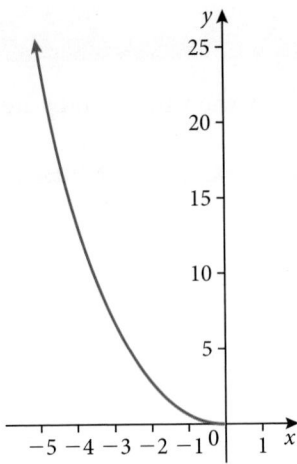

**Figure 2.17** The graph of $g(x) = x^2$ with domain $x \leqslant 0$

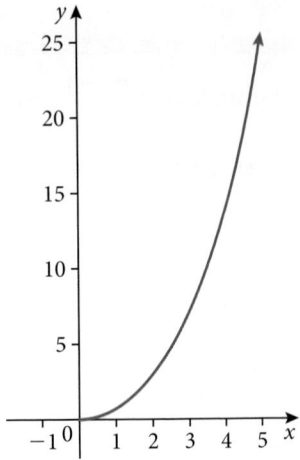

**Figure 2.18** The graph of $h(x) = x^2$ with domain $x \geqslant 0$

No horizontal line can pass through the graph of a one-to-one function at more than one point.

A function for which at least one element $y$ in the range is the image of more than one element $x$ in the domain is called a many-to-one function. Examples of many-to-one functions that we have already encountered are $y = x^2$, $x \in \mathbb{R}$ and $y = |x|$, $x \in \mathbb{R}$

As Figure 2.19 illustrates for $y = |x|$, a horizontal line exists that intersects a many-to-one function at more than one point. Thus, the inverse of a many-to-one function will not be a function.

**Figure 2.19** Graph of $y = |x|$; an example of a many-to-one function

## Finding the inverse of a function

If a function $f$ is always increasing or always decreasing in its domain (i.e. it is monotonic), then $f$ has an inverse $f^{-1}$.

### Example 2.9

The function $f$ is defined for $x \in \mathbb{R}$ by $f(x) = 4x - 8$

(a) Find an expression for the inverse function, $f^{-1}$

(b) Verify the result by showing that $(f \circ f^{-1})(x) = x$ and $(f^{-1} \circ f)(x) = x$

(c) Graph $f$ and its inverse function $f^{-1}$ on the same set of axes.

### Solution

(a) To find the equation for $f^{-1}$, start by switching the domain $(x)$ and range $(y)$ since the domain of $f$ becomes the range of $f^{-1}$ and the range of $f$ becomes the domain of $f^{-1}$. Also, recall that $y = f(x)$

$$f(x) = 4x - 8$$

$y = 4x - 8$      write $y = f(x)$

$x = 4y - 8$      interchange $x$ and $y$ (i.e. switching the domain and range)

$4y = x + 8$      solve for $y$ (dependent variable) in terms of $x$ (independent variable)

$y = \dfrac{x}{4} + 2$

$f^{-1}(x) = \dfrac{x}{4} + 2$      resulting equation is $y = f^{-1}(x)$

(b) Verify that $f$ and $f^{-1}$ are inverses by showing that:

$$f(f^{-1}(x)) = x \text{ and } f^{-1}(f(x)) = x$$

$$f\left(\frac{1}{4}x + 2\right) = 4\left(\frac{1}{4}x + 2\right) - 8 = x + 8 - 8 = x$$

$$f^{-1}(4x - 8) = \frac{1}{4}(4x - 8) + 2 = x - 2 + 2 = x$$

This confirms that $y = 4x - 8$ and $y = \dfrac{1}{4}x + 2$ are inverses of each other.

(c)

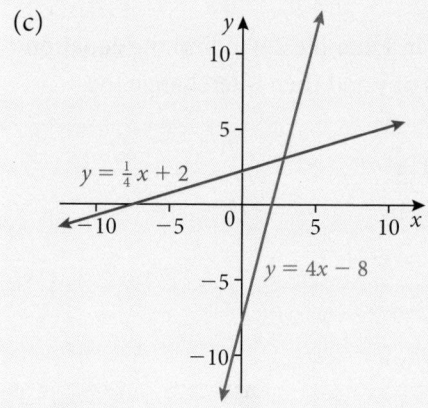

The method of interchanging domain ($x$) and range ($y$) to find the inverse function used in Example 2.9 also gives us a way for obtaining the graph of $f^{-1}$ from the graph of $f$. Given the reversing effect that a pair of inverse functions have on each other, if $f(a) = b$ then $f^{-1}(b) = a$. Hence, if the ordered pair $(a, b)$ is a point on the graph of $y = f(x)$, then the 'reversed' ordered pair $(b, a)$ must be on the graph of $y = f^{-1}(x)$. Figure 2.20 shows that the point $(b, a)$ can be found by reflecting the point $(a, b)$ about the line $y = x$. Therefore, as the figure illustrates, the statement on the right can be made about the graphs of a pair of inverse functions.

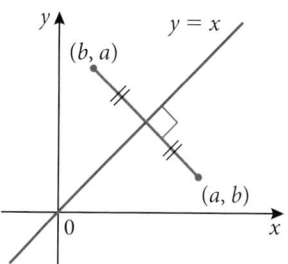

**Figure 2.20** The point $(b, a)$ is a reflection over the line $y = x$ of the point $(a, b)$

The graph of $f^{-1}$ is a reflection of the graph of $f$ about the line $y = x$

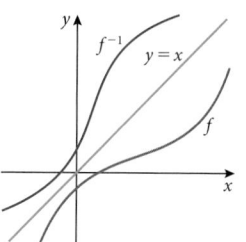

### Example 2.10

The function $f$ is defined for $x \in \mathbb{R}$ by $f(x) = \dfrac{x^2 + 3}{x^2 + 1}$

(a) Explain why $f$ does not have an inverse $f^{-1}$

The function $g$ is defined for $x \in \mathbb{R}$ by $g(x) = \dfrac{x^2 + 3}{x^2 + 1}, \, x \leqslant a$

(b) Find the greatest value of $a$ in order for the inverse function $g^{-1}$ to exist.

(c) Using the value of $a$ found in part (b),
  (i) find an expression for $g^{-1}$
  (ii) sketch graphs of $g$ and its inverse $g^{-1}$ on the same set of axes.

### Solution

(a) A graph of $f$ produced on a GDC reveals that it is not monotonic over its domain $]-\infty, \infty[$. It is increasing for $]-\infty, 0[$, and decreasing for $]0, \infty[$. Therefore, $f$ does not have an inverse $f^{-1}$ for $x \in \mathbb{R}$

(b) Using the graph, since $g$ is increasing for $]-\infty, 0[$, the greatest value of $a$ must be 0.

(c) (i) We use a method similar to that in Example 2.9 to find the equation for $g^{-1}$. First solve for $x$ in terms of $y$ and then interchange the domain $(x)$ and range $(y)$.

$$g(x) = \frac{x^2 + 3}{x^2 + 1} \Rightarrow y = \frac{x^2 + 3}{x^2 + 1}$$

$$\Rightarrow x^2 y + y = x^2 + 3$$

$$\Rightarrow x^2 y - x^2 = 3 - y$$

$$\Rightarrow x^2 = \frac{3 - y}{y - 1}$$

$$\Rightarrow x = \pm\sqrt{\frac{3 - y}{y - 1}} \Rightarrow y = \pm\sqrt{\frac{3 - x}{x - 1}}$$

Since the domain of $g$ is $x \in \, ]-\infty, 0[$, then the range of $g^{-1}$ will be $y \in \, ]-\infty, 0[$. Therefore, the inverse function is $g^{-1}(x) = -\sqrt{\dfrac{3 - x}{x - 1}}$

(ii) See GDC screen on the left.

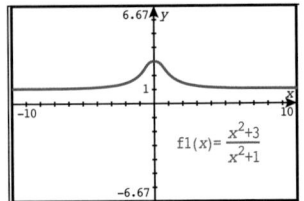

**Figure 2.21a** GDC screen for solution to Example 2.10 (a)

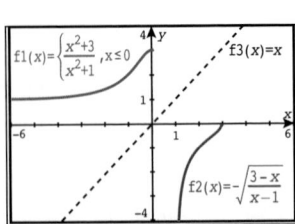

**Figure 2.21b** GDC screen for solution to Example 2.10 (c) (ii)

To find the inverse of a function $f$, use the following steps:

- Determine if the function is one-to-one; if not, restrict the domain so that it is.
- Replace $f(x)$ with $y$.
- Solve for $x$ in terms of $y$.
- Interchange $x$ and $y$.
- Replace $y$ with $f^{-1}(x)$.
- The domain of $f^{-1}$ is equal to the range of $f$; and the range of $f^{-1}$ is equal to the domain of $f$.

## Example 2.11

Consider the function $f(x) = \sqrt{x + 3}, x \geqslant -3$

(a) Determine the inverse function $f^{-1}$

(b) Find the domain of $f^{-1}$

### Solution

(a) Following the steps for finding the inverse of a function gives

$\quad\quad y = \sqrt{x + 3}$ $\quad\quad$ replace $f(x)$ with $y$

$\quad\quad y^2 = x + 3$ $\quad\quad$ solve for $x$ in terms of $y$; squaring both sides here

$\quad\quad x = y^2 - 3$ $\quad\quad$ solved for $x$

$\quad\quad y = x^2 - 3$ $\quad\quad$ interchange $x$ and $y$

$\quad\quad$ Therefore $f^{-1}(x) = x^2 - 3$ $\quad\quad$ replace $y$ with $f^{-1}(x)$

(b) The domain explicitly defined for $f$ is $x \geqslant -3$ and since the $\sqrt{\phantom{x}}$ symbol stands for the principal square root (positive), then the range of $f$ is all positive real numbers, i.e. $y \geqslant 0$. The domain of $f^{-1}$ is equal to the range of $f$, therefore the domain of $f^{-1}$ is $x \geqslant 0$

## Example 2.12

Consider the functions $f(x) = 2(x + 4)$ and $g(x) = \dfrac{1 - x}{3}$

(a) Find $g^{-1}$ and state its domain and range.

(b) Solve the equation $(f \circ g^{-1})(x) = 2$

### Solution

(a) $\quad\quad\quad y = \dfrac{1 - x}{3}$ $\quad\quad$ replace $g(x)$ with $y$

$\quad\quad\quad x = \dfrac{1 - y}{3}$ $\quad\quad$ interchange $x$ and $y$

$\quad\quad\quad 3x = 1 - y$ $\quad\quad$ solve for $y$

$\quad\quad\quad y = 1 - 3x$ $\quad\quad$ solved for $y$

$\quad\quad$ Therefore, $g^{-1}(x) = 1 - 3x$ $\quad\quad$ replace $y$ with $g^{-1}(x)$

$g$ is a linear function with domain $x \in \mathbb{R}$ and range $y \in \mathbb{R}$, therefore, for $g^{-1}$ the domain is $x \in \mathbb{R}$ and the range is $y \in \mathbb{R}$

(b) Find $(f \circ g^{-1})(x)$ first

$$(f \circ g^{-1})(x) = f(g^{-1}(x)) = f(1 - 3x) = 2(5 - 3x)$$

Then solve $(f \circ g^{-1})(x) = 2$

$$2(5 - 3x) = 2$$
$$5 - 3x = 1$$
$$3x = 4$$
$$x = \frac{4}{3}$$

## Exercise 2.4

1. Given that $f$ is a one-to-one function and:

   (a) $f(2) = -5$, write down $f^{-1}(-5)$

   (b) $f^{-1}(6) = 10$, write down $f(10)$

   (c) $f(-1) = 13$, write down $f^{-1}(13)$

   (d) $f^{-1}(b) = a$, write down $f(a)$

2. Given $g(x) = 3x - 7$, find $g^{-1}(5)$

3. Given $h(x) = x^2 - 8x, x \geqslant 4$, find $h^{-1}(-12)$

4. Find the inverse and domain of each of these functions.

   (a) $f(x) = 2x - 3$      (b) $f(x) = \dfrac{x + 7}{4}$

   (c) $f(x) = \sqrt{x}, x \geqslant 0$      (d) $f(x) = \dfrac{1}{x + 2}, x \neq -2$

   (e) $f(x) = 4 - x^2, x \geqslant 0$      (f) $f(x) = \sqrt{x - 5}, x \geqslant 5$

   (g) $f(x) = ax + b, a \neq 0$      (h) $f(x) = x^2 + 2x, x \leqslant -1$

   (i) $f(x) = \dfrac{x^2 - 1}{x^2 + 1}, x \leqslant 0$

5. Given $f(x) = x^3 + 1$

   (a) Find $f^{-1}$.

   (b) Sketch the graphs of $f$ and $f^{-1}$ on the same axes.

   (c) Write down the line of symmetry of the two graphs.

**6.** The graph of $y = f(x)$ is given. Sketch the graph of $f^{-1}$, labelling the new coordinates of any features labelled on the graph of $f$.

**(a)**

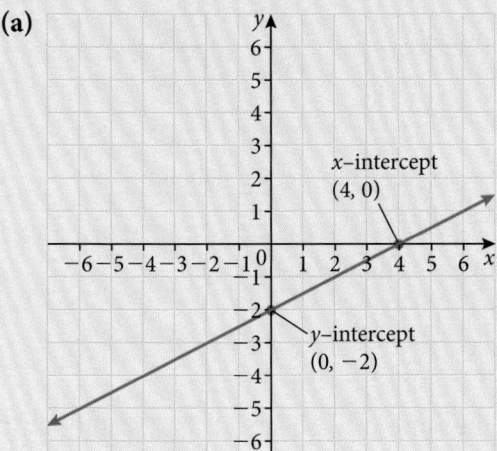

x-intercept (4, 0)

y-intercept (0, −2)

**(b)**

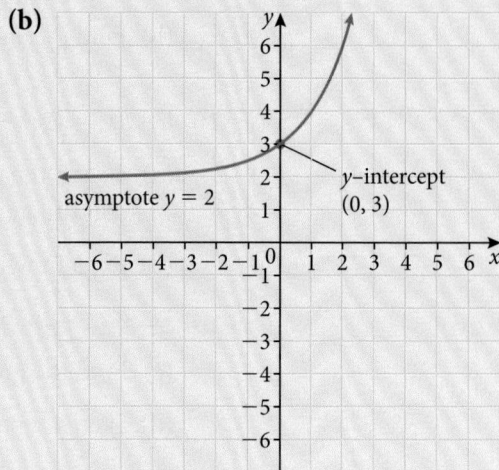

asymptote $y = 2$

y-intercept (0, 3)

**(c)**

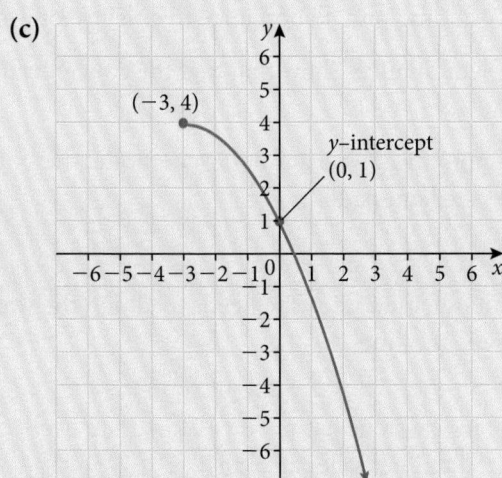

(−3, 4)

y-intercept (0, 1)

**7.** The graphs of functions $f$ and $g$ are shown.

Use the graphs to find:

**(a)** $g^{-1}(0)$

**(b)** $f^{-1}(4)$

**(c)** $(f \circ g^{-1})(1)$

**(d)** $(g^{-1} \circ f^{-1})(3)$

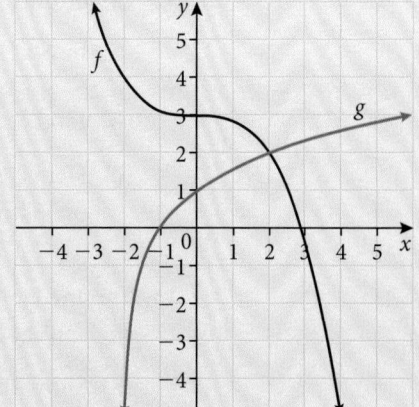

**8.** The functions $f$ and $g$ are defined by

$$f(x) = x^2 + 3, x \in \mathbb{R} \text{ and } g(x) = 2x + 1, x \in \mathbb{R}$$

**(a)** Find the function $f \circ g$ and state its range.

**(b)** Solve the inequality $f(x) < 14g^{-1}(x)$

9. The function $f(x) = 5\log_3(x + 2) - 4$, $x > a$

   (a) Write down the least possible value of $a$.

   (b) Find $f^{-1}$ and state its range.

10. Let $f(x) = 2x^3 - 3x^2 - 5x + 1$

    (a) Sketch the graph of $y = f(x)$ for $-4 \leqslant x \leqslant 4$, labelling the coordinates of any axes intercepts.

    (b) Explain why $f$ does not have an inverse function.

    Let $g(x) = 2x^3 - 3x^2 - 5x + 1$, $x \leqslant a$

    (c) Find the greatest value of $a$ for which $g^{-1}$ exists.

    (d) Sketch $y = g^{-1}(x)$ on the same axes as part (a).

11. Teresa makes lemonade by adding pure lemon juice to one litre of water.

    (a) Teresa adds 150 m$l$ of lemon juice. Show that that the concentration of the lemonade is 13.0%.

    (b) Teresa adds $x$ ml of lemon juice. Find a function to model the concentration, $C(x)$, of the lemonade.

    (c) Find an expression for $C^{-1}(x)$.

    (d) Hence find the amount of lemon juice she needs to add to the water to make a 40% lemonade mixture.

12. Giles participates in a race in which he has to run 10 km, then cycle 30 km. He runs at 20 km h$^{-1}$ and he cycles at $x$ km h$^{-1}$.

    (a) Show that a function to model his average speed, $S$ km h$^{-1}$, is given by $S(x) = \dfrac{80x}{x + 60}$

    (b) Find $S^{-1}(x)$

    (c) Hence find the speed he needs to cycle to achieve an average speed of 35 km h$^{-1}$.

13. Yvonne is training for a race. She runs at 15 km h$^{-1}$ for 20 minutes, then she runs at 22 km h$^{-1}$ for $t$ minutes.

    (a) Show that a function to model her average speed, $S$ km h$^{-1}$, is given by $S(t) = \dfrac{22t + 300}{t + 20}$

    (b) Find $S^{-1}(t)$

    (c) Hence find the value of $t$ if Yvonne's average speed is 20 km h$^{-1}$.

**14.** A model for the distance, $d$ km, to the horizon is given by $d(h) = \sqrt{h(h + 2r)}$, where $h > 0$ is the height in km above sea level and $r = 6371$ km is the radius of the earth.

   **(a)** Mount Everest is 8848 m high. Find the distance to the horizon from the summit.

   **(b)** Find $d^{-1}(h)$, stating its domain.

   **(c)** Hence, or otherwise, find the height above sea level when the distance to the horizon is 200 km.

## 2.5 Transformations of graphs

Even when you use your GDC to sketch the graph of a function, it is helpful to know what to expect in terms of the location and shape of the graph. In this section, we look at how certain changes to the equation of a function can affect, or transform, the location and shape of its graph. We will investigate three different types of transformations of functions that include how the graph of a function can be translated, reflected and stretched. Studying graphical transformations will help us to sketch and visualise many different functions efficiently.

Consider the functions $f_1(x) = x^2$, $f_2(x) = x^2 + 2$ and $f_3(x) = x^2 - 3$. We want to see how the graphs of $y = f_2(x)$ and $y = f_3(x)$ compare to the original function $y = f_1(x)$

We can sketch the graphs of the original and transformed functions, as shown in Figure 2.22.

All three graphs appear to have the same shape, but a different location.

The graph of $y = x^2 + k$ is found by transforming the graph of $y = x^2$ by a translation $\begin{pmatrix} 0 \\ k \end{pmatrix}$

In the column vector $\begin{pmatrix} a \\ b \end{pmatrix}$, the top number $a$ represents a horizontal translation of $a$ units and the bottom number $b$ represents a vertical translation of $b$ units.

Now consider the functions $f_2(x) = (x - 2)^2$ and $f_3(x) = (x + 3)^2$

We want to see how the graphs of $y = f_2(x)$ and $y = f_3(x)$ compare to the original function $y = f_1(x)$.

We can sketch the graphs of the original and transformed functions, as shown in Figure 2.23.

All three graphs appear to have the same shape, but a different location.

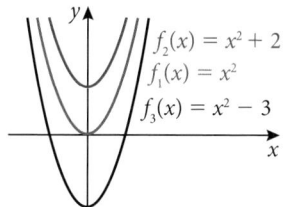

Figure 2.22 Graphs of the original and transformed functions

**Vertical translation of a function.**

The graph of $y = f(x) + k$ is obtained by translating the graph of $y = f(x)$ by $\begin{pmatrix} 0 \\ k \end{pmatrix}$

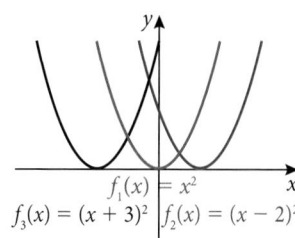

Figure 2.23 Graphs of the original and transformed functions

45

**Horizontal translation of a function**

The graph of $y = f(x - h)$ is obtained by translating the graph of $y = f(x)$ by $\begin{pmatrix} h \\ 0 \end{pmatrix}$

The graph of $y = (x - h)^2$ is found by transforming the graph of $y = x^2$ by a translation of $\begin{pmatrix} h \\ 0 \end{pmatrix}$

For the graph $y = f_2(x)$, $h = 2$ and the graph moved to the right.

For the graph $y = f_3(x)$, $h = -3$ and the graph moved to the left.

Now consider the function $f_2(x) = -x^2$

What is the transformation of the original function, $f_1(x) = x^2$?

We can sketch the graphs of the original and transformed functions, as shown in Figure 2.24.

Note that the point on the $x$-axis has not changed (i.e. it is invariant.)

**Reflection of a function in the $x$-axis**

The graph of $y = -f(x)$ is obtained by reflecting the graph of $y = f(x)$ in the $x$-axis.

The graph of $y = -x^2$ is found by transforming the graph of $y = x^2$ by a reflection in the $x$-axis.

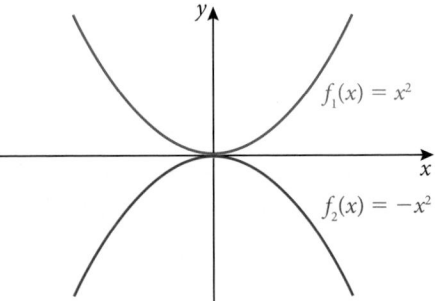

**Figure 2.24** Graphs of the original and transformed functions

Now consider the function $f_1(x) = x^3 + 2x^2 - 8x$ as the original function to be transformed. We will look at related curves of the form $y = a(x^3 + 2x^2 - 8x)$ where $a$ can take any real value. What is the transformation of the original $y = x^3 + 2x^2 - 8x$ curve when the value of $a$ is altered?

We can sketch the graphs of the original and transformed functions, as shown in Figure 2.25.

Note that the points on the $x$-axis have not changed (i.e. they are invariant.)

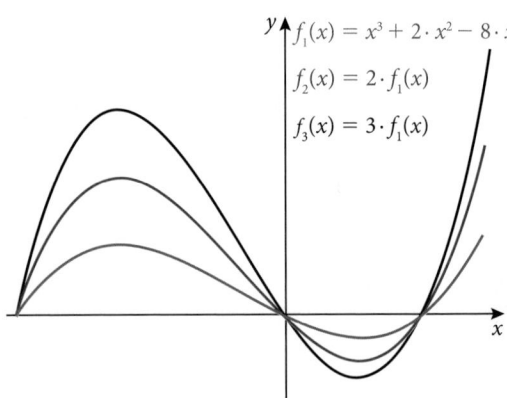

$f_1(x) = x^3 + 2 \cdot x^2 - 8 \cdot x$
$f_2(x) = 2 \cdot f_1(x)$
$f_3(x) = 3 \cdot f_1(x)$

**Vertical stretch of a function**

The graph of $y = a \cdot f(x)$ is obtained by stretching the graph of $y = f(x)$ by a scale factor of $a$, with the $x$-axis invariant.

The graph of $y = a(x^3 + 2x^2 - 8x)$ is found by transforming the graph of $y = f_1(x)$ by a stretch with scale factor $a$ and $x$-axis invariant.

**Figure 2.25** Graphs of the original and transformed functions

We continue to use the function $f_1(x) = x^3 + 2x^2 - 8x$ as the original function. We will now look at related curves of the form $y = (bx)^3 + 2(bx)^2 - 8(bx)$ where $b$ can take any real value. What is the transformation of the original $y = x^3 + 2x^2 - 8x$ curve when the value of $b$ is altered?

We can sketch the graphs of the original and transformed functions, as shown in Figure 2.26.

Points on the $x$-axis have moved, except for the origin, which is also a point on the $y$-axis. So the $y$-axis is invariant.

Points on the graph $y = f_1(2x)$ have moved parallel to the $x$-axis and are now half the distance from the $y$-axis.

Points on the graph $y = f_1(3x)$ have moved parallel to the $x$-axis and are now one third the distance from the $y$-axis.

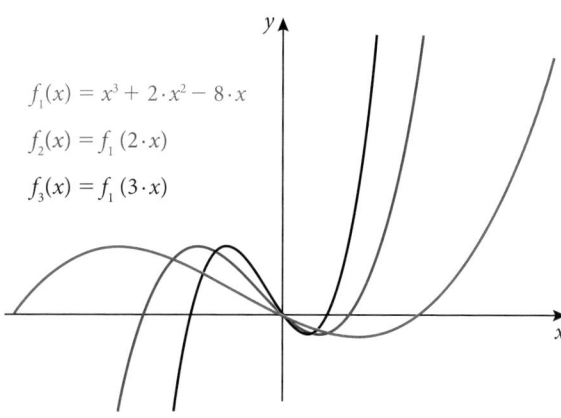

**Figure 2.26** Graphs of the original and transformed functions

The graph of $y = (bx)^2 - 2(bx) - 3$ is found by transforming the graph of $y = f_1(x)$ by a stretch with scale factor $\dfrac{1}{b}$ and $y$-axis invariant.

We continue to use the function $f_1(x) = x^3 + 2x^2 - 8x$ as the original function. We will look at the related curve $y = (-x)^3 + 2(-x)^2 - 8(-x)$

What is the transformation of the original $y = x^3 + 2x^2 - 8x$ curve?

We can sketch the graphs of the original and transformed functions, as shown in Figure 2.27.

Points on the $x$-axis have changed sign, except for the origin, which is unchanged.

The graph of $y = f(-x)$ is found by transforming the graph of $y = f(x)$ by a reflection in the $y$-axis.

It is possible to combine any of the above transformations to get a composite transformation of a graph.

**Horizontal stretch of a function**

The graph of $y = f(bx)$ is obtained by stretching the graph of $y = f(x)$ by a scale factor of $\dfrac{1}{b}$, with the $y$-axis invariant.

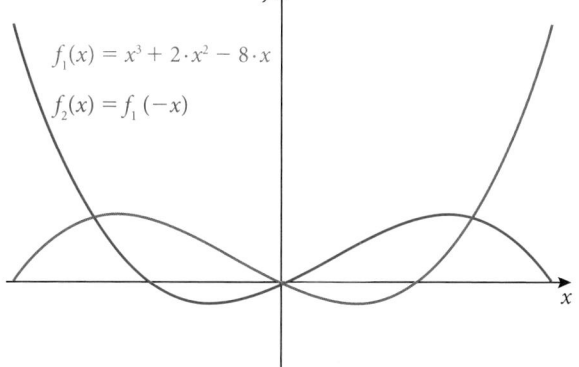

**Figure 2.27** Graphs of the original and transformed functions

**Reflection of a function in the $y$-axis**

The graph of $y = f(-x)$ is obtained by reflecting the graph of $y = f(x)$ in the $y$-axis.

### Example 2.13

The function $f(x) = x^2 - 3x$

(a) The graph of $y = g(x)$ is obtained from the graph of $y = f(x)$ by a translation of $\begin{pmatrix} 2 \\ -3 \end{pmatrix}$. Find an expression for the function $g(x)$.

(b) The graph of $y = h(x)$ is obtained from the graph of $y = f(x)$ by a stretch of scale factor 2, with the $y$-axis invariant. Find an expression for the function $h(x)$.

(c) The graph of $y = j(x)$ is obtained from the graph of $y = f(x)$ by a stretch of scale factor 2, with the $x$-axis invariant and a reflection in the $y$-axis. Find an expression for the function $j(x)$.

## Solution

(a) We can think of the translation $\begin{pmatrix} 2 \\ -3 \end{pmatrix} = \begin{pmatrix} 2 \\ 0 \end{pmatrix} + \begin{pmatrix} 0 \\ -3 \end{pmatrix}$

The first part $\begin{pmatrix} 2 \\ 0 \end{pmatrix}$ corresponds to a translation 2 units to the right, so we want $f(x - 2)$

The second part $\begin{pmatrix} 0 \\ -3 \end{pmatrix}$ corresponds to a translation 3 units down, so we want $f(x) - 3$

Combining these gives

$$g(x) = f(x - 2) - 3 = (x - 2)^2 - 3(x - 2) - 3$$

Expanding the brackets and simplifying gives

$$g(x) = x^2 - 7x + 7$$

The answer can be confirmed by graphing and checking that the second function is a translation of the first by $\begin{pmatrix} 2 \\ -3 \end{pmatrix}$

(b) A stretch of scale factor 2, with the $y$-axis invariant can be represented as $f\left(\frac{1}{2}x\right)$. So $h(x) = f\left(\frac{1}{2}x\right) = f\left(\frac{x}{2}\right) = \left(\frac{x}{2}\right)^2 - 3\left(\frac{x}{2}\right) = \frac{x^2}{4} - \frac{3x}{2}$

The answer can be confirmed by graphing and checking that the second function is a stretch scale factor 2, with $y$-axis invariant.

(c) A stretch of scale factor 2, with the $x$-axis invariant can be represented as $2f(x)$

A reflection in the $y$-axis can be represented as $f(-x)$

Combining these gives

$$j(x) = 2f(-x) = 2((-x)^2 - 3(-x)) = 2(x^2 + 3x) = 2x^2 + 6x$$

The answer can be confirmed by graphing and checking that the second function is a stretch of scale factor 2, with the $x$-axis invariant and a reflection in the $y$-axis.

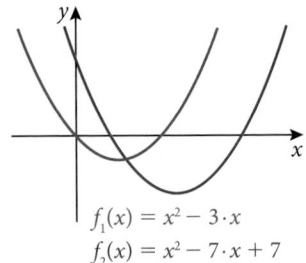

$f_1(x) = x^2 - 3 \cdot x$
$f_2(x) = x^2 - 7 \cdot x + 7$

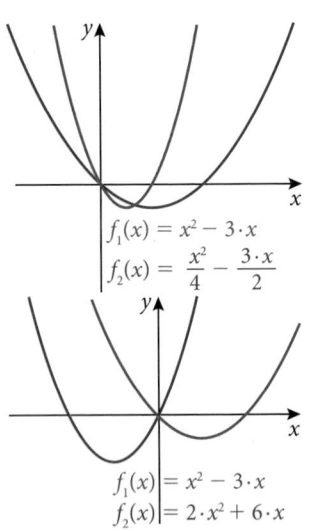

$f_1(x) = x^2 - 3 \cdot x$
$f_2(x) = \dfrac{x^2}{4} - \dfrac{3 \cdot x}{2}$

$f_1(x) = x^2 - 3 \cdot x$
$f_2(x) = 2 \cdot x^2 + 6 \cdot x$

**Figure 2.28** Graphs of the original and transformed functions for Example 2.13

It is often necessary to consider the order of transformations carefully, when finding an equation for a transformed function. For example, let $f(x) = x^2$ and consider the transformations

**A** A reflection in the $y$-axis, followed by a translation $\begin{pmatrix} 2 \\ 0 \end{pmatrix}$

**B** A translation $\begin{pmatrix} 2 \\ 0 \end{pmatrix}$ followed by a reflection in the $y$-axis.

**Transformations A**

- a reflection in the $y$-axis gives $g(x) = f(-x) = (-x)^2 = x^2$

- then a translation of $\begin{pmatrix} 2 \\ 0 \end{pmatrix}$ gives $g(x - 2) = (x - 2)^2$

So the final function is $h(x) = (x - 2)^2$

**Transformations B**

- a translation of $\begin{pmatrix} 2 \\ 0 \end{pmatrix}$ gives $g(x) = f(x - 2) = (x - 2)^2$

- then a reflection in the $y$-axis gives
  $g(-x) = (-x - 2)^2 = (-1)^2 (x + 2)^2 = (x + 2)^2$

So the final function is $h(x) = (x + 2)^2$

Note that the final function for transformations A does not equal the final function for transformations B. Therefore it is important to consider a sequence of transformations in the correct order.

## Exercise 2.5

1. Figure 2.29 shows $y = f(x)$

   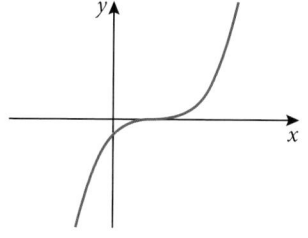

   **Figure 2.29** $y = f(x)$

   (a) Sketch $y = f(x) + 2$
   (b) Sketch $y = f(x + 2)$
   (c) Sketch $y = -f(x)$
   (d) Sketch $y = f(x - 2) - 2$
   (e) Sketch $y = 2f(x)$
   (f) Sketch $y = f(2x)$
   (g) Sketch $y = f(-x)$
   (h) Sketch $y = -2f(x + 2)$

2. The function $f$ is defined by $f(x) = 3x^2 - 2x$

   The graph of $y = f(x)$ is transformed to give the graph of $y = g(x)$

   Find the equation of $y = g(x)$ for each transformation below.
   Give each answer in the form $g(x) = ax^2 + bx + c$

   (a) A stretch of scale factor 4 and the $x$-axis invariant.

   (b) A stretch of scale factor 2 and the $y$-axis invariant.

   (c) A translation by $\begin{pmatrix} -1 \\ 3 \end{pmatrix}$

   (d) A translation by $\begin{pmatrix} 2 \\ -1 \end{pmatrix}$

   (e) A reflection in the $x$-axis.

   (f) A reflection in the $y$-axis.

3. The function $f$ is defined by $f(x) = x^3 - 4x$

   The graph of $y = f(x)$ is transformed by a reflection in the $y$-axis, followed by a translation of $\begin{pmatrix} 2 \\ 0 \end{pmatrix}$. Find the equation of the transformed graph.

4. The function $f$ is defined by $f(x) = x^3 - 4x$

   The graph of $y = f(x)$ is transformed by a reflection in the $x$-axis, followed by a translation of $\begin{pmatrix} 0 \\ 2 \end{pmatrix}$. Find the equation of the transformed graph.

5. In each case below, the function $g$ is obtained from function $f$ by a sequence of transformations. Describe fully the sequence of transformations.

(a) $g(x) = -2f(x + 4)$

(b) $g(x) = f(x - 3) - 2$

(c) $g(x) = f(-2x + 4)$

(d) $g(x) = 2f(x) + 4$

(e) $g(x) = 6 - 3f(x)$

6. David throws a ball. The height, $h$ metres above the ground, of the ball after it has travelled a horizontal distance $d$ metres can be modelled by the function

$$h(d) = 1 + 4d - 2d^2$$

(a) Find the distance travelled by the ball when it reaches the ground.

(b) Find the maximum height reached by the ball.

David throws the ball so that it reaches the same maximum height but travels twice as far.

(c) Find a function for $h(d)$

7. The number of people, $N$, infected with a virus can be modelled by the function $N(t) = \dfrac{200}{1 + 49\,(4)^{-t}}$, where $t \geqslant 0$, is the time in days.

(a) Find the initial number of infected people.

(b) Sketch a graph of $y = N(t)$, for $0 \leqslant t \leqslant 20$

(c) Write down the equation of the horizontal asymptote of the graph.

(d) On the same axes sketch the graphs of

(i) $y = \dfrac{1}{2}N(t)$

(ii) $y = N(2t)$

(iii) $y = 3N\left(\dfrac{t - 4}{2}\right)$

(e) For each function in part (d), write down:

(i) the initial number of infected people.

(ii) the equation of the horizontal asymptote.

(f) Show that $3N\left(\dfrac{t - 4}{2}\right) = \dfrac{a}{1 + b\,(c)^{-t}}$, where $a$, $b$ and $c$ are integers to be found.

8. A government charges income tax, $\$T$, based on income, $\$x$, according to the following model

$$T(x) = \begin{cases} 0 & x \leqslant 6000 \\ \dfrac{(x - 6000)^2}{40\,000} & 6000 < x \leqslant 10\,000 \\ \dfrac{x}{10} - 600 & 10\,000 < x \end{cases}$$

(a) Find the income tax for someone earning an income of:

    **(i)** $8000        **(ii)** $20 000

The government is considering three options to increase income tax.

Option 1: Increase income tax by $200, then by a factor of 1.2

Option 2: Increase income tax using the model $T(x + 1000)$

Option 3: Increase income tax using the model $T(2x + 1000)$

**(b)** Find a function to model the income tax for each option.

**(c)** For each option, find the income tax for someone earning an income of:

    **(i)** $8000        **(ii)** $20 000

9. Rob is investigating the cooling of a cup of hot green tea. He makes a cup and puts it on the table. The temperature, $T°C$, of the tea, at time $t$ minutes, can be modelled by the function $T(t) = 63(0.85)^t + 22$

  **(a)** Write down the initial temperature of the tea.

  **(b)** Sketch a graph of $T$ against $t$ for $0 \leqslant t \leqslant 40$

  **(c)** Find the temperature of the tea after 12 minutes.

  **(d)** Find the temperature of the tea after a long period of time.

Rob makes a second cup of tea, using an insulated cup. He finds that this cup cools half as fast as the first cup.

  **(e)** Find a model for the temperature $T_2$ of this second cup.

  **(f)** Verify that $T_2(24)$ equals $T(12)$

Rob makes a third cup of tea, and places it under a fan. He finds that this cup cools three times faster than the first cup.

  **(g)** Find a model for the temperature $T_3$ of this third cup.

  **(h)** Find how long it takes for this third cup to reach 30°C.

## Chapter 2 practice questions

1. A bank exchanges 600 EUR (Euros) into 2850 MYR (Malaysian Ringgit).

  **(a)** Find a linear function to convert Euros into Malaysian Ringgit.

  **(b)** Kathryn want to exchange 170 EUR. Find the amount of Malaysian Ringgit she receives.

  **(c)** Kenneth wants to exchange 3500 MYR. Find the amount he receives, in Euros.

2. A function $f$ is defined by $f(x) = \dfrac{2x - 5}{3x - 1}, x \in \mathbb{R}, x \neq \dfrac{1}{3}$

Find an expression for $f^{-1}(x)$

3. Karen is training for a race. She runs at $12\,\text{km}\,\text{h}^{-1}$ for 15 minutes, then she runs at $18\,\text{km}\,\text{h}^{-1}$ for $x$ kilometres.

   (a) Show that a function to model her average speed, $S\,\text{km}\,\text{h}^{-1}$, is given by $S(x) = \dfrac{36x + 108}{2x + 9}$

   (b) Find $S^{-1}(x)$

   (c) Find the value of $x$ if Karen's average speed is $16\,\text{km}\,\text{h}^{-1}$.

4. Gary throws a ball. The height, $h$ metres above the ground, of the ball after it has travelled a horizontal distance $d$ metres can be modelled by the function

   $$h(d) = 2 + 3d - \frac{1}{2}d^2$$

   (a) Find the distance travelled by the ball when it reaches the ground.

   (b) Find the maximum height reached by the ball.

   Gary throws the ball so that it reaches the same maximum height but travels half the distance.

   (c) Find a function for $h(d)$

5. The volume of liquid, $v\,\text{cm}^3$ in a glass is a function of the height, $h\,\text{cm}$ of the liquid, given by

   $$v(h) = \frac{2\pi}{3}\left(\frac{1}{4}h^3 - 3h^2 + 15h\right), \text{ where } 0 \leqslant h \leqslant 10$$

   (a) Sketch a graph of $y = v(h)$

   (b) Find the greatest volume of liquid that the glass can contain.

   (c) Find the height of liquid in the glass when it is

       (i) half full        (ii) one-third full.

   Giovanni pours pure orange juice into the glass until it reaches a height of 3 cm. He then pours pure water into the glass.

   (d) Find a function to model the concentration of the orange juice in the glass, for $3 \leqslant h \leqslant 10$

   (e) Find the concentration of the orange juice if Giovanni pours a total volume of $200\,\text{cm}^3$ into the glass.

6. (a) The graph represents a function $y = f(x)$, where $-3 \leqslant x \leqslant 5$
   The function has a maximum at $(3, 1)$ and a minimum at $(-1, -1)$.

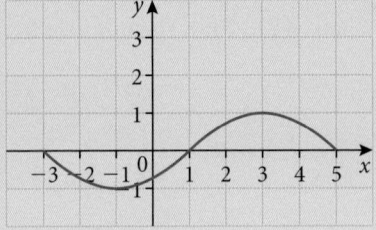

The functions $u$ and $v$ are defined as $u(x) = x - 3$ and $v(x) = 2x$ where $x \in \mathbb{R}$.

   **(i)** State the range of the function $u \circ f$

   **(ii)** State the range of the function $u \circ v \circ f$

   **(iii)** Find the largest possible domain of the function $f \circ v \circ u$

**(b) (i)** Explain why $f$ does not have an inverse.

   **(ii)** The domain of $f$ is restricted to define a function $g$ so that it has an inverse $g^{-1}$. State the largest possible domain of $g$.

   **(iii)** Sketch a graph of $y = g^{-1}(x)$, showing clearly the $y$-intercept and stating the coordinates of the endpoints.

**(c)** Consider the function defined by $h(x) = \dfrac{2x - 5}{x + d}, x \neq -d$ and $d \in \mathbb{R}$

   **(i)** Find an expression for the inverse function $h^{-1}(x)$

   **(ii)** Find the value of $d$ such that $h$ is a self-inverse function.

For this value of $d$, there is a function $k$ such that

$$h \circ k(x) = \frac{2x}{x + 1}, x \neq -1$$

   **(iii)** Find $k(x)$

**7.** The graphs of functions $f$ and $g$ are shown.

   **(a)** Use the graphs to find

      **(i)** $f^{-1}(-3)$

      **(ii)** $(f^{-1} \circ g)(1)$

      **(iii)** $(f^{-1} \circ f^{-1})(-4)$

   **(b)** Explain why $g$ does not have an inverse.

   **(c)** Solve $g(x) = (g \circ g)(3)$

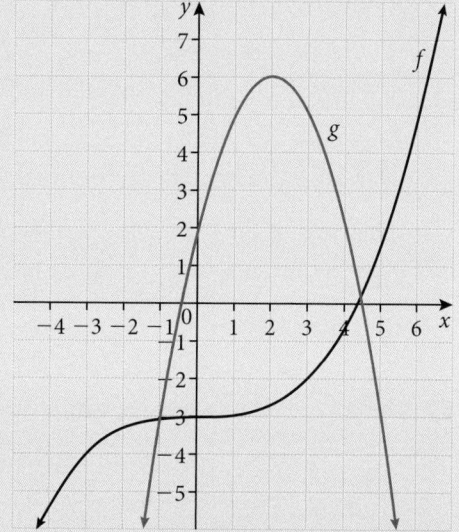

**8.** Let $f(x) = 4x - x^3 + 2x^2 - 3$

   **(a)** Sketch the graph of $y = f(x)$ for $-2 \leqslant x \leqslant 3$, labelling the coordinates of any axes intercepts.

   **(b)** Explain why $f$ does not have an inverse function.

Let $g(x) = 4x - x^3 + 2x^2 - 3, a \leqslant x \leqslant 2$

   **(c)** Find the least value of $a$ for which $g^{-1}$ exists.

   **(d)** Find $g^{-1}(3)$

9. (a) State the set of values of $a$ for which the function $f(x) = \log_a x$ exists, for all $x \in \mathbb{R}^+$

   (b) Given that $\log_x y = 9 \log_y x$, find all the possible expressions of $y$ as a function of $x$.

10. The activity, $A$ becquerel (Bq), of a radioactive substance can be modelled by the function $A(t) = \dfrac{80}{1.2^t}$, $t \geqslant 0$, where $t$ is the time in days.

    (a) Sketch the graph of $A$ against $t$, for $0 \leqslant t \leqslant 25$

    (b) Find the initial activity of the substance.

    (c) Find the time it takes for the activity of the substance to reach 25 Bq

    (d) Write down the equation of the horizontal asymptote of the graph of $A$.

    (e) Describe what happens to the activity of the substance after a long period of time.

    The activity of a second radioactive substance can be modelled by the function $B(t) = \dfrac{50}{1.1^t}$, $t \geqslant 0$, where $t$ is the time in days.

    (f) Find the time when the two substances have equal activity.

    (g) Find the time when $B = 2A$

11. The function $f$ is defined by $f(x) = \begin{cases} 1 - 2x, & x \leqslant 2 \\ \dfrac{3}{4}(x-2)^2 - 3, & x > 2 \end{cases}$

    (a) Determine whether or not $f$ is continuous.

    The graph of the function $g$ is obtained by applying, to the graph of $f$, a reflection in the $y$-axis followed by a translation by the vector $\begin{pmatrix} 2 \\ 0 \end{pmatrix}$

    (b) Find $g(x)$

12. The cost of a taxi ride, $\$C$, can be modelled by

    $$C(d) = \begin{cases} 3 + 0.8d, & 0 \leqslant d \leqslant 10 \\ x + 0.6d, & d > 10 \end{cases}$$

    where $d$ is the distance travelled in kilometres.

    (a) Given that $C$ is a continuous function, find the value of $x$.

    (b) Find the cost of a taxi ride of distance 16 km.

13. A species of rare bird is introduced into a wilderness area in the hope that the bird population will grow. The size of the population, $P$, can be modelled by

    $$P(t) = \dfrac{130}{1 + 4^{3-t}} \text{ where } t \text{ is the time in years.}$$

(a) Sketch the graph of $P$ against $t$, for $0 \leqslant t \leqslant 10$

(b) Find the initial number of birds that are introduced into the wilderness area.

(c) Find the time it takes for the number of birds to reach 50.

(d) Write down the equation of the horizontal asymptote of the graph of $P$.

(e) Describe what happens to the population of birds after a long period of time.

14. The diagram shows a sketch of the graph of $y = f(x)$

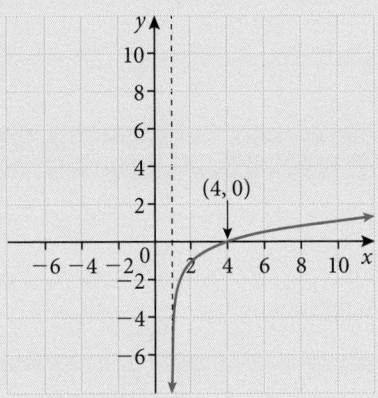

(a) Sketch the graph of $y = f^{-1}(x)$

(b) State the range of $f^{-1}$

(c) Given that $f(x) = \ln(ax + b)$, $x > 1$, find the value of $a$ and the value of $b$.

15. Consider the following functions:

$$f(x) = \frac{2x^2 + 3}{75}, x \geqslant 0 \text{ and } g(x) = \frac{|3x - 4|}{10}, x \in \mathbb{R}$$

(a) State the range of $f$ and of $g$.

(b) Find an expression for the composite function $f \circ g(x)$ in the form $\dfrac{ax^2 + bx + c}{3750}$, where $a$, $b$ and $c \in \mathbb{Z}$

(c) (i) Find an expression for the inverse function $f^{-1}(x)$

   (ii) State the domain and range of $f^{-1}$

16. A company wants to find the minimum cost of building an oil pipeline from point $A$ to point $B$ across two types of soil. The first type of soil is made of clay and the cost of building the pipeline through this type of soil is \$6 million per km. The second type of soil is sandy and the cost of building the pipeline is \$2 million per km. The boundary $XY$ between the two types of soil is a straight line running from east to west and point $A$ lies 50 km directly north of it. Point $B$ is 120 km to the east of point $A$ and 40 km directly south of the boundary. This is shown in the diagram along with a possible route $APB$, where $P$ is $x$ km east of $X$.

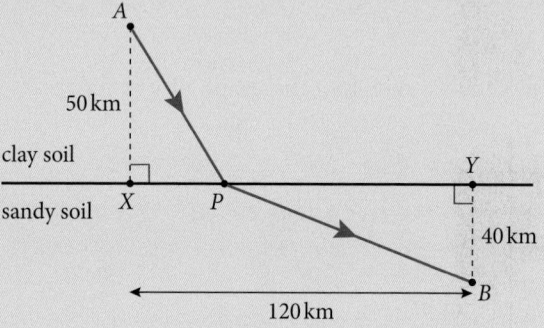

   (a) Find a function to model the cost of the pipeline in terms of $x$.

   (b) Find the minimum cost of building the pipeline.

17. Let $f(x) = \ln x$. The graph of $f$ is transformed into the graph of the function $g$ by a translation of $\begin{pmatrix} 3 \\ -2 \end{pmatrix}$, followed by a reflection in the $x$-axis. Find an expression for $g(x)$, giving your answer as a single logarithm.

18. The function $f$ is defined by $f(x) = 2x^2 - 3x + 1$

   The graph of $y = f(x)$ is transformed by a reflection in the $x$-axis, followed by a translation of $\begin{pmatrix} 1 \\ 2 \end{pmatrix}$, followed by a horizontal stretch with scale factor 2. Find the equation of the transformed graph. Give your answer in the form $f(x) = ax^2 + bx + c$

# Sequences and series

3

# 3 Sequences and series

## Learning objectives

By the end of this chapter, you should be familiar with…
- using the formulae for the $n$th term for arithmetic and geometric sequences
- using the formulae for the sum of $n$ terms for arithmetic and geometric series
- using sigma notation for the sums of sequences
- applications, including simple interest, compound interest, population growth, annual depreciation and modelling
- solving problems involving amortisation and annuities using technology.

The heights of consecutive bounces of a ball, compound interest, and Fibonacci numbers are just a few of the applications of sequences and series that you may be familiar with from previous courses. In this chapter you will review these concepts, consolidate your understanding and take them one step further.

## 3.1 Sequences

Look at the pattern in Figure 3.1

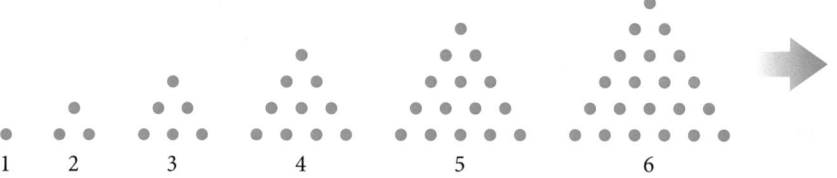

1        2        3        4        5        6

**Figure 3.1** Triangular pattern of dots

The first unit represents one dot, the second represents three dots, etc.
This pattern can be represented as a list of numbers written in a definite order:
$a_1 = 1, a_2 = 3, a_3 = 6, \ldots$

Here are some more examples of sequences:

16, 12, 18, 24, 30

$3, 9, 27, \ldots, 3^k, \ldots$

$\left\{ \dfrac{1}{i^2} ; i = 1, 2, 3, \ldots, 10 \right\}$

$\{b_1, b_2, \ldots, b_n, \ldots\}$, sometimes used with an abbreviation $(b_n)$

The first and third sequences are **finite** and the second and fourth are **infinite**. Note that in the second and third sequences, we are able to define a rule that yields the $n$th number in the sequence (called the $n$th term) as a function of $n$, the term's number. In this sense you can think of a sequence as a **function** that assigns a **unique** number $a_n$ to each positive integer $n$.

The pattern can also be described, for example, in function notation:
$f(1) = 1, f(2) = 3,$
$f(3) = 6,$ etc., where the domain is $\mathbb{Z}^+$

58

## Example 3.1

Find the first five terms and the 50th term of the sequence $(u_n)$ given that
$$u_n = 2 - \frac{1}{n^2}$$

### Solution

Since we know an explicit expression for the $n$th term as a function of its number $n$, we only need to find the value of that function for the required terms:

$$u_1 = 2 - \frac{1}{1^2} = 1; \ u_2 = 2 - \frac{1}{2^2} = \frac{7}{4}; \ u_3 = 2 - \frac{1}{3^2} = \frac{17}{9};$$

$$u_4 = 2 - \frac{1}{4^2} = \frac{31}{16}; \ u_5 = 2 - \frac{1}{5^2} = \frac{49}{25};$$

and $u_{50} = 2 - \frac{1}{50^2} = \frac{4999}{2500}$

So, informally, a sequence is an ordered set of real numbers. That is, there is a first number, a second, and so forth. The way we defined the function in Example 3.1 is called the **explicit** definition of a sequence. Another way of defining a sequence is the **recursive** (or **inductive**) definition. Example 3.2 shows how this is used.

**Notation for the terms of a sequence**
$u_1$ = first term
$u_2$ = second term
.
.
.
$u_n$ = $n$th term

## Example 3.2

Find the first five terms and the 20th term of the sequence $(b_n)$ given that
$b_n = 2(b_{n-1} + 3)$ and $b_1 = 5$

### Solution

The defining formula for this sequence is recursive. It allows us to find the $n$th term $b_n$ if we know the preceding term $b_{n-1}$. So, we can find the second term from the first, the third from the second, and so on. Since we know the first term $b_1 = 5$, we can calculate the the 2nd, 3rd, 4th and 5th terms:

$$b_2 = 2(b_1 + 3) = 2(5 + 3) = 16$$

$$b_3 = 2(b_2 + 3) = 2(16 + 3) = 38$$

$$b_4 = 2(b_3 + 3) = 2(38 + 3) = 82$$

$$b_5 = 2(b_4 + 3) = 2(82 + 3) = 170$$

Recursive sequences are introduced in this chapter to help clarify the underlying concepts. However, note that they do not appear explicitly in this section of the syllabus. One of the main applications of recursive sequences in this course is to solve differential equations using Euler's method, which you will study in chapter 20.

Be aware that not all sequences have formulae, either recursive or explicit. Some sequences are given only by listing their terms. Among the many kinds of sequences that there are, two types are of particular interest to us: arithmetic and geometric sequences, which we will discuss in the next two sections.

So, the first five terms of this sequence are 5, 16, 38, 82, and 170. Although, to find the 20th term, we must first find all 19 preceding terms. This is one of the drawbacks of this type of definition, unless we can change the definition into explicit form. However, we can find a term in a recursive sequence easily using a GDC, as shown below.

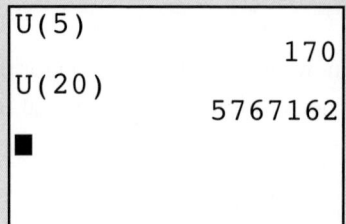

```
Plot1 Plot2 Plot3
 nMin=1
∴U(n)◼2(u(n−1)+3
)
 U(nMin)◼5◼
∴V(n)=
 V(nMin)=
∴W(n)=
```

```
U(5)
                    170
U(20)
              5767162
◼
```

## Exercise 3.1

1. Find the first five terms of each infinite sequence.

   (a) $a_n = 2n - 3$

   (b) $b_n = 2^n - 3$

   (c) $c_n = 3(2)^{-n}$

   (d) $\begin{cases} d_1 = 5 \\ d_n = d_{n-1} + 3, \quad \text{for } n > 1 \end{cases}$

   (e) $e_n = (-1)^n(2^n) + 3$

   (f) $\begin{cases} f_1 = 3 \\ f_n = f_{n-1} + 2n, \quad \text{for } n > 1 \end{cases}$

2. Find the first five terms and the 50th term of each infinite sequence.

   (a) $a_n = 2 - 5n$

   (b) $b_n = 2 \times 3^{n-1}$

   (c) $u_n = (-1)^{n-1}\dfrac{2n}{n^2 + 2}$

   (d) $a_n = n^{n-1}$

   (e) $a_n = 2a_{n-1} + 5$ and $a_1 = 3$

   (f) $u_{n+1} = \dfrac{3}{2u_n + 1}$ and $u_1 = 0$

   (g) $b_n = 3b_{n-1}$ and $b_1 = 2$

   (h) $a_n = a_{n-1} + 2$ and $a_1 = -1$

3. Find a recursive definition for each sequence.

   (a) $\dfrac{1}{3}, \dfrac{1}{12}, \dfrac{1}{48}, \dfrac{1}{192}, \ldots$

   (b) $\dfrac{a}{2}, \dfrac{2a^3}{3}, \dfrac{8a^5}{9}, \dfrac{32a^7}{27}, \ldots$

   (c) $a - 5k, 2a - 4k, 3a - 3k, 4a - 2k, 5a - k, \ldots$

4. Find an explicit formula that gives the $n$th term of each sequence.

   (a) $4, 7, 12, 19, \ldots$

   (b) $2, 5, 8, 11, \ldots$

   (c) $1, \dfrac{3}{4}, \dfrac{5}{9}, \dfrac{7}{16}, \dfrac{9}{25}, \ldots$

   (d) $\dfrac{1}{4}, \dfrac{3}{5}, \dfrac{5}{6}, 1, \dfrac{9}{8}, \ldots$

**5.** The Fibonacci sequence $F_n$ is defined by

$$F_1 = 1, F_2 = 1 \text{ and } F_n = F_{n-1} + F_{n-2}, n > 2$$

A second sequence $a_n$ is defined by

$$a_n = \frac{F_{n+1}}{F_n}, n \geqslant 1,$$

where $F_n$ is a member of the Fibonacci sequence.

**(a)** Use a spreadsheet to find the first 30 terms of both $F_n$ and $a_n$

**(b)** Hence, write down the limit of $a_n$ as $n$ tends to infinity, $\lim\limits_{n\to\infty} a_n$

**6.** The sequence $G_n$ is defined by $G_1 = G_2 = 1$ and $G_n = 2G_{n-1} + G_{n-2}, n > 2$

A second sequence $b_n$ is defined by $b_n = \frac{G_{n+1}}{G_n}, n \geqslant 1$

**(a)** Use a spreadsheet to find the first 30 terms of both $G_n$ and $b_n$

**(b)** Hence, write down the limit of $b_n$ as $n$ tends to infinity, $\lim\limits_{n\to\infty} b_n$

## 3.2 Arithmetic sequences

Here are three sequences and the most likely recursive formula for each one.

7, 14, 21, 28, 35, 42, ...        $a_1 = 7$ and $a_n = a_{n-1} + 7$, for $n > 1$

2, 11, 20, 29, 38, 47, ...        $a_1 = 2$ and $a_n = a_{n-1} + 9$, for $n > 1$

48, 39, 30, 21, 12, 3, $-6$, ...        $a_1 = 48$ and $a_n = a_{n-1} - 9$, for $n > 1$

Note that in each case, each term is formed by adding a constant number (which can be negative) to the preceding term. Sequences formed in this manner are called **arithmetic sequences**.

So, for the sequences above, 7 is the common difference for the first, 9 is the common difference for the second, and $-9$ is the difference for the third.

This description gives us the recursive definition of the arithmetic sequence. It is possible, however, to find the explicit definition of the sequence.

Applying the recursive definition repeatedly helps us to see the expression:

$$a_2 = a_1 + d$$

$$a_3 = a_2 + d = a_1 + d + d = a_1 + 2d$$

$$a_4 = a_3 + d = a_1 + 2d + d = a_1 + 3d$$

and so on.

So you can get to the $n$th term by adding $d$ to $a_1$, $(n - 1)$ times.

A sequence $a_1, a_2, a_3, ...$ is an **arithmetic sequence** if there is a constant $d$ for which

$$a_n = a_{n-1} + d$$

for all integers $n > 1$, where $d$ is the **common difference** of the sequence, and $d = a_n - a_{n-1}$ for all integers $n > 1$

# 3 Sequences and series

The **nth term** of an arithmetic sequence, $a_n$, with first term $a_1$ and common difference $d$, may be expressed explicitly as

$$a_n = a_1 + (n-1)d$$

This result is useful for finding any term of the sequence without knowing all the previous terms.

The arithmetic sequence can be looked at as a linear function as explained in the introduction to this chapter. i.e., for every increase of one unit in $n$, the value of the sequence will increase by $d$ units. As the first term is $a_1$, the point $(1, a_1)$ belongs to this function. The constant increase $d$ can be considered to be the gradient (slope) of this linear model, hence the $n$th term, the dependent variable in this case, can be found by using the point-slope form of the equation of a line:

$$y - y_1 = m(x - x_1)$$
$$a_n - a_1 = d(n-1) \Leftrightarrow a_n = a_1 + (n-1)d$$

## Example 3.3

For the sequence 2, 11, 20, 29, 38, 47, …, find:

(a) a formula for the $n$th term

(b) the 50th term.

### Solution

(a) This is an arithmetic sequence whose first term is 2 and common difference is 9. Therefore,

$$a_n = a_1 + (n-1)d = 2 + 9(n-1) = 9n - 7$$

(b) $a_{50} = 9(50) - 7 = 443$

## Example 3.4

(a) Find the recursive and the explicit forms of the sequence 13, 8, 3, −2, …

(b) Calculate the value of the 25th term.

### Solution

(a) This is clearly an arithmetic sequence, since −5 is the common difference.

The recursive definition is    $a_1 = 13$
$$a_n = a_{n-1} - 5, n > 1$$

The explicit definition is    $a_n = 13 - 5(n-1) = 18 - 5n$

(b) $a_{25} = 18 - 5(25) = -107$

## Example 3.5

Find the 20th term of the arithmetic sequence with first term 5 and fifth term 11.

### Solution

Since the fifth term is given, using the explicit general form:

$$a_5 = a_1 + (5 - 1)d$$
$$\Rightarrow 11 = 5 + 4d$$
$$\Rightarrow d = \frac{3}{2}$$

This leads to the general term

$$a_n = 5 + \frac{3}{2}(n - 1)$$

Therefore, $a_{20} = 5 + \frac{3}{2}(19) = \frac{67}{2}$

Sometimes a real-life situation can be modelled by an arithmetic sequence, even though it does not follow the sequence exactly. In such cases it is necessary to find an approximation for the common difference. One approach is to use an average of all the differences as an estimate for the common difference. An alternative approach is to use linear regression, which you will meet in Chapter 19. We can use a spreadsheet to calculate differences quickly and then average them.

## Example 3.6

An experiment was undertaken to investigate the relationship between the length of a spring and the mass hanging from it, as shown in the diagram.

The table shows the extension of the spring (cm) for each mass (g).

| Mass (g) | 10 | 20 | 30 | 40 | 50 | 60 | 70 | 80 |
|---|---|---|---|---|---|---|---|---|
| Extension (cm) | 2.4 | 4.8 | 7.0 | 9.5 | 11.8 | 14.2 | 16.4 | 18.9 |

It is believed that the data can be modelled by an arithmetic sequence, according to Hooke's law.

(a) Find an estimate of the common difference, by using an average of the differences.

(b) Find a model for the $n$th term $u_n$, where $n = \dfrac{\text{mass}}{10}$

(c) Use your model to predict the extension for a mass of
    (i) 100 g         (ii) 150 g

(d) In fact, further experiments found the extension at 150 g to be 30.9 cm. With reference to the spring, give a reason why the model does not give a good prediction in this case.

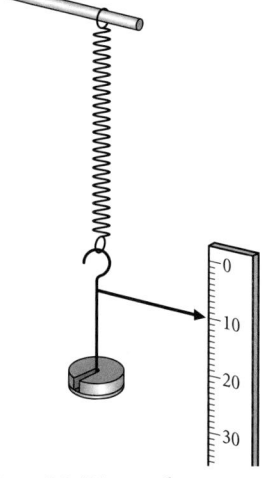

**Figure 3.2** Diagram for Example 3.6

### Solution

(a) Record the differences in an extra row in the table. We can do this quickly and efficiently using a GDC or spreadsheet.

| $n$ | 1 | 2 | 3 | 4 | 5 | 6 | 7 | 8 |
|---|---|---|---|---|---|---|---|---|
| **Extension (cm)** | 2.4 | 4.8 | 7.0 | 9.5 | 11.8 | 14.2 | 16.4 | 18.9 |
| **Difference (cm)** | – | 2.4 | 2.2 | 2.5 | 2.3 | 2.4 | 2.2 | 2.5 |

The average difference, $d = \dfrac{2.4 + 2.2 + 2.5 + 2.3 + 2.4 + 2.2 + 2.5}{7}$

$$= 2.36$$

(b) Since $u_1 = 2.4$, applying the formula gives $u_n = 2.4 + 2.36(n - 1)$
$$= 2.36n + 0.04$$

(c) (i) For 100 g, $u_{10} = 2.36 \times 10 + 0.04 = 23.6 \, \text{cm}$

   (ii) For 150 g, $u_{15} = 2.36 \times 15 + 0.04 = 35.4 \, \text{cm}$

(d) It is possible that this size mass has caused the spring to become almost fully extended, hence the model is no longer appropriate and cannot be extrapolated to this size mass.

## Simple interest

When we invest money in an account, we usually receive interest. When we borrow money then we usually pay interest, for example taking out a loan to buy a car, or a mortgage to buy a house, or using a credit card and not paying the balance off in full each month. Simple interest is interest calculated only on the initial investment. Suppose $2000 is invested in an account paying simple interest at a rate of 5% per year. How much money will there be in the account at the end of 4 years?

Table 3.1 shows how the amount can be calculated for each year.

| Time (years) | Amount in the account ($) |
|---|---|
| 0 | 2000 |
| 1 | $2000 + 2000 \times 0.05$ |
| 2 | $2000 + 2000 \times 0.05 + 2000 \times 0.05 = 2000 + 2000 \times 0.05 \times 2$ |
| 3 | $2000 + 2000 \times 0.05 \times 2 + 2000 \times 0.05 = 2000 + 2000 \times 0.05 \times 3$ |
| 4 | $2000 + 2000 \times 0.05 \times 3 + 2000 \times 0.05 = \mathbf{2000 + 2000 \times 0.05 \times 4}$ |

**Table 3.1** Simple interest calculations

This appears to be an arithmetic sequence with five terms (as both the beginning and the end of the first year are counted).

In general, if a **principal** of $P$ is invested in an account with a simple interest rate $r$ (expressed as a decimal) annually, then we can use the arithmetic sequence formula to calculate the **future value $A$,** which is accumulated after $t$ years.

If we repeat the steps above using the general terms, it becomes easier to develop the formula:

| Time (years) | Amount in the account ($) |
|---|---|
| 0 | $A_0 = P$ |
| 1 | $A_1 = P + Pr$ |
| 2 | $A_2 = A_1 + Pr = P + 2Pr$ |
| $\vdots$ | |
| $t$ | $A_t = P + Prt$ |

**Table 3.2** Developing a formula

Note that since we are counting from 0 to $t$, we have $t + 1$ terms, and hence using the arithmetic sequence formula,

$$a_n = a_1 + (n - 1)d \Rightarrow A_t = A_0 + (t)A_0 r = P + Prt$$

## Example 3.7

$3500 is invested in an account paying simple interest at a rate of 4.2% per year. Interest is added at the end of each year.

(a) Calculate the amount of money in the account after 6 years.

(b) Find the number of years it would take for the amount of money in the account to exceed $6000. (Assume no further money is invested or withdrawn from the account.)

## Solution

(a) The interest rate is 4.2%, so $r = 0.042$ and $P = 3500$. After 6 years:

Amount $= 3500 + 3500 \times 0.042 \times 6 = \$4382$

(b) After $n$ years, an expression for the amount in the account is:

Amount $= 3500 + 3500 \times 0.042 \times n = 3500 + 147n$

We need to solve the inequality:

$$3500 + 147n > 6000$$

This can be solved using a graph or table on the GDC, or algebraically.

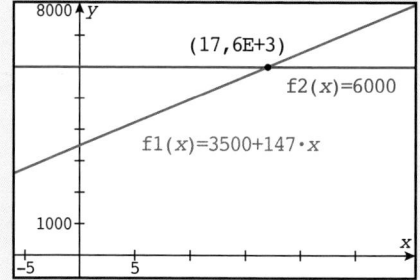
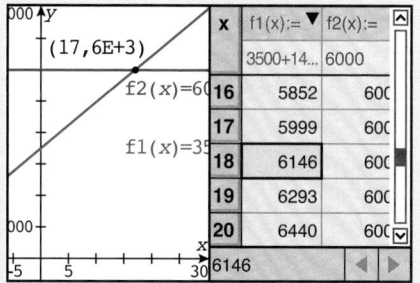

We see that $n > 17.6$, so the number of years is 18.

## Exercise 3.2

1. State whether each sequence is an arithmetic sequence. If yes, find the common difference and the 50th term. If not, state why not.

   (a) $a_n = 2n - 3$

   (b) $b_n = n + 2$

   (c) $c_n = c_{n-1} + 2$, and $c_1 = -1$

   (d) $2, 5, 7, 12, 19, \ldots$

   (e) $2, -5, -12, -19, \ldots$

2. For each arithmetic sequence in parts **(a)** to **(d)** find:

   (i)  the 8th term

   (ii)  an explicit formula for the $n$th term

   (iii)  a recursive formula for the $n$th term.

   (a) $-2, 2, 6, 10, \ldots$

   (b) $10.07, 9.95, 9.83, 9.71, \ldots$

   (c) $100, 97, 94, 91, \ldots$

   (d) $2, \dfrac{3}{4}, -\dfrac{1}{2}, -\dfrac{7}{4}, \ldots$

3. In an arithmetic sequence, $a_5 = 6$ and $a_{14} = 42$

   Find an explicit formula for the $n$th term of this sequence.

4. In an arithmetic sequence, $a_3 = -40$ and $a_9 = -18$

   Find an explicit formula for the $n$th term of this sequence.

5. For each finite sequence, the first 3 terms and the last term are given.

   Find the number of terms in each sequence.

   (a) $3, 9, 15, \ldots, 525$

   (b) $9, 3, -3, \ldots, -201$

   (c) $\dfrac{1}{3}, \dfrac{1}{2}, \dfrac{2}{3}, \ldots, 2\dfrac{5}{6}$

   (d) $1 - k, 1 + k, 1 + 3k, \ldots, 1 + 19k$

6. The 30th term of an arithmetic sequence is 147 and the common difference is 4. Find a formula for the $n$th term.

7. The first term of an arithmetic sequence is $-7$ and the common difference is 3. Determine if 9803 is a term in this sequence. If it is, state which one.

8. The first term of an arithmetic sequence is 9689 and the 100th term is 8996.

   (a) Show that the 110th term is 8926.

   (b) Determine if 1 is a term in this sequence. If it is, state which term.

9. The first term of an arithmetic sequence is 2 and the 30th term is 147. Determine if 995 is a term in this sequence. If it is, state which term.

**10.** An experiment was undertaken to investigate the relationship between the length of a spring and the mass hanging from it. The table shows the extension of the spring (cm) for each mass (g).

| Mass (g) | 10 | 20 | 30 | 40 | 50 | 60 | 70 | 80 |
|---|---|---|---|---|---|---|---|---|
| Extension (cm) | 4.4 | 8.7 | 13.2 | 17.4 | 21.7 | 26.1 | 30.3 | 34.7 |

It is believed that the data can be modelled by an arithmetic sequence, according to Hooke's law.

(a) Find an estimate of the common difference, by using an average of the differences.

(b) Find a model for the $n$th term $u_n$, where $n = \dfrac{\text{mass}}{10}$

(c) (i) Use your model to estimate the extension for a mass of 70 g.

(ii) Calculate the percentage error in the estimate found in part (i).

**11.** Marcus is planning to eat a tub of ice cream, one spoonful at a time. He believes that the mass of ice cream remaining can be modelled by an arithmetic sequence. He puts the tub on a set of measuring scales and collects the following data while he is eating.

| Number of spoonfuls | 1 | 2 | 3 | 4 | 5 | 6 |
|---|---|---|---|---|---|---|
| Mass remaining (g) | 280 | 255 | 235 | 200 | 175 | 145 |

(a) Find an estimate of the common difference, by using an average of the differences.

(b) Find a model for the $n$th term $u_n$

(c) (i) Use your model to estimate the original mass of ice cream in the tub.

(ii) Giving a reason to support your answer, explain if you think this is an overestimate or an underestimate.

(d) (i) Use your model to estimate the number of spoonfuls required to eat all of the ice cream in the tub.

(ii) Giving a reason to support your answer, explain if you think this is an overestimate or an underestimate.

**12.** $500 is invested in an account paying simple interest at a rate of 3.2% per year. Interest is paid at the end of each year.

(a) Calculate the amount of money in the account after 5 years.

(b) Find the number of years it would take for the amount of money in the account to exceed $2000. (Assume no further money is invested or withdrawn from the account.)

The value of an asset, such as a car, decreases over time due to wear and tear. This reduction in value of an asset over time is called depreciation.

13. The owner of a small company buys a company car for one of his employees. The purchase price of the car is \$16,500. For tax purposes the company can depreciate the value of the car by 10% of its purchase price each year.

   (a) Calculate the value of the car to the company after 4 years.

   (b) After the initial 4 years, the employee has the option to buy the car from his company by paying 50% of the purchase price. Determine the cost to the company if the employee buys the car.

14. An investment of \$450 is worth \$560 after 7 years of earning an unknown annual interest rate under the simple interest model. Find the interest rate to the nearest tenth of a percent.

15. Nanako takes out a loan of \$16,000 to buy a car. The loan has a simple interest rate of 8% per year. She pays back the loan at a fixed rate of \$250 per month. Find how long it will take for Nanako to repay the loan.

## 3.3 Geometric sequences

In this section we will consider sequences of the type:

1, 2, 4, 8, 16, …

4, 12, 36, 108, 324, …

600, 300, 150, 75, 37.5, …

3, −12, 48, −192, 768, …

In this type of sequence, each term is obtained by multiplying the previous term by a fixed constant. For example, in the second sequence, each term is 3 times the previous term. We can write this sequence recursively as

$$u_n = 3u_{n-1} \text{ and } u_1 = 4$$

Sequences of this type are called **geometric sequences.** Note that the ratio of consecutive terms equals a fixed constant. For example in the second sequence, we get

$$\frac{12}{4} = \frac{36}{12} = \frac{108}{36} = \frac{324}{108} = 3$$

The ratio of consecutive terms is a constant equal to 3. This is called the **common ratio** of the geometric sequence.

A sequence $u_1, u_2, u_3, \ldots$ is a **geometric sequence** if there is a constant $r$ for which $u_n = u_{n-1} \times r$ for all integers $n > 1$. The constant, $r$, is called the **common ratio** of the sequence, and $r = \dfrac{u_n}{u_{n-1}}$ for all integers $n > 1$.

## Example 3.8

For each sequence, state the common ratio and write a recursive formula.

(a) 1, 2, 4, 8, 16, …

(b) 600, 300, 150, 75, 37.5, …

(c) 3, −12, 48, −192, 768, …

### Solution

(a) The common ratio is 2 and a recursive formula is $u_n = 2u_{n-1}$, $u_1 = 1$

(b) The common ratio is $\frac{1}{2}$ and a recursive formula is $u_n = \frac{1}{2}u_{n-1}$, $u_1 = 600$

(c) The common ratio is −4 and a recursive formula is $u_n = -4u_{n-1}$, $u_1 = 3$

In practice, the recursive formula is not useful to obtain the terms of a sequence. For example, finding the 100th term in the sequences in Example 3.8 would be laborious, using the recursive formula. If we start with first term $u_1$ and common ratio $r$, then we can apply the recursive formula repeatedly, to find an explicit formula for the $n$th term.

$$u_2 = u_1 \times r$$

$$u_3 = u_2 \times r = u_1 \times r \times r = u_1 \times r^2$$

$$u_4 = u_3 \times r = u_1 \times r^2 \times r = u_1 \times r^3$$

and so on.

So, as you see, you can get to the $n$th term by multiplying $u_1$ by $r$, $(n-1)$ times.

The general ($n$th) term of a geometric sequence, $u_n$ with common ratio $r$ and first term $u_1$ may be expressed explicitly as

$$u_n = u_1 r^{n-1}$$

This result is useful in finding any term of the sequence without knowing all the previous terms.

## Example 3.9

(a) Find the 12th term in the sequence 3, −12, 48, −192, 768, …

(b) The third term of a geometric sequence is 5 and the seventh term is 405. Find the two possible values of the tenth term.

### Solution

(a) $u_1 = 3$ and $r = -4$, so $u_{12} = 3(-4)^{11} = -12\,582\,912$. Note that when the common ratio is negative, the sign of the terms in the sequence alternates.

(b) $u_3 = 5 \Rightarrow u_1 r^2 = 5$

$u_7 = 405 \Rightarrow u_1 r^6 = 405$

Dividing the second equation by the first gives

$$\frac{u_1 r^6}{u_1 r^2} = \frac{405}{5}$$

Cancelling common factors of $u_1$, $r^2$, and 5 gives

$$r^4 = 81 \Rightarrow r = \pm\sqrt[4]{81} \Rightarrow r = \pm 3$$

We can solve for $u_1$ using the first equation

$$u_1 = \frac{5}{r^2} = \frac{5}{9}$$

So,

$$u_{10} = u_1 r^9 = \frac{5}{9}(\pm 3)^9 = \pm 10\,935$$

The two possible values of the tenth term are $10\,935$ and $-10\,935$

## Compound interest

### Interest compounded annually

We have already looked at simple interest in section 3.2, which was calculated only on the initial investment. Compound interest is interest calculated both on the initial investment and also on interest already paid. Suppose \$2500 is invested in an account paying compound interest at a rate of 3% per year. How much money will there be in the account at the end of 4 years? It is important to note that the 3% interest is paid annually and is added to the account, so that in the following year it will also earn interest, and so on. Table 3.3 shows how the amount can be calculated for each year.

| Time (years) | Amount in the account (\$) |
|:---:|:---|
| 0 | 2500 |
| 1 | $2500 + 2500 \times 0.03 = 2500(1 + 0.03)$ |
| 2 | $2500(1 + 0.03) + (2500(1 + 0.03)) \times 0.03 = 2500(1 + 0.03)(1 + 0.03) = 2500(1 + 0.03)^2$ |
| 3 | $2500(1 + 0.03)^2 + (2500(1 + 0.03)^2) \times 0.03 = 2500(1 + 0.03)^2(1 + 0.03) = 2500(1 + 0.03)^3$ |
| 4 | $2500(1 + 0.03)^3 + (2500(1 + 0.03)^3) \times 0.03 = 2500(1 + 0.03)^3(1 + 0.03) = 2500(1 + 0.03)^4$ |

**Table 3.3** Compound interest

This appears to be a geometric sequence with 5 terms. Note that the number of terms is 5, as both the beginning and the end of the first year are counted.

In general, if a **principal** of \$P$ is invested in an account that has an interest rate $r$ (expressed as decimal) annually, and this interest is added to the principal at the end of each year, then we can use the geometric sequence formula to calculate the **future value $A$,** which is accumulated after $t$ years.

If we repeat the steps above, it becomes easier to develop the formula:

| Time (years) | Amount in the account ($) |
|---|---|
| 0 | $A_0 = P$ |
| 1 | $A_1 = P + P \times r = P(1 + r)$ |
| 2 | $A_2 = A_1(1 + r) = P(1 + r)^2$ |
| $\vdots$ | $\vdots$ |
| $t$ | $A_t = P(1 + r)^t$ |

**Table 3.4** Developing a formula

Note that since we are counting from 0 to $t$, we have $t + 1$ terms, and hence using the geometric sequence formula,

$$u_n = u_1 r^{n-1} \Rightarrow A_t = A_0(1 + r)^t$$

## Example 3.10

$1500 is invested in an account paying compound interest at a rate of 4.5% per year. Interest is added at the end of each year.

(a) Calculate the amount of money in the account after 6 years.

(b) Find the number of years it would take for the amount of money in the account to exceed $10,000. (Assume no further money is invested or withdrawn from the account.)

## Solution

(a) Each year the amount in the account increases by 4.5%. Calculate the percentage increase by multiplying by 1.045 for each year. After 6 years,

$$\text{Amount} = 1500 \times 1.045^6 = \$1953.39$$

(b) After $n$ years, an expression for the amount in the account is:

$$\text{Amount} = 1500 \times 1.045^n$$

We need to solve the inequality

$$1500 \times 1.045^n > 10\,000$$

This can be solved using a graph or table on a GDC, or algebraically using logarithms.

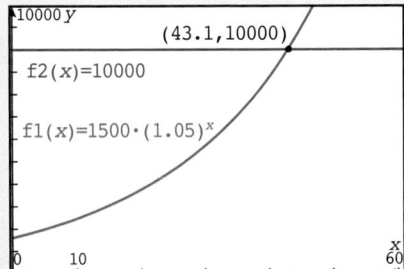
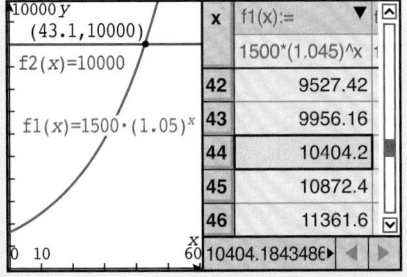

We see that $n > 43.1$, so the number of years is 44.

Example 3.10 can also be solved using the built-in financial package on a graphing calculator.

(a) Using the TVM solver, we make the following entries and solve for the future value, *FV*.

$N = 6$
$I\% = 4.5$
$PV = -1500$
$PMT = 0$
$FV = 1953.39$
$P/Y = 1$
$C/Y = 1$
*PMT: END*

The number of payments, *N*, is 6, because interest is compounded once per year for six years. The present value, *PV*, is negative, because the bank now has the money.

(b) Using the TVM solver, we make the following entries and solve to find the number of payments, *N*.

$N = 43.1$
$I\% = 4.5$
$PV = -1500$
$PMT = 0$
$FV = 10\,000$
$P/Y = 1$
$C/Y = 1$
*PMT: END*

A Time-Value-Money (TVM) solver is a built-in financial package on a GDC that allows a user to enter parameters of an investment and then calculate a missing value. The parameters of a TVM solver are as follows.

$N$ = number of payments made in total
$I\%$ = annual interest rate ($\times$ 100)
$PV$ = present value of payments
$PMT$ = value of the regular payment
$FV$ = future value of the payments
$P/Y$ = number of payments per year
$C/Y$ = number of compounding periods per year
*PMT: END   BEGIN*   payments made at beginning or end of year

## Interest compounded *n* times per year

Suppose that the principal *P* is invested as before but the interest is paid *n* times per year. Then $\frac{r}{n}$ is the interest paid every compounding period. Since every year we have *n* periods, therefore for *t* years, we have *nt* periods. The amount *A* in the account after *t* years is

$$A = P\left(1 + \frac{r}{n}\right)^{nt}$$

## Example 3.11

€1000 is invested in an account paying compound interest at a rate of 6%. Calculate the amount of money in the account after 10 years if

(a) the compounding is annual

(b) the compounding is quarterly

(c) the compounding is monthly.

## Solution

(a) The amount after 10 years compounding annually is

$$A = 1000(1 + 0.06)^{10} = €1790.85$$

(b) The amount after 10 years compounding quarterly is

$$A = 1000\left(1 + \frac{0.06}{4}\right)^{40} = €1814.02$$

(c) The amount after 10 years compounding monthly is

$$A = 1000\left(1 + \frac{0.06}{12}\right)^{120} = €1819.40$$

Example 3.11 can also be solved using the built-in financial package on a graphing calculator.

(a) Using the TVM solver, we make the following entries and solve to find the future value, FV.

$N = 10$
$I\% = 6$
$PV = -1000$
$PMT = 0$
$FV = 1790.85$
$P/Y = 1$
$C/Y = 1$
$PMT: END$

(b) Using the TVM solver, enter $N = 40$ and payments per year, $PpY = 4$

$N = 40$
$I\% = 6$
$PV = -1000$
$PMT = 0$
$FV = 1814.02$
$P/Y = 4$
$C/Y = 1$
$PMT: END$

(c) Using the TVM solver, enter $N = 120$ and payments per year, $PpY = 12$

$N = 120$
$I\% = 6$
$PV = -1000$
$PMT = 0$
$FV = 1819.40$
$P/Y = 2$
$C/Y = 1$
$PMT: END$

## Real rate of return

In Examples 3.10 and 3.11, the interest rate used to calculate the future value of an investment was the **nominal rate**. The nominal rate of interest is the interest rate before taking other factors, such as inflation, into account. In practice, the rate of inflation can affect the future value of an investment. In order to account for the rate of inflation, the nominal rate is adjusted to take inflation into account. This adjusted interest rate is called the **real rate of return.** Suppose the interest rate is 6% per year and the inflation rate is 2% per year. Then the **real rate of return** is only 4% per year.

**Inflation** means an overall rise in the cost of goods and services. This cost increase means that inflation reduces the purchasing power of each unit of currency.

73

### Example 3.12

$2500 is invested in an account paying an annual interest rate of 5.6%. The interest is compounded half-yearly. The inflation rate during the time of the investment is 2%.

(a) Find the real rate of return per year.

(b) Calculate the real value of the investment after 6 years.

### Solution

(a) The **nominal** rate of return per year is

$$\left(1 + \frac{0.056}{2}\right)^2 = 1.056784 \text{ or } 5.6784\%$$

The **real** rate of return per year is

$$5.6784 - 2 = 3.6784\%$$

(b) After 6 years the **real** value, $A$, of the investment is

$$A = 2500(1 + 0.036784)^6 = \$3105.06$$

## Exponential growth and decay

In the section on compound interest, the expression for the amount after $t$ years, $A_t = P(1 + r)^t$, is an example of exponential growth. There are several other applications of geometric sequences, linked to exponential growth or decay.

### Example 3.13

In 2016, the population growth rate in Singapore was 1.3% and the size of the population was 5.6 million (source: World Bank). Assuming the population growth rate remains constant, find:

(a) the size of the population in 2021

(b) the number of years until the population exceeds 7 million.

### Solution

(a) This can be modelled by a geometric sequence with first term 5.6 million and common ratio of 1.013

In 2021, population size $= 5.6 \times 1.013^5 = 5.97$ million

(b) We need to solve the inequality:

$$5.6 \times 1.013^t > 7$$

We can use a GDC to sketch a graph and/or create a table.

According to the model, the population will exceed 7 million after 17.3 years.

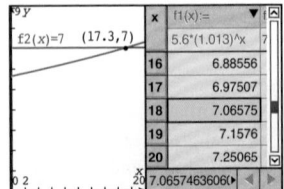

**Figure 3.3** Using a GDC to solve the inequality in Example 3.13 (b)

## Example 3.14

A company buys a machine for \$60,000. The expected lifetime of the machine is 6 years. At the end of 6 years, the company estimates it can sell the machine for \$8000. Calculate the annual rate of depreciation of the machine.

**Annual depreciation** is the reduction in value of an asset at a particular rate per year.

### Solution

The value of the machine can be modelled by a geometric sequence with $u_1 = 60\,000$ and $u_7 = 8000$. Substituting into $u_n = u_1 r^{n-1}$ gives

$$8000 = 60\,000 r^6$$

$$\Rightarrow r = \sqrt[6]{\dfrac{8000}{60\,000}} = 0.715$$

The annual rate of depreciation is $1 - 0.715 = 0.285$ or 28.5%

Example 3.14 can also be solved using the built-in financial package on a graphing calculator. Using the TVM solver, we make the following entries and solve for the interest rate, $I$.

$N = 6$
$I\% = -28.5$
$PV = -60\,000$
$PMT = 0$
$FV = 8000$
$P/Y = 1$
$C/Y = 1$
$PMT: END$

Note that the negative interest rate corresponds to depreciation.

## Exercise 3.3

In each of questions **1–15**:

**(a)** Determine whether the sequence is arithmetic, geometric, or neither.

**(b)** Find the common difference for the arithmetic sequences and the common ratio for the geometric sequences.

**(c)** Find the 10th term for each arithmetic or geometric sequence.

**1.** $3, 3^{a+1}, 3^{2a+1}, 3^{3a+1}, \ldots$

**2.** $a_n = 3n - 3$

**3.** $b_n = 2^{n+2}$

**4.** $c_n = 2c_{n-1} - 2$, and $c_1 = -1$

**5.** $u_n = 3u_{n-1}$ and $u_1 = 4$

**6.** $2, 5, 12.5, 31.25, 78.125, \ldots$

**7.** $2, -5, 12.5, -31.25, 78.125, \ldots$

**8.** $2, 2.75, 3.5, 4.25, 5, \ldots$

**9.** $18, -12, 8, -\dfrac{16}{3}, \dfrac{32}{9}, \ldots$

**10.** $52, 55, 58, 61, \ldots$

**11.** $-1, 3, -9, 27, -81, \ldots$

**12.** $0.1, 0.2, 0.4, 0.8, 1.6, 3.2, \ldots$

**13.** $3, 6, 12, 18, 21, 27, \ldots$

**14.** $6, 14, 20, 28, 34, \ldots$

**15.** $2.4, 3.7, 5, 6.3, 7.6, \ldots$

For each arithmetic or geometric sequence in questions **16–32**, find:

**(a)** the 8th term

**(b)** an explicit formula for the *n*th term

**(c)** a recursive formula for the *n*th term.

**16.** $-3, 2, 7, 12, \ldots$

**17.** $19, 15, 11, 7, \ldots$

**18.** $-8, 3, 14, 25, \ldots$

**19.** $10.05, 9.95, 9.85, 9.75, \ldots$

**20.** $100, 99, 98, 97, \ldots$

**21.** $2, \dfrac{1}{2}, -1, -\dfrac{5}{2}, \ldots$

**22.** $3, 6, 12, 24, \ldots$

**23.** $4, 12, 36, 108, \ldots$

**24.** $5, -5, 5, -5, \ldots$

**25.** $3, -6, 12, -24, \ldots$

**26.** $972, -324, 108, -36, \ldots$

**27.** $-2, 3, -\dfrac{9}{2}, \dfrac{27}{4}, \ldots$

**28.** $35, 25, \dfrac{125}{7}, \dfrac{625}{49}, \ldots$

**29.** $-6, -3, -\dfrac{3}{2}, -\dfrac{3}{4}, \ldots$

**30.** $9.5, 19, 38, 76, \ldots$

**31.** $100, 95, 90.25, \ldots$

**32.** $2, \dfrac{3}{4}, \dfrac{9}{32}, \dfrac{27}{256}, \ldots$

**33.** A geometric sequence has a first term of 3 and a common ratio of 5. Find the eighth term.

**34.** A geometric sequence has a third term of 7 and a common ratio of $\dfrac{1}{3}$ Find the first term.

**35.** A geometric sequence has a third term of 8 and a sixth term of 27. Find the tenth term.

**36.** A geometric sequence has a fifth term of 6 and a seventh term of 8.64 Find the possible values of the second term.

**37.** A geometric sequence has a second term of 2, a seventh term of $\dfrac{243}{512}$ and an *n*th term of $\dfrac{19\,683}{131\,072}$. Find the value of *n*.

**38.** John buys a car for $12,000. The value of the car depreciates by 12% each year. Find the value of the car after 11 years.

**39.** Sarah buys a car for $14,900. The car depreciates by a fixed rate each year, and after 8 years it is worth $2300. Find the annual rate of depreciation of the car.

**40.** In 2016, the UK government estimated the total greenhouse gas emissions in the UK at 483.0 million tonnes carbon dioxide equivalent. This was a decrease of 6.2% from the previous year.

   **(a)** Assuming that the percentage decrease remains constant at 6.2% per year, estimate the UK total greenhouse gas emissions in 2030.

   **(b)** Calculate the percentage decrease in UK total greenhouse gas emissions from 2016 to 2030.

**41.** Ashish invests $2500 in an account that pays 6.2% compound interest per year. He does not invest any further amount or withdraw any money from the account.

   **(a)** Find the amount in his account after 7 years.

   **(b)** Find the number of years until the amount in his account exceeds $5000.

**42.** At her daughter Jane's birth, Charlotte put $500 into a savings account. The interest she earned was 4% compounded quarterly. How much money will Jane have on her 16th birthday?

**43.** Stephen invests $4000 in an account that pays 4.7% interest compounded monthly. He does not invest any further amount or withdraw any money from the account.

   **(a)** Find the amount in his account after 5 years.

   **(b)** Find the number of months until the amount in his account doubles.

**44.** In 2016, the population growth rate in India was 1.2% and the population was 1.324 billion. The population growth rate in China was 0.5% and the population was 1.379 billion (source: World Bank). Assume the population growth rate in each country remains constant.

   **(a)** Find the size of the population in India in 2020.

   **(b)** Find the size of the population in China in 2020.

   **(c)** Find when the population of India will exceed that of China.

**45.** A bank pays interest at a rate of 4% compounded annually. If the inflation rate is 1.7%, calculate the real rate of return per year.

**46.** A bank pays interest at a rate of 3.5% compounded monthly. If the inflation rate is 2.1%, calculate the real rate of return per year.

## 3.4   Series

The word 'series' in common language implies much the same thing as 'sequence'. But in mathematics when we talk of a series, we are referring in particular to the sum of terms in a sequence. For example, for a sequence of values $a_n$, the corresponding series is the sequence $S_n$ with

$$S_n = a_1 + a_2 + \ldots + a_{n-1} + a_n$$

If the terms are in an arithmetic sequence we call the sums an **arithmetic series**.

## Sigma notation

Many of the series we consider in mathematics are **infinite** series. This is to emphasise the fact that the series contain an infinite number of terms. Any sum in the series $S_k$ will be called a **partial sum** and is given by:

$$S_k = a_1 + a_2 + \ldots + a_{k-1} + a_k$$

For convenience, this partial sum is written using the sigma notation:

$$S_k = \sum_{i=1}^{k} a_i = a_1 + a_2 + \ldots + a_{k-1} + a_k$$

Sigma notation is a concise and convenient way to represent long sums. Here, the symbol $\Sigma$ is the Greek capital letter sigma. So this expression means the sum of all the terms $a_i$ where $i$ takes the values from 1 to $k$. We can also write $\sum_{i=m}^{n} a_i$ to mean the sum of the terms $a_i$ where $i$ takes the values from $m$ to $n$. In such a sum, $m$ is called the lower limit and $n$ the upper limit.

### Example 3.15

Write out what is meant by:

(a) $\sum_{i=1}^{5} i^4$  (b) $\sum_{r=3}^{7} 3^r$  (c) $\sum_{j=1}^{n} x_j p_j$

### Solution

(a) $\sum_{i=1}^{5} i^4 = 1^4 + 2^4 + 3^4 + 4^4 + 5^4$  (b) $\sum_{r=3}^{7} 3^r = 3^3 + 3^4 + 3^5 + 3^6 + 3^7$

(c) $\sum_{j=1}^{n} x_j p_j = x_1 p_1 + x_2 p_2 + \ldots + x_n p_n$

### Example 3.16

Evaluate $\sum_{n=0}^{5} 2^n$

### Solution

$$\sum_{n=0}^{5} 2^n = 2^0 + 2^1 + 2^2 + 2^3 + 2^4 + 2^5 = 63$$

### Example 3.17

Write the sum $\dfrac{1}{2} - \dfrac{2}{3} + \dfrac{3}{4} - \dfrac{4}{5} + \ldots + \dfrac{99}{100}$ in sigma notation.

## Solution

The terms in the numerator and denominator are consecutive integers, so they take on the absolute value of $\dfrac{k}{k+1}$ or any equivalent form.

The signs of the terms alternate and there are 99 terms. To take care of the sign change, we use some power of $(-1)$ that will start with a positive value. If we use $(-1)^k$, then the first term will be negative, hence we can use $(-1)^{k+1}$ instead. We can therefore write the sum as:

$$(-1)^{1+1}\left(\frac{1}{2}\right) + (-1)^{2+1}\left(\frac{2}{3}\right) + (-1)^{3+1}\left(\frac{3}{4}\right) + \ldots + (-1)^{99+1}\left(\frac{99}{100}\right)$$

$$= \sum_{k=1}^{99}(-1)^{k+1}\left(\frac{k}{k+1}\right)$$

## Properties of the sigma notation

There are a number of useful results that we can obtain using sigma notation.

- For example, suppose we have a sum of constant terms

$$\sum_{i=1}^{5}2$$

What does this mean? If we write this out in full we get

$$\sum_{i=1}^{5}2 = 2 + 2 + 2 + 2 + 2 = 5 \times 2 = 10$$

In general, if we sum a constant $n$ times then we can write

$$\sum_{i=1}^{n}k = k + k + \ldots + k = n \times k = nk$$

- Suppose we have the sum of a constant times $i$. What does this give us? For example,

$$\sum_{i=1}^{5}5i = 5 \times 1 + 5 \times 2 + 5 \times 3 + 5 \times 4 + 5 \times 5 = 5(1 + 2 + 3 + 4 + 5)$$
$$= 75$$

However, this can also be interpreted as follows.

$$\sum_{i=1}^{5}5i = 5 \times 1 + 5 \times 2 + 5 \times 3 + 5 \times 4 + 5 \times 5 = 5(1 + 2 + 3 + 4 + 5)$$
$$= 5\sum_{i=1}^{5}i$$

which implies that

$$\sum_{i=1}^{5}5i = 5\sum_{i=1}^{5}i$$

In general, we can say

$$\sum_{i=1}^{n}ki = k \times 1 + k \times 2 + \ldots + k \times n$$
$$= k \times (1 + 2 + \ldots + n)$$
$$= k\sum_{i=1}^{n}i$$

- Suppose we need to consider the sum of two different functions, such as:

$$\sum_{k=1}^{n}(k^2 + k^3) = (1^2 + 1^3) + (2^2 + 2^3) + \dots + (n^2 + n^3)$$

$$= (1^2 + 2^2 + \dots + n^2) + (1^3 + 2^3 + \dots + n^3)$$

$$= \sum_{k=1}^{n}k^2 + \sum_{k=1}^{n}k^3$$

In general:

$$\sum_{k=1}^{n}(f(k) + g(k)) = \sum_{k=1}^{n}f(k) + \sum_{k=1}^{n}g(k)$$

## Arithmetic series

In an arithmetic series, we add the terms of an arithmetic sequence. It is very helpful to be able to find an explicit expression for the partial sum of an arithmetic series.

For example, find the partial sum for the first 50 terms of the series:

$$3 + 8 + 13 + 18 + \dots$$

We can express $S_{50}$ in two different ways and then add them together:

$$S_{50} = \quad 3 + \quad 8 + \quad 13 + \dots + 248$$

$$S_{50} = 248 + 243 + 238 + \dots + \quad 3$$

$$2S_{50} = 251 + 251 + 251 + \dots + 251$$

There are 50 terms in this sum, and hence

$$2S_{50} = 50 \times 251 \Rightarrow S_{50} = 6275$$

This reasoning can be extended to any arithmetic series in order to develop a formula for the $n$th partial sum $S_n$.

Let $(a_n)$ be an arithmetic sequence with first term $a_1$ and a common difference $d$. We can construct the series in two ways: forwards, by adding $d$ to $a_1$ repeatedly and backwards by subtracting $d$ from $a_n$ repeatedly. We get the following two expressions for the sum:

$$S_n = a_1 + (a_1 + d) + (a_1 + 2d) + \dots + (a_1 + (n-1)d),$$

and

$$S_n = = a_n + (a_n - d) + (a_n - 2d) + \dots + (a_n - (n-1)d)$$

By adding, term by term vertically, we get:

$$S_n = a_1 + (a_1 + d) + (a_1 + 2d) + \dots + (a_1 + (n-1)d)$$

$$S_n = a_n + (a_n - d) + (a_n - 2d) + \dots + (a_n - (n-1)d)$$

$$2S_n = (a_1 + a_n) + (a_1 + a_n) + (a_1 + a_n) + \dots + (a_1 + a_n)$$

Since we have $n$ terms, we can reduce the expression to $2S_n = n(a_1 + a_n)$

Dividing both sides by 2 gives $S_n = \frac{n}{2}(a_1 + a_n)$, which in turn can be rearranged to give an interesting perspective

of the sum, i.e., $S_n = n\left(\dfrac{a_1 + a_n}{2}\right)$ is $n$ times the average of the first and last terms.

If we substitute $a_1 + (n - 1)d$ for $a_n$ then we arrive at an alternative formula for the sum.

The sum, $S_n$, of $n$ terms of an arithmetic series with common difference $d$, first term $a_1$ and $n$th term $a_n$ can be expressed explicitly as

$$S_n = \frac{n}{2}(a_1 + a_n) \text{ or } S_n = \frac{n}{2}(2a_1 + (n - 1)d)$$

## Example 3.18

Find the partial sum for the first 50 terms of the series $3 + 8 + 13 + 18 + \ldots$

### Solution

Using the second formula for the sum we get:

$$S_{50} = \frac{50}{2}(2 \times 3 + 5(50 - 1)) = 25 \times 251 = 6275$$

Alternatively, to use the first formula we need to know the $n$th term.
So, $a_{50} = 3 + 49 \times 5 = 248$

$\Rightarrow S_{50} = 25(3 + 248) = 6275$

## Geometric series

A geometric series is the sum of the terms in a geometric sequence. As is the case with arithmetic series, it is desirable to find a general expression for the $n$th partial sum of a geometric series.

For example, find the partial sum for the first 20 terms of the series

$$3 + 6 + 12 + 24 + \ldots$$

Express $S_{20}$ in two different ways and subtract them:

$$\begin{aligned}
S_{20} &= 3 + 6 + 12 + \ldots + 1\,572\,864 \\
2S_{20} &= \phantom{3 + } 6 + 12 + \ldots + 1\,572\,864 + 3\,145\,728 \\
\hline
-S_{20} &= 3 \phantom{+ 6 + 12 + \ldots + 1\,572\,864} - 3\,145\,728 \\
\Rightarrow S_{20} &= 3\,145\,725
\end{aligned}$$

This reasoning can be extended to any geometric series in order to develop a formula for the $n$th partial sum $S_n$.

Let $(a_n)$ be a geometric sequence with first term $a_1$ and a common ratio $r \neq 1$.

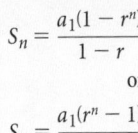

The sum, $S_n$, of $n$ terms of a geometric series with common ratio $r \neq 1$ and first term $a_1$ can be expressed explicitly as

$$S_n = \frac{a_1(1 - r^n)}{1 - r}$$

or

$$S_n = \frac{a_1(r^n - 1)}{r - 1}$$

It is better to use the first version of the formula when $r < 1$ and the second version when $r > 1$. This avoids having too many negative numbers in calculations.

We can construct the series in two ways as before, and using the definition of the geometric sequence, i.e., $a_n = a_{n-1} \times r$, we have:

$$S_n = a_1 + a_2 + a_3 + \ldots + a_{n-1} + a_n$$
$$rS_n = ra_1 + ra_2 + ra_3 + \ldots + ra_{n-1} + ra_n$$
$$= a_2 + a_3 + \ldots + a_{n-1} + a_n + ra_n$$

Now, subtract the first and last expressions to get

$$S_n - rS_n = a_1 - ra_n \Rightarrow S_n(1 - r) = a_1 - ra_n \Rightarrow S_n = \frac{a_1 - ra_n}{1 - r}, r \neq 1$$

This expression however, requires that $r$, $a_1$, as well as $a_n$ be known in order to find the sum. However, using the $n$th term expression developed earlier, we can simplify this formula for the sum:

$$S_n = \frac{a_1 - ra_n}{1 - r} = \frac{a_1 - ra_1 r^{n-1}}{1 - r} = \frac{a_1(1 - r^n)}{1 - r}, r \neq 1$$

### Example 3.19

Find the partial sum for the first 20 terms of the series $3 + 6 + 12 + 24 + \ldots$

### Solution

$$S_{20} = \frac{3(1 - 2^{20})}{1 - 2} = \frac{3(1 - 1\,048\,576)}{-1} = 3\,145\,725$$

### Example 3.20

In a geometric sequence the second term is 5 and the fifth term is 135. Find the sum of the first 12 terms in the sequence.

### Solution

$$u_2 = 5 \Rightarrow u_1 r = 5$$
$$u_5 = 135 \Rightarrow u_1 r^4 = 135$$

Dividing the second equation by the first equation gives

$$\frac{u_1 r^4}{u_1 r} = \frac{135}{5} \Rightarrow r^3 = 27 \Rightarrow r = 3$$

Substituting $r$ into the first equation gives

$$u_1 = \frac{5}{3}$$

Since $r > 1$, it is better to use $S_n = \frac{a_1(r^n - 1)}{r - 1}$

$$S_{12} = \frac{\frac{5}{3}(3^{12} - 1)}{3 - 1} = \frac{5}{6}(3^{12} - 1) = \frac{1\,328\,600}{3}$$

## Example 3.21

In a geometric sequence the sum of the first three terms is 304 and the sum of the first six terms is 1330. Find the sum of the first eight terms.

**Solution**

$$S_3 = 304 \Rightarrow \frac{u_1(r^3 - 1)}{r - 1} = 304$$

$$S_6 = 1330 \Rightarrow \frac{u_1(r^6 - 1)}{r - 1} = 1330$$

Dividing the second equation by the first equation:

$$\frac{r^6 - 1}{r^3 - 1} = \frac{1330}{304} = \frac{35}{8} \Rightarrow 8(r^6 - 1) = 35(r^3 - 1)$$

Rearranging:

$$8r^6 - 35r^3 + 27 = 0$$

Solving on a GDC gives the roots of the polynomial as

$r = 1$ and $r = \dfrac{3}{2}$

So $r = \dfrac{3}{2}$, since $r \neq 1$

```
polyRoots(8·x⁶-35·x³+27,x)
                    {1, 3/2}
```

Substituting $r$ into the first equation:

$$u_1 = \frac{304\left(\dfrac{3}{2} - 1\right)}{\left(\left(\dfrac{3}{2}\right)^3 - 1\right)} = 64$$

Finally,

$$S_8 = \frac{64\left(\left(\dfrac{3}{2}\right)^8 - 1\right)}{\dfrac{1}{2}} = 3152.5$$

We will now consider an application of geometric series to the compound interest model. It is often the case that an investor makes regular equal payments into an investment, all of which accumulate interest at an equal annual rate. Such an investment is called an **annuity**. We will consider the case in which equal payments are made annually, and interest is compounded annually at an equal rate at the end of the year. The general case in which payments and interest compounding periods do not coincide will be considered in the next section.

**Annuities** are equal regular payments that earn an annual interest rate.

When an annuity is paid at the beginning of the period, it is called **annuity due**.

### Example 3.22

Christine invests in a retirement plan in which equal payments of $1200 are made at the beginning of each year. Interest is earned on each payment at an annual rate of 2% per year, compounded annually.

(a) How much is the investment worth at the end of 30 years?

(b) How much interest has been earned on the investment in 30 years?

### Solution

(a) Note that in 30 years, Christine has invested $30 \times \$1200 = \$36,000$. However, each of the payments earns compound interest at the rate of 2% per year. The investment made in the first year has a value of $1200(1.02)^{30}$ at the end of the investment period; the investment in the second year has a value of $1200(1.02)^{29}$ and so on. The last payment earns interest for 1 year and has the value $1200(1.02)$. As such, the value of the annuity is the sum of 30 terms of the geometric series with first term $1200(1.02)$ and common ratio 1.02.

$$1200(1.02) + 1200(1.02)^2 + 1200(1.02)^3 + \ldots + 1200(1.02)^{30}$$
$$= 1200(1.02)[1 + 1.02 + 1.02^2 + \ldots + 1.02^{29}]$$
$$= 1200(1.02)\left[\frac{1.02^{30} - 1}{1.02 - 1}\right]$$
$$= \$49,655.33$$

(b) The value of the payments made is $36,000, thus the interest earned is $13,655.33

Example 3.22 (a) can also be solved using the built-in financial package on a graphing calculator. Using the TVM solver, we make the following entries:

$N = 30$
$I\% = 2$
$PV = 0$
$PMT = -1200$
$FV = 49\,655.33$
$P/Y = 1$
$C/Y = 1$
$PMT: \quad BEGIN$

The present value, $PV = 0$, since there is no money invested until the first payment of $1200 is made. The payment is made at the beginning of each year, so set the $PmtAt = BEGIN$. Then solve for future value, $FV$.

## Sum to infinity of a geometric series

In this section we consider if it is possible to calculate the sum of infinitely many terms in a geometric series.

First consider the infinite geometric series with first term 1 and common ratio 2.

$$\sum_{k=1}^{\infty} 2^{k-1} = 1 + 2 + 4 + 8 + 16 + \ldots$$

We can calculate the partial sums for 10, 50 and 100 terms.

$$\sum_{k=1}^{10} 2^{k-1} = \frac{2^{10} - 1}{2 - 1} = 1023$$

$$\sum_{k=1}^{50} 2^{k-1} = \frac{2^{50} - 1}{2 - 1} = 1\,125\,899\,906\,842\,623$$

$$\sum_{k=1}^{100} 2^{k-1} = \frac{2^{100} - 1}{2 - 1} = 1\,267\,650\,600\,228\,229\,401\,496\,703\,205\,375$$

We observe that as the number of terms in the partial sum increases, the sum also increases. It would appear that the sum to infinity, $\sum_{k=1}^{\infty} 2^{k-1}$, approaches infinity.

Now consider the infinite geometric series with first term 1 and common ratio $\frac{1}{2}$

$$\sum_{k=1}^{\infty} \left(\frac{1}{2}\right)^{k-1} = 1 + \frac{1}{2} + \frac{1}{4} + \frac{1}{8} + \frac{1}{16} + \ldots$$

We can calculate the partial sums for 10, 20 and 50 terms.

$$\sum_{k=1}^{10} \left(\frac{1}{2}\right)^{k-1} = \frac{1 - \left(\frac{1}{2}\right)^{10}}{1 - \frac{1}{2}} = 1.998046875$$

$$\sum_{k=1}^{20} \left(\frac{1}{2}\right)^{k-1} = \frac{1 - \left(\frac{1}{2}\right)^{20}}{1 - \frac{1}{2}} = 1.9999980926514$$

$$\sum_{k=1}^{50} \left(\frac{1}{2}\right)^{k-1} = \frac{1 - \left(\frac{1}{2}\right)^{50}}{1 - \frac{1}{2}} = 1.9999999999999982236$$

We observe that as the number of terms in the partial sum increases, the sum also increases, but it approaches a limiting value of 2. It would appear that the sum to infinity, $\sum_{k=1}^{\infty} \left(\frac{1}{2}\right)^{k-1}$, equals 2.

Using the language of limits, we define the sum to infinity, $S_\infty$ as the limit as $n$ tends to infinity of $S_n$. We write this as

$$S_\infty = \lim_{n \to \infty} S_n$$

For the above example we have

$$S_\infty = \lim_{n \to \infty} \sum_{k=1}^{n} \left(\frac{1}{2}\right)^{k-1} = \lim_{n \to \infty} \frac{1 - \left(\frac{1}{2}\right)^{n}}{1 - \frac{1}{2}} = \frac{1}{\frac{1}{2}} = 2 \text{, since } \lim_{n \to \infty} \left(\frac{1}{2}\right)^{n} = 0$$

If $S_\infty = L$, where $L$ is finite, then we say that the infinite series **converges** to $L$.

So this particular infinite series converges to 2. The first series, $\sum_{k=1}^{\infty} 2^{k-1}$, does not have a limit and we say it **diverges**.

We are now ready to develop a general rule for finding the sum to infinity of a geometric series.

$$S_{\infty} = \lim_{n \to \infty} S_n = \lim_{n \to \infty} \frac{u_1(1 - r^n)}{1 - r}, r \neq 1$$

If $|r| < 1$, then $\lim_{n \to \infty} r^n = 0$ and $S_{\infty} = \frac{u_1(1 - 0)}{1 - r} = \frac{u_1}{1 - r}$

### Example 3.23

Find the sum to infinity of the geometric series $45 - 15 + 5 - \frac{5}{3} + \frac{5}{9} - \ldots$

**Solution**

$u_1 = 45$ and $r = -\frac{1}{3}$

$$S_{\infty} = \frac{45}{1 - \left(-\frac{1}{3}\right)} = \frac{45}{\frac{4}{3}} = \frac{135}{4}$$

### Example 3.24

The sum of the first three terms of a geometric series is 140 and the sum to infinity is 160. Find the tenth term.

**Solution**

$$S_3 = 140 \Rightarrow \frac{u_1(1 - r^3)}{1 - r} = 140$$

$$S_{\infty} = 160 \Rightarrow \frac{u_1}{1 - r} = 160$$

Substituting the second equation into the first:

$$160(1 - r^3) = 140 \Rightarrow 1 - r^3 = \frac{140}{160} = \frac{7}{8}$$

Rearranging:

$$r^3 = \frac{1}{8} \Rightarrow r = \frac{1}{2}$$

Substituting $r$ into the second equation: $u_1 = 80$

Finally,

$$u_{10} = 80\left(\frac{1}{2}\right)^9 = \frac{5}{32}$$

## Example 3.25

If a ball has elasticity such that it bounces up 80% of its previous height, find the total vertical distance travelled down and up by this ball when it is dropped from a height of 3 metres. Ignore friction and air resistance.

### Solution

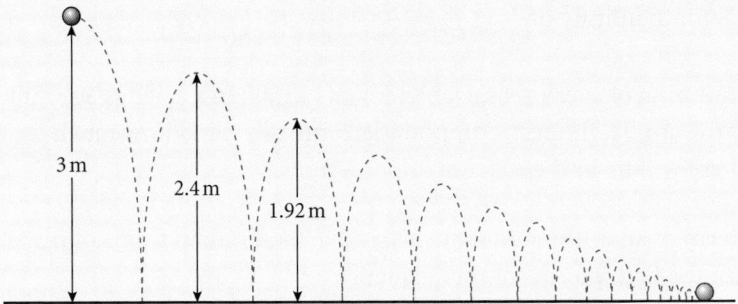

After the ball is dropped the initial 3 m, it bounces up and down a vertical distance of 2.4 m. Each bounce after the first bounce, the ball travels 0.8 times the previous height twice – once upwards and once downwards. So, the total vertical distance is given by

$$h = 3 + 2(2.4 + (2.4 \times 0.8) + (2.4 \times 0.8^2) + \dots) = 3 + 2 \times l$$

The amount in parentheses is an infinite geometric series with $a_1 = 2.4$ and $r = 0.8$. The value of that quantity is

$$l = \frac{2.4}{1 - 0.8} = 12$$

Hence the total distance required is

$$h = 3 + 2(12) = 27 \text{ m}$$

## Exercise 3.4

1. Find the sum of the arithmetic series $11 + 17 + \dots + 365$

2. Find the value of $9 + 13 + 17 + \dots + 85$

3. Find the value of $8 + 14 + 20 + \dots + 278$

4. Find the value of $155 + 158 + 161 + \dots + 527$

5. Find the value of $120 + 24 + \dfrac{24}{5} + \dfrac{24}{25} + \dots + \dfrac{24}{78\,125}$

6. Find the value of $2 - 3 + \dfrac{9}{2} - \dfrac{27}{4} + \ldots - \dfrac{177\,147}{1024}$

7. Find the value of $\displaystyle\sum_{k=0}^{13}(2 - 0.3k)$

8. Find the sum to infinity of $2 - \dfrac{4}{5} + \dfrac{8}{25} - \dfrac{16}{125} + \ldots$

9. Find the sum to infinity of $\dfrac{1}{3} + \dfrac{\sqrt{3}}{12} + \dfrac{1}{16} + \dfrac{\sqrt{3}}{64} + \dfrac{3}{256} + \ldots$

10. At the beginning of every year, Daniel invests \$1500 in a savings account that pays 6% annual interest, compounded annually. Find how much money there will be in the account after 15 years.

11. The $k$th term of an arithmetic sequence is $2 + 3k$. Find, in terms of $n$, the sum of the first $n$ terms of this sequence.

12. Find the least number of terms required for the series $17 + 20 + 23 + \ldots$ to exceed 678.

13. Find the least number of terms required for the series $-18 - 11 - 4 - \ldots$ to exceed 2335.

14. Consider the arithmetic sequence $3, 7, 11, \ldots, 999$

    (a) Find the sum of this sequence.

    (b) Every third term of the sequence is removed (i.e. $u_3, u_6, u_9, \ldots$ are removed). Find the sum of the terms of the remaining sequence.

15. The sum of the first 10 terms of an arithmetic sequence is 235, and the sum of the second 10 terms is 735. Find the first term and the common difference.

16. Use your GDC or a spreadsheet to evaluate each sum.

    (a) $\displaystyle\sum_{k=1}^{20}(k^2 + 1)$

    (b) $\displaystyle\sum_{i=3}^{17}\left(\dfrac{1}{i^2 + 3}\right)$

    (c) $\displaystyle\sum_{n=1}^{100}(-1)^n\dfrac{3}{n}$

**17.** A ball is dropped from a height of 16 m. Every time it hits the ground it bounces 81% of its previous height.

   **(a)** Find the maximum height it reaches after the 10th bounce.

   **(b)** Find the total distance travelled by the ball until it comes to rest. (Assume no friction and no loss of elasticity.)

**18.**

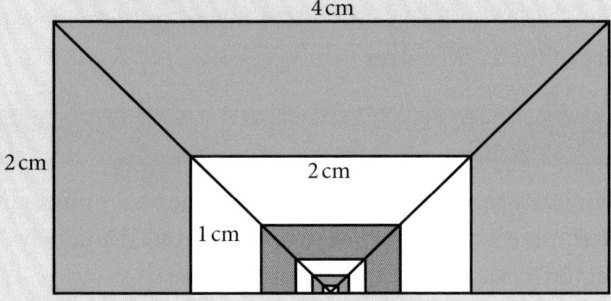

The largest rectangle has dimensions 4 by 2, as shown; another rectangle is constructed inside it with dimensions 2 by 1. The process is repeated. The region surrounding every other inner rectangle is shaded, as shown.

   **(a)** Find the total area for the three regions shaded already.

   **(b)** If the process is repeated indefinitely, find the total area of the shaded regions.

**19.** The sides of a square are 16 cm long. A new square is formed by joining the midpoints of the adjacent sides and two of the resulting triangles are coloured as shown.

   **(a)** If the process is repeated 6 more times, determine the total area of the shaded region.

   **(b)** If the process is repeated indefinitely, find the total area of the shaded region.

# 3 Sequences and series

## 3.5 Annuities and amortisation

In Example 3.22, we applied geometric series to the calculation of annuities, which are equal regular payments that earn an annual interest rate. We also showed how the problem could be solved using the built-in financial package on a graphing calculator. Example 3.26 compares the calculation of an annuity using geometric series to the use of a TVM solver, and provides an introduction to the use of the TVM solver.

### Example 3.26

A retirement plan requires a regular payment of $800 at the beginning of every year for 25 years. Interest is earned at 2% per year, compounded annually. Calculate the value of the investment at the end of the 25 years using

(a) a geometric series      (b) a TVM solver.

**Solution**

(a)  $800(1.02) + 800(1.02)^2 + \ldots + 800(1.02)^{25}$

$= 800(1.02)[1 + 1.02 + 1.02^2 + \ldots + 1.02^{24}]$

$= 800(1.02)\left[\dfrac{1.02^{25} - 1}{1.02 - 1}\right]$

$= \$26{,}136.72$

(b)  Using the TVM solver, we make the following entries:

$N = 25$

$I\% = 2$

$PV = 0$

$PMT = -800$

$FV = 26\,136.72$

$P/Y = 1$

$C/Y = 1$

$PMT:\quad BEGIN$

Note that because the $800 is being paid, it has a negative sign. However, the value of the annuity at the end of the 25 years is the future value (FV) and is positive, as the investor receives this money upon completion of the payments. Since payments are made at the beginning of the year, BEGIN is highlighted for the PMT option.

The TVM solver is very helpful in comparing financial instruments without having to create and evaluate a geometric series. Example 3.27 demonstrates how the annuity in Example 3.26 can be analysed using the TVM solver.

## Example 3.27

Consider the annuity from Example 3.26, in which $800 is invested each year at 2% per year, compounded annually. Using a TVM solver, answer the following questions.

(a) Find the value of the investment after 25 years if the monthly payment is changed to $810.

(b) Find the value of the monthly payment if the value of the investment at the end of the 25 years is to be $30,000.

(c) In Example 3.26, the annuity had a value of $26,136.72 after 25 years. If the payment per year is reduced to $761, find how much longer it takes for the investment to reach a value of $26,136.72

## Solution

(a) The following entries are input into the TVM solver:

$N = 25$

$I\% = 2$

$PV = 0$

$PMT = -810$

$FV = 26\,463.43$

$P/Y = 1$

$C/Y = 1$

$PMT:$   $BEGIN$

This means that within 25 years, an additional $250 is invested in payments, but an increase in $326.71 in future value is obtained due to the accumulation of interest.

(b) In this case, we solve for the unknown payment. The entries in the TVM solver are as follows:

$N = 25$

$I\% = 2$

$PV = 0$

$PMT = -918.25$

$FV = 30\,000$

$P/Y = 1$

$C/Y = 1$

$PMT:$   $BEGIN$

Thus, the payments need to be increased to $918.25 per year to have $30,000 after 25 years.

(c) In this example, we must find the number of payments. Since $P/Y=1$, the number of payments is given by $N$.

$N = 26$
$I\% = 2$
$PV = 0$
$PMT = -761$
$FV = 26\,136.72$
$P/Y = 1$
$C/Y = 1$
$PMT:\quad BEGIN$

Thus, reducing the yearly payment to $761 results in an additional year of investment in order to obtain the same future value of $26,136.72

So far, we have considered examples in which money is invested over time and accumulates interest according to the compound interest model. We now consider the situation in which one borrows an amount of money and repays it in regular payments that earn interest. We call this sum of money a **loan** and the process of repaying the loan is called **amortisation**. Example 3.28 demonstrates the amortisation of loans and the use of the TVM solver in comparing loan amortisation schemes.

### Example 3.28

In order to pay for a university course, a student takes a loan of $20,000. The unpaid balance on the loan has an interest rate of 3% per year, compounded annually. Use a TVM solver to answer the following questions about the amortisation of the loan.

(a) The loan is to be repaid in payments of $1200 per year. Find how long it takes to repay the loan, if payments are made at the end of each year.

(b) Hence, calculate how much has been paid in total in amortising the loan.

(c) If the loan is to be amortised in 10 years, find the annual payment, and how much has been paid in total.

### Solution

(a) Use the TVM solver to find the value of $N$.

$N = 23.4$
$I\% = 3$
$PV = 20000$
$PMT = -1200$
$FV = 0$
$P/Y = 1$
$C/Y = 1$
$PMT: END$

The present value is positive, as the student receives this money from an institution. As the payments are outgoing, they are negative in value. The future value is zero, as the student wants to pay off the loan. According to the TVM solver, the loan is amortised in about 23 years with payments of $1200 per year.

(b) $1200 \times 23.4 = \$28,080$

Thus, the student pays $28,080 in amortising the loan.

(c) Using $N = 10$, the inputs for the TVM solver are as follows:

$N = 10$

$I\% = 3$

$PV = 20\,000$

$PMT = -2344.61$

$FV = 0$

$P/Y = 1$

$C/Y = 1$

$PMT: END$

To amortise the loan in 10 years, the student must pay $2344.61 annually. However, the total amount paid in 10 years would be $2344.61 \times 10 = \$23,446.10$

Example 3.28 shows that if the annual payment increases, the time taken to amortise a loan decreases, and consequently, the total amount paid to amortise the loan decreases.

We will now use the TVM solver to analyse a loan in which the accumulation of interest does not coincide with the regular payments.

**Example 3.29**

A young family takes a loan of $100,000 to buy a house. Interest on the loan accumulates at the rate of 1.5% per year, compounded semi-annually. The family will amortise the loan in monthly payments, paid at the beginning of each month.

(a) The loan is to be amortised in 25 years. Find the monthly payment.

(b) Calculate how much the family has paid in total after 25 years.

(c) Explain the benefits to the family of increasing the monthly payment by $75. Justify your answer.

**Solution**

(a) Since payments are made monthly, $N = 25 \times 12 = 300$ and $P/Y = 12$
    In addition, since interest is compounded semi-annually, $C/Y = 2$
    Entering the values into the TVM solver:

$$N = 300$$
$$I\% = 1.5$$
$$PV = 100\,000$$
$$PMT = -399.22$$
$$FV = 0$$
$$P/Y = 12$$
$$C/Y = 2$$
$$PMT: \quad BEGIN$$

The monthly payment is $399.22

(b) At the end of 25 years, the family has paid $399.22 \times 300 = \$119{,}766$

(c) Using $474.22 for payment in the TVM Solver and solving for $N$ gives
    $N = 244.4$, so about 245 months in total.

    Thus, the family amortises the loan 55 months earlier than in part (a).
    Using $N = 245$, the total amount paid in amortising the loan is
    $245 \times \$475 = \$116{,}375$. If the family can afford to pay an additional
    $75 per month, they finish paying the loan more than four and a half
    years earlier, and pay $3391 less in total.

## Exercise 3.5

1. An investment of $300 per year is made for 15 years, earning 1% per
   year, compounded annually. Find the value of the investment using:

   **(a)** geometric series          **(b)** a TVM solver.

2. Explain the financial calculation being completed in each set of the
   TVM solver parameters.

   **(a)** $N = 30$                    **(b)** $N = 15$
   $I\% = 1.5$                       $I\% = 3$
   $PV = 0$                          $PV = 0$
   $PMT = -60$                       $PMT = -52.20$
   $FV = 2286.11$                    $FV = 1000$
   $P/Y = 1$                         $P/Y = 1$
   $C/Y = 1$                         $C/Y = 1$
   $PMT: \quad BEGIN$                $PMT: \quad BEGIN$

**(c)** $N = 21.76$

$I\% = 1.2$

$PV = 0$

$PMT = -100$

$FV = 2500$

$P/Y = 1$

$C/Y = 1$

$PMT: \quad BEGIN$

**(d)** $N = 30$

$I\% = 4$

$PV = 0$

$PMT = -50$

$FV = 1579.38$

$P/Y = 12$

$C/Y = 2$

$PMT: \quad BEGIN$

3. At the beginning of each year Peter invests $1800 in a savings account that pays 5% annual interest, compounded annually. Find how many years it will take until Peter has $50,000 in his account.

4. At the beginning of each year Anjali invests $2000 in a savings account that pays annual interest, compounded annually. After 15 years she has $60,000 in her account. Find the annual interest rate.

5. **(a)** An investment of $720 is made at the beginning of every year for 12 years. Interest accumulates at the rate of 3% per year, compounded annually. Use the TVM solver to find the value of the investment at the end of the 12 years.

   **(b)** A second investor invests $60 at the beginning of every month for 12 years, and interest accumulates at the rate of 3% per year, compounded annually. Use the TVM solver to find the value of the investment at the end of the 12 years. Compare your answer to your answer in part **(a)**.

6. **(a)** A teacher invests $50 at the beginning of every month into an investment fund that earns 2% per year, compounded semi-annually. Find the value of the investment at the end of 30 years.

   **(b)** The teacher is given the option to invest $75 per month instead at a rate of 1.7% per year, compounded semi-annually. Compare the value of this investment at the end of the 30 years to the answer in part **(a)**.

   **(c)** Another teacher would like to have $28,000 at the end of the 30 years. She invests monthly at a rate of 2% per year, compounded semi-annually. Calculate how much she must invest per month.

7. A family has taken out a loan of $250,000 to purchase a house. They agree to pay the bank $2100 at the end of every month to amortise the loan, and interest accumulates on the balance at a rate of 1.3% per year, compounded annually.

   **(a)** Find how long it takes to pay back the loan, in years and months.

   **(b)** Calculate how much have they paid in total in amortising the loan.

   **(c)** The family considers increasing the monthly payment to $2300. By completing appropriate calculations with a TVM solver, justify this decision.

### Chapter 3 practice questions

1. The seventh, third and first terms of an arithmetic sequence form the first three terms of a geometric sequence. The arithmetic sequence has first term $a$ and non-zero common difference $d$.

   (a) Show that $d = \dfrac{a}{2}$

   The seventh term of the arithmetic sequence is 3. The sum of the first $n$ terms in the arithmetic sequence exceeds the sum of the first $n$ terms in the geometric sequence by at least 200.

   (b) Find the least value of $n$ for which this occurs.

2. (a) Find the sum of all integers, between 10 and 200, which are divisible by 7.

   (b) Express the sum found in part (a) using sigma notation.

   (c) An arithmetic sequence has first term 1000 and common difference $-6$. The sum of the first $n$ terms of this sequence is negative. Find the least value of $n$.

3. The sum of the first two terms of a geometric series is 10 and the sum of the first four terms is 30.

   (a) Show that the common ratio $r$ satisfies $r^2 = 2$

   (b) Given $r = \sqrt{2}$
   
      (i) find the first term       (ii) find the sum of the first ten terms.

4. The fourth term in an arithmetic sequence is 34 and the tenth term is 76.

   (a) Find the first term and the common difference.

   (b) The sum of the first $n$ terms exceeds 5000. Find the least possible value of $n$.

5. A geometric sequence has first term $a$, common ratio $r$ and sum to infinity 76. A second geometric sequence has first term $a$, common ratio $r^3$ and sum to infinity 36. Find $r$.

6. (a) The arithmetic sequence $(u_n: n \in \mathbb{Z}^+)$ has first term $u_1 = 1.6$ and common difference $d = 1.5$. The geometric sequence $(v_n: n \in \mathbb{Z}^+)$ has first term $v_1 = 3$ and common ratio $r = 1.2$
   Find an expression for $u_n - v_n$ in terms of $n$.

   (b) Determine the set of values of $n$ for which $u_n > v_n$

   (c) Determine the greatest value of $u_n - v_n$
   Give your answer correct to 3 significant figures.

7. A metal rod 1 metre long is cut into 10 pieces, the lengths of which form a geometric sequence. The length of the longest piece is 8 times the length of the shortest piece. Find, to the nearest millimetre, the length of the shortest piece.

8. **(a)** Each time a ball bounces, it reaches 95% of the height reached on the previous bounce. Initially, it is dropped from a height of 4 metres.

   What height does the ball reach after its fourth bounce?

   **(b)** How many times does the ball bounce before it no longer reaches a height of 1 metre?

   **(c)** What is the total vertical distance travelled by the ball?

9. **(a)** A geometric sequence, $u_1, u_2, u_3, \ldots$ has $u_1 = 27$ and sum to infinity $\dfrac{81}{2}$

   Find the common ratio of the geometric sequence.

   **(b)** An arithmetic sequence $v_1, v_2, v_3, \ldots$ is such that $v_2 = u_2$ and $v_4 = u_4$.

   Find the greatest value of $N$ such that $\displaystyle\sum_{n=1}^{N} v_n > 0$

10. The sum, $S_n$, of the first $n$ terms of a geometric sequence, whose $n$th term is $u_n$, is given by

$$S_n = \frac{7^n - a^n}{7^n} \text{ where } a > 0$$

   **(a)** Find an expression for $u_n$

   **(b)** Find the first term and common ratio of the sequence.

   **(c)** Consider the sum to infinity of the sequence.

   **(i)** Determine the values of $a$ such that the sum to infinity exists.

   **(ii)** Find the sum to infinity when it exists.

11. A rope of length 81 metres is cut into $n$ pieces of increasing lengths that form an arithmetic sequence with a common difference of $d$ metres. Given that the lengths of the shortest and longest pieces are 1.5 metres and 7.5 metres respectively, find the values of $n$ and $d$.

12. **(a)** A bank offers loans of $\$P$ at the beginning of a particular month at a monthly interest rate of $I$. The interest is calculated at the end of each month and added to the amount outstanding. A repayment of $\$R$ is required at the end of each month. Let $S_n$ denote the amount outstanding, in dollars, immediately after the $n$th monthly repayment.

   **(i)** Find an expression for $S_1$ and show that:

$$S_2 = P\left(1 + \frac{I}{100}\right)^2 - R\left(1 + \left(1 + \frac{I}{100}\right)\right)$$

   **(ii)** Determine a similar expression for $S_n$

   Hence show that:

$$S_n = P\left(1 + \frac{I}{100}\right)^n - \frac{100R}{I}\left(\left(1 + \frac{I}{100}\right)^n - 1\right)$$

**(b)** Sue borrows $5000 at a monthly interest rate of 1% and plans to repay the loan in 5 years (60 months).

    **(i)** Calculate the required monthly repayment, giving your answer correct to 2 decimal places.

    **(ii)** After 20 months, Sue inherits some money and she decides to repay the loan completely at that time. Giving your answer correct to the nearest dollar, how much will she have to repay?

**13.** Phil takes out a bank loan of $150,000 to buy a house, at an annual interest rate of 3.5%. The interest is calculated at the end of each year and added to the amount outstanding.

To pay off the loan, Phil makes annual deposits of $P at the end of every year in a savings account, paying an annual interest rate of 2%. He makes his first deposit at the end of the first year after taking out the loan.

David visits a different bank and makes a single deposit of $Q into a new savings account, the annual interest rate being 2.8%.

**(a)** Find the amount Phil would owe the bank after 20 years. Give your answer to the nearest dollar.

**(b)** Show that the total value of Phil's savings after 20 years is:

$$\frac{(1.02^{20} - 1)P}{(1.02 - 1)}$$

**(c)** Given that Phil's aim is to own the house after 20 years, find the value for $P$ to the nearest dollar.

**(d) (i)** David wishes to withdraw $5000 at the end of each year for a period of $n$ years. Show that an expression for the minimum value of $Q$ is

$$\frac{5000}{1.028} + \frac{5000}{1.028^2} + \ldots + \frac{5000}{1.028^n}$$

    **(ii)** Hence or otherwise, find the minimum value of $Q$ that would permit David to withdraw annual amounts of $5000 indefinitely. Give your answer to the nearest dollar.

# Geometry and trigonometry 1

**4**

Calculating measurements of planar and 3-dimensional objects is important in the application of mathematics. This chapter revises and extends topics that may have been introduced in earlier years. Geometry and trigonometry go well together as they provide visual models for analyses.

## 4.1 Coordinate geometry in a plane – lengths of segments and midpoints

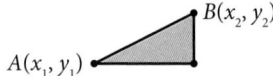

Figure 4.1 Coordinate points $A$ and $B$

Given the Cartesian coordinates of two points $A(x_1, y_1)$ and $B(x_2, y_2)$, the length $AB$ of the segment $[AB]$ is the hypotenuse of a right-angled triangle whose side lengths are the differences in the **abscissae** ($x$-measurements) and **ordinates** ($y$-measurements).

$AB = \sqrt{(x_2 - x_1)^2 + (y_2 - y_1)^2}$

### Example 4.1

Find the distance between each pair of points.

(a) $B(8, -1), N(-4, 4)$

(b) $C(12, 6), P(-3, -2)$

(c) $E(\sqrt{3}, 5), R(3\sqrt{3}, 3)$

### Solution

(a) $BN = \sqrt{(8 + 4)^2 + (-1 - 4)^2} = 13$

(b) $CP = \sqrt{(12 + 3)^2 + (6 + 2)^2} = 17$

(c) $ER = \sqrt{(\sqrt{3} - 3\sqrt{3})^2 + (5 - 3)^2} = 4$

The coordinates of the **midpoint** $C(x, y)$ of $[AB]$ are the **mean values** of the abscissae ($x$-coordinates) and the **mean values** of the ordinates ($y$-coordinates) of the points $A$ and $B$.

$$C(x, y) = C\left(\frac{x_1 + x_2}{2}, \frac{y_1 + y_2}{2}\right)$$

### Example 4.2

Find the coordinates of the midpoint of the line segment joining each pair of points in Example 4.1

(a) $B(8, -1), N(-4, 4)$

(b) $C(12, 6), P(-3, -2)$

(c) $E(\sqrt{3}, 5), R(3\sqrt{3}, 3)$

**Solution**

(a) $\left(\dfrac{8 - 4}{2}, \dfrac{-1 + 4}{2}\right) = \left(2, \dfrac{3}{2}\right)$

(b) $\left(\dfrac{12 - 3}{2}, \dfrac{6 - 2}{2}\right) = \left(\dfrac{9}{2}, 2\right)$

(c) $\left(\dfrac{\sqrt{3} + 3\sqrt{3}}{2}, \dfrac{5 + 3}{2}\right) = (2\sqrt{3}, 4)$

## Coordinate geometry in a plane – lines and intersections

Given two lines in a plane, three possibilities exist:

- The lines have the same gradient and the same $y$-intercept; hence, they are the same line (coincident lines).
- The lines have the same gradient, but different $y$-intercepts; hence, they are parallel and do not intersect.
- The lines have different gradients; hence, they intersect at one point.

**Same line**

Graph the first function, $f(x) = 2x - 3$, and graph the line passing through $(1, -1)$ and $(4, 5)$.
It overlaps the first function.
They are coincident lines.

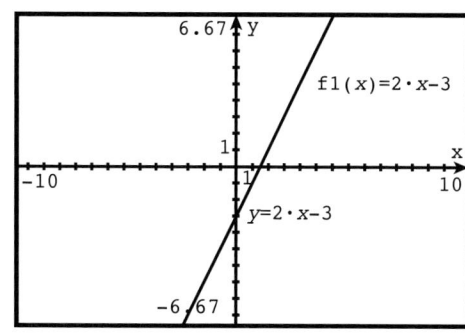

**Figure 4.2** Same line (coincident lines)

**Parallel lines**

The first function is $f(x) = 2x - 3$ in blue.

The second function is $f(x) = 2x + 4$ in red.

Their gradients are the same, but their $y$-intercepts are not.
The two lines are parallel.

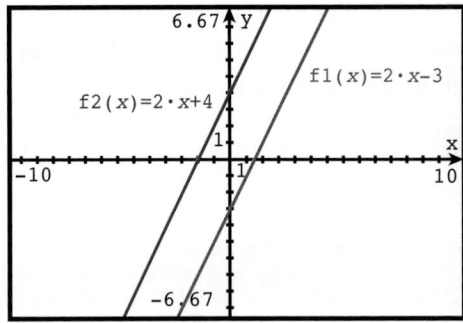

**Figure 4.3** Parallel lines

**Intersecting lines**

The first function is $f(x) = 2x - 3$ in blue.

The second function is $f(x) = -3x + 2$ in red.

Their gradients are not the same.
They are intersecting lines.

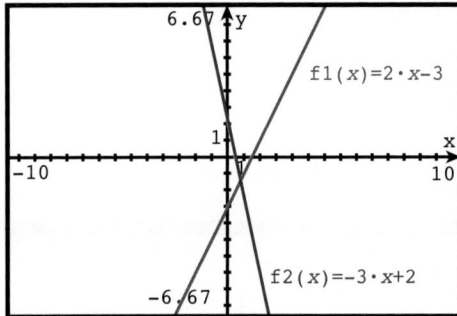

**Figure 4.4** Intersecting lines

---

### Example 4.3

By inspection, determine if each pair of lines are parallel, coincident, or intersecting.

(a) $y = 2x - 3$
   $y = -2x + 3$

(b) $y = 2x - 3$
   $2y = 4x - 6$

(c) $y = 2x - 3$
   $2x - y = 3$

(d) $y = 2x - 3$
   $2x - y = 6$

(e) $y = 2x - 3$
   $x = -2y + 3$

(f) $2x - 6y = 4$
   $x - 3y = 8$

---

**Solution**

(a) Intersecting: their gradients are different.

(b) Coincident: the second is a scalar multiple of the first.
   They are the same line.

(c) Coincident: they are the same line.

(d) Parallel: their gradients are the same, but their intercepts are different.

(e) Intersecting: their gradients are different.

(f) Parallel: their gradients are the same, but their intercepts are different.

---

## Coordinate geometry in a plane – perpendicular lines

If two lines are perpendicular to each other, one is said to be **normal** to the other at the point of intersection.

Recall that when two lines $l_1$ and $l_2$, with gradients $m_1$ and $m_2$ respectively, are perpendicular to each other, then:

$$m_1 \cdot m_2 = -1$$

and one gradient is the negative reciprocal of the other:

$$m_1 = -\frac{1}{m_2}$$

## Example 4.4

List three lines that are perpendicular to $y = \frac{2}{3}x + 6$

### Solution

The negative reciprocal of the gradient $\frac{2}{3}$ is $-\frac{3}{2}$, so consider the family of lines whose gradient

is $-\frac{3}{2}$, i.e. $y = -\frac{3}{2}x + c$

Choose three lines with different intercepts such as

$$y = -\frac{3}{2}x - 3, \ y = -\frac{3}{2}x \text{ and } y = -\frac{3}{2}x + 6$$

We can show this on a GDC. Plot the graph of $y = \frac{2}{3}x + 6$

Add these equations to the original plot. Their graphs are all perpendicular to the original.

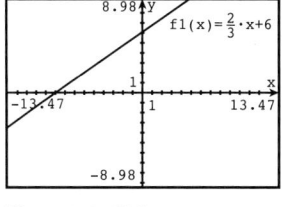

**Figure 4.5** GDC screen showing the graph of $y = \frac{2}{3}x + 6$

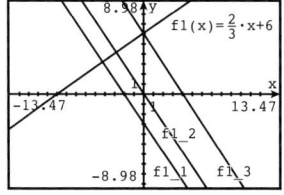

**Figure 4.6** Lines perpendicular to $y = \frac{2}{3}x + 6$

## Example 4.5

Find the equation of the perpendicular bisector of the segment $[AB]$, given the coordinates of $A$ and $B$ are $(-3, 6)$ and $(5, -2)$ respectively.

### Solution

The gradient of $AB$ is $\dfrac{6 + 2}{-3 - 5} = -1$, so the gradient of the perpendicular to

$[AB]$ is 1. The midpoint of $[AB]$ is $\left(\dfrac{-3 + 5}{2}, \dfrac{6 - 2}{2}\right) = (1, 2)$

Therefore, $y - 2 = (x - 1)$ or $y = x + 1$

With a GDC, the problem is easier to visualise:

Insert the points on a graph and (optionally) draw the segment $[AB]$

The perpendicular bisector of a line segment $[AB]$ is the normal to $[AB]$ at its midpoint.

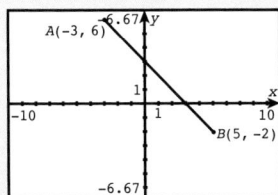 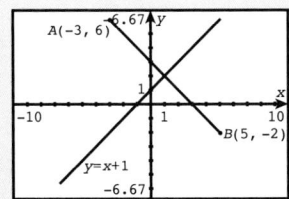

Draw the perpendicular bisector of the segment, then find its equation.

103

## Exercise 4.1

1. Find the distance between each pair of points.
   (a) $V(4, -3)$, $W(-2, 5)$
   (b) $K(3, -1)$, $L(-1, 7)$
   (c) $T(-6, 7)$, $G(2, -8)$

2. Find the midpoint of the segment between each pair of points.
   (a) $V(4, -3)$, $W(-2, 5)$
   (b) $K(3, -1)$, $L(1, 7)$
   (c) $T(-6, 7)$, $G(2, -8)$

3. Determine the gradient of each line and state the gradient of a line perpendicular to it.
   (a) $y = 4x + 1$
   (b) $y = -2x + 3$
   (c) $y = -\frac{2}{3}x - 1$

4. Find the coordinates of the points of intersection of each pair of lines.
   (a) $y = 3x - 1$
   $y = x + 7$
   (b) $y = 5x - 6$
   $y = 2x + 3$
   (c) $x - 2y = 11$
   $2x - y = 16$
   (d) $y = 3x + 2$
   $y = 6x + 1$
   (e) $y = \frac{1}{3}x + 13$
   $y = x - 7$
   (f) $x + y = -1$
   $3x - 2y = 12$

5. Determine the equation of the line parallel to $y = \frac{1}{2}x + 6$ that passes through the point $(2, -4)$.

6. Determine the equation of the line parallel to $y = -\frac{2}{3}x + 4$ that passes through the point $(-6, -1)$.

7. Determine the equation of the line perpendicular to $y = 2x - 6$ that passes through the point $(4, 2)$.

8. Determine the equation of the line perpendicular to $y = -\frac{2}{3}x + 4$ that passes through the point $(4, 5)$.

9. Determine the equation of the perpendicular bisector of the segment $[PQ]$, given the points $P(-5, 12)$ and $Q(7, 6)$.

10. Determine the equation of the perpendicular bisector of the isosceles trapezium, the coordinates of whose vertices are $A(-3, 8)$, $B(1, 10)$, $C(-4, 4)$, and $D(5, 9)$.

11. (a) The points $T(1, 3)$, $X(4, 6)$, $G(6, 4)$, and $C(8, 8)$ are coordinates on a treasure map.

    Find the buried treasure, which is at the intersection of the perpendicular bisector of the segment between the coconut tree ($T$) and the big $X$, with the line drawn between the green rock ($G$) and the opening of the cave ($C$).

**(b)** It appears that over time, the $X$ drawn on the ground has been redrawn and moved! Fortunately, the mark was moved, but always in line with the original two points $T$ and $X$. Can the treasure still be found along the line defined by $[GC]$?

# 4.2 Trigonometry

Trigonometry is the study of angles. Basic definitions relate to the use of a right-angled triangle, but any triangle will suffice. Trigonometry is used in many fields that involve angles and linear measurement, for instance, in surveying, navigating and GPS triangulation.

## Basic definitions

Given a right-angled triangle $ABC$, the sides opposite angles $A$, $B$, and $C$ are designated $a$, $b$, and $c$ respectively by mathematical convention (Figure 4.7).

### Inverses and reciprocals

If $\sin A = \dfrac{a}{c}$, its **inverse** is expressed as $A = \arcsin\left(\dfrac{a}{c}\right)$, often noted on a GDC as $\sin^{-1}\left(\dfrac{a}{c}\right)$. Be careful not to confuse this calculator notation with the reciprocal $\dfrac{1}{\sin A}$. The reciprocal, should one be needed, is known as a **cosecant** and expressed as $\csc A = \dfrac{c}{a}$. The other inverses and reciprocals are shown in Table 4.1.

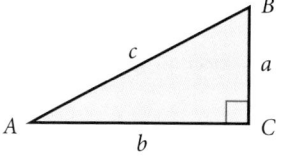

**Figure 4.7** Right-angled triangle $ABC$

| Ratio | Inverse | Reciprocal | |
|-------|---------|------------|--|
| $\sin A = \dfrac{a}{c}$ | $A = \arcsin\left(\dfrac{a}{c}\right)$ | $\csc A = \dfrac{c}{a}$ | (cosecant) |
| $\cos A = \dfrac{b}{c}$ | $A = \arccos\left(\dfrac{b}{c}\right)$ | $\sec A = \dfrac{c}{b}$ | (secant) |
| $\tan A = \dfrac{a}{b}$ | $A = \arctan\left(\dfrac{a}{b}\right)$ | $\cot A = \dfrac{b}{a}$ | (cotangent) |

**Table 4.1** Inverse and reciprocal trigonometric functions

## Example 4.6

Given the right-angled triangle $RST$, with $ST = 8$ cm and $RT = 15$ cm and $\hat{T} = 90°$ find the exact ratios and angles to the nearest degree.

(a) $\sin R$  

(b) $\cos R$

(c) $\tan R$  

(d) $\arcsin\left(\dfrac{r}{t}\right)$

(e) $\arctan\left(\dfrac{r}{s}\right)$  

(f) $\arccos\left(\dfrac{s}{t}\right)$

**Solution**

(a) $\dfrac{8}{17}$    (b) $\dfrac{15}{17}$    (c) $\dfrac{8}{15}$

(d) $28°$    (e) $28°$    (f) $62°$

A GDC provides all three ratios, but just one should be enough when solving a problem analytically using the Pythagorean theorem. The tangent ratio is particularly useful if you compare it to the gradient of a line.

## Modelling problems with trigonometry

Trigonometry is often used to measure distances and angles when the situation can be modelled with a 2-dimensional right-angled triangle. Sighting objects above the horizontal line of vision, such as when one looks upwards to the top of a tree, produces an **angle of elevation** relative to the horizontal; whereas, sighting objects below the horizontal line of vision, such as when one looks downwards from a higher position than the object, results in an **angle of depression** relative to the horizontal.

### Example 4.7

Standing 10 m at $A$ from the base of a flagpole $[TB]$ on level ground, and using an upside-down protractor, the angle of elevation of the top of the flagpole is found to be $57°$. How much higher than eye level is the flagpole?

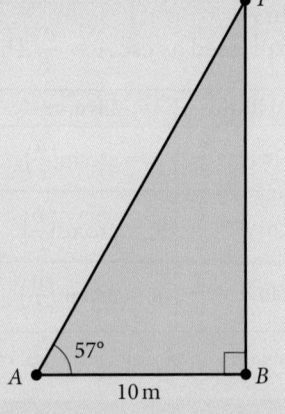

**Solution**

In this diagram, $[AB]$ is the horizontal line of vision parallel to level ground. The length $AB = 10$ m and $\widehat{A} = 57°$. Since point $A$ is at eye level, $[TB]$ is the height of the flagpole above eye level.

$$\tan A = \frac{TB}{10} \Rightarrow TB = 10 \cdot \tan 57° \Rightarrow TB \approx 15.4 \text{ m}$$

Trigonometry is also used extensively in map reading and navigation. By convention, angles are measured from north, in a clockwise direction, and expressed as a three digit number, called a **bearing**. Hence, east is 090°, south is 180°, and west is 270°, returning to north after a 360° rotation.

## Example 4.8

Orienteering is a sport that requires participants to cover terrain quickly using a map and a compass. If the bearing required for the next 50 m is 300°, how far west of the present position is the next position?

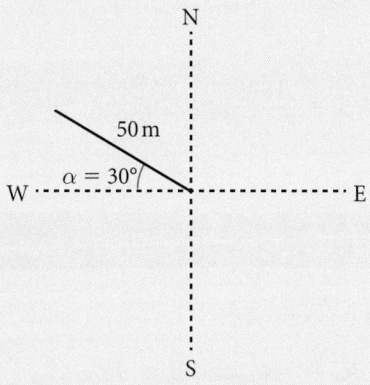

### Solution

In the diagram, since a bearing of 300° means an addition 30° beyond a westerly heading, the westward component is $50 \cos 30° \approx 43.4$ m

Airport runways are marked at both ends with large numbers that indicate the bearings rounded to the first two digits. A plane directed to runway 14L should be on a bearing of approximately 140° and landing on the left runway of two parallel runways. If the wind conditions change and the plane should be approaching the same runway from the opposite direction, the marking there will show 32R, requiring the pilot to maintain an approach with a bearing of approximately 320°. This runway is designated 14L-32R.

## Example 4.9

The construction of a new runway parallel to runway 06, which has an actual bearing of 62°, is being considered. If federal instrument landing regulations require that the runway centres be 1035 m apart, how much farther east will runway 06R be to the present one?

**Solution**

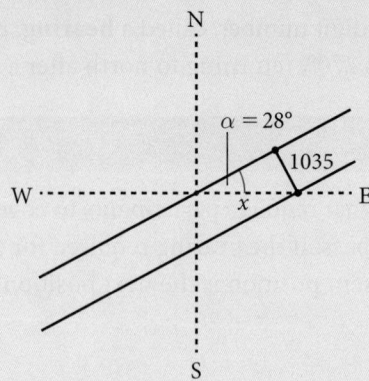

Since $\dfrac{1035}{x} = \sin 28° \Rightarrow x = \dfrac{1035}{\sin 28°} \approx 2205\,\text{m}$

## Exercise 4.2

1. In the right-angled triangle $ABC$,

   $$AC = 25\text{cm}, BC = 7\text{cm}, \text{ and } \widehat{B} = 90°$$

   Find the following ratios and angles to the nearest degree.

   **(a)** $\cos A$      **(b)** $\sin C$      **(c)** $\tan A$

   **(d)** $\arcsin\left(\dfrac{c}{b}\right)$      **(e)** $\arctan\left(\dfrac{a}{c}\right)$      **(f)** $\arccos\left(\dfrac{c}{b}\right)$

2. Determine the size of the smallest positive angle $\theta$ which satisfies:

   **(a)** $\sin\theta = \dfrac{4}{5}$      **(b)** $\cos\theta = \dfrac{8}{17}$      **(c)** $\tan\theta = 1$

3. A woman, standing 3 m from a streetlight, casts a 2-metre shadow on level ground. She is 1.62 m tall. What is the height of the streetlight? Solve this problem with and without trigonometry.

4. A kite is at the end of a 60 m string, hovering at an angle of elevation of 54°. Assume that the string is perfectly taut. How far above the ground is the kite? Give your answer to the nearest metre.

5. Determine the height of an isosceles triangle with a vertex angle of 72° and equal sides of length 2 m.

6. From a point on the ground 50 m from the base of a building, a communication antenna is visible on the top edge of the building. The base of the antenna has an angle of inclination of 39°, while the angle of inclination of the top of the antenna is 50°.

(a) Find, to the nearest metre:

  (i) the height of the building

  (ii) the height of the top of the antenna above ground

  (iii) the height of just the antenna.

(b) Why was it not possible to use the difference between the two angles of inclination of 11° to find the height of the antenna?

7. The angle of inclination to the top of a taller building from the top of a shorter one is 30°, while the angle of depression to the base of the taller building is 60°. If the difference in the buildings' heights is 16 m, how far apart are they?

8. A building 30 m high sits on top of a hill. The angles of elevation of the top and bottom of the building from the same spot at the base of the hill are measured to be 55° and 50° respectively. Relative to the base, how high is the hill?

9. A 30 m cable connects the top of an 18 m totem pole to a point on the ground. Why is it unnecessary to find the exact value of the angle of inclination to find the distance along the ground to the totem pole?

10. From a 50-metre lighthouse on the shoreline, a boat is sighted at an angle of depression of 4° moving directly towards the lighthouse at a constant speed. Five minutes later, the angle of depression of the boat is 12°. What is the speed of the boat in knots, given that a knot is defined to be $1.852 \, \text{km} \, \text{h}^{-1}$?

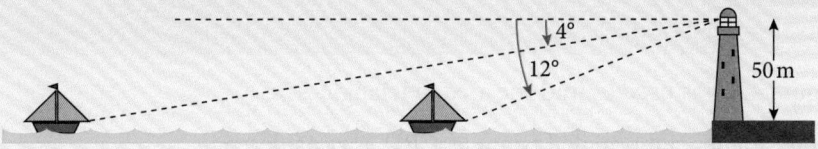

## 4.3  Areas of triangles, sine rule and cosine rule

The equation for the area of a triangle is Area $= \frac{1}{2}bh$, where the area is half the base ($b$) multiplied by the perpendicular height ($h$). Now, by using trigonometry, we have another possibility.

The base, $AC$, has length $b$ and the perpendicular height, $BD$, has length $h$ (Figure 4.8).

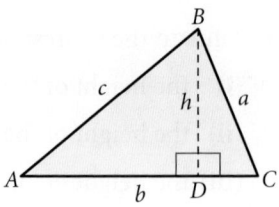

Figure 4.8  The area of triangle $ABC$

In the right-angled triangle $ABD$,

$$\sin A = \frac{h}{c} \Rightarrow c \sin A = h$$

So, if Area $= \frac{1}{2}bh$

$$\Rightarrow \text{Area} = \frac{1}{2}b(c \sin A)$$

This formula uses two sides and the included angle, and is thus known by the acronym SAS.

### Example 4.10

Find the area of $\triangle LMN$, given that $\widehat{M} = 60°$, $LM = 4\,\text{cm}$, and $MN = 10\,\text{cm}$

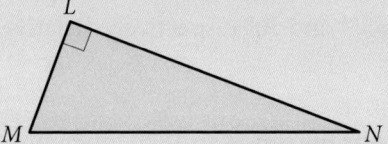

#### Solution

In this instance, $l = 10\,\text{cm}$, $n = 4\,\text{cm}$, and the included angle, $\widehat{M} = 60°$

$$\text{Area} = \frac{1}{2}l(n \sin M) = \frac{1}{2} \cdot 10 \cdot 4 \cdot \sin 60 = 20 \cdot \frac{\sqrt{3}}{2} = 10\sqrt{3}\ \text{cm}^2$$

### Sine rule

Figure 4.8 was used to find the area of triangle $ABC$.

If the smaller right-angled triangle, $CBD$, is considered,

$$\sin C = \frac{h}{a} \Rightarrow a \sin C = h$$

In which case, Area $= \frac{1}{2}b(a \sin C)$

As the height of triangle $ABC$ is the same regardless of whether one considers the right-angled triangle $ABD$ or $CBD$,

$$a \sin C = c \sin A$$

$$\Rightarrow \frac{a}{\sin A} = \frac{c}{\sin C}$$

The sine rule states:
$$\frac{a}{\sin A} = \frac{b}{\sin B} = \frac{c}{\sin C}$$

And with a simple rotation of the triangle, it can also be shown that $\dfrac{a}{\sin A} = \dfrac{b}{\sin B}$

This is known as the **sine rule**.

## Example 4.11

In the triangle $ABC$, $\widehat{A} = 42°$, $a = 6$ cm, and $\widehat{B} = 63°$

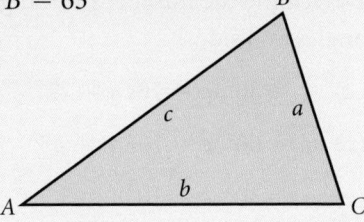

Find:

(a)  angle $C$        (b)  side $b$        (c)  side $c$.

## Solution

Note that one angle and its opposite side are known, and that a second angle is also given.

(a)  As two angles are known, the missing angle is the supplement of the sum of the two that are known.

$$\widehat{A} = 42° \text{ and } \widehat{B} = 63° \Rightarrow \widehat{C} = 180° - (42° + 63°) = 75°$$

(b)  Using the sine rule:

$$\frac{6}{\sin 42°} = \frac{b}{\sin 63°} \Rightarrow b = \frac{6 \sin 63°}{\sin 42°} \Rightarrow b \approx 7.99 \text{ cm}$$

(c)  Using the sine rule again:

Since the value of $b$ is an approximation to 3 significant figures, while other known measurements are exact, the exact measurements are preferred.

$$\frac{6}{\sin 42°} = \frac{c}{\sin 75°} \Rightarrow c = \frac{6 \sin 75°}{\sin 42°} \Rightarrow c \approx 8.66 \text{ cm}$$

In Example 4.11, two angles and one side (AAS) were known. However, using the sine rule, when only one angle is known, but two sides are known (SSA), does not guarantee a solution. There are two possible cases.

**Case 1**

The side opposite the known angle is too short to produce a triangle:

For an SSA diagram to produce exactly one solution, the length $a$ must be exactly the same length as the height or it must be longer than side $c$. Since the height of the triangle is $h = c \sin A$, this requires $a = c \sin A$ or $a > c$

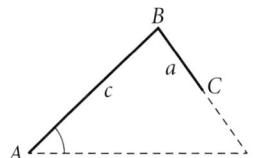

**Figure 4.9**  Case 1

**Case 2**

The side opposite the known angle produces two possible solutions:

The two possible triangles are $\triangle ABC$ and $\triangle ABC'$

Two triangles are produced if $a$ is longer than the height and shorter than side $c$. Hence, two triangles are produced only if $c \sin A < a < c$

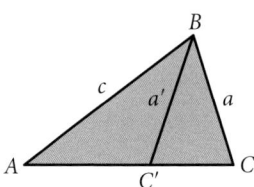

**Figure 4.10**  Case 2

**Example 4.12**

Determine the number of triangles that are formed with each group of angles and sides.

(a) $\hat{A} = 30°$, $c = 10$, $a = 12$        (b) $\hat{A} = 30°$, $c = 10$, $\hat{B} = 70°$

(c) $\hat{C} = 60°$, $c = 10$, $\hat{A} = 45°$       (d) $\hat{C} = 73°$, $c = 10$, $a = 12$

(e) $\hat{A} = 30°$, $c = 10$, $a = 5$         (f) $\hat{A} = 60°$, $c = 10$, $a = 9$

**Solution**

(a)

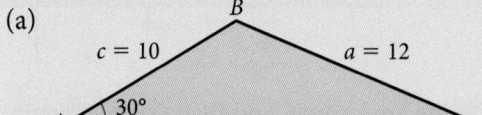

Drawing a diagram will show that side $a$ is longer than side $c$; hence there is only 1 triangle possible.

(b) 1      (c) 1      (d) 0      (e) 1      (f) 2

## Revisiting the area of a triangle with the sine rule

Recall that the area of a triangle can be expressed as Area $= \dfrac{1}{2}b(c \sin A)$

Using the sine rule, we know that $\dfrac{b}{\sin B} = \dfrac{c}{\sin C} \Rightarrow b = \dfrac{c \sin B}{\sin C}$

Substituting for $b$ in the area formula: Area $= \dfrac{1}{2}b(c \sin A)$

$$\Rightarrow \text{Area} = \dfrac{1}{2}\left(\dfrac{c \sin B}{\sin C}\right)(c \sin A)$$

$$= \dfrac{1}{2}\left(\dfrac{c^2 \sin A \sin B}{\sin C}\right)$$

**Example 4.13**

Find the area of $\triangle STU$, given $\hat{T} = 20°$, $t = 10$ cm, and $\hat{U} = 130°$

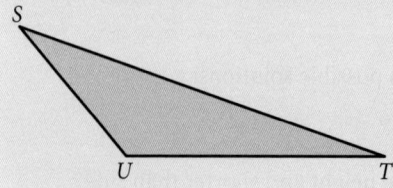

## Solution

Work out the size of the third angle: $\hat{S} = 180° - (20° + 130°) = 30°$

With the length of one side given, this diagram now shows an ASA situation, so use the alternative area formula:

$$\text{Area} = \frac{1}{2}\left(\frac{t^2 \sin S \sin U}{\sin T}\right)$$

$$= \frac{1}{2} \cdot \frac{10^2 \cdot \sin 30° \cdot \sin 130°}{\sin 20°} \simeq 56.0 \, \text{cm}^2$$

## Cosine rule

Consider the triangle $ABC$ again with the side $b$ divided into two parts at $D$ by the height $h$.

Let the lengths of the two parts be $AD = x$ and $DC = b - x$

Now, consider the two smaller right-angled triangles formed by the height $h$.

In the right-angled triangle $ABD$, note that

$$c^2 = h^2 + x^2 \Rightarrow c^2 - x^2 = h^2, \text{ and } \cos A = \frac{x}{c} \Rightarrow c \cos A = x$$

Consequently, in the right-angled triangle $CBD$

$$a^2 = h^2 + (b-x)^2 \Rightarrow a^2 = h^2 + b^2 - 2bx + x^2$$
$$\Rightarrow a^2 = (c^2 - x^2) + b^2 - 2bx + x^2$$
$$\Rightarrow a^2 = c^2 + b^2 - 2b(c \cos A)$$

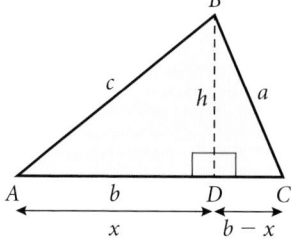

**Figure 4.11** Side $b$ is divided into two parts at $D$

### Example 4.14

Find the length of the side opposite angle $\hat{L}$ when the following triangle measurements are known.

(a) $j = 6$ cm, $k = 10$ cm, $\hat{L} = 120°$

(b) $j = 6\sqrt{2}$ cm, $k = 12$ cm, $\hat{L} = 135°$

## Solution

(a) $l^2 = 6^2 + 10^2 - 2 \cdot 6 \cdot 10 \cos 120°$

$= 36 + 100 - 120 \cdot \left(\frac{-1}{2}\right)$

$= 196$

$\Rightarrow l = \sqrt{196} = 14 \, \text{cm}$

(b) $l^2 = (6\sqrt{2})^2 + 12^2 - 2 \cdot 6\sqrt{2} \cdot 12 \cos 135°$

$= 72 + 144 - 144\sqrt{2} \cdot \left(\frac{-\sqrt{2}}{2}\right)$

$= 360$

$\Rightarrow l = \sqrt{360} = 6\sqrt{10} \, \text{cm}$

## Example 4.15

The distance from $A$ to $B$ cannot be measured due to a pond that is between them.
However, from point $C$, the distance $AC = 20$ m and $BC = 32$ m and $\widehat{C} = 60°$
Determine the distance $AB$.

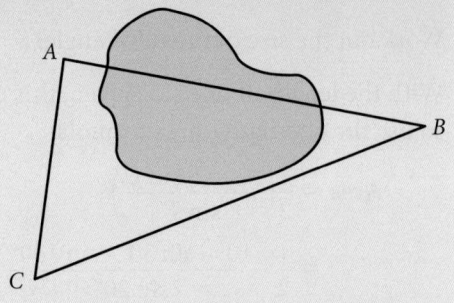

### Solution

Using the cosine rule, $AB = \sqrt{20^2 + 32^2 - 2 \cdot 20 \cdot 32 \cos 60°}$

$$= \sqrt{400 + 1024 - 1280 \cdot \frac{1}{2}}$$

$$\Rightarrow AB = \sqrt{784} = 28 \text{ m}$$

## Exercise 4.3

1. For each of **(a)** to **(d)**:

   **(i)** draw the triangle

   **(ii)** indicate which formula(e) and steps would be used to find the missing angles and/or sides and its area.

   **(a)** an SAS triangle        **(b)** an ASA triangle

   **(c)** an AAS triangle        **(d)** an SSS triangle

2. For each triangle $ABC$, find the missing lengths to 3 significant figures and missing angles to the nearest degree.

   **(a)** $\widehat{A} = 30°$, $c = 10$ cm, and $\widehat{B} = 72°$

   **(b)** $\widehat{A} = 36°$, $a = 8$ cm, and $\widehat{B} = 72°$

   **(c)** $a = 6$ cm, $\widehat{C} = 60°$, and $b = 10$ cm

   **(d)** $a = 12$ cm, $b = 18$ cm, and $c = 15$ cm

3. Find the areas of the triangles in question **2**.

4. Triangle $PQR$ has $\widehat{P} = 30°$, $r = 14$ cm, and $p = 10$ cm

   **(a)** Verify that triangle $PQR$ is an ambiguous case that produces two possible figures.

   **(b)** Find the missing side(s) and angle(s).

5. Determine the areas of the two possible triangles produced in question **4**.

## Measurements in 3 dimensions

## Volumes and surface areas

You should already be familiar with these formulae for volume and surface area.

| Polyhedron | Volume | Surface area | Where: |
|---|---|---|---|
| Cuboid | $V = l \cdot w \cdot h$ | $S = 2(l \cdot w + l \cdot h + w \cdot h)$ | $l$ = length<br>$w$ = width<br>$h$ = height |
| Sphere | $V = \frac{4}{3}\pi r^3$ | $S = 4\pi r^2$ | $r$ = radius |
| Prism | $V = A_{\text{base}} \cdot h$ | $S = 2 \cdot A_{base} + \sum_{i=1}^{n} A_{\text{lateral face}}$ | $A_{\text{base}}$ = area of polygonal base<br><br>$\sum_{i=1}^{n} A_{\text{lateral face}}$ = sum of the $n$ lateral faces, each of which is a rectangle |
| Pyramid | $V = \frac{1}{3} A_{\text{base}} \cdot h$ | $S = A_{\text{base}} + \sum_{i=1}^{n} A_{\text{lateral face}}$ | $A_{\text{base}}$ = area of polygonal base<br><br>$\sum_{i=1}^{n} A_{\text{lateral face}}$ = sum of the $n$ lateral faces, each of which is a triangle |
| Cylinder | $V = \pi r^2 h$ | $S = 2\pi r^2 + 2\pi rh$ or<br>$= 2\pi r(r + h)$ | $r$ = radius<br>$h$ = height |
| Cone | $V = \frac{1}{3}\pi r^2 h$ | $S = \pi r^2 + \pi rl$<br>where $l = \sqrt{r^2 + h^2}$,<br>so alternatively,<br>$S = \pi r(r + \sqrt{r^2 + h^2})$ | $r$ = radius<br>$h$ = height<br>$l$ = lateral height |

**Table 4.2**  Volume and surface area formulae

### Example 4.16

A triangular pyramid sits on top of a triangular prism. The prism has a height of $h$. How tall would the pyramid have to be for the prism and pyramid to have the same volume?

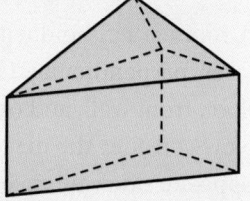

### Solution

Since the shapes have the same base, their volumes depend only on their heights. Hence, three times as tall.

### Example 4.17

A sphere has of radius $r$ and a cone has a radius of $r$ and a height of $2r$.

(a) How much bigger is the volume of the sphere than the volume of the cone?

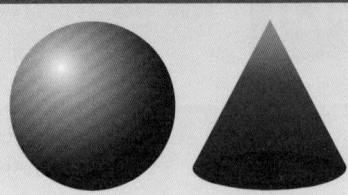

(b) What is the ratio of the surface area of the cone to that of the sphere?

### Solution

(a) The volume of the sphere is $V = \dfrac{4}{3}\pi r^3$

The volume of the cone is $V = \dfrac{1}{3}\pi r^2 h$
Since $h = 2r$

$$\Rightarrow V = \frac{1}{3}\pi r^2 h = \frac{1}{3}\pi r^2(2r) = \frac{2}{3}\pi r^3$$

(b) The surface area of the sphere is $S = 4\pi r^2$

while the surface area of the cone is $S = \pi r(r + \sqrt{r^2 + h^2})$

Since $h = 2r$

$$\Rightarrow S = \pi r(r + \sqrt{r^2 + (2r)^2}) = \pi r(r + \sqrt{5r^2})$$
$$= (1 + \sqrt{5})\pi r^2$$

The ratio of the surface area of the cone to that of the sphere is

$$1 : \frac{1 + \sqrt{5}}{4}$$

## Distances in 3 dimensions

We can find the distance between two points in 3-dimensional space by the extended use of the Pythagorean theorem.

A room is a cuboid with every pair of opposite faces parallel, and intersecting faces, perpendicular. Consider the point at the corner of the room where the floor, front wall, and one side wall meet (Figure 4.12). Call this point $A(x_1, y_1, z_1)$. Now consider the diagonally opposite point where the ceiling, back wall, and opposite side wall meet. Call this point $B(x_2, y_2, z_2)$.

On any wall, floor, or ceiling, a diagonal drawn is the hypotenuse of a right-angled triangle.

All points on the floor are mapped by the $x$- and $y$-axes. Consider the point on the floor directly below point $B$ and call it $B'$.

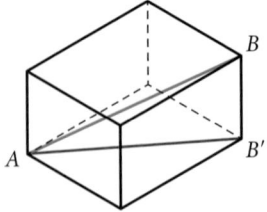

**Figure 4.12** Cuboid room

The diagonal drawn from point $A$ to point $B'$ is the hypotenuse of a right-angled triangle, and its length is $AB' = \sqrt{(x_2 - x_1)^2 + (y_2 - y_1)^2}$

Now visualise a screen, erected perpendicularly to the floor along $[AB']$
The diagonal line drawn from $A$ to $B$ on this screen is the hypotenuse of another right-angled triangle whose sides are $AB'$ along the floor and the height of the room, which is the difference of the $z$-values, $(z_2 - z_1)$

Hence, $(AB)^2 = (AB')^2 + (z_2 - z_1)^2$

$$= (x_2 - x_1)^2 + (y_2 - y_1)^2 + (z_2 - z_1)^2$$

$$\Rightarrow AB = \sqrt{(x_2 - x_1)^2 + (y_2 - y_1)^2 + (z_2 - z_1)^2}$$

Given two points in space, $A(x_1, y_1, z_1)$ and $B(x_2, y_2, z_2)$, the distance $AB$ is found by taking the square root of the sum of the squares of the differences in each dimension.

## Example 4.18

Find the distance between each pair of points.

(a) $A(1, 3, 5)$, $M(2, 1, 3)$

(b) $C(12, 7, -2)$, $P(8, 5, -2)$

(c) $F(7, 1, 3\sqrt{6})$, $S(4, 3, \sqrt{6})$

## Solution

(a) $AM = \sqrt{(2 - 1)^2 + (3 - 1)^2 + (3 - 5)^2} = 3$

(b) $CP = \sqrt{(8 - 12)^2 + (5 - 7)^2 + (-2 + 2)} = 2\sqrt{5}$

(c) $FS = \sqrt{(4 - 7)^2 + (3 - 1)^2 + (\sqrt{6} - 3\sqrt{6})^2} = 8$

## Example 4.19

A zip line is to be constructed across a rectangular, sloped piece of land from a point up a tree ($T$) in one corner to a mound ($M$) at the diagonally opposite corner on the ground. The field is 30 m × 60 m and the difference in elevation between the starting and ending points of the zip line is 20 m. Disregarding the sag of the zip line, what length of line would be required?

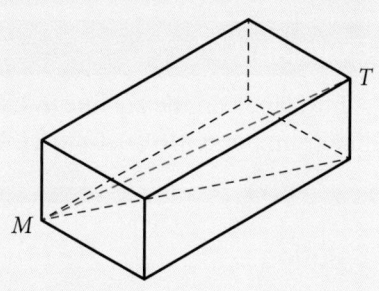

**Solution**

If the mound is considered to be at $M(0, 0, 0)$, then the starting point up the tree is $T(30, 60, 20)$. So $MT = \sqrt{30^2 + 60^2 + 20^2} = \sqrt{100(9 + 36 + 4)}$

$\Rightarrow MT = 70\,\text{m}$

## Midpoints of line segments in 3-dimensional space

The coordinates of $C$, the midpoint of the segment $AB$, where $A$ and $B$ have the coordinates $A(x_1, y_1, z_1)$ and $B(x_2, y_2, z_2)$, have the **mean value** of the $x$-, $y$-, and $z$-coordinates: $C(x, y, z) = C\left(\dfrac{x_1 + x_2}{2}, \dfrac{y_1 + y_2}{2}, \dfrac{z_1 + z_2}{2}\right)$

### Example 4.20

Find the midpoint of the line segment defined by each pair of points.

(a)  $A(1, 3, 5)$, $M(2, 1, 3)$

(b)  $E(3, -1, 2)$, $R(-3, 2, 4)$

(c)  $F(7, 1, 3\sqrt{6})$, $S(4, 3, \sqrt{6})$

**Solution**

(a) $\left(\dfrac{1 + 2}{2}, \dfrac{3 + 1}{2}, \dfrac{5 + 3}{2}\right) = \left(\dfrac{3}{2}, 2, 4\right)$

(b) $\left(\dfrac{3 - 3}{2}, \dfrac{-1 + 2}{2}, \dfrac{2 + 4}{2}\right) = \left(0, \dfrac{1}{2}, 3\right)$

(c) $\left(\dfrac{7 + 4}{2}, \dfrac{1 + 3}{2}, \dfrac{3\sqrt{6} + \sqrt{6}}{2}\right) = \left(\dfrac{3}{2}, 2, 2\sqrt{6}\right)$

Using similar triangles, we can determine the coordinates of any portion of an entire length if the measurements along the three axes are known.

### Example 4.21

A bird lands on the zip line in Example 4.19, two-thirds of the way along the line from point $M(0, 0, 0)$ to $T(30, 60, 20)$. What are its coordinates?

**Solution**

Take two-thirds of the measurements in the difference of the $x$-, $y$-, and $z$-coordinates. Hence, the bird is at $B\left(30 \cdot \dfrac{2}{3}, 60 \cdot \dfrac{2}{3}, 20 \cdot \dfrac{2}{3}\right) = B\left(20, 40, \dfrac{40}{3}\right)$

1. A cuboid measuring $10\,\text{cm} \times 12\,\text{cm} \times 20\,\text{cm}$ is cut into a rectangular pyramid, keeping the face that is $12\,\text{cm} \times 20\,\text{cm}$ as its base and $10\,\text{cm}$ as its height.

   (a) What is the volume of the pyramid?

   (b) What is the volume if the $10\,\text{cm} \times 12\,\text{cm}$ face is used as the base instead?

2. One official size 5 FIFA football is $22\,\text{cm}$ in diameter. The cover is made of 32 identical curved panels. What is the area of each panel?

3. The simplest Japanese *kokeshi* (wooden doll) is formed from a sphere (head) on top of a cylinder (body). Such a doll can be made from plastic blanks with the head being the same diameter as the body, then painted. The blanks are cylinders with a diameter of $5\,\text{cm}$ and a height of $15\,\text{cm}$.

   (a) What is the volume of the doll?

   (b) How much of the blank is wasted?

4. Using the plastic blanks in question **3**, a model of an ice cream cone is produced by creating a hemisphere on the base of a cone. The cone must be no longer than three times the height of the hemisphere. What percentage of the $5\,\text{cm} \times 15\,\text{cm}$ cylindrical blank is used in making this model of an ice cream cone?

5. A sphere is carved out of a cube. What percentage of the original volume is wasted?

6. A drone is flown along the edges of a 3-dimensional grid measured in metres, from $O(0, 0, 0)$ to $P(10, 20, 20)$, then directly back to point $O$.

   (a) What is the total distance flown?

   (b) What are its coordinates when it has covered three-quarters of the total distance?

## Chapter 4 practice questions

1. Find the distance between each pair of points.

   (a) $A(11, 2), B(-1, -3)$

   (b) $C(9, -3), D(3, 5)$

   (c) $E(5, 3\sqrt{5}), F(9, \sqrt{5})$

2. Find the coordinates of the midpoint of the line segment joining each pair of points.

   (a) $P(6, -1), Q(-4, 5)$

   (b) $R(-7, -2), S(-1, 8)$

   (c) $T(1, -4), U(8, 0)$

3. Determine if each pair of lines is parallel, coincident, or intersecting.

   **(a)** $2y = 4x - 2$
      $2x - y = 2$

   **(b)** $y = 2x - 3$
      $x = 2y - 3$

   **(c)** $y = 2x - 4$
      $3x - \dfrac{3}{2}y = 6$

4. Determine the coordinates of the points of intersection of each pair of lines.

   **(a)** $y = \dfrac{3}{2}x - 1$
      $y = x + 3$

   **(b)** $y = x - 6$
      $y = 2x + 4$

   **(c)** $x + 2y = 3$
      $2x - 3y = -8$

5. Determine the equation of the line parallel to

   $$y = -\frac{2}{3}x + 1 \text{ that passes through the point } (-3, 0)$$

6. Determine the equation of the line perpendicular to

   $$y = \frac{2}{3}x + 1 \text{ that passes through the point } (-3, -1)$$

7. Determine the equation of the line perpendicular to

   $$y = -\frac{1}{2}x - 5 \text{ that passes through the point } (-1, 5)$$

8. Determine the equation of the perpendicular bisector of the segment $[RS]$, given the coordinates of the endpoints are $R(7, -10)$ and $S(1, 2)$.

9. A family of isosceles triangles is built upon the line segment $[PQ]$ as the base where the coordinates of the endpoints are $P(-3, 6)$ and $Q(5, 10)$. Find the equation of the line that passes through the apex of all such isosceles triangles.

10. In the isosceles triangle $ABC$, $AB = AC = 41$ cm, $BC = 18$ cm Find the following ratios and angles, to the nearest degree.

    **(a)** $\cos B$

    **(b)** $\sin C$

    **(c)** $\tan\left(\dfrac{A}{2}\right)$

    **(d)** $B$

    **(e)** $A$

11. Two students who can run at the same speed decide to run from the corner of a 70 m × 105 m football pitch to the diagonally opposite corner. One runs along the diagonal, while the other runs along the perimeter. To the nearest metre, how much of a head start should the second runner have in order for them to arrive at the opposite corner at the same time?

12. A regular pentagon has equal sides of 24 cm. What is the distance from the centre to any vertex?

13. A rhombus has sides of length 20 cm. The two opposite acute angles are 36°. What is the length of the longer diagonal?

14. A tent flysheet is to be tethered with string to a nearby tree on level ground, 1 m away. The edge of the flysheet is to be 1.4 m off the ground. At what height on the tree should the string be attached to maintain an angle of depression of 30°?

15. From the top of a 40 m cliff, the angle of depression of the near bank of the stream below is 72° while that of the far bank is 60°. To the nearest metre, what is the width of the stream?

16. When viewed from a distance of 50 m, the top of a tree has an angle of elevation of 22°. As one approaches the tree, the angle of elevation changes to 40°. To the nearest metre, how much closer to the tree is the second view point compared to the initial one?

17. For each triangle $ABC$, find all missing lengths to 3 significant figures and missing angles to the nearest degree.
    (a) $\widehat{C} = 45°$, $a = 10$ cm, and $\widehat{B} = 30°$
    (b) $\widehat{A} = 110°$, $a = 8$ cm, and $\widehat{B} = 40°$
    (c) $c = 12$ cm, $\widehat{B} = 36°$, and $a = 18$ cm
    (d) $a = 12$ cm, $b = 6$ cm, and $c = 8$ cm

18. Find the areas of the triangles in question 17.

19. Shape $RST$ has a right angle at $T$, with $\widehat{R} = 36°$, $t = 14$ cm and $r = 6$ cm. At present, it is not a triangle. Determine the two possible changes necessary to the length of $r$ to form exactly one triangle.

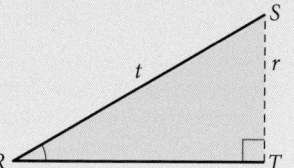

20. The following diagram shows a triangle $ABC$ and a sector $BDC$ of a circle with centre $B$ and radius 6 cm. The points $A$, $B$, and $D$ are on the same line.

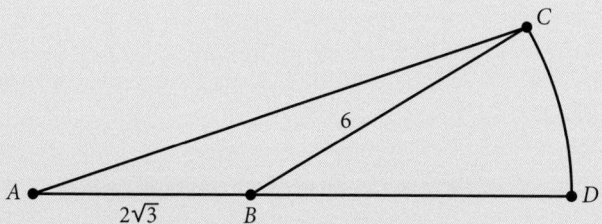

$AB = 2\sqrt{3}$ cm, $BC = 6$ cm, area of triangle $ABC = 3\sqrt{3}$ cm², $A\widehat{B}C$ is obtuse.
    (a) Find $A\widehat{B}C$
    (b) Find the exact area of the sector $BCD$.

**21.** In triangle $ABC$, $AB = 6$ cm and $AC = 8$ cm
The area of the triangle is $16$ cm².

(a) Find the two possible values for $\widehat{A}$.

(b) Given that $\widehat{A}$ is obtuse, find $BC$.

**22.** The triangle $ABC$ is equilateral of side 3 cm. The point $D$ lies on $[BC]$ such that $BD = 1$ cm. Find $\cos D\widehat{A}C$.

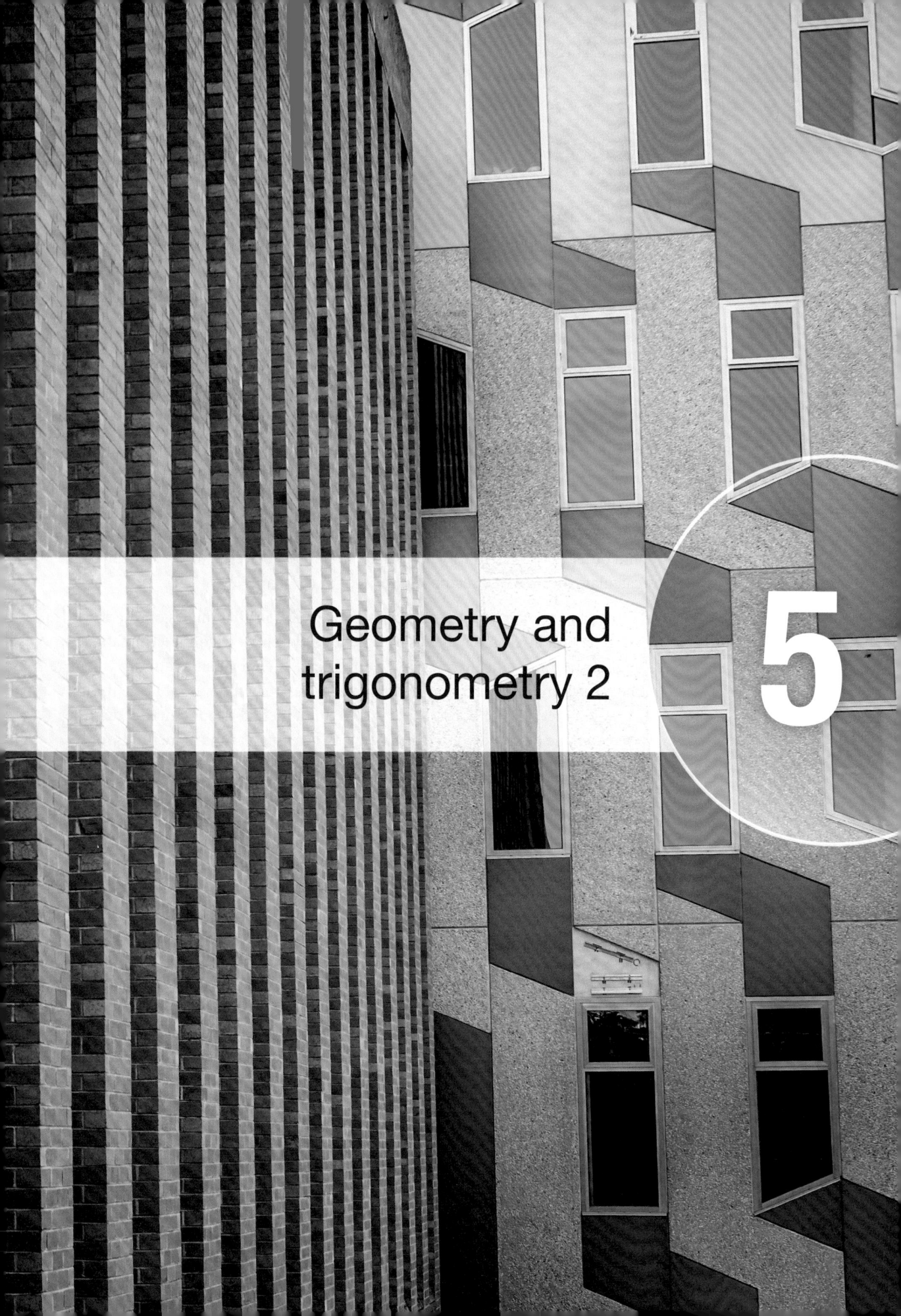

# Geometry and trigonometry 2

**5**

## Learning objectives

By the end of this chapter, you should be familiar with…

- calculating sector areas and arc lengths
- understanding radian measure and converting between degrees and radians
- understanding the definitions of $\sin \theta$, $\cos \theta$, $\tan \theta$ in the unit circle and using the trigonometric functions for all real numbers
- using the Pythagorean identity, $\tan \theta = \dfrac{\sin \theta}{\cos \theta}$, and graphical methods of solving trigonometric equations in a finite interval
- using and creating Voronoi diagrams, including: terminology (sites, vertices, edges, cells), adding a site to an existing Voronoi diagram using the incremental algorithm, nearest neighbour interpolation, and applications of Voronoi diagrams to distance/area and function interpolation.

You have already learned how to find many areas and lengths, in triangles and other polygons. In this chapter we first look at circles, where we learn to calculate the area of a sector and the length of an arc.

In the trigonometry part of this chapter, we will consider angles of rotation with the unit circle to enable us to generalise the three trigonometric ratios of sine, cosine, and tangent to apply to any angle. Then, by using ratios to calculate the measurement of arcs and sectors in a circle, we will develop the definition of radian measure. This also leads to two fundamental trigonometric identities. We conclude our study of trigonometry by learning graphical methods for solving trigonometric equations.

## 5.1 Arc length and area of a sector

A sector can be thought of as a pizza slice. This is an intuitive way to begin to understand how to calculate the area of a sector.

### Example 5.1

Suppose you have a pizza with a total area of 120 cm². The pizza is cut into 8 equal slices as shown in the diagram.

(a) What is the area of one slice?

(b) What is the total area of 3 slices?

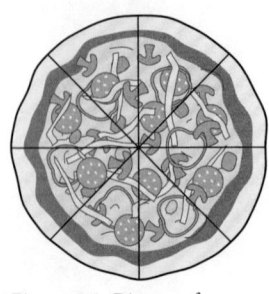

**Figure 5.1** Diagram for Example 5.1

### Solution

Use a ratio to solve the problem.

(a) Since we have cut the pizza into 8 equal slices, one slice is $\frac{1}{8}$ of the total area. Therefore, the area of one slice is $\frac{1}{8}(120) = 15\,\text{cm}^2$

(b) Now we have 3 slices, hence $\frac{3}{8}$ of the total area: $\frac{3}{8}(120) = 45\,\text{cm}^2$

## Area of a sector

Mathematically, a pizza slice is a **sector** of a circle. We are often interested in calculating the area of sectors, and we can use the method of Example 5.1 to calculate the area of any sector.

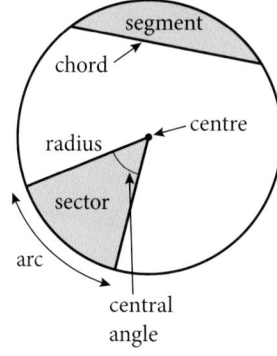

Figure 5.2 The parts of a circle

### Example 5.2

On golf courses, the grass is carefully watered to make sure that the grass receives neither too much nor too little water. On a certain area of a golf course, a lawn sprinkler is set to spray an arc of 150° with a maximum spray distance of 3 metres (Figure 5.3).

Calculate the area of grass that is watered by this sprinkler.

### Solution

Since the sprinkler sprays an arc of 150°, the watered area is $\frac{150}{360}$ of the total circle area. The area, $A$, of a circle is given by: $A = \pi r^2$

Therefore, the watered area $= \frac{150}{360}\pi(3)^2 = 3.75\pi \approx 11.8\,\text{m}^2$

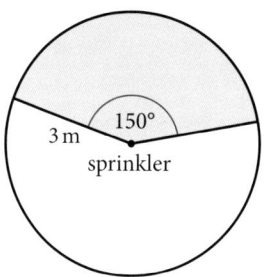

Figure 5.3 Diagram for Example 5.2

The area of a sector can be found using a fraction of the area of a circle. The numerator of the fraction is the central angle; the denominator is the whole circle (360°):

$$\text{Area of sector} = \frac{\theta}{360}A$$

This can be written as:

$$A_{\text{sector}} = \frac{\theta}{360}\pi r^2$$

where $r$ is the radius of the circle and $\theta$ is the measure of the sector's central angle in degrees.

Remember that a central angle is an angle with its vertex on the centre of a circle. The diagram to the right shows two central angles, with measurements of 70° and 290°.

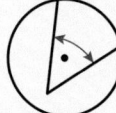 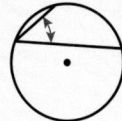 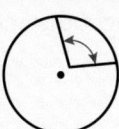

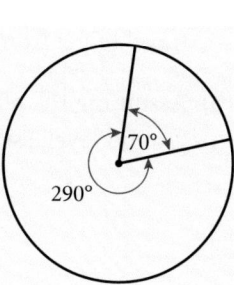

Beware! This method only works for central angle measurements! The angles shown above are not central angles.

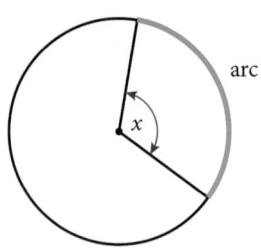

**Figure 5.4** An arc and its central angle

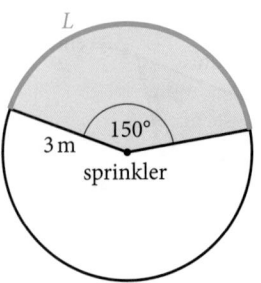

**Figure 5.5** Diagram for Example 5.3

## Length of an arc

Sometimes we are interested in the length of an **arc**.

We use the same logic we used for the area of a sector: we need the size of the central angle for the arc, and then we calculate the portion of the circumference using a ratio.

### Example 5.3

The golf course in Example 5.2 needs to add a boundary along the edge of the sprinkler arc. How long is the arc ($L$)?

### Solution

Since the arc measures 150°, we use that fraction of the circumference:

$$L = \frac{150}{360}(2\pi \times 3) = \frac{5\pi}{2} = 7.85 \text{ m (3 s.f.)}$$

Arc lengths can be used to calculate the shortest distance between two points on Earth, called the great circle route. When the great circle route passes over the North or South Pole, we can use the latitude of the two places to find the distance between them.

### Example 5.4

The great circle route from Dubai, UAE, to Seattle, USA, comes very close to the North Pole. In this case, the distance of the flight can be estimated by the arc length.

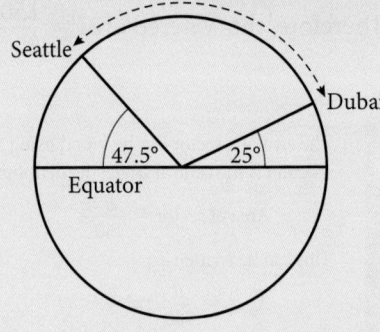

Dubai's airport is at 25° N and Seattle's airport is at 47.5° N as shown in the diagram, and the radius of the Earth is approximately 6370 km.

(a) Calculate the distance along the Earth's surface that an aeroplane must travel to get from Seattle to Dubai.

(b) Assuming the plane travels at a constant elevation of 10 km above ground level, estimate the actual flight distance.

## Solution

(a) First we must calculate the arc measure. Since the semi-circle formed by the equator measures 180°, the arc measure must be 180 − 47.5 − 25 = 107.5°. We are given that the radius of the Earth is 6370 km, so the distance along the Earth's surface is

$$\frac{107.5}{360}(2\pi)(6370) \approx 11\,952\,\text{km} = 12\,000\,\text{km (3 s.f.)}$$

(b) The new radius will be 6370 + 10 = 6380 km. Therefore we can estimate the actual flight distance to be

$$\frac{107.5}{360}(2\pi)(6380) \approx 11\,970\,\text{km} = 12\,000\,\text{km (3 s.f.)}$$

Note that our estimate of flight distance, to 3 significant figures, has not changed.

## Exercise 5.1

1. Find the length of the arc $s$ in each figure.

(a)
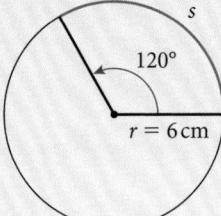

(b)
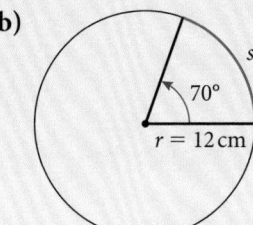

2. Find the area of the sector and the arc length determined by each given radius and central angle $\theta$. Give answers in terms of $\pi$.

(a) $\theta = 30°, r = 10\,\text{cm}$      (b) $\theta = 45°, r = 8\,\text{m}$

(c) $\theta = 52°, r = 180\,\text{mm}$      (d) $\theta = n°, r = 15\,\text{cm}$

3. Find the size of angle $\theta$ in the diagram.

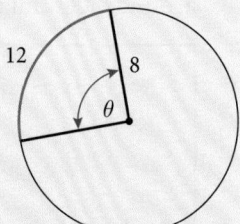

4. A bicycle with tires 70 cm in diameter is travelling such that its tires complete one and a half revolutions every second. That is, the **angular velocity** of each wheel is 1.5 revolutions per second.

   (a) What is the angular velocity of the wheel in degrees per second?

   (b) At what speed is the bicycle travelling along the ground, in $\text{km}\,\text{h}^{-1}$?

5. A sector of a circle with radius 4 has area $\dfrac{16}{5}\pi$

   Find the size of the central angle of the sector.

6. In the circle in this diagram, the value of the area of the shaded sector is equal to the value of the length of arc *l* (ignoring units). Find the radius of the circle.

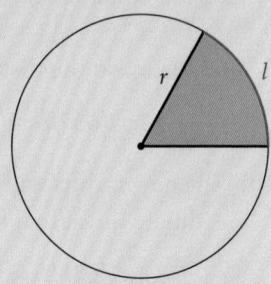

7. Many of the streets in Sun City, Arizona, USA, are formed by concentric circles. Maria and Norbert each go for a walk. Maria walks the path shown by the green arc, while Norbert walks along the path shown by the blue arc.

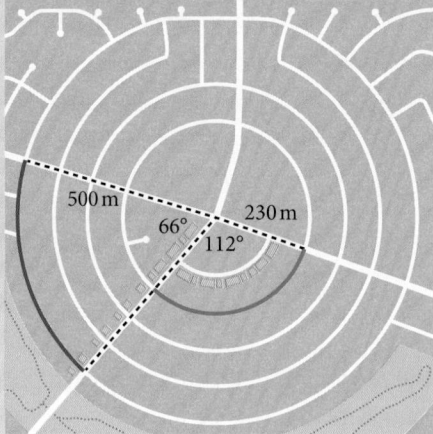

   Who walks further, and by how much?

8. A circular irrigation system consists of a 400-metre pipe that is rotated around a central pivot point.

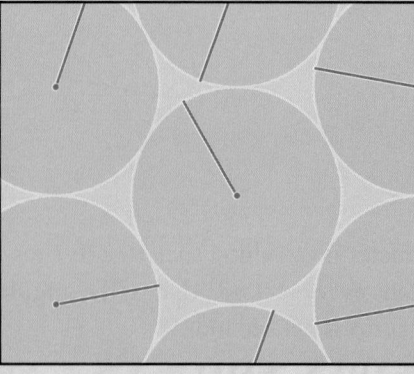

   The pipe makes one full revolution around the pivot point in a day. How much area, in square metres, is irrigated each hour?

9. Use the information given in Example 5.2 to answer these questions.

(a) In an arid climate, a golf course may require 2.5 mm of water per square metre of grass, per day. Calculate the volume of water required by the sector of grass in cm³.

The sprinkler delivers water at a rate of 800 cm³ min⁻¹.

(b) Calculate how long the sprinkler should run in order to provide 2.5 mm of water per square metre.

The sprinkler is now adjusted to spray in a 100° arc.

(c) Assuming the same flow rate and spray radius, calculate how long the sprinkler should run in order to provide 2.5 mm of water per square metre.

10. A nautical mile is the distance on the Earth's surface formed by a great-circle arc that measures 1 minute $\left(\frac{1}{60}\text{ of a degree}\right)$.

(a) Given that the radius of the Earth is 6370 km, find the length of 1 nautical mile in kilometres.

(b) Find the number of kilometres from the Equator to the South Pole.

11. A goat is tied to the outside corner of a rectangular building, using a 10 m leash. The building measures 5 m² by 8 m². Find the total area in which the goat can graze.

## 5.2 Angles of rotation and radian measure

In this section we formalise our understanding of angles greater than 180°, define positive and negative angles, and introduce a new method for measuring angles.

**Figure 5.6** Angles are not only in triangles – in this section we study angles of rotation, which can be used to describe this snowboarder's 360° jump

## Angles

Remember than an angle in a plane is made by rotating a ray about its endpoint, called the **vertex** of the angle. For the snowboarder in Figure 5.6, we can think of the snowboard as the ray. Since the snowboard does one complete revolution, the measure of the angle of rotation is 360°.

The starting position of the ray is called the **initial side** and the position of the ray after rotation is called the **terminal side** of the angle (Figure 5.7).

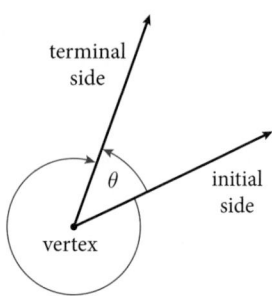

**Figure 5.7** Components of an angle

 If it seems strange that anti-clockwise angles are positive, remember that positive angles follow the numbering of the four quadrants (I, II, III, IV).

On a coordinate plane, an angle having its vertex at the origin and its initial side lying on the positive $x$-axis is said to be in **standard position** (Figure 5.8). A **positive angle** is produced when a ray is rotated in an anti-clockwise direction, and a **negative angle** when a ray is rotated in a clockwise direction.

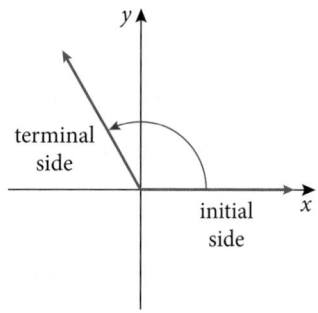

**Figure 5.8** Angle in standard position

Two angles in standard position that have the same terminal sides – regardless of the direction or number of rotations – are called **coterminal angles** (Figure 5.9). Greek letters are often used to represent angles, and the direction of rotation is indicated by an arc with an arrow at its endpoint. The $x$ and $y$ axes divide the coordinate plane into four quadrants (numbered with Roman numerals). Figure 5.9 shows a positive angle $\alpha$ and a negative angle $\beta$ that are coterminal in quadrant III.

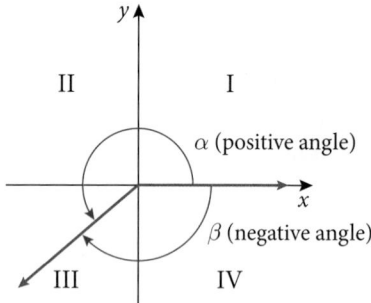

**Figure 5.9** Coterminal angles

## Measuring angles: degree measure and radian measure

Perhaps the most natural unit for measuring large angles is the **revolution**. For example, most cars have an instrument called a tachometer that indicates the number of revolutions per minute (rpm) at which the engine is operating. However, to measure smaller angles we need a smaller unit. In section 5.1 we used **degrees** to measure angles. There are 360 degrees in one revolution. Hence, the unit of one degree (1°) is defined to be one 360th of one anti-clockwise revolution about the vertex.

There is another method of measuring angles that is more natural. Instead of dividing a full revolution into an arbitrary number of equal divisions (e.g. 360), consider a central angle and its associated arc or the circle. Figure 5.10 shows three circles with radii of different lengths ($r_1 < r_2 < r_3$) and the same central angle $\theta$ subtending (intercepting) the arc lengths $s_1$, $s_2$ and $s_3$. Regardless of the size of the circle (i.e. length of the radius), the ratio of arc length, $s$, to radius, $r$, for a given circle will be constant. For the angle $\theta$ in Figure 5.10, $\frac{s_1}{r_1} = \frac{s_2}{r_2} = \frac{s_3}{r_3}$.

Note that this ratio is an arc length divided by another length (radius), so it is a number without units.

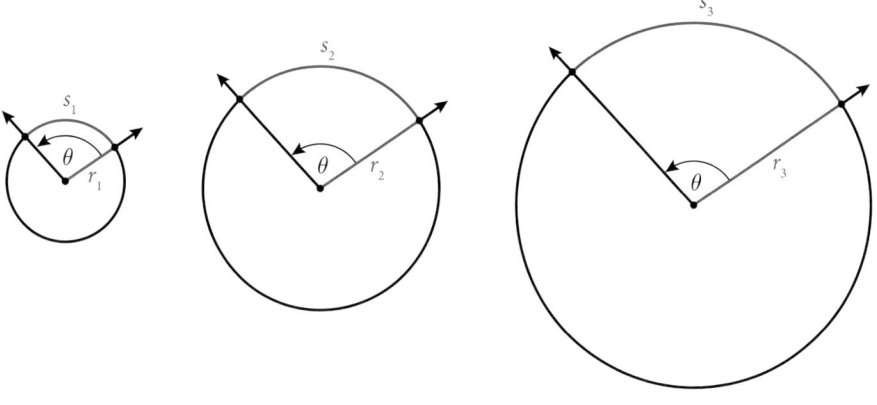

**Figure 5.10** The ratio of arc length to radius remains constant

**Major and minor arcs**
If a central angle is less than 180°, then the subtended arc is referred to as a minor arc. If a central angle is greater than 180°, then the subtended arc is referred to as a major arc.

The ratio $\frac{s}{r}$ indicates how many radius lengths, $r$, fit into the length of the arc $s$. For example, if $\frac{s}{r} = 2$, then the length of $s$ is equal to two radius lengths. Hence, this method of measuring arcs is given the name **radian** and leads to the following definition.

**Radian measure**
One **radian** is the measure of a central angle $\theta$ of a circle that subtends an arc $s$ of the circle that is exactly the same length as the radius $r$ of the circle. As shown in Figure 5.11, when $\theta = 1$, then arc length = radius.

When the measure of an angle is, for example, 5 radians, the word 'radians' does not indicate units (as when writing centimetres, seconds or degrees) but indicates the method of angle measurement. If the measure of an angle is in degrees, we must indicate this by word or symbol. For example, $\theta = 30$ degrees, or $\theta = 30°$. However, when radian measure is used it is customary to write no units or symbol. For example, a central angle $\theta$ that subtends an arc equal to 5 radius lengths (radians) is simply given as $\theta = 5$. However, radians is often given for clarity.

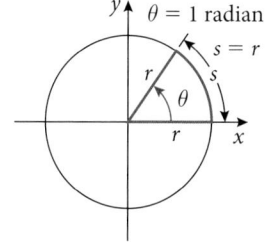

**Figure 5.11** When the radius $r$ is equal to the arc length $s$, the measure of the angle is 1 radian

## The unit circle

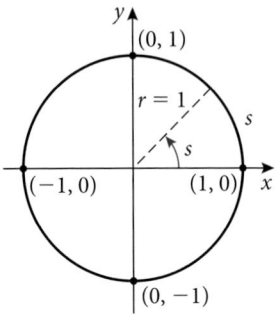

**Figure 5.12** The unit circle

When an angle is measured in radians it makes sense to draw it, or visualise it, so that it is in standard position. It follows that the angle will be a central angle of a circle whose centre is at the origin as shown in Figure 5.12. As Figure 5.11 illustrated, it makes no difference what size circle is used. The most practical circle to use is the circle with a radius of one unit, as shown in Figure 5.12.

The circle with a radius of one unit and centre at the origin (0, 0) is called the **unit circle**. The equation for the unit circle is $x^2 + y^2 = 1$. Using the familiar equation for the circumference of a circle $C = 2\pi r$, the circumference of the unit circle must be $C = 2\pi(1) = 2\pi$. Therefore, if an angle has a degree measure of 360° (i.e., the entire circle) then its radian measure is exactly $2\pi$. It follows that an angle of 180° has a radian measure of exactly $\pi$. This fact can be used to convert between degree measure and radian measure and vice versa.

**Conversion between degrees and radians**

Because $180° = \pi$ radians, $1° = \dfrac{\pi}{180}$ radians, and 1 radian $= \dfrac{180°}{\pi}$. An angle with a radian measure of 1 has a degree measure of approximately 57.3° (to 3 significant figures).

### Example 5.5

(a) Convert 30° and 45° to radian measure and sketch the corresponding arc on the unit circle.

(b) Use these results to convert 60° and 90° to radian measure.

### Solution

(a)

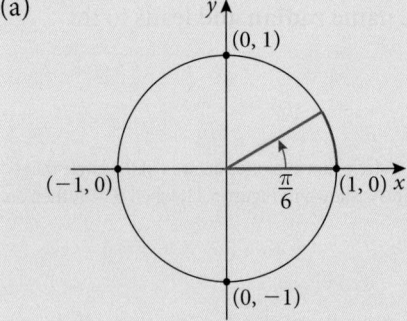

Note that the 'degree' units cancel.

$$30° = 30° \left(\frac{\pi}{180°}\right) = \frac{\pi}{6} \qquad 45° = 45° \left(\frac{\pi}{180°}\right) = \frac{\pi}{4}$$

(b) $60° = 2(30°)$ and $30° = \dfrac{\pi}{6}$, so $60° = 2\left(\dfrac{\pi}{6}\right) = \dfrac{\pi}{3}$

Similarly, $90° = 2(45°)$ and $45° = \dfrac{\pi}{4}$, so $90° = 2\left(\dfrac{\pi}{4}\right) = \dfrac{\pi}{2}$

It is very helpful to be able to quickly recall the four conversions from Example 5.5, as they are often used in trigonometry:

$$30° = \frac{\pi}{6}, 45° = \frac{\pi}{4}, 60° = \frac{\pi}{3} \text{ and } 90° = \frac{\pi}{2}$$

Of course, not all angles are multiples of 30° or 45° when expressed in degrees, and not all angles are multiples of $\frac{\pi}{6}$ or $\frac{\pi}{4}$ when expressed in radians. However, these 'special' angles often appear in problems and applications. Knowing these four facts can help you to quickly convert between degrees and radians for many common angles. For example, to convert 225° to radians, use the fact that 225° = 5(45°). Since 45° = $\frac{\pi}{4}$, then 225° = 5(45°) = $5\left(\frac{\pi}{4}\right) = \frac{5\pi}{4}$

## Example 5.6

(a) Convert the following radian measures to degrees. Express exactly, if possible. Otherwise express accurate to 3 significant figures.

(i) $\frac{4\pi}{3}$    (ii) $-\frac{3\pi}{2}$    (iii) 5    (iv) 1.38

(b) Convert the following degree measures to radians. Express exactly.

(i) 135°    (ii) −150°    (iii) 175°    (iv) 10°

## Solution

(a) (i) $\frac{4\pi}{3} = 4\left(\frac{\pi}{3}\right) = 4(60°) = 240°$

(ii) $-\frac{3\pi}{2} = -\frac{3}{2}(\pi) = -\frac{3}{2}(180°) = -270°$

(iii) $5\left(\frac{180°}{\pi}\right) \approx 286.479° = 286°$ (3 s.f.)

(iv) $1.38\left(\frac{180°}{\pi}\right) \approx 79.068° = 79.1°$ (3 s.f.)

(b) (i) $135° = 3(45°) = 3\left(\frac{\pi}{4}\right) = \frac{3\pi}{4}$

(ii) $-150° = -5(30°) = -5\left(\frac{\pi}{6}\right) = -\frac{5\pi}{6}$

(iii) $175°\left(\frac{\pi}{180°}\right) = \frac{35\pi}{36}$

(iv) $10°\left(\frac{\pi}{180°}\right) = \frac{\pi}{18}$

All GDCs will have a **degree mode** and a **radian mode**. Before doing any calculations with angles on your GDC, be certain that the mode setting for angle measurement is set correctly. Although you may be more familiar with degree measure, as you progress further in mathematics – and especially in calculus – radian measure is far more useful.

Because $2\pi$ is approximately 6.28 (to 3 significant figures), there are a little more than six radius lengths in one revolution, as shown in Figure 5.13.

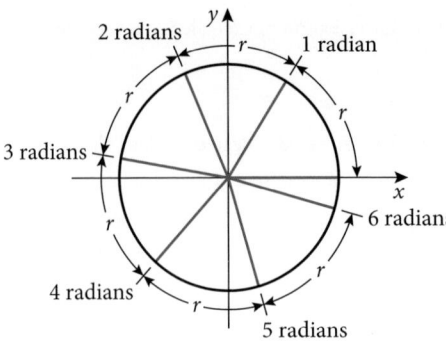

**Figure 5.13** Arcs with length of the radius placed along the circumference of a circle

**Figure 5.14** Degree measure and radian measure for common angles

Figure 5.14 shows all of the angles between 0° and 360° that are multiples of 30° or 45, and their equivalent radian measure. You will benefit by being able to quickly convert between degree measure and radian measure for these common angles.

## Arc length

For any angle $\theta$, its radian measure is given by $\theta = \dfrac{s}{r}$

Simple re-arrangement of this formula leads to a simple formula for arc length.

> Be careful! This arc length formula only works when the angle measure is in radians!

For a circle of radius $r$, a central angle $\theta$ subtends an arc of the circle of length $s$ given by $s = r\theta$ where $\theta$ is measured in radians.

### Example 5.7

A circle has a radius of 10 centimetres. Find the length of the arc of the circle subtended by a central angle of 150°.

### Solution

To use the formula $s = r\theta$, we must first convert 150° to radian measure.

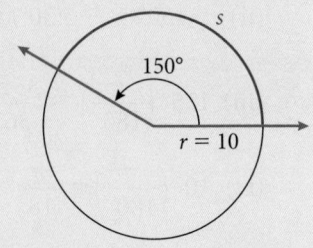

$$150° = 150° \left(\frac{\pi}{180°}\right) = \frac{150\pi}{180} = \frac{5\pi}{6}$$

Given that the radius, $r$, is 10 cm, substituting into the formula gives

> The units of the product $r\theta$ are equal to the units of $r$ because in radian measure $\theta$ has no units.

$$s = r\theta \Rightarrow s = 10\left(\frac{5\pi}{6}\right) = \frac{25\pi}{3} \approx 26.17994$$

The length of the arc is 26.2 cm (3 s.f.)

## Example 5.8

The diagram shows a circle with centre $O$ and radius $r = 6$ cm

Angle $AOB$ subtends the minor arc $AB$ such that the length of the arc is 10 cm. Find the measure of angle $AOB$ in degrees to 3 significant figures.

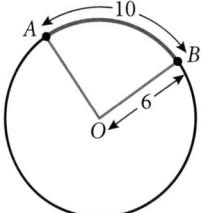

**Figure 5.15** Diagram for Example 5.8

### Solution

From the arc length formula $s = r\theta$, we can state that $\theta = \frac{s}{r}$. Remember that the result for $\theta$ will be in radian measure. Therefore, angle $AOB = \frac{10}{6} = \frac{5}{3}$ radians. Now, we convert to degrees: $\frac{5}{3}\left(\frac{180°}{\pi}\right) \approx 95.49297°$

The degree measure of angle $AOB$ is $95.5°$ (3 s.f.)

## Area of a sector

As in section 5.1, the area of a sector is a fraction of the area of the entire circle (a 'pizza slice'). With radian measure, calculating the area of a sector is a bit simpler than with degrees. Remember that the measure of the entire circle in radians in $2\pi$. We can use the logic from section 5.1 and write the area of a sector as

$$A = \frac{\theta}{2\pi}(\pi r^2) = \frac{1}{2}\theta r^2$$

where $\theta$ is given in radians.

In a circle of radius $r$, the area of a sector with a central angle $\theta$ measured in radians is

$$A = \frac{1}{2}\theta r^2$$

Again, be careful! This area of a sector formula works only when the angle measure is in radians.

## Example 5.9

A circle of radius 9 cm has a sector whose central angle has radian measure $\frac{2\pi}{3}$

Find the exact values of:

(a) the length of the arc subtended by the central angle.

(b) the area of the sector.

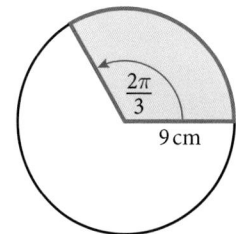

**Figure 5.16** Diagram for Example 5.9

### Solution

(a) $s = r\theta \Rightarrow s = 9\left(\frac{2\pi}{3}\right) = 6\pi$

The length of the arc is exactly $6\pi$ cm.

(b) $A = \frac{1}{2}r^2\theta \Rightarrow A = \frac{1}{2}(9)^2\left(\frac{2\pi}{3}\right) = 27\pi$

The exact area of the sector is $27\pi$ cm².

**Exercise 5.2**

In questions **1–9**, convert each angle to radians.

**1.** 60°   **2.** 150°   **3.** −270°

**4.** 36°   **5.** 135°   **6.** 50°

**7.** −45°   **8.** 400°   **9.** −480°

In questions **10–18**, convert each angle to degrees. If possible, express your answer exactly, otherwise express it accurate to 3 significant figures.

**10.** $\dfrac{3\pi}{4}$   **11.** $-\dfrac{7\pi}{2}$   **12.** 2

**13.** $\dfrac{7\pi}{6}$   **14.** −2.5   **15.** $\dfrac{5\pi}{3}$

**16.** $\dfrac{\pi}{12}$   **17** 1.57   **18.** $\dfrac{8\pi}{3}$

In questions **19–27**, the size of an angle in standard position is given. Find two angles – one positive and one negative – that are coterminal with the given angle. If no units are given, assume the angle is in radian measure.

**19.** 30°   **20.** $\dfrac{3\pi}{2}$   **21.** 175°

**22.** $-\dfrac{\pi}{6}$   **23.** $\dfrac{5\pi}{3}$   **24.** 3.25

**25.** 900°   **26.** $\dfrac{19\pi}{3}$   **27.** −51π

In questions **28–29**, find the length of the arc, $s$. Angles are in radians.

**28.**
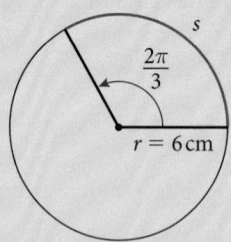

**29.**
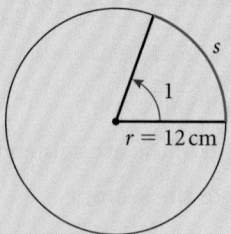

**30.** Find the angle $\theta$ in both radians and degrees.
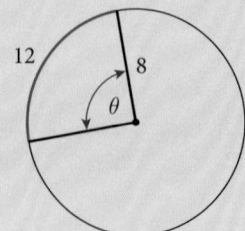

**31.** Find the radius $r$ of the circle.

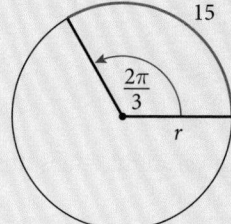

In questions **32–33**, find the area of the shaded sector.

**32.**

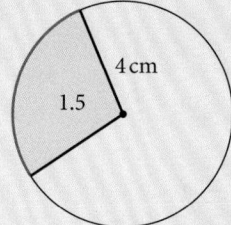

**33.**

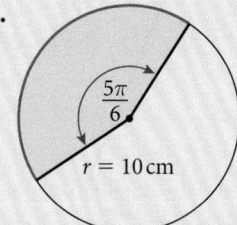

**34.** An arc of length 60 cm subtends a central angle $\alpha$ in a circle of radius 20 cm. Find the measure of $\alpha$ in both degrees and radians, accurate to 3 significant figures.

**35.** Find the length of an arc that subtends a central angle of 2 radians in a circle of radius 16 cm.

**36.** The area of a sector of a circle with a central angle of 60° is 24 cm². Find the radius of the circle.

**37.** Find the area of the sector and the arc length determined by each given radius and central angle $\theta$.

Give your answer in exact form or correct to 3 significant figures.

(a) $\theta = \dfrac{\pi}{3}$, $r = 20$ mm

(b) $\theta = -\dfrac{\pi}{6}$, $r = 70$ m

(c) $\theta = 2$, $r = 3$ km

(d) $\theta = 5$, $r = 3$ cm

**38.** Find the exact angle, in radians, of a sector with these measurements:

(a) Area $25\pi$ cm² and radius 5 cm

(b) Area $30\pi$ m² and radius 2 m

(c) Area 36 cm² and radius 6 m

(d) Area 40 m² and radius 2 m

(e) Arc length $75\pi$ km and radius 25 km

**(f)** Arc length $\frac{\pi}{3}$ m and radius 2 m

**(g)** Arc length 15 cm and radius 5 cm

**(h)** Arc length 100 mm and radius 20 mm

**39.** A cone is formed from a sector as shown so that the straight edges of the sector touch exactly without any overlap (not to scale):

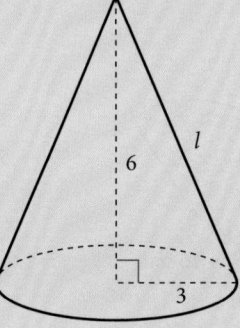 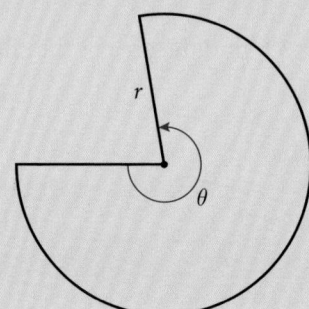

All measurements are in cm. Find:

**(a)** the slant length $l$

**(b)** the radius $r$

**(c)** the arc length of the sector

**(d)** the angle $\theta$ in radians.

<table>
<tr><td>**5.3**</td><td>**The unit circle and trigonometric functions**</td></tr>
</table>

So far you have learned about the three trigonometric ratios (sine, cosine, and tangent) in the context of triangles, for angles less than 180°. In this section, we will generalise the definitions of the trigonometric functions for any angle.

While generalising is an important part of mathematics, we have some very good reasons to do it: as you will see, trigonometric functions work very well for describing **periodic** behaviour – that is, any phenomenon that repeats itself with a fixed interval. Trigonometric functions can be used to describe tides, hours of daylight during a year, the motion of a person on a Ferris wheel, sound and light waves, etc.

Before you can use the trigonometric functions to model real-life phenomena, we need to evaluate sine, cosine, and tangent for any angle.

## The trigonometric functions in the unit circle

To generalise the definition of the trigonometric ratios, consider an angle of 45° in standard position in the unit circle, as shown in Figure 5.17.

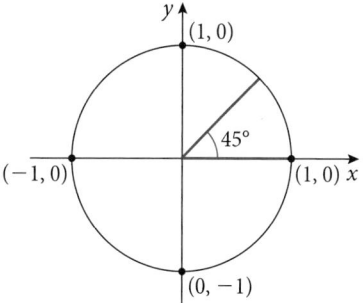

**Figure 5.17** The unit circle with a 45° angle

Now, suppose we want to find the sine of this 45° angle. Our normal approach would be to draw a right-angled triangle. We do the same here, as shown in Figure 5.18. The hypotenuse will be a radius of the unit circle and the terminal side of the angle; one leg will be a segment on the $x$ axis; the second leg will be perpendicular to the $x$ axis. Recall that the sides of a 45°-45°-90° triangle are in the ratio $1:1:\sqrt{2}$, but since the hypotenuse of our triangle is a radius of the unit circle, we must scale the ratio to make the hypotenuse have length of 1 unit. To do this we simply divide the side length by the length of the hypotenuse, $\sqrt{2}$. Hence the sides of the triangle are $\dfrac{1}{\sqrt{2}}, \dfrac{1}{\sqrt{2}}$, and $\dfrac{\sqrt{2}}{\sqrt{2}}$. We rationalise and simplify to obtain side lengths of $\dfrac{\sqrt{2}}{2}, \dfrac{\sqrt{2}}{2}$, and 1, as shown in Figure 5.18.

Now, we calculate the sine ratio as before. From Figure 5.18 we see that

$$\sin(45°) = \frac{\text{opposite}}{\text{hypotenuse}} = \frac{\dfrac{\sqrt{2}}{2}}{1} = \frac{\sqrt{2}}{2}$$

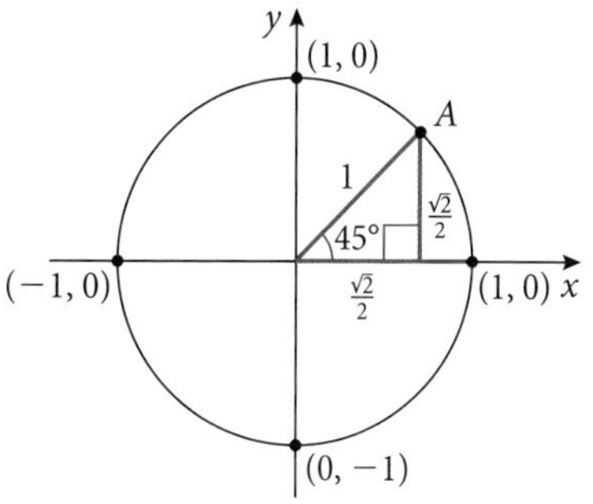

**Figure 5.18** A 45°-45°-90° triangle embedded in the unit circle

Note also that the coordinates of point $A$ must be $\left(\dfrac{\sqrt{2}}{2}, \dfrac{\sqrt{2}}{2}\right)$. In fact, we see that since the length of the hypotenuse is equal to 1, the value of the sine ratio will always be equal to the $y$ coordinate when we draw a right-angled triangle in this manner.

The same logic applies for the cosine, so that $\cos(45°) = \dfrac{\text{adjacent}}{\text{hypotenuse}} = \dfrac{\dfrac{\sqrt{2}}{2}}{1} = \dfrac{\sqrt{2}}{2}$ which implies that the value of the cosine ratio is always equal to the $x$ coordinate. What about the tangent ratio? Since the tangent ratio is $\dfrac{\text{opposite}}{\text{adjacent}}$, we see that it is equal to $\dfrac{y}{x}$ so that $\tan(45°) = 1$. In this sense, the tangent is analogous to the **gradient** of the terminal side. The reasoning we have used here is not specific to a 45°-45°-90° triangle.

We can now generalise the tangent ratios into the three trigonometric functions.

**Definition of the trigonometric functions**

Let $\theta$ be an angle in standard position and $(x, y)$ the point where the terminal side of angle $\theta$ intersects the unit circle. Then the trigonometric functions can be defined as

$$\sin \theta = y \qquad \cos \theta = x$$

$$\tan \theta = \frac{y}{x}, \quad x \neq 0$$

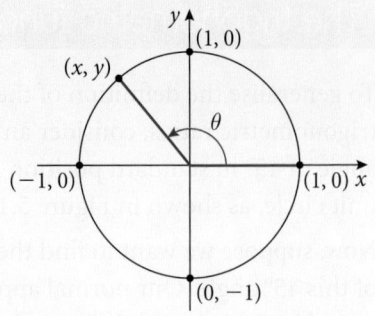

Now we consider angles in other quadrants, like the other three angles that are 45° from an axis: 135°, 225°, and 315°. These angles don't make sense in a right-angled triangle, but with our unit-circle definition of the trigonometric functions we can now calculate them. In Figure 5.19, we see these angles shown. By symmetry, we know the coordinates of the points on the unit circle from our discussion at the beginning of this chapter.

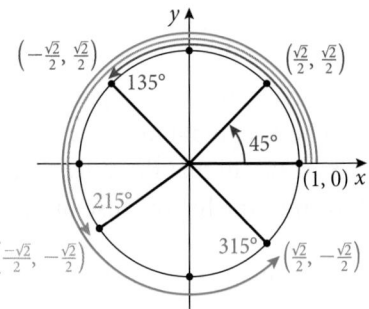

**Figure 5.19** Calculating values of trigonometric functions from coordinates of points on the unit circle

Symmetry is crucial to finding the values of the trigonometric ratios in the unit circle. Any angle which is a reflection across the $x$ or $y$ axis to a known angle can be found using this method.

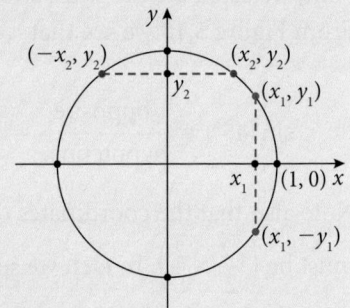

## Example 5.10

Evaluate the sine, cosine and tangent functions for $\theta = 45°$, and then use the results to evaluate the same functions for $\theta = 135°$, $\theta = 225°$ and $\theta = 315°$

### Solution

From the discussion at the beginning of the chapter, we know that $\sin(45°) = \frac{\sqrt{2}}{2}$, $\cos(45°) = \frac{\sqrt{2}}{2}$, and $\tan(45°) = 1$. Using the symmetry shown in Figure 5.19, we can then evaluate the values for the other three angles:

$$\sin(135°) = \frac{\sqrt{2}}{2} \qquad \sin(225°) = -\frac{\sqrt{2}}{2} \qquad \sin(315°) = -\frac{\sqrt{2}}{2}$$

$$\cos(135°) = -\frac{\sqrt{2}}{2} \qquad \cos(225°) = -\frac{\sqrt{2}}{2} \qquad \cos(315°) = \frac{\sqrt{2}}{2}$$

$$\tan(135°) = -1 \qquad \tan(225°) = 1 \qquad \tan(315°) = -1$$

Note that whether the value of a trigonometric function for a given angle is positive or negative depends on which quadrant the terminal side of the angle is in, because the signs of $x$, $y$, or $\frac{y}{x}$ change in each quadrant. You don't need to memorise this, because you already know where $y$ and $x$ are positive and negative, which tells you when sine and cosine are positive and negative. And, if you remember that tangent is $\frac{y}{x}$, i.e. the gradient of the terminal side, then it's easy to know when tangent is positive or negative as well.

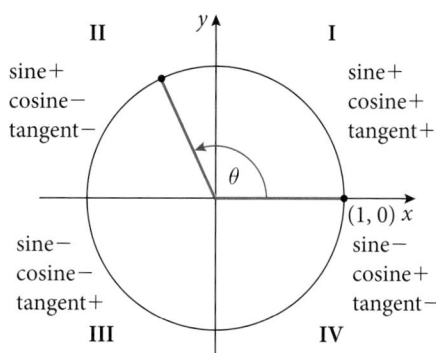

## Trigonometric functions of real numbers

Now, we want to move away from degrees and think of the trigonometric functions as having a domain of real numbers. This will be more powerful for modelling real-life phenomena. For this purpose, we think of 'wrapping' the real number line around the unit circle. We start by placing the zero at the point $(1, 0)$ and then bending the real number line around the unit circle, as shown in Figure 5.20.

Conveniently, each real number gets mapped to a point on the unit circle with an arc length equal to the number. For example, the number $t = \pi$ gets mapped to the point $(-1, 0)$ because $\pi$ is half of the circumference of the unit circle. Thus, the arc length formed by an angle in standard position, measuring $\pi$ radians, would also be $\pi$, and would have a terminal side intersecting at the point $(-1, 0)$. This is important, because we can now evaluate the trigonometric functions for radian measure as well.

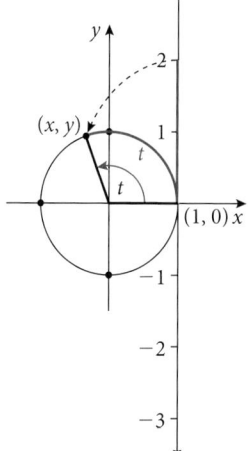

**Figure 5.20** 'Wrapping' the real number line around the unit circle

### Example 5.11

Evaluate the three trigonometric functions for each value of $t$.

(a) $t = 0$

(b) $t = \frac{\pi}{2}$

(c) $t = \pi$

(d) $t = \frac{3\pi}{2}$

(e) $t = 2\pi$

## Solution

Evaluating the trigonometric functions for any value of $t$ involves finding the coordinates of the point on the unit circle where the arc of length $t$ will 'wrap to' (or terminate) starting at the point $(1, 0)$. It is useful to remember that an arc of length $\pi$ is equal to one-half of the circumference of the unit circle. All of the values for $t$ in this example are positive, so the arc length will wrap along the unit circle in an anti-clockwise direction.

(a) An arc of length $t = 0$ has no length so it 'terminates' at the point $(1, 0)$. By definition:

$$\sin 0 = y = 0 \qquad \cos 0 = x = 1 \qquad \tan 0 = \frac{y}{x} = \frac{0}{1} = 0$$

(b) An arc of length $t = \frac{\pi}{2}$ is equivalent to one-quarter of the circumference of the unit circle, so it terminates at the point $(0, 1)$.

By definition:

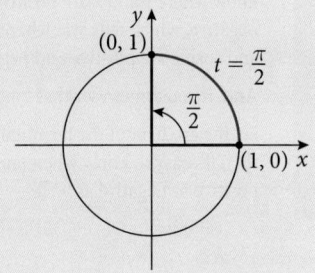

$$\sin \frac{\pi}{2} = y = 1 \qquad \cos \frac{\pi}{2} = x = 0$$

$$\tan \frac{\pi}{2} = \frac{y}{x} = \frac{1}{0} \text{ is undefined}$$

(c) An arc of length $t = \pi$ is equivalent to one-half of the circumference of the unit circle, so it terminates at the point $(-1, 0)$.

By definition:

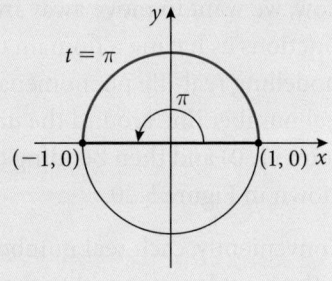

$$\sin \pi = y = 0 \qquad \cos \pi = x = -1$$

$$\tan \pi = \frac{y}{x} = \frac{0}{-1} = 0$$

(d) An arc of length $t = \frac{3\pi}{2}$ is equivalent to three-quarters of the circumference of the unit circle, so it terminates at the point $(0, -1)$.

By definition:

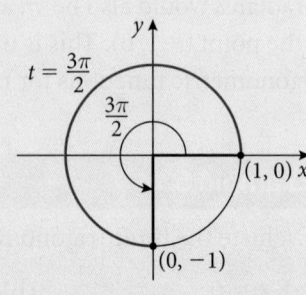

$$\sin \frac{3\pi}{2} = y = -1 \qquad \cos \frac{3\pi}{2} = x = 0$$

$$\tan \frac{3\pi}{2} = \frac{y}{x} = \frac{-1}{0} \text{ is undefined}$$

(e) An arc of length $t = 2\pi$ terminates at the same point as arc of length $t = 0$, so the values of the trigonometric functions are the same as found in part (a):

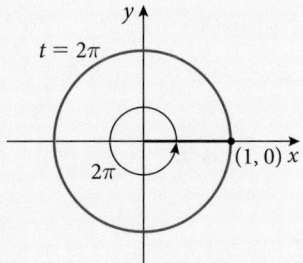

$$\sin 0 = y = 0 \qquad \cos 0 = x = 1$$

$$\tan 0 = \frac{y}{x} = \frac{0}{1} = 0$$

In practice, in this course, we will usually use a calculator to find the values of trigonometric functions. However, understanding how the values of the trigonometric functions are defined with the unit circle will help you create models and understand applications of trigonometric functions.

## Trigonometric identities

Finally, there are two fundamental **trigonometric identities** you should know. The first one is derived from the Pythagorean Theorem.

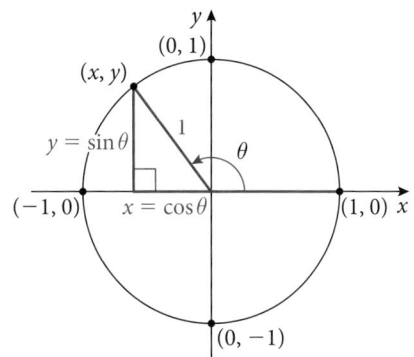

Remember that an **identity** is an equation that is true for all values of the variables.

Refer to Figure 5.21. Since the triangle has a radius of the unit circle as a hypotenuse, it must be equal to 1. From the Pythagorean Theorem, we know that $x^2 + y^2 = 1$ since $x$ and $y$ are the lengths of the legs of the triangle. But, since $x = \cos \theta$ and $y = \sin \theta$, we have

**Figure 5.21** The Pythagorean Identity

$$x^2 + y^2 = 1$$
$$(\cos \theta)^2 + (\sin \theta)^2 = 1$$
$$\cos^2 \theta + \sin^2 \theta = 1$$

This is known as the **Pythagorean Identity**.

The second identity is an alternate definition of the **tangent** function.

We know that the tangent is defined as $\tan \theta = \dfrac{y}{x}$

Since $\sin \theta = y$ and $\cos \theta = x$, we can substitute to obtain $\tan \theta = \dfrac{\sin \theta}{\cos \theta}$

The Pythagorean Identity and Tangent Identity are often useful when solving equations. Also, the Pythagorean Identity allows us to find the value of the sine function given the cosine function, or vice-versa.

**The Pythagorean Identity**

For any angle $\theta$:
$\cos^2 \theta + \sin^2 \theta = 1$

**The Tangent Identity**

For any angle $\theta$ where $\cos \theta \neq 0$, $\tan \theta = \dfrac{\sin \theta}{\cos \theta}$

### Example 5.12

Given that $\sin \theta = \dfrac{3}{4}$, and $\dfrac{\pi}{2} < \theta < \pi$, find the values of $\cos \theta$ and $\tan \theta$ without finding the size of angle $\theta$.

143

## Solution

Since $\cos^2 \theta + \sin^2 \theta = 1$

$$\cos^2 \theta + \left(\frac{3}{4}\right)^2 = 1$$

$$\cos^2 \theta = \frac{7}{16}$$

$$\cos \theta = \pm \frac{\sqrt{7}}{4}$$

However, since $\frac{\pi}{2} < \theta < \pi$, the angle $\theta$ must be in Quadrant II. The $x$-axis is negative in Quadrant II, so cosine must be negative, hence $\cos \theta = -\frac{\sqrt{7}}{4}$.

Since $\tan \theta = \frac{\sin \theta}{\cos \theta}$, we have $\tan \theta = \dfrac{\dfrac{3}{4}}{-\dfrac{\sqrt{7}}{4}} = -\frac{3}{\sqrt{7}} = -\frac{3\sqrt{7}}{7}$

Tangent is negative in Quadrant II, so this agrees with our answer.

## Exercise 5.3

In questions **1–3**, state the exact value (if possible) of the sine, cosine and tangent of the given angle.

1. **(a)** $\theta = 30°$    **(b)** $\theta = 150°$    **(c)** $\theta = 210°$    **(d)** $\theta = 330°$

2. **(a)** $\theta = 60°$    **(b)** $\theta = 120°$    **(c)** $\theta = 240°$    **(d)** $\theta = 300°$

3. **(a)** $\theta = 0°$    **(b)** $\theta = 90°$    **(c)** $\theta = 180°$    **(d)** $\theta = 270°$

4. Given $\sin \theta = \frac{4}{5}$ and $0 < \theta < \frac{\pi}{2}$, find the values of $\cos \theta$ and $\tan \theta$.

5. Given $\sin \theta = -\frac{2}{3}$ and $\pi < \theta < \frac{3\pi}{2}$, find the values of $\cos \theta$ and $\tan \theta$.

6. Given $\cos \theta = \frac{3}{4}$ and $\pi < \theta < 2\pi$, find the values of $\sin \theta$ and $\tan \theta$.

In questions **7–15**, $t$ is the length of an arc on the unit circle starting from $(1, 0)$.

**(a)** State the quadrant in which the terminal point of the arc lies.

**(b)** Find the coordinates of the terminal point $(x, y)$ on the unit circle. Give exact values for $x$ and $y$ if possible, otherwise approximate to 3 significant figures.

7. $t = \frac{\pi}{6}$      8. $t = \frac{5\pi}{3}$      9. $t = \frac{7\pi}{4}$

**10.** $t = \dfrac{3\pi}{2}$      **11.** $t = 2$      **12.** $t = -\dfrac{\pi}{4}$

**13.** $t = -1$      **14.** $t = -\dfrac{3\pi}{4}$      **15.** $t = 3.52$

In questions **16–24**, state the exact value (if possible) of the sine, cosine and tangent of the given real number.

**16.** $\dfrac{\pi}{3}$      **17.** $\dfrac{5\pi}{6}$      **18.** $-\dfrac{3\pi}{4}$

**19.** $\dfrac{\pi}{2}$      **20.** $-\dfrac{4\pi}{3}$      **21.** $3\pi$

**22.** $\dfrac{3\pi}{2}$      **23.** $-\dfrac{7\pi}{6}$      **24.** $1.25\pi$

In questions **25–28**, use the periodic properties of the sine and cosine functions to find the exact value of $\sin x$ and $\cos x$.

**25.** $x = \dfrac{13\pi}{6}$    **26.** $x = \dfrac{10\pi}{3}$    **27.** $x = \dfrac{15\pi}{4}$    **28.** $x = \dfrac{17\pi}{6}$

# 5.4   Graphical analysis of trigonometric functions

Graphical analysis is a fundamental part of mathematics, as you have seen in Chapter 2 (Functions). Therefore, we need to examine the graphs of sine, cosine, and tangent.

As we shall see in the course, many real-life applications involve solving trigonometric equations. In this section we will examine a graphical method for solving trigonometric equations.

## Graphs of the sine and cosine functions

Recall that the sine, cosine, and tangent functions have a domain of real numbers, based on the 'wrapping function' shown in Figure 5.22.

Now, however, we are going to 'unwrap' the function from the unit circle and graph them in a normal coordinate plane. All three trigonometric functions are periodic, i.e., their values repeat in a regular manner as we complete a move around the circle. The sine and cosine functions repeat every $2\pi$, so we say that the **period** of sine and cosine is $2\pi$.

Therefore, we only need to graph one period from $0 \leqslant t \leqslant 2\pi$ and then we know the graph will repeat itself.

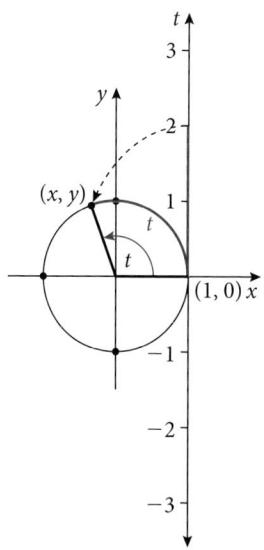

**Figure 5.22** The wrapping function for trigonometric functions

145

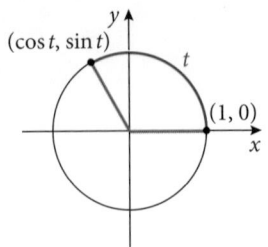

**Figure 5.23** $\sin t$ is the $y$-coordinate of the terminal point on the unit circle corresponding to the real number $t$

> The sine and cosine functions have a period of $2\pi$ because that is one revolution around the unit circle.

We know from the previous section that $\sin t$ is the $y$-coordinate of the terminal point on the unit circle corresponding to the real number $t$. (Figure 5.23).

In order to generate one period of the graph of $y = \sin t$, we need to record the $y$-coordinates of a point on the unit circle and the corresponding value of $t$ as the point travels anti-clockwise one revolution (one period) starting from the point $(1, 0)$. These values are then plotted on a graph with $t$ on the horizontal axis and $y$ ($\sin t$) on the vertical axis. Figure 5.24 illustrates this process in a sequence of diagrams.

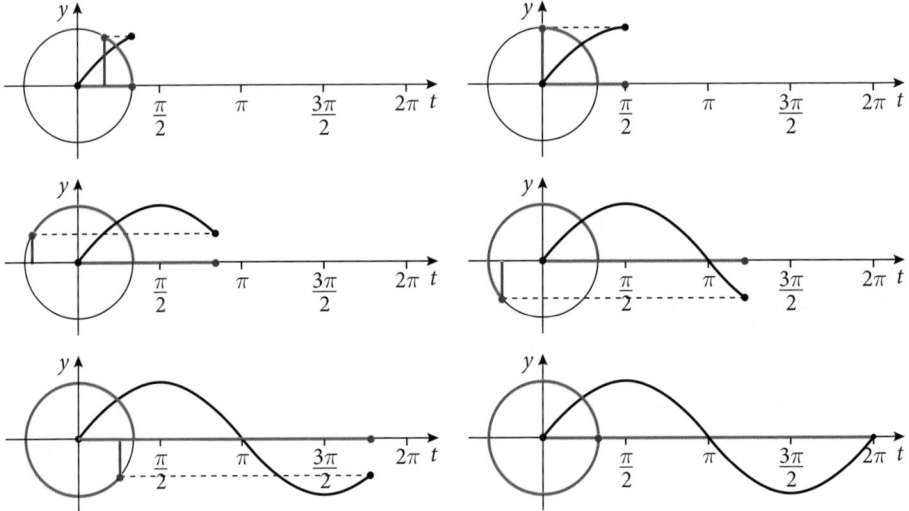

**Figure 5.24** Graph of the sine function for $0 \leqslant t \leqslant 2\pi$ generated from a point travelling around the unit circle

As the point $(\cos t, \sin t)$ travels along the unit circle, the $x$-coordinate ($\cos t$) goes through the same cycle of values as the $y$-coordinate ($\sin t$) does. The only difference is that the $x$-coordinate begins at a different value in the cycle. When $t = 0$, $y = 0$, but $x = 1$. The result is that the graph of $y = \cos t$ is the exact same shape as $y = \sin t$ but has simply been shifted to the left $\dfrac{\pi}{2}$ units. The graph of $y = \cos t$ for $0 \leqslant t \leqslant 2\pi$ is shown in Figure 5.25.

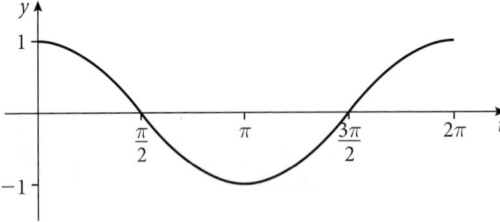

**Figure 5.25** Graph of $y = \cos t$ for $0 \leqslant t \leqslant 2\pi$

The convention is to use the letter $x$ to denote the variable in the domain of the function. Hence, from here on we will use the letter $x$ rather than $t$ and write the trigonometric functions as $y = \sin x$, $y = \cos x$ and $y = \tan x$.

Since the period for both the sine function and cosine function is $2\pi$, to graph $y = \sin x$ and $y = \cos x$ for wider intervals of $x$ we simply repeat the shape of the graph that we generated from the unit circle for $0 \leqslant x \leqslant 2\pi$ (Figures 5.24 and 5.25). Figure 5.26 shows graphs of $y = \sin x$ and $y = \cos x$ for $-4\pi \leqslant x \leqslant 4\pi$

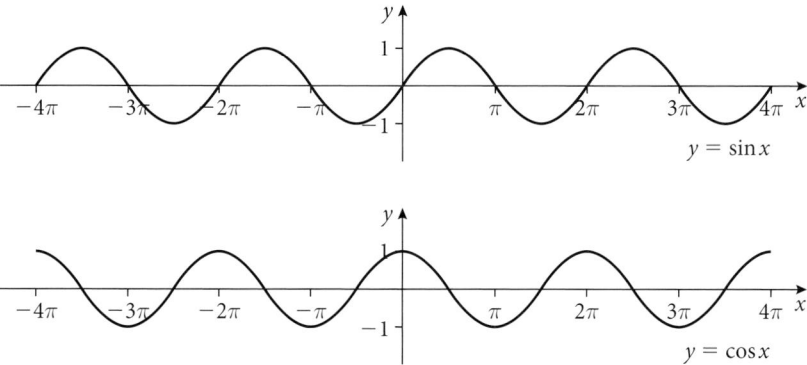

**Figure 5.26**  $y = \sin x$ and $y = \cos x$, $-4\pi \leqslant x \leqslant 4\pi$

Aside from their periodic behaviour, these graphs reveal further properties of the graphs of $y = \sin x$ and $y = \cos x$. Note that the sine function has a maximum value of $y = 1$ for all $x = \dfrac{\pi}{2} + k \cdot 2\pi, k \in \mathbb{Z}$, and has a minimum value of $y = -1$ for all $x = k \cdot 2\pi, k \in \mathbb{Z}$

The cosine function has a maximum value of $y = 1$ for all $x = k \cdot 2\pi, k \in \mathbb{Z}$, and has a minimum value of $y = -1$ for all $x = \pi + k \cdot 2\pi, k \in \mathbb{Z}$. We also see that both functions have a domain of all real numbers and a range of $-1 \leqslant y \leqslant 1$. (This makes sense based on our definitions from the unit circle: the coordinates will never be outside the radius of 1 unit.)

## Graph of the tangent function

The tangent function is defined as $\tan \theta = \dfrac{y}{x}, x \neq 0$ or $\tan \theta = \dfrac{\sin \theta}{\cos \theta}, \cos \theta \neq 0$

Therefore, we can consider the graphs of $y = \sin x$ and $y = \cos x$ to generate the graph of $y = \tan x$. We know that $y = \tan x$ will be zero whenever $\sin x = 0$. Likewise, we know that $y = \tan x$ will be undefined whenever $\cos x = 0$. As $x$ approaches these values where $\cos x = 0$, the value of $\tan x$ will become very large − either in the negative or in the positive direction. Thus, the graph of $y = \tan x$ has vertical asymptotes at $x = \dfrac{\pi}{2} + k\pi, k \in \mathbb{Z}$

By considering the sign of $\sin x$ before and after these values we can begin to see what the graph of the tangent function should look like. Finally, remember that the graph of the tangent function will be $\pm 1$ whenever $\sin x$ and $\cos x$ are equal or opposite $\left( \text{at } x = \dfrac{\pi}{4} + k\left(\dfrac{\pi}{2}\right) \right)$. With this logic, we can construct the graph of the tangent function, shown in Figure 5.27.

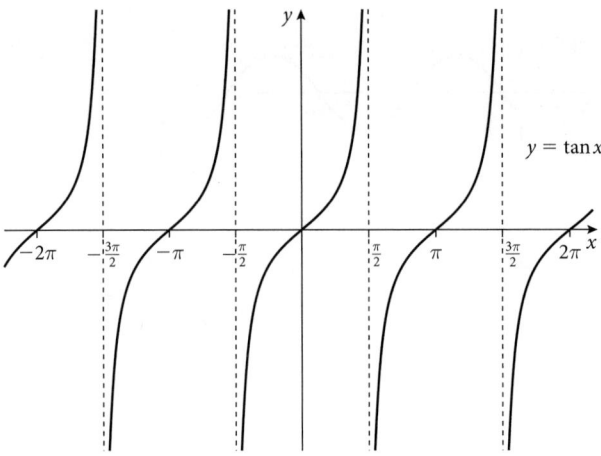

**Figure 5.27** $y = \tan x$ for $-2\pi < x < 2\pi$

## Graphical solutions to trigonometric equations

Now that we have the graphs of the trigonometric functions, we can use these to find solutions to trigonometric equations. In general, there can be an infinite number of solutions to trigonometric equations, but we will only consider equations over a finite interval of the real numbers.

In this chapter, we will not focus on how to draw graphs of trigonometric functions or how to create models using trigonometric functions. Although you learned many of the concepts you need in Chapter 4, we will revisit this topic in Chapter 9 when we look at sinusoidal models. So, for now, we will assume that you will use technology or be given the graph of the function.

In fact, we use graphical solutions all the time to solve complex trigonometric equations in real-life applications. For example, suppose the function

$h(t) = 2 \sin\left(\dfrac{2\pi t}{12}\right) + 3$ models the tide height for a certain location, where $t$ is

hours after midnight. Finding the times of high and low tide seems daunting. However, by graphing this function (Figure 5.28), we see that the high tides occur at 03:00 and 15:00, while the low tides occur at 09:00 and 21:00.

Tides don't really have a 12-hour period; it is actually about 12 hours and 25.2 minutes. This is the time it takes for the moon to pass from directly overhead to directly underfoot for a given location.

In Chapter 9, we will see how to refine this model for the actual tidal period.

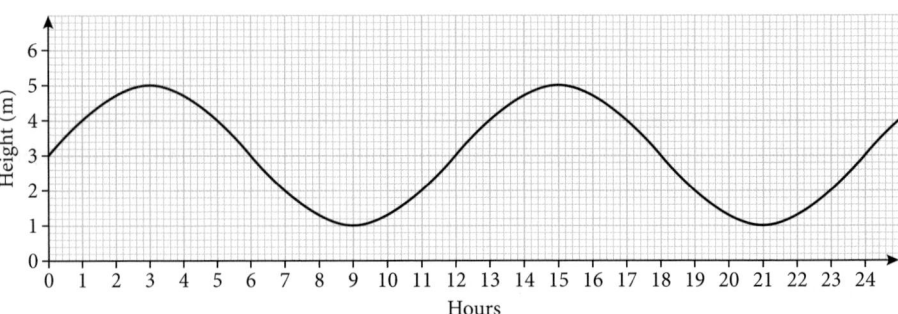

**Figure 5.28** A model of tide height in hours after midnight

## Example 5.13

Use the tide model $h(t) = 2 \sin\left(\frac{2\pi t}{12}\right) + 3$ and its graph in Figure 5.28 to answer the following questions.

(a) Write down the time(s) before noon (12:00) when the tide height is 4 m.

(b) Suppose a certain boat needs at least 2 m of water to safely pass this area. Write down the intervals that are not safe for this boat.

(c) Calculate the change in tidal height between 05:00 and 09:00

## Solution

(a) Reading across from the vertical axis, we see the tide is at 4 m at 01:00 and 05:00. (It is also at 4 m at times 13:00 and 17:00, but these are after noon.)

(b) The tide height is less than 2 m between 07:00 and 11:00, and also between 19:00 and 23:00.

(c) At 05:00, the tide is at 4 m. At 09:00, the tide is at 1 m. Therefore, the tide has decreased by 3 m.

If a graph is not given, then we can use our GDC to generate a graph, as shown in the next example.

## Example 5.14

The Vienna *Riesenrad* is a large Ferris wheel originally built in 1897 by the Englishman Walter Basset.

The height above ground of a passenger (in feet) on the *Riesenrad* after $t$ minutes can be modelled by the function $h(t) = 100 \cos\left(\frac{\pi}{5}(t - 5)\right) + 112$

Use your GDC to answer the following questions.

(a) Find the maximum height reached by a passenger on the *Riesenrad*.

(b) Find the height of a passenger 3 minutes after the ride has started.

(c) Find the number of minutes required for the wheel to make one complete revolution.

(d) Calculate the number of minutes that a passenger is more than 175 feet above the ground.

## Solution

Before answering the questions, we will enter the function in our GDC to obtain a useful graph. Since this is a real-number function, make sure your calculator is in radians mode.

For trigonometric models, unless otherwise specified, your GDC should be in radian mode.

149

Then we should enter the function to obtain a graph. Note that you will need to change the default window settings. Here the window is set to to $[-1, 11]$ on the $x$-axis and $[-10, 300]$ on the $y$-axis. (You can use your GDC's zoom in/out and zoom fit commands to find the viewing window if you don't know a good window setting.)

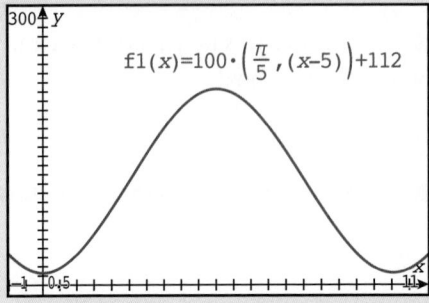

(a) To find the maximum height, we use the GDC **Maximum** command:

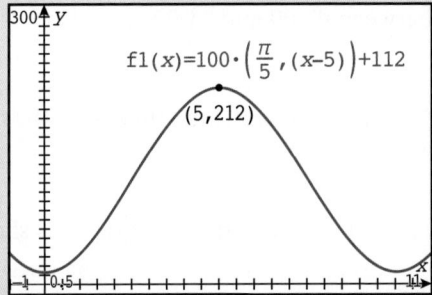

Thus, the maximum height reached is 212 feet (after 5 minutes).

(b) To find a passenger's height 3 minutes after the ride has started, evaluate $h(3)$. Use the function already entered into the GDC:

$$f1(3) \qquad\qquad 142.902$$

Thus the height of a passenger after 3 minutes is 143 feet (3 s.f.).

(c) We can infer the length of time required for one complete revolution by looking at the time difference between two minimum points:

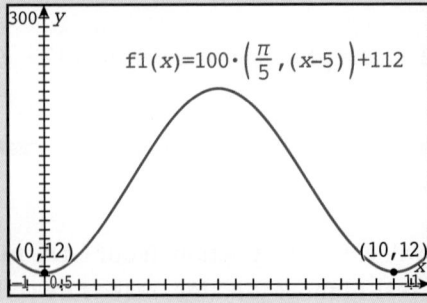

Since the first minimum is at the start of the ride (0 minutes) and the second minimum is 10 minutes later, it must take 10 minutes for one complete revolution.

(d) To find the number of minutes that a passenger is more than 175 feet above the ground, we must solve the inequality

$$100 \cos\left(\frac{\pi}{5}(x - 5)\right) + 112 > 175$$

We can use a GDC to find the boundary points graphically, by entering a second function $f(x) = 175$ and looking for intersection points:

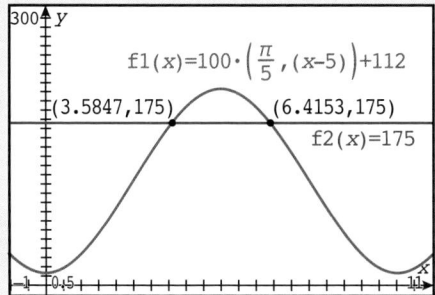

Thus we see that the passenger would be above 175 feet between $6.4153 - 3.5847 = 2.83$ minutes.

The GDC is a powerful tool for finding numerical solutions to equations in general, as we see in the next example.

! Learn how to use full precision to avoid inaccurate answers by storing results on your GDC rather than retyping answers. Consult your calculator instructions to find how to store values for later use and how to use values from the graph without retyping them

## Example 5.15

Solve the equation $\sin 2x = \cos x$ on the interval $0 \leqslant x < \pi$

### Solution

Again, we turn to our GDC for help. Since we are given a specific interval, we can set our $x$-axis to $[0, \pi]$ and the use the **zoom fit** function to get an appropriate viewing window. Also, we can enter the left-hand side of the equation as one function, and the right-hand side as a second function:

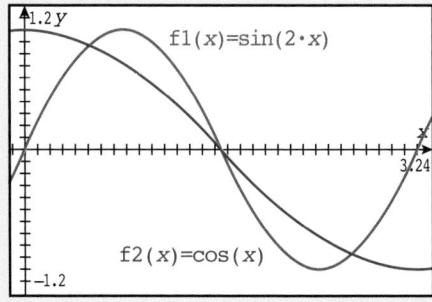

We can see there are three intersections of these functions on the given interval, so there are three solutions to the equation on the given interval:

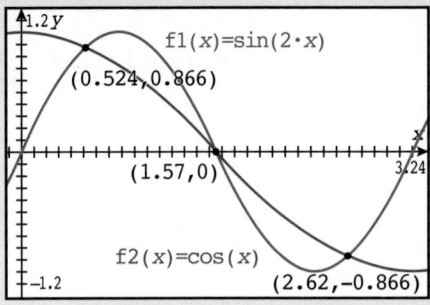

Therefore the three solutions are $x \in \{0.524, 1.57, 2.62\}$

Note that the $y$-coordinates are not meaningful in this context; our original equation had only $x$.

## Exercise 5.4

Use your GDC in questions **1–4** to solve the each equation on the given interval.

1. $\cos(2x) = \sin(x), \quad 0 \leqslant x < 2\pi$

2. $\dfrac{x}{2} = \cos(2x), \quad -\pi \leqslant x \leqslant \pi$

3. $6\sin\left(\dfrac{\pi}{30}(x - 15)\right) = 4, \quad 0 \leqslant x \leqslant 48$

4. $\cos^2(2x) + \sin^2(x) = 1.5, \quad 0 \leqslant x \leqslant \pi$

5. Without graphing, explain why $\cos^2 x + \sin^2 x = 2$ has no solution.

6. Solve each equation on the domain $0 \leqslant x \leqslant 2\pi$
   (a) $\sin(x - 2) = 0.715$ \qquad\qquad (b) $\sin^2 x = 1$
   (c) $\sin^2 x + -2\sin x = 1$ \qquad (d) $2\sin^2 x + 3\sin x - 2 = 0$

7. Given $\sin(2x) + k = 3.5, 0 \leqslant x \leqslant \pi$, find the values of $k$ such that
   (a) the equation has exactly one solution
   (b) the equation has exactly two solutions
   (c) the equation has exactly three solutions
   (d) the equation has no solutions.

8. Given $3\cos(x) + 2k = 1, 0 \leqslant x \leqslant 2\pi$, find the values of $k$ such that
   (a) the equation has exactly two solutions
   (b) the equation has exactly one solution
   (c) the equation has no solutions.

**Voronoi diagrams** have practical applications as varied as describing soap bubbles and animal dominance, analysing geographic markets, and partitioning regions for phone coverage.

Emergency services can sometimes locate individuals based on their smartphone signal. The first step in this process may be to locate the individual based on the nearest telecommunications tower. A map to identify service regions for each tower would look something like Figure 5.29.

Each red dot indicates a telecoms tower, and the blue lines are the boundaries of each region served by that tower.

The map in Figure 5.29 is an example of a **Voronoi diagram**. The polygons on the map, each of which shows the service area for a particular telecoms tower, are called **cells**. The red dots represent telecoms towers and are called **sites**. The line segment boundaries between each region are called **edges** and places where edges intersect are **vertices**. Note that the some of the edges of cells around the outermost sites are actually rays, not line segments.

A **Voronoi diagram** divides a plane into a number of regions called **cells**. Cells are divided by boundaries called **edges**. Each cell contains exactly one **site**, such that every point in a given cell is closer to that cell's site than any other site. Points on edges are equidistant from two or more sites, **vertices** formed by edge intersections are equidistant from at least three sites.

Geometrically, we know that the set of points that are equidistant from two points lies on the perpendicular bisector of the line segment between those two points. Therefore, every edge in a Voronoi diagram is a perpendicular bisector of a line segment connecting adjacent sites. We can use this fact to generate the equations of edges algebraically and to reason about points on the edges of a Voronoi diagram.

### Example 5.16

(a) Using the Voronoi diagram shown, identify the site closest to (30, 20).

(b) Give the coordinates of a vertex equidistant to four sites.

(c) Estimate the coordinates of a point equidistant to sites $A$, $E$, and $K$.

(d) Explain why the diagram cannot give a point equidistant to sites $C$, $F$, $I$, and $L$.

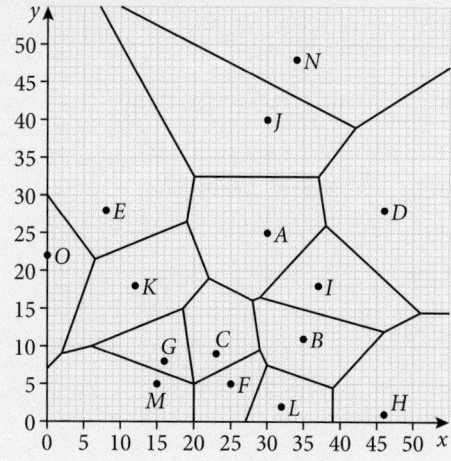

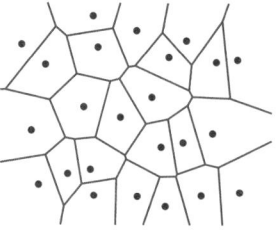

**Figure 5.29** Voronoi diagram

Recall these geometric objects:

A **line** goes through two points and has no endpoints.

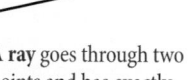

A **ray** goes through two points and has exactly one endpoint.

A **line segment** goes through two points and has exactly two endpoints.

Recall that the set of points equidistant to $A$ and $B$ is the perpendicular bisector of the line segment with endpoints $A$ and $B$. In other words, any point on the perpendicular bisector of segment $AB$ is equidistant to points $A$ and $B$.

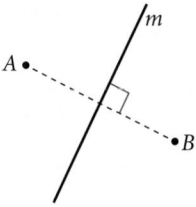

Edges in Voronoi diagrams mark points that are equidistant from at least two adjacent sites.

(e) The point $(a, a)$ is equidistant to sites $J$ and $N$. Find the value of $a$.

(f) Find the equation of the edge between sites $E(8, 28)$ and $K(12, 18)$ in the form $y = mx + c$

---

**Solution**

(a) The point $(30, 20)$ is within the cell containing $A$, so $A$ must be the closest site.

(b) The vertex located at $(20, 5)$ is formed by the intersection of 4 edges. Therefore, it must be equidistant to four sites, namely, $C$, $G$, $M$, and $F$.

(c) The vertex located at approximately $(19, 25.5)$ is equidistant to sites $A$, $E$, and $K$ since it is on the edges of the cells for all three sites.

(d) In this diagram, the edges of cells for sites $C$, $F$, $I$, and $L$ do not share a common vertex. Therefore, this diagram cannot give us a point equidistant to those four sites.

(e) Points equidistant from $J$ and $N$ must be on the edge separating the cells of $J$ and $N$. On that edge, there is only one point with equal $x$- and $y$-coordinates: the point $(40, 40)$. Therefore, $a = 40$.

> Try drawing the line $y = x$ on the diagram, since it contains all points with equal $x$ and $y$ coordinates.

(f) To find the equation of the edge, we must find the midpoint of segment $\overline{EK}$ and the gradient of a line perpendicular to $\overline{EK}$. First, the midpoint:

$$\text{Midpoint}_{EK} = \left(\frac{8 + 12}{2}, \frac{28 + 18}{2}\right) = (10, 23)$$

Next, we find the gradient of $\overline{EK}$:

$$m_{EK} = \frac{28 - 18}{8 - 12} = \frac{10}{-4} = -\frac{5}{2}$$

The gradient of the edge is therefore $\frac{2}{5}$

Finally, use point-slope form to generate the equation of the line:

$$y - 23 = \frac{2}{5}(x - 10)$$

$$\Rightarrow y = \frac{2}{5}x + 19$$

## Constructing Voronoi diagrams

There are a variety of methods (algorithms) to construct a Voronoi diagram from a list of sites. We will first look at how to **add** a site to an existing Voronoi diagram, and then use this algorithm to build a Voronoi diagram.

**Incremental Insertion Algorithm**

To add a new site $p$ to an existing Voronoi diagram, we use the following process:

1. Find the nearest existing site. Call this site $s$. (The nearest site is quickly found because the new site $p$ must be in its cell! If the new site is on an edge, start with either adjacent site.)

2. Construct the perpendicular bisector of the segment joining $p$ and $s$.

3. On this perpendicular bisector, create a segment with endpoints on the edges of the cell containing $s$. This segment is the first edge of the new cell containing $p$.

4. Build the remaining edges of the cell containing $p$ by repeating the following:
   a. At each endpoint of a new edge, divide the adjacent cell using a new perpendicular-bisector between the new site and the site for the adjacent cell.
   b. On the new perpendicular bisector, create a segment with endpoints on existing edges. (If only one existing edge is intersected, create a ray with its endpoint on that edge. If no existing edges are intersected, the entire perpendicular bisector becomes the new boundary).
   c. Continue until there are no unused new-edge endpoints.

5. Discard the edges inside the newly-created cell containing $p$.

## Example 5.17

Show the process for adding the site shown in red to the Voronoi diagram.

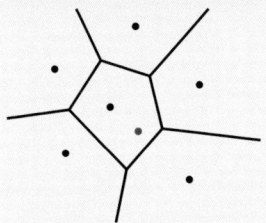

### Solution

**Step 1**

We identify the nearest neighbour.

**Step 2**

Construct a perpendicular bisector.

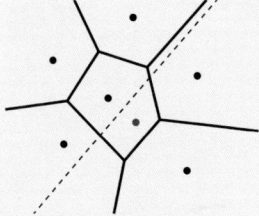

**Step 3**

Create a line segment with endpoints on existing cell boundaries (in red). This is the first edge of the new cell.

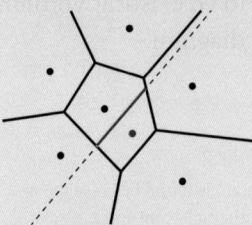

**Step 4a**

Continue on to the next cell adjacent to the newly-added line segment and create a new perpendicular bisector.

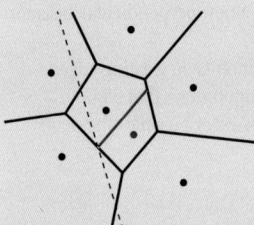

**Step 4b**

Create the next new cell edge (in red).

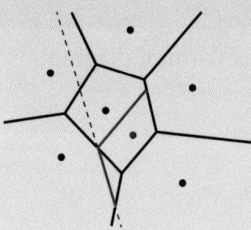

**Step 4a**

Repeat for the next adjacent cell.

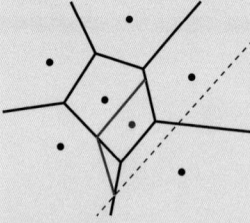

**Step 4b**

Continue creating new line segments to form the edges of the new site's cell.

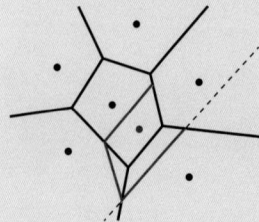

## Step 4a/b

Repeat until there are no more adjacent cells.

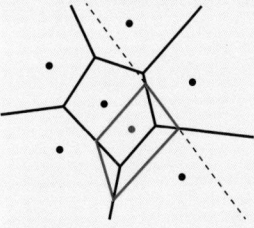

## Step 4c

Stop creating new edges when there are no more unused new-edge endpoints.

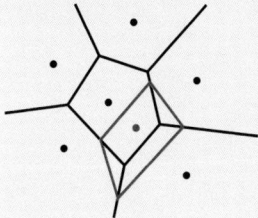

## Step 5

Remove any edges inside the new cell.

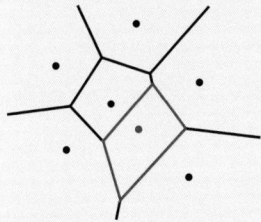

Note that in Step 4b, it may happen that a new edge is a ray instead of a line segment. This is the case when the new site is on the outer edge of the Voronoi diagram, and then the new cell may be unbounded – two of the edges will be rays instead of lines segments.

## Example 5.18

Show the steps to add site $W$ to the Voronoi diagram shown in the diagram.

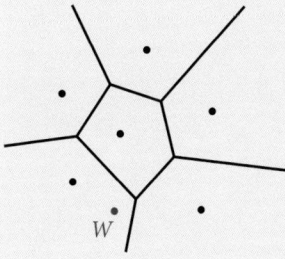

---

**Solution**

Find the perpendicular bisector between *W* and the containing cell's site.

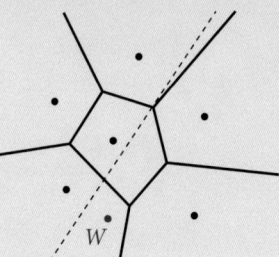

Add a ray along the perpendicular bisector.

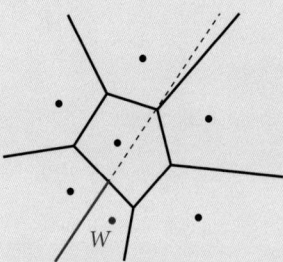

Find the perpendicular bisector between *W* and next adjacent site.

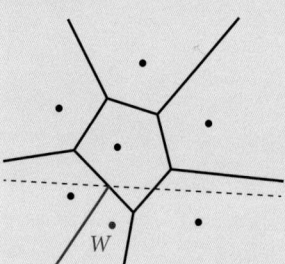

Add a line segment along the perpendicular bisector.

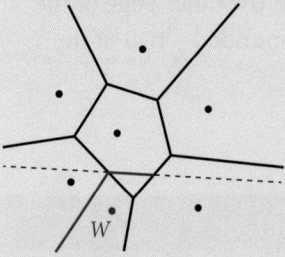

Find the perpendicular bisector between *W* and the next adjacent site.

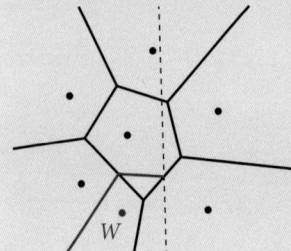

Add a ray along the perpendicular bisector.

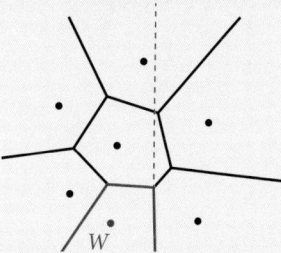

Remove edges inside $W$'s new cell.

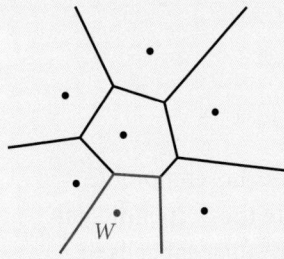

## Creating a new Voronoi diagram

To create a new Voronoi diagram, we apply the incremental insertion algorithm repeatedly.

To create a new Voronoi diagram from a list of sites $S_1, S_2, ..., S_n$:

1. Construct the perpendicular bisector between sites $S_1$ and $S_2$. (This creates the cells for $S_1$ and $S_2$.)
2. For each additional site $S_1, S_2, ..., S_n$ repeat the incremental insertion algorithm until there are no more vertices to insert.

It doesn't matter which two sites you start the Voronoi diagram with, nor does it matter in what order the remaining sites are added. When showing this process, however, it's usually clearer to start with sites near to each other and then work on the next-closest site.

### Example 5.19

Create a Voronoi diagram for the points $A(2, 0)$, $B(6, 6)$, $C(6, 0)$, and $D(15, 3)$.

### Solution

Begin by plotting the points and choosing two points to begin.

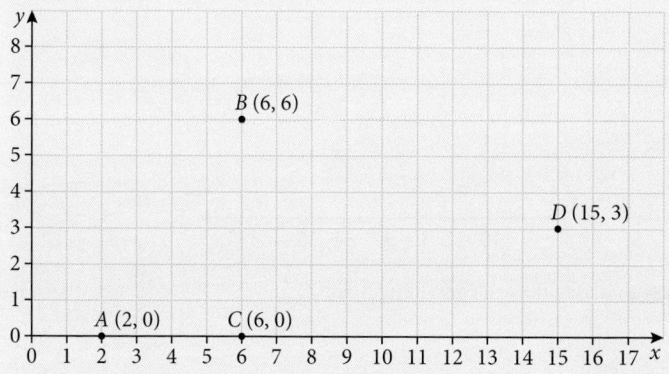

Since it doesn't matter which pair of points we begin with, we will start with $A$ and $B$ for simplicity. We first construct the perpendicular bisector of segment $AB$:

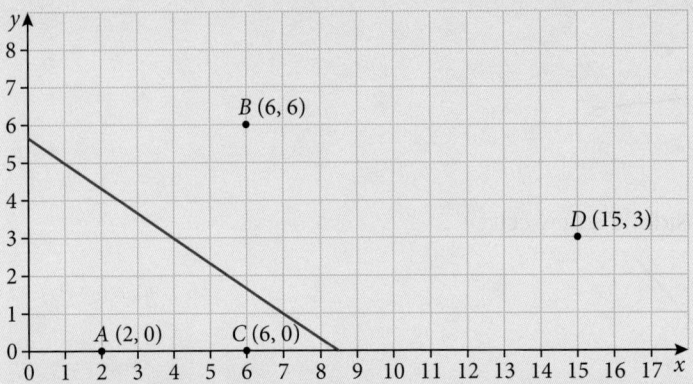

Then, we choose the next point to insert; here we have chosen point $C$. As per the Incremental Insertion Algorithm, we create the perpendicular bisector with $A$ and then work until there are no more adjacent cells to divide:

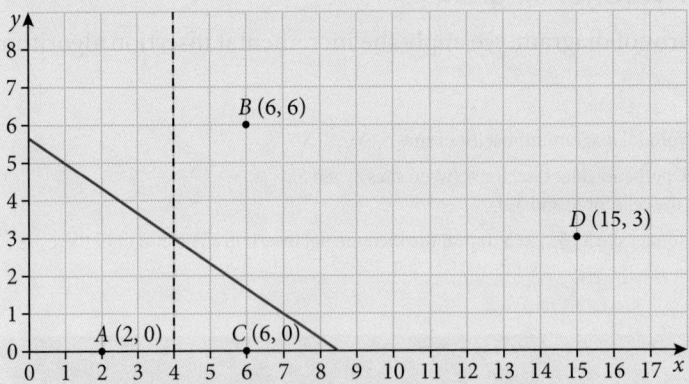

We divide the cell containing $A$ and $C$

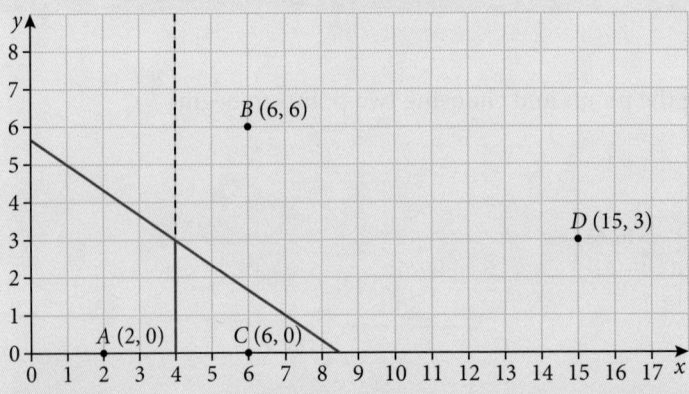

Add a line segment or ray to divide the cell containing $A$ and $C$

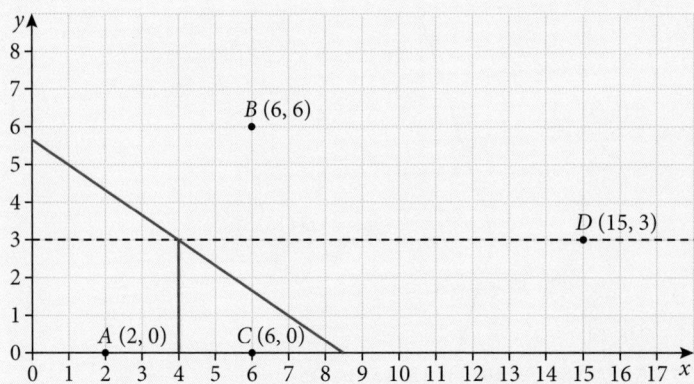

Add a perpendicular bisector to divide the next adjacent cell (the cell of site $B$ in this case)

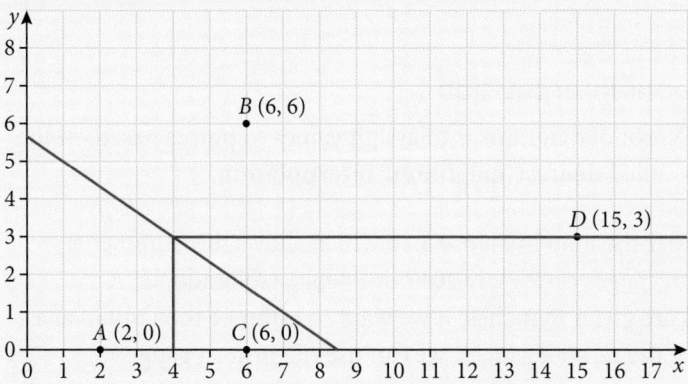

Remove edges inside the new cell

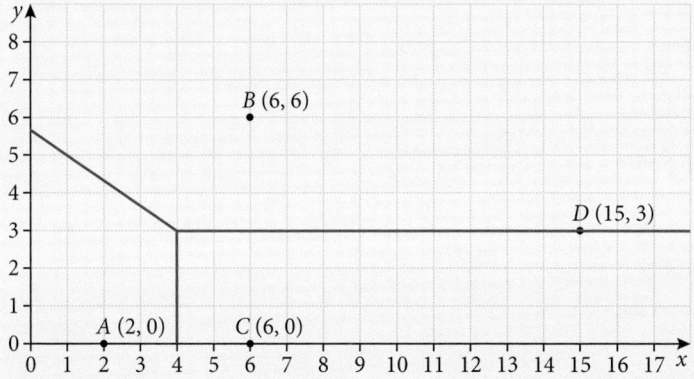

We have finished inserting site $C$, now we can insert site $D$

The process for inserting side $D$ is the same as for site $C$. Note that since $D$ is on an existing edge, we can choose to start by dividing the cell for either $B$ or $C$. Again, we follow the steps for the Incremental Insertion Algorithm to obtain the finished Voronoi diagram:

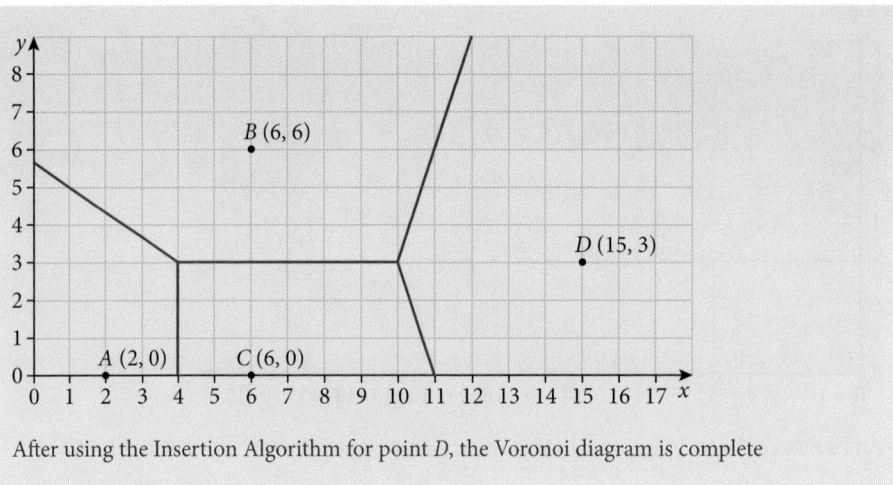

After using the Insertion Algorithm for point $D$, the Voronoi diagram is complete

### Nearest-neighbour interpolation

Using a Voronoi diagram to assign values to points is a **function** in the general sense, just like the functions you learned about in Chapter 2. In this case, the domain is a set of points and the range is the set of values or labels. Every point is assigned to exactly one value or label based on which site is closest to that point..

A common use of Voronoi diagrams is to assign values to points based on the nearest site. This is called **nearest-neighbour interpolation**.

For example, consider the map in Figure 5.30 which shows measurements from rain-collection gauges on 27 July 2017 in Pueblo, Colorado, USA. The measurements are given in inches. If we want to estimate the rainfall at a location near one of the rainfall gauges, we can use the nearest gauge.

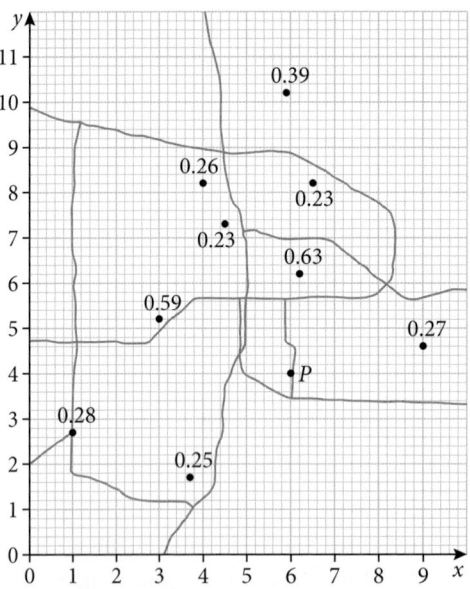

**Figure 5.30** A map showing measurements of rain-collection gauges in inches from Pueblo, Colorado, USA. The grey lines are major roads.

### Example 5.20

Use the map of rainfall measurements in Figure 5.25 to estimate the rainfall at the coordinates $P(6, 4)$ using nearest-neighbour interpolation.

## Solution

To estimate the rainfall at $P(6, 4)$, we will use the **nearest-neighbour** method. To determine the nearest neighbour, we will first build a Voronoi diagram using each point as a site. This is the resulting Voronoi diagram.

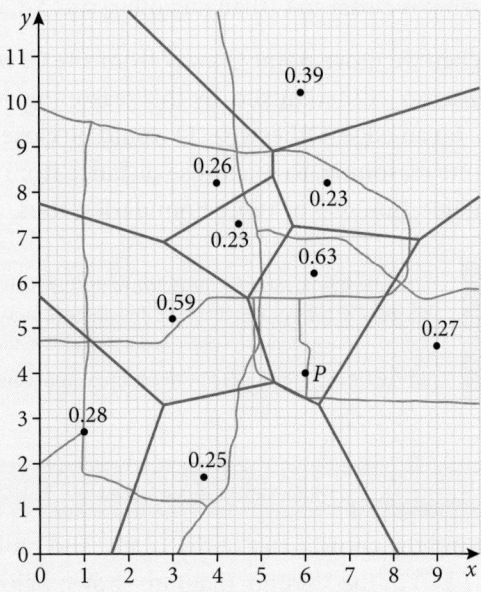

The map of rainfall gauges after creating a Voronoi diagram.

Now, to determine the nearest neighbour to point $P(6, 4)$, we simply look at which site the point is contained within. We can see from the diagram that the nearest site has a reading of 0.63 inches. Therefore, we estimate that the rainfall at $P(6, 4)$ is 0.63 inches.

You might argue that 0.63 inches is not the best estimate for the rainfall at point $P$. Wouldn't it be better to try to make some sort of average of the rainfall gauges surrounding $P$? Indeed, there are different methods for interpolating unknown values. A common method that uses an area-weighted average of the nearest sites is called **natural neighbour** interpolation, but it is beyond the scope of this text.

## Largest empty circle

One interesting application of Voronoi diagrams concerns finding the point(s) that are farthest from any site. For example, consider a Voronoi diagram where the sites are towns and you are looking for a place to put a toxic waste dump. Of course, no one wants the toxic waste dump in their backyard, so it makes sense to find the location farthest from any of the towns. This is why the **largest empty circle (LEC) problem** is sometimes called the **toxic waste dump problem**.

On a more positive note, suppose that a restaurant is looking for a new location within a city. They might be interested to see which location is farthest from competing restaurants, but still within the city. So, they draw a Voronoi diagram where the competing restaurants in the city are sites, as shown in Figure 5.31.

Visually, it appears that the largest area without a restaurant, that is, the point farthest from any existing restaurant, must be in the area between sites $B$, $C$, $F$, and $K$, shown in green in Figure 5.32.

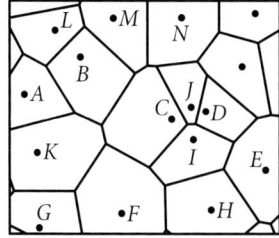

**Figure 5.31** A map of existing competitor restaurants in a city

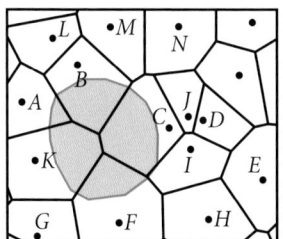

**Figure 5.32** General area for a new restaurant location shown in green

163

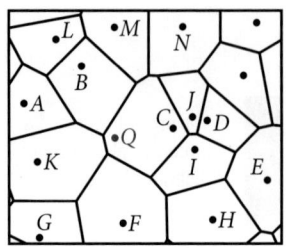

**Figure 5.33** Point $Q$ is one possible location. Is point $Q$ the farthest point from any existing restaurant?

The largest empty circle (LEC) that is contained within a Voronoi diagram must be centred on a vertex.

But where exactly should the restaurant go? Consider this: we could put it at point $Q$, shown in Figure 5.33. But, since $Q$ is located inside the cell of site $C$, then it is closer to $C$ than any other site. So, we should move it away from site $C$.

When we move point $Q$ away from site $C$, we need to be careful not to get closer to a different site. By this reasoning, it should be apparent that $Q$ must be on an edge. Otherwise, $Q$ would be closer to one site than another. By the same reasoning, to maximise distance from all existing sites, $Q$ must be on an intersection of 3 or more boundaries: thus, $Q$ must be on a vertex.

> Actually, it is possible for the centre of the LEC to be just on a boundary, not a vertex, if it is near the edge of the Voronoi diagram. To handle this case, we would need to learn about something called a convex hull. This topic is not included in the IB syllabus, so we will only look at cases where we want the LEC to be inside the Voronoi diagram.

### Example 5.21

Find the coordinates of the centre and radius of the largest empty circle (LEC) in the Voronoi diagram shown below.

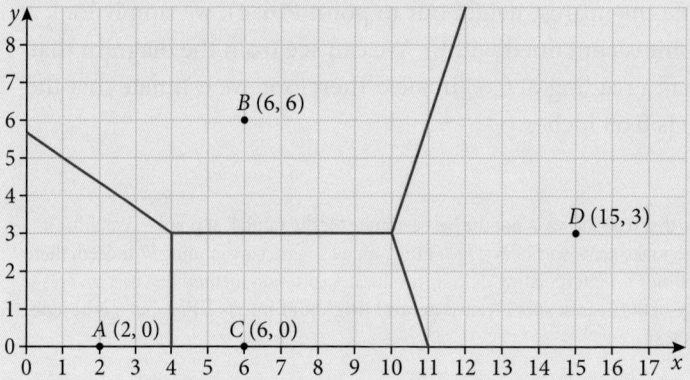

### Solution

Since the centre of the LEC must be on a vertex of the Voronoi diagram, we only have to consider two points: (4, 3) and (10, 3). Although it is clear visually that (10, 3) must be the centre of the LEC, we can check algebraically by using the distance formula.

For (4, 3), it is equidistant to sites $A$, $B$, and $C$, so we can choose any of those coordinates to find the radius. Using site $A(2, 0)$, we have
$$d = \sqrt{(4-2)^2 + (3-0)^2} = \sqrt{4+9} = \sqrt{13}$$

For (10, 3), it is equidistant to sites $B$, $C$, and $D$, so we can choose any of those coordinates to find the radius. Using site $C(6, 0)$, we have
$$d = \sqrt{(10-6)^2 + (3-0)^2} = \sqrt{16+9} = \sqrt{25} = 5$$

Since $5 > \sqrt{13}$, the point (10, 3) must be the centre of the LEC, with radius 5.

1. A new site named $W$ will be placed at $(15, 15)$ in the Voronoi diagram shown.

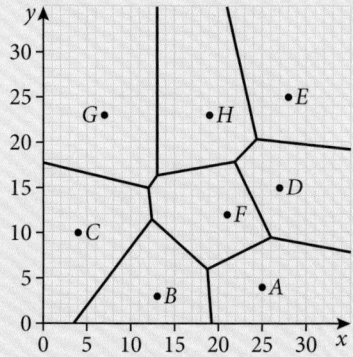

(a) Write down the letter of the existing site's cell which will contain $W$.

(b) Find the equation of the edge separating $W$ from site $G$ $(7, 23)$.

(c) Sketch the perpendicular bisectors required to adjust the diagram after site $W$ is included. Then write down the letters of the sites whose cells will be changed after the Voronoi diagram is adjusted for site $W$.

(d) An additional site $X$ is to be placed at $(5, 5)$. Will every edge of $X$'s cell be a line segment?

2. Copy the Voronoi diagram below onto graph paper. Show the steps to add a new site at $(5, 5)$.

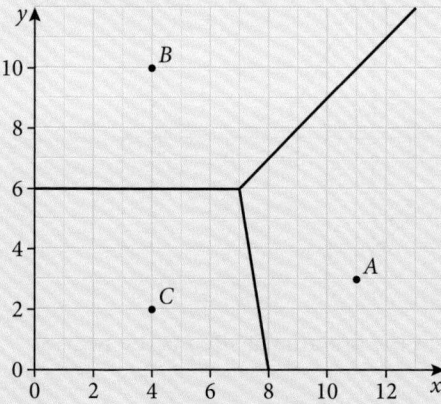

3. The sites $S$, $P$, $Q$ (not shown), and $R$ in the Voronoi diagram represent the locations of telecommunications towers. The grid units are in kilometres. Use the Voronoi diagram to answer the following questions.

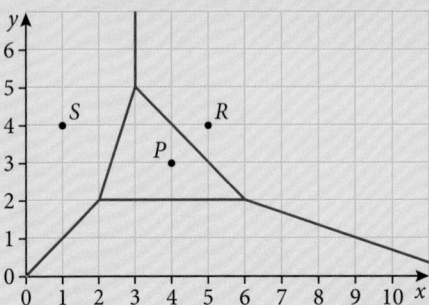

(a) Site $Q$ is missing from the diagram. Write down its coordinates.

(b) Calculate the area covered by the telecoms tower located at site $P$.

(c) Find the distance from point $X(9, 2)$ to the nearest telecoms tower.

4. In less populated areas, internet service providers may use WiMAX antennas to provide service. The Voronoi diagram below represents four WiMAX antennas located at sites $J$, $K$, $L$, and $M$. The axis units are in kilometres.

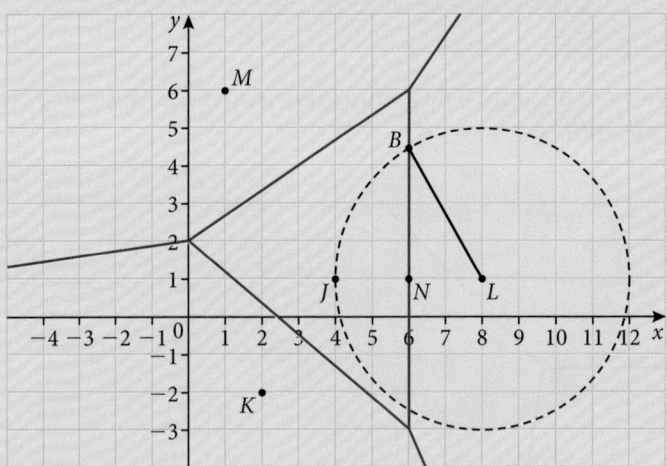

(a) Calculate the area for which the antenna at site $J$ is the nearest antenna.

(b) Calculate the angle $BLN$.

(c) The antenna at site $L$ has a maximum range of 4 km (shown by the shaded area). Calculate the area for which antenna $L$ is the nearest antenna that is also within the range of 4 km.

5. A soil scientist takes samples of the soil in a certain area and records the following data:

| Site | Location | Soil type |
|------|----------|-----------|
| A | (1, 2) | clay |
| B | (6, 3) | sand |
| C | (2, 4) | silt |
| D | (4, 5) | chalk |

Assume the units are metres.

(a) Beginning with sites $A$ and $B$, and continuing alphabetically, draw a Voronoi diagram for sites $A$, $B$, $C$, and $D$, using the incremental insertion algorithm. Show your method by drawing:

   (i)   the Voronoi diagram after sites $A$ and $B$ are added

   (ii)  the Voronoi diagram after site $C$ is added

   (iii) the finished Voronoi diagram.

(b) Use nearest neighbour interpolation to determine the likely soil type at point $M(3.5, 2.5)$.

(c) Give a reason why it is not possible to use nearest-neighbour interpolation to determine the likely soil type at point $N(5, 4)$.

The soil scientist takes an additional sample at location $E(6, 0)$ and determines that the soil type at $E$ is loam.

(d) If site $E$ is added to the Voronoi diagram, which cells would be divided? Give a reason for each cell that would be divided.

(e) Given the new sample at site $E$, does your answer to (b) change? Give a reason why or why not.

(f) The study area is bounded by the line formed by the lines $x = 0$, $y = 0$, $x = 7$, and $y = 6$
Determine the area of the study area that is likely to be loam.

6. Using the following Voronoi diagram, find the centre and radius of the largest empty circle.

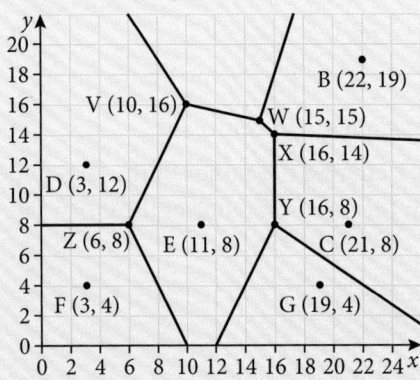

7. In the following Voronoi diagram, each labelled point is the location of a bank branch in a city.

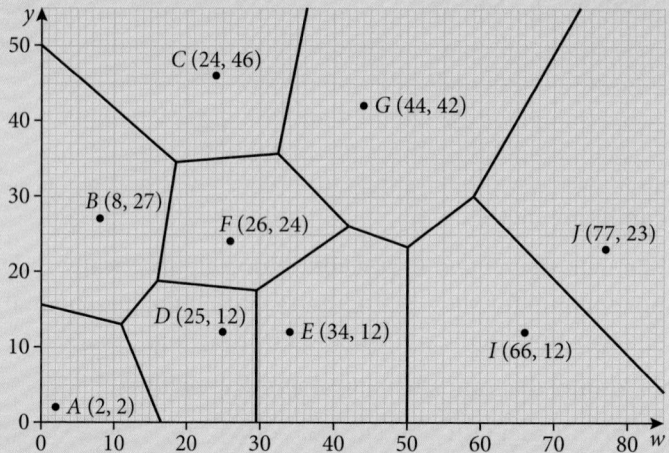

The bank decides to construct a new branch at $H(66, 28)$.

(a) Add this branch as a site and modify the Voronoi diagram accordingly.

(b) The bank wants to add one more branch in the area bounded by branches $E$, $G$, $H$, and $I$. Find the optimal location for a new branch based on the largest empty circle.

## Chapter 5 practice questions

1. (a) The graph of $y = \sin\left(x + \dfrac{\pi}{9}\right)$ intersects the $x$-axis at point $A$.

   Find the $x$-coordinate of $A$, where $0 < x < \pi$

   (b) Solve the equation $\sin\left(x + \dfrac{\pi}{9}\right) = -\dfrac{1}{2}$, for $0 \leq x \leq \pi$

2. Solve the equation $2\cos x = \sin 2x$, for $0 < x < 3\pi$

3. Figure 5.34 shows two concentric circles with centre $O$.

   The radius of the smaller circle is 8 cm and the radius of the larger circle is 10 cm. Points $A$, $B$ and $C$ are on the circumference of the larger circle such that $A\widehat{O}B$ is $\dfrac{\pi}{3}$ radians.

   (a) Find the length of the arc $ACB$.

   (b) Find the area of the shaded region.

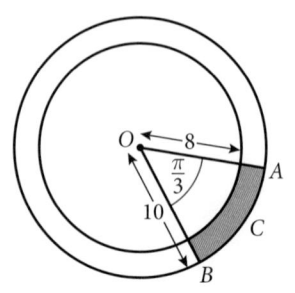

**Figure 5.34** Diagram for quesstion 3

**4.** The diagram shows a circle with centre $O$ and radius $r$. The shaded sector $OACB$ has an area of 27 cm². Angle $A\widehat{O}B = \theta = 1.5$ radians.

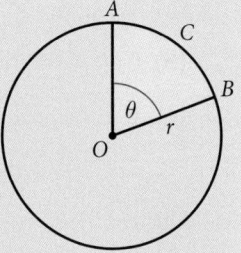

   **(a)** Find the radius of the circle.

   **(b)** Calculate the length of the minor arc $ACB$.

**5.** The diagram shows a circle with centre $O$ and radius $r$. The length of arc $ABC$ is $3\pi$ cm and $A\widehat{O}C = \dfrac{2\pi}{9}$

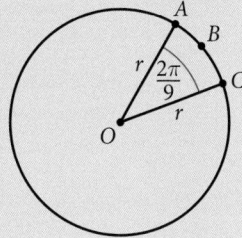

   **(a)** Find the value of $r$.

   **(b)** Find the perimeter of sector $OABC$.

   **(c)** Find the area of sector $OABC$.

**6.** The diagram shows a circle with centre $O$ and radius $r$.

Points $A$, $B$ and $C$ are on the circle and $A\widehat{O}C = \theta$

The area of sector $OABC$ is $\dfrac{4}{3}\pi$ and the

length of arc $ABC$ is $\dfrac{2}{3}\pi$.

Find the value of $r$ and of $\theta$.

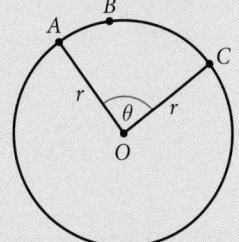

**7.** The diagram shows a circle with centre $O$ and radius $r$.

The angle $A\widehat{O}B = \theta$

The length of the arc $AB$ is 24 cm.
The area of the sector $OAB$ is 180 cm².
Find the value of $r$ and of $\theta$.

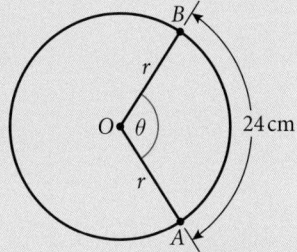

**8.** The graph below models the height of the tide at a certain beach, with time shown as hours after midnight.

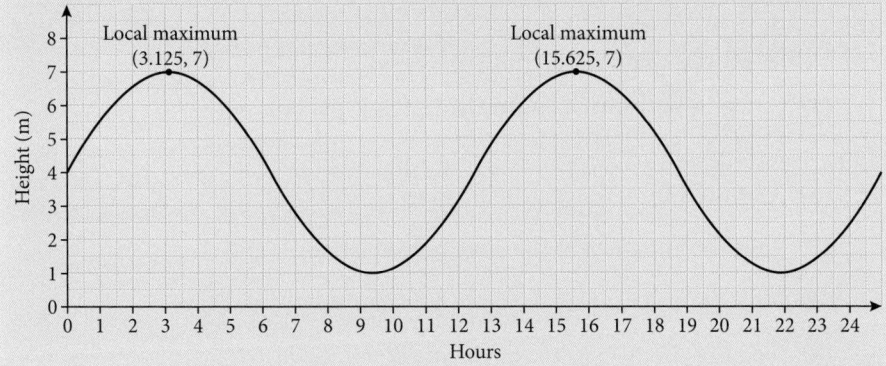

169

**(a)** Write down the period of the tide.

**(b)** Calculate the amplitude of the tide.

**(c)** Use the graph to estimate the amount of time in this 24-hour period that the tide height is at least 6 m.

9. Consider the function $y = 3 \cos(2x) + 1$

   **(a)** Sketch the graph of this function for $0° \leq x \leq 180°$

   **(b)** Write down the period of the function.

   **(c)** Using your GDC, find the smallest possible value of $x$, $0° \leq x \leq 180°$, for which $3 \cos(2x) + 1 = 2$

10. The graph of the function $f(x) = 2 \cos(4x) - 1$, where $0° \leq x \leq 90°$, is shown below.

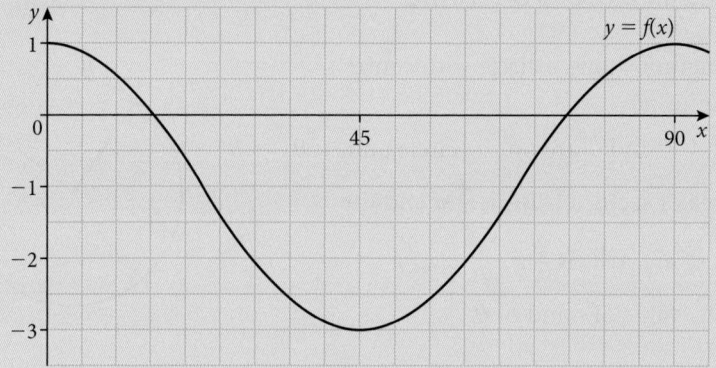

   **(a)** Copy the diagram and draw the graph of the function $g(x) = \sin(2x) - 2$ for $0° \leq x \leq 90°$

   **(b)** Write down the number of solutions to the equation $f(x) = g(x)$, for $0° \leq x \leq 90°$

   **(c)** Write down one value of $x$ for which $f(x) > g(x)$, for $0° \leq x \leq 90°$

   **(d)** Given that $f(x) < g(x)$ in the interval $a < x < b$, use your GDC to find the value of:
   
   **(i)** $a$          **(ii)** $b$

11. For this question, use a grid with $0 \leq x \leq 18$ and $0 \leq y \leq 12$.

   **(a)** Create a Voronoi diagram with sites at $A(7, 0)$, $B(3, 4)$, $C(11, 8)$, and $D(15, 4)$.

   **(b)** Add the site $X(7, 4)$ to the diagram.

   **(c)** Find the equation of the edge between sites $X$ and $A$.

   **(d)** Find the equation of the edge between sites $X$ and $C$.

   **(e)** Add the site $Y(7, 8)$ to the diagram.

**(f)** Find the centre of the largest empty circle with coordinates $(a, b)$ such that $3 \leq a \leq 15$ and $0 \leq b \leq 8$

**(g)** Add site $Z$ at $(a, b)$.

**(h)** Find the equation of the edge between sites $Z$ and $D$. Write your answer in the $ax + by = c$ where $a, b, c \in \mathbb{Z}$

**12.** For this question, use a grid with $-10 \leq x \leq 10$ and $-10 \leq y \leq 4$

**(a)** Create a Voronoi diagram with sites at $A(0, 1)$, $B(-7, -2)$, $C(3, -2)$, and $D(6, -5)$.

**(b)** Add the site $X(0, -2)$ to the diagram.

**(c)** Find the equation of the edge between sites $X$ and $C$.

**(d)** Add the site $Y(3, -6)$ to the diagram.

**(e)** Find the equation of the edge between sites $X$ and $Y$. Write your answer in the form $ax + by = c$ where $a, b, c \in \mathbb{Z}$

**13.** In the Voronoi diagram each site is the location of a petrol station in a city.

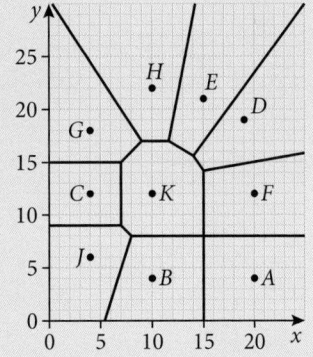

**(a)** Given that a new petrol station is to be located as far as possible from existing petrol stations,

**(i)** give a reason why the new location should be on a vertex of the Voronoi diagram

**(ii)** give a reason why $(15, 8)$ is the best location for the new station.

**14.** Use the Voronoi diagram shown to answer the questions.

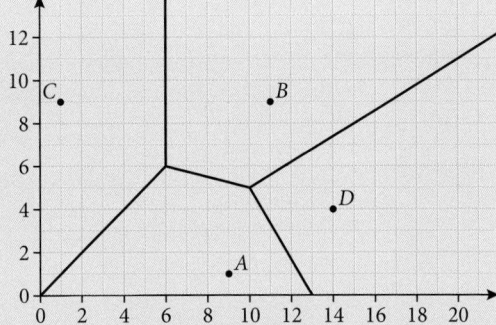

**(a)** Find the centre and radius of the largest empty circle. Give two justifications for your answer.

**(b)** Insert a new site $X$ at the centre of the largest empty circle.

**15.** Sites $A$, $B$, $C$, and $D$ are collinear. Explain why the Voronoi diagram for $A$, $B$, $C$, and $D$ has no vertices.

16. Voronoi diagrams of regular polygons:

   (a) Sites $A$, $B$, and $C$ are located at the vertices of an equilateral triangle. Sketch the Voronoi diagram for these sites.

   (b) Sites $A$, $B$, $C$, and $D$ are located at the vertices of a square. Sketch the Voronoi diagram for these sites.

   (c) Give a description for the Voronoi diagram for a regular $n$-gon including the position of the edges and the number of vertices.

Complex numbers

6

By the end of this chapter, you should be familiar with…

- the properties of complex numbers
- complex number calculations
- complex numbers as vectors in the complex plane
- powers and roots of complex numbers
- the use of complex numbers in STEM applications.

**Real numbers** are those that can be represented as points on a number line, such as integer, rational, or irrational quantities. **Complex numbers** cannot be represented on a number line, but are analytic solutions to equations whose solutions are not real numbers. As these complex numbers start with the acceptance of the **imaginary** unit, it gives the mistaken notion that a solution may be merely a figment of mathematical imagination; however, the relevance of non-real solutions to equations has been known by mathematicians for centuries. The many applications in electronics and with fractal images are more current uses you may already be familiar with.

## 6.1 Imaginary numbers

For any real number $a$, its square is guaranteed to be a positive real number. If $a$ is positive, $a^2$ is positive; if $a$ is negative, $a^2$ is positive again. So, what happens mathematically when you need a solution to $x^2 = -1$? Since real numbers will not give a solution, an imaginary number is required. This number is called the imaginary unit i, where $i^2 = -1$, which implies that $i = \sqrt{-1}$. Like other well-known constants, such as $\pi$, $\phi$, and e, this constant i has many applications, some of which will be presented later in this chapter.

> There is a reason why we do not define $i = \sqrt{-1}$. It is the convention in mathematics that when we write $\sqrt{9}$ then we mean the non-negative square root of 9, namely 3. We do not mean $-3$.
>
> i does not belong to this category since we cannot say that i is the positive square root of $-1$, i.e., $i > 0$. If we do, then $-1 = i \cdot i > 0$, which is false, and if we say $i < 0$, then $-i > 0$, and $-1 = -i \cdot -i > 0$, which is also false. Actually $-i$ is also a square root of $\sqrt{-1}$ because $-i \cdot -i = i^2 = -1$
>
> With this in mind, we can use a convention which calls i the **principal** square root of $-1$ and write $i = \sqrt{-1}$

When we isolate the factor of $-1$ when there are square roots in an expression, then we can express all negative radicands using the imaginary unit i.

## Example 6.1

Express each value using the imaginary unit i.

(a) $\sqrt{-16}$  (b) $\sqrt{-18}$  (c) $\sqrt{-27}$  (d) $\sqrt{-4} \cdot \sqrt{-9}$

### Solution

(a) $\sqrt{-16} = \sqrt{16} \cdot \sqrt{-1} = 4i$

(b) $\sqrt{-18} = \sqrt{18} \cdot \sqrt{-1} = 3\sqrt{2}\,i$

(c) $\sqrt{-27} = \sqrt{27} \cdot \sqrt{-1} = 3\sqrt{3}\,i$

(d) $\sqrt{-4} \cdot \sqrt{-9} = \sqrt{4} \cdot \sqrt{-1} \cdot \sqrt{9} \cdot \sqrt{-1}$
$\qquad\qquad = 2i \cdot 3i = 6i^2 = -6$

 Be very careful with the last example.

Given $i = \sqrt{-1}$ and $i^2 = -1$, consider further powers of i:

The first four:

$i = \sqrt{-1}$

$i^2 = -1$

$i^3 = i^2 \cdot i = -i$

$i^4 = i^2 \cdot i^2 = -1 \cdot -1 = 1$

The next four:

$i^5 = i^4 \cdot i = 1 \cdot i = i$

$i^6 = i^4 \cdot i^2 = 1 \cdot i^2 = -1$

$i^7 = i^4 \cdot i^3 = 1 \cdot i^3 = -i$

$i^8 = i^4 \cdot i^4 = 1 \cdot 1 = 1$

Note that every power $k$ that is an integer multiple of 4 produces $i^k = 1$ and the pattern repeats.

## Example 6.2

Express each in its simplest form.

(a) $i^{31}$  (b) $i^{125}$

### Solution

Divide 31 and 125 by 4 and look for the remainder.

(a) $i^{31} = i^3 = -i$

(b) $i^{125} = i^1 = i$

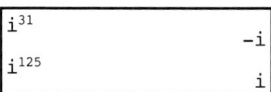

**Figure 6.1** GDC screen for the solution to Example 6.2

Complex numbers add another dimension to real numbers represented by points on a number line. This new dimension contains values that are not real, and contains all real numbers as well as imaginary numbers. This super-set of numbers is called the set of **complex numbers.** A complex number has the form $z = a + bi$ where $a$ is the real component and $b$ is the imaginary component.

If $z = a + bi$, then $a$ is called the real part of $z$ and written as Re $z$, and $b$ is called the imaginary part of $z$ and written as Im $z$.

The zeros of a function are the solutions to $y = 0$. Graphically, the zeros of a function are the $x$-intercepts, where $y = 0$.

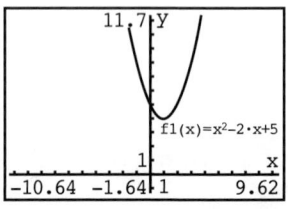

**Figure 6.2** GDC screen for the solution to Example 6.3

---

**Example 6.3**

Find the zeros of $y = x^2 - 2x + 5$

---

**Solution**

As the graph of $y = x^2 - 2x + 5$ shows, there are no real zeros.

Using the quadratic formula:

$$x = \frac{-b \pm \sqrt{b^2 - 4ac}}{2a} = \frac{2 \pm \sqrt{(-2)^2 - 4 \cdot 1 \cdot 5}}{2}$$

$$\Rightarrow x = \frac{2 \pm \sqrt{4 - 20}}{2} = \frac{2 \pm \sqrt{-16}}{2}$$

$$\Rightarrow x = \frac{2 \pm 4i}{2} = 1 \pm 2i$$

---

**Exercise 6.1**

1. Express each value using the imaginary unit i.

   (a) $\sqrt{-36}$    (b) $\sqrt{-12}$    (c) $\sqrt{-63}$    (d) $\sqrt{-8} \cdot \sqrt{-18}$

2. Express each number as a complex number in the form $a + bi$

   (a) $4 + \sqrt{-9}$    (b) $-3 - \sqrt{-4}$    (c) $-18 + \sqrt{-18}$

   (d) $4\sqrt{2} - \sqrt{-8}$    (e) $\sqrt{-4}$    (f) $12 + \sqrt{-12}$

   (g) $-\sqrt{49}$    (h) $(2 + \sqrt{-4})i$

3. Express each number in its simplest form.

   (a) $i^7$    (b) $i^{66}$    (c) $i^{721}$    (d) $i^{-24}$

4. Find the zeros of each function.

   (a) $y = x^2 + 2x + 10$    (b) $y = x^2 - 4x + 7$

   (c) $y = 2x^2 + 4x + 5$    (d) $y = x^2 - 2x + 10$

   (e) $y = x^2 + 6x + 10$    (f) $y = x^2 + 3x + 3$

5. $x^k - 1 = 0$ has $k$ roots. For the value of $k$ indicated, find them by factorising $x^k - 1$. Knowing that $(x - 1)$ is a factor may be useful. There will be an alternative strategy presented later in this chapter.

   (a) $k = 4$    (b) $k = 3$

## 6.2   Operations with complex numbers

Since a complex number has the form $z = a + bi$ that separates the real component $a$ from the imaginary component $b$, the addition and subtraction of complex numbers is very straightforward: combine the components separately.

### Example 6.4

Find each sum.

(a) $(1 + 3i)$ and $(4 + i)$           (b) $(5 - 2i)$ and $(-8 + 5i)$

(c) $(2\sqrt{3} - 4i)$ and $(\sqrt{3} + i)$

### Solution

(a) $\begin{cases} 1 + 4 = 5 \\ 3i + i = 4i \end{cases}$        (b) $\begin{cases} 5 - 8 = -3 \\ -2i + 5i = 3i \end{cases}$

     The sum is $5 + 4i$            The sum is $-3 + 3i$

(c) $\begin{cases} 2\sqrt{3} + \sqrt{3} = 3\sqrt{3} \\ -4i + i = -3i \end{cases}$

     The sum is $3\sqrt{3} - 3i$

```
1+3i+4+i
                        5+4i
5-2i+-8+5i
                       -3+3i
2√3-4i+√3+i
                       3√3-3i
```

**Figure 6.3** GDC output for the solutions to Example 6.4

The product of complex numbers is nothing more than the product of two binomials, found in much the same way as the product $(3x - 4y)(5x + 2y)$, for instance. We can find the product $(3 - 4i)(5 + 2i)$ in a similar way.

$$(3 - 4i)(5 + 2i) = 3 \cdot 5 + 3 \cdot 2i + (-4i) \cdot 5 + (-4i) \cdot (2i)$$
$$= 15 + 6i - 20i - 8i^2$$
$$= 15 + 8 + (6 - 20)i \quad \text{since } i^2 = -1$$
$$= 23 - 14i$$

Before we divide complex numbers, we need a reminder on the simplification of rational expressions containing irrational denominators.

### Example 6.5

Express each fraction with a rational number denominator.

(a) $\dfrac{6}{\sqrt{2}}$          (b) $\dfrac{12}{3 - \sqrt{5}}$

### Solution

(a) $\dfrac{6}{\sqrt{2}} \cdot \left(\dfrac{\sqrt{2}}{\sqrt{2}}\right) = \dfrac{6\sqrt{2}}{2} = 3\sqrt{2}$

(b) We need the conjugate of $3 - \sqrt{5}$ which is $3 + \sqrt{5}$, so

$$\frac{12}{3 - \sqrt{5}} = \frac{12}{3 - \sqrt{5}} \cdot \left(\frac{3 + \sqrt{5}}{3 + \sqrt{5}}\right)$$

$$= \frac{12(3 + \sqrt{5})}{9 - 5} = \frac{12(3 + \sqrt{5})}{4} = 3(3 + \sqrt{5}) = 9 + 3\sqrt{5}$$

By considering any complex number $a + bi$ as $a + b\sqrt{-1}$, it essentially has the same form as the binomial $3 - \sqrt{5}$ above. The **complex conjugate** of $z = a + bi$ is called $z^* = a - bi$ and serves a similar purpose to the binomials. Consider the product of $z = a + bi$ and its complex conjugate, $z^* = a - bi$:

$$(a + bi)(a - bi)$$
$$= a^2 - (bi)^2$$
$$= a^2 - b^2i^2$$
$$= a^2 + b^2$$

Just as binomial conjugates produce rational products, the multiplication of a complex number by its complex conjugate produces an entirely real value.

---

### Example 6.6

Express each rational expression in the form $a + bi$

(a) $\dfrac{6}{1 + i}$      (b) $\dfrac{50}{-4 + 3i}$      (c) $\dfrac{9}{2 + \sqrt{5}i}$      (d) $\dfrac{2 - 3i}{3 + 2i}$

---

```
6
---
1+i
             3-3i
50
----
-4+3i
             -8-6i
```

**Figure 6.4** GDC output for the solutions to Example 6.6 (a) and (b)

### Solution

(a) $\dfrac{6}{1 + i} = \dfrac{6}{1 + i} \cdot \left(\dfrac{1 - i}{1 - i}\right) = \dfrac{6(1 - i)}{1 + 1} = \dfrac{6(1 - i)}{2} = 3(1 - i) = 3 - 3i$

(b) $\dfrac{50}{-4 + 3i} = \dfrac{50}{-4 + 3i} \cdot \left(\dfrac{-4 - 3i}{-4 - 3i}\right) = \dfrac{50(-4 - 3i)}{16 + 9} = \dfrac{50(-4 - 3i)}{25}$

$\quad\quad = 2(-4 - 3i) = -8 - 6i$

(c) $\dfrac{9}{2 + \sqrt{5}i} = \dfrac{9}{2 + \sqrt{5}i} \cdot \left(\dfrac{2 - \sqrt{5}i}{2 - \sqrt{5}i}\right) = \dfrac{9(2 - \sqrt{5}i)}{4 + 5} = 2 - \sqrt{5}i$

(d) $\dfrac{2 - 3i}{3 + 2i} = \dfrac{2 - 3i}{3 + 2i} \cdot \left(\dfrac{3 - 2i}{3 - 2i}\right) = \dfrac{(2 - 3i)(3 - 2i)}{9 + 4} = \dfrac{-13i}{13} = -i$

---

Consider the roots, $x = 1 \pm 2i$, of the equation $y = x^2 - 2x + 5$ in Example 6.3 which we found using the quadratic formula. They are complex conjugates of each other. We can generate the original equation from the roots. In general:

- when a quadratic polynomial in $x$ has zeros $r_1$ and $r_2$ then

$$(x - r_1) \text{ and } (x - r_2) \text{ are its factors}$$

- when $(x - r_1)$ and $(x - r_2)$ are its factors then the quadratic is

$$(x - r_1)(x - r_2) = x^2 - (r_1 + r_2)x + r_1 r_2$$

In other words, the **negative of the sum of the zeros** is the coefficient of $x$, and the **product of the zeros** is the constant term.

## Example 6.7

Show how the original function $y = x^2 - 2x + 5$ can be found when only its zeros, $x = 1 \pm 2i$, are known.

### Solution

Since the complex conjugates $(1 + 2i)$ and $(1 - 2i)$ have a sum of 2 and a product of 5, the original function would be $y = x^2 - 2x + 5$

A polynomial has **zeros**, a polynomial equation has **roots**.

## Example 6.8

A quadratic equation of the form $ax^2 + bx + c = 0$ has one known root, $x = 3 - 4i$

If $a$, $b$, and $c$ are all real numbers, then what is the other root?

### Solution

In Example 6.7, it was shown that the sum of the roots would be $-b$ and the product of the roots would be $c$. For $b$ and $c$ to be real numbers, the other root must be the complex conjugate of the root given.

Hence, the other root is $x = 3 + 4i$

## Example 6.9

Find the equation with the roots given in Example 6.8

### Solution

Given that $3 - 4i$ and $3 + 4i$ are the roots, their sum is 6 and their product is 25. If we assume that $a = 1$, then the required equation is $x^2 - 6x + 25 = 0$. However, as there is no specific value given for $a$, the other coefficients are multiples of $a$, and the general solution is $ax^2 - 6ax + 25a = 0$

 Remember that the coefficient of the quadratic term need not be 1.

**Exercise 6.2**

1. Find the sum of each pair of complex numbers.

   **(a)** $(2 - 4i)$ and $(-3 + 2i)$      **(b)** $(-1 + i)$ and $(3 - 2i)$

   **(c)** $(2\sqrt{2} + i)$ and $(-\sqrt{2} - 2i)$

2. Find each product.

   **(a)** $(1 - 4i)(1 + 4i)$      **(b)** $(2\sqrt{3} + i)(-2\sqrt{3} + i)$

   **(c)** $(3 + 4i)(3 - 4i)$

3. Find the product of each given complex number and its conjugate.

   **(a)** $4 - 3i$      **(b)** $-5 + 12i$      **(c)** $-4 - 2\sqrt{5}i$

4. Express each quotient in $a + bi$ form.

   **(a)** $\dfrac{5}{2 - i}$      **(b)** $\dfrac{1 - 2i}{1 + 2i}$      **(c)** $\dfrac{2 - 4i}{-3 + 2i}$

5. A quadratic function $y = ax^2 + bx + c$ has real coefficients $a$, $b$, and $c$. Find the function when one of the zeros is:

   **(a)** $1 - i$      **(b)** $-7 + i$      **(c)** $-2\sqrt{3} + \sqrt{3}\,i$

6. Find the quadratic function whose zeros are:

   **(a)** $x = 2 + \sqrt{3}, x = 2 - \sqrt{3}$      **(b)** $x = \dfrac{1 \pm \sqrt{5}}{2}$

   **(c)** $x = -1 \pm 2i$      **(d)** $x = \dfrac{3}{2} \pm \dfrac{5}{2}i$

   **(e)** $x = \dfrac{1 \pm \sqrt{5}\,i}{2}$      **(f)** $x = -2\sqrt{3} \pm \sqrt{3}\,i$

7. Find the quadratic function of the form $y = x^2 + bx + c$ whose zeros are:

   **(a)** $(5 + 2i)$ and $(3 - i)$      **(b)** $(3 + 2i)$ and $(-3 - 2i)$

   **(c)** $(3 + \sqrt{2}i)$ and $(-3 - \sqrt{2}i)$

8. Let $z = a + bi$. Find the values of $a$ and $b$ if $(2 + 3i) \cdot z = 7 + i$

9. $(2 + yi)(x + i) = 1 + 3i$, where $x$ and $y$ are real numbers. Solve for $x$ and $y$.

10. Consider the complex number $z = 1 + \sqrt{3}\,i$

    **(a)** Evaluate $z^3$

    **(b)** Prove that $z^{6n} = 8^{2n}$, where $n \in \mathbb{Z}^+$

    **(c)** Hence, find $z^{48}$

11. Consider the complex number $z = -\sqrt{2} + \sqrt{2}i$

    (a) Evaluate $z^2$

    (b) Prove that $z^{4k} = (-16)^k$, where $k \in \mathbb{Z}^+$

    (c) Hence, find $z^{46}$

12. Given that $z$ is a complex number such that $|z + 4i| = 2|z + i|$ find the value of $|z|$.

13. Write the complex number $z = 3 + \dfrac{2i}{2 - \sqrt{2}\,i}$ in the form $a + bi$

14. Find the values of the two real numbers $x$ and $y$ if
    $(x + yi)(4 - 7i) = 3 + 2i$

15. Find the complex number $z$ and write it in the form $a + bi$:

    (a) $(z + 1)i = 3z - 2$            (b) $z = 3 - 4i$

## 6.3 The complex plane

Real numbers can be found on a number line, but imaginary numbers cannot. However, purely imaginary numbers are ordered in the same way as real numbers.

For example, with real numbers: $\;1 < \sqrt{2} < 2 < e < 3 < \pi$

and with imaginary numbers: $\quad i < \sqrt{2}i < 2i < ei < 3i < \pi i$

There are operations such as those in section 6.2 where we take two purely imaginary numbers or two complex numbers and produce a real number.

If we use a separate number line to show the imaginary component of a complex number, and put it at right angles to the real number line, we have the **complex plane** (Figure 6.5). This is similar to the $x$- and $y$-axes we use for graphs of functions. Complex numbers of the form $a + bi$ can be represented as a coordinate point measuring $a$ units along the horizontal $x$-axis, and $b$ units along the vertical $y$-axis. This complex plane is called the **Argand plane** (also known as an Argand diagram).

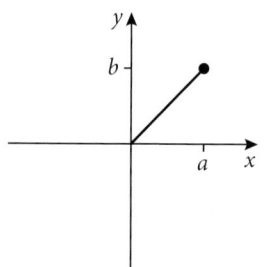

**Figure 6.5** $a + bi$ in the complex plane

Since this plane is essentially the Cartesian plane, there are many ways we can use this representation. The distance of a number $z$ from the origin on the Argand plane is known as the **absolute value** or **modulus** of the complex number, $|z|$.

## Example 6.10

Consider the complex number $z_1 = -3 + 4i$

(a) Sketch it and find its distance from the origin.

(b) Compare this distance to $\sqrt{z_1 \cdot z_1^*}$

## Solution

(a) The complex number $z_1 = -3 + 4i$ is shown in the Argand diagram. Its distance from the origin is the length of the hypotenuse of the right-angled triangle that could be drawn, and is 5 units long.

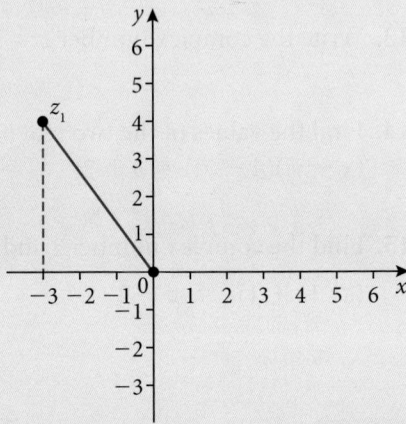

(b) As $z_1 = -3 + 4i$, its conjugate is $z_1^* = -3 - 4i$, and their product is $9 + 16 = 25$

In general, $z \cdot z^* = |z|^2$

Hence, $\sqrt{z_1 \cdot z_1^*} = 5$ and is exactly the same as the answer above.

## Example 6.11

Given the complex numbers $z_1 = -3 + 4i$ and $z_2 = 5 + i$:

(a) plot them on the complex plane

(b) find the modulus of each number (the distance from the origin)

(c) find their sum, and call it $z_3$

(d) plot $z_3$ in relation to $z_1$ and $z_2$. What do you notice?

## Solution

(a)

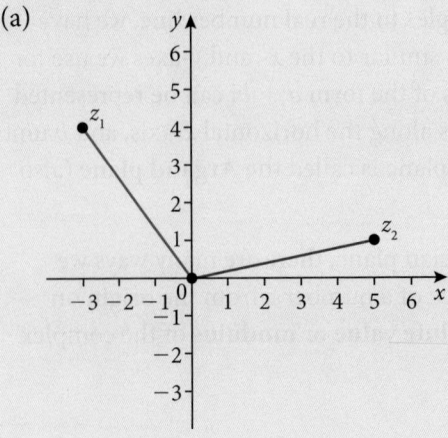

(b) $|z_1| = \sqrt{(-3)^2 + 4^2} = 5$

$|z_2| = \sqrt{5^2 + 1^2} = \sqrt{26}$

(c) $z_3 = z_1 + z_2$

$= (-3 + 4i) + (5 + i) = 2 + 5i$

(d)

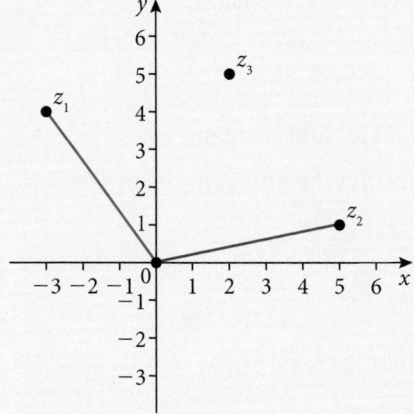

The real and imaginary parts of $z_3$ are the sums of the real and imaginary parts of $z_1$ and $z_2$. This produces a parallelogram when plotted.

Also, the modulus of $z_3$ is not the sum of the moduli of $z_1$ and $z_2$.

Compare $3 - 4i$ and $-3 + 4i$

Is one bigger than the other? As complex numbers are represented by points representing two different components, it should be easy to see why complex numbers are not ordered.

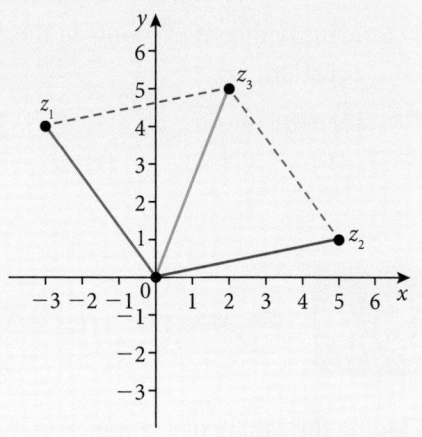

## Exercise 6.3

1. Graph these complex numbers in the same complex plane:
$3 + 4i, 4 + 3i, -3 + 4i, -4 + 3i, 5i, -5$

   (a) What is common to all of them?

   (b) Name another complex number in quadrant 3 with the same property.

   (c) Does $-2\sqrt{3} + \sqrt{13}i$ have the same property? Why?

   (d) Name both complex numbers whose imaginary part is $3\sqrt{2}$ that fit the pattern.

2. Find the modulus of each complex number.

   (a) $1 + i$      (b) $\sqrt{3} + i$      (c) $2i$

   (d) $-2i$      (e) $-5 - 12i$      (f) $-5 + 12i$

   (g) $-21 + 20i$      (h) $2\sqrt{3} + 4\sqrt{6}i$

3. Consider the two complex numbers $z_1 = 1 + 2i$ and $z_2 = 5 + i$

   (a) Plot these points on an Argand diagram.

   (b) Find their moduli.

   (c) Plot the point $z_3$, given that it is the sum of $z_1$ and $z_2$

   (d) Draw a quadrilateral by connecting the points in the order $Oz_1z_3z_2$ then back to $O$.

   (e) What is the significance of each of the two diagonals in the quadrilateral drawn?

4. Use an Argand diagram to show that $|z_1 + z_2| \leqslant |z_1| + |z_2|$

5. The complex numbers $z_1 = 2\sqrt{3} - 2i$, $z_2 = 2 + 2i$, and $z_3 = (2\sqrt{3} - 2i)(2 + 2i)$ represent the vertices of a triangle in an Argand diagram. Find its area.

6. Identify the set of points in the complex plane that correspond to each equation.

   (a) $|z| = 3$      (b) $z^* = -z$      (c) $z + z^* = 8$

   (d) $|z - 3| = 2$      (e) $|z - 1| + |z - 3| = 2$

## 6.4 Powers and roots of complex numbers

Finding the square of a complex number, such as $1 + \sqrt{3}i$, takes only a little effort, since by binomial multiplication:

$$(1 + \sqrt{3}i)^2 = (1 + \sqrt{3}i)(1 + \sqrt{3}i)$$
$$= 1 + \sqrt{3}i + \sqrt{3}i + 3i^2$$
$$= -2 + 2\sqrt{3}i$$

However, it is much more difficult to expand $(1 + \sqrt{3}i)^6$, unless we use some trigonometry.

### The polar form of complex numbers

Consider the complex number $z = a + bi$ in the Argand diagram.

By drawing a segment from the origin to the complex number represented as a point, a right-angled triangle is identified with an angle, its adjacent side, its opposite side, and hypotenuse (Figure 6.6).

The angle, shown in red, is the **argument** or **position angle**, $\theta$, and the distance $r$ from the origin to the point is its absolute value, otherwise known as its **modulus** or **magnitude**. This description is called its **modulus-argument** or **polar form**.

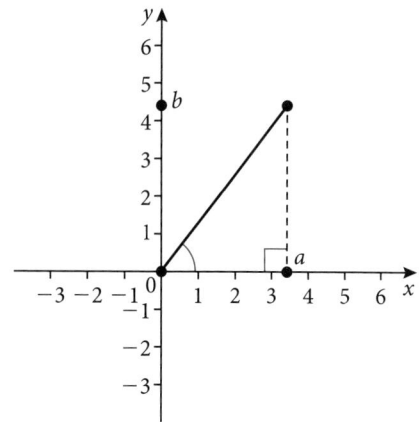

**Figure 6.6** A right-angled triangle is identified

In the right-angled triangle with (position) angle $\theta$ and hypotenuse $r$,
$$a = r\cos\theta \quad \text{and} \quad b = r\sin\theta$$
$$a + b\mathrm{i} = r\cos\theta + r\sin\theta \cdot \mathrm{i} = r(\cos\theta + \mathrm{i}\cdot\sin\theta)$$
This expression is often written in shorthand as $r\,\mathrm{cis}\,\theta$

Although the addition and subtraction of complex numbers is simple in $z = a + b\mathrm{i}$ form, the polar form is often easier for multiplication and division, when we use trigonometric identities. Consider the multiplication of two complex numbers $z_1 = a_1 + b_1\mathrm{i}$ and $z_2 = a_2 + b_2\mathrm{i}$

In polar form, they are $z_1 = r_1\,\mathrm{cis}\,\theta_1$ and $z_2 = r_2\,\mathrm{cis}\,\theta_2$

Their product is

$$z_1 z_2 = (r_1\,\mathrm{cis}\,\theta_1)(r_2\,\mathrm{cis}\,\theta_2)$$
$$= r_1(\cos\theta_1 + \mathrm{i}\sin\theta_1)\cdot r_2(\cos\theta_2 + \mathrm{i}\sin\theta_2) + \mathrm{i}^2\sin\theta_1\sin\theta_2)$$
$$= r_1 r_2(\cos\theta_1\cos\theta_2 + \mathrm{i}\cdot\cos\theta_1\sin\theta_2 + \mathrm{i}\cdot\sin\theta_1\cos\theta_2, \text{ but } \mathrm{i}^2 = -1, \text{ so}$$
$$z_1 z_2 = r_1 r_2[(\cos\theta_1\cos\theta_2 - \sin\theta_1\sin\theta_2) + \mathrm{i}(\cos\theta_1\sin\theta_2 + \sin\theta_1\cos\theta_2)]$$

Since $\cos\theta_1\cos\theta_2 - \sin\theta_1\sin\theta_2 = \cos(\theta_1 + \theta_2)$ and

$\cos\theta_1\sin\theta_2 + \sin\theta_1\cos\theta_2 = \sin(\theta_1 + \theta_2)$

$$z_1 z_2 = r_1 r_2[\cos(\theta_1 + \theta_2) + \mathrm{i}\sin(\theta_1 + \theta_2)] = r_1 r_2[\mathrm{cis}(\theta_1 + \theta_2)]$$

Simply put, the product of two complex numbers in polar form contains the product of their moduli $(r_1 r_2)$ and the sum of their arguments $(\theta_1 + \theta_2)$ Geometrically, when $z_1 = r_1\mathrm{cis}\theta_1$ is multiplied by $z_2$, the complex number $z_1$ is rotated by $\theta_2$ and stretched by the scalar $r_2$.

## Example 6.12

Find the products of the complex numbers in polar form.

(a) $z_1 = 2\,\mathrm{cis}\,\dfrac{\pi}{6}$ and $z_2 = 4\,\mathrm{cis}\,\dfrac{5\pi}{6}$

(b) $z_1 = 6\,\mathrm{cis}\,\dfrac{\pi}{3}$ and $z_2 = 10\,\mathrm{cis}\,\dfrac{\pi}{6}$

185

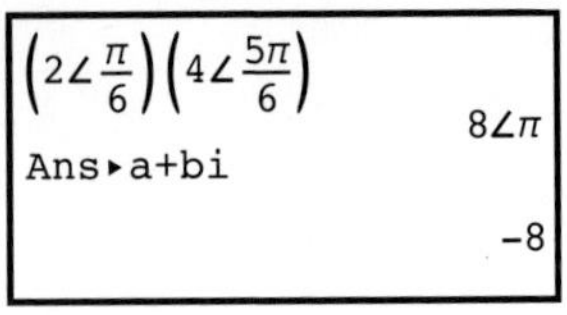

$$\left(2\angle\frac{\pi}{6}\right)\left(4\angle\frac{5\pi}{6}\right)$$

$$8\angle\pi$$

Ans▸a+bi

$$-8$$

**Figure 6.7** GDC output for the solution to Example 6.12 (a)

## Solution

(a) $z_1 \cdot z_2 = \left(2 \operatorname{cis}\frac{\pi}{6}\right) \cdot \left(4 \operatorname{cis}\frac{5\pi}{6}\right)$

$$= 2 \cdot 4 \operatorname{cis}\left(\frac{\pi}{6} + \frac{5\pi}{6}\right)$$

$$= 8 \operatorname{cis} \pi = 8(\cos \pi + i \sin \pi)$$

$$= -8$$

(b) $z_1 \cdot z_2 = \left(6 \operatorname{cis}\frac{\pi}{3}\right) \cdot \left(10 \operatorname{cis}\frac{\pi}{6}\right)$

$$= 60 \operatorname{cis}\frac{\pi}{2} = 60\left(\cos\frac{\pi}{2} + i \sin\frac{\pi}{2}\right)$$

$$= 60i$$

Now, consider the quotient of two complex numbers $z_1 = r_1 \operatorname{cis} \theta_1$ and $z_2 = r_2 \operatorname{cis} \theta_2$

Their quotient is

$$\frac{z_1}{z_2} = \frac{r_1 \operatorname{cis} \theta_1}{r_2 \operatorname{cis} \theta_2}$$

$$= \frac{r_1(\cos \theta_1 + i\sin \theta_1)}{r_2(\cos \theta_2 + i\sin \theta_2)}$$

Multiply both numerator and denominator by $z_2{}^*$

$$\frac{z_1}{z_2} = \frac{r_1(\cos \theta_1 + i\sin \theta_1)}{r_2(\cos \theta_2 + i\sin \theta_2)} \cdot \frac{(\cos \theta_2 - i\sin \theta_2)}{(\cos \theta_2 - i\sin \theta_2)}$$

$$= \frac{r_1}{r_2} \cdot \frac{(\cos \theta_1 \cos \theta_2 - i\cos \theta_1 \sin \theta_2 + i\sin \theta_1 \cos \theta_2 - i^2 \sin \theta_1 \sin \theta_2)}{\cos^2 \theta_2 - i^2 \sin^2 \theta_2}$$

but $i^2 = -1$,

so $\dfrac{z_1}{z_2} = \dfrac{r_1}{r_2} \cdot \dfrac{(\cos \theta_1 \cos \theta_2 + \sin \theta_1 \sin \theta_2) + i(\sin \theta_1 \cos \theta_2 - \cos \theta_1 \sin \theta_2)}{\cos^2 \theta_2 + \sin^2 \theta_2}$

Since $\cos \theta_1 \cos \theta_2 + \sin \theta_1 \sin \theta_2 = \cos(\theta_1 - \theta_2)$, and

$\sin \theta_1 \cos \theta_2 - \cos \theta_1 \sin \theta_2 = \sin(\theta_1 - \theta_2)$, while

$\cos^2 \theta_2 + \sin^2 \theta_2 = 1$

$$\frac{z_1}{z_2} = \frac{r_1}{r_2} \cdot [\cos(\theta_1 - \theta_2) + i\sin(\theta_1 - \theta_2)] = \frac{r_1}{r_2} \cdot \operatorname{cis}(\theta_1 - \theta_2)$$

The quotient of two complex numbers in polar form uses the quotient of their moduli $\dfrac{r_1}{r_2}$ and the difference of their arguments $(\theta_1 - \theta_2)$. Intuitively, the quotient should involve the opposite operations from the product.

## Example 6.13

Find the quotient $\dfrac{z_1}{z_2}$. State your answer in $a + b$i form.

(a) $z_1 = 12 \operatorname{cis} \dfrac{2\pi}{3}$ and $z_2 = 4 \operatorname{cis} \dfrac{\pi}{6}$

(b) $z_1 = 8 \operatorname{cis} \dfrac{5\pi}{6}$ and $z_2 = 2 \operatorname{cis} \dfrac{\pi}{6}$

## Solution

(a) $\dfrac{z_1}{z_2} = \dfrac{12 \operatorname{cis} \dfrac{2\pi}{3}}{4 \operatorname{cis} \dfrac{\pi}{6}} = \dfrac{12}{4} \operatorname{cis}\left(\dfrac{2\pi}{3} - \dfrac{\pi}{6}\right) = 3 \operatorname{cis}\dfrac{\pi}{2} = 3$i

$$\left(12\angle\dfrac{2\pi}{3}\right) \div \left(4\angle\dfrac{\pi}{6}\right)$$
$$3\angle\dfrac{1}{2}\pi$$
$$\texttt{Ans} \blacktriangleright \texttt{a+bi}$$
$$3\texttt{i}$$

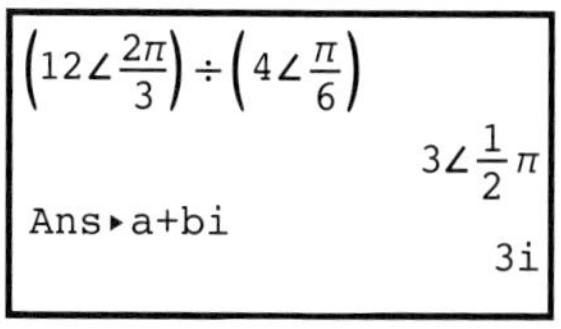

**Figure 6.8** GDC output for the solution to Example 6.13 (a)

(b) $\dfrac{z_1}{z_2} = \dfrac{8 \operatorname{cis} \dfrac{5\pi}{6}}{2 \operatorname{cis} \dfrac{\pi}{6}} = 4 \operatorname{cis}\dfrac{2\pi}{3} = -2 + 2\sqrt{3}$i

## Powers of complex numbers

Consider the square of $z = a + b$i $= r \operatorname{cis} \theta$:

$$(a + b\text{i})^2 = (r \operatorname{cis} \theta)^2 = (r \operatorname{cis} \theta)(r \operatorname{cis} \theta) = r^2 \operatorname{cis} 2\theta$$

Exploring higher powers in this fashion,

$$(a + b\text{i})^3 = (a + b\text{i})^2 \cdot (a + b\text{i}) = (r^2 \operatorname{cis} 2\theta) \cdot (r \operatorname{cis} \theta) = r^3 \operatorname{cis} 3\theta$$

To generalise, $(a + b\text{i})^n = r^n \operatorname{cis} n\theta$. This is known as **de Moivre's theorem**.

## Example 6.14

For each expression, raise the complex number to the power indicated and state the result in $a + b$i form.

(a) $(1 + \sqrt{3}\,\text{i})^6$          (b) $(2 + 2\text{i})^4$

## Solution

It is helpful to show the number on an Argand diagram.

(a)

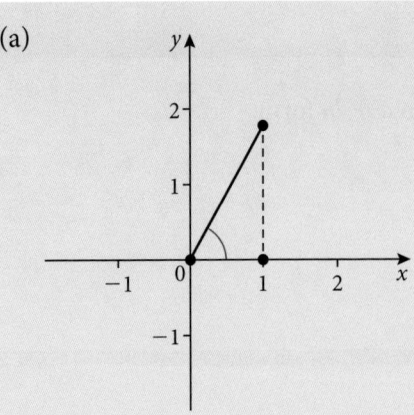

$1 + \sqrt{3}\,i$: $r = \sqrt{1^2 + \sqrt{3}^2} = 2$ and $\theta = \arctan\dfrac{\sqrt{3}}{1} = 60°$ or $\dfrac{\pi}{3}$ radians, so

$$(1 + \sqrt{3}\,i)^6 = \left(2\,\text{cis}\dfrac{\pi}{3}\right)^6 = 2^6\,\text{cis}\,2\pi$$

$$= 64(\cos 2\pi + i \sin 2\pi) = 64$$

(b)

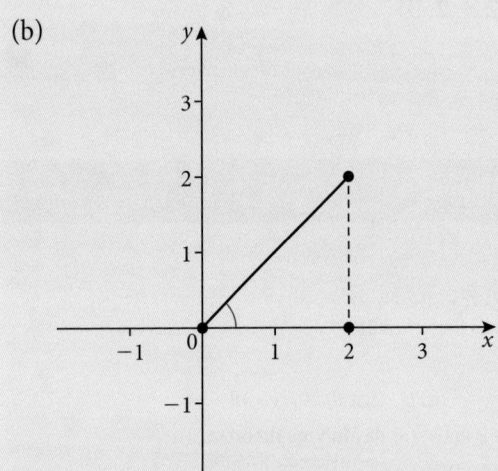

$2 + 2i$: $r = \sqrt{2^2 + 2^2} = 2\sqrt{2}$ and

$\theta = \arctan 1 = 45°$ or $\dfrac{\pi}{4}$ radians, so

$$(2 + 2i)^4 = \left(2\sqrt{2}\,\text{cis}\dfrac{\pi}{4}\right)^4 = (2\sqrt{2})^4\,\text{cis}\dfrac{4\pi}{4}$$

$$= 64(\cos \pi + i \sin \pi) = -64$$

## Roots of complex numbers

Now, take a close look at part (a) of Example 6.14

If $(1 + \sqrt{3}\,i)^6 = 64$ and $(2 + 2i)^4 = -64$, then $\sqrt[6]{64} = 1 + \sqrt{3}\,i$
and $\sqrt[4]{-64} = 2 + 2i$

We can use de Moivre's theorem to verify that the results are correct. Apply de Moivre's theorem, which is assumed to be true for rational numbers $n = \dfrac{1}{k}$ as well, to verify that $\sqrt[6]{64} = 1 + \sqrt{3}\,i$

64 as a complex number is $64 + 0i$ which is merely a point on the $x$-axis.

$$\Rightarrow r = 64 \text{ and } \theta = 0, \text{ so } \sqrt[6]{64} = (64\operatorname{cis}0)^{\frac{1}{6}} = 64^{\frac{1}{6}}(\cos 0 + i \sin 0) = 2$$

This is disappointing. What happened to $\sqrt[6]{64} = 1 + \sqrt{3}\,i$?

A little bit of trigonometry will help. Recall that $\sin \theta$ and $\cos \theta$ are periodic functions. Adding a multiple of 360° or $2\pi$ just produces another co-terminal angle.

Thus, $\sin \theta = \sin(\theta + 2k\pi)$ and $\cos \theta = \cos(\theta + 2k\pi)$ where $k \in \mathbb{Z}$

$$\sqrt[6]{64} = (64\operatorname{cis}0)^{\frac{1}{6}} = (64\operatorname{cis}(0 + 2k\pi))^{\frac{1}{6}} = 64^{\frac{1}{6}}\left(\cos\frac{0 + 2k\pi}{6} + i\sin\frac{0 + 2k\pi}{6}\right)$$

$$= 2\left(\cos\frac{k\pi}{3} + i\sin\frac{k\pi}{3}\right) \text{ for } k \in \{0, 1, 2, 3, 4, 5\} \text{ which produces six}$$

possibilities

$$= \{2, 1 + \sqrt{3}i, -1 + \sqrt{3}\,i, -2, -1 - \sqrt{3}\,i, 1 - \sqrt{3}i\}$$

We can also analyse $\sqrt[4]{-64}$ in a similar way.

$-64$ as a complex number is $-64 + 0i$ which is also a point on the $x$-axis in the opposite direction, but the distance is still $64 \Rightarrow r = 64$ and $\theta = \pi$

$$\sqrt[4]{-64} = 64\operatorname{cis}\pi^{\frac{1}{4}} = (64\operatorname{cis}(\pi + 2k\pi))^{\frac{1}{4}}$$

$$= 64^{\frac{1}{4}}\left(\cos\frac{\pi + 2k\pi}{4} + i\sin\frac{\pi + 2k\pi}{4}\right)$$

$$= 2\sqrt{2}\left(\cos\frac{\pi + 2k\pi}{4} + i\sin\frac{\pi + 2k\pi}{4}\right) \text{ for } k \in \{0, 1, 2, 3\} \text{ which}$$

produces

$$= \left\{ \begin{array}{l} 2\sqrt{2}\left(\cos\dfrac{\pi}{4} + i\sin\dfrac{\pi}{4}\right), 2\sqrt{2}\left(\cos\dfrac{3\pi}{4} + i\sin\dfrac{3\pi}{4}\right), \\ 2\sqrt{2}\left(\cos\dfrac{5\pi}{4} + i\sin\dfrac{5\pi}{4}\right), 2\sqrt{2}\left(\cos\dfrac{7\pi}{4} + i\sin\dfrac{7\pi}{4}\right) \end{array} \right\}$$

$$= \{2 + 2i, -2 + 2i, -2 - 2i, 2 - 2i\}$$

## The Euler form of complex numbers

There is yet another form that complex numbers can take: the **Euler form**. Every complex number can be expressed as $z = r \cdot e^{i\theta}$. Unfortunately, the change from $r\operatorname{cis}\theta$ to $r \cdot e^{i\theta}$ requires some understanding of infinite series, which is not part of this course.

This topic is not required for exams. It is mentioned here as an application of what you learned.

Euler form is also known as Exponential form.

If a complex number is in Euler form, a GDC will readily convert it into Cartesian form. Consider the complex number $z = 1 + i$

A quick mental sketch should show an isosceles right-angled triangle with $r = \sqrt{2}$ and $\theta = \dfrac{\pi}{4}$; hence, $z = \sqrt{2}\,e^{\frac{i\pi}{4}}$

On a GDC, enter $z = \sqrt{2}\,e^{\frac{i\pi}{4}}$, then press ENTER:

$$\boxed{\begin{array}{l} \sqrt{2}e^{\,i\pi/4} \\ \hspace{3cm} 1+1\,i \end{array}}$$

Now, consider $z = \left(\sqrt{2}\,e^{\frac{i\pi}{4}}\right)^3$

$$= 2\sqrt{2}\,e^{\frac{3i\pi}{4}} \quad \text{but a GDC is faster}$$

By default, your GDC is set to accept input in $r \cdot e^{i\theta}$ form and produce results in $a + bi$ form.

$$\boxed{\begin{array}{l} \left(\sqrt{2}e^{\,i\pi/4}\right)^3 \\ \hspace{3cm} -2+2\,i \end{array}}$$

If you wish to go from $a + bi$ form to Euler form quickly, some GDCs allow you to change the settings to make this possible.

---

### Example 6.15

(a) Convert $1 + \sqrt{3}i$ to the forms: $r\operatorname{cis}\theta$ and $r \cdot e^{i\theta}$

(b) Use the Euler form to find $(1 + \sqrt{3}i)^6$

---

#### Solution

First of all, an Argand diagram is useful.

$$r = \sqrt{1^2 + \sqrt{3}^2} = 2$$

$$\theta = \arctan\frac{\sqrt{3}}{1} = 60° \text{ or } \frac{\pi}{3} \text{ radians}$$

(a)  $1 + \sqrt{3}i = 2\operatorname{cis}\dfrac{\pi}{3}$ or $2e^{\frac{i\pi}{3}}$

(b)  $(1 + \sqrt{3}i)^6 = \left(2e^{\frac{i\pi}{3}}\right)^6 = 64e^{2\pi i} = 64$

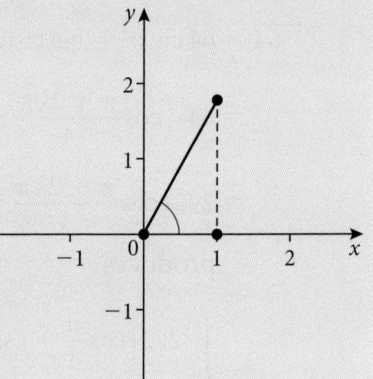

Consider how much easier it would be to explain the multiplication and division of complex numbers when $r$ and $\theta$ are known. No trigonometric identities are required, just basic exponent rules.

$$\text{Given} \begin{cases} z_1 = r_1 \operatorname{cis}\theta_1 = r_1\,e^{i\theta_1} \\ z_2 = r_2 \operatorname{cis}\theta_2 = r_2\,e^{i\theta_2} \end{cases}, \text{ their product and quotient respectively are}$$

$$z_1 z_2 = (r_1\,e^{i\theta_1})(r_2\,e^{i\theta_2}) = r_1 r_2 \cdot e^{i(\theta_1+\theta_2)} \text{ and } \frac{z_1}{z_2} = \frac{r_1\,e^{i\theta_1}}{r_2\,e^{i\theta_2}} = \frac{r_1}{r_2} \cdot e^{i(\theta_1-\theta_2)}$$

1. Express each complex number in polar form.

   (a) $3 + 3i$

   (b) $-3 + 3i$

   (c) $3 - 3i$

   (d) $-3 - 3i$

   (e) $5 + 5\sqrt{3}i$

   (f) $5\sqrt{3} + 5i$

   (g) $-5 + 5\sqrt{3}i$

   (h) $-5\sqrt{3} - 5i$

2. Find the products of each pair of complex numbers. State your answer in $a + bi$ form.

   (a) $z_1 = 5 \operatorname{cis} \dfrac{\pi}{3}$ and $z_2 = 3 \operatorname{cis} \dfrac{\pi}{6}$

   (b) $z_1 = 4 \operatorname{cis} \dfrac{2\pi}{3}$ and $z_2 = 2 \operatorname{cis} \dfrac{2\pi}{3}$

3. Find the quotient $\dfrac{z_1}{z_2}$. State your answer in $a + bi$ form.

   (a) $z_1 = 6 \operatorname{cis} \dfrac{\pi}{2}$ and $z_2 = 2 \operatorname{cis} \dfrac{\pi}{6}$

   (b) $z_1 = 16 \operatorname{cis} \dfrac{3\pi}{2}$ and $z_2 = 4 \operatorname{cis} \dfrac{\pi}{6}$

   (c) $z_1 = 8 \operatorname{cis} \dfrac{\pi}{4}$ and $z_2 = 2 \operatorname{cis} \dfrac{\pi}{2}$

   (d) $z_1 = 16 \operatorname{cis} \dfrac{\pi}{6}$ and $z_2 = 2 \operatorname{cis} \dfrac{\pi}{3}$

4. For each expression, raise the complex number to the power indicated and state the result in $a + bi$ form.

   (a) $(-1 + i)^5$

   (b) $(-1 - i)^4$

   (c) $(-\sqrt{3} + i)^4$

   (d) $(-\sqrt{3} - i)^6$

5. Find each root.

   (a) $\sqrt[3]{8}$

   (b) $\sqrt{(1 + \sqrt{3}\,i)}$

   (c) $\sqrt[4]{-16}$

6. Write each complex number in Euler form.

   (a) $-6 + 6i$

   (b) $2\sqrt{3} + 6i$

   (c) $1 - \sqrt{3}i$

7. Find the product of the complex numbers in Euler form.

   (a) $z_1 = 3e^{\frac{\pi i}{8}}$ and $z_2 = 2e^{\frac{3\pi i}{8}}$

   (b) $z_1 = 4e^{\frac{3\pi i}{4}}$ and $z_2 = 3e^{\frac{\pi i}{2}}$

8. Find the quotient $\dfrac{z_1}{z_2}$ of these complex numbers and give the answer in $a + bi$ form.

   (a) $z_1 = 12e^{\frac{3\pi i}{4}}$ and $z_2 = 3e^{\frac{\pi i}{8}}$

   (b) $z_1 = 16e^{\frac{3\pi i}{2}}$ and $z_2 = 2e^{\frac{\pi i}{4}}$

9. This chapter started with $i = \sqrt{-1}$. This time, consider $\sqrt{i}$.

   (a) How many roots should there be?

   (b) Find them in $a + bi$ form.

   (c) Are these roots the negative of the roots of $\sqrt{-i}$? Explain your answer.

10. Starting with the complex number $z = -1 + 0i$, show how Euler's formula $e^{i\pi} + 1 = 0$ can be found by expressing in Euler form.

## 6.5  Applications of complex numbers

The application of complex numbers within mathematics can be found in topics such as differential equations and eigenvalues; however, the discussion of these topics requires an understanding of mathematics beyond what has been covered in the HL syllabus to this point. Complex numbers are used in fluid dynamics, control theory, quantum mechanics, and Fourier transforms, for example. We will look at the use of complex numbers in vectors and alternating current (AC) electrical circuits. Vectors are covered in chapter 8. We will consider impedance in AC circuits here.

Consider an AC waveform, commonly known as a **sinusoidal** curve (Figure 6.9).

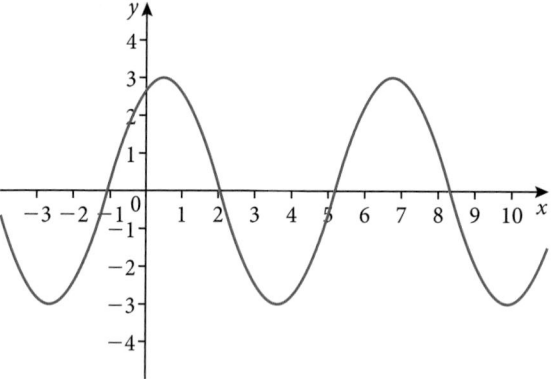

**Figure 6.9**  Sinusoidal curve

All AC waveforms have three distinguishing features. Each waveform:

- is periodic, repeating after a cycle is completed, over a span of $n$ degrees or radians. The number of cycles per second (frequency) is given in hertz (Hz).

- has an **amplitude**, which is the measure of its potential difference or current

- has a specific starting point, called the **phase**, typically measured in degrees relative to its basic waveform which is generally set at the origin.

The frequency for AC circuits is constant and is generally set at 50 Hz or 60 Hz, and can be taken out of consideration. However, in circuits, both the potential difference/current and phase do vary. Since there are then two components to be considered, AC circuits can and are modelled well by complex numbers.

Consider the potential difference given as a sinusoidal function,
$V = 10 \cos\left(\omega t + \dfrac{\pi}{4}\right)$ whose potential difference (amplitude) is 10, with a phase

angle of $\dfrac{\pi}{4}$. These characteristics can be illustrated by the Argand diagram shown.

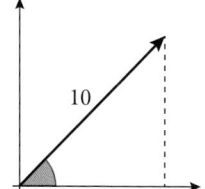

**Figure 6.10** Potential difference in the complex plane

In electrical notation, this is stated as $V = 10\angle 45°$ where the phase angle is always measured in degrees. In Euler form, it is $V = 10e^{45°i}$. As you can see from the simple right-angled triangle diagram above, the real component is $5\sqrt{2}$ as is the imaginary component. The potential difference with phase shift is described by the complex number $5\sqrt{2} + 5\sqrt{2}i$

## Multiple sinusoidal functions

When more than one AC source is placed in a series circuit, the difference in phase between them can be resolved by considering the complex number representation of each source, then added.

### Example 6.16

Three AC sources are placed in series, with potential differences $V_1 = \cos \omega t$,
$V_2 = 4\cos\left(\omega t + \dfrac{\pi}{4}\right)$, and $V_3 = 6\left(\cos \omega t + \dfrac{\pi}{2}\right)$
Find the components of the total, $V = V_1 + V_2 + V_3$

### Solution

In electrical notation, $V_1 = 1\angle 0°$, $V_2 = 4\angle 45°$, and $V_3 = 6\angle 90°$

$$\Rightarrow V = 1 + (2\sqrt{2} + 2\sqrt{2}i) + 6i = (1 + 2\sqrt{2}) + (6 + 2\sqrt{2})i$$

$$\Rightarrow |V| = \sqrt{(1 + 2\sqrt{2})^2 + (6 + 2\sqrt{2})^2} \approx 9.62$$

$$\text{and } \arctan\left(\frac{6 + 2\sqrt{2}}{1 + 2\sqrt{2}}\right) \approx 66.6°$$

$$\therefore V = 9.62\angle 66.6° \text{ or } 9.62e^{66.6°i}$$

To convert your answer to $a + bi$ form, use your GDC.

In $a + bi$ form, $V \approx 3.83 + 8.83i$

```
π*i/180→C
                 0.0174532925i
R
                  9.622784407
T
                  66.55614454
Re^TC
   3.828427124+8.828427125i
```

# 6 Complex numbers

## Impedance – complex variables used in electrical theory

**Resistance** ($R$) is a measure used in direct current (DC) circuits. In AC circuits, **impedance** ($Z$) is the measure of resistance that includes reactance due to capacitance ($X_C$) and inductance ($X_L$). The formula $V = IR$ becomes $V = IZ$ with AC circuits.

Potential difference across a resistor is noted as $V_R$, across a capacitor, $V_C$, and across an inductor, $V_L$. Both $V_C$ and $V_L$ are considered imaginary components, with the symbol j used to denote the imaginary unit in electrical theory.

Potential difference across a resistor $V_R$ is in phase with the current; however, potential difference across a capacitor $V_C$ lags, and potential difference across an inductor $V_L$ leads. Hence, impedance creates a shift measured in degrees, named the **phase angle**, with $\theta = \arctan\dfrac{X_L - X_C}{R}$. Note how $\theta$ is calculated as $\arctan\left(\dfrac{\text{imaginary component}}{\text{real component}}\right)$ just as for complex numbers in the complex plane. In electrical notation, $Z$ represents impedance, and $Z = R + j(X_L - X_C)$. This is basically the form $z = a + bi$

### Example 6.17

A particular AC circuit has a resistor of $6\,\Omega$, a reactance across an inductor of $11\,\Omega$ and a reactance across a capacitor of $3\,\Omega$.

(a) Express the impedance of the circuit as a complex number in Cartesian form.

(b) Express the impedance in Euler form, with $\theta$ given in degrees, correct to 3 significant figures.

### Solution

(a) $R = 6$, $X_L = 11$, $X_C = 3$, so $Z = 6 + j(11 - 3) = 6 + 8j$ (this compares to $z = 6 + 8i$)

(b) $|Z| = \sqrt{6^2 + 8^2} = 10$ and $\theta = \arctan\dfrac{8}{6} \approx 53.1°$. Hence, $Z = 10\,e^{53.1°i}$

This angle measurement may not work with some calculators which expect the angle measurement in radians. Here is a possible conversion:

Store a simple conversion factor into your GDC to convert from degrees to radians. The imaginary value was stored in C and can be reused.

```
π*i/180→C
            0.0174532925i
```

As a simple example, take $1 + i$ which has $r = \sqrt{2}$ and $\theta = 45°$, so $z = \sqrt{2}e^{45°i}$

```
π*i/180→C
            0.0174532925i
.................................
√2e^{45C}
                    1+1i
```

## Example 6.18

A resistor, an inductor, and a capacitor are connected in series in an AC circuit, with potential differences across them of 8.0 V, 10.5 V, and 4.5 V respectively. What is the potential difference (EMF) of the source?

### Solution

$$V_R = 8, \ V_L = 10.5, \text{ and } V_C = 4.5$$

Since $V = V_R + j(V_L - V_C)$

$$= 8.0 + (10.5 - 4.5)j = 8 + 6j$$

The magnitude of the source potential difference is $|V| = \sqrt{8^2 + 6^2} = 10$, that is, 10 V.

## Example 6.19

The current in a given AC circuit is $4.1 - 5.3j$ A and the impedance is $6.2 + 2.3j \ \Omega$. What is the magnitude of the potential difference?

### Solution

Since $V = IZ$, $|V| = |I| \cdot |Z| = \sqrt{4.1^2 + 5.3^2} \cdot \sqrt{6.2^2 + 2.3^2} \approx 44.3$ V

## Impedance in parallel circuits

The work with complex numbers becomes much more useful when **parallel** circuits are analysed. The resistance of three resistors connected in parallel is given by: $\dfrac{1}{R} = \dfrac{1}{R_1} + \dfrac{1}{R_2} + \dfrac{1}{R_3}$. So, in AC circuits, impedance is defined as

$$\frac{1}{Z} = \frac{1}{Z_1} + \frac{1}{Z_2} + \frac{1}{Z_3}$$

Consider when there are two impedances in a parallel circuit, $Z_1$ and $Z_2$:

$$\frac{1}{Z} = \frac{1}{Z_1} + \frac{1}{Z_2} \Rightarrow \frac{1}{Z} = \frac{Z_1 + Z_2}{Z_1 Z_2} \Rightarrow Z = \frac{Z_1 Z_2}{Z_1 + Z_2}$$ which is the product

of two complex numbers divided by their sum. The addition of complex numbers is easiest in $a + bi$ form, but their multiplication would be simpler in polar or Euler form.

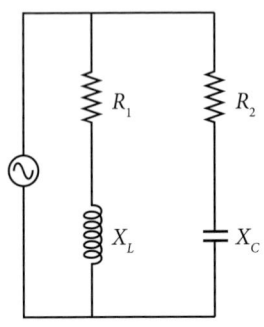

**Figure 6.11** Circuit diagram for Example 6.20

## Example 6.20

A parallel AC circuit has two loops. The first has a resistor with $R_1 = 80\,\Omega$ and an inductor with $X_L = 60\,\Omega$, while the second has a resistor with $R_2 = 12\,\Omega$ and a capacitor with $X_C = 5\,\Omega$.

Find the combined impedance.

### Solution

In the first loop, $Z_1 = 80 + 60j$, and in the second, $Z_2 = 12 - 5j$

The combined impedance is $Z = \dfrac{Z_1 Z_2}{Z_1 + Z_2}$, so we need to find their product and their sum. Again, since the sum is much easier in $a + bi$ form, start with the denominator. $Z_1 + Z_2 = 92 + 55j$

Next convert $Z_1$, $Z_2$, and $Z_1 + Z_2$ with your GDC into Euler form:

Change the output mode to Euler form, then enter the impedances.

Note that the angle is in radians, but that setting is kept until the final conversion.

```
80+60i→A
                100e^0.6435011088i
12-5i→B
              13e^-039479111197i
92+55i→C
        107.1867529e^0.53881950016i
(A*B)/C
        12.12836442e^-0.2901095125i
```

Now, reset the mode to its default to produce an answer in $a + bi$ form.

```
12-5i→B
              13e^-039479111197i
92+55i→C
        107.1867529e^0.53881950016i
(A*B)/C
        12.12836442e^-0.2901095125i
(A*B)/C
        11.62155105-3.469405518i
```

To 3 significant figures, this complex number is $11.6 - 3.47i$
i.e. $Z = 11.6 - 3.47j$

## Exercise 6.5

1. A series AC circuit has a resistor of $R = 12\,\Omega$, a reactance across an inductor of $X_L = 12\,\Omega$ and a reactance across a capacitor of $X_C = 3\,\Omega$.

   (a) Express the impedance of the circuit as a complex number in Cartesian form.

   (b) Express the impedance in Euler form, with $\theta$ given in degrees to 3 significant figures.

2. Find the impedance of a series AC circuit with $R = 4\,\Omega$, $X_L = 2\,\Omega$, and $X_C = 5\,\Omega$ in Cartesian form.

3. A resistor, an inductor, and a capacitor in a series AC circuit have potential differences across the of 9.0 V, 15.0 V, and 3.0 V respectively. What is the potential difference of the source?

4. The potential differences across a resistor ($V_R = 6$ V), an inductor ($V_L = 11.5$ V), and capacitor ($V_C = 3.5$ V) are individually measured in a series circuit. Find the potential difference at the source.

5. The current in a given AC circuit is $6 - 3j$ A and the impedance is $8 + 4j\,\Omega$. What is the magnitude of the potential difference? [Remember $|V| = |I| \cdot |Z|$]

6. The potential difference across a given AC circuit is 100 V and the impedance is $4 - 3j\,\Omega$. What is the magnitude of the current?

7. A parallel AC circuit has two loops. The first has a resistor with $R_1 = 12\,\Omega$ and an inductor with $X_L = 5\,\Omega$, while the second has a resistor with $R_2 = 8\,\Omega$ and a capacitor with $X_C = 6\,\Omega$

   Find the combined impedance.

8. In the circuit in question 7, $R_1 = R_2 = 8\,\Omega$, $X_L = 6\,\Omega$ and $X_C = 6\,\Omega$

   What is the impedance?

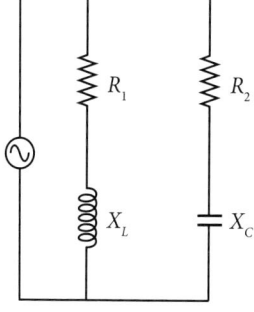

**Figure 6.12** Circuit diagram for question 7

## Chapter 6 practice questions

1. Express each value using the imaginary unit i.

   (a) $\sqrt{-49}$        (b) $\sqrt{-18}$

   (c) $\sqrt{-9} \cdot \sqrt{-1}$        (d) $\sqrt{-12} \cdot \sqrt{-27}$

2. Express each as a complex number in the form $a + bi$

   (a) $\sqrt{-16}$        (b) $25 + \sqrt{-25}$        (c) $5 + \sqrt{-25}$

   (d) $-3\sqrt{2} + \sqrt{-18}$        (e) $2\sqrt{3} - \sqrt{-12}$        (f) $-\sqrt{i^2}$

   (g) $-\sqrt{i^4}$        (h) $(-3 + \sqrt{-9})\,i$

3. Express each value in simplest form.

(a) $i^{22}$          (b) $i^{21}$          (c) $i^{20}$          (d) $i^{19}$

4. Find the zeros of each function.

(a) $y = x^2 + 4x + 8$          (b) $y = x^2 - 6x + 10$

(c) $y = x^2 - 8x + 25$          (d) $y = x^2 - 4x + 8$

(e) $y = x^2 - 10x + 29$          (f) $y = x^2 + 8x + 32$

5. Find the sums of the complex numbers.

(a) $(5 - i)$ and $(-4 + 3i)$          (b) $(4 + 2i)$ and $(-2 + i)$

(c) $(\sqrt{5} - 3i)$ and $(2\sqrt{5} + i)$

6. Find the products of the complex numbers.

(a) $(6 - 4i)(6 + 4i)$          (b) $(7 + 2i)(-7 + 2i)$

(c) $(3\sqrt{3} + i)(3\sqrt{3} - i)$

7. Find the product of each complex number and its conjugate.

(a) $9 + 12i$          (b) $6 - 8i$          (c) $3 - 3\sqrt{2}i$

8. Express each rational expression in $a + bi$ form.

(a) $\dfrac{10}{3 + i}$          (b) $\dfrac{14}{\sqrt{3} - 2i}$          (c) $\dfrac{1 - 3i}{-3 + 9i}$

9. A quadratic function $y = ax^2 + bx + c$ has real coefficients $a$, $b$, and $c$. Find the function if one of the zeros is known.

(a) $1 + 2i$          (b) $4 + 3i$          (c) $-3\sqrt{2} + \sqrt{2}i$

10. Find the quadratic function of the form $y = x^2 + bx + c$ with the given zeros.

(a) $(-1 - i)$ and $(-1 + i)$          (b) $(2 + \sqrt{2}i)$ and $(2 - \sqrt{2}i)$

(c) $(3 + 2i)$ and $(4 - i)$

11. Plot a complex number in each of the four quadrants, each with a modulus of 10.

12. If $2\sqrt{2}$ is the imaginary component of a complex number also with a modulus of 10, what are the possible values of its real component?

13. On the complex plane, connect the origin $O$ to any two points $a + bi$ and $c + di$ then construct a parallelogram with sides parallel to those segments. What is the significance of the two diagonals of the parallelogram?

14. Express each complex number in polar form.

(a) $2 - 2i$

(b) $-2 - 2i$

(c) $2 + 2i$

(d) $-2 + 2i$

(e) $2 + 2\sqrt{3}i$

(f) $2 - 2\sqrt{3}i$

(g) $-2 + 2\sqrt{3}i$

(h) $-2 - 2\sqrt{3}i$

15. Find the products of these complex numbers. State your answer in the form $a + bi$

(a) $z_1 = 6\operatorname{cis}\dfrac{\pi}{2}$ and $z_2 = 2\operatorname{cis}\dfrac{3\pi}{4}$

(b) $z_1 = 8\operatorname{cis}\dfrac{\pi}{3}$ and $z_2 = 2\operatorname{cis}\dfrac{\pi}{6}$

16. Find the quotient $\dfrac{z_1}{z_2}$ of these complex numbers and give your answer in the form $a + bi$

(a) $z_1 = 9\operatorname{cis}\dfrac{3\pi}{4}$ and $z_2 = 3\operatorname{cis}\dfrac{\pi}{4}$

(b) $z_1 = 10\operatorname{cis}\dfrac{5\pi}{4}$ and $z_2 = 2\operatorname{cis}\dfrac{\pi}{2}$

(c) $z_1 = 2\operatorname{cis}\dfrac{\pi}{3}$ and $z_2 = 4\operatorname{cis}\dfrac{\pi}{6}$

(d) $z_1 = 12\operatorname{cis}\dfrac{\pi}{4}$ and $z_2 = 3\operatorname{cis}\dfrac{\pi}{2}$

17. Raise the complex number to the power indicated and state the result in the form $a + bi$

(a) $(1 + i)^4$

(b) $(1 - i)^4$

(c) $(-1 + \sqrt{3}\,i)^3$

(d) $(2\sqrt{3} + 2i)^4$

18. Find the indicated roots.

(a) $\sqrt[3]{-8}$

(b) $\sqrt[3]{64}$

(c) $\sqrt{(-3 + 3\sqrt{3}i)}$

19. Write each complex number in Euler form.

(a) $1 - i$

(b) $-2 - 2i$

(c) $-\sqrt{3} + 3i$

20. Find the products of these complex numbers and give the answer in the form $a + bi$

(a) $z_1 = 2e^{\frac{\pi i}{6}}$ and $z_2 = 5e^{\frac{3\pi i}{2}}$

(b) $z_1 = 2e^{\frac{2\pi i}{3}}$ and $z_2 = 4e^{\frac{\pi i}{6}}$

21. Find the quotients $\dfrac{z_1}{z_2}$ of these complex numbers and give the answer in the form $a + bi$

(a) $z_1 = 10e^{\frac{3\pi i}{2}}$ and $z_2 = 2e^{\frac{\pi i}{4}}$

(b) $z_1 = 12e^{\frac{\pi i}{2}}$ and $z_2 = 3e^{\frac{3\pi i}{4}}$

22. In a series AC circuit there is a resistor of $R = 12\,\Omega$, a reactance across an inductor of $X_L = 3\,\Omega$ and a reactance across a capacitor of $X_C = 8\,\Omega$.

(a) Express the impedance of the circuit as a complex number in the form $a + bi$

(b) Express the impedance in Euler form, with $\theta$ given in radians to 3 significant figures.

23. Determine the impedance in $a + bi$ form of a series AC circuit if it contains a resistor, inductor, and capacitor with $R = 4\,\Omega$, $X_L = 7\,\Omega$, and $X_C = 3\,\Omega$.

24. A resistor, an inductor, and a capacitor in a series AC circuit have potential differences across them of 12.0 V, 4.5 V, and 9.5 V respectively. What is the potential difference of the source?

25. The potential differences across a resistor, an inductor, and capacitor are individually measured in a series circuit and found to be $V_R = 8$ V, $V_L = 4.5$ V, and $V_C = 10.5$ V. Find the potential difference at the source.

26. The current in a given AC circuit is $I = 2 + j$ A. If the impedance is $X_L = 4 + 2j\ \Omega$, what is the magnitude of the potential difference?

27. The potential difference across a given AC circuit is 65 V and the impedance is $12 - 5j\ \Omega$. What is the magnitude of the current?

28. Consider the parallel circuit shown in the diagram. A 6 Ω resistor and 8 Ω inductor are in the first loop, and a 6 Ω resistor and a 3 Ω capacitor are in the second loop. Find their combined impedance.

29. Consider the complex numbers $u = 2 + 3i$ and $v = 3 + 2i$

    (a) Given that $\dfrac{1}{u} + \dfrac{1}{v} = \dfrac{10}{w}$, express $w$ in the form $a + bi$, $a, b \in \mathbb{R}$

    (b) Find $w^*$ and express it in the form $r \cdot e^{i\theta}$

30. (a) Find three distinct roots of the equation $8z^3 + 27 = 0$, $z \in \mathbb{C}$ giving your answers in modulus-argument form.

    (b) The roots are represented by the vertices of a triangle in an Argand diagram. Show that the area of the triangle is $\dfrac{27\sqrt{3}}{16}$

31. Let $w = \cos\dfrac{2\pi}{7} + i\sin\dfrac{2\pi}{7}$

    (a) Verify that $w$ is a root of the equation $z^7 - 1 = 0$, $z \in \mathbb{C}$

    (b) (i) Expand $(w - 1)(1 + w + w^2 + w^3 + w^4 + w^5 + w^6)$
        (ii) Hence deduce that $1 + w + w^2 + w^3 + w^4 + w^5 + w^6 = 0$

    (c) Write down the roots of the equation $z^7 - 1 = 0$, $z \in \mathbb{C}$ in terms of $w$ and plot these roots on an Argand diagram.

    Consider the quadratic equation $z^2 + bz + c = 0$ where $b, c \in \mathbb{R}$, $z \in \mathbb{C}$. The roots of this equation are $\alpha$ and $\alpha^*$ where $\alpha^*$ is the complex conjugate of $\alpha$.

    (d) (i) Given that $\alpha = w + w^2 + w^4$, show that $\alpha^* = w^6 + w^5 + w^3$
        (ii) Find the value of $b$ and the value of $c$.

    (e) Using the values for $b$ and $c$ obtained in part (d) (ii), find the imaginary part of $\alpha$, giving your answer in surd form.

32. One root of the equation $x^2 + ax + b = 0$ is $2 + 3i$ where $a, b \in \mathbb{R}$ Find the value of $a$ and the value of $b$.

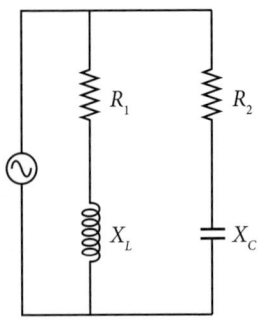

**Figure 6.13** Circuit diagram for question 28

# Matrix algebra

**7**

Matrices have been, and remain, significant mathematical tools. Uses of matrices span several areas, from simply solving systems of simultaneous linear equations to describing atomic structure, designing computer game graphics, analysing relationships, coding, and operations research. If you have ever used a spreadsheet program, or have created a table, then you have used a matrix. Matrices make the presentation of data understandable and help make calculations easy to perform. For example, your teacher's grade book may look something like this:

| Student | Quiz 1 | Quiz 2 | Test 1 | Test 2 | Homework | Grade |
|---------|--------|--------|--------|--------|----------|-------|
| **Tim** | 70 | 80 | 86 | 82 | 95 | A |
| **Maher** | 89 | 56 | 80 | 60 | 55 | C |
| $\vdots$ | $\vdots$ | $\vdots$ | $\vdots$ | $\vdots$ | $\vdots$ | $\vdots$ |

**Table 7.1** Example of teacher's grade book

If we want to know Tim's grade on Test 2, we simply follow along the row 'Tim' to the column 'Test 2' and find that he achieved a mark of 82. Take a look at the matrix below about the number of cameras sold at shops in four cities.

| | Venice | Rome | Budapest | Prague |
|---|--------|------|----------|--------|
| **Digital compact** | 153 | 98 | 74 | 56 |
| **Digital standard** | 211 | 120 | 57 | 29 |
| **DSLR** | 82 | 31 | 12 | 5 |
| **Other** | 308 | 242 | 183 | 107 |

**Table 7.2** Number of cameras sold in four cities

If we want to know how many digital standard cameras were sold in the Budapest shop, we follow along the row 'Digital standard' to the column 'Budapest' and find that 57 digital standard cameras were sold.

## What is a matrix?

A matrix is a rectangular array of elements. The elements can be symbolic expressions or numbers.

Matrix $A$ is denoted by

$$A = \begin{pmatrix} a_{11} & a_{12} & \cdots & a_{1n} \\ a_{21} & a_{22} & \cdots & a_{2n} \\ \vdots & \vdots & \vdots & \vdots \\ a_{m1} & a_{m2} & \cdots & a_{mn} \end{pmatrix} \left. \begin{array}{l} \leftarrow \\ \leftarrow \\ \vdots \\ \leftarrow \end{array} \right\} m \text{ rows}$$

$$\underbrace{\uparrow \quad \uparrow \quad \cdots \quad \uparrow}_{n \text{ columns}}$$

Row $i$ of $A$ has $n$ elements and is $(a_{i1} \quad a_{i2} \quad \cdots \quad a_{in})$

Column $j$ of $A$ has $m$ elements and is $\begin{pmatrix} a_{1j} \\ a_{2j} \\ \vdots \\ a_{mj} \end{pmatrix}$

The number of rows and columns of a matrix defines its size (order). So, a matrix that has $m$ rows and $n$ columns is said to have an $m \times n$ ($m$ by $n$) order. A matrix $A$ with $m \times n$ order is sometimes denoted as $[A]_{m \times n}$ or $[A]_{mn}$ to show that $A$ is a matrix with $m$ rows and $n$ columns. (Sometimes $[a_{ij}]$ is used to represent a matrix.) The camera sales matrix has a $4 \times 4$ order. When $m = n$, the matrix is said to be a square matrix with order $n$, so the camera sales matrix is a square matrix of order 4.

Every entry in a matrix is called an **entry** or **element** of the matrix and is denoted by $a_{ij}$, where $i$ is the row number and $j$ is the column number of that element. The ordered pair $(i, j)$ is also called the **address** of the element. So, in the grade book matrix example, the entry $(2, 4)$ is 60, the student Maher's grade on Test 2, while $(2, 4)$ in the camera sales matrix example is 29, the number of digital standard cameras sold in the Prague shop.

A vector is a matrix that has only one row or one column. There are two types of vector: row vectors and column vectors.

## Row vector

If a matrix has one row, it is called a row vector.

$B = (b_1 \quad b_2 \quad \ldots \quad b_m)$ is a row vector with **dimension** $m$.

$B = (1 \quad 2)$ could represent the position of a point in a plane and is an example of a row vector of dimension 2.

## Column vector

If a matrix has one column, it is called a column vector.

$C = \begin{pmatrix} c_1 \\ c_2 \\ \vdots \\ c_n \end{pmatrix}$ is a column vector with dimension $n$.

$C = \begin{pmatrix} 1 \\ 2 \end{pmatrix}$ again could represent the position of a point in a plane and is an example of a column vector of dimension 2.

Vectors can be represented by row or column matrices.

## Submatrix

If some row(s) and/or column(s) of a matrix $A$ are deleted, the remaining matrix is called a **submatrix** of $A$.

For example, if we are interested in the sales of only the three main types of camera and only in Italian cities, we can represent them with the following submatrix of the original matrix:

$$\begin{pmatrix} 153 & 98 \\ 211 & 120 \\ 82 & 31 \end{pmatrix}$$
Submatrix

$$\begin{pmatrix} 153 & 98 & 74 & 56 \\ 211 & 120 & 57 & 29 \\ 82 & 31 & 12 & 5 \\ 308 & 242 & 183 & 107 \end{pmatrix}$$
Original matrix

## Zero matrix

A zero matrix is one for which all entries are equal to zero, ($a_{ij} = 0$ for all $i$ and $j$)

Some zero matrix examples: $(0 \quad 0)$ $\begin{pmatrix} 0 & 0 \\ 0 & 0 \end{pmatrix}$ $\begin{pmatrix} 0 & 0 & 0 \\ 0 & 0 & 0 \end{pmatrix}$ $\begin{pmatrix} 0 \\ 0 \end{pmatrix}$

## Diagonal matrix

In a square matrix, the entries $a_{11}$, $a_{22}$, ..., $a_{nn}$ are called the **diagonal elements** of the matrix. Sometimes the diagonal of the matrix is also called the **principal** or **main diagonal** of the matrix.

What is the diagonal in our camera sales matrix?

Here $a_{11} = 153$, $a_{22} = 120$, $a_{33} = 12$, and $a_{44} = 107$

## Triangular matrix

You can use a matrix to show distances between different cities.

$$\begin{pmatrix} 153 & 0 & 0 & 0 \\ 0 & 120 & 0 & 0 \\ 0 & 0 & 12 & 0 \\ 0 & 0 & 0 & 107 \end{pmatrix}$$

|  | Graz | Salzburg | Innsbruck | Linz |
|---|---|---|---|---|
| **Vienna** | 191 | 298 | 478 | 185 |
| **Graz** |  | 282 | 461 | 220 |
| **Salzburg** |  |  | 188 | 135 |
| **Innsbruck** |  |  |  | 320 |

**Table 7.3** Distance (in km) between Austrian cities.

The data in Table 7.3 can be represented by a triangular matrix. It is an upper triangular matrix, in this case.

$$\begin{pmatrix} 191 & 298 & 478 & 185 \\ 0 & 282 & 461 & 220 \\ 0 & 0 & 188 & 135 \\ 0 & 0 & 0 & 320 \end{pmatrix}$$

In a triangular matrix, the entries on one side of its diagonal are all zero.

A triangular matrix is a square matrix with order $n$ for which $a_{ij} = 0$ when $i > j$ (upper triangular) or alternatively when $i < j$ (lower triangular).

Another way of representing the distance data is given by the following matrix.

|  | Vienna | Graz | Salzburg | Innsbruck | Linz |
|---|---|---|---|---|---|
| **Vienna** | 0 | 191 | 298 | 478 | 185 |
| **Graz** | 191 | 0 | 282 | 461 | 220 |
| **Salzburg** | 298 | 282 | 0 | 188 | 135 |
| **Innsbruck** | 478 | 461 | 188 | 0 | 320 |
| **Linz** | 185 | 220 | 135 | 320 | 0 |

Again, the data in the table can be represented by a matrix called a **symmetric** matrix.

In such matrices, $a_{ij} = a_{ji}$ for all $i$ and $j$. All symmetric matrices are square.

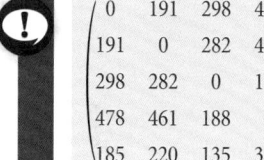

# Matrix operations

## Equal matrices

Two matrices $A$ and $B$ are equal if the orders of $A$ and $B$ are the same (number of rows and columns are the same for $A$ and $B$) and $a_{ij} = b_{ij}$ for all $i$ and $j$.

For example, $\begin{pmatrix} 2 & 3 \\ 5 & 7 \end{pmatrix}$ and $\begin{pmatrix} 2 & x \\ x^2 - 4 & 7 \end{pmatrix}$ are equal only if $x = 3$ and $x^2 - 4 = 5$

which can only be true if $x = 3$

## Adding and subtracting matrices

We can add two matrices $A$ and $B$ only if they are the same size. If $C$ is the sum of the two matrices, then $C = A + B$ where $c_{ij} = a_{ij} + b_{ij}$, so we add corresponding terms, one by one.

For example

$$\begin{pmatrix} 2 & 3 \\ 5 & 7 \end{pmatrix} + \begin{pmatrix} x & y \\ a & b \end{pmatrix} = \begin{pmatrix} 2 + x & 3 + y \\ 5 + a & 7 + b \end{pmatrix}$$

We carry out subtraction in a similar way

$$\begin{pmatrix} 2 & 3 & 1 \\ 5 & 7 & 0 \end{pmatrix} - \begin{pmatrix} x & y & 8 \\ a & b & 2 \end{pmatrix} = \begin{pmatrix} 2-x & 3-y & -7 \\ 5-a & 7-b & -2 \end{pmatrix}$$

The operations of addition and subtraction of matrices obey all rules of algebraic addition and subtraction.

## Multiplying a matrix by a scalar

A scalar is any object that is not a matrix. You multiply each term of the matrix by the scalar.

$A$ is an $m \times n$ matrix, and $c$ is a scalar. The scalar product of $c$ and $A$ is another matrix $B = cA$, such that every entry $b_{ij}$ of $B$ is a multiple of its corresponding entry in $A$. So, for every entry in $B$, we have $b_{ij} = c \times a_{ij}$

## Matrix multiplication

At first glance, the following definition may seem unusual. You will see later, however, that this definition of the product of two matrices has many practical applications.

> It is often convenient to rewrite the scalar multiple $cA$ by factoring $c$ out of every entry in the matrix. For instance, in the matrix below, the scalar $\frac{1}{2}$ has been factored out of the matrix.
> $$\begin{pmatrix} \frac{1}{2} & -\frac{3}{2} \\ \frac{5}{2} & \frac{1}{2} \end{pmatrix} = \frac{1}{2}\begin{pmatrix} 1 & -3 \\ 5 & 1 \end{pmatrix}$$

$A = [a_{ij}]$ is an $m \times n$ matrix and $B = [b_{ij}]$ is an $n \times p$ matrix. The product $AB$ is an $m \times p$ matrix $AB = [c_{ij}]$ where

$$c_{ij} = \sum_{k=1}^{n} a_{ik}b_{kj} = a_{i1}b_{1j} + a_{i2}b_{2j} + \ldots + a_{in}b_{nj}$$

for each $i = 1, 2, \ldots, m$ and $j = 1, 2, \ldots, n$

For the product of two matrices to be defined, the number of columns in the first matrix must be the same as the number of rows in the second matrix.

$$\underset{m \times n}{A} \quad \underset{n \times p}{B} = \underset{m \times p}{AB}$$

This definition means that each entry with an address $ij$ in the product $AB$ is obtained by multiplying the entries in the **$i$th row** of $A$ by the corresponding entries in the **$j$th column** of $B$ and then adding the results:

$$c_{ij} = (a_{i1} \quad a_{i2} \quad \ldots \quad a_{in})\begin{pmatrix} b_{1j} \\ b_{2j} \\ \vdots \\ b_{nj} \end{pmatrix} = a_{i1}b_{1j} + a_{i2}b_{2j} + \ldots + a_{in}b_{nj}$$

### Example 7.1

Find $C = AB$ when $A = \begin{pmatrix} 3 & -5 & 2 \\ 2 & 1 & 7 \end{pmatrix}$ and $B = \begin{pmatrix} 3 & -2 & 1 & 5 \\ 5 & 8 & -4 & 0 \\ -9 & 10 & 5 & 3 \end{pmatrix}$

## Solution

**A** is a 2 × 3 matrix, **B** is a 3 × 4 matrix, so the product will be a 2 × 4 matrix. Every entry in the product is the result of multiplying the entries in the rows of **A** and columns of **B**. For example

$$c_{12} = \sum_{k=1}^{3} a_{1k}b_{k2} = (a_{11} \quad a_{12} \quad a_{13})\begin{pmatrix} b_{12} \\ b_{22} \\ b_{32} \end{pmatrix} = (3 \quad -5 \quad 2)\begin{pmatrix} -2 \\ 8 \\ 10 \end{pmatrix}$$

$$= 3 \times (-2) - 5 \times 8 + 2 \times 10 = -26$$

and

$$c_{23} = \sum_{k=1}^{3} a_{2k}b_{k3} = (a_{21} \quad a_{22} \quad a_{23})\begin{pmatrix} b_{13} \\ b_{23} \\ b_{33} \end{pmatrix} = (2 \quad 1 \quad 7)\begin{pmatrix} 1 \\ -4 \\ 5 \end{pmatrix}$$

$$= 2 \times 1 + 1 \times (-4) + 7 \times 5 = 33$$

Repeat the operation for each entry in the solution matrix to get:

$$C = AB = \begin{pmatrix} -34 & -26 & 33 & 21 \\ -52 & 74 & 33 & 31 \end{pmatrix}$$

We can also use our GDC to find the product.

```
[A][B]
[[-34 -26 33 21...
 [-52 -74 33 31...
```

**Figure 7.1** Using a GDC to find a matrix product

Here are some examples of matrix multiplication. Multiplying a 2 × 3 matrix by a 3 × 2 matrix results in a 2 × 2 product matrix.

$$\underset{2 \times 3}{\begin{pmatrix} 5 & 0 & 3 \\ -2 & 1 & 2 \end{pmatrix}} \underset{3 \times 2}{\begin{pmatrix} -2 & 4 \\ 1 & -1 \\ 3 & -2 \end{pmatrix}} = \underset{2 \times 2}{\begin{pmatrix} -1 & 14 \\ 11 & -13 \end{pmatrix}}$$

When matrices are the same size, the product is the same size.

$$\underset{2 \times 2}{\begin{pmatrix} 4 & -5 \\ 1 & 7 \end{pmatrix}} \underset{2 \times 2}{\begin{pmatrix} 1 & 0 \\ 0 & 1 \end{pmatrix}} = \underset{2 \times 2}{\begin{pmatrix} 4 & -5 \\ 1 & -7 \end{pmatrix}}$$

$$\underset{3 \times 3}{\begin{pmatrix} 5 & 0 & 3 \\ -2 & 1 & 2 \\ 2 & 1 & 3 \end{pmatrix}} \underset{3 \times 3}{\begin{pmatrix} -\dfrac{1}{7} & -\dfrac{3}{7} & \dfrac{3}{7} \\ -\dfrac{10}{7} & -\dfrac{9}{7} & \dfrac{16}{7} \\ \dfrac{4}{7} & \dfrac{5}{7} & -\dfrac{5}{7} \end{pmatrix}} = \underset{3 \times 3}{\begin{pmatrix} 1 & 0 & 0 \\ 0 & 1 & 0 \\ 0 & 0 & 1 \end{pmatrix}}$$

When a matrix of order 2 is multiplied by the matrix $\begin{pmatrix} 1 & 0 \\ 0 & 1 \end{pmatrix}$, the product is the original matrix. The matrix $\begin{pmatrix} 1 & 0 \\ 0 & 1 \end{pmatrix}$ is called the **identity** matrix of order 2.

The **identity matrix** of order $n$ is a diagonal matrix where $a_{ii} = 1$

207

Two further identity matrices are $\begin{pmatrix} 1 & 0 & 0 \\ 0 & 1 & 0 \\ 0 & 0 & 1 \end{pmatrix}$ and $\begin{pmatrix} 1 & 0 & 0 & 0 \\ 0 & 1 & 0 & 0 \\ 0 & 0 & 1 & 0 \\ 0 & 0 & 0 & 1 \end{pmatrix}$

Sometimes, the identity matrix is denoted simply by $I$, or by $I_n$, where $n$ is the order. So, the identity matrix with three rows and columns is $I_3$, and the identity matrix with four rows and columns is $I_4$.

**Example 7.2**

Let $A = (2 \quad -1 \quad 3)$ and $B = \begin{pmatrix} 2 \\ 5 \\ 4 \end{pmatrix}$

Calculate:

(a) $AB$          (b) $BA$

**Solution**

(a) $(2 \quad -1 \quad 3)\begin{pmatrix} 2 \\ 5 \\ 4 \end{pmatrix} = 2 \times 2 + (-1) \times 5 + 3 \times 4 = 11$

(b) $\begin{pmatrix} 2 \\ 5 \\ 4 \end{pmatrix}(2 \quad -1 \quad 3) = \begin{pmatrix} 2 \times 2 & 2 \times (-1) & 2 \times 3 \\ 5 \times 2 & 5 \times (-1) & 5 \times 3 \\ 4 \times 2 & 4 \times (-1) & 4 \times 3 \end{pmatrix} = \begin{pmatrix} 4 & -2 & 6 \\ 10 & -5 & 15 \\ 8 & -4 & 12 \end{pmatrix}$

Note that the order of multiplication affects the product. Matrix multiplication, in general, is **not commutative**. It is usually not true that $AB = BA$.

Let $A = \begin{pmatrix} 3 & 6 \\ 5 & 2 \end{pmatrix}$ and $B = \begin{pmatrix} -2 & 3 \\ 1 & 5 \end{pmatrix}$, then $AB = \begin{pmatrix} 3 & 6 \\ 5 & 2 \end{pmatrix}\begin{pmatrix} -2 & 3 \\ -1 & 5 \end{pmatrix} = \begin{pmatrix} -0 & 39 \\ -8 & 25 \end{pmatrix}$

but $BA = \begin{pmatrix} -2 & 3 \\ 1 & 5 \end{pmatrix}\begin{pmatrix} 3 & 6 \\ 5 & 2 \end{pmatrix} = \begin{pmatrix} 9 & -6 \\ 28 & 16 \end{pmatrix} \Rightarrow AB \neq BA$

However, there are some special cases where matrix multiplication is commutative. For example

$A = \begin{pmatrix} 3 & 6 \\ 5 & 2 \end{pmatrix}$ and $B = \begin{pmatrix} 2 & 6 \\ 5 & 1 \end{pmatrix}$, then $AB = \begin{pmatrix} 3 & 6 \\ 5 & 2 \end{pmatrix}\begin{pmatrix} 2 & 6 \\ 5 & 1 \end{pmatrix} = \begin{pmatrix} 36 & 24 \\ 20 & 32 \end{pmatrix}$ and

$BA = \begin{pmatrix} 2 & 6 \\ 5 & 1 \end{pmatrix}\begin{pmatrix} 3 & 6 \\ 5 & 2 \end{pmatrix} = \begin{pmatrix} 36 & 24 \\ 20 & 32 \end{pmatrix} \Rightarrow AB = BA$

Multiplying by an identity matrix is also commutative.

$$\begin{pmatrix} a & b & c \\ d & e & f \\ g & h & i \end{pmatrix}\begin{pmatrix} 1 & 0 & 0 \\ 0 & 1 & 0 \\ 0 & 0 & 1 \end{pmatrix} = \begin{pmatrix} a & b & c \\ d & e & f \\ g & h & i \end{pmatrix}$$

$$\begin{pmatrix} 1 & 0 & 0 \\ 0 & 1 & 0 \\ 0 & 0 & 1 \end{pmatrix}\begin{pmatrix} a & b & c \\ d & e & f \\ g & h & i \end{pmatrix} = \begin{pmatrix} a & b & c \\ d & e & f \\ g & h & i \end{pmatrix}$$

## Example 7.3

Use the information given in the table to set up a matrix to find the camera sales in each city.

|  | Venice | Rome | Budapest | Prague |
|---|---|---|---|---|
| Digital compact | 153 | 98 | 74 | 56 |
| Digital standard | 211 | 120 | 57 | 29 |
| DSLR | 82 | 31 | 12 | 5 |
| Other | 308 | 242 | 183 | 107 |

The average selling price for each type of camera is as follows:

Digital compact €1200; Digital standard €1100; DSLR €900; Other €600

### Solution

We set up a matrix multiplication in which the individual camera sales are multiplied by the corresponding price. Since the rows represent the sales of the different types of camera, create a row matrix of the different prices and perform the multiplication.

$$(1200 \quad 1100 \quad 900 \quad 600)\begin{pmatrix} 153 & 98 & 74 & 56 \\ 211 & 120 & 57 & 29 \\ 82 & 31 & 12 & 5 \\ 308 & 242 & 183 & 107 \end{pmatrix}$$

$$= (674\,300 \quad 422\,700 \quad 272\,100 \quad 167\,800)$$

So, the sales (in euros) from each city are:

|  | Venice | Rome | Budapest | Prague |
|---|---|---|---|---|
| Sales | 674 300 | 422 700 | 272 100 | 167 800 |

Remember that we are multiplying a $1 \times 4$ matrix with a $4 \times 4$ matrix and hence we get a $1 \times 4$ matrix.

## Exercise 7.1

1. Consider the matrices

$$A = \begin{pmatrix} -2 & x \\ y-1 & 3 \end{pmatrix} \qquad B = \begin{pmatrix} x+1 & -3 \\ 4 & y-2 \end{pmatrix}$$

$$C = \begin{pmatrix} 1 & 2x & -1 \\ 2 & 3 & 0 \end{pmatrix} \qquad D = \begin{pmatrix} 1 & 2 \\ 2x & 3 \\ -1 & 0 \end{pmatrix}$$

(a) Evaluate:

   (i) $A + B$     (ii) $3A - B$     (iii) $A + C$

(b) Find $x$ and $y$ such that $A = B$

(c) Find $x$ and $y$ such that $A + B$ is a diagonal matrix.

(d) Find $AB$ and $BA$

(e) Find $x$ and $y$ such that $C = D$

2. Solve for the variables:

(a) $\begin{pmatrix} 3 & 0 \\ 4 & 2 \end{pmatrix}\begin{pmatrix} x \\ y \end{pmatrix} = \begin{pmatrix} 6 \\ -12 \end{pmatrix}$

(b) $\begin{pmatrix} 2 & p \\ 3 & q \end{pmatrix}\begin{pmatrix} 4 \\ 5 \end{pmatrix} = \begin{pmatrix} 18 \\ -8 \end{pmatrix}$

(c) $\begin{pmatrix} 3 & -6 \\ 5 & 7 \end{pmatrix} + \begin{pmatrix} a & b \\ c & d \end{pmatrix} = \begin{pmatrix} 0 & 2 \\ 6 & -4 \end{pmatrix}$

3. The diagram shows the major highways connecting some European cities: Vienna (V), Munich (M), Frankfurt (F), Stuttgart (S), Zurich (Z), Milan (L), and Paris (P).

The partially completed matrix below shows the number of direct routes between these cities.

(a) Use the diagram to copy and complete the matrix.

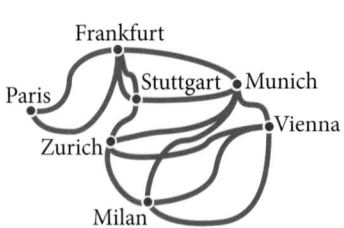

**Figure 7.2** Diagram for question 3

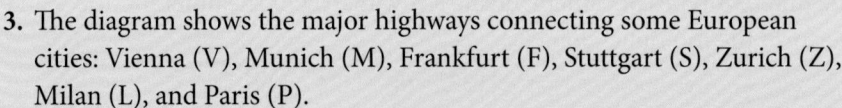

|   | V | M | F | S | Z | L | P |
|---|---|---|---|---|---|---|---|
| V | 0 | 1 | 0 | 0 | 1 | 2 | 0 |
| M |   |   |   |   |   |   |   |
| F |   |   |   |   |   |   |   |
| S |   |   |   |   |   |   |   |
| Z |   |   |   |   |   |   |   |
| L |   |   |   |   |   |   |   |
| P |   |   |   |   |   |   |   |

(b) Multiply the matrix from part (a) by itself and interpret what it signifies.

4. Consider the matrices

$$A = \begin{pmatrix} 2 & 5 & 1 \\ 0 & -3 & 2 \\ 7 & 0 & -1 \end{pmatrix} \qquad B = \begin{pmatrix} m & -2 \\ 3m & -1 \\ 2 & 3 \end{pmatrix}$$

$$C = \begin{pmatrix} x-1 & 5 & y \\ 0 & -x & y+1 \\ 2x+y & x-3y & 2y-x \end{pmatrix}$$

(a) Find $A + C$

(b) Find $AB$

(c) Find $BA$

(d) Solve for $x$ and $y$ if $A = C$

(e) Find $B + C$

(f) Solve for $m$ if $3B + 2\begin{pmatrix} -1 & m^2 \\ -5 & 2 \\ 1 & -1 \end{pmatrix} = \begin{pmatrix} 7 & 12 \\ 17 & 1 \\ 2m+2 & 7 \end{pmatrix}$

5. Find $a$, $b$, and $c$ so that the following equation is true.

$$2\begin{pmatrix} a-1 & b \\ c+2 & 3 \end{pmatrix} + \begin{pmatrix} 3 & -1 \\ 0 & 5 \end{pmatrix} = \begin{pmatrix} -5 & 5 \\ 8 & c+9 \end{pmatrix}$$

6. Find $x$ and $y$ so that the following equation is true.

$$\begin{pmatrix} 2 & -3 \\ -5 & 7 \end{pmatrix}\begin{pmatrix} x-11 & 1-x \\ -5 & x+2y \end{pmatrix} = \begin{pmatrix} 1 & 0 \\ 0 & 1 \end{pmatrix}$$

7. Find $m$ and $n$ so that the following equation is true.

$$\begin{pmatrix} m^2-1 & m+2 \\ 5 & -2 \end{pmatrix} = \begin{pmatrix} 3 & n+1 \\ 5 & n-5 \end{pmatrix}$$

8. There are two shops in your area. Your shopping list consists of 2 kg of tomatoes, 500 g of meat, and 3 litres of milk. Prices differ between the different shops, and it is difficult to switch between shops to make certain you are paying the least amount of money. A better strategy is to check where you pay less on average. The prices of the different items are given in the table. Which shop should you go to?

| Product | Price in shop A | Price in shop B |
| --- | --- | --- |
| Tomatoes | €1.66/kg | €1.58/kg |
| Meat | €2.55/100 g | €2.6/100 g |
| Milk | €0.90/litre | €0.95/litre |

**9.** Consider the matrices

$$A = \begin{pmatrix} 2 & 0 \\ -5 & 1 \end{pmatrix} \qquad B = \begin{pmatrix} 3 & -1 \\ 1 & 4 \end{pmatrix} \qquad C = \begin{pmatrix} -3 & 5 \\ 2 & 7 \end{pmatrix}$$

(a) Find $A + (B + C)$ and $(A + B) + C$

(b) Make a conjecture about the addition of $2 \times 2$ matrices observed in part (a) and prove it.

(c) Find $A(BC)$ and $(AB)C$

(d) Make a conjecture about the multiplication of $2 \times 2$ matrices observed in part (c) and prove it.

**10.** A company sells air conditioning units, electric heaters and humidifiers. Row matrix $A$ represents the number of units sold of each appliance last year, and matrix $B$ represents the profit margin for each unit. Find $AB$ and describe what this product represents.

$$A = (235 \quad 562 \quad 117) \qquad B = \begin{pmatrix} €120 \\ €95 \\ €56 \end{pmatrix}$$

**11.** Find $r$ and $s$ such that $rA + B = A$ is true, where

$$A = \begin{pmatrix} 2 & 3 \\ 5 & 7 \end{pmatrix} \qquad B = \begin{pmatrix} -12 & -18 \\ s - 8 & -42 \end{pmatrix}$$

**12.** Let $A = \begin{pmatrix} 1 & 1 \\ 0 & 1 \end{pmatrix}$

(a) Find:

    (i) $A^2$        (ii) $A^3$        (iii) $A^4$        (iv) $A^n$

Let $B = \begin{pmatrix} 3 & 3 \\ 0 & 3 \end{pmatrix}$

(b) Find:

    (i) $B^2$        (ii) $B^3$        (iii) $B^4$        (iv) $B^n$

**13.** Solve for $x$ and $y$ such that $AB = BA$ when

$$A = \begin{pmatrix} 2 & 3 \\ 4 & 1 \end{pmatrix} \text{ and } B = \begin{pmatrix} x & 2 \\ y & 3 \end{pmatrix}$$

**14.** Solve for $x$ and $y$ such that $AB = BA$ when

$$A = \begin{pmatrix} 3 & x \\ -2 & 1 \end{pmatrix} \text{ and } B = \begin{pmatrix} 5 & 2 \\ y & 1 \end{pmatrix}$$

**15.** Solve for $x$ such that $AB = BA$ when

$$A = \begin{pmatrix} 1 & 2 & 3 \\ x & 2 & -3 \\ 1 & 0 & 4 \end{pmatrix} \text{ and } B = \begin{pmatrix} -8 & x+3 & 12 \\ 23 & x-6 & -18 \\ 2 & -2 & 8 \end{pmatrix}$$

**16.** Solve for $x$ and $y$ such that $AB = BA$ when

$$A = \begin{pmatrix} y & 2 & y+2 \\ x & 2 & -3 \\ 1 & y-1 & 4 \end{pmatrix} \text{ and } B = \begin{pmatrix} -8 & x+3 & 12 \\ 23 & x-6 & -18 \\ 2 & -2 & 8 \end{pmatrix}$$

## 7.2  Applications to systems

There is a wide range of applications of matrices in solving systems of equations.

Recall from algebra that the equation of a straight line can take the form

$ax + by = c$ where $a$, $b$, and $c$ are constants, and $x$ and $y$ are variables.

We say this is a linear equation in two variables. Similarly, the equation of a plane in three-dimensional space has the form

$ax + by + cz = d$ where $a$, $b$, $c$, and $d$ are constants, and $x$, $y$, and $z$ are variables.

We say that this is a linear equation in three variables.

A solution of a linear equation in $n$ variables (in this case 2 or 3) is an ordered set of real numbers $(x_0, y_0, z_0)$ so that the equation in question is satisfied when these values are substituted for the corresponding variables. For example, the equation $x + 2y = 4$ is satisfied when $x = 2$ and $y = 1$

Some other solutions are:  $x = -4$ and $y = 4$

$x = 0$   and $y = 2$

$x = -2$ and $y = 3$

The set of all solutions of a linear equation is its solution set, and when this set is found, the equation is said to have been solved. To describe the entire solution set we often use a **parametric representation**, as illustrated in the following examples.

## Example 7.4

Solve the linear equation $x + 2y = 4$

### Solution

To find the solution set of an equation in two variables, we solve for one variable in terms of the other. For instance, if we solve for $x$, we obtain

$$x = 4 - 2y$$

In this form, $y$ is free, as it can take on any real value, while $x$ is not free, since its value depends on that of $y$. To represent this solution set in general terms, we introduce a third variable, for example $t$, called a parameter, and by letting $y = t$ we represent the solution set as

$$x = 4 - 2t, y = t, t \text{ is any real number}$$

Particular solutions can then be obtained by assigning values to the parameter $t$. For instance, $t = 1$ yields the solution $x = 2$ and $y = 1$, and $t = 3$ yields the solution $x = -2$ and $y = 3$

Note that the solution set of a linear equation can be represented parametrically in several ways. For instance, in Example 7.4, if we solve for $y$ in terms of $x$, the parametric representation would take the form:

$$x = m, y = 2 - \frac{1}{2}m, m \text{ is a real number}$$

Also, by choosing $m = 2$, one particular solution is $(x, y) = (2, 1)$, and when $m = -2$, another particular solution is $(-2, 3)$.

## Example 7.5

Solve the linear equation $3x + 2y - z = 3$

### Solution

Choosing $x$ and $y$ as the free variables, we solve for $z$.

$$z = 3x + 2y - 3$$

Letting $x = p$ and $y = q$, we obtain the parametric representation:

$$x = p, y = q, z = 3p + 2q - 3, \text{ where } p \text{ and } q \text{ are any real numbers}$$

A particular solution is $(x, y, z) = (1, 1, 2)$

Parametric representation is very important when we study vectors and lines later on in the book.

## Systems of linear equations

A system of $k$ equations in $n$ variables is a set of $k$ linear equations in the same $n$ variables. For example

$$2x + 3y = 3$$
$$x - y = 4$$

is a system of two linear equations in two variables, while

$$x - 2y + 3z = 9$$
$$x - 3y = 4$$

is a system with two equations and three variables, and

$$x - 2y + 3z = 9$$
$$x - 3y = 4$$
$$2x - 5y + 5z = 17$$

is a system with three equations and three variables.

A solution of a system of equations is an ordered set of numbers $x_0, y_0, \ldots$ which satisfy every equation in the system. For example $(3, -1)$ is a solution of

$$2x + 3y = 3$$
$$x - y = 4$$

Both equations in the system are satisfied when $x = 3$ and $y = -1$ are substituted into the equations. However, $(0, 1)$ is not a solution of the system; it satisfies the first equation, but it does not satisfy the second.

In this chapter, we will use matrix methods to solve systems of equations.

Taking our example above, we can write the system of equations in matrix form:

$$\begin{cases} 2x + 3y = 3 \\ x - y = 4 \end{cases} \Rightarrow \begin{pmatrix} 2 & 3 \\ 1 & -1 \end{pmatrix}\begin{pmatrix} x \\ y \end{pmatrix} = \begin{pmatrix} 3 \\ 4 \end{pmatrix}$$

The representation of the system of equations this way enables us to use matrix operations in solving systems of equations. This matrix equation can be written as

$$\begin{pmatrix} 2 & 3 \\ 1 & -1 \end{pmatrix}\begin{pmatrix} x \\ y \end{pmatrix} = \begin{pmatrix} 3 \\ 4 \end{pmatrix} \Rightarrow AX = C$$

where $A$ is the coefficient matrix, $X$ is the variable matrix, and $C$ is the constant matrix. However, to solve this equation, the inverse of a matrix has to be defined as the solution of the system in the form

$$X = A^{-1}C$$

where $A^{-1}$ is the inverse of the matrix $A$.

# 7  Matrix algebra

## Matrix inverse

To solve the equation $2x = 6$ for $x$, we need to multiply both sides of the equation by $\frac{1}{2}$:

$$\frac{1}{2} \times 2x = \frac{1}{2} \times 6 \Rightarrow x = 3 \qquad \text{This is so, because } \frac{1}{2} \times 2 = 2 \times \frac{1}{2} = 1$$

> A square matrix $B$ is the inverse of a square matrix $A$ if $AB = BA = I$ where $I$ is the identity matrix.

$\frac{1}{2}$ is the multiplicative inverse of 2. The inverse of a matrix is defined in a similar manner and plays a similar role in solving a matrix equation, such as

$$AX = C$$

The notation $A^{-1}$ is used to denote the inverse of a matrix $A$. Thus, $B = A^{-1}$

> Note that only square matrices can have multiplicative inverses.

### Example 7.6

Are the matrices $A = \begin{pmatrix} 7 & 5 \\ 4 & 3 \end{pmatrix}$ and $B = \begin{pmatrix} 3 & -5 \\ -4 & 7 \end{pmatrix}$ multiplicative inverses?

### Solution

$$AB = \begin{pmatrix} 7 & 5 \\ 4 & 3 \end{pmatrix}\begin{pmatrix} 3 & -5 \\ -4 & 7 \end{pmatrix} = \begin{pmatrix} 21 - 20 & -35 + 35 \\ 12 - 12 & -20 + 21 \end{pmatrix} = \begin{pmatrix} 1 & 0 \\ 0 & 1 \end{pmatrix}$$

$$BA = \begin{pmatrix} 3 & -5 \\ -4 & 7 \end{pmatrix}\begin{pmatrix} 7 & 5 \\ 4 & 3 \end{pmatrix} = \begin{pmatrix} 21 - 20 & 15 - 15 \\ -28 + 28 & -20 + 21 \end{pmatrix} = \begin{pmatrix} 1 & 0 \\ 0 & 1 \end{pmatrix}$$

So $A$ and $B$ are multiplicative inverses.

We can also find the inverse using a GDC.

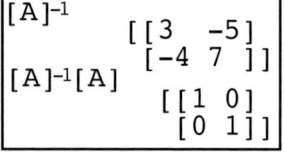

```
[A]⁻¹
            [[3  -5]
             [-4  7 ]]
[A]⁻¹[A]
            [[1  0]
             [0  1]]
```

**Figure 7.3** GDC screen for the solution to Example 7.6

We will now find the general form for the inverse of a matrix.

Let $A = \begin{pmatrix} a & b \\ c & d \end{pmatrix}$ and assume $A^{-1} = \begin{pmatrix} e & f \\ g & h \end{pmatrix}$ and then solve the following

matrix equation for $e, f, g,$ and $h$ in terms of $a, b, c,$ and $d$.

$$\begin{pmatrix} a & b \\ c & d \end{pmatrix}\begin{pmatrix} e & f \\ g & h \end{pmatrix} = \begin{pmatrix} 1 & 0 \\ 0 & 1 \end{pmatrix} \Rightarrow \begin{pmatrix} ae + bg & af + bh \\ ce + dg & cf + dh \end{pmatrix} = \begin{pmatrix} 1 & 0 \\ 0 & 1 \end{pmatrix}$$

Now we can set up two systems to solve for the required variables:

$$\begin{pmatrix} ae + bg & af + bh \\ ce + dg & cf + dh \end{pmatrix} = \begin{pmatrix} 1 & 0 \\ 0 & 1 \end{pmatrix}$$

$$\left.\begin{matrix} ae + bg = 1 \\ ce + dg = 0 \end{matrix}\right\} \Rightarrow \left.\begin{matrix} dae + dbg = d \\ bce + bdg = 0 \end{matrix}\right\} \Rightarrow e = \frac{d}{ad - bc}, g = \frac{-c}{ad - bc}$$

$$\left.\begin{matrix} af + bh = 0 \\ cf + dh = 1 \end{matrix}\right\} \Rightarrow \left.\begin{matrix} daf + dbh = 0 \\ bcf + bdh = b \end{matrix}\right\} \Rightarrow f = \frac{-b}{ad - bc}, h = \frac{a}{ad - bc}$$

In a matrix $A = \begin{pmatrix} a & b \\ c & d \end{pmatrix}$, if $ad - bc \neq 0$, then its inverse $A^{-1} = \begin{pmatrix} \dfrac{d}{ad-bc} & \dfrac{-b}{ad-bc} \\ \dfrac{-c}{ad-bc} & \dfrac{a}{ad-bc} \end{pmatrix}$

or $A^{-1} = \dfrac{1}{ad-bc}\begin{pmatrix} d & -b \\ -c & a \end{pmatrix}$

## Example 7.7

Find the inverse of $A = \begin{pmatrix} 4 & 7 \\ 3 & 5 \end{pmatrix}$

### Solution

Here $a = 4$, $b = 7$, $c = 3$, and $d = 5$, so $ad - bc = -1$

Thus $A^{-1} = \dfrac{1}{ad-bc}\begin{pmatrix} d & -b \\ -c & a \end{pmatrix} = \dfrac{1}{-1}\begin{pmatrix} 5 & -7 \\ -3 & 4 \end{pmatrix} = \begin{pmatrix} -5 & 7 \\ 3 & -4 \end{pmatrix}$

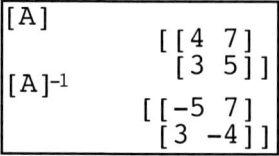

**Figure 7.4** GDC screen for the solution to Example 7.7

The number $ad - bc$ is called the **determinant** of the $2 \times 2$ matrix

$$A = \begin{pmatrix} a & b \\ c & d \end{pmatrix}$$

The notation we will use for this number is **det $A$** or **$|A|$**, so we write this as:

$$\det A = |A| = ad - bc$$

The determinant plays an important role in determining whether or not a matrix has an inverse.

When the determinant is zero ($ad - bc = 0$), the matrix does not have an inverse. A matrix that does not have an inverse is called a **singular matrix**; a matrix that does have an inverse is called a **non-singular** matrix.

## Example 7.8

Solve the system of equations using matrices.

$$2x + 3y = 3$$
$$x - y = 4$$

### Solution

In matrix form, the system can be written as

$$\begin{pmatrix} 2 & 3 \\ 1 & -1 \end{pmatrix}\begin{pmatrix} x \\ y \end{pmatrix} = \begin{pmatrix} 3 \\ 4 \end{pmatrix}$$

Write the equation in the form $X = A^{-1}C$

$$\begin{pmatrix} x \\ y \end{pmatrix} = \begin{pmatrix} 2 & 3 \\ 1 & -1 \end{pmatrix}^{-1}\begin{pmatrix} 3 \\ 4 \end{pmatrix}$$

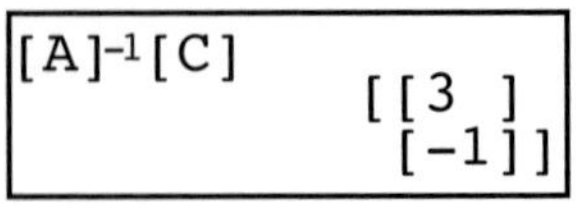

Find $A^{-1}$, then substitute into the equation and simplify

$$\Rightarrow \begin{pmatrix} x \\ y \end{pmatrix} = -\frac{1}{5}\begin{pmatrix} -1 & -3 \\ -1 & 2 \end{pmatrix}\begin{pmatrix} 3 \\ 4 \end{pmatrix}$$

$$\Rightarrow \begin{pmatrix} x \\ y \end{pmatrix} = -\frac{1}{5}\begin{pmatrix} -15 \\ 5 \end{pmatrix} = \begin{pmatrix} 3 \\ -1 \end{pmatrix}$$

```
[A]⁻¹[C]
           [[3 ]
            [-1]]
```
**Figure 7.5** GDC screen for the solution to Example 7.8

In general, a system of equations can be written in matrix form as $AX = B$

There is a solution to the system when $A$ is non-singular, which is $X = A^{-1}B$

If $B = 0$, the system is **homogeneous**. A homogeneous system will always have a solution, called the **trivial** solution, $X = 0$ when $A$ is non-singular. When $A$ is singular then the system has an infinite number of solutions.

We use a similar procedure to solve systems of equations in three variables. However, we will use a GDC to find the inverse of a $3 \times 3$ matrix. As in the case of a $2 \times 2$ matrix, the existence of an inverse for a $3 \times 3$ matrix depends on the value of its determinant.

There are two methods for calculating the determinant of a $3 \times 3$ matrix $A$:

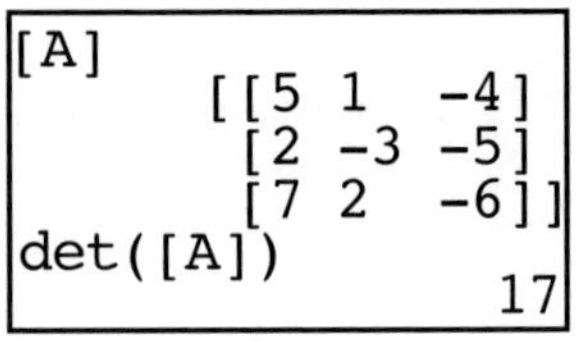

**Method 1**

```
[A]
      [[5 1  -4]
       [2 -3 -5]
       [7 2  -6]]
det([A])
             17
```
**Figure 7.6** Method 1

$$A = \begin{pmatrix} a & b & c \\ d & e & f \\ g & h & i \end{pmatrix} \Rightarrow \det A = a(ei - fh) - b(di - fg) + c(dh - eg)$$

For example, if $A = \begin{pmatrix} 5 & 1 & -4 \\ 2 & -3 & -5 \\ 7 & 2 & -6 \end{pmatrix}$

then $\det A = 5(18 + 10) - 1(-12 + 35) - 4(4 + 21) = 17$

**Method 2**

Use a special set up as follows:

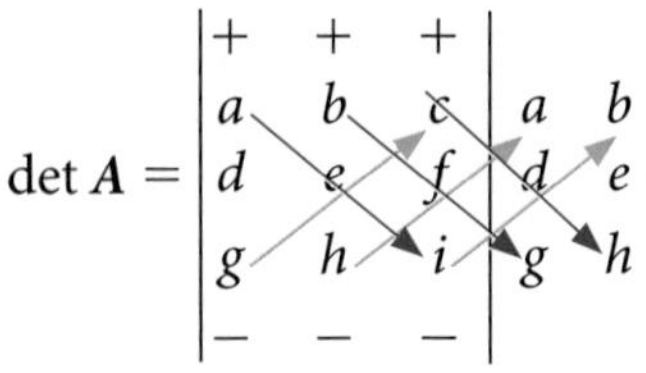

$$\det A = \begin{vmatrix} a & b & c \\ d & e & f \\ g & h & i \end{vmatrix}\begin{matrix} a & b \\ d & e \\ g & h \end{matrix} = aei + bfg + cdh - gec - hfa - idb$$

218

This is done by copying the first two columns and adding them to the end of the matrix, multiplying down the main diagonals and adding the products, and then multiplying up the second diagonals and subtracting them from the previous product as shown. For example:

$$
\begin{array}{ccccc}
+ & + & + & & \\
5 & 1 & -4 & 5 & 1 \\
2 & -3 & -5 & 2 & -3 \\
7 & 2 & 6 & 7 & 2 \\
- & - & - & &
\end{array}
$$

$$
= 5 \cdot (-3)(-6) + 1 \cdot (-5) \cdot 7 + (-4) \cdot 2 \cdot 2 - 7(-3)(-4)
$$
$$
- 2(-5) \cdot 5 - (-6) \cdot 2 \cdot 1
$$
$$
= 90 - 35 - 16 - 84 + 50 + 12 = 152 - 135 = 17
$$

This arrangement is a re-ordering of the calculations involved in the first method.

## Example 7.9

Solve the system of equations

$$
\begin{aligned}
5x + y - 4z &= 5 \\
2x - 3y - 5z &= 2 \\
7x + 2y - 6z &= 5
\end{aligned}
$$

### Solution

We write this system in matrix form:

$$
\begin{pmatrix} 5 & 1 & -4 \\ 2 & -3 & -5 \\ 7 & 2 & -6 \end{pmatrix} \begin{pmatrix} x \\ y \\ z \end{pmatrix} = \begin{pmatrix} 5 \\ 2 \\ 5 \end{pmatrix}
$$

Since det $A = 17 \neq 0$, we can find the solution in the same way we did for the $2 \times 2$ matrix:

$$
\begin{pmatrix} 5 & 1 & -4 \\ 2 & -3 & -5 \\ 7 & 2 & -6 \end{pmatrix} \begin{pmatrix} x \\ y \\ z \end{pmatrix} = \begin{pmatrix} 5 \\ 2 \\ 5 \end{pmatrix} \Rightarrow \begin{pmatrix} x \\ y \\ z \end{pmatrix} = \begin{pmatrix} 5 & 1 & -4 \\ 2 & -3 & -5 \\ 7 & 2 & -6 \end{pmatrix}^{-1} \begin{pmatrix} 5 \\ 2 \\ 5 \end{pmatrix}
$$

To check our work, using a GDC, we can store the answer matrix as **D** and then substitute the values into the system

$$
\begin{pmatrix} 5 & 1 & -4 \\ 2 & -3 & -5 \\ 7 & 2 & -6 \end{pmatrix} \begin{pmatrix} 3 \\ -2 \\ 2 \end{pmatrix} = \begin{pmatrix} 15 - 2 - 8 \\ 6 + 6 - 10 \\ 21 - 4 - 12 \end{pmatrix} = \begin{pmatrix} 5 \\ 2 \\ 5 \end{pmatrix}
$$

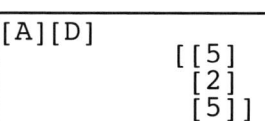

```
[A]⁻¹[C]
                [[3  ]
                 [-2 ]
                 [2  ]]

[A][D]
                [[5]
                 [2]
                 [5]]
```

**Figure 7.7** GDC screens for the solution to Example 7.9

219

## Area of a triangle

An interesting application of determinants that you may find helpful is finding the area of a triangle whose vertices are given as points in a coordinate plane.

### Example 7.10

Find the area of triangle $ABC$ whose vertices are

$A(1, 3)$, $B(5, -1)$ and $C(-2, 5)$.

**Solution**

We let $(x_1, y_1) = (1, 3)$, $(x_2, y_2) = (5, -1)$, and $(x_3, y_3) = (-2, 5)$

To find the area, we evaluate the determinant

$$\begin{vmatrix} x_1 & y_1 & 1 \\ x_2 & y_2 & 1 \\ x_3 & y_3 & 1 \end{vmatrix} = \begin{vmatrix} 1 & 3 & 1 \\ 5 & -1 & 1 \\ -2 & 5 & 1 \end{vmatrix} = -4$$

Using this value, we can conclude that the area of the triangle is

$$\text{Area} = \left| \frac{1}{2} \begin{vmatrix} 1 & 3 & 1 \\ 5 & -1 & 1 \\ -2 & 5 & 1 \end{vmatrix} \right| = \left| \frac{1}{2} \cdot -4 \right| = 2$$

## Lines in plane

What happens when the three points are collinear? The triangle becomes a line segment and the area becomes zero. This fact allows us to develop two techniques that are very helpful in dealing with questions of collinearity and equations of lines.

For example, consider the points $A(-2, -3)$, $B(1, 3)$ and $C(3, 7)$. Find the area of 'triangle' $ABC$.

$$\text{Area} = \left| \frac{1}{2} \begin{vmatrix} -2 & -3 & 1 \\ 1 & 3 & 1 \\ 3 & 7 & 1 \end{vmatrix} \right| = \left| \frac{1}{2} \cdot 0 \right| = 0$$

This result can be stated in general as a test for collinearity.

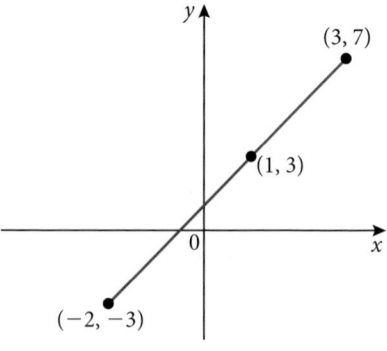

**Figure 7.8** Three collinear points

## Example 7.11

Determine if the points $(-2, 3)$, $(2, 5)$ and $(5, 7)$ lie on the same line.

### Solution

Set up the matrix as given above:

$$\begin{vmatrix} -2 & 3 & 1 \\ 2 & 5 & 1 \\ 5 & 7 & 1 \end{vmatrix} = 2 \neq 0$$

The points cannot lie on a line because the value of the determinant is not equal to zero.

## Two-point equation of a line

The test for collinearity leads us to a method for finding the equation of a line containing two points. Consider two points $(x_1, y_1)$, $(x_2, y_2)$ which lie on a given line. To find the equation of the line through these two points, we introduce a general point $(x, y)$ on the line. These three points $(x_1, y_1)$, $(x_2, y_2)$ and $(x, y)$ are collinear, and hence they satisfy the determinant equation

$$\begin{vmatrix} x & y & 1 \\ x_1 & y_1 & 1 \\ x_2 & y_2 & 1 \end{vmatrix} = 0$$

which gives us the equation of the line in the form:

$$(y_1 - y_2)x + (x_2 - x_1)y + (x_1 y_2 - y_1 x_2) = 0$$

which in turn is of the form: $Ax + By + C = 0$

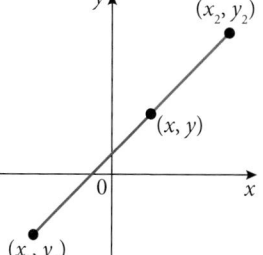

**Figure 7.9** $(x_1, y_1)$, $(x_2, y_2)$ and $(x, y)$ are collinear

## Example 7.12

Find the equation of the line through $(-2, 3)$ and $(3, 7)$

### Solution

Apply the determinant formula for the equation of a line.

$$\begin{vmatrix} x & y & 1 \\ -2 & 3 & 1 \\ 3 & 7 & 1 \end{vmatrix} = (3 - 7)x + (3 + 2)y + (-14 - 9) = 0$$

The equation of the line is $-4x + 5y - 23 = 0$

We can use column vectors instead of row vectors, and the calculation in Example 7.12 becomes:

$$\begin{vmatrix} x & -2 & 3 \\ y & 3 & 7 \\ 1 & 1 & 1 \end{vmatrix} = (3-7)x + (-14-9) + (3+2)y = 0$$

$$-4x + 5y - 23 = 0$$

The matrix $A^t = \begin{pmatrix} x & -2 & 3 \\ y & 3 & 7 \\ 1 & 1 & 1 \end{pmatrix}$ is called the transpose of the matrix $A = \begin{pmatrix} x & y & 1 \\ -2 & 3 & 1 \\ 3 & 7 & 1 \end{pmatrix}$

where each row of the matrix becomes a column of the transpose. The transpose has the same determinant as the matrix itself.

## Coding and decoding messages: cryptography

Data encryption is essential in applications such as online banking. Encryption uses encoding-decoding methods in which matrices play a very important role.

The methods included here are not secure enough to use for applications such as internet banking, but they result in codes that are not easy to break and provide a good introduction to the ideas of encryption.

The process of encryption is called cryptography. In cryptography, a message that has not yet been encrypted is called plaintext, after the encryption process the encrypted message is called ciphertext. The process of converting the plaintext to ciphertext is called enciphering and the reverse process where the ciphertext is converted to plaintext is called deciphering. One such method, called Hill's method, involves dividing the plaintext message into sets of $n$ letters, each of which is replaced by $n$ cipher letters. This is called a polygraphic system. Hill-ciphers require a matrix based polygraphic system. A system of cryptography in which the plaintext is divided into sets of $n$ letters, each of which is replaced by a set of $n$ cipher letters, is called a polygraphic system. For example, {*mathematics is a great subject*} = {math emat ics⊔ is⊔a ⊔gre at⊔s ubje ct⊔⊔}. In this case we used blocks of 4 letters. [⊔ stands for space]

**Modular/clock arithmetic**

In order to work efficiently with cryptography, some basic knowledge of modular arithmetic is helpful.

Two integers, $a$ and $b$ are said to be congruent modulo $n$, written as $a \equiv b \bmod n$, if they leave the same remainder when divided by $n$. For example $41 \equiv 5 \bmod 12$ or $15 \equiv 3 \bmod 4$. Alternatively, $a \equiv b \bmod n$ also means that $n \mid (a - b)$. Note that $41 - 5 = 36$, and $12 \mid 36$

In calculations, using the same modulus, you can replace any integer by any integer congruent to it. For example, in mod 4, $19 \times 3 \equiv 1 \bmod 4$ because you can write it as $3 \times 3 \equiv 9 \equiv 1 \bmod 4$, or alternatively $19 \times 3 = 57$ which leaves a remainder of 1 when divided by 4. Thus $19 \times 3 = 57 \equiv 1 \bmod 4$.

When replacing numbers by their equivalents, it is a good idea to either add or subtract multiples of the modulus until you reach a number less than the mod. Remember that when dividing by $n$ the possible remainders are $0, 1, …, n - 1$. For example, the closest multiple of 4 to 57 is 56.

Thus $57 - 56 = 1$, which explains why $57 \equiv 1 \bmod 4$.

Let us describe the process with an example.

Say you want to send the message 'listen to me please!'

The process is as follows:

1. Choose a code table similar to the one below. The table depends on how many letters/symbols you need and what language you use. We are using English here, so we need a table to cater for the whole alphabet at least.

| | A | B | C | D | E | F | G | H | I | J | K | L | M | N |
|---|---|---|---|---|---|---|---|---|---|---|---|---|---|---|
| 0 | 1 | 2 | 3 | 4 | 5 | 6 | 7 | 8 | 9 | 10 | 11 | 12 | 13 | 14 |
| O | P | Q | R | S | T | U | V | W | X | Y | Z | ! | ? | . |
| 15 | 16 | 17 | 18 | 19 | 20 | 21 | 22 | 23 | 24 | 25 | 26 | 27 | 28 | 29 |

**Table 7.4** Code table

The first cell is for a space.

2. Then, translate the text message into codes from the table.

| L | I | S | T | E | N | | T | O | | M | E | | P | L | E | A | S | E | ! |
|---|---|---|---|---|---|---|---|---|---|---|---|---|---|---|---|---|---|---|---|
| 12 | 9 | 19 | 20 | 5 | 14 | 0 | 20 | 15 | 0 | 13 | 5 | 0 | 16 | 12 | 5 | 1 | 19 | 5 | 27 |

3. Choose a non-singular coding matrix of any order of your choice. We will use a $3 \times 3$ matrix. Also, for convenience, we will choose it to have a determinant of 1. For example, we will use the matrix

$$\begin{pmatrix} 1 & 0 & 1 \\ -1 & 1 & 0 \\ 0 & 1 & 2 \end{pmatrix}$$

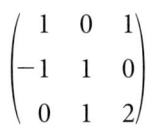 An efficient way of getting a matrix with determinant 1 is to start with an identity matrix and then add or subtract rows or multiples of rows (or columns)

4. Subdivide your codes into columns (or rows) of 3 each, including spaces. If you don't have enough letters to fill the columns, use a space. (This is so, because our coding matrix is of order 3.)

$$\begin{pmatrix} 12 & 20 & 0 & 0 & 0 & 5 & 5 \\ 9 & 5 & 20 & 13 & 16 & 1 & 27 \\ 19 & 14 & 15 & 5 & 12 & 19 & 0 \end{pmatrix}$$

5. Now, multiply the coding matrix by the code matrix.

$$\begin{pmatrix} 1 & 0 & 1 \\ -1 & 1 & 0 \\ 0 & 1 & 2 \end{pmatrix}\begin{pmatrix} 12 & 20 & 0 & 0 & 0 & 5 & 5 \\ 9 & 5 & 20 & 13 & 16 & 1 & 27 \\ 19 & 14 & 15 & 5 & 12 & 19 & 0 \end{pmatrix}$$

This is a $3 \times 3$ matrix multiplied by a $3 \times 7$ matrix, thus the result is a $3 \times 7$ matrix.

$$\begin{pmatrix} 31 & 34 & 15 & 5 & 12 & 24 & 5 \\ -3 & -15 & 20 & 13 & 16 & -4 & 22 \\ 47 & 33 & 50 & 23 & 40 & 39 & 27 \end{pmatrix}$$

6. Before we give out the ciphered message, we need to replace the numbers in pink with their congruent numbers mod 30 (This is so, because we are using 30 codes).

The message will now be:

$$\begin{pmatrix} 1 & 4 & 15 & 5 & 12 & 24 & 5 \\ 27 & 15 & 20 & 13 & 16 & 26 & 22 \\ 17 & 3 & 20 & 23 & 10 & 9 & 27 \end{pmatrix}$$

This is equivalent to the message A!QDOCOTTEMWLPJXZIEV!

The GDC output is as shown.

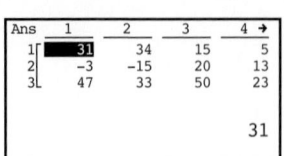

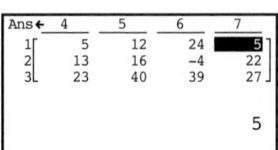

**Figure 7.10** GDC output

7. The receiver of the message will decipher the message by multiplying the inverse of the coding matrix by the message matrix. In this case, the inverse of the coding matrix is:

$$\begin{pmatrix} 2 & 1 & -1 \\ 2 & 2 & -1 \\ -1 & -1 & 1 \end{pmatrix}$$

$$\begin{pmatrix} 2 & 1 & -1 \\ 2 & 2 & -1 \\ -1 & -1 & 1 \end{pmatrix} \begin{pmatrix} 1 & 4 & 15 & 5 & 12 & 24 & 5 \\ 27 & 15 & 20 & 13 & 16 & 26 & 22 \\ 17 & 3 & 20 & 23 & 10 & 9 & 27 \end{pmatrix}$$

$$= \begin{pmatrix} 12 & 20 & 30 & 0 & 30 & 65 & 5 \\ 39 & 35 & 50 & 13 & 46 & 91 & 27 \\ -11 & -16 & -15 & 5 & -18 & -41 & 0 \end{pmatrix}$$

Now, replace the numbers in pink with their congruent counterparts mod 30

$$\begin{pmatrix} 12 & 20 & 0 & 0 & 0 & 5 & 5 \\ 9 & 5 & 20 & 13 & 16 & 1 & 27 \\ 19 & 14 & 15 & 5 & 12 & 19 & 0 \end{pmatrix}$$

Which is the matrix for the original message.

---

### Example 7.13

You receive the following message.

UHTUWWE??SCVPMALU!TJ.ZMYFLL

You also know from your sender that the coding matrix is

$$\begin{pmatrix} 3 & 2 & 2 \\ 1 & 1 & 1 \\ 1 & 1 & 2 \end{pmatrix}$$

Use Table 7.4 to decode the message.

## Solution

We first use the code table to write out the matrix corresponding to the coded message:

$$\begin{pmatrix} 21 & 21 & 5 & 19 & 16 & 12 & 20 & 26 & 6 \\ 8 & 23 & 28 & 3 & 13 & 21 & 10 & 13 & 12 \\ 20 & 23 & 28 & 22 & 1 & 27 & 29 & 25 & 12 \end{pmatrix}$$

Next, we multiply the inverse of the coding inverse by this matrix

$$\begin{pmatrix} 3 & 2 & 2 \\ 1 & 1 & 1 \\ 1 & 1 & 2 \end{pmatrix}^{-1} \begin{pmatrix} 21 & 21 & 5 & 19 & 16 & 12 & 20 & 26 & 6 \\ 8 & 23 & 28 & 3 & 13 & 21 & 10 & 13 & 12 \\ 20 & 23 & 28 & 22 & 1 & 27 & 29 & 25 & 12 \end{pmatrix}$$

$$= \begin{pmatrix} 5 & -25 & -51 & 13 & -10 & -30 & 0 & 0 & -18 \\ -9 & 48 & 79 & -29 & 35 & 45 & -9 & 1 & 30 \\ 12 & 0 & 0 & 19 & -12 & 6 & 19 & 12 & 0 \end{pmatrix}$$

Now, we replace the numbers that are less than 0 or over 30 with their congruent counterparts mod 30.

$$\begin{pmatrix} 5 & 5 & 9 & 13 & 20 & 0 & 0 & 0 & 12 \\ 21 & 18 & 19 & 1 & 5 & 15 & 21 & 1 & 0 \\ 12 & 0 & 0 & 19 & 18 & 6 & 19 & 12 & 0 \end{pmatrix}$$

Now, replacing the ciphers with letters

EULER IS MASTER OF US ALL

> If numbers are to be included in messages, then you can extend the code table by 10 to represent the integers from 0 to 9. Your mod will then be 40.

---

## Exercise 7.2

1. Consider the matrix $M$ which satisfies the matrix equation

$$\begin{pmatrix} 3 & 7 \\ -4 & -9 \end{pmatrix} M = \begin{pmatrix} 2 & 1 \\ 3 & 5 \end{pmatrix}$$

   (a) Find the inverse of matrix $\begin{pmatrix} 3 & 7 \\ -4 & -9 \end{pmatrix}$

   (b) Hence, write $M$ as a product of two matrices.

   (c) Evaluate $M$.

   (d) Now consider the equation containing the matrix $N$:

$$N \begin{pmatrix} 3 & 7 \\ -4 & -9 \end{pmatrix} = \begin{pmatrix} 2 & 1 \\ 3 & 5 \end{pmatrix}$$

   (i) Express $N$ as a product of two matrices.

   (ii) Evaluate $N$.

   (e) Write a short paragraph describing your work on this problem.

225

**2.** Find the matrix $E$ in the following equation.

$$\begin{pmatrix} 1 & 3 \\ 3 & 4 \end{pmatrix} = \begin{pmatrix} 1 & 0 \\ 3 & 1 \end{pmatrix} E \begin{pmatrix} 1 & 0 \\ 0 & -5 \end{pmatrix}$$

**3. (a)** Prove that the matrix $A = \begin{pmatrix} 2 & -3 & 1 \\ 1 & 1 & -3 \\ 3 & -2 & -3 \end{pmatrix}$ should have an inverse.

**(b)** Write out $A^{-1}$.

**(c)** Hence, solve the system of equations

$$\begin{cases} 2x - 3y + z = 4.2 \\ x + y - 3z = -1.1 \\ 3x - 2y - 3z = 2.9 \end{cases}$$

**4.** Find the inverse for each matrix:

**(a)** $A = \begin{pmatrix} \dfrac{\sqrt{3}}{2} & -\dfrac{1}{2} \\ \dfrac{1}{2} & \dfrac{\sqrt{3}}{2} \end{pmatrix}$

**(b)** $B = \begin{pmatrix} a & 1 \\ a+2 & \dfrac{3}{a}+1 \end{pmatrix}$

**5.** For what values of $x$ is the following matrix singular?

$$A = \begin{pmatrix} x+1 & 3 \\ 3x-1 & x+3 \end{pmatrix}$$

**6.** Find $n$ such that $\begin{pmatrix} 2 & -1 & 4 \\ 2n & 2 & 0 \\ 2 & 1 & 4n \end{pmatrix}$ is the inverse of $\begin{pmatrix} -2 & -3 & 4 \\ 1 & 2 & -2 \\ 3n & 2 & -5n \end{pmatrix}$

**7.** Consider the two matrices $A = \begin{pmatrix} 4 & 2 \\ 0 & -3 \end{pmatrix}$ and $B = \begin{pmatrix} 2 & 1 \\ 3 & 5 \end{pmatrix}$

**(a)** Find $X$ such that $XA = B$

**(b)** Find $Y$ such that $AY = B$

**(c)** Is $X = Y$? Explain.

**8.** Consider the two matrices

$$P = \begin{pmatrix} 2 & 0 & -1 \\ 3 & 5 & 4 \\ 1 & 0 & -1 \end{pmatrix} \text{ and } Q = \begin{pmatrix} 3 & -1 & 1 \\ 4 & 0 & 0 \\ 1 & -2 & -1 \end{pmatrix}$$

**(a)** Find $PQ$ and $QP$.

**(b)** Find:

**(i)** $P^{-1}$          **(ii)** $Q^{-1}$          **(iii)** $P^{-1}Q^{-1}$

**(iv)** $Q^{-1}P^{-1}$          **(v)** $(PQ)^{-1}$          **(vi)** $(QP)^{-1}$

**(c)** Write a few sentences about your observations in parts **(a)** and **(b)**.

9. Consider the matrices

$$A = \begin{pmatrix} 3 & -2 & 1 \\ -4 & 1 & -3 \\ 1 & -5 & 1 \end{pmatrix} \quad B = \begin{pmatrix} -29 \\ 37 \\ -24 \end{pmatrix}$$

**(a)** Find the matrix $C$ where $AC = B$

**(b)** Solve the system of equations

$$\begin{cases} 3x - 2y + z = -29 \\ 4x - y + 3z = -37 \\ -x + 5y - z = 24 \end{cases}$$

10. Solve the matrix equation:

$$\begin{pmatrix} 2 & 2 + x \\ 5 & 4 + x \end{pmatrix} \begin{pmatrix} 3 & x \\ x - 4 & 2 \end{pmatrix} = \begin{pmatrix} 3 & x \\ x - 4 & 2 \end{pmatrix} \begin{pmatrix} 2 & 2 + x \\ 5 & 4 + x \end{pmatrix}$$

11. Consider the matrices $A$ and $B$. Find $x$ and $y$ such that $AB = BA$

**(a)** $A = \begin{pmatrix} 2 & 1 \\ 5 & 3 \end{pmatrix}$          and     $B = \begin{pmatrix} 2 - x & 1 \\ 5x & y \end{pmatrix}$

**(b)** $A = \begin{pmatrix} 3 & 1 \\ -5 & 2 \end{pmatrix}$          and     $B = \begin{pmatrix} 1 - x & x \\ 5x & y \end{pmatrix}$

**(c)** $A = \begin{pmatrix} 3 + x & 1 \\ -5 & 2 \end{pmatrix}$     and     $B = \begin{pmatrix} y - x & x \\ 5x - y + 1 & y + x \end{pmatrix}$

12. Use matrix methods to find an equation of a line that contains the given points.

**(a)** $A(-5, -6), B(3, 11)$          **(b)** $A(5, -2), B(3, -2)$

**(c)** $A(-5, 3), B(-5, 8)$

13. Find the area of the parallelogram with the given points as three of its vertices.

**(a)** $A(-5, -6), B(3, 11), C(8, 1)$          **(b)** $A(3, -5), B(3, 11), C(8, 11)$

**(c)** $A(4, -6), B(-3, 9), C(7, 7)$

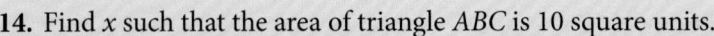

**14.** Find $x$ such that the area of triangle $ABC$ is 10 square units.

  (a) $A(x, -6)$, $B(3, 11)$, $C(8, 3)$

  (b) $A(-5, x)$, $B(3, x+2)$, $C(x^2+2x - 3, 1)$

**15.** Find the value of $k$ such that the points $P$, $Q$, and $R$ are collinear.

  (a) $P(2, -5)$, $Q(4, k)$, $R(5, -2)$      (b) $P(-6, 2)$, $Q(-5, k)$, $R(-3, 5)$

$f(x)$ is called the characteristic polynomial of $A$.

**16.** Consider the matrix $A = \begin{pmatrix} 2 & 7 \\ 5 & 5 \end{pmatrix}$. Define $f(x) = \det(xI - A)$ where $x$ is any real number and $I$ is the identity matrix.

  (a) Find $\det(A)$.

  (b) Expand $f(x)$ and compare the constant term to your answer in (a).

  (c) How is the coefficient of $x$ in the expansion of $f(x)$ related to $A$?

  (d) Find $f(A)$ and simplify it.

  (e) Now repeat parts (a)–(d) with matrix $B = \begin{pmatrix} a & b \\ c & d \end{pmatrix}$

**17.** Consider the matrix $A = \begin{pmatrix} 2 & 7 & 1 \\ -1 & 3 & 2 \\ 5 & 5 & -4 \end{pmatrix}$

  Define $f(x) = \det(xI - A)$, where $x$ is any real number and $I$ is the identity matrix.

  (a) Find $\det(A)$.

  (b) Expand $f(x)$ and compare the constant term to your answer in (a).

  (c) How is the coefficient of $x^2$ in the expansion of $f(x)$ related to $A$?

  (d) Find $f(A)$ and simplify it.

  (e) Now repeat parts (a)–(d) with matrix $B = \begin{pmatrix} a & b & c \\ d & e & f \\ g & h & i \end{pmatrix}$

**18.** (a) Use Table 7.4 to decode the following message, given that ∪ stands for a space.

    S.TPEHZO?WPOSWYSFPV!∪LWPUO∪

    The coding matrix is $\begin{pmatrix} 2 & 2 & 3 \\ 2 & 3 & 2 \\ 1 & 1 & 2 \end{pmatrix}$

  (b) Choose an appropriate matrix of your own, code your answer and decode it.

## 7.3 Further properties and applications

In question 8 of Exercise 7.2, you were asked to make some observations concerning the answers to parts (a) and (b). The question shows some properties of inverse matrices.

You should have found out that:

$$P^{-1}Q^{-1} = \begin{pmatrix} -2 & 2 & -1 \\ \dfrac{23}{5} & -\dfrac{22}{5} & \dfrac{12}{5} \\ -4 & \dfrac{15}{4} & -2 \end{pmatrix} \qquad Q^{-1}P^{-1} = \begin{pmatrix} -\dfrac{7}{20} & \dfrac{1}{20} & \dfrac{11}{20} \\ \dfrac{17}{5} & -\dfrac{1}{5} & -\dfrac{26}{5} \\ \dfrac{109}{20} & -\dfrac{7}{20} & -\dfrac{157}{20} \end{pmatrix}$$

$$(PQ)^{-1} = \begin{pmatrix} -\dfrac{7}{20} & \dfrac{1}{20} & \dfrac{11}{20} \\ \dfrac{17}{5} & -\dfrac{1}{5} & -\dfrac{26}{5} \\ \dfrac{109}{20} & -\dfrac{7}{20} & -\dfrac{157}{20} \end{pmatrix} \qquad (QP)^{-1} = \begin{pmatrix} -2 & 2 & -1 \\ \dfrac{23}{5} & -\dfrac{22}{5} & \dfrac{12}{5} \\ -4 & \dfrac{15}{4} & -2 \end{pmatrix}$$

So $(PQ)^{-1} \neq P^{-1}Q^{-1}$, but $(PQ)^{-1} = Q^{-1}P^{-1}$

and $(QP)^{-1} = P^{-1}Q^{-1}$

This leads to the following general result.

When $A$ and $B$ are non-singular matrices of order $n$, then $AB$ is also non-singular and
$$(AB)^{-1} = B^{-1}A^{-1}$$
The proof of this theorem is straightforward:

To show that $B^{-1}A^{-1}$ is the inverse of $AB$, we need only show that it conforms to the definition of an inverse matrix. That is,
$$(AB)(B^{-1}A^{-1}) = (B^{-1}A^{-1})(AB) = I$$
Now, $(AB)(B^{-1}A^{-1}) = A(B\,B^{-1})A-1 = A(I)A^{-1} = AA^{-1} = I$

Similarly, $(B^{-1}A^{-1})(AB) = B^{-1}(A^{-1}A)B = B-1(I)B = B-1B = I$

Hence, $AB$ is non-singular and its inverse is $B^{-1}A^{-1}$

Non-singular matrices also have these properties:
$$(A^{-1})^{-1} = A$$
$$(cA)^{-1} = \frac{1}{c}A^{-1}; c \neq 0$$
$$\det(AB) = \det A \cdot \det B$$
$$\det A^{-1} = \frac{1}{\det A}$$

We can prove the last property using the third property.

Since $AA^{-1} = I$, then

$$\det(AA^{-1}) = \det I \Rightarrow \det A \cdot \det A^{-1} = 1 \Rightarrow \det A^{-1} = \frac{1}{\det A}$$

In Section 7.2, we solved a system of equations using inverse matrices. The method works only when the system has a unique solution. In many cases, there is either an infinite number of solutions or the system is inconsistent. We can use another method of solution.

## Some terminology

In Section 7.2 we learned how to solve a system of equations by writing the system in matrix form. When the system has a unique solution then it can be solved. However, the method is limited and it has a strict constraint. If we use a slightly different arrangement, we can use matrices to find the solution whether it is unique, there are an infinite number of solutions, or simply no solution. We write the system as follows.

$$\begin{pmatrix} 2 & 3 & -4 & | & 8 \\ 0 & 2 & 4 & | & -3 \\ 1 & 0 & -2 & | & 4 \end{pmatrix}$$

This called the **augmented** matrix of the system. It is customary to put a bar between the coefficients and the answers. However, this bar is not necessary and we will not be using it in this book. Just remember that the last column is the answers' column.

## Gauss-Jordan elimination

The idea behind this method is very simple. We apply certain simple operations to the system of equations to reduce them into a special form that is easy to solve. We keep applying the operations until we have a form that is easy to solve. The operations are called **elementary row operations** and they can be applied to the system without changing the solution to the system. That is, the solution to the reduced system (**reduced row echelon form**) is the same as that for the original system. We can apply the operations either to the system itself or to its augmented matrix. As it is easier to work with the augmented matrix, we recommend that you first write the augmented matrix, reduce it, and then write the equivalent system to read the solution from.

There are three types of elementary row operations:

• multiply any row by a non-zero real number

• interchange any two rows

• add a multiple of one row to another row.

We will demonstrate the method with an example.

Note that the order in which we apply the operations is not unique.

### Example 7.14

Consider the following system with its augmented matrix and simplify it to its reduced row echelon form.

$$\begin{cases} 2x + y - z = 2 \\ x + 3y + 2z = 1 \\ 2x + 4y + 6z = 6 \end{cases} \Leftrightarrow \begin{pmatrix} 2 & 1 & -1 & | & 2 \\ 1 & 3 & 2 & | & 1 \\ 2 & 4 & 6 & | & 6 \end{pmatrix}$$

Switch row 1 and row 2:

$$\begin{cases} x + 3y + 2z &= 1 \\ 2x + y - z &= 2 \\ 2x + 4y + 6z &= 6 \end{cases} \Leftrightarrow \left(\begin{array}{ccc|c} 1 & 3 & 2 & 1 \\ 2 & 1 & -1 & 2 \\ 2 & 4 & 6 & 6 \end{array}\right)$$

Multiply row 3 by $\frac{1}{2}$:

$$\begin{cases} x + 3y + 2z &= 1 \\ 2x + y - z &= 2 \\ x + 2y + 3z &= 3 \end{cases} \Leftrightarrow \left(\begin{array}{ccc|c} 1 & 3 & 2 & 1 \\ 2 & 1 & -1 & 2 \\ 1 & 2 & 3 & 3 \end{array}\right)$$

Multiply row 1 by $-2$ and add it to row 2, and multiply row 1 by $-1$ and add it to row 3 (we replace the second row with the result):

$$\begin{cases} x + 3y + 2z &= 1 \\ -5y - 5z &= 0 \\ -y + z &= 2 \end{cases} \Leftrightarrow \left(\begin{array}{ccc|c} 1 & 3 & 2 & 1 \\ 0 & -5 & -5 & 0 \\ 0 & -1 & 1 & 2 \end{array}\right)$$

Note that row 1 did not change and rows 2 and three were replaced with the result of the elementary operation.

Multiply row 2 by $-\frac{1}{5}$

$$\begin{cases} x + 3y + 2z &= 1 \\ y + z &= 0 \\ -y + z &= 2 \end{cases} \Leftrightarrow \left(\begin{array}{ccc|c} 1 & 3 & 2 & 1 \\ 0 & 1 & 1 & 0 \\ 0 & -1 & 1 & 2 \end{array}\right)$$

Now, add row 2 to row 3, and multiply row 2 by $-3$ and add it to row 1:

$$\begin{cases} x - z &= 1 \\ y + z &= 0 \\ 2z &= 2 \end{cases} \Leftrightarrow \left(\begin{array}{ccc|c} 1 & 0 & -1 & 1 \\ 0 & 1 & 1 & 0 \\ 0 & 0 & 2 & 2 \end{array}\right)$$

Now multiply row 3 by $\frac{1}{2}$:

$$\begin{cases} x - z &= 1 \\ y + z &= 0 \\ z &= 1 \end{cases} \Leftrightarrow \left(\begin{array}{ccc|c} 1 & 0 & -1 & 1 \\ 0 & 1 & 1 & 0 \\ 0 & 0 & 1 & 1 \end{array}\right)$$

Lastly, add row 3 to row 1, and multiply row 3 by $-1$ and add it to row 2:

$$\begin{cases} x &= 2 \\ y &= -1 \\ z &= 1 \end{cases} \Leftrightarrow \left(\begin{array}{ccc|c} 1 & 0 & 0 & 2 \\ 0 & 1 & 0 & -1 \\ 0 & 0 & 1 & 1 \end{array}\right)$$

As you can see, this last system is very easy to read the solution from.
You can verify that this solution is also the solution to the original system.

[A]
```
[[2  1  -1  2]
 [1  3   2  1]
 [2  4   6  6]]
```

rref([A])
```
[[1  0  0   2]
 [0  1  0  -1]
 [0  0  1   1]]
```

**Figure 7.11** Carrying out the operation on a GDC

The simplified matrix is in its reduced row echelon form.

Of course, when we do the work, we do not have to show the processes in parallel. We just perform the operation on the matrix and then translate it into the equation form at the end.

You can carry out this whole operation easily using a GDC.

---

**Example 7.15**

Solve the system of equations:
$$\begin{cases} x + y + 2z = 1 \\ x \quad\; + z = 2 \\ \quad y + z = 0 \end{cases}$$

---

**Solution**

The augmented matrix is

$$\begin{cases} x + y + 2z = 1 \\ x \quad\; + z = 2 \\ \quad y + 2z = 0 \end{cases} \Leftrightarrow \begin{pmatrix} 1 & 1 & 2 & 1 \\ 1 & 0 & 1 & 2 \\ 0 & 1 & 1 & 0 \end{pmatrix}$$

Multiply row 1 by $-1$ and add to row 2:

$$\begin{cases} x + y + 2z = 1 \\ \;\;- y - z = 1 \\ \quad y + z = 0 \end{cases} \Leftrightarrow \begin{pmatrix} 1 & 1 & 2 & 1 \\ 0 & -1 & -1 & 1 \\ 0 & 1 & 1 & 0 \end{pmatrix}$$

Add row 2 to row 1 and row 2 to row 3:

$$\begin{cases} x \quad\; + z = 2 \\ \;\;- y - z = 1 \\ \qquad\quad 0 = 1 \end{cases} \Leftrightarrow \begin{pmatrix} 1 & 0 & 1 & 2 \\ 0 & -1 & -1 & 1 \\ 0 & 0 & 0 & 1 \end{pmatrix}$$

At this stage, work can stop because if you write the last row as an equation, then it reads

$$0x + 0y + 0z = 1$$

This statement cannot be true for any value, and hence the system is inconsistent.

You can also use a GDC.

[B]
```
[[1  1  2  1]
 [1  0  1  2]
 [0  1  1  0]]
```

rref([A])
```
[[1  0  1  0]
 [0  1  0  0]
 [0  0  0  1]]
```

**Figure 7.12** GDC screens for the solution to Example 7.15

**Example 7.16**

Solve the system of equations:

$$\begin{cases} 2x + y - z &= 4 \\ x + 3y + 7z &= 7 \\ 2x + 4y + 8z &= 10 \end{cases}$$

**Solution**

The augmented matrix is:

$$\begin{cases} 2x + y - z &= 4 \\ x + 3y + 7z &= 7 \\ 2x + 4y + 8z &= 10 \end{cases} \Leftrightarrow \begin{pmatrix} 2 & 1 & -1 & 4 \\ 1 & 3 & 7 & 7 \\ 2 & 4 & 8 & 10 \end{pmatrix}$$

Swap rows 1 and 2:

$$\begin{cases} x + 3y + 7z &= 7 \\ 2x + y - z &= 4 \\ 2x + 4y + 8z &= 10 \end{cases} \Leftrightarrow \begin{pmatrix} 1 & 3 & 7 & 7 \\ 2 & 1 & -1 & 4 \\ 2 & 4 & 8 & 10 \end{pmatrix}$$

Multiply row 2 by $-1$ and add to row 3; multiply row 1 by $-2$ and add to row to 2:

$$\begin{cases} x + 3y + 7z &= 7 \\ -5y - 15z &= -10 \\ 3y + 9z &= 6 \end{cases} \Leftrightarrow \begin{pmatrix} 1 & 3 & 7 & 7 \\ 0 & -5 & -15 & -10 \\ 0 & 3 & 9 & 6 \end{pmatrix}$$

Multiply row 2 by $-\dfrac{1}{5}$; multiply row 3 by $\dfrac{1}{3}$:

$$\begin{cases} x + 3y + 7z &= 7 \\ y + 3z &= 2 \\ y + 3z &= 2 \end{cases} \Leftrightarrow \begin{pmatrix} 1 & 3 & 7 & 7 \\ 0 & 1 & 3 & 2 \\ 0 & 1 & 3 & 2 \end{pmatrix}$$

Multiply row 2 by $-1$ and add to row 3; multiply row 2 by $-3$ and add to row 1

$$\begin{cases} x + 3y + 7z &= 7 \\ y + 3z &= 2 \\ 0 &= 0 \end{cases} \Leftrightarrow \begin{pmatrix} 1 & 0 & -2 & 1 \\ 0 & 1 & 3 & 2 \\ 0 & 0 & 0 & 0 \end{pmatrix}$$

Since the last row is all zeros, there is not much that we can do. The conclusion is that this last row is true for any choice of values for the variables. Now we are left with a system of two equations and three variables.

$$\begin{cases} x & -2z & = & 1 \\ & y+3z & = & 2 \end{cases}$$

We need to solve for two of the variables in terms of the third. A wise choice here would be to solve for $x$ and $y$ in terms of $z$. That is

$$x = 1 + 2z, y = 2 - 3z$$

This means that for every choice of a value for $z$, we have a corresponding solution for the system. For example, if $z = 0$, then the solution would be $(1, 2, 0)$, for $z = 2$, the solution is $(5, -4, 2)$, and so on. This means that we have an infinite number of solutions. So we present the solution in terms of a parameter such as $t$. We let $z = t$, and our general solution would then be

$$(1 + 2t, 2 - 3t, t)$$

## Reduced row echelon form

A matrix is in **reduced row echelon form** when it satisfies the following properties:

- If there are any rows consisting entirely of 0s, they appear at the bottom of the matrix.
- In any non-zero row, the first non-zero entry is 1. This entry is called the **pivot** of the row.
- For any consecutive rows, the pivot of the lower row must be to the right of the pivot of the preceding row.
- Any column that contains a pivot, has zeros everywhere else.

Matrix $A$ is in reduced row echelon form, but matrix $B$ is not.

$$A = \begin{pmatrix} \boxed{1} & 0 & 3 & 0 & 5 & 8 \\ 0 \to & \boxed{1} & 4 & 0 & 4 & 2 \\ 0 & 0 \to & 0 \to & \boxed{1} & 5 & 2 \\ 0 & 0 & 0 & 0 & 0 & 0 \end{pmatrix}$$

$$B = \begin{pmatrix} 1 & 0 & 0 & 2 & 3 & 4 & 5 \\ 0 & 0 & 0 & 1 & 3 & 6 & 7 \\ 0 & 0 & 1 \leftarrow & 0 & 2 & 2 & 1 \\ 0 & 0 & 0 & 0 & 0 & 0 & 0 \\ 0 & 0 & 0 & 0 & 0 & 0 & 0 \end{pmatrix}$$

## Curve fitting

Another application of matrices (systems) is to help fit specific models to sets of points.

234

## Example 7.17

Fit a quadratic model to pass through the points $(-1, 10)$, $(2, 4)$, and $(3, 14)$

### Solution

The problem is to find parameters $a$, $b$, and $c$ that will force the curve representing the function $f(x) = ax^2 + bx + c$ to contain the given points. This means $f(-1) = 10, f(2) = 4$, and $f(3) = 14$

Since we need to find the three unknown parameters, we need three equations which are offered by the conditions above.

$$f(x) = ax^2 + bx + c$$
$$f(-1) = a - b + c = 10$$
$$f(2) = 4a + 2b + c = 4$$
$$f(3) = 9a + 3b + c = 14$$

This is clearly a system of three equations which can be solved using matrix methods.

Using the reduced row echelon form, we get the following result

$$\begin{pmatrix} 1 & -1 & 1 & | & 10 \\ 4 & 2 & 1 & | & 4 \\ 9 & 3 & 1 & | & 14 \end{pmatrix} \Leftrightarrow \begin{pmatrix} 1 & 0 & 0 & | & 3 \\ 0 & 1 & 0 & | & -5 \\ 0 & 0 & 1 & | & 2 \end{pmatrix}$$

Which means that $a = 3$, $b = -5$, and $c = 2$, so the function is

$$f(x) = 3x^2 - 5x + 2$$

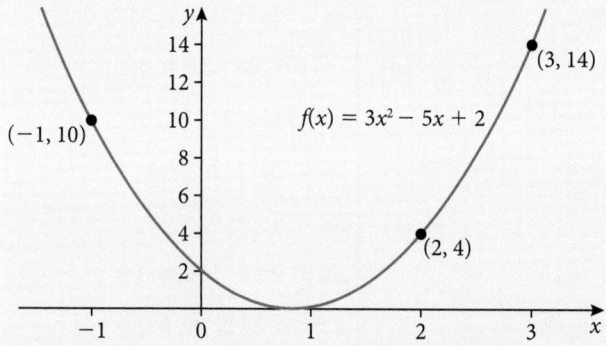

Equivalently, we can use the inverse matrix directly:

$$\begin{pmatrix} 1 & -1 & 1 \\ 4 & 2 & 1 \\ 9 & 3 & 1 \end{pmatrix} \begin{pmatrix} a \\ b \\ c \end{pmatrix} = \begin{pmatrix} 10 \\ 4 \\ 14 \end{pmatrix} \Leftrightarrow \begin{pmatrix} a \\ b \\ c \end{pmatrix} = \begin{pmatrix} 1 & -1 & 1 \\ 4 & 2 & 1 \\ 9 & 3 & 1 \end{pmatrix}^{-1} \begin{pmatrix} 10 \\ 4 \\ 14 \end{pmatrix} = \begin{pmatrix} 3 \\ -5 \\ 2 \end{pmatrix}$$

You can also use a GDC.

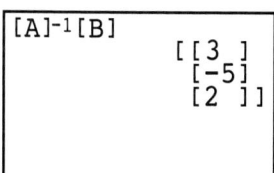

**Figure 7.13** GDC screens for the solution to Example 7.17

# 7 Matrix algebra

## Exercise 7.3

1. Let the matrix $A = \begin{pmatrix} 5 & 6 \\ -1 & 0 \end{pmatrix}$. Find the values of the real numbers $m$ such that $\det(A - mI) = 0$, where $I$ is the $2 \times 2$ multiplication identity matrix.

2. **(a)** Find the values of $a$ and $b$ given that the inverse of the matrix

$$A = \begin{pmatrix} a & -4 & -6 \\ -8 & 5 & 7 \\ -5 & 3 & 4 \end{pmatrix} \text{ is the matrix } B = \begin{pmatrix} 1 & 2 & -2 \\ 3 & b & 1 \\ -1 & 1 & 3 \end{pmatrix}$$

   **(b)** For the values of $a$ and $b$ found in part **(a)**, solve the system of linear equations

$$\begin{cases} x + 2y - 2z = 5 \\ 3x + by + z = 0 \\ -x + y - 3z = a - 1 \end{cases}$$

3. Find the value(s) of $m$ so that the matrix $\begin{pmatrix} 1 & m & 1 \\ 3 & 1-m & 2 \\ m & -3 & m-1 \end{pmatrix}$ is singular.

4. Solve each system of equations. If a solution does not exist, justify why not.

   **(a)** $\begin{cases} 4x - y + z = -5 \\ 2x + 2y + 3z = 10 \\ 5x - 2y + 6z = 1 \end{cases}$
   **(b)** $\begin{cases} 4x - 2y + 3z = -2 \\ 2x + 2y + 5z = 16 \\ 8x - 5y - 2z = 4 \end{cases}$

   **(c)** $\begin{cases} 5x - 3y + 2z = 2 \\ 2x + 2y - 3z = 3 \\ x - 7y + 8z = -4 \end{cases}$
   **(d)** $\begin{cases} 3x - 2y + z = -29 \\ -4x + y - 3z = 37 \\ x - 5y + z = -24 \end{cases}$

   **(e)** $\begin{cases} 2x + 3y + 5z = 4 \\ 3x + 5y + 9z = 7 \\ 5x + 9y + 17z = 13 \end{cases}$
   **(f)** $\begin{cases} 2x + 3y + 5z = 4 \\ 3x + 5y + 9z = 7 \\ 5x + 9y + 17z = 1 \end{cases}$

   **(g)** $\begin{cases} -x + 4y - 2z = 12 \\ 2x - 9y + 5z = -25 \\ -x + 5y - 4z = 10 \end{cases}$
   **(h)** $\begin{cases} x - 3y - 2z = 8 \\ -2x + 7y + 3z = -19 \\ x - y - 3z = 3 \end{cases}$

5. **(a)** Find the values of $k$ such that the matrix $A$ is not singular.

$$A = \begin{pmatrix} 1 & 1 & k-1 \\ k & 0 & -1 \\ 6 & 2 & -3 \end{pmatrix}$$

**(b)** Find the value(s) of $k$ such that $A$ is the inverse of $B$, where

$$B = \begin{pmatrix} k-3 & -3 & k \\ 3 & k+2 & -1 \\ -2 & -4 & 1 \end{pmatrix}$$

**(c)** For the value of $k$ found in part **(b)**, apply elementary row operations to reduce the matrix $\begin{pmatrix} 1 & 1 & k-1 & 1 & 0 & 0 \\ k & 0 & -1 & 0 & 1 & 0 \\ 6 & 2 & -3 & 0 & 0 & 1 \end{pmatrix}$ into

$\begin{pmatrix} 1 & 0 & 0 & a & b & c \\ 0 & 1 & 0 & d & e & f \\ 0 & 0 & 1 & g & h & i \end{pmatrix}$ where $a, b, c, \ldots i$ are to be determined.

**6. (a)** Find the values of $k$ such that the matrix $A$ is not singular.

$$A = \begin{pmatrix} \dfrac{2}{5} & \dfrac{-17}{5} & \dfrac{k+9}{5} \\ \dfrac{-1}{5} & \dfrac{21}{5} & \dfrac{-13}{5} \\ k-2 & 3 & -2 \end{pmatrix}$$

**(b)** Find the value(s) of $k$ such that $A$ is the inverse of $B$, where

$$B = \begin{pmatrix} k+1 & 1 & k \\ 2 & k+2 & -3 \\ 3 & 6 & -5 \end{pmatrix}$$

**(c)** For the value of $k$ found in part **(b)**, apply elementary row operations to reduce the matrix $\left( \begin{array}{ccc|ccc} 2 & -17 & k+9 & 1 & 0 & 0 \\ -1 & 21 & -13 & 0 & 1 & 0 \\ 5(k-2) & 15 & -10 & 0 & 0 & 1 \end{array} \right)$

into $\left( \begin{array}{ccc|ccc} 1 & 0 & 0 & a & b & c \\ 0 & 1 & 0 & d & e & f \\ 0 & 0 & 1 & g & h & i \end{array} \right)$ where $a, b, c, \ldots i$ are to be determined.

**7.** Use elementary row operations to transform the matrix $[A \,|\, I]$ to a matrix in the form $[I \,|\, B]$. Comment on the relationship between $A$ and $B$ and support your conclusion.

**(a)** $\left( \begin{array}{ccc|ccc} 2 & 0 & 3 & 1 & 0 & 0 \\ -1 & 1 & 1 & 0 & 1 & 0 \\ 2 & -2 & 1 & 0 & 0 & 1 \end{array} \right)$

**(b)** $\left( \begin{array}{ccc|ccc} 1 & 4 & 6 & 1 & 0 & 0 \\ 2 & -3 & 1 & 0 & 1 & 0 \\ -1 & 18 & 16 & 0 & 0 & 1 \end{array} \right)$

8. Determine the function $f$ so that the curve representing it contains the indicated points.

There is more than one curve.

(a) $f(x) = ax^2 + bx + c$ to contain $(-1, 5)$, $(2, -1)$, and $(4, 35)$

(b) $f(x) = ax^2 + bx + c$ to contain $(-1, 12)$ and $(2, -3)$

9. Consider this system of equations.
$$\begin{cases} 2x + y + 3z = -5 \\ 3x - y + 4z = 2 \\ 5x + 7z = m - 5 \end{cases}$$
Find the value(s) of $m$ for which this system is consistent. For the value of $m$ found, find the most general solution of the system.

10. Consider this system of equations.
$$\begin{cases} -3x + 2y + 3z = 1 \\ 4x - y - 5z = -5 \\ x + y - 2z = m - 3 \end{cases}$$
Find the value(s) of $m$ for which this system is consistent. For the value of $m$ found, find the most general solution of the system.

11. Consider the matrix $A = \begin{pmatrix} 3 & -4 & -6 \\ -8 & 5 & 7 \\ -5 & 3 & 4 \end{pmatrix}$

(a) Find $\det(A)$.

(b) Add a multiple of one row to another row to transform the matrix $A$ into matrix $B$ in triangular form.

(c) Find $\det(B)$.

(d) Use a GDC to find $\det(C)$ for $C = \begin{pmatrix} 2 & 1 & -3 & 5 \\ 4 & 3 & -4 & -6 \\ 6 & -8 & 5 & 7 \\ -6 & -5 & 3 & 4 \end{pmatrix}$

(e) Repeat parts (b) and (c) for $C$.

## 7.4 Eigenvectors and eigenvalues

Recall that if a matrix has one column, it is called a column vector. We usually denote vectors with an arrow as shown. (You will learn more about vectors in Chapter 9.)

$$\vec{v} = \begin{pmatrix} a \\ b \end{pmatrix} \text{ or } \vec{u} = \begin{pmatrix} a \\ b \\ c \end{pmatrix}$$

If we multiply an $n \times n$ matrix by an $n \times 1$ vector we will get a new $n \times 1$ vector. In other words, $A\vec{v} = \vec{w}$

However, instead of just getting a brand new vector out of the multiplication, we would like to know if it is possible to instead get:

$$A\vec{v} = \lambda\vec{v}$$

In other words, is it possible, for certain $\lambda$ and $\vec{v}$, that matrix multiplication is the same as multiplying the vector by a constant? The answer is yes, it is possible for this to happen, but it won't happen for just any value of $\lambda$ and $\vec{v}$. There is a particular value of $\lambda$ and $\vec{v}$ for which this works. They always occur in pairs and $\lambda$ is an **eigenvalue** of $A$ and $\vec{v}$ is an **eigenvector** of $A$.

Coverage in this course is limited to $2 \times 2$ matrices only.

In order to see how we can develop a method for finding eigenvectors and eigenvalues, we start at the original equation $A\vec{v} = \lambda\vec{v}$

Firstly, we must make sure that $\vec{v} \neq \vec{0}$. If $\vec{v} = \vec{0}$, then $A\vec{v} = \lambda\vec{v}$ will be true for any value of $\lambda$.

Now, rewriting the original equation and simplifying:

$$A\vec{v} = \lambda\vec{v}$$
$$A\vec{v} - \lambda\vec{v} = A\vec{v} - \lambda I\vec{v} = (A - \lambda I)\vec{v} = \vec{0}$$

Note that before we factored out the $\vec{v}$ we added the identity matrix $I$. This is equivalent to multiplying by 1. We needed to do this because without it we would have had the difference of a matrix, $A$, and a constant, $\lambda$, and this can't be done. We now have the difference of two matrices of the same size which can be done.

In order to find the eigenvectors for a matrix we will need to solve a homogeneous system. If a system is written in matrix form as you recall from section 7.3, we will either have the trivial solution, $\vec{v} = \vec{0}$, or we will have an infinite number of solutions.

A homogeneous solution can have an infinite number of solutions if the coefficient matrix is singular.

So, to solve the equation $(A - \lambda I)\vec{v} = \vec{0}$, the matrix $A - \lambda I$ must be singular, and hence $\det(A - \lambda I) = 0$. This is called the **characteristic equation**. $\det(A - \lambda I)$ is called the **characteristic polynomial**.

## Example 7.18

Find the eigenvalues and eigenvectors of the matrix $A = \begin{pmatrix} 2 & 7 \\ -1 & -6 \end{pmatrix}$

### Solution

The first step is to find the eigenvalues.

$$A - \lambda I = \begin{pmatrix} 2 & 7 \\ -1 & -6 \end{pmatrix} - \lambda\begin{pmatrix} 1 & 0 \\ 0 & 1 \end{pmatrix} = \begin{pmatrix} 2 - \lambda & 7 \\ -1 & -6 - \lambda \end{pmatrix}$$

In particular we need to determine where the value(s) of $\lambda$ for which the determinant of this matrix is zero.

$$\det(A - \lambda I) = (2 - \lambda)(-6 - \lambda) + 7 = \lambda^2 + 4\lambda - 5$$

Thus, the characteristic polynomial is $\lambda^2 + 4\lambda - 5$

Now, finding the zeros of this polynomial will give us the eigenvalues

$$\lambda^2 + 4\lambda - 5 = (\lambda + 5)(\lambda - 1) = 0$$
$$\Rightarrow \lambda_1 = -5, \lambda_2 = 1$$

Each of these eigenvalues will generate an eigenvector.

For $\lambda_1 = -5$:

$$(A - \lambda I)\vec{v} = \vec{0} \Rightarrow \begin{pmatrix} 7 & 7 \\ -1 & -1 \end{pmatrix} \vec{v} = \vec{0}$$

Let $\vec{v} = \begin{pmatrix} a \\ b \end{pmatrix} \Rightarrow \begin{pmatrix} 7 & 7 \\ -1 & -1 \end{pmatrix}\begin{pmatrix} a \\ b \end{pmatrix} = \begin{pmatrix} 0 \\ 0 \end{pmatrix} \Rightarrow \begin{cases} 7a + 7b = 0 \\ -a - b = 0 \end{cases}$

We can either use simple addition of 6 times the second equation and add it to the first, or use Gauss-Jordan method:

$$a + b = 0 \Rightarrow b = -a \Rightarrow \vec{v} = \begin{pmatrix} a \\ -a \end{pmatrix}$$

This is a general eigenvector, but in fact we usually need specific vectors, and hence any value for $a$ will suffice, say $a = 1$ here. Therefore $\vec{v} = \begin{pmatrix} 1 \\ -1 \end{pmatrix}$ is a first eigenvector. Note that for any other choice of $a$, the resulting vector will be parallel to this one.

For $\lambda_2 = 1$:

$$(A - \lambda I)\vec{v} = \vec{0} \Rightarrow \begin{pmatrix} 1 & -7 \\ -1 & -7 \end{pmatrix} \vec{v} = \vec{0}$$

$$\Rightarrow \begin{pmatrix} 1 & 7 \\ -1 & -7 \end{pmatrix}\begin{pmatrix} a \\ b \end{pmatrix} = \begin{pmatrix} 0 \\ 0 \end{pmatrix} \Rightarrow \begin{cases} a + 7b = 0 \\ -a - 7b = 0 \end{cases}$$

Again, this will lead to

$$a + 7b = 0 \Rightarrow a = -7b \Rightarrow \vec{v} = \begin{pmatrix} -7b \\ b \end{pmatrix}$$

Using $b = 1$ (or any other value, not zero, of your choice) we have

$$\vec{v} = \begin{pmatrix} -7 \\ 1 \end{pmatrix} \text{ as a second eigenvector.}$$

Note that $\begin{pmatrix} 2 & 7 \\ -1 & -6 \end{pmatrix}\begin{pmatrix} -1 \\ 1 \end{pmatrix} = \begin{pmatrix} 5 \\ -5 \end{pmatrix} = -5\begin{pmatrix} -1 \\ 1 \end{pmatrix}$ as required, and

$$\begin{pmatrix} -2 & 7 \\ -1 & -6 \end{pmatrix}\begin{pmatrix} -7 \\ 1 \end{pmatrix} = \begin{pmatrix} -7 \\ 1 \end{pmatrix} = 1\begin{pmatrix} -7 \\ 1 \end{pmatrix} \text{ as required.}$$

If $A$ is a square matrix, and if $\vec{v}$ is an eigenvector, then any scalar multiple $k\vec{v}$ is also an eigenvector.

If $A$ is a square matrix and $\vec{v}$ and $\vec{w}$ are eigenvectors, then $\vec{u} + \vec{w}$ is an eigenvector.

If $A$ is a square matrix, and if $\lambda$ is an eigenvalue and $\vec{v}$ is a corresponding eigenvector, then $\lambda^n$ is an eigenvalue and $\vec{v}$ is an eigenvector of $A^n$.

## Example 7.19

Find the eigenvalues and eigenvectors of the matrix $A = \begin{pmatrix} 1 & -1 \\ 2 & 4 \end{pmatrix}$ and the matrix $A^4$.

### Solution

The first step is to find the eigenvalues of $A$.

$$A - \lambda I = \begin{pmatrix} 1 & -1 \\ 2 & 4 \end{pmatrix} - \lambda \begin{pmatrix} 1 & 0 \\ 0 & 1 \end{pmatrix} = \begin{pmatrix} 1 - \lambda & -1 \\ 2 & 4 - \lambda \end{pmatrix}$$

The determinant of this matrix must be zero.

$$\det \begin{pmatrix} 1 - \lambda & -1 \\ 2 & 4 - \lambda \end{pmatrix} = 0$$

$$\Rightarrow \lambda^2 - 5\lambda + 6 = 0$$

$$\Rightarrow \lambda_1 = 2, \lambda_2 = 3$$

For $\lambda_1 = 2$:

$$(A - \lambda I)\vec{v} = \vec{0} \Rightarrow \begin{pmatrix} -1 & -1 \\ 2 & 2 \end{pmatrix}\vec{v} = \vec{0}$$

$$\Rightarrow \begin{pmatrix} -1 & -1 \\ 2 & 2 \end{pmatrix}\begin{pmatrix} a \\ b \end{pmatrix} = \begin{pmatrix} 0 \\ 0 \end{pmatrix} \Rightarrow \begin{cases} -a - b = 0 \\ 2a + 2b = 0 \end{cases}$$

Solving for $a$ and $b$ will yield an infinite number of solutions such that $b = -a$

Thus, any vector of the form $\vec{v} = \begin{pmatrix} t \\ -t \end{pmatrix}$ is an eigenvector. Just for demonstration purposes, we have:

$$A\vec{v} = \begin{pmatrix} 1 & -1 \\ 2 & 4 \end{pmatrix}\begin{pmatrix} t \\ -t \end{pmatrix} = \begin{pmatrix} 2t \\ -2t \end{pmatrix} = 2\begin{pmatrix} t \\ -t \end{pmatrix}$$

For $\lambda_2 = 3$:

$$(A - \lambda I)\vec{v} = \vec{0} \Rightarrow \begin{pmatrix} -2 & -1 \\ 2 & 1 \end{pmatrix}\vec{v} = \vec{0}$$

$$\Rightarrow \begin{pmatrix} -2 & -1 \\ 2 & 1 \end{pmatrix}\begin{pmatrix} a \\ b \end{pmatrix} = \begin{pmatrix} 0 \\ 0 \end{pmatrix} \Rightarrow \begin{cases} -2a - b = 0 \\ 2a + b = 0 \end{cases}$$

Solving for $a$ and $b$ will yield an infinite number of solutions such that $b = -2a$.

Thus, anyvector of the form $\vec{v} = \begin{pmatrix} t \\ -2t \end{pmatrix}$ is an eigenvector.

Also, for demonstration purposes, we have:

$$A\vec{v} = \begin{pmatrix} 1 & -1 \\ 2 & -4 \end{pmatrix} \begin{pmatrix} t \\ -2t \end{pmatrix} = \begin{pmatrix} 3t \\ -6t \end{pmatrix} = 3 \begin{pmatrix} t \\ -2t \end{pmatrix}$$

For $A^4$ the eigenvalues are $2^4$ and $3^4$ with the same eigenvectors as before.

$$A^4 = \begin{pmatrix} -49 & -65 \\ 130 & 146 \end{pmatrix} \text{ and } \begin{pmatrix} -49 & -65 \\ 130 & 146 \end{pmatrix} \begin{pmatrix} t \\ -t \end{pmatrix} = 16 \begin{pmatrix} t \\ -t \end{pmatrix}$$

$$\text{and } \begin{pmatrix} -49 & -65 \\ 130 & 146 \end{pmatrix} \begin{pmatrix} t \\ -2t \end{pmatrix} = \begin{pmatrix} 81t \\ -2t \end{pmatrix}$$

### Example 7.20

Find the eigenvalues and eigenvectors of the matrix $A = \begin{pmatrix} 1 & 1 \\ 6 & 2 \end{pmatrix}$

### Solution

$$A - \lambda I = \begin{pmatrix} 1 & 1 \\ 6 & 2 \end{pmatrix} - \lambda \begin{pmatrix} 1 & 0 \\ 0 & 1 \end{pmatrix} = \begin{pmatrix} 1 - \lambda & 1 \\ 6 & 2 - \lambda \end{pmatrix}$$

$$\det(A - \lambda I) = (1 - \lambda)(2 - \lambda) - 6 = \lambda^2 - 3\lambda - 4 = 0$$

$$\Rightarrow \lambda_1 = -1, \lambda_2 = 4$$

For $\lambda_1 = -1$:

$$(A - \lambda I)\vec{v} = \vec{0} \Rightarrow \begin{pmatrix} 2 & 1 \\ 6 & 3 \end{pmatrix} \vec{v} = \vec{0}$$

$$\Rightarrow \begin{pmatrix} 2 & 1 \\ 6 & 3 \end{pmatrix} \begin{pmatrix} a \\ b \end{pmatrix} = \begin{pmatrix} 0 \\ 0 \end{pmatrix} \Rightarrow \begin{cases} 2a + b = 0 \\ 6a + 3b = 0 \end{cases}$$

Solving for $a$ and $b$ will yield an infinite number of solutions such that $b = -2a$

Thus, any vector of the form $\vec{v} = \begin{pmatrix} t \\ -2t \end{pmatrix}$ is an eigenvector. Also, for demonstration purposes, let us choose $t = 1$

$$A\vec{v} = \begin{pmatrix} 1 & 1 \\ 6 & 2 \end{pmatrix} \begin{pmatrix} 1 \\ -2 \end{pmatrix} = \begin{pmatrix} -1 \\ 2 \end{pmatrix} = (-1) \begin{pmatrix} 1 \\ -2 \end{pmatrix}$$

For $\lambda_2 = 4$:

$$(A - \lambda I)\vec{v} = \vec{0} \Rightarrow \begin{pmatrix} -3 & 1 \\ 6 & -2 \end{pmatrix}\vec{v} = \vec{0}$$

$$\Rightarrow \begin{pmatrix} -3 & 1 \\ 6 & -2 \end{pmatrix}\begin{pmatrix} a \\ b \end{pmatrix} = \begin{pmatrix} 0 \\ 0 \end{pmatrix} \Rightarrow \begin{cases} -3a + b = 0 \\ 6a - 2b = 0 \end{cases}$$

Solving for $a$ and $b$ will yield an infinite number of solutions such that $b = 3a$

Thus, any vector of the form $\vec{v} = \begin{pmatrix} t \\ 3t \end{pmatrix}$ is an eigenvector. Also, for demonstration purposes, let us choose $t = 1$

$$A\vec{v} = \begin{pmatrix} 1 & 1 \\ 6 & 2 \end{pmatrix}\begin{pmatrix} 1 \\ 3 \end{pmatrix} = \begin{pmatrix} 4 \\ 12 \end{pmatrix} = 4\begin{pmatrix} 1 \\ 3 \end{pmatrix}$$

To find eigenvalues and eigenvectors of a square matrix $A$:

form the matrix $A - \lambda I$

- solve the equation $\det(A - \lambda I) = 0$; the real solutions are the eigenvalues of $A$
- for each eigenvalue $\lambda_0$, form the matrix $A - \lambda_0 I$
- solve the homogeneous system $(A - \lambda_0 I)\vec{X} = \vec{0}$

When the characteristic equation gives distinct real eigenvalues, the corresponding eigenvectors are said to be linearly independent.

The proof is straightforward.

$$\text{Let } A = \begin{pmatrix} a & b \\ 0 & c \end{pmatrix}$$

**Eigenvalues of triangular matrices**
The eigenvalues of a triangular matrix are its diagonal entries.

Then, $\det(A - \lambda I) = \det\begin{pmatrix} a - \lambda & b \\ 0 & c - \lambda \end{pmatrix} = 0 \Rightarrow (a - \lambda)(c - \lambda) = 0 \Rightarrow \lambda = a$
or $\lambda = c$

## Diagonalisation

Why is this useful? Suppose you wanted to find $A^3$. If $A$ can be diagonalised, then:

$$A^3 = (PDP^{-1})^3 = (PDP^{-1})(PDP^{-1})(PDP^{-1})$$

$$= PDP^{-1}PDP^{-1}PDP^{-1} = PD(P^{-1}P)D(P^{-1}P)DP^{-1}$$

But $P^{-1}P = I$, so,

$$A^3 = PD(P^{-1}P)D(P^{-1}P)DP^{-1} = PDDDP^{-1} = PD^3P^{-1}$$

A square matrix $A$ can be diagonalised if it has the property that there exists a matrix $P$ that has an inverse and a diagonal matrix $D$ such that $A = PDP^{-1}$.

In general, when $A = PDP^{-1}$, then $A^n = PD^nP^{-1}$

The connection between eigenvectors, eigenvalues and diagonalisation is given by the diagonalistion theorem which states that an $n \times n$ matrix $A$ is diagonalisable if and only if $A$ has $n$ linearly independent eigenvectors.

If $v_1, v_2, \cdots, v_n$ are linearly independent eigenvectors of $A$ and $\lambda_1, \lambda_2, \cdots, \lambda_n$ are their corresponding eigenvalues, then $A = PDP^{-1}$, where

$P = (v_1, v_2, \cdots, v_n)$, i.e. $P$ is the matrix whose columns are the eigenvectors, and

$$D = \begin{pmatrix} \lambda_1 & 0 & 0 & 0 \\ 0 & \lambda_2 & 0 & 0 \\ 0 & 0 & \ddots & 0 \\ 0 & 0 & 0 & \lambda_n \end{pmatrix}$$

## Example 7.21

Diagonalise the matrix $A = \begin{pmatrix} 1 & 1 \\ 6 & 2 \end{pmatrix}$

### Solution

This is the same matrix as in Example 7.20, so the eigenvalues are $-1$ and 4. The eigenvectors are

$$\vec{v_1} = \begin{pmatrix} 1 \\ -2 \end{pmatrix} \text{ and } \vec{v}_2 = \begin{pmatrix} 1 \\ 3 \end{pmatrix}, \text{ and so,}$$

$$P = \begin{pmatrix} 1 & 1 \\ -2 & 3 \end{pmatrix} \text{ and } D = \begin{pmatrix} -1 & 0 \\ 0 & 4 \end{pmatrix}$$

Therefore

$$A = \begin{pmatrix} 1 & 1 \\ 6 & 2 \end{pmatrix} = \begin{pmatrix} 1 & 1 \\ -2 & 3 \end{pmatrix} \begin{pmatrix} -1 & 0 \\ 0 & 4 \end{pmatrix} \begin{pmatrix} 1 & 1 \\ -2 & 3 \end{pmatrix}^{-1}$$

To verify that the equation above holds, we perform the multiplication on the right-side of the equation:

$$\begin{pmatrix} 1 & 1 \\ -2 & 3 \end{pmatrix} \begin{pmatrix} -1 & 0 \\ 0 & 4 \end{pmatrix} \begin{pmatrix} 1 & 1 \\ -2 & 3 \end{pmatrix}^{-1} = \begin{pmatrix} -1 & 4 \\ 2 & 12 \end{pmatrix} \begin{pmatrix} \frac{3}{5} & \frac{-1}{5} \\ \frac{2}{5} & \frac{1}{5} \end{pmatrix} = \begin{pmatrix} 1 & 1 \\ 6 & 2 \end{pmatrix}$$

```
Mat Amat B(Mat A)⁻¹
              [1 1]
              [6 2]
```

**Figure 7.14** GDC screen for the solution to Example 7.21

## Example 7.22

Given that $A = \begin{pmatrix} 1 & 1 \\ 6 & 2 \end{pmatrix}$, calculate $A^4$

### Solution

From Example 7.20:

$$A = \begin{pmatrix} 1 & 1 \\ 6 & 2 \end{pmatrix} = \begin{pmatrix} 1 & 1 \\ -2 & 3 \end{pmatrix} \begin{pmatrix} -1 & 0 \\ 0 & 4 \end{pmatrix} \begin{pmatrix} 1 & 1 \\ -2 & 3 \end{pmatrix}^{-1}$$

So

$$A^4 = \begin{pmatrix} 1 & 1 \\ -2 & 3 \end{pmatrix} \begin{pmatrix} -1 & 0 \\ 0 & 4 \end{pmatrix}^4 \begin{pmatrix} 1 & 1 \\ -2 & 3 \end{pmatrix}^{-1}$$

$$= \begin{pmatrix} 1 & 1 \\ -2 & 3 \end{pmatrix} \begin{pmatrix} (-1)^4 & 0 \\ 0 & 4^4 \end{pmatrix} \begin{pmatrix} \dfrac{3}{5} & \dfrac{-1}{5} \\ \dfrac{2}{5} & \dfrac{1}{5} \end{pmatrix}$$

$$= \begin{pmatrix} 1 & 1 \\ -2 & 3 \end{pmatrix} \begin{pmatrix} 1 & 0 \\ 0 & 256 \end{pmatrix} \begin{pmatrix} \dfrac{3}{5} & \dfrac{-1}{5} \\ \dfrac{2}{5} & \dfrac{1}{5} \end{pmatrix} = \begin{pmatrix} 1 & 256 \\ -2 & 768 \end{pmatrix} \begin{pmatrix} \dfrac{3}{5} & \dfrac{-1}{5} \\ \dfrac{2}{5} & \dfrac{1}{5} \end{pmatrix}$$

$$= \begin{pmatrix} 103 & 51 \\ 306 & 154 \end{pmatrix}$$

## Example 7.23

Given that $A^t$ is the transpose of a matrix $A$, prove that $A$ and $A^t$ have the same eigenvalues.

### Solution

Recall that the eigenvalues of $A$ are the solutions to its characteristic equation

$$\det (A - \lambda I) = 0$$

In order to find the eigenvalues for $A^t$, we find its characteristic equation. However, remembering that a matrix and its transpose have equal determinants, we have

$$\det (A - \lambda I) = \det (A - \lambda I)^t = \det (A^t - \lambda I^t) = \det (A^t - \lambda I)$$

Which means that both $A$ and $A^t$ have the same characteristic polynomials. Therefore, $A$ and $A^t$ have the same eigenvalues.

## Example 7.24

Let $A$ be the matrix $A = \begin{pmatrix} 1 & 1 & -4 \\ 2 & 0 & -4 \\ -1 & 1 & -2 \end{pmatrix}$

Find a diagonalisation of $A$.

## Solution

The characteristic equation of $A$ is

$$\det(A - \lambda I) = \det\begin{pmatrix} 1 - \lambda & 1 & -4 \\ 2 & -\lambda & -4 \\ -1 & 1 & -2 - \lambda \end{pmatrix}$$

$$= -(\lambda + 1)(\lambda + 2)(\lambda - 2) = 0$$

And so, the eigenvalues of $A$ are $\lambda_1 = -1, \lambda_2 = -2, \lambda_3 = 2$

We use the Gauss-Jordan method to find the eigenvectors:

$$\vec{v}_1 = \begin{pmatrix} 1 \\ 2 \\ 1 \end{pmatrix}, \vec{v}_2 = \begin{pmatrix} 1 \\ 1 \\ 1 \end{pmatrix}, \vec{v}_3 = \begin{pmatrix} 1 \\ 1 \\ 0 \end{pmatrix}$$

Thus, $D = \begin{pmatrix} -1 & 0 & 0 \\ 0 & -2 & 0 \\ 0 & 0 & 2 \end{pmatrix}$ and $P = \begin{pmatrix} 1 & 1 & 1 \\ 2 & 1 & 1 \\ 1 & 1 & 0 \end{pmatrix} \Rightarrow P^{-1} = \begin{pmatrix} -1 & 1 & 0 \\ 1 & -1 & 1 \\ 1 & 0 & -1 \end{pmatrix}$

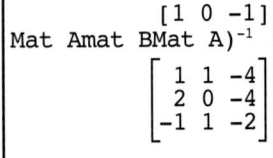

We have the factorisation:

$$\begin{pmatrix} 1 & 1 & -4 \\ 2 & 0 & -4 \\ -1 & 1 & -2 \end{pmatrix} = \begin{pmatrix} 1 & 1 & 1 \\ 2 & 1 & 1 \\ 1 & 1 & 0 \end{pmatrix}\begin{pmatrix} -1 & 0 & 0 \\ 0 & -2 & 0 \\ 0 & 0 & 2 \end{pmatrix}\begin{pmatrix} -1 & 1 & 0 \\ 1 & -1 & 1 \\ 1 & 0 & -1 \end{pmatrix}$$

You can use a GDC.

**Figure 7.15** GDC screen for the solution to Example 7.24

To diagonalise a square matrix $A$:

- form the matrix $A - \lambda I$ and find the eigenvalues and eigenvectors as before
- use the eigenvectors found as column vectors to form matrix $P$
- check that $P$ is non-singular (has an inverse).
- form the diagonal matrix $D$ where the diagonal entries are the eigenvalues found earlier.
  Now, $A = PDP^{-1}$ (equivalently, $D = P^{-1}AP$)

## Markov chains

This is an introduction to Markov chains. More details will be given in Chapter 15.

### Example 7.25

In a rural town, a survey indicated that 85% of the children of university educated parents went to university, while only 35% of the children of parents that do not have university education went to university. What percentage of the second generation went to university if initially 30% of the town parents were university educated?

## Solution

The situation described is a part of a sequence of experiments in which the outcomes and their associated chances depend on the outcomes of the preceding experiments. So, for this case, the percentage of university-oriented children, depends on whether their parents went to university, and for those in turn, their parents and so on.

Such a process is called a **Markov process**. The outcome of any experiment is called the **state** of the experiment.

A tree diagram helps to show the situation. Let C stand for university educated person and N for not-university educated.

The first generation has 30% university educated, and hence 70% are not. In order to find the percentages for the third generation, we first establish the state of the second generation.

The university educated children, C, in this generation could come from university educated parents $0.30 \times 0.85$, or from not-university educated parents $0.70 \times 0.35$

The proportion of university educated children in the next generation is:
$0.30 \times 0.85 + 0.70 \times 0.35 = 0.50$

The children without a university education, N, in this generation could come from university educated parents $0.30 \times 0.15$, or from not-university educated parents $0.70 \times 0.65$

The proportion of children without a university education for the next generation is: $0.30 \times 0.15 + 0.70 \times 0.65 = 0.50$.

For the third generation we will have to expand every branch with two new ones, i.e. eight separate possible outcomes! You can imagine how the tree will look if you want more generations.

It is very useful to describe the Markov process by a **transition matrix**. The transition matrix gives the proportion going from one state to another. For example, in this question.

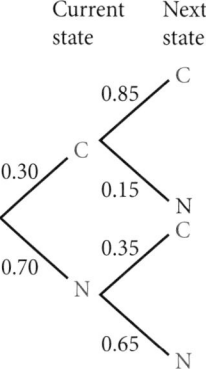

**Figure 7.16** Tree diagram for the solution to Example 7.25

Current state

$$\begin{array}{cc} \phantom{x}C & \phantom{x}N \end{array}$$
$$\begin{pmatrix} 0.85 & 0.35 \\ 0.15 & 0.65 \end{pmatrix} \begin{array}{l} C \\ N \end{array} \text{Next state}$$

The initial state is

$$\begin{pmatrix} x_0 \\ y_0 \end{pmatrix} = \begin{pmatrix} 0.30 \\ 0.70 \end{pmatrix}$$

And thus, the next generation is

$$\begin{pmatrix} 0.85 & 0.35 \\ 0.15 & 0.65 \end{pmatrix} \begin{pmatrix} 0.30 \\ 0.70 \end{pmatrix} = \begin{pmatrix} 0.85 \times 0.30 + 0.35 \times 0.70 \\ 0.15 \times 0.30 + 0.65 \times 0.70 \end{pmatrix} = \begin{pmatrix} 0.50 \\ 0.50 \end{pmatrix}$$

7      Matrix algebra

For the second generation, we just multiply the transition matrix by this result

$$\begin{pmatrix} 0.85 & 0.35 \\ 0.15 & 0.65 \end{pmatrix} \begin{pmatrix} 0.50 \\ 0.50 \end{pmatrix} = \begin{pmatrix} 0.85 \times 0.50 + 0.35 \times 0.50 \\ 0.15 \times 0.50 + 0.65 \times 0.50 \end{pmatrix} = \begin{pmatrix} 0.60 \\ 0.40 \end{pmatrix}$$

Or, put differently

$$\begin{pmatrix} 0.85 & 0.35 \\ 0.15 & 0.65 \end{pmatrix}^2 \begin{pmatrix} 0.30 \\ 0.70 \end{pmatrix} = \begin{pmatrix} 0.50 \\ 0.50 \end{pmatrix}$$

We can summarise the steps as follows.

In a two-state Markov process, we let the proportions of moving from one state to the other be given by a transition matrix

$$T = T_0 \qquad \overset{\text{From}}{\begin{pmatrix} p_{1,1} & p_{2,1} \\ p_{1,2} & p_{2,2} \end{pmatrix}}$$

where $p_{i,j}$ represents the proportion moving from state $i$ to state $j$. (from C to N or from N to N….
in the example). Note that the sum of the entries in each column must be 1.
We also let the initial state be represented by

$$X_0 = \begin{pmatrix} x_0 \\ y_0 \end{pmatrix}, \text{ then the state after } n \text{ experiments will have a state represented by}$$

$$X_n = T^n X_0$$

For more state cases, the matrices will be of an appropriate order. For example, in a 3-state process

$$T = \begin{pmatrix} p_{1,1} & p_{2,1} & p_{3,1} \\ p_{1,2} & p_{2,2} & p_{3,2} \\ p_{1,3} & p_{2,3} & p_{3,3} \end{pmatrix} \text{ and } X_0 = \begin{pmatrix} x_0 \\ y_0 \\ z_0 \end{pmatrix}$$

This process can also be done with horizontal vectors. That is

$$X_n = X_0 T^n = (x_0, y_0, z_0) \begin{pmatrix} p_{1,1} & p_{1,2} & p_{1,3} \\ p_{2,1} & p_{2,2} & p_{2,3} \\ p_{3,1} & p_{3,2} & p_{3,3} \end{pmatrix}^n$$

This means that the sum of entries in each row must be 1.

## Example 7.26

Each year 3% of the population living in a certain city will move to the suburbs and 6% of the population living in the suburbs will move into the city. At present 975 000 people live in the city itself, and 525 000 live in the suburbs. Assuming that the total population of the area does not change, find the distribution of the population:

(a)  1 year from now

(b)  10 years from now.

## Solution

(a) The transition matrix is:

Current state

$$
\begin{array}{cc}
C & S
\end{array}
$$
$$
\begin{pmatrix} 0.97 & 0.06 \\ 0.03 & 0.94 \end{pmatrix} \begin{array}{l} C \\ S \end{array} \text{Next state}
$$

The initial state is:

$$
\begin{pmatrix} x_0 \\ y_0 \end{pmatrix} = \begin{pmatrix} 975\,000 \\ 525\,000 \end{pmatrix}
$$

One year from now, is one stage above the initial one

$$
X_1 = TX_0 \Rightarrow \begin{pmatrix} 0.97 & 0.06 \\ 0.03 & 0.94 \end{pmatrix} \begin{pmatrix} 975\,000 \\ 525\,000 \end{pmatrix} = \begin{pmatrix} 977\,250 \\ 522\,750 \end{pmatrix}
$$

(b) 10 years from now:

$$
X_{10} = TX_0 \Rightarrow \begin{pmatrix} 0.97 & 0.06 \\ 0.03 & 0.94 \end{pmatrix}^{10} \begin{pmatrix} 975\,000 \\ 525\,000 \end{pmatrix} \approx \begin{pmatrix} 990\,264 \\ 509\,736 \end{pmatrix}
$$

## Example 7.27

A taxi company divides the city into three zones: I, II, and III. The company determined from previous records that 60% of passengers picked up in zone I stay in zone I, 30% go to zone II and 10% to zone III. Of those picked up in zone II, 40% go to zone I, 30% to zone II, and 30% to zone III. Of those picked up in zone III, 30% go to zone I, 30% to zone II, and 40% to zone III.

At the beginning of the day, 80% of the taxis are in zone I, 15% are in zone II, and 5% are in zone III.

(a) What is the distribution of the taxis after they have each made one trip?

(b) On average, all taxis make 20 trips per day. What is the distribution by the end of the day?

## Solution

$$
T = \begin{pmatrix} 0.6 & 0.4 & 0.3 \\ 0.3 & 0.3 & 0.3 \\ 0.1 & 0.3 & 0.4 \end{pmatrix}, X_0 = \begin{pmatrix} 0.80 \\ 0.15 \\ 0.05 \end{pmatrix}
$$

(a) $X_1 = TX_0 = \begin{pmatrix} 0.6 & 0.4 & 0.3 \\ 0.3 & 0.3 & 0.3 \\ 0.1 & 0.3 & 0.4 \end{pmatrix} \begin{pmatrix} 0.80 \\ 0.15 \\ 0.05 \end{pmatrix} = \begin{pmatrix} 0.555 \\ 0.300 \\ 0.145 \end{pmatrix}$

(b) $X_{10} = T^{10}X_0 = \begin{pmatrix} 0.6 & 0.4 & 0.3 \\ 0.3 & 0.3 & 0.3 \\ 0.1 & 0.3 & 0.4 \end{pmatrix}^{10} \begin{pmatrix} 0.80 \\ 0.15 \\ 0.05 \end{pmatrix} \approx \begin{pmatrix} 0.471 \\ 0.300 \\ 0.229 \end{pmatrix}$

Eigenvalues and eigenvectors can be used to evaluate the powers of transition matrices. The approach is similar to what we did with diagonalising matrices earlier. In Example 7.26, we can diagonalise the transition matrix

$$\begin{pmatrix} 0.97 & 0.06 \\ 0.03 & 0.94 \end{pmatrix} = \begin{pmatrix} 2 & 1 \\ 1 & -1 \end{pmatrix} \begin{pmatrix} 1 & 0 \\ 0 & 0.91 \end{pmatrix} \frac{1}{3} \begin{pmatrix} 1 & 1 \\ 1 & -2 \end{pmatrix}$$, which will enable us to find its powers more easily.

## Exercise 7.4

1. For each matrix:
   **(i)** find the characteristic polynomial
   **(ii)** find the eigenvalues and eigenvectors
   **(iii)** diagonalise the matrix.

   **(a)** $A = \begin{pmatrix} 3 & -1 \\ 2 & 0 \end{pmatrix}$
   **(b)** $A = \begin{pmatrix} 1 & 4 \\ 2 & -1 \end{pmatrix}$
   **(c)** $A = \begin{pmatrix} 2 & 0 \\ -1 & 3 \end{pmatrix}$

   **(d)** $A = \begin{pmatrix} 0 & -1 \\ -3 & 2 \end{pmatrix}$
   **(e)** $A = \begin{pmatrix} 1 & 2 \\ 2 & 1 \end{pmatrix}$
   **(f)** $A = \begin{pmatrix} 1 & 0 \\ 0 & 2 \end{pmatrix}$

   **(g)** $A = \begin{pmatrix} 2 & 3 \\ 1 & 4 \end{pmatrix}$
   **(h)** $A = \begin{pmatrix} 7 & -1 \\ 6 & 2 \end{pmatrix}$
   **(i)** $A = \begin{pmatrix} 4 & 9 \\ 2 & 7 \end{pmatrix}$

   **(j)** $A = \begin{pmatrix} 1 & 1 \\ 1 & 0 \end{pmatrix}$
   **(k)** $A = \begin{pmatrix} a & 0 \\ 0 & b \end{pmatrix}$
   **(l)** $A = \begin{pmatrix} 3 & 2 \\ 2 & 3 \end{pmatrix}$

   **(m)** $A = \begin{pmatrix} 0 & a^2 \\ b^2 & 0 \end{pmatrix}$
   **(n)** $A = \begin{pmatrix} 1 & 3 \\ 4 & 2 \end{pmatrix}$
   **(o)** $A = \begin{pmatrix} 3 & 7 \\ 5 & 10 \end{pmatrix}$

   **(p)** $A = \begin{pmatrix} 15 & -8 \\ 18 & -15 \end{pmatrix}$
   **(q)** $A = \begin{pmatrix} -5 & 4 \\ -8 & 7 \end{pmatrix}$
   **(r)** $A = \begin{pmatrix} 8 & 6 \\ -15 & -11 \end{pmatrix}$

2. The transition matrix for a Markov process is given by

   $$\begin{array}{cc} \text{State} & \\ \begin{array}{cc} 1 & 2 \end{array} & \\ \begin{pmatrix} 0.3 & 0.6 \\ 0.7 & 0.4 \end{pmatrix} & \begin{array}{c} \text{State 1} \\ \text{State 2} \end{array} \end{array}$$

   **(a)** What does the entry 0.3 represent?
   **(b)** The initial-state distribution vector is given by

   $$X_0 = \begin{array}{c} \text{State 1} \\ \text{State 2} \end{array} \begin{pmatrix} 0.4 \\ 0.6 \end{pmatrix}$$

   Find the distribution of the system after one observation.

3. Kevin is either happy or sad. If he is happy one day, then he is happy the next day four times out of five. If he is sad one day, then he is sad the next day one time out of three. Over the long term, what are the chances that Kevin is happy on any given day?

4. Three grocery chains serve a large area in a certain country. During the year, grocery A expects to retain 80% of its customers, 5% are lost to grocery B, and 15% to grocery C. Grocery B expects to retain 90% of its customers, and loses 5% to each of the other two groceries. Grocery C expects to retain 75% of its customers, and loses 10% to grocery A and 15% to grocery B.

   (a) Construct the transition matrix for the Markov chain that describes the change in the market share.

   (b) Currently the market share is 0.4 for grocery A, 0.3 for grocery B and 0.3 for grocery C. What share of the market is held by each grocery after 1 year?

   (c) Assuming the trend continues, what share does each grocery hold after 2 years?

5. TG Polling conducted a poll 6 months before elections in a country in which a liberal and a conservative were running for president. TG found that 60% of the voters intended to vote for the conservative and 40% for the liberal. In a poll conducted 3 months later, TG found that 70% of those who had earlier stated a preference for the conservative candidate still maintained that preference, whereas 30% of those voters now preferred the liberal candidate. Of those who earlier preferred the liberal, 80% still maintained their preference, whereas 20% switched to the conservative.

   (a) If elections were held at this time (after 3 months), who would win?

   (b) If the trend continues, which candidate will win the election?

6. Three truck manufacturers A, B, and C share the domestic market in a certain country. Their current market shares are 60%, 30% and 10% respectively. Market studies show that manufacturer A retains 75% of its customers, and loses 15% to manufacturer B and 10% to manufacturer C. Of the customers who buy from manufacturer B, 90% would keep their preference, while 5% go to each of manufacturers A and C. Of the customers who buy from manufacturer C, 85% are retained, while 5% would buy from manufacturer A and 10% from manufacturer B.

   (a) Assuming that these sentiments reflect the buying habits of customers in the future, determine the market share that will be held by each manufacturer after 2 years.

   (b) Under the same conditions, determine the market share that will be held by each manufacturer after 5 years.

7. By reviewing its donation records, the alumni office of a university finds that 80% of its alumni who contribute to the annual fund one year will also contribute the next year, and 30% of those who do not contribute one year will contribute the next. Consider a new graduate who did not give a donation in the initial year after graduation.

(a) Construct the probable future donation record for three years of such new graduates who did not give a donation in the initial year after graduation.

(b) Consider the situation on the 11th year and conjecture a pattern for the long term.

8. A car rental agency has three rental locations, Zurich ($Z$), Geneva ($G$), and Basel ($B$). A customer may rent a car from any of the three locations and return the car to any of the three locations. The manager finds that customers return the cars to the various locations according to the following probabilities:

Rented from location

$$\begin{array}{ccc} Z & G & B \end{array}$$
$$\begin{pmatrix} 0.8 & 0.3 & 0.2 \\ 0.1 & 0.2 & 0.6 \\ 0.1 & 0.5 & 0.2 \end{pmatrix} \begin{array}{l} Z \\ G \\ B \end{array} \begin{array}{l} \text{Returned} \\ \text{to} \\ \text{location} \end{array}$$

Previous records show that 40% of the fleet are rented in Geneva, 35% in Zurich, and 25% in Basel. Find the long-term trend in terms of percentage of cars present at each location.

9. 200 000 people live in a certain city and 25 000 people live in its suburbs. The Regional Planning Commission determines that each year 5% of the city population moves to the suburbs and 3% of the suburban population moves to the city.

(a) Assuming that the total population remains constant, make a table that shows the populations of the city and its suburbs over a five-year period (round to the nearest integer).

(a) Over the long term, how will the population be distributed between the city and its suburbs?

## 7.5 Matrices and geometric transformations

In Chapter 3 you learned about function transformations: Reflecting in the $x$-axis, or $y$-axis, stretching vertically or horizontally, or in both directions or combinations of those. In this chapter we focus on transformations of the plane.

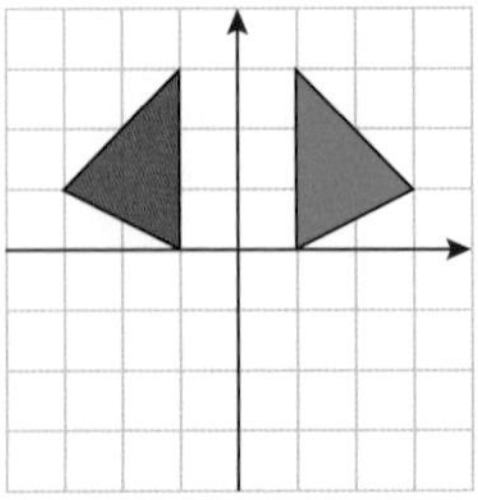

The green triangle in the diagram in Figure 7.17, for example, is reflected in the $y$-axis to get the red triangle. As the diagram shows, only the position has changed. All angles and sides still have the same measures.

Take a look at the coordinates of the points making the triangles:

**Figure 7.17** The green triangle is reflected in the $y$-axis to get the red triangle

Green triangle: $(1, 0)$, $(3, 1)$, and $(1, 3)$

Red triangle: $(-1, 0)$, $(-3, 1)$, and $(-1, 3)$

That means, any point with coordinates $(x, y)$ is reflected into a point $(-x, y)$.

This process can be achieved by matrix multiplication. Consider the matrix $M = \begin{pmatrix} -1 & 0 \\ 0 & 1 \end{pmatrix}$ and construct a matrix, $T$ with the coordinates of the vertices of the green triangle as its columns $T = \begin{pmatrix} 1 & 3 & 1 \\ 0 & 1 & 3 \end{pmatrix}$

Now multiply $M$ by $T$: $MT = \begin{pmatrix} -1 & 0 \\ 0 & 1 \end{pmatrix}\begin{pmatrix} 1 & 3 & 1 \\ 0 & 1 & 3 \end{pmatrix} = \begin{pmatrix} -1 & -3 & -1 \\ 0 & 1 & 3 \end{pmatrix}$

The columns of the resulting matrix are the coordinates of the red triangle.

The method we can use to find the matrix representing a transformation is given by the following theorem.

**Matrix Basis Theorem**

Let $T$ be a transformation represented by a matrix $M$. Then if

$$T: \begin{pmatrix} 1 \\ 0 \end{pmatrix} \mapsto \begin{pmatrix} a \\ c \end{pmatrix}, \text{ and } T: \begin{pmatrix} 0 \\ 1 \end{pmatrix} \mapsto \begin{pmatrix} b \\ d \end{pmatrix}, \text{ then } M = \begin{pmatrix} a & b \\ c & d \end{pmatrix}$$

Here is a list of major transformations that you are familiar with and their corresponding matrices.

1.  **Reflection in the $x$-axis:** $\begin{pmatrix} 1 \\ 0 \end{pmatrix} \mapsto \begin{pmatrix} 1 \\ 0 \end{pmatrix}, \begin{pmatrix} 0 \\ 1 \end{pmatrix} \mapsto \begin{pmatrix} 0 \\ -1 \end{pmatrix} \Rightarrow M = \begin{pmatrix} 1 & 0 \\ 0 & -1 \end{pmatrix}$

2.  **Reflection in the $y$-axis:** $\begin{pmatrix} 1 \\ 0 \end{pmatrix} \mapsto \begin{pmatrix} -1 \\ 0 \end{pmatrix}, \begin{pmatrix} 0 \\ 1 \end{pmatrix} \mapsto \begin{pmatrix} 0 \\ 1 \end{pmatrix} \Rightarrow M = \begin{pmatrix} -1 & 0 \\ 0 & 1 \end{pmatrix}$

3.  **Reflection in $y = x$:** $\begin{pmatrix} 1 \\ 0 \end{pmatrix} \mapsto \begin{pmatrix} 0 \\ 1 \end{pmatrix}, \begin{pmatrix} 0 \\ 1 \end{pmatrix} \mapsto \begin{pmatrix} 1 \\ 0 \end{pmatrix} \Rightarrow M = \begin{pmatrix} 0 & 1 \\ 1 & 0 \end{pmatrix}$

4.  **Reflection in $y = -x$:** $\begin{pmatrix} 1 \\ 0 \end{pmatrix} \mapsto \begin{pmatrix} 0 \\ -1 \end{pmatrix}, \begin{pmatrix} 0 \\ 1 \end{pmatrix} \mapsto \begin{pmatrix} -1 \\ 0 \end{pmatrix} \Rightarrow M = \begin{pmatrix} 0 & -1 \\ -1 & 0 \end{pmatrix}$

5.  **Horizontal dilation** by a constant $k$ (stretch/shrink):

    $\begin{pmatrix} 1 \\ 0 \end{pmatrix} \mapsto \begin{pmatrix} k \\ 0 \end{pmatrix}, \begin{pmatrix} 0 \\ 1 \end{pmatrix} \mapsto \begin{pmatrix} 0 \\ 1 \end{pmatrix} \Rightarrow M = \begin{pmatrix} k & 0 \\ 0 & 1 \end{pmatrix}$

    For example, triangle $(1, 1)$, $(2, 2)$, $(1, 4)$ will be transformed into

    $\begin{pmatrix} k & 0 \\ 0 & 1 \end{pmatrix}\begin{pmatrix} 1 & 2 & 1 \\ 1 & 2 & 4 \end{pmatrix} = \begin{pmatrix} k & 2k & k \\ 1 & 2 & 4 \end{pmatrix}$

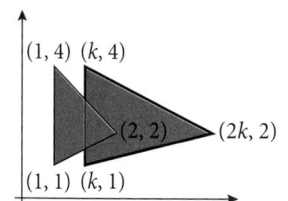

**Figure 7.18** Horizontal dilation

6. **Vertical dilation** by a constant $m$ (stretch/shrink):

$$\begin{pmatrix}1\\0\end{pmatrix} \mapsto \begin{pmatrix}1\\0\end{pmatrix}, \begin{pmatrix}0\\1\end{pmatrix} \mapsto \begin{pmatrix}0\\m\end{pmatrix} \Rightarrow M = \begin{pmatrix}1 & 0\\0 & m\end{pmatrix}$$

For example, triangle (1, 1), (2, 2), (1, 4) will be transformed into

$$\begin{pmatrix}1 & 0\\0 & m\end{pmatrix}\begin{pmatrix}1 & 2 & 1\\1 & 2 & 4\end{pmatrix} = \begin{pmatrix}1 & 2 & 1\\m & 2m & 4m\end{pmatrix}$$

7. The matrix $M = \begin{pmatrix}k & 0\\0 & m\end{pmatrix}$ represents a **horizontal dilation of magnitude $k$ and a vertical dilation of magnitude $m$.** When $k = m$, this is called scaling.

For example, triangle (1, 1), (2, 2), (1, 4) will be transformed into

$$\begin{pmatrix}k & 0\\0 & m\end{pmatrix}\begin{pmatrix}1 & 2 & 1\\1 & 2 & 4\end{pmatrix} = \begin{pmatrix}k & 2k & k\\m & 2m & 4m\end{pmatrix}$$

### Example 7.28

The triangle $ABC$ has vertices $A(2, -3)$, $B(3, 1)$, and $C(-1, 4)$.

(a) Find the area of $ABC$.

(b) Find the coordinates of the image of $ABC$ under each of the following transformations:

　(i) reflection in the $x$-axis

　(ii) reflection in the line $y = x$

(c) Find the coordinates of the image of $ABC$ under each of the following transformations, find the area of the image, find the determinant of the transformation matrix, and make a conjecture about the relationship between the three quantities.

　(i) dilation of magnitude 2 in the horizontal direction

　(ii) horizontal dilation of magnitude $k$ and vertical dilation of magnitude $m$.

### Solution

(a) From section 7.2, the area is given by

$$A = \frac{1}{2}\begin{vmatrix}x_1 & y_1 & 1\\x_2 & y_2 & 1\\x_3 & y_3 & 1\end{vmatrix} = \frac{1}{2}\begin{vmatrix}2 & -3 & 1\\3 & 1 & 1\\-1 & 4 & 1\end{vmatrix} = \frac{19}{2}$$

(b) (i) Reflection in the $x$-axis: $\begin{pmatrix} 1 & 0 \\ 0 & -1 \end{pmatrix} \begin{pmatrix} 2 & 3 & -1 \\ -3 & 1 & 4 \end{pmatrix}$

$$= \begin{pmatrix} 2 & 3 & -1 \\ 3 & -1 & -4 \end{pmatrix}$$

that is $(2, 3), (3, -1), (-1, -4)$

(ii) Reflection in the line $y = x$: $\begin{pmatrix} 0 & 1 \\ 1 & 0 \end{pmatrix} \begin{pmatrix} 2 & 3 & -1 \\ -3 & 1 & 4 \end{pmatrix}$

$$= \begin{pmatrix} -3 & 1 & 4 \\ 2 & 3 & -1 \end{pmatrix}$$

that is $(-3, 2), (1, 3), (4, -1)$

(c) (i) Dilation of magnitude 2 in the horizontal direction:

$$\begin{pmatrix} 2 & 0 \\ 0 & 1 \end{pmatrix} \begin{pmatrix} 2 & 3 & -1 \\ -3 & 1 & 4 \end{pmatrix} = \begin{pmatrix} 4 & 6 & -2 \\ -3 & 1 & 4 \end{pmatrix}$$

So, image has the vertices $(4, -3), (6, 1), (-2, 4)$

$$\text{area} = \frac{1}{2} \begin{Vmatrix} 4 & -3 & 1 \\ 6 & 1 & 1 \\ -2 & 4 & 1 \end{Vmatrix} = 19 \text{ and determinant is } \begin{vmatrix} 2 & 0 \\ 0 & 1 \end{vmatrix} = 2$$

Thus, area of image = determinant $\times$ Area of pre-image. In fact, it must be the absolute value of the determinant.

(ii) Dilations in both directions:

$$\begin{pmatrix} k & 0 \\ 0 & m \end{pmatrix} \begin{pmatrix} 2 & 3 & -1 \\ -3 & 1 & 4 \end{pmatrix} = \begin{pmatrix} 2k & 3k & -k \\ -3m & m & 4m \end{pmatrix}$$

So, image has the vertices

$$(2k, -3m), (3k, m), (-k, 4m),$$

$$\text{area} = \frac{1}{2} \begin{Vmatrix} 2k & -3m & 1 \\ 3k & m & 1 \\ -k & 4m & 1 \end{Vmatrix} = \frac{19}{2}|km|$$

Thus, area of image = |determinant| $\times$ Area of object (pre-image)

 If a matrix $M$ is a transformation matrix, then Area of image = $|\det M| \times$ area of object.

The above transformations are called **affine transformations**. Affine transformations are known to preserve points, (for example, a triangle is mapped into a triangle), also, sets of parallel lines and planes remain parallel after an affine transformation. An affine transformation does not necessarily preserve angles between lines or distances between points, though it does preserve ratios of distances between points lying on a straight line.

Two more transformations worth mentioning, but may not be examined without guidance are **horizontal and vertical shearing**:

A horizontal shearing of magnitude $k$ is given by $\begin{pmatrix} 1 & k \\ 0 & 1 \end{pmatrix}\begin{pmatrix} x \\ y \end{pmatrix} = \begin{pmatrix} x + ky \\ y \end{pmatrix}$ and a vertical shearing

of magnitude $m$ is given by $\begin{pmatrix} 1 & 0 \\ m & 1 \end{pmatrix}\begin{pmatrix} x \\ y \end{pmatrix} = \begin{pmatrix} x \\ y + mx \end{pmatrix}$

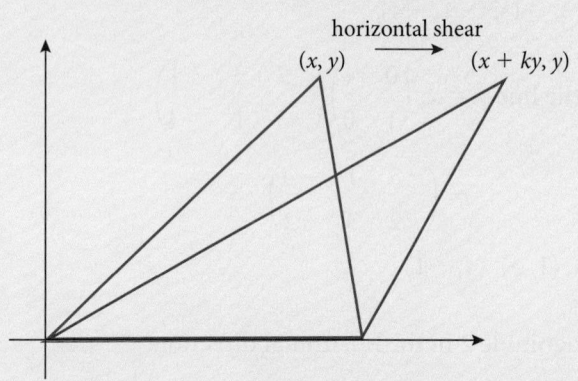

Two more transformations, that are a special type of affine transformation, are called **isometries**: **translations** and **rotations**. These transformations preserve angles and distances. That is, in a translation, a triangle is mapped into a congruent triangle. Similarly, for rotations.

### Example 7.29

Consider the matrix $A = \begin{pmatrix} \dfrac{\sqrt{3}}{2} & -\dfrac{1}{2} \\ \dfrac{1}{2} & \dfrac{\sqrt{3}}{2} \end{pmatrix}$

and the unit square with vertices $(0, 0)$, $(1, 0)$, $(1, 1)$, and $(0, 1)$.

(a) Perform the transformation, sketch graphs showing the unit square and its image, and conjecture what this transformation represents.

(b) What do $A^2$ and $A^3$ represent?

### Solution

(a) The transformation mapped

$$\begin{pmatrix} 0 & 1 & 1 & 0 \\ 0 & 0 & 1 & 1 \end{pmatrix} \text{ to } \begin{pmatrix} 0 & \dfrac{\sqrt{3}}{2} & \dfrac{\sqrt{3}-1}{2} & -\dfrac{1}{2} \\ 0 & \dfrac{1}{2} & \dfrac{\sqrt{3}+1}{2} & \dfrac{\sqrt{3}}{2} \end{pmatrix}$$

As the figure implies, we have a rotation through an angle of 30°. This is confirmed by observing that the image of $\begin{pmatrix} 1 \\ 0 \end{pmatrix}$ is $\begin{pmatrix} \frac{\sqrt{3}}{2} \\ \frac{1}{2} \end{pmatrix}$ which is $\begin{pmatrix} \cos 30° \\ \sin 30° \end{pmatrix}$.

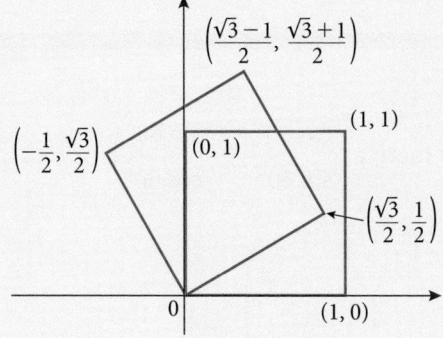

(b) $A^2 = \begin{pmatrix} \frac{1}{2} & -\frac{\sqrt{3}}{2} \\ \frac{\sqrt{3}}{2} & \frac{1}{2} \end{pmatrix}$ which implies that the image of $\begin{pmatrix} 1 \\ 0 \end{pmatrix}$ is $\begin{pmatrix} \frac{1}{2} \\ \frac{\sqrt{3}}{2} \end{pmatrix} = \begin{pmatrix} \cos 60° \\ \sin 60° \end{pmatrix}$

which in turn implies a rotation of 60°.

$A^3 = \begin{pmatrix} 0 & -1 \\ 1 & 0 \end{pmatrix}$ implying a rotation of 90°.

We can generalise the above discussion for rotations around the origin through any angle.

As the diagram shows, the rotation maps $\begin{pmatrix} 1 \\ 0 \end{pmatrix}$ to $\begin{pmatrix} \cos\theta \\ \sin\theta \end{pmatrix}$ and $\begin{pmatrix} 0 \\ 1 \end{pmatrix}$ to $\begin{pmatrix} -\sin\theta \\ \cos\theta \end{pmatrix}$, thus we can state the following result:

The matrix $\begin{pmatrix} \cos\theta & -\sin\theta \\ \sin\theta & \cos\theta \end{pmatrix}$ represents a **rotation** of angle $\theta$ around the origin.

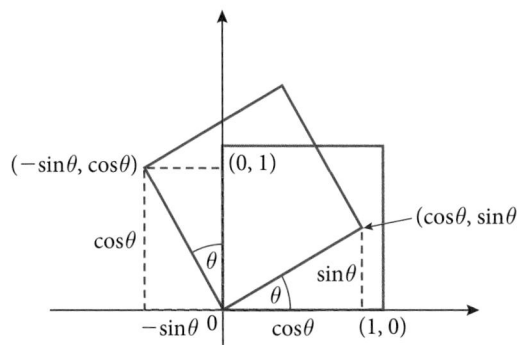

### Example 7.30

Consider the unit square in Example 7.29. Perform the following two rotations on the square and draw a sketch of the result:

(a)  rotation of 60° around the origin

(b)  rotation of 180° around the origin.

### Solution

(a)  We multiply $\begin{pmatrix} 0 & 1 & 1 & 0 \\ 0 & 0 & 1 & 1 \end{pmatrix}$ by matrix $\begin{pmatrix} \cos 60° & -\sin 60° \\ \sin 60° & \cos 60° \end{pmatrix}$

$$= \begin{pmatrix} \dfrac{1}{2} & -\dfrac{\sqrt{3}}{2} \\ \dfrac{\sqrt{3}}{2} & \dfrac{1}{2} \end{pmatrix} \text{ to get } \begin{pmatrix} 0 & \dfrac{1}{2} & \dfrac{1-\sqrt{3}}{2} & -\dfrac{\sqrt{3}}{2} \\ 0 & \dfrac{\sqrt{3}}{2} & \dfrac{\sqrt{3}+1}{2} & \dfrac{1}{2} \end{pmatrix}$$

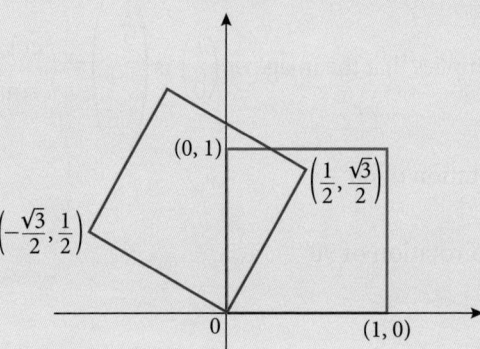

(b)  We multiply by

$$\begin{pmatrix} \cos 180° & -\sin 180° \\ \sin 180° & \cos 180° \end{pmatrix}$$

$$= \begin{pmatrix} -1 & 0 \\ 0 & -1 \end{pmatrix} \text{ to get}$$

$$\begin{pmatrix} 0 & -1 & -1 & 0 \\ 0 & 0 & -1 & -1 \end{pmatrix}$$

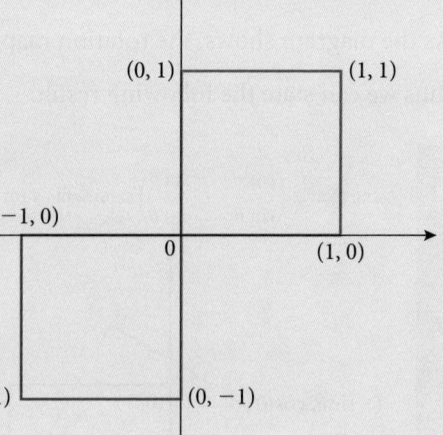

The matrix $\begin{pmatrix} -1 & 0 \\ 0 & -1 \end{pmatrix}$ represents a reflection in the origin.

In fact, this is nothing but a reflection in the origin.

**Translation**

A translation of $h$ units horizontally and $k$ units vertically is simply achieved by mapping $\begin{pmatrix} x \\ y \end{pmatrix}$ into $\begin{pmatrix} x + h \\ y + k \end{pmatrix}$. This is called a translation of $\begin{pmatrix} h \\ k \end{pmatrix}$

As is, we cannot perform translation using matrix multiplication. However, we can follow a procedure similar to what we did in using determinants to find the area of a triangle. We introduce a 'new' set of 3D coordinates $(x, y, 1)$, called **homogeneous coordinates** to represent the 2D coordinates $(x, y)$. Translations can then be performed with matrix multiplication in the following manner.

$$\begin{pmatrix} 1 & 0 & h \\ 0 & 1 & k \\ 0 & 0 & 1 \end{pmatrix} \begin{pmatrix} x \\ y \\ 1 \end{pmatrix} = \begin{pmatrix} x + h \\ y + k \\ 1 \end{pmatrix}$$

Rotation in this system can also be performed by considering the rotation matrix as

$$\begin{pmatrix} \cos\theta & -\sin\theta & 0 \\ \sin\theta & \cos\theta & 0 \\ 0 & 0 & 1 \end{pmatrix}$$

In fact, all matrices introduced earlier can be written in this manner:

$$\begin{pmatrix} a & b & 0 \\ c & d & 0 \\ 0 & 0 & 1 \end{pmatrix}$$

## Composition of transformations

Let $S$ and $T$ be two transformations. The transformation $S \circ T$ is called the **composition** of the two transformations where $T$ is applied first. It is also written as $S(T(x, y))$

For example, if $S(x, y) = (-x, y)$ and $T(x, y) = (-y, x)$, then:

$$S \circ T(x, y) = S(T(x, y)) = S(-y, x) = (y, x)$$

and
$$T \circ S(x, y) = T(-x, y) = (-y, -x)$$

Composition of transformations can be achieved with matrix multiplication.

 If $M$ is the matrix representation of a transformation $M$, and $N$ is that of transformation $N$, then the product $MN$ is the matrix representation of $M \circ N$

Composition of transformations is not commutative. That is, in general $M \circ N \neq N \circ M$

### Example 7.31

Consider the unit square in Example 7.29. Perform the following transformations on the square.

(a) a translation of $\begin{pmatrix} 2 \\ 3 \end{pmatrix}$

(b) a translation of $\begin{pmatrix} 2 \\ 3 \end{pmatrix}$ followed by a rotation of 180°

(c) a rotation of 180° followed by a translation of $\begin{pmatrix} 2 \\ 3 \end{pmatrix}$

**Solution**

(a) $\begin{pmatrix} 0 & 1 & 1 & 0 \\ 0 & 0 & 1 & 1 \end{pmatrix} \Rightarrow \begin{pmatrix} 0+2 & 1+2 & 1+2 & 0+2 \\ 0+3 & 0+3 & 1+3 & 1+3 \end{pmatrix} = \begin{pmatrix} 2 & 3 & 3 & 2 \\ 3 & 3 & 4 & 4 \end{pmatrix}$

It can be done by $\begin{pmatrix} 1 & 0 & 2 \\ 0 & 1 & 3 \\ 0 & 0 & 1 \end{pmatrix} \begin{pmatrix} 0 & 1 & 1 & 0 \\ 0 & 0 & 1 & 1 \\ 1 & 1 & 1 & 1 \end{pmatrix} = \begin{pmatrix} 2 & 3 & 3 & 2 \\ 3 & 3 & 4 & 4 \\ 1 & 1 & 1 & 1 \end{pmatrix}$

(b) Translation is done as above. Applying rotation to the result

$$\begin{pmatrix} -1 & 0 \\ 0 & -1 \end{pmatrix} \begin{pmatrix} 2 & 3 & 3 & 2 \\ 3 & 3 & 4 & 4 \end{pmatrix} = \begin{pmatrix} -2 & -3 & -3 & -2 \\ -3 & -3 & -4 & -4 \end{pmatrix}$$

This can be done by the following matrix multiplication

$$\begin{pmatrix} -1 & 0 & 0 \\ 0 & -1 & 0 \\ 0 & 0 & 1 \end{pmatrix} \begin{pmatrix} 1 & 0 & 2 \\ 0 & 1 & 3 \\ 0 & 0 & 1 \end{pmatrix} \begin{pmatrix} 0 & 1 & 1 & 0 \\ 0 & 0 & 1 & 1 \\ 1 & 1 & 1 & 1 \end{pmatrix} = \begin{pmatrix} -2 & -3 & -3 & -2 \\ -3 & -3 & -4 & -4 \\ 1 & 1 & 1 & 1 \end{pmatrix}$$

(c) Rotation: $\begin{pmatrix} -1 & 0 \\ 0 & -1 \end{pmatrix} \begin{pmatrix} 0 & 1 & 1 & 0 \\ 0 & 0 & 1 & 1 \end{pmatrix} = \begin{pmatrix} 0 & -1 & -1 & 0 \\ 0 & 0 & -1 & -1 \end{pmatrix}$, applying

translation to the result $\begin{pmatrix} 0+2 & -1+2 & -1+2 & 0+2 \\ 0+3 & 0+3 & -1+3 & -1+3 \end{pmatrix}$

$$= \begin{pmatrix} 2 & 1 & 1 & 2 \\ 3 & 3 & 2 & 2 \end{pmatrix}$$

This can be done by the following matrix multiplication

$$\begin{pmatrix} 1 & 0 & 2 \\ 0 & 1 & 3 \\ 0 & 0 & 1 \end{pmatrix} \begin{pmatrix} -1 & 0 & 0 \\ 0 & -1 & 0 \\ 0 & 0 & 1 \end{pmatrix} \begin{pmatrix} 0 & 1 & 1 & 0 \\ 0 & 0 & 1 & 1 \\ 1 & 1 & 1 & 1 \end{pmatrix} = \begin{pmatrix} 2 & 1 & 1 & 2 \\ 3 & 3 & 2 & 2 \\ 1 & 1 & 1 & 1 \end{pmatrix}$$

Note in Example 7.31 that a translation followed by a rotation is not the same as a rotation followed by a translation.

## Example 7.32

Describe the effect of each transformation on any point $(x, y)$ in the plane and on a circle with equation $x^2 + y^2 = 16$

(a) $\begin{pmatrix} 2 & 0 \\ 0 & 2 \end{pmatrix}$      (b) $\begin{pmatrix} 3 & 0 \\ 0 & 1 \end{pmatrix}$

**Solution**

(a) This is a simple dilation (scale in this case).

$$\begin{pmatrix} 2 & 0 \\ 0 & 2 \end{pmatrix}\begin{pmatrix} x \\ y \end{pmatrix} = \begin{pmatrix} 2x \\ 2y \end{pmatrix}$$

The new coordinates are

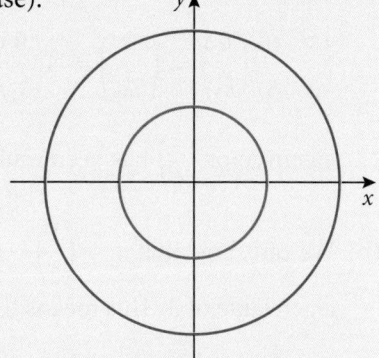

$$\begin{pmatrix} x_1 \\ y_1 \end{pmatrix} = \begin{pmatrix} 2x \\ 2y \end{pmatrix} \Rightarrow \begin{pmatrix} x \\ y \end{pmatrix} = \begin{pmatrix} \frac{x_1}{2} \\ \frac{y_1}{2} \end{pmatrix}$$

so, the equation of the image is

$x^2 + y^2 = 16 \Rightarrow \left(\frac{x_1}{2}\right)^2 + \left(\frac{y_1}{2}\right)^2 = 16 \Rightarrow x_1^2 + y_1^2 = 64$, which is another

circle with a larger radius (which you can write as $x^2 + y^2 = 64$). Note that the new circle has an area 4 times the original. The determinant of the matrix is 4.

(b) This is a dilation in the $x$-direction.

$$\begin{pmatrix} 3 & 0 \\ 0 & 1 \end{pmatrix}\begin{pmatrix} x \\ y \end{pmatrix} = \begin{pmatrix} 3x \\ y \end{pmatrix}$$

The new coordinates are

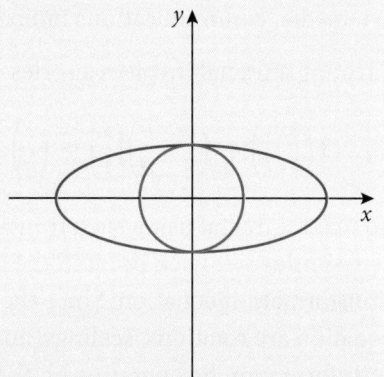

$$\begin{pmatrix} x_1 \\ y_1 \end{pmatrix} = \begin{pmatrix} 3x \\ y \end{pmatrix} \Rightarrow \begin{pmatrix} x \\ y \end{pmatrix} = \begin{pmatrix} \frac{x_1}{3} \\ y_1 \end{pmatrix}$$

so the equation of the image is

$$x^2 + y^2 = 16 \Rightarrow \left(\frac{x_1}{3}\right)^2 + y_1^2 = 16$$

$$\Rightarrow \frac{x_1^2}{144} + \frac{y_1^2}{16} = 1 \left(\text{or } \frac{x^2}{144} + \frac{y^2}{16} = 1\right) \text{ which is an ellipse.}$$

## Example 7.33

Consider each transformation matrix.
Find an eigenvector and interpret its meaning.

(a) $\begin{pmatrix} 3 & 0 \\ 0 & 5 \end{pmatrix}$      (b) $\begin{pmatrix} 3 & 0 \\ 8 & -1 \end{pmatrix}$

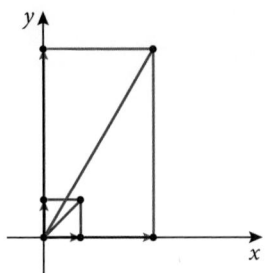

**Figure 7.19** Diagram for solution to Example 7.33 (a)

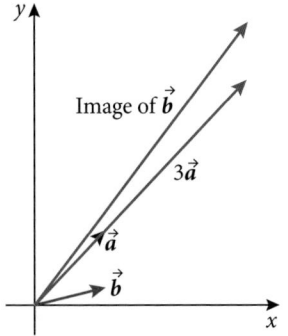

**Figure 7.20** Diagram for solution to Example 7.33 (b)

## Solution

(a) From section 7.4, you know that the characteristic equation is $(\lambda - 3)(\lambda - 5) = 0$ implying that the eigenvalues are 3 and 5 with corresponding eigenvectors $\begin{pmatrix} k_1 \\ 0 \end{pmatrix}$ and $\begin{pmatrix} 0 \\ k_2 \end{pmatrix}$. This means that vectors along the $x$-axis and along the $y$-axis will have their images multiples of themselves. i.e., $\begin{pmatrix} 3 & 0 \\ 0 & 5 \end{pmatrix}\begin{pmatrix} k_1 \\ 0 \end{pmatrix} = \begin{pmatrix} 3k_1 \\ 0 \end{pmatrix} = 3\begin{pmatrix} k_1 \\ 0 \end{pmatrix}$ and $\begin{pmatrix} 3 & 0 \\ 0 & 5 \end{pmatrix}\begin{pmatrix} 0 \\ k_2 \end{pmatrix} = \begin{pmatrix} 0 \\ 5k_2 \end{pmatrix} = 5\begin{pmatrix} 0 \\ k_2 \end{pmatrix}$. Note from the graph that an eigenvector $\begin{pmatrix} 1 \\ 0 \end{pmatrix}$ has been multiplied by 3 and $\begin{pmatrix} 0 \\ 1 \end{pmatrix}$ by 5.

(b) We can show that $\vec{a} = \begin{pmatrix} 1 \\ 2 \end{pmatrix}$ is an eigenvector corresponding to an eigenvalue of 3. This means that only vectors parallel to $\begin{pmatrix} 1 \\ 2 \end{pmatrix}$ are transformed into multiples of themselves. See the diagram where we show another vector with its image.

### Fractals

Fractal mathematics is a growing field of study. Fractals are applied in today's art, media, communications industry, and science.

Creating a fractal involves a series of affine transformations of the form

$$T\left(\begin{pmatrix} x \\ y \end{pmatrix}\right) = \begin{pmatrix} a & b \\ c & d \end{pmatrix}\begin{pmatrix} x \\ y \end{pmatrix} + \begin{pmatrix} e \\ f \end{pmatrix}, \text{ where } a, b, c, d, e \text{ and } f \text{ are scalars.}$$

To make a fractal using such transformations, we begin with an initial **self-similar** set. Each point in the set is transformed as dictated by the transformation equation. Since the only operations present in transformation equation are rotations, scalings, and translations, one can visually arrive at the transformation by operating on the whole set rather than one point at a time. Below is one of the most famous fractals, the **Sierpinski carpet**.

The 8 transformation equations used to generate this image are:

1. $T_1\left(\begin{pmatrix} x \\ y \end{pmatrix}\right) = \begin{pmatrix} \frac{1}{3} & 0 \\ 0 & \frac{1}{3} \end{pmatrix}\begin{pmatrix} x \\ y \end{pmatrix} + \begin{pmatrix} 0 \\ 0 \end{pmatrix}$

2. $T_2\left(\begin{pmatrix} x \\ y \end{pmatrix}\right) = \begin{pmatrix} \frac{1}{3} & 0 \\ 0 & \frac{1}{3} \end{pmatrix}\begin{pmatrix} x \\ y \end{pmatrix} + \begin{pmatrix} \frac{1}{3} \\ 0 \end{pmatrix}$

3. $T_3\left(\begin{pmatrix} x \\ y \end{pmatrix}\right) = \begin{pmatrix} \frac{1}{3} & 0 \\ 0 & \frac{1}{3} \end{pmatrix}\begin{pmatrix} x \\ y \end{pmatrix} + \begin{pmatrix} \frac{2}{3} \\ 0 \end{pmatrix}$

4. $T_4\left(\begin{pmatrix} x \\ y \end{pmatrix}\right) = \begin{pmatrix} \frac{1}{3} & 0 \\ 0 & \frac{1}{3} \end{pmatrix}\begin{pmatrix} x \\ y \end{pmatrix} + \begin{pmatrix} \frac{2}{3} \\ \frac{1}{3} \end{pmatrix}$

5. $T_5\left(\begin{pmatrix} x \\ y \end{pmatrix}\right) = \begin{pmatrix} \frac{1}{3} & 0 \\ 0 & \frac{1}{3} \end{pmatrix}\begin{pmatrix} x \\ y \end{pmatrix} + \begin{pmatrix} \frac{2}{3} \\ \frac{2}{3} \end{pmatrix}$     6. $T_6\left(\begin{pmatrix} x \\ y \end{pmatrix}\right) = \begin{pmatrix} \frac{1}{3} & 0 \\ 0 & \frac{1}{3} \end{pmatrix}\begin{pmatrix} x \\ y \end{pmatrix} + \begin{pmatrix} \frac{1}{3} \\ \frac{2}{3} \end{pmatrix}$

7. $T_7\left(\begin{pmatrix} x \\ y \end{pmatrix}\right) = \begin{pmatrix} \frac{1}{3} & 0 \\ 0 & \frac{1}{3} \end{pmatrix}\begin{pmatrix} x \\ y \end{pmatrix} + \begin{pmatrix} 0 \\ \frac{2}{3} \end{pmatrix}$     8. $T_8\left(\begin{pmatrix} x \\ y \end{pmatrix}\right) = \begin{pmatrix} \frac{1}{3} & 0 \\ 0 & \frac{1}{3} \end{pmatrix}\begin{pmatrix} x \\ y \end{pmatrix} + \begin{pmatrix} 0 \\ \frac{1}{3} \end{pmatrix}$

Start with the unit square, call it $U_0$, then apply the transformation equations to the square, you will get $U_1$. Notice how square 1 is the result of applying $T_1$, with no translation, $T_2$ has a translation of 1:3 in the horizontal direction, and so on. The square in the middle marked with white will be removed. Next, for each new image, apply the equations again, and we get $U_2$, the squares left in the centre of each square will be removed, and so on. Eventually, with many iterations, obviously done with software, it is possible to get something similar to $U_n$ and more!

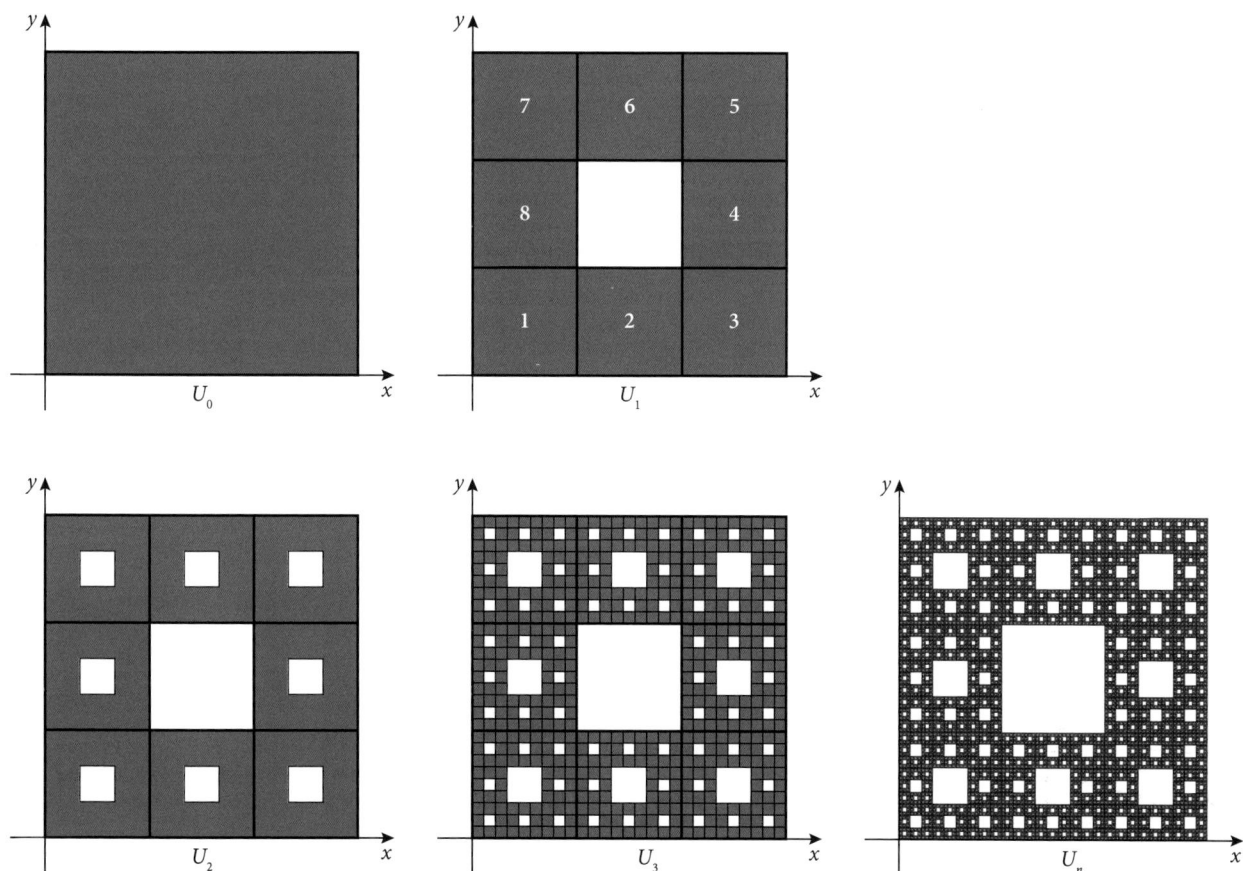

**Figure 7.21** Sierpinski carpet

263

**Example 7.34**

The **Koch curve** is constructed by removing the middle third of a line segment (say length 1 unit) and replacing it with two sides of an equilateral triangle. Here is the first iteration.

The Koch curve can be created with four transformations:

1. The first shrinks the whole segment to one third of its original size. This maps $AD$ to $AB$.

2. The second is a dilation of ratio 1:3 followed by a rotation of 60° and a horizontal translation of 1:3. This maps $AD$ to $BE$.

3. The third transformation is a dilation of ratio 1:3 followed by a rotation of 260° and a translation of $\frac{1}{2}$ in the horizontal direction and $\frac{1}{3}\left(\frac{\sqrt{3}}{2}\right) = \frac{\sqrt{3}}{6}$ in the vertical direction. This maps $AD$ to $EC$.

4. The fourth transformation is a dilation of ratio 1:3 followed by a translation of 2:3 in the horizontal direction. This maps $AD$ to $CD$.

(a) Find the matrices representing each transformation.

(b) Draw the first three iterations.

**Solution**

(a) 1. $\begin{pmatrix} \frac{1}{3} & 0 \\ 0 & \frac{1}{3} \end{pmatrix}\begin{pmatrix} x \\ y \end{pmatrix} + \begin{pmatrix} 0 \\ 0 \end{pmatrix}$

2. $\begin{pmatrix} \cos 60 & -\sin 60 \\ \sin 60 & \cos 60 \end{pmatrix}\begin{pmatrix} \frac{1}{3} & 0 \\ 0 & \frac{1}{3} \end{pmatrix}\begin{pmatrix} x \\ y \end{pmatrix} + \begin{pmatrix} \frac{1}{3} \\ 0 \end{pmatrix} = \begin{pmatrix} \frac{1}{6} & -\frac{\sqrt{3}}{6} \\ \frac{\sqrt{3}}{6} & \frac{1}{6} \end{pmatrix}\begin{pmatrix} x \\ y \end{pmatrix} + \begin{pmatrix} \frac{1}{3} \\ 0 \end{pmatrix}$

3. $\begin{pmatrix} \cos(-60) & -\sin(-60) \\ \sin(-60) & \cos(-60) \end{pmatrix}\begin{pmatrix} \frac{1}{3} & 0 \\ 0 & \frac{1}{3} \end{pmatrix}\begin{pmatrix} x \\ y \end{pmatrix} + \begin{pmatrix} \frac{1}{2} \\ \frac{\sqrt{3}}{6} \end{pmatrix} = \begin{pmatrix} \frac{1}{6} & \frac{\sqrt{3}}{6} \\ -\frac{\sqrt{3}}{6} & \frac{1}{6} \end{pmatrix}\begin{pmatrix} x \\ y \end{pmatrix} + \begin{pmatrix} \frac{1}{2} \\ \frac{\sqrt{3}}{6} \end{pmatrix}$

4. $\begin{pmatrix} \frac{1}{3} & 0 \\ 0 & \frac{1}{3} \end{pmatrix}\begin{pmatrix} x \\ y \end{pmatrix} + \begin{pmatrix} \frac{2}{3} \\ 0 \end{pmatrix}$

(b) For the second iteration, we apply the same procedure on each segment created in the first iteration:

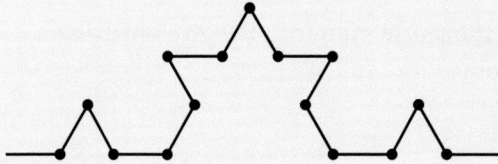

For the third iteration, we apply the same procedure on each segment created in the second iteration:

## Exercise 7.5

1. Find the image of $\triangle ABC$ with vertices $A\,(0, 0)$, $B\,(3, 0)$, and $C\,(3, 1)$ under each of the following transformations, then describe the effect of the transformation in words.

(a) $T(x, y) = \begin{pmatrix} -1 & 0 \\ 0 & 1 \end{pmatrix}$

(b) $T(x, y) = \begin{pmatrix} 2 & 0 \\ 0 & 2 \end{pmatrix}$

(c) $T(x, y) = \begin{pmatrix} 1 & 0 \\ 0 & -1 \end{pmatrix}$

(d) $T(x, y) = \begin{pmatrix} 0 & 1 \\ 1 & 0 \end{pmatrix}$

(e) $T(x, y) = \begin{pmatrix} 1 & 0 \\ 0 & 3 \end{pmatrix}$

(f) $T(x, y) = \begin{pmatrix} 0 & -1 \\ 1 & 0 \end{pmatrix}$

2. Describe the effect of applying the transformation $\begin{pmatrix} 0 & 2 \\ 2 & 0 \end{pmatrix}$ on objects in the plane.

3. Determine whether the matrix $A$ below may represent a rotation about the origin. Explain your answer: $A = \begin{pmatrix} \dfrac{\sqrt{3}}{2} & -\dfrac{1}{2} \\ -\dfrac{1}{2} & \dfrac{\sqrt{3}}{2} \end{pmatrix}$

4. Describe the effect upon the line with equation $3x + 2y = 6$ of the transformation:

(a) $\begin{pmatrix} 3 & 0 \\ 0 & 3 \end{pmatrix}$

(b) $\begin{pmatrix} 0 & 3 \\ 3 & 0 \end{pmatrix}$

5. Show that a reflection in the $x$-axis followed by a reflection in the $y$-axis is equivalent to a rotation of $180°$ about the origin.

6. Describe the effect of the transformation with the given matrix on the graph of the given equation:

   (a) $\begin{pmatrix} 2 & 0 \\ 0 & 2 \end{pmatrix}$; $x^2 + (y - 1)^2 = 9$

   (b) $\begin{pmatrix} 2 & 0 \\ 0 & -1 \end{pmatrix}$; $3x + 2y = 6$

7. Show that:

   (a) a reflection in the $x$-axis followed by a reflection in the line $y = x$ is equivalent to a rotation about the origin through $90°$

   (b) a reflection in the line $y = x$ followed by a reflection in the $x$-axis is equivalent to a rotation about the origin through $-90°$

8. Consider the unit square with vertices $(0, 0)$, $(1, 0)$, $(1, 1)$, and $(0, 1)$.

   Find the image of this square under the following transformations and sketch the graphs of the final images:

   (a) a translation of $\begin{pmatrix} 2 \\ 3 \end{pmatrix}$ followed by a dilation $\begin{pmatrix} 2 & 0 \\ 0 & 3 \end{pmatrix}$

   (b) a dilation $\begin{pmatrix} 2 & 0 \\ 0 & 3 \end{pmatrix}$ followed by a translation $\begin{pmatrix} 2 \\ 3 \end{pmatrix}$

9. Consider the transformation $\begin{pmatrix} 1 & 0 \\ 3 & 1 \end{pmatrix}$ and the line $3x + y = 6$

   Find the equation of the image of the line under the transformation. Then choose three points on the line and find their images and check the correctness of your equation.

10. Consider each of the transformation matrices, find the eigenvectors for each and interpret them in terms of transformations.

    (a) $\begin{pmatrix} 1 & 0 \\ 0 & -1 \end{pmatrix}$    (b) $\begin{pmatrix} 0 & 1 \\ 1 & 0 \end{pmatrix}$    (c) $\begin{pmatrix} 1 & 2 \\ 2 & 1 \end{pmatrix}$    (d) $\begin{pmatrix} 2 & 2 \\ 6 & 3 \end{pmatrix}$

11. A rotation about the origin of $\alpha°$ followed by a rotation of $\beta°$ is equivalent to a rotation of $\alpha° + \beta°$

    (a) Write down the matrix for a rotation of $\alpha° + \beta°$

    (b) By considering rotation of $\alpha°$ followed by a rotation of $\beta°$ as a composition of transformations, find the matrix for this transformation.

    (c) Compare your answers to (a) and (b).

**12.** Consider a Koch curve as described in Example 7.34, where the length of each side is 1 unit.

**(a)** Find an expression for the length of the $n$th iterated curve.

**(b)** Hence, find the length of the curve as it is iterated an infinite number of times.

If instead of starting with a line, you start with an equilateral triangle, you get a **Koch snowflake**.

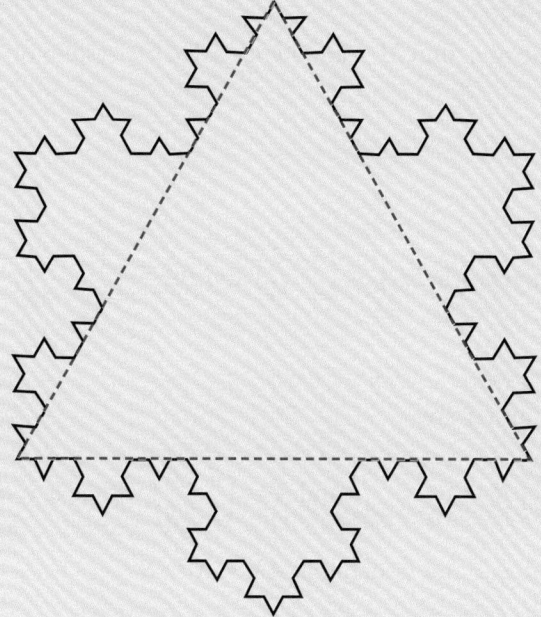

**(c)** Find an expression for the number of small sides after $n$ iterations.

**(d)** Find the length of each side after $n$ iterations.

**(e)** Find the perimeter after $n$ iterations, and hence the perimeter of the fractal.

**(f)** Find the number of smaller triangles (each missing one side).

**(g)** Find the area of each triangle.

**(h)** Find the total area of the triangles after $n$ iterations, and hence the area of the fractal.

## Chapter 7 practice questions

1. If $A = \begin{pmatrix} 2x & 3 \\ -4x & x \end{pmatrix}$ and det $A = 14$, find the value of $x$.

2. Let $M = \begin{pmatrix} a & 2 \\ 2 & -1 \end{pmatrix}$ where $a \in \mathbb{Z}$

   (a) Find $M^2$ in terms of $a$.

   (b) $M^2 = \begin{pmatrix} 5 & -4 \\ -4 & 5 \end{pmatrix}$ find the value of $a$.

   (c) Using this value of $a$, find $M^{-1}$ and hence solve the system of equations
   $$\begin{cases} -x + 2y = -3 \\ 2x - y = 3 \end{cases}$$

3. Two matrices are given, where $A = \begin{pmatrix} 5 & 2 \\ 2 & 0 \end{pmatrix}$ and $BA = \begin{pmatrix} 11 & 2 \\ 44 & 8 \end{pmatrix}$
   Find $B$.

4. The matrices $A$, $B$, and $X$ are:

   $$A = \begin{pmatrix} 3 & 1 \\ -5 & 6 \end{pmatrix} \quad B = \begin{pmatrix} 4 & 8 \\ 0 & -3 \end{pmatrix} \text{ and } X = \begin{pmatrix} a & b \\ c & d \end{pmatrix} \text{ with } a, b, c, d \in \mathbb{Q}$$

   Find the values of $a$, $b$, $c$ and $d$ such that $AX + X = B$

5. $A = \begin{pmatrix} 5 & -2 \\ 7 & 1 \end{pmatrix}$ is a $2 \times 2$ matrix.

   (a) Write out $A^{-1}$

   (b) (i) If $XA + B = C$, where $B$, $C$, and $X$ are $2 \times 2$ matrices, express $X$ in terms of $A^{-1}$, $B$, and $C$.

   (ii) Find $X$ when $B = \begin{pmatrix} 6 & 7 \\ 5 & -2 \end{pmatrix}$ and $C = \begin{pmatrix} -5 & 0 \\ -8 & 7 \end{pmatrix}$

6. Let $A = \begin{pmatrix} a & b \\ c & 1 \end{pmatrix}$ and $B = \begin{pmatrix} 1 & 2 \\ d & c \end{pmatrix}$

   (a) Find $A + B$

   (b) Find $AB$

7. (a) State the inverse of the matrix $A = \begin{pmatrix} 1 & -3 & 1 \\ 2 & 2 & -1 \\ 1 & -5 & 3 \end{pmatrix}$

**(b)** Hence solve the system of simultaneous equations

$$\begin{cases} x - 3y + z = 1 \\ 2x + 2y - z = 2 \\ x - 5y + 3z = 3 \end{cases}$$

**8.** Let $C = \begin{pmatrix} -2 & 4 \\ -1 & 7 \end{pmatrix}$ and $D = \begin{pmatrix} 5 & 2 \\ -1 & a \end{pmatrix}$

The matrix $Q$ is given such that $3Q = 2C - D$

**(a)** Find $Q$

**(b)** Find $CD$

**(c)** Find $D^{-1}$

**9. (a)** Find the values of $a$ and $b$ given that the matrix $A = \begin{pmatrix} a & -4 & -6 \\ -8 & 5 & 7 \\ -5 & 3 & -4 \end{pmatrix}$

is the inverse of the matrix $B = \begin{pmatrix} 1 & 2 & -2 \\ 3 & b & 1 \\ -1 & 1 & -3 \end{pmatrix}$

**(b)** For the values of $a$ and $b$ found in part **(a)**, solve the system of linear equations:

$$\begin{cases} x + 2y - 2z = 5 \\ 3x + by + z = 0 \\ -x + y - 3z = a - 1 \end{cases}$$

**10. (a)** Given matrices $A$, $B$, $C$ for which $AB = C$ and $\det A \neq 0$, express $B$ in terms of $A$ and $C$.

**(b)** Let $A = \begin{pmatrix} 1 & 2 & 3 \\ 2 & -1 & 2 \\ 3 & -3 & 2 \end{pmatrix}$ $D = \begin{pmatrix} -4 & 13 & -7 \\ -2 & 7 & -4 \\ 3 & -9 & 5 \end{pmatrix}$ and $C = \begin{pmatrix} 5 \\ 7 \\ 10 \end{pmatrix}$

    **(i)** Find the matrix $DA$

    **(ii)** Find $B$ when $AB = C$

**(c)** The following three equations represent three planes that intersect at a point. Find the coordinates of this point.

$$x + 2y + 3z = 5, \quad 2x - y + 2z = 7 \text{ and } 3x - 3y + 2z = 10$$

11. (a) Find the determinant of the matrix $\begin{pmatrix} 1 & 1 & 2 \\ 1 & 2 & 1 \\ 2 & 1 & 5 \end{pmatrix}$

(b) Find the value of $\lambda$ for which the following system of equations can be solved.

$$\begin{pmatrix} 1 & 1 & 2 \\ 1 & 2 & 1 \\ 2 & 1 & 5 \end{pmatrix} \begin{pmatrix} x \\ y \\ z \end{pmatrix} = \begin{pmatrix} 3 \\ 4 \\ \lambda \end{pmatrix}$$

(c) For this value of $\lambda$, find the general solution to the system of equations.

12. Find the eigenvalues and eigenvectors of the matrix $A = \begin{pmatrix} 2 & 2 \\ 5 & -1 \end{pmatrix}$

13. At a large private university, the student loan program evaluates the payment status of student loans. The loans are divided into three categories: loans paid within 14 days are up to date, D, loans paid within 15–60 days are considered late, L, and those paid after 60 days are labelled as problematic, P. Each year, some of the students change categories because they get behind in payments or catch up. The table shows the fraction of students that change from one category to another or stay in the same category.

|  |  | Move from category | | |
|---|---|---|---|---|
|  |  | D | L | P |
| Move to category | D | 0.86 | 0.62 | 0.17 |
|  | L | 0.08 | 0.29 | 0.37 |
|  | P | 0.06 | 0.09 | 0.46 |

One year, the fraction of students in each category was 0.8 in D, 0.11 in L, and 0.09 in P.

(a) Find the fraction in each category the next year.

(b) Find the fraction in each category three years later.

14. (a) Diagonalise the matrix $A = \begin{pmatrix} 1 & 0 \\ 6 & -1 \end{pmatrix}$

(b) Hence or otherwise find $A^8$

**15.** A country is divided into three demographic regions. It is found that each year 5% of the residents of region 1 move to region 2, and 5% move to region 3. Of the residents of region 2, 15% move to region 1 and 10% move to region 3. And of the residents of region 3, 10% move to region 1 and 5% move to region 2. What percentage of the population resides in each of the three regions after a long period of time?

**16.** Two competing television channels, channel 1 and channel 2, each have 50% of the viewer market at some initial point in time. Over each one-year period, channel 1 captures 10% of channel 2's share, and channel 2 captures 20% of channel 1's share.

  **(a)** What is each channel's market share after one year?

  **(b)** Track the market shares of channels 1 and 2 in part **(a)** over a five-year period.

  **(c)** If this trend continues, what is the market share of each station?

**17.** Sierpinski' s triangle is another fractal that can be constructed using transformations, starting with an equilateral triangle. Three transformations are required. The first three stages are shown.

At stage 1, we take a scale of ratio $\frac{1}{2}$ and create one triangle, which is positioned at the lower left corner of the original. Next, we create another triangle and translate it to fit the lower right corner, and similarly we fill the upper corner, thus cutting out the middle triangle as shown. The process is then repeated to each new shaded triangle, and so on.

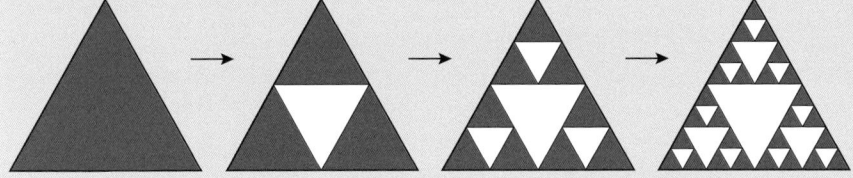

  **(a)** Count the number of shaded triangles at each stage (stage 0 has 1, stage 1 has 3). Predict the number of shaded triangles at stage 4 and stage 5.

  **(b)** What is the number of triangles at stage $n$?

  **(c)** Letting the area at stage 0 to be $a$, what is the total shaded area at each stage? As $n$ becomes large without bound, what happens to the shaded area?

  **(d)** Find the matrix transformations required to create the fractal.

**18.** A study in 2015 divided occupations in the United Kingdom into upper level, U, (executives and professionals), middle level, M, (supervisors and skilled manual workers), and lower level, L, (unskilled).

To determine the mobility across these levels in a generation, about two thousand men were asked, 'At which level are you, and at which level was your father when you were fourteen years old?' Here is a summary:

| | Father's occupation | | | |
|---|---|---|---|---|
| | U | M | L | |
| Son's occupation | 0.60 | 0.26 | 0.14 | U |
| | 0.29 | 0.37 | 0.34 | M |
| | 0.16 | 0.27 | 0.57 | L |

For example, a child of a middle-class worker has a 0.26 chance of moving into an upper class job.

With initial distribution of respondents' fathers given below, find the distributions for the next five generations.

Upper: 0.12, middle: 0.32, lower: 0.56

Vectors

8

By the end of this chapter, you should be familiar with…

- describing vectors using standard notation
- determining the vector equations of lines
- applying vector equations to kinematic models
- finding scalar and vector products
- calculating the angle between vectors in 3-dimensional space (3-space).

A **scalar** is a quantity, but a **vector** is a representation of a directed quantity with two components, its **magnitude** and **direction**. For example, mass is a scalar, but force is a vector.

In the same way that a complex number can have a real and an imaginary component, a vector can have a horizontal and a vertical component. A vector can be also expressed in trigonometric form, as a complex number can. However, unlike complex numbers, vectors are not limited to two dimensions. Indeed, a vector can represent a force in $n$ dimensions, although this chapter will consider just 2- and 3-dimensional vectors.

## 8.1 Vector representation

A 2-dimensional vector $u$ is often represented by an arrow in the Cartesian plane. It has a starting point and an ending point; thus, it has a magnitude. It has some specific heading suggested by the arrow; thus, it has a direction. However, without a set of axes and perhaps a grid as a background, you cannot verify the magnitude and the direction.

**Figure 8.1** 2D vector

If the vector is on axes and a grid, you can determine its properties. Note that a vector need not be drawn starting at the origin; it can lie anywhere, provided neither its magnitude nor direction are altered. Hence, two vectors are equal if their magnitudes and directions are equal, regardless of their location in the plane or space.

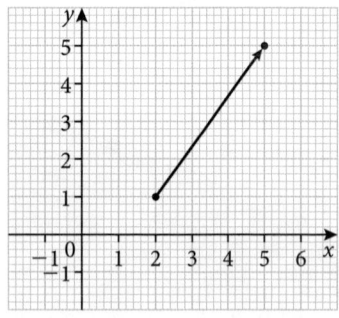

**Figure 8.2** Vector on a grid

Moving the vector in Figure 8.2 so that it starts at the origin should make it easier to spot a right-angled triangle with sides 3 and 4. Hence, its magnitude is 5 and its direction is $\theta = \arctan\left(\frac{4}{3}\right)$

This should look familiar: complex numbers are described in trigonometric form, $r(\cos\theta + i\sin\theta)$ and abbreviated as $r\operatorname{cis}\theta$, with a magnitude and direction.

A 2-dimensional vector can be described by its horizontal and vertical components, $\boldsymbol{u} = \begin{pmatrix} 3 \\ 4 \end{pmatrix}$

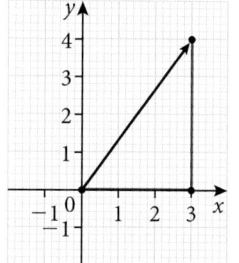

**Figure 8.3** Moving the vector so it starts at the origin

Note the use of **bold italic** font to indicate vectors.

A vector $\boldsymbol{v}$ in 3 dimensions can also be described by its magnitude and direction. The magnitude is the diagonal of a cuboid which you learned to calculate in chapter 4; however, the direction is considerably more difficult to specify, since its correct direction can only be given relative to each of the three axes, and requires that all three directions, $\theta_x$, $\theta_y$, and $\theta_z$ be found.

Should $\boldsymbol{v}$ be described by its measurements relative to the $x$-, $y$-, and $z$-axes, it would require no effort other than to list its components in order as $\boldsymbol{v} = \begin{pmatrix} x \\ y \\ z \end{pmatrix}$

Although representing a point or vector in 4 dimensions (or more) may present a daunting diagram to draw, adding another value to a column matrix is decidedly simpler.

## Vector properties

Any vector $\boldsymbol{w}$ can be represented by an arrow.

A second arrow with the same magnitude points in the opposite direction, and is $-\boldsymbol{w}$.

Another arrow twice as long points in the same direction: it is $2\boldsymbol{w}$.

When $\boldsymbol{w}$ is a 3-dimensional vector defined as $\boldsymbol{w} = \begin{pmatrix} x \\ y \\ z \end{pmatrix}$, then $-\boldsymbol{w} = \begin{pmatrix} -x \\ -y \\ -z \end{pmatrix}$

and $2\boldsymbol{w} = \begin{pmatrix} 2x \\ 2y \\ 2z \end{pmatrix}$

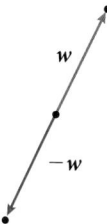

**Figure 8.4** $-\boldsymbol{w}$ has the same magnitude as $\boldsymbol{w}$ but points in the opposite direction

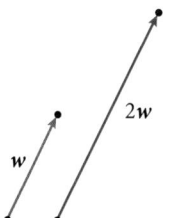

**Figure 8.5** $2\boldsymbol{w}$ is twice as long as $\boldsymbol{w}$

### Example 8.1

Let vector $\boldsymbol{u} = \begin{pmatrix} -2 \\ 14 \end{pmatrix}$

Determine the components of vector $\boldsymbol{v}$ when:

(a) $\boldsymbol{v} = 3\boldsymbol{u}$        (b) $\boldsymbol{v} = -2\boldsymbol{u}$        (c) $\boldsymbol{u} = 2\boldsymbol{v}$

## Solution

(a) $\mathbf{v} = 3\begin{pmatrix} -2 \\ 14 \end{pmatrix} = \begin{pmatrix} 3 \times -2 \\ 3 \times 14 \end{pmatrix} = \begin{pmatrix} -6 \\ 42 \end{pmatrix}$ (b) $\begin{pmatrix} 4 \\ -28 \end{pmatrix}$ (c) $\begin{pmatrix} -1 \\ 7 \end{pmatrix}$

## Vector addition and scalar multiplication

Just as complex numbers are added by summing the real and imaginary components separately, the measurements of vectors in the $x$-, $y$-, and $z$-directions are summed separately. Vectors with the same number of dimensions can be added and subtracted easily using their column matrix notation; vectors with different numbers of dimensions cannot. When vectors and their scalar multiples are combined, the result is called a **linear combination**.

### Example 8.2

Let vector $\mathbf{u} = \begin{pmatrix} -3 \\ 5 \\ 2 \end{pmatrix}$ and $\mathbf{v} = \begin{pmatrix} 1 \\ 0 \\ -2 \end{pmatrix}$

Determine the vector components of $\mathbf{w}$, when:

(a) $\mathbf{w} = \mathbf{u} + \mathbf{v}$ (b) $\mathbf{w} = \mathbf{u} - 2\mathbf{v}$ (c) $\mathbf{w} = 2\mathbf{u} + 3\mathbf{v}$

### Solution

The linear combinations of $\mathbf{u}$ and $\mathbf{v}$ produce the following vectors.

(a) $\mathbf{w} = \begin{pmatrix} -3 \\ 5 \\ 2 \end{pmatrix} + \begin{pmatrix} 1 \\ 0 \\ -2 \end{pmatrix} = \begin{pmatrix} -3+1 \\ 5+0 \\ 2-2 \end{pmatrix} = \begin{pmatrix} -2 \\ 5 \\ 0 \end{pmatrix}$ (b) $\begin{pmatrix} -5 \\ 5 \\ 6 \end{pmatrix}$ (c) $\begin{pmatrix} -3 \\ 10 \\ -2 \end{pmatrix}$

## Unit vectors

Unit vectors have a magnitude of 1. Assume that a vector $\mathbf{v}$ has components $\begin{pmatrix} a \\ b \end{pmatrix}$

Its magnitude, designated by the notation $|\mathbf{v}|$, is equal to 1 only if $\sqrt{a^2 + b^2} = 1$
By dividing each component by $\sqrt{a^2 + b^2}$, its magnitude becomes equal to 1.

### Example 8.3

Determine the components of vector $\mathbf{u}$, the unit vector of $\mathbf{v} = \begin{pmatrix} -6 \\ 8 \end{pmatrix}$, and find the components of the vector $\mathbf{w}$ in the same direction as $\mathbf{v}$ and with magnitude 7.

## Solution

Vector $\boldsymbol{u} = \dfrac{\boldsymbol{v}}{|\boldsymbol{v}|} = \dfrac{1}{\sqrt{6^2 + 8^2}} \begin{pmatrix} -6 \\ 8 \end{pmatrix} = \begin{pmatrix} -0.6 \\ 0.8 \end{pmatrix}$

Vector $\boldsymbol{w} = 7\boldsymbol{u} \Rightarrow \boldsymbol{w} = 7 \begin{pmatrix} -0.6 \\ 0.8 \end{pmatrix} = \begin{pmatrix} -4.2 \\ 5.6 \end{pmatrix}$

Unit vectors in three dimensions are found in the same manner.

## Example 8.4

Given vector $\boldsymbol{v} = \begin{pmatrix} 2 \\ 1 \\ -2 \end{pmatrix}$, find the components of a vector $\boldsymbol{w}$ in the same

direction as $\boldsymbol{v}$ with a magnitude of 9.

## Solution

Vector $\boldsymbol{w} = 9\boldsymbol{u} \Rightarrow \boldsymbol{w} = 9 \begin{pmatrix} \frac{2}{3} \\ \frac{1}{3} \\ -\frac{2}{3} \end{pmatrix} = \begin{pmatrix} 6 \\ 3 \\ -6 \end{pmatrix}$

In 2-dimensional space, the unit vectors along the $x$- and $y$-axes are called $\boldsymbol{i} = \begin{pmatrix} 1 \\ 0 \end{pmatrix}$ and $\boldsymbol{j} = \begin{pmatrix} 0 \\ 1 \end{pmatrix}$ respectively. Similarly, in 3-dimensional space, the unit vectors along the $x$-, $y$-, and $z$-axes are called $\boldsymbol{i} = \begin{pmatrix} 1 \\ 0 \\ 0 \end{pmatrix}, \boldsymbol{j} = \begin{pmatrix} 0 \\ 1 \\ 0 \end{pmatrix}$, and $\boldsymbol{k} = \begin{pmatrix} 0 \\ 0 \\ 1 \end{pmatrix}$ respectively. These three vectors are also known as the **base vectors** in 3-dimensional space.

## Vector operations illustrated

Consider the geometric view of vectors.

If $\boldsymbol{u} = \begin{pmatrix} a \\ b \end{pmatrix}$ and $\boldsymbol{v} = \begin{pmatrix} c \\ d \end{pmatrix}$, then $\boldsymbol{u} + \boldsymbol{v} = \begin{pmatrix} a + c \\ b + d \end{pmatrix}$

Since $\boldsymbol{v} + \boldsymbol{u} = \begin{pmatrix} a + c \\ b + d \end{pmatrix}$ as well, $\boldsymbol{u} + \boldsymbol{v} = \boldsymbol{v} + \boldsymbol{u}$

With two vectors starting at the same point, their sum is shown by the black vector (Figure 8.6) extending to the diagonally opposite corner of a parallelogram.

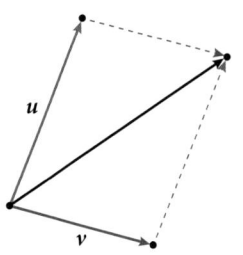

**Figure 8.6** The sum is shown by the black vector

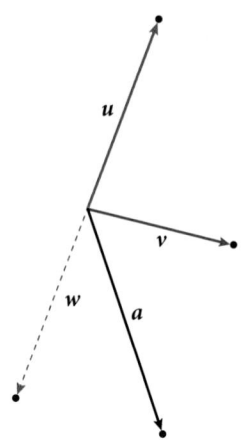

Now, let vector $w = -u$ shown in Figure 8.7 as a dashed blue line in the opposite direction to $u$.

Therefore, the black vector, labelled $a = v + w = w + v$

Since $w = -v$, $a = v - u = -u + v$

When vector $a$ is placed in the original diagram with vectors $u$ and $v$ (Figure 8.8), it should become evident that it forms the other diagonal of the parallelogram. How can you tell from the diagram that $a = v - u$ and not $u - v$?

The tip of $a$ is at the tip of $v$ while its bottom is at the tip of $u$. Or to move from the bottom of $a$ to its tip, move backwards through $u$, then forwards through $v$.

**Figure 8.7** $w = -u$

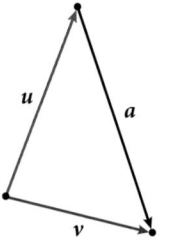

**Figure 8.8** How can you tell that $a = v - u$?

---

### Example 8.5

Consider vectors $u$, $v$, $w$, and $a$ drawn in an anti-clockwise sequence from point $P$.

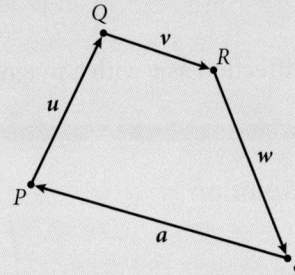

(a) Describe each vector sum as a single vector.

   (i) $u + v$

   (ii) $u + v + w$

(b) How does the sum $u + v + w$ compare to vector $a$?

(c) What is the sum of all four vectors?

---

### Solution

(a) (i) $u + v$ is a vector that starts at $P$ and ends at $R$. The vector is $\overrightarrow{PR}$.

   (ii) $u + v + w$ is a vector from $P$ to $S$, and can simply be written as $\overrightarrow{PS}$.

(b) $\overrightarrow{PS}$ is the opposite vector to $a$.

(c) The sum of all four vectors is the zero vector, $\mathbf{0}$.

---

### Exercise 8.1

1. $u = \begin{pmatrix} 5 \\ -2 \end{pmatrix}$. Determine the components of vector $v$ when:

   (a) $v = 2u$         (b) $v = -3u$         (c) $u = \dfrac{v}{3}$

2. $u = \begin{pmatrix} 6 \\ 0 \\ -3 \end{pmatrix}$ and $v = \begin{pmatrix} 2 \\ -1 \\ 2 \end{pmatrix}$. Determine the vector components of $w$ when:

   (a) $w = u - v$        (b) $w = u + 2v$       (c) $w = 3u - 4v$

**3.** Find the unit vector of each given vector.

(a) $u = \begin{pmatrix} 12 \\ -5 \end{pmatrix}$ 　　　(b) $u = \begin{pmatrix} -1 \\ 3 \end{pmatrix}$ 　　　(c) $u = \begin{pmatrix} 1 \\ 2 \\ 2 \end{pmatrix}$

**4.** Find $(x, y)$ so that the diagram is a parallelogram.

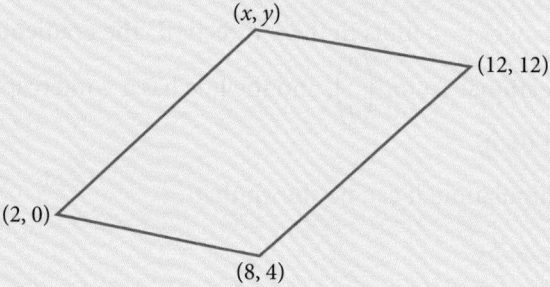

**5.** Find the values of $x$ and $y$ in the parallelogram.

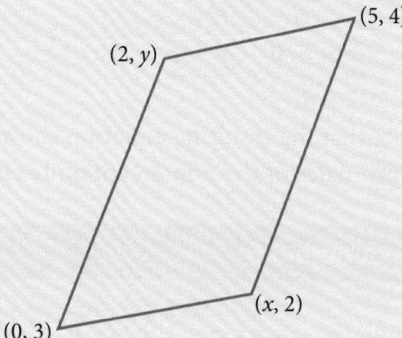

**6.** Find the value of scalars $r$ and $s$ such that $\begin{pmatrix} 8 \\ 46 \end{pmatrix} = r\begin{pmatrix} 1 \\ 9 \end{pmatrix} + s\begin{pmatrix} 1 \\ -4 \end{pmatrix}$

**7.** Write $\begin{pmatrix} 4 \\ 7 \end{pmatrix}$ as a linear combination of $\begin{pmatrix} 2 \\ 3 \end{pmatrix}$ and $\begin{pmatrix} 2 \\ 1 \end{pmatrix}$

**8.** Write $\begin{pmatrix} 5 \\ -5 \end{pmatrix}$ as a linear combination of $\begin{pmatrix} 1 \\ -1 \end{pmatrix}$ and $\begin{pmatrix} -1 \\ 1 \end{pmatrix}$

**9.** Write $\begin{pmatrix} -11 \\ 0 \end{pmatrix}$ as a linear combination of $\begin{pmatrix} 2 \\ 5 \end{pmatrix}$ and $\begin{pmatrix} 3 \\ 2 \end{pmatrix}$

**10.** Find vectors, each with a magnitude of 10 units, parallel to these vectors.

(a) $\begin{pmatrix} 4 \\ -3 \end{pmatrix}$ 　　　(b) $\begin{pmatrix} -5 \\ -12 \end{pmatrix}$ 　　　(c) $\begin{pmatrix} 2 \\ -1 \\ -2 \end{pmatrix}$

# 8 Vectors

## 8.2 Vector and parametric equations of lines

Lines require just two distinct points to be defined in both 2 and 3 dimensions. In 2 dimensions, a familiar description of a line may be in the form $y = mx + b$. Consider the gradient, $m$, to be equal to $\frac{q}{p}$, where for every non-zero $p$ units in the $x$-direction, the line changes by $q$ units in the $y$-direction. The gradient can then be described by a directional vector $\begin{pmatrix} p \\ q \end{pmatrix}$, and the line has a $y$-intercept, so it passes through the point $(0, b)$.

With simple algebraic manipulation, it can be shown that:

$$y = \frac{q}{p}x + b \Rightarrow y - b = \frac{q}{p}x$$

$$\Rightarrow \frac{y - b}{q} = \frac{x}{p}$$

$$\text{or } \frac{x - 0}{p} = \frac{y - b}{q}$$

Note that the resulting equation shows the vector components and the coordinates of the $y$-intercept quite clearly.

### Example 8.6

Find the equation of the line in the form $\frac{x - h}{p} = \frac{y - k}{q}$ that passes through the given point $A$ with gradient $m$.

(a) $A(5, -2)$, $m = \frac{2}{3}$

(b) $A(1, 2)$, $m = -\frac{1}{12}$

### Solution

(a) In this question, $h = 5$, $k = -2$, $p = 3$, and $q = 2$ hence, $\frac{x - 5}{3} = \frac{y + 2}{2}$

(b) $\frac{x - 1}{12} = \frac{y - 2}{-1}$

Now, consider a line in 3 dimensions that contains a point $A(x_0, y_0, z_0)$ with a gradient described by the vector $\begin{pmatrix} p \\ q \\ s \end{pmatrix}$

The addition of another dimension merely adds another displacement and another denominator $\frac{x - x_0}{p} = \frac{y - y_0}{q} = \frac{z - z_0}{s}$

This is commonly known as the Cartesian form of a line.

## Example 8.7

Find the equation of the line in 3-space that passes through the given point $A$ with the gradient described by the direction vector $\boldsymbol{u}$ with components $\begin{pmatrix} p \\ q \\ s \end{pmatrix}$

(a) $A(1, 2, 3)$, $\boldsymbol{u} = \begin{pmatrix} 4 \\ 5 \\ 6 \end{pmatrix}$

(b) $A(6, -2, 4)$, $\boldsymbol{u} = \begin{pmatrix} 1 \\ -2 \\ 3 \end{pmatrix}$

## Solution

(a) $\dfrac{x - 1}{4} = \dfrac{y - 2}{5} = \dfrac{z - 3}{6}$

(b) $x - 6 = \dfrac{y + 2}{-2} = \dfrac{z - 4}{3}$

Consider the point $A(x_0, y_0, z_0)$ in relation to the origin, $O(0, 0, 0)$. The segment $[OA]$ has a length, but no direction. Now, should this segment start at $O$ and end at $A$, then it would be a vector, called the **position vector** $\overrightarrow{OA} = \begin{pmatrix} x_0 \\ y_0 \\ z_0 \end{pmatrix} = \boldsymbol{r}_0$

Now, with the point $A$ and the direction vector $\boldsymbol{v} = \begin{pmatrix} p \\ q \\ s \end{pmatrix}$, both given as vectors, the equation of the line containing point $A$ is then simply the vector sum of the position vector $\overrightarrow{OA} = \begin{pmatrix} x_0 \\ y_0 \\ z_0 \end{pmatrix}$ and the direction vector $\begin{pmatrix} p \\ q \\ s \end{pmatrix}$, the scalar multiple of which, $k \begin{pmatrix} p \\ q \\ s \end{pmatrix}$, would ensure it is a line, not a segment. The vector equation of a line is given in the key fact box on the right.

You may find it easier to work with the **parametric form** of this equation:

$\begin{pmatrix} x \\ y \\ z \end{pmatrix} = \begin{pmatrix} x_0 \\ y_0 \\ z_0 \end{pmatrix} + k \begin{pmatrix} p \\ q \\ s \end{pmatrix}$ implies that $x = \begin{cases} x = x_0 + kp \\ y = y_0 + kq \\ z = z_0 + ks \end{cases}$ by matrix addition.

Simplifying the equation will give us

$x - x_0 = kp \Rightarrow \dfrac{x - x_0}{p} = k$, and similarly, $\dfrac{y - y_0}{q} = k$, as well as $\dfrac{z - z_0}{s} = k$

Thus, you can see that the parametric form will lead to the Cartesian form of the equation of the line

$k = \dfrac{x - x_0}{p} = \dfrac{y - y_0}{q} = \dfrac{z - z_0}{s}$

---

In terms of notation, vector $\overrightarrow{OA}$ is also noted simply as $\boldsymbol{a}$, vector $\overrightarrow{OB}$ as $\boldsymbol{b}$, and so on.

The **vector equation** of the line $\boldsymbol{r} = \boldsymbol{r}_0 + k\boldsymbol{v}$

$= \begin{pmatrix} x_0 \\ y_0 \\ z_0 \end{pmatrix} + k \begin{pmatrix} p \\ q \\ s \end{pmatrix}$

$\boldsymbol{r}$ is the position vector of any point on the line, while $\boldsymbol{r}_0$ is the position vector of a fixed point ($A$ in this case) on the line and $\boldsymbol{v}$ is the vector parallel to the given line.

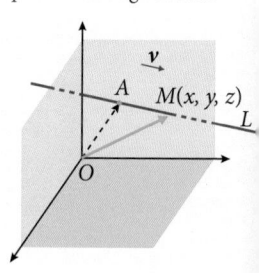

# 8 | Vectors

## Example 8.8

Find the equation of the line in vector form which passes through the given point $A$ with direction vector $\boldsymbol{u}$.

(a) $A(5, 2, -3)$, $\boldsymbol{u} = \begin{pmatrix} 4 \\ 7 \\ -6 \end{pmatrix}$

(b) $A(1, 0, -4)$, $\boldsymbol{u} = \begin{pmatrix} 2 \\ 1 \\ 8 \end{pmatrix}$

### Solution

(a) $\boldsymbol{r} = \begin{pmatrix} 5 \\ 2 \\ -3 \end{pmatrix} + k \begin{pmatrix} 4 \\ 7 \\ -6 \end{pmatrix}$

(b) $\boldsymbol{r} = \begin{pmatrix} 1 \\ 0 \\ -4 \end{pmatrix} + k \begin{pmatrix} 2 \\ 1 \\ 8 \end{pmatrix}$

## Exercise 8.2

1. Find the equation of the line in the form $\dfrac{x - h}{p} = \dfrac{y - k}{q}$ that passes through the given point $P$ with gradient $m$.

    (a) $A(1, -3)$, $m = \dfrac{3}{4}$

    (b) $A(6, -1)$, $m = \dfrac{1}{4}$

    (c) $A(-9, 5)$, $m = -\dfrac{3}{2}$

    (d) $A(7, 1)$, $m = -\dfrac{2}{5}$

2. Find the equation of the line in 3-space that passes through the given point $A$ with the gradient described by the direction vector $\boldsymbol{u}$.

    (a) $A(6, 7, 0)$, $\boldsymbol{u} = \begin{pmatrix} 1 \\ -2 \\ 2 \end{pmatrix}$

    (b) $A(-3, -2, 9)$, $\boldsymbol{u} = \begin{pmatrix} -3 \\ 3 \\ 6 \end{pmatrix}$

3. Find the equation of the line in vector form which passes through the given point $A$ with direction vector $\boldsymbol{u}$.

    (a) $A(1, 0, 2)$, $\boldsymbol{u} = \begin{pmatrix} 3 \\ -4 \\ 5 \end{pmatrix}$

    (b) $A(-2, 3, 0)$, $\boldsymbol{u} = \begin{pmatrix} 1 \\ -1 \\ 2 \end{pmatrix}$

4. Name the coordinates of the point on the given line whose $x$-value is 6.

    (a) $\boldsymbol{r} = \begin{pmatrix} 3 \\ 2 \end{pmatrix} + k \begin{pmatrix} 1 \\ 3 \end{pmatrix}$

    (b) $\boldsymbol{r} = \begin{pmatrix} -2 \\ 4 \end{pmatrix} + k \begin{pmatrix} 2 \\ -1 \end{pmatrix}$

    (c) $\boldsymbol{r} = \begin{pmatrix} 8 \\ 0 \\ -1 \end{pmatrix} + k \begin{pmatrix} 2 \\ 2 \\ -3 \end{pmatrix}$

    (d) $\boldsymbol{r} = \begin{pmatrix} 5 \\ -2 \\ 7 \end{pmatrix} + k \begin{pmatrix} 2 \\ 2 \\ -4 \end{pmatrix}$

5. Find the vector equation of each line given its Cartesian form.

(a) $\dfrac{x-3}{2} = \dfrac{y+2}{-2} = \dfrac{z-1}{4}$

(b) $\dfrac{x}{3} = \dfrac{y-3}{-2} = z+1$

(c) $\dfrac{x+3}{\frac{1}{2}} = \dfrac{y}{5} = \dfrac{z+2}{-2}$

(d) $\dfrac{x-a}{m} = \dfrac{y-b}{n} = \dfrac{z-c}{p}$

6. Find the vector equation of the line passing through $P(3, -1, -5)$, parallel to each line.

(a) $r = \begin{pmatrix} 1 \\ -6 \\ 0 \end{pmatrix} + k\begin{pmatrix} 2 \\ -2 \\ 4 \end{pmatrix}$

(b) $r = \begin{pmatrix} -1 \\ -5 \\ 3 \end{pmatrix} + k\begin{pmatrix} 3 \\ 2 \\ -4 \end{pmatrix}$

(c) $\dfrac{x+6}{3} = \dfrac{y-8}{-1} = \dfrac{z+1}{5}$

(d) $\dfrac{x+9}{7} = y+2 = \dfrac{z+11}{-2}$

7. A triangle $ABC$ has coordinates $A(2, 4, 6)$, $B(4, -1, 7)$, and $C(-2, -3, 1)$. Determine the vector components of $a$, $b$, and $c$ where $a$ is $\overrightarrow{OA}$, $b$ is $\overrightarrow{OB}$ and $c$ is $\overrightarrow{OC}$.

8. Find the vector equation of the line passing through each pair of points.

(a) $A(-3, 7)$ and $B(-1, 11)$

(b) $C(2, -5)$ and $D(-2, 1)$

(c) $E(8, -2, 1)$ and $F(-3, 10, 7)$

(d) $G(0, -6, -3)$ and $H(7, -1, 0)$

(e) $J(3, -4, 5)$ and $K(5, -2, 5)$

(f) $L(-7, -4, 2)$ and $M(2, -4, 12)$

9. Express each of the lines in question 8 in Cartesian form.

10. Find the point of intersection of the vector equations of the two lines:

$$r_1 = \begin{pmatrix} 5 \\ 1 \end{pmatrix} + k\begin{pmatrix} 3 \\ -2 \end{pmatrix} \text{ and } r_2 = \begin{pmatrix} -2 \\ 2 \end{pmatrix} + t\begin{pmatrix} 4 \\ 1 \end{pmatrix}$$

11. Show that lines $\dfrac{x-2}{1} = \dfrac{y-2}{3} = \dfrac{z-3}{1}$ and $\dfrac{x-2}{1} = \dfrac{y-3}{4} = \dfrac{z-4}{2}$ intersect and find the coordinates of $P$, the point of intersection.

## 8.3 Kinematics

Objects in motion can be analysed with vectors, provided that the motion is constant. When rates of change are involved, the analyses require the use of differential calculus which will be presented in Chapter 13.

Constant motion directly towards or away from an observer is the simplest.

## Example 8.9

A drone is hovering directly above a fixed point $P$. It then flies at a constant altitude in a direction described by the vector $v = \begin{pmatrix} 6 \\ 8 \end{pmatrix}$, moving 6 m to the east for every 8 m to the north, every second. Determine the distance to the drone from point $P$ after:

(a) 1 second                                (b) 3.2 seconds

(c) 15 seconds                          (d) $t$ seconds

## Solution

The distance flown every second by the drone is the magnitude of the vector $|v| = \sqrt{6^2 + 8^2} = 10$ m. The diagram represents a plan view of the motion of the drone, rendered in 2D-space.

(a) $|v| = 10$ m                       (b) $3.2|v| = 32$ m

(c) $15|v| = 150$ m                (d) $t|v| = 10t$ m

**Figure 8.9** Diagram for solution to Example 8.9

Now, consider a situation when the drone does not fly directly towards or away from a fixed point.

## Example 8.10

A drone starts from a position 10 m to the west of a fixed point $O$ and flies along the vector $v = \begin{pmatrix} 6 \\ 8 \end{pmatrix}$

Find the distance of the drone from $O$ after:

(a) 1 second                            (b) 2 seconds.

## Solution

(a) Consider the point $O$ to be at the origin $(0, 0)$. The drone starts at $S(-10, 0)$. After $t$ seconds, with the vector $v$ moving the drone by $\begin{pmatrix} 6t \\ 8t \end{pmatrix}$, it will be at the point $P(-10 + 6t, 0 + 8t)$.

The position vector is $\overrightarrow{OP} = \begin{pmatrix} -10 + 6t \\ 8t \end{pmatrix}$

Its magnitude is the distance from the origin, and is $|\overrightarrow{OP}| = \sqrt{(-10 + 6t)^2 + (8t)^2}$

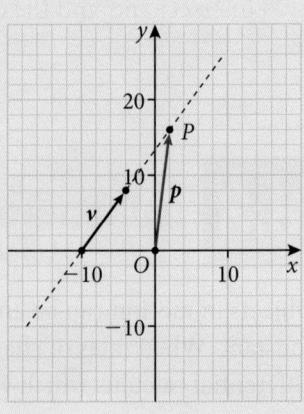

At $t = 1$

$$\left|\overrightarrow{OP}\right| = \sqrt{(-10 + 6)^2 + (8)^2} = \sqrt{80} = 4\sqrt{5}\,\text{m}$$

(b) At $t = 2$

$$\left|\overrightarrow{OP}\right| = \sqrt{(-10 + 12)^2 + (16)^2} = \sqrt{260} = 2\sqrt{65}\,\text{m}$$

## Minimum distance between a point and an object in motion

The distance to the drone is not constant, as shown in Figure 8.10. Note that the expression under the square root sign is a quadratic expression with a positive leading coefficient; hence, an expression with a minimum.

Expand the brackets in the expression:

$$\left|\overrightarrow{OP}\right| = \sqrt{(-10 + 6t)^2 + (8t)^2} = \sqrt{100 - 120t + 36t^2 + 64t^2}$$

$$= \sqrt{100 - 120t + 100t^2}$$

Now, let $y = 100 - 120t + 100t^2$. As every quadratic function of the form $y = ax^2 + bx + c$ has a vertex at $\left(-\dfrac{b}{2a}, c - \dfrac{b^2}{4a}\right)$ and with $a > 0$, $y$ has a minimum at $c - \dfrac{b^2}{4a} = 64$

So the minimum distance of the point $P$ from $O$ is $\left|\overrightarrow{OP}\right| = \sqrt{y} = \sqrt{64} = 8$ m

## Minimum distance between two objects in motion

Both the drone and the point of observation are moving along vectors.

While the drone flies along the path $d = \begin{pmatrix} -10 + 6t \\ 8t \end{pmatrix}$, the observer travels from $O(0, 0)$ along a road defined by the vector $u = \begin{pmatrix} 4 \\ 10 \end{pmatrix}$

The observer's path is defined as $u = \begin{pmatrix} 4t \\ 10t \end{pmatrix}$

Disregarding the drone's altitude by using a 2-dimensional map, the distance to the drone can be expressed using vectors.

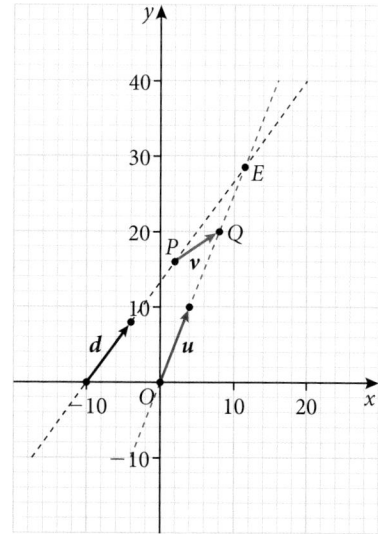

Note that although the paths cross at the point labelled $E$ in the diagram, the observer and the drone reach it at different times.

**Figure 8.10** Vector from the drone to the observer

285

The vector $\overrightarrow{PQ} = \overrightarrow{OQ} - \overrightarrow{OP} = \begin{pmatrix} 4t \\ 10t \end{pmatrix} - \begin{pmatrix} -10 + 6t \\ 8t \end{pmatrix} = \begin{pmatrix} -2t + 10 \\ 2t \end{pmatrix}$ and the

magnitude, similar to the earlier question, is $\left|\overrightarrow{PQ}\right| = \sqrt{(-2t + 10)^2 + (2t)^2}$

$$= \sqrt{4t^2 - 40t + 100 + 4t^2}$$

Let $y = 8t^2 - 40t + 100 \Rightarrow c - \dfrac{b^2}{4a} = 100 - \dfrac{(-40)^2}{4 \cdot 8} = 50 \text{ m}$, and the minimum

distance between the two moving points is $\left|\overrightarrow{PQ}\right| = \sqrt{50} = 5\sqrt{2}\,\text{m}$

## Minimum distance between two moving objects in 3 dimensions

Finding the distance between objects in 3 dimensions requires the use of the same process as noted above. Vectors allow the flexibility to change the number of dimensions using the same process.

### Example 8.11

A drone starts on the ground at the point $A(0, 0, 0)$ and flies at a constant

velocity along the vector $\boldsymbol{u} = \begin{pmatrix} 6 \\ 8 \\ 2 \end{pmatrix}$ each second, while another drone, already

hovering at the point $B(24, 12, 18)$ starts to descend at the same time in the

direction $\boldsymbol{v} = \begin{pmatrix} -2 \\ 4 \\ -4 \end{pmatrix}$. Find the minimum distance between the two drones.

### Solution

Since the first drone travels along the directional vector $\boldsymbol{u} = \begin{pmatrix} 6 \\ 8 \\ 2 \end{pmatrix}$ starting

from its initial position $A(0, 0, 0)$, after $t$ seconds, its position can be indicated by the vector

$$\boldsymbol{a} = \begin{pmatrix} 0 \\ 0 \\ 0 \end{pmatrix} + t \cdot \begin{pmatrix} 6 \\ 8 \\ 2 \end{pmatrix}$$

Similarly, the second drone, starting from $B(24, 12, 18)$ with a directional

$\boldsymbol{v} = \begin{pmatrix} -2 \\ 4 \\ -4 \end{pmatrix}$ will have a position that is indicated by $\boldsymbol{b} = \begin{pmatrix} 24 \\ 12 \\ 18 \end{pmatrix} + t \cdot \begin{pmatrix} -2 \\ 4 \\ -4 \end{pmatrix}$

As in the example in 2 dimensions, $\boldsymbol{a} = \begin{pmatrix} 6t \\ 8t \\ 2t \end{pmatrix}$ and $\boldsymbol{b} = \begin{pmatrix} 24 - 2t \\ 12 + 4t \\ 18 - 4t \end{pmatrix}$

$$\Rightarrow \overrightarrow{AB} = \begin{pmatrix} 24 - 8t \\ 12 - 4t \\ 18 - 6t \end{pmatrix} \Rightarrow \left|\overrightarrow{AB}\right| = \sqrt{(24 - 8t)^2 + (12 - 4t)^2 + (18 - 6t)^2}$$

$$= \sqrt{116(9 - 6t + t^2)}$$

As the expression under the square root sign is a quadratic with a leading positive coefficient, it will have a minimum distance of

$$c - \frac{b^2}{4a} = 9 - \frac{6^2}{4 \cdot 1} = 0 \text{ when } t = -\frac{b}{2a} = \frac{6}{2} = 3$$

The two drones will in fact crash at $t = 3$ seconds as the distance between them will be 0.

Did you notice that at $t = 4.5$ seconds, the second drone would have hit the ground?

## Exercise 8.3

1. A robot vacuum cleaner, set in motion on an empty gymnasium floor, moves in the direction $u = \begin{pmatrix} 1.2 \\ 2 \end{pmatrix}$, representing a path 1.2 m to the right for every 2 m forwards, every minute. Determine the distance travelled after:

   (a) 1 minute    (b) 2 minutes    (c) 10 minutes    (d) $t$ minutes

2. The vacuum cleaner in question **1** maintains movement along its set direction. An observer is 12 m to the right of the vacuum cleaner's starting position. Find the distance of the vacuum cleaner from the observer after:

   (a) 1 minute    (b) 2 minutes.

3. Another robot vacuum cleaner starts at the same time as the first one in question **1**, but exactly 10 m to the right of the observer and in the direction of $v = \begin{pmatrix} -2 \\ 2 \end{pmatrix}$

   Determine the closest distance between the two devices.

4. Consider a drone moving in the direction $u = \begin{pmatrix} 2 \\ 1 \\ -2 \end{pmatrix}$, given in m s$^{-1}$, when observed at the point $P(24, 16, 48)$.

   (a) How far is it from $O(0, 0, 0)$ when observed?

   (b) How far will it move in the next 10 seconds?

   (c) How far will it be from $O(0, 0, 0)$ after 10 seconds?

5. The drone in question **4** is being tracked by a sensor located at $Q(10, 0, 2)$.

   (a) Write an expression for the displacement from the sensor to the drone after $t$ seconds.

   (b) Find the minimum distance between the drone and sensor to 3 significant figures.

6. Two projectiles travelling through space have directional vectors, given in m s$^{-1}$, of $u = \begin{pmatrix} 20 \\ 20 \\ 20 \end{pmatrix}$ and $v = \begin{pmatrix} 30 \\ -20 \\ -10 \end{pmatrix}$

287

The first projectile, with directional vector **u**, is at $P(1300, 2800, 1000)$ when the second, with directional vector **v**, is at $Q(1000, 4000, 2000)$.

(a) Assume you are looking down on the projectiles from above. At what $x$- and $y$- coordinates do the paths of the projectiles intersect?

(b) Determine the time it would take each projectile to reach that $(x, y)$ location.

(c) Hence, determine whether the two projectiles appear to collide when viewed from above.

(d) What is the closest that they will come to each other? Give an answer to 3 significant figures.

(e) At what time does this occur?

## 8.4 Scalar and vector products

### The scalar product

The **scalar product** of two 2-dimensional vectors $u = \begin{pmatrix} a \\ b \end{pmatrix}$ and $v = \begin{pmatrix} d \\ e \end{pmatrix}$ is given by:

$$u \cdot v = ad + be$$

Similarly, the scalar product of two 3-dimensional vectors $u = \begin{pmatrix} a \\ b \\ c \end{pmatrix}$ and $v = \begin{pmatrix} d \\ e \\ f \end{pmatrix}$ is given by:

$$u \cdot v = ad + be + cf$$

Note that due to the use of the dot symbol for this operation, the scalar product is also known as the **dot product**.

### Example 8.12

Find the scalar product of each pair of vectors.

(a) $u = \begin{pmatrix} 2 \\ 3 \end{pmatrix}, v = \begin{pmatrix} -1 \\ 2 \end{pmatrix}$
(b) $u = \begin{pmatrix} 4 \\ 5 \end{pmatrix}, v = \begin{pmatrix} 5 \\ -4 \end{pmatrix}$

(c) $u = \begin{pmatrix} -2 \\ 1 \\ 2 \end{pmatrix}, v = \begin{pmatrix} 2 \\ 10 \\ -3 \end{pmatrix}$
(d) $u = \begin{pmatrix} 7 \\ 0 \\ -1 \end{pmatrix}, v = \begin{pmatrix} 2 \\ 1 \\ -2 \end{pmatrix}$

### Solution

(a) $u \cdot v = 2 \cdot (-1) + 3 \cdot 2 = 4$

(b) $u \cdot v = 4 \cdot 5 + 5 \cdot (-4) = 0$

(c) $u \cdot v = (-2) \cdot 2 + 1 \cdot 10 + 2 \cdot (-3) = 0$

(d) $u \cdot v = 7 \cdot 2 + 0 \cdot 1 + (-1) \cdot (-2) = 16$

Describing vectors by their directional components, for example, $v = \begin{pmatrix} x_0 \\ y_0 \\ z_0 \end{pmatrix}$,

facilitates the calculation of the scalar product. However, vectors can also be described by their magnitudes and directions. As the directions of vectors in 3-space require three separate measurements, $\theta_x$, $\theta_y$, and $\theta_z$, calculating the angle between vectors by starting with their polar forms is not practical.

When the magnitudes and the angle between vectors are known, you can calculate their scalar product as

$u \cdot v = |u| \cdot |v| \cos \theta$

Let $u$ and $v$ be drawn from the same point as shown.

Then $|u - v|^2 = (u - v) \cdot (u - v) = u^2 + v^2 - 2u \cdot v$

$\qquad = |u|^2 + |v|^2 - 2u \cdot v$

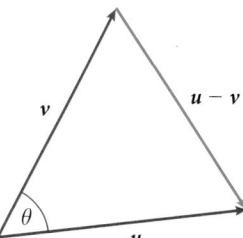

Also, using the law of cosines,

$|u - v|^2 = |u|^2 + |v|^2 - 2|u| \cdot |v| \cdot \cos \theta$

By comparing the two results above:

**Figure 8.11** The difference of two vectors

$|u|^2 + |v|^2 - 2u \cdot v = |u|^2 + |v|^2 - 2|u| \cdot |v| \cdot \cos \theta$

$-2u \cdot v = -2|u| \cdot |v| \cdot \cos \theta$

$u \cdot v = |u| \cdot |v| \cos \theta$

## Example 8.13

Find the scalar product for each pair of vectors, where $\theta$ is the angle between them.

(a) $|u| = 4$, $|v| = 8$, $\theta = 60°$        (b) $|u| = 8\sqrt{3}$, $|v| = 6$, $\theta = 30°$

(c) $|u| = 10$, $|v| = 4\sqrt{2}$, $\theta = 45°$      (d) $|u| = 12$, $|v| = 6$, $\theta = 120°$

## Solution

(a) $u \cdot v = 4 \cdot 8 \cdot \cos 60° = 16$       (b) $u \cdot v = 8\sqrt{3} \cdot 6 \cdot \cos 30° = 72$

(c) $u \cdot v = 10 \cdot 4\sqrt{2} \cdot \cos 45° = 40$     (d) $u \cdot v = 12 \cdot 6 \cdot \cos 120° = -36$

The **work** done by any force is defined as the product of the force multiplied by the distance it moves a certain object in the direction of the force. In other words, it is the product of the force multiplied by the displacement of the object. As such, work is then the dot product between the vectors representing the force ($F$) and displacement ($D$), $W = F \cdot D$

The SI unit of work is the joule (J), defined as the work expended by a force of one newton through a displacement of one metre.

## Example 8.14

Find the work done by the force $F = \begin{pmatrix} 400 \\ -50 \end{pmatrix}$ N in moving an object between points $M(2, 3)$ and $N(12, 43)$ given in metres. Your answer should be in joules (J).

## The vector product

The **vector product** of two vectors is another vector at right angles to the first two.

Using the properties of determinants, the vector (cross) product is equivalent to

$$\boldsymbol{u} \times \boldsymbol{v} = \begin{vmatrix} i & j & k \\ a & b & c \\ d & e & f \end{vmatrix}$$

Whereas the **scalar product** of two vectors yields a scalar quantity, the **vector product** of two vectors is another vector, at right angles to the two given vectors. The components of the vector

product of two vectors $\boldsymbol{u} = \begin{pmatrix} a \\ b \\ c \end{pmatrix}$ and $\boldsymbol{v} = \begin{pmatrix} d \\ e \\ f \end{pmatrix}$ are $\boldsymbol{u} \times \boldsymbol{v} = \begin{pmatrix} bf - ce \\ cd - af \\ ae - bd \end{pmatrix}$

Note that with the use of the symbol $\times$, the vector product is also known as the **cross product**. The resultant vector is orthogonal to the two original vectors and follows the **right hand rule**.

$(\boldsymbol{u} \times \boldsymbol{v})$ is a vector perpendicular to both $\boldsymbol{u}$ and $\boldsymbol{v}$ and obeying the right-hand rule, and has the magnitude: $|\boldsymbol{u} \times \boldsymbol{v}| = |\boldsymbol{u}||\boldsymbol{v}|\sin\theta$

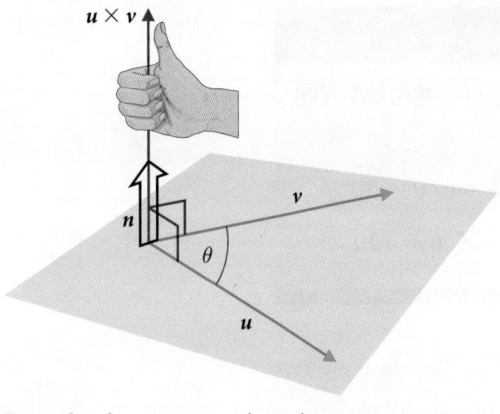

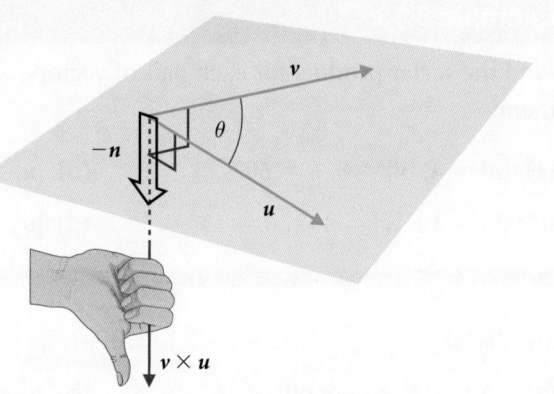

Remember that $\boldsymbol{u} \times \boldsymbol{v} = -(\boldsymbol{v} \times \boldsymbol{u})$

The cross product $\boldsymbol{v} \times \boldsymbol{u}$ gives a vector in the opposite direction to $\boldsymbol{u} \times \boldsymbol{v}$

You can use a GDC to find the vector product. Look for a function with a reference to 'cross products'. The screenshot shows an example.

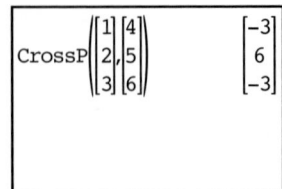

**Figure 8.12** Using a GDC to find a vector product

### Example 8.15

Find the vector product $\boldsymbol{u} \times \boldsymbol{v}$ of each pair of vectors.

(a) $\boldsymbol{u} = \begin{pmatrix} 1 \\ 0 \\ 0 \end{pmatrix}$, $\boldsymbol{v} = \begin{pmatrix} 0 \\ 1 \\ 0 \end{pmatrix}$ 

(b) $\boldsymbol{u} = \begin{pmatrix} 1 \\ 0 \\ 0 \end{pmatrix}$, $\boldsymbol{v} = \begin{pmatrix} 0 \\ 0 \\ 1 \end{pmatrix}$

(c) $\boldsymbol{u} = \begin{pmatrix} -1 \\ 2 \\ -2 \end{pmatrix}$, $\boldsymbol{v} = \begin{pmatrix} 3 \\ 1 \\ 2 \end{pmatrix}$ 

(d) $\boldsymbol{u} = \begin{pmatrix} 2 \\ 2 \\ -1 \end{pmatrix}$, $\boldsymbol{v} = \begin{pmatrix} 1 \\ 2 \\ -2 \end{pmatrix}$

## Solution

(a) $u \times v = \begin{pmatrix} 0 \times 0 - 0 \times 1 \\ 0 \times 0 - 1 \times 0 \\ 1 \times 1 - 0 \times 0 \end{pmatrix} = \begin{pmatrix} 0 \\ 0 \\ 1 \end{pmatrix}$     (b) $u \times v = \begin{pmatrix} 0 \\ -1 \\ 0 \end{pmatrix}$

(c) $u \times v = \begin{pmatrix} 6 \\ -4 \\ -7 \end{pmatrix}$     (d) $u \times v = \begin{pmatrix} -2 \\ 3 \\ 2 \end{pmatrix}$

## Exercise 8.4

1. Find the scalar product of each pair of vectors.

   (a) $u = \begin{pmatrix} 4 \\ -3 \end{pmatrix}, v = \begin{pmatrix} 3 \\ 3 \end{pmatrix}$     (b) $u = \begin{pmatrix} 6 \\ -2 \end{pmatrix}, v = \begin{pmatrix} -2 \\ -6 \end{pmatrix}$

   (c) $u = \begin{pmatrix} 3 \\ -1 \\ 4 \end{pmatrix}, v = \begin{pmatrix} -1 \\ 5 \\ 2 \end{pmatrix}$     (d) $u = \begin{pmatrix} -7 \\ 4 \\ 3 \end{pmatrix}, v = \begin{pmatrix} -1 \\ -4 \\ 3 \end{pmatrix}$

2. Find the scalar product, accurate to 3 significant figures, of each pair of vectors where $\theta$ is the angle between them.

   (a) $|u| = 6, |v| = 9, \theta = 30°$     (b) $|u| = 12, |v| = 8, \theta = 45°$

   (c) $|u| = 12, |v| = 3, \theta = 23°$     (d) $|u| = 10, |v| = 13, \theta = 13°$

3. Find the work done by the force $F = \begin{pmatrix} 30 \\ 150 \end{pmatrix}$ N in moving an object between points $A(0, 30)$ and $B(15, 70)$ given in metres. State your answer in joules.

4. Find the vector products $u \times v$ and $v \times u$ for each pair of vectors.

   (a) $u = \begin{pmatrix} 0 \\ 1 \\ 0 \end{pmatrix}, v = \begin{pmatrix} 0 \\ 0 \\ 1 \end{pmatrix}$     (b) $u = \begin{pmatrix} 0 \\ 0 \\ 1 \end{pmatrix}, v = \begin{pmatrix} 1 \\ 0 \\ 0 \end{pmatrix}$

   (c) $u = \begin{pmatrix} 3 \\ 4 \\ -2 \end{pmatrix}, v = \begin{pmatrix} 4 \\ -1 \\ 4 \end{pmatrix}$     (d) $u = \begin{pmatrix} -7 \\ 2 \\ -3 \end{pmatrix}, v = \begin{pmatrix} -2 \\ 2 \\ 6 \end{pmatrix}$

## 8.5 Angles between vectors

Using the scalar product, the angle between two vectors $u$ and $v$ can be given by
$$\theta = \arccos\left(\frac{u \cdot v}{|u||v|}\right)$$

As there are two formulae for the scalar product, we can work out the angle between vectors, regardless of whether the vectors are in 2 or 3 dimensions. Since vectors are unchanged if their magnitudes and directions are unchanged, by considering any two vectors that share a common starting point, they can be drawn to be coplanar, and the angle between them identified.

### Example 8.16

Find the angle, to the nearest degree, between each pair of vectors.

(a) $u = \begin{pmatrix} 5 \\ 12 \end{pmatrix}$, $v = \begin{pmatrix} 4 \\ -3 \end{pmatrix}$

(b) $u = \begin{pmatrix} 8 \\ 6 \end{pmatrix}$, $v = \begin{pmatrix} 8 \\ 15 \end{pmatrix}$

(c) $u = \begin{pmatrix} 2 \\ 2 \\ -1 \end{pmatrix}$, $v = \begin{pmatrix} 1 \\ 2 \\ -2 \end{pmatrix}$

(d) $u = \begin{pmatrix} -2 \\ 1 \\ 2 \end{pmatrix}$, $v = \begin{pmatrix} 6 \\ -2 \\ -3 \end{pmatrix}$

### Solution

(a) $u \cdot v = ad + be$
$= 5 \times 4 + 12 \times (-3) = -16$
$|u| = 13, |v| = 5$
$u \cdot v = |u||v| \cos \theta$
$|u||v| \cos \theta = -16$
$\therefore \theta = \arccos\left(\frac{-16}{13 \cdot 5}\right) \approx 104°$

(b) $u \cdot v = 154, |u| = 10, |v| = 17$
$\therefore \theta = \arccos\left(\frac{154}{10 \cdot 17}\right) \approx 25°$

(c) $u \cdot v = 8, |u| = 3, |v| = 3$
$\therefore \theta = \arccos\left(\frac{8}{3 \cdot 3}\right) \approx 27°$

(d) $u \cdot v = -20, |u| = 3, |v| = 7$
$\therefore \theta = \arccos\left(\frac{-20}{3 \cdot 7}\right) \approx 162°$

### Exercise 8.5

1. Find the angle between each pair of vectors, to the nearest degree.

(a) $u = \begin{pmatrix} -3 \\ 1 \end{pmatrix}$, $v = \begin{pmatrix} 2 \\ 6 \end{pmatrix}$

(b) $u = \begin{pmatrix} -5 \\ 4 \end{pmatrix}$, $v = \begin{pmatrix} 2 \\ 3 \end{pmatrix}$

(c) $u = \begin{pmatrix} -2 \\ 1 \\ -2 \end{pmatrix}$, $v = \begin{pmatrix} 3 \\ -2 \\ -2 \end{pmatrix}$

(d) $u = \begin{pmatrix} 3 \\ -2 \\ 2 \end{pmatrix}$, $v = \begin{pmatrix} 4 \\ -3 \\ 9 \end{pmatrix}$

2. Determine whether $\boldsymbol{u}$ is orthogonal, parallel to $\boldsymbol{v}$, or neither.

(a) $\boldsymbol{u} = \begin{pmatrix} -\frac{1}{2} \\ 2 \end{pmatrix}$, $\boldsymbol{v} = \begin{pmatrix} -2 \\ \frac{1}{2} \end{pmatrix}$
   
   (b) $\boldsymbol{u} = \begin{pmatrix} 8 \\ 4 \end{pmatrix}$, $\boldsymbol{v} = \begin{pmatrix} 6 \\ -12 \end{pmatrix}$

(c) $\boldsymbol{u} = \begin{pmatrix} 2\sqrt{3} \\ 2 \end{pmatrix}$, $\boldsymbol{v} = \begin{pmatrix} 1 \\ -\sqrt{3} \end{pmatrix}$

3. Find the interior angles of the triangle $ABC$, the coordinates of whose vertices are given.

   (a) $A(1, 2)$, $B(3, 4)$, $C(2, 5)$

   (b) $A(3, 4)$, $B(-1, -7)$, $C(-8, -2)$

   (c) $A(3, -5)$, $B(1, -9)$, $C(-7, -9)$

4. Find a vector perpendicular to $\boldsymbol{u} = \begin{pmatrix} 3 \\ 5 \end{pmatrix}$

In question 4, your answer will not be unique.

5. Determine if triangle $ABC$ is right-angled by using the scalar product, where the coordinates of its vertices are $A(1, -3)$, $B(2, 0)$, and $C(6, -2)$

6. For what non-zero value(s) of $b$ are the vectors $\begin{pmatrix} -6 \\ b \end{pmatrix}$ and $\begin{pmatrix} b \\ b^2 \end{pmatrix}$ perpendicular?

7. Two vectors $\begin{pmatrix} 3 \\ 4 \end{pmatrix}$ and $\begin{pmatrix} x \\ 1 \end{pmatrix}$ have an angle of 30° between them. Find the possible values of $x$.

8. Use the scalar product to prove that the diagonals of a rhombus are perpendicular to each other.

9. Given the points $A$ and $B$, use the scalar product to find the equation of a circle whose diameter is $[AB]$.

   (a) $A(1, 2)$, $B(3, 4)$
   
   (b) $A(3, 4)$, $B(-1, -7)$

## Chapter 8 practice questions

1. Vector $\boldsymbol{u} = \begin{pmatrix} 8 \\ -2 \end{pmatrix}$

   Determine the components of vector $\boldsymbol{v}$ when:

   (a) $\boldsymbol{v} = -\boldsymbol{u}$
   
   (b) $\boldsymbol{v} = \frac{\boldsymbol{u}}{2}$
   
   (c) $3\boldsymbol{u} = 2\boldsymbol{v}$

2. Vector $\boldsymbol{u} = \begin{pmatrix} 6 \\ -2 \\ 7 \end{pmatrix}$ and $\boldsymbol{v} = \begin{pmatrix} -2 \\ 8 \\ -3 \end{pmatrix}$

   Determine the vector components of $\boldsymbol{w}$ when:

   (a) $\boldsymbol{w} = -\boldsymbol{u} + \boldsymbol{v}$
   
   (b) $\boldsymbol{w} = \boldsymbol{u} - \boldsymbol{v}$
   
   (c) $\boldsymbol{w} = 5\boldsymbol{u} - 2\boldsymbol{v}$

3. Find the equation of each line in the form $\dfrac{x-h}{p} = \dfrac{y-k}{q}$ that passes through the given point $P$ with gradient $m$.

   **(a)** $P(-2, 2)$, $m = \dfrac{3}{5}$             **(b)** $P(4, -1)$, $m = -\dfrac{2}{3}$

   **(c)** $P(0, 4)$, $m = \dfrac{3}{2}$              **(d)** $P(11, 7)$, $m = -\dfrac{4}{3}$

4. Find the equation of the line in 3-space that passes through the given point $P$ with the gradient described by the direction vector $u$.

   **(a)** $P(3, 0, -2)$, $u = \begin{pmatrix} 2 \\ -3 \\ -6 \end{pmatrix}$        **(b)** $P(-4, 4, 0)$, $u = \begin{pmatrix} 5 \\ 0 \\ -2 \end{pmatrix}$

5. Find the equation of the line in vector form which passes through the given point $P$ with direction vector $u$.

   **(a)** $P(-1, 3, -2)$, $u = \begin{pmatrix} 2 \\ -3 \\ 6 \end{pmatrix}$        **(b)** $P(-9, 3, -3)$, $u = \begin{pmatrix} 2 \\ 0 \\ -2 \end{pmatrix}$

6. A radio-controlled boat in a pond moves in a direction described by the vector $u = \begin{pmatrix} -1 \\ 2 \end{pmatrix}$, moving 1 m to the west for every 2 m to the north, every minute. Determine, accurate to 3 significant figures, the distance covered by the boat in:

   **(a)** 1 minute      **(b)** 2.25 minutes      **(c)** 10 minutes      **(d)** $t$ minutes

7. The boat in question **6** starts from a position 20 m to the east of the person controlling it and moves along the same vector $u$. Find the distance from the container to the boat after:

   **(a)** 1 minute      **(b)** $t$ minutes.

8. A second boat, starting at the same time and 25 m to the west of the boat in question **7**, moves in a direction described by the vector $v = \begin{pmatrix} 2 \\ 1 \end{pmatrix}$, moving 2 m to the east for every 1 m to the north, every minute.

   **(a)** Determine the distance between the two boats after:
       **(i)** 1 minute        **(ii)** $t$ minutes

   **(b)** Would the two boats collide?

9. Find the scalar product for each pair of vectors.

   **(a)** $u = \begin{pmatrix} -5 \\ 2 \end{pmatrix}$, $v = \begin{pmatrix} 2 \\ 1 \end{pmatrix}$           **(b)** $u = \begin{pmatrix} -3 \\ -6 \end{pmatrix}$, $v = \begin{pmatrix} -6 \\ 3 \end{pmatrix}$

   **(c)** $u = \begin{pmatrix} 8 \\ 2 \\ -7 \end{pmatrix}$, $v = \begin{pmatrix} -1 \\ 4 \\ 0 \end{pmatrix}$        **(d)** $u = \begin{pmatrix} -2 \\ 2 \\ -1 \end{pmatrix}$, $v = \begin{pmatrix} 2 \\ -3 \\ 6 \end{pmatrix}$

**10.** Find the scalar product for each pair of vectors, where $\theta$ is the angle between them.

(a) $|u| = 7$, $|v| = 11$, $\theta = 60°$      (b) $|u| = 11.2$, $|v| = 5$, $\theta = 120°$

(c) $|u| = 9$, $|v| = 9$, $\theta = 45°$      (d) $|u| = 13$, $|v| = 6$, $\theta = 23°$

**11.** Find each vector product: $u \times v$

(a) $u = \begin{pmatrix} 0 \\ 1 \\ 0 \end{pmatrix}$, $v = \begin{pmatrix} 1 \\ 0 \\ 0 \end{pmatrix}$      (b) $u = \begin{pmatrix} -1 \\ 0 \\ 0 \end{pmatrix}$, $v = \begin{pmatrix} 0 \\ -1 \\ 0 \end{pmatrix}$

(c) $u = \begin{pmatrix} 1 \\ 2 \\ 2 \end{pmatrix}$, $v = \begin{pmatrix} 2 \\ 1 \\ 2 \end{pmatrix}$      (d) $u = \begin{pmatrix} 5 \\ 2 \\ -2 \end{pmatrix}$, $v = \begin{pmatrix} 2 \\ 1 \\ 6 \end{pmatrix}$

**12.** Find the angle between each pair of vectors, correct to the nearest degree.

(a) $u = \begin{pmatrix} -9 \\ 12 \end{pmatrix}$, $v = \begin{pmatrix} 4 \\ 3 \end{pmatrix}$      (b) $u = \begin{pmatrix} 13 \\ -12 \end{pmatrix}$, $v = \begin{pmatrix} -15 \\ -8 \end{pmatrix}$

(c) $u = \begin{pmatrix} 2 \\ 1 \\ 2 \end{pmatrix}$, $v = \begin{pmatrix} 1 \\ -2 \\ -2 \end{pmatrix}$      (d) $u = \begin{pmatrix} -1 \\ 2 \\ 2 \end{pmatrix}$, $v = \begin{pmatrix} 3 \\ 6 \\ -2 \end{pmatrix}$

**13.** Ryan and Jack have model aeroplanes which take off from level ground. Jack's aeroplane takes off after Ryan's.

The position of Ryan's aeroplane $t$ seconds after it takes off is given by

$$r = \begin{pmatrix} 5 \\ 6 \\ 0 \end{pmatrix} + t \begin{pmatrix} -4 \\ 2 \\ 4 \end{pmatrix} \text{m}$$

(a) Find the speed of Ryan's aeroplane.

(b) Find the height of Ryan's aeroplane after two seconds.

The position of Jack's aeroplane $s$ seconds after it takes off is given by:

$$r = \begin{pmatrix} -39 \\ 44 \\ 0 \end{pmatrix} + s \begin{pmatrix} 4 \\ -6 \\ 7 \end{pmatrix} \text{m}$$

(c) Show that the paths of the aeroplane are perpendicular.

The two aeroplane collide at the point $(-23, 20, 28)$.

(d) How long after Ryan's aeroplane takes off does Jack's aeroplane take off?

**14.** A line $L$ passes through points $A(-2, 4, 3)$ and $B(-1, 3, 1)$.

(a) (i) Show that $\overrightarrow{AB} = \begin{pmatrix} 1 \\ -1 \\ -2 \end{pmatrix}$      (ii) Find $|\overrightarrow{AB}|$

(b) Find a vector equation for $L$.

The diagram shows the line $L$ and the origin $O$.

The point $C$ also lies on $L$.

Point $C$ has position vector $\begin{pmatrix} 0 \\ y \\ -1 \end{pmatrix}$

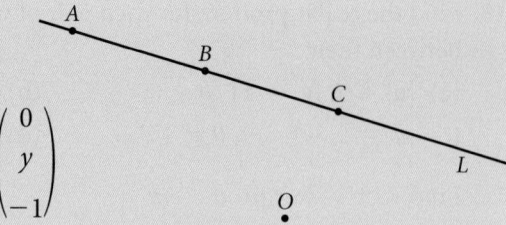

(c) Show that $y = 2$

(d) (i) Find $\overrightarrow{OC} \cdot \overrightarrow{AB}$

    (ii) Hence, write down the size of the angle between $OC$ and $L$.

(e) Hence or otherwise, find the area of triangle $OAB$.

15. Let $\mathbf{u} = -3\mathbf{i} + \mathbf{j} + \mathbf{k}$ and $\mathbf{v} = m\mathbf{j} + n\mathbf{k}$, where $m, n \in \mathbb{R}$. Given that $\mathbf{v}$ is a unit vector perpendicular to $\mathbf{u}$, find the possible values of $m$ and of $n$.

16. The points $A$ and $B$ lie on a line $L$, and have position vectors $\begin{pmatrix} -3 \\ -2 \\ 2 \end{pmatrix}$ and $\begin{pmatrix} 6 \\ 4 \\ -1 \end{pmatrix}$ respectively. Let $O$ be the origin.

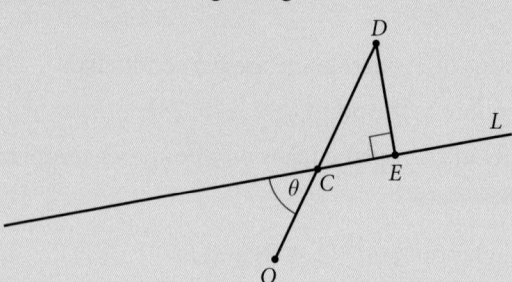

(a) Find $\overrightarrow{AB}$

The point $C$ also lies on $L$, such that $\overrightarrow{AC} = 2\overrightarrow{CB}$

(b) Show that $\overrightarrow{OC} = \begin{pmatrix} 3 \\ 2 \\ 0 \end{pmatrix}$

Let $\theta$ be the angle between $\overrightarrow{AB}$ and $\overrightarrow{OC}$.

(c) Find $\theta$.

Let $D$ be a point such that $\overrightarrow{OD} = k\,\overrightarrow{OC}$, where $k > 1$. Let $E$ be a point on $L$ such that $C\hat{E}D$ is a right angle.

(d) (i) Show that $\left|\overrightarrow{DE}\right| = (k - 1)\left|\overrightarrow{OC}\right|\sin\theta$

    (ii) The distance from $D$ to line $L$ is less than 3 units.
        Find the possible values of $k$.

# Modelling real-life phenomena

9

By the end of this chapter, you should be familiar with…

- modelling linear, quadratic, cubic, exponential, direct/inverse variation, trigonometric, and logistic phenomena, including piecewise combinations

- developing and fitting models; recognising the context, choosing an appropriate model, determining a reasonable domain and range, using technology to find parameters

- testing and reflecting upon models; commenting on appropriateness and reasonableness of a model, justifying the choice of a model

- using models; reading, interpreting, and making predictions, avoiding dangerous extrapolation.

Mathematical models help us to describe the world around us. In this chapter, we will look at several different kinds of mathematical models. We will examine how to choose, develop, test, apply, and extend a model. Here are some examples of the different kinds of models we see around us:

- On a long flight, if the airspeed of an aeroplane is constant, so the distance remaining to the destination can be described by a **linear** model.

- In a situation where the price to manufacture $x$ units of some product decreases **linearly**, the revenue from selling $x$ units can be described by a **quadratic** model.

- The volume of a balloon relative to its diameter can be described by a **cubic** model.

- The spread of algae in a polluted lake can be described by an **exponential** model.

- A DJ charges a fixed amount to provide music for a party. The cost is spread equally among everyone who attends the party. The cost per person can be described by an **inverse variation** model.

- A population growing exponentially at first but then approaching a maximum can be described by a **logistic** model.

- The price of electricity is often billed per kilowatt-hour (kWh), so the cost of powering the lights of a stadium relative to the time the lights are on can be described by a **direct variation** model.

- The height of a person above the ground on a Ferris wheel can be described by a **trigonometric** model.

The process of mathematical modelling is a design process and is illustrated in Figure 9.1.

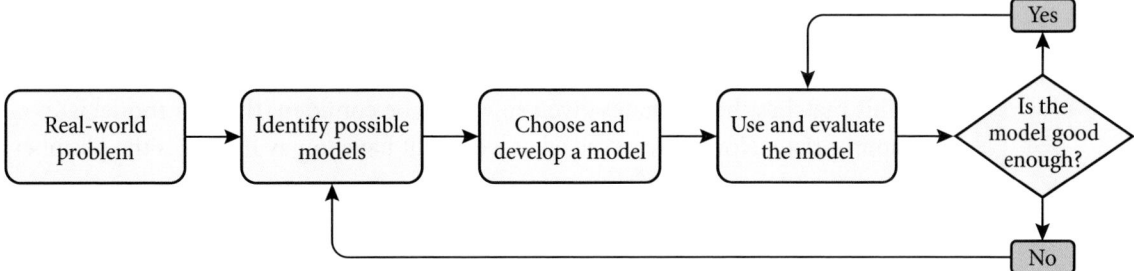

**Figure 9.1** The process of mathematical modelling

In this chapter we will learn how to develop models both by hand and by using technology. We will discuss how to test a model and then how to reflect upon or analyse the validity of the model. Finally, we will talk about how to use and, if needed, extend or revise a model.

# 9.1 Polynomial functions

**Linear models** are used to describe situations where one quantity (the **dependent variable**) increases at a fixed rate relative to another quantity (the **independent variable**).

## Developing and testing a linear model

Suppose that you need a plumbing repair in your home. You call a plumber to ask about how much it will cost. Of course, the plumber cannot give you an exact cost but gives the estimates shown in Table 9.1.

Since we know that the cost must depend on the time required, **cost** is the dependent variable and **time required** is the independent variable.

| Time required (hours) | Cost of repair (€) |
|:---:|:---:|
| 1 | 110 |
| 2 | 185 |
| 3 | 260 |

**Table 9.1** Repair cost estimates

To decide if this situation can be described by a linear model, we need to see if the rate of change is constant. To do this, check the gradient (in this context, the cost per hour) between two or more pairs of points:

$$\frac{185 - 110}{2 - 1} = 75 \quad \text{and} \quad \frac{260 - 185}{3 - 2} = 75$$

Since the cost per hour is constant, a linear model is appropriate for this situation. In addition, we have discovered that the cost per hour is €75. However, there seems to be another part to the cost. We can use the gradient-intercept form of an equation to find a linear model. We will use $y$ for the dependent variable (cost), and $x$ for the independent variable (time):

$$y - y_1 = m(x - x_1) \quad \Rightarrow \quad y - 110 = 75(x - 1) \quad \Rightarrow \quad y = 75x + 35$$

We can see that the plumber's hourly rate is €75/hr, and he adds a fixed amount of €35. This is probably to compensate him for travelling to your home! Finally, it's a good idea to test the model to make sure it describes the situation:

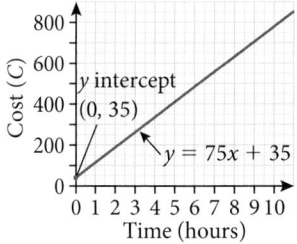

**Figure 9.2** The graph of a linear model of the cost of hiring a particular plumber to come to your home.

299

Remember that in a graph the independent variable is usually placed on the horizontal axis, and the dependent variable is placed on the vertical axis.

Linear models describe situations where the rate of change (gradient) is constant.

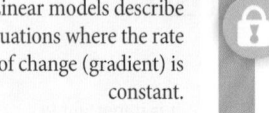

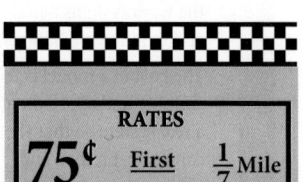

**Figure 9.3** New York City taxi rates from the late 1970s. Linear models are very well suited to this sort of situation

For 1 hour: $y = 75(1) + 35 = 110$

For 2 hours: $y = 75(2) + 35 = 185$

This matches the estimates given so we can be confident that our model is appropriate. Now, we can see what we might have to pay if it takes the plumber a whole 8-hour day to fix our problem:

For 8 hours $y = 75(8) + 35 = €635$

We hope that our plumbing repair is not that serious!

## Extending and revising models

Sometimes we need to revise our initial model to make it more useful. Here is an example.

Consider the taxi costs shown in Figure 9.3. Because the cost increases at a fixed rate relative to the distance driven (10 cents for each $\frac{1}{7}$ mile driven), a linear model is likely to be suitable. Since the cost depends on the number of miles travelled, we will make cost the dependent variable and distance travelled the independent variable. It often helps to make a table showing the independent and dependent variables:

| Miles driven | Calculation | Cost ($) |
|---|---|---|
| $\frac{1}{7}$ | $0.75 + 0$ | 0.75 |
| $\frac{2}{7}$ | $0.75 + 0.10(1)$ | 0.85 |
| $\frac{3}{7}$ | $0.75 + 0.10(2)$ | 0.95 |
| $\frac{4}{7}$ | $0.75 + 0.10(3)$ | 1.05 |

**Table 9.2** Taxi costs

By explicitly showing our calculations, we get a good idea of how to develop our model. From the table, it appears that an appropriate model is Cost $= 0.75 + 0.10x$. But, be careful! In this case, what is $x$? Note that the number we are multiplying by 0.10 is not the number of miles driven – it is the number of $\frac{1}{7}$ miles after the first $\frac{1}{7}$ mile.

Since most people don't think in terms of sevenths of miles, it would be useful if our independent variable was simply distance in miles. Let's revise our table:

| Miles driven | The number of $\frac{1}{7}$ miles after the first $\frac{1}{7}$ mile | Calculation | Cost ($) |
|---|---|---|---|
| $\frac{1}{7}$ | $\frac{1}{7} - \frac{1}{7} = 0$ | $0.75 + 0.10(0)$ | 0.75 |
| $\frac{2}{7}$ | $7\left(\frac{2}{7} - \frac{1}{7}\right) = 1$ | $0.75 + 0.10(1)$ | 0.85 |
| $\frac{3}{7}$ | $7\left(\frac{3}{7} - \frac{1}{7}\right) = 2$ | $0.75 + 0.10(2)$ | 0.95 |
| $\frac{4}{7}$ | $7\left(\frac{4}{7} - \frac{1}{7}\right) = 3$ | $0.75 + 0.10(3)$ | 1.05 |
| $m$ | $7\left(m - \frac{1}{7}\right)$ | $0.75 + 0.10x$ | $C$ |

**Table 9.3** Revised table for taxi costs

To change 'miles driven' into 'the number of $\frac{1}{7}$ miles after the first $\frac{1}{7}$ mile', we need to subtract $\frac{1}{7}$ (representing the first $\frac{1}{7}$ mile) and then multiply by 7 (so that each $\frac{1}{7}$ mile is counted as one unit). So now we know that

$C = 0.75 + 0.10x$ and $x = 7\left(m - \frac{1}{7}\right)$ where $m$ is the number of miles.

By substituting the equation for $x$ into the equation for $C$, we get

$$C = 0.75 + 0.10\left(7\left(m - \frac{1}{7}\right)\right) \;\Rightarrow\; C = 0.7m + 0.65$$

A graph of this function is shown in Figure 9.4.

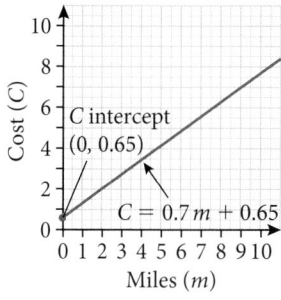

**Figure 9.4** A graph of the linear model for the cost of hiring a New York taxi in the 1970s

## Models don't often capture reality perfectly

If we look at the graph of our model from the taxi cost example, we see that the model suggests that for a journey of 0 miles, we will pay $0.65. However, we know that we will always pay at least $0.75. What went wrong?

The problem here is that the model assumes that the incremental cost (represented by the gradient of the function) is continuous – that is, that the taxi will charge us for any increment of a mile. However, we know that the taxi will charge for each $\frac{1}{7}$ of a mile. To make this point clear, consider what happens if we drive 0.5 miles. The model suggests that the cost would be $C = 0.7(0.5) + 0.65 = \$1.00$

But we know that 0.5 miles is more than $\frac{3}{7}$ of a mile and less than $\frac{4}{7}$ so we would actually get charged

$$C = 0.7\left(\frac{4}{7}\right) + 0.65 = \$1.05$$

That is, our model works as long as we round the miles up to the nearest $\frac{1}{7}$ of a mile. As stated in the introduction, all models are a simplification of reality. This model can help us see how the cost of the ride relates to the length of the ride, but we need to be careful in using it to predict exact costs.

### Example 9.1

The I&T Fitness Centre charges a one-time joining fee of $100 and then charges $2 per visit.

(a) Develop a model for this situation.

(b) Use your model to find the cost of 20 visits.

(c) Draw a graph of your model.

### Solution

(a) Since the rate is already given ($2 per visit), plus a one-time cost of $100, we can write the model directly:

$$C = 100 + 2v$$

where $C$ is the total cost and $v$ is the number of visits.

(b) Using our model, 20 visits would cost $C = 100 + 2(20) = \$140$

(c)

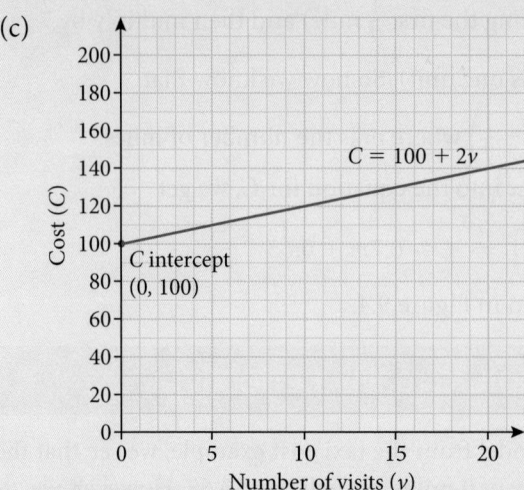

The graph in Example 9.1 has been drawn with a continuous line, even though it is only possible to visit a whole number of times. For convenience, we will often draw a graph as a continuous line even though that may not strictly represent reality.

## Example 9.2

A visitor to the I&T Fitness Centre decides not to buy a membership and instead just pays the daily rate. The first visit costs her $12. At the end of the month, she has visited 5 more times and paid a total of $60 for those five visits.

(a) Develop a linear model for this the total cost after $v$ visits.

(b) Use your model to predict the cost of 20 visits.

(c) After how many visits is it better value to buy the membership given in Example 9.1?

## Solution

(a) One visit costs $12, and five visits cost $60. To be sure there are no extra fees, check that the cost for five visits is the same ratio as the cost for one visit. For five visits, the cost per visit is $\frac{60}{5} = \$12$

Therefore, the rate of change is constant and we can write the model:

$C = 12v$ where $C$ is the total cost and $v$ is the number of visits.

(b) Using our model, 20 visits would cost $C = 12(20) = \$240$

This is much more expensive than the membership plan in Example 9.1!

(c) To find when the membership plan in Example 9.1 is better value, we will first find when the two plans are equal. We can do this algebraically or graphically.

Algebraically, we are looking for the number of visits ($v$) that produce the same cost. Therefore,

$$C = 100 + 2v = 12v \implies 10 = v$$

Therefore, the two plans cost the same for ten visits. Since we know that the membership plan costs only $2 per visit, we know that it will be cheaper for any number of visits more than 10.

Graphically, we can graph both models and look for the intersection:

The intersection point of the two graphs tells us that the two plans will cost the same ($120) for 10 visits.

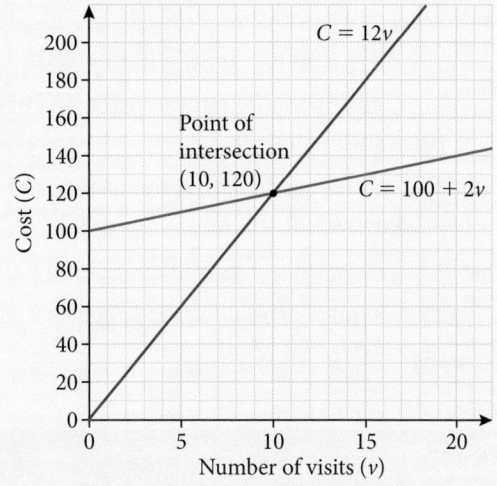

## Interpreting and evaluating linear models

It is important to be able to recognise the structure of a linear model and interpret its meaning. Also, we must be careful to recognise the limitations of linear models.

### Example 9.3

The number of items that a clothing store sells can be modelled by the function $N = 1000 - 5p$ where $N$ is the number of items sold and $p$ is price of a pair of jeans in euros.

(a) Use the model to predict the number of pairs of jeans sold when the jeans are priced at €100.

(b) Interpret the value of the gradient and $N$-intercept in context.

(c) Interpret the value of the $p$-intercept in context.

(d) Use the model to predict the number of pairs of jeans sold when the jeans are priced at €500 per pair. Give a reason why this prediction is not reasonable.

### Solution

(a) The model predicts that the number of pairs of jeans sold when they are priced at €100 is:

$$N = 1000 - 5(100) = 500 \text{ pairs of jeans.}$$

(b) The gradient of $-5$ means that for each euro that the price increases, the number of pairs of jeans sold decreases by 5. The $N$-intercept of $(0, 1000)$ suggests that the number of pairs of jeans sold when the jeans are free will be 1000 – probably not realistic!

Example 9.3 shows that linear models often become nonsensical for certain extreme values of the independent variable. For this reason, it's best to give a limitation on the domain of the model. In the model above, it is sensible to limit the domain of the model to $1 \leqslant p \leqslant 200$. The upper bound of $p = 200$ is when the model predicts that $N$ will be zero.

(c) The $p$-intercept occurs when $N = 0$. Therefore, we must solve

$$0 = 1000 - 5p \text{ to obtain } p = 200$$

This tells us that when the price of a pair of jeans is €200, the number of pairs sold will be 0.

(d) The model predicts that the number of pairs of jeans sold when they are priced at €500 will be:

$$N = 1000 - 5(500) = -1500 \text{ pairs}$$

This is not reasonable as it suggests that they will sell a negative number of jeans!

## Quadratic models

Quadratic models appear frequently in many real-world situations including area, economics, projectile motion, and falling objects. In this section we will look primarily at quadratic models of the form $y = ax^2 + bx + c$

Quadratic models are used in situations where the rate of change in the dependent variable changes linearly with respect to the independent variable. For example:

- Objects falling due to gravity and projectile motion: the acceleration is constant, the velocity follows a linear model, and the displacement (position) follows a quadratic model.

- Revenue models: the number of items sold of some item based on the price follows a linear model; the revenue from selling the number of items follows a quadratic model.

Quadratic models also have some geometric properties that make them well-suited to designing satellite dishes and to model bridge spans.

Key properties of quadratic functions:

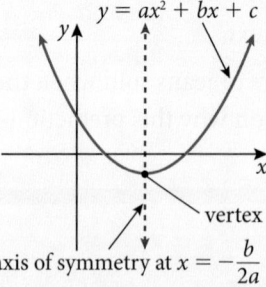

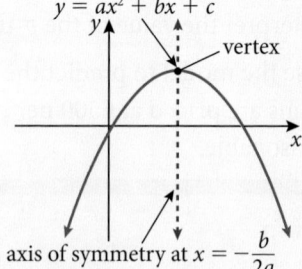

If $a > 0$ then the parabola is concave up     If $a < 0$ then the parabola is concave down

For a quadratic function of the form $y = ax^2 + bx + c$, where $a \neq 0$:

- The graph of a quadratic function is roughly U-shaped and is called a **parabola**.

- Concavity: The graph of the function is **concave up** if and only if $a > 0$
  It is **concave down** if and only if $a < 0$

- Symmetry: The graph of the function is symmetrical about the vertical line with equation
  $x = -\dfrac{b}{2a}$. This line is called the **axis of symmetry**.

- Maximum/minimum: The function has a **maximum** (when concave down) or **minimum** (when concave up) where the graph intersects the axis of symmetry. This point is called the **vertex** of the parabola. The $x$-coordinate of the vertex is therefore given by $x = -\dfrac{b}{2a}$
- The **$y$-intercept** of the graph is $(0, c)$
- The **$x$-intercepts** of the graph, also called the **zeros** of the function, can be found using the quadratic formula: $x = \dfrac{-b \pm \sqrt{b^2 - 4ac}}{2a}$

  Remember that a quadratic graph may have zero, one, or two $x$-intercepts.

## Example 9.4

A ball is thrown upwards from the top of a building. The height of the ball from ground level can be modelled by the function $h(t) = -4.9t^2 + 11t + 50$ where $h(t)$ is the height of the ball in metres and $t$ is the time in seconds after the ball was thrown.

(a) Sketch a graph of the function.

(b) Write down the height of the building.

(c) Find the time when the ball reaches its maximum height.

(d) Find the maximum height reached by the ball.

(e) Find the time when the ball hits the ground.

(f) Describe a reasonable domain and range for this model.

### Solution

We can solve this problem by using either an algebraic approach (by hand) or using a GDC to analyse the graph. We will show both approaches here.

**Algebraic approach**

(a) A sketch of the function is shown in the diagram.

We know the ball travels upwards before coming back down and we know the parabola is concave down since $a < 0$

(b) The height of the building is the initial height of the object which is the height of the ball when $t = 0$: in other words, the $y$ intercept. Therefore, the height of the building is 50 m.

(c) The time when the ball reaches its maximum height is the $t$-coordinate of the vertex of the parabola. Find the $t$ coordinate, which is also the location of the axis of symmetry:

$$t = -\frac{b}{2a} = -\frac{11}{2(-4.9)} = 1.12 \text{ s (3 s.f.)}$$

(d) To find the maximum height reached by the ball, substitute the $t$ value from (c) back into the function:

$$h(1.12) = -4.9(1.12)^2 + 11(1.12) + 50 = 56.2 \text{ m (3 s.f.)}$$

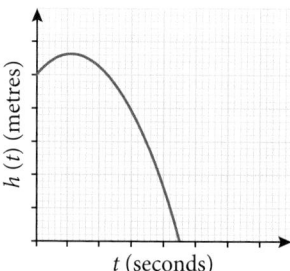

**Figure 9.5** Sketch the graph of $h$

(e) The ball hits the ground when the height is zero.

Therefore, we must solve the equation $0 = -4.9t^2 + 11t + 50$

Using the quadratic formula:

$$t = \frac{-b \pm \sqrt{b^2 - 4ac}}{2a} = \frac{-11 \pm \sqrt{11^2 - 4(-4.9)(50)}}{2(-4.9)} = 4.51 \text{ s (3 s.f.)}$$

(f) We should limit the domain to positive values of $t$, from $t = 0$ (when the ball is thrown) until $t = 4.51$ (when the ball hits the ground). Therefore a reasonable domain for this model is $0 \leqslant t \leqslant 4.51$(s) The range indicates the possible heights of the ball. Here, $0 \leqslant h(t) \leqslant 56.2$(m) makes sense since we know the maximum height of the ball and we presume the ball does not go underground.

### GDC approach

(a) Here we use a GDC to obtain a graph. We need to be careful to adjust the viewing window appropriately.

On a GDC, we need to use $x$ in place of $t$. It is important to note that this graph is height versus time – it is not the trajectory (path) of the object. The horizontal axis represents time, not distance.

(b) The Graph Trace feature can be used to evaluate the function at $x = 0$ to find the initial height.

(c) We can use a GDC to find the maximum height.

We see that the $x$-coordinate is 1.12, so the ball reaches a maximum height at 1.12 s.

(d) Using the same point from part (c), we see that the $y$-coordinate is 56.2, so the ball reaches a maximum height at 56.2 m.

(e) We can use our GDC to find the positive zero of the function.

Therefore, the ball hits the ground after 4.51 s.

(f) We use the same logic as in the algebraic approach to conclude that a reasonable domain is $0 \leqslant t \leqslant 4.51$(s) and a reasonable range is $0 \leqslant h(t) \leqslant 56.2$(m)

In the next few examples we will look at building a quadratic function.

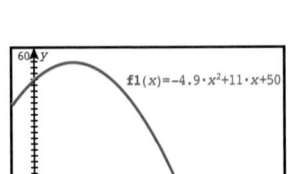

**Figure 9.6** GDC screen for solution to Example 9.4(a)

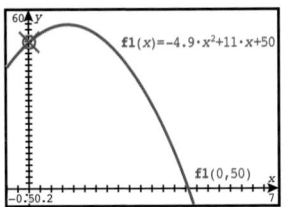

**Figure 9.7** GDC screen for solution to Example 9.4(b)

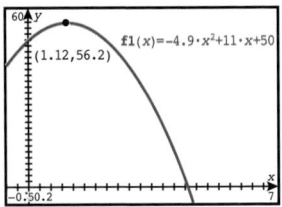

**Figure 9.8** GDC screen for solution to Example 9.4(c)

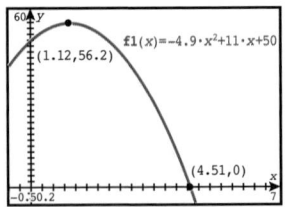

**Figure 9.9** GDC screen for solution to Example 9.4(e)

### Example 9.5

The demand for a certain style of jeans is modelled by the function $D = 1000 - 5p$ where $D$ is the number of items sold and $p$ is price of the jeans in euros.

(a) Develop a model for the revenue earned from selling these jeans based on the selling price $p$.

(b) Use your model to find the price the jeans should be sold at in order to maximise revenue.

(c) Find the maximum revenue predicted by your model.

(d) Give a reasonable domain and range for your model.

## Solution

(a) In general,

revenue = selling price × number of units sold

Therefore, we can develop a model for revenue, $R$, by multiplying the selling price, $p$, by the expression for the demand:

$$R = p(1000 - 5p) \Rightarrow R = -5p^2 + 1000p$$

(b) Since this is a quadratic model with a concave-down graph $(a < 0)$, we know there will be a maximum at the vertex. In this case, the $p$-coordinate of the vertex is the selling price and the $R$-coordinate is the revenue. To find the price, we use the formula for the $p$-coordinate of the vertex:

$$p = -\frac{b}{2a} = -\frac{1000}{2(-5)} = 100 \text{ euros}$$

(c) To find the maximum revenue, we need to evaluate the model for 100 pairs of jeans (from part (b)):

$$R = -5(100)^2 + 1000(100) = 50\,000 \text{ euros}$$

(d) Clearly, our model doesn't make much sense if the price $p$ is less than zero. But what about a maximum price? Since we know that this is a concave-down quadratic function with a vertex above the $p$-axis, we know there will be two $p$-intercepts. Solving algebraically, we have:

$$0 = -5p^2 + 1000p \Rightarrow p = 0, p = 200$$

Therefore the $p$-intercepts are 0 and 200.
So $0 \leqslant p \leqslant 200$ is a reasonable domain.
The range is given by the minimum and maximum values of the function on this domain, $0 \leqslant R \leqslant 50\,000$

It's often helpful to generate a graph of the model. In this case, we can clearly see the maximum revenue and can use a GDC to verify our results in parts (c) and (d).

Note that value 5E+4 for the $y$-coordinate of the vertex is the GDC's version of scientific notation (or standard form). i.e.

$$5E+4 = 5 \times 10^4 = 50\,000$$

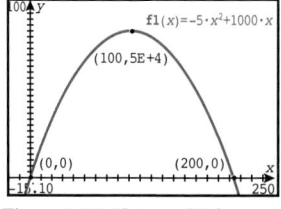

**Figure 9.10** Using a GDC to verify results

When using a GDC, it's important to take care when setting the viewing window. The default view is often set at $-10 \leqslant x \leqslant 10$ and $-10 \leqslant y \leqslant 10$ or less. For example, the display for the model in Example 9.5 shows how this default view can be very misleading.

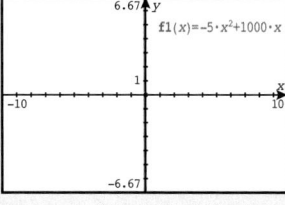

The graph is there but with the current settings it is almost indistinguishable from the $y$-axis. We can use our knowledge of the function to choose more appropriate settings.

After setting the $x$-axis to a suitable domain, in this case $0 \leqslant x \leqslant 300$, and using the Zoom Fit feature to scale the $y$-axis to fit the function, we get the result on the right. The negative part of the graph isn't very useful, but you can use the Zoom Box function to examine the graph more closely.

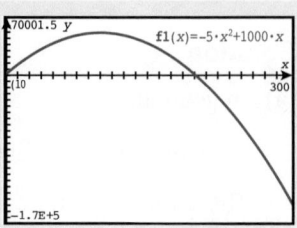

## Example 9.6

The Sydney Harbour Bridge is supported by two spans that can be modelled by quadratic functions.

The lower span is approximately 503 m wide and 118 m tall at its highest point.

Develop a quadratic model for the lower span such that one end of the span is positioned at $(0, 0)$.

### Solution

We can start by drawing our axes and locating the vertex at the maximum point of the lower span. Since the vertex must be halfway along the length of the bridge, its $x$-coordinate must be 251.5

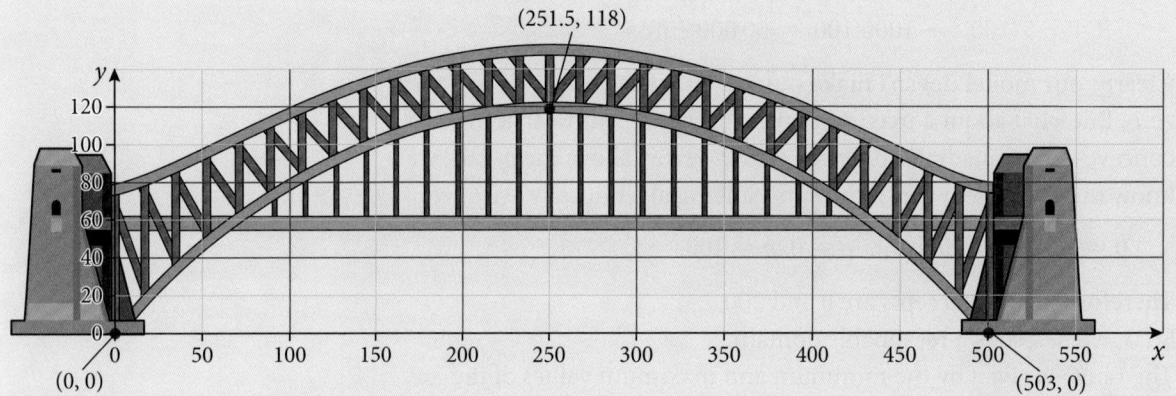

Then, we can use the general quadratic equation $y = ax^2 + bx + c$ and some algebra to find the model for the lower span. First, since the graph must pass through $(0, 0)$, it must be true that

$$0 = a(0)^2 + b(0) + c \Rightarrow c = 0$$

So

$$y = ax^2 + bx$$

Next, since the graph must also pass through $(251.5, 118)$ and $(503, 0)$, we can generate a system of equations by substituting each coordinate pair into the model:

$$\begin{cases} 118 = a(251.5)^2 + b(251.5) \\ 0 = a\,503^2 + b(503) \end{cases} \Rightarrow \begin{cases} 63\,252.25a + 251.5b = 118 \\ 25\,53009a + 505b = 0 \end{cases}$$

We can solve this system of equations (algebraically, or by using a GDC) to obtain:

$$a = -0.001866$$

$$b = 0.9384$$

Therefore, our model is $y = -0.001866x^2 + 0.9384x$

If we graph this function with appropriate axes, we can see that it fits the lower span very well.

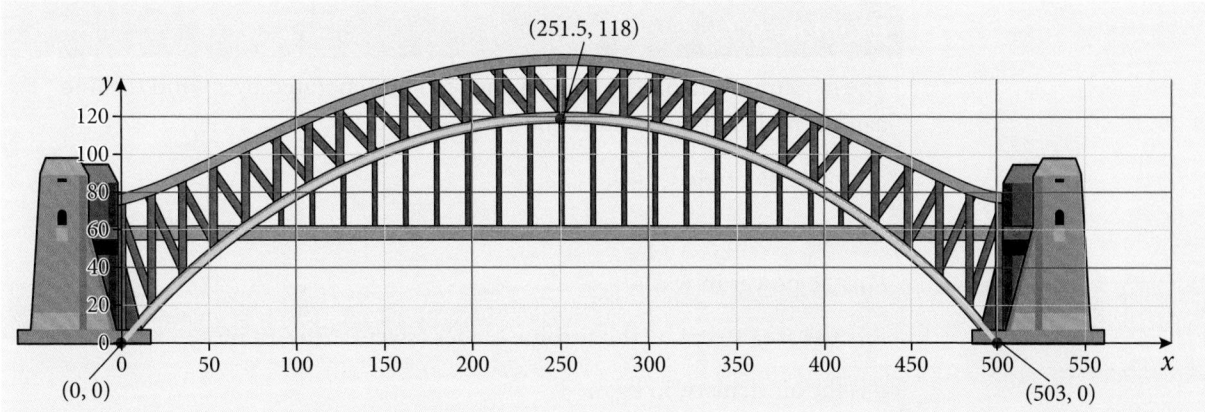

(251.5, 118)

(0, 0)

(503, 0)

Note that most GDCs have a quadratic regression feature that can also find the model.

Note that the values for $a$ and $b$ calculated by the GDC agree with our values.

| | A x | B y | C | D | E |
|---|---|---|---|---|---|
| = | | | | =QuadReg( | |
| 2 | 0 | 0 | Title | Quadratic... | |
| 3 | 251.5 | 118 | RegEqn | a*x^2+b*x... | |
| 4 | 503 | 0 | a | −0.001866 | |
| 5 | | | b | 0.93837 | |
| 6 | | | c | 0. | |
| E4 | | | | | |

Remember that the graph may appear differently depending on the viewing window chosen on your GDC. For example, if the scale of the units on the $x$ and $y$ axes is not 1:1 the function appears taller than it should. Using the GDC's Zoom Square function will fix this.

## Cubic models

Cubic models are slightly more complex than quadratic models. We often encounter cubic models when we are dealing with quantities based on volume (such as optimising the volume or surface area of a package or calculating forces from wind or water). Cubic models are also widely used in computer graphics, in everything from modelling the curves of the letters in this textbook to smoothing computer-generated effects and animation. In the section, we will examine cubic models of the form

$$f(x) = ax^3 + bx^2 + cx + d$$

Note that a simple cubic model may have $b$, $c$, and $d$, equal to zero. In that case, it can also be considered a direct variation model, which we will study later in this chapter.

## Example 9.7

The maximum theoretical power that can be generated by a wind turbine can be modelled by the function

$$P = 0.297AdV^3$$

where

$P$ is the power in watts

$A$ is the area swept by the turbine blades (swept area), in $m^2$

$d$ is the air density, in $kg\,m^{-3}$

$V$ is the wind speed in $m\,s^{-1}$

A certain wind turbine has a swept area of $80\,m^2$ and is located at sea level where the air density is $1.225\,kg\,m^{-3}$

(a) Find the cubic model for this wind turbine.

(b) Use your model to calculate the maximum theoretical power generated when the wind speed is $10\,m\,s^{-1}$

(c) Given that wind speeds above $20\,m\,s^{-1}$ are strong enough to cause damage, determine if this turbine could produce 300 000 watts of power.

(d) Determine a reasonable domain for this model.

### Solution

(a) Substitute the known values to find the model for this particular turbine. Therefore, the cubic model for this turbine is

$$P = 0.297AdV^3$$
$$= 0.297(80)(1.225)V^3 = 29.1V^3$$

(b) The maximum theoretical power generated when the wind speed is $10\,m\,s^{-1}$ is

$$P = 29.1V^3 = 29.1(10)^3 = 29\,100 \text{ watts (3 s.f.)}$$

(c) Using the model, we can find the wind speed required to produce 300 000 watts:

$$P = 29.1V^3 \Rightarrow 300\,000 = 29.1V^3 \Rightarrow V = \sqrt[3]{10309} = 21.8\,m\,s^{-1}$$

Since this is greater than the wind speed that will cause damage, it is not reasonable to expect this turbine to generate 300 000 watts.

(d) Since wind speeds above $20\,m\,s^{-1}$ are strong enough to cause damage to the turbine, that should be the upper limit for our domain, and since negative wind speeds are unrealistic, a reasonable domain is $0 \leqslant V \leqslant 20$

We often use models to find **optimal** solutions. That is, we are interested in maximising or minimising a quantity. Example 9.8 examines a classic problem.

## Example 9.8

The dimensions of a piece of A4 paper, to the nearest centimetre, are $21 \times 30$ cm. It is possible to create an open box by cutting out square corners and folding the remaining flaps up, as shown in the diagram.

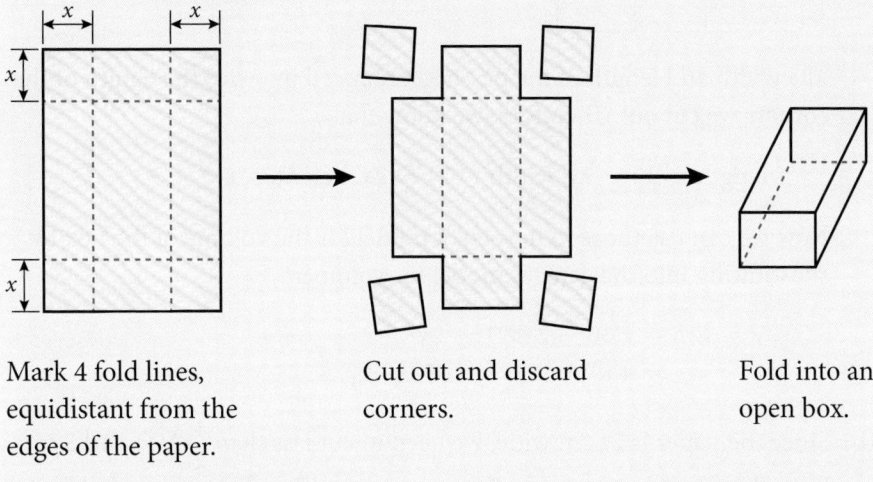

Mark 4 fold lines, equidistant from the edges of the paper.

Cut out and discard corners.

Fold into an open box.

(a) Develop a model in the form $V = ax^3 + bx^2 + cx + d$ for the volume of the open box.

(b) Determine a reasonable domain for your model.

(c) Find the dimensions of the open box with the largest volume that can be created using this method.

(d) Calculate the maximum volume of the open box.

### Solution

(a) The first step in solving this problem is to find an appropriate model. Since we are interested in the volume of the box, it seems appropriate to start with a model for volume:

$$V = lwh$$

Looking at the second step in the diagram, we can see that part of the width and length of the paper becomes the height for the box.

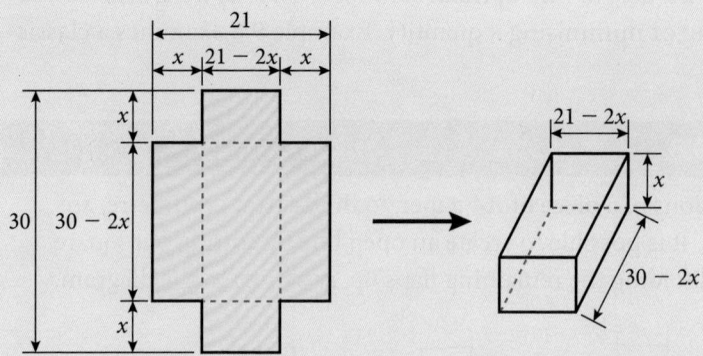

The width and length of the paper get reduced by twice the length of the corners we cut out. Therefore, we know that

$$\text{length} = 30 - 2x, \text{width} = 21 - 2x, \text{height} = x$$

Now we can use those to develop a model for the volume of the box by substituting into the general model for volume:

$$V = lwh = x(30 - 2x)(21 - 2x)$$
$$= 4x^3 - 102x^2 + 630x$$

(b) Since the paper is 21 cm wide, we need to cut less than $\frac{1}{2}(21) = 10.5$ cm from the edge in order to make an open box. Therefore, a reasonable domain is $0 \leqslant x < 10.5$

(c) Our goal is to find the maximum volume. To do this, we can use a GDC to graph the model and look for a maximum value.

```
☐ f1(x)=(30–2x)(21–2x)(x)          ≡
```

From the graph shown, we can conclude that the value of $x$ that produces the maximum volume is 4.06 cm. To find the dimensions of the box, we need to go back to our expressions for the length, height, and width of the box and we can use these to obtain the missing dimensions:

$$\text{height} = x = 4.06 \text{ cm}$$
$$\text{length} = 30 - 2x = 21.9 \text{ cm}$$
$$\text{width} = 21 - 2x = 12.9 \text{ cm}$$

(d) The volume of the open box is given in GDC's output, since the volume is the $y$-coordinate in our GDC; $V = 1140 \text{ cm}^3$.

---

When we expand the right hand side of the volume equation, we can verify that this is a cubic model. However, since we are going to use a GDC to analyse this model, there is no need to expand the model.

Remember, when using a GDC to analyse a model, it is important to choose the viewing window carefully. Since the $x$-axis represents the distance of each fold from the edge of the paper (which is also equal to the size of each square we cut out), we are only interested in positive $x$ values. Also, because our paper is only 21 cm wide, $x$ must be less than $\frac{21}{2} = 10.5$

Therefore, set your window to $0 \leqslant x < 10.5$ and use your GDC's Zoom fit to scale the $y$-axis accordingly.

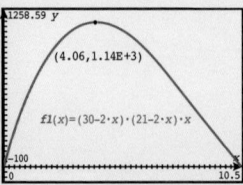

Remember that
$1.14E + 3 = 1.14 \times 10^3$
$= 1140 \text{ cm}^3$

## Piecewise models

Piecewise functions were introduced in Chapter 3. Piecewise functions are often used to model situations that display behaviour that fits with one or more type of function. Although we will introduce the idea of piecewise models in this section, piecewise models are not limited to polynomial functions.

---

### Example 9.9

The position of a certain Formula One race car on a straight track can be modelled by the piecewise function

$$s(t) = \begin{cases} 7.2\,t^2, & 0 \leqslant t < 4.5 \\ 3.47\,t^2 + 55.6t + a, & 4.5 \leqslant t < 6 \\ 97.2(t - 6) + b, & t \geqslant 6 \end{cases}$$

where $s(t)$ is the position in metres at time $t$ in seconds.

(a) Find the value of the constants $a$ and $b$, where $a, b \in \mathbb{R}$

(b) Find the time it will take for the car to travel

    (i)  500 m         (ii)  1 km

(c) Plot a graph of this model on your GDC showing the position of the car during the first 10 s.

(d) By examining the graph, estimate the interval when the car is travelling fastest and hence calculate its maximum speed.

---

### Solution

(a) The position at time $t = 4.5$ must be the same for the first two functions. Therefore, it must be true that $7.2t^2 = 3.47t^2 + 55.6t + a$ when $t = 4.5$ Substituting and solving:

$$7.2(4.5)^2 = 3.47(4.5)^2 + 55.6(4.5) + a \implies a = -175$$

Likewise, the position at time $t = 6$ must be the same for the second and third functions, so we have:

$$3.47(6)^2 + 55.6(6) - 175 = 97.2(6 - 6) + b \implies b = 284$$

(b) (i) To find the time taken to travel 500 m, we need to find the position of the car at the end of each function of the piecewise function. For the first function, the car will be at position $7.2(4.5)^2 = 146$ m

At the end of the second function, the car will be at $3.47(6)^2 + 55.6(6) - 175 = 284$ m

Therefore, the car will reach 500 m during the third function.

Now we can solve: $97.2(t - 6) + 284 = 500 \implies t = 8.22$ s

(ii) Likewise, the car will travel 1 km in
$97.2(t - 6) + 284 = 1000 \implies t = 13.4$ s

(c)

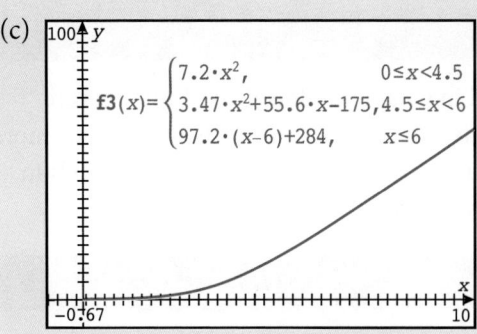

$$f3(x)=\begin{cases} 7.2 \cdot x^2, & 0 \le x < 4.5 \\ 3.47 \cdot x^2 + 55.6 \cdot x - 175, & 4.5 \le x < 6 \\ 97.2 \cdot (x-6) + 284, & x \le 6 \end{cases}$$

(d) It appears that the speed is greatest (the graph is steepest) for the third part of the piecewise function, after $t \ge 6$ s. Since the third part of the function is linear, we can calculate the rate of change directly.

$$s(t) = 97.2(t - 6) + 284 = 97.2t - 299$$
$$\Rightarrow \text{ the speed is } 97.2 \text{ m s}^{-1} \text{ (350 km h}^{-1})$$

## Exercise 9.1

1. An aeroplane is currently 3000 km from its destination, travelling at a constant speed of 900 km h$^{-1}$

   (a) Develop a linear model for the distance $d$ remaining after $t$ hours of travel.

   (b) Interpret the $d$-intercept of your model in context.

   (c) Interpret the $t$-intercept of your model in context.

   (d) State a reasonable domain and range for your model.

2. An aeroplane is currently 5000 km from its destination. 1.5 hours later, it is 3800 km from its destination.

   (a) Develop a linear model for the distance $d$ remaining after $t$ hours of travel.

   (b) Interpret the gradient of your model in context.

   (c) Interpret the $d$-intercept of your model in context.

   (d) Interpret the $t$-intercept of your model in context.

   (e) State a reasonable domain and range for your model.

3. A table showing EU and USA shoe sizes is given.

   (a) Develop a linear model to find the EU shoe sizes given the US shoe sizes.

   (b) Use your model to predict the EU shoe size for a USA Men's shoe size of 12.

   (c) Use your model to predict the USA Men's shoe size for an EU shoe size of 44.

   (d) Interpret the gradient of your model in context.

   (e) Given that USA Men's shoe sizes typically run from 6 to 16, find a reasonable domain and range for your model.

| USA (Men's) | EU |
|---|---|
| 7 | 40 |
| 8 | 41 |
| 9 | 42 |

**Table 9.4** Shoe size data for question 3

**4.** Julie has collected data on how long it takes her to read books, based on the number of pages. The data she collected are shown in the table.

| Number of pages ($n$) | 340 | 290 | 500 |
|---|---|---|---|
| Time to read (minutes) | 490 | 420 | 714 |

(a) Develop a linear model for the time $t$ required to read $n$ pages.

(b) Use your model to predict the time required to read 1000 pages. Give your answer to the nearest 10 minutes.

(c) Interpret the gradient and $t$-intercept of your model in context.

(d) State a reasonable domain and range for your model.

**5.** Given that $68°F = 20°C$ and $212°F = 100°C$,

(a) Develop a linear model for degrees Fahrenheit ($F$) in terms of degrees Celsius ($C$).

(b) Explain, in context, what the gradient of your model represents.

(c) Interpret the $F$-intercept of your model in context.

(d) Interpret the $C$-intercept of your model in context.

(e) Use your model to convert 10°C into °F.

(f) Use your model to find the numerical value in °C that is the same in °F.

(g) Given that absolute zero (the lowest possible temperature) is $-273°C$, calculate a reasonable domain and range for your model.

**6.** Two plastic cup factories, Cups R Us and Cupomatic, can produce cups printed with the image of your choice. At Cups R Us, the mandatory setup and design cost is ZAR350 and the cost per cup is ZAR8.50

(a) Develop a linear model for the cost $C$ of an order at Cups R Us based on the number of cups $n$.

(b) Write down a reasonable domain and range for your model.

(c) Use your model to calculate the cost of an order of:

    (i) 100 cups     (ii) 200 cups     (iii) 400 cups.

(d) Calculate the average cost per cup for:

    (i) 100 cups     (ii) 200 cups     (iii) 400 cups.

(e) Hence, give a reason why, in general, it is more cost-effective to order more cups.

Cupomatic charges ZAR2150 for 200 cups and ZAR3750 for 400 cups.

(f) Develop a linear model for the cost $D$ of an order at Cupomatic based on the number of cups $n$.

(g) Write down a reasonable domain and range for your model.

(h) Interpret the gradient of your model in context.

(i) Use your linear model to predict the cost of 600 cups.

(j) For orders of more than $x$ cups, it is less expensive to order from Cupomatic. Find the value of $x$.

Cups R Us will waive the setup and design cost if a customer orders at least 500 cups.

(k) Develop a piecewise model for the cost $C$ of an order at Cups R Us based on the number of cups $n$.

(l) The main competitor of Cups R Us is Cupomatic. You are given that the model for the cost $D$ of ordering $n$ cups from Cupomatic is
$$D = 8n + 550$$

It is less expensive to order from Cupomatic if $x$, the number of cups ordered, is in the interval $a \leqslant x < b$ or $x > k$

Find the values of $a$, $b$, and $k$.

7. As of 2018, taxi cabs for working hours in London, England, are:

For the first 234.8 m or 50.4 s (whichever is reached first) there is a minimum charge of £2.60

For each additional 117.4 m or 25.2 s (whichever is reached first), or part thereof, if the distance travelled is less than 9656.1 m there is a charge of £0.20

Once the distance has reached 9656.1 m there is a charge of £0.20 for each additional 86.9 m or 18.7 s (whichever is reached first), or part thereof.

(a) Develop a piecewise linear model for the cost $C$ of a taxi ride based on the distance travelled $m$ in metres.

*Ignore the 'part thereof' and interpret as a linear rate.*

(b) Find the cost of 0.2, 5, and 15 km rides.

(c) Develop a piecewise linear model for the cost $D$ of a taxi ride based on the time taken $t$ in seconds, ignoring distance.

(d) Find the cost of 0.5, 5, and 15 minute rides.

(e) Given that the actual taxi fare is always the greater of the two models, find:

(i) the cost of a ride that takes 20 minutes to go 4 km

(ii) the cost of a ride that takes 5 minutes to go 4 km.

**8.** On Earth, the position of a falling object can be modelled by the function $h(t) = -4.9t^2 + v_0 t + h_0$ where $h(t)$ is the height in metres after $t$ seconds, $v_0$ is the initial velocity and $h_0$ is the initial height.

   **(a)** Formulate a model for the height of a ball thrown upwards with an initial velocity of $5\,\text{m s}^{-1}$ from the roof of a 60 m tall building.

   **(b)** Use your model to find:

      **(i)** the maximum height of the ball

      **(ii)** the time until the ball hits the ground

      **(iii)** the interval of time for which the ball is more than 50 m above the ground.

**9.** A rock falls off the top of a cliff. Let $h$ be its height above ground in metres, after $t$ seconds. The table on the right gives values of $h$ and $t$.

Jane thinks that the function $f(t) = -0.25t^3 - 2.32t^2 + 1.93t + 106$ is a suitable model for the data. Use Jane's model to:

   **(a)** write down the height of the cliff

   **(b)** find the height of the rock after 4.5 s

   **(c)** find after how many seconds the height of the rock is 30 m.

Kevin thinks that the function $g(t) = -5.2t^2 + 9.5t + 100$ is a better model for the data.

   **(d)** Use Kevin's model to find the value of $t$ when the rock hits the ground.

   **(e)** Create graphs of $f$, $g$, and the data given. By comparing the graphs of $f$ and $g$ with the plotted data, explain which function is a better model for the height of the falling rock.

| $t$ (s) | $h$ (m) |
|---|---|
| 1 | 105 |
| 2 | 98 |
| 3 | 84 |
| 4 | 60 |
| 5 | 26 |

**Table 9.5** Data for question 9

**10.** The diagram shows two ships $A$ and $B$. At noon, ship $A$ was 15 km due north of ship $B$. Ship $A$ was moving south at $15\,\text{km h}^{-1}$ and ship $B$ was moving east at $11\,\text{km h}^{-1}$.

Find the distance between the ships:

   **(a)** at 13:00

   **(b)** at 14:00.

Let $s(t)$ be the distance between the ships $t$ hours after noon, for $0 \leqslant t \leqslant 4$

   **(c)** Show that $s(t) = \sqrt{346\,t^2 - 450t + 225}$

   **(d)** Plot the graph of $s(t)$.

   **(e)** Due to poor weather, the captain of ship $A$ can only see another ship if they are less than 8 km apart. Explain why the captain cannot see ship $B$ between noon and 16:00.

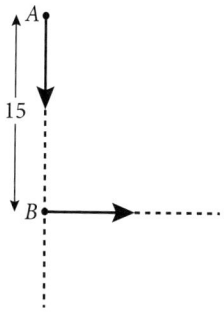

**Figure 9.11** Diagram for question 10

**11.** A small manufacturing company makes and sells $x$ machines each month. The monthly cost $C$, in dollars, of making $x$ machines is given by

$$C(x) = 2600 + 0.4x^2$$

The monthly income $I$, in dollars, obtained by selling $x$ machines is given by $I(x) = 150x - 0.6x^2$

(a) Show that the company's monthly profit can be calculated using the quadratic function $P(x) = -x^2 + 150x - 2600$

(b) The maximum profit occurs at the vertex of the function $P(x)$. How many machines should be made and sold each month for a maximum profit?

(c) If the company does maximise profit, what is the selling price of each machine?

(d) Given that $P(x) = (x - 20)(130 - x)$ find the smallest number of machines the company must make and sell each month in order to make a positive profit.

12. A base jumper leaping from a bridge falls according to the model $s(t) = -4.9t^2 + 300$ where $s(t)$ is the distance in metres above the ground at time $t$, in seconds, until she opens her parachute. After she opens her parachute, she falls at a constant rate of $7 \text{ m s}^{-1}$.

(a) Write down the height of the bridge she is jumping from.

(b) On her first jump, the base jumper opens her parachute after 4 s of free-fall.

   (i) Find the height of the base jumper at the instant she opens her parachute.

   The piecewise model for the height of the base jumper can be written as

   $$h(t) = \begin{cases} -4.9t^2 + 300, & t < t_0 \\ -7t + k, & t \geq t_0 \end{cases}$$

   (ii) Write down the value of $t_0$.

   (iii) Find the value of $k$.

   (iv) Find the height of the base jumper at 1, 2, and 8 s.

   (v) For how many seconds is the base jumper at least 100 m above the ground?

   (vi) Find the time when the base jumper touches the ground.

   (vii) Write down a suitable domain and range for this piecewise model.

(c) On a second jump, the base jumper now opens her parachute when she is 100 m from the ground.

   (i) The base jumper opened her parachute after $x$ s of free-fall. Find the value of $x$.

   (ii) Develop a piecewise model for the height $g$ of the parachutist after $t$ s.

   (iii) Find the time when the base jumper touches the ground.

**(d)** On a third jump, the base jumper opens her parachute 2 s before she would have otherwise hit the ground.

    **(i)** Find the time at which she opens her parachute.

    **(ii)** At the time the base jumper opens her parachute, she is $d$ m above the ground. Find the value of $d$.

    **(iii)** Develop a piecewise model for the height $f$ of the base jumper after $t$ s.

    **(iv)** Find the time when the base jumper touches the ground.

# 9.2   Exponential and logarithmic models

Exponential models arise in situations where the rate of change is a constant **factor**, that is, when the next value is found by multiplying by a constant factor. This sort of change can produce surprising results.

## Just how fast is exponential growth?

A classic fable about the game of chess goes something like this:

A great long time ago a wise man invented the game of chess. This king of this land was so pleased with this new game that he offered the wise man the riches of his kingdom as a gift. The wise man replied, "My needs are modest. Instead of your riches, please place one grain of rice on the first square of the chessboard, two grains of rice on the second square, four grains of rice on the third square, and so on, doubling the number of grains of rice for each of the remaining squares."

The king replied, "What a silly man! I offer him the riches of my kingdom and all he asks for is a few grain of rice!"

How much rice did the wise man ask for? To investigate this, let's build a table. Since we are multiplying by 2 each time, we can write the multiplication using exponents.

| Square | Grains of rice | Grains of rice |
|:---:|:---:|:---:|
| 1 | 1 | $1 = 2^0$ |
| 2 | 2 | $1 \times 2 = 2^1$ |
| 3 | 4 | $1 \times 2 \times 2 = 2^2$ |
| 4 | 8 | $1 \times 2 \times 2 \times 2 = 2^3$ |

**Table 9.6** Grains of rice on the first four squares

Now, we could keep multiplying by 2 each time until we get to the 64th square, but that seems like a lot of work. If we look carefully, we can see a pattern: the exponent of 2 is equal to one less than the number of the square. Now we can add a few more rows to our table.

According to the Food and Agriculture Organisation of the United Nations, the estimated worldwide production of rice in 2017 was 759.6 million tonnes, or $7.596 \times 10^{11}$ kg of rice. Therefore, the amount of rice on the last square alone is $\dfrac{1.84 \times 10^{14}}{7.596 \times 10^{11}} \approx 240$ times more than the entire worldwide harvest of rice in 2017!

| Square | Grains of rice | Grains of rice |
|--------|----------------|----------------|
| 1 | 1 | 1 |
| 2 | 2 | $2^1$ |
| 3 | 4 | $2^2$ |
| 4 | 8 | $2^3$ |
| ... | ... | ... |
| $n$ | | $2^{n-1}$ |
| 64 | | $2^{63}$ |

**Table 9.7** Adding a few more rows to the table

So, we can conclude that the king must place $2^{63}$ grains of rice on the last square. How much rice is that? If we estimate that one kilogram of rice contains approximately 50 000 grains of rice, then we have $\dfrac{2^{63}}{50000} = 1.84 \times 10^{14}$ kg of rice.

To see this more clearly, let's graph the model we developed.

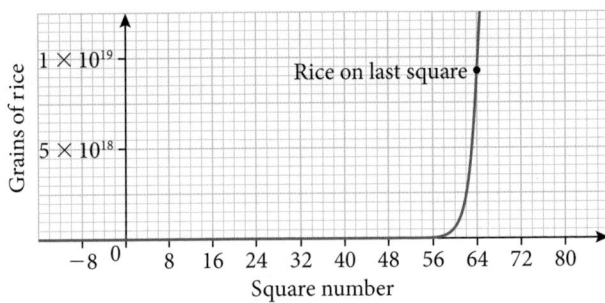

**Figure 9.12** Graphing the model

Most calculators can't even graph numbers this large. At the scale of this graph, it looks like there are almost no grains of rice until somewhere around the 56th square. Only if we zoom in to the very first few squares can we see some of the initial growth.

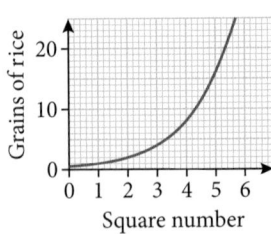

**Figure 9.13** Zooming in on the graph

So, both visually and numerically, we can see that exponential functions can describe very rapid change. We will need to keep this in mind when we consider the reasonableness of our models.

Of course, doubling the number of grains of rice on each square of a chessboard might seem like an extreme example. However, many real-life phenomena double in a fixed interval. For example, a single bacterium in ideal conditions can divide itself into two bacteria every 15 minutes! Furthermore, any amount that increases by a fixed fraction will also double in a fixed interval. For example, the mean university tuition fee increase in the USA is 4.2% per annum. This doesn't sound like much, but it means that tuition rates double every 17 years (approximately).

You can quickly calculate an estimate of doubling time by the 'Rule of 70':

$$\text{Doubling time} \approx \frac{70}{\% \text{ rate}}$$

For example, for something that is increasing at 7% per month, it will take $\frac{70}{7} \approx 10$ months to double. The actual value is about 10.2 months (3 s.f.). You will learn how to calculate the doubling time precisely in this section.

**Warning:** Do not use this method to calculate your final answer in any exercise as it is only an approximation. Use it only to check if your answer is reasonable.

## Developing exponential models

Exponential models can sometimes be developed from examining a table, as in the case of the grains-of-rice-on-a-chessboard example. Other times we know about or can hypothesise a percentage growth, doubling time, or growth factor. Any of these can be used to develop an exponential model. Example 9.10 looks at when the growth rate is known.

### Example 9.10

Suppose that a population of wombats is increasing at 7% per annum. The population at the beginning of 2018 is recorded to be 240 individuals. Assume that this rate of growth remains constant.

(a)  Develop a model for the population of wombats over time.

(b)  Use your model to predict the number of wombats at the beginning of 2025.

(c)  How long will it take the population of wombats to double?

(d)  An ecologist estimates that the region the wombats inhabit can sustain at most 1200 wombats. In what year will the population reach 1200 wombats?

It is common to think of the starting date as time 0. This could be the starting hour, day, month, or year. Doing so can simplify the calculations in our models. However, we have to be careful to convert back to actual times or dates when interpreting our results.

### Solution

(a)  To develop a model, we will start by making a table and see if we can see a pattern. We start by simply adding 10% for each year. However, to simplify our model, we will start with year 0 as 2018.

| Year | Years since 2018 | Calculation of number of wombats | Number of wombats |
|------|------------------|----------------------------------|-------------------|
| 2018 | 0 | 240 | 240.0 |
| 2019 | 1 | $240 + 240 \times 0.07$ | 256.8 |
| 2020 | 2 | $256.8 + 256.8 \times 0.07$ | 274.8 |
| 2021 | 3 | $274.8 + 274.8 \times 0.07$ | 294.0 |

There doesn't seem to be an obvious pattern. However, if we make a change to how we calculate the number of wombats we might have better luck.

Although it doesn't make sense to have 256.8 wombats, we keep the decimal value in order to make our subsequent calculations more accurate. On your GDC, you should store the previous results to maintain full precision in your calculations, as shown:

| | |
|---|---|
| $24 + 240 \cdot 0.07$ | $256.8$ |
| $256.8 + 256.8 \cdot 0.07$ | $274.776$ |
| $274.776 + 274.776 \cdot 0.07$ | $294.01$ |

The key insight relies on a clever factorisation of our calculation:

| | |
|---|---|
| Original expression | $240 + 240(0.07)$ |
| Using the distributive property | $= 240(1 + 0.07)$ |
| Simplify the sum | $= 240(1.07)$ |

Now we rewrite our table multiplying the previous result by 1.07 each time.

| Year | Years since 2018 | Calculation of number of wombats | Number of wombats |
|---|---|---|---|
| 2018 | 0 | 240 | 240.0 |
| 2019 | 1 | 240(1.07) | 256.8 |
| 2020 | 2 | 240(1.07)(1.07) | 274.8 |
| 2021 | 3 | 240(1.07)(1.07)(1.07) | 294.0 |

We can see that we are multiplying each row by a factor of 1.07. This means that the exponent of the 1.07 factor is equal to the number of years since 2018, which allows us to write an exponential model as a function

$$P(n) = 240(1.07)^n$$

where $P(n)$ is the population of wombats and $n$ is the number of years since 2018.

(b) Since 2025 is 7 years since 2018, our model predicts that the population of wombats will be $P(7) = 240(1.07)^7 = 385$ wombats.

(c) To work out the time taken for the wombat population to double, we need to solve the equation $P(n) = 480 = 240(1.07)^n$
We will show an algebraic and GDC approach here.

### Algebraic approach

From the equation $480 = 240(1.07)^n$, simplifying and taking the logarithm of both sides, we obtain $n = 10.2$ (3 s.f.). Thus, the population will double after 10.2 years.

### GDC approach

To use a GDC, graph the left-hand side and the right-hand side of the equation $480 = 240(1.07)^n$ as two separate functions. We will need to think carefully about a suitable viewing window. For the $x$-axis, a domain of $0 \leqslant x \leqslant 20$ is a good start. Since the $y$-axis is the number of wombats, and we know we are starting with 240 and looking for the point when the population is 480, we could choose $200 \leqslant y \leqslant 600$
Then using the GDC's Intersect feature, we find that the population of wombats will reach 480 after 10.2 years. Therefore, the population will double every 10.2 years.

*Since the growth rate of 7% is constant, the population of wombats will double every 10.2 years.*

*We could also use the numerical solver on a GDC:*

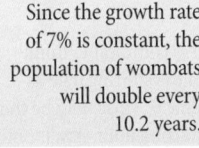

$\text{nSolve}\left(480 = 240 \cdot (1.07)^n, n\right) \quad 10.2448$

*If you know that an equation has only one solution then using the numerical solver feature on a GDC is a quick way to find the solution. (You can tell numerical solvers to look for a solution close to a starting value that you guess, but it can be difficult deciding where to start!) Make sure that you are confident that there is, in fact, only one solution to the equation you are trying to solve. Otherwise, a numerical solver can give you a solution that doesn't make sense in the context of the equation.*

(d) To find the year during which the population reaches 1200 wombats, solve $P(n) = 1200 \Rightarrow 1200 = 240(1.07)^n$. As before, simplify then take the logarithm of both sides to obtain $n = 23.8$ (3 s.f.). Therefore, the population of wombats will reach 1200 after 23.8 years, during the year 2041.

The technique we used for developing the model for exponential growth is general. (In fact, it even works when there is a decrease that is a constant factor.) If we look back at Example 9.10, we can see the general form of an exponential model.

In the general model, the fixed period we refer to may be days, hours, years, generations, chessboard squares, etc. Remember that $p$ is the fraction of the previous value that is added each time. For example, if a quantity is doubling every period, we are adding 100% so $p = 1$. A quantity that is increases by 10% every period has $p = 0.10$. A quantity that decreases by 15% every period has $p = -0.15$. When $p$ is negative the model exhibits **exponential decay**.

Sometimes, it is necessary to add a constant $c$ to the exponential model to obtain $f(x) = k(1 + p)^x + c$

The value of $c$ represents the asymptotic value of the quantity. We will see this in the next example.

A quantity that is changing at a fixed fractional rate $p$ per fixed period can be modelled by the exponential function $f(x) = k(1 + p)^x$ where $k$ is the initial value of the quantity.

## Interpreting exponential models

As with any model, we need to be able to understand what an exponential model can tell us and how it might be useful. Example 9.11 looks at how to interpret an exponential model.

### Example 9.11

The value of a certain new car decreases according to the model
$V(t) = 39\,000(0.72)^t + 1000$ where $V(t)$ represents the value in euros of the car $t$ years after is it purchased.

(a) Find the original purchase price of the car.

(b) Interpret the meaning of the value 1000 in the model.

(c) Calculate when the value of the car will be half its original value.

#### Solution

(a) The original purchase price is
$V(0) = 39\,000(0.72)^0 + 1000 = 40\,000$ euros.

(b) The expression $39\,000(0.72)^t$ tends towards zero as $t$ increases. Therefore, the value 1000 in the model represents the eventual or residual value of the car. (This is probably the value of the car as scrap!)

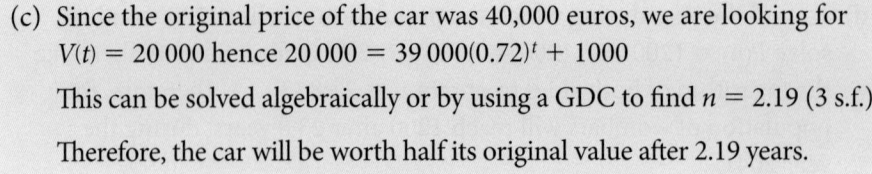

**Figure 9.14** GDC screen for the solution to Example 9.11(c)

If a quantity is approaching a constant value $c$ from an initial value $k + c$, and the difference $k - c$ is changing at a fixed fractional rate $p$ per fixed period, the growth or decay in the quantity can be modelled by the exponential function $f(x) = k(1 + p)^x + c$ where $k + c$ is the initial value of the quantity.

(c) Since the original price of the car was 40,000 euros, we are looking for $V(t) = 20\,000$ hence $20\,000 = 39\,000(0.72)^t + 1000$

This can be solved algebraically or by using a GDC to find $n = 2.19$ (3 s.f.)

Therefore, the car will be worth half its original value after 2.19 years.

Note that when using an exponential model of the form $f(x) = k(1 + p)^x + c$ the value $k$ is no longer the initial value.

The generic statement of the model makes sense: If we consider Example 9.11, the original purchase price of €40,000 is not decreasing by a fixed percentage. Rather, it is the amount above the residual value of €1000 that is decreasing at a fixed percentage per period, as shown in Table 9.8.

| Years since purchase | Value (EUR) | Percentage change in total value of car | Value (EUR) above residual value of 1000 EUR | Percentage change in value above residual value of car |
|---|---|---|---|---|
| 0 | 40,000 | | 39,000 | |
| 1 | 29,080 | −27.3% | 28,080 | −28.0% |
| 2 | 21,218 | −27.0% | 20,218 | −28.0% |
| 3 | 15,557 | −26.7% | 14,557 | −28.0% |
| 4 | 11,481 | −26.2% | 10,481 | −28.0% |

**Table 9.8** The amount above the residual value changes by a constant percentage

It's common in science applications to use a different form of the exponential model. Instead of changing the base of the exponent by adding the fractional rate $p$ to 1, we add a parameter $r$ as a factor in the exponent and use the special base e, so $f(x) = ke^{rx}$

The value of $r$ is often determined through experimental evidence. It turns out that the two models are equivalent, as the Example 9.12 shows.

| Time (minutes) | Temperature (°C) |
|---|---|
| 0 | 95.0 |
| 1 | 79.0 |
| 2 | 66.4 |
| 3 | 56.5 |
| 4 | 48.7 |

**Table 9.9** Data for Example 9.12

## Example 9.12

A student measures the temperature of a cup of coffee at regular intervals in a room where the temperature is 20°C. Her data are shown in the table.

(a) Develop an exponential model of the form $f(x) = ke^{rx} + c$, where $k$, $r$, and $c$ are constants to be determined.

(b) Use your model to find the temperature of the coffee after 8 minutes.

(c) Use your model to predict when the temperature of the coffee will reach room temperature, to 3 significant figures.

(d) Find the rate of decrease in the difference between the coffee temperature and room temperature as a percentage per minute.

## Solution

(a) We know the coffee is cooling, so our model must be a decreasing exponential function. We also know that a decreasing exponential function will tend towards zero unless a constant is added, so the expression $ke^{rx}$ must represent the (decreasing) difference between the room temperature and the coffee, while $c$ represents room temperature (20°C). Therefore, $c = 20$

Since $f(0) = 95$, we have: $95 = ke^{r(0)} + 20 \Rightarrow k = 75$

Thus, so far we have $f(x) = 75e^{rx} + 20$

Using the data value from the first minute, we know $f(1) = 79.0$ so we can substitute our known values and solve for $r$:

$$79.0 = 75e^{r(1)} + 20 \Rightarrow r = -0.240$$

Therefore, the model is $f(x) = 75e^{-0.240x} + 20$

(b) After 8 minutes, the coffee temperature is

$$f(8) = 75e^{-0.240(8)} + 20 = 31.0°C$$

(c) Theoretically speaking, the coffee cup will never reach room temperature according to the model. However, with three significant figures of accuracy, it will be equal to room temperature when the temperature is less than 20.05°C. So, we are looking for the time when $f(x) = 20.05$

### Algebraic approach

As before, we substitute known values and use logarithms to solve

$$20.05 = 75\,e^{-0.240x} + 20 \Rightarrow x = 30.5$$

Therefore the coffee temperature will be within 3 significant figures of room temperature after 30.5 minutes.

### GDC approach

We can directly enter the equation into the numeric solver:

nSolve$(20.05 = 75 \cdot e^{-0.24 \cdot x} + 20, x)$

$30.4718$

(d) We need to read the question carefully: 'Find the rate of decrease in the difference between the coffee temperature and room temperature.' If we calculate each difference and then find the percentage change, we get the following table.

| Time (minutes) | Temperature (°C) | Difference between coffee temperature and room temperature | Percentage change from previous minute |
|---|---|---|---|
| 0 | 95.00 | 75.00 | |
| 1 | 79.00 | 59.00 | −21.3% |
| 2 | 66.41 | 46.41 | −21.3% |
| 3 | 56.51 | 36.51 | −21.3% |
| 4 | 48.72 | 28.72 | −21.3% |

Therefore, the percentage change per minute in the difference between the coffee temperature and room temperature is −21.3%

Note that in part (d), what we are doing is removing the effect of the room temperature and considering the cooling rate of the coffee. Algebraically, that means we are considering the expression $75e^{-0.240x}$ If we are looking for a percentage rate, we could convert this expression into a percentage-decrease model by using the law of exponents $x^{ab} = (x^a)^b$:

$75e^{-0.240x} = 75(e^{-0.240})^x$
$\qquad = 75(0.787)^x$

Since $1 - 0.213 = 0.787$, we confirm our result that the coffee temperature difference is decreasing by a fixed rate of 21.3%.

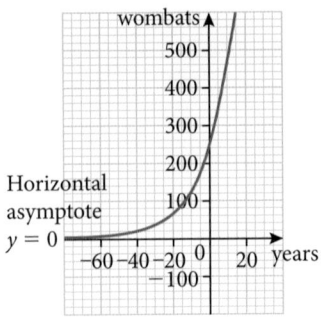

**Figure 9.15** Graph of wombat population

Half-life refers to a process of radioactive decay. The half-life of a substance is the average amount of time it take for the activity of a substance to decrease to half its previous value. The activity is the number of nuclei that decay per second or becquerel (Bq).

For example, the half-life of fermium-253 is 3 days. If the activity of a sample of fermium-253 is 100 Bq, then after 3 days the average activity will decrease to 50 Bq. After another 3 days, the average activity will be 25 Bq.

All living things on Earth continuously take in carbon. Since a tiny fraction of the carbon on Earth is the naturally occurring radioactive isotope carbon-14, the fraction of carbon-14 in all living things is constant. However, once an organism dies, it stops taking in carbon and the carbon-14 in its body begins to decay. Because of this, scientists can use carbon-14 dating to determine how many years have passed since an organism died.

## Graphical interpretation

It's worth looking at the graphs of the last few models. Note that in each case, the location of the horizontal asymptotes is given by the value of $c$ in the model $f(x) = k(1 + p)^x + c$

For the wombat example (Figure 9.15), $c = 0$ so the asymptote is at $y = 0$. In this context, the horizontal asymptote is not meaningful because it occurs in negative years since 2018, so there doesn't seem to be a useful interpretation of it.

For the car example (Figure 9.16), $c = 1000$ so the asymptote is at $y = 1000$. In this context, it represents the eventual (residual) value of the car.

For the coffee example (not shown), $c = 20$ so the asymptote is at $y = 20$. In this context, it represents the room temperature, which is the eventual temperature of the coffee.

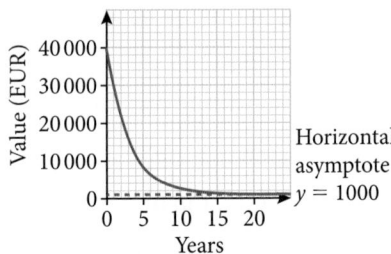

**Figure 9.16** Graph of car value

For an exponential model of the form
$$f(x) = k(1 + p)^x + c \quad \text{or} \quad f(x) = ke^{rx} + c$$
the graph of $f$ has a horizontal asymptote at $y = ic$

A common use of exponential functions is half-life calculations.

### Example 9.13

A model for carbon-14 dating is $A(t) = ke^{-0.000121t}$, where $A(t)$ is the activity of carbon-14 remaining from an initial activity $k$, after $t$ years.

(a) Show that the half-life of carbon 14 is 5730 years, to 3 significant figures.

(b) A tissue sample from a body discovered on the border between Austria and Italy has an average activity of 6.92 Bq. A sample of live tissue of the same size has an average activity of 12 Bq. Determine the age of the sample.

### Solution

(a) A half-life is the time taken for the activity of carbon-14 to halve. Therefore, we can set $k = 2$ and $A(t) = 1$

Substituting into $A(t) = ke^{-0.000121t}$
$$1 = 2e^{-0.000121t} \quad \Rightarrow \quad t = 5728.5$$

Therefore, to 3 significant figures, the half-life of carbon-14 is 5730 years.

(b) Using the model, we obtain
$$6.92 = 12e^{-0.000121t} \quad \Rightarrow \quad 0.577 = e^{-0.000121t} \quad \Rightarrow \quad t = 4550$$

Therefore, the age of the tissue sample is 4550 years.

1. When a skydiver is falling towards the Earth, she will accelerate until the force due to gravity becomes equal to the force due to air resistance. The velocity of the skydiver at this point is called terminal velocity. A skydiver records the difference between her velocity and terminal velocity every 5 s and obtains the data shown in the table.

| Free-fall time (s) | 0 | 5 | 10 | 15 | 20 |
|---|---|---|---|---|---|
| Difference between velocity and terminal velocity (m s$^{-1}$) | 56 | 23.4 | 9.79 | 4.10 | 1.71 |

   (a) Develop an exponential model for this data.

   (b) Find the difference between velocity and terminal velocity at 7 s.

   (c) When is the skydiver first within 5 m s$^{-1}$ of terminal velocity?

   (d) Predict the time when the skydiver will be within 1 m s$^{-1}$ of terminal velocity.

2. The number of bacteria in two colonies, A and B, starts increasing at the same time. The number of bacteria in colony A after $t$ hours is modelled by the function $A(t) = 12e^{0.4t}$

   (a) Find the number of bacteria in colony A after four hours.

   (b) How long does it take for the number of bacteria in colony A to reach 400?

   The number of bacteria in colony B after $t$ hours is modelled by the function $B(t) = 24e^{kt}$

   After four hours, there are 60 bacteria in colony B.

   (c) Find the value of $k$.

   (d) The number of bacteria in colony A first exceeds the number of bacteria in colony B after $n$ hours, where $n \in \mathbb{Z}$. Find the value of $n$.

3. Jose takes medication. After $t$ minutes, the concentration of medication remaining in his bloodstream is given by $A(t) = 10(0.5)^{0.014t}$ where A is in milligrams per litre.

   (a) Write down $A(0)$.

   (b) Find the concentration of medication remaining in his bloodstream after 50 minutes.

   (c) At 13:00, when there is no medication in Jose's bloodstream, he takes his first dose of medication. He can take his medication again when the concentration of medication reaches 0.395 milligrams per litre. What time will Jose be able to take his medication again?

4. A large lizard known as a Gila Monster is about 16 cm long at birth. For the first 8 years of its life, the Gila Monster's length increases by about 8% each year.

   (a) Formulate a function to model the length $L$ of a Gila Monster $t$ years after birth.

   (b) Estimate the length of a 3-year-old Gila Monster.

   (c) Estimate the age of a 25-cm long Gila Monster.

5. DDT is a toxic insecticide that was widely used in the past. The function $A = A_0 e^{kt}$ can be used to describe the amount $A$ of DDT left in an area $t$ years after an initial application of $A_0$ units.

   (a) Given that half-life of DDT is about 15 years, find the value of $k$.

   (b) Find the amount of DDT left after 2 years when 50 units were initially applied to an area.

   (c) A sample of soil is tested and 35 units of DDT are found. It is known that the last application of DDT was 20 years ago. Find the initial amount of DDT applied to this soil.

   (d) In one area, 120 units of DDT are applied. Given that the safe level of DDT is 40 units, for how long will the area be unsafe?

6. In certain soil conditions, the half-life of the pesticide glyphosate is about 45 days. A scientist studying the how this chemical decays wrote in his notes that the data in an experiment could be modelled with the function $A(t) = 500(0.5)^t$

   (a) Find $A(0)$ and interpret its value in context.

   (b) Explain what the variable $t$ represents in this context.

   (c) Find $A(1)$ and interpret in context.

7. Sejah placed a baking tin, that contained cake mix, in a preheated oven in order to bake a cake. The temperature in the centre of the cake mix, $T$, in degrees Celsius (°C) is given by

   $$T(t) = 150 - a \times (1.1)^{-t}$$

   where $t$ is the time, in minutes, since the cake was placed in the oven. The graph of $T$ is shown in the diagram.

   (a) Write down what the value of 150 represents in the context of the question.

   (b) The temperature in the centre of the cake mix was 18°C when placed in the oven. Find the value of $a$.

   (c) The baking tin is removed from the oven 15 minutes after the temperature in the centre of the cake mix has reached 130°C. Find the total time that the baking tin is in the oven.

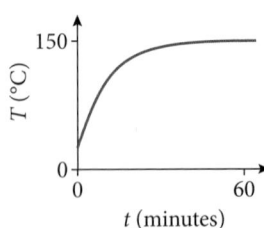

**Figure 9.17** Diagram for question 7

8. The number of fish, $N$, in a pond is decreasing according to the model
$$N(t) = ab^{-t} + 40, \ t \geqslant 0$$
where $a$ and $b$ are positive constants, and $t$ is the time in months since the number of fish in the pond was first counted. The fish are first counted in January. Some of the data collected is shown in the table.

(a) Find the value of $a$ and of $b$.

(b) Use your model to estimate the number of fish in September.

(c) Use your model to find the first month when there were less than 50 fish.

(d) The number of fish in the pond will not decrease below $p$. Write down the value of $p$.

| Month | Fish |
|---|---|
| January | 850 |
| May | 100 |
| September | ? |

Table 9.10 Data for question 8

9. A cup of boiling water is placed in a room to cool. The temperature of the room is 20 °C. This situation can be modelled by the exponential function $T = a + b(k^{-m})$, where $T$ is the temperature of the water, in °C, and $m$ is the number of minutes for which the cup has been placed in the room. A sketch of the situation is given in the diagram.

(a) Explain why $a = 20$

Some data collected on the temperature of the water is given in the table.

(b) Find the value of $b$ and of $k$.

(c) Find the temperature of the water 5 minutes after it has been placed in the room.

(d) Find the total time needed for the water to reach a temperature of 35°C. Give your answer in minutes and seconds, correct to the nearest second.

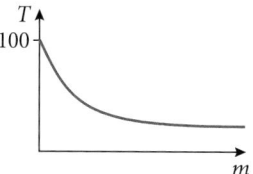

Figure 9.18 Diagram for question 9

| Time (minutes) | Temperature (°C) |
|---|---|
| 0 | 100 |
| 1 | 85 |
| 5 | ? |
| ? | 35 |

Table 9.11 Data for question 9

10. A potato is placed in an oven heated to a temperature of 200°C. The temperature of the potato, in °C, is modelled by the function $p(t) = 200 - 190(0.97)^t$, where $t$ is the time, in minutes, that the potato has been in the oven.

(a) Write down the temperature of the potato at the moment it is placed in the oven.

(b) Find the temperature of the potato half an hour after it has been placed in the oven.

(c) After the potato has been in the oven for $k$ minutes, its temperature is 40°C. Find the value of $k$.

11. The amount of electrical charge, $C$, stored in a smartphone battery is modelled by $C(t) = 2.5 - 2^{-t}$, where $t$, in hours, is the time for which the battery has been charged.

(a) Write down the amount of electrical charge in the battery at $t = 0$

(b) The line $L$ is the horizontal asymptote to the graph. Write down the equation of $L$.

(c) To download a game to the smartphone, an electrical charge of 2.4 units is needed. Find the time taken to reach this charge. Give your answer correct to the nearest minute.

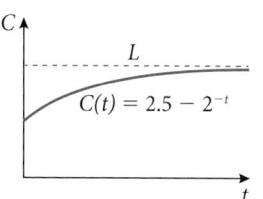

Figure 9.19 Diagram for question 11

12. The number of cells, $C$, in a culture is given by the equation $C = p \times 2^{0.5t} + q$ where $t$ is the time in hours measured from 12:00 on Monday and $p$ and $q$ are constants. The number of cells in the culture at 12:00 on Monday is 47. The number of cells in the culture at 16:00 on Monday is 53.

(a) Write down two equations in $p$ and $q$.

(b) Calculate the value of $p$ and of $q$.

(c) Find the number of cells in the culture at 22:00 on Monday.

## 9.3 Trigonometric models

Trigonometric models are well-suited to describing repetitive phenomena: tides, seasonal temperatures, the motion of wheels, etc. In this section, we will examine the models $y = a \sin(b(x - c)) + d$ and $y = a \cos(b(x - c)) + d$

### Exploration

To be able to use the sine and cosine functions to model effectively, you need to understand the difference between the graphs of sine and cosine and how the parameters $a$, $b$, and $d$ affect them. This is best done by using a graphing application such as *GeoGebra* that allows you to observe carefully the effect each parameter has on the shape of the graph.

#### What does $a$ do?

Make sure you work in **radian** mode.

Use your graphing application to generate a graph of $y = a \sin x$. Experiment with different values of $a$ to see how the graph changes. Make sure to set $a$ to at least the following four values: $a = 1, \frac{1}{2}, 2, -2$

You should see that $|a|$ is equal to the **amplitude** of the graph, which is half of the **wave height**, as shown in Figure 9.20. Also, if $a$ is negative, then the graph is reflected in the $x$-axis.

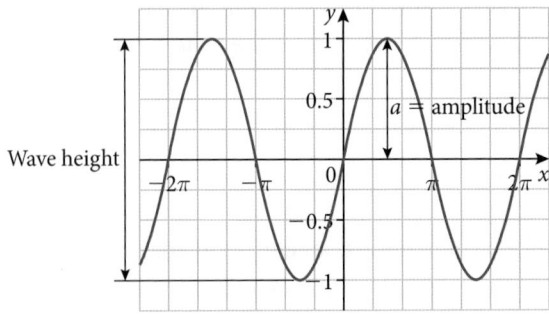

**Figure 9.20** Amplitude is half of wave height

## What does *b* do?

Use your graphing application to generate a graph of $y = \sin(bx)$. Experiment with different values of $b$ to see how the graph changes. Make sure to set $b$ to at least the following four values: $b = 1, \frac{1}{2}, 2, -2$

You should see that $b$ determines the **wavelength** of the graph. In mathematics and science, we refer to this as the **period** of the function. The period is the distance it takes for the graph to repeat itself – the distance between two adjacent local maxima or local minima, for the sine and cosine functions. But, how exactly does $b$ affect the period? If we write down the values of $b$ and the length of the period, we get Table 9.12. What does it tell us? First that the sign of $b$ does not seem to make a difference to the period since the period is the same for both 2 and $-2$. Next, that as the absolute value of $b$ increases, the period gets shorter. This suggests that the length of the period and the value of $b$ vary inversely. We can also see from the table that the product of $|b|$ and the period has a constant value of $2\pi$. Therefore, we can conclude that

$$|b|p = 2\pi \;\Rightarrow\; p = \frac{2\pi}{|b|}$$ as shown in Figure 9.21.

Also, note that if $b$ is negative, then the graph is reflected across the $y$-axis.

| Value of *b* | Length of period (*p*) |
|:---:|:---:|
| $-2$ | $\pi$ |
| $\frac{1}{2}$ | $4\pi$ |
| $1$ | $2\pi$ |
| $2$ | $\pi$ |

Table 9.12  Values of *b* and the length of the period

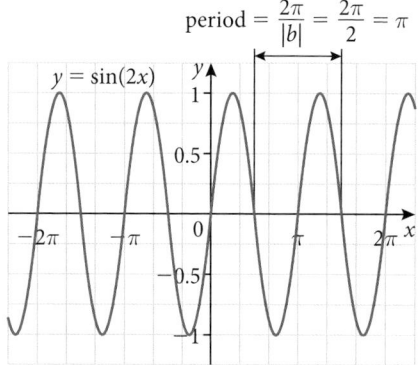

Figure 9.21  Relationship between period and *b*

## What does *c* do?

Use your graphing application to generate a graph of $y = \sin(x - c)$. Experiment with different values of $c$ to see how the graph changes.

Make sure to set $c$ to at least the following four values: $c = -\pi, 0, \frac{\pi}{3}, \pi$

You should see that $c$ produces a horizontal translation or **phase shift**, as shown in Figure 9.22.

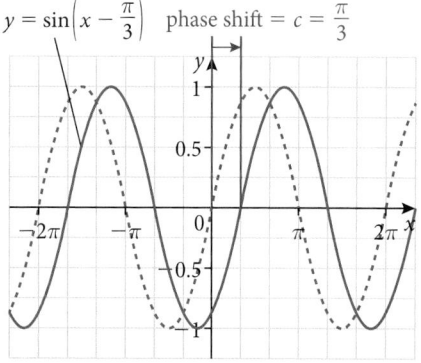

Figure 9.22  Phase shift of a periodic function

## What does *d* do?

Use your graphing application to generate a graph of $y = \sin(x) + d$. Experiment with different values of $d$ to see how the graph changes. Make sure to set $d$ to at least the following four values: $d = 0, \frac{1}{2}, 2, -2$

You should see that $d$ controls the vertical position of the graph. Specifically, the value of $d$ gives us the position of the **principal axis** of the function. The principal axis is a horizontal line with equation $y = d$ (Figure 9.23).

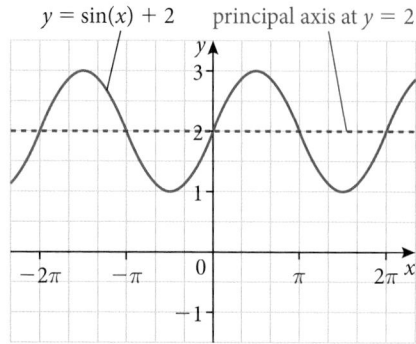

Figure 9.23  The principal axis of a periodic function

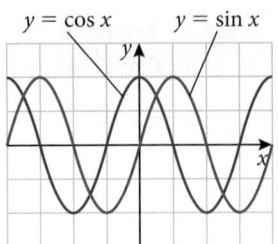

$y = \cos x$    $y = \sin x$

**Figure 9.24** Comparison of sine and cosine graphs

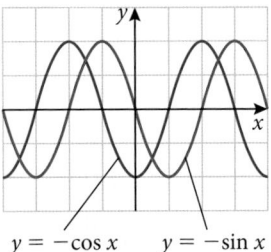

$y = -\cos x$    $y = -\sin x$

**Figure 9.25** When $a < 0$, the graphs of sine and cosine are reflected across the $x$-axis

## The difference between the graphs of sine and cosine

Note that we have only used sine so far. This is because the graphs of sine and cosine are very similar – the properties of $a$, $b$, $c$, and $d$ are the same for both. However, there is an important difference between the graphs of sine and cosine. Graph the functions $y = \cos x$ and $y = \sin x$

Look at the behaviour of both functions when $x = 0$, as shown in Figure 9.24. At $x = 0$ cosine has a maximum that decreases as $x$ increases. At $x = 0$ sine begins on the principal axis and then increases as $x$ increases. Is this always true? No – remember how the parameters $a$, $b$, and $d$ affect the graphs. Because the value of $a$ can cause the graph to be reflected across the $x$-axis, when $a < 0$ cosine will begin at a minimum and then increase, and sine will begin on the principal axis and decrease at first. These graphs are shown in Figure 9.25.

For the trigonometric models $y = a\sin(b(x - c)) + d$ and $y = a\cos(b(x - c)) + d$
- amplitude $= |a|$
- The amplitude is equal to half the wave height.
- For $a < 0$, the graph is reflected across the $x$-axis.
- period $= \dfrac{2\pi}{|b|}$
- For $b < 0$, the graph is reflected across the $y$-axis.
- The principal axis is at $y = d$
- Use a sine model when you want the initial value of the model to be equal to the location of the principal axis (half-way between the maximum and minimum values). If the model should initially increase, let $a > 0$. If the model should initially decrease, let $a < 0$
- Use a cosine model when you want the initial value of the model to be a maximum (let $a > 0$) or minimum (let $a < 0$); or, use a sine model with a phase shift.

## Developing trigonometric models

Now that we have an understanding of the parameters of trigonometric models, we can model some real-life periodic phenomena.

### Example 9.14

A reflector is attached to a bicycle wheel at a point 17 cm from the centre of the wheel. The radius of the wheel is 30 cm. The wheel rotates clockwise and takes 4 s to complete one revolution.

(a) Assuming that the reflector is at the topmost position at time $t = 0$, develop a model for the height of $h$, in cm, of the reflector above the ground at any time $t$ in seconds.

(b) Revise your model so that the reflector is at the bottommost position at time $t = 0$

(c) Revise your model so that the reflector is at the rightmost position at time $t = 0$

(d) Revise your model for a point on the outer edge of the wheel. Assume that the point is at the leftmost position at time $t = 0$

## Solution

(a) Since the radius of the wheel is 30 cm, and the reflector is 17 cm from the centre of the wheel, we know that the minimum height of the reflector is $30 - 17 = 13$ cm. At the topmost position, the reflector will be $30 + 17 = 47$ cm from the ground. Since the amplitude is half of the wave height, it must be $\dfrac{47 - 13}{2} = 17$ cm, so $a = 17$ or $a = -17$

That makes sense since the reflector is located 17 cm from the centre of the wheel! Also, the principal axis must be halfway between the minimum and maximum values, so it must be at $\dfrac{47 + 13}{2} = 30$ cm

hence $d = 30$. That makes sense because the centre of the wheel is 30 cm above the ground. Since it takes 4 seconds for the wheel to complete one revolution, the period of our model must equal 4. We can then find $|b|$:

$$\text{Period} = \frac{2\pi}{|b|} \Rightarrow 4 = \frac{2\pi}{|b|} \Rightarrow |b| = \frac{2\pi}{4} = \frac{\pi}{2}$$

So $b = \dfrac{\pi}{2}$ or $b = -\dfrac{\pi}{2}$

Finally, we need to decide if a sine or cosine model is better suited for this situation. Since we are told that the reflector begins at the topmost point, our model must begin at a maximum. Therefore, we choose the cosine function, and we keep both $a$ and $b$ positive since no reflection across the $x$ or $y$-axis is needed.

Thus, we have $a = 17$, $b = \dfrac{\pi}{2}$, and $d = 30$, and our model is

$$h = 17 \cos\left(\frac{\pi t}{2}\right) + 30$$

We can check by graphing two complete periods ($0 \leqslant t \leqslant 8$), as shown.

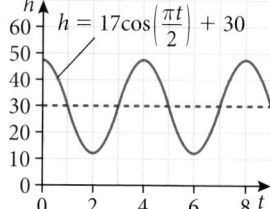

**Figure 9.26** The graph appears to fit the behaviour we expect

(b) For the reflector to start at the bottommost position, the model must have a minimum as an initial value. In this case, we simply let $a$ be negative, and revise our model accordingly:

$$h = -17 \cos\left(\frac{\pi t}{2}\right) + 30 \text{ as shown in the graph.}$$

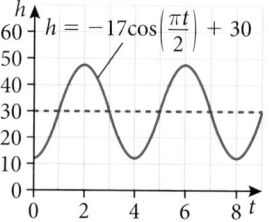

**Figure 9.27** Diagram for solution to Example 9.14(b)

(c) For the reflector to start at the rightmost position, the model must have an initial value equal to the location of the principal axis. Therefore, we can either use a sine model or apply a phase shift to our current model.

### Sine model

We are told that the wheel rotates clockwise, which will cause the height of the reflector to initially decrease. Therefore, we must let $a < 0$ and so our model is:

$$h = -17 \sin\left(\frac{\pi t}{2}\right) + 30$$

333

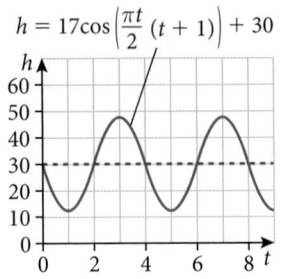

$$h = 17\cos\left(\frac{\pi t}{2}(t+1)\right) + 30$$

**Figure 9.28** The sine model above produces the same graph

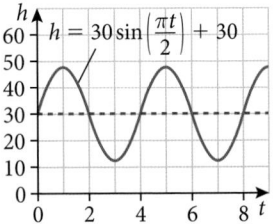

**Figure 9.29** Diagram for solution to Example 9.14(d)

**Cosine model with phase shift**

To use a cosine model with a phase shift, recall that the reflector initially started at the topmost position and rotated clockwise. Therefore, after one-quarter of a period, it would be at the right-most position. So, we simply need to turn the 'wheel' by adding a quarter-period to $t$.

Since each period is 4 seconds, we add $\frac{1}{4}(4) = 1$ second to $t$:

$$h = 17\cos\left(\frac{\pi}{2}(t+1)\right) + 30 \text{ as shown in the graph.}$$

(d) A point on the outer edge of the wheel will have a similar model, but the amplitude of the model is now equal to the radius of the wheel. Also, since the point is in the leftmost position at time $t = 0$, the model must increase initially. Therefore, we will use a sine model with $a = +30$

$$h = 30\sin\left(\frac{\pi t}{2}\right) + 30 \text{ as shown in the graph.}$$

## Example 9.15

The distance from the ground of a person on Vienna's famous *Riesenrad* ferris wheel can be modelled by the function $h(t) = a\sin\left(\frac{\pi}{5}\left(t - \frac{5}{2}\right)\right) + d$ where $h(t)$ is the height in feet at time $t$ in minutes.

(a) Given that the diameter of the *Riesenrad* is 200 feet, deduce the value of $a$.

(b) The boarding platform at the lowest point of the *Riesenrad* is 12 feet above the ground. Find the value of $d$.

(c) Calculate the number of minutes until a person riding the *Riesenrad* reaches the top of the wheel.

(d) Calculate the time required for one complete rotation.

(e) Determine a reasonable domain and range for this model if each ride is exactly one complete revolution.

(f) In one 10 minute ride, during what interval of time will a person on the *Riesenrad* be at least 100 feet above the ground?

(g) Comment on the reasonableness of this model.

## Solution

(a) Since the diameter of the *Riesenrad* is 200 feet, the wave height of the model must be 200, so the amplitude $a = 100$

(b) Since the minimum value of the function is 12 feet, the principal axis must be located at $12 + 100 = 112$. Therefore $d = 112$

(c) By graphing the model, we obtain the display shown.
    We can see that a local maximum is at $t = 5$ minutes.

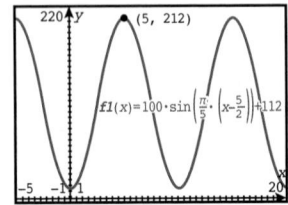

**Figure 9.30** GDC screen for solution to Example 9.15 (c)

(d) Since the first maximum is reached 5 minutes after the first minimum (when the person board the *Riesenrad*, we can conclude that one complete rotation takes 10 minutes. This agrees with the graph.

(e) Since one complete revolution takes 10 minutes, the domain for this model should be $0 \leqslant t \leqslant 10$

In that time, the person will travel from 12 feet to 212 feet, so the range is $12 \leqslant h(t) \leqslant 212$

(f) Again, we can use a GDC to solve this, by adding a second function with the constant value of 100.

Then, by using the Intersection command, we can find the times when a person goes above and below 100 feet.

We see that a person would be at least 100 feet above the ground for $2.31 \leqslant t \leqslant 7.69$ minutes.

(g) Although this model seems reasonable, it assumes that the rotation speed of the wheel is constant. In reality, the wheel would need to stop at regular intervals to allow passengers to get on and off.

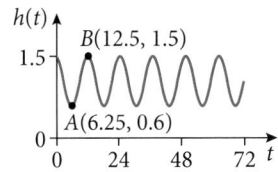

**Figure 9.31** GDC screen for solution to Example 9.15 (f)

We do not consider the next period, since the question asked only above the first ten minutes.

## Exercise 9.3

1. At Grande Anse Beach the height of the water, in metres, is modelled by the function $h(t) = p \cos(q \times t) + r$ where $t$ is the number of hours after 21:00 hours on 10 December 2017. The diagram shows the graph of $h$, for $0 \leqslant h \leqslant 72$

The point $A(6.25, 0.6)$ represents the first low tide and $B(12.5, 1.5)$ represents the next high tide.

(a) How much time is there between the first low tide and the next high tide?

(b) Find the difference in height between low tide and high tide.

(c) Find the value of $p$.

(d) Find the value of $q$.

(e) Find the value of $r$.

(f) There were two high tides on 12 December 2017. At what time did the second high tide occur?

(g) Calculate the number of minutes where the height of the water is at most 0.8 m in the first 24 hours after 21:00 hours on 10 December 2017.

2. The depth, $d$ metres, of water in a port is modelled by the function $d(t) = p \cos(qt) + 7.5$ for $0 \leqslant t \leqslant 12$, where $t$ is the number of hours after high tide. At high tide, the depth is 9.7 m. At low tide, which is 7 hours later, the depth is 5.3 m.

(a) Find the value of $p$.

(b) Find the exact value of $q$.

(c) Use the model to find the depth of the water 10 hours after high tide.

**Figure 9.32** Diagram for question 1

3. The height, $h$ metres, of a seat on a Ferris wheel after $t$ minutes is given by $h(t) = -15\cos(1.2t) + 17$, for $t \geq 0$

   (a) Find the height of the seat when $t = 0$

   (b) The seat first reaches a height of 20 m after $k$ minutes. Find the value of $k$.

   (c) Calculate the time needed for the seat to complete a full rotation, giving your answer correct to one decimal place.

4. The diagram represents a large Ferris wheel, with a diameter of 100 m.

   Let $P$ be a point on the wheel. The wheel starts with $P$ at the lowest point, at ground level. The wheel rotates at a constant rate, in an anticlockwise direction. One revolution takes 20 minutes.

   (a) Write down the height of $P$ above ground level after

      (i) 10 minutes        (ii) 15 minutes.

   (b) Let $h(t)$ metres be the height of $P$ above ground level after $t$ minutes.

      (i) Given that $h$ can be expressed in the form $h(t) = a\cos bt + c$, find the values of $a$, $b$, and $c$.

      (ii) The height $h$ could be expressed using a sine function instead of a cosine function. Given that $h$ can be expressed in the form $h(t) = a\sin(b(t - d)) + c$, find the value of $d$.

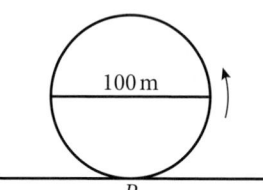

**Figure 9.33** Diagram for question 4

5. The diagram shows a waterwheel with a bucket. The wheel rotates at a constant rate in an anticlockwise direction.

   The diameter of the wheel is 8 m. The centre of the wheel, $A$, is 2 m above the water level. After $t$ s, the height of the bucket above the water level is given by $h = a\sin bt + 2$

   (a) Find the value of $a$.

   (b) The wheel turns at a rate of one rotation every 30 s. Find the value of $b$.

   (c) Calculate the number of seconds the bucket is underwater during one rotation.

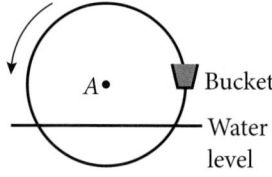

**Figure 9.34** Diagram for question 5

6. The diagram represents a large Ferris wheel at an amusement park. The points $P$, $Q$ and $R$ represent different positions of a seat on the wheel.

   The wheel has a radius of 50 m and rotates clockwise at a rate of one revolution every 30 minutes. A seat starts at the lowest point $P$, when its height is one metre above the ground.

   (a) Find the height of a seat above the ground after 15 minutes.

   (b) After six minutes, the seat is at point $Q$. Find its height above the ground at $Q$.

   (c) The height of the seat above ground after $t$ minutes can be modelled by the function $h(t) = a\sin(b(t - c)) + d$. Find the values of $a$, $b$, $c$ and $d$.

   (d) Hence find the value of $t$ the first time the seat is 96 m above the ground.

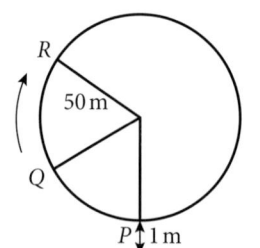

**Figure 9.35** Diagram for question 6

7. The temperature in degrees Celsius during a 24-hour period is shown on the graph and is given by the function $f(t) = a \cos(bt) + c$ where $a$, $b$, and $c$ are constants, $t$ is the time in hours, and $(bt)$ is measured in degrees.

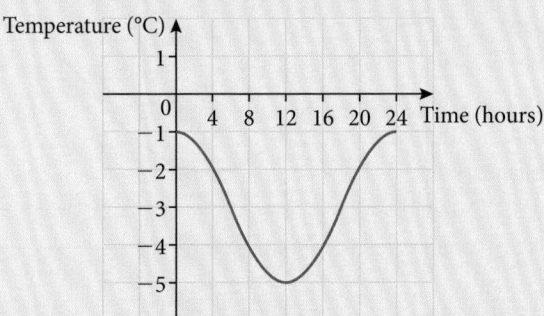

(a) Write down the value of $a$.

(b) Find the value of $b$.

(c) Write down the value of $c$.

(d) Write down the interval of time during which the temperature is increasing from $-4°C$ to $-2°C$.

8. The height, $h$, in centimetres, of a bicycle pedal above the ground at time, $t$ seconds, is a cosine function of the form $h(t) = A \cos(bt) + C$, where $(bt)$ is measured in degrees. The graph of this function for $0 \leqslant t \leqslant 4.3$ is shown on the right.

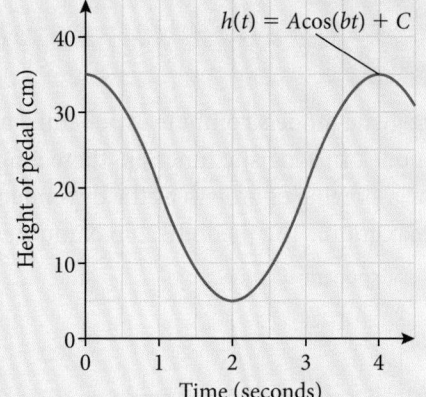

(a) Write down the maximum height of the pedal above the ground.

(b) Write down the minimum height of the pedal above the ground.

(c) Find the amplitude of the function.

(d) Hence or otherwise, find the value of $A$ and of $C$.

(e) Write down the period of the function $h(t)$, including units.

(f) Hence find the value of $b$.

(g) Calculate the first value of $t$ for which the height of the pedal above the ground is 30 cm.

(h) Calculate the number of times the pedal rotates in one minute.

9. The depth, $d$, in metres, of water in a harbour is given by the function $d = 4 \sin(0.5t) + 7$, where $t$ is in minutes, $0 \leqslant t < 1440$

(a) Write down the amplitude of $d$.

(b) Find the maximum value of $d$.

(c) Find the period of $d$. Give your answer in hours.

(d) On Tuesday, the minimum value of $d$ occurs at 14:00. Find when the next maximum value of $d$ occurs.

**10.** The graph represents the temperature ($T°C$) in Washington measured at midday during a period of thirteen consecutive days starting at day 0. These points also lie on the graph of the function $T(x) = a + b\cos(cx)$, $0 \leqslant x \leqslant 12$, where $a, b, c \in \mathbb{R}$.

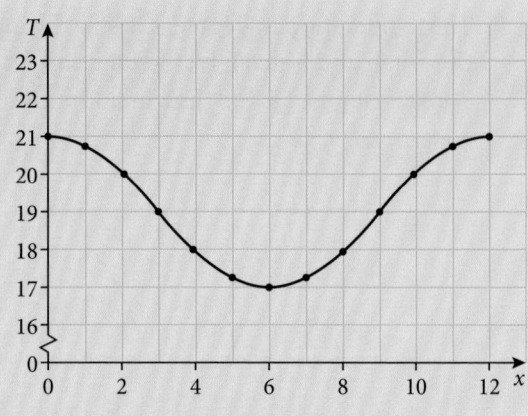

(a) Find the value of $a$ and of $b$.

(b) Find the value of $c$.

(c) Using the graph, or otherwise, write down the part of the domain for which the midday temperature is less than 18.5°C.

Logistic models combine exponential and bounded exponential growth into one model. In a logistic function, there is a period of exponential growth, followed by a slowing of growth as the dependent variable approaches a theoretical maximum. The theoretical maximum is shown as a horizontal asymptote.

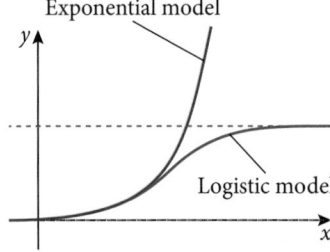

**Figure 9.36** The logistic function has a characteristic S-curve and a horizontal asymptote

The characteristic S-curve of the logistic function is also known as a **sigmoid curve**.

Logistic models are useful for modelling situations where growth is limited in some way.

For example, suppose we have a class of 20 students. If the teacher tells one student that there will be a test in two weeks, how will this news spread through the class, assuming that each student who knows about the test will tell one other student each day? Well, at first, each student who knows will tell one other person every day and it is likely that the person they tell will not yet know about the test. However, after a few days, more students will have heard the news already, so the chance that a student will tell someone new decreases, which causes the rate at which the news spreads to slow down. Eventually, all twenty students will have heard the news, and no further 'growth' is possible. The table shows how this news could spread, where the students are named A-T.

| | A | B | C | D | E | F | G | H | I | J | K | L | M | N | O | P | Q | R | S | T | Total number of students who know about test |
|---|---|---|---|---|---|---|---|---|---|---|---|---|---|---|---|---|---|---|---|---|---|
| **Day 1** | A | | | | | | | | | | | | | | | | | | | | **1** |
| tells | | B | | | | | | | | | | | | | | | | | | | |
| **Day 2** | A | B | | | | | | | | | | | | | | | | | | | **2** |
| tells | H | R | | | | | | | | | | | | | | | | | | | |
| **Day 3** | A | B | | | | | | H | | | | | | | | | | R | | | **4** |
| tells | J | F | | | | | | Q | | | | | | | | | | N | | | |
| **Day 4** | A | B | | | | F | | H | | J | | | | N | | | Q | R | | | **8** |
| tells | Q | H | | | | Q | | Q | | G | | | | D | | | P | L | | | |
| **Day 5** | A | B | | D | | F | G | H | | J | | L | | N | | P | Q | R | | | **12** |
| tells | Q | P | | A | | E | A | I | | M | | C | | G | | H | C | T | | | |
| **Day 6** | A | B | C | D | E | F | G | H | I | J | | L | M | N | | P | Q | R | | T | **17** |
| tells | C | O | G | I | S | I | E | N | S | E | | J | P | Q | | L | H | M | | A | |
| **Day 7** | A | B | C | D | E | F | G | H | I | J | | L | M | N | O | P | Q | R | S | T | **19** |
| tells | K | L | M | T | I | T | H | L | J | S | | M | B | J | I | F | G | Q | E | O | |
| **Day 8** | A | B | C | D | E | F | G | H | I | J | K | L | M | N | O | P | Q | R | S | T | **20** |
| tells | D | S | P | N | B | L | N | C | S | H | M | T | O | T | L | N | T | N | P | E | |
| **Day 9** | A | B | C | D | E | F | G | H | I | J | K | L | M | N | O | P | Q | R | S | T | **20** |
| tells | C | F | T | R | Q | E | L | C | M | P | E | N | I | H | I | C | J | H | F | P | |

**Table 9.13** The spread of news about a test. A purple cell shows when a student is first told about the test; an orange cell shows when they are told again.

On day 1, student A tells student B. On day 2, student A tells student H, and student B tells student R. Students continue to spread the news, but already on day 4 we see that student A tells student Q (but student Q already knew). Likewise, students B, F, and H spread the news to other students who have already been told (H, Q, and Q, respectively). By day 7, almost all of the students in the class have learned the news, and on day 8 the last remaining student learns of the news. No further growth is possible. If we graph the number of students who know about the test against the day number, we obtain Figure 9.37.

In the graph we can see evidence of the *S*-curve of the logistic model. Naturally, we would like to fit a logistic model to this data. The logistic model is given by

$$f(x) = \frac{L}{1 + Ce^{-kx}}$$

where $C, k > 0$, so there are three parameters ($L$, $C$, and $k$) that we need to find.

An analytical method for finding the parameters for the logistic model based on data will be discussed at in Chapter 20. Here we will look at the effect of the parameters and learn how to solve for them in simplified examples or using a GDC (check your GDC for logistic regression). When we enter the data into a GDC, we can use it to estimate the logistic model for this data as shown in the screenshot.

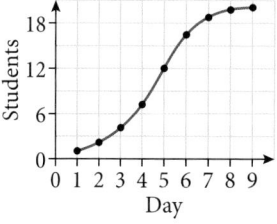

**Figure 9.37** How news spreads in a group of 20 students

You will learn more about the logistic model, including its derivation, in Chapter 20.

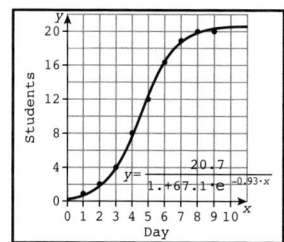

**Figure 9.38** The logistic model found by a GDC for the student example

339

To help us determine $C$ and $k$, it is helpful to realise that the logistic curve has rotational symmetry around a point with $y$-coordinate equal to $y = \dfrac{L}{2}$ (this point is also a point of inflection). Generally it is also possible to identify this point on a graph. Therefore, given $L$, we can determine a relationship between $C$ and $k$. Since the curve must pass through $\left(x, \dfrac{L}{2}\right)$ we can substitute and simplify as follows:

$$f(x) = \frac{L}{1 + Ce^{-kx}} \implies \frac{L}{2} = \frac{L}{1 + Ce^{-kx}}$$

$$2 = 1 + Ce^{-kx}$$

$$\frac{1}{C} = e^{-kx}$$

$$\ln\left(\frac{1}{C}\right) = -kx$$

$$x = \frac{\ln C}{k}$$

The equation $x = \dfrac{\ln C}{k}$ also gives us some insight into the location of the point of inflection. Recall that $\ln C < 0$ when $0 < C < 1$, which implies that the point of inflection has a negative $x$-coordinate for $0 < C < 1$. Likewise, $\ln C = 0$ when $C = 1$, so the $x$-coordinate of the point of inflection will be zero. Finally, for $C > 1$, $\ln C > 0$, so the point of inflection has a positive $x$-coordinate.

You will learn more about points of inflection in Chapter 14.

The logistic function and its parameters are shown in Figure 9.39.

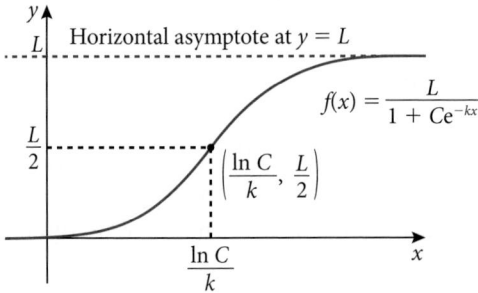

**Figure 9.39** Graphical properties of the logistic function

The logistic model is given by

$$f(x) = \frac{L}{1 + Ce^{-kx}}, \quad C, k > 0 \text{ where}$$

$L$ is the limiting value; the function has a horizontal asymptote at $y = L$

$C$ determines the horizontal position of the growth part of the curve, such that for

- $0 < C < 1$ the point of inflection has a negative $x$-coordinate;
- $C = 1$ the point of inflection is on the $y$-axis;
- $C > 1$ the point of inflection has a positive $x$-coordinate.

$k$ determines the growth rate (how steep the growth part of the curve is).

The location of the point of inflection is $\left(\dfrac{\ln C}{k}, \dfrac{L}{2}\right)$

There is also a horizontal asymptote at $y = 0$. Here, we are interested in the non-zero asymptote.

Example 9.16 shows another way of finding a logistic model.

## Example 9.16

In a class of 30 students, one student comes to school sick one day. The infection of the remaining students follows a logistic model of the form $S(d) = \dfrac{L}{1 + Ce^{-kd}}$ where $d$ is the number of days since the initial infection.

(a) Write down the value of $L$.

(b) Find the value of $C$ given that the first student is sick on day 0.

(c) Given that half of the students are infected after 5 days, find the value of $k$.

(d) Sketch a graph of the logistic model, showing clearly the point of inflection and any asymptotes.

### Solution

(a) Since the maximum number of students that can be infected is 30, we know $L = 30$

(b) Using the fact that $S(0) = 1$:

$$S(0) = 1 = \frac{30}{1 + Ce^{-k \times 0}} \quad \Rightarrow \quad C = 29$$

(c) If half the students are infected after 5 days, we know that the point of inflection must be at $(5, 15)$. Hence,

$$\frac{\ln C}{k} = 5 \quad \Rightarrow \quad \ln 29 = 5k \quad \Rightarrow \quad k = \frac{\ln 29}{5} = 0.673$$

(d) We know the model has a horizontal asymptote at $y = 30$ and a point of inflection at $(5, 15)$.

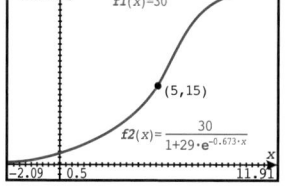

**Figure 9.40** GDC Screen for solution to Example 9.16 (d)

Logistic models can be useful in predicting the limiting (maximum) value.

## Example 9.17

The population curve of bacteria in a closed environment is known to follow a logistic model. A student begins an experiment by placing 10 bacteria in a Petri dish. The student records the following data:

| Time (hours) | 0 | 1 | 2 | 3 |
|---|---|---|---|---|
| Number of bacteria | 10 | 150 | 1100 | 6700 |

(a) Construct a logistic model for the population of bacteria in terms of hours since the experiment started.

(b) Write down the maximum population of the bacteria according to your model.

(c) Find the minimum number of whole hours until the population of bacteria will be within 5% of the maximum population.

341

(d) Since the logistic model is asymptotic to the maximum value, the maximum will never be reached. However, at some point in time the number of bacteria will get close enough to the maximum for any practical purpose. Find the minimum number of hours until the population of bacteria, given in 3 significant figures, is equal to the maximum population.

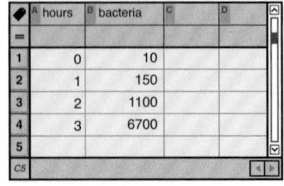

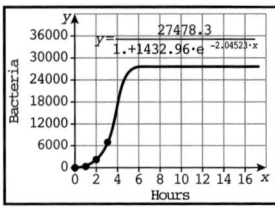

**Figure 9.41** GDC screens for solution to Example 9.17 (a)

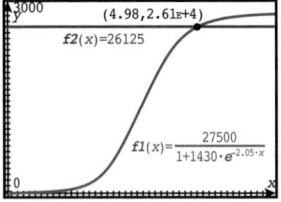

**Figure 9.42** GDC Screen for solution to Example 9.17 (c)

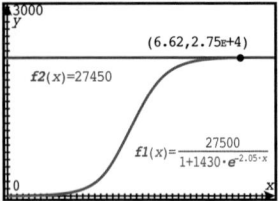

**Figure 9.43** GDC Screen for solution to Example 9.17 (d)

## Solution

(a) The logistic model is of the form $P(t) = \dfrac{L}{1 + Ce^{-kt}}$ where $P(t)$ is the population $t$ hours after the experiment started.
Use a GDC to find the logistic model.

Therefore, our model (to 3 significant figures) is $P(t) = \dfrac{27\,500}{1 + 1430e^{-2.05t}}$

(b) According to our model, the maximum bacteria population is $L = 27\,500$ bacteria.

(c) We want to find when the population will reach 95% of the maximum, i.e., $0.95 \times 27\,500 = 26\,125$ bacteria. We can use a GDC to graph the logistic model alongside the line $y = 26\,125$ and then find the intersection point:

Therefore, the minimum number of whole hours required for the population to be within 5% of its maximum is 5 hours.

(d) Since the maximum number of bacteria is 27 500, any integer between 27 450 and 27 549 is equal to 27 500 when 3 significant figures are used. Thus, we follow the same procedure as part (c), but this time we look for an intersection with the line $y = 27\,450$

Therefore, the bacteria population is within 3 significant figures of the maximum at 6.62 hours.

### Example 9.18

A population of white-tailed deer in a certain park is known to follow a logistic model of the form $P(t) = \dfrac{L}{1 + Ce^{-kt}}$ with a logistic growth rate of $k = 0.24$. The population is monitored and it is noted that at the start of 2020 there are 1000 deer. Three years later, there are 2000 deer.

(a) Find the values of $C$ and $L$ for this population of deer.

(b) Predict the maximum possible population of deer for this park.

(c) Deer give birth every spring. Calculate the number of whole years required for the population to reach half of its maximum.

## Solution

(a) Using a logistic model of the form $P(t) = \dfrac{L}{1 + Ce^{-kt}}$, we are given
$k = 0.24$ and we will let $t$ equal the number of years since 2020.
Therefore, the function must pass through $(0, 1000)$ and $(3, 2000)$.
Substituting:

$$1000 = \frac{L}{1 + Ce^{-0.24 \times 0}} \quad \text{and} \quad 2000 = \frac{L}{1 + Ce^{-0.24 \times 3}}$$

$$L = 1000(1 + C) \qquad\qquad L = 2000(1 + Ce^{-0.72})$$

Therefore:

$$1000(1 + C) = 2000(1 + Ce^{-0.72})$$

Solving for $C$: $C = \dfrac{1}{1 - 2e^{-0.72}} = 37.7$

Solving for $L$: $L = 1000(1 + C) = 38\,700$

(b) According to the logistic model, the maximum population of deer in this park is $L = 38\,700$ deer.

(c) The population will reach half its maximum at the inflection point, at

$t = \dfrac{\ln C}{k} = \dfrac{\ln 37.7}{0.24} = 15.1$ years. However, since the population will not

increase until the spring, we must wait until later in that year. Therefore, 16 whole years are required.

---

## Exercise 9.4

1. The velocity of a peregrine falcon's high-speed attack dive can be described by a logistic model. Researchers captured the data shown for one such dive:

   (a) Use your GDC to find a logistic model of the form $P(t) = \dfrac{L}{1 + Ce^{-kt}}$

   (b) Hence, predict the maximum velocity this falcon could attain.

   (c) How many seconds will it take the falcon to be within 1% of its maximum velocity?

   | Time in dive (s) | Velocity (m s⁻¹) |
   |---|---|
   | 2 | 10 |
   | 4 | 53 |
   | 6 | 90 |

   Table 9.14 Data for question 1

2. On a farm with 500 cows, one cow is determined to have a serious disease. The infection of the remaining cows follows a logistic model of the form
   $B(d) = \dfrac{L}{1 + Ce^{-kd}}$, where $B(d)$ is the number of cows infected after $d$ days.

   (a) Write down the value of $L$.

   (b) Find the value of $C$.

   (c) Given that half of the cows are infected after 10 days, find the value of $k$.

**(d)** Sketch a graph of the logistic model, showing clearly the point of inflection and any asymptotes.

**(e)** After many days will 90% of the cows be infected?

**3.** The population of rats on an island is known to follow a logistic model of the form $P(t) = \dfrac{L}{1 + Ce^{-kt}}$, where $P(t)$ is the population after $t$ years, with a logistic growth rate of $k = 0.2$. The population is monitored, and it is noted that at the start of 2000 there are 1000 rats. At the start of 2005, there are 2500 rats.

**(a)** Find the vales of $C$ and $L$ for this population of rats.

**(b)** Predict the maximum possible population of rats on this island.

**(c)** Calculate the number of years for the population to reach half of its maximum.

# **9.5** Direct and inverse variation

Direct variation and inverse variation are two types of models that occur so often in real life that they deserve a special look.

## Direct variation

In general, **direct variation** occurs when one quantity increases along with another quantity. **Inverse variation** occurs when one quantity decreases along with another quantity.

Suppose a light-weight perforated ball is dropped from the roof of a building. Since air resistance will not be negligible, the standard constant acceleration due to gravity ($g = 9.81$ m s$^{-1}$) does not apply. Although the standard falling-object model does not apply, the distance the ball has fallen varies directly with the square of the time since it was dropped. Symbolically, we write $d = at^2$ where $d$ is the distance travelled, $t$ is the time elapsed, and $a$ is a **constant of variation** that we will determine.

At the instant the ball is dropped, it has travelled 0 m. One second later, it has travelled 4 m.

Note that we cannot use the point $(0, 0)$ to find the constant of variation, since it produces the equation $0 = 0$.

By two seconds, it has travelled 16 m. By three seconds, it has travelled 36 m. We can express this data as ordered pairs: $(0, 0)$, $(1, 4)$, $(2, 16)$, $(3, 36)$

In common usage, direct variation is often used only to refer to linear direct variation models, where $y = ax$. In this text, and in the IB syllabus, we use a broader definition, as shown in the key fact box.

How can we find the constant of variation? We simply need to substitute one of the ordered pairs into the model: $4 = a(1)^2 \Rightarrow a = 4$

Therefore the direct variation model is $d = 4t^2$

Direct variation models are simplified polynomial models of the form
$$y = ax^n \text{ where } n \in \mathbb{Z}, n > 0$$
The constant $a$ is sometimes called **the constant of variation**.

## Example 9.19

The cost $C$ of a mobile phone call varies directly with the length of the call $m$ in minutes. A recent 5-minute phone call cost $0.60

(a) Formulate a direct variation model for this situation.

(b) Use your model to predict the cost of a 7-minute phone call.

(c) Find the length of a call that costs $1.56

(d) Write down the cost per minute for mobile phone calls.

### Solution

(a) Since a 5-minute phone call cost $0.60, we have

$$C = am \implies 0.60 = a(5) \implies a = 0.12$$

therefore, the model is $C = 0.12m$

(b) A 7-minute phone call would cost $C = 0.12(7) = \$0.84$

(c) A call that cost $1.56 must have $1.56 = 0.12m \implies m = 13$ minutes

(d) From the model, we see that $a = 0.12$

Therefore, calls cost $0.12 per minute.

## Example 9.20

The volume of a sphere varies directly with the cube of the radius of the sphere. You are given that a sphere with a radius of 14 cm has a volume of 11 500 cm³ (3 s.f.).

(a) Find a direct variation model.

(b) Use your model to find the volume of a sphere with a radius of 7 cm to 3 significant figures.

(c) Using 6 significant figures, find the percentage error between your model and the volume given by the formula $V = \frac{4}{3}\pi r^3$

### Solution

(a) Since the volume varies directly with the cube of the radius, we have

$$V = ar^3$$

Given that the volume of a sphere with radius 14 cm has a volume of 11 500 cm³, we can solve for the constant of variation:

$$11\,500 = a(14)^3 \implies a = 4.19 \text{ (3 s.f.)}$$

Therefore, our model is $V = 4.19r^3$

(b) For a sphere of radius 7 cm, we have

$$V = 4.19(7)^3 = 1440 \text{ cm}^3 \text{ (3 s.f.)}$$

(c) The theoretical volume of a 7 cm sphere is $V = \frac{4}{3}\pi(7)^3 = 1436.76 \text{ cm}^3$

The percentage error is therefore $\dfrac{|1436.76 - 1437.17|}{1437.76} \times 100 = 0.0285\%$

## Inverse variation

Suppose you want to hire a DJ for a party. The cost of the DJ is $500 for the night. To cover the cost, you will sell tickets. What price should you set for the tickets, based on how many you think you will sell? This is inverse variation, because the more tickets you sell, the lower the ticket price can be. Consider a few data points: if you only sell one ticket, the ticket price will need to be $500 to cover costs (but it would be a lonely party!). If you can sell two tickets, then each one can be $250.

If you can sell 10 tickets, then each one is $\frac{500}{10} = \$50$

For 20 tickets, each one is $\frac{500}{20} = \$25$

We have already developed the model: to cover costs, the ticket price is the total cost divided by the number of tickets you can sell. Symbolically:

$$P = \frac{500}{n}$$

where $P$ is the ticket price and $n$ is the number of tickets you must sell, $n > 0$

This produces an interesting graph, shown in Figure 9.44.

Note that in Figure 9.44 the graph approaches – but does not intersect – the horizontal axis. This makes sense, because as we sell more and more tickets, we could reduce the price per ticket. But, if we give away the tickets for free, we can't possibly cover the cost of the DJ.

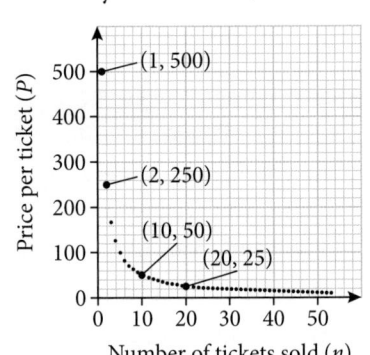

**Figure 9.44** Price per ticket varies inversely with the number of tickets sold

In this example, we can rewrite our model using a negative exponent:

$$P = \frac{500}{n} \implies P = 500n^{-1}$$

This leads us to a definition of inverse variation models (see key fact box).

Note that the only difference between direct and inverse variation models is the sign of the exponent.

In Figure 9.44, we have drawn the graph as a series of dots but have not connected them. Why not? It's not possible to sell 1.5 tickets – we can only sell whole tickets. So, we plot the graph as a series of points rather than a smooth function. As noted earlier, we will often draw the graph as a smooth function even though that may not strictly represent reality.

Inverse variation models are of the form

$$y = ax^{-n}$$

where $n \in \mathbb{Z}, n < 0$

The constant $a$ is sometimes called the constant of variation.

Remember that in scientific terms, weight is a force and is measured in newtons.

### Example 9.21

The weight of an object (near Earth) varies inversely with the square of the distance from the centre of the Earth. At sea level (6370 km from the centre of the Earth), an astronaut weighs 600 N.

(a) Find a model for the weight of the astronaut based on the distance from the centre of the Earth.

(b) The International Space Station orbits at an altitude of 408 km. Find the weight of the astronaut while on board the International Space Station.

(c) Find the required distance from the centre of the Earth for the astronaut to weigh half what she weighs at sea level.

## Solution

(a) Since the weight varies inversely with the square of the distance, we can write $w = \dfrac{a}{d^2}$ where $w$ is the weight in N, $d$ is the distance from the centre of the Earth, in km, and $a$ is the constant of variation that we will determine. We find $a$ by substituting known values:

$$w = \frac{a}{d^2} \implies 600 = \frac{a}{6370^2} \implies a = 2.43 \times 10^{10}$$

Therefore, our model is $w = \dfrac{2.43 \times 10^{10}}{d^2}$

(b) At an altitude of 408 km, the distance from the centre of the Earth is $6370 + 408 = 6778$ km. Therefore, the weight of the astronaut would be

$$w = \frac{2.43 \times 10^{10}}{6778^2} = 530 \text{ N}$$

(c) We need to solve

$$300 = \frac{2.43 \times 10^{10}}{d^2} \implies 300\, d^2 = 2.43 \times 10^{10}$$

$$\implies d = \sqrt{\frac{2.43 \times 10^{10}}{300}} = 9010 \text{ km (3 s.f.)}$$

The answer to (b) might seem counter-intuitive: If the astronaut still weighs 530 N on the International Space Station, why do we always see videos of astronauts floating? The answer is that the space station and the astronauts inside it are actually falling towards the Earth at the same rate, so the astronauts seem to be floating inside the space station.

## Exercise 9.5

1. Choose always/sometimes/never:

   (a) Direct variation models always/sometimes/never pass through the origin $(0, 0)$.

   (b) Direct variation models are always/sometimes/never a type of polynomial model.

   (c) Direct variation models are always/sometimes/never a type of linear model.

   (d) Direct variation models are always/sometimes/never a type of exponential model.

2. Choose always/sometimes/never:

   (a) Inverse variation models always/sometimes/never pass through the origin $(0, 0)$.

   (b) Inverse variation models are always/sometimes/never a type of polynomial model.

   (c) Inverse variation models are always/sometimes/never a type of linear model.

   (d) Inverse variation models are always/sometimes/never a type of exponential model.

3. Given that $y$ varies directly with $x$, and $y = 462$ when $x = 11$, find:

   (a) $y$ when $x = 5$          (b) $x$ when $y = 672$

4. Given that $y$ varies directly with the square of $x$, and $y = 10$ when $x = 5$, find:

   (a) $y$ when $x = 20$          (b) $x$ when $y = 40$

5. Given that $y$ varies directly with the cube of $x$, and $y = 250$ when $x = 5$, find:

   (a) $y$ when $x = 8$          (b) $x$ when $y = 128$

6. Given that $y$ varies inversely as $x$, and $y = 10$ when $x = 5$, find:

   (a) $y$ when $x = 20$          (b) $x$ when $y = 0.5$

7. Given that $y$ varies inversely as the square of $x$, and $y = 10$ when $x = 5$, find:

   (a) $y$ when $x = 20$          (b) $x$ when $y = 2.5$

8. Given that $y$ varies inversely as the cube of $x$, and $y = 54$ when $x = 5$, find:

   (a) $y$ when $x = 15$          (b) $x$ when $y = 250$

9. The height of an image produced by a projector varies directly with the distance from the screen. The image is 1.5 m tall when the projector is 2 m from the screen.

   (a) Find a direct variation model for the height of the image $S$ given the distance $d$ from the screen.

   (b) Use your model to predict the height of the image for a projector 7 m from the screen.

   (c) Use your model to find the distance required to project an image 10 m tall.

10. The velocity of a falling object varies directly with the amount of time it has been falling.

    (a) Given that an object that has been falling for two seconds has a velocity of 19.6 m s$^{-1}$, calculate the constant of variation.

    (b) Calculate the velocity of an object that has been falling for 4 s.

    (c) Terminal velocity for a skydiver is approximately 200 km h$^{-1}$. Calculate the approximate number of seconds it takes a skydiver to reach terminal velocity according to this model.

**11.** The position of a falling object varies directly with the square of the number of seconds it has been falling. Assume the initial velocity is zero.

    **(a)** Given that an object that has been falling for 10 s towards the surface of the moon travels 162 m, find the constant of variation.

    **(b)** Calculate the distance an object falls in 5 s.

    **(c)** Calculate the time required for an object to fall 200 m.

**12.** The radius of a satellite's orbit around the Earth varies inversely with the square of the velocity of the satellite.

    **(a)** Given that a satellite that travels at a velocity of 7700 m s$^{-1}$ has an orbital radius of $6.75 \times 10^6$ m, calculate the constant of variation.

    **(b)** Find the orbital velocity of a satellite with an orbital radius of $7.0 \times 10^6$ m.

    **(c)** Find the orbital radius of a satellite with a velocity of 8000 m s$^{-1}$.

**13.** The volume of a dodecahedron varies directly with the cube of the length of an edge $a$, as shown in the diagram.

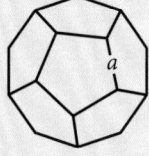

    **(a)** Given that the volume of a dodecahedron is 958 cm$^3$ with the edge length $a = 5$, find the constant of variation.

    **(b)** Find the volume of a dodecahedron when the edge length is 8 cm.

    **(c)** Find the edge length of a dodecahedron with volume 100 m$^3$.

**14.** The power a wind turbine can generate varies directly with the cube of the wind speed. A certain turbine can generate 314 W when the wind speed is 8 m s$^{-1}$.

    **(a)** Find the constant of variation.

    **(b)** Find the power generated when the wind speed is 12 m s$^{-1}$.

    **(c)** Find the wind speed necessary to generate 2000 W.

# 9.6    Further modelling skills

In the real world, it's crucial to be able to identify, develop, evaluate, and use models. In this chapter we have examined different models and touched on many of the uses and limitation of models. However, it's also important to be able to choose an appropriate model. Consider the following example of the misuse of a model.

One of the authors recorded the mass of his young baby at several points after she was born. The data is shown in the table, and a graph is shown in Figure 9.45.

| Days since birth | 0 | 4 | 17 | 45 | 73 |
|---|---|---|---|---|---|
| Mass (kg) | 3.36 | 3.4 | 3.69 | 5.29 | 6.15 |

**Table 9.15** Baby mass data since birth

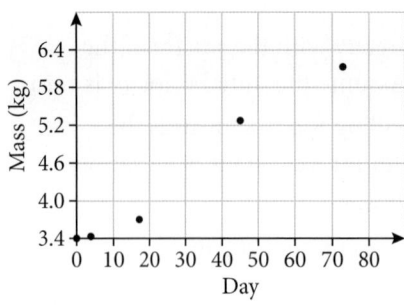

**Figure 9.45** Mass versus days since birth for a young child

The data appears approximately linear, so the author used technology to generate a linear model for this data:

$$m = 0.0408d + 3.24,$$ where $m$ is mass in kilograms $d$ days after birth.

The model suggests that she is gaining 0.0408 kg per day, with an initial mass of 3.24 kg. Graphically, the model appears to fit the data reasonably well.

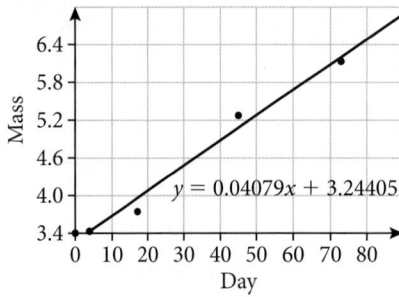

**Figure 9.46** The model appears to fit the data reasonably well

Of course, the author was very interested in the eventual mass of his young baby girl. What would her mass be when she is 1 year old? Five years old? Ten years old? He used the model to make some predictions, shown in the table below (he even added leap days to be sure):

| Days since birth | 365 (1 year) | 1826 (5 years) | 3652 (10 years) |
|---|---|---|---|
| Mass (kg) | 18.1 | 77.7 | 152 |

**Table 9.16** Mass predictions

Are these results reasonable? A mass of 77.7 kg for a five-year-old girl is alarming, and a mass of 152 kg for a ten-year-old girl is definitely a cause for concern.

What went wrong? The mathematical mechanics are perfectly correct.

There are two flaws in our reasoning. One, our choice of model (linear model) assumes that the rate of change is constant (in this case, 40.8 g per day), which is not correct: we know that the growth rate will decrease as the child gets older. Perhaps a logistic model would be a better choice. Two, we **extrapolated** beyond known data. We will discuss both of these issues in this section.

## Choosing a model

One of the most challenging parts of modelling in the real world is choosing an appropriate model. Often, we try to use our understanding of a situation when we are choosing and developing a model. Other times we may look at the shape of the data (as seen on a graph) to try to help us choose a model.

Here are some guidelines to help, with examples from the introduction to this chapter:

**Consider the rate of change.**

- Is it constant? Try a linear model.

  **Example**: On a long flight, the airspeed of an aeroplane is constant, so the distance remaining to the destination can be described by a linear model.

- Is the quantity increasing/decreasing by a fixed percentage or ratio? Try an exponential model.

  **Examples:** The number of algae in a polluted lake doubles every 3 days; a bank account earns 5% quarterly interest, compounded annually; the value of a car is decreasing by 25% per year; what will the activity of a radioisotope be after some time.

- Is the quantity increasing/decreasing at a linearly increasing/decreasing rate? Try a quadratic model.

  **Examples:** The price to manufacture $x$ units of some product decreases linearly, the revenue from selling $x$ units can be described by a quadratic model; the velocity of a falling object changes linearly, the position follows a quadratic model.

**Consider the nature of the phenomenon.**

- Does it relate volume to a linear quantity? Try a cubic model.

  **Examples:** The volume of a balloon relative to its diameter can be described by a cubic model; electricity generated by a wind turbine based on wind speed.

- Is it cyclical/periodic/repeating? Try a trigonometric model.

  **Examples:** Tide height; the position of a person on Ferris wheel; seasonal average temperatures.

**Consider the shape of the data as shown in Figures 9.47–9.53**

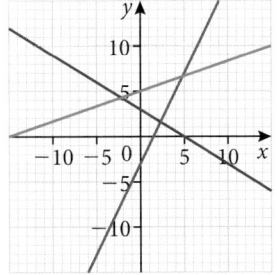

**Figure 9.47** Linear models

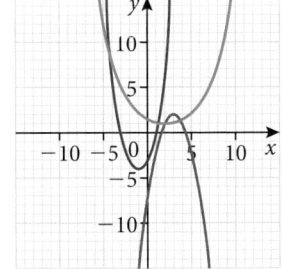

**Figure 9.48** Quadratic models

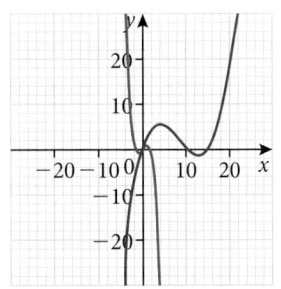

**Figure 9.49** Cubic models

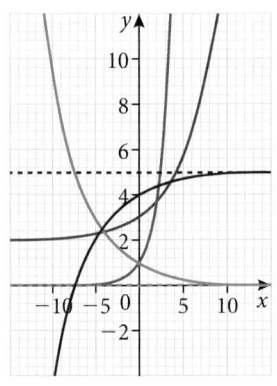

**Figure 9.50** Exponential models

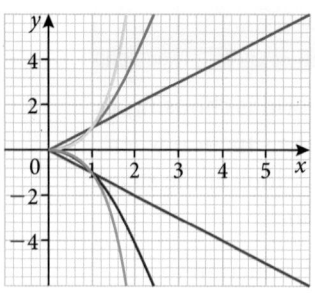

**Figure 9.51** Direct variation models

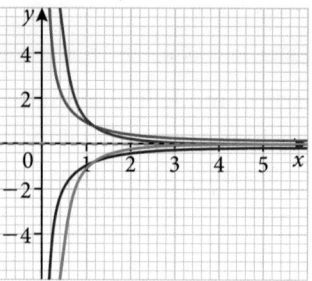

**Figure 9.52** Inverse variation models

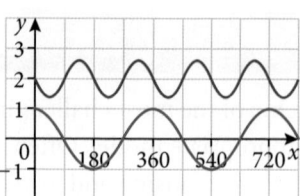

**Figure 9.53** Trigonometric models

### Example 9.22

For each of the following examples choose an appropriate model and give a reason for your choice.

(a) Modelling a person's mass as a function of their height.

(b) The number of hours of day-light in Tokyo during each day of the year.

(c) The population of bacteria in a Petri dish over time.

(d) The resale value of a computer over time.

(e) The distance from Earth of a Voyager space probe traveling away at a constant rate.

(f) The height of the golf ball Alan Shepard hit on the moon in 1971.

### Solution

(a) Since weight is based on volume, and volume varies with the cube of height, we should use a cubic model of the form $W = kh^3$

(b) Number of hours of day light each day is a periodic phenomenon, so a trigonometric model is appropriate.

(c) The population of bacteria in a Petri dish will grow exponentially (until the maximum population for the dish is reached), so an exponential model is appropriate.

(d) A computer's value will depreciate exponentially, so an exponential model is appropriate.

(e) Since the Voyager is travelling away from Earth at a constant rate, a linear model is appropriate.

(f) The golf ball will follow a quadratic falling-object model as it is still subject to gravity – just less than on Earth.

## Testing a model

Once a model is developed (using the techniques in this chapter), it is important to test it. In fact, in real life, it is often the case that after testing a model you decide to develop an alternative model or to revise your existing model. How do we test a model? We can test a model by looking at the fit of the model for the known data. If the model does not seem to fit well, it can be a sign that we do not fully understand the phenomenon we are trying to model. In that case, we may decide to try a different model, or perhaps more research is necessary.

Developing useful models is the work of many people: economists, doctors, aid workers, actuaries, and many other researchers. This chapter is a basic introduction to the art and science of modelling.

It is important to understand that, in the real world, there is often no 'right' model. Instead, in real-life applications we often judge a model by its usefulness. For example, what does it tell us about the data we have? Does it tell us something about the nature of the phenomena we are modelling? Does it allow us to make meaningful predictions? Are the predictions surprising but plausible? What assumptions are we making?

Asking these sorts of questions is a crucial skill for the practicing mathematical modeller.

### Example 9.23

Anna is attempting to model the temperature of a pond at her school. She has collected 12 points of data, measuring the temperature of the pond every 2 hours. Based on her data, she believes a quadratic model may be useful, so she has used the vertex and $y$-intercept to fit a model. The graph shows her data points and model.

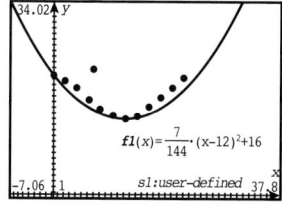

Figure 9.54 Graph for Example 9.23

(a) Give two reasons why Anna's choice of model may not be appropriate.

(b) Suggest a better model and give a reason why it would be more appropriate.

### Solution

(a) Anna's model doesn't seem to fit the data very well. Also, the temperature of the pond probably varies during the day and night in a periodic way, which would not suggest a quadratic model.

(b) Since the temperature of the pond probably varies periodically, a trigonometric model would be more appropriate.

## Interpolation versus extrapolation

As in the example at the start of the section, we can quickly get into trouble when we extrapolate. But what exactly is extrapolation? How can we tell when we are using a model appropriately and when we are extrapolating? One way to check is to look at the range of known data, as in Example 9.24.

### Example 9.24

A used-car dealer has recently sold five of the same kind of car, called the Canyonero. All five were in similar condition, but the ages of the cars varied. The data are shown in the table.

| Age of car (years) | 2 | 3 | 4 | 6 | 10 |
|---|---|---|---|---|---|
| Sale price ($) | 20 000 | 14 500 | 12 000 | 6100 | 2000 |

The dealer decides to use the model $P = 20\,000\,(0.75)^{\,t-2}$ to describe the price of the Canyonero ($P$ represents the price and $t$ represents the age in years).

(a) Use the model to predict the price of a new Canyonero. Comment on the usefulness of your prediction.

(b) Use the model to predict the price of an 8-year-old Canyonero. Comment on the usefulness of your prediction.

(c) Use the model to predict the price of a 15-year-old Canyonero. Comment on the usefulness of your prediction.

### Solution

(a) According to the model, the price of a new Canyonero would be $P = 20\,000\,(0.75)^{0-2} = \$35,600$ (3 s.f.). Although this may seem like a reasonable prediction, our data only includes cars from 2 to 10 years old, so we cannot be confident in this prediction. Therefore, it is not a useful prediction.

(b) According the model, the price of an 8-year-old Canyonero would be $P = 20\,000\,(.75)^{8-2} = \$3560$ (3 s.f.). Since 8 years is within the range of the age of cars in our data, we can be reasonably confident in this prediction. Therefore, this is a useful prediction.

(c) According the model, the price of a 15-year-old Canyonero would be $P = 20\,000\,(.75)^{15-2} = \$475$ (3 s.f.). Not only does this value seem like a small amount, but it is not within the range of the age of cars in our data. Therefore, it is not a useful prediction.

As you can see in Example 9.24, it is important to note that we decide whether a prediction is interpolation or extrapolation based on the range of our independent variable. This is easier to see when we look at a graph (Figure 9.55).

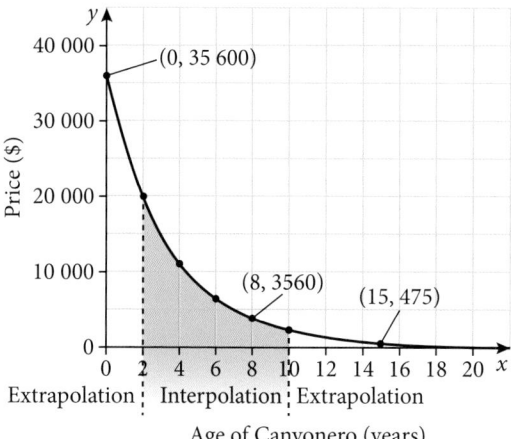

Figure 9.55 Interpolation is based on the range of the independent variable

As shown in Figure 9.55, any predictions based on values within the range of the independent variable are considered **interpolations**. Therefore, any predictions for cars between 2 and 10 years old would be interpolation and therefore relatively safer and more useful.

On the other hand, any predictions based on values outside the range of the independent variable are considered **extrapolations**. Therefore, any predictions for cars newer than 2 years old or older than 10 years old are considered extrapolation. Extrapolated values are less reliable and therefore less useful, since we simply can't be sure what the value of the car might be outside the range that we have observed.

Although it is tempting to look at the value of a prediction to decide whether a prediction is useful or not, it is critical to consider first whether the value of the independent variable is within the range of known values of the independent variable. That is, we can decide whether a value is interpolation or extrapolation before we make a prediction. Prediction based on extrapolation should always be used with caution!

**Interpolations** are predictions based on values within the range of known values of the independent variable. **Extrapolations** are predictions based on values outside the range of known values of the independent variable.

Given a model $f(x)$ based on data with minimum $x$-value $a$ and a maximum $x$-value $b$, and domain $a \leqslant x \leqslant b$

- If $a \leqslant c \leqslant b$, then $f(c)$ is an **interpolated value**.
- If $c < a$ or $c > b$, then $f(c)$ is an **extrapolated value**.

**Exercise 9.6**

1. Identify appropriate models for the following situations.

   (a) The mass of an elephant from birth to death.

   (b) The acceleration produced by exerting a constant force on various masses.

   (c) The total cost of fuel purchased at a fixed price per litre.

   (d) A population of rats on an island.

   (e) The area of a rectangle where the length is twice as large as the width.

   (f) Average monthly temperature over a year in Tokyo.

   (g) The volume of a cube as a function of its side length.

   (h) The velocity of a skydiver jumping from an aeroplane.

   (i) The value of a savings account growing at a fixed percentage rate, compounded monthly.

   (j) The height of the tide on a beach.

   (k) The value of a new car over time.

   (l) The cost of hiring a bus, per person.

   (m) The cost of a wedding as a function of the number of guests.

2. For each data set, identify the valid domain for interpolation. Assume that the first row in the table is the independent variable.

   (a)

   | Time in dive (s) | 2 | 4 | 6 |
   |---|---|---|---|
   | Velocity (m s$^{-1}$) | 10 | 53 | 90 |

   (b)

   | $x$ | 41 | 33 | 100 | 76 | 11 |
   |---|---|---|---|---|---|
   | $y$ | 33.4 | 31.8 | 52.3 | 47.3 | 16.6 |

   (c)

   | $n$ | 8.2 | 3.9 | 5.5 | 1.3 | 8.5 |
   |---|---|---|---|---|---|
   | $P(n)$ | 15.5 | 11.5 | 12.3 | 6.0 | 15.2 |

3. In general, what are the key feature(s) that we might see to suggest that we should use each model:

   (a) linear

   (b) quadratic

   (c) trigonometric

   (d) exponential

   (e) logistic

   (f) inverse

1. A city is concerned about pollution, and decides to look at the number of people using taxis. At the end of the year 2000, there were 280 taxis in the city. After $n$ years the number of taxis, $T$, in the city is given by
$T = 280 \times 1.12^n$

   (a) Find the number of taxis in the city at the end of 2005.

   (b) Find the year in which the number of taxis is double the number of taxis there were at the end of 2000.

   At the end of 2000 there were 25 600 people in the city who used taxis. After $n$ years the number of people, $P$, in the city who used taxis is given by $P = \dfrac{2\,560\,000}{10 + 90e^{-0.1n}}$

   (c) Find the value of $P$ at the end of 2005, giving your answer to the nearest whole number.

   (d) After seven complete years, will the value of $P$ be double its value at the end of 2000? Justify your answer.

   Let $R$ be the ratio of the number of people using taxis in the city to the number of taxis. The city will reduce the number of taxis if $R < 70$

   (e) Find the value of $R$ at the end of 2000.

   (f) After how many complete years will the city first reduce the number of taxis?

2. The profit $(P)$ in Swiss Francs made by three students selling homemade lemonade is modelled by the function $P = -\dfrac{1}{20}x^2 + 5x - 30$ where $x$ is the number of glasses of lemonade sold.

   (a) Copy and complete the table below.

   | $x$ | 0 | 10 | 20 | 30 | 40 | 50 | 60 | 70 | 80 | 90 |
   |---|---|---|---|---|---|---|---|---|---|---|
   | $P$ | | 15 | | | 90 | | | 75 | 50 | |

   (b) Draw the graph of $P$ against $x$ by plotting the points given in the table. Label your graph.

   (c) Use your graph to find:

   (i) the maximum possible profit

   (ii) the number of glasses that need to be sold to make the maximum profit

   (iii) the number of glasses that need to be sold to make a profit of 80 Swiss Francs

   (iv) the amount of money initially invested by the three students.

   The three students Baljeet, Jane and Fiona share the profits in the ratio of $1:2:3$ respectively. Given that they sold 40 glasses of lemonade

   (v) calculate Fiona's share of the profits.

3. The function $Q(t) = 0.003t^2 - 0.625t + 25$ represents the amount of energy in a battery after $t$ minutes of use.

   (a) Write down the amount of energy held by the battery immediately before it was used.

   (b) Calculate the amount of energy available after 20 minutes.

   (c) Given that $Q(10) = 19.05$, find the average amount of energy lost per minute for the interval $10 \leqslant t \leqslant 20$

   (d) Calculate the number of minutes it takes for the energy to reach zero.

4. Jashanti is saving money to buy a car. The price of the car, in US Dollars (USD), can be modelled by the equation $P = 8500(0.95)^t$

   Jashanti's savings, in USD, can be modelled by the equation $S = 400t + 2000$

   In both equations $t$ is the time in months since Jashanti started saving for the car.

   (a) Write down the amount of money Jashanti saves per month.

   (b) Use your graphic display calculator to find how long it will take for Jashanti to have saved enough money to buy the car.

   (c) Jashanti does not want to wait too long and wants to buy the car two months after she started saving. She decides to ask her parents for the extra money that she needs. Calculate how much extra money Jashanti needs.

5. A building company has many rectangular construction sites, of varying widths, along a road. The area, $A$, of each site is given by the function $A(x) = x(200 - x)$ where $x$ is the width of the site in metres and $20 \leqslant x \leqslant 180$

   (a) Site $S$ has a width of 20 m. Write down the area of $S$.

   (b) Site $T$ has the same area as site $S$, but a different width. Find the width of $T$.

   (c) When the width of the construction site is $b$ metres, the site has a maximum area.

      (i) Write down the value of $b$.

      (ii) Write down the maximum area.

   (d) The range of $A(x)$ is $m \leqslant A(x) \leqslant n$
   Hence write down the value of $m$ and of $n$.

6. Water has a lower boiling point at higher altitudes. The relationship between the boiling point of water ($T$) and the height above sea level ($h$) can be described by the model $T = -0.0034h + 100$ where $T$ is measured in degrees Celsius (°C) and $h$ is measured in metres from sea level.

   (a) Write down the boiling point of water at sea level.

   (b) Use the model to calculate the boiling point of water at a height of 1.37 km above sea level.

   (c) At the top of Mt. Everest, water boils at 70°C. Use the model to calculate the height above sea level of Mt. Everest.

7. The temperature in degress Celsius (°C) of a pot of water removed from a cooker is given by $T(m) = 20 + 70 \times 2.72^{-0.4m}$ where $m$ is the number of minutes after the pot is removed from the cooker.

   (a) Show that the temperature of the water when it is removed from the cooker is 90°C.

   (b) Calculate the temperature of the water after 10 minutes.

   (c) Calculate how long it takes for the temperature of the water to reach 56°C.

   (d) Write down the temperature approached by the water after a long time. Justify your answer.

   Consider the function $S(m) = 20m - 40$ for $2 \leqslant m \leqslant 6$

   The function $S(m)$ represents the temperature of soup in a pot placed on the cooker two minutes after the water has been removed. The soup is then heated.

   (e) Comment on the meaning of the constant 20 in the formula for $S(m)$ in relation to the temperature of the soup.

   (f) Solve the equation $S(m) = T(m)$ and interpret the solution in context.

   (g) Hence describe by using inequalities the set of values of $m$ for which $S(m) > T(m)$

8. Shiyun bought a car in 1999. The value of the car $V$, in USD, is depreciating according to the exponential model $V = 25\,000 \times 1.5^{-0.2t}$, $t \geqslant 0$, where $t$ is the time, in years, that Shiyun has owned the car.

   (a) Write down the value of the car when Shiyun bought it.

   (b) Calculate the value of the car three years after Shiyun bought it. Give your answer correct to two decimal places.

   (c) Calculate the time for the car to depreciate to half of its value since Shiyun bought it.

9. In an experiment it is found that a culture of bacteria triples in number every four hours. There are 200 bacteria at the start of the experiment.

    **(a)** Write down the number of the bacteria after 8 hours.

    **(b)** Calculate how many bacteria there will be after one day.

    **(c)** Find how long it will take for there to be two million bacteria.

10. The cost per person, in euros, when $x$ people hire an aeroplane can be determined by the function $C(x) = x + \dfrac{200}{x}$

    **(a)** Calculate the cost per person when 40 people hire the aeroplane.

    **(b)** When the number of people who hire the aeroplane is $a$ or $b$, the cost per person is 33 euros. Find the value of $a$ and of $b$.

    **(c)** When $n$ people hire the aeroplane, the cost per person is the minimum possible value.

    **(i)** Find the value of $n$.

    **(ii)** Find the minimum cost per person, to the nearest 0.01 euro.

11. A Ferris wheel with diameter 122 m rotates clockwise at a constant speed. The wheel completes 2.4 rotations every hour. The bottom of the wheel is 13 m above the ground.

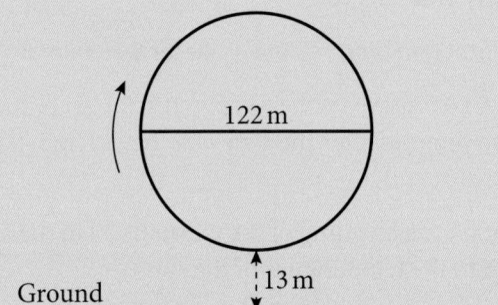

Ground ⎮13 m

A seat starts at the bottom of the wheel.

**(a)** Find the maximum height above the ground of the seat.

After $t$ minutes, the height $h$ m above the ground of the seat is given by $h = 74 + a \cos bt$

**(b)** Show that the period of $h$ is 25 minutes.

**(c)** Write down the exact value of $b$.

**(d)** Find the value of $a$.

**(e)** Sketch the graph of $h$, for $0 \leqslant t \leqslant 50$

**(f)** In one rotation of the wheel, find the probability that a randomly selected seat is at least 105 m above the ground.

# Descriptive statistics

10

Statistics are used daily in newspapers, on television, on the internet, in advertisements, and in ordinary conversations. Consider the following examples.

- In a certain large city, people complain that police take a long time to respond to calls for help. A news reporter collected some data over a month and discovered the following: for 1200 victims of a major assault, it took an average of 5 minutes and 20 seconds for the victim to call the police. Once the police were called, she found out that on average a police car was on the scene within 2 minutes and 40 seconds.

- A person sent 10 job applications with his original name and 10 applications with a popular name. The applications with the original name received no response. Those with the popular name received 8 responses.

- *Business Insider* (19 July 2016) reported that just four countries are home to over 60% of the world's high net worth individuals, according to the World Wealth Report by the consulting firm Capgemini.

- The UK Office for National Statistics reported that households without children spent a higher proportion of their total spending on transport than households with children. (Statistical Bulletin, 18 January 2018.)

- Antibacterial soaps are no better than regular soap. This is the result of a study carried out in a large city with 448 households where 224 were given antibacterial soaps and the other 224 were given normal liquid soap. The families used the soap for a year. All participants' hands were cultured for bacteria at the beginning and the end of the study. By the end of the year, tests revealed that they had the same number of bacteria regardless of which soap they used.

Each of these news clips involves information in the form of numbers – length of time, number of responses, percent of wealth, etc. In each of the news items, conclusions can be drawn from the data presented. Statistics is a way to get information from data.

Statistics is gathering, organising, and drawing conclusions from data.

# 10.1 Data and variables

Data are numerical evidences. To learn from data, we need more than just the numbers. The numbers in a medical study, for example, mean little without some knowledge of the goals of the study and of what blood pressure, heart rate, and other measurements contribute to those goals. That is, data are numbers with a context, and we need to understand the context if we are to make sense of the numbers. On the other hand, measurements from the study's several hundred subjects are of little value to even the most knowledgeable medical expert until you organise, display, and summarise them. We start with examining data. This branch of statistics is called **descriptive statistics**.

## Example 10.1

A new advertisement for an ice-cream in May last year resulted in a 35% increase in sales. Therefore, the advertisement was effective. Do you agree?

### Solution

A major flaw in this argument is looking at the numbers only. A 35% increase is good. However, knowing the purpose and background is also important. In the northern hemisphere, general sales of ice-cream increase in the months of June, July, and August regardless of advertisements.

## Variables

Any set of data contains information about some group of individuals. The information is organised in variables.

Take for example part of the school records by the end of school year, which may look like Figure 10.1.

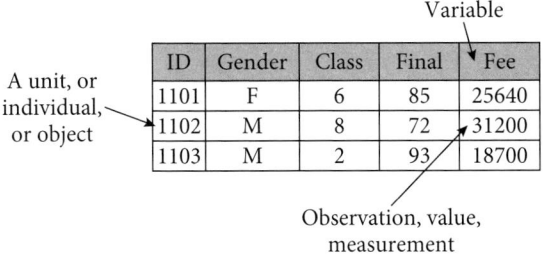

Figure 10.1 A sample of school records

**Individuals** are the objects described in a set of data. Individuals are sometimes people. When the objects that we want to study are not people, we often call them **cases**. Units, objects and other names are frequently used too.

A **variable** is any characteristic of an individual. A variable can take different values for different individuals.

The data set is constructed to keep track of some information about the students. The individuals are the students in the class. There are five variables in this data set. These include an identifier for each student – ID, gender, class,

A **qualitative (categorical** or **attribute) variable** places an individual into one of two or more groups or categories.

A **quantitative variable** takes numerical values representing counts or measurements for which arithmetic operations such as adding and averaging make sense. Quantitative variables can be split into two types.

**Discrete variables** result from a finite (countable) number of possible values. Number of children in a family, number of customers waiting to be served, the number of eggs a hen can lay per week, etc. are discrete variables.

**Continuous variables** can have infinitely many possible values that can be associated with points on a continuous scale in such a way that there are no gaps or interruptions. The time it takes to finish your homework, the thickness of a hard disk, the amount of milk a cow can produce per day, etc. are continuous variables.

The **distribution** of a variable tells us what values it takes and how often it takes these values.

final score, and fee. There are no units of measurement for ID, gender or class. The other variables both have a unit. For example, Final could be a number of marks and Fee is probably a currency. Some variables, like gender and class, simply place individuals into categories. Others, like Final and Fee, take numerical values for which we can do arithmetic and find averages.

It makes sense to give an average grade or fee, but it does not make sense to give an average gender. We can, however, count the numbers of female and male students and do arithmetic with these counts.

### Example 10.2

Identify each variable as qualitative or quantitative, and if quantitative, whether it is discrete or continuous.

(a) A cigarette contains 16.13 mg of tar.

(b) There were 15 females and 21 males in the statistics course last term.

(c) Major earthquakes happened in 1952, 1960, 1964, 2004, and 2011.

(d) The average rainfall in Rio de Janeiro for the first 6 months of the year is: 114 mm, 105 mm, 103 mm, …

(e) Upon completion of training for his upcoming bicycle race, Kevin's mass was 3.26 kg less than when he started.

(f) Zip codes (post codes) of a country.

### Solution

(a) Quantitative, continuous

(b) Qualitative

(c) Quantitative, discrete

(d) Quantitative, continuous

(e) Quantitative, continuous

(f) Qualitative (There is no numerical meaning, they are just numbers and sometimes letters.)

### Populations and samples

Descriptive statistics is a range of methods for planning the collection of, collecting, organising, summarising, and presenting data. There are also tools to analyse, interpret, and draw conclusions based on the collected data. This is called **inferential** statistics. At the core of these two branches we have the concepts of **population** and **sample**.

**Population** is the complete collection of all elements (scores, people, measurements, etc.) to be studied.

A **sample** is a sub-collection of elements drawn from a population.

If we are interested in predicting which party will win the next elections, then our population will be all registered voters. An opinion poll organisation will choose a certain number of voters and ask them about their preferences. This is a sample.

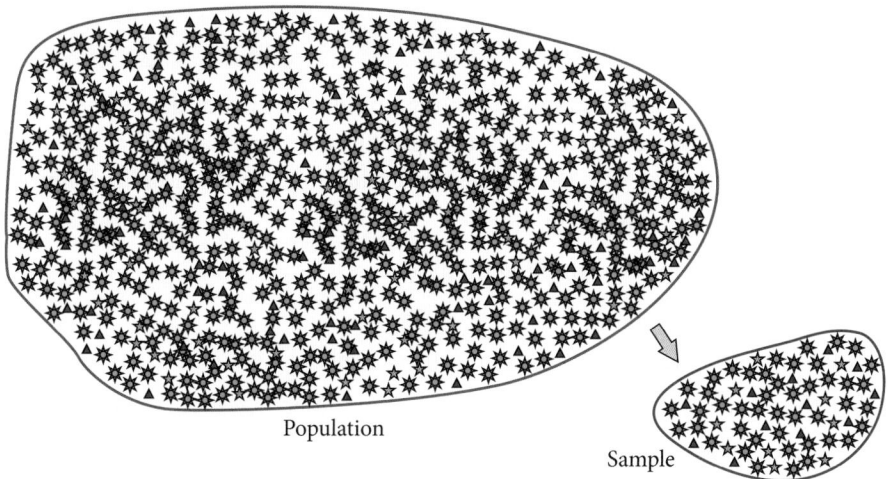

Population

Sample

**Figure 10.2** Population and sample

A computer company buys a shipment of hard disks from a supplier. The contract with the supplier states that less than 1% of the disks may be defective. The company cannot test all the shipment. It is time-consuming, costly, and also partly damages the units. They perform what is called acceptance sampling by taking a sample, say of 100 disks, and if more than 1 defective disk is found, the shipment is returned. The whole shipment is the population and the 100 disks selected is a sample.

A **parameter** is a numerical characteristic of the population. A **statistic** is a numerical characteristic of the sample.

Information contained in the sample is usually used to make an inference concerning a **parameter**, which is a numerical characteristic of the population. For instance, in predicting the winning party in elections, we are interested in the proportion of voters choosing that party on election day. That is a parameter. If we are interested in the average time commuters spend on using public transport, then that is a parameter. A parameter is estimated by computing a similar characteristic of the sample. This is called a **statistic**.

## Example 10.3

150 000 European Union students attend university in the UK. To estimate the average yearly expenditure of these students a service provider plans to conduct a study by interviewing 500 students.

What is the population, the parameter, the sample and the statistic?

### Solution

The population is the 150 000 EU students in the UK.

The parameter is the average yearly expenditure.

The 500 students who will be interviewed constitute a sample.

The average expenditure of this sample is a statistic that could be used to estimate the parameter of interest.

## Sampling

As you know by now, any study concerning populations needs data to be collected. Usually we do not collect data from the entire population. For statistical studies, data from samples is used. Schemes similar to the one below are followed:

- Specify the population of interest.
- Choose an appropriate sampling method.
- Collect the sample data.
- Analyse the pertinent information in the sample.
- Use the results of the sample analysis to make an inference about the population.
- Provide a measure of the inference's reliability.

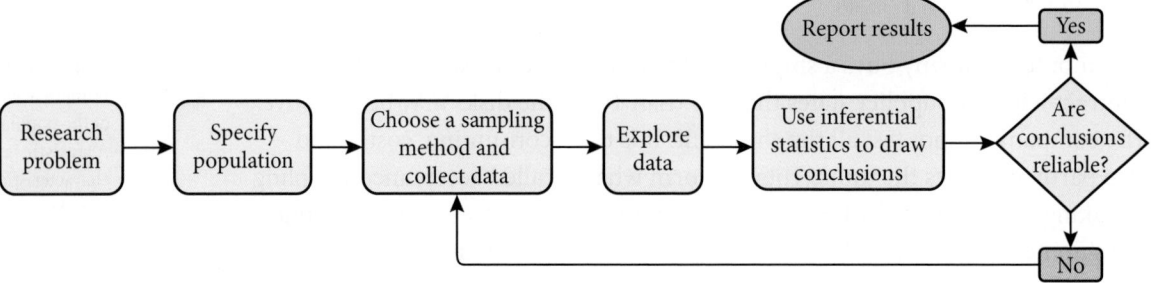

**Figure 10.3** Sampling process

### Reasons for sampling

Taking a sample instead of conducting a census offers several advantages.

- The sample can save money and time. If an eight-minute interview is being undertaken, conducting the interviews with a sample of 100 people rather than with a population of 100 000 is obviously less expensive. In addition to the cost savings, the significantly smaller number of interviews usually requires less total time.

- For given resources, the sample can broaden the scope of the study. With fixed resources, more detailed information can be gathered by taking a sample than gathering information from the whole population. Concentrating on fewer individuals or items, the study can be broadened in scope to allow for more specialised questions.

 A sample that represents the characteristics of the population as closely as possible is called a **representative sample**.

- Some research processes are destructive to the product or item being studied. For example, if light bulbs are being tested to determine how long they burn or if candy bars are being taste tested to determine whether the taste is acceptable, the product is destroyed.

- If accessing the entire population is impossible, a sample is the only option. If sampling is deemed to be appropriate, it must be decided how to select a sample. Since the sample will be employed to draw conclusions about the entire population, it is crucial that the sample is **representative** – it should reflect as closely as possible the relevant parameter of the population under consideration.

## Random and non-random sampling

The two main types of sampling are random and non-random. In **random sampling** every unit of the population has the same probability of being selected into the sample. Random sampling implies that chance enters into the process of selection.

In **non-random sampling** not every unit of the population has the same probability of being selected into the sample.

Sometimes random sampling is called **probability sampling** and non-random sampling is called **nonprobability sampling**. Because every unit of the population is not equally likely to be selected in non-random sampling, assigning a probability of occurrence is impossible. The statistical methods presented and discussed in your syllabus assume that the data comes from random samples.

Non-random sampling methods are not appropriate techniques for gathering data to be analysed by most of the statistical methods presented in this book. However, several non-random sampling techniques are described in this section, primarily to alert you to their characteristics and limitations.

## Random sampling

Three basic random sampling techniques – simple random sampling (SRS), stratified random sampling, and systematic random sampling – are discussed here. Each technique offers advantages and disadvantages. Some techniques are simpler to use, some are less costly, and others show greater potential for reducing sampling error.

**Sampling error** occurs when, by chance, the sample does not represent the population.

### Sampling or chance error

Generally, all samples selected from the same population will give different results because they contain different elements of the population. Additionally, the results obtained from any one sample will not be exactly the same as the ones obtained from a census. The difference between a sample result and the result we would have obtained by conducting a census is called the **sampling error**, assuming that the sample is random and no non-sampling error has been made.

Non-sampling errors can occur both in a sample survey and in a census. Such errors occur because of human mistakes and not chance.

### Simple random sampling

The most elementary random sampling technique is **simple random sampling**. Simple random sampling can be viewed as the basis for the other random sampling techniques. With simple random sampling, each unit of the frame is numbered from 1 to $N$ (where $N$ is the size of the population). Next, a random number generator (or a table of random numbers, which is an outdated technique) is used to select $n$ items into the sample.

Suppose it has been decided to interview 20 out of 659 high school students in a school to form an understanding of their views of a new block-scheduling the school wants to adopt.

We number the students from 001 (or simply 1) to 659 and have a random generator choose 20 of them. The numbers chosen will be your sample. The GDC output gives us a list of chosen numbers. The screenshot (Figure 10.4) shows five of them.

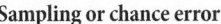

```
RanInt#(1,659,20)
{217,100,191,518,252▶
```

**Figure 10.4** GDC random number generator

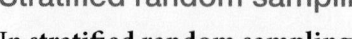

## Stratified random sampling

In **stratified random sampling**, the population is divided into non-overlapping subpopulations called strata. The researcher then extracts a sample from each of the subpopulations. The main reason for using stratified random sampling is that it has the potential for reducing sampling error.

With stratified random sampling, the potential to match the sample closely to the population is greater than it is with simple random sampling because portions of the total sample are taken from different population subgroups. However, stratified random sampling is generally more costly than simple random sampling because each unit of the population must be assigned to a stratum before the random selection process begins.

Strata selection is usually based on available information. Such information may have been obtained from previous censuses or surveys. Stratification benefits increase as the strata differ more. Internally, a stratum should be relatively homogeneous; externally, strata should contrast with each other.

For example, in FM radio markets, age of listener is an important determinant of the type of programming used by a station. Figure 10.5 contains a stratification by age with three strata, based on the assumption that age makes a difference in preference of programming.

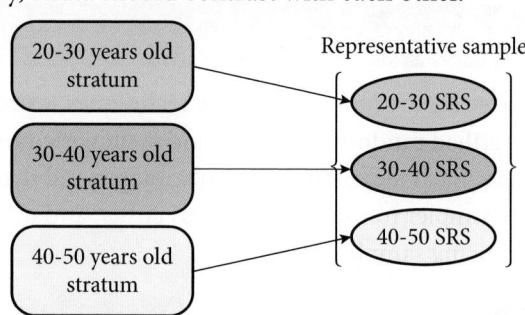

**Figure 10.5** Stratified random sampling

This stratification implies that listeners of 20 to 30 years of age tend to prefer the same type of programming, which is different from that preferred by listeners of 30 to 40 and of 40 to 50 years of age. Within each age subgroup (stratum), there is homogeneity; between each pair of subgroups there is a difference, or heterogeneity. A sample from each stratum is taken and together they constitute a representative sample of the whole population.

## Systematic sampling

Unlike stratified random sampling, **systematic sampling** is not done in an attempt to reduce sampling error. Rather, systematic sampling is used because of its convenience and relative ease of administration. With systematic sampling, every $k$th item is selected to produce a sample of size $n$ from a population of size $N$. The value of $k$, sometimes called the sampling cycle, can be determined using

$$k = \frac{N}{n}$$

If $k$ is not an integer value, then a whole-number value should be used.

For example, suppose we need a sample of 20 students from the 659 high school students using systematic sampling. $n = 20$ and $N = 659$, so:

$$k = \frac{659}{20} \approx 32$$

From the list of 659 students we randomly choose a starting number between 1 and 32, say 11 for example, and then after that we choose every 32nd number, i.e., $11 + 32 = 43, 75, 107$

Besides convenience, systematic sampling has other advantages. Because systematic sampling is evenly distributed across the population, a knowledgeable person can easily determine whether a sampling plan has been followed in a study.

## Non-random sampling

Sampling techniques used to select elements from the population by any mechanism that does not involve a random selection process are called **non-random sampling techniques**. Because chance is not used to select items for the samples, these techniques are non-probability techniques and are not desirable for use in gathering data to be analysed by standard methods of inferential statistics. Sampling error cannot be determined objectively for these sampling techniques. Two non-random sampling techniques are presented here: convenience sampling, and quota sampling.

### Convenience sampling

In **convenience sampling**, elements for the sample are selected for the convenience of the researcher. The researcher typically chooses elements that are readily available, nearby, or willing to participate. The sample tends to be less variable than the population because in many environments the extreme elements of the population are not readily available. The researcher will select more elements from the middle of the population. For example, a convenience sample of homes for door-to-door interviews might include houses where people are at home, houses with no dogs, houses near the street, first-floor apartments, and houses with friendly people. In contrast, a random sample would require the researcher to gather data only from houses and apartments that have been selected randomly, no matter how inconvenient or unfriendly the location. If research is carried out in a mall, a convenience sample might be selected by interviewing only shoppers who pass the location and look friendly.

### Quota sampling

**Quota sampling** appears to be similar to stratified random sampling. However, instead of selecting a sample from each stratum, we use a non-random sampling method to gather data from one stratum until the desired quota of samples is filled. Quotas are described by setting the sizes of the samples to be obtained from the subgroups. Generally, a quota is based on the proportions of the subclasses in the population.

For example, a company is test marketing a new soft drink and is interested in how different age groups react to it. The researchers go to a shopping mall and interview shoppers aged 16–20, for example, until enough responses are obtained to fill the quota. In quota sampling, an interviewer would begin by asking a few filter questions; if the respondent represents a subclass whose quota has been filled, the interviewer would stop the interview.

Quota sampling can be useful if no previous information is available for the population. For example, suppose we want to stratify the population into cars using different types of winter tyres but we do not have lists of users of a particular brand of tyres. Through quota sampling, we would proceed by interviewing all car owners and rejecting all users who do not use the particular brand until the quota of users of the particular brand is filled.

Quota sampling is less expensive than most random sampling techniques because it essentially is a technique of convenience. Another advantage of quota sampling is the speed of data gathering. We do not have to call back or send out a second questionnaire if we do not receive a response; we just move on to the next element.

The problem with quota sampling is that it is a non-random sampling technique. Some researchers believe that a solution to this issue can be achieved if the quota is filled by randomly selecting elements and discarding those not from a stratum. This way quota sampling is essentially a version of stratified random sampling. The object is to gain the benefits of stratification without the high costs of stratification. However, it remains a nonprobability sampling method.

## Exercise 10.1

1. Identify the experimental units, sensible population and sample on which each variable is measured. Then indicate whether the variable is quantitative or qualitative.

   (a) Gender of a student.

   (b) Number of errors on a final exam for 10th grade students.

   (c) Height of a newly born child.

   (d) Eye colour for children aged less than 14.

   (e) Amount of time it takes to travel to work.

   (f) Rating of a country's leader: excellent, good, fair, poor.

   (g) Country of origin of students at international schools.

2. For each situation:

   (i) identify the population

   (ii) give two examples of how an appropriate sample of the population can be obtained.

   (a) A large high school with students in grades 10–12 has 42 classes with a total of 1176 students. Information about sizes of students are needed for the school to stock enough sweatshirts for those interested in purchasing one. The sample size required is 30.

**(b)** In a machine parts factory, bolts are produced on an assembly line in three shifts and are collected in large containers. For quality control purposes, bolts are to be tested for their size and strength using a digital sensor.

3. Identify each variable as numerical (quantitative) or categorical (qualitative) data. If the data is numerical, classify it as discrete or continuous.

   **(a)** The blood type of a group of blood donors.

   **(b)** The number of cars each family in a community has.

   **(c)** The length of a fish caught from a pond.

   **(d)** The amount of time spent per evening studying mathematics.

   **(e)** The volume of liquid in a canned drink.

   **(f)** The number of languages spoken in a given community.

   **(g)** 100-metre race times.

   **(h)** Colours used by maths teachers to write on their whiteboards.

   **(i)** The rating of trumpet solos as superior, excellent or good.

4. Group the set of situations as categorical or numerical. For the numerical data, identify each as discrete or continuous.

   **(a)** Volume of paint in a can.

   **(b)** Number of children in a family.

   **(c)** Time taken to finish a marathon.

   **(d)** The length of an average children's film.

   **(e)** The height of the students in an IB Mathematics class.

   **(f)** The average carbon dioxide emissions in a large city.

   **(g)** The religion(s) practiced in each household of a city.

5. Classify each as an example of descriptive or inferential statistics.

   **(a)** Gathering data to determine if the average mass of 16-year-old football players in the UK is 60 kg.

   **(b)** Gathering data to determine a basketball player's free throw percentage for a season of 10 games.

6. State whether inferential or descriptive statistics would be needed for the following.

   **(a)** Finding the average mass of all 12-year-old boys in Spain.

   **(b)** Finding the average amount of money you spend on entertainment for the next 3 months.

   **(c)** Finding the number of heads that show after flipping a coin 100 times.

   **(d)** Determining if the proportion of women who wear seat belts is greater than the proportion of men who wear seat belts.

   **(e)** Determining if the average IB Mathematics score for boys is greater than the average IB Mathematics score for girls.

7. A professor wanted to select 20 students from his class of 300 students to collect detailed information on their profiles. He used his knowledge and expertise to select these 20 students.

    (a) Is this sample a random or a non-random sample? Explain.

    (b) Is this an SRS, a systematic sample, a stratified sample, a convenience sample, a judgment sample, or a quota sample? Explain your answer.

    (c) What kind of error, if any, will be made with this kind of sample? Explain.

    Suppose the professor enters the names of all the students enrolled in his class on a computer. He then selects a sample of 20 students at random using a statistical software package.

    (d) Explain whether this sample is random or non-random.

    (e) Explain what kind of sample it is.

    (f) Do you think any error will be made in this case? Explain your answer.

8. A company has 1000 employees, of whom 58% are men and 42% are women. The research department at the company wanted to conduct a quick survey by selecting a sample of 100 employees and asking them about their opinions on an issue. They divided the population of employees into two groups, men and women, and then selected 58 men and 42 women from these respective groups. Explain what kind of sample it is.

9. A large university would like to determine the views of its students on an increase in fees. They would also like to compare the faculties of business and arts and sciences as well as the graduate school. Describe a sampling plan to achieve this goal.

10. An opinion poll is to be given to a sample of 90 members of a local gym. The members are first divided into men and women, and then a simple random sample of 45 men and a separate simple random sample of 45 women are taken. What sampling method has been used? Explain.

## 10.2 Displaying distributions using graphs

Statistical tools and ideas help us examine data in order to describe their main features. This examination is called **exploratory data analysis.** Like an explorer crossing unknown lands, we want first to simply describe what we see.

One efficient approach is to begin with a graph or graphs. Then add numerical summaries of specific aspects of the data. This section presents methods for describing a single variable.

There are several ways of summarising and describing data. Among them are tables and graphs and numerical measures.

I just purchased a bag of milk chocolate sweets that have six different colours. There are 55 sweets in the bag: 16 brown, 13 red, 9 yellow, 7 green, 4 blue, and 6 orange. Table 10.1 shows these counts.

| Colour | Frequency |
|--------|-----------|
| Brown | 16 |
| Red | 13 |
| Yellow | 9 |
| Green | 7 |
| Blue | 4 |
| Orange | 6 |

**Table 10.1**  Colours of sweets

This table is called a frequency table and it describes the distribution of sweet colour frequencies. Not surprisingly, this kind of distribution is called a **frequency distribution**. Often a frequency distribution is shown graphically as in Figure 10.6.

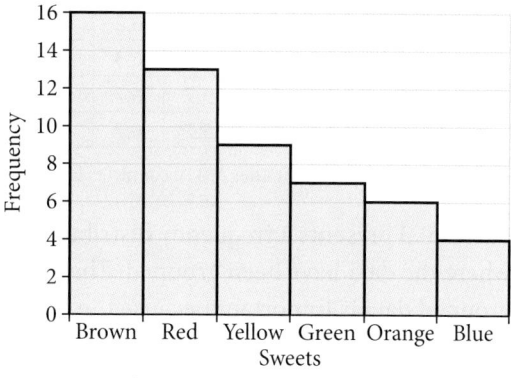

**Figure 10.6**  A bar graph of colours of sweets

The distribution shown in Figure 10.6 concerns just one bag of sweets. What about the distribution of colours for all sweets?

The manufacturer of the sweets provides some information about this matter, but they do not say exactly how many sweets of each colour they have ever produced. Instead, they report proportions rather than frequencies. Figure 10.7 shows these proportions.

Since every sweet is one of the six familiar colours, the six proportions shown in the figure add to 100. We call Figure 10.7 a relative frequency distribution (or probability distribution) because if you choose a sweet at random, the probability of getting, say, a brown sweet is equal to the proportion of sweets that are brown (0.30).

**Figure 10.7**  Relative frequency distribution

Note that the distributions in Figures 10.6 and 10.7 are not identical. Figure 10.6 portrays the distribution in a **sample** of 55 of the sweets. Figure 10.7 shows the proportions for all of the sweets. Chance factors involving the machines used by the manufacturer introduce random variation into the different bags produced. Some bags will have a distribution of colours that is close to Figure 10.7; others will be further away.

Graphs similar to the ones in Figures 10.6 and 10.7 are mainly used for qualitative variables and are called **bar graphs**.

Bar graphs and similar charts help an audience grasp a distribution quickly. They are, however, of limited use for data analysis because it is usually easy to understand data on a single categorical variable, such as highest frequency, without a graph. We will move on to quantitative variables, where graphs are essential tools.

Raw data, or data that have not been summarised in any way, are sometimes referred to as ungrouped data. Table 10.2 contains 110 data points of raw data of the unemployment rates for 110 countries for 2017 as reported by the International Monetary Fund (IMF). Data that have been organised into a frequency distribution are called **grouped data**.

| 0.7 | 3.1 | 4.0 | 4.4 | 5.1 | 5.8 | 6.7 | 7.7 | 9.0 | 11.3 | 14.6 |
| 1.0 | 3.2 | 4.0 | 4.4 | 5.2 | 5.8 | 6.7 | 8.0 | 9.1 | 11.7 | 15.3 |
| 1.1 | 3.2 | 4.0 | 4.5 | 5.4 | 5.8 | 6.8 | 8.0 | 9.3 | 11.8 | 16.5 |
| 2.0 | 3.4 | 4.0 | 4.6 | 5.4 | 6.0 | 6.9 | 8.1 | 9.4 | 12.2 | 17.2 |
| 2.2 | 3.4 | 4.0 | 4.7 | 5.5 | 6.0 | 6.9 | 8.3 | 9.4 | 12.2 | 18.9 |
| 2.2 | 3.7 | 4.2 | 4.9 | 5.6 | 6.1 | 6.9 | 8.4 | 9.8 | 12.2 | 19.6 |
| 2.8 | 3.7 | 4.2 | 5.0 | 5.6 | 6.2 | 7.1 | 8.7 | 10.1 | 12.2 | 20.5 |
| 2.9 | 3.8 | 4.2 | 5.0 | 5.7 | 6.3 | 7.1 | 8.7 | 10.2 | 12.5 | 21.5 |
| 2.9 | 3.8 | 4.2 | 5.0 | 5.7 | 6.7 | 7.2 | 8.9 | 11.0 | 12.8 | 22.5 |
| 3.0 | 3.9 | 4.4 | 5.0 | 5.7 | 6.7 | 7.4 | 9.0 | 11.3 | 13.9 | 27.5 |

**Table 10.2** Unemployment rates of 110 countries for 2017 as reported by IMF

Table 10.3 presents a frequency distribution for the data displayed in Table 10.2, where the data have been grouped. The distinction between ungrouped and grouped data is important because the calculation of statistics differs between the two types of data.

When grouping data, there are some steps to be followed.

**Step 1** Determine the range of the raw data. The range is often defined as the difference between the largest and smallest numbers. The range for the data in Table 10.2 is 26.8 (27.5 − 0.7)

**Step 2** Determine how many classes it will contain. A good general rule is to select between 5 and 15 classes. If the frequency distribution contains too few classes, the data summary may be too general to be useful. Too many classes may result in a frequency distribution that does not aggregate the data enough to be helpful. The final number of classes is arbitrary.

By examining the range we can determine a number of classes that will span the range adequately and also be meaningful to the user. The data in Table 10.2 were grouped into 9 classes for Table 10.3.

**Step 3** Determine the width of the class interval. An approximation of the class width can be calculated by dividing the range by the number of classes.

For this data, this approximation would be $\dfrac{26.8}{9} = 2.98$

| Rate | Frequency |
|------|-----------|
| $0 \leqslant x < 3$ | 9 |
| $3 \leqslant x < 6$ | 44 |
| $6 \leqslant x < 9$ | 26 |
| $9 \leqslant x < 12$ | 14 |
| $12 \leqslant x < 15$ | 8 |
| $15 \leqslant x < 18$ | 3 |
| $18 \leqslant x < 21$ | 3 |
| $21 \leqslant x < 24$ | 2 |
| $24 \leqslant x < 27$ | 0 |
| $27 \leqslant x < 30$ | 1 |

**Table 10.3** Distribution for grouped unemployment rates

Normally, the number is rounded up to the next whole number, which in this case is 3. The frequency distribution must start at a value equal to or lower than the lowest number of the ungrouped data and end at a value equal to or higher than the highest number. The lowest unemployment rate is 0.7 and the highest is 27.5, so we start the frequency distribution at 0 and end it at 30. Table 10.3 contains the completed frequency distribution for the data in Table 10.2. Class endpoints are selected so that no value of the data can fit into more than one class.

The midpoint of each class interval is called the **class midpoint** and is sometimes referred to as the **class mark**. It is the value halfway across the class interval and can be calculated as the average of the two class endpoints. For example, in the distribution of Table 10.3, the midpoint of the class interval $3 \leqslant x < 6$ is 4.5, or $\dfrac{3 + 6}{2}$

The class midpoint is important, because it becomes the representative value for each class in most grouped statistics calculations.

**Relative frequency** is the proportion of the total frequency that is in any given class interval in a frequency distribution. Relative frequency is the individual class frequency divided by the total frequency. For example, from Table 10.3, the relative frequency for the class interval $3 \leqslant x < 6$ is $\dfrac{44}{110} = 0.4$ (Table 10.4).

**Cumulative frequency** is a running total of frequencies through the classes of a frequency distribution. The cumulative frequency for each class interval is the frequency for that class interval added to the preceding cumulative total. In Table 10.4 the cumulative frequency for the first class is the same as the class frequency: 9. The cumulative frequency for the second class interval is the frequency of that interval (44) plus the frequency of the first interval (9), which gives a cumulative frequency of 53. This process continues through the last interval, at which point the cumulative total should equal the sum of the frequencies (110). Table 10.4 gives cumulative frequencies for the data in Table 10.3.

| Rate | Frequency | Class midpoint | Relative frequency | Cumulative frequency |
|---|---|---|---|---|
| $0 \leqslant x < 3$ | 9 | 1.5 | 0.0818 | 9 |
| $3 \leqslant x < 6$ | 44 | 4.5 | 0.4000 | 53 |
| $6 \leqslant x < 9$ | 26 | 7.5 | 0.2364 | 79 |
| $9 \leqslant x < 12$ | 14 | 10.5 | 0.1273 | 93 |
| $12 \leqslant x < 15$ | 8 | 13.5 | 0.0727 | 101 |
| $15 \leqslant x < 18$ | 3 | 16.5 | 0.0273 | 104 |
| $18 \leqslant x < 21$ | 3 | 19.5 | 0.0273 | 107 |
| $21 \leqslant x < 24$ | 2 | 22.5 | 0.0182 | 109 |
| $24 \leqslant x < 27$ | 0 | 25.5 | 0.0000 | 109 |
| $27 \leqslant x < 30$ | 1 | 28.5 | 0.0091 | 110 |

$$0.0818 = \frac{9}{110}$$

$$0.2364 = \frac{26}{110}$$

**Table 10.4** Cumulative frequency distribution for unemployment data

## Example 10.4

The table shows the scores of 80 students in an IB class achieved in an assessment, out of a total of 100 marks.

| | | | | | | | | | |
|---|---|---|---|---|---|---|---|---|---|
| 42 | 50 | 61 | 70 | 81 | 90 | 57 | 65 | 76 | 47 |
| 41 | 50 | 61 | 71 | 82 | 92 | 57 | 65 | 78 | 47 |
| 42 | 50 | 61 | 71 | 83 | 93 | 57 | 66 | 79 | 48 |
| 42 | 52 | 61 | 72 | 83 | 94 | 57 | 67 | 86 | 49 |
| 45 | 52 | 62 | 72 | 83 | 98 | 57 | 67 | 88 | 49 |
| 45 | 53 | 62 | 72 | 84 | 98 | 57 | 67 | 89 | 59 |
| 45 | 53 | 63 | 74 | 84 | 45 | 58 | 67 | 89 | 69 |
| 45 | 56 | 64 | 75 | 84 | 46 | 58 | 68 | 59 | 69 |

(a) Develop a frequency and relative frequency table for the data.

(b) Develop a cumulative and relative cumulative frequency table of the data.

## Solution

(a) Since the lowest potential grade is 40 and the highest is 100, we will choose 6 classes. We may choose another number too. However, separating grades into classes of multiples of 10 is a sensible choice:

$$\frac{98 - 41}{6} \approx 10$$

| Scores achieved ($x$) | Number of students (Frequency) | Relative frequency | Cumulative frequency | Relative cumulative frequency |
|---|---|---|---|---|
| $40 \leqslant x < 50$ | 15 | 18.7% | 15 | 18.7% |
| $50 \leqslant x < 60$ | 18 | 22.5% | 33 | 41.2% |
| $60 \leqslant x < 70$ | 18 | 22.5% | $33 + 18 = 51$ | 63.7% |
| $70 \leqslant x < 80$ | 11 | 13.8% | 62 | 77.5% |
| $80 \leqslant x < 90$ | 12 | 15.0% | 74 | 92.5% |
| $90 \leqslant x \leqslant 100$ | 6 | 7.5% | 80 | 100% |
| **Total** | **80** | **100%** | | |

Each cell in the relative frequency column is the ratio of the cell's frequency to the total number. For example, the first cell is

$$\frac{15}{80} = 0.187 = 18.7\% \text{ and the last cell is } \frac{6}{80} = 0.075 = 7.5\%$$

(b) Each cell in the cumulative frequency column is the sum of the cumulative frequency of the previous cell and the frequency of the interval itself. The last column shows relative cumulative frequency.

## Histograms

Although frequency distribution tables are useful for analysing large sets of data, their table format may not be as visually informative as a graph. If a frequency distribution has been developed from a quantitative variable, a **histogram** can be constructed directly from the frequency distribution. A histogram is a graph that consists of vertical bars constructed on a horizontal line that is marked off with intervals for the variable being displayed. The intervals correspond to those in a frequency distribution table. The area of each bar is proportional to the number of observations in that interval. In many cases, the histogram enables the data to be interpreted more easily.

You cannot use histograms for qualitative variables. However, you can use a bar chart instead.

A histogram shows three general types of information:

1. It provides a visual indication of where the approximate centre of the data is.

2. We can gain an understanding of the degree of spread (or variation) in the data. The more the data cluster around the centre, the smaller the variation in the data. If the data are spread out from the centre, the data exhibits greater variation.

3. We can observe the shape of the distribution. Is it reasonably flat, is it skewed to one side or the other, is it balanced around the centre, or is it bell shaped?

We can construct a histogram using the data in Table 10.3 using technology (Figure 10.8).

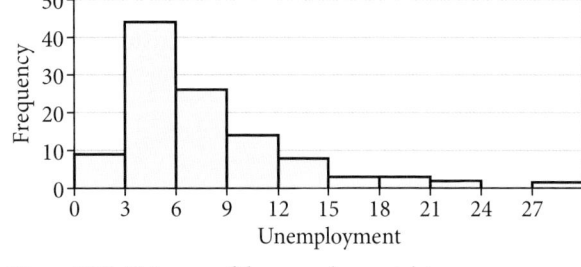

**Figure 10.8** Histogram of the unemployment data

Each class is represented by a rectangle with a height that corresponds to the frequency. For example, the first class has a frequency of 9, and so, the height of the rectangle is 9.

Figure 10.8 shows that the data are skewed to the right: there are fewer countries with high unemployment. We can also see that very few countries have an unemployment rate less than 3% and that a majority of the countries have rates between 3% and 9%.

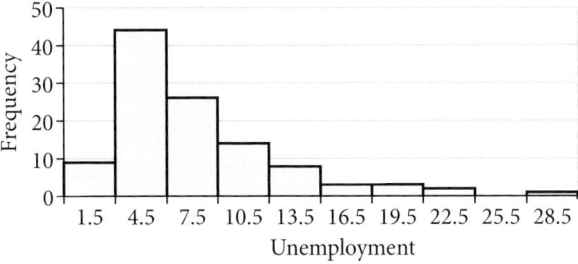

**Figure 10.9** Histogram for unemployment data using midpoints

GDCs can also produce histograms of relatively good quality. The screenshot in Figure 10.10 shows an example where class boundaries have been entered in one list and the frequencies in another list. The GDC then draws the histogram.

Histograms produced using technology can be presented in different forms. In Figure 10.8, we used the standard convention for histograms where the class boundaries separate the classes with the lowest boundary included in the class while the upper one is included in the next class.

Some software presents the histogram where the rectangles are placed directly above the class midpoints (Figure 10.9).

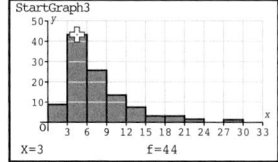

**Figure 10.10** GDC histogram

## Cumulative and relative frequency histograms and graphs

For many practical situations, it is more efficient to look at histograms representing the relative frequencies instead of frequencies. This is especially true when we compare two samples of different sizes. The shape of the histogram will not change.

For example, with the same unemployment data, Figure 10.11 shows a relative frequency histogram. Note that the shape is still the same but the numbers on the vertical axis shows percent (relative frequency) instead.

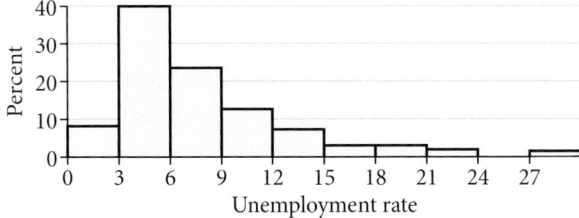

**Figure 10.11** Relative frequency histogram for unemployment data

A **frequency graph** (also called frequency polygon) shows the same information as the histogram with the rectangles replaced by points at the midpoint of each class (Figure 10.12).

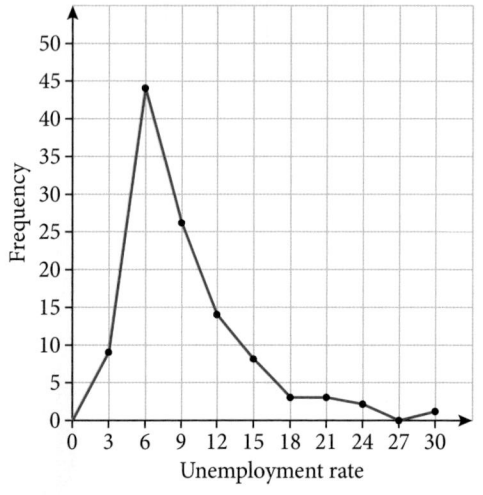

**Figure 10.12** Frequency graph for unemployment data

**Figure 10.13** Cumulative frequency graph the same data

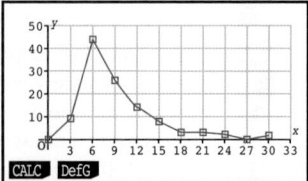

**Figure 10.14** GDC frequency graph

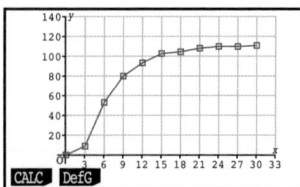

**Figure 10.15** GDC cumulative frequency graph

A **cumulative frequency graph** (polygon) for each class. It has to be drawn at the upper limit for each class because its height represents the cumulative frequency up to that point. For example, in Figure 10.13 the height of the graph corresponding to 6 is 44, which is the cumulative frequency up to 6.

Figures 10.14 and 10.15 show two graphs produced by a GDC.

## Example 10.5

At a busy intersection, the speed of passing cars is recorded and the information is represented in a cumulative frequency graph.

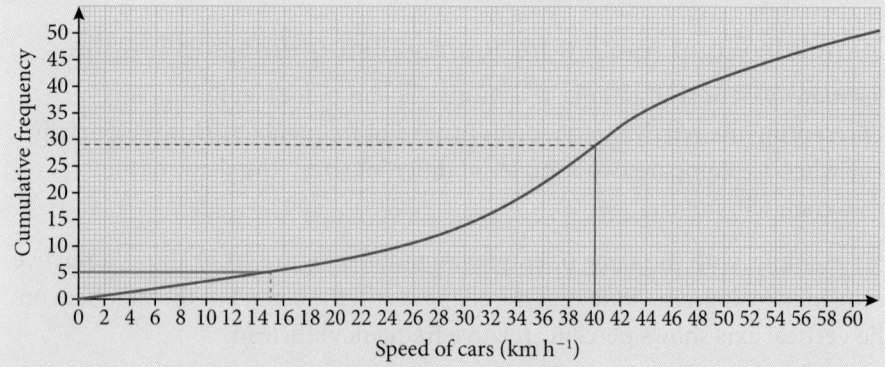

(a) How many cars were included in the sample?

(b) The speed limit at the intersection is 40 km h$^{-1}$. How many cars are driving at or below the speed limit? What percentage of the cars is that?

(c) If a car is travelling more than 10 km h$^{-1}$ above the speed limit, they receive a speeding ticket. How many cars would receive a speeding ticket?

(d) The slowest 10% of all drivers are seen as a slow safety hazard, and are issued a fine. What is the maximum speed considered to be a slow safety hazard?

## Solution

(a) From the graph, the speed of 50 cars were measured in the sample.

(b) Drawing a vertical line to the graph at 40 km h$^{-1}$ (red solid line) we can project horizontally onto the cumulative frequency axis (red dashed line) and we find that the slowest 29 cars out of 50 drive at or below the speed limit. This is 58% of the cars.

(c) 10 km h$^{-1}$ above the limit is 50 km h$^{-1}$. Drawing a vertical line to the graph at 50 km h$^{-1}$, we can project horizontally onto the cumulative frequency axis and we find that 41 cars out of 50 drive at or below 50 km h$^{-1}$, thus, 9 cars are potential recipients of a ticket.

(d) Calculating 10% of 50 gives 5, meaning that the 5 slowest cars are driving at a hazardously slow speed. Drawing a horizontal line to the graph for a cumulative frequency of 5 (green solid line) and projecting to the speed axis (green dashed line) gives the maximum speed of 15 km h$^{-1}$ for hazardously slow driving.

58% of the data are at 40 or below, so we call 40 km h$^{-1}$ the 58th **percentile**.

82% of the data are at 50 or below, so we call 50 km h$^{-1}$ the 82nd **percentile**.

 When $p$% of the data lie on $P$ or below, we call $P$ the $p$th **percentile**.

## Example 10.6

Data is collected for the number of text messages that 50 randomly selected high school students sent during one day. The results are listed below.

8, 52, 38, 48, 42, 9, 15, 36, 36, 53, 10, 8, 46, 46, 9, 11, 12, 24, 49, 34, 10, 11, 9, 11, 45, 25, 25, 37, 14, 16, 20, 22, 12, 43, 36, 23, 23, 26, 27, 16, 21, 29, 29, 38, 30, 47, 34, 39, 48, 46

(a) Set up a frequency and relative cumulative frequency table for the data with 7 classes.

(b) Draw a frequency histogram of the data.

(c) Draw a relative cumulative frequency graph.

(d) Estimate the 10th as well as the 90th percentiles of the data from your graph in part (c).

## Solution

(a) Since the minimum value is 8 and the maximum value is 53, then

the class size would be $\dfrac{53 - 8}{7} = 6.4$

Round this to 7 and start the table with the minimum 8.

| Class | Class midpoint | Frequency | Cumulative frequency | Relative cumulative frequency |
|---|---|---|---|---|
| $8 \leqslant t < 15$ | 11.5 | 13 | 13 | $13/50 = 0.26$ |
| $15 \leqslant t < 22$ | 18.5 | 5 | $13 + 5 = 18$ | 0.36 |
| $22 \leqslant t < 29$ | 25.5 | 8 | 26 | 0.52 |
| $29 \leqslant t < 36$ | 32.5 | 5 | 31 | 0.62 |
| $36 \leqslant t < 43$ | 39.5 | 8 | $31 + 8 = 39$ | 0.78 |
| $43 \leqslant t < 50$ | 46.5 | 9 | 48 | 0.96 |
| $50 \leqslant t < 57$ | 53.5 | 2 | 50 | 1.00 |

(b)

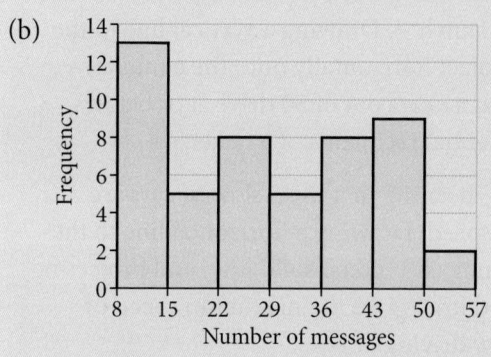

(c)

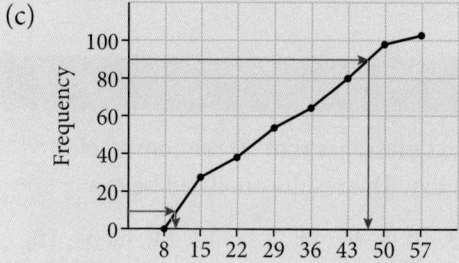

(d) As shown in the diagram, the 10th percentile is at 9 or 10 messages, while the 90th percentile is at 46 or 47 messages.

## Exercise 10.2

1. A group of college students were selected at random and their ages were recorded. The table shows a summary.

| Age | 17 | 18 | 19 | 20 | 21 | 22 | 23–25 | Over 25 |
|---|---|---|---|---|---|---|---|---|
| Number of students | 3 | 72 | 62 | 31 | 12 | 9 | 5 | 6 |
| Cumulative frequency | 3 | 75 | 137 | $x$ | 180 | 189 | 194 | $y$ |

(a) What are the values of $x$ and $y$?

(b) How many students are younger than 21?

(c) Find the value of the 25th percentile.

2. The heights (in metres) of 30 students randomly chosen in a large school are recorded in the table.

| 0.92 | 1.32 | 1.76 | 1.63 | 1.79 | 1.28 | 1.77 | 1.62 | 1.611 | 1.85 |
| 1.26 | 1.67 | 1.77 | 1.78 | 1.93 | 1.73 | 1.55 | 1.52 | 1.89 | 1.59 |
| 1.78 | 1.73 | 1.15 | 1.76 | 1.69 | 1.72 | 1.04 | 1.53 | 1.58 | 2.00 |

(a) Choose an appropriate interval width and create a frequency table to represent the data. Is the data discrete or continuous?

(b) Draw a frequency histogram representing the data.

(c) Draw a cumulative frequency graph for the data.

(d) Using your graphs, summarise the data in a short paragraph.

3. The table gives the frequency and cumulative frequency of the amount of time spent by 60 students, in hours, studying per evening.

| Time spent studying ($h$) | Frequency | Cumulative frequency |
| --- | --- | --- |
| $0 \leqslant h < 1$ | 7 | 7 |
| $1 \leqslant h < 2$ | $p$ | 18 |
| $2 \leqslant h < 3$ | 28 | $q$ |
| $3 \leqslant h \leqslant 4$ | $r$ | 60 |

(a) State the values of $p$, $q$ and $r$.

(b) Draw a frequency histogram and cumulative frequency graph for the data.

(c) Summarise the data in a short paragraph, using your graphs and tables.

4. The frequency bar graph in Figure 10.16 shows IB assessment scores for a group of students.

(a) How many student scores were included in the sample?

(b) What percentage of students achieved a score of 6?

(c) How many scores below 3 were there and what percentage of the total does this represent?

(d) The bar chart is a good representation of a cohort of 800 students. How many students in the cohort would you expect to receive a grade of 7?

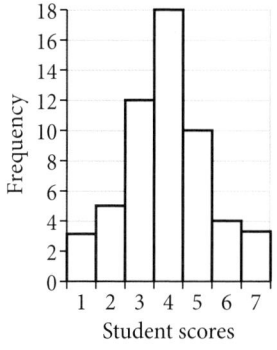

**Figure 10.16** Graph for question 4

5. The cumulative frequency graph shows the information about the mass (in kg) of a group of students selected at random from a large public school.

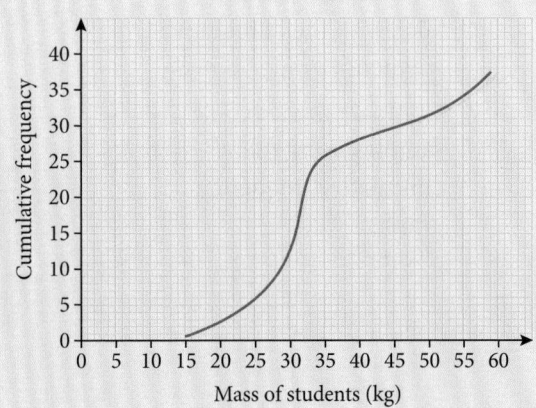

**(a)** Copy and complete the frequency table.

| Mass of students, $M$ (kg) | Frequency |
|---|---|
| $15 \leqslant M < 30$ | |
| $30 \leqslant M < 45$ | |
| $45 \leqslant M \leqslant 60$ | |

**(b)** How many students were included in the data sample?

**(c)** For which interval of masses does the frequency increase the most?

**(d)** How many students have a mass under 40 kg?

**(e)** Students in the bottom 25% by mass are given the option to participate in a specialised nutritional program. What is the maximum mass a student could have to be eligible for the program?

**6.** During an ecological survey, 100 pike are caught from a lake, and their lengths (in cm) are measured. The data is shown in the cumulative frequency graph.

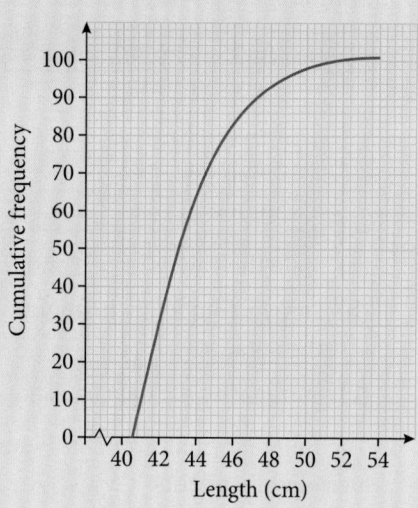

**(a)** What is the range of lengths that were observed in the survey?

**(b)** How many pike had a length smaller than 42 cm?

**(c)** How many pike had a length greater than 47 cm?

**(d)** Pike that are longer than 50 cm are considered big. What percentage of the pike are big?

**(e)** A second similar survey is completed at a different location in the same lake, with the cumulative frequency graph shown.

Compare the two pike samples using the cumulative frequency graphs.

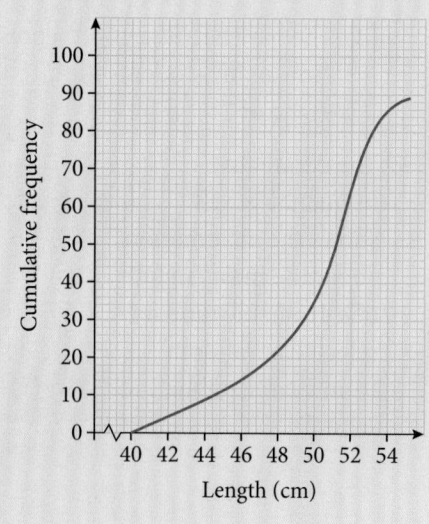

7. A large company has a report for days employees are absent per year. Table 10.5 shows a summary of absences.

   (a) Set up a cumulative frequency table.

   (b) Draw a cumulative frequency graph.

| Absences (days) | Frequency |
|---|---|
| $0 \leqslant x \leqslant 5$ | 30 |
| $6 \leqslant x \leqslant 11$ | 62 |
| $12 \leqslant x \leqslant 17$ | 61 |
| $18 \leqslant x \leqslant 23$ | 30 |
| $24 \leqslant x \leqslant 29$ | 17 |

Table 10.5 Data for question 7

8. Grade point averages (GPA) in several colleges are on a scale of 0 to 4. Here are the GPAs of 45 students at a certain college.

| 1.8 | 1.9 | 1.9 | 2.0 | 2.1 | 2.1 | 2.1 | 2.2 | 2.2 | 2.3 | 2.3 | 2.4 | 2.4 | 2.4 | 2.5 |
|---|---|---|---|---|---|---|---|---|---|---|---|---|---|---|
| 2.5 | 2.5 | 2.5 | 2.5 | 2.5 | 2.6 | 2.6 | 2.6 | 2.6 | 2.6 | 2.7 | 2.7 | 2.7 | 2.7 | 2.7 |
| 2.8 | 2.8 | 2.8 | 2.9 | 2.9 | 2.9 | 3.0 | 3.0 | 3.0 | 3.1 | 3.1 | 3.1 | 3.2 | 3.2 | 3.4 |

   Prepare a frequency histogram, a relative frequency histogram, and a cumulative frequency graph. Describe the data in two to three sentences.

9. The table shows the grades of an IB course with 40 students on a 100-point test.

| 61 | 62 | 93 | 94 | 91 | 92 | 86 | 87 | 55 | 56 |
|---|---|---|---|---|---|---|---|---|---|
| 63 | 64 | 86 | 87 | 82 | 83 | 76 | 77 | 57 | 58 |
| 94 | 95 | 89 | 90 | 67 | 68 | 62 | 63 | 72 | 73 |
| 87 | 88 | 68 | 69 | 65 | 66 | 75 | 76 | 84 | 85 |

   Use graphical methods you learned so far to describe the grades.

10. The length of time in months between repeated speeding violations of 50 young drivers are given in the table.

| 2.1 | 1.3 | 9.9 | 0.3 | 32.3 | 8.3 | 2.7 | 0.2 | 4.4 | 7.4 |
|---|---|---|---|---|---|---|---|---|---|
| 9 | 18 | 1.6 | 2.4 | 3.9 | 2.4 | 6.6 | 1 | 2 | 14.1 |
| 14.7 | 5.8 | 8.2 | 8.2 | 7.4 | 1.4 | 16.7 | 24 | 9.6 | 8.7 |
| 19.2 | 26.7 | 1.2 | 18 | 3.3 | 11.4 | 4.3 | 3.5 | 6.9 | 1.6 |
| 4.1 | 0.4 | 13.5 | 5.6 | 6.1 | 23.1 | 0.2 | 12.6 | 18.4 | 3.7 |

   (a) Construct a histogram for the data.

   (b) Would you describe the shape as symmetric?

   (c) The law in this country requires that the driving license be taken away if the driver repeats the violation within a period of 10 months. Use a cumulative frequency graph to estimate the fraction of drivers who may lose their license.

11. To decide on the number of counters needed to be open during rush hours in a supermarket, the management collected data from 60 customers

| 3.6 | 0.7 | 5.2 | 0.6 | 1.3 | 0.3 | 1.8 | 2.2 | 1.1 | 0.4 |
|---|---|---|---|---|---|---|---|---|---|
| 1 | 1.2 | 0.7 | 1.3 | 0.7 | 1.6 | 2.5 | 0.3 | 1.7 | 0.8 |
| 0.3 | 1.2 | 0.2 | 0.9 | 1.9 | 1.2 | 0.8 | 2.1 | 2.3 | 1.1 |
| 0.8 | 1.7 | 1.8 | 0.4 | 0.6 | 0.2 | 0.9 | 1.8 | 2.8 | 1.8 |
| 0.4 | 0.5 | 1.1 | 1.1 | 0.8 | 4.5 | 1.6 | 0.5 | 1.3 | 1.9 |
| 0.6 | 0.6 | 3.1 | 3.1 | 1.1 | 1.1 | 1.1 | 1.4 | 1 | 1.4 |

   for the time they spent waiting to be served. The times in minutes are given in the table.

   (a) Construct a relative frequency histogram for the times.

   (b) Construct a cumulative frequency graph and estimate the number of customers who have to wait 2 minutes or more.

12. The cumulative frequency graph shows the speeds of cars ($km\,h^{-1}$) passing through an intersection.

    (a) State the minimum speed of a car travelling through the intersection.

    (b) What percentage of cars drive at a speed higher than $55\,km\,h^{-1}$ through the intersection?

    (c) Given that 40% of the cars travelling through the intersection have a speed higher than $k\,km\,h^{-1}$, what is the value of $k$?

    (d) Find the 60th percentile.

| Time, $t$ (s) | Frequency |
|---|---|
| $t < 60$ | 12 |
| $60 \leqslant t < 120$ | 15 |
| $120 \leqslant t < 180$ | 42 |
| $180 \leqslant t < 240$ | 105 |
| $240 \leqslant t < 300$ | 66 |
| $300 \leqslant t < 360$ | 45 |
| $t \geqslant 360$ | 15 |

**Table 10.6** Data for question 13

13. The waiting time, $t$, in seconds, that it takes 300 customers at a supermarket cash register are recorded in Table 10.6.

    (a) Draw a histogram of the data.

    (b) Construct a cumulative frequency graph of the data.

    (c) Use the cumulative frequency graph to estimate the waiting time that is exceeded by 25% of the customers.

    (d) Find the 75th percentile.

14. The bar graph shows the number of days in hospital spent by heart patients in Austrian hospitals in the 2017–2018 period.

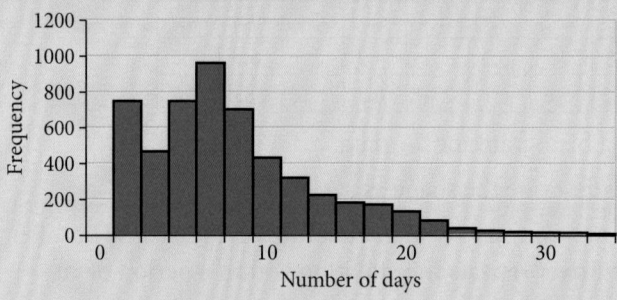

    (a) Describe the data in a few sentences.

    (b) Draw a cumulative frequency graph for the data.

    (c) What percent of the patients stayed less than 6 days?

| Speed, $s$ ($km\,h^{-1}$) | Frequency |
|---|---|
| $60 \leqslant s < 75$ | 70 |
| $75 \leqslant s < 90$ | 110 |
| $90 \leqslant s < 105$ | 150 |
| $105 \leqslant s < 120$ | 70 |
| $120 \leqslant s < 135$ | 40 |
| $s \geqslant 135$ | 10 |

**Table 10.7** Data for question 15

15. Radar devices are installed at several locations on a main highway. Speeds, in $km\,h^{-1}$ of 400 cars travelling on that highway are measured and summarised in Table 10.7.

    (a) Construct a frequency table for the data.

    (b) Draw a histogram to illustrate the data.

    (c) Draw a cumulative frequency graph for the data.

    (d) The speed limit in this country is $130\,km\,h^{-1}$. Use your graph from part (c) to estimate the percentage of the drivers driving faster than this limit.

## 10.3 Measures of central tendency and spread

> ### Help Wanted
>
> Falsead, a fast-growing company, is looking for a lab assistant in its research division.
>
> Minimum requirement:
>
> Master's degree and 2 years of experience.
>
> All benefits.          Average salary is €36,000
>
> **Call 01-234 567**

**Figure 10.17**  Is it true that the average salary is €36,000?

Catherine was very interested in this advertisement and decided to find out more. After some research she found last year's financial report and the salary structure given in Table 10.8. It is true that the average salary is €36,000. The total salary earned by the 9 employees of the company is €108,000.

So, the average salary is $\frac{324\,000}{9} = 36\,000$. But she also discovered that the salaries are bunched up at the low end of the scale. More than half of the employees earn under €16,000. The salary of €36,000 quoted in the advertisement is not at all typical of the pay at Falsead.

In this section, we will discuss how to describing characteristics of data like this one in more meaningful ways.

| Function | Salary | Number |
|---|---|---|
| CEO | €180,000 | 1 |
| Manager | €24,000 | 1 |
| Lab coordinator | €21,000 | 2 |
| Assistants | €15,600 | 5 |

**Table 10.8**  Salary structure

## Measures of location and centre

One of the first things we want to know about a data set such as salaries, prices of items, number of visitors, and so on is 'about how much?' or 'about how many?'

About how much do people earn at Falsead? About how much is the price of a mountain bike, about how many viewers of a TV program, etc.

When we ask, 'about how much?' we probably want to capture with a single number what is typical of the data.

What single number is most representative of an entire list of numbers? We will study three common measures of location: the **mean**, the **median**, and the **mode**. The mean, median and mode are all 'most representative,' but for different, related notions of representativeness.

The two most common measures of centre are the mean and the median. The mean is the 'average value' and the median is the 'middle value.' These are two different ideas for 'centre,' and the two measures behave differently.

385

## Mean

The average or mean of a data set is usually simple to find. Just add all the values and divide by the number of values.

For example, the average salary at Falsead is

$$\frac{180\,000 + 24\,000 + 2 \times 21\,000 + 5 \times 15\,600}{9} = \frac{324\,000}{9} = 36\,000$$

The **arithmetic mean** or **average** of a set of $n$ measurements (data set) is equal to the sum of the measurements divided by $n$.

$$\text{Mean} = \bar{x} = \frac{x_1 + x_2 + \cdots + x_n}{n} = \frac{\sum_{i=1}^{n} x_i}{n} = \frac{\int x}{n}$$

```
1-Variable
x̄=36000
Σx=324000
Σx²=3.5074ᴇ+10
σx=51001.9607
Sx=54095.7484
n=9                    ↓
```

**Figure 10.18** GDC mean value

In practice, you can key the data into your GDC which will give you the value of the mean.

In the IB syllabus, all data is considered as a population. Thus, the mean of the data set is a parameter. Usually, the mean of the population is called $\mu$. The calculation of $\mu$ is the same as before, $\mu = \dfrac{x_1 + x_2 + \cdots + x_N}{N} = \dfrac{\sum_{i=1}^{N} x_i}{N}$ where $N$ is the population size.

The salaries in Table 10.8 are misleading since they contain the salary of the CEO as an employee. This salary is an outlier, and does not belong to the same category of employee salaries. If we exclude the CEO's salary, the average salary in Falsead is

$$\bar{x} = \frac{24\,000 + 2 \times 21\,000 + 5 \times 15\,600}{8} = \frac{144\,000}{8} = 18\,000$$

This illustrates an important weakness of the mean as a measure of centre: the mean is sensitive to the influence of a few extreme observations. These may be outliers, but a skewed distribution that has no outliers will also pull the mean towards its long tail. Because the mean cannot resist the influence of extreme observations, we say that it is not a **resistant measure** of centre.

A measure that is resistant does more than limit the influence of outliers. Its value does not respond strongly to changes in a few observations, no matter how large those changes may be. The mean fails this requirement because we can make the mean as large as we wish by making a large enough increase in just one observation.

## Median

A second measure of central tendency is the median, which is the value in the middle position when the measurements are ordered from smallest to largest. The median of this data can only be calculated if we first sort them in ascending order.

The **median, $M$** is the midpoint of a distribution. Half the observations are smaller than the median and the other half are larger than the median. Here is a rule for finding the median:

1. Arrange all observations in order of size, from smallest to largest.

2. If the number of observations $n$ is odd, the median $M$ is the centre observation in the ordered list. Find the location of the median by counting $\frac{(n+1)}{2}$ observations up from the bottom of the list. For example, if $n = 5$, then $\frac{(5+1)}{2} = 3$. That is, the median is the third point. In general, if $n$ is odd, then it can be written as $n = 2k + 1$ for some integer $k$, and thus $\frac{n+1}{2} = \frac{2k+1+1}{2} = k+1$. This implies that when we arrange data in ascending order, we get something similar to the set-up below:

$$\underbrace{x_1, x_2, x_3, \cdots, x_k,}_{k} \underset{\underset{\text{Median}}{\Uparrow}}{\overset{\overset{M}{\overbrace{x_{k+1}}}}{}} \underbrace{, x_{k+2}, x_{k+3}, \cdots, x_n}_{k}$$

3. If the number of observations, $n$, is even, the median $M$ is the mean of the two centre observations in the ordered list. The location of the median is again $\frac{(n+1)}{2}$ from the bottom of the list. For example, if $n = 6$, then $\frac{(6+1)}{2} = 3.5$. That is, the median is the point with position between 3 and 4. In general, if $n$ is even, then it can be written as $n = 2k$ for some integer $k$, and thus $\frac{n+1}{2} = \frac{2k+1}{2} = k + \frac{1}{2}$. In this case, the median is the average of the two middle points. This implies that when we arrange data in ascending order, we get something similar to the set-up below:

$$\underbrace{x_1, x_2, x_3, \cdots, x_k,}_{k} \underset{\underset{\text{Median}}{\Uparrow}}{\overset{\overset{\frac{x_k + x_{k+1}}{2}}{M}}{}} \underbrace{, x_{k+1}, x_{k+2}, \cdots, x_n}_{k}$$

## Example 10.7

The following are the five closing prices of NASDAQ's stock for the first business week in October 2018: 19.96, 20.08, 20.74, 21.12, 21.04. Find the median and the mean stock price for that week.

### Solution

There are five values, so the position is given by $\frac{(n+1)}{2} = \frac{(5+1)}{2} = 3$

The median is the 3rd value.

19.96   20.08   **20.74**   21.04   21.12

                  ⇑

            Median

$$\text{Mean} = \bar{x} = \frac{19.96 + 20.08 + 20.74 + 21.12 + 21.04}{5} = 20.59$$

Medians require little arithmetic, so they are easy to find by hand for small sets of data. Arranging even a moderate number of observations in order takes a long time, so that finding the median by hand for larger sets of data is tedious. You can use your GDC to find both the mean and the median.

```
Mean(List 1)
                20.588
Median(List 1)
                 20.74
```

# 10 Descriptive statistics

In the previous example we calculated the sample median by finding the third measurement to be in the middle position. Let us demonstrate the case of even numbers with the following.

Let us assume that you took six tests last term and your marks were, in ascending order,

$$52 \quad 63 \quad \underbrace{74 \quad 78}_{\Uparrow} \quad 80 \quad 89$$

There are two 'middle' observations here. To find the median, choose a value halfway between the two middle observations:

$$m = \frac{74 + 78}{2} = 76$$

Although both the mean and median are good measures for the centre of a distribution, the median is less sensitive to extreme values or outliers. For example, the value 52 in the six tests example is lower than all your test scores and is the only failing score you have. The median, 76, is not affected by this outlier even if it were much lower than 52. Assume, for example that your lowest score is 12 rather than 52, the median's calculation

$$12 \quad 63 \quad \underbrace{74 \quad 78}_{\Uparrow} \quad 80 \quad 89$$

still gives the same median of 76. If we were to calculate the mean of the original set, we would get

$$\bar{x} = \frac{\sum x}{6} = \frac{436}{6} \approx 73$$

While the new mean with 12 as lowest score is

$$\bar{x} = \frac{\sum x}{6} = \frac{396}{6} = 66$$

Clearly, the low outlier 'pulled' the mean towards it while leaving the median untouched. However, because the mean depends on every observation and uses all the information in the data, it is generally, wherever possible, the preferred measure of central tendency.

## Mode

A third way to locate the centre of a distribution is to look for the value of $x$ that occurs most. This measure of the centre is called the **mode**.

**Mode** is a French word that means *fashion* – an item that is most popular or common.

The **mode** is the value that occurs with the highest frequency in a data set.

### Example 10.8

The following data give the speeds (in km h$^{-1}$) of eight cars that were stopped for speeding violations on a street with speed limit of 50 km h$^{-1}$:

67, 72, 64, 71, 69, 64, 64, 78

Find the mode.

In this data set, 64 occurs three times, and each of the remaining values occurs only once. Because 64 occurs with the highest frequency, it is the mode.

In the case of a tie for the most frequently occurring value, two modes are listed. Then the data are said to be **bimodal**. Data sets with more than two modes are referred to as **multimodal**.

In applications, the concept of mode is often used in determining sizes. As an example, manufacturers who produce cheap rubber bands might only produce them in one size in order to save on machine setup costs. In determining the one size to produce, the manufacturer would most likely produce bands in the modal size. The mode is an appropriate measure of central tendency for nominal-level data.

We cannot say for sure which of the three measures of central tendency is a better measure overall. Each of them may be better under different situations. Probably the mean is the most-used measure of central tendency, followed by the median. The mean has the advantage that its calculation includes each value of the data set. The median is a better measure when a data set includes outliers. The mode is simple to locate, but it is not of much use in many practical applications.

## Relationships between the mean, median and mode

Some histograms and frequency distributions can be symmetric and others can be skewed.

Knowing the values of the mean, median, and mode can give us some idea about the shape of a frequency distribution curve.

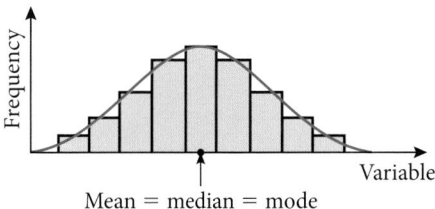

Mean = median = mode

**Figure 10.19** Symmetric distribution

For a symmetric histogram and frequency distribution curve with one peak, the values of the mean, median, and mode are the same, and they lie at the centre of the distribution.

For a histogram and a frequency distribution curve skewed to the right (positively skewed), the value of the mean is the largest, that of the mode is the smallest, and the value of the median lies between these two. Note that the mode always occurs at the peak point. The value of the mean is the largest in this case because it is sensitive to extreme values that occur in the right tail. These extremes pull the mean to the right.

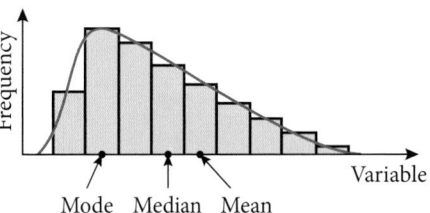

Mode   Median   Mean

**Figure 10.20** Positively skewed distribution

When a histogram and a frequency distribution curve are skewed to the left (negatively skewed), the value of the mean is the smallest and that of the mode is the largest, with the value of the median lying between these two. In this case, the extreme values in the left tail pull the mean to the left.

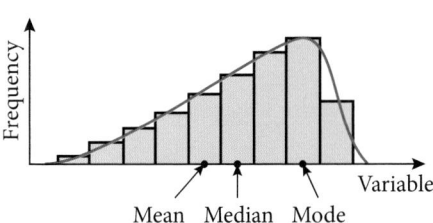

Mean   Median   Mode

**Figure 10.21** Negatively skewed distribution

389

## Measures of variability and spread

Measures of location summarise what is typical of elements of a list, but not every element is typical. Are all the elements close to each other? Are most of the elements close to each other? What is the biggest difference between elements? On average, how far are the elements from each other? The answer lies in the measures of spread or variability.

It is possible that two data sets have the same mean, but the individual observations in one set could vary more from the mean than the observations in the second set do. It takes more than the mean alone to describe data. Measures of variability (also called measures of dispersion or spread) which include the range, the variance, the standard deviation, and interquartile range help to summarise the data.

For example, Bryan's and Jim's test scores in their mathematics class are:

      Bryan: 60, 65, 70, 75, 80

      Jim: 45, 55, 70, 85, 95

To compare the performance of these students, we calculate their means and find out that both have a mean grade of 70. The question is, which 70 is more typical of each student's performance?

As we can see from Figure 10.22, Bryan's 70 is more typical of his performance than Jim's because his test scores are closer to the 70 average than Jim's.

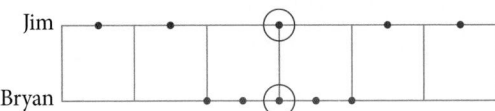

**Figure 10.22** Comparison of scores

The simplest useful numerical description of a distribution consists of both a measure of centre and a measure of spread.

## Range and quartiles

Finding the range for ungrouped data: range = largest value − smallest value.

The **range** is the simplest measure of dispersion to calculate. It is obtained by taking the difference between the largest and the smallest values in a data set.

The range for Jim's test scores is 95 − 45 = 50 while Bryan's is 80 − 60 = 20

With a range of 20 in comparison to 50, Bryan's grade of 70 is more typical for his performance than Jim's.

The range shows the full spread of the data, but it depends on only the smallest and largest observations, which may be outliers. We can improve our description of dispersion by also looking at the spread of the middle half of the data. The **quartiles** mark out the middle half of the data.

**Quartiles** are the summary measures that divide ranked data sets into four equal parts. Three measures will divide any data set into four equal parts. These three measures are the **first quartile** ($Q_1$), the **second quartile** ($Q_2$), and the **third quartile** ($Q_3$). The data should be ranked in increasing order before the quartiles are determined. The quartiles are defined as follows.

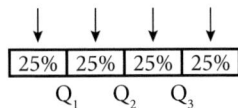

Note that the parts do not have to have the same length, but each should contain 25% of the data points.

**Quartiles** are three summary measures that divide a ranked data set into four equal parts. The second quartile is the same as the median of a data set. The first quartile is the value of the middle term among the observations that are less than the median, and the third quartile is the value of the middle term among the observations that are greater than the median.

Approximately 25% of the values in a ranked data set are less than $Q_1$ and about 75% are greater than $Q_1$. The second quartile, $Q_2$, divides a ranked data set into two equal parts; hence, the second quartile and the median are the same. Approximately 75% of the data values are less than $Q_3$ and about 25% are greater than $Q_3$.

The difference between the third and the first quartiles gives the **interquartile range**; that is, IQR = interquartile range = $Q_3 - Q_1$
IQR describes the middle half of the data.

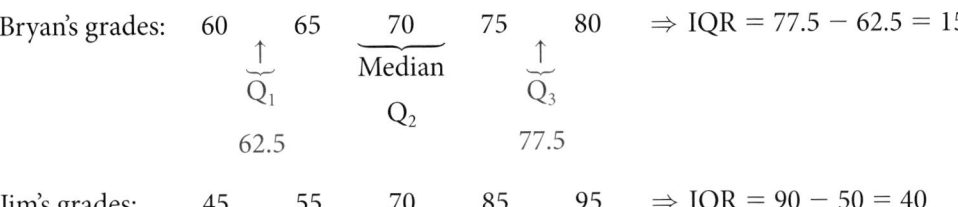

## Example 10.9

The table gives the 2017 profits of the top 10 companies in the world in US$ billions.

| Company | Profit |
|---|---|
| Walmart, US | 486 |
| State Grid, China | 315 |
| Sinopec, China | 268 |
| China National Petroleum | 263 |
| Toyota, Japan | 255 |
| Shell, UK-NL | 240 |
| VW, Germany | 240 |
| Berkshire Hathaway, US | 224 |
| Apple, US | 216 |
| GE, US | 205 |

(a) (i)  Find the values of the three quartiles.

    (ii)  Where does the 2017 profit of Apple fall in relation to these quartiles?

(b) Find the interquartile range.

The value of $Q_1 = \$224$ billion indicates that 25% of the companies in this sample had profits less than \$224 billion and 75% of the companies had profits higher than \$224 billion. Similarly, we can state that half of these companies had profits less than \$247.5 billion and the other half had profits greater than \$247.5 billion since the second quartile is \$247.5 billion. The value of $Q_3 = \$268$ billion indicates that 75% of the companies had profits less than \$268 billion and 25% had profits greater than this value.

### Solution

(a) (i)  First, we rank the given data in increasing order. Then we calculate the three quartiles.

Values less than the median

| 205 | 216 | 224 | 240 | 240 |
|-----|-----|-----|-----|-----|

$Q_1$

$$\begin{array}{c} 247.5 \\ \text{Median} = Q_2 \\ \overbrace{\dfrac{240 + 255}{2}} \end{array}$$

Values greater than the median

| 255 | 263 | 268 | 315 | 486 |
|-----|-----|-----|-----|-----|

$Q_3$

The value of $Q_2$, which is also the median, is given by the value of the middle term in the ranked data set. For the data of this example, this value is the average of the 5th and 6th terms. Consequently, $Q_2$ is \$247.5 billion.

The value of $Q_1$ is given by the value of the middle term of the five values that fall below the median (or $Q_2$). So, $Q_1$ is \$224 billion.

The value of $Q_3$ is given by the value of the middle term of the five values that fall above the median. For the data of this example, $Q_3$ is \$268 billion.

(ii)  The profit of Apple is \$216 billion, which is less than $Q_1$. So the profit of Apple is in the bottom 25% of the profits for 2017.

(b)  The interquartile range is given by the difference between the values of the third and the first quartiles. Thus,

$$\text{IQR} = \text{interquartile range} = Q_3 - Q_1 = 268 - 224 = \$44 \text{ billion}$$

The first quartile $Q_1$ is the median of the observations whose position in the ordered list is to the left of the location of the overall median.

The third quartile $Q_3$ is the median of the observations whose position in the ordered list is to the right of the location of the overall median.

You can calculate the values in Example 10.9 by entering the data into a GDC as a list and then finding the descriptive statistics for them.

```
1-Variable
x̄=271.2
Σx=2712
Σx²=795496
σx=77.4606997
Sx=81.6507467
↓n=10
```

```
1-Variable
↑MinX=205
Q1=224
Med=247.5
Q3=268
MaxX=486
↓Mod=240
```

**Figure 10.23** Calculating the values in Example 10.14 on a GDC

### Five-number summary – box plot

To get a quick summary of both centre and spread for a data set, combine all five numbers.

The **five-number summary** of a set of observations consists of the smallest observation, the first quartile, the median, the third quartile, and the largest observation, written in order from smallest to largest. In symbols, the five-number summary is:

$$\text{minimum} < Q_1 < M < Q_3 < \text{maximum}$$

As you have seen in Example 10.9, these five numbers offer a reasonably complete description of centre and spread.

The five-number summary leads to another visual representation of a distribution, the **box plot**. Figure 10.24 shows box plots for both Bryan's and Jim's grades.

The diagram shows clearly that both samples have the same centre, and both are symmetric around the centre, however, Jim's grades show more spread around the median.

We can put the information from the five-number summary together in one graphical display called a **box plot**, or a **box-and-whisker** plot.
It has upper and lower fences which are usually at 1.5 times the interquartile range from the upper and lower quartiles, respectively.
An **outlier** is a value which is lower than the lower fence or higher than the upper fence.

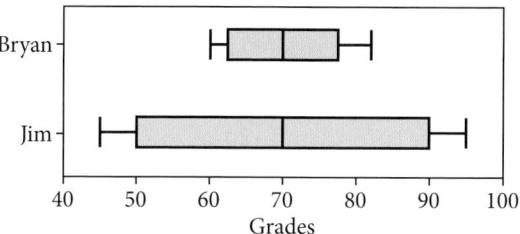

**Figure 10.24** Box plot of grades

## Example 10.10

A consumer agency is interested in studying weekly food and living expenses of college students. A survey of 80 students yielded the following expenses to the nearest euro.

| 38 | 50 | 55 | 60 | 46 | 51 | 58 | 64 | 50 | 49 | 48 | 65 | 58 | 61 | 65 | 53 |
| 39 | 51 | 56 | 61 | 48 | 53 | 59 | 65 | 54 | 54 | 54 | 59 | 65 | 66 | 47 | 49 |
| 40 | 51 | 56 | 62 | 47 | 55 | 60 | 63 | 60 | 59 | 59 | 50 | 46 | 45 | 54 | 47 |
| 41 | 52 | 57 | 64 | 50 | 53 | 58 | 67 | 67 | 66 | 65 | 58 | 54 | 52 | 55 | 52 |
| 44 | 52 | 57 | 64 | 51 | 55 | 61 | 68 | 67 | 54 | 55 | 48 | 57 | 57 | 66 | 66 |

(a) Find the five-number summary.  (b) Draw a box plot.

## Solution

(a) Using a GDC, we find the five numbers:
Minimum = 38, $Q_1$ = 50.5, Median = 55, $Q_3$ = 61, and Maximum = 68

(b) Draw an axis spanning the range of the data; mark the numbers corresponding to the median, minimum, maximum, and the lower and upper quartiles.

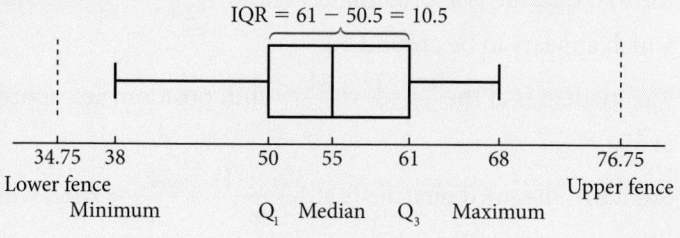

An outlier is an unusual observation. It lies at an abnormal distance from the rest of the data. There is no unique way of describing what an outlier is. A common practice is to consider any observation that is further than 1.5 IQR from the first quartile or the third quartile an outlier. Outliers are important in statistical analysis. Outliers may contain important information not shared with the rest of the data. Statisticians look very carefully at outliers because of their influence on the shapes of distributions and their effect on the values of the other statistics.

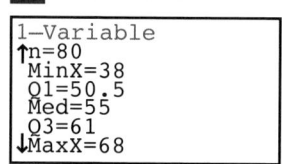

**Figure 10.25** GDC screen for the solution to Example 10.10

Draw a rectangle with lower end at $Q_1$ and upper end at $Q_3$, as shown.

To help us consider outliers, mark the points corresponding to lower and upper fences. Mark them with a dashed line since they are not part of the box. The fences are constructed at the following positions:

Lower fence: $Q_1 - 1.5 \times IQR$ (in this case: $50.5 - 1.5\,(10.5) = 34.75$)
Upper fence: $Q_3 + 1.5 \times IQR$  (in this case: $61 + 1.5\,(10.5) = 76.75$)

Any point beyond the lower or upper fence is considered an **outlier**.

Extend horizontal lines called 'whiskers' from the ends of the box to the smallest and largest observations that are not outliers. In this case these are 38 and 68.

If the data has an outlier, mark it with an asterisk ($*$) on the graph.

To demonstrate this point, consider what would happen if our maximum was 120 and not 68. Since the whisker is 76.75 and $120 > 76.75$, then 120 is an outlier as is clear by the box plots.

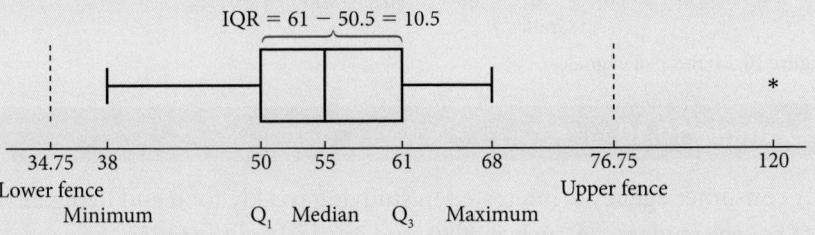

Here is the output of a GDC for both cases.

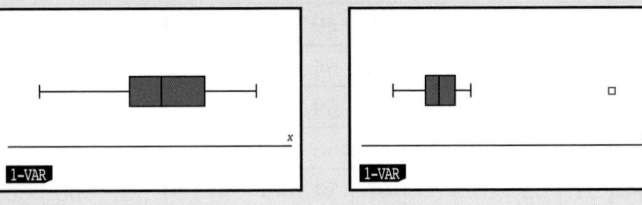

As you see, the box contains the middle 50% of the data. The width of the box is nothing but the IQR! Now we know that the middle 50% of the students' expenditure is €10.50

The box plot is not the only graph we use to explore data. You can also use the cumulative frequency polygon or ogive. Here is the ogive for the expenses data.

Note how we locate the first quartile. Since there are 80 observations, the first quartile is approximately at the $\frac{n+1}{4} = \frac{81}{4} \approx$ 20th position, which appears to be around 50.

The median is at the $\frac{n+1}{2} = \frac{81}{2} \approx$ 40th position, i.e., approximately at 55.

Similarly, the third quartile is at $\frac{3(n+1)}{4} = \frac{243}{4} \approx$ 61st, which happens here at approximately 61.

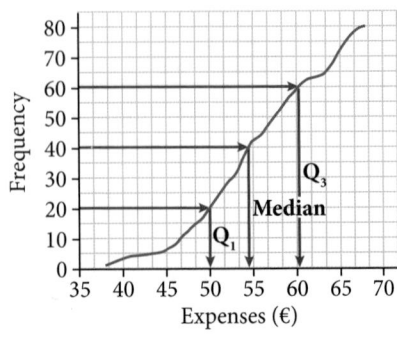

**Figure 10.26** Ogive for expense data

## Example 10.11

The data below shows the heart rates of randomly chosen Females and Males taken from a large group of college students:

Females: 78, 70, 52, 55, 68, 60, 66, 90, 71, 87

Males: 70, 64, 68, 80, 77, 71, 100, 89, 40, 55

(a) Find the five-number summaries.    (b) Find the range and IQR for each.

(c) Draw box plots of both data sets.    (d) Compare the two sets.

## Solution

(a) Using a GDC here are the summaries:

Females:

Minimum = 52, $Q_1$ = 60

Median = 69, $Q_3$ = 78, and Maximum = 90

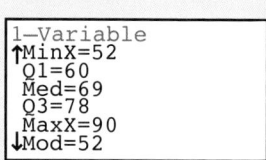

Males:

Minimum = 40, $Q_1$ = 64

Median = 70.5, $Q_3$ = 80

and Maximum = 100

(b) Females' range = 90 − 52 = 38

IQR = 78 − 60 = 18

Males' range = 100 − 40 = 60

IQR = 80 − 64 = 16

(c) Box plots are shown in the screenshot.

(d) The Females' data have a smaller range, a lower median rate, but a larger IQR than the non-smokers. Neither data set appears to have outliers.

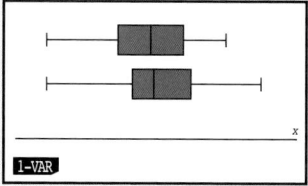

## Variance and standard deviation

The range, like the mean, has the disadvantage of being influenced by outliers. Look at the example of students' expenses. In the original data, the range was 68 − 38 = 30, while when we replaced 68 by 120, the range became 120 − 38 = 82. Just one extreme value almost tripled the range. Moreover, another disadvantage of using the range as a measure of dispersion is that its calculation is based on two values only: the largest and the smallest. All other values in a data set are ignored when calculating the range. Thus, it is not a very satisfactory measure of dispersion.

The **standard deviation** is the most frequently used measure of dispersion. The value of the standard deviation tells us how closely the values of a data set are clustered around the mean. In general, a lower value of the standard deviation for a data set indicates that the values of that data set are spread over a relatively smaller range around the mean, Conversely, a higher value of standard deviation indicates that the values are spread out more around the mean.

The standard deviation measures spread by looking at how far the observations are from their mean.

The standard deviation is obtained by taking the positive square root of the **variance**. The variance calculated for population data is denoted by $\sigma^2$ (read as sigma squared), and the variance calculated for sample data is denoted by $s^2$. Consequently, the standard deviation calculated for population data is denoted by $\sigma$ and the standard deviation calculated for sample data is denoted by **s**.

The basic formulae that are used to calculate the variance are:
$$\sigma^2 = \frac{\sum(x - \mu)^2}{N} \text{ and } s^2 = \frac{\sum(x - \bar{x})^2}{n}$$
The quantity $x - \mu$ or $x - \bar{x}$ is called the deviation of the $x$ value from the mean.
The sum of the deviations of the $x$ values from the mean is always zero; that is
$$\sum(x - \mu) = 0 \text{ and } \sum(x - \bar{x}) = 0$$
We will use the sample notation for the mean and standard deviation.

For example, in Bryan's and Jim's test scores in their mathematics class:

Bryan: 60, 65, 70, 75, 80

Jim: 45, 55, 70, 85, 95

In both cases, $\bar{x} = 70$

Bryan: $\sum(x - \bar{x}) = (60 - 70) + (65 - 70) + (70 - 70) + (75 - 70) + (80 - 70)$
$$= -10 - 5 + 0 + 5 + 10 = 0$$

Jim: $\sum(x - \bar{x}) = (45 - 70) + (55 - 70) + (70 - 70) + (85 - 70) + (95 - 70)$
$$= -25 - 15 + 0 + 15 + 25 = 0$$

For this reason, we square the deviations to calculate the variance and standard deviation.

Thus, for Bryan,

$$s^2 = \frac{\sum(x - \bar{x})^2}{5}$$

$$= \frac{(60 - 70)^2 + (65 - 70)^2 + (70 - 70)^2 + (75 - 70)^2 + (80 - 70)^2}{5}$$

$$= \frac{250}{5} = 50$$

and $s = \sqrt{50} \approx 7.07$

For Jim,

$$s^2 = \frac{\sum(x - \bar{x})^2}{5}$$

$$= \frac{(45 - 70)^2 + (55 - 70)^2 + (70 - 70)^2 + (85 - 70)^2 + (95 - 70)^2}{5}$$

$$= \frac{1700}{5} = 340$$

and $s = \sqrt{340} \approx 18.44$

We can observe from this that the spread in Jim's scores is more than double that of Bryan.

You can use your GDC to calculate these values, as shown in Figure 10.27.

If you are interested in the standard deviation only, most GDCs have a command for it.

**Very important**
In all GDCs, the standard deviation we need so far in this course is $\sigma x$ and not $Sx$.

More important than the details of hand calculation are the properties that determine the usefulness of the standard deviation:

- $s$ measures spread about the mean and should be used only when the mean is chosen as the measure of centre.

- $s = 0$ only when there is no spread. This happens only when all observations have the same value. Otherwise, $s$ is greater than zero. As the observations become more spread out about their mean, $s$ gets larger.

- $s$ has the same units of measurement as the original observations. For example, if you measure wages in dollars per hour, $s$ is also in dollars per hour.

- Like the mean $\bar{x}$, $s$ is not resistant. Strong skewness or a few outliers can greatly increase $s$.

## Measures of centre and spread for grouped data

The calculation of the mean and variance for grouped data is essentially the same as for raw data. The difference lies in the use of frequencies to save typing (writing) all numbers. Table 10.9 shows a comparison.

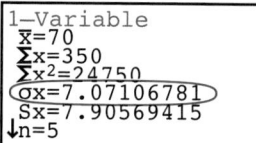

**Figure 10.27** Using a GDC to calculate standard deviation

The five-number summary is usually better than the mean and standard deviation for describing a skewed distribution or a distribution with extreme outliers. Use $\bar{x}$ and $s$ only for reasonably symmetric distributions that are free of outliers.

| Statistic | Raw data | Grouped data | Grouped data with intervals |
|---|---|---|---|
| $\bar{x}$ | $\bar{x} = \dfrac{\sum\limits_{all\,x} x}{n}$ | $\bar{x} = \dfrac{\sum\limits_{all\,x} x_i \times f(x_i)}{\sum f(x_i)}$ | $\bar{x} = \dfrac{\sum\limits_{all\,m} m_i \times f(m_i)}{\sum f(m_i)}$ <br><br> First find the midpoint of each class and then multiply the midpoints by the frequencies of the corresponding classes. The sum of these products gives an approximation for the sum of all values. To find the value of the mean, divide this sum by the total number of observations. |
| $s^2$ | $s^2 = \dfrac{\sum\limits_{all\,x}(x_i - \bar{x})^2}{n}$ | $s^2 = \dfrac{\sum\limits_{all\,x}(x_i - \bar{x})^2 \times f(x_i)}{\sum f(x_i)}$ | $s^2 = \dfrac{\sum\limits_{all\,m}(m_i - \bar{x})^2 \times f(m_i)}{\sum f(m_i)}$ |

**Table 10.9** Grouped data calculations

When we estimate the mean of a data set from its frequency table, answers may differ from the real mean. In this example, the real mean is 55.475. The reason is that we are replacing the real values with estimates. For example, in the interval $35 \leqslant x < 45$, we replace the real values with 40. This way, we are assuming that all numbers in this interval are equal to 40.

For example, when we group the students' expense data into a frequency table, we get the data shown in Table 10.10.

| Expenses | Midpoint $m$ | Number of students |
|---|---|---|
| $35 \leqslant x < 45$ | 40 | 5 |
| $45 \leqslant x < 55$ | 50 | 32 |
| $55 \leqslant x < 65$ | 60 | 30 |
| $65 \leqslant x < 75$ | 70 | 13 |

**Table 10.10** Frequency table for grouped expense data

$$\bar{x} \approx \frac{\sum\limits_{all\ m} m_i \times f(m_i)}{n}$$

$$= \frac{40 \times 5 + 50 \times 32 + 60 \times 30 + 70 \times 13}{80} = 56.375$$

Your GDC will give you the result in the same way as in the non-grouped data. The difference here is that you need to organise your data in two lists, one for the class midpoints and one for the frequency.

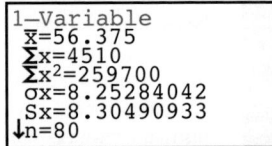

```
1-Variable
x̄=56.375
Σx=4510
Σx²=259700
σx=8.25284042
Sx=8.30490933
↓n=80
```

|  | List 1 | List 2 |
|---|---|---|
| SUB |  |  |
| 1 | 40 | 5 |
| 2 | 50 | 32 |
| 3 | 60 | 30 |
| 4 | 70 | 13 |

```
TOOL  EDIT  DELETE  DEL-ALL  INSERT
```

**Figure 10.28** You should organise your data into two lists on your GDC

### Effect of constant changes to the original data on centre and spread measures

Consider Bryan's test scores again. The teacher decides to add 5 marks to each score at the end of the term. What are the new measures for Bryan's scores?

Bryan's new scores are: 65, 70, 75, 80, 85. Table 10.11 gives both the old and new measurements.

| Measure | Old | New |
|---|---|---|
| Median | 70 | The middle observation is 75. Median is 75. |
| Mean | 70 | $\bar{x} = \dfrac{65 + 70 + 75 + 80 + 85}{5} = 75$ |
| Quartiles | $Q_1 = 62.5, Q_3 = 77.5$ | $Q_1 = 67.5, Q_3 = 82.5$ |
| IQR | 15 | $82.5 - 67.5 = 15$ |
| St. deviation | 7.07 | 7.07 |

**Table 10.11** Effect of adding 5 to each score

Note that 5 is added to each of the median, mean, and quartiles, but the IQR and standard deviation are not affected. This is so because, as you observe from the box plots, we moved the data by 5 units to the right, keeping the spread as it was.

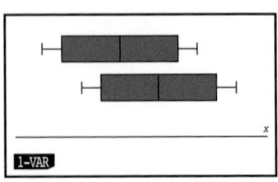

**Figure 10.29** New box plot shifted to the right by 5 units

## Example 10.12

Five students took a trial test on paper 1 of an IB exam.

Their scores, out of 120 are 111, 96, 87, 72, 60

(a) Find the median, IQR, mean and standard deviation of the scores.

(b) The scores must be scaled to be out of a total of 40 instead.
Find the new median, IQR, mean and standard deviation of the scores.

### Solution

| Measure | Old | New |
|---|---|---|
| Median | 87 | 29 |
| Mean | 85.2 | 28.4 |
| Quartiles | $Q_1 = 66, Q_3 = 103.5$ | $Q_1 = 22, Q_3 = 34.5$ |
| IQR | 37.5 | 12.5 |
| St. deviation | 17.86 | 5.95 |

Note that the new measures are equal to one third of the old ones.

When data values are multiplied by a constant $k$, then all centre and spread statistics are multiplied by $k$.

## Example 10.13

Instruments that measure blood sugar level (glucose) may use one of two systems of measurement: mmol/L or mg/dL.
Every mmol/L is equivalent to 18 mg/dL.

Tim keeps track of his blood sugar level. Last month's readings are as follows.

The average reading was 5.2 mmol/L, the standard deviation was 2.3 mmol/L, and the range was 3.2 mmol/L.

Find Tim's readings in mg/dL.

### Solution

Since every 1 mmol/L = 18 mg/dL, then Tim's readings in mg/dL are:

Average = $5.2 \times 18 = 93.6$ mg/dL

Standard deviation = $2.3 \times 18 = 41.4$ mg/dL

Range = $3.2 \times 18 = 57.6$ mg/dL

## Exercise 10.3

1. The number of visitors per week (thousands) for a book exhibition were

   9, 7, 8, 11, 9, 6, 10, 8, 12, 6, 8, 13, 7, 9, 10, 9, 10, 11, 12, 8, 7, 13, 10, 7, 7

   The data has been organised in the frequency table.

   | Number of visitors | 6 | 7 | 8 | 9 | 10 | 11 | 12 | 13 |
   |---|---|---|---|---|---|---|---|---|
   | Frequency | 2 | 5 | $n$ | 4 | 4 | 2 | 2 | 2 |

   (a) Write down the value of $n$.

   (b) Calculate the mean number and standard deviation of visitors per week, from the raw data and from the table.

   (c) What percentage of the weeks had more than 10 000 visitors?

   (d) What is the modal number of visitors?

2. The following data give the number of shoplifters apprehended during 9 weeks preceding Christmas 2017 at a large department store.

   1, 1, 8, 3, 1, 1, 7, 26, 51

   (a) Find the mean, median, and mode for these data.

   (b) Calculate the deviations of the data values from the mean. What is the sum of these deviations?

   (c) Calculate the range, variance, and standard deviation.

   (d) Draw a box plot. Can you explain why no whisker appears on the lower side?

   (e) Decide if there are any outliers. Explain.

3. The following data give the numbers of car thefts that occurred in a large city in the past 12 months.

   60, 30, 70, 10, 140, 30, 80, 70, 20, 60, 90, 10

   (a) Find the mean, median, and mode for these data.

   (b) Calculate the range and standard deviation.

   (c) Draw a box plot.

   (d) Decide if there are any outliers. Explain.

4. The pulse rates of 15 patients chosen at random from visitors of a local clinic are 72, 80, 67, 68, 80, 68, 80, 56, 76, 68, 71, 76, 60, 79, 71

   (a) Calculate the mean and standard deviation of the pulse rate of the patients at the clinic.

   (b) Draw a box plot of the data and indicate the values of the different parts of the box.

   (c) Check if there are any outliers.

**5.** The number of passengers on 50 flights from Washington to London on a commercial airline is given in the table.

| 165 | 173 | 158 | 171 | 177 | 156 | 178 | 210 | 160 | 164 |
| 141 | 127 | 119 | 146 | 147 | 155 | 187 | 162 | 185 | 125 |
| 163 | 179 | 187 | 174 | 166 | 174 | 139 | 138 | 153 | 142 |
| 153 | 163 | 185 | 149 | 154 | 154 | 180 | 117 | 168 | 182 |
| 130 | 182 | 209 | 126 | 159 | 150 | 143 | 198 | 189 | 218 |

(a) Calculate the mean and standard deviation of the number of passengers on this airline between the two cities.

(b) Develop a cumulative frequency graph. Estimate the median, first and third quartiles. Draw a box plot.

(c) Find the IQR and use it to check whether there are any outliers.

**6.** The table shows the frequency distribution of the daily commuting times (in minutes) from home to work for all 50 employees of a company.

| Daily commuting time (minutes) | Number of employees |
| --- | --- |
| $0 \leqslant t < 10$ | 8 |
| $10 \leqslant t < 20$ | 18 |
| $20 \leqslant t < 30$ | 12 |
| $30 \leqslant t < 40$ | 8 |
| $40 \leqslant t < 50$ | 4 |

(a) Set up a cumulative frequency table and graph.

(b) Estimate the median, first and third quartiles as well as IQR.

(c) Find the mean and standard deviation.

**7.** The table gives the frequency distribution of the amounts of telephone bills for October 2018 for a sample of 50 families.

| Amount of telephone bill (€) | Number of families |
| --- | --- |
| $40 \leqslant m < 70$ | 9 |
| $70 \leqslant m < 100$ | 11 |
| $100 \leqslant m < 130$ | 16 |
| $130 \leqslant m < 160$ | 10 |
| $160 \leqslant m < 190$ | 4 |

(a) Find the average value of a telephone bill for families in this community.

(b) Find the standard deviation.

(c) The amount of every bill contains a €24 flat monthly rate plus the cost of telephone calls. Find the average cost of calls and the standard deviation.

8. Every year, the faculty of a major university's business school chooses 10 students from the current graduating class that they feel will be most likely to succeed. Five years later, they evaluate their choices by looking at the annual incomes of the group of 10. The following is the list of the annual incomes of the class of 2007 (thousands of US$):

    59, 68, 84, 78, 107, 382, 56, 74, 97, 60

    (a) Find the mean, median, standard deviation, and IQR.

    (b) Does this data set contain any outlier? If yes, ignore the outlier and recalculate the statistics. Which of these measures changes by a greater amount? Explain.

    (c) Which measures provide a better summary for these data? Explain.

9. Students in one section of an SL maths class scored the following marks on a trial exam paper:

    5, 5, 4, 6, 3, 7, 7, 3, $x$

    (a) The section's average score is 5. What is the value of $x$?

    (b) Find the median and standard deviation of this group.

    (c) One student was absent on the day of the trial exam. What is the minimum score required of that student for the average score of the whole group to be 6?

    (d) The student from part (c) received a mark of 5. Another section of the class has 12 students, and their average score was 4.5. What is the average of both sections?

10. For restaurants, the time customers linger over coffee and dessert negatively affects profit. To learn more about this variable, a sample of 14 days was observed, and the average time spent by different groups was recorded (to the nearest minute):

    26, 21, 28, 28, 56, 45, 40, 32, 32, 29, 30, 27, 20, 25

    (a) Find a detailed summary of the data.

    (b) What do the results tell you about the amount of time spent in this restaurant?

11. The following data represent the ages in years of a sample of 25 jockeys from a local race track:

    31, 43, 56, 23, 49, 42, 33, 61, 44, 28, 48, 38, 44,
    35, 40, 64, 52, 42, 47, 39, 53, 27, 36, 35, 20

    (a) Find the 5-number summary, and IQR.

    (b) Construct a box plot for the ages and identify any extreme values. What does the box plot tell you about the distribution of the data?

    (c) Construct a relative frequency distribution for the data, using five class intervals and the value 20 as the lower limit of the first class.

    (d) Construct a relative frequency histogram for the data. What does the histogram tell you about the distribution of the data?

1. The temperatures in °C, at midday in a German city during summer of 2017, were measured for eight days and the results are recorded below.

    21, 12, 15, 12, 24, $T$, 30, 24

    The mean temperature was found to be 21°C.

    (a) Find the value of $T$.

    (b) Write down the mode.

    (c) Find the median.

2. The age (in months) when a child starts to walk is observed for a random sample of children from a town in France. The results are:

    14.3, 11.6, 12.2, 14.0, 20.4, 13.4, 12.9, 11.7, 13.1

    (a) (i) Find the mean of the ages of these children.

    (ii) Find the standard deviation of the ages of these children.

    (b) Find the median age.

3. The following data are listed in ascending order: 2, $b$, 3, $a$, 6, 9, 10, 12

    The mean is 6 and the median is 5. Find:

    (a) the value of $a$

    (b) the value of $b$.

4. The table shows the frequency distribution of the number of dental fillings for a group of 50 children.

| Number of fillings | 0 | 1 | 2 | 3 | 4 | 5 |
|---|---|---|---|---|---|---|
| Frequency | 4 | 3 | 8 | $x$ | 4 | 1 |

    (a) Find the value of $x$.

    (b) Find:

    (i) the mean number of fillings

    (ii) the median number of fillings

    (iii) the standard deviation of the number of fillings.

    (c) The first row in the table was entered by mistake. It should have been 2, 3, 4, 5, 6, and 7 instead.

    Without going through the calculations required for part (b), find the mean, median and standard deviation.

5. During the first 240 games played in 2017 in a National football league the number of goals scored are given in the table.

| Number of goals | 0 | 1 | 2 | 3 | 4 | 5 |
|---|---|---|---|---|---|---|
| Number of games | 48 | 66 | 57 | 51 | 3 | 15 |

    (a) Find the mean number of goals scored per game.

    (b) Find the median number of goals scored per game.

**6.** The masses, in kg, of 80 adult males were collected and are summarised in the box-and-whisker plot.

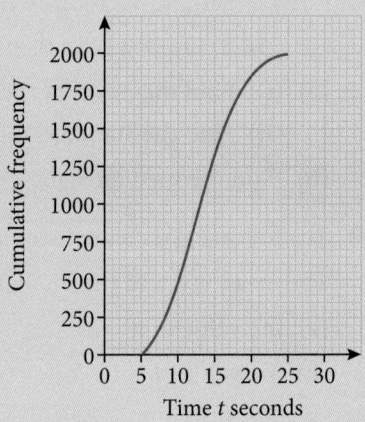

40  45  50  55  60  65  70  75  80
Weight (kg)

   **(a)** Write down the median mass of the males.

   **(b)** Calculate the interquartile range.

   **(c)** Estimate the number of males who have a mass between 61 kg and 66 kg.

   **(d)** Estimate the mean mass of the lightest 40 males.

**7.** The diagram shows the cumulative frequency graph for the time $t$ taken to perform a certain task by 2000 men.

   **(a)** Use the diagram to estimate:

      **(i)** the median time

      **(ii)** the upper quartile and the lower quartile

      **(iii)** the interquartile range.

   **(b)** Find the number of men who take more than 11 seconds to perform the task.

   **(c)** 55% of the men took less than $p$ seconds to perform the task. Find $p$.

| Time | Frequency |
|------|-----------|
| $5 \leqslant t < 10$ | 500 |
| $10 \leqslant t < 15$ | 850 |
| $15 \leqslant t < 20$ | $a$ |
| $20 \leqslant t < 25$ | $b$ |

   **(d)** Write down the value of:

      **(i)** $a$       **(ii)** $b$

   **(e)** Find an estimate of:

      **(i)** the mean time

      **(ii)** the standard deviation of the time.

Everyone who performs the task in less than one standard deviation below the mean will receive a bonus. Pedro takes 9.5 seconds to perform the task.

   **(f)** Does Pedro receive the bonus? Justify your answer.

8. The heights of 200 students are recorded in the table.

| Height ($h$) in cm | Frequency |
|---|---|
| $140 \leqslant h < 150$ | 2 |
| $150 \leqslant h < 160$ | 28 |
| $160 \leqslant h < 170$ | 63 |
| $170 \leqslant h < 180$ | 74 |
| $180 \leqslant h < 190$ | 20 |
| $190 \leqslant h < 200$ | 11 |
| $200 \leqslant h < 210$ | 2 |

(a) Write down the modal group.

(b) Calculate an estimate of the mean and standard deviation of the heights.

The cumulative frequency curve for this data is shown.

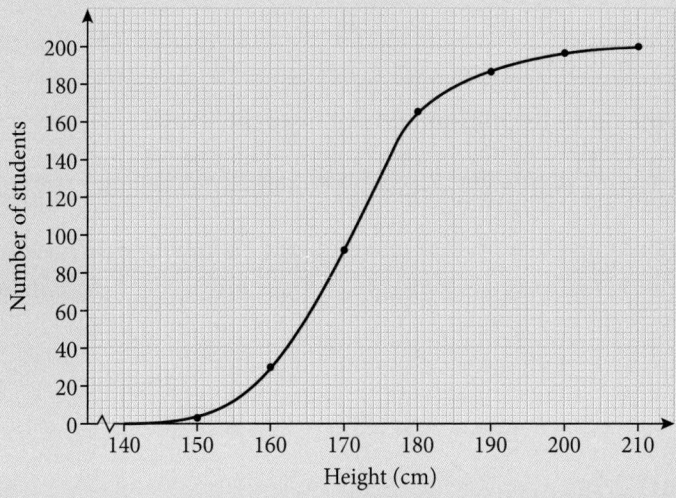

(c) Write down the median height.

(d) The upper quartile is 177.3 cm. Calculate the interquartile range.

(e) Find the percentage of students with heights less than 165 cm.

9. 120 Mathematics students in a school sat an examination. Their scores (given as a percentage) were summarised on a cumulative frequency diagram, as shown.

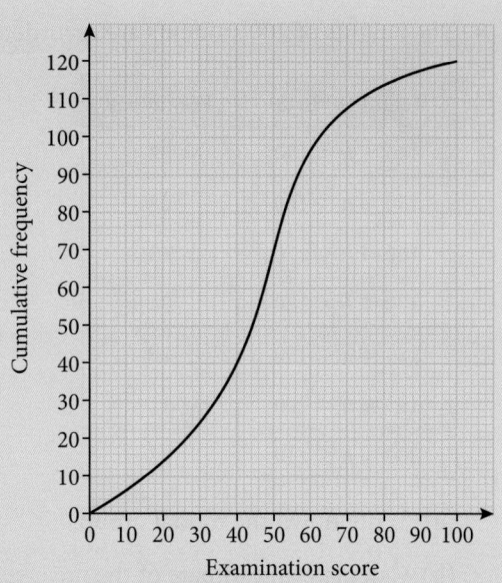

(a) Copy and complete the grouped frequency table for the students.

| Examination score $x$ (%) | $0 \leqslant x < 20$ | $20 \leqslant x < 40$ | $40 \leqslant x < 60$ | $60 \leqslant x < 80$ | $80 \leqslant x < 100$ |
|---|---|---|---|---|---|
| Frequency | 14 | 26 | | | |

(b) Write down the mid-interval value of the $40 \leqslant x < 60$ interval.

(c) Calculate an estimate of the mean examination score of the students.

Probability of events

11

### Learning objectives

By the end of this chapter, you should be familiar with...

- concepts of trial, outcome, equally likely outcomes, relative frequency, sample space ($U$) and event
- complementary events
- expected number of occurrences
- combined events and mutually exclusive events
- conditional probability and independent events
- Venn diagrams, tree diagrams, sample space diagrams and tables of outcomes to calculate probabilities.

Uncertainty plays an important role in our daily lives and activities as well as in business, science, and almost all fields. Investors cannot be sure which stocks will deliver the better growth over the following year, engineers try to reduce the likelihood that a machine breaks down, publishers cannot be sure of how sales of a new book are going to be over its life cycle. In each of these cases, chance is involved. In this chapter we will develop some tools to tackle situations that involve chance or uncertainty.

In Chapter 10, we learned how to describe a data set. In Chapter 17 we will consider how to use sample data to draw conclusions about the population from which we drew our sample. The techniques we use in drawing conclusions are part of what we call inferential statistics, which uses **probability** as one of its tools. To use this tool properly, we must first understand how it works. This chapter introduces you to the language and basic tools of probability.

The variables we discussed in Chapter 10 can now be redefined as random **variables**, whose values depend on the chance selection of the elements in the sample. Using probability as a tool, you will be able to create **probability distributions** in Chapter 15 that serve as models for random variables. You can then describe these using a mean and a standard deviation as you did in Chapter 10.

## 11.1 Concepts and definitions

Probability is the likelihood of something happening in the future. Consider playing a game where it is necessary to roll a 6 on a 6-sided dice. Each roll of the dice can be considered as a **trial** in which there are six possible outcomes: (1, 2, 3, 4, 5 or 6). We are interested in the event 'roll a 6'. If the dice is unbiased, then each of the six outcomes is equally likely, so the probability of rolling a 6 is $\frac{1}{6}$. If the dice is biased, then we could carry out multiple trials and use the outcomes to calculate the relative frequency of rolling a 6. For example, if the dice is rolled 100 times and a 6 occurs 11 times, then the relative frequency of rolling a 6 is $\frac{11}{100}$

A trial is said to be random when there is uncertainty concerning which of two or more possible outcomes will result.

A description of a random phenomenon in the language of mathematics is called a **probability model**. For example, when we flip a coin, the description of coin flipping has two parts:

- a list of possible outcomes
- a probability for each outcome.

This two-part description is the starting point for a probability model.

The sample space of a random trial (or experiment) is the set of all possible outcomes.

For example, for one flip of a coin, the sample space is

$U = \{\text{Heads, Tails}\}$, or simply $U = \{\text{H, T}\}$

> The IB notation for sample space is $U$, but it is possible to use any other letter. $U$ denotes the universal set, i.e. the set containing all objects.

## Example 11.1

Flip a coin twice (or two coins once) and record the results. Write down the sample space.

### Solution

$U = \{\text{HH, HT, TH, TT}\}$

A **simple event** is a unique possible outcome of a trial (or experiment).

An **event** is an outcome or a **set of outcomes** of a random trial (or experiment.)

We can also look at the event as a subset of the sample space or as a collection of simple events.

The probability of an event $= \dfrac{\text{number of outcomes of the event}}{\text{total number of outcomes}}$

This can be written as $P(A) = \dfrac{n(A)}{n(U)}$ where $P(A)$ denotes the probability of event $A$ occurring, $n(A)$ denotes the number of outcomes of event $A$ and $n(U)$ denotes the total number of outcomes in the sample space.

## Example 11.2

Three fair coins are flipped and the number of heads is counted. Find the probability of observing exactly two heads.

The sample space is

$$U = \{HHH, HHT, HTH, THH, HTT, THT, TTH, TTT\}$$

Three of the eight possible outcomes have exactly two heads,

so the probability $= \dfrac{3}{8}$

Example 11.2 shows the probability of observing exactly two heads when three coins are flipped. Using a computer, we can simulate what happens if we repeatedly flip three coins. Table 11.1 shows the results of 30 trials.

| Trial | Coin 1 | Coin 2 | Coin 3 | Number of Heads | Cumulative number of exactly 2 Heads | Relative frequency of exactly 2 Heads |
|---|---|---|---|---|---|---|
| 1 | H | H | T | 2 | 1 | 1.000 |
| 2 | T | T | T | 0 | 1 | 0.500 |
| 3 | H | T | T | 1 | 1 | 0.333 |
| 4 | T | T | T | 0 | 1 | 0.250 |
| 5 | H | H | H | 3 | 1 | 0.200 |
| 6 | T | H | T | 1 | 1 | 0.167 |
| 7 | H | T | T | 1 | 1 | 0.143 |
| 8 | T | T | T | 0 | 1 | 0.125 |
| 9 | T | H | H | 2 | 2 | 0.222 |
| ⋮ | ⋮ | ⋮ | ⋮ | ⋮ | ⋮ | ⋮ |
| 28 | T | T | T | 0 | 10 | 0.357 |
| 29 | H | H | H | 3 | 10 | 0.345 |
| 30 | H | H | T | 2 | 11 | 0.367 |

**Table 11.1**  Results of 30 coin flips

It is helpful to graph the relative frequency to see what happens over a period of time, as shown in Figure 11.1.

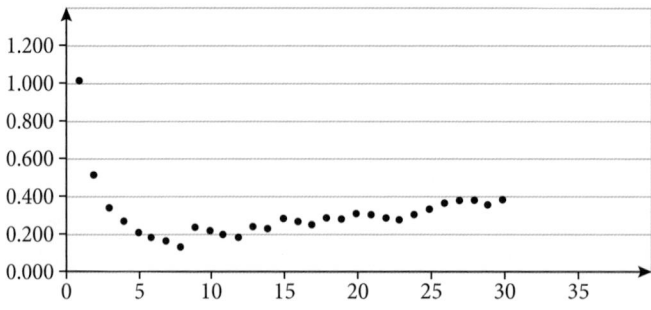

**Figure 11.1**  Relative frequency of obtaining exactly 2 heads when three coins are flipped 30 times

As the number of trials increases, the relative frequency appears to approach a value just under 0.4, which agrees with the result of 0.375 found in Example 11.2. Simulations provide a quick way to model random events. Figure 11.2 shows a second simulation of flipping three coins, but this time with 200 trials.

After a large number of trials, the relative frequency appears to approach a value around 0.375

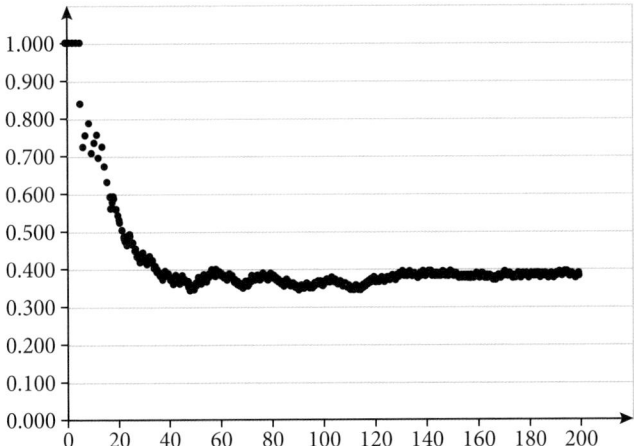

**Figure 11.2** Relative frequency of obtaining exactly 2 heads when three coins are tossed 200 times

In some problems it can be helpful to write down the outcomes in a table or grid.

## Example 11.3

Two dice are rolled, and the sum of the numbers is found.
Find the probability that the sum is greater than 8.

### Solution

We can represent the sample space in a table.

| | | Dice 1 | | | | | |
|---|---|---|---|---|---|---|---|
| | | **1** | **2** | **3** | **4** | **5** | **6** |
| | **1** | 2 | 3 | 4 | 5 | 6 | 7 |
| | **2** | 3 | 4 | 5 | 6 | 7 | 8 |
| **Dice 2** | **3** | 4 | 5 | 6 | 7 | 8 | 9 |
| | **4** | 5 | 6 | 7 | 8 | 9 | 10 |
| | **5** | 6 | 7 | 8 | 9 | 10 | 11 |
| | **6** | 7 | 8 | 9 | 10 | 11 | 12 |

There are 10 outcomes greater than 8 (coloured blue in the table) and 36 outcomes altogether. So the probability $= \dfrac{10}{36} = \dfrac{5}{18}$

In the examples of flipping a coin or rolling a dice, the outcomes are equally likely, which enables us to calculate the theoretical probability. In the theory of equally likely outcomes, probability has to do with symmetries and the indistinguishability of outcomes. If a given experiment or trial has $n$ possible outcomes, among which there is no preference, then they are **equally likely**.

The probability of each outcome is then $\dfrac{1}{n}$

In many events, the outcomes are not equally likely, so we can use experimental results to find the probability instead. Suppose you were planning a trip to Faro, Portugal and you wanted to find the probability of it raining during the day. Historically, during the month of June, Faro has an average of 5 rainy days.

So the experimental probability of rain $= \dfrac{5}{30} = \dfrac{1}{6}$

The **relative frequency** of an event is the ratio of the number of times the event occurs to the total number of trials.

Regardless of whether we are using theoretical or experimental probability, there are a number of rules that apply.

- Any probability is a number between 0 and 1, i.e. the probability $P(A)$ of any event $A$ satisfies $0 \leqslant P(A) \leqslant 1$. If an event is impossible, then its probability is 0. Likewise, if the event is certain to happen, then its probability is 1.

- The sample space $U$ contains all possible outcomes, so $P(U) = 1$

  Suppose that the probability that you receive a 7 on your IB exam is 0.2 Then the probability of not receiving a 7 on the exam is 0.8. The event that contains the outcomes not in $A$ is called the **complement** of $A$, and is denoted by $A'$.

  $$P(A') = 1 - P(A), \text{ or } P(A) = 1 - P(A')$$

- If two events have no outcomes in common, the probability that one or the other occurs is the sum of their individual probabilities. Two events that have no outcomes in common, and hence can never occur together are called **mutually exclusive events** (or disjoint events).

  $$P(A \text{ or } B) = P(A) + P(B)$$

  This is the **addition rule for mutually exclusive events**.

For example, when flipping three coins, the events of getting exactly two heads or exactly two tails are mutually exclusive.

The expected number of occurrences of an event is the product of the total number of trials and the probability of the event.

We can make use of either the theoretical or experimental probability to calculate how many times we expect a random event to occur in a given number of trials.

---

### Example 11.4

Data from the transport accident commission, below, shows the probability of road fatalities in the state of Victoria, Australia for the year 2017.

| Age group | Below 21 years | 21–39 years | 40–59 years | Over 60 years |
|---|---|---|---|---|
| Probability | 0.12 | 0.31 | 0.27 | $k$ |

(a) Find the value of $k$.

(b) Find the probability that a road fatality involves someone younger than 40.

(c) 260 road fatalities are recorded. Find the expected number involving someone younger than 40.

## Solution

(a) The sum of all the probabilities must equal 1:
$$k = 1 - 0.12 - 0.31 - 0.27 = 0.3$$

(b) The probability that a road fatality involves
someone younger than $40 = 0.12 + 0.31 = 0.43$

(c) The expected number $= 260 \times 0.43 = 112$

In Chapter 10, we calculated the mean value from a frequency table using

$$\mu = \frac{\sum_{i=1}^{k} f_i x_i}{n}$$

We can use a similar approach to calculate the mean or expected value
of a probability distribution if we know the relative frequency of each
event. Multiplying each outcome by its relative frequency and then
adding all these together gives the expected (mean) value, as shown
in the next example.

### Example 11.5

A car insurance company needs to estimate how much it will pay out in
claims during a year, in order to calculate how much to charge customers to
buy a policy. Based on past claims, they have collected the data shown below.

| Type of accident | Relative frequency | Average cost of claim ($) |
|---|---|---|
| Car is written off (it is not reparable) | 0.02 | 15 000 |
| Serious | 0.06 | 8 000 |
| Minor | 0.12 | 3 000 |
| No accident | 0.80 | 0 |

(a) Find the expected cost of a claim for each customer.

(b) For each policy sold, the company needs to cover the expected cost of
a claim, and in addition, cover the expenses of running the company.
The company charges 70% more than the expected cost of a claim
for each policy. Find the cost of buying a policy.

### Solution

(a) From the table we can see that 2% of accidents result in the car being
written off, which costs the company an average of $15,000 each time.
The expected cost of this type of accident is $0.02 \times 15\,000 = \$300$

We can calculate the expected cost for each of the other types of
accident in a similar way.

| Type of accident | Expected cost |
|---|---|
| Car is written off | $0.02 \times 15\,000 = \$300$ |
| Serious | $0.06 \times 8000 = \$480$ |
| Minor | $0.12 \times 3000 = \$360$ |
| No accident | $\$0$ |

Adding up the expected cost for each type of accident gives the expected cost of a claim for each customer as $300 + 480 + 360 = \$1140$

(b) Since the company charges 70% more than the expected cost of a claim, we can multiply the answer in part (a) by 1.7 to get $1140 \times 1.7 = \$1938$

You will learn more about calculating the expected value of a probability distribution in Chapter 15.

## Exercise 11.1

1. A fair coin is flipped three times and the number of heads is recorded. Find the probability of obtaining at most one head.

2. Two fair, six-sided dice are rolled and the difference between the two numbers is recorded.
   (a) Find the probability that the difference is:
       (i) equal to 1
       (ii) greater than 3.
   (b) The two dice are rolled 500 times. Find the expected number of times the difference is equal to 1.

3. Twenty identical balls numbered from 1 to 20 are placed into a bag. One ball is selected at random. Find the probability that the ball is:
   (a) a multiple of 5      (b) a prime number      (c) a factor of 60.

4. Two fair, six-sided dice are rolled and the product of the two numbers is recorded.
   (a) Find the probability that the product is:
       (i) a prime number            (ii) a square number
       (iii) a triangular number     (iv) at least 25
       (v) less than 36.
   (b) The two dice are rolled 200 times. Find the expected number of times the product is a prime number.

5. In a biased dice the probability of rolling a 6 is three times the probability of rolling any other number. Find the probability of rolling:
   (a) a 6                    (b) an odd number.

6. Given that $P(A) = 0.27$, find $P(A')$.

**7.** A regular deck of playing cards contains 52 cards. A card is selected at random. Find the probability that the card is:

(a) a club

(b) a jack, queen or king

(c) not the ace of spades.

**8.** On Monday morning, my class wanted to know how many hours students spent studying during the weekend. They stopped schoolmates at random as they arrived at school and asked each: 'How many hours did you study during the weekend?' Here are the data they collected:

| Number of hours | 0 | 1 | 2 | 3 | 4 | 5 |
|---|---|---|---|---|---|---|
| Number of students | 4 | 12 | 8 | 3 | 2 | 1 |

(a) Estimate the probability that a student spent less than three hours studying during the weekend.

(b) Estimate the probability that a student studied for two or three hours.

(c) Estimate the probability that a student studied less than six hours.

**9.** In 2017, a survey of 4318 households in the UK gave the results shown.

(a) A household is randomly chosen from the survey.
Find the probability that:

(i) there is only one person in the household

(ii) there are at least four people in the household.

| Number of persons in household (all ages) | Frequency |
|---|---|
| 1 | 1221 |
| 2 | 1439 |
| 3 | 698 |
| 4 | 611 |
| 5 | 262 |
| 6 or more | 87 |

(b) There were 27.2 million households in the UK in 2017. Assuming that the survey is representative of all UK households, find the expected number of UK households with:

(i) less than three persons      (ii) three or four persons.

**10.** Genevieve bought a biased dice. She rolled the dice 60 times and recorded the results in this frequency table:

| Score | 1 | 2 | 3 | 4 | 5 | 6 |
|---|---|---|---|---|---|---|
| Frequency | 4 | 18 | 3 | 15 | 6 | 14 |

(a) For this dice, find the relative frequency of scoring:

(i) a 3                          (ii) an even number.

(b) Genevieve rolls the dice 50 more times.
Find the expected number of times she scores a 6.

**11.** A hospital monitors the effectiveness of two different drugs used to treat a particular condition. They record the number of patients observed to either make an improvement or not after taking the drug.

|  | Drug A | Drug B |
|---|---|---|
| **No improvement** | 12 | 7 |
| **Improvement** | 75 | 62 |

(a) Find the relative frequency of improvement for each drug.

(b) Based on this trial, state which drug appears to be more effective.

(c) In a second trial, 200 further patients are treated with drug A. Find the expected number of patients who make an improvement.

**12.** An estimate of the world population in 2018 is given below. Numbers are in millions.

| Age | Male | Female | Total |
|---|---|---|---|
| **14 years or younger** | 964 | 899 | 1863 |
| **15–24** | 611 | 572 | 1183 |
| **25–54** | 1523 | 1488 | 3011 |
| **55–64** | 307 | 322 | 629 |
| **65 and over** | 284 | 352 | 636 |
| **Total** | 3689 | 3633 | 7322 |

A person is chosen at random. Find the probability that this person is:

(a) a female                    (b) a 15–24 year old male

(c) at least 65 years old.

**13.** The table shows the number of people who work at home and are over the age of 15 in Thailand.

|  | Male | Female |
|---|---|---|
| **Municipal area** | 39 943 | 75 892 |
| **Non-municipal area** | 62 782 | 261 634 |

A person is selected at random. Find the probability that the person is:

(a) male

(b) female and lives in a non-municipal area.

**14.** In 2016, a survey in Thailand recorded the number of people aged 6 or over using the internet by region. The results are shown in the table.

| Region | Population aged 6 or over | Number using internet |
|---|---|---|
| **Bangkok** | 10 801 | 7474 |
| **Central** | 24 227 | 12 840 |
| **North** | 14 391 | 5958 |
| **Northeast** | 23 243 | 8367 |
| **South** | 11 218 | 5194 |
| **Whole kingdom** | 83 880 | 39 833 |

(a) Calculate the relative frequency of internet users for each region and for the whole kingdom.

(b) Bangkok is the capital of Thailand. Give a reason why it has the highest relative frequency of internet users.

(c) In 2016 the population aged 6 or over of the whole kingdom of Thailand was 62.8 million. Find the expected number of internet users in Thailand.

15. An international health insurance company needs to estimate how much it will pay out in claims during a year, in order to calculate how much to charge customers to buy a policy. Based on past claims, they have collected the following data.

| Type of claim | Relative frequency | Average cost of claim ($) |
|---|---|---|
| Back problem | 0.002 | 10 000 |
| Respiratory infection (includes influenza) | 0.0015 | 3000 |
| Gastro or abdominal problems | 0.001 | 2500 |
| Cancer | 0.0004 | 100 000 |
| Other | 0.003 | 4000 |

(a) Find the expected cost of a claim for each customer.

(b) For each policy sold, the company needs to cover the expected cost of a claim, and in addition, cover the expenses of running the company. The company charges 60% more than the expected cost of a claim for each policy. Find the cost of buying a policy.

# 11.2 Representing the sample space

## Venn diagrams

We will now explore some additional methods of representing the sample space which can help to solve further problems in probability. We can use Venn diagrams to represent probability. We will use a rectangle to represent the sample space and closed curves to represent events. Figure 11.3 shows two events $A$ and $B$ and the sample space $U$.

For example, suppose event $A$ is rolling an odd number on a regular dice and event $B$ is rolling a prime number. We can represent the six outcomes in the sample space in a Venn diagram, as shown in Figure 11.4.

The outcomes 3 and 5 are both odd numbers and prime numbers so they belong to both the events $A$ and $B$.

In Figure 11.4 we see that $A \cap B = \{3, 5\}$

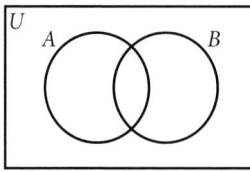

**Figure 11.3** A Venn diagram for two events $A$ and $B$

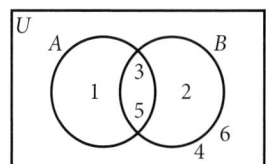

**Figure 11.4** A Venn diagram showing the outcomes of events $A$ and $B$

The intersection of two events $A$ and $B$, denoted by $A \cap B$, is the event containing all outcomes common to $A$ and $B$.

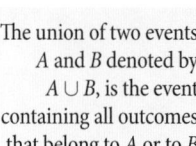

The union of two events $A$ and $B$ denoted by $A \cup B$, is the event containing all outcomes that belong to $A$ or to $B$ or to both.

If we want to find the probability of rolling a number that is both odd and prime, then we want to find the probability of both $A$ and $B$ occurring. We can calculate this using

$$P(A \text{ and } B) = P(A \cap B) = \frac{n(A \cap B)}{n(U)} = \frac{2}{6} = \frac{1}{3}$$

If we want to find the probability of rolling a number that is odd or prime, then we want to find the probability of $A$ or $B$ occurring. We cannot use the rule for the addition of mutually exclusive events, because we can see from the Venn diagram that events $A$ and $B$ are not mutually exclusive. They have the outcomes 3 and 5 in common. To find this answer, we need to define another operation, called union.

In Figure 11.4 we see that $A \cup B = \{1, 2, 3, 5\}$

One approach to finding the probability of $A$ or $B$ occurring is to use

$$P(A \text{ or } B) = P(A \cup B) = \frac{n(A \cup B)}{n(U)} = \frac{4}{6} = \frac{2}{3}$$

An alternative approach is to recognise from the Venn diagram that if we count the outcomes in $A$ and the outcomes in $B$, then we have counted the outcomes in the intersection twice. From this we see that:

$$n(A \cup B) = n(A) + n(B) - n(A \cap B)$$

Using this, we can derive a further probability rule for combined events:

$$P(A \cup B) = \frac{n(A \cup B)}{n(U)} = \frac{n(A) + n(B) - n(A \cap B)}{n(U)}$$

$$= \frac{n(A)}{n(U)} + \frac{n(B)}{n(U)} - \frac{n(A \cap B)}{n(U)}$$

$$= P(A) + P(B) - P(A \cap B)$$

For any two events $A$ and $B$,
$P(A \text{ or } B) = P(A \cup B)$
$= P(A) + P(B) - P(A \cap B)$

Note that this general rule for combined events also applies to the case of mutually exclusive events. Suppose events $A$ and $B$ are mutually exclusive, so that they have no outcomes in common.

Then $n(A \cap B) = 0$ and therefore $P(A \cap B) = 0$

Then this simplifies to:

$$P(A \cup B) = P(A) + P(B) - P(A \cap B)$$

$$= P(A) + P(B) - 0$$

$$= P(A) + P(B)$$

If events $A$ and $B$ are mutually exclusive, then $P(A \cap B) = 0$

which is the same as the rule for the addition of mutually exclusive events.

### Example 11.6

In a group of 80 students, 62 students play basketball, 47 students play football and 11 students play neither sport. Find the probability that a student selected at random from the class plays:

(a) both basketball and football

(b) only basketball.

## Solution

Let $B$ and $F$ be the events that a student plays basketball and football respectively. We can represent this information using a Venn diagram.

Since 11 students play neither sport, we can deduce that
$n(B \cup F) = 80 - 11 = 69$

Using $n(B \cup F) = n(B) + n(F) - n(B \cap F)$ gives
$$69 = 62 + 47 - n(B \cap F)$$

So $n(B \cap F) = 109 - 69 = 40$

We can now draw a Venn diagram.

Using the Venn diagram gives

(a) $P(B \cap F) = \dfrac{40}{80} = \dfrac{1}{2}$

(b) $P(B \cap F') = \dfrac{22}{80} = \dfrac{11}{40}$

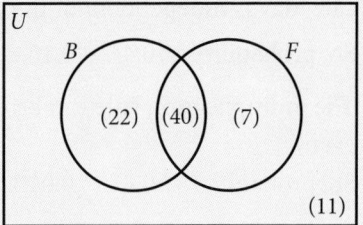

When using Venn diagrams you can either list the actual outcomes in each part of the diagram, or you can indicate the number of outcomes, as shown in Example 11.6.

In some problems, it is useful to apply the multiplication rule for probabilities.

Consider the following situation. In a large school, 55% of the students are male. It is also known that the percentage of musicians among males and females in this school are the same, 22%. A male student was selected to perform a solo in a concert, but another student complained that this decision was biased due to gender. What is the probability of selecting a student at random from this school who is a male musician?

We can think of the problem in the following manner. Since the proportion of musicians is the same in both groups, music and gender are independent of each other in the sense that knowing that the student is a male does not influence the probability that he is a musician.

The chance we pick a male student is 55%. From those 55% of the population, we know that 22% are musicians, so by simple arithmetic the probability that we select a male musician is $0.22 \times 0.55 = 0.121$. By comparison, the probability that we pick a female student is 45% and from those we know 22% are musicians. So the chance of selecting a female musician is $0.22 \times 0.45 = 0.099$. Comparing the two results, we can see that if a student was randomly selected to perform the solo at the concert, then there is a greater chance of selecting a male.

These are examples of the multiplication rule for independent events.

Two events $A$ and $B$ are independent if knowing that one of them occurs does not change the probability that the other occurs.

The multiplication rule for independent events states that if two events $A$ and $B$ are independent, then $P(A \cap B) = P(A)P(B)$

419

## Example 11.7

Every day, the probability that Andrew cycles to school is 0.4

Find the probability that Andrew cycles to school:

(a) two days in a row

(b) five days in a row.

### Solution

(a) Since the probability that Andrew cycles to school is always 0.4, each day is independent of the previous one.

So, probability $= 0.4^2 = 0.16$

(b) The multiplication rule can be extended to more than two independent events.

So, probability $= 0.4^5 = 0.01024$

## Example 11.8

Computers sometimes require repairs. It is estimated that 17% of one brand of computers require one repair during the first year after purchase, 7% will need repairs twice during the first year, and 4% require three or more repairs.

(a) Find the probability that a computer chosen at random from this company will need:

(i) some repair during the first year after its purchase

(ii) no repairs during the first year after its purchase

(iii) no more than one repair during the first year after its purchase.

(b) Anjali buys two of these computers. Find the probability that:

(i) neither will require repair during the first year after purchase

(ii) both will need repair during the first year after purchase.

### Solution

(a) Since all of the events listed are mutually exclusive, we can use the addition rule.

(i) P(some repairs) = P(one or two or three or more repairs)
$$= 0.17 + 0.07 + 0.04 = 0.28$$

(ii) P(no repairs) = $1 -$ P(some repairs) $= 1 - 0.28 = 0.72$

(iii) P(no more than one repair) = P(no repairs or one repair)
$$= 0.72 + 0.17 = 0.89$$

(b) Since repairs on the two computers are independent of each other, we can use the multiplication rule and the answers from part (a).

(i) P(neither will need repair) $= (0.72)(0.72) = 0.5184$

(ii) P(both will need repair) $= (0.28)(0.28) = 0.0784$

## Tree diagrams

To solve problems involving independent events, it is often helpful to draw a tree diagram. In a tree diagram, the possible outcomes of each event are shown as branches. For example, suppose event $A$ has three outcomes and event $B$ has two outcomes. Then we can represent the sample space using a tree diagram, as shown in Figure 11.5.

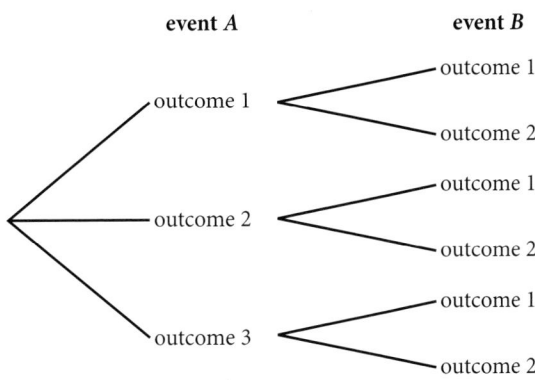

**Figure 11.5** A tree diagram representing the outcomes of two events $A$ and $B$

The probability of each outcome can then be written on its branch.

### Example 11.9

Peter applies for a new car insurance policy. His insurance company estimates that the chance of him having a serious accident is 0.002 and the chance of a minor accident is 0.009. Dan also applies for a new car insurance policy, and his company estimates that the chance of him having an accident is 0.006. Find the probability that:

(a) Peter has a serious accident and Dan has an accident

(b) exactly one of them has an accident.

#### Solution

We can represent the sample space using the tree diagram, where:

$S$ = serious accident

$M$ = minor accident

$A$ = accident and no accident

$A'$ = no accident

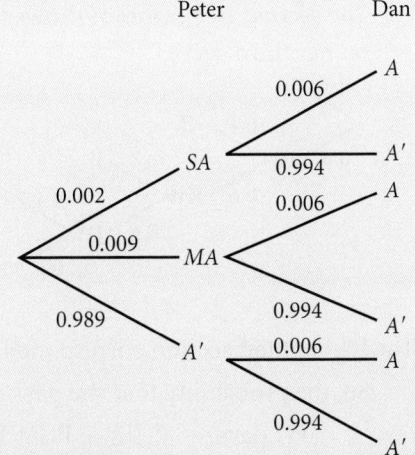

(a) P(Peter has a serious accident and Dan has an accident)

$= P(SA, A)$

$= (0.002)(0.006) = 0.000012$

(b) P(Exactly one of them has an accident)

$= P(SA, A') + P(MA, A') + P(A', A)$

$= (0.002)(0.994) + (0.009)(0.994) + (0.989)(0.006) = 0.0169$

### Example 11.10

Jane and Kate frequently play tennis together. Whoever serves first wins 60% of the time. They play by alternating the serve and usually play for a prize, which is a chocolate bar. The first one who loses on her serve will have to buy the chocolate. Jane serves first.

(a) Find the probability that Jane pays on her second serve.

(b) Find the probability that Jane pays for the chocolate.

(c) Find the probability that Kate pays for the chocolate.

### Solution

A tree diagram can help in solving this problem. Let JW stand for Jane winning her serve, JL for Jane losing her serve and hence paying for the chocolate. KW and KL are defined similarly.

The highlighted section corresponds to the branches followed in part (a).

(a) For Jane to pay on her second serve, she should win her first serve; Kate must also win her first serve, and then Jane loses her second serve (see the tree diagram). The probability that this happens is:

$$P(JW) \cdot P(KW) \cdot P(JL) = 0.6 \times 0.6 \times 0.4 = 0.4 \times (0.6)^2 = 0.144$$

(b) For Jane to pay, she needs to be the first one to lose on her serve. This means she loses on her first serve or her second or her third, and so on. The second tree diagram shows the sequence of events for Jane to lose on her third serve.

The highlighted section corresponds to the branches followed in part (b).

So, the probability that she pays is

$$P(J \text{ pays}) = P(JL) + P(JW) \times P(KW) \times P(JL) + P(JW) \times$$
$$P(KW) \times P(JW) \times P(KW) \times P(JL) + \cdots$$
$$= 0.4 + 0.4 \times (0.6)^2 + 0.4 \times (0.6)^4 + \cdots$$

This appears to be the sum of an infinite geometric series with $(0.6)^2$ as common ratio, hence

$$P(\text{J pays}) = 0.4 + 0.4 \times (0.6)^2 + 0.4 \times (0.6)^4 + \cdots = \frac{0.4}{1 - (0.6)^2}$$
$$= 0.625$$

(c) $P(\text{K pays}) = 1 - P(\text{J pays}) = 1 - 0.625 = 0.375$

This also gives us the opportunity to look at it differently. For Kate to pay, she needs to lose on her first serve, i.e., $0.6 \times 0.4$, or on her second, third etc.

$$P(\text{K pays}) = 0.6 \times 0.4 + 0.6 \times (0.6)^2 \times 0.4 + 0.6 \times (0.6)^4 \times 0.4$$
$$+ \cdots = \frac{0.6 \times 0.4}{1 - (0.6)^2} = 0.375$$

1. Events $A$ and $B$ are such that $P(A) = \frac{3}{4}$, $P(A \cup B) = \frac{4}{5}$ and $P(A \cap B) = \frac{3}{10}$. Find $P(B)$.

2. Events $A$ and $B$ are such that
   $P(A) = \frac{7}{10}$, $P(A \cup B) = \frac{9}{10}$ and $P(A \cap B) = \frac{3}{10}$
   Find:
   (a) $P(B)$      (b) $P(A' \cap B)$      (c) $P(A \cup B')$
   (d) $P(A' \cap B')$      (e) $P(A' \cup B')$

3. Events $A$ and $B$ are such that $P(A) = \frac{1}{3}$ $P(A \cup B) = \frac{4}{9}$ and $P(B) = \frac{2}{9}$.
   Show that $A$ and $B$ are neither independent nor mutually exclusive.

4. Events $A$ and $B$ are such that $P(A) = \frac{3}{7}$ and $P(A \cap B) = \frac{3}{10}$.
   If $A$ and $B$ are independent, find $P(A \cup B)$

5. In a certain city, the percentage of new drivers passing their driving test the first time is 60%. If a driver fails the first test, there is a chance of passing it on a second try, two weeks later. 75% of the second-chance drivers pass the test. Otherwise, the driver has to retrain and take the test after 6 months. Find the probability that a randomly chosen new driver will pass the test without needing to retrain.

6. People with O-negative type blood are universal donors. i.e., they can donate blood to individuals with any blood type. Only 8% of people have O-negative type blood.
   (a) Write down the probability that a randomly chosen person does not have O-negative type blood.
   (b) Two people arrive independently at a donor centre to give blood. Find the probability that:
      (i) both have O-negative type blood
      (ii) at least one of them has O-negative type blood
      (iii) only one of them has O-negative type blood.
   (c) Eight people arrive independently to give blood. Find the probability that at least one of them has O-negative type blood.

7. PINs for smartphones usually consist of four digits (0 to 9 inclusive) that are not necessarily different.

   (a) Find the total number of possible PINs.

   (b) Find the probability that a PIN chosen at random does not start with a zero.

   (c) Find the probability that a PIN contains at least one zero.

8. An urn contains six red balls and two blue balls. A ball is chosen at random and its colour is recorded. The ball is replaced, and a second ball is randomly chosen. Find the probability that exactly one of the balls is red.

9. Two dice are rolled and the numbers on the top face are observed.

   (a) Find the probability that at least one dice shows a 6.

   (b) Find the probability that the sum of the two numbers is at most 10.

   (c) Find the probability that a dice shows 4 or the sum is 10.

10. The table shows the number of students in a large school categorised by grade and gender.

| Gender \ Grade | Grade 9 | Grade 10 | Grade 11 | Grade 12 | Total |
|---|---|---|---|---|---|
| Male | 180 | 170 | 230 | 220 | 800 |
| Female | 200 | 130 | 190 | 180 | 700 |

   (a) Find the probability that a student chosen at random from the school:

   (i)   is a female

   (ii)  is a male grade 12 student

   (iii) is a grade 12 student

   (iv)  is a grade 12 or female student.

   (b) Find the probability that a grade 12 student chosen at random is a male.

   (c) Are gender and grade independent of each other at this school? Explain your answer.

11. Some young people do not like to wear glasses. A survey considered a large number of students as to whether they needed glasses to correct their vision and whether they used the glasses when they needed to. The table shows the results.

| | | Used glasses when needed | |
|---|---|---|---|
| | | Yes | No |
| Need glasses for correct vision | Yes | 0.41 | 0.15 |
| | No | 0.04 | 0.40 |

Find the probability that a randomly chosen person from this group:

    **(i)** needs glasses to correct their vision

    **(ii)** needs glasses but does not use them.

**(b)** A student is chosen at random from those that need glasses. Find the probability that the student does not use them.

**(c)** Explain whether the events of using and needing glasses are independent.

**12.** In a large graduating class there are 100 students taking the IB examination. 60 students are taking Mathematics SL, 35 students are taking Physics SL, and 12 are taking both.

**(a)** A student is chosen at random. Find the probability that this student is taking neither Mathematics SL nor Physics SL.

**(b)** Determine whether taking Physics SL and Mathematics SL are independent.

**13.** A shop in Germany accepts only Mastercard and Visa. It estimates that 21% of its customers use Mastercard, 57% use Visa, and 13% use both cards.

**(a)** Find the probability that a customer uses a Mastercard or Visa card.

**(b)** Find the probability that a customer uses neither card.

**(c)** Find the probability that a customer uses exactly one of these cards.

**14.** 132 of 300 patients at a hospital are signed up for a special exercise program, which consists of a swimming class and an aerobics class. Each of these 132 patients takes at least one of the two classes. There are 78 patients in the swimming class and 84 in the aerobics class. Find the probability that a randomly chosen patient at this hospital is:

**(a)** not in the exercise program     **(b)** enrolled in both classes.

**15.** An ordinary unbiased 6-sided dice is rolled three times. Find the probability of rolling:

**(a)** three 2s                     **(b)** at least one 2

**(c)** exactly one 2.

**16.** An athlete is shooting arrows at a target. She has a record of hitting the centre 30% of the time. Find the probability that she hits the centre:

**(a)** for the first time with her second shot

**(b)** exactly once with her first three shots

**(c)** at least once with her first three shots.

17. Two unbiased dodecahedral dice with faces numbered 1–12 are rolled. The score is the number on the top face. Find the probability that:

    (a) at least one dice scores 12      (b) the sum of the scores is 12

    (c) the sum of the scores is at least 20.

    Two events are defined as:

    $A = \{$at least one dice scores 10$\}$

    $B = \{$the sum of the scores is at most 15$\}$

    (d) Find the probability of each of the following events:

        (i) $A \cap B$      (ii) $A \cup B$      (iii) $(A \cap B)'$      (iv) $A' \cap B'$

18. George and Kassanthra play a game in which they take turns rolling two unbiased 6-sided dice. The first one who rolls a sum of 6 wins. Kassanthra rolls the dice first. Find the probability that:

    (a) Kassanthra wins on her second roll

    (b) George wins on his second roll

    (c) Kassanthra wins.

19. In 2017, a survey in the UK asked people where they watched films on any device. The results are shown in the table.

    | Location | Bedroom | Kitchen | Garden | Bathroom |
    | --- | --- | --- | --- | --- |
    | Percentage | 51 | 16 | 9 | 9 |

    (a) Two people are selected at random. Find the probability that:

        (i) both watch films in their bedroom

        (ii) at least one of them watches films in their bedroom.

    (b) In 2017, the UK population was 66.2 million. Find the expected number of people who watch films in the bathroom.

20. The results of people taking their driving test in Singapore, from July 2017 to June 2018, are shown in the table.

    | Driving centre | Number taking test | Number passing test |
    | --- | --- | --- |
    | Singapore Safety Driving Centre | 3678 | 1947 |
    | Bukit Batok Driving Centre | 6418 | 3846 |
    | Comfort Driving Centre | 7345 | 3931 |

    (a) Find which driving centre has the highest pass rate.

    Two candidates arrive independently at the Singapore Safety Driving Centre.

    (b) Find the probability that:

        (i) they both pass the test

        (ii) at least one of them passes the test.

Selvan takes his test at the Bukit Batok Driving Centre and Pauline takes her test at the Comfort Driving Centre.

(c) Find the probability that:

   (i) they both pass the test      (ii) exactly one of them passes the test.

In the following year, 18 200 people are anticipated to take their driving test at one of these three centres.

(d) Find the number of people expected to pass their test.

21. A rabbit has two genes for fur colour, one from its father and one from its mother. These genes can be the same or different. If the genes are different, then the dominant gene determines the fur colour. The gene for black fur, B, is dominant over the gene for brown fur, b. This means that if a rabbit has genes Bb, then it will have black fur. Parents pass one of their genes to their children at random.

(a) A male rabbit has genes Bb, and a female rabbit has genes Bb. The possible genes of their kittens (children) are shown in the table (Punnett square).

|         |   | Father |    |
|---------|---|--------|----|
|         |   | B      | b  |
| Mother  | B | BB     | Bb |
|         | b | Bb     | bb |

   Find the probability that the kitten has:

   (i) brown fur                    (ii) one gene of each type.

(b) The rabbits in part (a) have 20 kittens. Find the expected number with black fur.

(c) Another male rabbit has genes BB, and another female rabbit has genes Bb. Find the probability that their kitten has:

   (i) brown fur                    (ii) one gene of each type.

A rabbit also has two genes for fur thickness. It can have thick fur, T, or thin fur, t. The gene for thick fur is dominant over the gene for thin fur. The genes for fur thickness are independent of the genes for fur colour.

(d) A male rabbit has genes BbTt, and a female rabbit has genes BbTt. Find the probability that their kitten has:

   (i) thick brown fur       (ii) thin black fur

   (iii) thick black fur.

(e) The rabbits in part (d) have 32 kittens. Find the expected number with thin brown fur.

(f) A rabbit breeder has 80 rabbits. 62 have black fur and 55 have thick fur. 4 rabbits have thin, brown fur. A rabbit is selected at random. Find the probability that it has:

   (i) thick black fur       (ii) thin black fur.

## 11.3 Conditional probability

In probability, conditioning means incorporating new restrictions on the outcome of an experiment: updating probabilities to take into account new information. This section describes conditioning, and how conditional probability can be used to solve complicated problems.

### Example 11.11

A public health department is concerned about the adverse health effects of eating sugar. They interviewed 768 students from grades 10–12 and asked them what they had eaten the previous day. They categorised the students into the following three categories: high sugar consumers, moderate sugar consumers and low sugar consumers. The results are summarised in the table.

| | High sugar consumer | Moderate sugar consumer | Low sugar consumer | Total |
|---|---|---|---|---|
| **Male** | 127 | 73 | 214 | 414 |
| **Female** | 99 | 66 | 189 | 354 |
| **Total** | 226 | 139 | 403 | 768 |

A student is selected at random from this study. Find the probability that the student is

(a) a female     (b) a male high sugar consumer     (c) a low sugar consumer.

#### Solution

(a) $P(\text{female}) = \frac{354}{768} = 0.461$

So, 46.1% of the sample are females.

(b) There are 127 males categorised as high sugar consumers, so the chance of selecting a male high sugar consumer is

$P(\text{male high sugar consumer}) = \frac{127}{768} = 0.165$

(c) There are a total of 403 low sugar consumers, so the probability is

$P(\text{low sugar consumer}) = \frac{403}{768} = 0.525$

In Example 11.11, suppose we know that the selected student is a female. Does that influence the probability that the selected student is a low sugar consumer? Yes, it does.

Knowing that the selected student is a female changes our choices. The revised sample space is not made up of all students anymore. It is only the female students. The chance of finding a low sugar consumer among the females is $\frac{189}{354} = 0.534$. In other words, 53.4% of the females are low sugar consumers as compared to the 52.5% of low sugar consumers in the whole population.

This probability is called a conditional probability, we write this as

$$P(\text{low sugar consumer}|\text{female}) = \frac{189}{354}$$

We read this as probability of selecting a low sugar consumer given that we have selected a female.

The conditional probability of $A$ given $B$, $P(A|B)$, is the probability of event $A$, updated on the basis of the knowledge that event $B$ occurred. Suppose that $A$ is an event with probability $P(A) = p \neq 0$, and that $A \cap B = \varnothing$ (i.e. $A$ and $B$ are mutually exclusive). If we learn that $B$ occurred, we know $A$ did not occur, so we should revise the probability of $A$ to be zero. So $P(A|B) = 0$ (the conditional probability of $A$ given $B$ is zero).

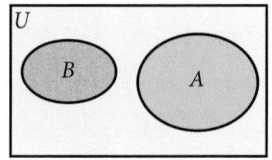

**Figure 11.6** A Venn diagram representing a mutually exclusive event

On the other hand, suppose that $A \cap B = B$ ($B$ is a subset of $A$, so $B$ implies $A$). Then if we learn that $B$ occurred, we know $A$ must have occurred as well, so we should revise the probability of $A$ to be 100%, $P(A|B) = 1$ (the conditional probability of $A$ given $B$ is 100%).

Remember that the probability we assign to an event can change if we know that some other event has occurred. This idea is the key to understanding conditional probability.

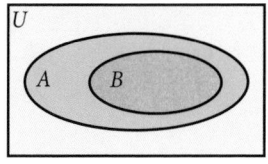

**Figure 11.7** A Venn diagram representing $A \cap B = B$

Imagine the following scenario.

You are playing cards and your opponent is about to give you a card. What is the probability that the card you receive is a queen?

There are 52 cards in the deck, 4 of which are queens. So, assuming that the deck was thoroughly shuffled, the probability of receiving a queen is

$$P(\text{queen}) = \frac{4}{52} = \frac{1}{13}$$

This calculation assumes that you know nothing about any cards already dealt from the deck.

Suppose now that you are looking at the five cards you have in your hand, and one of them is a queen. You know nothing about the other 47 cards except that exactly three queens are among them. The probability of being given a queen as the next card, given what you know is

$$P(\text{queen}|1 \text{ queen in hand}) = \frac{3}{47} \neq \frac{1}{13}$$

So, knowing that there is one queen among your five cards changes the probability of the next card being a queen.

Consider Example 11.11 again. We can express the original table of frequencies as relative frequencies or probabilities (Table 11.2).

| | High sugar consumer | Moderate sugar consumer | Low sugar consumer | Total |
|---|---|---|---|---|
| Male | 0.165 | 0.095 | 0.279 | 0.539 |
| Female | 0.129 | 0.086 | 0.246 | 0.461 |
| Total | 0.294 | 0.181 | 0.525 | 1.000 |

**Table 11.2** Relative frequencies of sugar consumption

To find the probability of selecting a student at random and finding that student is a low sugar consumer female, we look at the intersection of the female row with the non-smoking column and find that this probability is 0.246.

Looking at this calculation from a different perspective, we can think about it in the following way.

We know that the percentage of females in our sample is 46.1, and among those females, we found that 53.4% of those are low sugar consumer. So, the percentage of female low sugar consumer in the population is the 53.4% of those 46.1% females, i.e., $(0.534)(0.461) = 0.246$

In terms of events, this can be read as:

P(low sugar consumer|female) × P(female) = P(female and low sugar consumer)
$$= P(\text{female} \cap \text{low sugar consumer})$$

This is an example of the multiplication rule of any two events $A$ and $B$:

Given any events $A$ and $B$, the probability that both events happen is given by
$$P(A \cap B) = P(B)\,P(A|B)$$

It is helpful to visualise this formula using a tree diagram. Figure 11.8 shows that the probability of event $A$ occurring or not, depends upon whether or not event $B$ occurs. To find the probability of both $A$ and $B$ occurring, we multiply the probabilities along the relevant branches, i.e. $P(A \cap B) = P(B)\,P(A|B)$

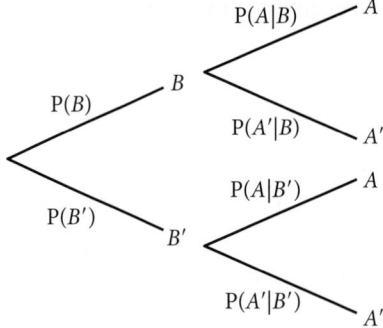

**Figure 11.8** A tree diagram illustrating conditional probabilities

If $P(A \cap B) = P(B)\,P(A|B)$, and if $P(B) \neq 0$, we can rearrange the multiplication rule to produce a definition of the conditional probability $P(A|B)$ in terms of the 'unconditional' probabilities $P(A \cap B)$ and $P(B)$.

When $P(B) \neq 0$, the conditional probability of $A$ given $B$ is:
$$P(A|B) = \frac{P(A \cap B)}{P(B)}$$

Why does this formula make sense?

First of all, note that it does agree with the intuitive answers we found previously.

If $A \cap B = \emptyset$, then $P(A \cap B) = 0$, so $P(A|B) = \dfrac{0}{P(B)} = 0$

and if $A \cap B = B$, $P(A|B) = \dfrac{P(B)}{P(B)} = 1$

Now, if we learn that $B$ occurred, we can restrict our attention to just those outcomes that are in $B$, and disregard the rest of $U$, so we have a new sample space that is just $B$ (Figure 11.9).

For $A$ to have occurred in addition to $B$ requires that $A \cap B$ occurred, so the conditional probability of $A$ given $B$ is $\dfrac{P(A \cap B)}{P(B)}$, just as we defined it above.

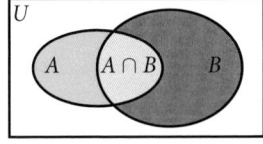

**Figure 11.9** A Venn diagram illustrating conditional probability

## Example 11.12

In an experiment to study the phenomenon of colour-blindness, researchers collected information concerning 1000 people in a small town and categorised them according to colour-blindness and gender. The table shows a summary of the findings.

|  | Male | Female | Total |
|---|---|---|---|
| **Colour-blind** | 40 | 2 | 42 |
| **Not colour-blind** | 470 | 488 | 958 |
| **Total** | 510 | 490 | 1000 |

Find the probability that a person is colour-blind, given that the person is a female.

### Solution

To answer this question, we notice that we do not have to search the whole population for this event. We limit our search to the 490 females. As we only need to consider females, then when we search for colour-blindness, we only look for the females who are colour-blind, i.e., the intersection. Here there are only two females. Therefore the chance we get a colour-blind person, given the person is a female is:

$$P(C|F) = \frac{P(C \cap F)}{P(F)} = \frac{n(C \cap F)}{n(F)} = \frac{2}{490} = 0.004$$

where $C$ is for colour-blind and $F$ is for female.

Notice that in Example 11.12 we used the frequency rather than the probability. However these are equivalent since dividing by $n(S)$ will transform the frequency into a probability.

$$\frac{n(C \cap F)}{n(F)} = \frac{\dfrac{n(C \cap F)}{n(S)}}{\dfrac{n(F)}{n(S)}} = \frac{P(C \cap F)}{P(F)} = P(C|F)$$

### Example 11.13

A box contains 14 dark and 6 milk chocolates. Sarah randomly selects three chocolates to eat. Find the probability that Sarah selects:

(a) at least one milk chocolate

(b) three milk chocolates, given that she selects at least one milk chocolate.

### Solution

(a) Find the probability of selecting no milk chocolates and subtract from 1.

$$P(\text{at least one milk chocolate}) = 1 - \frac{14}{20} \cdot \frac{13}{19} \cdot \frac{12}{18} = \frac{194}{285}$$

In the IB exam, probabilities can be with or without replacement.

(b) $$P(3 \text{ milk} \mid \text{at least one milk}) = \frac{\dfrac{6}{20} \cdot \dfrac{5}{19} \cdot \dfrac{4}{18}}{\dfrac{194}{285}} = \frac{5}{194}$$

### Example 11.14

A life insurance company estimates the probability that an applicant will die in the next ten years, in order to determine the cost of a life insurance policy. David applies for a policy, and based on the details in his application, the company estimates the chance of him developing heart disease is 0.03 and the chance of him having a stroke is 0.02. Based on their previous records, the company estimates that if an individual develops heart disease, the chance of dying is 0.6; if an individual has a stroke, the chance of dying is 0.45 and the chance of dying from another cause is 0.015.

Find the probability that in the next ten years

(a) David dies

(b) David developed heart disease, given that he dies

(c) David had a stroke, given that he dies.

### Solution

(a) We can represent this problem using a tree diagram, where:

$H$ = heart disease

$S$ = stroke

$D$ = death

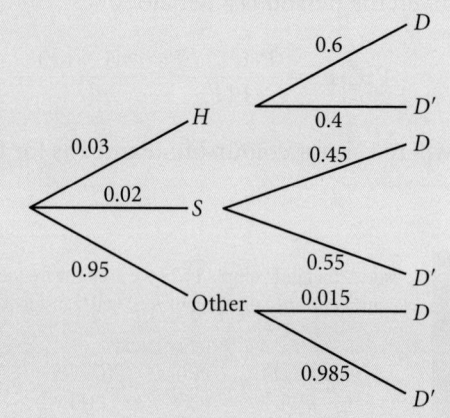

We can calculate the probability that David dies using

$$P(D) = P(H, D) + P(S, D) + P(Other, D)$$
$$= (0.03)(0.6) + (0.02)(0.45) + (0.95)(0.015)$$
$$= 0.04125$$

(b) We can write the probability that David developed heart disease, given that he dies as $P(H|D)$. Using the rule for conditional probability, we can calculate this using:

$$P(H|D) = \frac{P(H \cap D)}{P(D)}$$

Using the tree diagram and the result from part (a):

$$= \frac{(0.03)(0.6)}{0.04125}$$
$$= 0.436$$

(c) Similarly, we can calculate the probability that David had a stroke, given that he dies, using

$$P(S|D) = \frac{P(S \cap D)}{P(D)}$$

Using the tree diagram and the result from part (a):

$$= \frac{(0.02)(0.45)}{0.04125}$$
$$= 0.218$$

## Example 11.15

For a particular airline, the probability that a regularly scheduled flight departs on time is $P(D) = 0.83$, the probability that it arrives on time is $P(A) = 0.92$, and the probability that it arrives and departs on time is $P(A \cap D) = 0.78$. Find the probability that a flight:

(a) arrives on time given that it departed on time

(b) departs on time given that it arrived on time.

### Solution

(a) The probability that a flight arrives on time given that it departed on time is:

$$P(A|D) = \frac{P(A \cap D)}{P(D)} = \frac{0.78}{0.83} = 0.94$$

(b) The probability that a flight departs on time given that it arrived on time is:

$$P(D|A) = \frac{P(D \cap A)}{P(A)} = \frac{0.78}{0.92} = 0.85$$

## Example 11.16

Humans and chimpanzees are closely related genetically. Chimpanzees can be susceptible to catching human diseases. Research has shown that some chimpanzees in Gombe Stream National Park in Tanzania have become infected with a parasite common to humans that causes respiratory disease. Scientists studying the chimpanzees have developed a test to help determine whether a chimpanzee is infected with the parasite. The test is 98% accurate, which means it gives a correct result 98% of the time, and an incorrect result 2% of the time. There is evidence to indicate that 5% of the chimpanzees in Gombe Stream National Park are infected with the parasite.

(a) Find the probability that a randomly chosen chimpanzee in the park:

    (i)    tests positive for the parasite

    (ii)    tests negative for the parasite

    (iii)   is infected with the parasite, given that it tests negative for the parasite

    (iv)   is not infected with the parasite, given that it tests positive for the parasite.

(b) The result in part (a) (iii) is called a false negative, because the test did not correctly identify the chimpanzee as having the parasite. Comment on the significance of this result.

(c) The result in part (a) (iv) is called a false positive, because the test incorrectly identified a healthy chimpanzee as having the parasite. Comment on the significance of this result.

## Solution

We can represent this problem using a tree diagram.

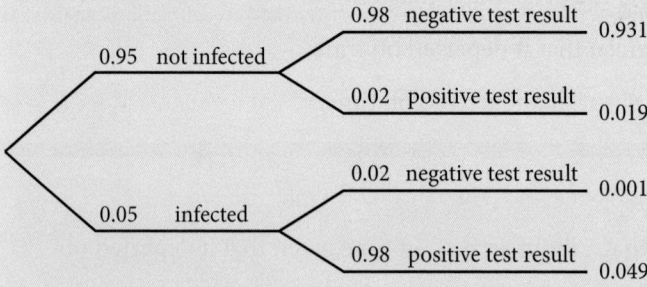

(a) (i)    $P(\text{positive test result}) = 0.049 + 0.019 = 0.068$

    (ii)   $P(\text{negative test result}) = 0.931 + 0.001 = 0.932$

    (iii) $P(\text{infected}|\text{negative test result}) = \dfrac{0.001}{0.932} = 0.00107$

    (iv) $P(\text{not infected}|\text{positive test result}) = \dfrac{0.019}{0.068} = 0.279$

(b) The false negative result of 0.00107 is a very small probability. This is a good thing, because if the test does not identify an infected chimpanzee, then it will not receive medical treatment, which could lead to its condition worsening.

(c) The false positive result of 0.279 is quite a significant probability. It means that around 28 out of every 100 chimpanzees that test positive are not actually infected with the parasite. If the treatment for the infection is expensive, then it would be a costly waste to treat chimpanzees that are not infected. The treatment might have negative side effects, so again treating chimpanzees without the infection is not desirable. Depending on the costs and/or side effects, it might be possible to develop a second test to improve the probability that a positive test result indicates that a chimpanzee actually has the infection.

## Independence

Two events are independent if learning that one occurred does not affect the chance that the other occurred. That is, when $P(A|B) = P(A)$, and vice versa.

This means that if we apply our definition to the general multiplication rule, then

$$P(A \cap B) = P(A|B)\, P(B) = P(A)\, P(B)$$

which is the multiplication rule for independent events we studied earlier.

These results provide some helpful tools in checking the independence of events:

Two events are independent if and only if either
$$P(A \cap B) = P(A)\, P(B), \text{ or } P(A|B) = P(A)$$

### Example 11.17

Take another look at Example 11.15. Are the events of arriving on time $(A)$ and departing on time $(D)$ independent?

#### Solution

We can answer this question in two different ways:

Given that $P(A) = 0.92$, and we found that $P(A|D) = 0.94$. Since the two values are not the same, then we can say that the two events are not independent.

Alternatively,

$$P(A \cap D) = 0.78 \text{ and } P(A)\, P(D) = (0.92)(0.83) = 0.76 \neq P(A \cap D)$$

## Exercise 11.3

1. A chocolate manufacturer is developing new products and trials them on visitors to one of its shops. Each visitor is asked to taste three products and choose their favourite one. The results of the trials are shown in the table.

|  | Male | Female |
|---|---|---|
| **Product A** | 42 | 33 |
| **Product B** | 23 | 46 |
| **Product C** | 57 | 61 |

A visitor from the trial is chosen at random. Find the probability that the visitor:

(a) is female and prefers product B

(b) is female, given that the visitor prefers product B

(c) prefers product B, given that she is female

(d) prefers product A

(e) prefers product A, given that he is male.

2. A car manufacturer tests airbags from two different companies by simulating accidents in its factory on test dummies. The results of the tests are shown in the table.

|  | Survives with minor injury or no injury | Survives with major injury | Does not survive |
|---|---|---|---|
| **Airbag A** | 34 | 12 | 4 |
| **Airbag B** | 29 | 15 | 2 |

A test dummy is chosen at random. Find the probability that:

(a) the dummy survives with major injury

(b) the dummy does not survive, given that airbag B was tested

(c) airbag A was tested, given that the dummy survives with minor injury or no injury.

100 further tests are carried out on airbag A.

(d) Find the expected number of test dummies that survive.

3. Blood type varies by nationality. The percentage of citizens with blood types $O^+$, $A^+$, $B^+$ or other are shown in the table for Peru, Portugal and the Philippines. The population of each country (in millions) is 32, 10 and 105 respectively.

| Country | $O^+$ | $A^+$ | $B^+$ | Other |
|---|---|---|---|---|
| **Peru** | 70 | 18 | 8 | 4 |
| **Portugal** | 36 | 40 | 7 | 17 |
| **Philippines** | 37 | 29 | 27 | 7 |

(a) Find the number of people in the Philippines with blood type $A^+$

Find the probability that a randomly chosen citizen from one of the three countries:

(b) has blood type $A^+$

(c) has blood type $B^+$, given that the citizen is from the Philippines

(d) is from Peru, given that the citizen has blood type $O^+$

(e) is from Portugal, given that the citizen does not have blood type $O^+$.

4. In 1951, Solomon Asch conducted an experiment to examine the impact of group pressure on a person to conform. He asked participants to answer straightforward multiple-choice questions. Unknown to the participant, the room also contained several actors who all gave an incorrect answer before it was the participant's turn. Asch also had a control group of participants who were asked the questions without anyone else present. The results are shown in the table.

| | Correct response | Incorrect response |
|---|---|---|
| Control group | 218 | 2 |
| Peer pressure group | 408 | 192 |

A participant from the trial is chosen at random.

(a) Find the probability that they gave an incorrect response, given that they are in the control group.

(b) Find the probability that they gave an incorrect response, given that they are in the peer pressure group.

(c) Based on your answers to parts (a) and (b), explain whether or not Asch's experiment indicates that group pressure has an impact on a person to conform.

5. Two regular, unbiased six-sided dice are thrown. Given that the sum of the two numbers is 10, find the probability that both dice show a 5.

6. The owner of a coffee shop records sales each day. On a particular day there were 157 customers. 125 customers ordered coffee and 43 ordered pastry. 21 customers ordered neither coffee nor pastry.

Find the probability that a customer ordered:

(a) coffee and pastry

(b) pastry, given that they ordered coffee

(c) coffee, given that they did not order pastry.

7. In a group of 90 students, 54 students do service with young children and 47 students do service with elderly people. 15 students do service with both young children and elderly people.

A student is selected at random from the group. Find the probability that:

(a) the student does service with elderly people, given that the student does service with young children

**(b)** the student does service with young children, given that the student does not do service with elderly people.

Three students are selected at random from the group.

**(c)** Find the probability that all three students do service with young children, given that at least one student does service with young children.

8. The tree diagram shows part of the calculation a car insurance company uses to estimate the probability of a customer having an accident, based on the number of accidents they had in the previous year.

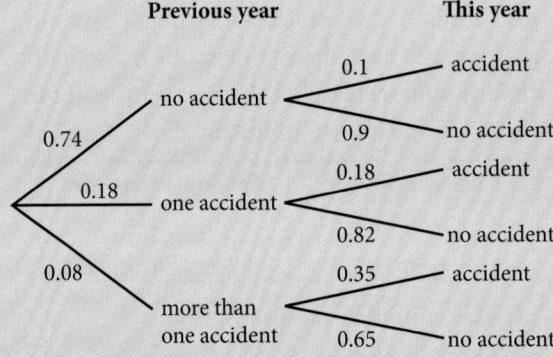

**(a)** Write down the probability that a customer:
   **(i)** has an accident, given that they had one accident the previous year
   **(ii)** has no accident, given that they had no accident the previous year.

**(b)** The insurance company has 12 000 customers. Find the expected number of customers who have an accident this year.

**(c)** Find the probability that a customer had more than one accident in the previous year, given that they have an accident this year.

9. A box contains 12 blue balls and 8 red balls.

**(a)** Ashish takes a ball at random and replaces it, then Jessica takes a ball at random. Find the probability that:
   **(i)** Jessica took a red ball, given that Ashish took a red ball
   **(ii)** at least one person took a red ball
   **(iii)** Ashish took a red ball, given that at least one person took a red ball.

**(b)** Ashish takes a ball at random and keeps it, then Jessica takes a ball. Find the probability that:
   **(i)** Jessica took a red ball, given that Ashish took a red ball
   **(ii)** at least one person took a red ball
   **(iii)** Ashish took a red ball, given that at least one person took a red ball.

**10.** In a game, a player spins a spinner which gives the result *A*, *B*, or *C* with probability 0.3, 0.5, and 0.2, respectively. The player then chooses a token at random from a bag with the matching label *A*, *B*, or *C*. Some of the tokens are prize tokens. The probability of choosing a prize token from bag *A*, *B* and *C* is 0.2, 0.1 and 0.5, respectively. Find the probability that:

(a) the player wins a prize

(b) the player chose a token from bag *A*, given that he won a prize.

**11.** Stephen lives in Nottingham. The probability of a rainy day in Nottingham is 0.27. On a rainy day, the probability that Stephen cycles to school is 0.2. Otherwise, the probability that Stephen cycles to school is 0.7. Find the probability that:

(a) Stephen cycles to school

(b) it is a rainy day in Nottingham, given that Stephen cycles to school.

**12.** An enzyme-linked immunosorbent assay (ELISA) is a test that detects and measures antibodies in your blood. An ELISA test can be used to test for HIV infection. The test is accurate 99% of the time. The USA has a population of 326 million and about 1.1 million people are known to be infected with HIV. A randomly selected person from the USA is given the ELISA test.

(a) Find the probability that they
  (i)   test positive for HIV
  (ii)  test negative for HIV
  (iii) are infected with HIV, given that they test negative for HIV
  (iv)  are not infected with HIV, given that they test positive for HIV.

(b) The result in part **(a) (iii)** is called a false negative, because the test did not correctly identify the infected person as having HIV. Comment on the significance of this result.

(c) The result in part **(a) (iv)** is called a false positive, because the test incorrectly identified a healthy person as having HIV. Comment on the significance of this result.

**13.** A significant proportion of car accidents in the Algarve region of Portugal are caused by drunk drivers. In order to reduce the number of drunk drivers, the local police randomly stop drivers and test the blood alcohol concentration (BAC) of a driver. The BAC limit for driving in Portugal is 0.5 g of alcohol in the body for every litre of blood. Police records show that, late at night, about 7% of drivers have a BAC greater than 0.5 g/litre of blood. One method the police use for testing BAC is a saliva strip test. This test is accurate 98% of the time. If a driver tests positive (i.e. the test indicates that their BAC is greater than 0.5 g/litre of blood) then the police will suspend their driving license and fine them.

(a) Find the probability that a driver stopped at random has:
  (i) a positive test result
  (ii) a negative test result
  (iii) a false negative result (i.e. their BAC is greater than 0.5 g/litre of blood, given that they test negative)
  (iv) a false positive result (i.e. their BAC is less than 0.5 g/litre of blood, given that they test positive).

During a particular week, the police test 1000 drivers.

(b) Comment on the implications of the result in part (a) (iii).
(c) Comment on the implications of the result in part (a) (iv).

14. Let $A$ and $B$ be two events with $P(A) = 0.6$ and $P(B) = 0.7$
   Find $P(A \cup B)$ given that:
   (a) $A$ and $B$ are independent events
   (b) $P(A|B) = 0.8$

15. Let $A$ and $B$ be two events with $P(A) = 0.4$, $P(B) = 0.55$
   and $P(A|B) = 0.7$ Find $P(A|B')$

16. Let $A$ and $B$ be two events with $P(A) = 0.3$, $P(B|A) = 0.5$ and $P(B|A') = 0.7$
   (a) Find $P(B)$
   (b) Find $P(A|B)$

17. Let $A$ and $B$ be two events with $P(A) = 0.6$ and $P(A \cup B) = 0.8$
   Find $P(B)$ given that:
   (a) $A$ and $B$ are independent events
   (b) $P(B|A) = 0.25$
   (c) $P(A|B') = 0.2$

## Chapter 11 practice questions

1. Two independent events $A$ and $B$ are given such that
   $P(A) = k$, $P(B) = k + 0.5$, and $P(A \cap B) = 0.14$
   (a) Find the value of $k$
   (b) Find $P(A \cup B)$
   (c) Find $P(A'|B')$

2. Two events $A$ and $B$ are such that
   $P(A|B) = 0.4$, $P(B|A) = 0.6$, and $P(A \cap B) = 0.24$
   (a) Find $P(B)$
   (b) Explain why events $A$ and $B$ are not mutually exclusive.
   (c) Determine whether $A$ and $B$ are independent.
   (d) Find $P(B \cap A')$

3. Two events $A$ and $B$ are such that $P(A) = \frac{9}{16}$, $P(B) = \frac{3}{8}$, and $P(A|B) = \frac{1}{4}$
   Find the probability that:
   (a) both events will happen
   (b) only one of the events will happen
   (c) neither event will happen.

4. For events $A$ and $B$, the probabilities are $P(A) = \frac{3}{11}$ and $P(B) = \frac{4}{11}$
   Calculate the value of $P(A \cap B)$ when:
   (a) $P(A \cup B) = \frac{6}{11}$
   (b) events $A$ and $B$ are independent.

5. Given that events $A$ and $B$ are independent with $P(A \cap B) = 0.3$ and $P(A \cap B') = 0.3$, find $P(A \cup B)$.

6. Two ordinary, six-sided dice are rolled. Find the probability of getting:
   (a) at least one six
   (b) two sixes, given that at least one dice shows a six.

7. In a survey of 200 people, 90 of whom were female, it was found that 60 people were unemployed, including 20 males.
   (a) Using this information, copy and complete the table.

   |  | Male | Female | Total |
   |---|---|---|---|
   | Unemployed |  |  |  |
   | Employed |  |  |  |
   | Total |  |  | 200 |

   (b) A person is selected at random from this group of 200.
   Find the probability that this person is:
   (i) an unemployed female
   (ii) a male, given that the person is employed.

8. Communications satellites are difficult to repair when something goes wrong. One satellite works on solar energy and it has two systems that provide electricity. The main system has a probability of failure of 0.002, and the back-up system, that works independently of the main one, has a failure rate of 0.01. The satellite will continue to function as long as one system provides electricity. Find the probability that:
   (a) both systems fail
   (b) at least one system does not fail
   (c) the main system failed, given that at least one system failed.

**9.** Maegan wants to research what her classmates do for exercise. She surveys the 125 students in her year and finds that 82 like cycling and 55 like inline skating. Four students like neither of these activities.

(a) Find the probability that a randomly selected student:

(i) likes either cycling or inline skating

(ii) likes either cycling or inline skating, but not both.

(b) Given that a student likes cycling, find the probability that the student likes inline skating.

**10.** In a school of 88 boys, 36 study Economics ($E$), 29 study History ($H$) and 37 do not study either subject.

(a) A boy is selected at random.

(i) Calculate the probability that he studies both Economics and History.

(ii) Given that he studies Economics, calculate the probability that he does not study History.

(b) A group of three boys is selected at random from the school.

(i) Calculate the probability that none of these boys studies Economics.

(ii) Calculate the probability that at least one of these boys studies Economics.

**11.** In a factory producing parts for computers, there are three machines that work independently to produce one of the components. In any production process, machines are not 100% fault free. The production after one batch from these machines is given in the table.

| | Defective | Non-defective |
| --- | --- | --- |
| **Machine I** | 6 | 120 |
| **Machine II** | 4 | 80 |
| **Machine III** | 10 | 150 |

(a) A component is chosen at random from this production run. Find the probability that the chosen component is:

(i) from machine I

(ii) a defective component from machine II

(iii) non-defective or from machine I

(iv) from machine I, given that it is defective.

(b) Explain whether or not the quality of the component depends on the machine used.

**12.** The market share of smartphone sales in India in 2017 is shown below:

| Company | Market share (%) |
|---------|------------------|
| Samsung | 24.7 |
| Xiaomi | 20.9 |
| Vivo | 9.4 |
| Lenovo | 7.8 |
| OPPO | 7.5 |

(a) Calculate the combined market share of all the other companies selling smartphones in India.

(b) Two smartphone users are selected at random in India. Find the probability that:

(i) they both use Xiaomi

(ii) at least one uses Samsung

(iii) exactly one uses Lenovo.

(c) Three smartphone users are selected at random in India. Find the probability that at least one uses Vivo.

(d) In 2017, there were an estimated 300 million smartphone users in India. Find the expected number using OPPO.

**13.** Papua New Guinea is the most linguistically diverse country in the world, with nearly 850 languages spoken. It has a population of 7.6 million people, and the number of people speaking its four official languages is shown in the table.

| Language | Number who can speak it |
|----------|-------------------------|
| Tok Pisin | 4 000 000 |
| Hiri Motu | 120 000 |
| English | 114 000 |
| Papua New Guinean sign language | 30 000 |

(a) Find the probability that a randomly chosen citizen can speak Hiri Motu.

(b) Two citizens are chosen at random from Papua New Guinea. Find the probability that:

(i) they both speak Tok Pisin

(ii) at least one uses Papua New Guinean sign language

(iii) they both speak the same official language

(iv) they both speak Hiri Motu, given that they both speak the same official language.

(c) A visiting tourist wants to find someone who can speak English. He approaches a group of five local citizens. Find the probability that at least one of the citizens speaks English.

**14.** A car insurance company needs to estimate how much it will pay out in claims during a year, in order to calculate how much to charge customers to buy a policy. Based on past claims, they have collected the following data.

| Type of accident | Relative frequency | Average cost of claim ($) |
|---|---|---|
| Car is written off | 0.01 | 16 000 |
| Serious | 0.055 | 7000 |
| Minor | 0.09 | 3500 |
| No accident | 0.845 | 0 |

(a) Find the expected cost of a claim for each customer.

(b) For each policy sold, the company needs to cover the expected cost of a claim, and in addition cover the expenses of running the company. The company charges 50% more than the expected cost of a claim for each policy. Find the cost of buying a policy.

(c) Find the probability that the car is written off, given that the customer has an accident.

**15.** An international school in Singapore randomly tests its high school students for drug use. If a student is using drugs, the test gives a positive result 99.5% of the time. If a student is not using drugs, the test gives a negative result 99% of the time. Data from the Singapore government suggests the proportion of high school students using drugs is 0.0006

(a) A student is randomly selected for a drug test. Find the probability that this student:
   (i) tests positive for drugs
   (ii) is not using drugs, given that they test positive
   (iii) tests negative for drugs
   (iv) is using drugs, given that they test negative.

(b) Based on the result in (a) (ii), the high school principal decides that any student who tests positive needs to take a second test. If both tests show positive results, then disciplinary action is taken. Under this new scenario, find the probability that a randomly chosen student:
   (i) faces disciplinary action
   (ii) is not using drugs, given that they face disciplinary action.

**16.** Bea applies for a new car insurance policy. The insurance company estimates that the chance of her having an accident in the next year is 0.003. Sam also applies for a new car insurance policy, and the company estimates that the chance of him having an accident in the next year is 0.005. Find the probability that:

(a) both drivers have an accident

(b) exactly one of them has an accident

(c) Bea has an accident, given that at least one of them has an accident.

**17.** An airport authority tests prospective employees for drug use. The test is 98% accurate. Each job applicant takes this test twice. The tests are done at separate times and are designed to be independent of each other. Find the probability that:

    **(a)** a drug user is detected (i.e. the applicant fails at least one test)

    **(b)** a drug user passes both tests

    **(c)** a non-user fails both tests.

**18.** People in North America were asked to state the genre for the last film they watched. The results of the survey are shown in the table.

| Genre | Frequency |
|---|---|
| Adventure | 58 |
| Action | 43 |
| Drama | 35 |
| Comedy | 33 |
| Thriller | 18 |
| Horror | 10 |
| Romantic comedy | 9 |
| Musical | 4 |
| Other | 3 |

    **(a)** A person in North America is selected at random.
        Find the probability that the last movie they watched was:

        **(i)** an adventure movie

        **(ii)** a drama or a thriller.

    **(b)** Two people in North America are selected at random.
        Find the probability that:

        **(i)** the last movie they each watched was a comedy

        **(ii)** the last movie they each watched was a comedy, given that at least one of them watched a comedy.

    **(c)** Ten people in North America are selected at random.
        Find the probability that:

        **(i)** the last movie they each watched was an action movie

        **(ii)** among the last movies they each watched, at least one of them watched an action movie.

**19.** A new blood test has been shown to be effective in the early detection of a disease. The probability that the blood test correctly identifies someone with this disease is 0.99, and the probability that the blood test correctly identifies someone without the disease is 0.95. The incidence of this disease in the general population is 0.0001. A doctor administered the blood test to a patient and the test result indicated that this patient had the disease. Find the probability that the patient has the disease.

20. Jack and Jill play a game, by throwing one dice in turn. If the dice shows a 1, 2, 3 or 4, the player who threw that dice wins the game. If the dice shows a 5 or 6, the other player has the next throw. Jack plays first and the game continues until there is a winner.

    (a) Write down the probability that Jack wins on his first throw.

    (b) Calculate the probability that Jill wins on her first throw.

    (c) Calculate the probability that Jack wins the game.

21. A bag contains 15 green counters and 7 black counters.

    (a) Jasmine takes a counter at random and replaces it, then Shihui takes a counter. Find the probability that:

        (i) Shihui took a green counter, given that Jasmine took a black counter

        (ii) at least one person took a black counter

        (iii) Jasmine took a green counter, given that at least one person took a green counter.

    (b) Jasmine takes a counter at random and keeps it, then Shihui takes a counter. Find the probability that:

        (i) Shihui took a green counter, given that Jasmine took a black counter

        (ii) at least one person took a black counter

        (iii) Jasmine took a green counter, given that at least one person took a green counter.

22. The results of people taking their driving theory test in the UK, from April 2017 to March 2018, are shown in the table for different driving test centres.

| Test centre | Number of males taking the test | Number of males passing the test | Number of females taking the test | Number of females passing the test |
|---|---|---|---|---|
| Aberdeen | 5892 | 3027 | 5805 | 3249 |
| Bangor | 3112 | 1412 | 3017 | 1461 |
| Cheltenham | 3332 | 1751 | 3130 | 1829 |
| Doncaster | 8506 | 3682 | 8072 | 3938 |

    (a) Find which test centre has the highest overall pass rate.

    (b) Given that a female candidate passes her test at one of these test centres, find the probability she took it at the Doncaster test centre.

    Two female candidates arrive independently at the Aberdeen test centre.

    (c) Find the probability that:

        (i) they both pass the test

        (ii) they both pass the test, given that at least one of them passes.

Louise takes her test at the Bangor test centre and James takes his test at the Cheltenham test centre.

(d) Find the probability that:

    (i) they both pass the test

    (ii) exactly one of them passes the test.

In the following year, 6000 males are anticipated to take their theory test at the Aberdeen test centre.

(e) Find the expected number of males who pass their test.

23. The crest on top of the head of a chicken is called a comb. One type of comb is a pea comb. A chick has two genes for a pea comb, one from its father and one from its mother. These genes can be the same or different. If the genes are different, then the dominant gene determines whether or not the chick will have a pea comb. The gene for a pea comb, P, is dominant over the gene for lack of a pea comb, p. This means that if a chick has genes Pp, then it will have a pea comb. Parents pass one of their genes to their chicks at random.

(a) A male chicken has genes Pp, and a female chicken has genes pp.
The possible genes of their chicks are shown in the Punnett diagram.

|        |   | Father |    |
|--------|---|--------|----|
|        |   | P      | p  |
| Mother | p | Pp     | pp |
|        | p | Pp     | pp |

Find the probability that their chick will have:

    (i) a pea comb

    (ii) one gene of each type.

(b) The chickens in part (a) have 10 chicks. Find the expected number with a pea comb.

(c) Another male chicken has genes Pp, and another female chicken has genes Pp. Find the probability that their chick will have:

    (i) a pea comb

    (ii) one gene of each type.

A chicken also has two genes for feather structure. It can have normal feather structure, H, or a woolly appearance, h. The gene for normal feather structure is dominant over the gene for a woolly appearance. The genes for a pea comb are independent to the genes for feather structure.

(d) A male chicken has genes PpHh, and a female chicken has genes ppHh. Find the probability that their chick will have:

    (i) a woolly appearance

    (ii) a pea comb and normal feather structure

    (iii) a pea comb and a woolly appearance.

(e) The chickens in part (d) have 16 chicks. Find the expected number with no pea comb and normal feather structure.

(f) A farmer has 120 chickens. 87 have a pea comb and 38 have a woolly appearance. 55 chickens have a pea comb and a normal feather structure. If a chicken is selected at random, find the probability that it has:

(i) a pea comb and a woolly appearance

(ii) no pea comb and normal feather structure

(iii) no pea comb, given that it has a normal feather structure.

24. In Canada, the rate of incidence of prostate cancer amongst men is estimated at 129 per 100 000 men. The PSA test is a blood test that can help diagnose prostate cancer. Clinical trials have shown that 65% of positive test results are false positives (i.e. the man did not actually have prostate cancer.) Assume that the accuracy of the test is the same for both detecting cancer and detecting non-cancer.

(a) Find the accuracy of the test.

(b) During one year, 800 000 men are given the PSA test. Find the expected number who receive a false negative result.

# Graph theory

**12**

By the end of this chapter, you should be familiar with…

- working with graphs; vertices, edges, adjacent vertices, adjacent edges, completing graphs, weighted graphs
- using directed graphs; in degree and out degree of a directed graph
- recognising subgraphs and trees, and using tree and cycle algorithms with undirected graphs
- setting up and using adjacency matrices and weighted adjacency tables; recognising walks, and finding the number of $k$-length walks between vertices
- constructing transition matrices for a strongly connected graph
- working with walks, trails, paths, circuits, cycles; recognising Eulerian trails and circuits and Hamiltonian paths and cycles
- finding minimum spanning tree (MST) using Kruskal's and Prim's algorithms
- using the Route Inspection Problem, and algorithm for solution, to determine the shortest route around a weighted graph with up to 4 odd vertices, going along each edge at least once
- using the travelling salesman problem to determine the Hamiltonian cycle of least weight in a weighted complete graph.

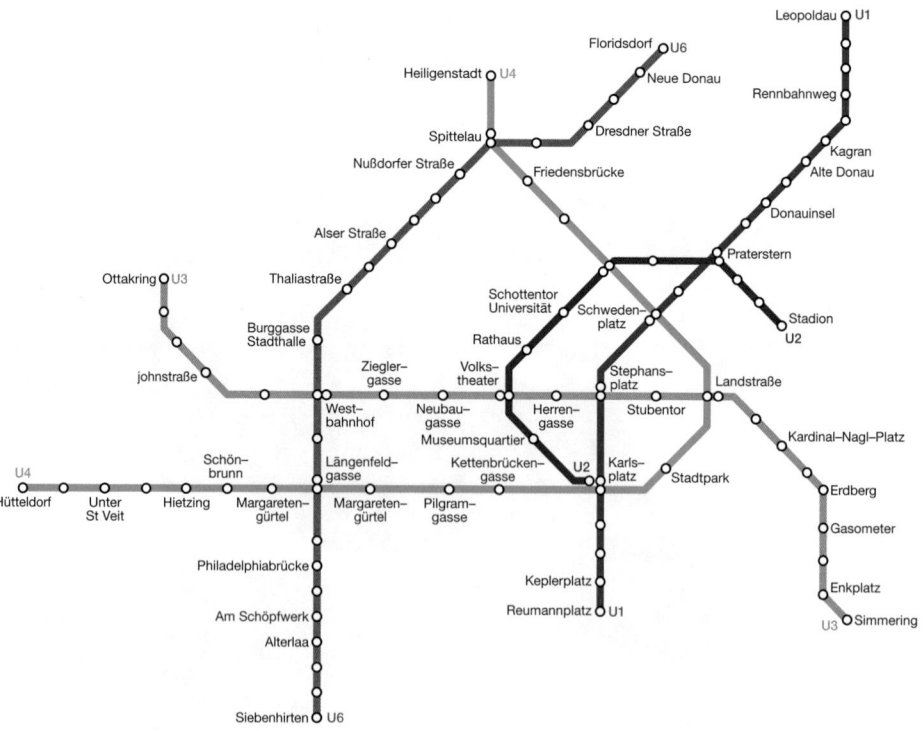

**Figure 12.1** Vienna metro map

Figure 12.1 shows a map of Vienna's metro. Maps like this one do not generally correspond to the real geographic sites in the city but rather the way in which the different stations are organised. A passenger using the underground can use the map to plan a route from one station to the other. The map is simply a diagrammatic means of representing how the stations are interconnected.

# 12.1 Graphs: definitions

When using a map, we are more concerned with seeing how to get from one point to another using the routes available. Consequently, we are dealing with two sets of objects: locations and routes. Such situations involving two sets give rise to relations between the elements of the sets. If $V$ denotes the set of vertices (also called nodes or points) and $E$ denotes the set of edges (routes, lines) made up of unordered pairs of vertices, then a graph $G$ is the non-empty set consisting of vertices and edges such as demonstrated in Figure 12.2.

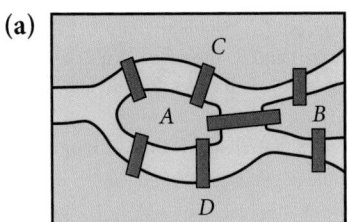 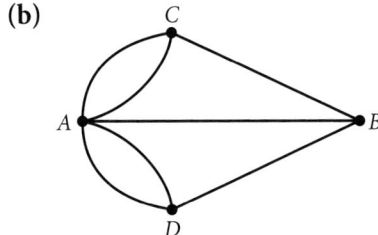

**Figure 12.2** Königsberg bridge problem

Figure 12.2 shows the Königsberg bridge problem. The Pregel river passes through the city of Königsberg (now Kaliningrad) and divides it into two banks and two islands in the middle. There were seven bridges that connect the four land areas of the city. Residents of the city had a popular problem – namely to determine whether it was possible to take a walk through the city using each of the bridges exactly once.

Figure 12.3 represents a computer network. Each computer is connected to the network by one cable. In this network, there is at most one cable between any two computers and there is no cable that connects a computer to itself. This network can be modelled by a **simple graph**, which consists of vertices that represent the computers and undirected edges that represent the cables. Each edge connects two different vertices and no two edges connect the same pair of vertices.

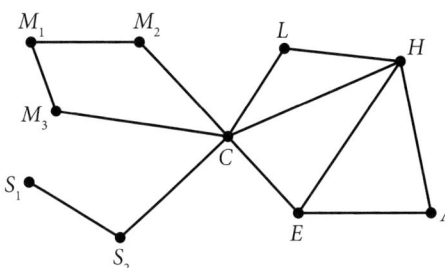

**Figure 12.3** Representation of a computer network

The Königsberg problem inspired Euler to find a solution which appeared in a paper published in 1736. Euler's idea was that the physical layout of land, water and bridges could be modelled by the graph shown in Figure 12.2(b). The land is represented by points $A$, $B$, $C$, and $D$, and the bridges by lines (edges) which could be curved. By means of such a graph, the real problem is transformed into a mathematical one. Given the graph in Figure 12.2(b), is it possible to choose a vertex, traverse the edges one after the other and come back to the starting vertex using every edge only once? Euler showed that it was impossible. We will visit this problem later in the chapter.

 A **graph** $G(V, E)$ consists of two sets $V$, a nonempty set of **vertices** (**nodes** or **points**), and $E$, a set of unordered pairs of different elements of $V$ called **edges** (**arcs**, or **sides**).

In this book, all graphs are assumed to be **finite graphs**, which means that they consist of a finite number of vertices and edges.

Edges in a graph are allowed to cross each other without intersecting at a vertex. See Figure 12.4.

A graph, with no direction assigned to its edges is **undirected**.

Vertices are denoted by single letters or by numbers, so we can say vertex $A$ or $a$, or 1, and edges connecting two vertices $u$ and $v$ by either $(u, v)$, $u$-$v$, $uv$, or by a single variable such as $e_1$.

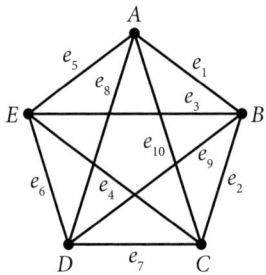

**Figure 12.4** A finite graph with five vertices and ten edges

A graph where all pairs of adjacent vertices are connected by only one edge is a simple graph. The graph shown in Figure 12.4 is a simple graph.

In graph theory we do not concern ourselves with the shape of edges or position of the vertices. What is important is which vertices are connected by which edges. The graphs in Figure 12.5 are the same, but represented in different ways. We consider the two graphs to be equivalent.

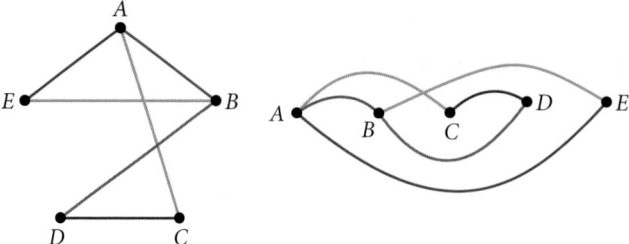

**Figure 12.5** Equivalent graphs

Two vertices $A$ and $B$ in an undirected graph $G$ are called **adjacent** (or neighbours) if $u = \{A, B\}$ is an edge in $G$. The edge $u$ is said to be **incident** with vertices $A$ and $B$. The edge $u$ is also said to connect $A$ and $B$. $A$ and $B$ are also called the endpoints of $\{A, B\}$. Two edges are said to be adjacent if they have a vertex in common.

If an edge has only one endpoint, then the edge joins the vertex to itself and is called a **loop**.

If two edges have the same endpoints, they are called multiple edges or parallel edges.

The degree of a vertex in an undirected graph is the number of edges incident with it. The loop, however, contributes two degrees to the vertex it is incident with. The degree of a vertex $a$ is denoted by $\deg(a)$. A vertex with degree 0, is said to be **isolated** and a vertex with degree 1 is **pendant**. Vertices with odd degrees are called odd vertices and those with even degrees are even vertices.

## Example 12.1

Identify the elements of the two graphs.

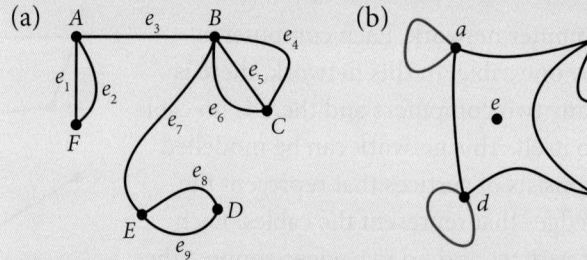

A **simple graph** $G = (V, E)$ is a graph that contains no loops or parallel edges. If there is more than one edge adjacent to two vertices the graph is called a **multiple graph** or a **multigraph**. For instance, the graphs in Example 12.1 are multigraphs while the graphs in Figures 12.3, 12.4 and 12.5 are simple. A graph is called a **connected graph** if there is a path between any two vertices.

### Solution

(a) $A$ is adjacent to $B$ and $F$, while $F$ and $B$ are not. $B$ is adjacent to $C$ and $E$ but not to $D$. $e_1$ is incident with $F$ and $A$, and so is $e_2$. $e_1$ and $e_2$ are multiple (parallel) edges. Also, $e_4$, $e_5$ and $e_6$ are multiple (parallel) edges and $e_8$ and $e_9$ are as well. There are no loops in the first graph. $\deg(A) = 3$, $\deg(B) = 5$, and $\deg(E) = 3$. $A$, $B$, $C$ and $E$ are odd, while $F$ and $D$ are even. $e_1$ and $e_3$ are adjacent since they have $A$ as a common vertex. $e_6$ and $e_7$ are also adjacent.

(b) $a$ and $d$ have loops incident with them. $\deg(a) = 4$, with 2 degrees from the loop! Edges $cd$ and $cb$ are adjacent since they have vertex $c$ in common. Vertex $e$ with $\deg(e) = 0$ is isolated while vertex $f$ with $\deg(f) = 1$ is pendant.

**The handshaking theorem**

Let $G = (V, E)$ be a graph with $e$ edges, i.e., $|E| = e$. Then the sum of all degrees of the vertices in $V$ is twice the number of edges.

That is $\sum_{v \in V} \deg(v) = 2e$

This applies even if the graph is a multigraph.

## Proof of the handshaking theorem

Every edge contributes 2 to the sum of the degrees of the vertices since every edge is incident with exactly two vertices (may be equal!). So, by adding all the vertex degrees we count each edge twice.

For instance, in Example 12.1,
Graph (a) has 9 edges and $3 + 5 + 3 + 2 + 3 + 2 = 18$ degrees
Graph (b) has 7 edges and $4 + 3 + 2 + 4 + 0 + 1 = 14$ degrees

This is called the Handshaking Theorem, because of the resemblance between an edge having two endpoints and a handshake involving two hands.

### Example 12.2

In a graph with 4 vertices $a$, $b$, $c$ and $d$, the degrees are as follows:
$\deg(a) = 4$, $\deg(b) = \deg(d) = 5$, and $\deg(c) = 2$. Is this graph possible? If yes draw a representation, and if not, justify why not?

#### Solution

Since the sum of the degrees is 16, there is a possible graph with $\dfrac{16}{2} = 8$ edges

Figure 12.6 shows a possible graph.

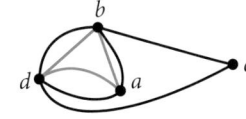

**Figure 12.6** Diagram for solution to Example 12.2

The handshaking theorem gives rise to another important theorem (see margin box note).

In Figure 12.3, the graph has two odd vertices, $S_1$ and $E$; in Figure 12.5, $A$ and $B$ are odd vertices; in Example 12.1, $A$, $B$, $C$, and $E$ are odd vertices in graph (a), while $b$ and $f$ are odd in graph (b); and finally, in Example 12.2, $b$ and $d$ are the odd vertices.

Figure 12.7 shows subgraphs of Figure 12.4. The subgraphs are coloured to distinguish them from the parent graph.

An undirected graph $G = (V, E)$ can only have an even number of odd vertices.

Given that $G = (V, E)$ is a graph then $G_1 = (V_1, E_1)$ is called a **subgraph** of $G$ if $V_1 \subseteq V$ and $E_1 \subseteq E$ and $V_1 \neq \varnothing$

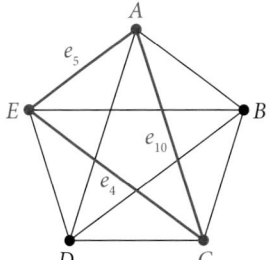
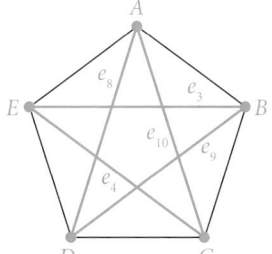
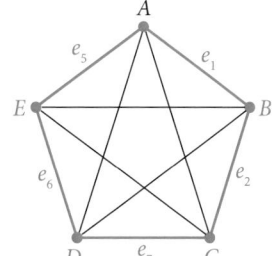
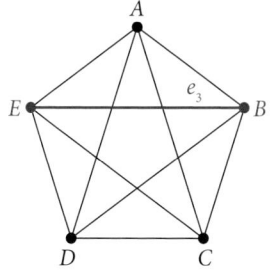

**Figure 12.7** Subgraphs of Figure 12.4

### Example 12.3

Find the union of the graphs $G_1$ and $G_2$.

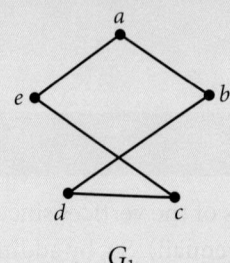

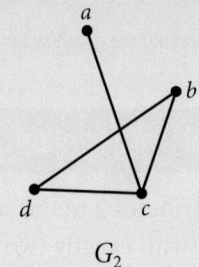

$G_1$ $G_2$

### Solution

The vertex set of the union $G_1 \cup G_2$ is the union of the two vertex sets. So $E = E_1 \cup E_2 = \{a, b, c, d, e\}$

The edge set is the union of the two edge sets, i.e., $V = V_1 \cup V_2 = \{ae, ab, ac, bc, bd, cd, ce\}$

The union is displayed on the right.

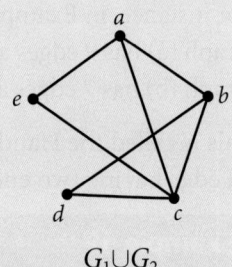

$G_1 \cup G_2$

## Some special graphs

So far, we have only considered undirected graphs. Adding direction to edges gives a slightly different graph, the directed graph or simply the digraph. The difference from the previous discussion is that edges in a directed graph have directions. For example, the edge $ab$ is not the same as the edge $ba$.

Figure 12.8 shows a digraph. Note each edge $e_i$ is represented by an arrow rather than simply an arc.

$G$ consists of 4 vertices $a$, $b$, $c$, and $d$; and seven arcs: $e_1 = (b, a)$, $e_2 = (b, a)$, $e_3 = (a, d)$, $e_4 = (d, b)$, $e_5 = (d, c)$, $e_6 = (c, b)$, and $e_7 = (b, b)$. Each directed arc has an initial vertex and a terminal vertex. So, $e_3$ has $a$ as its initial point and $d$ as its terminal point.

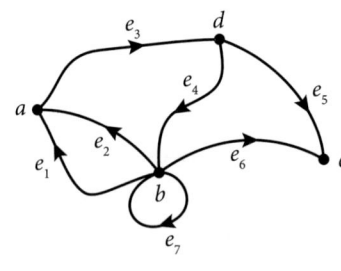

**Figure 12.8** A sample digraph

$e_7$ is a loop with the same initial and terminal vertex $b$. $e_1$ and $e_2$ are called parallel edges since they have the same initial vertex $b$ and terminal vertex $a$.

According to the definition, a loop contributes one in-degree and one out-degree for the vertex.

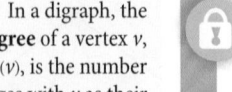

In Figure 12.8, for example, $\deg^-(a) = 2$, and $\deg^+(a) = 1$

Also, $\deg^-(b) = 2$ (one degree from $e_4$ and one from $e_7$), while $\deg^+(b) = 4$

Moreover, $\deg^-(c) = 2$, and $\deg^+(c) = 0$

**Theorem 3**

In a digraph $G(V, E)$, $|E| = \sum_{v \in V} \deg^+(v) = \sum_{v \in V} \deg^-(v)$

**Proof**

Since each edge has an initial vertex and a terminal vertex, the sum of the in-degrees is the same as the number of edges. The same is true for the out-degrees.

A simple graph $G = (V, E)$ is called a **complete graph** if for all $a, b \in V$ there is an edge $\{a, b\}$. A complete graph with $n$ vertices is denoted by $K_n$. Here are the graphs of $K_n$ where $n = 1, 2, 3, 4$ and $5$

The number of edges in a complete graph $K_n$ is given by $|K_n| = \dfrac{n(n-1)}{2}$

**Proof**

The number of vertices is $n$ and each edge connects two vertices, therefore there are $\dbinom{n}{2} = \dfrac{n(n-1)}{2}$ edges.

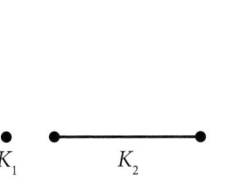 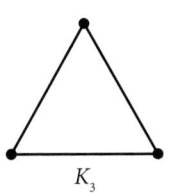 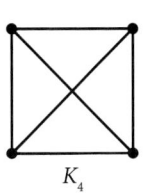 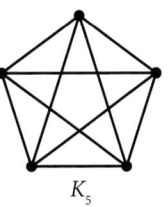

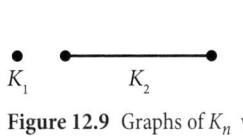

**Figure 12.9** Graphs of $K_n$ where $n = 1, 2, 3, 4$ and $5$

When dealing with sets, the complement of a set $A$ is the set containing the elements of the universal set $U$ that are not in the given set itself. The complete graphs here play a similar role to the universal set. The complement of $G$ which has $n$ vertices is the subgraph of $K_n$ consisting of the $n$ vertices in $G$ and all the edges that are not in $G$. So, two vertices are adjacent in $G'$ if and only if they are not adjacent in $G$.

Figure 12.10 shows the graphs of $G$ and $G'$.

Let $G = (V, E)$ be a simple graph. The **complement** of $G$, denoted by $G'$ is a graph that contains the same set of vertices as the graph $G$ and contains all the edges that are not in $G$.

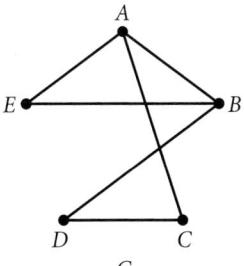 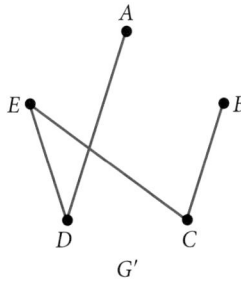 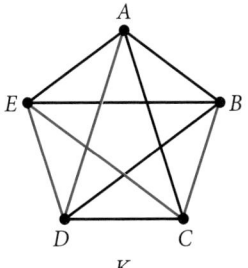

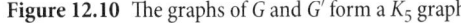

**Figure 12.10** The graphs of $G$ and $G'$ form a $K_5$ graph

Note that those two graphs together form a $K_5$ graph. In some books it is said that those two graphs complement each other to form a complete graph.

Another similarity with the complement of a set can be seen here when we look for the complement of $K_n$. The complement of $K_n$ consists of all the vertices and no edges and it is called a **null graph**. This is similar to the case when we look for the complement of $U$. It is the empty set.

Here are some examples of complete bipartite graphs.

A simple graph $G = (V, E)$ is said to be a **bipartite graph** if the vertex set $V$ can be separated into two subsets $V_1$ and $V_2$ such that $V_1 \cup V_2 = V$ and $V_1 \cap V_2 = \varnothing$, often called a **partition**, and all the edges for the set $E$ are of the form $\{X, Y\}$ such that $X \in V_1, Y \in V_2$ (no edge in $G$ connects either 2 vertices in $V_1$ nor 2 vertices in $V_2$). A bipartite graph is said to be a **complete bipartite graph** when every vertex from $V_1$ is adjacent to every vertex from $V_2$. The most common notation of a complete bipartite graph is $K_{m,n}$ where $|V_1| = m$ and $|V_2| = n$

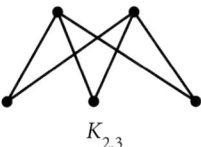 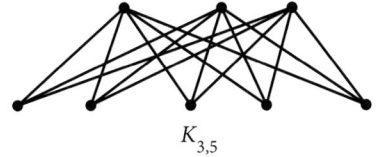

**Figure 12.11** Examples of bipartite graphs

Figure 12.12 shows a bipartite graph. The vertices can be split into two disjoint sets and no edge connects two vertices from the same set. If we simply colour vertices with different colours (red and blue) we can see that no blue vertex is adjacent to a blue vertex, therefore two possible partitions are $V_1 = \{A, C, E\}$ and $V_2 = \{B, D, F\}$

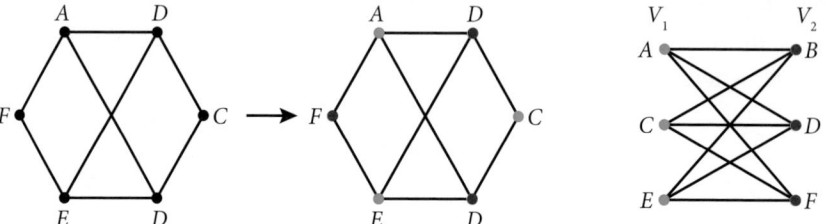

**Figure 12.12** No blue vertex is adjacent to another blue vertex

This can be made clearer by re-arranging the graph without changing the way the vertices are connected. With this, it becomes obvious that we have a bipartite graph.

## Example 12.4

Which of the following graphs are bipartite?

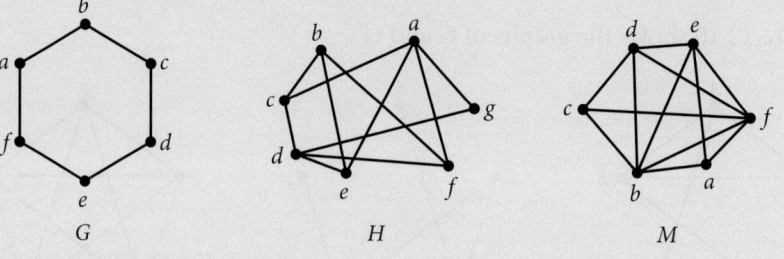

### Solution

G:  We can colour the vertices with two different colours without any two adjacent vertices having the same colour. By re-arranging the vertices, we can clearly see that we are able to separate them into two sets. So, $G$ is bipartite.

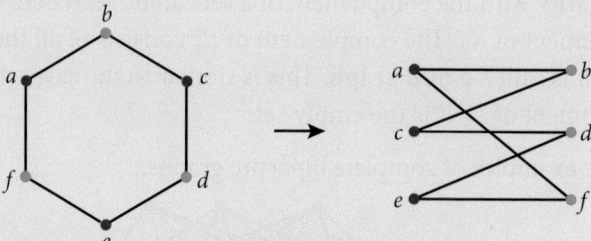

*H:* Doing the same thing for *H* will also yield a bipartite graph.

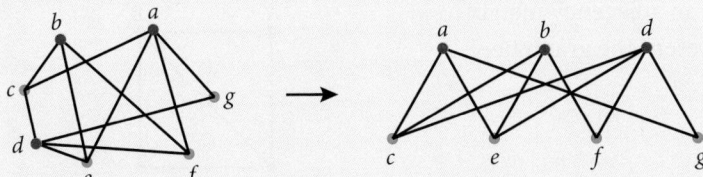

*M:* *M* cannot be bipartite. If you consider vertex *b* and vertex *f*, they cannot be in the same subset as they are adjacent. So, they should be in different subsets. Now, *a* can either be in the subset containing *b*, but it cannot since the two are adjacent. *a* should be in the subset containing *f*, but that cannot happen either.

## Graph representation

Diagrams are sometimes the best way to represent and understand graphs. However, there are other methods of representing graphs that may, at times, be more convenient. In this section we will see how we can represent graphs in different ways.

### Adjacency matrices

For any graph, we can store information about the number of edges connecting each pair of vertices in matrix form. Consider the graph and matrix in Figure 12.13.

Every row corresponds to a vertex and every column corresponds to a vertex. The entries in each row correspond to the number of edges connecting that vertex to the vertices represented by the columns.

For example, in row 1 of the matrix there is a 1 in column 2. This is because there is one edge connecting vertex 1 to vertex 2.

In row 3 of the matrix there is a 2 in column 4. This is because there are 2 edges connecting vertices 3 and 4.

Note that only row 5 has a non-zero entry (1) in the main diagonal because there is a loop in the graph at vertex 5.

The following definitions formalise the idea and introduce some notation.

For a multigraph the definition can be adjusted to reflect the fact there could be more than one edge between two vertices. So, for a multigraph, we can say that the adjacency matrix has the property

$$a_{i,j} = \begin{cases} k(i,j) & k = \text{number of edges between } v_i \text{ and } v_j \\ 0 & \text{otherwise} \end{cases}$$

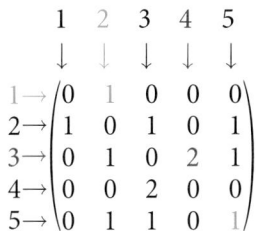

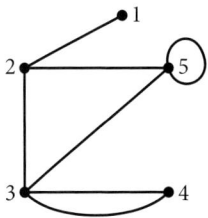

**Figure 12.13** Every row and every column of the matrix corresponds to a vertex in the graph

The **adjacency matrix** $A_G$ of a simple graph $G = (V, E)$ with *n* vertices is an $n \times n$ matrix containing 1 or 0 in such a way that any entry of the matrix

$$a_{i,j} = \begin{cases} 1 & \text{if } \{v, v_j\} \text{ is an edge from } E \\ 0 & \text{otherwise} \end{cases}$$

457

## Example 12.5

(a) Use an adjacency matrix to represent the graph here.

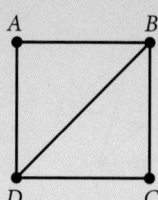

(b) Draw a graph represented by the adjacency matrix $B = \begin{pmatrix} 0 & 1 & 1 & 0 \\ 1 & 0 & 0 & 1 \\ 1 & 0 & 0 & 0 \\ 0 & 1 & 0 & 0 \end{pmatrix}$

### Solution

(a)
$$A = \begin{pmatrix} 0 & 1 & 0 & 1 \\ 1 & 0 & 1 & 1 \\ 0 & 1 & 0 & 1 \\ 1 & 1 & 1 & 0 \end{pmatrix}$$

(b)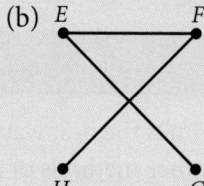

Note that in a simple graph, all the entries on the main diagonal of its adjacency matrix are 0. This is the case since there are no loops in a simple graph. The matrix is also symmetric about its main diagonal since the simple graph is not a digraph. Thus, when there is an edge between $v_i$ and $v_j$ it contributes 1 to the $(i, j)$ entry. Similarly this edge is between $v_j$ and $v_i$ and it contributes 1 to the $(j, i)$ entry.

Note that all the entries in the adjacency matrices of complete graphs are 1, except on the main diagonal where they are all 0.

For example, the adjacency matrix of $K_3$ is
$$A_{K_3} = \begin{pmatrix} 0 & 1 & 1 \\ 1 & 0 & 1 \\ 1 & 1 & 0 \end{pmatrix}$$

The adjacency matrices of complementary graphs each have the main diagonal 0, but all the other entries are complementary. This means whenever there is a 1 in one matrix it is 0 in the other matrix and vice versa, apart from the main diagonal, of course. When we add them, we obtain an adjacency matrix of a complete graph.

## Example 12.6

Use an adjacency matrix to represent the multigraph.

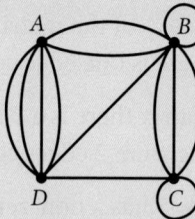

### Solution

Take the first row for example. Since there is no loop at $A$, the entry $(1, 1) = 0$. There 2 edges from $A$ to $B$, thus entry $(1, 2) = 2$. $A$ and $C$ are not adjacent, so entry $(1, 3) = 0$. And finally, there are 4 edges between $A$ and $D$ making entry $(1, 4) = 4$

$$A = \begin{pmatrix} 0 & 2 & 0 & 4 \\ 2 & 1 & 3 & 1 \\ 0 & 3 & 1 & 1 \\ 4 & 1 & 1 & 0 \end{pmatrix}$$

## Example 12.7

Write the adjacency matrices of $G$ and $G'$.

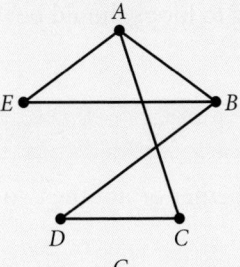

$G$

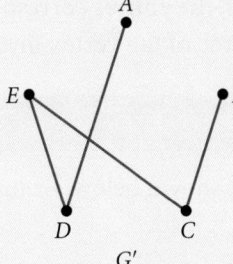

$G'$

### Solution

$$G = \begin{pmatrix} 0 & 1 & 1 & 0 & 1 \\ 1 & 0 & 0 & 1 & 1 \\ 1 & 0 & 0 & 1 & 0 \\ 0 & 1 & 1 & 0 & 0 \\ 1 & 1 & 0 & 0 & 0 \end{pmatrix} \text{ and } G' = \begin{pmatrix} 0 & 0 & 0 & 1 & 0 \\ 0 & 0 & 1 & 0 & 0 \\ 0 & 1 & 0 & 0 & 1 \\ 1 & 0 & 0 & 0 & 1 \\ 0 & 0 & 1 & 1 & 0 \end{pmatrix}$$

## Incidence matrices

Another way that can be helpful in comparing different graphs to check if they have a similar structure is the **incidence matrix**. The incidence matrix consists of $n$ rows corresponding to the vertices that a graph has, and $k$ columns corresponding to the edges that this graph has. The matrix will have a 1 in the entry $(i, j)$ if the edge $e_j$ is incident with the vertex $v_i$.

The **incidence matrix** $I_G$ of a simple graph $G = (V, E)$ with $n$ vertices and $k$ edges is an $n \times k$ matrix containing 1 or 0 in such a way that any entry of the matrix

$$a_{i,j} = \begin{cases} 1 & \text{if } e_j \text{ is incident with } v_i \\ 0 & \text{otherwise} \end{cases}$$

## Example 12.8

Represent the graph with an incidence matrix.

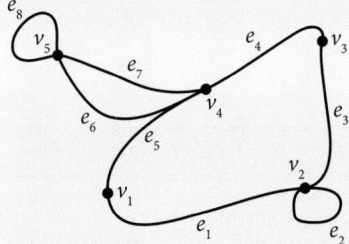

### Solution

$$\begin{array}{c} \\ v_1 \\ v_2 \\ v_3 \\ v_4 \\ v_5 \end{array} \begin{array}{cccccccc} e_1 & e_2 & e_3 & e_4 & e_5 & e_6 & e_7 & e_8 \\ \left( \begin{array}{cccccccc} 1 & 0 & 0 & 0 & 1 & 0 & 0 & 0 \\ 1 & \boxed{1} & 1 & 0 & 0 & 0 & 0 & 0 \\ 0 & 0 & 1 & 1 & 0 & 0 & 0 & 0 \\ 0 & 0 & 0 & 1 & 1 & 1 & 1 & 0 \\ 0 & 0 & 0 & 0 & 0 & 1 & 1 & \boxed{1} \end{array} \right) \end{array}$$

Note how multiple edges are represented by columns with identical entries while loops are the only columns with exactly one entry equal to 1.

In the case of simple graphs, the row totals give the degree of each vertex of the graph. In multigraphs however, the entries corresponding to loops should be multiplied by 2 to give the degree of the vertex involved.

---

### Example 12.9

Consider the graphs $G$ and $H$ shown below. Examine whether or not the two graphs are the same.

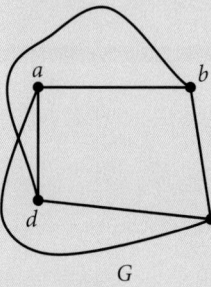 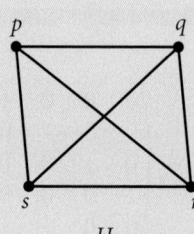

G          H

---

**Solution**

Set up the adjacency matrices for the two graphs:

$$
\begin{array}{c c} 
\begin{array}{c} 
\phantom{a} \\ a \\ b \\ c \\ d 
\end{array}
\begin{array}{cccc} 
a & b & c & d \\
\left(\begin{array}{cccc} 
0 & 1 & 1 & 1 \\
1 & 0 & 1 & 1 \\
1 & 1 & 0 & 1 \\
1 & 1 & 1 & 0 
\end{array}\right)
\end{array}
& = &
\begin{array}{c} 
\phantom{p} \\ p \\ q \\ r \\ s 
\end{array}
\begin{array}{cccc} 
p & q & r & s \\
\left(\begin{array}{cccc} 
0 & 1 & 1 & 1 \\
1 & 0 & 1 & 1 \\
1 & 1 & 0 & 1 \\
1 & 1 & 1 & 0 
\end{array}\right)
\end{array}
\end{array}
$$

> The **degree sequence** of a graph is the list of degrees of the vertices of the graph, listed from smallest (largest) degree to largest (smallest).

Since the graphs are simple, then the column/row totals are the degrees of each vertex. If we list the degrees in ascending (descending) order, we see that they are the same. When we rearrange the matrices, the adjacency matrices of both are identical.

---

### Example 12.10

Consider the two graphs and examine if they are the same.

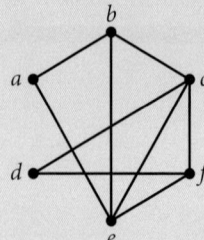 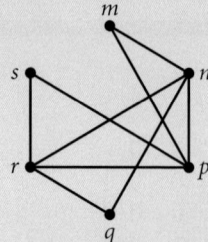

## Solution

Set up the adjacency matrices for both graphs:

$$
\begin{array}{c|cccccc|c}
 & a & b & c & d & e & f & \text{Totals} \\
\hline
a & 0 & 1 & 0 & 0 & 1 & 0 & 2 \\
b & 1 & 0 & 1 & 0 & 1 & 0 & 3 \\
c & 0 & 1 & 0 & 1 & 1 & 1 & 4 \\
d & 0 & 0 & 1 & 0 & 0 & 1 & 2 \\
e & 1 & 1 & 1 & 0 & 0 & 1 & 4 \\
f & 0 & 0 & 1 & 1 & 1 & 0 & 3 \\
\end{array}
\quad \text{and} \quad
\begin{array}{c|cccccc|c}
 & m & n & p & q & r & s & \text{Totals} \\
\hline
m & 0 & 1 & 1 & 0 & 0 & 0 & 2 \\
n & 1 & 0 & 1 & 1 & 1 & 0 & 4 \\
p & 1 & 1 & 0 & 0 & 1 & 1 & 4 \\
q & 0 & 1 & 0 & 0 & 1 & 0 & 2 \\
r & 0 & 1 & 1 & 1 & 0 & 1 & 4 \\
s & 0 & 0 & 1 & 0 & 1 & 0 & 2 \\
\end{array}
$$

Since these graphs are simple, the column/row totals are the degrees of each vertex. We can see that the degree sequence of the first graph is 2, 2, 3, 3, 4, 4 while the second graph is 2, 2, 2, 4, 4, 4. This means that the two graphs are not the same.

## Exercise 12.1

1. For each graph write down:
   (i) the number of vertices
   (ii) the number of edges
   (iii) the degree of each vertex.

(a) 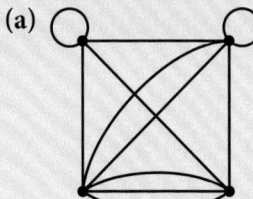  (b) 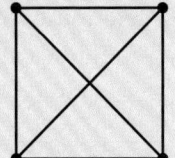  (c)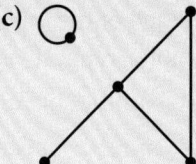

2. Consider a group of five people at a party. Is it possible for each of them to chat with:

   (a) 3 other people from the group

   (b) 4 other people from the group?

   If possible, represent the solution in a form of a graph.

3. What is the minimum number of edges a simple connected graph with $n$ vertices can have?

4. A graph has $n$ vertices. What is number of edges when the graph is complete?

5. Find the number of vertices and edges for each graph.

   (a) $K_{3,4}$          (b) $K_{13,17}$          (c) $K_{m,n}$

6. A complete bipartite graph $K_{m,n}$ has 24 vertices and 128 edges. Find the number of vertices in each partition.

 A graph is called a **connected graph** if there is a path between any two vertices.

7. A graph is **r-regular** if all the vertices have the same degree $r$.

   (a) How many vertices does a 3-regular graph have when it has 12 edges?

   (b) Is it possible to have a regular simple graph with 14 edges? Explain your solution.

   (c) How many regular simple graphs there are with $p$ edges, where $p$ is a prime number?

   (d) If the number of edges in a graph is $e$ and vertices $v$, show that, if the graph is simple and connected, then $v - 1 \leqslant e \leqslant \dfrac{v(v-1)}{2}$

8. Show that in a simple connected graph there are at least two vertices of the same degree.

9. Prove that any subgraph of a bipartite graph must be bipartite.

10. Explain which of the following graphs are bipartite.

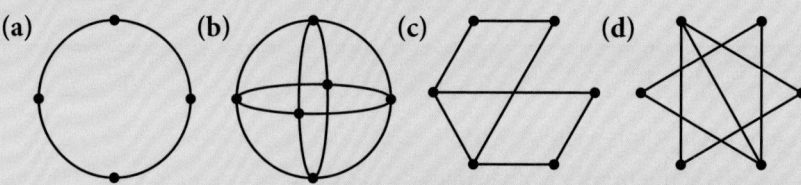

(a)        (b)        (c)        (d)

11. A graph with $v = 7$, has the following vertex degrees: 2, 3, 3, 3, 4, 4, 5
    What is the number of edges of this graph?

12. In each of the following, determine whether it is possible to have a simple graph. If yes, draw it; if not, explain why not.

   (a) Number of vertices, $v = 5$, vertex degrees: 1, 3, 3, 4, 4

   (b) Number of vertices, $v = 6$, vertex degrees: 1, 3, 3, 4, 4, 5

   (c) Number of vertices, $v = 6$, vertex degrees: 1, 2, 2, 3, 3, 3

13. For each graph in question **1**, write down its adjacency matrix.

14. Draw the graph for each adjacency matrix. Which graphs are the same?

(a) $\begin{pmatrix} 0 & 1 & 1 & 1 \\ 1 & 0 & 1 & 0 \\ 1 & 1 & 0 & 1 \\ 1 & 0 & 1 & 0 \end{pmatrix}$ (b) $\begin{pmatrix} 1 & 0 & 0 & 1 \\ 0 & 1 & 1 & 0 \\ 0 & 1 & 1 & 0 \\ 1 & 0 & 0 & 1 \end{pmatrix}$ (c) $\begin{pmatrix} 0 & 1 & 0 & 1 \\ 1 & 0 & 1 & 1 \\ 0 & 1 & 0 & 1 \\ 1 & 1 & 1 & 0 \end{pmatrix}$

(d) $\begin{pmatrix} 0 & 1 & 0 & 1 \\ 1 & 0 & 1 & 0 \\ 0 & 1 & 0 & 1 \\ 1 & 0 & 1 & 0 \end{pmatrix}$ (e) $\begin{pmatrix} 1 & 0 & 1 & 0 \\ 0 & 1 & 0 & 1 \\ 1 & 0 & 1 & 0 \\ 0 & 1 & 0 & 1 \end{pmatrix}$ (f) $\begin{pmatrix} 1 & 1 & 1 & 1 \\ 1 & 1 & 1 & 1 \\ 1 & 1 & 1 & 1 \\ 1 & 1 & 1 & 1 \end{pmatrix}$

15. Write down the adjacency matrix for each graph and show that they are the same graphs.

(a)        (b)

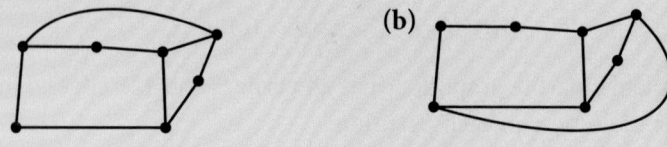

**16.** Determine which graphs are the same.

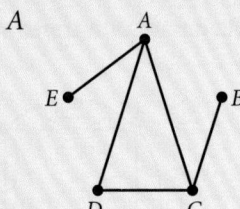

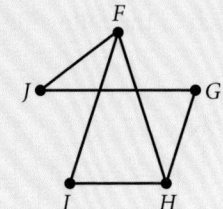

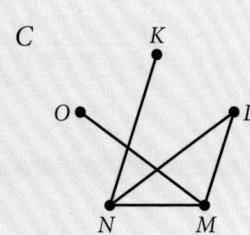

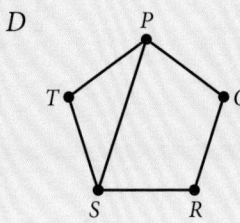

Many of the applications of graph theory are to do with paths formed by travelling along the edges of graphs. The example of the Königsberg bridges is one of the oldest. Network links and how messages travel between different nodes, post routes, garbage collection, etc., are some of the current applications.

Let $G = (V, E)$ be a graph. A **walk** is a sequence of alternating vertices and edges that starts and ends with a vertex and where each edge is adjacent to its neighbouring vertices. Stated slightly differently, an $v_0 - v_n$ walk in $G$ is a finite alternating sequence

$$v_0, e_1, v_1, e_2, v_2, \ldots, e_{n-1}, v_{n-1}, e_n, v_n$$

of vertices and edges starting at vertex $v_0$ and ending at vertex $v_n$ and involving the $n$ edges $e_i = \{v_{i-1}, v_i\}$, where $1 \leq i \leq n$. $v_0$ and $v_n$ do not have to be different.

The **length of a walk** $n$, is the number of edges used in the sequence.

A walk may repeat both edges and vertices.

Consider the graph in Figure 12.14.

The blue coloured **walk** is the walk *abdcbef*. Note that vertex *b* has been visited twice. The length of this walk is 6. No edge has been visited more than once.

The walk *abdcedb* has a length of 6 and uses the edge *bd* twice and the vertices *b* and *d* are used twice.

A walk like the first one is known as a **trail**.

A **trail** is a walk in which no edge appears more than once. A trail (like *abcbda*) which begins and ends at the same vertex is called a **circuit**.

A walk (like *abef*) where no vertex is visited more than once is called a **path**. A path (like *abceda*) which begins and ends at the same vertex is called a **cycle**.

Like several things in graph theory, there is still no unique way of labelling walks. For example, if a graph $G$ has the following set of vertices $V = \{a, b, c, \ldots\}$, then a walk can be described as

    $a, \{a, b\}, b, \{b, c\}, \ldots$

or simply as

    $\{a, b\}, \{b, c\}, \ldots$

or as

    $a, b, c, \ldots$

or as

    *abc*…

Example 12.11 introduces slight variations to this definition.

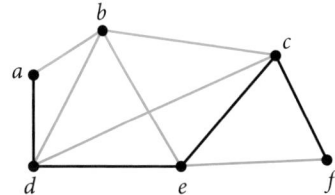

**Figure 12.14** Vertex *b* has been visited twice in the blue walk

463

## Example 12.11

Determine which of the following sequences is a walk, a path or a trail.

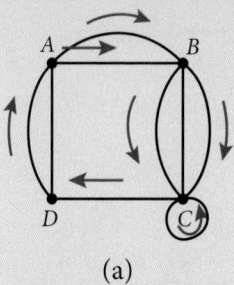
(a)

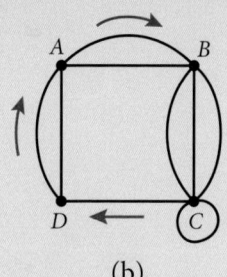
(b)

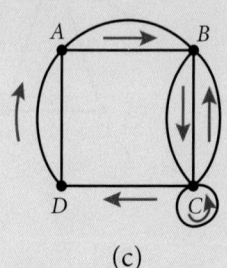
(c)

### Solution

(a) $A, \{A,B\}_{lower}, B, \{B,C\}, C, \{C,C\}_{loop}, C, \{C,D\}, D, \{D,A\}, A, \{A,B\}_{upper}, B$.

This is a trail since no edge has been repeated. Starting at vertex $A$ to vertex $B$ we use the lower edge, while at the end of the sequence again from vertex $A$ to vertex $B$ we used the upper edge. This sequence cannot be a path since vertices $C$, $A$ and $B$ have been repeated.

(b) $C, \{C,D\}, D, \{D,A\}, A, \{A,B\}_{upper}, B$.

This is a path since no vertex has been repeated.

(c) $C, \{C,C\}_{loop}, C, \{C,D\}, D, \{D,A\}, A, \{A,B\}_{upper}, B, \{B,C\}_{middle}, C, \{C,B\}_{middle}, B$.

This is just a walk as it cannot be a trail because the middle edge from $B$ to $C$ has been repeated twice.

Every path is a trail, while a trail can be a path only in a simple graph.

## Adjacency matrices and walks

Adjacency matrices are useful in determining the number of possible walks in a graph. Let's take a $K_3$ graph and its adjacency matrix.

The adjacency matrix also represents walks of length 1.

### How many different walks of length 2 can we have in $K_3$?

Since this graph is regular, all the vertices will be treated equally. Start walking from $A$ and work out where can we arrive after travelling through two edges.

$$A, \{A,B\}, B, \{B,C\}, C \qquad A, \{A,B\}, B, \{B,A\}, A$$
$$A, \{A,C\}, C, \{C,B\}, B \qquad A, \{A,C\}, C, \{C,A\}, A$$

Two walks of length 2 will end up back in $A$, while only one walk of length 2 will end up in $B$ or $C$.

$$A_{K_3} = \begin{pmatrix} 0 & 1 & 1 \\ 1 & 0 & 1 \\ 1 & 1 & 0 \end{pmatrix}$$

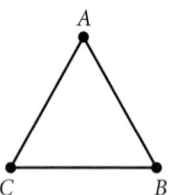

**Figure 12.15** $K_3$ and its adjacency matrix

A regular graph is a graph where all vertices have the same degree.

Now look at the square of the adjacency matrix.

$$A_{K_3}^2 = A_{K_3} \cdot A_{K_3} = \begin{pmatrix} 0 & 1 & 1 \\ 1 & 0 & 1 \\ 1 & 1 & 0 \end{pmatrix} \cdot \begin{pmatrix} 0 & 1 & 1 \\ 1 & 0 & 1 \\ 1 & 1 & 0 \end{pmatrix} = \begin{pmatrix} 2 & 1 & 1 \\ 1 & 2 & 1 \\ 1 & 1 & 2 \end{pmatrix}$$

The entries are the number of walks of the length 2 in $K_3$. Two walks from each vertex back to the same vertex and one walk from each vertex to each of the other 2.

Consider the multigraph in Figure 12.16.

Its adjacency matrix is $A_G = \begin{pmatrix} 0 & 2 & 0 & 2 \\ 2 & 0 & 3 & 0 \\ 0 & 3 & 1 & 1 \\ 2 & 0 & 1 & 0 \end{pmatrix}$

and the square of the matrix is $A_G^2 = \begin{pmatrix} 8 & 0 & 8 & 0 \\ 0 & 13 & 3 & 7 \\ 8 & 3 & 11 & 1 \\ 0 & 7 & 1 & 5 \end{pmatrix}$

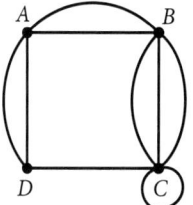

**Figure 12.16** Multigraph

Here for example, the matrix suggests that there are 8 walks of length 2 from $A$ to $C$. We will not list them, we will just explain how to find them. There are 2 edges from $A$ to $B$ and then 3 edges to get from $B$ to $C$. Therefore by the counting principle there are $3 \times 2 = 6$ walks from $A$ to $C$ through $B$. On the other hand there are 2 edges from $A$ to $D$ and only 1 edge from $D$ to $C$. Therefore there are 2 ways from $A$ to $C$ through $D$. Now, the total number of walks from $A$ to $C$ is then $6 + 2 = 8$ which is suggested by the matrix. On the other hand it looks like there are many walks of length 2 from $C$ back to itself. There are 3 edges to $B$ and 3 edges back, therefore 9 walks through $B$ altogether. There is only one walk to $D$ and back. At the end there is a loop at $C$, therefore if we go through it twice we have 11 walks.

 Let $G$ be a graph containing $v$ vertices and $A_G$ be its adjacency matrix. The number of walks of length $n$ from vertex $v_i$ to $v_j$ is given by the $(i, j)$th entry of $A_G^n$, $n \in \mathbb{Z}^+$

## Example 12.12

Determine which of these sequences is a closed walk, a cycle or a circuit.

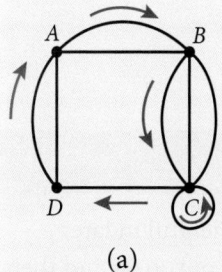

(a)

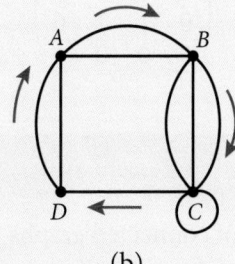

(b)

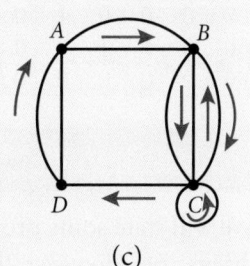

(c)

**A**   $C, \{C,C\}_{\text{loop}}, C, \{C,D\}, D, \{D,A\}, A, \{A,B\}_{\text{upper}}, B, \{B,C\}_{\text{left}}, C.$

**B**   $D, \{D,A\}, A, \{A,B\}_{\text{upper}}, B, \{B,C\}_{\text{right}}, C, \{C,D\}, D.$

**C**   $A, \{A,B\}_{\text{lower}}, B, \{B,C\}_{\text{middle}}, C, \{C,B\}_{\text{right}}, B, \{B,C\}_{\text{middle}}, C, \{C,C\}_{\text{loop}}, C, \{C,D\}, D, \{D,A\}, A.$

465

## Solution

(a) The sequence is a circuit since it is closed and no edge has been repeated. This sequence cannot be a cycle because of the loop at *C*.

(b) The sequence is a cycle since it is closed and no vertex has been repeated.

(c) The sequence is a closed walk since it cannot be a circuit because the middle edge from *B* to *C* has been repeated twice.

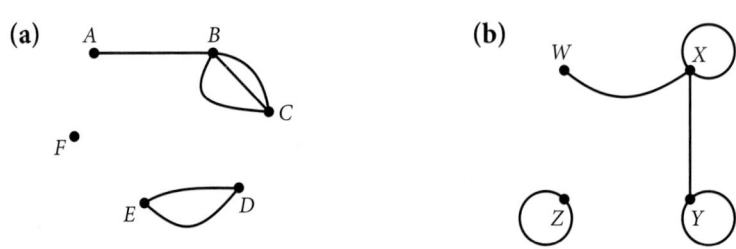

**Figure 12.17** Disconnected graphs

All the graphs presented so far are connected. However, graphs (a) and (b) in Figure 12.17 are not connected since both contain vertices, or even subgraphs, that are isolated. Note that in the case of the vertex *Z*, even though it is isolated, the degree of it is not equal to zero, unlike the vertex *F*.

The adjacency matrix of a graph that is not connected contains only zero in a row and a column of the isolated vertex or contains only one 1 at the diagonal position in that row or column. On the other hand disconnected subgraphs can be shown as diagonal matrices where all the other entries are zero.

Graphs *G* and *H* have the following adjacency matrices,

$$\begin{pmatrix} 0 & 1 & 0 & 0 & 0 & 0 \\ 1 & 0 & 3 & 0 & 0 & 0 \\ 0 & 3 & 0 & 0 & 0 & 0 \\ 0 & 0 & 0 & 0 & 1 & 0 \\ 0 & 0 & 0 & 1 & 0 & 0 \\ 0 & 0 & 0 & 0 & 0 & 0 \end{pmatrix} \quad \text{and} \quad \begin{pmatrix} 0 & 1 & 0 & 0 \\ 1 & 1 & 1 & 0 \\ 0 & 1 & 1 & 0 \\ 0 & 0 & 0 & 1 \end{pmatrix}$$

## Properties of connected graphs

We will state some properties of connected graphs that are helpful in later discussion. However, they are not required for examination purposes and their proofs are not supplied in this book.

**Property 1:** Let $G = (V, E)$ be a simple connected graph, and let *a* and *b* be two vertices in *G* that are not adjacent. If a graph $G_1$ is formed by adding the edge *ab* to *G*, then $G_1$ has a cycle that contains the edge *ab*.

**Property 2:** When an edge is removed from a cycle in a connected graph, the result is a graph that is still connected.

# Eulerian graphs

## Example 12.12

Which of the undirected graphs have an Eulerian circuit? Which only have an Eulerian trail?

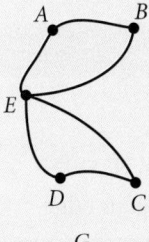

$G$

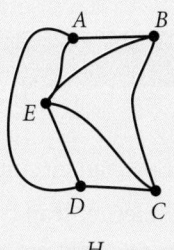

$H$

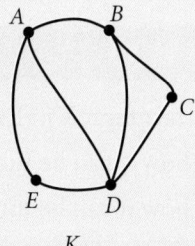

$K$

### Solution

Graph $G$ has an Eulerian circuit. Look at *AECDEBA* for example.

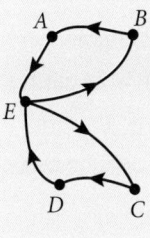

$G$

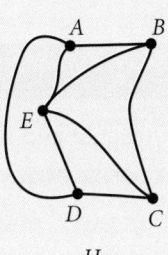

$H$

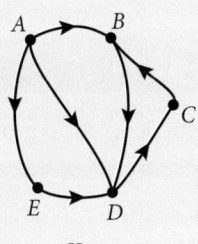

$K$

Graph $H$ has neither an Eulerian circuit nor trail. You will be able to confirm this later in the chapter.

Graph $K$ does not have an Eulerian circuit, but it has an Eulerian trail *AEDCBDAB*.

Look back at the figures in example 12.12. Graph $G$ has deg($A$) = deg($B$) = deg($C$) = deg($D$) = 2, and deg($E$) = 4. That is why $G$ is Eulerian.

Graph $H$ has deg($A$) = deg($B$) = deg($C$) = deg($D$) = 3, and deg($E$) = 4 Only one of the vertices is even while the rest are all odd. Thus it cannot be Eulerian.

Graph $K$ has deg($C$) = deg($E$) = 2, deg($D$) = 4, while deg($A$) = deg($B$) = 3 This is why it does not have an Eulerian circuit. We know however that it has an Eulerian trail. This can be confirmed using the following theorem.

### An informal approach to Eulerian trails and circuits

Consider an Eulerian walk $W$, which we can consider as a sequence of edges $e_1, e_2, e_3, ..., e_n$. Now consider a vertex $v$. Each edge incident with $v$ is used exactly once in the walk. Say $v$ is not the first or last vertex of the walk. Walk along $W$. Each time we arrive at $v$, say along edge $e_i$, we must leave along edge $e_{i+1}$.

Each time we visit $v$ we use two edges. Say the number of times we visit $v$ is $k$. Then $v$ has degree $2k$, an even number. What if $v$ is the first or last vertex? Then the same reasoning applies except for the first or last edge in the walk. If the walk is closed (circuit), then the first and last edges both visit $v$ and we still have an even number. If the walk is open (trail), then either the first or last edge visits $v$, but not both and we see that $v$ has odd degree. Thus the first and last vertices of $W$ have odd degree and we have two vertices of odd degree.

### Example 12.13

Consider the graph $K$ in Example 12.12.

(a) Show how it can be turned into an Eulerian circuit by adding an edge.

(b) Show how it can be turned into an Eulerian trail by removing an edge.

#### Solution

(a) By adding an edge $BA$, we are able to have the circuit $AEDCBDABA$.

(b) By removing the edge $BA$, we get the trail $AEDCBDAB$.

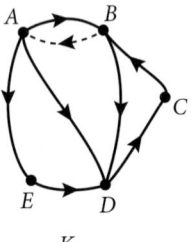

**Figure 12.18** Diagram for solution to Example 12.13

### Example 12.14

Consider the Königsberg bridges problem again and try to solve it.

#### Solution

Note that $\deg(B) = \deg(C) = \deg(D) = 3$, and $\deg(A) = 5$

With reference to the Key Fact boxes on page 467, no Eulerian circuit is possible in such a graph, nor an Eulerian trail.

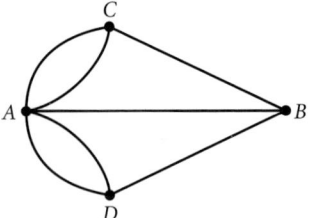

**Figure 12.19** Diagram for solution to Example 12.14

Example 12.15 shows how an Eulerian circuit can be constructed in an Eulerian graph.

### Example 12.15

The vertices in the graph (Figure 12.20) are the roads connecting several cities that you want to visit. You don't want to use the same road twice and you want to return home to city $a$. Find a route for your trip.

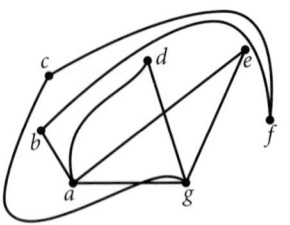

**Figure 12.20** Diagram for Example 12.15

#### Solution

This question is asking you to find an Eulerian circuit for the given graph.

This is an Eulerian graph since the degree at every vertex is even.

First construct a circuit $C$ beginning with $a$. $adga$ is such a circuit. Since it does not include all the edges, it is not Eulerian. Next, look for a vertex that is adjacent to a non-used edge. $a$ and $g$ are such vertices. Beginning with $g$ for example, construct a circuit using unused edges. $geabfcg$ is such a circuit.

Since no more solid edges remain, the procedure stops here. To combine the two circuits, join them at vertex *g* where the second circuit started.

Thus the Eulerian circuit for the graph is *adgeabfcga*, as shown in Figure 12.21.

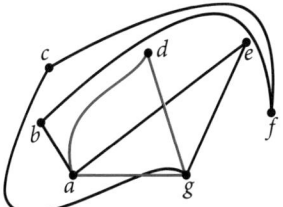

**Figure 12.21** Join the two circuits here

### Example 12.16

In which of these graphs it is possible to find an Eulerian trail or an Eulerian circuit? When it is possible, find an example of the trail or circuit and when it is not possible, explain why not.

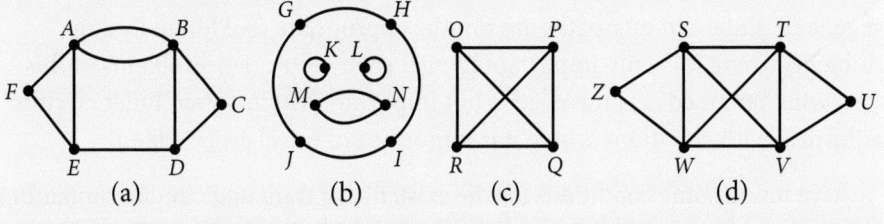

**Figure 12.22** The Eulerian circuit for the graph is *adgeabfcga*

(a)    (b)    (c)    (d)

### Solution

(a)  Looking at vertex degrees, we have:

$\deg(A) = \deg(B) = 4$, $\deg(C) = \deg(F) = 2$ and $\deg(D) = \deg(E) = 3$

Since two vertices have odd degrees, it is possible to find a trail.

We need to start from a vertex of an odd degree so one possible Eulerian trail would be

$D$, $\{D,C\}$, $C$, $\{C,B\}$, $B$, $\{B,D\}$, $D$, $\{D,E\}$, $E$, $\{E,F\}$, $F$, $\{F,A\}$, $A$, $\{A,B\}_{\text{upper}}$, $B$, $\{B,A\}_{\text{lower}}$, $A$, $\{A,E\}$, $E$.

(b)  Even though all vertices are of an even degree (2) the graph is not connected therefore it is not possible to find an Eulerian trail or an Eulerian circuit.

(c)  All the vertices are of the same degree (3), so it is not possible to find an Eulerian trail or an Eulerian circuit.

(d)  Looking at vertex degrees we have:

$\deg(S) = \deg(T) = \deg(V) = \deg(W) = 4$ and $\deg(U) = \deg(Z) = 2$

Thus it is possible to find a circuit. We can start from any vertex so one of the possible Eulerian circuits would be $STVWSVUTWZS$.

If we apply the algorithm presented in Example 12.15, we can start with a circuit $SZWS$ for example. Then $WTVW$, and lastly $VUTSV$. Now we join the first two at $W$ getting a new circuit $SZWTVWS$. Lastly, we join this circuit with the third one at $V$, and thus getting $SZWTVUTSVWS$ as our Eulerian circuit.

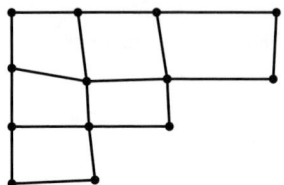

**Figure 12.23** The vertices represent postal box locations

## Hamiltonian graphs

Figure 12.23 shows a graph where the vertices represent locations of postal boxes where mail has to be collected every day. The postal service must find a route so that mail can be collected from each of these boxes. Would an Euler circuit suffice for this job?

The answer is no. An Euler circuit would not provide a good solution since the primary goal is simply visiting each vertex rather than travelling each edge. In this problem it would be very inefficient to require each edge to be travelled since this would force multiple visits to the same vertex.

In general, Euler circuits/paths are not the appropriate tool for analysing problems where it is only important to visit each vertex. For problems of this type, whether an edge is travelled is not important. (Remember, Euler circuits/paths deal with situations where it is important to travel every edge.)

We have found some conditions for the existence of trails and circuits containing all the edges of a graph only once. Can we do a similar task with vertices? Is it possible to find a path or a cycle that contain all the vertices in a given graph?

Let $G = (V, E)$ be a connected graph. A path that contains all vertices of $G$ is called a **Hamiltonian path**.

A cycle that contains all vertices of $G$ is called a **Hamiltonian cycle**.

A connected graph that contains a Hamiltonian cycle is called a **Hamiltonian graph**.

### Example 12.17

In which of these graphs it is possible to find a Hamiltonian path or a Hamiltonian cycle? When possible find an example of the path or cycle and when not, explain why not.

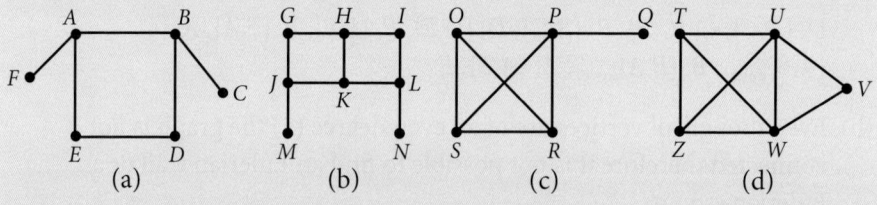

(a)      (b)      (c)      (d)

### Solution

(a) Two vertices have a degree of 1, so if we leave any of the two vertices we cannot come back to it, therefore it is not possible to find a cycle. A possible Hamiltonian path would be

$F, \{F,A\}, A, \{A,E\}, E, \{E,D\}, D, \{D,B\}, B, \{B,C\}, C$

(b) It is not possible to find a Hamiltonian circuit because there are two vertices of degree 1. Neither is it possible to find a Hamiltonian path since at the end there are two non-adjacent vertices that we need to connect.

(c) There is only one vertex of a degree of 1, therefore it is not possible to find a cycle. A possible Hamiltonian path would be *QPOSR*.

(d) It is possible to find a Hamiltonian cycle. We can start from any vertex so one such possible cycle would be *VUZTWV*.

Unlike the situation with Eulerian trails and circuits, there is no well-known test, or listing of requisites, that can be employed to establish whether a graph contains a Hamiltonian path or cycle. In its place, there are some negative tests, which can explain that a certain graph cannot contain such a cycle of path. There are some theorems that establish either necessary conditions or sufficient conditions for a graph to have a Hamiltonian cycle or path. We will examine some of these in the following pages. When faced with certain graphs, however, we will time and again resort to trial and error.

Unfortunately, these two theorems give us only **sufficient** conditions, **not necessary** conditions for the statement. Also, once we know the existence of a Hamiltonian cycle there is no guidance for finding Hamiltonian paths or cycles.

**Optional**

Let $G = (V, E)$ be a simple connected graph. If $|V| = n$, $n \geqslant 3$ and for each vertex $A \in V$, $\deg(A) \geqslant \dfrac{n}{2}$ then the graph $G$ has a Hamiltonian cycle. This is known as **Dirac's theorem**.

Let $G = (V, E)$ be a simple connected graph. If $|V| = n$, $n \geqslant 3$ and for each pair of nonadjacent vertices $A, B \in V$, $\deg(A) + \deg(B) \geqslant n$ then the graph $G$ has a Hamiltonian circuit. This fact is known as **Ore's theorem** and is a generalisation of Dirac's theorem.

## Example 12.18

In which of these bipartite graphs is it possible to find a Hamiltonian path or a Hamiltonian cycle? If possible then give an example and if not give a reason why not.

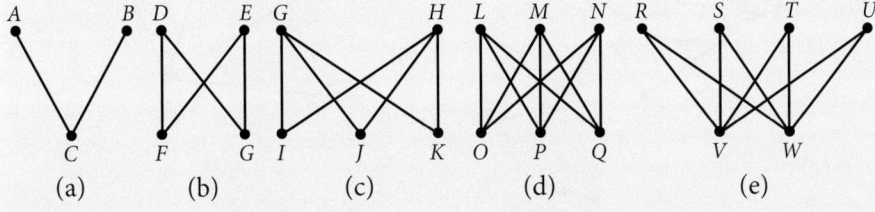

### Solution

(a) There is a Hamiltonian path $A$, $\{A,C\}$, $C$, $\{C,B\}$, $B$, but no cycle. We can see that the vertices don't satisfy the conditions of Dirac's or Ore's theorems.

(b) There is a Hamiltonian cycle. One such possible cycle would be

$D$, $\{D,F\}$, $F$, $\{F,E\}$, $E$, $\{E,G\}$, $G$, $\{G,D\}$, $D$.

All four vertices have a degree of 2 and they satisfy the conditions of Dirac's and Ore's theorems.

(c) There is a Hamiltonian path but no cycle. To find one such path we need to start from a vertex of degree 2 not to repeat a vertex before we travel through all of them. One possible path is $J$, $\{J,G\}$, $G$, $\{G,I\}$, $I$, $\{I,H\}$, $H$, $\{H,K\}$, $K$. The vertices don't satisfy the condition of the theorems since vertices $I$, $J$ and $K$ have a degree of 2 which is less than 2.5. Also taken two at a time, the sum of their degrees is 4 which is less than 5.

(d) There is a Hamiltonian cycle. One such possible cycle would be

$L$, $\{L,O\}$, $O$, $\{O,M\}$, $M$, $\{M,P\}$, $P$, $\{P,N\}$, $N$, $\{N,Q\}$, $Q$, $\{Q,L\}$, $L$

All four vertices have a degree of 3 and they satisfy the conditions of Dirac's and Ore's theorems.

(e) There is neither a Hamiltonian path nor cycle. The problem is that every time we visit a 2-degree vertex, we need to leave it and re-visit a 4-degree vertex, hence there is no Hamiltonian cycle.

Example 12.18 points to two possible negative tests.

**Bipartite graphs – negative test**

If $G$ is a bipartite graph with $V_1$ and $V_2$ subsets of vertices. Let subset 1 have $m$ vertices and subset 2, $n$ vertices.
- If $m \neq n$, $G$ cannot have a Hamiltonian cycle.
- If $m$ and $n$ differ by 2 or more, there is no Hamiltonian path.

# 12 Graph theory

## Exercise 12.2

1. Explain why both of these graphs are Eulerian and find an Eulerian circuit for each.

   (a) 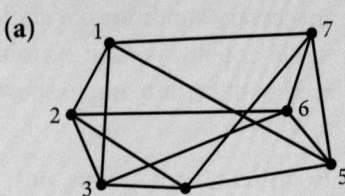   (b)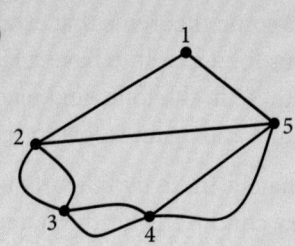

2. In each of the graphs either find an Eulerian circuit or explain why no Eulerian circuit exists.

   (a) 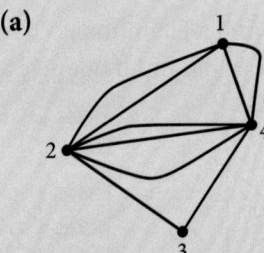   (b)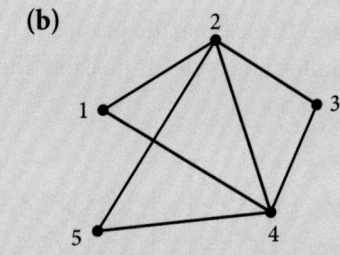

   (c)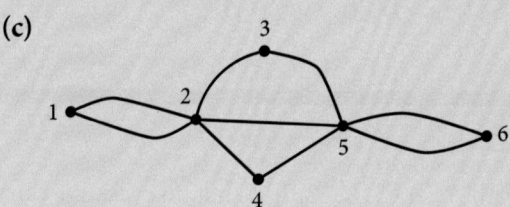

3. Under what conditions would each of these be Eulerian? Justify your answer.

   (a) $K_n$   (b) $K_{m,n}$

4. Are the graphs in questions **1** and **2** Hamiltonian? If one is not Hamiltonian but has a Hamiltonian path, find it.

5. Consider the three graphs in Figure 12.24 of an infinite sequence of graphs which we call $T_n$.

   (a) Find an Eulerian circuit where possible, or justify why one does not exist.

   (b) Find a Hamiltonian cycle where possible, or justify why one does not exist.

   (c) When is $T_n$ Eulerian? When is it Hamiltonian?

6. How many paths of length 1, 2, 3, or 4 are there between $a$ and $e$ in the simple graph in Figure 12.25?

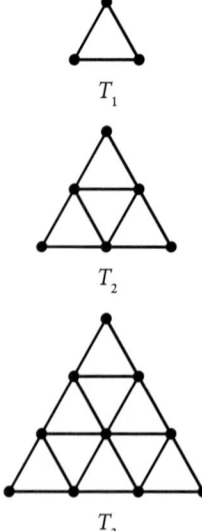

**Figure 12.24** Diagrams for question 5

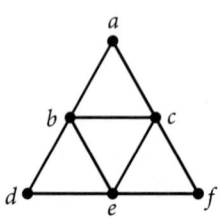

**Figure 12.25** Diagram for question 6

**7.** Find the number of paths of length $x$ between the vertices in in $K_5$ when $x$ is

   **(a)** 4          **(b)** 5          **(c)** 6

**8.** Consider the graph $K_{3,4}$. Let $a$ and $b$ be two vertices in the subset of 3 non-adjacent vertices. Find the number of paths of length $x$ between these vertices when $x$ is:

   **(a)** 4      **(b)** 5      **(c)** 6      **(d)** 7

**9.** Determine if each graph below has a Hamiltonian cycle. If it does, find one such cycle. If it does not, justify why not. For the graphs that do not have a cycle, do any of them have a Hamiltonian path? If yes find it and if not, justify why not.

**(a)**
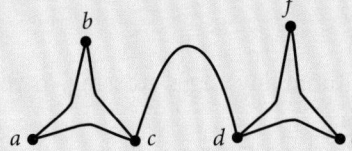

**(b)**
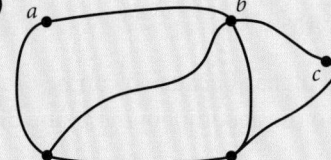

**(c)**
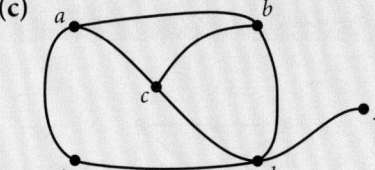

**(d)**
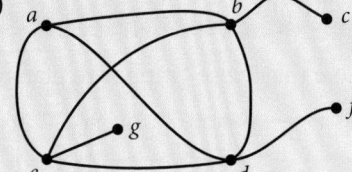

**(e)**
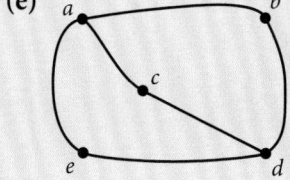

**(f)**
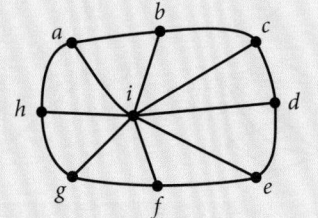

One of the applications of graph theory is in the design of electronic components. In computer chips, electronic components are assembled using 'printed' circuits where the conducting strips are printed onto boards of insulating material. The conducting strips may not cross, as that would lead to malfunction of the component because of short circuits. Complex circuits where crossing strips are unavoidable have to be printed on several boards which are then packed together. Naturally, manufacturers want to print circuits onto the minimum number of boards for obvious reasons. This is an application where graphs, representing components of circuits have to be planar.

A **planar graph** is a graph that can be represented by a diagram in which no edges cross. Such a diagram is called a plane diagram (also known as planar representation or embedding). For example, $K_4$ is a planar graph.

For example, Figure 12.26 shows three diagrams of $K_4$. The first is not a plane diagram, while the second and third are.

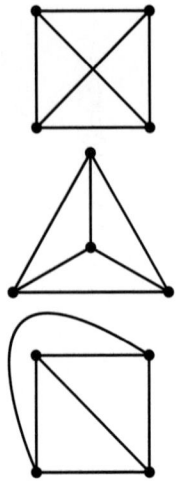

**Figure 12.26** The second and third diagrams are plane diagrams

### Example 12.19

Is the graph known as 3-cube, $Q_3$ shown below planar?

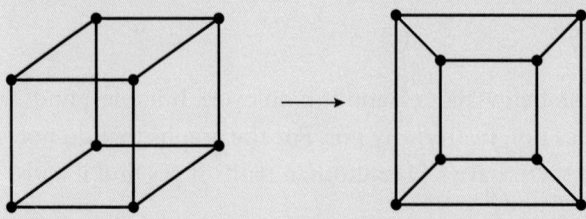

### Solution

$Q_3$ is planar because it can be drawn without any edges crossing, as you can see in the accompanying plane diagram.

### Example 12.20

Show that these graphs are planar.

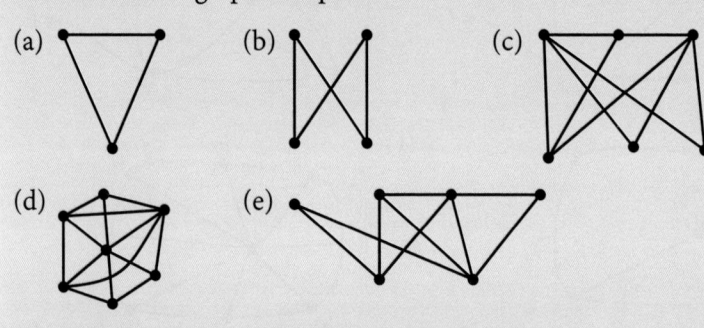

(a)    (b)    (c)    (d)    (e)

### Solution

The graphs are redrawn to show that no two edges in any of the graphs cross. Hence they are planar.

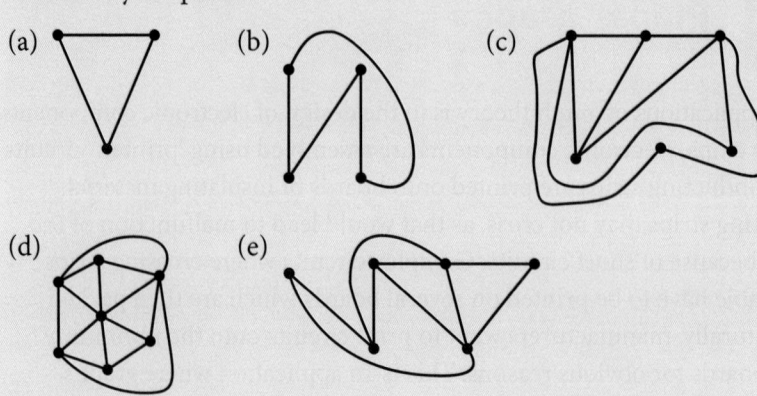

(a)    (b)    (c)    (d)    (e)

## Example 12.21

Investigate which of the complete graphs $K_n$ and complete bipartite graphs $K_{m,n}$ are planar.

### Solution

It is obvious that the following complete graphs are planar: $K_1$, $K_2$, $K_3$, $K_{2,1}$ and $K_{2,2}$ (as shown in Example 12.20). The planar embedding for $K_4$ and $K_{3,2}$ are easily found and are shown in the diagram.

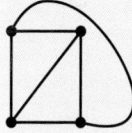 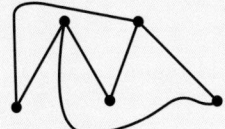

Whether $K_5$ and $K_{3,3}$ are planar needs to be investigated. Start with $K_5$. After drawing the pentagon and all diagonals from one vertex, proceed by drawing one edge at the time.

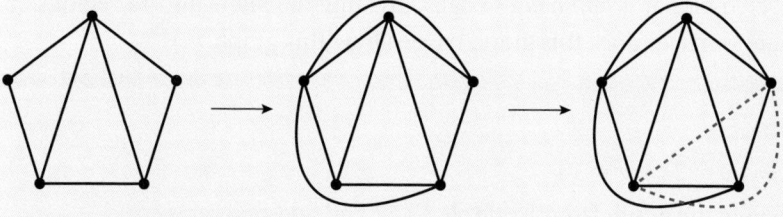

It becomes clear that in order to draw the last edge we must cross one of the previously drawn edges, therefore it is not possible to find a planar representation of $K_5$.

Apply a similar approach to find a plane diagram of $K_{3,2}$

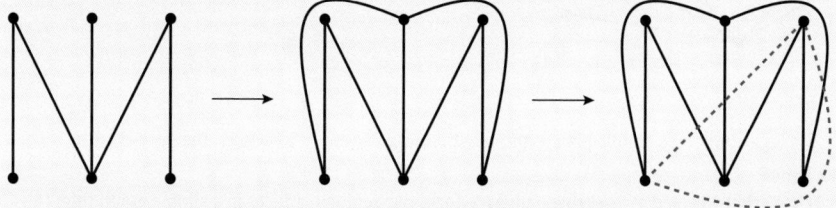

You can see that before reaching the last edge, that there is no way to draw any edge left without crossing some other edge. Thus $K_{3,2}$ is not planar.

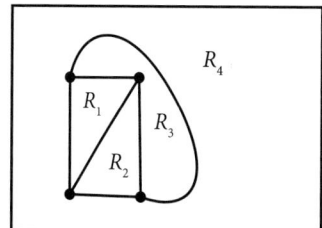

**Figure 12.27** The plane is split into 4 regions

## Euler's formula

A planar representation of a graph partitions the plane into separate regions. For example, the graph of $K_4$ is shown in Figure 12.27, and it splits the plane into 4 regions or faces. Euler showed that all graph diagrams of the same graph partition the plane into the same number of regions. He accomplished this by finding a relationship between the number of regions, the number of edges and the number of vertices of a planar graph.

**Euler's formula**
Let $G = (V, E)$ be a connected planar simple graph (multigraph) where $|V| = v$, $|E| = e$ and $f$ is the number of faces or regions this graph's planar embedding establishes in the plane, then
$v - e + f = 2$

## Example 12.22

Verify Euler's formula for the connected planar graph.

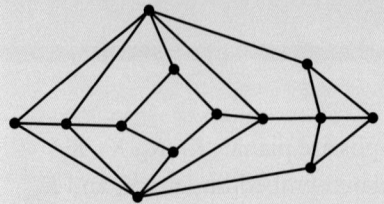

### Solution

The graph has 13 vertices 22 edges and 11 regions.

So, $13 - 22 + 11 = 24 - 22 = 2$

## Example 12.23

A connected planar graph has 24 edges, dividing the plane into 12 regions. How many vertices does this graph have? Draw this graph.

### Solution

$v - 24 + 12 = 2 \Rightarrow v = 14$

You can take the graph from Example 12.22 and add one vertex.

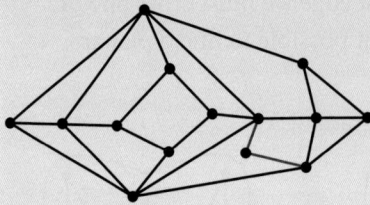

 If $G$ is a connected simple planar graph with $e$ edges and $v > 2$ vertices, then $e \leqslant 3v - 6$

**Proof**

Given that we need at least three edges to form two regions or faces in a simple graph then $2e \geqslant 3f$. By using the Euler's formula we obtain:

$$\left.\begin{array}{l} 2 + e - v = f \\ 2e \geqslant 3f \end{array}\right\} \Rightarrow 2e \geqslant 3(2 + e - v) \Rightarrow 2e \geqslant 6 + 3e - 3v \Rightarrow e \leqslant 3v - 6$$

## Example 12.24

Show that $K_5$ is not planar.

### Solution

$K_5$ is a simple connected graph with $e = 10$ and $v = 5$. If it were planar, then $e = 10 \leqslant 3v - 6 = 15 - 6 = 9$, which is not true. Thus $K_5$ is not planar.

If $G$ is a connected simple planar graph with $e$ edges and $v > 2$ vertices and no circuits of length 3, then $e \leqslant 2v - 4$.

**Proof**

Since there are no circuits of degree 3, then we need at least 4 edges to form two regions. Hence $2e \geqslant 4f$. Thus

$$\left. \begin{array}{l} 2 + e - v = f \\ 2e \geqslant 4f \end{array} \right\} \Rightarrow 2e \geqslant 4(2 + e - v) \Rightarrow 2e \geqslant 8 + 4e - 4v \Rightarrow 2e \leqslant 4v - 8 \Rightarrow e \leqslant 2v - 4$$

## Example 12.25

Show that $K_{3,3}$ is not planar.

### Solution

$K_{3,3}$ is a simple connected graph with no circuit of length 3. $v = 6$ and $e = 9$. If it were planar, then $e = 9 \leqslant 2v - 4 = 12 - 4 = 8$ which is not true. Thus $K_{3,3}$ is not planar.

## Homeomorphic graphs

If we remove an edge, (for example $\{A,B\}$) from a graph and we add another vertex $C$ together with the edges $\{A,C\}$ and $\{B, C\}$ such an operation is called an elementary subdivision. Graphs are **homeomorphic** if they can be obtained from the same graph by a sequence of elementary subdivisions.

To understand this idea, consider the graphs in Figure 12.28.

Graph $H$ is obtained from $G$ by one elementary subdivision: remove edge $ac$ from $G$, then add the edges $ae$ and $ec$ to the graph. Graph $K$ is obtained from $G$ by two elementary subdivisions. Remove $ab$ and add $ag$ and $gb$, and remove $ad$ and add $af$ and $fd$. Thus $H$ and $K$ are homeomorphic.

Kuratowski's theorem is a useful result of this discussion.

**Kuratowski's theorem**

A graph $G = (V, E)$ is not a planar graph if and only if it contains a subgraph homeomorphic to $K_5$ or $K_{3,3}$

## Example 12.26

Is this graph planar?

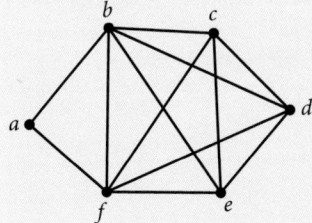

Since $K_{3,3}$ is a simple connected graph, if we were to apply theorem 2, then we have

$e = 9 \leqslant 3v - 6 = 12$

which is true! It would be an error to conclude that $K_{3,3}$ is planar. This is using the converse of the theorem without proving it. Unfortunately, the theorem we proved is necessary but not sufficient. That is, if the graph is planar, then the relation is true.

Since $K_5$ and $K_{3,3}$ are not planar it is obvious that all the graphs containing $K_5$ or $K_{3,3}$ as subgraphs are not planar either. Moreover, all the graphs that contain a subgraph that can be obtained from $K_5$ or $K_{3,3}$ using certain permitted operations are not planar.

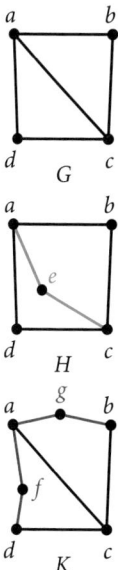

**Figure 12.28** Homeomorphic graphs

### Solution

The graph is not planar since $K_5$ ($bcdef$) is a subgraph.

---

## Transition matrix for a strongly connected graph

### Example 12.27

Are the directed graphs $A$ and $B$ strongly connected or weakly connected?

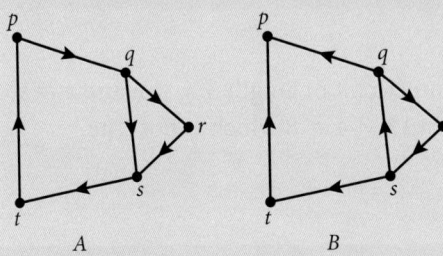

A directed graph is **strongly connected** if there is a path from $a$ to $b$ and from $b$ to $a$ whenever $a$ and $b$ are vertices in the graph.

A directed graph is **weakly connected** if there is a path between every two vertices in the underlying undirected graph.

### Solution

$A$ is strongly connected because there is a path between any two vertices in this directed graph (verify this). Hence, $A$ is also weakly connected. Graph $B$ is not strongly connected. There is no directed path from $p$ to $q$ in this graph. However, $B$ is weakly connected, because there is a path between any two vertices in the underlying undirected graph of $H$ (verify this).

If we set up adjacency matrices of the graphs in Example 12.27, we get:

$$A = \begin{pmatrix} 0 & 1 & 0 & 0 & 0 \\ 0 & 0 & 1 & 1 & 0 \\ 0 & 0 & 0 & 1 & 0 \\ 0 & 0 & 0 & 0 & 1 \\ 1 & 0 & 0 & 0 & 0 \end{pmatrix} \qquad B = \begin{pmatrix} 0 & 0 & 0 & 0 & 0 \\ 1 & 0 & 1 & 0 & 0 \\ 0 & 0 & 0 & 1 & 0 \\ 0 & 1 & 0 & 0 & 1 \\ 1 & 0 & 0 & 0 & 0 \end{pmatrix}$$

A graph is strongly connected if starting at any vertex $a$ we can reach any different vertex $b$. In terms of matrices, this means that if there is a positive integer $n$ such that the matrix
$$M = I + G + G^2 + \ldots + G^n$$
is positive (where all entries are greater than zero), then the graph is strongly connected. In other words, if there is at least one path from vertex $a$ to vertex $b$ of length at most $n$, then we can travel from $a$ to $b$. If this happens for all vertices, then the graph is strongly connected.

Note that the matrix for $B$ has one row of zeros, which tells us that there is no edge starting at $p$ and going to any other vertex. Thus $B$ is not strongly connected.

Recall that for an adjacency matrix $G$, $G^k$ entries give you the number of paths of length $k$ between different vertices. This gives us a tool to use adjacency matrices to check whether a graph is strongly connected.

For graph $A$ in Example 12.27, we have

$$M = I + A + A^2 + A^3 + A^4 = \begin{pmatrix} 2 & 1 & 1 & \boxed{2} & 2 \\ 2 & 2 & 1 & 2 & 2 \\ 1 & 1 & 1 & 1 & 1 \\ 1 & 1 & 1 & 2 & 1 \\ 1 & 1 & 1 & 2 & 2 \end{pmatrix}$$

This means that given any vertex, you can reach all other vertices. For example, we can get from vertex $p$ to vertex $s$ in two ways, each of length at most equal to 4. If we take

$$N = I + A + A^2 + A^3 = \begin{pmatrix} 1 & 1 & 1 & 2 & 2 \\ 2 & 1 & 1 & 2 & 2 \\ 1 & 1 & \boxed{0} & 1 & 1 \\ 1 & 1 & 1 & 1 & 1 \\ 1 & 1 & 1 & 2 & 1 \end{pmatrix}$$

from vertex $r$ back to itself, there is no path up to length of 3.

If we try to raise $B$ to any power, its first row will always be zero. This means that we cannot go from $p$ to any other vertex. Thus, it is not strongly connected.

The **transition matrix $A$** associated to a directed graph is defined as follows. If there is an edge from $a$ to $b$ and the out-degree of vertex $a$ is $d$, then on column

$a$ and row $b$ we put $\dfrac{1}{d}$. Otherwise we mark column $a$, row $b$ with zero. Note that we first look at the column, then at the row. We usually write $\dfrac{1}{d}$ on the edge going from vertex $a$ to an adjacent vertex $b$, thus obtaining a weighted graph.

Consider the graph $A$ in Figure 12.29 with the weights on its edges as described.

## Example 12.28

Interpret the graph in Figure 12.29 to represent webpages where the vertices are the pages and the edges between vertices to be the links from each page to the other(s).

The transition matrix models the behaviour of a random visitor to a webpage. The visitor chooses a page at random, then follows its links to other web pages for as long as they wish.

At each step the probability that the person moves from vertex $a$ to vertex $b$ is zero if there is no link from $a$ to $b$ and $\dfrac{1}{d}$ otherwise. Recall that $d$ is the out degree of vertex.

The initial stage is that the visitor chooses at random which page they will visit. Given the links above, what is the long-term probability that each page will be visited?

### Solution

The initial state can be modelled by a vector $v = \begin{pmatrix} 0.2 \\ 0.2 \\ 0.2 \\ 0.2 \\ 0.2 \end{pmatrix}$ because the visitor is

choosing randomly between 5 pages. Thus the probability of visiting one page is $\dfrac{1}{5} = 0.2$. At the first step, the probability of each page (vertex) to be

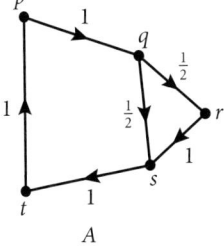

**Figure 12.29** Directed graph with weights on its edges

The transition matrix associated with this graph is

$$\begin{pmatrix} 0 & 0 & 0 & 0 & 1 \\ 1 & 0 & 0 & 0 & 0 \\ 0 & \frac{1}{2} & 0 & 0 & 0 \\ 0 & \frac{1}{2} & 1 & 0 & 0 \\ 0 & 0 & 0 & 1 & 0 \end{pmatrix}$$

Note that the sum of the entries in each column is 1. This is similar to the transition matrix you experienced in Markov chains in Chapter 7.

To find this stationary vector, consider large powers of the transition matrix:

$$A^{10}\cdot v = \begin{pmatrix} 0.2 \\ 0.2 \\ 0.125 \\ 0.25 \\ 0.225 \end{pmatrix}$$

$$A^{100}\cdot v = \begin{pmatrix} 0.222 \\ 0.222 \\ 0.111 \\ 0.222 \\ 0.222 \end{pmatrix}$$

$$p = \begin{pmatrix} \frac{2}{9} \\ \frac{2}{9} \\ \frac{1}{9} \\ \frac{2}{9} \\ \frac{2}{9} \end{pmatrix}$$

The vector $p$ is called the **page rank vector**.

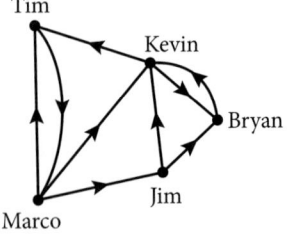

**Figure 12.30** Diagram for Example 12.29

visited after one click is $A\cdot v = \begin{pmatrix} 0 & 0 & 0 & 0 & 1 \\ 1 & 0 & 0 & 0 & 0 \\ 0 & \frac{1}{2} & 0 & 0 & 0 \\ 0 & \frac{1}{2} & 1 & 0 & 0 \\ 0 & 0 & 0 & 1 & 0 \end{pmatrix} \cdot \begin{pmatrix} 0.2 \\ 0.2 \\ 0.2 \\ 0.2 \\ 0.2 \end{pmatrix} = \begin{pmatrix} 0.2 \\ 0.2 \\ 0.1 \\ 0.3 \\ 0.2 \end{pmatrix}$

After two clicks the probability is $A^2\cdot v = \begin{pmatrix} 0 & 0 & 0 & 0 & 1 \\ 1 & 0 & 0 & 0 & 0 \\ 0 & \frac{1}{2} & 0 & 0 & 0 \\ 0 & \frac{1}{2} & 1 & 0 & 0 \\ 0 & 0 & 0 & 1 & 0 \end{pmatrix}^2 \cdot \begin{pmatrix} 0.2 \\ 0.2 \\ 0.2 \\ 0.2 \\ 0.2 \end{pmatrix} = \begin{pmatrix} 0.2 \\ 0.2 \\ 0.1 \\ 0.2 \\ 0.3 \end{pmatrix}$

The probability of a page being visited after step $n$ is thus $A^n v$.

Knowing that Markov chains usually converge to a stationary probability distribution vector $p = \begin{pmatrix} p_1 \\ p_2 \\ \vdots \\ p_k \end{pmatrix}$ where $p_i$ represents the probability that a visitor visits page $x$ at any given time.

### Example 12.29

In studies of group behaviour, it is not surprising to observe that certain people can influence others. A directed graph called an **influence graph** can be used to model this behaviour. Each person of the group is represented by a vertex. There is a directed edge from vertex $a$ to vertex $b$ when the person represented by vertex $a$ can influence the person represented by vertex $b$. Figure 12.30 shows the relationship described. On a topic of equal interest, rank the influence of each after some good exchange of opinions on a social media platform.

### Solution

The transition matrix modelling the situation is

$$A = \begin{array}{c} \\ B \\ J \\ K \\ M \\ T \end{array} \begin{array}{c} \begin{array}{ccccc} B & J & K & M & T \end{array} \\ \begin{pmatrix} 0 & \frac{1}{2} & \frac{1}{2} & 0 & 0 \\ 0 & 0 & 0 & \frac{1}{3} & 0 \\ 1 & \frac{1}{2} & 0 & \frac{1}{3} & 0 \\ 0 & 0 & 0 & 0 & 1 \\ 0 & 0 & \frac{1}{2} & \frac{1}{3} & 0 \end{pmatrix} \end{array}$$

Recall that the entry representing connection from $a$ to $b$ is $\frac{1}{d}$ where $d$ is the out-degree of $a$.

Since the topic is of equal interest, their influence is equally distributed among the five people. So, the probability is $\frac{1}{5} = 0.20$ and the initial state vector is $v = \begin{pmatrix} 0.2 \\ 0.2 \\ 0.2 \\ 0.2 \\ 0.2 \end{pmatrix}$

The influence after the first round of discussion is $Av$.

$$\begin{pmatrix} \dfrac{1}{5} \\[6pt] \dfrac{1}{15} \\[6pt] \dfrac{11}{30} \\[6pt] \dfrac{1}{5} \\[6pt] \dfrac{1}{6} \end{pmatrix}$$

The stationary probability distribution vector is given by $A^k v$ for large values of $k$:

$$\begin{pmatrix} 0.1852 \\ 0.0741 \\ 0.2963 \\ 0.2222 \\ 0.2222 \end{pmatrix}$$

Apparently, Kevin stands to be the most influential, followed by Marco, Tim, Bryan and Jim.

## Exercise 12.3

1. For each graph, decide whether the graph is planar. If it is, give reason for your decision and draw a planar presentation; if not, justify why not.

(a)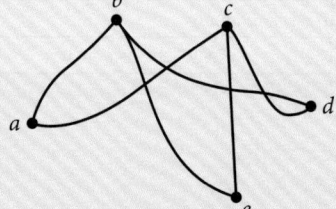

(b)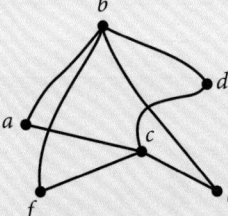

(c)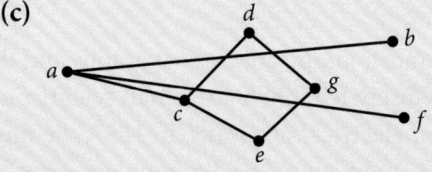

(d)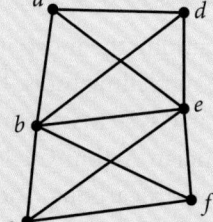

2. A connected planar graph contains 10 vertices and partitions the plane into seven regions. What is the number of edges in the graph?

3. Find the maximum number of edges in a simple connected planar graph with:

   **(a)** 7 vertices             **(b)** 8 vertices.

4. Find the minimum number of vertices in a simple connected planar graph with:

   **(a)** 14 edges             **(b)** 21 edges.

5. A connected planar graph has 8 vertices with 3 degrees each.
   How many regions are created by a planar embedding of this graph?

6. Determine whether or not these graphs are planar.

   **(a)** 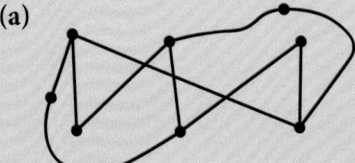      **(b)**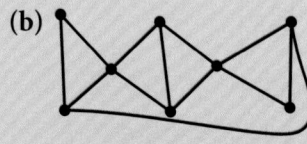

7. The table shows links available on four webpages that connect to each other.

   **(a)** Draw a graph modelling the situation.

   **(b)** Set up a transition matrix.

   **(c)** Find the page rank vector of the 4 webpages assuming that a surfer starts with any of the pages at random.

| https://www.pearsonglobalschools.com | | |
|---|---|---|
| inScholastics.com | | |

| https://www.inscholastics.com | | |
|---|---|---|
| www.pearsonglobalschools.com | www.ismtf.org | www.wazir-garry-math.org |

| https://www.ismtf.org | |
|---|---|
| www.pearsonglobalschools.com | www.wazir-garry-math.org |

| https://www.wazir-garry-math.org | |
|---|---|
| www.pearsonglobalschools.com | inScholastics.com |

## 12.4    Trees

Trees are among the most, if not the most, important class of graphs and they make fine modelling tools. In 1847, Gustav Kirchhof, a German scientist used them to solve systems of equations for electrical networks. In 1857, the English mathematician Arthur Cayley used them to count the different isomers of saturated hydrocarbons. Today, trees are widely used in mathematics, computer science, and many other fields including social sciences.

For example, a common representation of genealogical charts of a family is called a family tree. In the form of a graph, vertices represent the family members, while edges represent the parent-child relationship. Here is a family tree that represents the ancestors of the Austrian emperor Franz Joseph I.

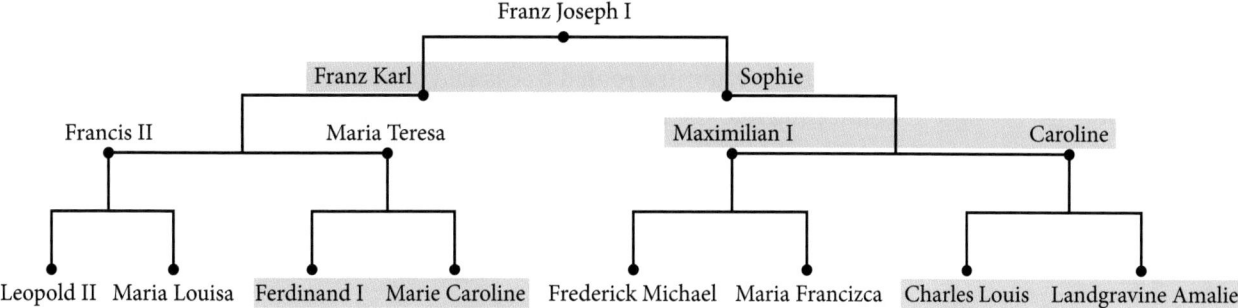

**Figure 12.31** Family tree

You should be familiar with trees in graph theory. We have already discussed several instances for connected graphs that do not contain cycles. These are trees. As in graph theory, tree terminology is unfortunately not standard. We will use the IB terminology in this book.

For example, propane $C_3H_8$ has the structure shown in Figure 12.32.

The structure has no cycle, so it is a tree.

Let $T = (V, E)$ be a simple connected graph. If $T$ contains no cycles it is a **tree**. A **subtree** is a subgraph of a tree that is a tree itself.

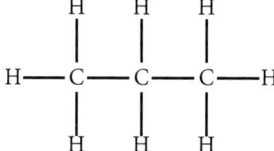

**Figure 12.32** Structure for propane

## Example 12.30

Which of these graphs are trees? Give reasons for your answers.

(a) 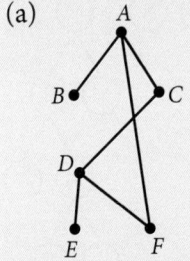    (b) 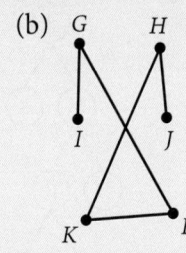    (c) 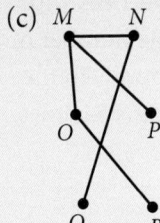    (d)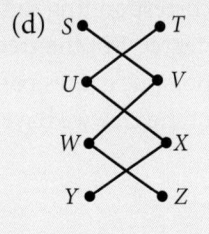

### Solution

Graphs (b) and (c) are trees. Graph (a) contains a cycle *ACDFA*, while graph (d) is not connected.

A graph $T = (V, E)$ is a tree if and only if there is a **unique simple path** between any pair of vertices.

In many applications of trees, such as the family tree in Figure 12.31, organisational trees, computer file systems, networks, etc., a vertex is designated as the **root**. Since there is a unique path from the root to each vertex of the tree, we direct each edge away from the root in a manner described by Figure 12.33. A tree with its root, produces a graph called a rooted tree.

We can change any tree into a rooted tree by the choice of the root.

In a rooted tree, the starting vertex is the root while the other vertices are called parent, child, siblings, ancestors and descendants. A vertex of a tree with no children is called a leaf. Vertices that have children are called internal vertices.

Let $T = (V, E)$ be a tree. Let $v_i$ be a vertex such that every edge is directed away from it, the $T$ is called a **rooted tree**.

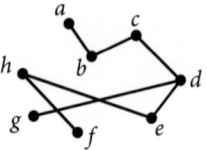

'unrooted' tree

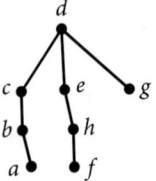

With root d

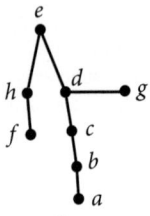

With root e

**Figure 12.33** Rooted trees

A tree $T = (V, E)$ with $n$ vertices has $n - 1$ edges.

In Figure 12.33, for the tree with root $d$, $b$ is a parent of $a$ and $a$ is a child of $b$. Vertices $c$, $e$ and $g$ are siblings, since they have the same parent $d$. Ancestors of $f$ are $d$, $e$ and $h$, whereas $a$ and $f$ have no descendants and therefore each of them is a leaf. We can say that all the vertices in the tree are descendants of the root. An internal vertex in a rooted tree is said to be at a level $i$ when the path connecting it to the root is $i$. For example, in the tree with root $d$: $c$, $e$ and $g$ are at level 1, while $a$ and $f$ are at level 3. In the tree with root $e$: $h$ and $d$ are at level 1, while $a$ is at level 4.

> All vertices in a rooted tree have each a degree at least 2, except for the leaves. Each leaf has a degree of 1.

### Example 12.30

Marco and Roberto play a tennis game. They agree that whoever wins a total of three games first or two games in a row will be declared the winner. How many outcomes are possible, and what is the maximum number of games they will play?

### Solution

The situation can be represented by a tree. There could be 10 possible outcomes corresponding to the vertices with degree 1 in the tree. The number of possible games corresponds to the layers of the tree we have, that is 5 games.

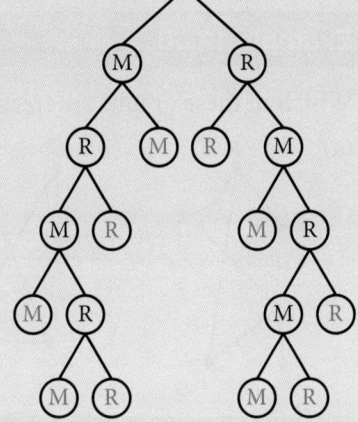

## Spanning trees

All connected graphs have trees that span them. Consider the following situation. In a small mountainous area, winter is harsh and snow makes it sometimes difficult to keep all the towns connected to the rest of the world. Because of the cost involved and the equipment needed for the task, authorities try to make sure that a minimum number of roads between the towns are kept clear by ploughing as few roads as possible. Graph $G$ in Figure 12.34 shows the road network and two possible networks of ploughed roads $T_1$ and $T_2$. These subgraphs of $G$ are called spanning trees of $G$.

- Let $G = (V, E)$ be a connected graph. A subgraph $H$ of $G$ is a **spanning tree** of $G$ if $H$ is a tree which contains every vertex of $G$.
- Every connected graph has a spanning tree.

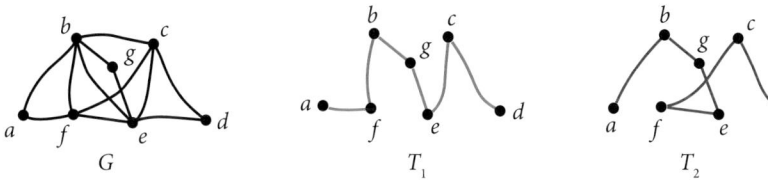

Figure 12.34  Road network, *G*, and two possible networks of ploughed roads.

## How to find a spanning tree?

Spanning trees can be constructed either by removing edges (vertices are not removed) that form cycles, or by building a tree one edge at a time.

## Method 1: Edge removal

Given that *G*(*V*, *E*) is a connected graph, remove edges one at a time in such a way that the resulting graph always remains connected. If this is done until no further edge can be removed then the resulting graph is a spanning tree.

## Method 2: Edge addition

Given that *G* = (*V*, *E*) is a connected graph, start with the subgraph containing all the vertices from the set *V*. Join the edges, one edge at a time in such a way that the resulting graph has no cycle. If this is done until no further edge can be added, then the resulting graph is a spanning tree.

There are a few algorithms for constructing spanning trees. Kruskal's algorithm makes use of the fact that every connected graph has a spanning tree. It proceeds by successively adding edges that have not already been used. We will consider a non-programming set of instructions for this algorithm.

### Example 12.31

Apply Kruskal's algorithm to find a spanning tree for graph *G* (Figure 12.35).

### Solution

Construct a spanning tree using the steps in Kruskal's algorithm. The steps are summarised in the table. Observe that the number of vertices is seven.

| Edge in *G* | Cycle formed? | Edges in tree | Number of edges in tree | Notes |
|---|---|---|---|---|
| *ab* | no | *ab* | 1 | |
| *bf* | no | *ab, bf* | 2 | |
| *fa* | yes | *ab, bf* | 2 | No edges added |
| *fe* | no | *ab, bf, fe* | 3 | |
| *eg* | no | *ab, bf, fe, eg* | 4 | |
| *gb* | yes | *ab, bf, fe, eg* | 4 | No edges added |
| *ec* | no | *ab, bf, fe, eg, ec* | 5 | |
| *ed* | no | *ab, bf, fe, eg, ec, ed* | 6 | Stop, *i* = 7 − 1 |

**Kruskal's algorithm**
Given a graph *G*(*V*, *E*) a simple, connected graph, and |*V*| = *n*. Find a spanning tree *T* for *G*.

**Algorithm**
Set the counter *i* = 0 (*i* is the number of edges of the tree. Every time we add an edge, we increase this number by 1)

**Step 1:** Select an edge, $e_1$. If $e_1$ does not create a cycle, add it to the tree and set *i* = 1, and add $e_1$ to the tree *T*.

**Step 2:** For $1 \leqslant i \leqslant n - 2$, if edges $e_1, e_2, ..., e_i$ have been selected, then select edge $e_{i+1}$ from the remaining edges so that the subgraph determined by $e_1, e_2, ..., e_{i+1}$ contains no cycles.

**Step 3:** Replace *i* by *i* + 1. If *i* = *n* − 1, the subgraph *T* determined by $e_1, e_2, ..., e_{i+1}$ is connected with *n* − 1 edges and *n* vertices, and hence is a spanning tree. If *i* < *n* − 1, return to step 2.

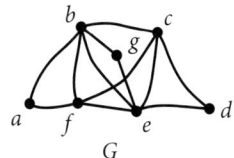

Figure 12.35  Graph for Example 12.31

The diagram shows an example of the spanning tree. Note though that this is not a unique tree and we could have created a different one.

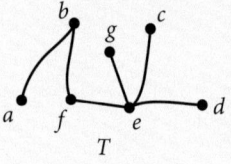

485

## Example 12.32

Find a spanning tree for this graph.

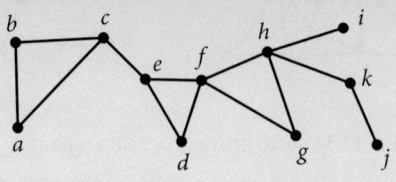

### Solution

Start at $f$, and note that $n = 11$

| Edge in G | Cycle formed? | Edges in tree | Number of edges in tree | Notes |
|---|---|---|---|---|
| fg | no | fg | 1 | |
| gh | no | fg, gh | 2 | |
| hi | no | fg, gh, hi | 3 | |
| hk | no | fg, gh, hi, hk | 4 | |
| hf | yes | fg, gh, hi, hk | 4 | No edges added |
| kj | no | fg, gh, hi, hk, kj | 5 | |
| ⋮ | ⋮ | ⋮ | ⋮ | |
| ba | no | fg, gh, hi, hk, kj, fd, de, ec, cb, ba | 10 | Stop, $i = 11 - 1$ |

Here is a possible spanning tree.

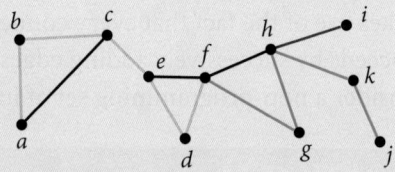

## Exercise 12.4

1. Consider the tree in the diagram.

   (a) List the leaves of this tree.

   (b) List the parents of 4, 8, and 15.

   (c) List the descendants of 3, 7, and 15.

   (d) List the siblings of 4, 7, and 9.

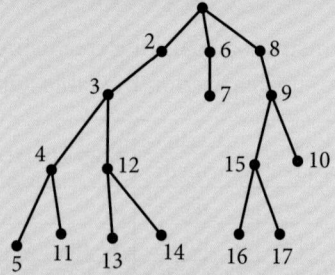

2. Let $T(u, e)$ and $S(v, f)$ be two trees, where $u$ and $v$ are the set of vertices and $e$ and $f$ are the sets of edges for the two trees. If $|e| = 17$, and $|v| = 2|u|$, find $|u|$, $|v|$ and $|f|$.

3. $G(V, E)$ is a connected undirected graph with $|E| = 30$. What is the maximum number of vertices?

4. $T(V, E)$ is a tree with $n$ vertices where $n \geq 2$, How many different paths are there in $T$?

**5. (a)** Find two different spanning trees for $K_{2,3}$. How many such trees are there?

**(b)** How many different spanning trees are there for $K_{2,n}$? $n \in \mathbb{Z}^+$.

**6.** Use an edge removal process to find a spanning tree for each graph.

**(a)** 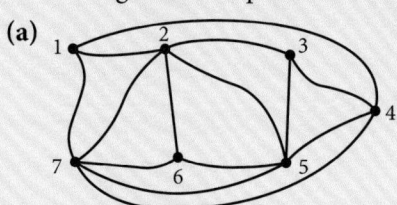  **(b)**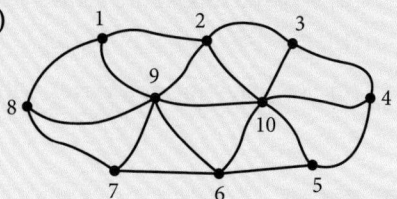

**(c)**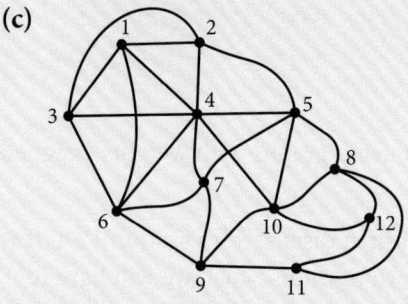

**7.** Use Kruskal's algorithm to find a spanning tree for each graph in question **6**.

**8.** Find a spanning tree for each graph. Start with 1 or *a*.

**(a)**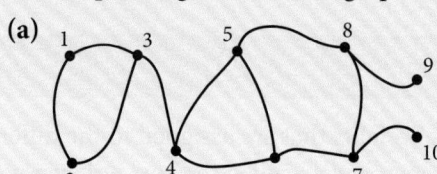

**(b)**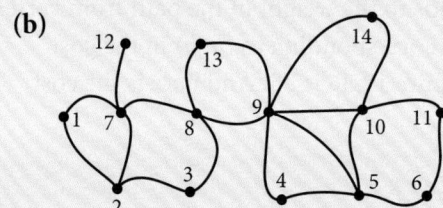

**(c)**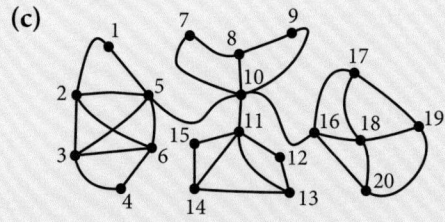

**(d)**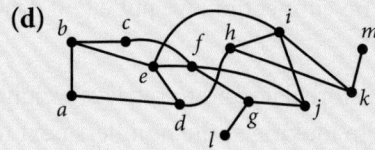

9. A **cycle** $C_n$, $n \geqslant 3$ is a graph in which every vertex has an order of 2. A **wheel** $W_n$, $n \geqslant 3$ is a graph consisting of a cycle $C_n$ and an additional point that is connected to all the vertices in the cycle.

Find a spanning tree for:

(a) $W_6$ starting at the centre vertex

(b) $K_5$

(c) $K_{3,4}$, starting at a vertex with degree 3.

## 12.5 Weighted graphs and greedy algorithms

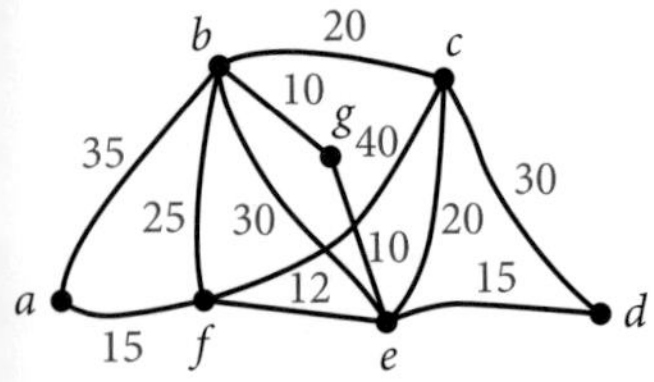

Several real situations can be modelled using graphs with weights assigned to their edges.

Consider the roads in the mountainous area discussed in section 12.4. If we know the distances between the towns, to minimise cost, we will have to minimise the total distance travelled. Airlines use such graphs to represent distances and times between different airports, networks utilise such graphs to represent the response time between different nodes, and many other applications. Such graphs are called weighted graphs.

**Figure 12.36** Weighted graph

Let $G = (V, E)$ be a graph. If a numerical value or a weight is assigned to every edge of $G$, then we say that $G$ is a **weighted graph**.

The **weight of a path** is the sum of all the weights of all the edges in that path.

### Representation

A convenient way of representing the weights that are assigned to the different edges is to use a special type of adjacency matrix called the **cost adjacency matrix** $C_G$. The entry $(i, j)$ corresponds to the weight of the path from vertex $i$ to vertex $j$. So for example, the entry corresponding to $(a, b)$ in the cost adjacency matrix for the graph in Figure 12.36 is 35. And here is the cost adjacency matrix for that graph:

$$C_G = \begin{array}{c|ccccccc} & a & b & c & d & e & f & g \\ \hline a & - & 35 & - & - & - & 15 & - \\ b & 35 & - & 20 & - & 30 & 25 & 10 \\ c & - & 20 & - & 30 & 20 & 40 & - \\ d & - & - & 30 & - & 15 & - & - \\ e & - & 30 & 20 & 15 & - & 12 & 10 \\ f & 15 & 25 & 40 & - & 12 & - & - \\ g & - & 10 & - & - & 10 & - & - \end{array}$$

We use the convention that where there is no connection, we put a $(-)$ dash. You can also use a 0.

The cost adjacency matrix is a good tool to store data and to retrieve weights of edges when needed without getting lost in looking at the numbers next to each edge.

Weighted graphs are associated with spanning trees that have a minimum weight. In the example in this section, we are interested in finding a spanning tree with minimum weight. Such trees are called **minimal** (or **minimum**) **spanning trees**. There are a few algorithms that help us find such trees. These are called greedy algorithms. Two of these will be discussed in this section: Kruskal's algorithm and Prim's algorithm.

Kruskal's algorithm for minimal spanning trees is an extension of his algorithm for spanning trees introduced earlier. In this new one, we keep track of the weight of the edge.

**Kruskal's algorithm (for a weighted graph)**
Given that a graph $G(V, E)$ is a simple, weighted, connected graph, and $|V| = n$. Find a spanning tree $T$ for $G$.

**Algorithm**
Set the counter $i = 0$. ($i$ is the number of edges of the sought tree. Every time we add an edge, we increase this number by 1)

**Step 1:** Select an edge, $e_1$, where $e_1$ does not create a cycle and it has the smallest possible weight, add it to the tree, and set $i = 1$, and add $e_1$ to the tree $T$.

**Step 2:** For $1 \leq i \leq n - 2$, if edges $e_1, e_2, ..., e_i$ have been selected, then select edge $e_{i+1}$ from the remaining edges so that the subgraph determined by $e_1, e_2, ..., e_{i+1}$ contains no cycles and the weight of $e_{i+1}$ is the smallest possible.

**Step 3:** Replace $i$ by $i + 1$. If $i = n - 1$, the subgraph $T$ determined by $e_1, e_2, ..., e_{i+1}$ is connected with $n - 1$ edges and $n$ vertices, and hence is a spanning tree. If $i < n - 1$, return to step 2.

## Example 12.33

Apply Kruskal's algorithm to find a minimal spanning tree for the graph.

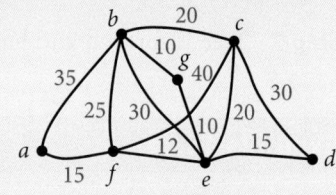

### Solution

**Step 1** Select edge $bg$ as it has the lowest weight ($ge$ too), $i = 1$, the weight is 10

**Step 2** Now select $ge$ with smallest possible weight of 10, no cycle is formed, add it to the tree, the weight is 20, $i = 2$

**Step 3** Now select $fe$ with weight 12, no cycle is formed, add it to $T$, the weight is 32, $i = 3$

**Step 4** Select $af$, then $ed$, add it to $T$, the weight is $32 + 15 + 15 = 62$, $i = 3 + 1 + 1 = 5$

**Step 5** Select $bc$ (or $ed$), add it to $T$, the weight is $62 + 20 = 82$, $i = 6$ STOP

The tree is $T\{bg, ge, fe, af, ed, bc\}$ with minimal weight of 82.

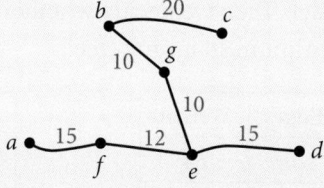

You can also use a table to summarise the steps.

## Example 12.33

Find a minimal spanning tree for the network shown in Figure 12.37

### Solution

We will arrange the weights in a non-decreasing order of weights to make the choice of the edges to be added easy. This method of applying Kruskal's algorithm is very helpful especially in graphs with a relatively small number of edges.

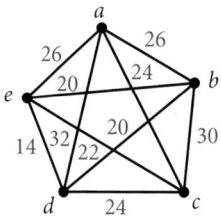

**Figure 12.37** Spanning tree for Example 12.33

| Weight | 14 | 20 | 20 | 22 | 24 | 24 | 26 | 26 | 30 | 32 |
|--------|----|----|----|----|----|----|----|----|----|----|
| Edge | *de* | *db* | *eb* | *ec* | *dc* | *ac* | *ea* | *ab* | *bc* | *ad* |

**Step 1**   Select *de*, weight 14, add to $T$, $i = 1$

**Step 2**   Select *db*, no cycle is formed, weight 20, add to $T$, $i = 2$

**Step 3**   Select *eb*, cycle is formed, reject

**Step 4**   Select *ec*, no cycle is formed, weight 22, add to $T$, $i = 3$

**Step 5**   Select *dc*, cycle is formed, reject

**Step 6**   *ac*, no cycle is formed, weight 24, add to $T$, $i = 4$. STOP

**Step 7**   Tree is formed and has a weight of $14 + 20 + 22 + 24 = 80$

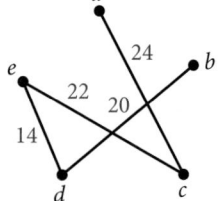

**Figure 12.38** Diagram for solution to Example 12.33

## Example 12.34

Use Kruskal's algorithm to find a minimum spanning tree for this graph.

Find the weight of the minimum spanning tree.

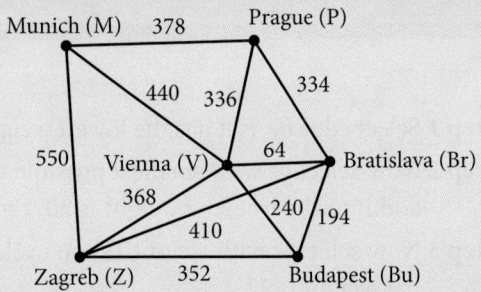

### Solution

We will list all the edges in a table and then sort them into non-descending order. Then we decide whether or not we are going to include them in the minimum spanning tree.

| Edge | Weight |
|------|--------|
| V-Br | 64 |
| V-Bu | 240 |
| V-P | 336 |
| V-Z | 368 |
| V-M | 440 |
| Br-Bu | 194 |
| Bu-Z | 352 |
| Z-M | 550 |
| M-P | 378 |
| P-Br | 334 |
| Br-Z | 410 |

| Edge | Weight | Decision |
|------|--------|----------|
| V-Br | 64 | Yes, $i = 1$ |
| Br-Bu | 194 | Yes, $i = 2$ |
| V-Bu | 240 | No, the cycle V-Br-Bu-V |
| P-Br | 334 | Yes, $i = 3$ |
| V-P | 336 | No, the cycle V-Br-P-V |
| Bu-Z | 352 | Yes, $i = 4$ |
| V-Z | 368 | No, the cycle V-Bu-P-V |
| M-P | 378 | Yes, $i = 5$, STOP |
| Br-Z | 410 | |
| V-M | 440 | |
| Z-M | 550 | |

The three remaining edges form a cycle with the edges already included in the spanning tree, hence they are not included. Also, after we have included the fifth edge we stop since a tree with six vertices contains five edges. We know that any additional edge to the tree will form a cycle with some of the already existing edges.

By Kruskal's algorithm the minimum spanning tree is shown.

The minimum spanning tree has a weight of
$64 + 194 + 334 + 352 + 378 = 1322$

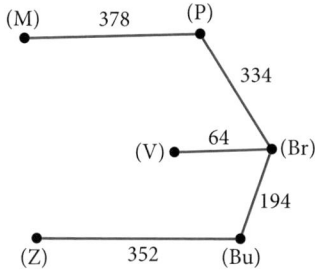

**Figure 12.39** Diagram for solution to example 12.34

## Example 12.35

Use Kruskal's algorithm to find the weight of a minimum spanning tree in this graph.

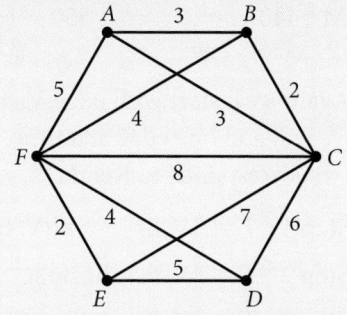

### Solution

| Edge | Weight |
|------|--------|
| {A,B} | 3 |
| {A,C} | 3 |
| {A,F} | 5 |
| {B,C} | 2 |
| {B,F} | 4 |
| {C,D} | 6 |
| {C,E} | 7 |
| {C,F} | 8 |
| {D,F} | 4 |
| {D,E} | 5 |
| {E,F} | 2 |

| Edge | Weight | Decision |
|------|--------|----------|
| {B,C} | 2 | Yes, $i = 1$ |
| {E,F} | 2 | Yes, $i = 2$ |
| {A,B} | 3 | Yes, $i = 3$ |
| {A,C} | 3 | No, creates cycle $BCAB$ |
| {B,F} | 4 | Yes, $i = 4$ |
| {D,F} | 4 | Yes, $i = 5$. STOP |
| {A,F} | 5 | |
| {D,E} | 5 | |
| {C,D} | 6 | |
| {C,E} | 7 | |
| {C,F} | 8 | |

So the minimum spanning tree has a weight of $2 + 2 + 3 + 4 + 4 = 15$

It is also possible that instead of the edge $\{A, B\}$ we include the edge $\{A,C\}$.

Note that edge $\{E, F\}$ was added, even though it was not adjacent to any existent edge in the tree. The algorithm will guarantee that the tree will eventually be formed by focusing on $n - 1$ edges with no cycles.

Prim's algorithm is similar to Kruskal's with the exception that it requires the added edges to be adjacent to existent edges of the tree.

## Example 12.36

Use Prim's algorithm to find a minimum spanning tree in the graph in Example 12.34

### Solution

The data from the diagram can be stored into a cost adjacency matrix.

|     | V   | Br  | Bu  | Z   | M   | P   |
|-----|-----|-----|-----|-----|-----|-----|
| V   | −   | 64  | 240 | 368 | 440 | 336 |
| Br  | 64  | −   | 194 | 410 | −   | 334 |
| Bu  | 240 | 194 | −   | 352 | −   | −   |
| Z   | 368 | 410 | 352 | −   | 550 | −   |
| M   | 440 | −   | −   | 550 | −   | 378 |
| P   | 336 | 334 | −   | −   | 378 | −   |

Again we'll start with edge Vienna-Bratislava that has a length of 64 and then we will add one edge at a time. Once we reach 5 edges in the set we will stop. (wt corresponds to weight)

**Step 1** $T = \{\{V, Br\}\}$, $\quad$ wt$(\{V, Br\}) = 64$

**Step 2** $T = \{\{V, Br\}, \{Br, Bu\}\}$, $\quad$ wt$(\{Br, Bu\}) = 194$

**Step 3** $T = \{\{V, Br\}, \{Br, Bu\}, \{Br, P\}\}$, $\quad$ wt$(\{Br, P\}) = 334$

**Step 4** $T = \{\{V, Br\}, \{Br, Bu\}, \{Br, P\}, \{Bu, Z\}\}$, $\quad$ wt$(\{Bu, Z\}) = 352$

**Step 5** $T = \{\{V, Br\}, \{Br, Bu\}, \{Br, P\}, \{Bu, Z\}, \{P, M\}\}$, $\quad$ wt$(\{P, M\}) = 378$ $\quad$ STOP

So we have the same minimum spanning tree with a weight of 1322.

Note how in step 2, we added {Br, Bu} because it is adjacent to {V, Br} and in step 4 {Bu, Z} because it is adjacent to {Br, Bu}. This is not a requirement of Kruskal's algorithm. In this specific example, both algorithms happened to add the edges in the same order. This is not always the case. Note how in example 4 step 2, we added {E, F} even though it is not adjacent to {B, C} which is in the tree already. To show the difference between the two algorithms, the next example will apply Prim's algorithm to the same graph.

## Example 12.37

Apply Prim's algorithm to the graph given in example 12.35.
For demonstration purposes, the cost adjacency matrix is shown.

$$C_G = \begin{array}{c|cccccc} & A & B & C & D & E & F \\ \hline A & - & 3 & 3 & - & - & 5 \\ B & 3 & - & 2 & - & - & 4 \\ C & 3 & 2 & - & 6 & 7 & 8 \\ D & - & - & 6 & - & 5 & 4 \\ E & 2 & 2 & 7 & 5 & 2 & 2 \\ F & 5 & 4 & 8 & 4 & 2 & - \end{array}$$

## Solution

Since there are two edges with the same weight of 2, we can start with either of them. We will start with the edge {B,C}.

**Step 1**   $T = \{\{B, C\}\}$,   $\text{wt}(\{B, C\}) = 2$

**Step 2**   $T = \{\{B, C\},\{A, C\}\}$,   $\text{wt}(\{A, C\}) = 3$

Note: {A, C} is added as it is adjacent to {B, C}. Also note that at this stage in Kruskal's algorithm, we add {E, F} instead because it is the next "lightest" edge.

**Step 3**   $T = \{\{B, C\},\{A, C\},\{B, F\}\}$,   $\text{wt}(\{B, F\}) = 4$

**Step 4**   $T = \{\{B, C\},\{A, C\},\{B, F\},\{F, E\}\}$,   $\text{wt}(\{F, E\}) = 2$

**Step 5**   $T = \{\{B, C\},\{A, C\},\{B, F\},\{F, E\},\{F, D\}\}$,   $\text{wt}(\{F, D\}) = 4$     STOP

So the minimum spanning tree has the same weight of 15, but the process of adding edges to the tree had a different order.

Kruskal's algorithm appears to be the easier of the two. However, this is only true for small graphs. As the graph size increases, spotting a cycle in Kruskal's algorithm is more difficult than in Prim's algorithm.

---

## Example 12.38

Apply Kruskal's and Prim's algorithms to find a minimum spanning tree for this graph.

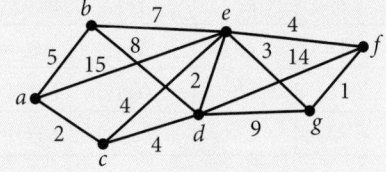

## Solution

In both cases, since we have 7 vertices, we will stop after finding 6 edges. We will set up a table of weights that will help us in finding the spanning trees we need.

| Weight | 1 | 2 | 2 | 3 | 4 | 4 | 4 | 5 | 7 | 8 | 9 | 14 | 15 |
|--------|----|----|----|----|----|----|----|----|----|----|----|----|----|
| Edge | fg | de | ac | eg | ef | cd | ce | ab | be | bd | dg | df | ae |

Kruskal's algorithm

| Weight | Edge | Cycle | Tree | Total weight | i |
|--------|------|-------|------|--------------|---|
| 1 | fg | No | fg | 1 | 1 |
| 2 | de | No | fg, de | 3 | 2 |
| 2 | ac | No | fg, de, ac | 5 | 3 |
| 3 | eg | No | fg, de, ac, eg | 8 | 4 |
| 4 | ef | Yes, reject | fg, de, ac, eg | 8 | 4 |
| 4 | cd | No | fg, de, ac, eg, cd | 12 | 5 |
| 4 | ce | Yes, reject | fg, de, ac, eg, cd | 12 | 5 |
| 5 | ab | No | fg, de, ac, eg, cd, ab | 17 | 6 |
| | | STOP | Tree found | 17 | |

493

The minimum spanning tree is shown in the diagram.

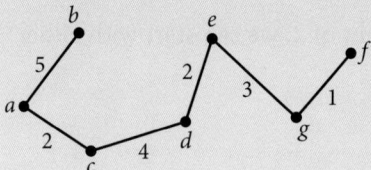

Prim's algorithm

| Weight | Edge | Adjacent | Cycle | Tree | Total weight | $i$ |
|---|---|---|---|---|---|---|
| 1 | fg | | No | fg | 1 | 1 |
| 2 | de, ac | No | | fg | 1 | 1 |
| 3 | eg | Yes | No | fg, eg | 4 | 2 |
| 2 | de | Yes | No | fg, eg, de | 6 | 3 |
| 2 | ac | No | | fg, eg, de | 6 | 3 |
| 4 | ef | Yes | Yes, reject | fg, eg, de | 6 | 3 |
| 4 | cd | Yes | No | fg, eg, de, cd | 10 | 4 |
| 2 | ac | Yes | No | fg, eg, de, cd, ac | 12 | 5 |
| 4 | ce | Yes | Yes, reject | fg, eg, de, cd, ac | 12 | 5 |
| 5 | ab | Yes | No | fg, eg, de, cd, ac, ab | 17 | 6 |
| | | | STOP | Tree found | 17 | |

Note that we found a minimum spanning tree with the same weight using both algorithms. In this example it turned out to be the same tree. However, this is often not the case. The only common result should be the weight of the tree. Also worth noting is that in Kruskal's algorithm, once you finish investigating a minimum weight you move to the next level, while in Prim's algorithm, if the adjacency test fails, then you need to revisit the level at a later stage as happened to edges *ac*, *de* (weight of 2) and *ce* (weight 4).

### Exercise 12.5

1. Use Kruskal's algorithm to find a minimum spanning tree for each weighted graph.

(a)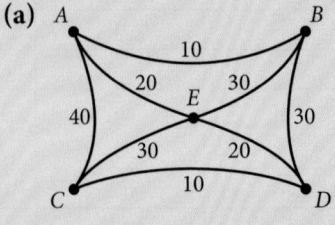

(b)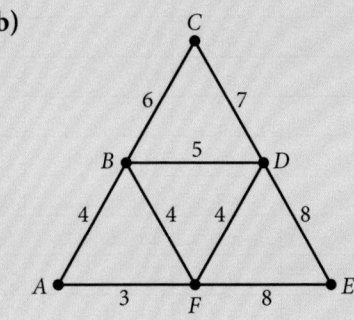

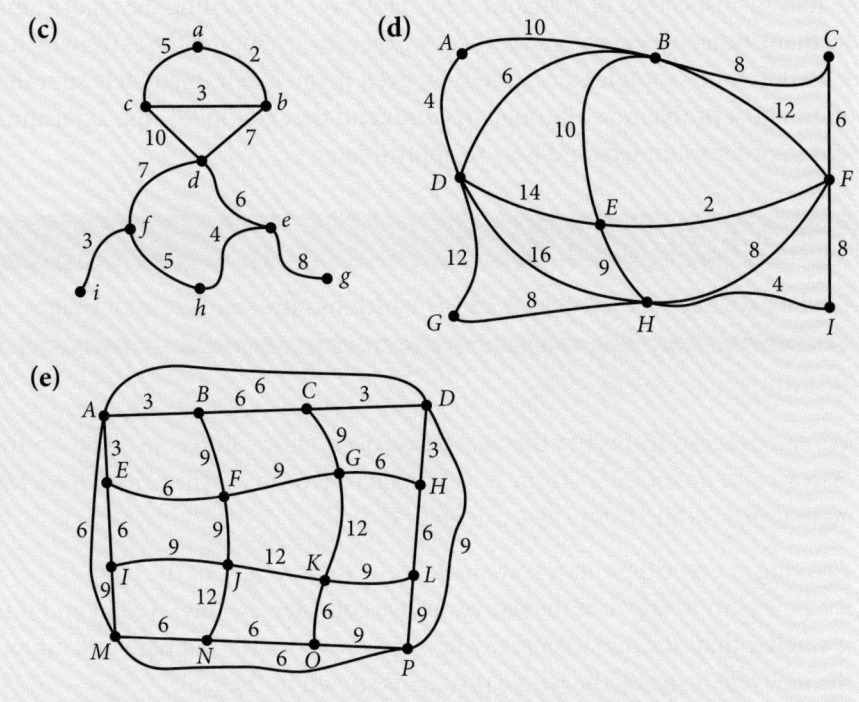

**(c)**

**(d)**

**(e)**

2. Use Prim's algorithm to find a minimum spanning tree for each weighted graph in question **1**.

3. Describe the differences between the results of questions **1** and **2**.

4. The diagram shows the network for a large bus company. To minimise cost, some routes must be discontinued. Find out which routes should be kept to ensure that transport between all these cities is still possible (not necessarily direct). Distances are given in hundreds of km.

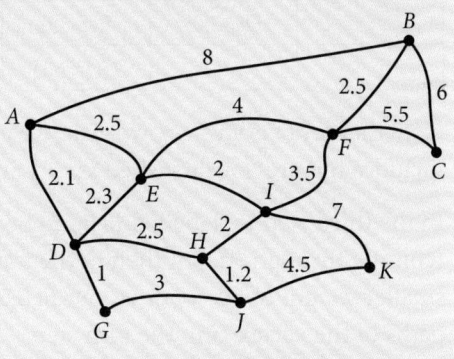

**Shortest path, route inspection and the travelling salesman problem**

A shortest path is a path from one vertex to another in a weighted graph using the smallest possible weight. As a path, no edges or vertices are visited more than once. The shortest path, especially in complex networks, is not always evident. That is why, Edsger Dijkstra, a Dutch mathematician created a way for finding the shortest path in his shortest-path algorithm. In this section, we will discuss the algorithm and apply it to a few situations. However, you want to keep in mind that in textbook examples, the solution may be more readily obvious by inspection or trial and error. However, by learning the algorithms, like many other aspects of graph theory, you are developing the skills which can be used in more complex situations. So, even if you can immediately spot the solution to a problem, we strongly recommend that you follow the algorithm's steps in order to understand how to apply it later.

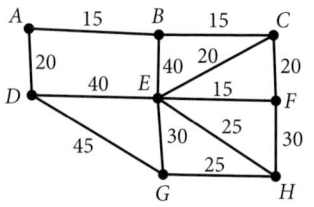

**Figure 12.40** Diagram for Example 12.39

### Example 12.39

Find a path between vertices $A$ and $H$ which has the smallest total weight.

### Solution

We can proceed from $A$ to the nearest vertices, taking into consideration the least weight possible. So, from $A$ we can go to $B$ or to $D$. Then from $B$ we can go to $C$ or $E$, while from $D$ we can go to $E$ or $G$.

Arriving at every new vertex, we look at the total weight of the path. If there is a new path to arrive at the old vertex we consider the total weight and if it is smaller than the one we already have we cross out the old path and adopt the new one instead. The whole process is shown in the table:

| Step 1 | Step 2 | Step 3 | Step 4 | Step 5 |
|--------|--------|--------|--------|--------|
| $A$ | $B(A,15)$ | $C(B,30)$ | $F(C,50)$ | |
| | | | $E(C,50)$ | |
| | $D(A,20)$ | ~~$E(B,55)$~~ | | $H(E,75)$ |
| | | $G(D, 65)$ | ~~$H(G,90)$~~ | |

Note that for every vertex we visit, we give it a temporary label which includes the previously visited vertex and the total weight so far.

In the third step we labelled $E(B,55)$ because so far this is the smallest weight (coming through $B$), but then in the fourth step, once we reached $E$ with a smaller weight of 50, we cross out $E(B,55)$. The same thing happens to the paths of the vertex $H$ in the fifth step.

So the path with the smallest weight is $ABCEH$.

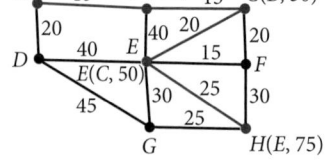

**Figure 12.41** Diagram for the solution to Example 12.39

Example 12.39 demonstrates the general rule used in Dijkstra's algorithm. It proceeds by finding the shortest path from $A$ to its adjacent vertices, then the shortest path to a second level set of vertices, and so on until the length of the shortest path to $H$ is found.

The algorithm performs a sequence of iterations. A key set of vertices is assembled by adding one vertex at each iteration. A labelling process is executed at each iteration. In this labelling process, a vertex $w$ is labelled with the length of a shortest path from $a$ to $w$ that contains only vertices from the key set. The vertex added to the set is one with minimal label among those vertices not already members of the set.

**Dijkstra's algorithm**

To find a shortest path from vertex $a$ to vertex $z$ in a weighted graph, proceed as follows:

**Step 1** Set $v_1 = a$ and assign to this vertex the label $(-, 0)$. Assign every other vertex a temporary label of $\infty$, where $\infty$ is reckoned to be larger than any real number.

**Step 2** Until $z$ has been assigned a permanent label, do the following:

   **i** Use the notation $L(v)$ to represent the shortest path from the source, $a$, to the present vertex $v$.

   **ii** Take the vertex $v_i$ that most recently acquired a permanent label, say $d$. For each vertex that is adjacent to $v_i$ and has not yet received a permanent label, if $d + w(v_iv) < L(v)$, the current temporary label of $v$, update $L(v)$ to $d + w(v_iv)$.

   **iii** Take a vertex $v$ that has a temporary label smallest among all temporary labels in the graph and make its temporary label permanent. If there are several vertices $v$ that tie for the smallest temporary label, make any choice.

   **iv** To keep track of the successive steps, use the following convention: Each vertex, $v$, is labelled with an ordered pair $(x, l)$, where $x$ represents the vertex just preceding $v$ and $l$ is the shortest length of the path from $a$. All labels are temporary, until the algorithm identifies their path as shortest, then they are changed into permanent labels, which we will denote by circling the vertex. Any temporary label that does not become permanent will be crossed out.

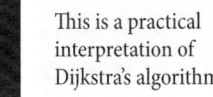

This is a practical interpretation of Dijkstra's algorithm.

## Example 12.40

Use Dijkstra's algorithm to find the shortest path between $P$ and $W$ in this graph.

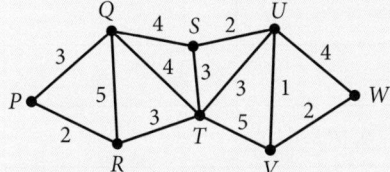

## Solution

Note that in this example we will draw the graph at different stages. You do not have to do this when performing the algorithm. In the diagram we have used the convention that if a vertex is not labelled, then it has the label $(-, \infty)$

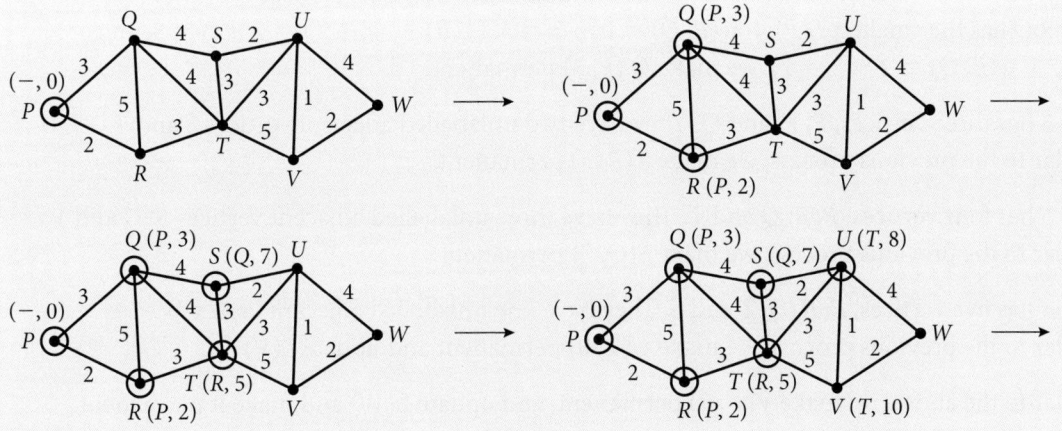

497

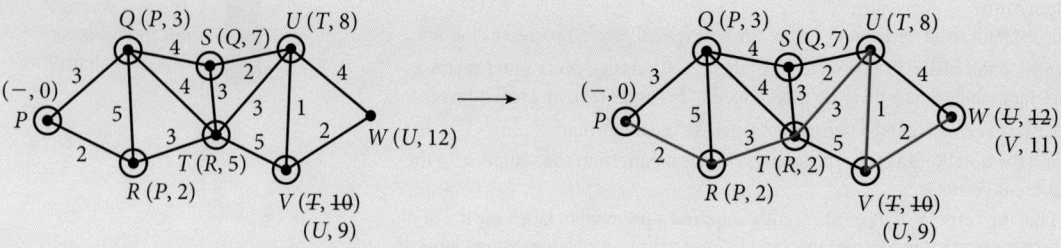

The table shows the steps. Each cell contains the length of the path and the preceding vertices. The highlighted cells are the ones describing the shortest path. Each cell lists also the path lengths that are calculated at this stage.

| Step 1 | Step 2 | Step 3 | Step 4 | Step 5 | Step 6 | Step 7 |
|---|---|---|---|---|---|---|
| $L(P) = 0$ | | | | | | |
| $L(R) = \infty$ | $L(R) = 2, \{P\}$ | | | | | |
| $L(Q) = \infty$ | $L(Q) = \infty$ | $L(Q) = 3, \{P\}$ | | | | |
| $L(T) = \infty$ | $L(T) = \infty$ | $L(T) = 5, \{P,R\}$ | | | | |
| $L(S) = \infty$ | $L(S) = \infty$ | $L(S) = \infty$ | $L(S) = 7, \{P,Q\}$ | | | |
| $L(U) = \infty$ | $L(U) = \infty$ | $L(U) = \infty$ | $L(U) = \infty$ | $L(U) = 8, \{P,R,T\}$ | | |
| $L(V) = \infty$ | $L(V) = \infty$ | $L(V) = \infty$ | $L(V) = \infty$ | $L(V) = 10, \{P,R,T\}$ | $L(V) = 9, \{P,R,T,U\}$ | |
| $L(W) = \infty$ | $L(W) = \infty$ | $L(W) = \infty$ | $L(W) = \infty$ | $L(W) = \infty$ | $L(W) = 12, \{P,R,T,U\}$ | $L(W) = 11, \{P,R,T,U,V\}$ |

**Step 1**  We start by labelling $P(-, 0)$ since there is no vertex to precede it. Make it permanent.

**Step 2**  From $P$ there are two unlabelled vertices, $Q$ and $R$. Since $L(P) = 0$, Vertex $R$ gives the smallest $L(P) + w(P, R) = 0 + 2$, then we label $R(P, 2)$ and we add it to the path $S$. Make it permanent.

**Step 3**  Now $S$ has two vertices, $P$ and $R$. They have two unlabelled adjacent vertices $Q$ and $T$. Vertex $Q$ has the smallest $L(P) + w(P, Q) = 0 + 3 = 3$, $(L(R) + w(R, T) = 2 + 3 = 5$, and $(L(R) + w(R, Q) = 2 + 5 = 7)$, we make $Q (P, 3)$ permanent.

**Step 4**  Now $S$ has three vertices, $P$, $R$, and $Q$. They have two unlabelled adjacent vertices $S$ and $T$. Similar to the previous process; we make $T(R, 5)$ permanent.

**Step 5**  Now $S$ has four vertices, $P$, $R$, $Q$ and $T$. They have three unlabelled adjacent vertices $S$, $U$ and $V$. Similar to the previous process; we make $S(Q, 7)$ permanent.

**Step 6**  Now $S$ has five vertices, $P$, $R$, $Q$, $T$ and $S$. They have one unlabelled adjacent vertex $W$. Similar to the previous process; we make $U(T, 8)$ permanent and update $L(V)$.

**Step 7**  Similar to the above, we make $V(U, 9)$ permanent, and update $L(W)$ and make it permanent.

So the shortest path is $PRTUVW$ and it has a length of 11.

## Example 12.41

Find a shortest path from $a$ to $z$ in this graph.

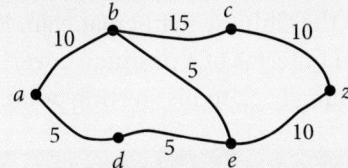

### Solution

We will follow the algorithm by labelling the graph without a table this time.

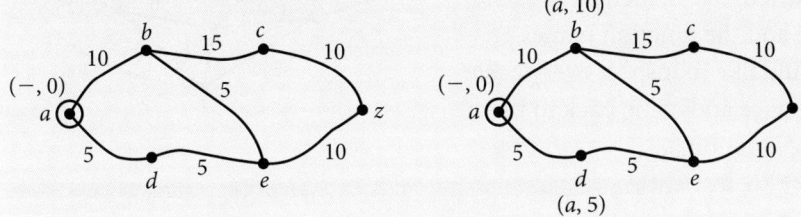

First we label and make $a$ permanent. Next we label vertices $d$ with $(a, 5)$ to indicate the length of the path and that it is visited through $a$. Similarly we label $b(a, 10)$.

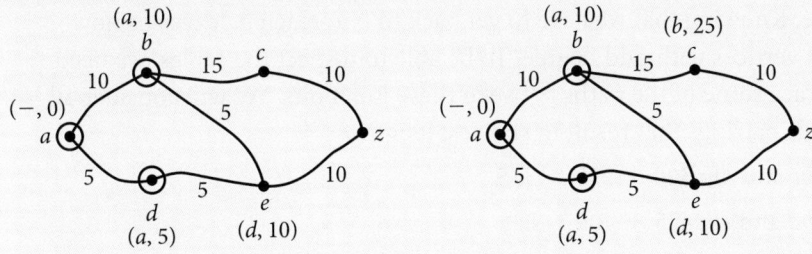

Next we make $d$ permanent and update vertex $e$. Then we make $b$ permanent and update vertices $c$ and $e$ (no change in $e$).

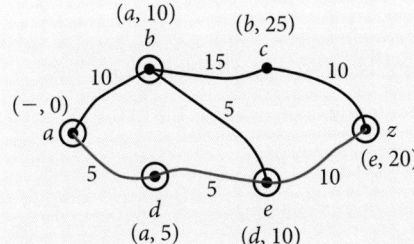

Next we make $e$ permanent and update $z$. At this point, we can make the label at $z$ permanent, a shortest path has been found, namely *adez*.

Note that in this example, it is not necessary to change the label of a vertex $v$ if $d + w(v_iv) \not< L(v)$, and that it is also unnecessary to make all vertices in the graph permanent as long as they don't contribute towards a shortest path.

## The route inspection problem

In 1962 the Chinese mathematician, Kwan Mei-Ko posed an inspection problem in terms of a postman covering each road of a network exactly once and coming back to his starting point.

### Example 12.42

A cable network has to be inspected for possible faulty wires. The diagram represents a sketch of the wires along with the length , in metres, of each section and the junction names. We would like to inspect every cable at least once and come back to the starting junction, $a$.

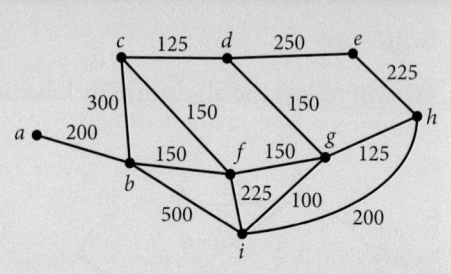

### Solution

The problem is similar to finding an Eulerian circuit. However, this is not possible since we have four vertices with odd degree: $a$, $c$, $d$, and $h$.

Since we are starting at $a$, $ab$ has to be retraced. This makes $b$ also with odd degree. Knowing that we have to get back to $b$ to reach $a$, leaves us now with 4 vertices with odd degree. To be able to inspect the cables, we need to retrace some of the paths between these junctions. We will consider all possible pairings that result in shortest lengths:

$$bc \text{ and } dh: 300 + 275 = 575$$

$$bd \text{ and } ch: 425 + 400 = 825$$

$$bh \text{ and } cd: 425 + 125 = 550$$

So, $bh$ and $cd$ is the shortest, and hence we will retrace these paths.

The original network has 2850 m, and we will retrace

$$ab = 200 \text{ and } bh + cd = 550$$

giving a total length of 3600 m. Such a route is: *abifcdgihgfbcdehgfba*.

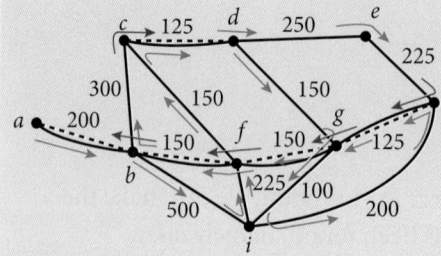

**Route Inspection algorithm**

Step 1 Find all vertices of odd degree.

Step 2 For each pair of odd vertices, find the path of shortest length.

Step 3 Pair up all odd vertices from step 2, so that the sum of the lengths is minimum.

Step 4 In the original graph, duplicate the shortest-length paths found in step 3.

Step 5 Find an Eulerian circuit containing all edges of the new graph.

As you can see, when the number of edges to be inspected is large, the process is tedious to follow. The Route Inspection algorithm proposed by Kuan Mei-Ko makes the process more systematic.

## Example 12.43

A guard patrols the campus of a large school given by the graph. The weights of the edges are distances given in metres. If he must pass through each street at least once during his shift, find the minimum distance the guard will cover.

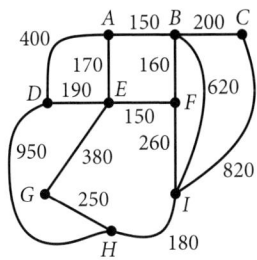

**Figure 12.42** Diagram for Example 12.43

### Solution

The sum of all the distances in the graph is 4880 metres.

There are four odd vertices: $A$, $D$, $F$ and $H$. We need to investigate all the possible pairings and then choose the shortest paths between pairs of vertices.

| Pairing | Shortest path | Distance |
|---------|---------------|----------|
| A,D | A E D | 360 |
| A,F | A B F | 310 |
| A,H | A B F I H | 750 |
| D,F | D E F | 360 |
| D,H | D E G H | 820 |
| F,H | F I H | 440 |

Now we need to look at the pairings that will include all four vertices and give us the minimum sum of the distances. The pairings are $A,D$ and $F,H$ and the paths that we will repeat are $AED$ and $FIH$ with their distances of 360 and 440 m.

So the minimum distance the guard will cover in one shift is $4880 + 360 + 440 = 5680$ m. One path is: $ADHGEDEFBIHIFICBAEA$.

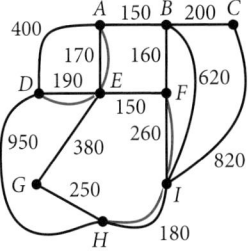

**Figure 12.43** Diagram for solution to Example 12.43

## Example 12.44

A truck collecting trash has to visit a neighbourhood with the street network shown in Figure 12.44. Give a possible plan that minimises the distance it has to travel. Distances are in kilometres.

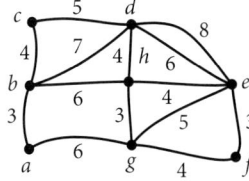

**Figure 12.44** Diagram for Example 12.44

### Solution

Vertices $d$ and $e$ are odd. So, we first duplicate the shortest path between them which is 6, and then try to find the minimum distance to be travelled.

Since all vertices are even, the graph is Eulerian. We can use the algorithm developed in Example 10 of chapter 3, or any other method to find the circuit.

If we start at $b$, we can create a cycle $bcdb$, which can be joined at $b$ with $hdefgh$, which can be joined at $h$ with $edegab$. Our route is then $bcdbhdefghedegab$ with length of $68 + 6$ (retracing $de$) $= 74$ km

This is not a unique solution. You can find other circuits with the same minimum length of 74 km.

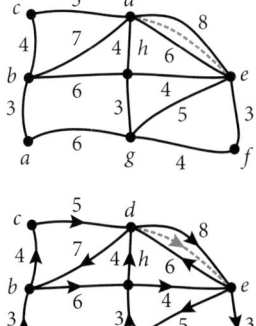

**Figure 12.45** Diagrams for solution to Example 12.44

## The travelling salesman problem

Given a set of cities and the cost of travel between each pair of them, the travelling salesman problem (TSP), is an attempt to find the cheapest way of visiting all of the cities and returning to your starting point.

The simplicity of the statement of the problem is misleading. The TSP is one of the most considered problems in computational mathematics and yet no successful solution method is known for the general case.

The TSP naturally arises in many transportation and logistics applications, for example, practical uses for the TSP include routing trucks for package pickups and material handling in warehouses. Other applications involve the scheduling of service calls at communications businesses, and many others.

Although transportation applications are the most natural setting for the TSP, the simplicity of the model has led to many interesting applications in other areas. A classic example is the scheduling of a machine to drill holes in a circuit board. In this case the holes to be drilled are the cities, and the cost of travel is the time it takes to move the drill head from one hole to the next. The technology for drilling varies from one industry to another, but whenever the travel time of the drilling device is a significant portion of the overall manufacturing process then the TSP can play a role in reducing costs.

Basically, the travelling salesman problem is related to the search for Hamiltonian cycles in a graph.

### Example 12.45

A salesman lives in Vienna. On a business trip by car, he will visit Prague, Munich, Zagreb, Budapest and Bratislava. The diagram shows the distances between the cities, in kilometres. (Not all routes have been included in the diagram). Find the length of the shortest cycle.

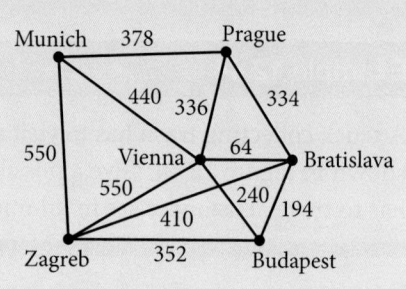

### Solution

There are several possible Hamiltonian cycles and for each one we calculate the total distance travelled.

The shortest cycle is the second one from Vienna to Prague, Munich, Zagreb, Budapest, Bratislava and back to Vienna, which has the total distance of 1874 km. Since every cycle can have two directions it is possible to visit all the cities in the reverse order.

| Cycle | Distance (km) |
|---|---|
| V Br P M Z Bu V | 1918 |
| V P M Z Bu Br V | 1874 |
| V M Z Bu Br P V | 2206 |
| V Z Bu Br P M V | 2066 |
| V Bu Br P M Z V | 2064 |

**Table 12.1** Table for solution to Example 12.45

The solution presented in the example is a trial and error approach. Are there other approaches?

Remembering that a Hamiltonian cycle is a cycle that visits every vertex in a connected graph exactly once, we see that the classical TSP is a Hamiltonian cycle with minimum length. However, as with the route inspection problem, we allow vertices to be visited more than once.

If you can inspect all possible routes involved in the TSP, you will be able to find the minimum total weight. However, as the number of vertices increases, checking all possibilities becomes impractical. There is an assumption that the graph under consideration is complete, and as such, theoretically the number of possible routes to inspect for a graph with $n$ vertices will be $\dfrac{(n-1)!}{2}$ (considering routes in reverse order). For example, if you have 5 cities, then the number of routes to be inspected will be 12 and if you have 10 cities the number will jump to 181 440. If there are 20 cities, then number will be $6.0 \times 10^{17}$. Even if you have a fast computer that can calculate 1 000 000 routes per second, it would take such a computer approximately 19 years to finish the task! So far, there is no known solution to the general TSP. Mathematicians resort to finding near-minimum-weight solutions. Many algorithms have been developed. The nearest neighbour algorithm, nearest insertion algorithm, cutting-plane methods, and branch-and-cut methods are a few such algorithms. The IB syllabus does not require you to use such algorithms and thus we will not discuss these concepts in detail. We will just demonstrate the use of one of the algorithms without requiring you to do it.

We will consider complete graphs with at least three vertices. Such graphs will have a Hamiltonian cycle. Moreover, since the number of vertices is finite, then the number of Hamiltonian cycles will also be finite. Thus there must be at least one with minimum weight.

Also, since the weights of the edges in the complete graphs represent the shortest distances between nodes of the original route network, the complete graph must satisfy the **triangle inequality**. The sum of two sides of a triangle must be larger than or equal to the third side. Thus for every choice of three vertices, $v_i$, $v_j$, and $v_k$ the following must be true

$$w(v_i, v_j) + w(v_i, v_k) \geqslant w(v_j, v_k)$$

where $w(v_i, v_j)$ is the weight of the corresponding edge.

**The nearest neighbour algorithm**
**Step 1** Choose a starting vertex.
**Step 2** Consider the edge of smallest weight incident to this vertex. If the other end of this edge is not visited yet, add it to the tour.
**Step 3** Repeat step 2 until all vertices have been visited.
**Step 4** Add the edge connecting the last visited vertex to the starting vertex.

This algorithm will sometimes produce a minimal Hamiltonian cycle, but in general, it may produce cycles with a weight that is considerably greater than the minimum weight.

We will use the complete graph in Figure 12.46 to demonstrate the algorithm. The salesman is to start and end at $A$.

Starting at $A$, the first edge is $AF$ since 6 is the minimum among 6, 8, 15, 18, and 20.

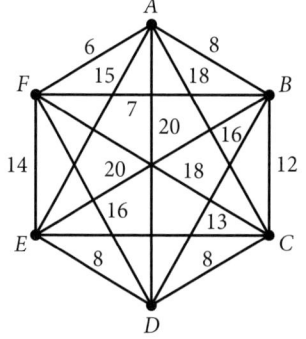

**Figure 12.46** The salesperson is to start and end at $A$

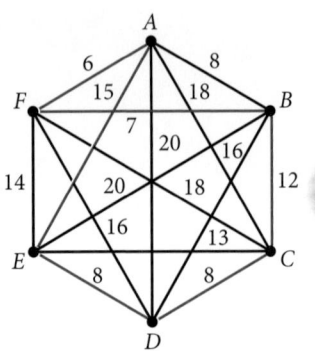

**Figure 12.47** The whole route has a total weight of 56

With the same reasoning, *FB* is chosen, with weight 7. *BC* is next with a weight of 12, followed by *CD* with 8, *DE* with 8 and finally, we get back to *A* with 15. The whole route has a total weight of 56, see Figure 12.47.

**The nearest insertion algorithm**

This algorithm creates a cycle in the graph and then enlarges it to include a vertex which is closest to the given cycle.

**Step 1** Choose a starting vertex, $u_1$.

**Step 2** Consider the edge of smallest weight incident to this vertex. Add it to the cycle C. The vertex at the other end of this edge is added to C, call it $u_2$.

**Step 3** Select an edge with minimum weight that joins a vertex in C to one not in C, call the new vertex *v*.

**Step 4** Next, we enlarge the cycle to include the new vertex *v*. Now we consider the expression
$$x = w(u_i, v) + w(u_j, v) - w(u_i, u_j)$$
We choose the pair of vertices $u_i$ and $u_j$ for which *x* is minimum, we then include the edges $(u_i, v)$ and $(u_j, v)$ and we remove $(u_i, u_j)$. [*x* represents the increase in the weight of the cycle when we add *v*].

**Step 5** Repeat steps 3 and 4 until we include all vertices in the cycle.

Let's apply the nearest insertion algorithm to Figure 12.46.

We start with *AF* as it is the smallest, then we add *B*. Now we have a cycle *AFB* as shown in the first diagram in Figure 12.48.

Now consider all possible cycle expansions by comparing the *x* values for adding any of the remaining vertices. Here are the values:

Consider vertex *C*: $AC + CF - AF = 18 + 18 - 6 = 30$,
$AC + CB - AB = 18 + 12 - 8 = 22$, $BC + CF - FB = 18 + 12 - 7 = 23$

Consider vertex *D*: $20 + 16 - 8 = 28$, $16 + 20 - 6 = 30$, $16 + 16 - 7 = 25$

Consider vertex *E*: $14 + 15 - 6 = 23$, $15 + 20 - 8 = 27$, $14 + 20 - 7 = 27$

So, 22 is the minimum, and since it corresponds to the connection of *C* and *A* and *B*, we add *AC* and *BC* and remove *AB*. Now the cycle is *AFBCA* and is given in the second diagram.

Repeat the same steps for the new cycle:

Consider vertex *D*: $8 + 16 - 12 = 12$, $8 + 20 - 18 = 10$, $16 + 20 - 6 = 30$, $16 + 16 - 7 = 25$

Consider vertex *E*: $14 + 15 - 6 = 23$, $14 + 20 - 7 = 27$, $15 + 13 - 18 = 10$, $20 + 13 - 12 = 21$

Thus *DC* and *DA* are added and *AC* removed. (Note that we could have added *E* at this stage instead of *D*).

Lastly, consider *E*: $15 + 8 - 20 = 3$, $14 + 20 - 7 = 27$, $14 + 15 - 6 = 23$, $20 + 13 - 12 = 21$, $13 + 8 - 8 = 13$. Thus we add *ED* and *DA* and remove *DA*. The route now has a weight of 56 as before.

The routes created by these two algorithms are not always equal. And neither of them will definitely produce a minimum weight Hamiltonian cycle.

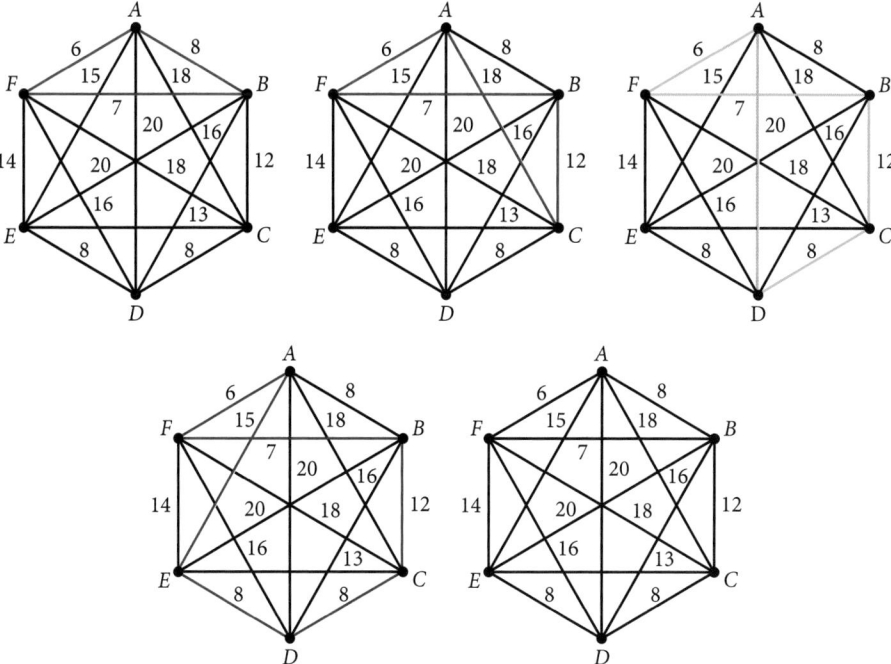

**Figure 12.48** Graphs for nearest insertion algorithm

---

### Example 12.46

Consider the graph in Figure 12.49a and use both algorithms to find a minimum TSP tour.

#### Solution

##### Nearest neighbour algorithm

Start at $A$, the next vertex must be $F$. From $F$, the edge with smallest weight leads to $C$, then similarly from $C$ to $D$, then to $B$, to $E$, and finally back to $A$. The total weight is 55.

##### Nearest insertion algorithm

The first cycle could be $AFE$ with weight of 30. See the first diagram in Figure 12.50. Considering $x$ values for possible expansion, we find that can be achieved by adding vertex $D$ with $x = 0$. We add $AD$ and $DE$ and remove $AE$ (second diagram in Figure 12.50). The weight so far is 30. Applying the algorithm again, we can add $C$ to the cycle by adding $FC$ and $CE$ and removing $FE$ (third diagram). The cycle $AFCEDA$ has a weight of 39 so far. Finally, we expand the cycle to include $B$ by adding $BD$ and $BE$ the have an $x$ value of 10, and removing $ED$ (fourth diagram). So, the cycle now is $AFCEBDA$ with a total weight of 49, which is less than what we achieved with the nearest neighbour algorithm.

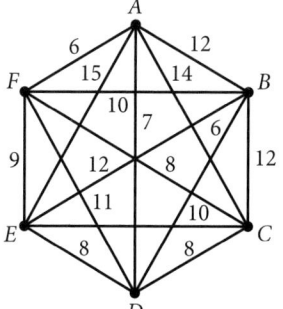

**Figure 12.49a** Diagram for Example 12.46

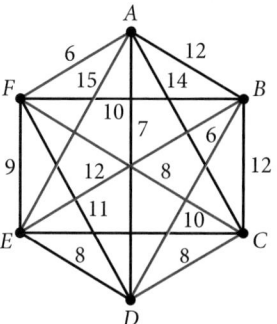

**Figure 12.49b** Diagram for the nearest neighbour algorithm solution to Example 12.46

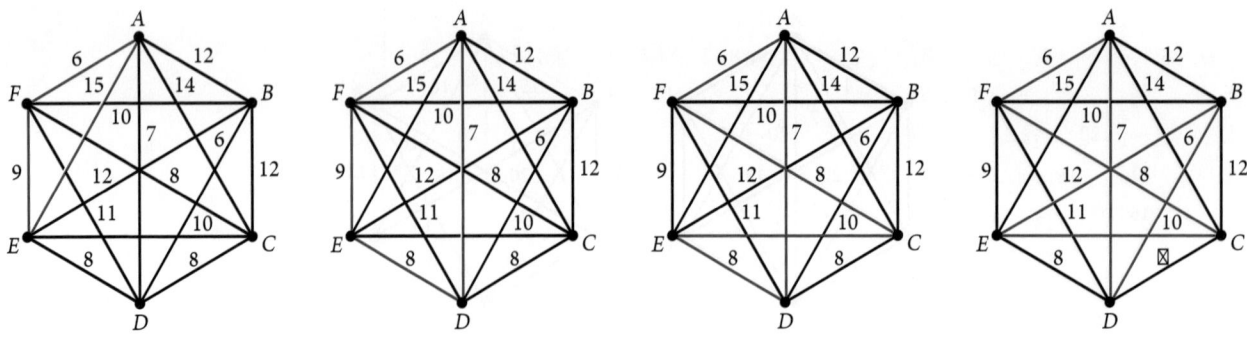

**Figure 12.50** Diagram for the nearest insertion algorithm solution to Example 12.46

## Lower and upper bounds

As Example 12.46 shows, the nearest neighbour algorithm did not lead us to a Hamiltonian cycle with minimum possible weight. As you observed, we were able to find an improved cycle. How far can we go?

A lower bound can be found by using algorithms that help us find minimal spanning trees. The argument is as follows. If we have a minimum weight Hamiltonian cycle in a complete graph, then we can remove one vertex $v$ and all edges incident to it. Then we have a minimal spanning tree passing through the rest of the vertices. The weight of the Hamiltonian cycle is the weight of this minimal spanning tree plus the total weight of the edges we just removed.

This argument leads to the following lower bound algorithm.

**Step 1** Choose a vertex $v$ in the complete graph and find the total of the two smallest edge weights incident to $v$.

**Step 2** Find the total weight of a minimum spanning tree going through all the remaining vertices.

**Step 3** The sum of the two totals is a lower bound.

Consider the graph in Example 12.46, for instance. Remove $A$ and its incident edges from the graph.

A minimum spanning tree for the remaining vertices is marked in green and has a weight of 30. The two edges with minimum total weight are $AF$ and $AD$ with a weight of 13. Hence a lower bound for the cycle is 43, which is less than the smallest we found, 49.

So, now we can say that the minimum weight for a Hamiltonian cycle lies between 43 and 49.

We used the weight of the Hamiltonian cycle we found earlier as an upper bound. There are a few ways of looking at an upper bound. One is to say the upper bound is the length of any cycle you manage to find, or, in general, is twice the length of a minimal spanning tree. The reason for this is a worst case scenario. That is, the travelling salesman would visit every city and return, that way tracing each edge of the spanning tree twice.

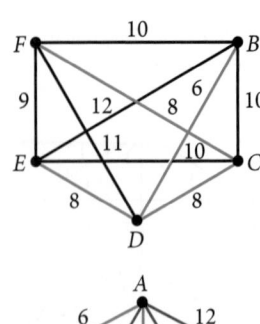

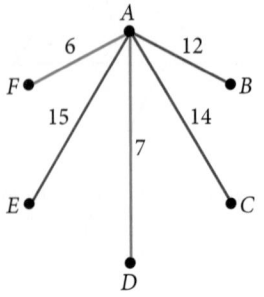

**Figure 12.51** Lower and upper bounds

## Example 12.47

We will try to find a lower bound, an upper bound, and a possible shortest route for the salesman in Vienna.

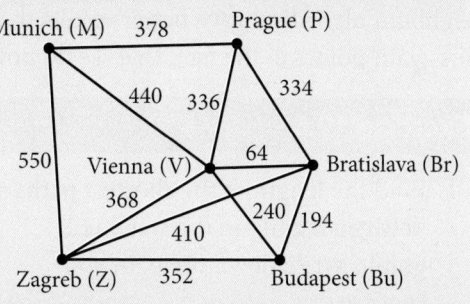

## Solution

The graph is not complete. However, in the TSP we are allowed to add new edges which represent the minimum weight between two vertices that are not adjacent in the original graph. For example, Budapest and Prague are not directly connected; however, a path of minimum length of $334 + 194 = 528$ km through Bratislava can be added. Similarly, Budapest–Munich can have an extra edge of length $440 + 240 = 680$ km added as well as Prague–Zagreb with 704 and Munich–Bratislava with 504 km. The new complete graph is shown in Figure 12.52.

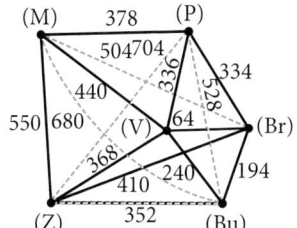

**Figure 12.52** Diagram for solution to Example 12.47

To find a lower bound, remove Vienna, for example, and all edges incident to it, and then find a minimum spanning tree for the rest. The tree weight is 1258 km.

The minimum total weight of two of the edges from Vienna is $64 + 240 = 304$ km and together with the minimum spanning tree give us a lower bound of 1562 km. Note that if we remove another city, we may receive a different lower bound! An upper bound may be the route weight we found earlier of 1874 km. Thus we are confident that our minimal Hamiltonian cycle would be between 1562 km and 1874 km.

Apply the nearest insertion algorithm. You will expand the cycles, starting with V, Br, Bu and you will get the sequence shown in Figure 12.53.

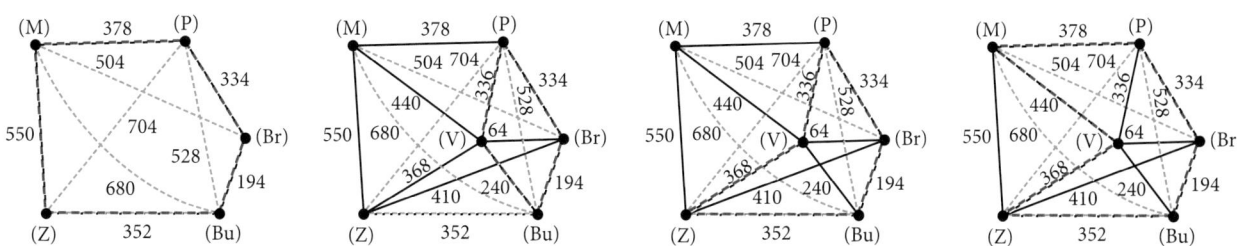

**Figure 12.53** Nearest insertion algorithm

Unfortunately, the algorithm here did not yield the best results. The length of the route is 2066 km, which is greater than the upper bound.

Applying the nearest neighbour algorithm yields Figure 12.54

507

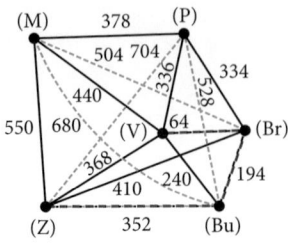

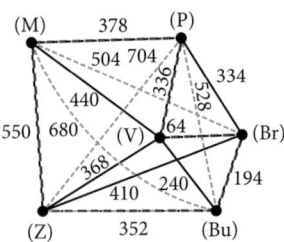

**Figure 12.54** Nearest neighbour algorithm

The total weight of this route is 1874 km, the same as the method we used at the start of this section and which we used as a lower bound. Note that the nearest neighbour algorithm gave better results than the nearest insertion algorithm. This again points to the fact, that we do not have a unique solution to the TSP.

## Exercise 12.6

1. Find the length of the shortest path between $a$ and $f$ in the weighted graph. Write down the path you suggest.

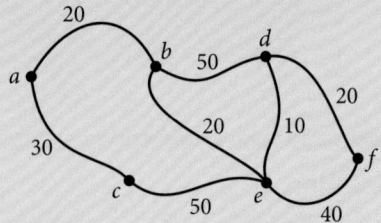

2. Find the length of the shortest path between $A$ and $H$ in the following weighted graph. Write down the path you find.

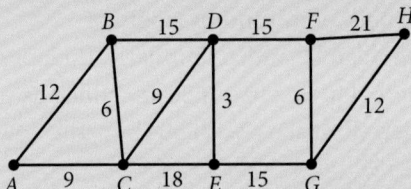

3. A circuit board has the following sub-network. The weight on each edge represents the time (in microseconds) that it takes a DC signal to flow through it. Find the minimum time it takes a signal to go from $a$ to $u$. Write down the path that gives this time.

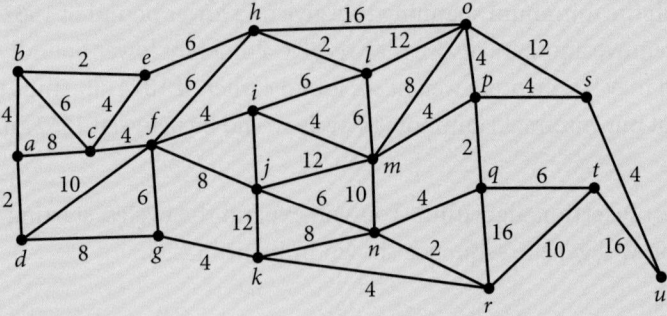

4. Find the shortest route between $a$ and $d$ in question **1**.

5. Find the shortest route between $A$ and $F$, and between $B$ and $H$ in question **2**.

6. Solve the TSP for this graph.

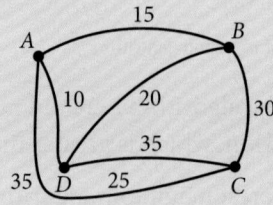

**7.** Solve the TSP for this graph.

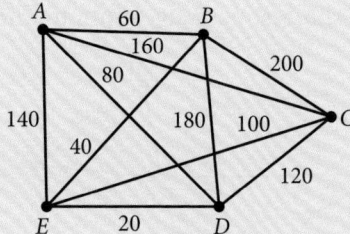

**8.** Find the route with the least total cost if a tourist wants to visit each of the cities once. The weight on each edge is the cheapest possible return flight between the two cities. The prices are in Euros.

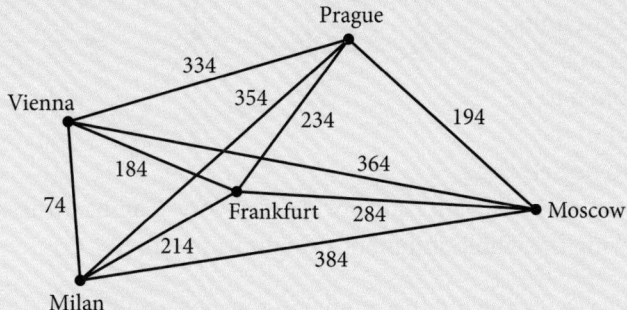

**9.** Find the route with the least total cost if a tourist wants to visit each of the cities once. The weight on each edge is the cheapest possible return flight between the two cities. The prices are in Euros.

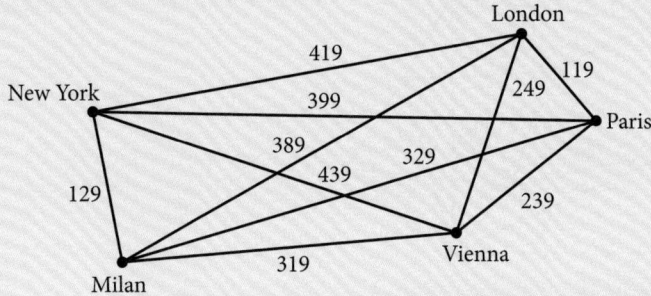

**10.** The nodes $A$, $B$, $C$, $D$, and $E$ in a network have to be connected with the minimum length of cable. The distances between the nodes are given in the table. Find the most efficient set up.

|   | A | B | C | D | E |
|---|---|---|---|---|---|
| **A** |   | 100 | 90 | 80 | 110 |
| **B** | 100 |   | 130 |   | 120 |
| **C** | 90 | 130 |   | 120 |   |
| **D** | 80 |   | 120 |   | 130 |
| **E** | 110 | 120 |   | 130 |   |

11. Use a shortest path algorithm to find the shortest route from *a* to *e*.

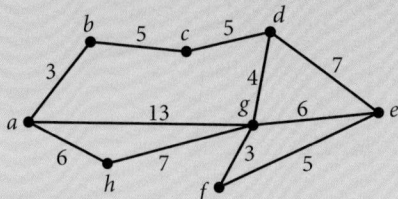

12. The graph is a network of a transport company where the weights of the edges are distances in 10 km units. A shipment has to be transported from *a* to *l*. However, a part of the shipment has to be delivered to *f* first. Find the most efficient route for this delivery. Compare the result to delivering the whole shipment straight from *a* to *l*.

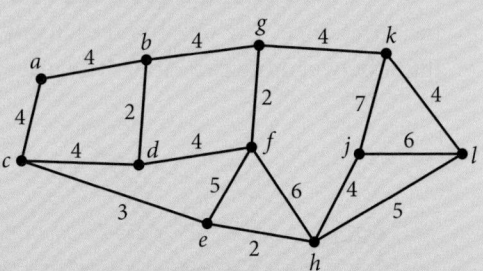

13. You are in charge of organising the campaign tour for a politician. The map has distances between the different cities he must visit. He is based in *E* and needs to get back there at the end of the tour. Find a suitable tour of minimum length.

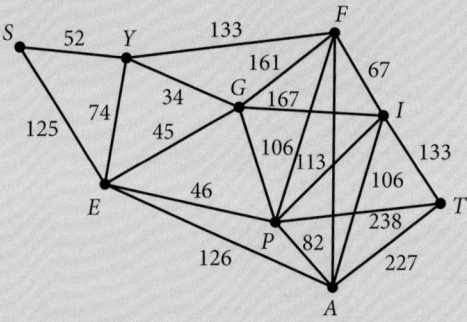

14. A road sweeper truck is required to sweep all the streets in a block of the city for which a map is supplied. Distances are in hundreds of metres. Find a route of minimum total length.

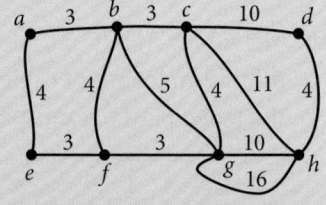

15. A local telephone network has to be inspected for possible defects. Find the shortest possible inspection tour to ensure that all cables have been checked. The sketch gives the length of each cable in hundreds of metres.

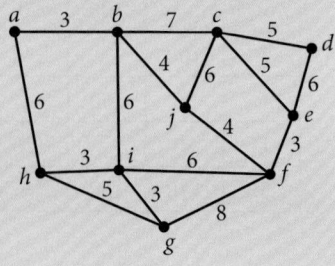

1. Use Dijkstra's algorithm to find the shortest path between the vertices $B$ and $F$.

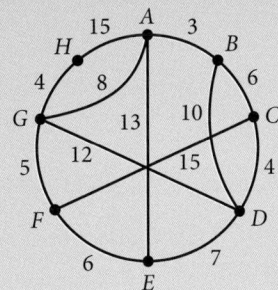

2. The bus routes connecting various cities and the cost of the tickets in dollars are given in the table.

| City | A | B | C | D | E | F |
|------|---|---|---|---|---|---|
| A | – | 25 | 42 | – | 55 | 28 |
| B |   | – | 15 | 63 | – | 17 |
| C |   |   | – | 12 | 20 | – |
| D |   |   |   | – | 22 | 40 |
| E |   |   |   |   | – | 10 |
| F |   |   |   |   |   | – |

   (a) Draw the weighted graph that represents all the routes between the cities.

   (b) Jerry would like to travel from $A$ to $D$. Determine the cheapest route and find how much will Jerry pay for his travel.

3. Ravi has a band and he needs to post the posters in his neighbourhood for the upcoming concert in a club. The graph represents the plan of the posts in the neighbourhood and Ravi's home is denoted by the vertex $A$. The distances between the posts are given in kilometres.

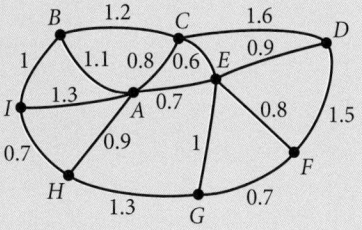

   Find the shortest distance Ravi will need to travel in order to put the posters on all the posts before coming back home.

4. Jenny collects frequent flyer points and she has earned over $230\,000$ points with a particular airline. The cost of the tickets in points between the cities is given in the matrix below. Each entry represents thousands of points.

$$C_G = \begin{bmatrix} 0 & 0 & 10 & 0 & 20 & 25 & 10 \\ 0 & 0 & 10 & 18 & 54 & 0 & 0 \\ 10 & 10 & 0 & 8 & 0 & 50 & 0 \\ 0 & 18 & 8 & 0 & 0 & 0 & 45 \\ 20 & 54 & 0 & 0 & 0 & 28 & 32 \\ 25 & 0 & 50 & 0 & 28 & 0 & 16 \\ 10 & 0 & 0 & 45 & 32 & 16 & 0 \end{bmatrix}$$

(a) Draw a weighted graph representing the possible flights between the cities with the corresponding cost in points.

(b) Jenny would like to make a round trip and visit all the cities. What is the cheapest route, and will she have enough points for such a trip or she will need to buy some additional points to pay for the trip?

5. Jack is a night guard and during the night shift he must patrol through every single corridor of a warehouse. The plan of the corridors is given in this graph. The time needed to patrol each corridor is given in minutes.

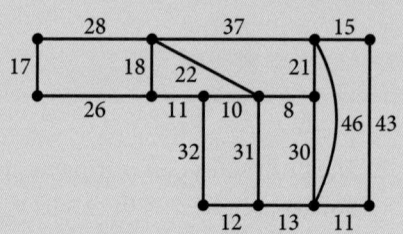

Is it possible for Jack to patrol the whole warehouse during his night shift from 10 p.m. to 6 a.m.? If yes, how many minutes will he have for a break? If not, how much longer he would need to stay in order to fulfil his duty?

6. Cycle $C_n$, $n \geq 3$, is a graph in which every vertex has an order of 2. Wheel $W_n$, $n \geq 3$, is a graph that consists of a cycle $C_n$ and an additional point that is connected to all the vertices in the cycle. Here are some examples of cycles and wheels.

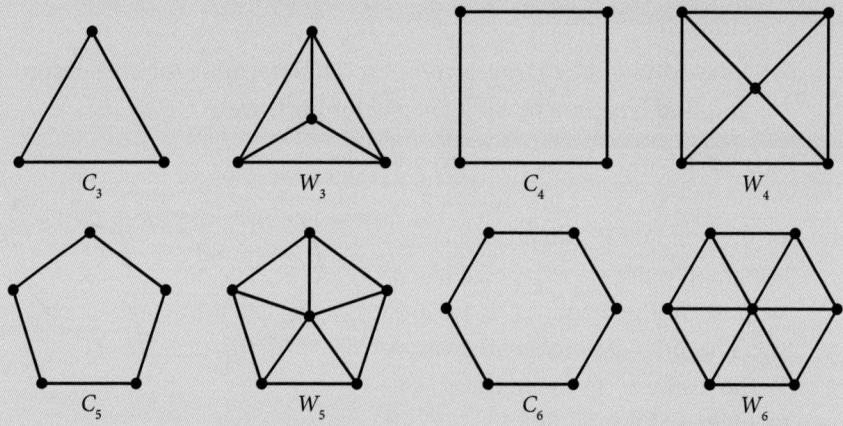

(a) Show that the number of edges in a wheel $W_n$ is twice the number of edges in a cycle $C_n$

(b) Show that in $C_4$ there are $2^{n-1}$ paths of length $n$ between

(i) adjacent vertices, when $n$ is odd

(ii) non-adjacent vertices, when $n$ is even.

7. Apply Kruskal's and Prim's algorithms to find the minimum spanning tree for this graph.

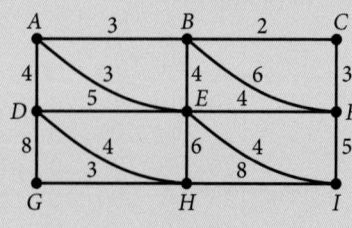

Show all the decision steps in both algorithms. Draw the minimum spanning tree and state its weight.

**8.** The information on the distances between some cities in a country is provided in the table. The distances are given in kilometres.

| City | P | Q | R | S | T | U |
|------|-----|-----|-----|-----|-----|-----|
| P | – | – | – | – | – | – |
| Q | 200 | – | – | – | – | – |
| R | 292 | 487 | – | – | – | – |
| S | 333 | 465 | 222 | – | – | – |
| T | 86 | 282 | 203 | 257 | – | – |
| U | 333 | 509 | 133 | 97 | 235 | – |

The government would like to construct a system of highways connecting all the cities. Determine which highways should be built so that the cost of the construction is minimised. Assume that the cost of a kilometre of highway is constant.

**9.** Peter needs to install sockets that will be connected by an optical cable in his apartment so that he can watch TV, use his phone and the internet in the rooms. The positions of the sockets are shown on the graph. The distances between the sockets are given in metres.

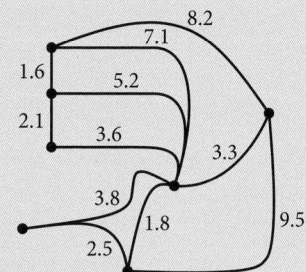

Given that the cost of optical cable is 70 cents per metre, find the minimum price Peter will pay for the cable.

**10.** A parent–teacher organisation (PTO) at an international school has 6 people working for it. They are Adam, Bernard, Cecile, Donatella, Eva and Flor. They can communicate in at least one language according to the table.

| Name | English | Spanish | French | German |
|------|---------|---------|--------|--------|
| Adam | ✓ | ✓ | ✓ | |
| Bernard | ✓ | | ✓ | ✓ |
| Cecile | | | ✓ | |
| Donatella | | ✓ | ✓ | |
| Eva | | | | ✓ |
| Flor | | ✓ | | |

**(a)** Draw a graph indicating which people can communicate with each other.

**(b)** Cecile ordinarily communicates with Flor with the help of Donatella. Unfortunately Donatella has gone to visit her mother. Can Cecile still communicate with Flor? Write how she can.

**(c)** Without whom is it impossible to communicate with all the members of the PTO? Give your reasons.

**11.** Let $k_n$ be the complete graph of order $n$ and $k_{m,n}$ be a bipartite graph of orders $m$ and $n$.

    **(a)** Explain the following, giving one example of each:

        **(i)** $k_5$, the complete graph of order 5

        **(ii)** a bipartite graph $k_{3,3}$

    **(b)** Show that $k_{3,3}$ has a Hamiltonian cycle, giving appropriate reasons.

**12.** The floor plan shows the ground level of a new home. Is it possible to enter the house through the front door and exit through the rear door, going through each internal doorway exactly once? Give a reason for your answer.

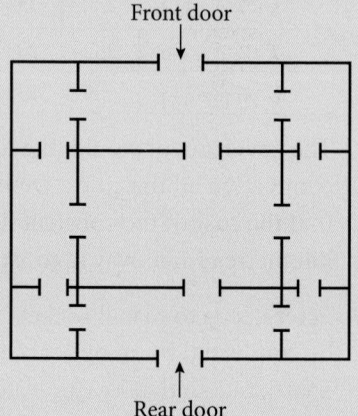

**13.** Apply Prim's algorithm to the weighted graph to obtain the minimal spanning tree starting with the vertex $A$.

Find the weight of the minimal spanning tree.

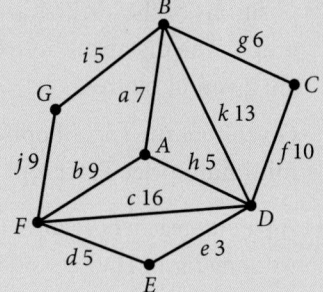

**14.** The network of cities served by an airline is shown in the diagram, with edges representing direct connections.

    **(a)** Copy and complete the table below which shows the least number of edges connecting each pair of cities in this network. (Such a table gives the least number of stops required between cities in this region.)

|   | A | B | C | D | E | F |
|---|---|---|---|---|---|---|
| **A** |   |   |   |   |   |   |
| **B** |   |   | 1 |   |   |   |
| **C** |   |   |   |   |   |   |
| **D** |   |   |   |   |   |   |
| **E** |   |   | 2 |   |   |   |
| **F** |   |   |   |   |   |   |

**(b)** The accessibility index is used to determine how easy it is to get to a particular city. To calculate the accessibility index for a given city, the total of each column is divided by the degree of the vertex representing it. The most accessible city is the one with the smallest accessibility index, and the least accessible city has the largest accessibility index.

    **(i)** Which city in this region is the most accessible and which city is the least accessible? Give your reasons.

    **(ii)** A new flight is added between cities $A$ and $C$. With this change, which city is the most accessible and which city is the least accessible? Show your working.

**15.** Graphs $U$ and $V$ with 8 vertices each are shown.

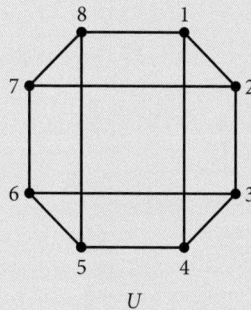 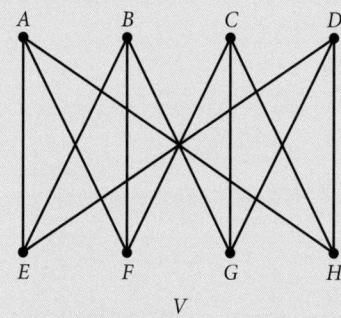

    $U$                                     $V$

**(a)** Set up an adjacency matrix for graph $U$.

**(b)** By setting up an appropriate adjacency matrix for $V$, show that the two graphs are the same.

**(c)** Determine whether $V$ is planar.

**16.** In an offshore drilling site for a large oil company, the distances between the planned wells are given below in metres.

|    | 1   | 2   | 3   | 4   | 5   | 6   | 7   | 8   | 9   | 10  |
|----|-----|-----|-----|-----|-----|-----|-----|-----|-----|-----|
| 2  | 30  |     |     |     |     |     |     |     |     |     |
| 3  | 40  | 60  |     |     |     |     |     |     |     |     |
| 4  | 90  | 190 | 130 |     |     |     |     |     |     |     |
| 5  | 80  | 200 | 10  | 160 |     |     |     |     |     |     |
| 6  | 70  | 40  | 20  | 40  | 130 |     |     |     |     |     |
| 7  | 60  | 120 | 50  | 90  | 30  | 60  |     |     |     |     |
| 8  | 50  | 140 | 90  | 70  | 140 | 70  | 40  |     |     |     |
| 9  | 40  | 170 | 140 | 60  | 50  | 90  | 50  | 70  |     |     |
| 10 | 200 | 80  | 150 | 110 | 90  | 30  | 190 | 90  | 100 |     |
| 11 | 150 | 30  | 200 | 120 | 190 | 120 | 60  | 190 | 150 | 200 |

**(a)** It is intended to construct a network of paths to connect the different wells in a way that minimises the sum of the distances between them.

    Use Prim's algorithm to find a network of paths of minimum total length that can span the whole site.

**(b)** Pipes are laid under water. Well 1 has the largest amount of oil to be pumped per day, and Well 11 is designed to be the main transportation hub. The only possible connections to be made between wells are shown in the diagram.

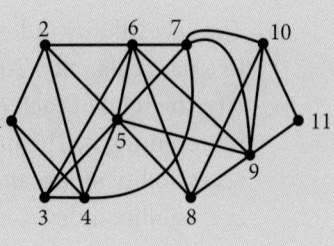

The associated cost for each pipe, in $100,000s, are given in the table. Use Dijkstra's algorithm to find the path with minimum cost that can transport oil from well 1 to well 11.

|    | 1 | 2 | 3 | 4 | 5 | 6 | 7 | 8 | 9 | 10 |
|----|---|---|---|---|---|---|---|---|---|----|
| 2  | 6 |   |   |   |   |   |   |   |   |    |
| 3  | 3 |   |   |   |   |   |   |   |   |    |
| 4  | 8 | 7 | 2 |   |   |   |   |   |   |    |
| 5  |   | 14| 12| 6 |   |   |   |   |   |    |
| 6  |   | 16| 19|   | 7 |   |   |   |   |    |
| 7  |   |   |   | 24| 20| 29|   |   |   |    |
| 8  |   |   |   |   | 23| 15|   |   |   |    |
| 9  |   |   |   |   | 56| 30| 41| 50|   |    |
| 10 |   |   |   |   |   |   | 42| 25| 40|    |
| 11 |   |   |   |   |   |   |   |   | 32| 22 |

**17.** Let *G* be the graph shown in Figure 12.55.

**(a)** Has *G* got an Eulerian circuit? Give a reason for your answer.

**(b)** What is the adjacency matrix of the graph *G*? Determine how many walks of length 2 are there from vertex *A* to vertex *C*.

**(c)** Use Kruskal's algorithm to find the minimum spanning tree for graph *G*.

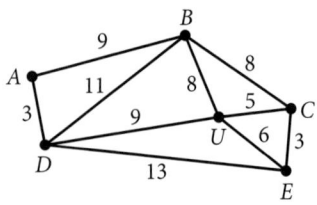

**Figure 12.55** Graph for question 17

**18.** The diagram shows a weighted graph.

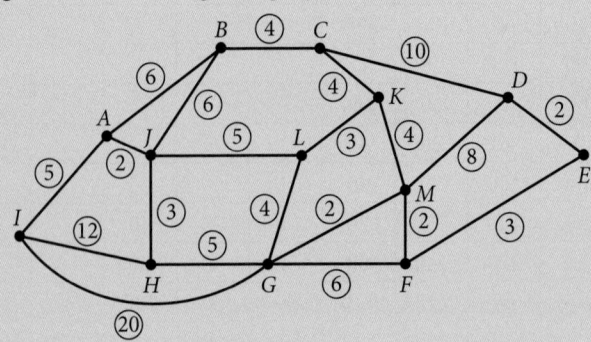

Use Kruskal's algorithm to find a minimum spanning tree as follows.

**(a)** State the major steps in the algorithm.

**(b)** Execute the algorithm showing the steps in your work.

**(c)** Sketch the minimum spanning tree found, and write down its weight.

**19.** Consider the two graphs $G$ and $H$.

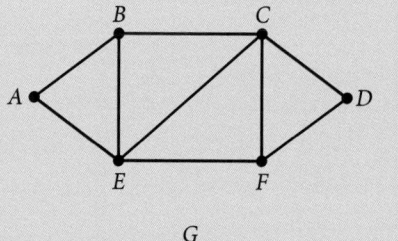 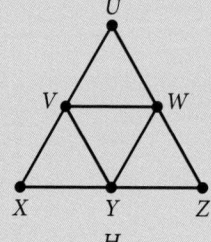

(a) Find an Eulerian trail for graph $G$ starting with vertex $B$.

(b) State a result which shows that the graph $H$ has an Eulerian circuit.

**20.** The diagram shows a weighted graph.

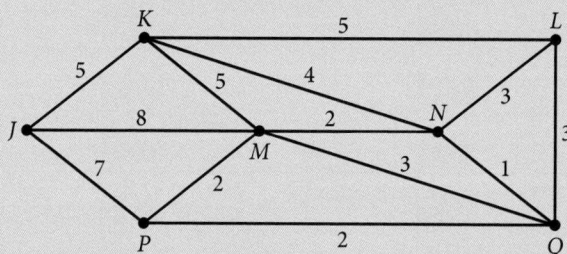

Use Prim's algorithm to find a minimal spanning tree, starting at $J$. Draw the tree, and find its total weight.

**21.** The floor plan of a certain building is shown. There are four rooms $A$, $B$, $C$, $D$ and doorways are indicated between the rooms and to the exterior $O$.

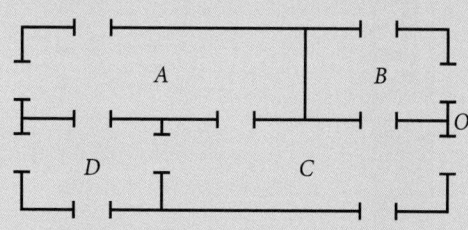

(a) Draw a graph by associating a vertex with each room using the letters $A$, $B$, $C$, $D$, and $O$. If there is a door between the two rooms, draw an edge joining the corresponding two vertices.

(b) Does the graph in part (a) possess an Eulerian trail? Give a reason for your answer. What does your answer mean about the floor plan?

(c) Does the graph in part (a) possess a Hamiltonian cycle? Give a reason for your answer. What does your answer mean about the floor plan?

**22.** Consider the matrix $M$, shown in Figure 12.56.

(a) Draw a planar graph $G$ with 5 vertices $A$, $B$, $C$, $D$, $E$ such that $M$ is its adjacency matrix.

(b) Give a reason why $G$ has an Eulerian circuit.

(c) Find an Eulerian circuit for $G$.

(d) Find a spanning tree for $G$.

$$M = \begin{array}{c c} & \begin{array}{c c c c c} A & B & C & D & E \end{array} \\ \begin{array}{c} A \\ B \\ C \\ D \\ E \end{array} & \begin{pmatrix} 0 & 1 & 1 & 1 & 1 \\ 1 & 0 & 0 & 2 & 1 \\ 1 & 0 & 0 & 2 & 1 \\ 1 & 2 & 2 & 0 & 1 \\ 1 & 1 & 1 & 1 & 0 \end{pmatrix} \end{array}$$

**Figure 12.56** Matrix for question 22

**23.** Let $G$ be a weighted graph with 6 vertices $L$, $M$, $N$, $P$, $Q$, and $R$. The weight of the edges joining the vertices is given in the table.

|   | $L$ | $M$ | $N$ | $P$ | $Q$ | $R$ |
|---|---|---|---|---|---|---|
| $L$ | – | 4 | 3 | 5 | 1 | 4 |
| $M$ | 4 | – | 4 | 3 | 2 | 7 |
| $N$ | 3 | 4 | – | 2 | 4 | 3 |
| $P$ | 5 | 3 | 2 | – | 3 | 4 |
| $Q$ | 1 | 2 | 4 | 3 | – | 5 |
| $R$ | 4 | 7 | 3 | 4 | 5 | – |

For example, the weight of the edge joining the vertices $L$ and $N$ is 3.

(a) Use Prim's algorithm to draw a minimum spanning tree starting at $M$.

(b) What is the total weight of the tree?

**24.** Let $G_1$ and $G_2$ be two graphs whose adjacency matrices are shown here.

$G_1$

|   | $A$ | $B$ | $C$ | $D$ | $E$ | $F$ |
|---|---|---|---|---|---|---|
| $A$ | 0 | 2 | 0 | 2 | 0 | 0 |
| $B$ | 2 | 0 | 1 | 1 | 0 | 1 |
| $C$ | 0 | 1 | 0 | 1 | 2 | 1 |
| $D$ | 2 | 1 | 1 | 0 | 2 | 0 |
| $E$ | 0 | 0 | 2 | 2 | 0 | 2 |
| $F$ | 0 | 1 | 1 | 0 | 2 | 0 |

$G_2$

|   | $a$ | $b$ | $c$ | $d$ | $e$ | $f$ |
|---|---|---|---|---|---|---|
| $a$ | 0 | 1 | 3 | 0 | 1 | 2 |
| $b$ | 1 | 0 | 1 | 3 | 2 | 0 |
| $c$ | 3 | 1 | 0 | 2 | 1 | 3 |
| $d$ | 0 | 3 | 2 | 0 | 2 | 0 |
| $e$ | 1 | 2 | 1 | 2 | 0 | 1 |
| $f$ | 2 | 0 | 3 | 0 | 1 | 0 |

(a) Which one does not have an Eulerian trail? Give a reason for your answer.

(b) Find an Eulerian trail for the other graph.

**25.** The matrix here is the adjacency matrix of a graph $H$ with 6 vertices $A$, $B$, $C$, $D$, $E$, $F$.

(a) Show that $H$ is not planar.

(b) Find a planar subgraph of $H$ by deleting one edge from it.

(c) Show that any subgraph of $H$ (excluding $H$ itself) is planar.

$$\begin{array}{c} \quad\; A \; B \; C \; D \; E \; F \\ \begin{array}{c} A \\ B \\ C \\ D \\ E \\ F \end{array} \left(\begin{array}{cccccc} 0 & 1 & 0 & 1 & 1 & 0 \\ 1 & 0 & 1 & 0 & 0 & 1 \\ 0 & 1 & 0 & 1 & 1 & 0 \\ 1 & 0 & 1 & 0 & 0 & 1 \\ 1 & 0 & 1 & 0 & 0 & 1 \\ 0 & 1 & 0 & 1 & 1 & 0 \end{array}\right) \end{array}$$

**26.** Let $G$ be the graph in Figure 12.57.

(a) Find the total number of Hamiltonian cycles in $G$, starting at vertex $A$. Explain your answer.

(b) (i) Find a minimum spanning tree for the subgraph obtained by deleting $A$ from $G$.

(ii) Hence, find a lower bound for the travelling salesman problem for $G$.

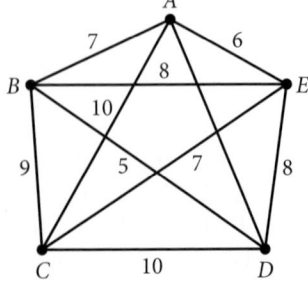

**Figure 12.57** Graph for question 26

**(c)** Give an upper bound for the travelling salesman problem for the graph above.

**(d)** Show that the lower bound you have obtained is not the best possible for the solution to the travelling salesman problem for $G$.

**27.** A graph $G$ has this adjacency matrix:

$$
\begin{array}{c@{\ }ccccccc}
 & A & B & C & D & E & F \\
A & 0 & 1 & 0 & 1 & 0 & 0 \\
B & 1 & 0 & 1 & 0 & 2 & 0 \\
C & 0 & 1 & 0 & 1 & 0 & 0 \\
D & 1 & 0 & 1 & 0 & 1 & 1 \\
E & 0 & 2 & 0 & 1 & 0 & 1 \\
F & 0 & 0 & 0 & 1 & 1 & 0 \\
\end{array}
$$

**(a)** Draw the graph $G$.

**(b)** Explain why $G$ has an Eulerian circuit. Find such a circuit.

**(c)** Find a Hamiltonian path.

**28.** Consider this graph.

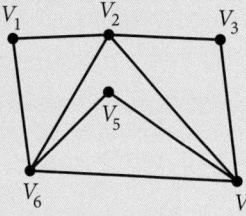

**(a)** Show that this graph has

    **(i)** an Eulerian circuit     **(ii)** a Hamiltonian cycle.

**(b)** The edge joining $V_2$ and $V_6$ is removed. Does the graph still have an Eulerian circuit and a Hamiltonian cycle? Give reasons for your answers.

**(c)** Replace the edge joining $V_2$ and $V_6$, and remove the edge joining $V_1$ and $V_2$

    **(i)** Find an Eulerian trail.     **(ii)** Find a Hamiltonian path.

**29.** Christian plays a computer game in which he must pass certain rooms and in each room he collects some points. The points in the first level of the game are given in the following matrix.

$$
C_G = \begin{pmatrix}
0 & 2 & 3 & 4 & 0 & 0 & 2 & 0 & 0 \\
2 & 0 & 3 & 2 & 3 & 0 & 0 & 0 & 0 \\
3 & 3 & 0 & 0 & 0 & 4 & 0 & 0 & 3 \\
4 & 2 & 0 & 0 & 2 & 0 & 4 & 0 & 0 \\
0 & 3 & 0 & 2 & 0 & 5 & 0 & 4 & 0 \\
0 & 0 & 4 & 0 & 5 & 0 & 0 & 0 & 2 \\
2 & 0 & 0 & 4 & 0 & 0 & 0 & 6 & 0 \\
0 & 0 & 0 & 0 & 4 & 0 & 6 & 0 & 5 \\
0 & 0 & 3 & 0 & 0 & 2 & 0 & 5 & 0 \\
\end{pmatrix}
$$

In order to advance to the next level of the game he must visit all the rooms in the shortest possible time. Find the maximum possible points Christian can collect at the first level.

**30.** The weights of the edges of a graph with vertices $A$, $B$, $C$, $D$ and $E$ are given in the table.

|   | A | B | C | D | E |
|---|---|---|---|---|---|
| **A** | – | 10 | 15 | 11 | 16 |
| **B** | 10 | – | 12 | 19 | 13 |
| **C** | 15 | 12 | – | 18 | 14 |
| **D** | 11 | 19 | 18 | – | 17 |
| **E** | 16 | 13 | 14 | 17 | – |

(a) Use any method to find an upper bound for the travelling salesman problem for this graph.

(b) (i) Use Kruskal's algorithm to find and draw a minimum spanning tree for the subgraph obtained by removing the vertex $E$ from the graph.

(ii) State the total weight of this minimum spanning tree and hence find a lower bound for the travelling salesman problem for this graph.

**31.** The weights of the edges of a complete graph $G$ are shown in the following table.

|   | A | B | C | D | E | F |
|---|---|---|---|---|---|---|
| **A** | – | 5 | 4 | 7 | 6 | 2 |
| **B** | 5 | – | 6 | 3 | 5 | 4 |
| **C** | 4 | 6 | – | 8 | 1 | 6 |
| **D** | 7 | 3 | 8 | – | 7 | 3 |
| **E** | 6 | 5 | 1 | 7 | – | 3 |
| **F** | 2 | 4 | 6 | 3 | 3 | – |

(a) Determine whether or not $G$ is planar.

(b) Starting at $B$, use Prim's algorithm to find and draw a minimum spanning tree for $G$. Your solution should indicate the order in which the vertices are added. State the total weight of your tree.

**32.** Consider the adjacency matrices for the graphs $G_1$ and $G_2$:

(a) Draw the graphs of $G_1$ and $G_2$.

(b) For each graph, giving a reason, determine whether or not it

(i) is simple

(ii) is connected

(iii) is bipartite

(iv) is a tree

(v) has an Eulerian trail, giving an example of a trail if one exists.

$$G_1: \begin{array}{c c} & \begin{matrix} p & q & r & s & t \end{matrix} \\ \begin{matrix} p \\ q \\ r \\ s \\ t \end{matrix} & \begin{pmatrix} 0 & 1 & 0 & 1 & 0 \\ 1 & 0 & 2 & 0 & 1 \\ 0 & 2 & 0 & 1 & 0 \\ 1 & 0 & 1 & 0 & 1 \\ 0 & 1 & 0 & 1 & 0 \end{pmatrix} \end{array}$$

$$G_2: \begin{array}{c c} & \begin{matrix} P & Q & R & S & T \end{matrix} \\ \begin{matrix} P \\ Q \\ R \\ S \\ T \end{matrix} & \begin{pmatrix} 0 & 0 & 0 & 1 & 1 \\ 0 & 0 & 0 & 1 & 0 \\ 0 & 0 & 0 & 1 & 0 \\ 1 & 1 & 1 & 0 & 0 \\ 1 & 0 & 0 & 0 & 0 \end{pmatrix} \end{array}$$

**33.** The weights of the edges in a simple graph $G$ are given in this table:

| Vertex | A | B | C | D | E | F |
|--------|---|---|---|---|---|---|
| A | – | 4 | 6 | 16 | 15 | 17 |
| B | 4 | – | 5 | 17 | 9 | 16 |
| C | 6 | 5 | – | 15 | 8 | 14 |
| D | 16 | 17 | 15 | – | 15 | 7 |
| E | 15 | 9 | 8 | 15 | – | 18 |
| F | 17 | 16 | 14 | 7 | 18 | – |

(a) Use Prim's algorithm, starting with vertex $F$, to find and draw the minimum spanning tree for $G$. Your solution should indicate the order in which the edges are introduced.

(b) Use your tree to find an upper bound for the travelling salesman problem for $G$.

**34.** Let $H$ be the weighted graph shown.

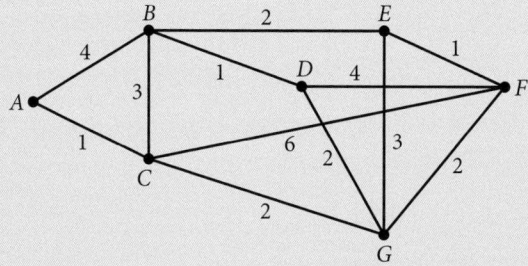

(a) (i) Name the two vertices of odd degree.

(ii) State the shortest path between these two vertices.

(iii) Using the route inspection algorithm, or otherwise, find a walk, starting and ending at $A$, of minimum total weight which includes every edge at least once.

(iv) Calculate the weight of this walk.

(b) Write down a Hamiltonian cycle in $H$.

**35.** The weighted graph $G$ is shown. Graph $G'$ is produced by deleting vertex $A$ from $G$.

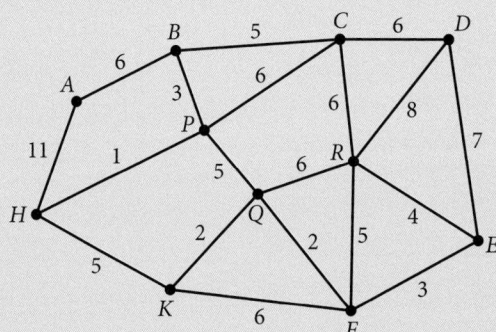

(a) Use Kruskal's algorithm to find the minimum spanning tree of graph $G'$ and state its weight.

(b) Hence find the weight of a lower bound for the Hamiltonian cycle in $G$ beginning at vertex $A$.

(c) For a complete graph with $n$ vertices, prove that no more than $\dfrac{(n-1)!}{2}$, $n \geqslant 3$, Hamiltonian cycles have to be examined to find the Hamiltonian cycle of least weight.

(d) How many cycles in $G$ would have to be examined to find the one with the least weight?

36. (a) The diagram shows the graph $G$.

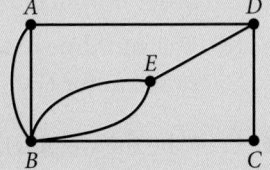

   (i) Explain briefly why $G$ has no Eulerian circuit.

   (ii) Determine whether or not $G$ is bipartite.

   (iii) Write down the adjacency matrix of $G$. Hence find the number of walks of length 4 beginning at $A$ and ending at $C$.

(b) The cost adjacency matrix of a graph with vertices $P, Q, R, S, T, U$ is shown.

Use Dijkstra's algorithm to find the length of the shortest path between the vertices $P$ and $S$. Show all the steps used by the algorithm and list the order of the vertices in the path.

|   | P | Q | R | S | T | U |
|---|---|---|---|---|---|---|
| P | – | 8 | – | – | – | 4 |
| Q | 8 | – | 7 | – | 2 | 3 |
| R | – | 7 | – | 6 | 3 | – |
| S | – | – | 6 | – | 9 | – |
| T | – | 2 | 3 | 9 | – | 7 |
| U | 4 | 3 | – | – | 7 | – |

37. The weights of the edges of a graph are given in the following table.

|   | A | B | C | D | E | F | G | H | I |
|---|---|---|---|---|---|---|---|---|---|
| A |   | 5 |   | 2 |   |   |   |   |   |
| B | 5 |   | 4 | 3 | 5 | 6 |   |   |   |
| C |   | 4 |   |   |   | 3 |   |   |   |
| D | 2 | 3 |   |   | 7 |   | 6 | 8 |   |
| E |   | 5 |   | 7 |   | 1 |   | 3 |   |
| F |   | 6 | 3 |   | 1 |   |   | 4 | 4 |
| G |   |   | 6 |   |   |   |   | 4 |   |
| H |   |   | 8 | 3 | 4 | 4 |   |   | 2 |
| I |   |   |   |   | 4 | 2 |   |   |   |

(a) Starting at $A$, use Prim's algorithm to find a minimum spanning tree for this graph.

(b) Draw this minimum spanning tree and state its weight.

**38.** A graph $G$ has $e$ edges and $n$ vertices.

    **(a)** Show that the sum of the degrees of the vertices is twice the number of edges.

    **(b)** Deduce that $G$ has an even number of vertices of odd degree.

    **(c) (i)** Graph $G$ is connected, planar and divides the plane into exactly four regions. If $(n-1)$ vertices have degree three and exactly one vertex has degree $d$, determine the possible values of $(n, d)$.

        **(ii)** For each possible $(n, d)$, draw a graph which satisfies the conditions described in **(i)**.

**39.** The graph $G$ has the following cost adjacency matrix.

|   | A | B | C | D | E |
|---|---|---|---|---|---|
| **A** | – | 9 | – | 8 | 4 |
| **B** | 9 | – | 7 | – | 2 |
| **C** | – | 7 | – | 7 | 3 |
| **D** | 8 | – | 7 | – | 5 |
| **E** | 4 | 2 | 3 | 5 | – |

    **(a)** Draw $G$ in a planar form.

    **(b)** Giving a reason, determine the maximum number of edges that could be added to $G$ while keeping the graph both simple and planar.

    **(c)** List all the distinct Hamiltonian cycles in $G$ beginning and ending at $A$, noting that two cycles each of which is the reverse of the other are to be regarded as identical. Hence determine the Hamiltonian cycle of least weight.

**40. (a)** The matrix here shows the distances between towns $A$, $B$, $C$, $D$ and $E$.

|   | A | B | C | D | E |
|---|---|---|---|---|---|
| **A** | – | 5 | 7 | 10 | 6 |
| **B** | 5 | – | 2 | 9 | – |
| **C** | 7 | 2 | – | 3 | 8 |
| **D** | 10 | 9 | 3 | – | – |
| **E** | 6 | – | 8 | – | – |

        **(i)** Draw the graph, in its planar form, that is represented by the matrix.

        **(ii)** Write down, with reasons, whether or not it is possible to find an Eulerian trail in this graph.

        **(iii)** Solve the route inspection problem with reference to this graph if $A$ is to be the starting and finishing point. Write down the walk and determine the length of the walk.

    **(b)** Show that a graph cannot have exactly one vertex of odd degree.

**41. (a) (i)** Let $M$ be the adjacency matrix of a bipartite graph. Show that the leading diagonal entries in $M^{37}$ are all zero.

        **(ii)** What does the $(i, j)$th element of $M + M^2 + M^3$ represent?

    **(b)** Prove that a graph containing a triangle cannot be bipartite.

    **(c)** Prove that the number of edges in a bipartite graph with $n$ vertices is less than or equal to $\dfrac{n^2}{4}$

# 12 | Graph theory

**42.**

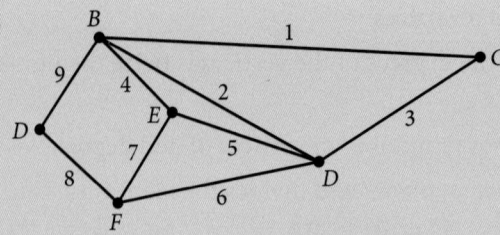

The diagram shows the weighted graph G.

(a) Determine whether or not G is bipartite.

(b) (i) Write down the adjacency matrix for G.

   (ii) Find the number of distinct walks of length 4 beginning and ending at A.

(c) Starting at A, use Prim's algorithm to find and draw the minimum spanning tree for G. Your solution should indicate clearly the way in which the tree is constructed.

# Introduction to differential calculus

## 13

# 13 Introduction to differential calculus

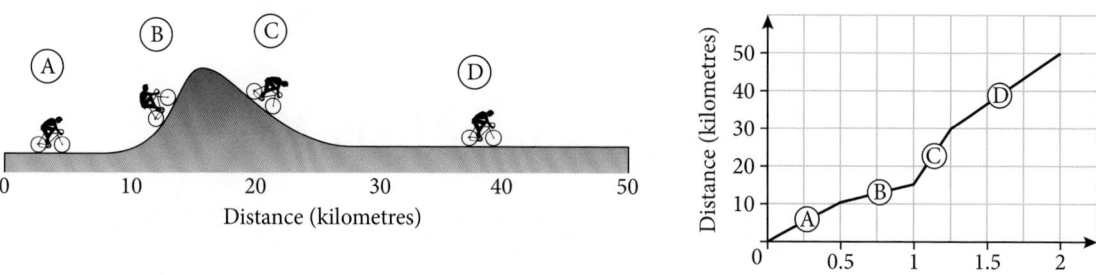

**Figure 13.1** The graph is a distance–time graph for a 50-kilometre bicycle ride that included going up and then down a hill. The cyclist's speed is least in interval B and greatest in interval C. The cyclist's speed is about the same in intervals A and D. The shape of the distance–time graph gives information about the speed of the cyclist during a certain interval and at a particular moment (instant) during the ride

Calculus is the branch of mathematics that was developed to analyse and model continuous change rather than discrete change. The change in the number of students graduating from your school from one year to the next is **discrete**: it only has integer values. The change in the distance you've travelled over any time interval while riding a bicycle can be an infinite number of values, depending on the precision of the distance measurements. In this case, the change in distance is **continuous**. Calculus is a tremendously useful tool for studying and modelling real-life phenomena which often display continuous change – and when we say change, we really mean a **rate of change**. For example, $20\,\mathrm{km\,h^{-1}}$ is a rate of change (change of distance with respect to time). The mathematical techniques that you have previously studied have limitations when it comes to modelling continuous change. For example, consider the curve in Figure 13.2 illustrating the motion of an object by indicating the distance ($y$ metres) travelled after a certain amount of time ($t$ seconds). Without calculus, we can compute the **average velocity** between two different times (Figure 13.3) but not the velocity at a specific time or instant; that is, the **instantaneous velocity**.

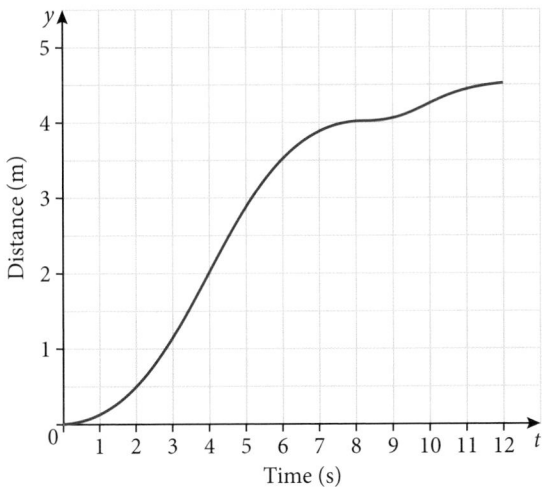

**Figure 13.2** Distance–time graph for an object's motion

**Figure 13.3** Computing average velocity from a distance–time graph

Calculus is divided into two main parts: **differential calculus** and **integral calculus**. In this chapter we will introduce differential calculus.

## 13.1 Limits and instantaneous rate of change

The limit of a function is a fundamental concept of calculus. Without limits, calculus would not exist.

### The idea of a limit

In this course, we will estimate the value of a limit by using technology to produce an appropriate table and/or graph.

### Example 13.1

Imagine that you are given 50 cm of wire and are told to use the wire to form a rectangle with the largest possible area. What dimensions should the rectangle have?

#### Solution

Let $b$ represent the base (or width) of the rectangle and $h$ represent the height (or length) of the rectangle. Since the perimeter of the rectangle is 50 cm, then $2b + 2h = 50$. It follows that $b + h = 25$ and $h = 25 - b$. Thus, the area, $A$, of the rectangle expressed as a function of $b$ is

$$A(b) = b(25 - b) = 25b - b^2$$

We can use technology (e.g. a spreadsheet, as in the screenshot to the right) to create a table showing values of $A$ determined by different values of $b$ to get an idea for the dimensions of a rectangle with maximum area.

|   | b | A(b) |
|---|---|---|
| = |   | =25*b–b^2 |
| 1 | 12 | 156 |
| 2 | 12.2 | 156.16 |
| 3 | 12.4 | 156.24 |
| 4 | 12.5 | 156.25 |
| 5 | 12.6 | 156.24 |
| 6 | 12.8 | 156.16 |
| 7 | 13 | 156 |

**Table 13.1** Spreadsheet data for solution to Example 13.1

527

When $f(x)$ becomes arbitrarily close to a unique fixed number $L$ as $x$ approaches $c$ from either side, then we say that the **limit** of $f(x)$ as $x$ approaches $c$ is $L$. In mathematical notation, this is written as $\lim\limits_{x \to c} f(x) = L$

It appears that the maximum area occurs when $b = 12.5$ cm. Using the terminology of limits, we can say that the limit of the area as $b$ approaches 12.5 cm is 156.25 cm².

When $b = 12.5$ cm the value of $h$ is also 12.5 cm. Thus, the rectangle of maximum area with a perimeter of 50 cm is a square with sides of 12.5 cm.

In Example 13.1, $A(b)$ got closer and closer to 156.25 as $b$ approached 12.5 from either direction (and was equal to 156.25 when $b = 12.5$). We can write this limit as $\lim\limits_{b \to 12.5} A(b) = 156.25$

Hence, in this example $\lim\limits_{x \to c} f(x) = f(c)$. However, for many limits we are not able to evaluate $f(c)$. That is, it's often not possible to evaluate the limit of a function as the input approaches some number $c$ by directly substituting $c$ into the functions, as Example 13.2 illustrates.

| $x$ | $f(x)$ |
|------|-----------|
| 2.90 | 5.90 |
| 2.99 | 5.99 |
| 3.00 | undefined |
| 3.01 | 6.01 |
| 3.10 | 6.10 |

**Table 13.2** Values of $f(x)$ in Example 13.2

### Example 13.2

Find $\lim\limits_{x \to 3} \dfrac{x^2 - 9}{x - 3}$

### Solution

Notice that the function $f(x) = \dfrac{x^2 - 9}{x - 3}$ is not defined at $x = 3$ (division by zero). However, the graph and table of values indicate that as $x$ approaches 3 from both the left and right, $f$ approaches 6. Therefore, we can write $\lim\limits_{x \to 3} f(x) = 6$

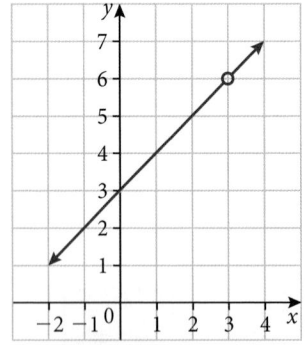

**Figure 13.4** Graph of $f$ in Example 13.2

The value of each of the limits in Examples 13.1 and 13.2 was a finite number (156.25 in Example 13.1, and 6 in Example 13.2). Also, each of these limits was evaluated as the variable in question approached a specific number (12.5 in Example 13.1, and 6 in Example 13.2). The value of some limits do not approach a finite number and instead increase without bound (goes to positive infinity) or decrease without bound (goes to negative infinity). Also, for some limits we need to evaluate (or estimate) the limit as the variable goes to positive or negative infinity. Another common feature to Examples 13.1 and 13.2 is that we were interested in the limit as the variable approached a specific number from both directions. Some of these scenarios are illustrated in Examples 13.3 and 13.4.

Infinity is not a number. The symbol $\infty$ is used to indicate that a quantity or measure increases indefinitely.

## Example 13.3

Find the value of each of the limits, if it exists.

(a) $\lim\limits_{x \to 3^+} \dfrac{x^2 + 9}{x - 3}$  (b) $\lim\limits_{x \to 3^-} \dfrac{x^2 + 9}{x - 3}$

The limits in Example 13.3 are left-hand and right-hand limits because we are interested in the value of the limit as $x$ gets closer to 3 coming from values less than 3 (from the left) or as $x$ gets closer to 3 coming from values greater than 3 (from the right).

### Solution

(a) Although the function $f(x) = \dfrac{x^2 + 9}{x - 3}$ is very similar to the function in the limit for Example 13.2, its graph is significantly different.

The graph of $f(x) = \dfrac{x^2 + 9}{x - 3}$, along with the vertical line $x = 3$ (dashed), is shown in the diagram. As $x$ approaches 3 from the positive side, the graph goes up indefinitely; getting closer and closer to the line $x = 3$ (a vertical asymptote). Thus, $\lim\limits_{x \to 3^+} \dfrac{x^2 + 9}{x - 3}$ increases without bound. The limit does not have a numerical value – it does not exist. Although not mathematically correct, it is acceptable to express this result as

$$\lim\limits_{x \to 3^+} \dfrac{x^2 + 9}{x - 3} = \infty$$

(b) As $x$ approaches 3 from the negative side, the graph goes down indefinitely; getting closer and closer to the line $x = 3$.

Thus, $\lim\limits_{x \to 3^-} \dfrac{x^2 + 9}{x - 3}$ decreases without bound (does not exist) which

we can write as $\lim\limits_{x \to 3^-} \dfrac{x^2 + 9}{x - 3} = -\infty$

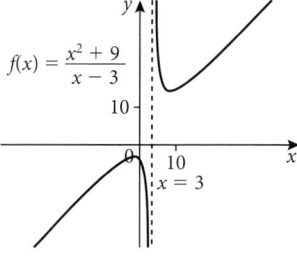

**Figure 13.5** Diagram for the solution to Example 13.3

## Example 13.4

Find $\lim\limits_{x \to \infty} \left(1 + \dfrac{1}{x}\right)^x$

### Solution

Surprisingly, the value of this limit appears to be a specific number that is approximately 2.718 (rounded to 4 significant figures). That is,

$$\lim\limits_{x \to \infty} \left(1 + \dfrac{1}{x}\right)^x \approx 2.718$$

You may recognise that this is the approximate value of the irrational number e which has been mentioned previously in Chapters 1 and 9.

| $x$ | $\left(1 + \dfrac{1}{x}\right)^x$ |
|---|---|
| 1. | 2. |
| 10. | 2.59374 |
| 25. | 2.66584 |
| 50. | 2.69159 |
| 100. | 2.70481 |
| 500. | 2.71557 |
| 1000. | 2.71692 |
| 10000. | 2.71815 |
| 100000. | 2.71827 |

**Table 13.3** Table for the solution to Example 13.4

## Instantaneous rate of change

What is the difference between average rate of change and instantaneous rate of change?

## Example 13.5

The distance between Lusaka and Livingstone in Zambia is approximately 500 km. On a vacation, it took one of the textbook authors 5 hours to drive from Lusaka to Livingstone. What was the author's average rate of change?

### Solution

In this context, average rate of change refers to the author's average speed, given by $\dfrac{\Delta \text{ distance}}{\Delta \text{ time}}$ for the 5-hour time interval.

So, the author's average speed is $\dfrac{500}{5} = 100 \text{ km h}^{-1}$

> The Greek letter $\Delta$ (capital delta) is commonly used to represent change. For example, the expression $\Delta x$ means 'the change in $x$'; $\Delta$ distance means 'the change in distance.'

Does that mean that the author was always driving at $100 \text{ km h}^{-1}$? Of course not, $100 \text{ km h}^{-1}$ represents an average rate of change. We are also interested in the instantaneous rate of change. In a car, you can see your instantaneous rate of change at any moment by simply looking at the speedometer. One of the fundamental applications of differential calculus is determining the instantaneous rate of change of a function.

## Example 13.6

Write down the (instantaneous) rate of change for the function $f(x) = 3x - 2$

### Solution

For a linear function, the average rate of change and instantaneous rate of change are the same constant value which is simply the gradient of the line. Since the gradient is the coefficient of the variable ($x$ in this case), the instantaneous rate of change is 3.

> Remember that gradient and slope mean the same thing.

The gradient of a line is constant, but what about the gradient of the graph of a function that is not a straight line?

To develop an approach that will help answer this question, consider an experiment where a toy car rolls down an inclined ramp (Figure 13.6).

> The word 'curve' is often quivalent to the word 'function', even if the function is linear.

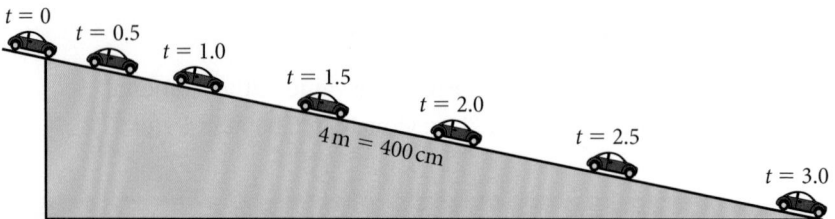

**Figure 13.6** Toy car rolling down an inclined ramp

It takes the toy car 3 seconds to travel the entire length of the 4-metre ramp. At intervals of half a second, the distance the car is from the top of the ramp is recorded (Table 13.4).

Speed is the rate of change of distance with respect to time.

Velocity is the rate of change of position with respect to time: velocity $= \dfrac{\Delta \text{ position}}{\Delta \text{ time}}$. We already know how to calculate the average velocity of the car. For example, the average velocity of the car

for the interval from $t = 0$ to $t = 3$ is $\dfrac{\Delta \text{ position}}{\Delta \text{ time}} = \dfrac{400 \text{ cm}}{3 \text{ s}} \approx 133 \text{ cm s}^{-1}$

But, how do we find the instantaneous velocity at $t = 1$?

| Time (s) | Distance (cm) |
|---|---|
| 0 | 0 |
| 0.5 | 28 |
| 1.0 | 69 |
| 1.5 | 127 |
| 2.0 | 203 |
| 2.5 | 294 |
| 3.0 | 400 |

**Table 13.4** Time and distance data for toy car rolling down an inclined ramp

You may have carried out this experiment in a science course. How does the average velocity for the toy car vary over different time intervals?

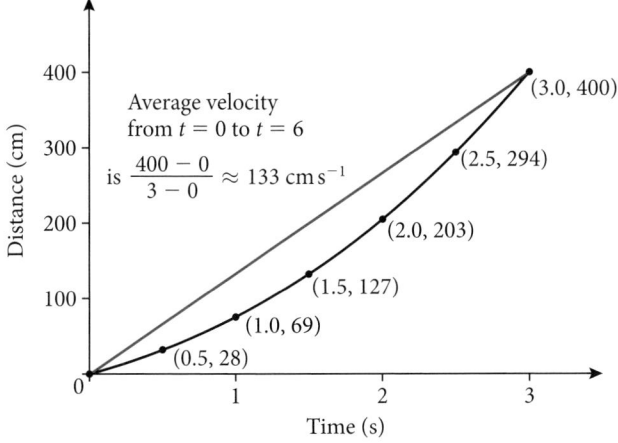

**Figure 13.7** Distance–time graph showing the average velocity from $t = 0$ to $t = 3$

Suppose that we want to find the instantaneous velocity at time $t = 1$

Figures 13.8 and 13.9 show secant lines whose gradients are the average velocity for the intervals $1 < t < 3$ and $1 < t < 2.5$ respectively. As we decrease the upper endpoint of the interval from $t = 3$ to $t = 2.5$, we move it closer to $t = 1$. We can continue this process by moving the upper endpoint even closer to $t = 1$ (Figure 13.10). As we do this, the gradient of the secant line (average velocity) approaches the gradient of the line that is tangent to the curve at $t = 1$ (Figure 13.11).

You should be familiar with expressing speed as $\dfrac{\Delta \text{ distance}}{\Delta \text{ time}}$. Here we use velocity and position instead of speed and distance because we want to allow negative values.

A secant (or chord) is a line or line segment that intersects the graph of a curve at two points.

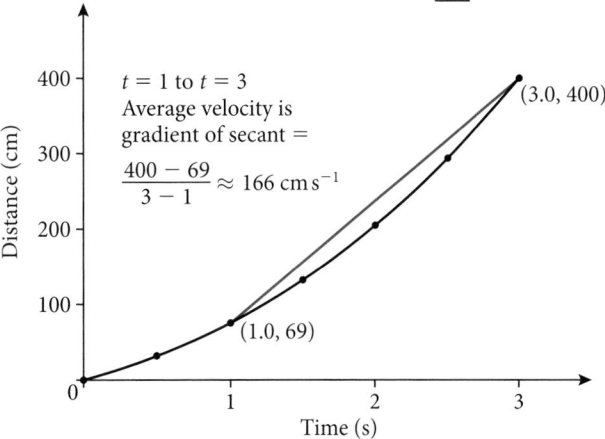

**Figure 13.8** Average velocity from $t = 1$ to $t = 3$

531

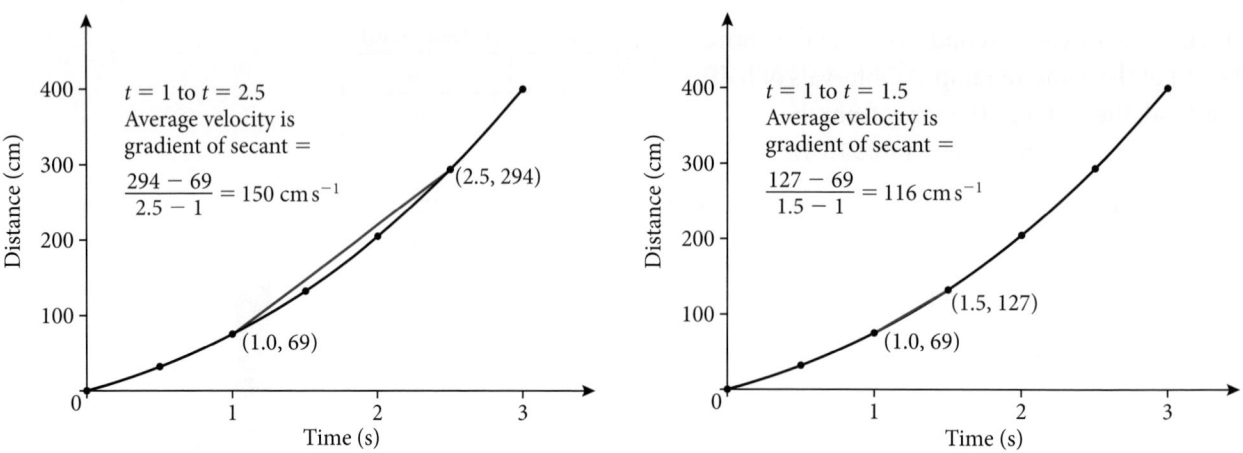

**Figure 13.9** Average velocity from $t = 1$ to $t = 2.5$

**Figure 13.10** Average velocity from $t = 1$ to $t = 1.5$

Using the language of limits, we can say that the gradient of the **tangent** to the distance–time graph at $t = 1$ (Figure 13.11) is the limit of the gradient of the secant line as the time interval gets narrower and narrower focusing on $t = 1$

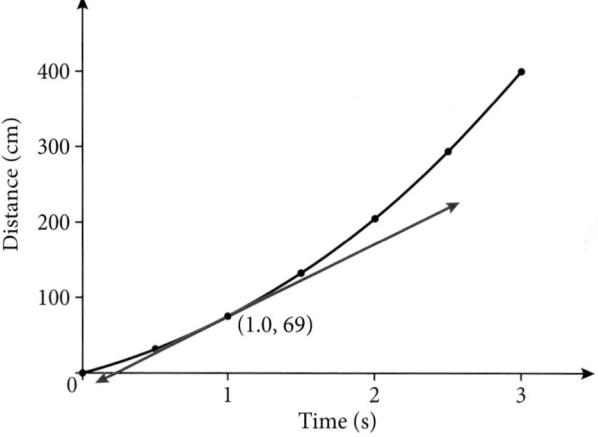

**Figure 13.11** Velocity at $t = 1$ is the limit of the average velocity computed over a narrowing time interval focused on $t = 1$

A line that is tangent to a curve just touches the curve at a point (the point of tangency) but may intersect with the curve at another point. Graphical illustration of a tangent:

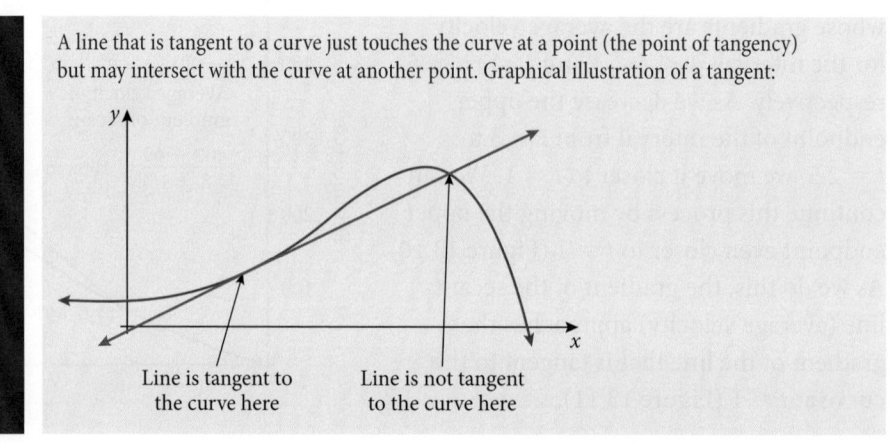

We are interested in finding the gradient of the tangent to the curve at $t = 1$. We can approximate this by finding the gradient of secants for progressively smaller intervals. However, the sequence of intervals will approximate the gradient only at $t = 1$. We want to find the instantaneous rate of change so we need a different sequence for each instant in time.

For a linear function, we can represent the gradient for all points on the line with a single number, i.e. the gradient of the line. For a non-linear function, the gradient can have many different values, depending on the point we are interested in. Therefore, we cannot represent the gradient for all the points on a graph that curves (non-linear) with a single number. Instead, we will need another function – called the **gradient function** or **derivative** – that will give us the gradient at any point. We will examine the gradient function in the next section. We can also estimate the gradient by drawing a tangent by eye and estimating the gradient of the tangent.

---

**Example 13.7**

The distance, $d$, in metres, covered by a child travelling along a horizontal waterslide after time $t$, in seconds, is shown in the graph.

(a) Describe the velocity of the child during the first two seconds of her journey.

(b) By drawing suitable tangents to the graph, estimate the velocity of the child at points $A$ and $B$.

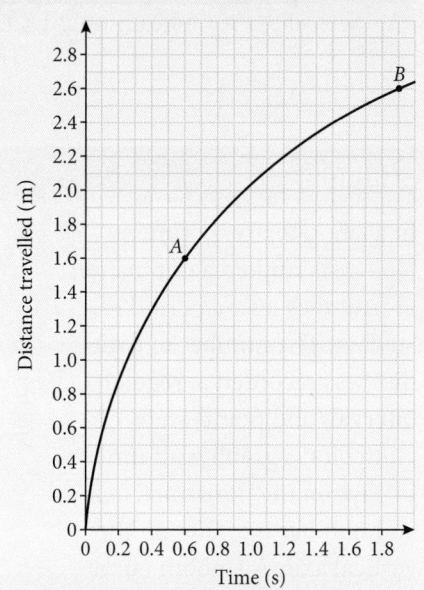

**Solution**

(a) The distance–time graph is steepest at $t = 0$, and its gradient decreases as $t$ increases. This can be seen using the tangents to the graph at $t = 0, 0.1, 0.25$ and $0.425$, shown in blue on the diagram. The child moves fastest upon touching the waterslide and is slowed down by friction throughout the duration of the ride.

(b) By drawing a suitable tangent at points $A$ and $B$ (in red) and calculating the gradient, the velocity of the child at $A$ is approximately $1.25 \text{ m s}^{-1}$ and at $B$ is approximately $0.5 \text{ m s}^{-1}$.

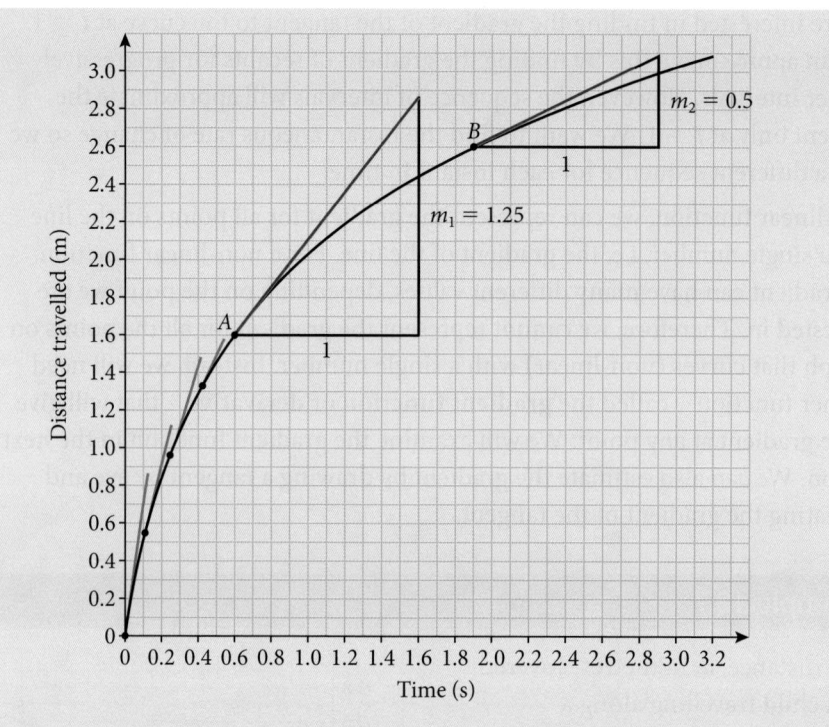

## Example 13.8

In a laboratory experiment, biologists studied a population of fruit flies over a period of 50 days. The number of fruit flies was recorded at regular intervals and points were plotted on a graph with time $t$ in days on the horizontal axis and number $n$ of flies on the vertical axis. A smooth curve was drawn through the plotted points and is shown in the graph.

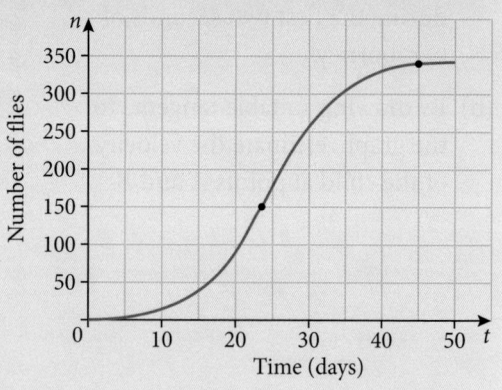

On the 23rd day of the experiment there were 150 flies and on the 45th day there were 340 flies.

(a) Compute the average rate of change of the number $n$ of flies from day 23 to day 45. Give your answer to 3 significant figures and include the units.

(b) Draw by eye a tangent to the graph at the point (23, 150). Use this tangent to estimate the instantaneous rate of change of the number of flies on day 23.

## Solution

(a)

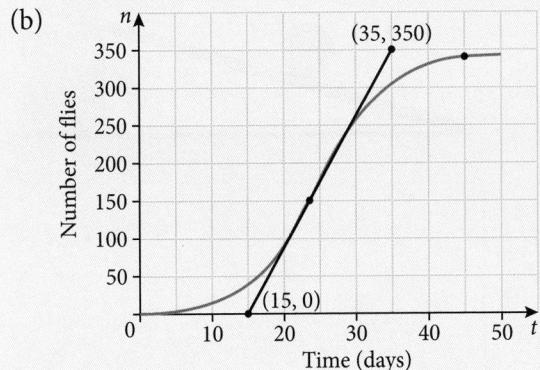

From day 23 to day 45, $\Delta n = 340 - 150 = 190$ and $\Delta t = 45 - 23 = 22$

Thus, the average rate of change in the number $n$ of fruit flies from day

23 to day 45 is $\dfrac{\Delta n}{\Delta t} = \dfrac{190}{22} \approx 8.64$ flies per day

(b)

A reasonably accurate tangent is shown in the diagram that passes through the points $(15, 0)$ and $(35, 350)$. Using these points gives the gradient for the tangent at the point $(23, 150)$.

$$\frac{350 - 0}{35 - 15} = 17.5 \text{ flies per day}$$

Thus, the instantaneous rate of change of the fly population on day 23 is approximately 17.5 flies per day. Answers will vary depending on where the tangent is drawn.

## Exercise 13.1

1. One of Zeno's paradoxes is the following:

   Achilles is in a footrace with a tortoise. Since Achilles knows he runs 10 times as fast as the tortoise, he allows the tortoise a 100 m head start. Then, Achilles begins. Achilles quickly runs the first 100 m, during which time the tortoise crawls 10 m.

The Tortoise and Achilles paradox is one of the best known paradoxes credited to the Greek philosopher Zeno of Elea, in southern Italy.

Achilles quickly runs the next 10 m, during which time the tortoise crawls 1 m. Achilles quickly runs the next 1 m, during which time the tortoise crawls 0.1 m. Achilles continues to chase the tortoise, but each interval he runs the tortoise crawls one-tenth of that distance, so the tortoise is always ahead of Achilles.

Can Achilles catch the tortoise? Use the logic of limits to argue that he can.

2. Using either a table or a graph on your GDC, evaluate each limit. Give approximate answers correct to 3 significant figures.

(a) $\lim_{x \to 2} \dfrac{x^2 - 4}{x - 2}$

(b) $\lim_{x \to -1} \dfrac{x^2 - 3x - 4}{x + 1}$

(c) $\lim_{x \to 9} \dfrac{9 - x}{3 - \sqrt{x}}$

(d) $\lim_{x \to 2^+} \dfrac{x^2 + 4}{x - 2}$

(e) $\lim_{x \to 1} \dfrac{x^3 - 1}{x - 1}$

(f) $\lim_{x \to 0} \dfrac{\sqrt{x + 2} - \sqrt{2}}{x}$

3. A ball rolls down an inclined ramp. The time and distance measurements are recorded for a 6-second interval. The table and graph show the measurements.

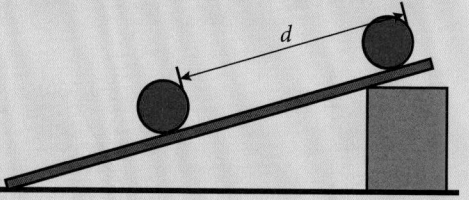

| Time (s) | Distance (cm) |
|---|---|
| 0 | 0 |
| 1 | 13 |
| 2 | 62 |
| 3 | 136 |
| 4 | 247 |
| 5 | 386 |
| 6 | 553 |

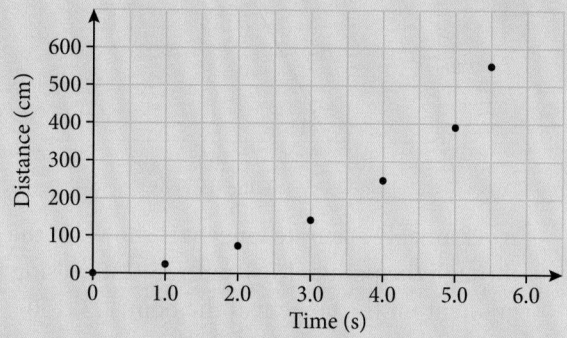

(a) Compute the average velocity of the ball during the time intervals:

(i) $t = 0$ to $t = 2$    (ii) $t = 2$ to $t = 4$    (iii) $t = 4$ to $t = 6$

Give answers with units to 3 significant figures.

(b) Estimate the instantaneous velocity of the ball at $t = 3$. Briefly explain your procedure for obtaining the estimate.

4. In each graph, the distance, $d$, in km, from a starting point for a cyclist is plotted against time, $t$, in hours.

(i) Calculate the average velocity of each cyclist between points $P$ and $Q$, by drawing an appropriate secant.

**(ii)** Estimate the instantaneous velocity at point *P* for each cyclist, by drawing an appropriate tangent.

**(iii)** Are any of the cyclists stationary at any time during their ride? If so, which cyclist and during what time interval? Justify your answer.

**(a)**

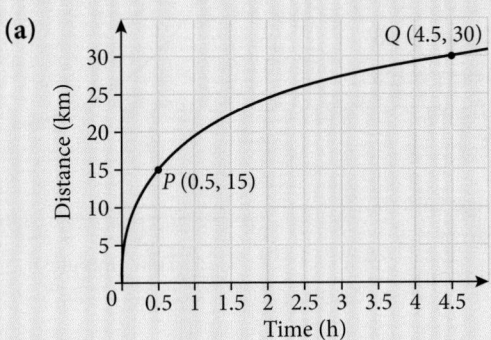

**(b)**

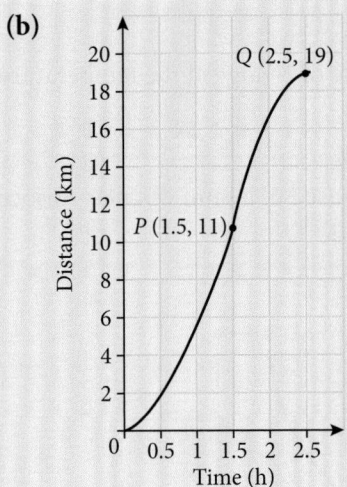

**(c)**

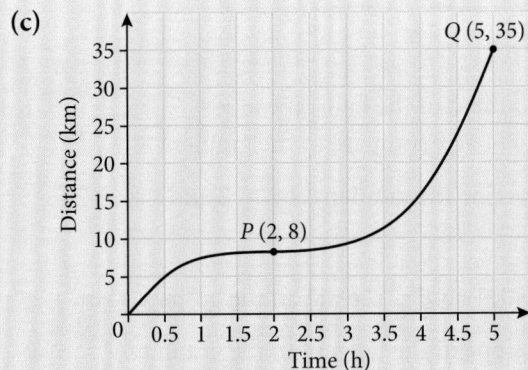

**5.** Boiling water is poured into a pan. The temperature (in degrees Celsius) of the water is measured regularly during the first seven minutes after the water was poured into the pan. These data points (time, temperature) are plotted on a graph and a smooth curve is drawn through these points. Give all answers to 3 significant figures.

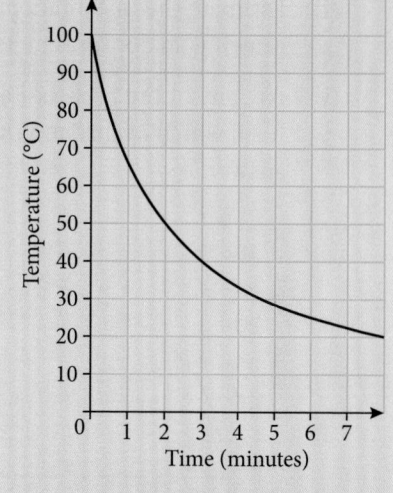

(a) Approximate the average rate of change of the temperature per minute for the interval from $t = 0$ to $t = 7$

(b) Approximate the instantaneous rate of change of the temperature per minute at $t = 1$

(c) Approximate the instantaneous rate of change of the temperature per minute at $t = 5$

**6.** A half-pipe is a skateboard ramp composed of a semi-circular surface as shown in the diagram.

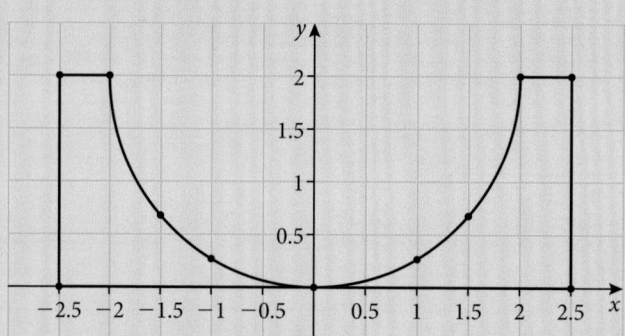

(a) Approximate the gradient of the half-pipe at $x = -1.5, -1, 0, 1, 1.5$ How does the gradient at $x$ compare to the gradient at $-x$?

(b) How would you describe the gradient at $x = 2$ and $x = -2$?

**7.** The points $A\left(-1, \frac{1}{e}\right)$, $B(0, 1)$ and $C(1, e)$ lie on the graph $y = e^x$, as shown. Estimate the slope of the tangent to the graph of $y = e^x$ at each of the points $A$, $B$ and $C$. Using your GDC find approximate values (to 3 significant figures) for $\frac{1}{e}$ and e. Comment on your results.

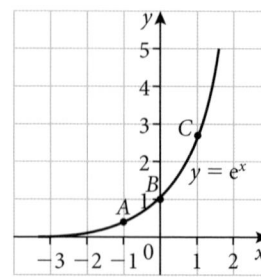

**Figure 13.12** Graph of $y = e^x$ for question 7

# 13.2 Derivative of a function

The **derivative** of a function, also known as the **gradient function**, is a function that gives us the gradient of the tangent to a curve at any point. We will now look at using and interpreting the derivative function.

There are many ways to denote the derivative of a function $y = f(x)$.

The most common notations are:

| Notation | Read as | Comment |
|---|---|---|
| $y'$ | 'y prime' | This notation is brief but does not name the independent variable. |
| $\dot{y}$ | 'y dot' | The derivative with respect to time; equivalent to $\dfrac{dy}{dt}$ |
| $\dfrac{dy}{dx}$ | 'dy dx' or 'the derivative of y with respect to x' | Names both dependent and independent variables and uses d for derivative. |
| $f'(x)$ | 'f prime of x' or 'the derivative of f of x' | Emphasises the name of the function and names the independent variable. |

**Table 13.5** Common notations for the derivative of a function $y = f(x)$

Given a function $y = f(x)$, we will refer to the **derivative function** with the notation $y'$, $\dot{y}$, $f'(x)$, or $\dfrac{dy}{dx}$. Remember, the derivative function is a rule for computing the gradient of the tangent to the graph of the original function $y = f(x)$ at a particular point, i.e. the instantaneous rate of change of $f(x)$ at a particular value of $x$.

The notation $\dfrac{dy}{dx}$ does not indicate d multiplied by $y$ or $x$. Instead, you can think of d$y$ and d$x$ as names for 'a small change in $y$' and 'a small change in $x$'. When we say, 'a small change' we mean an infinitesimally small number, but not equal to zero. This notation helps us remember that the derivative function is giving us the instantaneous rate of change in $y$ with respect to $x$. In this way, it is analogous to how the gradient of a line, $m$, is defined to be the change in $y$ compared to the change in $x$.

$$m = \frac{\Delta y}{\Delta x}$$

## Example 13.9

A ball is thrown upwards from the roof of a building. The vertical height of the ball from the ground in metres after $t$ seconds is given by the function
$h(t) = -4.9t^2 + 18t + 50$

The derivative function is given by $h'(t) = -9.8t + 18$

(a) Find the initial velocity of the ball.

(b) Find the time at which the velocity of the ball is zero. Interpret the meaning of this.

(c) Find the velocity of the ball at the moment it hits the ground.

### Solution

(a) Since $h(t)$ is the position function at a given time, then $h'(t)$ represents the change in position with respect to time, i.e., the velocity.

Therefore, the initial velocity of the ball is given at time $t = 0$, thus
$h'(0) = -9.8(0) + 18 = 18\,\mathrm{m\,s^{-1}}$

Remember that we sometimes use other variables instead of $x$ and $y$, and not all functions are named $f$. For example, you may see $h'(t)$, $P'(n)$, $\dfrac{dh}{dt}$, $\dfrac{dP}{dn}$

539

Note that you need to be careful in choosing when to use the derivative function and when to use the original function. Remember that the derivative function will generally give you a rate of change.

(b) The ball has zero velocity when $h'(t) = 0$

$$-9.8t + 18 = 0 \Rightarrow t = 1.84\,\text{s}$$

This is the time when the ball reaches its maximum height, since it is stationary for an instant between moving up and falling down.

(c) The ball hits the ground when $h(t) = 0$. Therefore, we must solve the quadratic equation $-4.9t^2 + 18t + 50 = 0$ to obtain $t = 5.52$ or $t = -1.85$

The negative solution is not meaningful in this context. Therefore, the ball hits the ground at time $t = 5.52\,\text{s}$, and has a velocity of $h'(5.52) = -9.8(5.52) + 18 = -36.1\,\text{m}\,\text{s}^{-1}$

## Interpretations of the derivative

It is important to be able to reason abstractly about the sign of the derivative. Specifically, what does the **sign of the derivative** tell us about the function? For this, it is sufficient to revisit linear functions and consider positive, zero, and negative gradients as shown in Figure 13.13.

When a gradient is positive, the function is increasing. When a gradient is zero, the function is constant (neither increasing nor decreasing). When a gradient is negative, the function is decreasing. This straightforward reasoning is no different for non-linear functions. We simply use the sign of the derivative to determine if a function is increasing, constant, or decreasing.

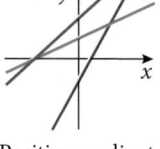

Positive gradient

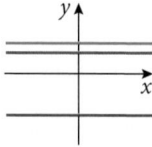

Zero gradient

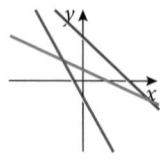

Negative gradient

**Figure 13.13** Lines with positive, zero and negative gradients

For a given function $f(x)$, when:
- $f'(x) > 0$, the gradient is positive, the function is increasing at $x$
- $f'(x) = 0$, the gradient is zero, the function is constant (neither increasing nor decreasing) at $x$
- $f'(x) < 0$, the gradient is negative, the function is decreasing at $x$.

### Example 13.10

Given the derivative $f'(x) = x^2 - x - 12$, identify intervals where the function $f$ is increasing and decreasing.

### Solution

#### Algebraic approach

This is a quadratic function that is concave up. So, if we find the $x$-intercepts, then we can reason about the sign of the derivative. The $x$-intercepts are found by setting the function equal to zero and solving.

$$f'(x) = x^2 - x - 12$$
$$0 = (x - 4)(x + 3)$$
$$x = 4 \text{ or } x = -3$$

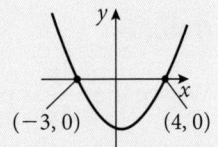

The derivative function is positive for $x < -3$ or $x > 4$ and negative for $-3 < x < 4$

Therefore, the function $f$ is increasing for $x < -3$ or $x > 4$ and decreasing for $-3 < x < 4$

### Graphical approach

If we graph the derivative using a GDC, we can quickly come to the same conclusion as before.

Looking at the graph, we see that $f'(x)$ is positive for $x < -3$ or $x > 4$ and negative for $-3 < x < 4$

Therefore, the function $f$ is increasing for $x < -3$ or $x > 4$ and decreasing for $-3 < x < 4$

**Figure 13.14** GDC screen for solution to Example 13.10

Remember, when you are looking at the graph of a derivative, you need to look only at whether the graph is positive (above the $x$-axis), zero (intersecting the $x$-axis), or negative (below the $x$-axis).

## Example 13.11

The rate of change in a population is given by the function $\dfrac{dP}{dt} = \dfrac{4}{t + 1}$

where $P$ is the population in thousands and $t$ is time in years, $t \geq 0$

(a) Find the initial rate of change in the population.

(b) Describe how the population changes as time progress.

## Solution

(a) The initial rate of change in the population is $\dfrac{dP}{dt} = \dfrac{4}{0 + 1} = 4$

The population is increasing by 4000 individuals per year.

(b) To see how the population changes, examine a graph of the derivative on a GDC, as shown.

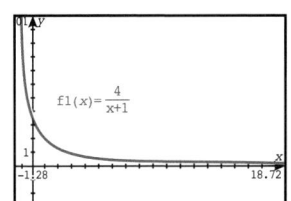

We see that the derivative is positive everywhere, so the population is growing. However, the derivative approaches zero, which means that the growth rate of the population is decreasing. Therefore, the population is levelling off as time progresses.

**Figure 13.15** GDC screen for solution to Example 13.11

## Example 13.12

Given the graph of the derivative $\dfrac{dy}{dx}$ shown, write down intervals where $y$ is increasing and decreasing.

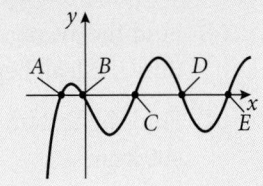

541

**Solution**

We see that $\dfrac{dy}{dx}$ is positive for $A < x < B$ and $C < x < D$ and $x > E$

Therefore, $y$ is increasing in those intervals. Likewise, $\dfrac{dy}{dx}$ is negative for

$x < A$ and $B < x < C$ and $D < x < E$, so $y$ is decreasing in those intervals.

One important use of the derivative is to help sketch graphs of functions. While we often use a GDC to generate graphs of functions, understanding the connection between the derivative of a function and the graph of the function is still important.

## Exercise 13.2

| $t$ | Distance travelled | Velocity |
|---|---|---|
| 0 | | |
| 1 | | |
| 2 | | |
| 3 | | |

**Table 13.6** Table for question 1

1. A toy car rolling down a 400-cm ramp travels a distance (cm) given by $d(t) = 35t^2 + 40t$ where $t$ is the time (s). The derivative function is given as $d'(t) = 70t + 40$

   (a) Find the time it takes the toy car to reach the end of the 400-cm ramp.

   (b) Calculate the average velocity of the toy car.

   (c) Copy and complete the table.

   (d) Give a reason why the distance at time $t = 3$ s does not make sense in the context of this question.

   (e) Write down the velocity of the car at the moment it reaches the end of the ramp.

   (f) Write down the velocity of the toy car the moment it begins rolling down the ramp.

   (g) Calculate the time when the car has travelled 200 cm.

   (h) Calculate the velocity of the car when it is half way down the ramp.

   (i) Calculate the time when the velocity is $130\,\text{cm s}^{-1}$.

2. An oil leak underneath a car creates a circular puddle. As the oil leaks, the area of the puddle is given by $A = \pi r^2$ where $r$ is the radius of the puddle in cm. The derivative is given by $\dfrac{dA}{dr} = 2\pi r$

   (a) Find the area of the puddle when the radius is 10 cm.

   (b) What are the units of $\dfrac{dA}{dr}$?

   (c) Find the average rate of change in the area of the puddle from the time the leak begins until the area is $400\pi\,\text{cm}^2$.

   (d) Find the instantaneous rate of growth of the puddle when the area is $400\pi\,\text{cm}^2$.

**3.** A runner goes for a run in the mountains covering 25 km in five hours. Her distance is described by the function

$d(t) = \frac{2}{3}t^3 - \frac{35}{6}t^2 + \frac{35}{2}t$, where $d(t)$ is in km and $t$ is in hours, $0 \leqslant t \leqslant 5$

The derivative function is given as $d'(t) = 2t^2 - \frac{35}{3}t + \frac{35}{2}$

(a) Find the distance covered after 2.5 hours.

(b) Verify that the runner completes 25 km at 5 hours.

(c) Find the runner's average speed for the 5 hour run.

(d) Give the units of $d'(t)$

(e) Find the runner's speed at the moment she starts and at the moment she stops.

(f) Find the time when the runner is going the fastest and the runner's maximum speed.

(g) Find the runner's minimum speed and the time when this occurs.

**4.** The distance, $s$ (m), travelled by a child sliding along a horizontal waterslide is modelled by the function $s = 2\sqrt{3t}$, where $0 \leqslant t \leqslant 3$ (s).

The derivative is given as $\frac{ds}{dt} = \sqrt{\frac{3}{t}}$

(a) Write down the units of $\frac{ds}{dt}$

(b) Find the average velocity of the child during the 3-second interval.

(c) Find the velocity of the child at 1, 2, and 3 seconds.

(d) Find the distance travelled by the child after 1, 2, and 3 seconds.

**5.** Consider the graph of $f$.

(a) Between which two consecutive points is the average rate of change of the function greatest?

(b) At what points is the instantaneous rate of change of $f$

(i) positive  (ii) negative  (iii) zero?

(c) For which two pairs of points is the average rate of change approximately equal?

**6.** Copy and complete Table 13.7 by matching each function with its derivative.

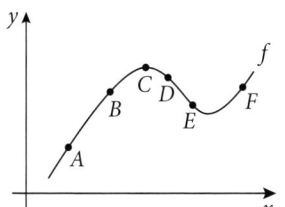

Figure 13.16  Graph for question 5

| Function | Derivative diagram |
|:---:|:---:|
| $f_1$ | |
| $f_2$ | |
| $f_3$ | |
| $f_4$ | |

Table 13.7  Table for question 6

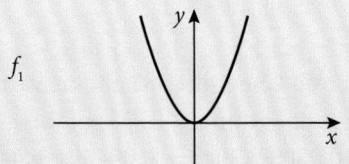

(a)

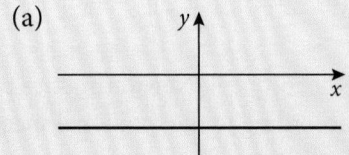

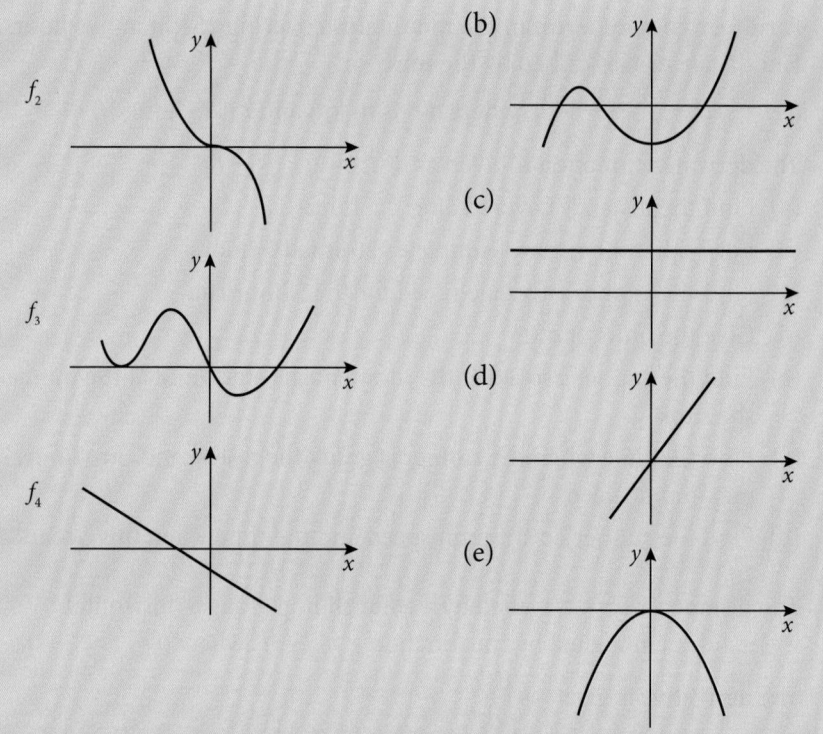

7. Identify the intervals where $f$ is increasing or decreasing.

(a)

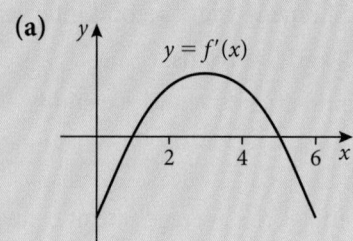

(b)

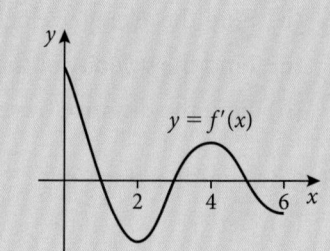

8. The function $g(x)$ is defined for $-3 \leqslant x \leqslant 3$
   The behaviour of $g'(x)$ is given in the table.

| $x$ | $-3 < x < -2$ | $-2$ | $-2 < x < 1$ | $1$ | $1 < x < 3$ |
|---|---|---|---|---|---|
| $g'(x)$ | negative | 0 | positive | 0 | negative |

(a) For what intervals is $g$ decreasing?

(b) Given that $g(0) = 0$, make a rough sketch of the graph of $g$.

9. The torque in Newton-metres produced by a Chevrolet 383 engine varies
   according to the rotational speed (rotations per minute, RPM) of the
   engine according to the model $T(r) = -0.00002720r^2 + 0.08459r + 440.0$
   where $T(r)$ is the torque in Nm and $r$ is revolutions $\text{min}^{-1}$ (RPM).
   Give your answers to 3 significant figures.

(a) Explain the meaning of a positive value for $T'(r)$

**(b)** Use your GDC to obtain a graph of $T'(r)$ for the interval $0 \leqslant r \leqslant 10\,000$. On what intervals is the torque increasing? Decreasing? Justify your answer using $T'(r)$.

**(c)** Use your GDC to determine the maximum torque and the RPM at which this occurs. What is the value of $T'(r)$ at this point?

**10.** The rate of change in the concentration of a certain drug in a patient's bloodstream is given by $\dfrac{dC}{dt} = (-0.646t^{1.15} + 36.8t^{0.15})(0.98^t)$

where $C$ is in nanograms per ml and $t$ is in minutes since the patient was given the drug, for $0 \leqslant t \leqslant 480$

**(a)** Use your GDC to generate a graph of $\dfrac{dC}{dt}$

**(b)** Find intervals where the concentration of the drug is increasing and decreasing.

**(c)** Find the time at which the concentration is decreasing the fastest and the rate at this time.

**(d)** Predict when the maximum concentration of the drug occurs.

**11.** The rate of change in the population of Windelberg can be modelled by the function $P'(t) = \dfrac{12\,000}{8500(0.82^t) + 1.22^t + 184}$ where $P$ is the population in thousands and $t$ is time in years since 1980.

**(a)** Give the units of $P'(t)$

**(b)** Use your GDC to generate a graph of this derivative.

**(c)** Give a reason why the population of Windelberg does not decrease.

**(d)** Describe the general pattern of population change in Windelberg.

**(e)** Find the instantaneous rate of change in the population in 1990.

**(f)** Find the interval when the population is increasing by at least $20\,000$ people per year.

**(g)** Find the time when the population is increasing the fastest and the rate of increase at this time.

# 13.3 Derivatives of functions of the form $f(x) = ax^n$

So far in this chapter, we have examined the uses and interpretations of the derivative. However, we can also find the derivative of a function. In this section, we will develop a rule for finding the derivative of a function in the form $f(x) = ax^n$ where $a \in \mathbb{R}$ and $n \in \mathbb{Q}$. This rule – called the **power rule** – will allow us to find the derivative of a variety of functions containing terms in the form $ax^n$, including any polynomial function.

The process of finding a derivative is called **differentiation**.

## Exploration of the power rule

Your GDC is able to graph the derivative of any function. Look up instructions for your particular GDC to learn how to do this.

### Example 13.13

Graph the derivative of each function. For each, attempt to find the derivative function by examining the graph.

(a) $y = 5x$      (b) $y = x^2$      (c) $y = 3x^2$

(d) $y = x^3$      (e) $y = \frac{1}{2}x^3$

Then, summarise your findings and make a conjecture.

**Solution**

(a)
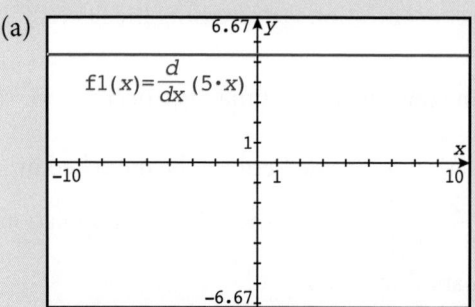

We see that $\frac{dy}{dx} = 5$, which agrees with our understanding that the gradient of linear functions of the form $y = mx + c$ is the value of $m$.

(b)
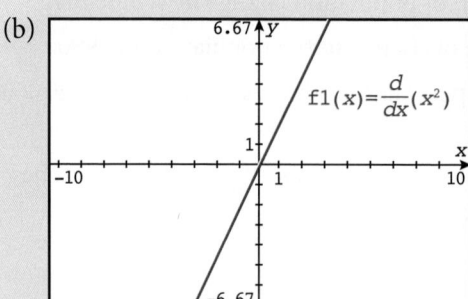

It seems that the derivative of the quadratic function (degree $= 2$) is a linear function (degree $= 1$). Look at the gradient of the line. It has a gradient of 2, so $\frac{dy}{dx} = 2x$

(c) The graph of the derivative of $y = 3x^2$ is shown here, but we can also compare it with the graph of the derivative from the previous question, $\frac{dy}{dx} = 2x$

The graph of the derivative $y = 3x^2$ is quite a bit steeper than $y = 2x$, the derivative of $y = x^2$. If you examine some points or a table of values, you should be able to see that the derivative of $y = 3x^2$ is exactly 3 times as large at the derivative of $y = x^2$ for a particular value of $x$. Hence, we have $\dfrac{dy}{dx} = 3(2x) = 6x$

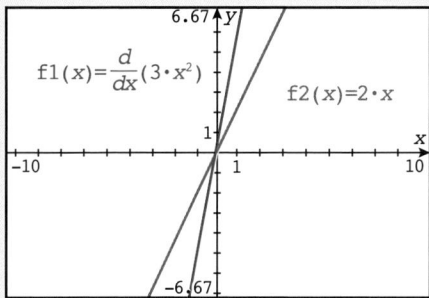

(d)

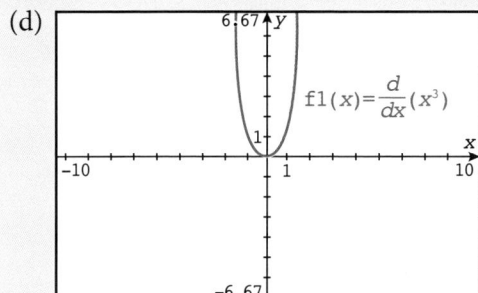

It appears that the derivative of a cubic function (degree = 3) is a quadratic function (degree = 2). Is $\dfrac{dy}{dx} = x^2$?

We can add $y = x^2$ to the graph and compare.

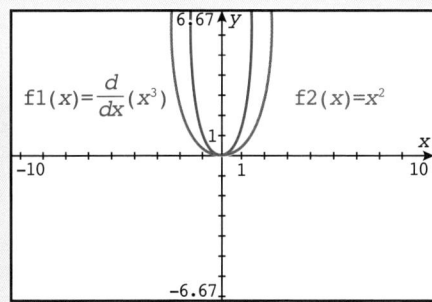

So the derivative of $y = x^3$ is not simply $\dfrac{dy}{dx} = x^2$

Again, by examining a table of values or checking some of the points on the derivative function, you should be able to convince yourself that the derivative of $y = x^3$ is actually $\dfrac{dy}{dx} = 3x^2$

(e) We can compare the graph of the derivative of $y = \frac{1}{2}x^3$ with the graph of $y = 3x^2$

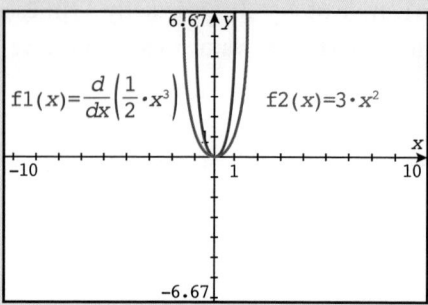

As in part (d), the coefficient of $\frac{1}{2}$ has changed the derivative as well.

If you examine it closely, you will see that the derivative is a scaled-down version of $y = 3x^2$. In fact, it appears to be scaled by exactly half.

Hence, we have $\frac{dy}{dx} = \frac{1}{2}(3x^2) = \frac{3}{2}x^2$

Table 13.8 summarises the findings.

| Function | Derivative |
|----------|-----------|
| $y = 5x$ | $\frac{dy}{dx} = 5$ |
| $y = x^2$ | $\frac{dy}{dx} = 2x$ |
| $y = 3x^2$ | $\frac{dy}{dx} = 6x$ |
| $y = x^3$ | $\frac{dy}{dx} = 3x^2$ |
| $y = \frac{1}{2}x^3$ | $\frac{dy}{dx} = \frac{3}{2}x^2$ |

**Table 13.8** Summary of results from exploring power rule for differentiation

The results of this example lead us to the power rule.

The **power rule** works on power functions; that is, functions of the form $f(x) = ax^n$ where $a \in \mathbb{R}, n \in \mathbb{Q}$

For a function of the form $f(x) = ax^n$, $n \in \mathbb{Q}$, the derivative is equal to $f'(x) = anx^{n-1}$
In words: to differentiate a power expression multiply by the exponent, then reduce the exponent by one.

Note that although we explored integer exponents to find the power rule, it applies to rational exponents as well.

The derivative of a sum (or difference) is the sum (or difference) of the derivatives. That is, if $h(x) = f(x) \pm g(x)$, then $h'(x) = f'(x) \pm g'(x)$

What about functions that have more than one term, like $f(x) = 4x^2 - 10x$? The derivative of a sum is the sum of the derivatives. So, since the derivative of $4x^2$ is $8x$ and the derivative of $-10x$ is $-10$, it must be true that $f'(x) = 8x - 10$

## Example 13.14

Use the power rule to find the derivative of each function.

(a) $y = x^4$

(b) $y = 10x^2$

(c) $y = 15$

(d) $y = \frac{3}{x^2}$

(e) $f(x) = -\frac{1}{3}x^3$

(f) $s = \frac{t - 5t^4}{t^3}$

(g) $g(x) = x^4 - \frac{1}{3}x^3 + 10x^2 + 15$

(h) $y = 5x^2 - 4x + \frac{8}{x}$

(i) $y = \sqrt{x}$

(j) $y = \frac{3x^2 + 2x}{\sqrt{x}}$

**Solution**

(a) $\dfrac{dy}{dx} = 4 \times x^{4-1} = 4x^3$

(b) $\dfrac{dy}{dx} = 10 \times 2x^{2-1} = 20x$

(c) For $y = 15$, remember that, by definition, a constant doesn't change. Therefore, the rate of change of a constant is always zero.

Hence, $\dfrac{dy}{dx} = 0$

Or, you can think of this as $y = 15x^0$ so that the derivative is

$\dfrac{dy}{dx} = 15 \times 0x^{0-1} = 0$

(d) Rewrite $y = \dfrac{3}{x^2}$ in the form $y = ax^n$

$y = 3x^{-2} \Rightarrow \dfrac{dy}{dx} = -2 \times 3x^{-2-1} = -6x^{-3} = -\dfrac{6}{x^3}$

> By convention, we usually write expressions with positive exponents, as shown for example in part (d).

(e) $f'(x) = -\dfrac{1}{3} \times 3x^2 = -x^2$

(f) First, simplify the function: $s = \dfrac{t - 5t^4}{t^3} = \dfrac{t}{t^3} - \dfrac{5t^4}{t^3} = t^{-2} - 5t$

$\dfrac{ds}{dt} = -2 \times t^{-2-1} - 5 \times t^{1-1} = -2t^{-3} - 5t^0 = -\dfrac{2}{t^3} - 5$

(g) We already know the derivative of each term of this polynomial function, we just need to put them together using the sum rule above:

$g'(x) = 4 \times x^{4-1} - 3 \times \dfrac{1}{3}x^{3-1} + 2 \times 10x^{2-1}$

$\quad\quad = 4x^3 - x^2 + 20x$

Note that the $+15$ disappears since its derivative is zero.

(h) $y = 5x^2 - 4x + \dfrac{8}{x}$

$\quad\quad = 5x^2 - 4x + 8x^{-1}$

$\dfrac{dy}{dx} = 5 \times 2x^{2-1} - 1 \times 4x^{1-1} + (-1) \times 8x^{-1-1}$

$\quad\quad = 10x - 4 - \dfrac{8}{x^2}$

(i) $y = \sqrt{x} = x^{\frac{1}{2}}$

$\dfrac{dy}{dx} = \dfrac{1}{2}x^{-\frac{1}{2}}$

$\quad\quad = \dfrac{1}{2\sqrt{x}}$

(j) $y = \dfrac{3x\,2 + 2x}{\sqrt{x}} = 3x^{\frac{3}{2}} + 2x^{\frac{1}{2}}$

$y' = 3 \times \dfrac{3}{2}x^{\frac{1}{2}} + 2 \times \dfrac{1}{2}x^{-\frac{1}{2}}$

$= \dfrac{9\sqrt{x}}{2} + \dfrac{1}{\sqrt{x}}$

Although the power rule is relatively straightforward, it can be applied to many functions.

### Example 13.15

Find the equation of the tangent to the graph of $h(x) = 4 - 2x - \dfrac{2}{3}x^2$ at the point where $x = -2$

Confirm your result by graphing $h$ and the tangent on your GDC.

### Solution

$h(-2) = 4 - 2(-2) - \dfrac{2}{3}(-2)^2 = 4 + 4 - \dfrac{8}{3} = \dfrac{24}{3} - \dfrac{8}{3} = \dfrac{16}{3}$

So, point of tangency is $\left(-2, \dfrac{16}{3}\right)$

$h'(x) = -2 - \dfrac{4}{3}x$

$h'(-2) = -2 - \dfrac{4}{3}(-2) = -\dfrac{6}{3} + \dfrac{8}{3} = \dfrac{2}{3}$

So, the gradient of the tangent at $x = -2$ is $\dfrac{2}{3}$

$y - \dfrac{16}{3} = \dfrac{2}{3}[x - (-2)] \Rightarrow y = \dfrac{2}{3}x + \dfrac{4}{3} + \dfrac{16}{3}$

Therefore, the equation of the tangent to $h$ at $\left(-2, \dfrac{16}{3}\right)$ is $y = \dfrac{2}{3}x + \dfrac{20}{3}$

This result is confirmed by graphing the function $h$ and the tangent, and then using the intersect command on the GDC to confirm the coordinates of the point of tangency.

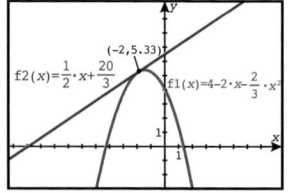

**Figure 13.17** GDC screen for solution to Example 13.15

### Exercise 13.3

1. For each function **(a)** to **(g)**

   **(i)** find the derivative

   **(ii)** compute the slope of the graph of the function at the indicated point.

   Use a GDC to confirm your results.

   **(a)** $y = 3x^2 - 4x$, at $(0, 0)$

   **(b)** $y = 1 - 6x - x^2$, at $(-3, 10)$

**(c)** $y = \dfrac{2}{x^3}$, at $(-1, 2)$

**(d)** $y = x^5 - x^3 - x$, at $(1, -1)$

**(e)** $y = (x + 2)(x - 6)$, at $(2, -16)$

**(f)** $y = 2x + \dfrac{1}{x} - \dfrac{3}{x^3}$, at $(1, 0)$

**(g)** $y = \dfrac{x^3 + 1}{x^2}$, at $(-1, 0)$

2. Find the coordinates of any points on the graph of the function where the slope is equal to the given value.

   **(a)** $y = x^2 + 3x$, slope $= 3$

   **(b)** $y = x^3$, slope $= 12$

   **(c)** $y = x^2 - 5x + 1$, slope $= 0$

   **(d)** $y = x^2 - 3x$, slope $= -1$

3. The slope of the curve $y = x^2 + ax + b$, where $a, b \in \mathbb{R}$ at the point $(2, -4)$ is $-1$. Find the value of $a$ and the value of $b$.

4. Use the power rule to show that the gradient of $y = mx + b$, where $m$ and $b$ are constants, is $m$.

5. Consider the function $g(x) = \sqrt[3]{x}$

   **(a)** Find the derivative function $g'(x)$

   **(b) (i)** Find the exact gradient of the line that is tangent to the graph of $g$ at the point where $x = 8$

   **(ii)** Hence, find the equation of the tangent to the graph of $g$ at the point where $x = 8$

   **(c) (i)** Find the gradient of the tangent to the graph of $g$ at the origin $(0, 0)$. Comment on your answer.

   **(ii)** Hence, find the equation of the tangent to the graph of $g$ at the origin.

6. Find the equation of the tangent to the graph of $y = 3\sqrt{x} - 6$ at $(9, 3)$

7. Boiling water is poured into a cup. The temperature of the water, $C$, in degrees Celsius, after $t$ minutes is given by $C(t) = 19 + \dfrac{182}{t^{\frac{3}{2}}}$, for $t \geqslant 1$
   Give your answers correct to 3 significant figures.

   **(a)** Find the average rate of change of the temperature from $t = 2$ to $t = 6$

   **(b)** Find the instantaneous rate of change of the temperature at $t = 4$

8. A car is parked with the windows and doors closed for five hours. The temperature inside the car, $C$, in degrees Celsius, $C$, is given by $C = 2\sqrt{t^3} + 17$ with $t$ representing the number of hours since the car was first parked.

(a) Find the average rate of change of the temperature from $t = 1$ to $t = 4$

(b) Find the function which gives the instantaneous rate of change of the temperature for any time $t$ where $0 < t < 5$

(c) Find the time $t$ at which the instantaneous rate of change of the temperature is equal to the average rate of change from $t = 1$ to $t = 4$

9. The cumulative number of HIV AIDS cases reported in the United States from 1983 to 1998 follows the cubic model $C = -222t^3 + 7260t^2 - 12\,700t + 13500$, where $C$ is the cumulative number of HIV AIDS cases, and $t$ is the number of years since 1983.

(a) Write down the units of $\dfrac{dC}{dt}$

(b) Find $\dfrac{dC}{dt}$

(c) Find the value of $\dfrac{dC}{dt}$ for 1990 and interpret in context.

(d) Find the number of new cases reported in 1990. Compare to your answer from (c): should these two values be the same? Explain.

(e) Within the years 1983 to 1998, during what intervals was the cumulative number of HIV AIDS cases increasing and decreasing. Justify your answer using $\dfrac{dC}{dt}$

10. A colony of bacteria is exposed to a drug that stimulates reproduction. The number of bacteria is given by the model $P = 1200 + 17t^2 - t^3$ where $P$ is the number of bacteria present $t$ minutes after the drug is introduced, $0 \le t \le 20$

(a) Find $\dfrac{dP}{dt}$

(b) Find intervals where the population of bacteria is increasing and decreasing. Justify your answer using $\dfrac{dP}{dt}$.

(c) What is the rate of change in the population of bacteria 10 minutes after the drug is introduced?

(d) Find the time when the bacteria are dying at a rate of 165 per minute.

(e) Use your GDC and the derivative to find when the population of bacteria is increasing the fastest. What is the rate of increase at this time?

11. A tennis player hits a ball straight up. The height of the ball above the ground is described by the model $h(t) = -4.9t^2 + 16t + 1$, where $h(t)$ is the height in metres at time $t$ seconds after the ball is hit.

(a) Find $h'(t)$ and interpret its meaning. Include units in your answer.

(b) Find the velocity of the tennis ball 1 s after it is hit.

(c) At what time is the tennis ball descending at a rate of $10 \, \mathrm{m \, s^{-1}}$?

(d) At what time is the velocity of the tennis ball equal to zero? What is special about this point?

(e) Find the maximum speed of the tennis ball and the time at which this occurs.

## 13.4 Derivatives of composite functions, products and quotients

### The chain rule

We know how to differentiate functions such as $f(x) = x^3 + 2x - 3$ and $g(x) = \sqrt{x}$, but how do we differentiate the composite function $g(f(x)) = \sqrt{x^3 + 2x - 3}$ where $f$ and $g$ are nested functions? In this example, $f$ is nested inside $g$. The answer is that we use the **chain rule**. As most functions that we encounter in applications are composite functions, it can be argued that the chain rule is the most important, and most widely used, rule of differentiation.

Put simply, the chain rule says that given two nested functions the derivative of their composite is the product of their derivatives. For example, the function $y = 8x + 6 = 2(4x + 3)$ is the composite of the functions $y = 2u$ and $u = 4x + 3$. Note that the function $y$ is in terms of $u$, and the function $u$ is in terms of $x$.

$$\frac{dy}{dx} = 8, \frac{dy}{du} = 2 \text{ and } \frac{du}{dx} = 4$$

How are the derivatives of these three functions related?

Since $8 = 2 \cdot 4$, the derivatives are related such that $\dfrac{dy}{dx} = \dfrac{dy}{du} \cdot \dfrac{du}{dx}$

In other words, the rate of change of $y$ with respect to $x$ is the product of the rate of change of $y$ with respect to $u$ and the rate of change of $u$ with respect to $x$.

### Example 13.16

The polynomial function $y = 16x^4 - 8x^2 + 1 = (4x^2 - 1)^2$ is the composite of $y = u^2$ and $u = 4x^2 - 1$

Use the chain rule to find $\dfrac{dy}{dx}$, the derivative of $y$ with respect to $x$.

### Solution

$$y = u^2 \Rightarrow \frac{dy}{du} = 2u$$

$$u = 4x^2 - 1 \Rightarrow \frac{du}{dx} = 8x$$

Applying the chain rule:

$$\frac{dy}{dx} = \frac{dy}{du} \cdot \frac{du}{dx} = 2u \cdot 8x$$
$$= 2(4x^2 - 1) \cdot 8x$$
$$= 64x^3 - 16x$$

In this particular case, we could have differentiated the function in expanded form by differentiating term-by-term rather than differentiating the factored form by the chain rule.

$$\frac{dy}{dx} = \frac{d}{dx}(16x^4 - 8x^2 + 1) = 64x^3 - 16x$$

It is not always easier to differentiate powers of polynomials by expanding and then differentiating term-by-term. For example, it is far better to find the derivative of $y = (3x + 5)^8$ by the chain rule.

Composite functions are often written using nested function notation. For example, the notation $f(g(x))$ denotes a function composed of functions $f$ and $g$ such that $g$ is the inside function and $f$ is the outside function. For the composite function $y = (4x^2 - 1)^2$ in Example 13.16, the inside function is $g(x) = 4x^2 - 1$ and the outside function is $f(u) = u^2$. Looking again at the solution for Example 13.16, we see that we can choose to express and work out the chain rule in function notation rather than in Leibniz notation.

For $y = f(g(x)) = (4x^2 - 1)^2$ and $y = f(u) = u^2$, $u = g(x) = 4x^2 - 1$

**Leibniz notation**

$$\frac{dy}{dx} = \frac{dy}{du} \cdot \frac{du}{dx} = 2u \cdot 8x$$
$$= 2(4x^2 - 1) \cdot 8x$$
$$= 64x^3 - 16x$$

**Function notation**

$$\frac{d}{dx}[f(g(x))] = f'(u) \cdot g'(x) = 2u \cdot 8x$$
$$= f'(g(x)) \cdot g'(x) = 2(4x^2 - 1) \cdot 8x$$
$$= 64x^3 - 16x$$

This leads us to formally state the chain rule in two different notations.

**The chain rule**
If $y = f(u)$ is a function in terms of $u$ and $u = g(x)$ is a function in terms of $x$, then the function $y = f(g(x))$ is differentiated as follows.
In Leibniz form:
$$\frac{dy}{dx} = \frac{dy}{du} \cdot \frac{du}{dx}$$
In function notation form:
$$\frac{dy}{dx} = \frac{d}{dx}[f(g(x))]$$
$$= f'(g(x)) \cdot g'(x)$$

The chain rule needs to be applied carefully. Consider the function notation form for the chain rule $\frac{d}{dx}[f(g(x))] = f'(g(x)) \cdot g'(x)$. Although it is the product of two derivatives, it is important to point out that the first derivative involves the function $f$ differentiated at $g(x)$ and the second is function $g$ differentiated at $x$. The chain rule is easy to remember in Leibniz form, $\frac{dy}{dx} = \frac{dy}{du} \cdot \frac{du}{dx}$, but you

The differentiation rule
$$\frac{dy}{dx} = \frac{dy}{du} \cdot \frac{du}{dx} \text{ is called}$$
the chain rule because it deals with a chain of functions, or a function of functions (i.e. a composite function).

should remember that they are **not** fractions. The expressions $\frac{dy}{dx}$, $\frac{dy}{du}$ and $\frac{du}{dx}$ are derivatives or, more precisely, limits, and although $du$ and $dx$ essentially represent very small changes in the variables $u$ and $x$, we cannot guarantee that they are non-zero.

The function notation form of the chain rule offers a very useful way of saying the rule in words, and thus, a very useful structure for applying it.

f is outside function      g is inside function

$$\frac{dy}{dx} = \frac{d}{dx}[f(g(x))] = f'(g(x)) \cdot g'(x)$$

derivative of outside function    ×    derivative of inside function

The chain rule in words:

$$\left(\begin{array}{c}\text{derivative of} \\ \text{composite}\end{array}\right) = \left(\begin{array}{c}\text{derivative of outside function} \\ \text{with inside function unchanged}\end{array}\right) \times \left(\begin{array}{c}\text{derivative of} \\ \text{inside function}\end{array}\right)$$

The chain rule is the most important rule of differentiation. Forgetting to apply the chain rule when necessary, or by applying it improperly, is a common source of errors in calculus computations. It is important to understand it, practice it, and master it.

## Example 13.17

Differentiate each function by applying the chain rule.

(a) $y = \sqrt{3x^2 + 5x}$     (b) $y = \dfrac{1}{3x^2 + x}$     (c) $y = \sqrt[3]{(7 - 5x)^2}$

### Solution

Start by decomposing the composite function into an outside function and inside function.

(a) $y = f(g(x)) = \sqrt{3x^2 + 5x}$

So, the outside function is $f(u) = \sqrt{u} = u^{\frac{1}{2}} \Rightarrow f'(u) = \dfrac{1}{2}u^{-\frac{1}{2}}$

and the inside function is $g(x) = 3x^2 + 5x$

$$\frac{dy}{dx} = f'(g(x)) \cdot g'(x) = \frac{1}{2}(3x^2 + 5x)^{-\frac{1}{2}} \cdot (6x + 5)$$

$$\frac{dy}{dx} = \frac{6x + 5}{2(3x^2 + 5x)^{\frac{1}{2}}} \text{ or } \frac{6x + 5}{2\sqrt{3x^2 + 5x}}$$

(b) $y = f(g(x)) = \dfrac{1}{3x^2 + x}$

So, the outside function is $f(u) = \dfrac{1}{u} = u^{-1} \Rightarrow f'(u) = -u^{-2}$

and the inside function is $g(x) = 3x^2 + x$

$$\frac{dy}{dx} = f'(g(x)) \cdot g'(x) = -(3x^2 + x)^{-2} \cdot (6x + 1)$$

$$\frac{dy}{dx} = -\frac{6x + 1}{(3x^2 + x)^2}$$

(c) First change from radical form to rational exponent form

$$y = \sqrt[3]{(7 - 5x)^2} = (7 - 5x)^{\frac{2}{3}}$$

$$y = f(g(x)) = (7 - 5x)^{\frac{2}{3}}$$

Aim to write a function in a way that eliminates any confusion regarding the argument of the function. For example, write $\sin(x^2)$ rather than $\sin x^2$; $1 + \ln x$ rather than $\ln x + 1$; $5 + \sqrt{x}$ rather than $\sqrt{x} + 5$; $\ln(4 - x^2)$ rather than $\ln 4 - x^2$.

So, the outside function $f(u) = u^{\frac{2}{3}} \Rightarrow f'(u) = \dfrac{2}{3}u^{-\frac{1}{3}}$

and the inside function $g(x) = 7 - 5x$

By the chain rule, $\dfrac{dy}{dx} = f'(g(x)) \cdot g'(x)$

$$= \frac{2}{3}(7 - 5x)^{-\frac{1}{3}} \cdot (-5)$$

$$= -\frac{10}{3(7 - 5x)^{\frac{1}{3}}} \text{ or } -\frac{10}{3(\sqrt[3]{7 - 5x})}$$

### Example 13.18

Find the derivative of the function $y = (2x + 3)^3$ by

(a) expanding the binomial and differentiating term-by-term

(b) the chain rule.

#### Solution

(a)  $y = (2x + 3)^3 = (2x + 3)(2x + 3)^2$

$\qquad = (2x + 3)(4x^2 + 12x + 9)$

$\qquad = 8x^3 + 24x^2 + 18x + 12x^2 + 36x + 27$

$\qquad = 8x^3 + 36x^2 + 54x + 27$

$\dfrac{dy}{dx} = 24x^2 + 72x + 54$

(b)  $y = f(g(x)) = (2x + 3)^3 \Rightarrow y = f(u) = u^3; u = g(x) = 2x + 3$

$\qquad\qquad\qquad\qquad\qquad \Rightarrow f'(u) = 3u^2; g'(x) = 2$

$\dfrac{dy}{dx} = \dfrac{dy}{du} \cdot \dfrac{du}{dx} = 3u^2 \cdot 2 = 6u^2$

$\qquad = 6(2x + 3)^2$

$\qquad = 6(4x^2 + 12x + 9)$

$\qquad = 24x^2 + 72x + 54$

## The product rule

With the differentiation rules that we have learned so far we can differentiate some functions that are products. For example, we can differentiate the function $f(x) = (x^2 + 3x)(2x - 1)$ by expanding and then differentiating the polynomial term-by-term.

$$f(x) = (x^2 + 3x)(2x - 1) = 2x^3 + 5x^2 - 3x$$

$$f'(x) = 2\frac{d}{dx}(x^3) + 5\frac{d}{dx}(x^2) - 3\frac{d}{dx}(x)$$

$$f'(x) = 6x^2 + 10x - 3$$

We know that the derivative of a sum or difference of two functions is the sum or difference of their derivatives. Is the derivative of the product of two functions the product of their derivatives? Let's try this with an example.

$$f(x) = (x^2 + 3x)(2x - 1)$$

$$f'(x) = \frac{d}{dx}(x^2 + 3x) \cdot \frac{d}{dx}(2x - 1)$$

$$f'(x) = (2x + 3) \cdot 2$$

$$f'(x) = 4x + 6$$

This is not the same result that we obtained earlier.

So, the derivative of a product of two functions is not the product of their derivatives. However, there are many products, such as $y = (4x - 3)^3(x - 1)^4$ and $f(x) = x^2 \sin x$, for which it is either difficult or impossible to write the function as a polynomial.

When $y$ is a function in terms of $x$ that can be expressed as the product of two functions $u$ and $v$ that are also in terms of $x$, then the product $y = uv$ can be differentiated using the **product rule**.

$$\frac{dy}{dx} = \frac{d}{dx}(uv)$$

$$= u\frac{dv}{dx} + v\frac{du}{dx}$$

or, equivalently, if $y = f(x) \cdot g(x)$, then

$$\frac{dy}{dx} = \frac{d}{dx}[f(x) \cdot g(x)]$$

$$= f(x) \cdot g'(x) + g(x) \cdot f'(x)$$

### Example 13.19

Find the derivative of the function $y = (x^2 + 3x)(2x - 1)$ by

(a) expanding the binomial and differentiating term-by-term

(b) using the product rule.

### Solution

(a) Expanding gives $y = (x^2 + 3x)(2x - 1) = 2x^3 + 5x^2 - 3x$

Therefore, $\dfrac{dy}{dx} = 6x^2 + 10x - 3$

(b) Let $u(x) = x^2 + 3x$ and $v(x) = 2x - 1$, then $y = u(x) \cdot v(x)$
or simply $y = uv$

Using the product rule (in Leibniz form),

$$\frac{dy}{dx} = \frac{d}{dx}(uv) = u\frac{dv}{dx} + v\frac{du}{dx} = (x^2 + 3x) \cdot 2 + (2x - 1) \cdot (2x + 3)$$

$$= (2x^2 + 6x) + (4x^2 + 4x - 3)$$

$$= 6x^2 + 10x - 3$$

The result obtained from using the product rule agrees with the derivative obtained from differentiating the expanded polynomial.

As with the chain rule, it is very helpful to remember the structure of the product rule in words.

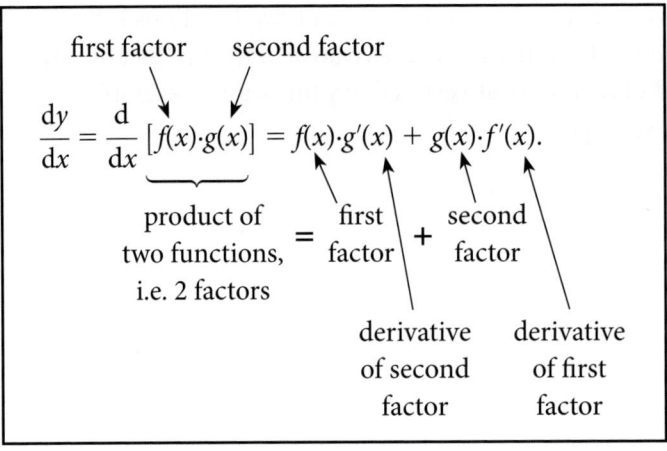

## The quotient rule

Just as the derivative of the product of two functions is not the product of their derivatives, the derivative of a quotient of two functions is not the quotient of their derivatives.

When $y$ is a function in terms of $x$ that can be expressed as the quotient of two functions $u$ and $v$ that are also in terms of $x$, then the quotient $y = \dfrac{u}{v}$ can be differentiated using the **quotient rule**:

$$\frac{dy}{dx} = \frac{d}{dx}\left(\frac{u}{v}\right) = \frac{v\dfrac{du}{dx} - u\dfrac{dv}{dx}}{v^2}$$

or, equivalently, if $y = \dfrac{f(x)}{g(x)}$, then

$$\frac{dy}{dx} = \frac{d}{dx}\left[\frac{f(x)}{g(x)}\right] = \frac{g(x) \cdot f'(x) - f(x) \cdot g'(x)}{[g(x)]^2}$$

Since order is important in subtraction, make sure you set up the numerator of the quotient rule correctly.

Note that we could have proved the quotient rule by writing the quotient $\dfrac{f(x)}{g(x)}$ as the product $f(x)[g(x)]^{-1}$ and apply the product rule and chain rule. As some of the examples here show, the derivative of a quotient can also be found by means of the product rule and/or the chain rule.

As with the chain rule and the product rule, it is helpful to recognise the structure of the quotient rule by remembering it in words.

$$\binom{\text{derivative of}}{\text{quotient}} = \frac{(\text{denominator}) \times \binom{\text{derivative of}}{\text{numerator}} - (\text{numerator})\binom{\text{derivative of}}{\text{denominator}}}{(\text{denominator})^2}$$

### Example 13.20

For each function, find its derivative (i) by the quotient rule, and (ii) by another method.

(a) $g(x) = \dfrac{5x - 1}{3x^2}$ 

(b) $h(x) = \dfrac{1}{2x - 3}$ 

(c) $f(x) = \dfrac{3x - 2}{2x - 5}$

## Solution

(a) (i) $g(x) = y = \dfrac{u}{v} = \dfrac{5x-1}{3x^2}$

$g'(x) = \dfrac{dy}{dx} = \dfrac{v\dfrac{du}{dx} - u\dfrac{dv}{dx}}{v^2}$

$= \dfrac{3x^2 \cdot 5 - (5x-1)\cdot 6x}{(3x^2)^2}$

$= \dfrac{-5x+2}{3x^3}$

(ii) Use algebra to split the numerator:

$g(x) = \dfrac{5x-1}{3x^2} = \dfrac{5x}{3x^2} - \dfrac{1}{3x^2} = \dfrac{5}{3x} - \dfrac{1}{3x^2} = \dfrac{5}{3}x^{-1} - \dfrac{1}{3}x^{-2}$

Now, differentiate term-by-term using the power rule.

$g'(x) = \dfrac{5}{3}\dfrac{d}{dx}(x^{-1}) - \dfrac{1}{3}\dfrac{d}{dx}(x^{-2})$

$= -\dfrac{5}{3x^2} + \dfrac{2}{3x^3}$

Note that the results for parts (i) and (ii) are equivalent:

$-\dfrac{5}{3x^2} + \dfrac{2}{3x^3} = -\dfrac{5x}{3x^3} + \dfrac{2}{3x^3} = \dfrac{-5x+2}{3x^3}$

(b) (i) $y = \dfrac{f(x)}{g(x)} = \dfrac{1}{2x-3} \Rightarrow f(x) = 1$ and $g(x) = 2x-3$

Using the quotient rule (function notation form),

$\dfrac{dy}{dx} = \dfrac{d}{dx}\left[\dfrac{f(x)}{g(x)}\right] = \dfrac{g(x)\cdot f'(x) - f(x)\cdot g'(x)}{[g(x)]^2}$

$= \dfrac{(2x-3)\cdot 0 - 1\cdot(2)}{(2x-3)^2}$

$= \dfrac{-2}{(2x-3)^2}$

(ii) $y = f(g(x)) = \dfrac{1}{2x-3} = (2x-3)^{-1}$

The outside function is $f(u) = u^{-1} \Rightarrow f'(u) = -u^{-2}$

The inside function is $g(x) = 2x-3$

Using the chain rule (function notation form):

$\dfrac{dy}{dx} = f'(g(x))\cdot g'(x) = -(2x-3)^{-2}\cdot 2$

$\dfrac{dy}{dx} = -\dfrac{2}{(2x-3)^2}$

(c) (i) $f(x) = y = \dfrac{u}{v} = \dfrac{3x-2}{2x-5}$

$$f'(x) = \frac{dy}{dx} = \frac{v\dfrac{du}{dx} - u\dfrac{dv}{dx}}{v^2}$$

$$= \frac{(2x-5)\cdot 3 - (3x-2)\cdot 2}{(2x-5)^2}$$

$$= \frac{-11}{(2x-5)^2}$$

(ii) Rewrite $f(x)$ as a product and apply the product rule (with chain rule embedded)

$$f(x) = y = \frac{3x-2}{2x-5} = (3x-2)(2x-5)^{-1}$$

for $y = uv$, $u = 3x - 2$ and $v = (2x-5)^{-1}$

$v = (2x-5)^{-1}$ is a composite function, so we need to use the chain rule to find $\dfrac{dv}{dx}$

$$f'(x) = \frac{d}{dx}(uv) = u\frac{dv}{dx} + v\frac{du}{dx}$$

$$= (3x-2)\cdot\frac{d}{dx}[(2x-5)^{-1}] + (2x-5)^{-1}\cdot 3$$

Applying the chain rule for $\dfrac{d}{dx}[(2x-5)^{-1}]$

$$= (3x-2)[-(2x-5)^{-2}\cdot 2] + 3(2x-5)^{-1}$$

$$= (-6x+4)(2x-5)^{-2} + 3(2x-5)^{-1}$$

Taking out a factor of $(2x-5)^{-2}$

$$= (2x-5)^{-2}[(-6x+4) + 3(2x-5)]$$

$$= (2x-5)^{-2}[-6x+4+6x-15]$$

$$f'(x) = \frac{-11}{(2x-5)^2}$$

The function
$h(x) = \dfrac{3x^2}{5x-1}$
looks similar to the function $g$ in Example 13.20 (a) (they are reciprocals). However, it is not possible to split the denominator and express as two fractions. Remember that $\dfrac{3x^2}{5x-1}$ is not equivalent to $\dfrac{3x^2}{5x} - \dfrac{3x^2}{1}$. Hence, in order to differentiate $h(x) = \dfrac{3x^2}{5x-1}$ we would apply either the quotient rule, or the product rule with the function rewritten as $h(x) = 3x^2(5x-1)^{-1}$, and using the chain rule to differentiate the factor $(5x-1)^{-1}$.

As Example 13.20 demonstrates, before differentiating a quotient it is worthwhile to consider if performing some algebra may allow other more efficient differentiation techniques to be used.

## Higher derivatives

If $y = f(x)$ is a function of $x$ then, in general, the derivative, $f'(x)$, will be some other function of $x$. For certain applications taking the derivative of the derivative – that is, the second derivative denoted by $f''(x)$ or $\dfrac{d^2y}{dx^2}$ – provides us with useful information. For example, the rate of change (derivative) of displacement is velocity and the rate of change (derivative) of velocity is acceleration. Hence, the second derivative of displacement is acceleration.

The notation $\dfrac{d^2y}{dx^2}$ comes from the idea that the first derivative is $\dfrac{d}{dx}(y) = \dfrac{dy}{dx}$, so the second

derivative is $\dfrac{d}{dx}\left(\dfrac{dy}{dx}\right) = \dfrac{d^2y}{dx^2}$

Notice that the $dx$ in the denominator is considered to be a single entity, so we don't write $d^2x^2$.

The $n$th derivative of $y$ with respect to $x$ is denoted by $\dfrac{d^ny}{dx^n}$. If the notation $f(x)$ is used, the first,

second and third derivatives are written as $f'(x), f''(x)$ and $f'''(x)$, respectively. Likewise, you may see $y', y'', y'''$ or $\dot{y}, \ddot{y}, \dddot{y}$ for first, second, and third derivatives, respectively. The fourth derivative and higher is denoted using a superscript number written in brackets. For example, $f^{(4)}(x)$ represents the fourth derivative of the function $f$ with respect to $x$.

## Example 13.21

The horizontal distance $s$ (m) travelled by a model aeroplane thrown in a straight line is given by the function $s(t) = \dfrac{t^2 - 4}{t}$ for $2 \leqslant t \leqslant 10$ (s).

(a) (i) By finding the first derivative of $s(t)$, determine the function $v(t)$ that expresses the velocity of the model aeroplane (in m s$^{-1}$).

   (ii) What is the sign of the aeroplane's velocity throughout the time interval $2 \leqslant t \leqslant 10$? What does this indicate about its motion?

(b) (i) By finding the second derivative of $s(t)$, determine the function $a(t)$ that expresses the acceleration of the model aeroplane (in m s$^{-2}$).

   (ii) What is the sign of the aeroplane's acceleration throughout the time interval $2 \leqslant t \leqslant 10$? What does this indicate about its motion?

## Solution

(a) (i) Applying the quotient rule:
$$s'(t) = v(t) = \frac{t(2t) - (t^2 - 4)(1)}{t^2} \Rightarrow v(t) = \frac{t^2 + 4}{t^2}$$

   (ii) The aeroplane's velocity is positive throughout the time interval $2 \leqslant t \leqslant 10$. A graph of $v(t) = \dfrac{t^2 + 4}{t^2}$ on a GDC confirms this.

   This means that during $2 \leqslant t \leqslant 10$, the aeroplane is continually moving further away from its initial position.

(b) (i) Applying the quotient rule again:
$$s''(t) = v'(t) = a(t) = \frac{t^2(2t) - (t^2 + 4)(2t)}{(t^2)^2} = \frac{2t^3 - 2t^3 - 8t}{t^4}$$
$$\Rightarrow a(t) = \frac{-8}{t^3}$$

   (ii) The aeroplane's acceleration is negative throughout the time interval $2 \leqslant t \leqslant 10$. A graph of $a(t) = \dfrac{-8}{t^3}$ on a GDC verifies this.

   Since velocity is positive and acceleration is negative, the aeroplane is continually slowing down on this interval.

## Exercise 13.4

1. Find the derivative of each function.

   (a) $y = (3x - 8)^4$

   (b) $y = \sqrt{1 - x}$

   (c) $y = \dfrac{3}{x^3}$

   (d) $y = \dfrac{x^3 + 1}{2x}$

   (e) $y = (x^2 + 4)^{-2}$

   (f) $y = \dfrac{x}{x + 1}$

   (g) $y = \dfrac{1}{\sqrt{x + 2}}$

   (h) $y = (2x^2 - 1)^3$

   (i) $y = x\sqrt{1 - x}$

   (j) $y = \dfrac{1}{3x^2 - 5x + 7}$

   (k) $y = \sqrt[3]{2x + 5}$

   (l) $y = (2x - 1)^3(x^4 + 1)$

   (m) $y = \sqrt{3x^2 - 2}$

   (n) $y = \dfrac{x^2}{x + 2}$

   (o) $y = \dfrac{x + 1}{x - 1}$

2. Given that $y = \dfrac{1}{x^4}$, find:

   (a) $\dfrac{dy}{dx}$

   (b) $\dfrac{d^2y}{dx^2}$

3. A curve has equation $y = (x - 1)(2x + 1)^2$

   (a) Find $\dfrac{dy}{dx}$ by first expanding the right side of the equation and applying the power rule.

   (b) Find $\dfrac{dy}{dx}$ by applying the product rule and chain rule.

4. Consider the function $f(x) = \dfrac{x^2 - 3x + 4}{(x + 1)^2}$

   (a) Show that $f'(x) = \dfrac{5x - 11}{(x + 1)^3}$

   (b) Show that $f''(x) = \dfrac{-10x + 38}{(x + 1)^4}$

5. Find the first and second derivatives of the function $f(x) = \dfrac{x - a}{x + a}$, $a \in \mathbb{R}$

6. Given that $y = x\sqrt{x + 1}$, show that $\dfrac{dy}{dx} = \dfrac{3x + 2}{2\sqrt{x + 1}}$

7. If $y = x^4 - 6x^2$, show that $y$, $\dfrac{dy}{dx}$, and $\dfrac{d^2y}{dx^2}$ are all negative on the interval $0 < x < 1$, but that $\dfrac{d^3y}{dx^3}$ is positive on the same interval.

8. Consider the right-angled triangle with sides $b$ and $b + 1$ and hypotenuse $h$.

   (a) Express $h$ as a function of $b$.

   (b) Show that $\dfrac{dh}{db} = \dfrac{2b + 1}{\sqrt{2b^2 + 2b + 1}}$

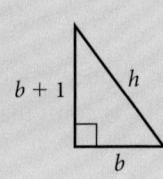

## 13.5 Derivatives of $\sin x$, $\cos x$, $\tan x$, $e^x$ and $\ln x$

For functions such as $\sin x$, $\cos x$, $\tan x$, $e^x$ and $\ln x$, we have to try different approaches to make sense of the derivative. While it is easy enough to look up the derivative of a given function in a formula booklet, it is worth trying to reason mathematically and use some of the methods we have explored to find unknown derivatives.

## Derivatives of trigonometric functions

The derivative is a rule that gives us the slope of the tangent to the graph of a function at a particular point. Thus, we can use a function's derivative to deduce the behaviour of its graph. Conversely, we can gain insight about the derivative of a function from the shape of its graph. Let's investigate the derivative of $\sin x$ by analysing its graph.

The graph of $y = \sin x$ (Figure 13.18) has a period of $2\pi$, so the same will be true of its derivative that gives the slope at each point on the graph. Therefore, it's only necessary for us to consider the portion of the graph in the interval $0 \leqslant x \leqslant 2\pi$

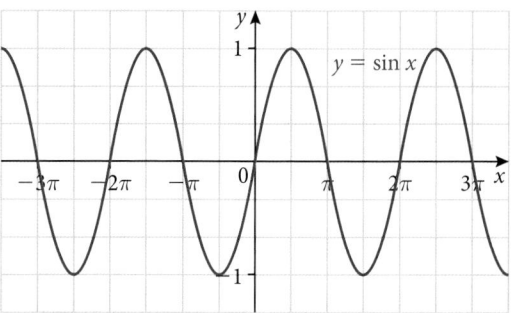

**Figure 13.18** Graph of $y = \sin x$

Figure 13.19 shows two pairs of axes having equal scales on the $x$- and $y$-axes and corresponding $x$-coordinates aligned vertically. On the top pair of axes $y = \sin x$ is graphed with tangents drawn at nine selected points. The points are chosen such that the slopes of the tangents at those points, in order, appear to be equal to

$$1, \frac{1}{2}, 0, -\frac{1}{2}, -1, -\frac{1}{2}, 0, \frac{1}{2}, 1$$

The values of these slopes are then plotted in the bottom graph with the $y$-coordinate of each point indicating the gradient of the curve for that particular $x$ value. Hence, the points in the bottom pair of axes should be on the graph of the derivative of $y = \sin x$

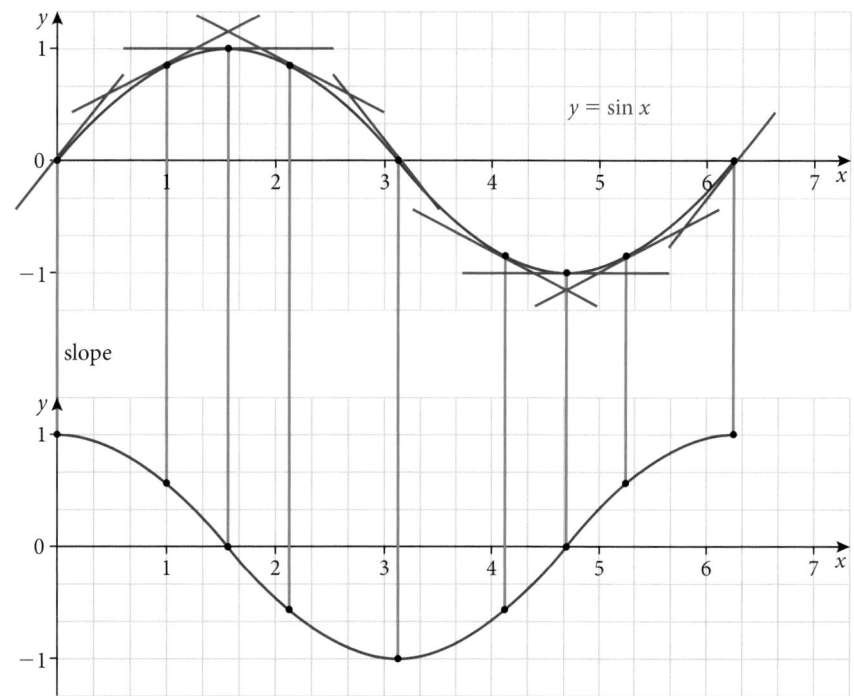

**Figure 13.19** Analysing the gradient of tangents to the graph of $y = \sin x$

 Note that the graphs in Figures 13.18 and 13.19 have $x$ in radians. We must use radian measure with trigonometric functions in calculus.

563

Clearly, the points representing the gradient of the tangents to $y = \sin x$ plotted in Figure 13.19 are tracing out the graph of $y = \cos x$. Therefore, the derivative of $\sin x$ is $\cos x$, that is, $\dfrac{d}{dx}(\sin x) = \cos x$. What about the derivative of $\cos x$?

We can use the GDC command that evaluates the derivative of a function at a specified point to graph the value of the derivative at all points on a graph. We used this feature of our GDC when forming a conjecture which led to the power rule for differentiation in section 13.3.

Figure 13.20 shows a graph of the derivative of $\cos x$; that is,

$$y = \dfrac{d}{dx}(\cos x)$$

The graph of $y = \dfrac{d}{dx}(\cos x)$ is the graph of $y = \sin x$ reflected about the $x$-axis.

Thus, $\dfrac{d}{dx}(\cos x) = -\sin x$

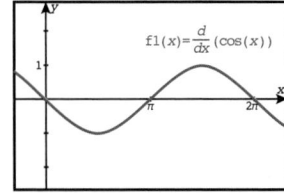

**Figure 13.20** Graph of the derivative of $\cos x$

Although we will use similar graphical approaches to conjecture the derivatives for $y = e^x$ and $y = \ln x$, it does not always work so smoothly. For example, let's analyse the graph of $y = \tan x$ in an attempt to guess its derivative.

The GDC screens in Figure 13.21 show the graph of $y = \tan x$ and the graph of the derivative of $\tan x$ (in bold) on the same set of axes. Although, in general, it is incorrect to graph a function and its derivative on the same axes (the units on the vertical axis will not be the same), it is helpful to see the connection between the graph of a function and that of its derivative.

The graph of the derivative of $\tan x$ is always above the $x$-axis meaning that the derivative is always positive. This agrees with the fact that the tangent function, except for where it is undefined, is always increasing (moving upwards) as the values of $x$ increase. Unfortunately, the shape of the graph does not suggest a rule for the derivative of $\tan x$. We will need another approach.

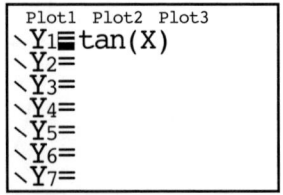

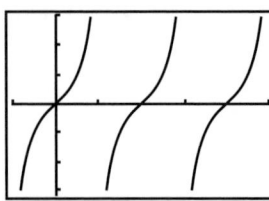

Let's write $\tan x$ as $\dfrac{\sin x}{\cos x}$ and use the quotient rule.

$$\frac{d}{dx}(\tan x) = \frac{d}{dx}\left(\frac{\sin x}{\cos x}\right) = \frac{\cos x \dfrac{d}{dx}(\sin x) - \sin x \dfrac{d}{dx}(\cos x)}{\cos^2 x}$$

$$= \frac{\cos x \cos x - \sin x(-\sin x)}{\cos^2 x}$$

$$= \frac{\cos^2 x + \sin^2 x}{\cos^2 x}$$

$$= \frac{1}{\cos^2 x}$$

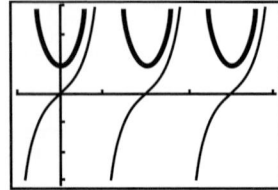

**Figure 13.21** GDC screens plotting $y = \tan x$ and its derivative

Therefore, $\dfrac{d}{dx}(\tan x) = \dfrac{1}{\cos^2 x}$

| f(x) | f'(x) |
|------|-------|
| $\sin x$ | $\cos x$ |
| $\cos x$ | $-\sin x$ |
| $\tan x$ | $\dfrac{1}{\cos^2 x}$ |

**Table 13.9** Derivatives of basic trigonometric functions

## Example 13.22

Find the derivative of each function.

(a) $y = \cos(\sqrt{x})$  (b) $y = \dfrac{x^3}{\sin x}$  (c) $y = x^2 \tan(3x)$

## Solution

(a) Applying the chain rule:

$$\frac{dy}{dx} = \frac{d}{dx}[\cos(\sqrt{x})] = -\sin(\sqrt{x}) \cdot \frac{d}{dx}(\sqrt{x})$$

$$= -\sin(\sqrt{x}) \cdot \frac{d}{dx}(x^{\frac{1}{2}})$$

Applying the power rule:

$$= -\sin(\sqrt{x}) \cdot \left(\frac{1}{2}x^{-\frac{1}{2}}\right)$$

Therefore, $\dfrac{dy}{dx} = -\dfrac{\sin(\sqrt{x})}{2(\sqrt{x})}$

(b) **Method 1 (quotient rule)**

Applying the quotient rule:

$$\frac{dy}{dx} = \frac{d}{dx}\left(\frac{x^3}{\sin x}\right) = \frac{\sin x \cdot \frac{d}{dx}(x^3) - x^3 \cdot \frac{d}{dx}(\sin x)}{\sin^2 x}$$

Therefore, $\dfrac{dy}{dx} = \dfrac{3x^2 \sin x - x^3 \cos x}{\sin^2 x}$

**Method 2 (product rule and chain rule)**

Rewriting as a product:

$$\frac{dy}{dx} = \frac{d}{dx}\left(\frac{x^3}{\sin x}\right) = \frac{d}{dx}[x^3 \cdot (\sin x)^{-1}]$$

Applying the product rule:

$$= x^3 \cdot \frac{d}{dx}[(\sin x)^{-1}] + (\sin x)^{-1} \cdot \frac{d}{dx}(x^3)$$

$$= x^3[-(\sin x)^{-2}\cos x] + (\sin x)^{-1}(3x^2)$$

Taking out the common factor of $(\sin x)^{-2}$:

$$= (\sin x)^{-2}[-x^3 \cos x + 3x^2 \sin x]$$

Therefore $\dfrac{dy}{dx} = \dfrac{3x^2 \sin x - x^3 \cos x}{\sin^2 x}$

(c) $\dfrac{dy}{dx} = \dfrac{d}{dx}[x^2 \tan(3x)] = x^2 \cdot \dfrac{d}{dx}(\tan(3x)) + \tan(3x) \cdot \dfrac{d}{dx}(x^2)$

Applying the product rule and the chain rule:

$$= x^2 \cdot \left(\frac{1}{\cos^2(3x)} \cdot 3\right) + \tan(3x) \cdot (2x)$$

$$= \frac{3x^2}{\cos^2(3x)} + 2x \tan(3x)$$

You may be tempted to find the derivative of $e^x$ by applying the power rule $\dfrac{d}{dx}(x^n) = nx^{n-1}$ but this only applies if a variable is raised to a constant power. An exponential function, such as $y = e^x$, is a constant raised to a variable power, so the power rule does not apply.

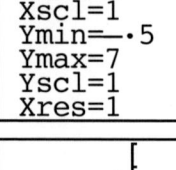

```
Plot1  Plot2  Plot3
\Y1◪e^(X)
\Y2=
\Y3=
\Y4=
\Y5=
\Y6=
\Y7=
```

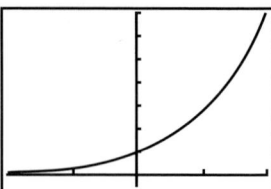

```
WINDOW
 Xmin=-2
 Xmax=2
 Xscl=1
 Ymin=-.5
 Ymax=7
 Yscl=1
 Xres=1
```

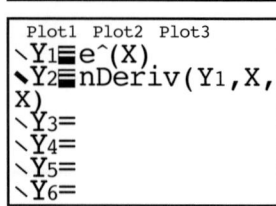

```
Plot1  Plot2  Plot3
\Y1◪e^(X)
\Y2◪nDeriv(Y1,X,
X)
\Y3=
\Y4=
\Y5=
\Y6=
```

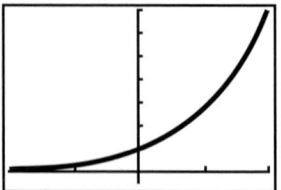

**Figure 13.23** GDC screens plotting $y = e^x$ and its derivative

When $f(x) = e^x$, then
$$f'(x) = e^x$$
In Leibniz notation
$$\frac{d}{dx}(e^x) = e^x$$

## Derivative of $e^x$

The graph of the exponential function $y = e^x$ passes through $(0, 1)$, has the $x$-axis as a horizontal asymptote, and is a continually increasing exponential growth curve (Figure 13.22).

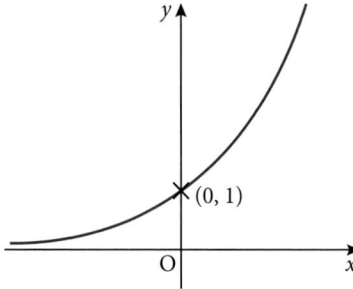

**Figure 13.22** The graph of $y = e^x$ is increasing over its entire domain

The exponential function is an important function for modelling exponential growth and decay. Example 13.4 showed that the number e is the limit of $\left(1 + \dfrac{1}{x}\right)^x$ as $x \to \infty$. As with the cosine function, let's try to guess the derivative of $e^x$ by using a GDC to graph its derivative.

The graph of the derivative of $e^x$ appears to be identical to $e^x$ itself. This is a very interesting result, but one which we see fits exactly with the nature of exponential growth and decay. The derivative of the exponential function is the exponential function. The gradient of the graph of $y = e^x$ at any point $(x, e^x)$ is equal to the $y$-coordinate of the point.

Look back at question 5 in Exercise 13.1 in connection to the derivative of $y = e^x$

### Example 13.23

Differentiate each function.

(a) $y = e^{2x + \ln x}$      (b) $y = \sqrt{x^2 + e^{4x}}$      (c) $y = \dfrac{e^x - e^{-x}}{e^x + e^{-x}}$

### Solution

(a) Because $e^{2x + \ln x} = e^{2x}e^{\ln x}$ and $e^{\ln x} = x$, then $e^{2x + \ln x} = xe^{2x}$
$$\frac{dy}{dx} = \frac{d}{dx}(e^{2x + \ln x}) = \frac{d}{dx}(e^{2x}e^{\ln x}) = \frac{d}{dx}(xe^{2x})$$
Applying the product rule and the chain rule
$$= x \cdot \frac{d}{dx}(e^{2x}) + e^{2x} \cdot \frac{d}{dx}(x)$$
Therefore, $\dfrac{dy}{dx} = 2xe^{2x} + e^{2x}$

(b) $\dfrac{dy}{dx} = \dfrac{d}{dx}(\sqrt{x^2 + e^{4x}}) = \dfrac{d}{dx}[(x^2 + e^{4x})^{\frac{1}{2}}]$

Applying the power rule and the chain rule

$$= \dfrac{1}{2}(x^2 + e^{4x})^{-\frac{1}{2}} \cdot \dfrac{d}{dx}(x^2 + e^{4x})$$

$$= \dfrac{2x + 4e^{4x}}{2\sqrt{x^2 + e^{4x}}}$$

Therefore, $\dfrac{dy}{dx} = \dfrac{x + 2e^{4x}}{\sqrt{x^2 + e^{4x}}}$

(c) Using the quotient rule:

$$\dfrac{dy}{dx} = \dfrac{d}{dx}\left(\dfrac{e^x - e^{-x}}{e^x + e^{-x}}\right)$$

$$= \dfrac{(e^x + e^{-x}) \cdot \dfrac{d}{dx}(e^x - e^{-x}) - (e^x - e^{-x}) \cdot \dfrac{d}{dx}(e^x + e^{-x})}{(e^x + e^{-x})^2}$$

$$= \dfrac{(e^x + e^{-x})(e^x + e^{-x}) - (e^x - e^{-x})(e^x - e^{-x})}{(e^x + e^{-x})^2}$$

$$= \dfrac{(e^{2x} + 2e^x e^{-x} + e^{-2x}) - (e^{2x} - 2e^x e^{-x} + e^{-2x})}{(e^x + e^{-x})^2}$$

$$= \dfrac{4e^x e^{-x}}{(e^x + e^{-x})^2}$$

Therefore, $\dfrac{dy}{dx} = \dfrac{4}{(e^x + e^{-x})^2}$

## Derivative of $\ln x$

Again, we will try to find the derivative of a function by analysing the shape of its graph along with producing a graph of the derivative on a GDC. We wish to determine the derivative for the natural logarithm function, $y = \ln x$

The graph of $y = \ln x$ is shown in Figure 13.24. Its $x$-intercept is $(1, 0)$, and since its domain is all positive real numbers, it has no $y$-intercept. It is asymptotic to the $y$-axis, and the graph rises steadily, though less steeply as $x \to \infty$. There is neither an upper nor a lower bound, so its range is all real numbers.

Using our GDC, we can view a graph of $y = \ln x$, a graph of its derivative, and a table of ordered pairs with $x$ and the value of the derivative at $x$ (as computed by the GDC).

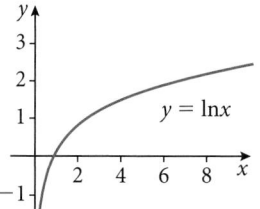

**Figure 13.24** The graph of $y = \ln x$ is increasing over its entire domain

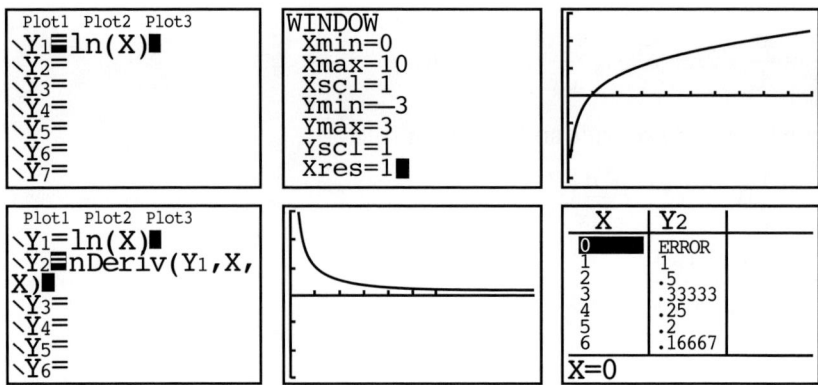

**Figure 13.25** GDC screens to show a graph of $y = \ln x$, a graph of its derivative, and a table of ordered pairs with $x$ and the value of the derivative at $x$

When $f(x) = \ln x$, then
$$f'(x) = \frac{1}{x}$$
In Leibniz notation
$$\frac{d}{dx}(\ln x) = \frac{1}{x}$$

In the table (bottom right screen in Figure 13.25), each value in the $Y_2$ column is the gradient of the curve (derivative) at the particular $x$ value for $y = \ln x$. From the graph of the derivative and especially from the table, we conjecture that the derivative of $\ln x$ is $\frac{1}{x}$

This agrees with the fact that for $x > 0$, the slope of the graph of $y = \ln x$ is always positive and as $x$ increases the gradient decreases.

---

### Example 13.24

(a) Given $y = \ln(1 - x^2)$, find $\dfrac{dy}{dx}$     (b) Given $f(x) = \ln\left(\dfrac{1 + x}{1 - x}\right)$, find $f'(x)$

---

**Solution**

(a) Applying $\dfrac{d}{dx}(\ln x) = \dfrac{1}{x}$ and the chain rule

$$\frac{dy}{dx} = \frac{1}{1 - x^2} \cdot \frac{d}{dx}(1 - x^2)$$

So $\dfrac{dy}{dx} = \dfrac{-2x}{1 - x^2}$ or $\dfrac{dy}{dx} = \dfrac{2x}{x^2 - 1}$

(b) Applying $\dfrac{d}{dx}(\ln x) = \dfrac{1}{x}$ and the chain rule

$$f'(x) = \frac{d}{dx}\left[\ln\left(\frac{1 + x}{1 - x}\right)\right] = \frac{1}{\frac{1+x}{1-x}} \cdot \frac{d}{dx}\left(\frac{1 + x}{1 - x}\right)$$

Applying the quotient rule

$$= \frac{1 - x}{1 + x} \cdot \frac{(1 - x)\dfrac{d}{dx}(1 + x) - (1 + x)\dfrac{d}{dx}(1 - x)}{(1 - x)^2}$$

$$= \frac{1 - x}{1 + x} \cdot \frac{1 - x + 1 + x}{(1 - x)^2}$$

$$= \left(\frac{1 - x}{1 + x}\right)\left(\frac{2}{(1 - x)^2}\right) = \frac{1}{1 + x} \cdot \frac{2}{1 - x}$$

So $f'(x) = \dfrac{2}{1 - x^2}$

## Exercise 13.5

1. Find the derivative of each function.

   (a) $y = x^2 e^x$

   (b) $y = \cos(4x)$

   (c) $y = e^{1-2x}$

   (d) $y = \dfrac{x}{1 + \sin x}$

   (e) $y = \dfrac{e^x}{x}$

   (f) $y = \dfrac{1}{2}\sin^2 2x$

   (g) $y = \dfrac{x}{e^x - 1}$

   (h) $y = \cos x \tan x$

   (i) $y = \sqrt{3 - \cos x}$

2. Find the equation of the tangent to the given curve at the specified value of $x$. Express the equation exactly in the form $y = mx + c$

   (a) $y = \ln\left(\dfrac{x}{3}\right), x = 6e$

   (b) $y = \ln(x^3 + 1), x = 1$

   (c) $y = \ln(\sin x), x = \dfrac{\pi}{4}$

   (d) $y = \ln\sqrt{x + 1}, x = 0$

   (e) $y = x \ln(x) - x, x = e$

   (f) $y = \dfrac{1}{\ln x}, x = e$

3. Consider the function $f(x) = e^x - x^3$

   (a) Find $f'(x)$ and $f''(x)$.

   (b) Find the $x$-coordinates (to 3 significant figures) for any points where $f'(x) = 0$.

   (c) Indicate the intervals for which $f(x)$ is increasing; and indicate the intervals for which $f(x)$ is decreasing.

4. (a) Show that the curves $y = e^{-x}$ and $y = e^{-x}\cos x$ are tangent at each point common to both curves.

   (b) Sketch the two curves over the interval $-\dfrac{\pi}{2} \leqslant x \leqslant \dfrac{3\pi}{2}$

5. A particle moves in a straight line such that its displacement, $s$, in metres, is given by $s(t) = 4\cos t - \cos 2t$ at $t$ seconds. The particle comes to rest after $T$ seconds, where $T > 0$

   Find the particle's acceleration at time $T$.

6. Given $h(x) = \dfrac{x^2 - 3}{e^x}$, find $h'(x)$

7. Find the second derivative, $\dfrac{d^2y}{dx^2}$, of each function.

   (a) $y = e^{-x}$

   (b) $y = \ln\left(\dfrac{1}{x}\right)$

   (c) $y = \sin^2 x$

   (d) $y = \sin(x^2)$

8. Consider the right-angled triangle shown.

   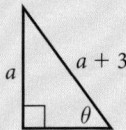

   (a) Show that $a = \dfrac{3\sin\theta}{\sin\theta - 1}$

   (b) Find $\dfrac{da}{d\theta}$

### Chapter 13 practice questions

1. A mass on the end of a spring moves up and down. Its height $h$ (in cm) above the ground is given by the function:

   $h(t) = t^3 - 7t^2 + 7t + 21$ for $0 \leqslant t \leqslant 5.5$ (s)

   (a) Find the two values of $t$ when the height of the mass is 6 cm above the ground.

   (b) Determine the function $v(t)$ that expresses the velocity of the mass.

   (c) Find the two values of $t$ when the velocity of the mass is zero (i.e. momentarily at rest).

   (d) Determine the function $a(t)$ that expresses the acceleration of the mass.

   (e) Find the acceleration (in $\text{cm s}^{-2}$) of the mass at

       (i) $t = 0$     (ii) $t = 5.5$

2. The function $f$ is defined as $f(x) = x^2$

   (a) Find the gradient of $f$ at the point $P$, where $x = 1.5$

   (b) Find an equation for the tangent to $f$ at the point $P$.

   (c) Draw a diagram to show clearly the graph of $f$ and the tangent at $P$.

   (d) The tangent in part (b) intersects the $x$-axis at the point $Q$, and the $y$-axis at the point $R$. Find the coordinates of $Q$ and $R$.

   (e) Verify that $Q$ is the midpoint of $[PR]$.

   (f) Find an equation, in terms of $a$, for the tangent to $f$ at the point $S(a, a^2)$, $a \neq 0$

   (g) The tangent in part (f) intersects the $x$-axis at the point $T$, and the $y$-axis at the point $U$. Find the coordinates of $T$ and $U$.

   (h) Prove that, whatever the value of $a$, $T$ is the midpoint of $[SU]$.

3. The curve $y = ax^3 - 2x^2 - x + 7$ has a gradient of 3 at the point where $x = 2$. Determine the value of $a$.

4. Consider the function $y = \dfrac{3x - 2}{x}$

   The graph of this function has a vertical and a horizontal asymptote.

   (a) Write down the equation of:

       (i) the vertical asymptote

       (ii) the horizontal asymptote.

   (b) Find $\dfrac{dy}{dx}$

   (c) Indicate the intervals for which the curve is increasing or decreasing.

5. Given the function $f(x) = x^2 - 3bx + (c + 2)$ where $b, c \in \mathbb{R}$, determine the values of $b$ and $c$ such that $f(1) = 0$ and $f'(3) = 0$

6. The displacement $s$ metres of a car, $t$ seconds after leaving a fixed point $A$, is given by the function $s(t) = 10t - \frac{1}{2}t^2$

   (a) Calculate the velocity of the car when $t = 0$

   (b) Calculate the value of $t$ when the velocity is zero.

   (c) Calculate the displacement of the car from $A$ when the velocity is zero.

7. Evaluate each limit.

   (a) $\lim\limits_{x \to \infty} \dfrac{2 - 3x + 5x^2}{8 - 3x^2}$ 
   (b) $\lim\limits_{x \to 0} \dfrac{\sqrt{x + 4} - 2}{x}$

   (c) $\lim\limits_{x \to 1} \dfrac{x^3 - 1}{x - 1}$ 
   (d) $\lim\limits_{x \to 1^-} \ln(1 - x)$

8. Find the derivative $f'(x)$ for each function.

   (a) $f(x) = \dfrac{x^2 - 4x}{\sqrt{x}}$ 
   (b) $f(x) = \dfrac{1}{x} + \dfrac{x}{2}$ 
   (c) $f(x) = \dfrac{7}{3x^{13}}$

9. An object moves along a straight line according to the position function $s(t) = t^3 - 9t^2 + 24t$

   Find the positions of the object when:

   (a) its velocity is zero 
   (b) its acceleration is zero.

10. In a controlled experiment, a tennis ball is dropped from the uppermost observation deck (447 m high) of the CN Tower in Toronto. The tennis ball's velocity is given by $v(t) = 66 - 66e^{-0.15t}$ where $v$ is in metres per second and $t \geq 0$ is in seconds.

    (a) Find the value of $v$ when:

    (i) $t = 0$ 
    (ii) $t = 10$

    (b) (i) Find an expression for the acceleration, $a$, as a function of $t$.
    (ii) Find the value of $a$ when $t = 0$

    (c) (i) As $t$ becomes large, what value does $v$ approach?
    (ii) As $t$ becomes large, what value does $a$ approach?
    (iii) Explain the relationship between the answers to parts (i) and (ii).

11. Consider the function $g(x) = 2 + \dfrac{1}{e^{3x}}$

    (a) (i) Find $g'(x)$.

    (ii) Explain briefly how this shows that $g(x)$ is a decreasing function for all values of $x$ (i.e. that $g(x)$ always decreases in value as $x$ increases).

    Let $P$ be the point on the graph of $g$ where $x = -\dfrac{1}{3}$

    (b) Find an expression in terms of e for

    (i) the $y$-coordinate of $P$

    (ii) the gradient of the tangent to the curve at $P$.

**12.** Differentiate each of these functions with respect to $x$.

(a) $y = \dfrac{1}{(2x + 3)^2}$     (b) $y = e^{\sin 5x}$     (c) $y = \tan^2(x^2)$

(d) $y = \dfrac{x}{e^x - 1}$     (e) $y = e^x \sin 2x$     (f) $y = (x^2 - 1)\ln(3x)$

**13.** The diagram shows the graph of a function $y = f(x)$

At which one of the five points on the graph:

(a) are $f'(x)$ and $f''(x)$ both negative

(b) is $f'(x)$ negative and $f''(x)$ positive

(c) is $f'(x)$ positive and $f''(x)$ negative?

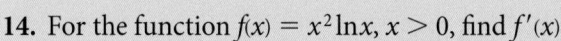

**14.** For the function $f(x) = x^2 \ln x$, $x > 0$, find $f'(x)$

**15.** For the function $f(x) = \dfrac{1}{2}\sin 2x + \cos x$, find the possible values of $\sin x$ for which $f'(x) = 0$

**16.** If $f(x) = \ln(2x - 1)$, $x > \dfrac{1}{2}$, find:

(a) $f'(x)$

(b) the value of $x$ where the gradient of $f(x)$ is equal to $x$.

**17.** Consider the function $y = \tan x - 8 \sin x$. Find:

(a) $\dfrac{dy}{dx}$

(b) the value of $\cos x$ for which $\dfrac{dy}{dx} = 0$

**18.** Let $y = \sin(kx) - kx \cos(kx)$, where $k$ is a constant.

Show that $\dfrac{dy}{dx} = k^2 x \sin(kx)$

**19.** If $y = \ln(2x - 1)$, find $\dfrac{d^2y}{dx^2}$

**20.** An experiment is carried out in which the number $n$ of bacteria in a liquid is given by the formula $n = 650e^{kt}$, where $t$ is the time in minutes after the start of the experiment and $k$ is a constant. The number of bacteria doubles every 20 minutes. Find:

(a) the exact value of $k$

(b) the rate of change of the number of bacteria when $t = 90$

**21.** The function $g$ is defined by $g(x) = \dfrac{2x}{x^2 + 6}$ for $x \geqslant b$ where $b \in \mathbb{R}$

(a) Show that $g'(x) = \dfrac{12 - 2x^2}{(x^2 + 6)^2}$

(b) Hence, find the smallest value of $b$ for which the inverse function $g^{-1}$ exists.

22. The graph in Figure 13.26 is of the function $y = f(x)$ for $a \leqslant x \leqslant b$

Make two copies of the graph shown and sketch the graphs of

(a) $y = f'(x)$                       (b) $y = f''(x)$

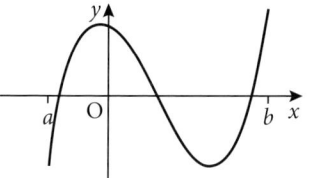

**Figure 13.26** Graph for question 22

23. The cost in thousands of Thai Baht (THB) to produce baseball caps is given by the function $c(x) = 2x^3 - 12x^2 + 30x$ where $x$ is in hundreds of caps. The derivative $c'(x)$ gives the marginal cost.

(a) Find the derivative $c'(x)$

(b) Find the marginal cost when 100 baseball caps are produced and interpret in context.

(c) Use $c'(x)$ to show that the cost function increases for all $x > 0$

The revenue in thousands of THB generated from selling $x$ baseball caps is given by the function $r(x) = 15x$ where $x$ is in hundreds of caps.

(d) Given that profit = revenue – cost, find a function $p(x)$ for the profit in THB from selling $x$ hundred baseball caps.

(e) Find the derivative $p'(x)$

(f) Find intervals where $p(x)$ is increasing and decreasing.

(g) The derivative $p'(x)$ gives the marginal profit. An optimum production level may be found when marginal profit is zero and $p(x)$ is positive. Find the optimum production level and the expected profit at this level.

24. The number of litres in a large tank of water $t$ minutes after a drain is opened is given by $Q(t) = 300(20 - t)^2, 0 \leqslant t \leqslant 20$

(a) Find the average rate at which the water is draining in the first 10 minutes.

(b) Find the rate that the water is draining at 10 minutes.

(c) Is the rate at which the water is draining increasing or decreasing? Speeding up or slowing down?

25. The spread of measles in a particular school is modelled by the function

$P(t) = \dfrac{200}{1 + 199(1.2)^{-t}}$ where $P(t)$ is the number of students who have

measles and $t$ is the number of days since the measles first appeared.

(a) Use your GDC to predict the maximum number of students infected.

(b) Use your GDC to obtain a graph of the derivative.

(c) Find the rate of infection 20 days after the first appearance.

(d) Find the average rate of infection for the first 20 days.

(e) Find the maximum rate of infection and the day at which this occurs.

(f) On which day after the maximum infection rate has passed does the infection rate drop to less than 1 person per day?

26. The value, $M$ (euros), of a motorcycle $t$ years after it is purchased can be modelled by the function $M(t) = \dfrac{4500}{t} + 750, t \geqslant 1$
    (a) Find $M'(t)$
    (b) Show that $M(t)$ is a decreasing function.
    (c) Find the time at which the value of the motorcycle is decreasing by €1000 per year.
    (d) Find the time, according to this function, at which the value of the motorcycle is decreasing the fastest.
    (e) Find the time at which the value of the motorcycle is decreasing by less than €50 year per year, and the value of the motorcycle at this point.

27. The profit $P(n)$ in thousands of dollars for a company producing t-shirts can be modelled by the function $P(n) = -2n^3 + 6n^2 + n$ where $n$ is the number of t-shirts produced (in thousands).
    (a) Find the interval(s) where the profit is increasing. Justify your response using $P'(n)$.
    (b) The optimum production level can be found by looking for when the marginal profit, given by $P'(n)$, changes from positive to negative. Find the optimum production level and the expected profit at this level.

28. The fuel efficiency in miles per gallon for vehicles in the USA can be modelled by the function $E(t) = -0.0007t^3 + 0.0278t^2 - 0.0843t + 12$ where $t$ is the number of years since 1970, $0 \leqslant t \leqslant 35$
    (a) Find $E'(t)$
    (b) Fuel efficiency was improving (increasing) between years $a$ and $b$. Find the values of $a$ and $b$.

29. The rate of change in the number of CD sales from 1991 to 2015 can be modelled by the function $S'(t) = 2.11t^2 - 74.4t + 498$ where $S(t)$ is the sales in millions per year at $t$ years since 1990, $1 \leqslant t \leqslant 25$
    (a) Describe the trend of CD sales during this time period.
    (b) Find the year in which the CD sales increased by 250 million.
    (c) In what year did CD sales first begin to decline?
    (d) In which year did the biggest decrease in CD sales happen, and how much did CD sales decrease?
    (e) Which graph below shows CD sales from 1991 to 2015?

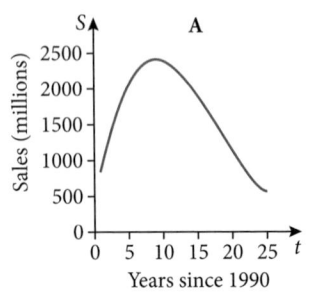
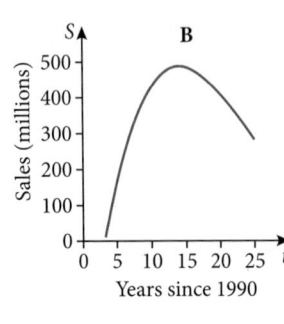
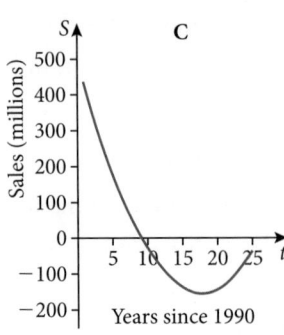
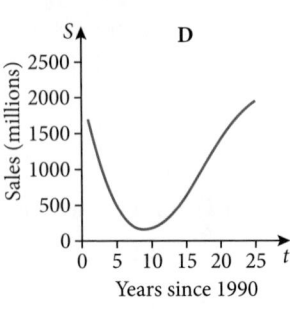

**Figure 13.27** Graphs for question 29

**30.** In an engine, a piston moves up and down according to the model $s = A\cos\left(\frac{\pi}{30}bt\right)$ where $s$ is the displacement in cm from the middle position, $A$ is the amplitude of the movement, $b$ is the frequency in cycles per minute, and $t$ is time in seconds. The distance that a piston travels in a Chevy 350 engine is 8.84 cm.

(a) Write down the value of $A$.

(b) Treating $b$ as a constant, find models for the velocity and acceleration of a Chevy 350 piston.

(c) Given the piston is moving at 550 cycles per minute, find the maximum velocity and acceleration of the piston.

(d) When the frequency of the piston increases by a factor of $m$, the acceleration increases by a factor of $k$. Find $k$ in terms of $m$.

(e) The frequency of a piston is equal to the revolutions per minute (RPM) of the engine. Use your answer for (d) to explain why engines in general cannot handle large RPMs.

**31.** The average daily high temperature in Stockholm, Sweden can be modelled by the function $T(d) = 11\cos\left(\frac{2\pi}{365}(d - 202)\right) + 11$ where $T(d)$ is the temperature in °C and $d$ is days since 1 January, $0 \leqslant d < 365$

(a) (i) On what day is the average temperature increasing the fastest?

(ii) What time of the year is this?

(iii) To the nearest tenth of a degree, how much warmer will the next day be, on average?

(b) (i) On what day is the average temperature decreasing the fastest?

(ii) What time of the year is this?

(iii) To the nearest tenth of a degree, how much colder will the next day be, on average?

(c) (i) On what days is the average temperature not changing?

(ii) What times of the year are these?

**32.** If a hemisphere-shaped bowl with a radius of 20 cm is filled with water to a depth of $x$ cm, the volume of water is given by $V = \pi x^2\left(20 - \frac{x}{3}\right)$ Find the rate of increase of volume per cm.

**33.** A paper manufacturer discharges pollutants into a nearby lake. A government agency requires that the paper manufacturer decrease their pollutant discharge by 25% each year. The concentration of pollutant in the lake can be modelled by the function $C = 1300(0.95)^n - 700(0.75)^n$ where $C$ is the concentration in parts per million (ppm) at year $n$, $n > 0$

(a) Pollutant concentrations above 700 ppm are considered dangerous. During what years is the concentration level dangerous?

(b) Find a function modelling the rate of change in the pollutant concentration per year.

(c) Hence, find intervals where the pollutant concentration is increasing and decreasing.

(d) During what years is the concentration increasing by more than 50 ppm per year?

(e) During what year is the concentration decreasing the fastest?

(f) Concentration levels less than 50 ppm are considered negligible.

   (i) Find the year when the concentration will first drop below 50 ppm.

   (ii) Give a reason why, according to this model, the pollutant concentration will remain less than 50 ppm. Give a reason for your answer.

34. Variable stars are those whose brightness increases and decreases in a periodic manner. The brightest variable star visible from Earth is Delta Cephei. The brightness of Delta Cephei varies from a magnitude of 4.35 to a magnitude of 3.65, with a cycle of 5.4 days.

(a) Formulate a model for the magnitude $M$ of the brightness of Delta Cephei based on time $t$ in days, where minimum magnitude is at time $t = 0$ in the form $M = a\cos(bt) + c$, where $a, b, c \in \mathbb{R}$ are constants to be determined.

(b) Find a model for the rate of change in the magnitude of the brightness of Delta Cephei.

(c) Hence, find intervals where the magnitude of the brightness is increasing and decreasing in the first week of the model.

35. The total number of cricket chirps during a given night can be modelled by the function $C = \frac{1}{\pi}\left(3\pi t - 12\cos\left(\frac{\pi}{12}t\right) + 12\right)$ where $C$ is the thousands of chirps and $t$ is in hours since sunset.

(a) Find a function for the number of cricket chirps per hour.

(b) During what time after sunset is the rate of chirps more than 3500 chirps per hour?

(c) Find the maximum number of chirps per hour and the time at which this occurs.

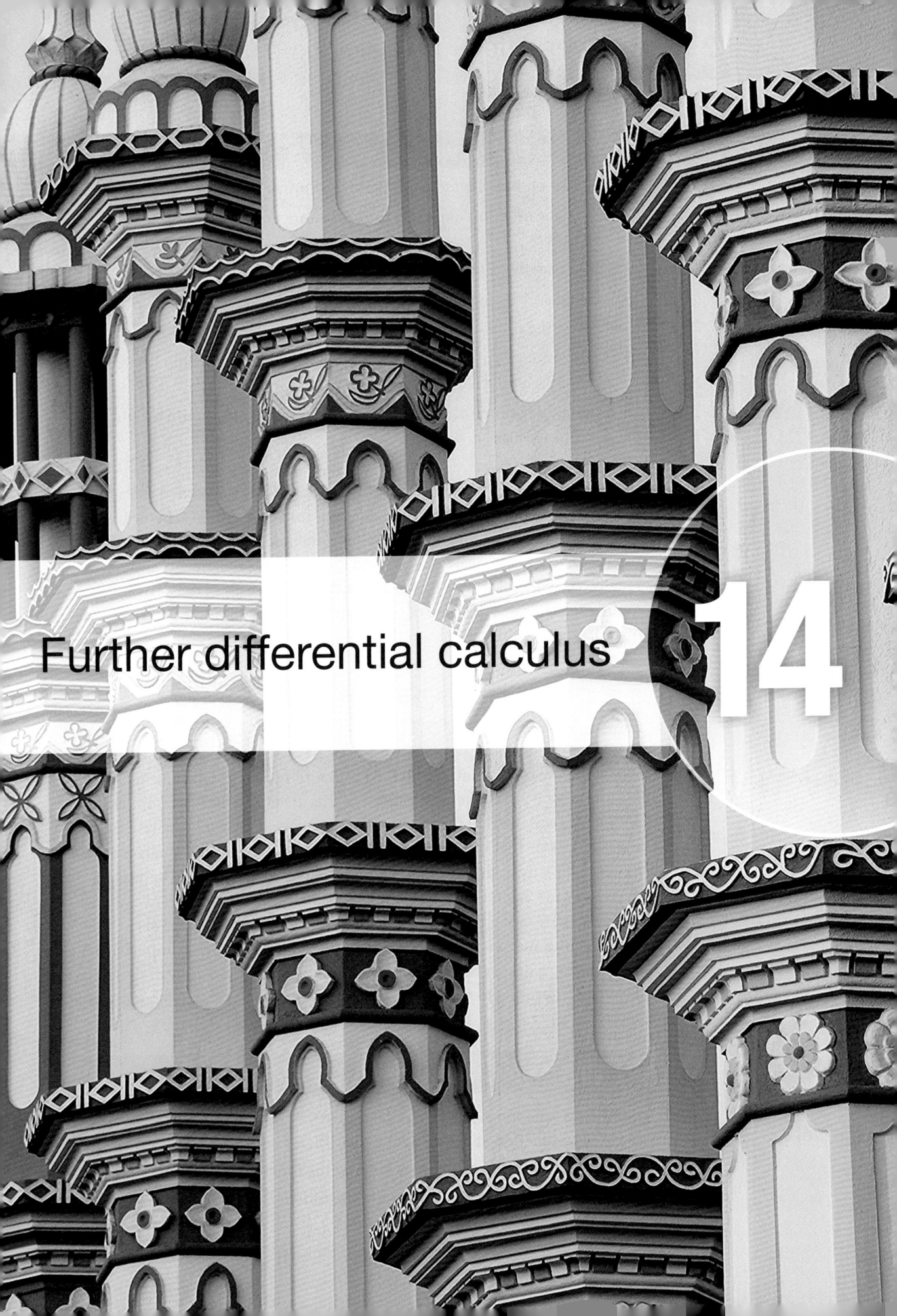

Further differential calculus

**14**

In Chapter 13, you were introduced to the derivative and learned rules for calculating the derivative for many different functions. In this chapter, we will apply that knowledge to analyse functions, including finding maxima and minima. We will also learn how to find the equations of tangents and normals. Finally, we will look at two important and powerful uses of the derivative: finding related rates (the rate of one quantity in relation to the rate of another quantity) and optimisation.

## 14.1 Minima, maxima and points of inflection

In real life, we often want to find the minimum or maximum value of a function. We will explore applications of this idea in Section 14.3. In this section, we will develop the calculus necessary to successfully analyse those situations. Note that in many situations, a GDC can be used to analyse a function directly to find maxima and minima using a **graphical** approach, as you saw in Chapter 9. When possible, there is no reason to avoid using a GDC for these cases; there is no need to use calculus for the sake of calculus. However, the **analytical** approach we develop in this section is important to help develop a deeper conceptual understanding of the meaning of derivatives as well as informing methods that you will use later in this chapter and this course in general.

## First derivative test

Consider a function $f(x)$. The first derivative tells us the instantaneous rate of change of a function at a particular value. Therefore:

- when $f'(x) > 0$, the gradient is positive; the function is increasing at $x$
- when $f'(x) = 0$, the gradient is zero; the function is neither increasing nor decreasing at $x$
- when $f'(x) < 0$, the gradient is negative; the function is decreasing at $x$.

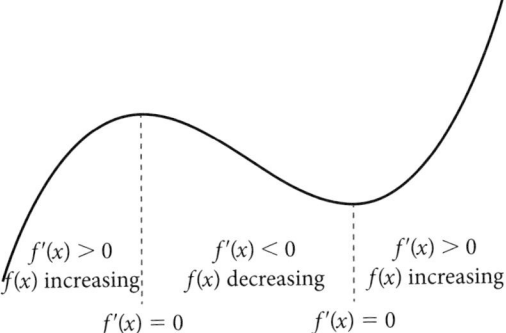

$f'(x) > 0$       $f'(x) < 0$      $f'(x) > 0$
$f(x)$ increasing   $f(x)$ decreasing   $f(x)$ increasing

$f'(x) = 0$        $f'(x) = 0$

**Figure 14.1** The sign of the first derivative tells us if the function is increasing, decreasing or neither

In this section, we are interested in using the first derivative to find **local extrema**. Local extrema can be either **maxima** or **minima**.

For a local maximum to occur, a continuous function must increase, then decrease. Likewise, for a local minimum to occur, the function must decrease, then increase. This is shown in Figure 14.2 as we move left to right along the graph of the function.

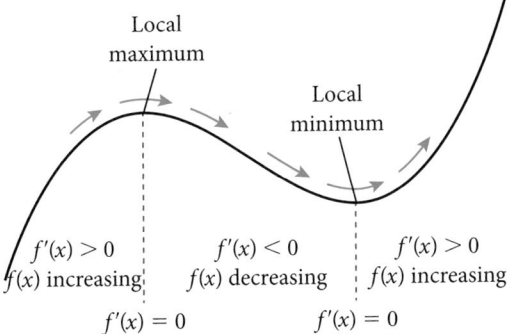

Local
maximum

Local
minimum

$f'(x) > 0$      $f'(x) < 0$      $f'(x) > 0$
$f(x)$ increasing   $f(x)$ decreasing   $f(x)$ increasing

$f'(x) = 0$        $f'(x) = 0$

**Figure 14.2** Local extrema occur on a continuous function when the first derivative changes sign

From Figure 14.2 we see that local extrema occur when there is a change in the sign of the first derivative. This leads to the **first derivative test**.

## Example 14.1

Find the coordinates of the local extremum of the function

$$f(x) = 2x^2 - 12x + 15$$

and classify it as a local maximum or local minimum.

### Solution

We could use our knowledge of quadratic functions to solve this problem, but we will apply our knowledge of calculus to reach the same result.

We need to find places where the first derivative changes sign. This can only happen if the first derivative is zero or undefined. We begin by finding the first derivative:

$$f'(x) = 2(2x) - 12 = 4x - 12$$

A local maximum (or minimum) is the greatest (or least) value of a function in a particular interval. Informally, we can think of these as the peaks or valleys of the graph of the function.

If a function is **continuous** on a **closed** interval, there will also be local extrema at the endpoints of the function.

As shown in Figure 14.2, for a continuous function $f$,

- a local **maximum** occurs at $c$ if $f'(x)$ changes sign from **positive** to **negative** at $x = c$.
- a local **minimum** occurs at $c$ if $f'(x)$ changes sign from **negative** to **positive** at $x = c$.

579

This is a linear function that is defined for all $x$, so we look for places where the first derivative is zero.

$$f'(x) = 0 \Rightarrow 4x - 12 = 0 \Rightarrow x = 3$$

Therefore, there is a possible extremum at $x = 3$. We find whether it is a maximum or minimum by applying the **first derivative test**. To do this, we must find the sign of the first derivative for intervals to the left and right of $x = 3$. For this it is helpful to make a **sign diagram**. A sign diagram simply shows the sign of the derivative on various intervals. We construct the sign diagram by first noting that the value of $f'(3) = 0$. Since 3 is the only value of $x$ where $f'(x) = 0$, it divides the number line into two intervals, greater than 3 and less than 3.

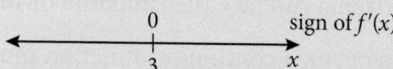

Next, we check the sign of $f'(x)$ by testing values of $x$ to the left and right of $x = 3$, so that we know whether $f(x)$ is increasing or decreasing on that interval. Pick values for $x$ that are easy to calculate:

To the left, $f'(0) = -12$, so $f'(x)$ is **negative**

To the right, $f'(10) = 28$, so $f'(x)$ is **positive**

Then we label our sign diagram accordingly.

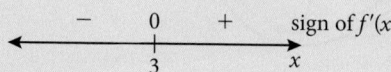

Now we can interpret the sign diagram to answer the question: Since the sign of $f'(x)$ changes from negative to positive at $x = 3$, there must be a **local minimum** at $x = 3$. The question asked us to find the coordinates of the point, so we evaluate $f(x)$ to find that $f(3) = -3$

Therefore, the local minimum occurs at $(3, -3)$.

It's important to consider values in the domain of a function where its derivative is undefined as well.

**Why is it necessary to test only one point in each interval?** The only way for the derivative to change sign is for it to pass through zero or to become undefined. Thus, we find the places where the derivative is zero or undefined, and then we need to test only one point in each interval bounded by those values.

### Example 14.2

Find the coordinates of all local extrema for the function $y = (x - 2)^{\frac{2}{3}} + 1$

### Solution

The first derivative is $\dfrac{dy}{dx} = \dfrac{2}{3}(x - 2)^{-\frac{1}{3}} = \dfrac{2}{3\sqrt[3]{x - 2}}$

There are no solutions to the equation $\dfrac{2}{3\sqrt[3]{x - 2}} = 0$, so the derivative is never equal to zero. However, it is undefined at $x = 2$. Therefore, we construct an appropriate sign diagram, by testing points to the left and right of $x = 2$

$-$ undefined $+$    sign of $f'(x)$

$x$

2

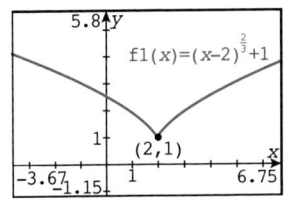

f1(x)=(x−2)$^{\frac{2}{3}}$+1

5.8 $y$

1

(2,1)

$x$

1    6.75

−3.67 −1.15

**Figure 14.3** GDC screen for the solution to Example 14.2

Therefore, we conclude there is a local minimum at $x = 2$

The coordinates of this point are $(2, 1)$.

This result can be confirmed by using a GDC to examine the graph (right) of $y = (x - 2)^{\frac{2}{3}} + 1$

Although the first derivative test can identify local maxima and minima, sometimes we are looking for the largest or smallest value of a function within some domain. In this case, we need to compare possible maxima and minima, including values at the endpoints of the domain. The values at the endpoints of the domain are always local maxima or minima. Suppose the domain endpoint is at $x = c$, then we can use the first derivative test in a one-sided manner.

For a function $f$ that is continuous on a closed interval with endpoint $c$, the endpoint is a local extremum.

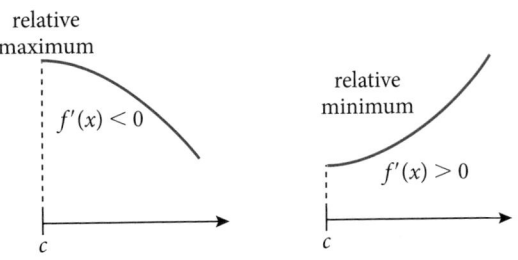

relative maximum

$f'(x) < 0$

relative minimum

$f'(x) > 0$

$c$    $c$

Left-hand endpoints (lower bounds) for $f(x)$ defined on $x \geqslant c$

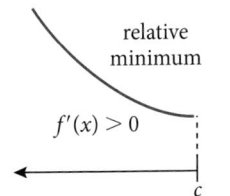

relative minimum

$f'(x) > 0$

relative maximum

$f'(x) < 0$

$c$    $c$

Right-hand endpoints (upper bounds) for $f(x)$ defined on $x \leqslant c$

Sometimes we are interested in finding the **absolute maximum** or **minimum**. To do this, we need to find all the local extrema and then compare them to see which one is the greatest or least value for the entire domain of $f$. In addition, the function may have more than one local extremum.

If a function $f$ is defined but not continuous for a closed interval, the interval endpoint(s) will still be local extrema. However, the first derivative test is not reliable for discontinuous functions.

## Example 14.3

Given $h(t) = 4t^3 + 3t^2 - 6t$ with domain $-\dfrac{3}{2} \leqslant t \leqslant 2$

(a) find the coordinates of all extrema and classify them as maxima or minima

(b) write down the coordinates of the absolute maximum and absolute minimum.

## Solution

It's a good idea to generate a graph of the function on a GDC (see right).

This helps us see likely extrema. In fact, we could use the GDC's maximum and minimum tools to find the local maximum and minimum directly. In this example, we will show the calculus approach.

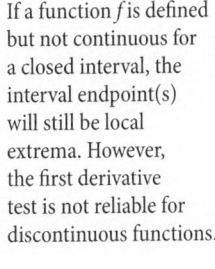

35.38 $y$

f1(x)=4·x³+3·x²−6·x

2

−3/2    0.1    2

−5.13

**Figure 14.4** GDC screen for the solution to Example 14.3

581

(a) Find the first derivative:

$$h'(t) = 12t^2 + 6t - 6$$

Then, find values of $t$ for which the first derivative is zero or undefined (but there are no undefined points on this derivative):

$$12t^2 + 6t - 6 = 0 \Rightarrow t = \left\{-1, \frac{1}{2}\right\}$$

Draw a sign diagram to visualise the intervals for the first derivative test.

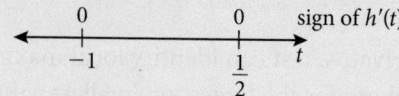

For example, we test $t = -2, 0, 1$ to obtain the sign of $h'(t)$ at each of those points: $h'(-2) = 30 > 0$, $h'(0) = -6 < 0$, and $h'(1) = 12 > 0$.

Remember that we don't need to find the exact value of $h'(t)$ – we just need to determine whether it is positive or negative.

Now, update the sign diagram:

There is a relative maximum at $t = -1$ and a relative minimum at $t = \frac{1}{2}$

However, we have not yet considered the endpoints. Since the first derivative is positive at both the lower and upper bound of the domain, the lower bound will be a relative minimum, while the upper bound will be a relative maximum. So, we have four relative extrema to consider.

Next, to find the coordinates, we must find the value of the function at each $t$ value:

| Relative **minimum** (lower bound of domain) | Relative **maximum** | Relative **minimum** | Relative **maximum** (upper bound of domain) |
|---|---|---|---|
| $h\left(-\frac{3}{2}\right) = \frac{9}{4}$ | $h(-1) = 5$ | $h\left(\frac{1}{2}\right) = -\frac{7}{4}$ | $h(2) = 32$ |

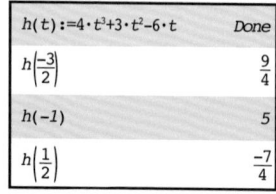

**Figure 14. 5** Most GDCs can define functions

Most GDCs can define functions, saving us repetitive calculations and reducing the chance of error (as shown on the left).

So, the coordinates and types of relative extrema are:

| Relative **minimum** (lower bound of domain) | Relative **maximum** | Relative **minimum** | Relative **maximum** (upper bound of domain) |
|---|---|---|---|
| $\left(-\frac{3}{2}, \frac{9}{4}\right)$ | $(-1, 5)$ | $\left(\frac{1}{2}, -\frac{7}{4}\right)$ | $(2, 32)$ |

(b) Now that we have done all the analysis of each relative extrema, it is easy to see that $(2, 32)$ is an absolute maximum and $\left(\dfrac{1}{2}, -\dfrac{7}{4}\right)$ is an absolute minimum over the given domain.

## Example 14.4

A toy rocket is launched upwards into the air. Its vertical position, $s$ metres, above the ground at $t$ seconds is given by $s(t) = -5t^2 + 18t + 1$

(a) Find the average velocity over the time interval from $t = 1$ to $t = 2$

(b) Find the instantaneous velocity at $t = 1$

(c) Find the maximum height reached by the rocket and the time at which this occurs.

Your GDC has built-in tools for solving polynomials (including quadratics). Learn to use them! Here is one example, for the quadratic in Example 14.3.

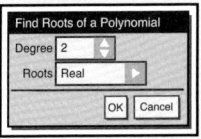

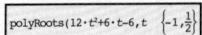

The solutions to this quadratic are $t = -1$ or $t = \dfrac{1}{2}$

## Solution

(a) $v_{\text{avg}} = \dfrac{s(2) - s(1)}{2 - 1} = 3 \, \text{m s}^{-1}$

(b) $s'(1) = 8 \, \text{m s}^{-1}$

(c) $s'(t) = 0 \Rightarrow t = 1.8$

Thus, $s$ has a stationary point at $t = 1.8$. We know $t$ must be positive and ranges from the time of launch ($t = 0$) to the time when the rocket hits the ground, i.e. $s = 0$. Since it doesn't make sense to include negative heights, we should check the position function and establish the domain of the function:

$s(t) = -5t^2 + 18t + 1 = 0$
$\Rightarrow t \approx -0.5472 \text{ or } t \approx 3.655$

So, the rocket hits the ground about $3.66 \, \text{s}$ after the time of launch. Hence, the domain for the position ($s$) and velocity ($v$) functions is $0 \leqslant t \leqslant 3.66$

Therefore, the function $s$ has three points we should check for a maximum: $t = 0$, $t = 1.8$ and $t \approx 3.66$

Apply the first derivative test and check the sign of the derivative $s'(t)$ for values on either side of $t = 1.8$, for example $t = 0$ and $t = 2$

$s'(0) = 18 > 0$ and $s'(2) = -2 < 0$

```
           +       0      −     sign of s'(t)
      ├──────────┼──────────┤
      0         1.8       3.66   t
```

Both of the domain endpoints, $t = 0$ and $t \approx 3.66$, are relative minima so we can ignore them. Since the function changes from increasing to decreasing at $t = 1.8$ and $s(1.8) = -5(1.8)^2 + 18(1.8) + 1 = 17.2$, the toy rocket reaches a maximum height of $17.2 \, \text{m}$, $1.8 \, \text{s}$ after it is launched.

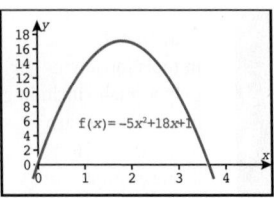

**Figure 14.6** Remember that most GDCs expect $x$ to be used as the independent variable for a graph, so we have replaced $t$ with $x$ here

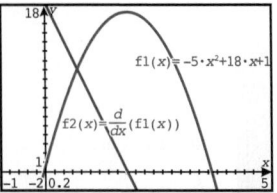

**Figure 14.7** GDC screen showing the derivative of the function

You will need to learn how to use your GDC to find the numerical derivative. It may have mathematical notation as shown here $\left(\dfrac{d}{dx}\right)$, or it may have a command like nDeriv(f1($x$),$x$). In either case, you will need to tell the calculator to differentiate with respect to a variable you specify. In this, we are differentiating with respect to $x$.

Be careful reading graphs with both a function and its derivative. Strictly speaking, we should not graph a function and its derivative on the same axes, because the units on the $y$ axes are different: remember that the $y$ value of the derivative is the rate of change in the original function.

Sometimes we may encounter functions that are difficult to differentiate algebraically. Your GDC can be a powerful tool for analysing these functions, as it has the capability to calculate **numerical** derivatives. Your GDC can tell you the value of the derivative at any point, and can therefore give a graph of the derivative. It cannot, however, give you an algebraic expression for the derivative.

For example, consider the position function in Example 14.4. We can enter the position function into a GDC and generate a graph (Figure 14.6).

Then, use the GDC to plot the derivative of that function (Figure 14.7).

We can then use the GDC to find the zero of the derivative function.

Now, we just need to interpret the results. Since $s'(1.8) = 0$, and the derivative is positive to the left and negative to the right of $t = 1.8$, we can conclude from the first derivative test that there must be a local maximum at $t = 1.8$. Furthermore, by looking at the graph of the position function (in blue), we can see that the position does indeed reach a maximum at $t = 1.8$. Since we have already entered the position function, we can also use the GDC to find the maximum height (Figure 14.8).

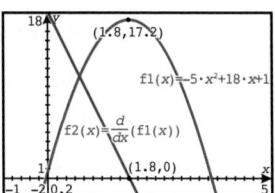

**Figure 14.8** Using the GDC to find the maximum height

This confirms the result in Example 14.4, showing that the toy rocket reaches a maximum height of 17.2 m, 1.8 s after launch.

Sometimes, we are looking for extrema in both the modelled quantity and in the rates of change. In these cases, we may use the second derivative to tell us about extrema in the first derivative.

### Example 14.5

After a drug is injected into a patient, the concentration $C$ of the drug in the bloodstream is given by the function

$$C = \frac{0.19t}{t^2 + 1}, t \geqslant 0$$

where $C$ represents the concentration in $\text{mg cm}^{-3}$, $t$ hours after injection.

(a) Find and interpret $\dfrac{dC}{dt}$

(b) Hence, find the time at which the bloodstream concentration is at a maximum and the bloodstream concentration at this time. Justify your answer using the first derivative test.

(c) Researchers are interested in the rate at which the bloodstream concentration is changing. In particular, they want to know when the bloodstream concentration is decreasing the fastest. Find the time at which the bloodstream concentration is decreasing the fastest and the rate of change in the bloodstream concentration at that time.

## Solution

(a) Using the quotient rule:

$$\frac{dC}{dt} = 0.19\left(\frac{1(t^2 + 1) - (2t)(t)}{(t^2 + 1)^2}\right) = 0.19\left(\frac{1 - t^2}{(t^2 + 1)^2}\right)$$

This function gives the instantaneous rate of change in the bloodstream concentration with respect to time, at time $t$.

(b) The derivative equals zero when the numerator of the rational derivative function equals zero, so $1 - t^2 = 0 \Rightarrow t = \pm 1$

Discard the negative value.

$$\begin{array}{ccccc} & + & 0 & - & \text{sign of } \frac{dC}{dt} \\ \vdash & & \vdash & \longrightarrow & \\ 0 & & 1 & & t \end{array}$$

Therefore, the maximum blood concentration occurs 1 hour after injection. The concentration at time $t = 1$ is

$$C = \frac{0.19(1)}{(1)^2 + 1} = 0.095 \, \text{mg cm}^{-3}$$

(c) To find when the rate of the bloodstream concentration is decreasing the fastest, we look for the minimum of the first derivative, which means the second derivative equals zero and has a sign change. We can use a GDC to find the minimum of the first derivative directly, or find the second derivative and look for zeros, or calculate by hand:

$$\frac{d^2C}{dt^2} = 0.19\left(\frac{(-2t)(-t^2 + 3)}{(t^2 + 1)^3}\right)$$

Solving to find possible extrema of the first derivative:

$$\frac{d^2C}{dt^2} = 0 \Rightarrow (-2t)(-t^2 + 3) = 0 \Rightarrow t = 0, \pm\sqrt{3} \text{ (reject negative solution)}$$

It makes sense that there is an extremum at time $t = 0$ since that is when the drug is first injected, so the bloodstream concentration is increasing quickly then. It appears that another extremum is at $t = \sqrt{3} = 1.73$ hours. We can check by analysing the sign of $\frac{d^2C}{dt^2}$

$$\begin{array}{ccccc} & - & 0 & + & \text{sign of } \frac{d^2C}{d^2t} \\ \vdash & & \vdash & \longrightarrow & \\ 0 & & \sqrt{3} & & t \end{array}$$

We conclude that there is a minimum in $\frac{dC}{dt}$ at $t = 1.73$ hours. Therefore the rate of change in the blood concentration is decreasing fastest at $t = 1.73$ hours. At this time the rate of change in bloodstream concentration is $\frac{dC}{dt} = -0.0238 \, \text{mg cm}^{-3} \, \text{h}^{-1}$

It is important to recognise when to use your GDC and when to find results by hand. In Example 14.5, it is quite possible to find the second derivative by hand, but doing so leads to a lot of algebra. Part (c), for example, could be done much more quickly by plotting the second derivative on a GDC.

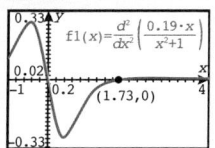

We can interpret this graph of the second derivative to see that $t = 1.73$ is a local minimum in the first derivative.

If you have the choice, it is usually more accurate and faster to use your GDC. In an exam, read the question carefully in case a certain method or exact answer is called for. Also, be sure to provide appropriate working or evidence for your results: sketching a graph, writing down values of a derivative, etc.

The first derivative test is especially useful in situations where we know a function that gives us a rate of change (a derivative), but not the original function.

### Example 14.6

A fruit distributor has noticed that the weekly change in the price of apples can be modelled by the function

$$PC = -\frac{3\pi}{208} \sin\left(\frac{\pi}{26}(x - 5)\right)$$

where $PC$ is the change in the price of a kilogram of apples and $x$ is the number of weeks since that start of the year.

(a) The fruit distributor would like to plan a sale when the price of apples will be at a minimum. What week should the sale be?

(b) Sales decrease when the price of apples approaches its maximum. In what week will the price of apples be the highest?

### Solution

(a) Since the given function predicts the change in the cost of apples, we can apply the first derivative test to find the minimum price of apples. Use a GDC to graph and locate the points where $PC = 0$

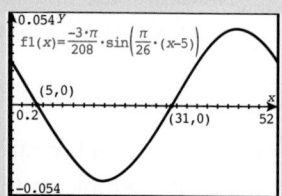

Since the price change (derivative) function is negative to the left of $x = 31$ and positive to the right, we conclude that the minimum price of apples is at week 31. The sale should be planned for week 31.

(b) With the same GDC graph, we can see that the price change (derivative) function is positive to the left of $x = 5$ and negative to the right, so we conclude that the maximum price of apples is at week 5.

## Second derivative test

In Chapter 13, you were introduced to the existence of a second derivative. How can that help us here? Consider how the first derivative is changing where $f'(x) = 0$:

- is the **first derivative** increasing (changing from negative to positive)
- or is the first derivative decreasing (changing from positive to negative)?

It stands to reason that if the first derivative tells us how the original function is changing, then the **second derivative** tells us how the first derivative is changing. So

- when the second derivative is positive, the first derivative is increasing
- when the second derivative is negative, the first derivative is decreasing.

Of course, the second derivative may not be positive or negative – it could be zero. In that case, it doesn't tell us anything useful about the first derivative, so we must revert to the first derivative test.

Put these facts together and you have the **second derivative test**.

**Second derivative test**

Given a function $f$ with $f'(c) = 0$:

When $f''(c) > 0$, $f'$ changes from negative to positive at $x = c$

    Hence, there is a local minimum at $x = c$

When $f''(c) < 0$, $f'$ changes from positive to negative at $x = c$

    Hence, there is a local maximum at $x = c$

When $f''(c) = 0$, the second derivative test is inconclusive; the first derivative test should be applied.

## Example 14.7

Given $f'(x) = x^2 - x - 6$, find the $x$-coordinates of all local extrema and identify them as maxima or minima.

### Solution

Find the values of $x$ where the first derivative is equal to zero or undefined. We will factorise the derivative function and solve algebraically:

$f'(x) = x^2 - x - 6 = (x - 3)(x + 2)$

so $f'(x) = 0 \Rightarrow x = 3$ or $x = -2$

Now find the second derivative:

$f''(x) = 2x - 1$

and evaluate at $x = 3$ and $x = -2$

$f''(3) = 5$ and $f''(-2) = -5$

Therefore:

since $f'(3) = 0$ and $f''(3) > 0$, there is a local minimum at $x = 3$

since $f'(-2) = 0$ and $f''(-2) < 0$, there is a local maximum at $x = -2$

**Figure 14.9** Concave up versus concave down

If you have trouble remembering what concave up and concave down look like, remember that up is happy, like a smile ☺, and down is sad, like a frown ☹.

The graph of $f(x)$ is **concave up** where $f'(x)$ is increasing and **concave down** where $f'(x)$ is decreasing. It follows that:
- When $f''(x) > 0$ for all $x$ in some interval of the domain of $f$, the graph of $f$ is concave up in the interval.
- When $f''(x) < 0$ for all $x$ in some interval of the domain of $f$, the graph of $f$ is concave down in the interval.

If, for some $x = c$, $f(c)$ exists, and the second derivative changes sign at $x = c$, then $(c, f(c))$ is a **point of inflection**.

Concavity is not defined for a line – it is neither concave up nor concave down.

We can get a bit more information from the second derivative. Since it tells us how the first derivative is changing, it tells us how the gradient is changing, which means it can tell us about **concavity**. In mathematics, concavity is whether a graph curves up (concave up) or curves down (concave down), as shown in Figure 14.9.

Since the second derivative is the rate of change in the first derivative, if the second derivative is positive then the graph must be curving up (concave up) since the first derivative is increasing. Likewise, if the second derivative is negative then the graph must be curving down (concave down) since the first derivative is increasing.

## Example 14.8

The function $f$ is continuous on the interval $0 \leqslant x \leqslant 4$, and the zeros of $f$ occur at $x = 1$ and $x = 4$. Information about the values of the derivatives of $f$ is given in the table.

587

| $x$ | $0 < x < 1$ | 1 | $1 < x < 2$ | 2 | $2 < x < 3$ | 3 | $3 < x < 4$ |
|---|---|---|---|---|---|---|---|
| $f'$ | − | 0 | − | − | − | 0 | + |
| $f''$ | + | 0 | − | 0 | + | + | + |

Use the information given in the table to:

(a)  state the $x$-coordinates of any extreme point(s)

(b)  state the $x$-coordinates any point(s) of inflection

(c)  sketch a possible graph of $f$.

## Solution

(a)  Extrema occur where $f'(x) = 0$ and there is a sign change in $f'(x)$ or the second derivative is positive or negative. Therefore, possible extrema are at $x = 1$ or $x = 3$

- At $x = 1$, $f''(x) = 0$ so we must examine the first derivative. There is no sign change in the first derivative so it is not an extremum.
- At $x = 3$, $f''(x) > 0$ so there is a local minimum.

(b)  Points of inflection occur when $f''$ changes sign.

This occurs when $x = 1$ or $x = 2$

(c)  It helps to start by sketching some arrows to represent the direction of the function at the known values.

| $x$ | $0 < x < 1$ | 1 | $1 < x < 2$ | 2 | $2 < x < 3$ | 3 | $3 < x < 4$ |
|---|---|---|---|---|---|---|---|
| $f'$ | ↘ | → | ↘ | ↘ | ↘ | → | ↗ |
| $f''$ | ↘ | 0 | ↘ | 0 | ↘ | ↘ | ↗ |

Then, we can sketch the curve.

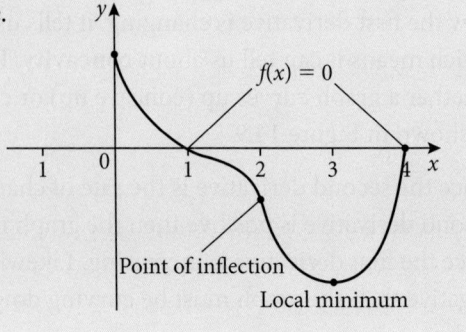

The second derivative is also very useful in kinematics, situations that involve position, velocity, and acceleration. Since velocity is the rate of change of displacement, we can find velocity via the first derivative. Likewise, since acceleration is the rate of change of velocity, we can find acceleration via the second derivative.

In kinematics, there is name and use for the third derivative of displacement, known as **jerk**, and occasionally even the fourth derivative, known as **jounce** or **snap**. The human body's physiological reaction time is limited; humans can only respond to changes in acceleration of a certain magnitude. For this reason, transportation engineers work hard to limit the maximum acceleration and jerk that passengers may be exposed to, even on everyday devices such as elevators.

## Example 14.9

Suppose the position of an object at time $t$ is given by the equation $s(t) = t^3 - 4t + 2$, $t \geqslant 0$, where $s$ is measured in metres and $t$ is measured in seconds. Find:

(a) the average velocity of the object from $t = 1$ to $t = 3$

(b) the instantaneous velocity of the object at $t = 3$

(c) the average acceleration of the object from $t = 1$ to $t = 3$

(d) the instantaneous acceleration of the object at $t = 3$

### Solution

(a) Average velocity is change in position over time, so average

velocity $= \dfrac{s(3) - s(1)}{3 - 1} = 9 \, \text{m s}^{-1}$

(b) Instantaneous velocity is given by $v(t) = s'(t) = 3t^2 - 4$, so
$v(3) = 3\,(3)^2 - 4 = 23 \, \text{m s}^{-1}$

(c) Average acceleration is change in velocity over time, so average

acceleration $= \dfrac{v(3) - v(1)}{3 - 1} = 12 \, \text{m s}^{-2}$

(d) Instantaneous acceleration is given by $a(t) = v'(t) = 6t$, so $a(3) = 18 \, \text{m s}^{-2}$

Recall that velocity and acceleration both have a direction. We usually consider up or forwards to be positive and down or backwards to be negative. This is why the leading coefficient in the standard falling object model for Earth is negative. It represents downward acceleration:

$s(t) = -4.9t^2 + v_0 t + s_0$

where $s(t)$ is the height in metres at time $t$ in seconds, $v_0$ is the initial velocity, and $s_0$ is the initial height.

But where does the 4.9 come from? It is half of the gravitational constant on Earth, as we can see if we find the second derivative:

$a(t) = \dfrac{d^2 s}{dt^2} = \dfrac{d^2}{dt^2}(-4.9t^2 + v_0 t + s_0) = \dfrac{d}{dt}(-9.8t + v_0) = -9.8 \, \text{m s}^{-2}$

However, we need to be careful when we talk about speed and speeding up or slowing down. Speed has no direction, but velocity does. Therefore, an object can be speeding up even when its acceleration is negative: an object in free-fall, with a negative velocity, is a perfect example. In other words, when the absolute value of an object's velocity is increasing, then the object is speeding up, otherwise it is slowing down.

When an object's velocity and acceleration have the same sign, the object is speeding up. When an object's velocity and acceleration have opposite signs, the object is slowing down.

## Example 14.10

Suppose an object is moving along a straight line so that its displacement relative to the origin at time $t$ is given by the equation $s(t) = 12t - t^3 + 1$, $t \geqslant 0$, where the position is measured in cm, and $t$ is the time in seconds.

(a) Find $s'(t)$, and give its meaning and its units.

(b) Find $s''(t)$, and give its meaning and its units.

(c) At time $t = 1$, is the object moving forwards or backwards? Is it speeding up or slowing down?

(d) When does the object change direction? Describe the position when this occurs.

(e) During the first 5 s, at what time is the object farthest from the origin?

### Solution

(a) $s'(t) = 12 - 3t^2$; this is the object's velocity at time $t$, in $\mathrm{m\,s}^{-1}$.

(b) $s''(t) = -6t$; this is the object's acceleration at time $t$, in $\mathrm{m\,s}^{-2}$.

(c) $v(1) = s'(1) = 12 - 3(1)^2 = 9\,\mathrm{m\,s}^{-1}$

The object is moving forward.

$a(1) = s''(1) = -6$

Since the sign of acceleration is opposite to the sign of the velocity, the object is slowing down.

(d) The object will change direction when velocity changes sign. So, we must find places where $v(t) = 0 \Rightarrow 12 - 3t^2 = 0 \Rightarrow t = \pm 2$; we discard the negative solution since $t \geqslant 0$

At $t = 2$, $a(t) = -6(2) = -12\,\mathrm{m\,s}^{-2}$, so velocity is changing from positive to negative. We also notice that this fulfils the second derivative test, so $t = 2$ is the location of a local maximum. In this context, this is the farthest the object is from the origin in that 'neighbourhood' of time.

(e) We know there is a local maximum at $t = 2$. At this time, we have a position of $s(2) = 17\,\mathrm{m}$. Now we simply need to check the endpoints of the domain to see if either of those are farther away. At $t = 0$, we have $s(0) = 1\,\mathrm{m}$, and at $t = 5$ we have $s(5) = -64\,\mathrm{m}$. Therefore, the maximum distance from the origin occurs at time $t = 5\,\mathrm{s}$

In supply chain analysis, distributors and logistics centres consider how fast demand is changing in order to adjust production and distribution. Example 14.11 shows how calculus is a useful tool in this situation.

## Example 14.11

The monthly sales of a new smartphone, in millions of units, are modelled by the function

$$S(t) = \frac{7t^4 + 8000t^2}{t^{4.9} + 110t^{2.9} + 3025t^{0.9}}, \text{ where } t \text{ is the time in months since release.}$$

(a) Calculate the value of $S'(2)$ and interpret it in context.

(b) Interpret the meaning of $S''(t)$ in context, with units.

(c) Find the time when the monthly sales begin to decline.

(d) The manufacturer wants to begin decreasing production as soon as the rate of increase in sales begin to slow down. At what time should the manufacturer begin to decrease production?

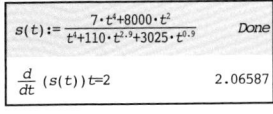

## Solution

(a) Using a GDC, we can see that $S'(2) = 2.07$. This tells us that 2 months after release, the monthly sales are increasing by 2.07 million units per month.

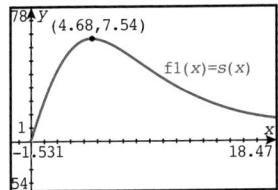

(b) $S''(t)$ is the rate of change in the change of sales, per month per month. Its units are millions of units per month$^2$.

(c) Using the monthly sales directly or finding the first derivative on a GDC, monthly sales reach a maximum at $t = 4.68$ months.

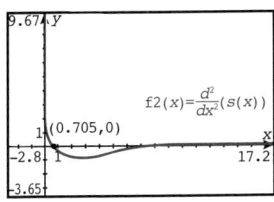

(d) The rate of change in monthly sales begins to decrease when the second derivative first becomes negative.

The second derivative becomes negative at $t = 0.705$ months, so the manufacturer should begin to decrease production at that time. This corresponds to a point of inflection in the monthly sales curve.

**Figure 14.10** GDC screens for the solution to Example 14.11

## Exercise 14.1

1. Find the vertex of the parabola using differentiation for each quadratic function.

   (a) $y = x^2 - 2x - 6$     (b) $y = 4x^2 + 12x + 17$

   (c) $y = -x^2 + 6x - 7$

2. For each function:

   (i)   find the coordinates of any points where $f'(x) = 0$

   (ii)  state, with reasons, whether each point from (i) is a minimum, maximum or neither

   (iii) sketch a graph of the equation and indicate the coordinates of each point from (i) on the graph.

   (a) $y = 2x^3 + 3x^2 - 72x + 5$     (b) $y = \frac{1}{6}x^3 - 5$

   (c) $y = -x(x - 3)^2$     (d) $y = x^4 - 2x^3 - 5x^2 + 6$

   (e) $y = x - \sqrt{x}$

3. For each function, find any local extrema and give the coordinates of each extremum.

(a) $f(x) = x^3 - 12x$

(b) $f(x) = \frac{1}{4}x^4 - 2x^2$

(c) $f(x) = x + \frac{4}{x}$

(d) $f(x) = -3x^5 + 5x^3$

(e) $f(x) = 3x^4 - 4x^3 - 12x^2 + 5$

4. For each of the following:

    (i) use your GDC to graph the derivative given.

    (ii) hence, find the locations of all local extrema and justify each using the first derivative test.

(a) $f'(x) = x^2 + 3x - 4$

(b) $h'(t) = t^2 - 7t + 10$

(c) $\dfrac{dy}{dx} = \dfrac{-x}{\sqrt{25 - x^2}}$

5. For each function below:

    (i) use your GDC to graph the derivative of the function

    (ii) hence, find the locations of all local extrema and justify each using the first derivative test. Verify your results by examining the graph of the function.

(a) $y = \sqrt{36 - x^2}$

(b) $f(x) = \sqrt{x^2 + 5} + \sqrt{25 - x^2}$

(c) $y = x^4 - 2x^3 - 5x^2 - 6$

6. The graphs of the derivative of a function $f$ are shown.

    (i) For what intervals is $f$ increasing or decreasing?

    (ii) At what value(s) of $x$ does $f$ have a local maximum or minimum?

(a)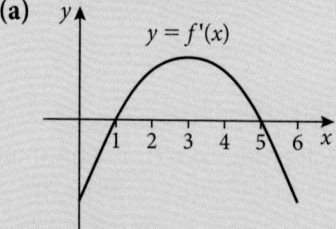

(b)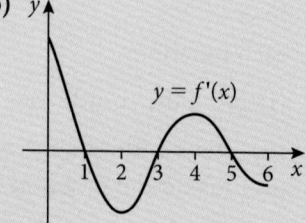

7. An object moves along a line such that its displacement $s$ metres from the origin $O$ is given by $s(t) = t^3 - 4t^2 + t$

(a) Find expressions for the object's velocity and acceleration in terms of $t$.

(b) For the interval $-1 \leqslant t \leqslant 3$, find the time at which the distance from zero is a maximum and find its value.

**(c)** For the interval $-1 \leqslant t \leqslant 3$, find the time at which the velocity is a minimum and find its value.

**(d)** In words, accurately describe the motion of the object during the interval $-1 \leqslant t \leqslant 3$

8. For each function, use calculus to find any relative extrema. State the coordinates of any such points and classify as maxima or minima. Use your GDC to verify graphically.

**(a)** $f(x) = -\dfrac{1}{2}x^4 + 4x^2 + 10$

**(b)** $f(x) = x^2 + \dfrac{16}{x}$

**(c)** $f(x) = x^2 + \dfrac{81}{x^2}$

**(d)** $f(x) = 2x^6 - 6x^2$

**(e)** $f(x) = \cos(x^2), 0 \leqslant x \leqslant 3$

**(f)** $f(x) = xe^x$

**(g)** $f(x) = x\sin(x), -1 \leqslant x \leqslant 6$

9. For each derivative, you are given that $f$ is a continuous function such that $c$ is the only value in the domain of $f$ where $f'(c) = 0$. Identify whether the point at $x = c$ is a local maximum, local minimum, neither, or cannot be determined.

**(a)** $f'(c - 1) > 0, f'(c) = 0, f'(c + 1) < 0$

**(b)** $f'(c - 1) < 0, f'(c) = 0, f'(c + 1) < 0$

**(c)** $f'(c - 1) < 0, f'(c) = 0, f'(c + 1) > 0$

**(d)** $f'(c) = 0, f''(c) > 0$

**(e)** $f'(c) = 0, f''(c) < 0$

**(f)** $f'(c) = 0, f''(c) = 0$

10. An object moves along a line such that its displacement $s$ metres from a fixed point $P$, at time $t$ seconds, is given by $s(t) = t(t - 3)(8t - 9)$

**(a)** Find the initial velocity and initial acceleration of the object.

**(b)** Find the velocity and acceleration of the object at $t = 3\,\text{s}$

**(c)** Find the values of $t$ for which the object changes direction. What significance do these times have in connection to the displacement of the object?

**(d)** Find the value of $t$ for which the object's velocity is a minimum. What significance does this time have in connection to the acceleration of the object?

11. The delivery cost per tonne of bananas, $D$ (in thousands of dollars), when $x$ tonnes of bananas are shipped is given by $D = 3x + \dfrac{100}{x}, x > 0$

**(a)** Find the value of $x$ for which the delivery cost per tonne of bananas is a minimum, and find the value of the minimum delivery cost.

(b) Use the second derivative test to justify that is value is in fact a minimum.

(c) Verify your results by checking the graph of the function.

12. The displacement $s$ metres of a car, $t$ seconds after leaving a fixed point $A$, is given by $s(t) = 10t - \frac{1}{2}t^2$

(a) Calculate the velocity when $t \geqslant 0$

(b) Calculate the value of $t$ when the velocity is zero.

(c) Calculate the displacement of the car from $A$ when the velocity is zero.

13. A ball is thrown vertically upwards from ground level such that its height $h$ in metres at $t$ seconds is given by $h = 14t - 4.9t^2$, $t \geqslant 0$

(a) State a function for the velocity of the ball.

(b) Find the maximum height the ball reaches and the time it takes to reach the maximum.

(c) At the moment the ball reaches its maximum height, what is the ball's velocity?

This problem is an extended version of one that appears in Chapter 9.

14. The diagram in Figure 14.11 represents a large Ferris wheel, with a diameter of 100 metres.

Let $P$ be a point on the wheel. The wheel starts with $P$ at the lowest point, at ground level. The wheel rotates at a constant rate, in an anti-clockwise direction. One revolution takes 20 minutes.

The height $h$ of point $P$ can be modelled by the function

$$h(t) = -50\sin\left(\frac{\pi}{10}(t + 5)\right) + 50$$

(a) Humans are more sensitive to vertical motion than horizontal motion. Show that the greatest vertical speed during the first rotation occurs at 5 minutes and 15 minutes.

(b) Vertical acceleration is what gives many amusement park rides their thrill factor. Find the times during the first rotation with the greatest absolute vertical acceleration.

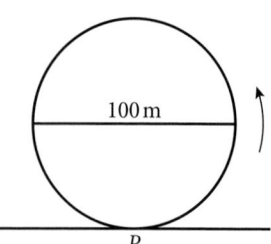

**Figure 14.11** Diagram for question 14

15. The cost in Chinese Renminbi to produce $x$ wooden toys can be modelled by the function $C(x) = 0.004x^2 + 5x + 2000$

Revenue from $x$ wooden toys can be modelled by the function $R(x) = -0.00016x^3 + 0.1x^2 + 20x$

Given that profit $P(x) = $ revenue $-$ cost, find the maximum profit and the number of wooden toys that should be produced to maximise profit.

16. The total cost, in euros, of producing $n$ bicycle wheels can be modelled by the function $C(n) = 20\,000 + 20n - n^2 + 0.005n^3$

(a) Find the number of bicycle wheels that should be produced to minimise the total cost and the minimum total cost.

(b) The average production cost per bicycle wheel can be expressed as $\dfrac{C(n)}{n}$. Find the minimum total production cost and the number of wheels that should be produced to achieve this minimum.

17. A chemical reaction is modelled by the function $R(t) = 100 - \dfrac{100}{1 + 60e^{-0.25t}}$, where $R(t)$ is the percentage of the reactant remaining at time $t$ in minutes, after a catalyst is introduced. Find the time at which the reaction rate is fastest.

18. The pH of a solution in an experiment can be modelled by the function $p(t) = \dfrac{8}{1 + 72e^{-0.09t}} + 3$, where $t$ is the time in seconds after the reaction began. At first, the reaction rate speeds up, then it slows down. Find the time when the reaction rate first begins to slow down.

19. The blood concentration of an injected medicine is modelled by the function $C(t) = \dfrac{0.13t}{t^2 + 1.5}$, where $C(t)$ is in mg cm$^{-3}$ and $t$ is the time in hours after injection.
    (a) Find $C'(t)$
    (b) Hence, find the time when the bloodstream concentration is at a maximum and the bloodstream concentration at this time. Justify your answer using the first derivative test.
    (c) Researchers are interested in the rate at which the bloodstream concentration is changing. In particular, they want to know when the bloodstream concentration is increasing the fastest and decreasing the fastest. Find the time at which the bloodstream concentration is changing the fastest and the rate of change in the bloodstream concentration at those times.

## 14.2   Tangents and normals

A **tangent** is a line that intersects a function and has the same gradient at the point of intersection. A **normal** is a line that intersects a function and is perpendicular to the tangent line at the point of intersection.

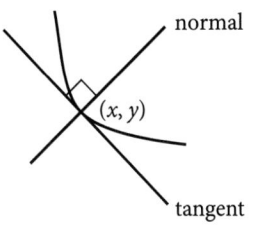

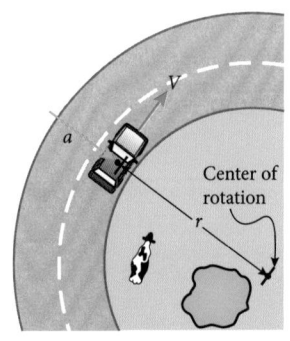

**Figure 14.12** When a car is travelling through a turn, its direction is changing, so it is accelerating. A force that is directed towards the centre of rotation accelerates the car so that it travels through the turn

We can use the derivative to find the gradient of the tangent (or normal) at a point on a function. Then, using the equation of a line, we can find the equations of tangents and normals.

### Example 14.12

Consider the function $g(x) = -x^2 + 8x - 14$

(a) Find the gradient of $g(x)$ at $x = 3$

(b) Hence write the equation of the tangent to $g(x)$ at $x = 3$

(c) Write the equation of the normal to $g(x)$ at $x = 3$

### Solution

(a) To find the value of the gradient, first find the derivative of $g(x)$

$$g'(x) = -2x + 8$$

Then, evaluate at $x = 3$ to obtain

$$g'(3) = 2$$

(b) Since the tangent must intersect $g(x)$ at the point of tangency, we can find the $y$-coordinate by evaluating $g(x)$ to obtain $g(3) = 1$. Therefore, $g(x)$ and the tangent both pass through $(3, 1)$. Now we can write the equation of the tangent, using point-slope form:

$$y - 1 = 2(x - 3)$$
$$y = 2x - 5$$

(c) Since a normal is perpendicular to a tangent, the product of their gradients must be $-1$

$$m_{normal} = \frac{-1}{m_{tangent}} = -\frac{1}{2}$$

The normal must also pass through the point of tangency $(3, 1)$, so

$$y - 1 = -\frac{1}{2}(x - 3)$$
$$y = -\frac{1}{2}x + \frac{5}{2}$$

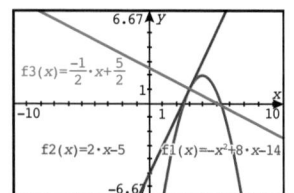

**Figure 14.13** The function (blue), the tangent (red), and the normal (green) for Example 14.12

We can graph the function, the tangent, and the normal to verify our work. Note that if the scales on the $x$ and $y$ axes are not the same, the lines will not appear perpendicular.

Sometimes we are interested in developing functions for which a tangent or normal has specific properties. For example, since the tangent has the same gradient as the curve, it can provide a smooth transition from one curve to another.

## Example 14.13

Lena is designing an archway for a building. The archway should be approximately triangular, but with a rounded top part defined by a parabola. She has sketched the design shown here.

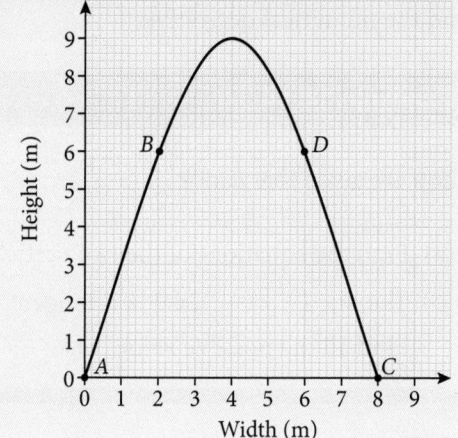

Height (m) / Width (m)

The line segments $AB$ and $DC$ are straight lines, while the curve is a parabola. The overall height of the archway is unknown.

The archway is symmetric about the line $x = 4$

(a) Write down the gradient of line segment $AB$.

(b) Write down the gradient of the curve at the maximum point of the archway.

(c) Given that the curve is modelled by the function $f(x) = ax^2 + bx + c$, find the values of $a$, $b$, and $c$, assuming that the gradient of $f(x)$ is equal to the gradients of $AB$ and $CD$ at the points where the straight lines and the curve intersect.

## Solution

(a) Line segment $AB$ has gradient $m_{AB} = \dfrac{6 - 0}{2 - 0} = 3$

(b) The gradient at the maximum point must be zero.

(c) Since the gradient of the curve must be zero at the maximum point, we know that $f'(4) = 0$. Likewise, to match the gradient of the line segment $AB$, we know that $f'(2) = 3$. The derivative of $f(x)$ can be written as $f(x) = 2ax + b$ and hence write:

$$f'(4) = 0 \Rightarrow 0 = 2a(4) + b \Rightarrow 0 = 8a + b$$

$$f'(2) = 3 \Rightarrow 3 = 2a(2) + b \Rightarrow 3 = 4a + b$$

Now we have a simultaneous linear system to solve:

Solving algebraically: $\begin{cases} 0 = 8a + b \\ 3 = 4a + b \end{cases} \Rightarrow b = 6, a = -\dfrac{3}{4}$

So far we have $f(x) = -\dfrac{3}{4}x^2 + 6x + c$

Now find the value of $c$. Since we know that the function must pass through $(2, 6)$, we can solve for $c$ using that point:

$$f(2) = 6$$
$$6 = -\frac{3}{4}(2)^2 + 6(2) + c$$
$$c = -3$$

Therefore, the equation of the curve must be $f(x) = -\dfrac{3}{4}x^2 + 6x - 3$

When we need to find the equation of a tangent or normal to a more complicated function, it is useful to know how to use your GDC to find the numerical derivative.

### Example 14.14

You are given that $h(x) = \dfrac{x^3 + 4x}{2x^2 - 3x}$

(a) Find the equations of the two vertical lines that are normal to $h$.

(b) The line $L_1: y = -2x + k$ is tangent to $h$.
Find all the possible values of $k$.

### Solution

(a) For a vertical line to be normal to $h$ the tangent must be horizontal. Hence, the value of $h'(x)$ at that point must be zero.
To start, plot the graph of $h$.

There are two points where the tangents are horizontal, at approximately $x = -1$ and $x = 4$

Graph the derivative function $h'$ to find where it is equal to zero.

It appears that $h'(x)$ has zeros at $x = -1$ and $x = 4$
Confirm this by using a GDC's Zero tool (twice) to find these zeros.

We are now confident that the normals to $h$ must pass through $x = -1$ and $x = 4$ and since these are vertical lines, the equations of the normals are also $x = -1$ and $x = 4$

(b) We need to find the points on $h$ where the gradient is $-2$, and then solve to find $k$.

Since we have already used a GDC to generate the graph of the derivative function $h'$, we can also use it to find the places where the gradient is equal to $-2$

We can do this by adding the line $y = -2$ to the graph and then using a GDC's Intersection tool (twice) to find where it intersects $h'(x)$.

From this display, we can see that $h'(x) = -2$ when $x = 0.382$ or $x = 2.62$ (3 s.f.)

Thus, the gradient of $h(x)$ is $-2$ at these $x$ values, and the tangents $y = -2x + k$ must intersect $h$ at these points. To find $k$, we must find the $y$-coordinates and then solve for $k$ in both cases.

We have already entered the function into a GDC, so we can use the GDC to calculate the $y$-values as $-1.85$ and $4.85$, respectively.

So we have two tangents; one at $(0.382, -1.85)$ and one at $(2.62, 4.85)$. Finally, solve for $k$, for each point.

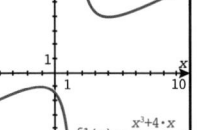

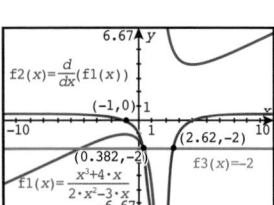

**Figure 14.14** GDC screens for the solution to Example 14.14

Learn how to use your GDC to store results from the graph, so that you avoid losing precision and do not have to retype long decimals. If you are retyping an approximate result into your GDC, you are not using it efficiently or effectively.

$$y = -2x + k$$
$$-1.85 = -2(0.382) + k$$
$$k = -1.09$$

$$y = -2x + k$$
$$4.85 = -2(2.62) + k$$
$$k = 10.1$$

So

$$y = -2x - 1.09 \text{ and } y = -2x + 10.1$$

We can verify this visually on a GDC.

Figure 14.15 shows the two lines we found, and we can see that they are tangent to $h(x)$.

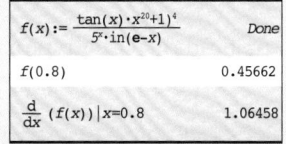

**Figure 14.15** $h(x)$ and the two lines with gradient $-2$ that are tangent to $h(x)$

Remember that your GDC is a powerful tool for finding derivatives numerically. Unless required to find exact answers, you should use your GDC when possible.

## Example 14.15

Find the equation of the normal to the curve $y = \dfrac{(\tan x)(x^{20} + 1)^4}{5^x \ln(e - x)}$ at the point where $x = 0.8$

Give the value of the gradient of the normal to 3 significant figures.

### Solution

It should be clear that we do not want to attempt to find the derivative of an equation of this type algebraically. Instead, we can use a GDC to find the numerical derivative and the $y$-coordinate at $x = 0.8$

$f(x) := \dfrac{\tan(x) \cdot x^{20}+1)^4}{5^x \cdot \text{in}(e-x)}$     *Done*

$f(0.8)$     0.45662

$\dfrac{d}{dx}(f(x))|x=0.8$     1.06458

The tangent passes through (0.8, 0.457) and has a gradient of 1.06. The normal, therefore, has a gradient of $-0.939$. So the equation of the normal is

$$y - 0.457 = -0.939(x - 0.8)$$
$$y = -0.939x + 1.21$$

Note that we can calculate the coefficients of the normal efficiently with a GDC.

The screenshots on the right show the efficient use of a GDC to find and verify the equation of the normal to a curve.

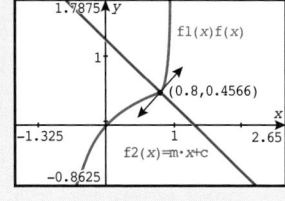

**Exercise 14.2**

1. For each function, find:
   **(i)** the equation of the tangent at the given point
   **(ii)** the equation of the normal at the given point.
   Use a GDC to confirm your results.
   **(a)** $y = 3x^2 - 4x$ at $(0, 0)$
   **(b)** $y = 1 - 6x - x^2$ at $(-3, 10)$
   **(c)** $y = x^2 + 2x + 1$ $x = -3$
   **(d)** $y = x^3 + x^2$ $x = -\dfrac{2}{3}$
   **(e)** $y = 3x^2 - x + 1$ $x = 0$
   **(f)** $y = 2x + \dfrac{1}{x}$ $x = \dfrac{1}{2}$
   **(g)** $y = \sin(2x)$ $x = \dfrac{\pi}{8}$
   **(h)** $y = e^x$ $x = 2$

2. For each function, find the coordinates of any points on the graph of the function where the tangent has the given gradient, $m$.
   **(a)** $y = x^2 + 3x$ $m = 3$
   **(b)** $y = x^3$ $m = 12$
   **(c)** $y = e^x$ $m = e^2$

3. Find the coordinates of any points of the graph of $y = \tan x$, $0 \leqslant x \leqslant 2\pi$, where the tangent has a gradient of 1.

4. Find the equations of the tangents to the curve $y = x^3 - 3x^2 + 2x$ at any point where the curve intersects the $x$-axis.

5. Find the equation of the tangent to the curve $y = x^2 - 2x$ which is perpendicular to the line $x - 2y = 1$

6. The line $y = 4x - 13$ is tangent to $y = x^2 + ax + b$ at the point $(3, -1)$, where $a, b \in \mathbb{R}$. Find the value of $a$ and the value of $b$.

7. The line $y = -\dfrac{1}{3}x + 3$ is normal to $y = x^2 + ax + b$ at the point $(3, 2)$, where $a, b \in \mathbb{R}$.
   Find the value of $a$ and the value of $b$.

8. Find the coordinates of the point on the graph of $y = x^2 - x$ at which the tangent is parallel to the line $y = 5x$

9. Find the equation of the tangent(s) to the curve $y = x^3 - 5x$ which are perpendicular to the line $x - 7y = 38$

10. Find the equation of the normal to the curve $y = x^2 + 4x - 2$ at the point where $x = -3$. Find the coordinates of the other point where this normal intersects the curve again.

11. Consider the function $g(x) = \dfrac{1 - x^3}{x^4}$

    (a) Use your GDC to find the value of the derivative when $x = 1$

    (b) Find the equation of both the tangent and the normal to the graph of $g$ at the point $(1, 0)$

12. Given the function $f(x) = x^3 + \dfrac{1}{2}x^2 + 1$, find:

    (a) the equation of the tangent to $f(x)$ at the point $\left(-1, \dfrac{1}{2}\right)$

    (b) the coordinates of another point on the graph of $f$ where the tangent is parallel to the tangent found in (a).

13. Find the equation of both the tangent and the normal to the curve $y = \sqrt{x}(1 - \sqrt{x})$ at the point where $x = 4$

14. (a) Using your GDC for assistance, make accurate sketches of the curves $y = x^2 - 6x + 20$ and $y = x^3 - 3x^2 - x$ on the same set of axes.

    The two curves have the same slope at an integer value for $x$ somewhere in the interval $0 \leqslant x \leqslant 7$

    (b) Find this value of $x$.

    (c) Find the equation for the tangent to each curve at this value of $x$.

15. The line $L$ is tangent to $f(x) = -x^2 - 6x - 4$ and passes through the origin.

    (a) Find the two possible points of tangency.

    (b) Find the corresponding two equations for $L$.

16. Find the equations of both lines that pass through the point $(2, -3)$ that are tangent to the parabola $y = x^2 + x$

17. Find the equation of the normal to the curve $y = \dfrac{e^x \cos x \, \ln(x + e)}{(x^{17} + 1)^5}$ at the point where $x = 0$. In your answer, give the value of the gradient of the normal to 3 significant figures.

18. Find an exact equation of the tangent to the curve $y = \sin x \cos 2x$ at the point where $x = \dfrac{\pi}{6}$

# 14.3   Optimisation

Optimisation is the process of finding values that will maximise or minimise another quantity. The following questions are typical optimisation problems.

- What production level or price should we choose to maximise profit?
- What production level should we choose to minimise costs?

- When is the maximum height or maximum velocity of a projectile reached?
- How can we make cuts and folds to a piece of paper to create an open box with maximum volume?
- How can we design a box to minimise packaging cost for a given volume?

Part of the process of solving optimisation problems is finding maximum and minimum values of functions, which you have already seen. However, this section will focus on the problem-solving skills necessary to get to that point. We start by revisiting an example from Chapter 9 in order to highlight some key steps.

### Example 14.16

**Finding a maximum volume (Open box problem)**
The dimensions of a piece of A4 paper, to the nearest cm, are $21 \times 30$ cm. It is possible to create an open box by cutting out square corners and folding the remaining flaps up, as shown in the diagram.

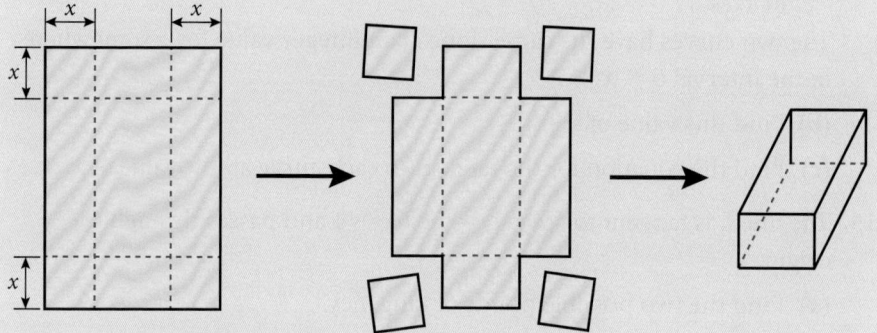

(a) Calculate the maximum volume of the open box.

(b) Find the dimensions of the open box with the largest volume that can be created using this method.

### Solution

(a) **Choose an appropriate model for the quantity to be optimised**

As discussed in Chapter 9, models come in many forms. In this section, our models often start with a simple, known formula. In this case, since we are interested in maximising the volume of the box, it seems appropriate to start with a model for the volume: $V = lwh$

**Draw an accurate diagram**

A diagram is helpful because it often uncovers relationships that will help us adapt our general model to the specific situation. In this case, we start with a piece of paper with dimensions of $21 \times 30$ cm. However, these are not the dimensions of the open box. If we look at the second step in the diagram above we see that part of the width and length of the paper becomes the height of the box.

Also, it is important to decide what the independent variable really is. In this case, it's the length of the side of the square we cut out from each corner. So, let's call that $x$ and then take the previous diagram and label it carefully as shown.

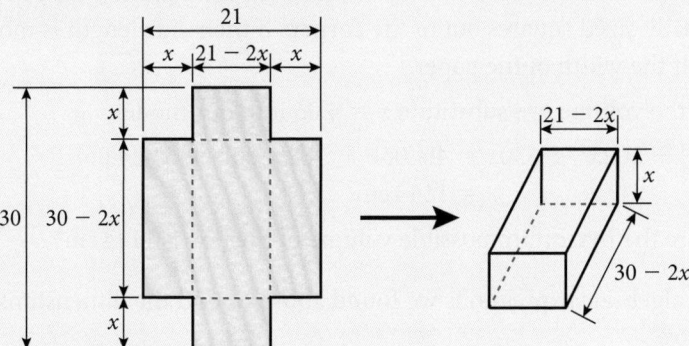

### Use the diagram to write algebraic relationships

When we label the dimensions, we can see that the width and length of the box are the width and length of the piece of paper reduced by twice the length of the corners we cut out. Therefore, we know that the length of the box is $30 - 2x$, the width of the box is $21 - 2x$ and the height of the box is $x$.

### Algebraically express the general model using a single variable

We can use the algebraic expressions we found above to develop a model for the volume of the box in terms of $x$ by substituting into the general model for volume:

$$V = lwh = (30 - 2x)(21 - 2x)x$$
$$= 4x^3 - 102x^2 + 630x$$

### Use calculus to find the extrema of the model

Finally, we can look for maxima by taking the derivative of the volume model:

$$V' = 12x^2 - 204x + 630$$

To find relative maxima, we look for value(s) where the first derivative is zero and changes from positive to negative:

$$12x^2 - 204x + 630 = 0$$

We can then use a GDC polynomial solver.

```
polyRoots(12·x²–204·x+630,x)
                    {4.0559,12.9441}
```

Hence $x = 12.9$ or $x = 4.06$

Then construct a sign diagram and test the intervals.

Therefore, it appears that a relative maximum for volume occurs when $x = 4.06$ (Note that values of $x \geqslant 10.5$ are nonsensical, since we can't cut two equally sized squares out of the corners if their side length is more than half the width of the paper.)

To find the volume, we substitute $x = 4.06$ into our model:

$$V = 4x^3 - 102x^2 + 630x = 4(4.06)^3 - 102(4.06)^2 + 630(4.06)$$
$$= 1144\,\text{cm}^3$$

Therefore the maximum possible volume of the box is $1144\,\text{cm}^3$

(b) Use the algebraic expressions we found above to find the dimensions of the box.

length $= 30 - 2x = 30 - 2(4.06) = 21.9\,\text{cm}$

width $= 21 - 2x = 21 - 2(4.06) = 12.9\,\text{cm}$

height $= x = 4.06\,\text{cm}$

The key steps for solving optimisation problems are as follows.

**Draw an accurate diagram of the situation**
A diagram can help us identify geometric and algebraic relationships we may have otherwise missed. These relationships are crucial for the next steps.
Choose an appropriate model for the quantity to be optimised.
This often starts with a general, well-known formula or model, or even a simple equation. More than one model or general formula may be needed.

**Decide what the independent variable is**
There is usually one independent variable that all the other variables are related to. It is your job to decide what that independent variable is, and then relate other variables to it in the next step.

**Write algebraic relationships**
Write algebraic relationships between unknown quantities so that you can express your general model in terms of the single independent variable you have chosen. Often, each relationship will express the unknown quantity in terms of the single independent variables, but sometimes additional substitution may be necessary.

**Express the general model in a single variable**
This step is a good one to set as your first goal, because it can help you decide on an independent variable. Also, unless you can express the general model in a single variable, you will not be able to use the calculus you have learned to optimise the quantity. Don't forget to consider the domain of the model, since some extrema may be nonsensical in the context of the problem.

**Use calculus or graphical analysis on your GDC to find the extrema of the model**
This step is usually the most straightforward piece of solving the problem (if you are confident in your calculus or GDC skills). Once you have a model for the quantity to be optimised in terms of a single independent variable, use the first derivative test, or your GDC, to find relative extrema. Be careful to consider endpoints of the domain as well, as you saw in section 14.2.

**Interpret your findings and answer the question in context**
Finally, don't forget to answer the question in context. Optimisation problems are sometimes long and you can forget the question you were trying to answer – make sure you answered the original question.
Note that these steps are not strictly in order – sometimes a model may be obvious from the beginning, sometimes it may be helpful to decide on the independent variable first, etc. These steps are the same regardless of context.

## Example 14.17

A factory capable of producing up to 1000 game consoles per day incurs the following daily costs, in US$: $1580 fixed costs and $120 per console for production costs. Equipment maintenance costs vary directly with the square of the number of consoles produced.

(a) Given that equipment maintenance costs are $11 per 10 consoles produced, find a function for equipment maintenance costs $m$ in terms of the number of consoles produced $x$.

(b) Find an expression for $C(x)$, the total cost of manufacturing $x$ game consoles per day.

(c) Show that total manufacturing costs per day increases as production increases for all $x > 0$

(d) The average cost per game console is given by the function $\overline{C}(x)$. Find an expression for $\overline{C}(x)$.

(e) Find the minimum average manufacturing cost and the number of game consoles which should be manufactured to minimise the average manufacturing cost.

## Solution

(a) Since equipment maintenance costs vary directly with the square of $x$, we have $m = kx^2$, so $11 = k(10)^2 \Rightarrow k = 0.11$. Therefore, equipment maintenance costs are given by $m = 0.11x^2$

(b) Total cost per day is a sum of all the costs, so our model is simply
$C(x) = 0.11x^2 + 120x + 1580$

(c) Use calculus to find the extrema – or, rather, to show that there are none within the domain. The first derivative is $C'(x) = 0.22x + 120$. This function is positive for all $x > 0$ so production costs are increasing for all $x > 0$

(d) Here we need a new model. Average costs per game console is the total cost divided by the number of units, hence
$$\overline{C}(x) = \frac{0.11x^2 + 120x + 1580}{x}$$

(e) Use a GDC to graph $\overline{C}'(x)$.
Costs are decreasing for $0 < x < 120$ and increasing for $120 < x < 1000$, so there must be a local minimum at $x = 120$. Therefore, average manufacturing cost is minimised when 120 units per day are produced. The average cost per console at this level is $\overline{C}(120) = \$146$

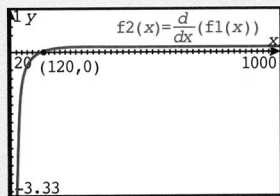

## Example 14.18

Two vertical posts, with heights of 7 m and 13 m, are secured by a rope going from the top of one post, to a point on the ground between the posts, and then to the top of the other post. The distance between the two posts is 25 m.

(a) How far from the base of the shorter post should the rope be anchored to the ground in order to minimise the length of the rope?

(b) What is the minimum length of rope necessary?

### Solution

(a) Start with an accurate diagram.

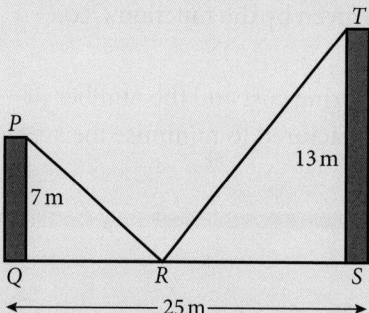

We have labelled the key points. Now, we need to think of some of the quantities in this diagram. Since the question asks for the length from the base of the shorter post ($Q$) to the anchor point ($R$), we will treat this as the independent variable and label it $x$. We are also trying to minimise the total length of the rope, so we will assign variables to each part of it as well, labelling them $l$ and $m$. Finally, since the posts are 25 m apart, we can label the horizontal distance from $R$ to the taller post as $25 - x$

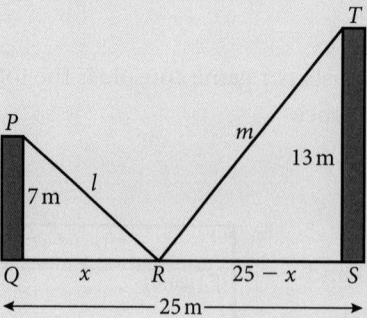

Now, use the diagram to try to find a general model. Since the quantity being optimised (the length of the rope) is the diagonal of a right-angled triangle, Pythagoras' theorem $c^2 = a^2 + b^2$ will probably be useful. Let's call the total length $T$. We know that $T = l + m$

Using Pythagoras' theorem and the equations $T = l + m$ and $RS = 25 - x$, we can write:

$$l^2 = x^2 + 7^2 \qquad\qquad m^2 = (25 - x)^2 + 13^2$$

$$l = \sqrt{x^2 + 49} \qquad\qquad m = \sqrt{(25 - x)^2 + 169}$$

which gives

$$T = \sqrt{x^2 + 49} + \sqrt{(25 - x)^2 + 169}$$

We now have a model for the quantity being optimised ($T$) expressed in a single variable ($x$). We are ready to use calculus to find the extrema, in this case the minimum value of $T$. Since this is a complicated expression, use a GDC to graph the derivative. To make the graph useful, consider the domain of our model. The length $x$ could be anywhere from 0 to 25, so scale the graph accordingly (set the $x$-axis from 0 to 25, then choose a reasonable $y$-axis to fit the derivative). Finally, use the GDC's Zero tool to find the $x$ value where $T' = 0$

The GDC tells us that $T' = 0$ when $x = 8.75$

Furthermore, we can see from the graph of the derivative that $T'$ is negative to the left of $x = 8.75$ and positive to the right of 8.75, so we conclude that $x = 8.75$ is a relative minimum for $T$.

Therefore, to minimise the length of the rope, it should be anchored 8.75 m from the base of the shorter post.

(b) Use the model to find the length of the rope. Hence,

$$T = \sqrt{x^2 + 49} + \sqrt{(25 - x)^2 + 169}$$
$$= \sqrt{(8.75)^2 + 49} + \sqrt{(25 - 8.75)^2 + 169}$$

We can use a GDC to calculate this without losing precision (remember to store the value from the graph page). Therefore, when $x = 8.75$ m, the length of the rope is $T = 32.0$ m

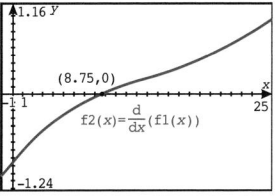

Figure 14.16 GDC screen for the solution to Example 14.18

By using a GDC to calculate and graph the first derivative, we were able to perform the first derivative test graphically.

Example 14.18 can also be solved without using calculus. Instead of using a GDC to graph the derivative function, we could analyse the length function directly to find the minimum. This method requires us to be more careful about considering the domain $0 \leqslant x \leqslant 25$ and setting the viewing window appropriately to obtain an appropriate display. Then, we can use the GDC's Minimum tool to find the minimum length and distance $x$ at the same time.

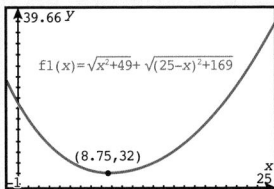

As before, we find that the rope should be anchored 8.75 m from the post, and that the minimum length of the rope is 32 m.

## Example 14.19

When a lifeguard on a beach sees a swimmer in distress, she must make a decision about the fastest way to get to the swimmer: should she run along the beach until she is as close to the swimmer as possible, then swim, or jump right in the water and swim? Of course, it depends on how fast she can swim relative to how fast she can run. (Ignore any currents in the water.)

Suppose that the lifeguard is at the point $L(0, 0)$ and a swimmer in distress is at the point $S(300, 50)$ (distances in metres). The edge of the water is along the $x$-axis. The lifeguard will leave point $L$, run along the water's edge (the $x$-axis) to point $C$, then swim to the swimmer at point $S$, as shown below.

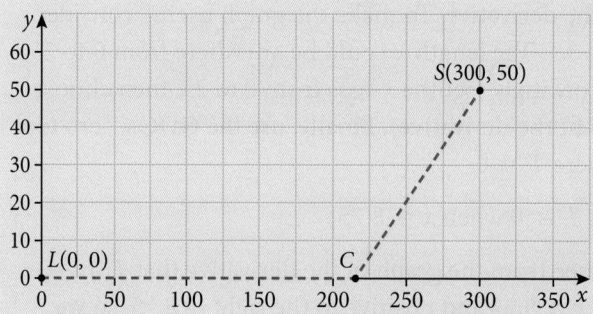

(a) Given that the lifeguard can run at a rate of 3 minutes and 30 seconds per km and swim at a rate of 1 minute per 100 m, where should point $C$ be located in order to reach the swimmer as fast as possible?

(b) What is the minimum time for the lifeguard to reach the swimmer?

## Solution

Since we are interested in minimising the total time, we should start with a very general model of $T = T_R + T_S$, where $T$ is the total time, $T_R$ is the running time, and $T_S$ is the swimming time. We will use seconds as the units for all times. Also, since we are interested in finding the point $C$, we should assign a variable to the $x$-coordinate (the $y$-coordinate will be zero). Let $C$ have coordinates $(c, 0)$; this is our independent variable.

Next, we have to consider how to find algebraic relationships for $T_R$ and $T_S$ in terms of $c$. Since both of these have to do with rate and time, we can use the model $\text{rate} = \dfrac{\text{distance}}{\text{time}}$, rearranged to $\text{time} = \dfrac{\text{distance}}{\text{rate}}$; we just need to be careful that the units of rate are the units of distance (metres) per time (seconds). The running part seems straightforward: wherever point $C$ is, the running distance will be exactly $c$ m. So, we can write:

$$T_R = \frac{c}{\frac{1000}{210}} = \frac{21c}{100}$$

Note that we converted the rate of 3 minutes 30 seconds per 1 km into 1000 m per 210 seconds.

Next, we need to find an expression for $T_S$. For this part, we can see that the lifeguard will swim the hypotenuse of a right-angled triangle, so we can add some more information to our diagram and we can use Pythagoras' theorem to model the distance the lifeguard will swim, $D_S$.

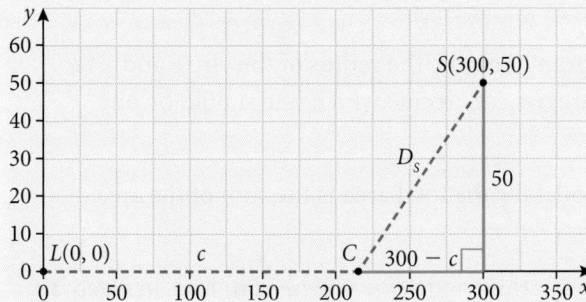

Therefore, an expression for $D_S$ is:

$$(D_S)^2 = (300 - c)^2 + 50^2 \Rightarrow D_S = \sqrt{(300 - c)^2 + 50^2}$$

Using the general distance–time model, we can now write:

$$T_S = \frac{\sqrt{(300 - c)^2 + 50^2}}{\frac{100}{60}} = \frac{3}{5}\sqrt{(300 - c)^2 + 50^2}$$

Substituting the expressions for $T_R$ and $T_S$ into our total-time model, we obtain an equation in a single variable:

$$T = T_R + T_S$$

$$T = \frac{21c}{100} + \frac{3}{5}\sqrt{(300 - c)^2 + 50^2}$$

Now we are ready to use calculus to find the value of $c$ which will minimise the total time it takes the lifeguard to reach the swimmer. For this expression, we will rely on a GDC to find the numerical derivative. Since $c$ can be anything in the interval $[0, 300]$, scale the $x$-axis accordingly and then use the GDC Zero tool to find the value of $c$ for which $T' = 0$

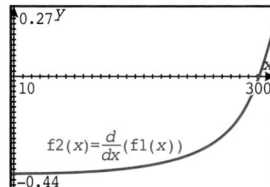

**Figure 14.17** GDC screen for the solution to Example 14.19

Now we just need to interpret the results. For part (a), the lifeguard can minimise the time it takes to reach the swimmer by running for 281 m and swimming the remaining distance. For part (b), to find the minimum time this will take, use a GDC to evaluate the function for total time at $x = 281$ to obtain 91.1 s. Therefore, it will take about 91 seconds for the lifeguard to reach the swimmer.

Again, we could use a GDC to analyse the function directly.

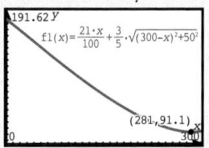

As before, the GDC tells us that the lifeguard should run for 281 m and the total time taken to reach the swimmer will be 91.1 seconds.

In other problems, we need to be careful to consider the endpoints of the domain as possible optimum solutions.

## Example 14.20

A supply of four metres of wire is to be used to form a square and a circle. How much of the wire should be used to make the square and how much should be used to make the circle in order to enclose the greatest amount of total area?

### Solution

Start with an accurate diagram. Let $r$ be the radius of the circle and $x$ be the side length of the square; we can decide which one should be our independent variable later.

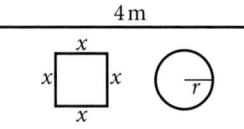

Figure 14.18 Diagram for the solution to Example 14.20

An appropriate general model for the total area is the sum of the area of a square and a circle, so $A = x^2 + \pi r^2$

Since we have two variables in this model, we will need to find an algebraic relationship to express the model in a single variable. Recall that the square and the circle are made from a single piece of wire that is 4 m long so we can write the algebraic relationship as

$4 = $ (perimeter of square) $+$ (circumference of circle)

$4 = 4x + 2\pi r$

This relationship can be solved for $r$, so that $x$ is the independent variable:

$$4 = 4x + 2\pi r \Rightarrow 2\pi r = 4 - 4x \Rightarrow r = \frac{2(1-x)}{\pi}$$

Substituting into our general model $A = x^2 + \pi r^2$, so that we now have a single variable:

$$A = x^2 + \pi r^2$$
$$\Rightarrow A = x^2 + \pi \left[ \frac{2(1-x)}{\pi} \right]^2 = x^2 + \frac{4(1-x)^2}{\pi}$$

What about the domain? Clearly, $x \geqslant 0$ since it is the length of the side of a square. Also, since the square's perimeter is $4x$, and the total length of wire is 4 m, we have $4x \leqslant 4$ so $x \leqslant 1$

So, the domain for our area model is $0 \leqslant x \leqslant 1$

Although this function could be written as a quadratic and solved by hand, we use a GDC again. Enter the function and ask the GDC to graph the numerical derivative.

Then, hide the area function, set the viewing window to the domain $0 \leqslant x \leqslant 1$, and find the zero of the derivative.

Now we need to interpret our results and answer the question. There is a local minimum for the area at $x = 0.560$ as the derivative is negative to the left and positive to the right. But, we want to maximise area. Recall the rules for the endpoints from section 14.1:

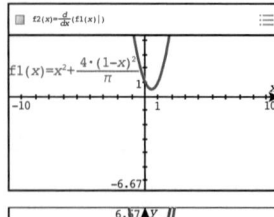

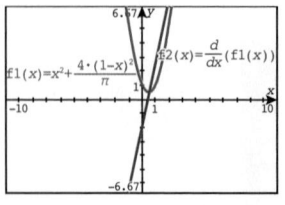

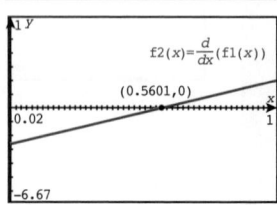

Figure 14.19 GDC screens for the solution to Example 14.20

- The left endpoint of the domain, at $x = 0$, has a negative derivative to the right, so it is a local maximum.

- The right endpoint of the domain, at $x = 1$, has a positive derivative to the left, so it is also a local maximum.

So, to find out which of the relative maxima is the absolute maximum, we need to check the value of the area at each:

- For $x = 0$, we have $A = (0)^2 + \dfrac{4(1-0)^2}{\pi} = \dfrac{4}{\pi} = 1.27\,\text{cm}^2$

- For $x = 1$, we have $A = (1)^2 + \dfrac{4(1-1)^2}{\pi} = 1\,\text{cm}^2$

Therefore, we conclude that the maximum area is enclosed when the side length of the square is 0, and the entire wire is used to make the circle.

What if we wanted to minimise the area enclosed? The side length of the square would be $x = 0.560\,\text{cm}$

## Exercise 14.3

1. Repeat Example 14.16 with a piece of US letter paper, which is 8.5 inches wide by 11 inches tall.

2. Repeat Example 14.18 with posts that are 8 m and 14 m tall and 30 m apart.

3. Repeat Example 14.19 where the lifeguard can run at 3 minutes and 40 seconds per 1 km and swim at 70 seconds per 100 m.

4. Find the dimensions of the rectangle with the maximum area that is inscribed in a semicircle with radius 1 cm. Two vertices of the rectangle are on the semicircle and the other two vertices are on the $x$-axis, as shown in the diagram.

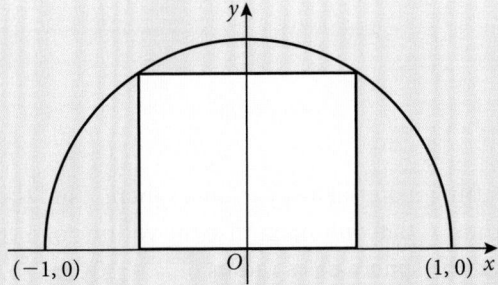

5. A rectangular piece of aluminium is to be rolled to make a cylinder with open ends (a tube). Regardless of the dimensions of the rectangle, the perimeter of the rectangle must be 40 cm. Find the dimensions of the rectangle that gives the maximum volume for the cylinder.

6. Find the minimum distance between the graph of the function $y = \sqrt{x}$ and the point $\left(\dfrac{3}{2}, 0\right)$

We can analyse the area function directly if we want to avoid using calculus. After obtaining a graph of the area function, we can see the relative minimum and two maxima at the endpoints of the domain.

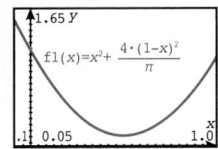

Use the Trace tool and type in '0' and '1' to find approximations for the area at each endpoint.

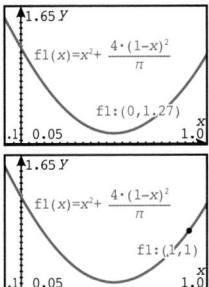

Again, the screenshots show that the area is maximised $(A = 1.27\,\text{cm}^2)$ when the side length of the square is zero.

7. The diagram consists of a rectangle $ABCD$ and two semicircles on either end. The rectangle has an area of $100\,\text{cm}^2$. Given that $x$ represents the length of the rectangle $AB$, find the value of $x$ that makes the perimeter of the entire figure a minimum.

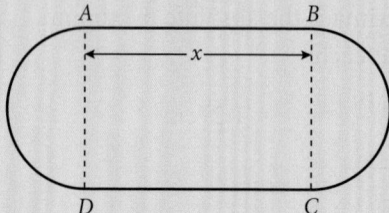

8. Charlie is walking from the wildlife observation tower (point $T$) to the Big Desert Park office (point $O$). The tower ($T$) is $7\,\text{km}$ due west and $10\,\text{km}$ due south of the office ($O$). There is a road that goes to the office that Charlie can get to if she walks $10\,\text{km}$ due north from the tower. Charlie can walk at a rate of $2\,\text{km}\,\text{h}^{-1}$ through the sandy terrain of the park, but she can walk at a faster rate of $5\,\text{km}\,\text{h}^{-1}$ on the road.

The point $A$ is the point Charlie should walk to, in order to minimise the total walking time from the tower to the office. Find the value of $x$ such that point $A$ is $x\,\text{km}$ from the office.

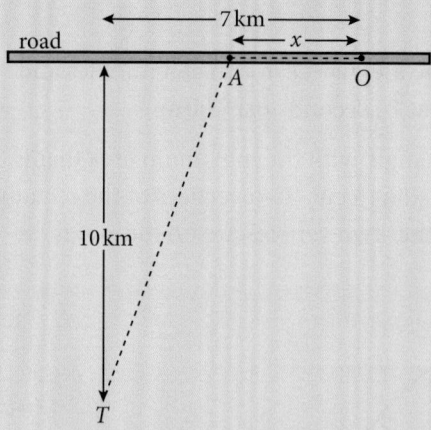

9. A cylinder is created by cutting material away from a sphere. Find the height, $h$, and the base radius, $r$, that will maximise the volume of a right circular cylinder, given that the sphere has a radius of $R = 10\,\text{cm}$.

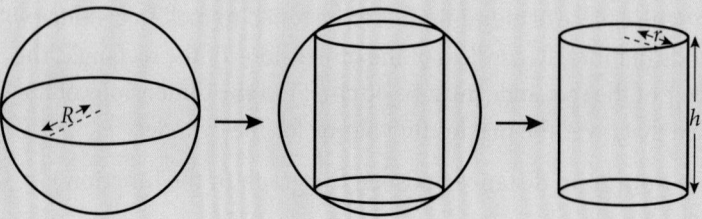

10. A manufacturer produces closed cylindrical cans of radius $r$ cm and height $h$ cm. Each can has a total surface area of $54\pi$ cm².

   (a) Solve for $h$ in terms of $r$, and hence find an expression for the volume, $V$ cm³, of each can in terms of $r$.

   (b) Find the value of $r$ for which the cans have their maximum possible volume.

11. The curve $y = ax^2 + bx + c$ has a maximum point at $(2, 18)$ and passes through the point $(0, 10)$. Find the values of $a$, $b$ and $c$.

12. A factory capable of producing up to 4000 GPS devices per day incurs the following daily costs, in US$: $2150 fixed costs, $85 per GPS device for production costs. Equipment maintenance costs vary directly with the square of the number of GPS devices produced.

   (a) Given that equipment maintenance costs are $4 per 5 GPS devices, find an expression for equipment maintenance costs in terms of the number of GPS devices produced, $x$.

   (b) Find an expression for $C(x)$, the total cost of manufacturing $x$ GPS devices per day.

   (c) Show that the total manufacturing costs per day increases as production increases for all $x > 0$

   (d) The average cost per GPS device is given by the function $\overline{C}(x)$ Find an expression for $\overline{C}(x)$

   (e) Find the minimum average manufacturing cost and the number of GPS devices which should be manufactured to minimise the average manufacturing cost.

13. A cone of height $h$ and radius $r$ is constructed from a circle with radius 10 cm by removing a sector $AOC$ of arc length $x$ cm and then connecting the edges $OA$ and $OC$. What arc length $x$ will produce the cone of maximum volume, and what is the volume?

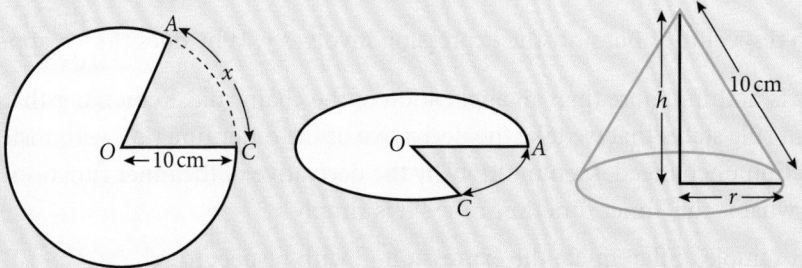

14. A ship sailing due south at 16 knots is 10 nautical miles north of a second ship going due west at 12 knots. Find the minimum distance between the two ships.

## 14.4 Related rates

The topic of related rates is one of the classic applications of differential calculus. Related rates problems are those that involve two or more variables that are changing with respect to time. For example:

- How fast is the radius of a balloon changing relative to the volume of air being pumped in?
- How fast is the angle of elevation changing as an aeroplane approaches a control tower?
- How fast is the area of an oil spill increasing relative to the rate of oil spilling out of a tank?

Solving a related rates problem requires careful application of the chain rule. Also, it is helpful to understand the basics of a technique known as implicit differentiation.

### Implicit differentiation

So far, all the functions we have found derivatives of have been in the form $y = \ldots$ or $f(x) = \ldots$. As we will see, this is not always the case, and implicit differentiation can help us work with expressions, equations, and relationships that are not written in function form. To start, consider these expressions and their derivatives.

| Expression | Result of differentiation with respect to $x$ |
|:---:|:---:|
| $3x + 5$ | 3 |
| $x^2$ | $2x$ |
| $\sin x$ | $\cos x$ |

| Expression | Result of differentiation with respect to $x$ |
|:---:|:---:|
| $3y + 5$ | $3 \cdot \dfrac{dy}{dx}$ |
| $y^2$ | $2y \cdot \dfrac{dy}{dx}$ |
| $\sin y$ | $\cos y \cdot \dfrac{dy}{dx}$ |

**Table 14.1** Expressions and their derivatives

In the left-hand table, we can see differentiation as usual: an expression containing $x$, with respect to $x$. In the right-hand table, we are still differentiating with respect to $x$, but now the expression involves $y$. Why does the $\dfrac{dy}{dx}$ appear?

This is nothing more than an application of the chain rule. Remember that the chain rule states that we take the derivative of the outer function with inside function unchanged, then multiply by the derivative of the inner function. But what is the 'inner function' of $y^2$? It is simply $y$.

For example, differentiate the expression $y^2$ with respect to $x$:

$$\frac{d}{dx}(y^2) = \underbrace{\frac{d}{dy}(y^2)}_{\substack{\text{derivative of} \\ \text{the outer function}}} \times \underbrace{\frac{dy}{dx}}_{\substack{\text{derivative of} \\ \text{the inner function}}} = 2y \times \frac{dy}{dx} = 2y\frac{dy}{dx}$$

Remember that $\dfrac{d}{dx}(y) = \dfrac{dy}{dx}$ because the rate of change in $y$, with respect to $x$, is by definition $\dfrac{dy}{dx}$. This logic applies when we differentiate with respect to any other variable. In this section, that variable will usually be $t$, representing time. As noted before, related rates problems involve quantities that change with respect to time.

## Example 14.21

Differentiate each equation with respect to $t$.

(a) $2y - 1 = x$ 

(b) $4y^3 - 8y^2 = 3x^2$ 

(c) $\dfrac{1}{y^2} + 2x^5 = \sqrt{t}$

(d) $A = \dfrac{4}{3}\pi r^3$ 

(e) $V = \dfrac{1}{3}Bh$ 

(f) $\sin\theta = \dfrac{x}{5}$

### Solution

(a) $\dfrac{d}{dt}(2y - 1) = \dfrac{d}{dt}(x)$

$\dfrac{d}{dt}(2y) + \dfrac{d}{dt}(-1) = \dfrac{dx}{dt}$

$2\dfrac{dy}{dt} + 0 = \dfrac{dx}{dt}$

$2\dfrac{dy}{dt} = \dfrac{dx}{dt}$

(b) $\dfrac{d}{dt}(4y^3 - 8y^2) = \dfrac{d}{dt}(3x^2)$

$12y^2\dfrac{dy}{dt} - 16y\dfrac{dy}{dt} = 6x\dfrac{dx}{dt}$

$(12y^2 - 16y)\dfrac{dy}{dt} = 6x\dfrac{dx}{dt}$

(c) $\dfrac{d}{dt}\left(\dfrac{1}{y^2} + 2x^5\right) = \dfrac{d}{dt}(\sqrt{t})$

$\dfrac{d}{dt}(y^{-2}) + \dfrac{d}{dt}(2x^5) = \dfrac{d}{dt}(t^{\frac{1}{2}})$

$-2y^{-3}\dfrac{dy}{dt} + 10x^4\dfrac{dx}{dt} = \dfrac{1}{2}t^{-\frac{1}{2}}$

$-\dfrac{2}{y^3}\dfrac{dy}{dt} + 10x^4\dfrac{dx}{dt} = \dfrac{1}{2\sqrt{t}}$

(d) $\dfrac{d}{dt}(A) = \dfrac{d}{dt}\left(\dfrac{4}{3}\pi r^3\right)$

$\dfrac{dA}{dt} = \dfrac{4}{3}\pi \times 3r^2\dfrac{dr}{dt}$

$\dfrac{dA}{dt} = 4\pi r^2\dfrac{dr}{dt}$

(e) Here we have a product of two variables, so use the product rule:

$\dfrac{d}{dt}(V) = \dfrac{d}{dt}\left(\dfrac{1}{3}Bh\right)$

$\dfrac{dV}{dt} = \dfrac{1}{3}\left(\dfrac{dB}{dt} \times h + B \times \dfrac{dh}{dt}\right)$

$\dfrac{dV}{dt} = \dfrac{1}{3}\left(h\dfrac{dB}{dt} + B\dfrac{dh}{dt}\right)$

(f) $\dfrac{d}{dt}(\sin\theta) = \dfrac{d}{dt}\left(\dfrac{x}{5}\right)$

$\cos\theta\dfrac{d\theta}{dt} = \dfrac{1}{5}\dfrac{dx}{dt}$

Note that:

- Unless the variables match, like $\frac{d}{dx}(x)$ or $\frac{d}{dt}(t)$, we always generate a derivative expression like $\frac{dy}{dx}$ or $\frac{dx}{dt}$. Otherwise, we use the fact that $\frac{d}{dx}(x) = 1$
- By convention, we write the derivative expression at the end of the term.
- You still need to remember all the normal rules of differentiation, like the product rule and quotient rule.

## Related rates

When a stone is thrown into a pond of water, a circular pattern of ripples is formed. We see an ever-widening circle moving across the water. As the circular ripple moves across the water, the radius $r$ of the circle, its circumference $C$, and its area $A$ all increase as a function of time $t$. Not only are these quantities (variables) functions of time, but their values at any particular time $t$ are related to one another by familiar formulae such as $C = 2\pi r$ and $A = \pi r^2$

Thus their rates of change are also related to one another.

### Example 14.22

A stone is thrown into a pond causing ripples in the form of concentric circles to move away from the point of impact at a rate of 20 cm per second. When a circular ripple has a radius of 50 cm, find:

(a) the rate of change of the circle's circumference

(b) the rate of change of the circle's area.

### Solution

In calculus, a derivative represents a rate of change of one variable with respect to another variable. If the circles are moving outward at a rate of $20\,\text{cm}\,\text{s}^{-1}$, then the rate of change of the radius is $20\,\text{cm}\,\text{s}^{-1}$. We can write this using calculus as $\frac{dr}{dt} = 20$

(a) Similar to the optimisation problems solved in Chapter 13, we must now identify a model to relate the variables we are interested in. The obvious choice to relate the circumference and radius is the familiar formula $C = 2\pi r$. Now, we use implicit differentiation to relate the rates to each other. Remember, we are differentiating with respect to time:

$$C = 2\pi r \Rightarrow \frac{d}{dt}(C) = \frac{d}{dt}(2\pi r)$$

$$\Rightarrow \frac{dC}{dt} = 2\pi \frac{dr}{dt}$$

Now, since we know that $\frac{dr}{dt} = 20$, we can substitute to find the rate of change of the circle's circumference: $\frac{dC}{dt} = 2\pi(20) = 40\pi\,\text{cm}\,\text{s}^{-1}$

(b) Again, identify a model that relates the variables we are interested in. We will use $A = \pi r^2$ and as before, we will use implicit differentiation:

$$A = \pi r^2 \Rightarrow \frac{d}{dt}(A) = \frac{d}{dt}(\pi r^2)$$

$$\Rightarrow \frac{dA}{dt} = 2\pi r \frac{dr}{dt}$$

This equation is telling us that the rate of change in area $\frac{dA}{dt}$ depends not only on the rate of change in the radius $\frac{dr}{dt}$ but also on the current value of the radius, which is 50 cm. Therefore, we have:

$$\frac{dA}{dt} = 2\pi r \frac{dr}{dt} = 2\pi(50)(20) = 2000\pi \, \text{cm}^2 \text{s}^{-1}$$

## Example 14.23

A 4-metre ladder stands upright against a vertical wall. The foot of the ladder is pulled away from the wall at a constant rate of $0.75 \, \text{m s}^{-1}$. Find how fast the top of the ladder is moving down the wall at the instant it is:

(a) 3 m above the ground

(b) 1 m above the ground.

Give your answers to 3 significant figures.

### Solution

Define the variables. Let $x$ and $y$ represent the distances of the foot and top of the ladder, respectively, from the bottom of the wall. Then, we need a model to relate these variables. Since $x$, $y$, and 4 are the sides of a right-angled triangle, Pythagoras' theorem applies, so we have:

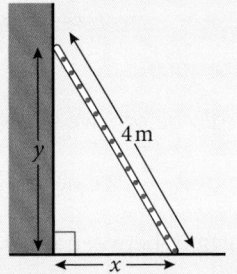

$$x^2 + y^2 = 4^2$$

What rates do we know already?

Given that the ladder is being pulled away at a rate of $0.75 \, \text{m s}^{-1}$, then

$$\frac{dx}{dt} = 0.75$$

So we know the rate $\frac{dx}{dt}$, and we need to find $\frac{dy}{dt}$ when $y = 3$ and when $y = 1$

Now we can use implicit differentiation:

$$\frac{d}{dt}(x^2 + y^2) = \frac{d}{dt}(16)$$

$$2x\frac{dx}{dt} + 2y\frac{dy}{dt} = 0$$

$$\frac{dy}{dt} = -\frac{x}{y}\frac{dx}{dt}$$

(a) We know $\dfrac{dx}{dt} = \dfrac{3}{4}$, and $y = 3\,\text{m}$, but what about $x$? Remember that in the original model, Pythagoras' theorem relates $x$ and $y$, so we use it to find the corresponding value for $x$:

$$x^2 + y^2 = 16 \Rightarrow x^2 + 3^2 = 16 \Rightarrow x = \sqrt{7} \ (x \text{ must be positive})$$

Now we can use the derivative equation:

$$\frac{dy}{dt} = -\frac{\sqrt{7}}{3} \cdot \frac{3}{4} = -\frac{\sqrt{7}}{4} \approx -0.661\,\text{m s}^{-1}$$

(b) For $y = 1$, we need to find the corresponding $x$ value and use the derivative equation again:

$$x^2 + y^2 = 16 \Rightarrow x^2 + 1^2 = 16 \Rightarrow x = \sqrt{15}$$

Then:

$$\frac{dy}{dt} = -\frac{\sqrt{15}}{1} \cdot \frac{3}{4} = -\frac{3\sqrt{15}}{4} \approx -2.90\,\text{m s}^{-1}$$

It makes sense that $\dfrac{dy}{dt}$ is negative because the distance $y$ decreases as the ladder slides down the wall.

### Example 14.24

In Example 14.23, how fast is the angle between the ladder and the ground changing when $y = 2\,\text{m}$?

**Solution**

From before, we know $\dfrac{dx}{dt} = \dfrac{3}{4}$ and we want to find $\dfrac{d\theta}{dt}$. We need a relationship, true at any instant, between the variables $\theta$ and $x$. Several trigonometric ratios could be used, but perhaps the most straightforward is

$$\cos\theta = \frac{x}{4} \Rightarrow x = 4\cos\theta$$

Now differentiate implicitly with respect to $t$:

$$\frac{d}{dt}(x) = \frac{d}{dt}(4\cos\theta)$$

$$\frac{dx}{dt} = -4\sin\theta\frac{d\theta}{dt}$$

When $y = 2$ we know that $\sin\theta = \dfrac{2}{4} = \dfrac{1}{2}$

Substitute the known quantities: $\dfrac{3}{4} = -4\left(\dfrac{1}{2}\right)\dfrac{d\theta}{dt} \Rightarrow \dfrac{d\theta}{dt} = -\dfrac{3}{8}$

Therefore the angle is decreasing at a rate of $\dfrac{3}{8}$ radians s$^{-1} \approx 21.5°\,\text{s}^{-1}$

Remember that in calculus we always work in radians. If degrees are needed, you can convert your final answer at the end of a problem.

The solution strategy used in Examples 14.23 and 14.24 is summarised here.

**Solving problems involving related rates**
1. Identify any rate(s) of change you know and the rate of change to be found.
2. Draw a diagram with all the important information clearly labelled.
3. Find a model that relates the variables whose rates of change are either known, or are to be found.
4. Differentiate the model (implicitly) with respect to time.
5. Substitute in all known values for any variables and any rates of change. Calculate the required rate of change. Be sure to include appropriate units with the result.

Only substitute a known variable before differentiating if it is actually a constant, that is, it is not changing over time. If you substitute variables before differentiating, then you will lose the derivative expressions you need to relate the rates in the problem.

Sometimes the model we choose will still leave us with variables that don't seem to be given in the problem. In these cases, we have to think carefully and find other models and relationships that can help us eliminate remaining variables.

## Example 14.25

Consider a conical tank as shown in the diagram.

The radius at the top is 4 m and its height is 8 m. The tank is being filled with water at a rate of $2\,\mathrm{m}^3\,\mathrm{min}^{-1}$. How fast is the water level rising when it is 5 m high?

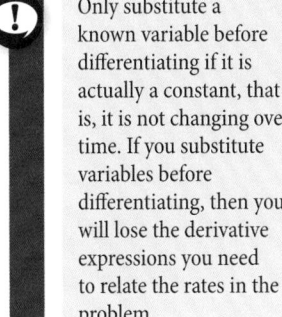

**Figure 14.20** Diagram for Example 14.25

### Solution

We know the rate of change of the volume with respect to time, that is, $\dfrac{\mathrm{d}V}{\mathrm{d}t} = 2\,\mathrm{m}^3\,\mathrm{min}^{-1}$ and we want to find the rate of change of the height of the water level with respect to time; we can call it $\dfrac{\mathrm{d}h}{\mathrm{d}t}$. We are also given that $h = 5$ for the particular instant we are interested in.

Not including $t$, there are three variables involved in this problem: $V$, $r$ and $h$. The formula for the volume of a cone will give us an equation that relates all these variables:

$$V = \frac{1}{3}\pi r^2 h$$

When we attempt to find the derivative, there are two variables on the right-hand side: $r$ and $h$. Therefore, we will have to use the product rule:

$$\frac{\mathrm{d}V}{\mathrm{d}t} = \frac{1}{3}\pi\left(2r\frac{\mathrm{d}r}{\mathrm{d}t} \times h + r^2 \times \frac{\mathrm{d}h}{\mathrm{d}t}\right)$$

$$\Rightarrow \frac{\mathrm{d}V}{\mathrm{d}t} = \frac{1}{3}\pi\left(2rh\frac{\mathrm{d}r}{\mathrm{d}t} + r^2\frac{\mathrm{d}h}{\mathrm{d}t}\right)$$

Now we have a bit of a problem. We are given values for $\dfrac{\mathrm{d}V}{\mathrm{d}t}$ and $h$, and we are solving for $\dfrac{\mathrm{d}h}{\mathrm{d}t}$, but what about $r$ and $\dfrac{\mathrm{d}r}{\mathrm{d}t}$? We need a way to find $r$ and $\dfrac{\mathrm{d}r}{\mathrm{d}t}$.

If we think carefully about a cone, we will realise that $r$ and $h$ are related as well by means of similar triangles.

Remember, don't substitute known values of variables before differentiation. For example, if we substitute $h = 5$ now (before differentiating), we will get $\dfrac{\mathrm{d}V}{\mathrm{d}t} = 0$, which is obviously wrong. You can substitute known values of constants that are not changing over time, but make sure that the value really is constant over time.

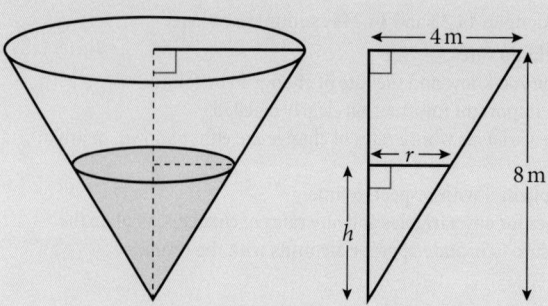

Therefore, we can write a second model and solve for $r$:

$$\frac{r}{h} = \frac{4}{8} \Rightarrow r = \frac{h}{2} = \frac{5}{2} \text{ since } h = 5$$

What about $\frac{dr}{dt}$? Again, we differentiate:

$$r = \frac{h}{2} \Rightarrow \frac{dr}{dt} = \frac{1}{2}\frac{dh}{dt}$$

Now we can substitute into our equation above, and solve for $\frac{dh}{dt}$:

$$\frac{dV}{dt} = \frac{1}{3}\pi\left(2rh\frac{dr}{dt} + r^2\frac{dh}{dt}\right)$$

$$\frac{dV}{dt} = \frac{1}{3}\pi\left(2\left(\frac{5}{2}\right)h\left(\frac{1}{2}\frac{dh}{dt}\right) + \left(\frac{5}{2}\right)^2\frac{dh}{dt}\right)$$

$$2 = \frac{1}{3}\pi\left(\frac{10}{2}(5)\left(\frac{1}{2}\frac{dh}{dt}\right) + \left(\frac{5}{2}\right)^2\frac{dh}{dt}\right)$$

$$2 = \frac{25\pi}{4}\frac{dh}{dt}$$

$$\frac{dh}{dt} = \frac{8}{25\pi} \approx 0.102 \text{ m min}^{-1} = 10.2 \text{ cm min}^{-1}$$

Therefore, when the water level is at 5 m it is rising at a rate of $10.2 \text{ cm min}^{-1}$.

Looking back, could we have saved ourselves some work if we had substituted for $r$ in the beginning? Let's try, starting from the volume of a cone, and substituting $r = \frac{h}{2}$

$$V = \frac{1}{3}\pi r^2 h = \frac{1}{3}\pi\left(\frac{h}{2}\right)^2 h$$

$$V = \frac{\pi}{12}h^3$$

The model is now much simpler. Finding the derivative yields:

$$\frac{dV}{dt} = \frac{\pi}{12} \times 3h^2\frac{dh}{dt}$$

$$2 = \frac{\pi}{12} \times 3(5)^2\frac{dh}{dt}$$

$$2 = \frac{25\pi}{4}\frac{dh}{dt} \Rightarrow \frac{dh}{dt} = \frac{8}{25\pi}$$

Therefore, where possible, remove extra variables by substituting before differentiating. However, take care to not substitute values into variables before differentiating.

We also have situations where more than one quantity is changing with respect to time, often at different rates.

## Example 14.26

At 12 noon, ship $A$ is 65 km due north of a second ship, $B$. Ship $A$ then sails south at a rate of $14\,\text{km}\,\text{h}^{-1}$, and ship $B$ sails west at a rate of $16\,\text{km}\,\text{h}^{-1}$.

(a) How fast are the two ships approaching each other at 13:30?

(b) At what time do the two ships stop approaching and begin moving away from each other?

### Solution

Let $a$ and $b$ be the distances that ships $A$ and $B$, respectively, are from the intersection of the ships' paths. Let $c$ be the distance between the two ships. Since $a$ is decreasing and $b$ is increasing, we know that

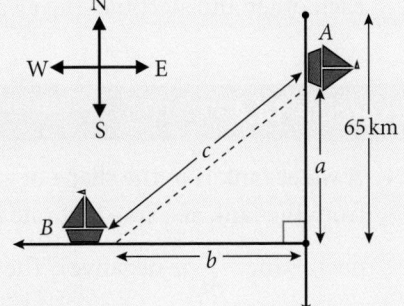

$$\frac{da}{dt} = -14\,\text{km}\,\text{h}^{-1} \text{ and } \frac{db}{dt} = 16\,\text{km}\,\text{h}^{-1}$$

(a) The three variables are related by the equation

$$c^2 = a^2 + b^2$$

Differentiating implicitly with respect to $t$ gives

$$2c\frac{dc}{dt} = 2a\frac{da}{dt} + 2b\frac{db}{dt}$$

The rate at which the ships are approaching is $\frac{dc}{dt}$. Solving for $\frac{dc}{dt}$,

$$\frac{dc}{dt} = \frac{a\dfrac{da}{dt} + b\dfrac{db}{dt}}{c}$$

Substituting $\frac{da}{dt} = -14$ and $\frac{db}{dt} = 16$

$$\frac{dc}{dt} = \frac{-14a + 16b}{c}$$

The distances $a$ and $b$ are both functions of time; thus, they can be written in terms of $t$ as $a = 65 - 14t$ and $b = 16t$

Evaluating these expressions when $t = 1.5$, gives $a = 44$, $b = 24$ and

$$c = \sqrt{44^2 + 24^2} \approx 50.12$$

Substituting these values into the expression for $\dfrac{dc}{dt}$ gives

$$\frac{dc}{dt} \approx \frac{-14(44) + 16(24)}{50.12} \approx -4.629$$

Therefore, at 13:30 the distance between the two ships is decreasing at a rate of approximately $4.63\,\text{km}\,\text{h}^{-1}$

(b) The time at which the two ships will stop approaching each other and begin to move away is when the value of $\dfrac{dc}{dt}$ changes from negative to positive. This can happen when $\dfrac{dc}{dt} = 0$

$$\frac{dc}{dt} = \frac{-14a + 16b}{c} = 0 \Rightarrow -14a + 16b = 0$$

Substituting in $a = 65 - 14t$ and $b = 16t$ gives

$$-14(65 - 14t) + 16(16t) = 0 \Rightarrow 452t - 910 = 0 \Rightarrow t = \frac{910}{452} \approx 2.013$$

Therefore, just moments after 15:30 the two ships will stop approaching each other and start moving away from each other.

## Exercise 14.4

1. A water tank is in the shape of an inverted cone. Water is being drained from the tank at a constant rate of $2\,\text{m}^3\,\text{min}^{-1}$ (since volume is decreasing, $\dfrac{dV}{dt}$ is negative). The height of the tank is $8\,\text{m}$, and the diameter of the top of the tank is $6\,\text{m}$. When the height of the water is $5\,\text{m}$, find, in units of $\text{cm}\,\text{min}^{-1}$:

   (a) the rate of change of the water level

   (b) the rate of change of the radius of the surface of the water.

2. A spherical balloon is being inflated at a constant rate of $240\,\text{cm}^3\,\text{s}^{-1}$ Find the rate at which the radius increasing:

   (a) when the radius is equal to $8\,\text{cm}$

   (b) after $5\,\text{seconds}$.

3. Oil is dripping from a car engine on to a garage floor, making a growing circular stain. The radius $r$ of the stain is increasing at a constant rate of $1\,\text{cm}\,\text{h}^{-1}$. When the radius is $4\,\text{cm}$, find:

   (a) the rate of change of the circumference of the stain

   (b) the rate of change of the area of the stain.

4. A hot air balloon is rising straight up from a level field at a constant rate of $50\,\text{m}\,\text{min}^{-1}$. An observer is standing $150\,\text{m}$ from the point on the ground where the balloon was launched. Let $\theta$ be the angle of elevation of the balloon from the observer. What is the rate of change of $\theta$ (in radians $\text{min}^{-1}$) when the height of the balloon is $250\,\text{m}$?

5. Kyoko is flying a kite at a constant height above level ground of 72 m. The wind carries the kite horizontally away at a rate of $6\,\text{m s}^{-1}$. How fast must Kyoko let out the string at the moment when the kite is 120 m away from her? (Note that the kite is 120 m away along the length of the string, not 120 m away along the ground.)

6. A 1.5 m tall girl is walking towards a 6 m tall lamppost at a constant rate of $2\,\text{m s}^{-1}$. The light from the lamppost causes the girl to cast a shadow. How fast is the tip of her shadow moving?

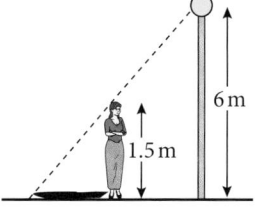

**Figure 14.21** Diagram for question 6

7. Two automobiles start from a point $A$ at the same time. One travels west at $60\,\text{km h}^{-1}$ and the other travels north at $35\,\text{km h}^{-1}$. How fast is the distance between them increasing 3 hours later?

8. A point moves along the curve $y = \sqrt{x^2 + 1}$ in such a way that $\dfrac{dx}{dt} = 4$. Find $\dfrac{dy}{dt}$ when $x = 3$

9. A horizontal trough is 4 m long, 1.5 m wide and 1 m deep. The trough's cross-section is an isosceles triangle. Water is flowing into the trough at a constant rate of $0.03\,\text{m}^3\,\text{s}^{-1}$. Find the rate at which the water level is rising 25 s after the water started flowing into the trough.

10. If the radius of a sphere is increasing at the constant rate of $3\,\text{mm s}^{-1}$, how fast is the volume changing when the surface area is 10 mm²? The surface area of a sphere with radius $r$ is $4\pi r^2$

11. Two roads, $A$ and $B$, intersect each other at an angle of 60°. Two cars, one on road $A$ travelling at $40\,\text{km h}^{-1}$ and the other on road $B$ travelling at $50\,\text{km h}^{-1}$, are approaching the intersection. If, at a certain moment, the two cars are both 2 km from the intersection, how fast is the distance between them changing?

12. If the diagonal of a cube is increasing at a rate of $8\,\text{cm s}^{-1}$, how fast is a side of the cube increasing?

13. A point $P$ is moving along a circle with equation $x^2 + y^2 = 100$ at a constant rate of $3\,\text{units s}^{-1}$. How fast is the projection of $P$ on the $x$-axis moving when $P$ is 5 units above the $x$-axis?

14. A jet is flying at a constant speed at an altitude of 10 000 m on a path that will take it directly over an observer on the ground. At a given instant the observer determines that the angle of elevation of the jet is $\dfrac{\pi}{3}$ radians and is increasing at a constant rate of $\dfrac{1}{60}$ radians sec⁻¹. Find the speed of the jet.

15. A television cameraman is filming an automobile race from a platform that is 40 m from the race track, following a car that is moving along a straight section of the track at $288\,\text{km h}^{-1}$. Find how fast, in degrees per second, the camera will be turning:

    (a) when the car is directly in front of the camera

    (b) half a second later.

    Give your answers to the nearest whole degree.

**16.** An aeroplane is flying due east at $640\,km\,h^{-1}$ and climbing vertically at a rate of $180\,m\,min^{-1}$. An airport control tower is tracking it. Determine how fast the distance between the aeroplane and the tower is changing when the aeroplane is 5 km above the ground over a point exactly 6 km due west of the tower. Express your answer in $km\,h^{-1}$.

### Chapter 14 practice questions

**1.** Tepees are cone-shaped dwellings that were traditionally used by nomadic tribes who lived on the Great Plains of North America. A tepee can be modelled as a cone with vertex $O$, radius $r$, height $h$, and slant height $l$.

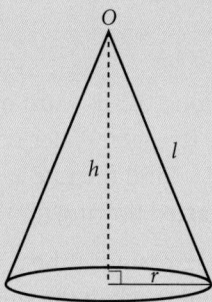

A model tepee is displayed at a Great Plains exhibition. The curved surface area of this tepee is covered by a piece of canvas that is $39.27\,m^2$ and has the shape of a semicircle, as shown in the diagram.

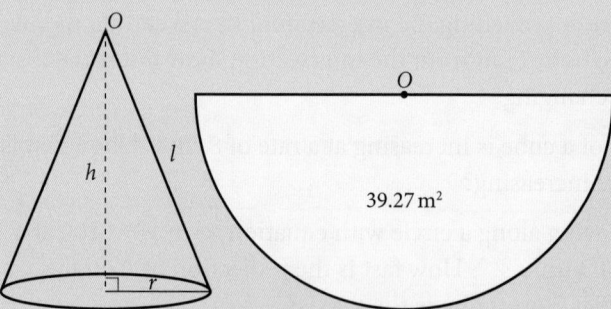

39.27 m²

**(a)** Show that the slant height $l$ is 5 m, correct to the nearest metre.

**(b) (i)** Find the circumference of the base of the cone.

**(ii)** Find the radius $r$ of the base.

**(iii)** Find the height $h$.

A company designs cone-shaped tents to resemble the traditional tepees. These cone-shaped tents come in a range of sizes such that the sum of the diameter and the height is equal to 9.33 m.

**(c)** Write down an expression for the height $h$ in terms of the radius $r$ of these cone-shaped tents.

**(d)** Show that the volume of the tent $V$ can be written as

$$V = 3.11\pi r^2 - \frac{2}{3}\pi r^3$$

**(e)** Find $\dfrac{dV}{dr}$

**(f) (i)** Determine the exact value of $r$ for which the volume is a maximum.

**(ii)** Find the maximum volume.

2. Let $f(x) = x^5$

   **(a)** Write down $f'(x)$.

   **(b)** Point $P(2, 32)$ lies on the graph of $f$. Find the gradient of the tangent to the graph of $y = f(x)$ at $P$.

   **(c)** Find the equation of the normal to the graph at $P$. Give your answer in the form $ax + by + d = 0$, where $a$, $b$ and $d$ are integers to be determined.

3. Consider the curve $y = x^3 + kx^2$, $k \in \mathbb{R}$

   **(a)** Write down $\dfrac{dy}{dx}$

   The curve has a local minimum at the point where $x = 3$

   **(b)** Find the value of $k$.

   **(c)** Find the value of $y$ at this local minimum.

4. Consider the graph of the function $f(x) = -x^3 - 3x^2 + 7$

   **(a)** Estimate, to 1 significant figure, the coordinates where the local maximum value of the function occurs.

   **(b)** Estimate, to 1 significant figure, the coordinates where the local minimum value of the function occurs.

   **(c)** Find the interval where $f'(x) > 0$

   **(d)** Find the equation of the tangent at $x = -3$

   **(e)** Sketch the graph, showing the tangent from **(d)**.

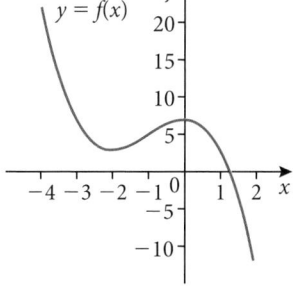

**Figure 14.22** Graph of $y = f(x)$ for question 4

5. A function is given as $f(x) = 3x^3 - 7x + \dfrac{4}{x} + 10$, $-5 \leqslant x \leqslant 5$, $x \neq 0$

   **(a)** Write down the derivative of the function.

   **(b)** Use your graphic display calculator to find the coordinates of the local minimum point of $f$ in the given domain.

   **(c)** Use your graphic display calculator to find the coordinates of the local maximum point of $f$ in the given domain.

6. A lobster trap is made in the shape of half a cylinder. It is constructed from a steel frame with netting pulled tightly around it. The steel frame consists of a rectangular base, two semi-circular ends and two further support rods, as shown in the diagram.

   The semi-circular ends each have radius $r$ and the support rods each have length $l$. Let $T$ be the total length of steel used in the frame.

(a) Write down an expression for $T$ in terms of $r$, $l$ and $\pi$.

The volume of the lobster trap is $0.75\,\text{m}^3$.

(b) Write down an equation for the volume of the lobster trap in terms of $r$, $l$ and $\pi$.

(c) Show that $T = (2\pi + 4)r + \dfrac{6}{\pi r^2}$

(d) Find $\dfrac{\mathrm{d}T}{\mathrm{d}r}$

The lobster trap is designed so that the length of steel used in its frame is a minimum.

(e) Show that the value of $r$ for which $T$ is a minimum is $0.719\,\text{m}$, correct to 3 significant figures.

(f) Calculate the value of $l$ for which $T$ is a minimum.

(g) Calculate the minimum value of $T$.

7. A parcel is in the shape of a rectangular prism. It has a length $l\,\text{cm}$, width $w\,\text{cm}$ and height $20\,\text{cm}$. The total volume of the parcel is $3000\,\text{cm}^3$.

(a) Express the volume of the parcel in terms of $l$ and $w$.

(b) Show that $l = \dfrac{150}{w}$

The parcel is tied up using a length of string that fits exactly around the parcel, shown in red in the diagram.

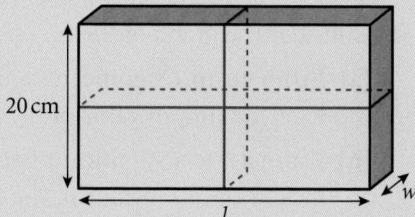

(c) Show that the length of string, $S\,\text{cm}$, required to tie up the parcel can be written as
$$S = 40 + 4w + \frac{300}{w}, 0 < w \leqslant 20$$

(d) Draw the graph of $S$ for $0 < w \leqslant 20$ and $0 < S \leqslant 500$, clearly showing the local minimum point. Use a scale of $2\,\text{cm}$ to represent 5 units on the horizontal axis $w$ (cm), and $2\,\text{cm}$ to represent 100 units on the vertical axis $S$ (cm).

(e) Find $\dfrac{\mathrm{d}S}{\mathrm{d}w}$

(f) Find the value of $w$ for which $S$ is a minimum.

(g) Write down the value $l$ of the parcel for which the length of string is a minimum.

(h) Find the minimum length of string required to tie up the parcel.

8. (a) Expand the expression $x(16x^3 - 27)$

(b) Differentiate $f(x) = x(16x^3 - 27)$

(c) Find the $x$-coordinate of the local minimum of the curve $y = f(x)$

9. Consider the function $f(x) = \dfrac{3}{4}x^4 - x^3 - 9x^2 + 20$

(a) Find $f(-2)$

(b) Find $f'(x)$

The graph of the function $f(x)$ has a local minimum at the point where $x = -2$

(c) Using your answer to part (b), show that there is a second local minimum at $x = 3$

(d) Sketch the graph of $f$ for $-5 \leqslant x \leqslant 5$ and $-40 \leqslant y \leqslant 50$

Indicate on your sketch the coordinates of the $y$-intercept.

(e) Write down the coordinates of the local maximum.

Let $T$ be the tangent to the graph of the function $f(x)$ at the point $(2, -12)$

(f) Find the gradient of $T$.

The line $L$ passes through the point $(2, -12)$ and is perpendicular to $T$.

$L$ has equation $x + by + c = 0$, where $b$ and $c \in \mathbb{Z}$

(g) Find:

    (i) the gradient of $L$

    (ii) the value of $b$ and the value of $c$.

10. Consider the function $f(x) = ax^3 - 3x + 5$, where $a \neq 0$

(a) Find $f'(x)$.

(b) Write down the value of $f'(0)$.

The function has a local maximum at $x = -2$

(c) Calculate the value of $a$.

11. The graph of the function $f(x) = \dfrac{14}{x} + x - 6$, for $1 \leqslant x \leqslant 7$ is shown.

(a) Calculate $f(1)$.

(b) Find $f'(x)$.

(c) Use your answer to part (b) to show that the $x$-coordinate of the local minimum point of the graph of $f$ is 3.7 correct to 2 significant figures.

(d) Find the range of $f$.

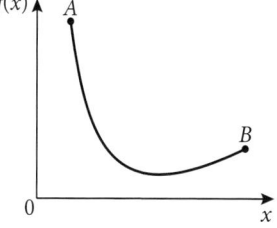

Figure 14.23 Graph for question 11

Points $A$ and $B$ lie on the graph of $f$. The $x$-coordinates of $A$ and $B$ are 1 and 7 respectively.

(e) Write down the $y$-coordinate of $B$.

(f) Find the gradient of the straight line passing through $A$ and $B$.

(g) $M$ is the midpoint of the line segment $AB$. Write down the coordinates of $M$.

$L$ is the tangent to the graph of the function $y = f(x)$ at the point on the graph with the same $x$-coordinate as $M$.

(h) Find the gradient of $L$.

(i) Find the equation of $L$. Give your answer in the form $y = mx + c$

12. The diagram shows an aerial view of a bicycle track. The track can be modelled by the quadratic function

$y = \dfrac{-x^2}{10} + \dfrac{27}{2}x$, where $x \geqslant 0$, $y \geqslant 0$,

$(x, y)$ are the coordinates of a point $x$ metres east and $y$ metres north of $O$, where $O$ is the origin $(0, 0)$. $B$ is a point on the track with coordinates $(100, 350)$.

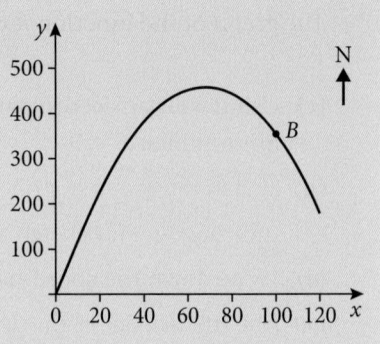

(a) The coordinates of point $A$ are $(75, 450)$. Determine whether point $A$ is on the track. Give a reason for your answer.

(b) Find the derivative of $y = \dfrac{-x^2}{10} + \dfrac{27}{2}x$

(c) Use the answer in part (b) to determine if $A$ $(75, 450)$ is the point furthest north on the track between $O$ and $B$. Give a reason for your answer.

(d) (i) Write down the midpoint of the line segment $OB$.

(ii) Find the gradient of the line segment $OB$.

Scott starts from a point $C(0, 150)$. He hikes along a straight road towards the bicycle track, parallel to the line segment $OB$.

(e) Find the equation of Scott's road. Express your answer in the form $ax + by = c$, where $a, b$ and $c \in \mathbb{R}$

(f) Use your graphic display calculator to find the coordinates of the point where Scott first crosses the bicycle track.

13. Given $f(x) = 5x^3 - 4x^2 + x$,

(a) find $f'(x)$

(b) find, using your answer to part (a), the $x$-coordinate of:

(i) the local maximum point

(ii) the local minimum point.

14. Consider the function $g(x) = bx - 3 + \dfrac{1}{x^2}$, $x \neq 0$

(a) Write down the equation of the vertical asymptote of the graph of $y = g(x)$

(b) Write down an expression for $g'(x)$.

The line $T$ is tangent to the graph of $y = g(x)$ at the point where $x = 1$.

The gradient of $T$ is 3.

(c) Show that $b = 5$

(d) Find the equation of $T$.

(e) Using your graphic display calculator, find the coordinates of the point where the graph of $y = g(x)$ intersects the $x$-axis.

**(f) (i)** Sketch the graph of $y = g(x)$ for $-2 \leqslant x \leqslant 5$ and $-15 \leqslant y \leqslant 25$ indicating clearly your answer to part **(e)**.

**(ii)** Draw the line $T$ on your sketch.

**(g)** Using your graphic display calculator, find the coordinates of the local minimum point of $y = g(x)$

**(h)** Write down the interval for which $g(x)$ is increasing in the domain $0 < x < 5$

**15.** The equation of a curve is given as $y = 2x^2 - 5x + 4$

**(a)** Find $\dfrac{dy}{dx}$

The equation of the line $L$ is $6x + 2y = -1$

**(b)** Find the $x$-coordinate of the point on the curve $y = 2x^2 - 5x + 4$ where the tangent is parallel to $L$.

**16.** Consider the function $f(x) = -\dfrac{1}{3}x^3 + \dfrac{5}{3}x^2 - x - 3$

**(a)** Sketch the graph of $y = f(x)$ for $-3 \leqslant x \leqslant 6$ and $-10 \leqslant y \leqslant 10$ showing clearly the axes intercepts and local maximum and minimum points. Use a scale of 2 cm to represent 1 unit on the $x$-axis, and a scale of 1 cm to represent 1 unit on the $y$-axis.

**(b)** Find the value of $f(-1)$.

**(c)** Write down the coordinates of the $y$-intercept of the graph of $f(x)$

**(d)** Find $f'(x)$.

**(e)** Show that $f'(-1) = -\dfrac{16}{3}$

**(f)** Explain what $f'(-1)$ represents.

**(g)** Find the equation of the tangent to the graph of $f$ at the point where $x$ is $-1$.

**(h)** Sketch the tangent to the graph of $f$ at $x = -1$ on your diagram.

$P$ and $Q$ are points on the curve such that the tangents to the curve at these points are horizontal. The $x$-coordinate of $P$ is $a$, and the $x$-coordinate of $Q$ is $b$, where $b > a$

**(i)** Write down the value of

**(i)** $a$      **(ii)** $b$

**(j)** Describe the behaviour of $f(x)$ for $a < x < b$

**17.** A shipping container is to be made with six rectangular faces, as shown in Figure 14.24.

The dimensions of the container are: length $2x$, width $x$, height $y$.

All the measurements are in metres. The total length of all twelve edges is 48 m.

**(a)** Show that $y = 12 - 3x$

**(b)** Show that the volume $V\,\mathrm{m}^3$ of the container is given by $V = 24x^2 - 6x^3$

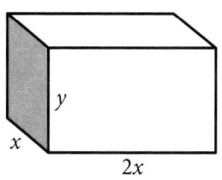

**Figure 14.24** Diagram for question 17

(c) Find $\dfrac{\mathrm{d}V}{\mathrm{d}x}$

(d) Find the value of $x$ for which $V$ is a maximum.

(e) Find the maximum volume of the container.

(f) Find the length and height of the container for which the volume is a maximum.

(g) The shipping container is to be painted. One litre of paint covers an area of $15\,\mathrm{m}^2$. Paint comes in tins containing four litres. Calculate the number of tins required to paint the shipping container.

18. In the sport of discus throwing, an athlete must throw the discus as far as possible from the centre of a 2.5 m diameter circle. The discus must land within a sector which has a central angle of 34.92°. Part of Kai's discus throw, before he releases the discus, can be modelled by the curve $y = -\sqrt{\dfrac{1}{2} - \dfrac{1}{4}x^2}$ where $x$ and $y$ are the coordinates, in metres, of the discus, with the origin placed at the centre of the throwing circle, as shown below. On one throw, Kai releases the discus at point $A$, and it travels according to the tangent to the curve $y$ at that point. The boundaries of the sector and throwing circle are shown in blue.

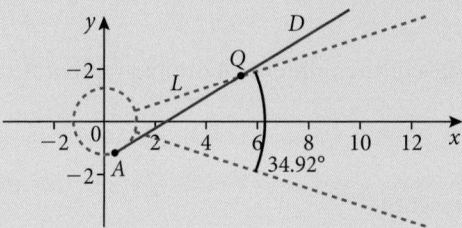

(a) The line $D$ is tangent to the curve at point $A$. Given that point $A$ has an $x$-coordinate of $x = 1$, find the equation of the line $D$ in the form $y = ax + b$, where $a, b \in \mathbb{R}$

(b) The boundary line $L$ passes through the centre of the throwing circle at the origin. Show that the equation of line $L$ is given by $y = 0.315x$

(c) Find the coordinates of point $Q$, the point at which the path of the discus leaves the sector.

(d) Kai's discus throw travels 20 m from the centre of the throwing circle before landing. For the throw to be valid, it must land within the sector. Is Kai's throw valid? Give a reason for your answer.

(e) The maximum distance Kai can throw (measured from the centre of the throwing circle) is 35 m. In order for Kai's throw to be valid, he must release the discus at point $R$ on the curve $y = -\sqrt{\dfrac{1}{2} - \dfrac{1}{4}x^2}$

(i) Find the coordinates of point $R$.

(ii) Find the equation of the tangent at point $R$ in the form $y = cx + d$, where $c, d \in \mathbb{R}$

**19.** The angle of repose of a material is the steepest possible angle in which a material can be piled. A particular type of dry sand has an angle of repose of 34°.

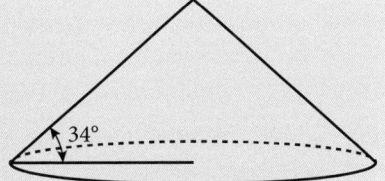

A quarry is depositing dry sand into a conical pile at a rate of $0.5\,\text{m}^3\,\text{h}^{-1}$. Four hours ago, the quarry began creating a new pile. Find the rate at which the diameter of the pile is growing.

**20.** An extendable ladder is sliding down a wall. The base of the ladder is sliding away from the wall at $0.5\,\text{m s}^{-1}$ while the length of the ladder is decreasing at twice that rate. How fast is the top of the ladder sliding down the wall when the bottom of the ladder is $1.2\,\text{m}$ from the base of the wall, and the ladder is $3.7\,\text{m}$ long?

**21.** Boyle's Law states that for a given mass of an ideal gas, volume is inversely proportional to pressure at a constant temperature.
A cylindrical piston with radius $4.5\,\text{cm}$ and an extended length of $9\,\text{cm}$ has a pressure of $80\,\text{kPa}$. The piston is compressed at a rate $4\,\text{cm s}^{-1}$. Find the rate at which the pressure is increasing when the length is $2\,\text{cm}$.

**22.** A woman in a rowing boat is $2\,\text{km}$ from the nearest point on the shore. The house that she wishes to reach is $6\,\text{km}$ from that point. She can row at a rate of $3\,\text{km h}^{-1}$ and walk at a rate of $5\,\text{km h}^{-1}$.

   **(a)** Show that the distance from the rowing boat to the house is $2\sqrt{10}\,\text{km}$.

   To get to the house in the minimum amount of time, she should row for $r\,\text{km}$ and walk for $w\,\text{km}$.

   **(b)** Find a function $r$ in terms of $w$.

   **(c)** Hence, find a function for the total time to reach the house in terms of $w$.

   **(d)** Hence or otherwise, find the values of $w$ and $r$.

   **(e)** Find the minimum amount of time required for her to reach the house.

**23.** A patient is given a hormone injection and the concentration of the hormone in her bloodstream is monitored. The rate of change of the bloodstream hormone concentration is given by
$$\frac{dC}{dt} = (-0.75t^{1.2} + 40t^{0.2})0.97^t \text{ where } \frac{dC}{dt} \text{ is in nanograms ml}^{-1}\text{min}^{-1}$$
and $t$ is the number of minutes since the injection was given.

   **(a)** Find the time when the hormone concentration is increasing the fastest and the rate of increase at this time.

   **(b)** Find the time when the maximum hormone concentration is present.

   **(c)** Find the time when the hormone concentration is decreasing the fastest and the rate of decrease at this time.

   **(d)** If the bloodstream concentration increases at more than $35$ nanograms ml$^{-1}$min$^{-1}$, it can be dangerous for the patient. Find the interval where the patient is at risk.

**24.** In a small town, the number of new customers for high-speed internet service provided by FibreLand can be modelled by the function

$$S'(t) = \frac{2000}{1200\,(0.65)^t + (1.25)^t + 250},$$ where $S'(t)$ is the number of new

subscribers in thousands, at time $t$, in years since 2000.

**(a)** Write down the units for $S'(t)$.

**(b)** In which year is the total number of subscribers increasing the fastest and at what rate?

**(c)** FibreLand has planned a peak growth rate of 15 new subscribers per day. During what time will this rate be exceeded? (Assume 365 days in a year.)

**(d)** Give a reason why the total number of subscribers will not decrease according to this model.

**25.** A barrel is made from strips of wood measuring 5 cm wide and 1 m long. A barrel maker has 40 strips with which to make a cylindrical barrel, including making the two ends of the barrel.

**(a)** Assuming that the wood can be entirely used (no wasted scraps), what are the dimensions of the barrel with maximum volume that can be made from these strips?

**(b)** What is the maximum volume, in litres?

# Probability distributions

# 15

In this chapter, you will be studying probability distributions by looking at the binomial, Poisson, and normal distributions. You will use your GDC to find many of the results. Its built-in menu of statistical options should prove to be very useful.

## 15.1 Random variables

Recall from Chapter 10 that data can be either **discrete** (countable) or **continuous** (measurable). Discrete quantities can be given by tallies that enumerate data such as the number of people, heads of lettuce, and motor vehicle accidents. Continuous data are measurements made of quantities such as length, area, mass, and time.

In order to describe both discrete and continuous data, we use a **random variable**, designated by $X$, where it can take on either discrete or continuous values over a pre-determined domain such as $x = \{1, 2, 3, 4, 5, 6\}$ or $1.2 \leqslant x \leqslant 8$ depending on whether your data are countable or measurable.

## Discrete data

Once data have been collected, the two descriptive statistics, mean and variance, need to be calculated, as you have done in Chapter 10. In a distribution, the mean is called the **expected value** of the random variable, and designated $\mathbf{E}(X)$. The variance is often written as $\mathbf{Var}(X)$.

The formulae for discrete data are $\mathrm{E}(X) = \mu = \sum_{i=1}^{n} x_i \cdot \mathrm{P}(X = x_i)$ and

$\mathrm{Var}(X) = \sum_{i=1}^{n}(x_i - \mu)^2 \cdot \mathrm{P}(X = x_i)$ or its computational equivalent $= \sum_{i=1}^{n} x_i^2 \cdot \mathrm{P}(X = x_i) - \mu^2$

## Example 15.1

A game that uses a spinner provides players with a chance to win money. The sectors of the spinner's circle are divided such that the probability of winning \$10 is $\frac{1}{20}$, the probability of winning \$2 is $\frac{1}{4}$, and the probability of winning nothing is $\frac{7}{10}$

What are the expected winnings per spin?

### Solution

Here, the random variable $X$ is the expected winnings. The possible values that the random variable can take are $x_1 = 10$, $x_2 = 2$, and $x_3 = 0$

Their probabilities are $P(X = x_1) = \frac{1}{20}$, $P(X = x_2) = \frac{1}{4}$, and $P(X = x_3) = \frac{7}{10}$

The expected value, $E(X)$ is found as the sum of the product of each prize value and its probability:

$$E(X) = \mu = 10 \cdot \frac{1}{20} + 2 \cdot \frac{1}{4} + 0 = \$1$$

You can do this on your GDC by entering the values for $x_i$ in one list and the probabilities as 'frequencies' in another.

The data values are entered in one list. Here, the list is named 'prize'. The probabilities corresponding to the data are entered in a second list. Here, the list is called 'probability' (Figure 15.1).

In a GDC, the expected value $E(X)$ is shown as $\bar{x}$, (Figure 15.2) while the standard deviation is given as $2.23607\ldots$, which incidentally is $\sqrt{5}$ since the variance is 5.

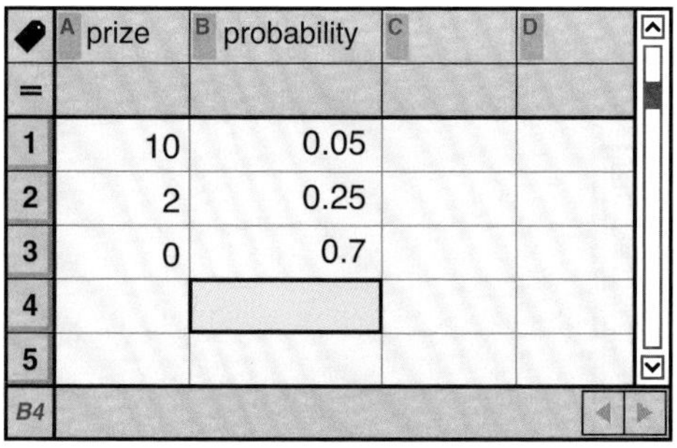

| A prize | B probability | C | D |
|---|---|---|---|
| = | | | |
| 1 | 10 | 0.05 | |
| 2 | 2 | 0.25 | |
| 3 | 0 | 0.7 | |
| 4 | | | |
| 5 | | | |

B4

**Figure 15.1**  A sample of lists in a GDC for Example 15.1

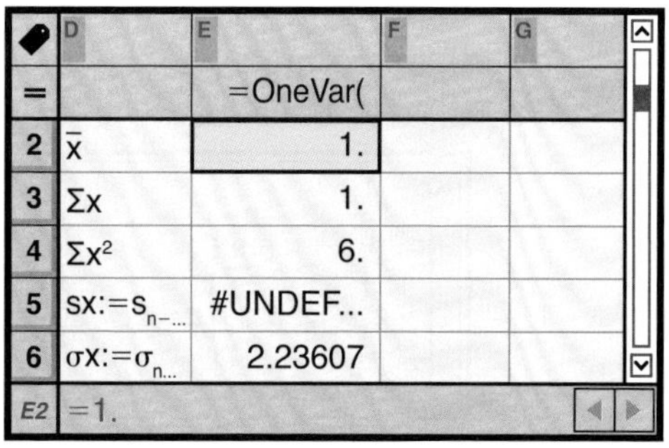

| D | E | F | G |
|---|---|---|---|
| = | | =OneVar( | |
| 2 | $\bar{x}$ | 1. | |
| 3 | $\Sigma x$ | 1. | |
| 4 | $\Sigma x^2$ | 6. | |
| 5 | sx:=s$_{n-\ldots}$ | #UNDEF... | |
| 6 | $\sigma$x:=$\sigma_{n\ldots}$ | 2.23607 | |

E2 =1.

**Figure 15.2**  The expected value $E(X)$ is shown as $\bar{x}$

Since the three sectors comprise the entire circle, it is easy to see that

$$0 \leqslant P(X = x_i) \leqslant 1 \quad \text{and} \quad \sum_{i=1}^{n} P(X = x_i) = 1$$

These are the conditions of any **probability distribution**.

## Continuous data

Continuous data require the use of an integral to describe the precise areas under a distribution curve. More detail and alternative approaches to integrals will be provided in Chapter 16, but you can calculate integrals using your GDC, with specific values for the lower bound and upper bound that match the prescribed domain.

In general, given a **probability density function** $f(x)$, the formulae for E(X) and Var(X) of continuous distributions are:

$$E(X) = \mu = \int_{-\infty}^{\infty} x \cdot f(x)\, dx$$

and

$$Var(X) = \int_{-\infty}^{\infty} (x - \mu)^2 \cdot f(x)\, dx$$

or its computational equivalent

$$= \int_{-\infty}^{\infty} x^2 \cdot f(x)\, dx - \mu^2$$

### Example 15.2

Waiting times in line are often modelled by exponential functions. In one situation, the probability density function (PDF) is modelled using the function:

$$f(x) = \begin{cases} \dfrac{1}{4}e^{-\frac{x}{4}}, & x \geqslant 0 \\[2mm] 0, & x < 0 \end{cases}$$

Find its mean and variance.

Note that $f(x)$ is always a positive value, and $\int_0^{\infty} \frac{1}{4}e^{-\frac{x}{4}}\, dx = 1$, so it serves well as a PDF.

### Solution

$$E(X) = \mu = \int_{-\infty}^{\infty} x \cdot \frac{1}{4}e^{-\frac{x}{4}}\, dx = \int_0^{\infty} x \cdot \frac{1}{4}e^{-\frac{x}{4}}\, dx = \lim_{b \to \infty} \left[ -(x+4) \cdot e^{-\frac{x}{4}} \right]_0^b = 4$$

Alternatively, as your GDC will not allow the use of limits or $\infty$, use 0 for the lower bound and pick a sufficiently large value for the upper bound. In this example, 100 is large enough, but if we choose 1000, we will find the result to be 4.

$$\int_0^{1000} \left( x \cdot \frac{1}{4} \cdot e^{\frac{-x}{4}} \right) dx \qquad 4.$$

The same approach can be used to find the variance.

$$Var(X) = \int_0^{\infty} x^2 \cdot \frac{1}{4}e^{-\frac{x}{4}}\, dx - 4^2 = \lim_{b \to \infty} \left[ -(x^2 + 8x + 32) \cdot e^{-\frac{x}{4}} \right]_0^b - 16 = 16$$

$$\int_0^{1000} \left( x^2 \cdot \frac{1}{4} \cdot e^{\frac{-x}{4}} \right) dx - 16 \qquad 16.$$

1. State whether or not the following measurements can be represented by a discrete or continuous random variable $X$.
   (a) The heights of students in your class
   (b) The number of songs you downloaded to your phone last month
   (c) The time you spend waiting for the bus
   (d) The number of earthquakes each year in Japan
   (e) How far you walked each day last week, according to your wrist device
   (f) The age of things that do not celebrate a birthday

2. There are a number of things that are considered countable but infinite that are still discrete and not continuous. Name some possibilities.

3. The number of points scored when a dice is rolled is the number showing on its upper face. Find the mean and variance of the points scored in the roll of a single, fair cubic (six-sided) dice.

4. The faces of a cubic dice are renumbered so that the faces have the numbers $\{1, 2, 3, 7, 8, 9\}$. Determine the mean and variance of the outcome of a single roll of this dice.

5. Due to a manufacturing defect, a cubic dice is made with the six faces $\{1, 1, 2, 3, 4, 5\}$. Determine the mean and variance of the outcome of a single roll of this dice.

6. A board game uses a dodecahedral (12-sided) dice, with faces numbered from 1 to 12. Find the mean and variance of the outcome of a single roll.

7. A random number generator produces digits 1, 2, or 3, but programmed so that $P(X = 1) = 0.4$, $P(X = 2) = a$, and $P(X = 3) = b$
   What is the relation between $a$ and $b$ when $E(X) = 1$?

8. A spinner is designed with the following outcomes and probabilities.

| $x$ | 2 | 4 | 6 | 8 |
|---|---|---|---|---|
| $P(X = x)$ | 0.4 | 0.3 | 0.2 | 0.1 |

   Find $E(X)$ and $Var(X)$.

9. A fair six-sided dice is rolled, but whenever the outcome is a 6, it is ignored and rolled again, thereby creating a random variable $X$ with five distinct equally likely outcomes that are recorded. Find $E(X)$ and $Var(X)$.

10. A fair coin is flipped four times and the number of heads recorded. Using the theoretical probabilities $P(X = x)$ in the table, where $x$ is the number of heads in four tosses, find $E(X)$ and $Var(X)$.

| $x$ | 0 | 1 | 2 | 3 | 4 |
|---|---|---|---|---|---|
| $P(X = x)$ | 0.0625 | 0.25 | 0.375 | 0.25 | 0.0625 |

**11.** Given the probability density function

$$f(x) = \begin{cases} \dfrac{2}{3}x, & 1 \leqslant x \leqslant 2 \\ 0, & \text{otherwise} \end{cases}$$

find:

(a) E(X) and Var(X)     (b) P(1.5 ≤ X ≤ 2)

**12. (a)** Explain why $f(x) = \begin{cases} \dfrac{1}{9}x^2, & 0 \leqslant x \leqslant 3 \\ 0, & \text{otherwise} \end{cases}$

would be appropriate as a probability density function.

(b) Find:

(i) E(X) and Var(X)     (ii) P(1 ≤ X ≤ 2)

**13.** Determine the value of $k$ such that $f(x) = \begin{cases} k \cdot x^2, & 1 \leqslant x \leqslant 3 \\ 0, & \text{otherwise} \end{cases}$

can be used as a probability density function.

**14. (a)** Determine the value of $k$ to 3 significant figures so that

$$f(x) = \begin{cases} k \cdot e^{-x^2}, & -3 \leqslant x \leqslant 3 \\ 0, & \text{otherwise} \end{cases}$$

can be used as a probability density function.

(b) Compare your results with the interval changed to:

(i) $-4 \leqslant x \leqslant 4$

(ii) $-\infty \leqslant x \leqslant \infty$

(c) Graph your function with your GDC. Does it look vaguely familiar?

The random variable for a binomial distribution is stated as $X \sim B(n, p)$ where $n$ is the number of independent Bernoulli trials and $p$ is the probability of success. The probability of exactly $r$ successes amongst the $n$ trials is expressed as
$$P(X = r)$$
$$= {}_nC_r \cdot p^r \cdot (1 - p)^{n-r}$$
where
$$r \in \{0, 1, 2, 3, \dots n\}$$
Your GDC has this calculation available as binompdf $(n, p, r)$.

The calculator function ${}_nC_r$ is also expressed as $\binom{n}{r}$, the number of **combinations** of $n$ items, taken $r$ at a time and is equal to $\dfrac{n!}{r!(n-r)!}$

# 15.2 Binomial distribution

Experiments whose outcomes are limited to one of two possibilities are called **Bernoulli trials**. Common examples of binomial outcomes include yes/no, success/failure, heads/tails, and for/against. Clearly, such outcomes must not be subject to interpretation, so tall/short would be considered unsuitable descriptors of outcomes unless there is a specific description given to the distinction between them.

The probabilities of the two outcomes need not be equal or even assumed to be equal; the only consideration is that there are two possibilities. As successive Bernoulli trials are conducted, the resulting distribution of outcomes is called a **binomial distribution**. Note that we are only considering **independent** events where the probability of the desired outcome (e.g. a success) remains unchanged in successive trials.

Since the expected number of successes after $n$ trials would be the average number of successes, if the probability of each success is $p$,

$$E(X) = \mu = np$$

The variance is found as

$$Var(X) = np(1 - p)$$

## Example 15.3

A biased coin lands with heads showing with a probability of $p = 0.6$. The coin is flipped 20 times. Find:

(a) the expected number of heads

(b) the variance

(c) the probability of exactly 13 heads in the 20 flips

(d) the probability of at least 13 heads.

### Solution

(a) $E(X) = 20(0.6) = 12$. On average, one would expect 12 heads in 20 flips.

(b) $Var(X) = 20(0.6)(0.4) = 4.8$

(c) $P(X = 13) = 20\,C_{13}\,(0.6)^{13} \cdot (0.4)^7 \approx 0.166$

(d) $P(X \geqslant 13) = 1 - [P(X = 0) + P(X = 1) + \dots + P(X = 12)] \approx 0.416$

```
binompdf(20,0,6,13)
               0.1658822656
1-binomcdf(20,0,6,12)
               0.4158929389
■
```

**Figure 15.3** GDC screen for the solution to Example 15.3

## Exercise 15.2

1. The random variable $X$ represents the number of heads in 20 flips of a fair coin.
   (a) Find $E(X)$ and $Var(X)$
   (b) Find $P(X = 10)$
   (c) Some people think that the result in part (b) should be much higher. With your GDC, determine the probability of all possible outcomes to justify your answer in part (b).

2. A fair coin is flipped 50 times and the random variable $X$ represents number of heads. Find:
   (a) $E(X)$ and $Var(X)$
   (b) $P(X = 25)$

3. A biased coin appears to land showing heads, on average, 80 times in 100 flips. Let the random variable $X$ represent the number of heads in the next 10 flips of this coin. Find:
   (a) $E(X)$ and $Var(X)$
   (b) $P(X = 8)$
   (c) $P(X \geqslant 1)$

4. Given a fair six-sided dice, you are interested only in rolling a 6. You roll the dice 60 times, and the random variable $X$ represents the number of times a 6 is rolled.

   (a) Find:

      (i) $E(X)$ and $Var(X)$      (ii) $P(X = 10)$

      (iii) $P(X = 0)$      (iv) $P(X = 60)$

   (b) Use your GDC to produce a bar chart of the distribution.

5. It is said that when a US penny is spun instead of flipped, it lands showing tails about four-fifths of the time. (All coins with a heavy relief on either side will exhibit this property, although the probabilities will vary.) Let the random variable $X$ represent the number of tails in 30 flips. Find:

   (a) $E(X)$ and $Var(X)$      (b) $P(X = 24)$

6. The random variable $X$ represents the number of times a 1 is rolled in six rolls of a biased six-sided dice. The probability of rolling a 1 with this dice is 0.3. Find:

   (a) $E(X)$ and $Var(X)$      (b) $P(X = 1)$      (c) $P(X < 6)$

7. The experiment in question **6** is modified to consider rolling the dice 60 times.

   (a) Calculate $P(X = 10)$ instead of $P(X = 1)$

      Did you get the same result as for $P(X = 1)$ in question **6**?

   (b) Find $E(X)$ and $Var(X)$ and compare your answers to those in question **6**.

   (c) How do their standard deviations compare?

8. A random number generator on a GDC is used to produce digits from 0 to 9. The random variable $X$ represents the number of times the number 8 appears in 20 digits generated.

   (a) Find $E(X)$ and $Var(X)$.

   (b) Find $P(X = 8)$, the probability that the digit 8 appears exactly 8 times in 20 digits.

   (c) Use the cumulative binomial function of your GDC to find $P(X \geq 8)$

9. The star striker on a football team has a lifetime probability of 0.9 for scoring a penalty kick in league play. Successive penalty kicks can be considered independent events, and the random variable $X$ represents the number of goals in the next ten penalty kicks taken. Find the probability distribution of $X$ as a histogram.

10. The probability of you passing a test without studying is 0.2, and $X$ represents the number of times you would pass without studying in the next four tests. Find $P(X > 1)$

**11.** As a dedicated cola drinker with a preference, you estimate that the probability of your detecting your cola between two brands in a blindfolded test is 0.8. Let $X$ represent the number of correct assessments in 10 trials. Find:

(a) $P(X \geqslant 8)$           (b) $P(X = 10)$

# 15.3 Poisson distribution

In independent events where the probability of a successful outcome is small, the **Poisson distribution** is an excellent estimator of binomial outcomes. Generally, however, for the Poisson distribution to be suitable, the rate of occurrence must be constant, and the number of times an event occurs within an interval of time should be proportional to the length of the interval. Unlike the binomial distribution, note that the emphasis with the Poisson is on the number of occurrences over a given period of time.

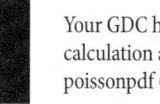 Your GDC has this calculation as poissonpdf $(m,r)$.

 The Poisson distribution is expressed as $X \sim \text{Po}(m)$ where $m$ is both the $E(X)$ and $\text{Var}(X)$, and the expected value, as in the binomial, is $E(X) = m = np$. Individual probabilities are found with the formula $P(X = r) = \dfrac{m^r \cdot e^{-m}}{r!}$ where $r \in \{0, 1, 2, 3, ...\}$

## The Poisson distribution as an estimator for small values of $p$

### Example 15.4

A spinner is designed to produce one of two outcomes, using a circle divided into two equal parts. Quite unexpectedly, it appears that the spinner needle lands on the dividing line between the two outcomes with a small probability of $p = 0.001$. What is the probability that in 3000 spins, the needle lands on the line 4 times instead of landing in one of the intended regions?

### Solution

As a binomial experiment, P(needle lands on the line 4 times) is calculated as $P(X = 4) = 3000C_4 \cdot (0.001)^4(0.999)^{2996}$. You can use your GDC to find this, as $P(X = 4) \approx 0.168$. As a Poisson approximation, first of all, the expected number of times the event would occur is
$E(X) = m = 3000(0.001) = 3$

$$\Rightarrow P(X = 4) = \frac{3^4 \cdot e^{-3}}{4!} \approx 0.168$$

As you might notice in the use of the binomial probability formula, before the use of GDCs, the computational impossibility made the Poisson distribution a viable alternative.

## The Poisson distribution as a probability model for the rate of occurrences

The Poisson distribution is generally used to model a distribution for the rate of occurrences of events over time.

Information about how the data were generated can help you decide whether the Poisson distribution fits. Assume that an interval is divided into a very large number of subintervals so that the probability of the occurrence of an event in any subinterval is very small. The Poisson distribution is based on four assumptions. We will use the term interval to refer to either a time interval or an area, depending on the context of the problem. There are four assumptions, as follows.

1. The probability of observing a single event over a small interval is approximately proportional to the size of that interval.

2. The probability of two events occurring in the same narrow interval is negligible.

3. The probability of an event within a certain interval does not change over different intervals.

4. The probability of an event in one interval is independent of the probability of an event in any other non-overlapping interval.

You should examine all of these assumptions carefully, especially the last two.

### Example 15.5

Your family spends time at a resort in an area subject to flooding on average every five years. What is the probability that in the next 10 years, it will flood three times?

### Solution

Let $X$ represent the number of floods affecting your resort. Since you expect a flood every 5 years, the average number of floods per year is 0.2

The expected number of floods in 10 years is

$$E(X) = m = 10(0.2) = 2 \Rightarrow P(X = 3) = \frac{2^3 \cdot e^{-2}}{3!} \approx 0.180$$

1. As you sit at the back of the class and appear to be engaged in each lesson, you've calculated the probability of being called upon to answer a question to be 0.01 on average. Find the probability that in the next 50 classes, you will be called upon:

   (a) exactly twice

   (b) exactly three times

   (c) not at all.

2. A new student in your class has heterochromia iridis, where the eyes are not of the same colour. The incidence of this genetic occurrence is 0.006. What is the probability that in a random collection of 1000 students, there are exactly 5 such students?

3. When playing darts, you think that the probability of hitting a bull's eye while blindfolded to be 0.001. Calculate the probability of doing so twice in 1000 attempts.

4. You have been purchasing a certain lottery ticket in every draw where the probability of winning is 0.0001

   What is the probability that in the next 100 lottery draws that you will win twice?

5. Researchers at the University of Leeds conducted a study which found that volcanic ash clouds occurred in northern Europe about once every 44 years. Determine the probability that in a century, there are three such occurrences.

6. When your father drives, you've noticed that he changes lanes, on average, every two minutes for no apparent reason. Find the probability that in a 20-minute drive, he will change lanes 10 times.

7. Earthquakes occur daily in Japan, with about 1500 earthquakes and tremors recorded every year. Find the probability that:

   (a) there are exactly 3 earthquakes recorded tomorrow

   (b) there are no earthquakes recorded tomorrow

   (c) there is at least one earthquake recorded tomorrow.

8. One of your teachers has been late for school just once in the last 100 days. What is the probability that she is late twice in the next 100 days?

9. One airline boasts of their on-time departures, having been behind schedule just once in 100 flights. What is the probability that there will be more than one late departure in the next 50 flights?

10. A lifeguard knows that the incidence of serious accidents at a particular pool is small, with records indicating one in 240 days on average. Find the probability that there are two serious accidents in the next 100 days.

## 15.4 Normal distribution

When the number of experiments grows in a binomial distribution, it can also be approximated by the **normal distribution**. Historically this has been practical, since the tabulation of probabilities for discrete outcomes was cumbersome and the binomial distribution for a large number of trials results in a block graph that begins to resemble a smooth curve. The smooth bell curve that depicts a normal distribution is often the model used in many applications. For instance, since the SAT examination marks are discrete values out of 800, there are 801 possible marks attainable; however, drawing 801 bars of a block graph is totally impractical.

The normal distribution, also commonly called the **Gaussian** distribution, is expressed as $X \sim N(\mu, \sigma^2)$ where $\mu$ is the mean and $\sigma^2$ is the variance. However, the random variable is continuous, so the probability of any one value among an infinite number of outcomes is predictably zero, but over an interval $(a, b)$, the probability can be expressed as a definite integral

$$P(a < X < b) = \frac{1}{\sqrt{2\pi}\,\sigma} \int_a^b e^{-\frac{(x-\mu)^2}{2\sigma^2}} \, dx$$

This is a rather daunting formula, but your GDC can carry out this calculation.

### Example 15.6

If test marks are normally distributed with mean $\mu = 100$ and standard deviation $\sigma = 16$, determine the probability of obtaining a mark within the interval (92, 117).

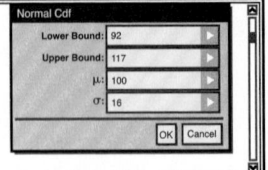

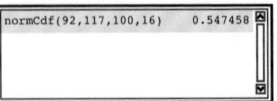

**Figure 15.4** GDC screens for the solution to Example 15.6

### Solution

The distribution is $X \sim N(100, 16^2)$

Enter the lower bound, upper bound, mean, and standard deviation into a GDC.

The answer is $P(92 < X < 117) \approx 0.547$

# The empirical rule

There is a quick rule of thumb that can be used to approximate the range of marks within 1, 2, or 3 standard deviation units from the mean, called the **empirical rule**, or simply the '68-95-99% rule'.

In every normal distribution, approximately:
- 68% of the distribution falls within 1 standard deviation unit of the mean, $P(\mu - \sigma < X < \mu + \sigma) \approx 0.68$
- 95% is within 2 standard deviation units, $P(\mu - 2\sigma < X < \mu + 2\sigma) \approx 0.95$, and
- 99.7% is within 3 standard deviation units, $P(\mu - 3\sigma < X < \mu + 3\sigma) \approx 0.997$

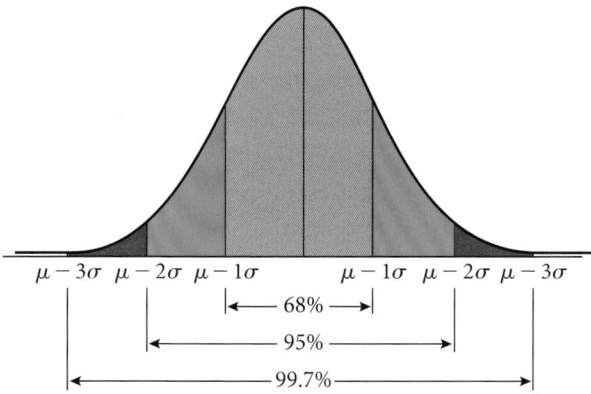

**Figure 15.5** The normal distribution

# The standardised score

Instead of centering the normal distribution about $\mu$ and labelling the points on the $x$-axis as $\mu + \sigma$, $\mu + 2\sigma$, $\mu + 3\sigma$, and so on, a more expedient system would be to start at 0 and count in units of $\sigma$, labelling the axis as 0, 1, 2, 3, and so on. This is called **standardisation** and uses the conversion $z = \dfrac{x - \mu}{\sigma}$, resulting in the distribution $Z \sim N(0, 1)$. We can restate the empirical rule.

In every standardised normal distribution,
- $P(-1 < Z < 1) \approx 0.68$
- $P(-2 < Z < 2) \approx 0.95$
- $P(-3 < Z < 3) \approx 0.997$

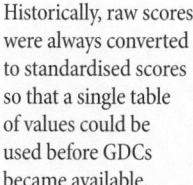

Historically, raw scores were always converted to standardised scores so that a single table of values could be used before GDCs became available.

# The inverse normal

Your GDC can also be used to find the raw score $x$ for a normally distributed random variable $X$ for which $P(X < x)$ is known, given its mean and standard deviation, through a feature called the **inverse normal**.

## Example 15.7

Suppose test scores are normally distributed with mean $\mu = 120$ and standard deviation $\sigma = 20$

Find the raw score $x$ such that $P(X < x) = 0.84$

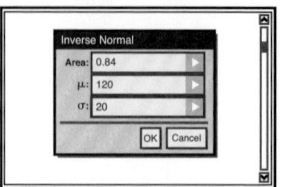

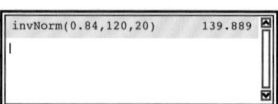

**Figure 15.6** GDC screens for the solution to Example 15.7

## Solution

Enter the values into your GDC following the pop-up menu. If the raw scores are discrete values, the 84th percentile score is 140. If the scores are discrete, always round up, regardless of the decimal value.

Now, think back to the empirical rule to consider an alternative view of this question. Since the normal PDF is symmetrical, this would mean that half of the 68% of the distribution would lie in the range $\mu < X < \mu + \sigma$

Therefore, 34% of the distribution would lie above the mean, at the 84th percentile score, $120 + 20 = 140$

## Exercise 15.4

1. Heights of young German men between 18 and 19 years of age follow a distribution that is approximately normal with mean 181 cm and standard deviation 8 cm. Describe this population of young men using the empirical rule.

2. A farmer finds that the masses of his eggs can be modelled by a normal distribution, with mean $\mu = 67$ g and standard deviation $\sigma = 6.4$ g If large eggs are considered to have masses in the range $63\,\text{g} \leqslant X \leqslant 73\,\text{g}$, what percentage of his eggs fall into this classification?

3. The results of a standardised test can be modelled by a standard distribution with mean $\mu = 60$ and standard deviation $\sigma = 6.2$ Determine the percentage of scores in the interval $[50, 65]$.

4. The heights in centimetres of students at a large secondary school are considered to be normally distributed, with $\mu = 165$ cm and $\sigma = 9.3$ cm. Determine the percentage of students expected to be taller than 180 cm.

5. The distance of a full marathon has varied over the years. The official distance is currently 42.195 km; however, this will vary as runners cannot follow the shortest path since other competitors get in the way. Suppose the distance covered can be modelled by a normal distribution with $X \sim N(43, 0.3^2)$

   (a) Determine the shortest distance covered by 5% of the runners.

   (b) Determine the longest distance covered by 95% of the runners.

6. Historical weather data for a city indicate that the highest temperature on 1 July can be considered to be normally distributed with mean 30°C and variance 6°C. Determine the percentage of the data that fall within one standard deviation of the mean.

646

7. Adult heights are often modelled by a normal distribution. Online data shows the heights of adult women in Spain can be modelled by $X \sim N(160, 7.1^2)$ given in centimetres. Determine the height of Spanish women at the 90th percentile.

8. Measurement errors are modelled by a normal distribution. Maintenance records indicate that the distribution of such errors can be given by $X \sim N(0.04, 0.1^2)$. Determine the percentage of measurement errors within the interval $[-0.1, 0.1]$

9. The time to failure of a motion sensor is found to be adequately modelled by a normal distribution with $X \sim N(2.8, 0.3^2)$, with the time given in years. What is the probability that the sensor will not fail for at least 3 years?

10. The NHS in the UK have conducted extensive studies on people's blood pressure, defining the ideal to be between 90/60 mm Hg and 120/80 mm Hg, with each pair of values being the systolic and diastolic pressures. If systolic blood pressure is modelled by $X \sim N(100, \sigma^2)$ mm Hg, and 31% of English males are considered to have high blood pressure, which is defined being consistently above 140 mm Hg, determine the value of the variance.

# 15.5 Transformations and combinations of data

Not all distributions lend themselves well to transformations and combinations; however, a few serve as illustrative examples of what is commonly seen in practice. Transformations and will be considered first, followed by linear combinations.

## Transformed data with the uniform and normal distributions

A uniform distribution is distinguished by a PDF that is a rectangle. A change to the outcomes by the addition of a constant does not change the shape of its PDF. The change would affect the value of the mean, but not the spread.

### Example 15.8

Consider the random variable $X$ representing the data set $x = \{1, 2, 3, 4, 5, 6\}$
Using your GDC, you can quickly find $E(X) = 3.5$ and as the standard deviation is shown to be approximately 1.71, the variance $Var(X) = \dfrac{35}{12}$ or approximately 2.92.
Determine the mean and variance if each datum is increased by 3.

### Solution

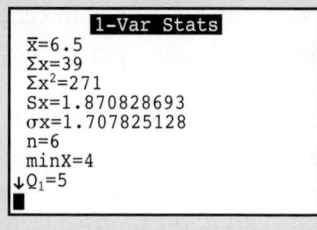

The data have been transformed into $x = \{4, 5, 6, 7, 8, 9\}$

Using a GDC, you can find that the mean increases by 3 to 6.5, but the standard deviation is shown to be unchanged; hence, the variance is also unchanged at $\dfrac{35}{12}$

Given $E(X) = \mu$ and $Var(X) = \sigma^2$, $E(X + c) = \mu + c$ while $Var(X + c) = \sigma^2$

Clearly, since each data value increased by 3, the mean increased by 3. However, increasing each data value by 3 did not change the differences from the new mean to the data points; hence, the variance is unchanged.

Now, let's consider changing the data values by taking a scalar multiple of each datum. What would you imagine the distribution to reveal?

### Example 15.9

Refer back to Example 15.8, and consider the random variable $X$ representing the data set $x = \{1, 2, 3, 4, 5, 6\}$. Recall that $E(X) = 3.5$ and the standard deviation was approximately 1.71, making $Var(X) = \dfrac{35}{12}$ or approximately 2.92.

Determine the mean and variance when each datum is multiplied by 2.

### Solution

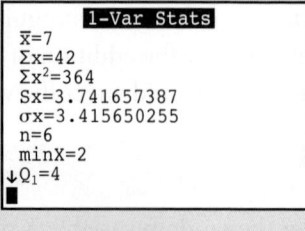

The data set has been transformed into $x = \{2, 4, 6, 8, 10, 12\}$

Using a GDC, the mean is found to be 7, and the standard deviation is found to be approximately 3.42, twice what it was before.

Clearly, as each value was doubled, the largest and smallest values were doubled, and the spread was doubled. The distribution would not become any taller as the frequencies would not change, but it would become twice as wide. Since the variance requires the **squares** of the differences between each data value and the mean, the variance would increase four-fold. The variance increased to $\dfrac{35}{3}$

Now, consider a normal distribution of data whose mean is $E(X) = \mu$ and variance $Var(X) = \sigma^2$ shown in blue in Figure 15.7. Translate the probability density function to the right by $c$ units to obtain the red graph and consider the change to the mean and variance.

It should be apparent that for the mean to increase by $c$, each datum must have increased by $c$. The PDF shows no change in spread, so the variance is unchanged.

On the other hand, if each of the data values were tripled, the range would triple, and the variance would increase 9-fold for the same reason as you've seen with the uniform distribution.

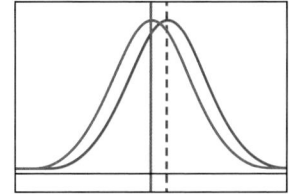

Given $E(X) = \mu$ and $Var(X) = \sigma^2$,
$E(aX) = a\mu$ while
$Var(aX) = a^2\sigma^2$

**Figure 15.7** PDF shifted by $c$ units

## Combinations of normal distributions

If two or more independent normal distributions are combined, the resulting distribution is also normal. This allows for the analyses of linear combinations of normally distributed data.

Given two normal distributions, $X_1$ and $X_2$, where $X_1 \sim N(\mu_1, \sigma_1^2)$ and
$X_2 \sim N(\mu_2, \sigma_2^2)$
$\Rightarrow X = X_1 + X_2$
$\sim N(\mu_1 + \mu_2, \sigma_1^2 + \sigma_2^2)$

### Example 15.10

The SAT exams were revised in 2016, with scores in the range 200–800 each in two areas: Math and ERW (Evidence-based Reading and Writing). The mean scores in 2017 were 527 in Math and 533 in ERW, with standard deviations of 107 and 100, respectively.

Universities use the combined score to determine admission.

What are the mean and variance of the combined score?

### Solution

Let $X_m$ be the random variable for the Math score and $X_v$ be the random variable for the ERW score. Since $X_m \sim N(527, 107^2)$ and $X_v \sim N(533, 100^2)$, $X \sim N(527 + 533, 107^2 + 100^2)$. The mean of the combined score is 1060; the variance is 21 449. (The standard deviation is approximately 146.)

## Combinations of Poisson distributions

If two or more independent Poisson distributions are combined, the resulting distribution is also a Poisson distribution.

### Example 15.11

A newly opened home improvement store has a lumber yard at one end and a gardening shop at the opposite end. Customers do not shop at both ends, although it is likely that they would shop in the middle of the store for other items such as tools and hardware. The store finds that the lumber yard has, on average, about 120 customers per day, while the gardening shop has 200 per day. In order to plan for a possible extension to the car park, the store needs to determine capacity.

(a) Explain why the number of shoppers at each end can be modelled by a Poisson distribution.

(b) What is the parameter for the combined distribution?

(c) Determine the expected number of customers in these two sections of the store in one hour of their 10-hour business day.

(d) Find the probability that in one hour, the total number of customers in these two sections of the store is 36.

### Solution

(a) The Poisson distribution suits the arrival of customers because it is based on a time interval. Let $X_1$ be the random variable for the number of shoppers at the lumber yard and $X_2$ represent the number of shoppers at the gardening shop. Given the 10-hour day, there are $E(X_1) = 12$ and $E(X_2) = 20$ customers per hour.

(b) Since $X_1 \sim \text{Po}(12)$ and $X_2 \sim \text{Po}(20)$, the sum of the two is $X = X_1 + X_2 \sim \text{Po}(32)$

(c) Independence is assumed, so the expected number is 32.

(d) $P(X = 36) = \dfrac{e^{-32} \cdot (32)^{36}}{36!} \approx 0.0522$

These transformations and combinations of data together are indicated in the IB Formula Booklet as follows.

$E(a_1 X_1) \pm E(a_2 X_2) = a_1 E(X_1) \pm a_2 E(X_2)$

$\text{Var}(a_1 X_1 \pm a_2 X_2) = a_1^2 \text{Var}(X_1) + a_2^2 \text{Var}(X_2)$

Note the $\pm$ sign on the left and the $+$ sign on the right due to the square of $a_2$

## Example 15.12

Test scores in a HL class are to be 'curved' as follows: every student will receive 5 marks, which are then added to twice the score on the test itself. Given that the test scores are normally distributed with an average of 35 and a standard deviation of 7 marks, find

(a) the mean and standard deviation of the 'curved' score

(b) the probability that a student receives a score of at least 65 after curving.

### Solution

(a) Let $X$ be the raw score on the test, and hence $Y = 2X + 5$ will be the curved score.

$$E(Y) = 2 \times 35 + 5 = 75$$
$$\sigma = \sqrt{\text{Var}(2X + 5)} = \sqrt{2^2\,\text{Var}(X)} = 2\sqrt{49} = 14$$

(b) $P(Y \geqslant 65) = 0.7625$

## Exercise 15.5

1. The uniform random variable $X$ has values $x = \{2, 4, 6, 8\}$

   (a) Find $E(X + 5)$ and $\text{Var}(X + 5)$

   (b) Find $E(3X)$ and $\text{Var}(3X)$

2. Given $X \sim N(100, 4^2)$, find the values of $E(X)$ and $\text{Var}(X)$ when:

   (a) each data value is incremented by 5

   (b) each data value is multiplied by 3.

3. Two independent random variables describe distributions that are both normal, with $X_1 \sim N(67, 12^2)$ and $X_2 \sim N(33, 5^2)$

   (a) Find $E(X)$ and $\text{Var}(X)$ where $X = X_1 + X_2$

   (b) Find $E(Y)$ and $\text{Var}(Y)$ where $Y = 2X_1 - X_2$

4. A local farm sells blueberries in 2-kg and 4-kg boxes. To facilitate the weighing and packaging, the 2-kg boxes are packed after weighing on one scale, while the 4-kg boxes are packed after weighing on another scale. The weights are normally distributed with the random variable $X_2 \sim N(2.3, 0.2^2)$ representing the mean and variance of the 2-kg boxes, and $X_4 \sim N(4.4, 0.2^2)$ representing the mean and variance of the 4-kg boxes. What is the probability that two 2-kg boxes contain more blueberries than one 4-kg box?

651

5. The manufacture of a certain tool requires two identical components made on a CNC machine, followed by a finishing process. The manufacturing times, in minutes, for the two processes are normally distributed with $X_1 \sim N(13, 0.4^2)$ and $X_2 \sim N(12, 0.3^2)$. Find the mean and variance of the distribution $X = 2X_1 + X_2$ then find $P(X \leqslant 40)$

6. Determine the mean of the sum of two independent Poisson distributions, given $X_1 \sim Po(3)$ and $X_2 \sim Po(4)$. Find the probability $P(X_1 + X_2 = 5)$

7. A telephone answering service looks after two unrelated small businesses: a travel agency and a used-car lot. The number of calls for the travel agency averages 40 per 8-hour day; the used car lot averages 20 per 8-hour day.

   (a) Express each as a Poisson distribution.

   (b) Determine:
      (i) the expected number of calls per hour for the travel agency
      (ii) the expected number of calls per hour for the used car business
      (iii) the expected number of calls answered per hour
      (iv) the probability of 10 calls answered in one hour.

8. It appears that birds come to feed randomly at a hummingbird feeder (containing nectar) and a bird feeder (containing seeds). In order to plan for an optimal restocking of feed, 'return visits' and 'visits by different birds' are counted as separate visits. The hummingbird feeder is visited 6 times in an hour, and the bird feeder is visited 12 times per hour. Find:

   (a) the probability that two hummingbirds come to feed in the next 10 minutes

   (b) the probability that two birds come to the bird feeder in the next 10 minutes

   (c) the probability that the feeders are visited a total of four times in the next 10 minutes. Compare your answer to your answers for parts (a) and (b)

   (d) the probability that the feeders are visited at most four times in the next 10 minutes.

# 15.6 Matrix applications (Markov chains)

When the probability distribution of possible events at discrete time intervals needs to be analysed, the use of matrices provides us with a mathematical model. Such events and transitions define a **Markov process**, where the next distribution or state depends only on the present state, not on previous history. Carried on repeatedly, any **stochastic** or random distribution of events is deterministic, and a Markov chain can produce long-term projections of a distribution that may lead to a steady state wherein the distribution becomes unchanged.

You studied Markov chains in Chapter 7.

The distribution and re-distribution of taxis in various garages, of planes at different hubs, the changing preferences in product choices such as food, cosmetics, cars, and clothing can be analysed if a **transition matrix** of changes, preferences, or states can be used to model changes over time.

Consider, for example, phone network operators and the preferences of consumers. Suppose that there are three phone network operators which we will call A, B, and C. At present, 20% of all subscribers use operator A, 30% use B, and 50% use C.

A tree diagram is the simplest way to keep track (Figure 15.8).

Now, suppose a survey of operator A subscribers showed that 40% of them would stay with A, and 25% would switch to B, while the remaining 35% would switch to C upon contract renewal. Similar surveys of operator B and C subscribers showed:

B→A: 25%, B→B: 45%, and B→C: 30%, and

C→A: 10%, C→B: 55%, and C→C: 35%

Let's update our tree diagram (Figure 15.9).

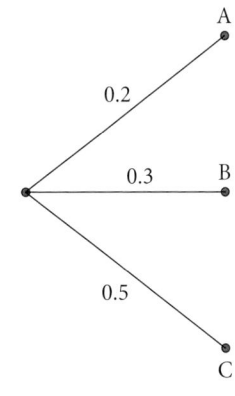

**Figure 15.8** Initial stage distribution

The original state had subscribers for A, B, and C in this distribution: (0.2  0.3  0.5)

Can you determine what the distribution of subscribers would be at the next state, given their preferences?

The next state for operator A (shown ending up in **red**):
$(0.2 \times 0.4) + (0.3 \times 0.25) + (0.5 \times 0.1) = 0.205$

The next state for operator B (shown ending up in **blue**):
$(0.2 \times 0.25) + (0.3 \times 0.45) + (0.5 \times 0.55) = 0.46$

The next state for operator C (shown ending up in **green**):
$(0.2 \times 0.35) + (0.3 \times 0.3) + (0.5 \times 0.35) = 0.335$

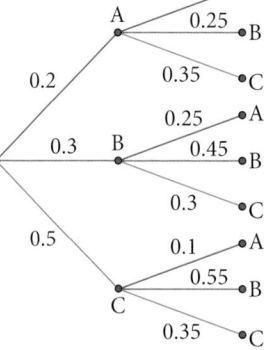

**Figure 15.9** Second stage distribution

Now, consider the calculations involved. You multiplied numbers and then added them to find the percentage of subscribers for each. If this looks vaguely familiar, it should be. When you multiplied matrices in Chapter 7, it's exactly the same set of calculations you used.

A system of equations such as
$$\begin{cases} 2x + y - z = 8 \\ x - y + 2z = -3 \\ 3x + 2y + z = 11 \end{cases}$$

translates into the matrix equation
$$\begin{pmatrix} 2 & 1 & -1 \\ 1 & -1 & 2 \\ 3 & 2 & 1 \end{pmatrix} \cdot \begin{pmatrix} x \\ y \\ z \end{pmatrix} = \begin{pmatrix} 8 \\ -3 \\ 11 \end{pmatrix}$$

Or
$$(x \quad y \quad z) \cdot \begin{pmatrix} 2 & 1 & 3 \\ 1 & -1 & 2 \\ -1 & 2 & 1 \end{pmatrix} = (8 \quad -3 \quad 11)$$

Similarly, the transition from the initial state to the next can be calculated as

$$(0.2 \quad 0.3 \quad 0.5) \cdot \begin{pmatrix} 0.4 & 0.25 & 0.35 \\ 0.25 & 0.45 & 0.3 \\ 0.1 & 0.55 & 0.35 \end{pmatrix} = (0.205 \quad 0.46 \quad 0.335)$$

The matrix equation is of the form $x_0 \cdot T = x_1$, where $x_0$ is the initial state, $T$ is the transition matrix, and $x_1$ is the first state after the transition. If the survey results that produced the transition matrix are reliable and stable, then it can be used to calculate $x_2, x_3, x_4, \ldots$ with each state depending on the previous distribution.

### Example 15.13

Using the transition matrix above and the distribution found as $x_1$, find $x_2$

### Solution

The use of a tree diagram is definitely not recommended! Instead, use your GDC to find $x_2 = x_1 \cdot T = (0.2305 \quad 0.4425 \quad 0.327)$

If you interpret the results, you will have noticed how operator C went from 50% market share to 33.5% after one transition, a significant drop, then to 32.7%. Is this an indication of steady declines to come?

Let's consider the next state. Of course, with a GDC, all that needs to be done is to hit 'ENTER' again to reveal $x_3 = x_2 \cdot T = (0.235525 \quad 0.4366 \quad 0.327875)$. The percentage market share is up for operator C, albeit slightly, to 32.7875%. In order to determine a long-term and perhaps **steady state**, you could keep hitting 'ENTER':

$$x_1 = x_0 \cdot T$$
$$x_2 = x_1 \cdot T = (x_0 \cdot T) \cdot T = x_0 \cdot T^2$$
$$x_3 = x_2 \cdot T = (x_0 \cdot T^2) \cdot T = x_0 \cdot T^3$$

However, since the pattern suggests that it would be more expedient to find the distribution at the $n$th state with the matrix calculation $x_n = x_0 \cdot T^n$

The 50th state, for instance, is simply

$$x_{50} = x_0 \cdot T^{50} = (0.2 \quad 0.3 \quad 0.5) \cdot \begin{pmatrix} 0.4 & 0.25 & 0.35 \\ 0.25 & 0.45 & 0.3 \\ 0.1 & 0.55 & 0.35 \end{pmatrix}^{50} \approx (0.236 \quad 0.436 \quad 0.328)$$

Incidentally, the calculation for the 100th state yields exactly the same results to ten decimal places. Your GDC makes these calculations possible. Just imagine a very, very large tree diagram as an alternative!

The term **deterministic** is sometimes applied to the Markov process. By investigating a different initial state, this can be seen.

## Example 15.14

In the investigation in Example 15.14, consider what would happen if the initial state had been $x_0 = (0.8 \quad 0.1 \quad 0.1)$

### Solution

At the 50th state, for instance,

$$x_{50} = x_0 \cdot T^{50} = (0.8 \quad 0.1 \quad 0.1) \cdot \begin{pmatrix} 0.4 & 0.25 & 0.35 \\ 0.25 & 0.45 & 0.3 \\ 0.1 & 0.55 & 0.35 \end{pmatrix}^{50}$$

$$\approx (0.236 \quad 0.436 \quad 0.328)$$

the same as above. The interesting outcome is that the long-term distribution is the same, regardless of the initial state.

Now, if you've observed that different initial states lead to the same, unchanging, **steady state distribution**, you might guess that what matters is the transition matrix $T$.

So, let's investigate a large power of $T$ shown in the screenshot in Figure 15.10.

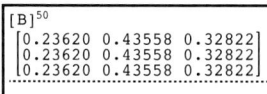

Figure 15.10 Do these values look familiar?

## Exercise 15.6

1. An electrical company is working on a large commercial construction project. At the end of each day, the workers typically leave equipment either on the current floor, on the roof, or in a storage room. The foreman observes that at the end of one day, the distribution of equipment was $x_0 = (0.7 \quad 0.2 \quad 0.1)$ and

he considers whether the transition matrix $T = \begin{pmatrix} 0.8 & 0.1 & 0.1 \\ 0.4 & 0.4 & 0.2 \\ 0.4 & 0.3 & 0.3 \end{pmatrix}$ might

model the distribution of equipment from day to day.

    **(a)** Determine the changing distribution states according to the model:

        **(i)**   $x_1$ (at the end of the next day)

        **(ii)**  $x_1, x_2, x_3$ (over the first 3 days)

        **(iii)** $x_{10}$ (at the end of 10 days).

    **(b)** Find the steady state, if there is one.

**2.** FAA regulations and airline management ensure that flight crews have the mandatory hours of rest between flights. Consequently, flight crews need to be rested and stationed in airport cities. Consider a small airline with four hubs A, B, C, and D between which they operate, but depending on the flight schedules, adequate rested crews must be available at each hub for the flights in any time period. Assuming the same daily departure and arrival schedules are repeated, the transition matrix is observed to be

$$T = \begin{pmatrix} 0.1 & 0.2 & 0.5 & 0.2 \\ 0.6 & 0 & 0.3 & 0.1 \\ 0.4 & 0.2 & 0.1 & 0.3 \\ 0.1 & 0.6 & 0.3 & 0 \end{pmatrix}$$

where, for instance in the top row,

$$\text{A} \rightarrow \text{A: } 0.1, \text{ A} \rightarrow \text{B: } 0.2, \text{ A} \rightarrow \text{C: } 0.5, \text{ and A} \rightarrow \text{D: } 0.2$$

    **(a)** If, on one particular day, the distribution of the flight crews is $x_0 = (0.25 \quad 0.25 \quad 0.25 \quad 0.25)$, determine:

        **(i)**   $x_1$ (at the end of the next day)

        **(ii)**  $x_1, x_2, x_3$ (over the first 3 days); is there a trend?

        **(iii)** $x_{10}$ (at the end of 10 days); is any trend you noted in part **(ii)** still there?

    **(b)** Find the steady state if there is one.

**3.** The IB Mathematics teachers in your district attended training that was available in one of two formats: face-to-face and online. Checking their experiences, it is found that 80% had the first type, 15% had the second, and 5% had none. Their initial state is described as $x_0 = (0.8 \quad 0.15 \quad 0.05)$

A survey reveals that for the following year, of those who had face-to-face training, 90% wanted the same again, 10% wanted no training, and nobody wanted online training. Of those who had online training, 80% wanted face-to-face training, 10% wanted online training, and 10% wanted no training. Of those who had no training, 80% wanted face-to-face training, 20% wanted online, and nobody opted for no training. Their preferences are given by the transition matrix:

$$T = \begin{pmatrix} 0.9 & 0 & 0.1 \\ 0.8 & 0.1 & 0.1 \\ 0.8 & 0.2 & 0 \end{pmatrix}$$

(a) Determine the changing distribution states according to the model:

(i) $x_1$ (after one year)

(ii) $x_1$, $x_2$, $x_3$ (over the first 3 years) and note if there is a trend.

(b) Find the steady state if there is one.

4. By surveying the owners of three different brands of ultralight laptops, A, B, and C, the transition matrix, given in the row/column order of A-B-C is found to be:

$$T = \begin{pmatrix} 0.5 & 0.3 & 0.2 \\ 0.3 & 0.3 & 0.4 \\ 0.1 & 0.5 & 0.4 \end{pmatrix}$$

(a) Without determining what the present distribution might be, explain how the transition matrix alone can determine the steady state.

(b) Which brand would become the preferred one eventually?

(c) The distribution in the $n$th state of this Markov chain was

$x_n = (0.3 \quad 0.35 \quad 0.35)$

Determine the distribution in its previous state, $x_{n-1}$

(d) Show that the distribution shown in the $n$th state will lead to the same steady state found in part **(b)**.

5. Three department stores serve a medium-sized city. During the year, store A expects to retain 80% of its customers, 5% are lost to store B, and 15% to store C. Store B expects to retain 90% of its customers and lose 5% to each of the other two stores. Store C expects to retain 75% of its customers, but lose 10% to A and 15% to B.

(a) Construct the transition matrix for the Markov chain that describes the change in the market share.

(b) Currently the market share of each store is 0.4 for A, 0.3 for B, and 0.3 for C. What share of the market is held by each store after 1 year?

(c) Assuming the trend continues, what share will each department store hold after 2 years?

6. TG Polling conducted a poll 6 months before elections in a country in which a liberal and a conservative were running for president. TG found that 60% of the voters intended to vote for the conservative and 40% for the liberal. In a poll conducted 3 months later, TG found that 70% of those who had earlier stated a preference for the conservative candidate still maintained that preference, whereas 30% of those voters now preferred the liberal candidate. Of those who earlier preferred the liberal, 80% still maintained their preference, whereas 20% switched to the conservative.

(a) If elections were held at this time (after 3 months), who would win?

(b) If the trend continues, which candidate will win the election?

7. Truck manufacturers A, B, and C share the domestic market in a certain country. Their current market shares are 60%, 30% and 10% respectively. Market studies show that manufacturer A retains 75% of its customers, and loses 15% to B and 10% to C. Of the customers who buy from B, 90% would keep their preference, while 5% go to each of A and C. Those who buy from C, 85% are retained, while 5% would buy from A and 10% from B.

   (a) Assuming that these sentiments reflect the buying habits of customers in the future, determine the market share that will be held by each manufacturer after the next 2 model years.

   (b) Under the same conditions, determine the market share that will be held by each manufacturer after the next 5 model years.

## Chapter 15 practice questions

1. A board game uses an octahedral dice with faces numbered 1 to 8 to determine the number of squares a player moves. Find the expected value of the number of squares a player moves on a single roll of the dice.

2. A discrete random variable $X$ can assume five possible values:

   12, 13, 15, 18, and 20

   Its probability distribution is shown below.

   | $x$ | 12 | 13 | 15 | 18 | 20 |
   |---|---|---|---|---|---|
   | $P(X = x)$ | 0.14 | 0.11 | | 0.26 | 0.23 |

   (a) Find $P(X = 15)$
   (b) Find the probability that $X$ equals 12 or 20
   (c) Find $P(X \leqslant 18)$
   (d) Find $E(X)$
   (e) Find $\text{Var}(X)$

3. A random number generator produces digits 1, 2, 3, 4, or 5, but is programmed so that the probability of each outcome is 0.05 higher than the probability of the outcome that is one less:

   $$P(X = 1) = a, P(X = 2) = a + 0.05, P(X = 3) = a + 0.1$$
   $$P(X = 4) = a + 0.15, \text{ and } P(X = 5) = a + 0.2$$

   Determine $E(X)$ and $\text{Var}(X)$

4. A store will give an instant discount of 40%, 20%, 10%, or 5% on customers' purchases during their sales promotion, depending on the outcome of a spinner. In the table, $x$ represents the possible percentage discounts and $P(X = x)$ represents the probability that the spinner will land on each of the percentage discounts.

| $x$        | 40   | 20   | 10   | 5    |
|------------|------|------|------|------|
| $P(X = x)$ | 0.05 | 0.1  | 0.25 | 0.6  |

Find $E(X)$ and $Var(X)$

5. $X$ has a probability distribution as shown in the table.

| $x$        | 5               | 10              | 15  | 20              | 25               |
|------------|-----------------|-----------------|-----|-----------------|------------------|
| $P(X = x)$ | $\dfrac{3}{20}$ | $\dfrac{7}{30}$ | $k$ | $\dfrac{3}{10}$ | $\dfrac{13}{60}$ |

   (a) Find the value of $k$
   (b) Find $P(X > 10)$
   (c) Find $P(5 < X \leqslant 20)$
   (d) Find the expected value and the standard deviation.

6. Determine the value of $k$ such that $f(x) = \begin{cases} k \cdot x^3, & 2 \leqslant x \leqslant 3 \\ \\ 0, & \text{otherwise} \end{cases}$

   can be used as a probability density function.

7. Find the value of $k$ such that the following represents a probability density function of a random variable $X$.

$$f(x) = \begin{cases} kx^2(2 - x), & 0 \leqslant x \leqslant 2 \\ \\ 0, & \text{otherwise} \end{cases}$$

8. The random variable $X$ has a PDF defined by:

$$f(x) = \begin{cases} k(-x^2 + 2x + 15), & 0 \leqslant x \leqslant 5 \\ \\ 0, & \text{otherwise} \end{cases}$$

   (a) Find the value of $k$.
   (b) Determine the mean and the mode.
   (c) Find the value $m$, the random variable which is larger than 50% of the population.

9. The function $f(x)$ is a PDF for a random variable $X$, defined by:

$$f(x) = \begin{cases} \dfrac{3}{4}x^2 (2 - x), & 0 \leqslant x \leqslant 2 \\ \\ 0, & \text{otherwise} \end{cases}$$

Find the variance and the standard deviation.

10. A 6 appears 40% of the time on a weighted six-sided dice. You roll this dice 60 times, and let the number of times it is a 6 be represented by the random variable $X$.

    (a) Find:
        (i) $P(X = 24)$
        (ii) $P(X = 22)$ and $P(X = 26)$
        (iii) $P(X = 20)$ and $P(X = 28)$

    (b) Conjecture a shape to this probability distribution.

11. A twenty-item multiple-choice test is taken by randomly guessing one of four possible answers to each question. If $X$ is the random variable representing the number of correct answers, find $P(X \geq 10)$ to calculate the probability of passing the test by guessing.

12. Houses in a large city are equipped with alarm systems to protect them from burglary. A company claims their system to be 98% reliable. That is, it will trigger an alarm in 98% of the cases. In a certain neighborhood, 10 houses equipped with this system experience an attempted burglary.

    (a) Find the probability that all the alarms work properly.
    (b) Find the probability that at least half of the houses trigger an alarm.
    (c) Find the probability that at most 8 alarms will work properly.

13. Sales records indicate that 40% of graphic novels were purchased by readers 30 years of age or older. If 15 readers are chosen at random, find the probability that:

    (a) at least ten of them are 30 years or older
    (b) ten of them are 30 or older
    (c) at most ten of them are younger than 30.

14. Owners of dogs in many countries buy health insurance for their dogs. In a large city, 3% of all dogs have health insurance. In a random sample of 100 dogs, find:

    (a) the expected number of dogs with health insurance
    (b) the probability that 5 of the dogs have health insurance
    (c) the probability that more than 10 dogs have health insurance.

15. The number of support phone calls coming into the central switchboard of a small computer company averages six per minute. Find the probability that:

    (a) no calls will arrive in a given one-minute period
    (b) at least two calls will arrive in a given one-minute period
    (c) at least two calls will arrive in a given two-minute period.

16. Passengers arrive at a security checkpoint in a busy airport at the rate of 8 passengers per 10-minute period. For the time between 8:00 and 8:10 on a given day, find the probability that:

    (a) exactly 8 passengers arrive

    (b) no more than 5 passengers arrive

    (c) at least 4 passengers arrive.

17. A certain website receives on average 0.2 hits per second. It is known that the number of hits on this site follows a Poisson distribution. Find the probability that:

    (a) no hits are registered during the next second

    (b) no hits are registered for the next 3 seconds.

18. There are an average of 4.4 faults per square metre in the knit of a certain fabric. It is assumed that the number of faults follows a Poisson distribution. Find the probability that:

    (a) one $1\,m^2$ piece of this fabric contains at least one fault

    (b) one $3\,m^2$ piece of this fabric contains at least one fault

    (c) three $1\,m^2$ pieces of this fabric contain one fault.

19. The results of a nationally conducted test can be modelled by a normal distribution with mean $\mu = 510$ and standard deviation $\sigma = 106$

    (a) Determine the percentage of scores in the interval [700, 800]

    (b) What is the score at the 90th percentile?

    (c) At what percentile is a score of 720?

20. A furnace thermocouple has a time to failure in years, modelled by a normal distribution with $X \sim N(6.8, 0.25^2)$. What is the probability that the thermocouple will last for at least 7 years?

21. Bottles of mineral water sold by a company are advertised to contain 1 litre of water. To ensure compliance with this quantity, the company adjusts its filling process to fill the bottles with an average of 1012 ml. The process follows a normal distribution with $\sigma = 5$ ml.

    (a) Find the probability that a randomly chosen bottle contains more than 1010 ml.

    (b) Find the probability that a randomly chosen bottle contains less than the advertised volume.

    (c) In a shipment of 10 000 bottles, what is the expected number of under-filled bottles?

22. Cholesterol plays a major role in a person's heart health. High blood cholesterol is a major risk factor for coronary heart disease and stroke. One of the measures of the level of cholesterol in the blood is the number of milligrams per deciliter of blood (mg/dL). According to the World Health Organisation, in general, less than 200 mg/dL is a desirable level and 200 to 239 mg/dL is borderline high. Above 240 mg/dL is a high risk level and the person with this level has more than twice the risk of heart disease as a person with less than 200 mg/dL.

    (a) What percentage do you expect to be classified as borderline high?
    (b) What percentage would be considered high risk?
    (c) Estimate the interquartile range of the cholesterol levels in this country.
    (d) Above what value are the highest 2% of adults' cholesterol levels in this country?

23. A manufacturer of car tyres claims that the tread life of its winter tyres can be described by a normal distribution with an average life of 52 000 km and a standard deviation of 4000 km.

    (a) If you buy a set of tyres from this manufacturer, is it reasonable for you to hope that they last more than 64 000 km?
    (b) What fraction of these tyres would you expect to last less than 48 000 km?
    (c) What fraction of these tyres would you expect to last between 48 000 km and 56 000 km?
    (d) What is the IQR of the tread life of this type of tyre?
    (e) The company wants to guarantee a minimum life for these tyres. They will refund customers whose tyres last less than a specific distance. What should their minimum life guarantee be so that they do not end up refunding more than 2% of their customers?

24. A family that runs a bookstore decided that their relatives should occupy a corner of the store to sell souvenirs. On average, during an 8-hour day, there are 80 people who arrive to shop for a book and 200 people to shop for souvenirs. What is the probability of having 40 people come to shop for a book or a souvenir in an hour?

25. A car rental company operates drop-off locations in three zones A, B, and C, to facilitate the return of cars rented on a daily basis. Their records, given by the transition matrix $T = \begin{pmatrix} 0.6 & 0.2 & 0.2 \\ 0.4 & 0.3 & 0.3 \\ 0.2 & 0.3 & 0.5 \end{pmatrix}$ indicate that a car rented in zone A is returned to zones A, B, and C with the probabilities 0.6, 0.2, and 0.2 respectively. The probabilities for cars from zone B are shown in the second row; from C, in the bottom row. The distribution of cars at the end of today is $x_0 = (0.35 \quad 0.35 \quad 0.3)$

(a) Determine:

    (i) $x_1$ (at the end of the next day)

    (ii) $x_1, x_2, x_3$ (over the first 3 days)

    (iii) $x_7$ (at the end of 1 week).

(b) Find the steady state, if there is one, to determine the optimal pre-distribution of cars.

26. Given the transition matrix $T = \begin{pmatrix} 0.75 & 0.25 \\ 0.5 & 0.5 \end{pmatrix}$ find the steady state. Show that the matrix is deterministic by starting with different initial states.

    (a) $x_0 = (0.5 \quad 0.5)$     (b) $x_0 = (0 \quad 1)$     (c) $x_0 = (1 \quad 0)$

27. A study suggests that the heights of the sons of short and tall men can be modelled by the transition matrix $T = \begin{pmatrix} 0.6 & 0.4 \\ 0.2 & 0.8 \end{pmatrix}$ where the probability that a short father's son will be short is 0.6, and that his son will be tall is 0.4, while the probability that a tall father has a short son is given as 0.2, while the probability that his son will also be tall is 0.8.

Find the steady state if there is one.

28. A forest has a large number of tall trees. The heights of the trees are normally distributed with a mean of 53 metres and a standard deviation of 8 metres. Trees are classified as giant trees if they are more than 60 metres tall.

    (a) A tree is selected at random from the forest.

        (i) Find the probability that this tree is a giant.

        (ii) Given that this tree is a giant, find the probability that it is taller than 70 metres.

    (b) Two trees are selected at random. Find the probability that they are both giants.

    (c) 100 trees are selected at random.

        (i) Find the expected number of these trees that are giants.

        (ii) Find the probability that at least 25 of these trees are giants.

29. A machine manufactures a large number of nails. The length, $L$ mm, of a nail is normally distributed, where $L \sim N(50, \sigma^2)$

    (a) Find $P(50 - \sigma < L < 50 + 2\sigma)$

    (b) The probability that the length of a nail is less than 53.92 mm is 0.975

    Show that $\sigma = 2.00$ (correct to 3 significant figures).

All nails with a length of at least $t$ mm are classified as large nails.

    (c) A nail is chosen at random. The probability that it is a large nail is 0.75 Find the value of $t$.

**(d) (i)** A nail is chosen at random from the large nails. Find the probability that the length of this nail is less than 50.1 mm.

**(ii)** Ten nails are chosen at random from the large nails. Find the probability that at least two nails have a length that is less than 50.1 mm.

30. The number of birds seen on a power line on any day can be modelled by a Poisson distribution with mean 5.84

    **(a)** Find the probability that during a certain seven-day week, more than 40 birds have been seen on the power line.

    **(b)** On Monday, there were more than 10 birds seen on the power line. Show that the probability of there being more than 40 birds seen on the power line from that Monday to the following Sunday, inclusive, can be expressed as

    $$\frac{P(X > 40) + \sum_{r=11}^{40} P(X = r) \cdot P(Y > 40 - r)}{P(X > 10)}$$

    where $X \sim Po(5.84)$ and $Y \sim Po(35.04)$

31. Students sign up at a desk for an activity during the course of an afternoon. The arrival of each student is independent of the arrival of any other student and the number of students arriving per hour can be modelled as a Poisson distribution with a mean of $\lambda$.

    The desk is open for 4 hours. If exactly five people arrive to sign up for the activity during that time, find the probability that exactly three of them arrived during the first hour.

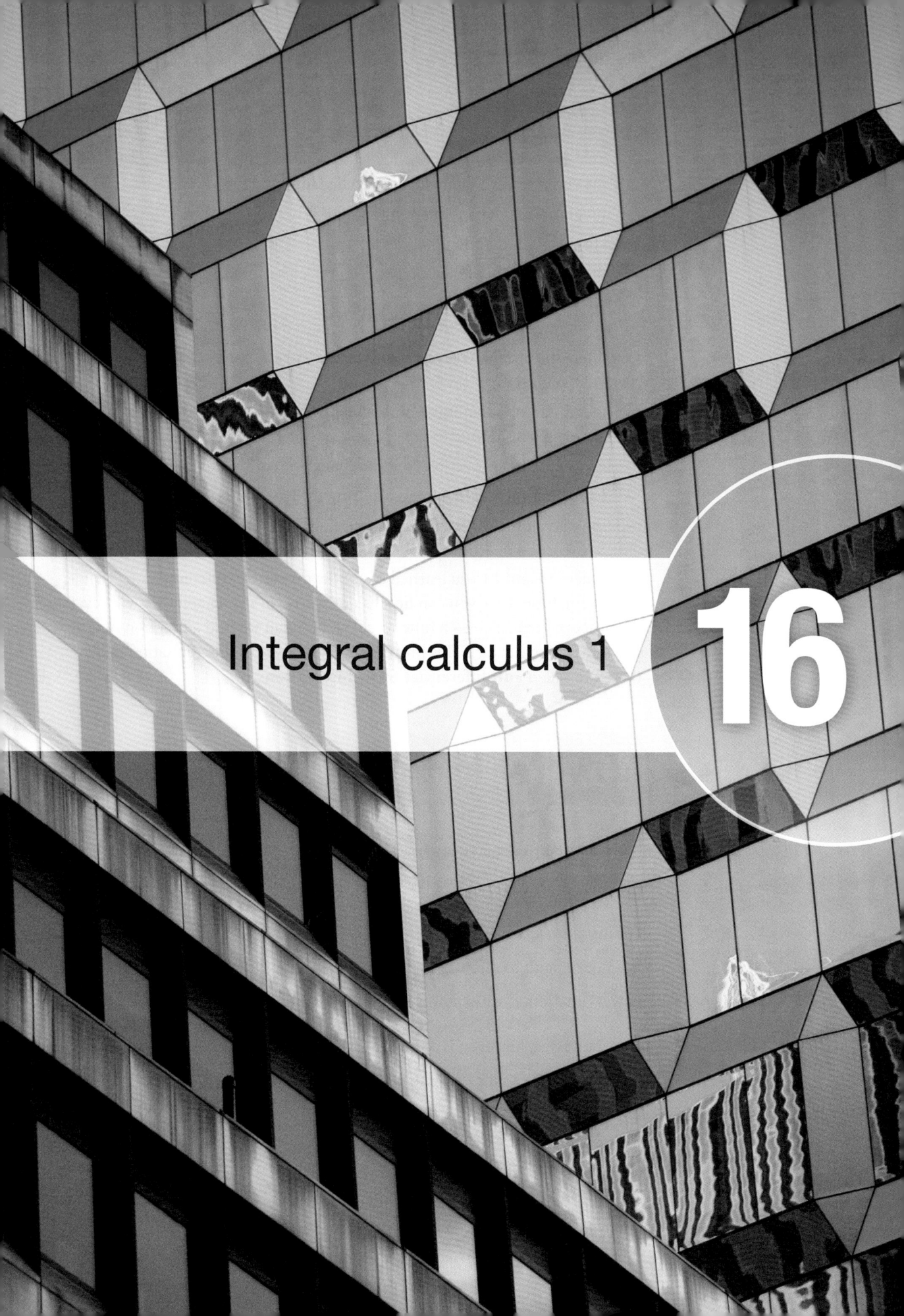

Integral calculus 1

16

## Learning objectives

By the end of this chapter, you should be familiar with...

- integration as the antidifferentiation of functions
- working with definite integrals
- finding areas under curves (between the curve and the $x$-axis), areas between curves
- antidifferentiation with a boundary value condition
- kinematic problems involving displacement $s$, velocity $v$, acceleration $a$ and total distance travelled
- integration of polynomial functions, trigonometric functions, and exponential functions
- integration by inspection (reverse chain rule) or by substitution
- approximating definite integrals using the trapezoidal rule
- finding volumes of revolution about the $x$-axis or $y$-axis.

In chapters 13 and 14 you learned about the process of differentiation. That is, given a function, how you can find its derivative. In this chapter we will reverse the process. That is, given a function, $f(x)$, how can we find a function $F(x)$ whose derivative is $f(x)$. This process is the opposite of differentiation and is therefore called **antidifferentiation or integration**.

## 16.1 Antiderivative

An **antiderivative** of the function $f(x)$ is a function $F(x)$ such that
$\frac{d}{dx} F(x) = F'(x) = f(x)$ wherever $f(x)$ is defined.

For instance, let $f(x) = x^2$.

It is not difficult to discover an antiderivative of $f(x)$. Keep in mind that this is a power function. If we examine the derivative of $x^3$ we find: $\frac{d}{dx}(x^3) = 3x^2$

This derivative, however, is 3 times $f(x)$. To compensate for the extra 3 we have to multiply by $\frac{1}{3}$ so that the antiderivative is now $\frac{1}{3}x^3$

Now, $\frac{d}{dx}\left(\frac{1}{3}x^3\right) = x^2$, and therefore, $\frac{1}{3}x^3$ is an antiderivative of $x^2$.

Table 16.1 shows some examples of functions, each paired with one of its antiderivatives.

Figure 16.1 shows the relationship between the derivative and the integral as opposite operations.

| Function $f(x)$ | Antiderivative $F(x)$ |
|---|---|
| 1 | $x$ |
| $x$ | $\frac{1}{2}x^2$ |
| $3x^2$ | $x^3$ |
| $x^4$ | $\frac{1}{5}x^5$ |
| $\cos x$ | $\sin x$ |
| $e^x$ | $e^x$ |
| $\sin x$ | $-\cos x$ |
| $\cos 2x$ | $\frac{1}{2}\sin 2x$ |

**Table 16.1** Examples of functions paired to an antiderivative

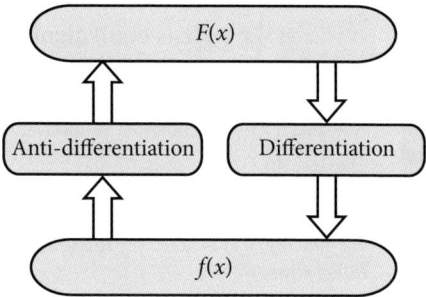

**Figure 16.1** Relationship between the derivative and the integral

## Example 16.1

Given the function $f(x) = 4x^3$

Find an antiderivative of $f(x)$

### Solution

$F(x) = x^4$ is such an antiderivative because $\frac{d}{dx}(F(x)) = 4x^3$

The following functions are also antiderivatives because the derivative of each one of them is also $4x^3$.

$$F_1(x) = x^4 + 27 \qquad F_2(x) = x^4 - \pi \qquad F_3(x) = x^4 + \sqrt{5}$$

Indeed, $F(x) = x^4 + c$ is an antiderivative of $f(x) = 4x^3$ for any choice of the constant $c$.

This is because $(F(x) + c)' = F'(x) + c' = F'(x) + 0 = f(x)$

Thus we can say that any single function $f(x)$ has many anti derivatives, whereas a function can have only one derivative.

If $F(x)$ is an antiderivative of $f(x)$, then so is $F(x) + c$ for any choice of the constant $c$.

## Notation

The notation

$$\int f(x)dx = F(x) + c$$

where $c$ is an arbitrary constant, means that $F(x) + c$ is an antiderivative of $f(x)$.

Equivalently, $F(x)$ satisfies the condition that

$$\frac{d}{dx}F(x) = F'(x) = f(x)$$

for all $x$ in the domain of $f(x)$

It is important to note that these two equations are just different notations to express the same fact. For example

$$\int x^2\,dx = \frac{1}{3}x^3 + c \text{ is equivalent to } \frac{d}{dx}\left(\frac{1}{3}x^3\right) = x^2$$

Note that if we differentiate an antiderivative of $f(x)$, we obtain $f(x)$.
Thus $\frac{d}{dx}\left(\int f(x)\,dx\right) = f(x)$

The expression $\int f(x)dx = F(x) + c$ is called an **indefinite integral** of $f(x)$

The function $f(x)$ is called the **integrand**, and the constant $c$ is called the **constant of integration**.

The integral symbol $\int$ is made like an elongated capital S. It is, in fact, a medieval S, used by Leibnitz as an abbreviation for the Latin word *summa* (sum).

We think of the combination $\int [\ ]dx$ as a single symbol; we fill in the blank with the formula of the function whose antiderivative we seek. We may regard the differential $dx$ as specifying the independent variable $x$ both in the function $f(x)$ and in its antiderivatives. This is true for any independent variable. For example, when the independent variable is $t$, the notation must be adjusted appropriately:

$\frac{d}{dt}\left(\int f(t)\,dt\right) = f(t)$ and $\int f(t)dt = F(t) + c$ are equivalent statements.

| Derivative formula | Equivalent integration formula |
|---|---|
| $\frac{d}{dx}(x^3) = 3x^2$ | $\int 3x^2\,dx = x^3 + c$ |
| $\frac{d}{dx}(\sqrt{x}) = \frac{1}{2\sqrt{x}}$ | $\int \frac{1}{2\sqrt{x}}\,dx = \sqrt{x} + c$ |
| $\frac{d}{dt}(\tan t) = \frac{1}{\cos^2 t} = \sec^2 t$ | $\int \frac{dt}{\cos^2 t} = \int \sec^2 t\,dt = \tan t + c$ |
| $\frac{d}{dv}\left(v^{\frac{3}{2}}\right) = \frac{3}{2}v^{\frac{1}{2}}$ | $\int \frac{3}{2}v^{\frac{1}{2}}\,dv = v^{\frac{3}{2}} + c$ |

**Table 16.2** Elementary integration formulae

The integral sign and differential serve as delimiters, adjoining the integrand on the left and right, respectively. In particular we do not write $\int dx\,f(x)$ when we mean $\int f(x)dx$

## Basic integration formulae

| | Derivation formula | Integration formula |
|---|---|---|
| 1 | $\frac{d}{dx}(x) = 1$ | $\int dx = x + c$ |
| 2 | $\frac{d}{dx}(x^{n+1}) = (n + 1)x^n, n \neq -1$ | $\int x^n\,dx = \frac{x^{n+1}}{n + 1} + c, n \neq -1$ |
| 3 | $\frac{d}{dx}(\sin x) = \cos x$ | $\int \cos x\,dx = \sin x + c$ |
| 4 | $\frac{d}{dv}(\cos v) = -\sin v$ | $\int \sin v\,dv = -\cos v + c$ |

**Table 16.3** Basic integration formulae

| 5 | $\dfrac{d}{dt}(\tan\theta) = \dfrac{1}{\cos^2\theta} = \sec^2\theta$ | $\displaystyle\int \dfrac{d\theta}{\cos^2\theta} = \int \sec^2\theta\, d\theta = \tan\theta + c$ |
|---|---|---|
| 6 | $\dfrac{d}{dv}(e^v) = e^v$ | $\displaystyle\int e^v\, dv = e^v + c$ |
| 7 | $\dfrac{d}{dx}(\ln|x|) = \dfrac{1}{x}$ | $\displaystyle\int \dfrac{1}{x}\, dx = \ln|x| + c$ |

**Table 16.3 cont.** Basic integration formulae

Formula 7 is a special case of the power rule.

If we are asked to integrate $\dfrac{1}{x}$, we may attempt to do so using the power rule:

$$\int \dfrac{1}{x}\, dx = \int x^{-1}\, dx = \dfrac{1}{(-1)+1}x^{(-1)+1} + c = \dfrac{1}{0}x^0 + c, \text{ which is undefined.}$$

However, if you recall from chapter 13:

$$\dfrac{d}{dx}(\ln x) = \dfrac{1}{x}, x > 0$$

which implies $\displaystyle\int \dfrac{1}{x}\, dx = \ln x + c, x > 0$

However, the function $\dfrac{1}{x}$ is differentiable for $x < 0$ too. So, we must be able to find its integral.

The solution lies in the chain rule.

If $x < 0$, then we can write $x = -u$ where $u > 0$. Then $dx = -du$, and
$\displaystyle\int \dfrac{1}{x}\, dx = \int \dfrac{1}{-u}(-du) = \int \dfrac{1}{u}\, du = \ln u + c, u > 0$, but $u = -x$, therefore
when $x < 0$

$$\int \dfrac{1}{x}\, dx = \ln u + c = \ln(-x) + c, \text{ and combining the two results gives}$$

$$\int \dfrac{1}{x}\, dx = \ln|x| + c, x \neq 0$$

---

### Example 16.2

Evaluate

(a) $\displaystyle\int 3\cos x\, dx$                        (b) $\displaystyle\int (x^3 + x^2)\, dx$

---

### Solution

(a) $\displaystyle\int 3\cos x\, dx = 3\int \cos x\, dx = 3\sin x + c$

(b) $\displaystyle\int (x^3 + x^2)\, dx = \int x^3\, dx + \int x^2\, dx = \dfrac{x^4}{4} + \dfrac{x^3}{3} + c$

Suppose that $f(x)$ and $g(x)$ differentiable functions and $k$ is a constant, then:

1. A constant factor can be moved through an integral sign; i.e.,
$$\int kf(x)\, dx = k\int f(x)\, dx$$

2. An antiderivative of a sum (difference) is the sum (difference) of the antiderivatives; i.e. $\int (f(x) \pm g(x))dx$
$$= \int f(x)\, dx \pm \int g(x)\, dx$$

# 16 Integral calculus 1

## Example 16.3

Evaluate

(a) $\int \dfrac{t^3 - 3t^5}{t^5}\,dt$

(b) $\int \dfrac{x + 5x^4}{x^2}\,dx$

### Solution

Sometimes it is useful to rewrite the integrand in a different form before performing the integration.

(a) $\int \dfrac{t^3 - 3t^5}{t^5}\,dt = \int \dfrac{t^3}{t^5}\,dt - \int \dfrac{3t^5}{t^5}\,dt = \int t^{-2}dt - \int 3\,dt = \dfrac{t^{-1}}{-1} - 3t + c = \dfrac{-1}{t} - 3t + c$

(b) $\int \dfrac{x + 5x^4}{x^2}\,dx = \int \dfrac{x}{x^2}\,dx + \int \dfrac{5x^4}{x^2}\,dx = \int \dfrac{1}{x}\,dx + \int 5x^2 dx = \ln|x| + 5\cdot\dfrac{x^3}{3} + c$

## Integration by simple substitution – change of variables

In this section we will study substitution, a technique which can often be used to transform complex integration problems into simpler ones.

The method of substitution depends on our understanding of the chain rule as well as the use of variables in integration. Two facts to recall:

1. When we find an antiderivative, we can use any other variable.

That is $\int f(u)\,du = F(u) + c$, where $u$ is a dummy variable in the sense that it can be replaced by any other variable.

2. Using the chain rule:

$\dfrac{d}{dx}(F(u(x))) = F'(u(x))\cdot u'(x)$

Which can be written in integral form as:

$\int F'(u(x))\cdot u'(x)\,dx = F(u(x)) + c$

Or equivalently, since $F(x)$ is an antiderivative of $f(x)$:

$\int f(u(x))\cdot u'(x)\,dx = F(u(x)) + c$

For our purposes it will be useful and simpler to let $u(x) = u$ and to write $\dfrac{du}{dx} = u'(x)$ in its differential form $du = u'(x)\,dx$ or simply $du = u'\,dx$

We can now write the integral as

$\int f(u(x))\cdot u'(x)\,dx = \int f(u)\,du = F(u(x)) + c = F(u) + c$

Example 16.4 shows how the method works.

## Example 16.4

Evaluate

(a) $\int 3x^2(x^3 + 2)^{10}\,dx$

(b) $\int \tan x\,dx;\ F(0) = 1$

(c) $\int \cos 5x\,dx$

(d) $\int \cos x^2 \cdot x\,dx$

(e) $\int e^{3x+1}\,dx$

670

## Solution

(a) To integrate this function, it is easiest to make the substitution $u = x^3 + 2$, and so $du = 3x^2 dx$. Now we can write the integral as

$$\int 3x^2(x^3 + 2)^{10} dx = \int u^{10} du = \frac{u^{11}}{11} + c = \frac{(x^3 + 2)^{11}}{11} + c$$

(b) Rewrite the integrand and then make the substitution:

$$\int \tan x \, dx = \int \frac{\sin x}{\cos x} dx = \int \frac{1}{\cos x} \cdot \sin x \, dx$$

We now let $u = \cos x \Rightarrow du = -\sin x \, dx$, so

$$\int \tan x \, dx = \int \frac{1}{\cos x} \cdot \sin x \, dx = \int \frac{1}{u}(-du) = -\int \frac{1}{u} du = -\ln|u| + c$$

This last result can be then expressed in two ways:

$$\int \tan x \, dx = -\ln|\cos x| + c, \text{ or}$$

$$\int \tan x \, dx = \ln\left|\frac{1}{(\cos x)}\right| + c = \ln|\sec x| + c$$

Since $F(0) = 1$, $\ln\left|\frac{1}{\cos 0}\right| + c = 1 \Rightarrow \ln 1 + c = 1 \Rightarrow c = 1$,

and $\int \tan x \, dx = \ln\left|\frac{1}{\cos x}\right| + 1$

(c) Let $u = 5x$, then $du = 5dx \Rightarrow dx = \frac{1}{5} du$, and so

$$\int \cos 5x \, dx = \int \cos u \cdot \frac{1}{5} du = \frac{1}{5} \int \cos u \, du = \frac{1}{5} \sin u + c = \frac{1}{5} \sin 5x + c$$

(d) By letting $u = x^2$, $du = 2x \, dx$ and so

$$\int \cos x^2 \cdot x \, dx = \frac{1}{2} \int \cos x^2 \cdot 2x \, dx = \frac{1}{2} \int \cos u \, du = \frac{1}{2} \sin u + c = \frac{1}{2} \sin x^2 + c$$

(e) $\int e^{3x+1} dx = \frac{1}{3} \int e^{3x+1} 3dx = \frac{1}{3} \int e^u du = \frac{1}{3} e^u + c = \frac{1}{3} e^{3x+1} + c$

> The main challenge in using the substitution rule is to think of an appropriate substitution. You should try to select $u$ to be a part of the integrand whose differential is also included (except for the constant). In Example 16.4(a), we selected $u$ to be $(x^3 + 2)$ knowing that $du = 3x^2 dx$. Then we compensated for the absence of 3. Finding the right substitution is a bit of an art, and you need to acquire it. It is quite usual that your first guess may not work.

In part (c), another method can be applied:
The substitution $u = 5x$ requires $du = 5dx$. As there is no factor of 5 in the integrand, and since 5 is a constant, we can multiply and divide by 5 so that we group the 5 and $dx$ to form the $du$ required by the substitution:

$\int \cos 5x \, dx = \frac{1}{5} \int \cos x \cdot 5 \, dx$
$= \frac{1}{5} \int \cos u \, du = \frac{1}{5} \sin u + c$
$= \frac{1}{5} \sin 5x + c$

In integration, multiplying by a constant inside the integral and compensating for that with the reciprocal outside the integral depends on formula 2 on page 670. That is
$\int kf(x) \, dx = k \int f(x) \, dx$
However, you cannot do this with a variable.
So, you cannot say, for example,
$\int \cos x^2 \, dx$
$= \frac{1}{2x} \int \cos x^2 \cdot 2x \, dx$

## Example 16.5

Evaluate

(a) $\int e^{-3x} dx$

(b) $\int \sin^2 x \cos x \, dx$

(c) $\int 2\sin(3x - 5) \, dx$

(d) $\int e^{mx+n} dx$

(e) $\int x\sqrt{x} \, dx; \quad F(1) = 2$

## Solution

(a) Let $u = -3x$, then $du = -3dx$
$$\int e^{-3x}dx = -\frac{1}{3}\int e^{-3x}(-3\,dx) = -\frac{1}{3}\int e^u\,du = -\frac{1}{3}e^u + c = -\frac{1}{3}e^{-3x} + c$$

(b) Let $u = \sin x \Rightarrow du = \cos x\,dx$, and hence
$$\int \sin^2 x \cos x\,dx = \int u^2\,du = \frac{1}{3}u^3 + c = \frac{1}{3}\sin^3 x + c$$

(c) Let $u = 3x - 5$, then $du = 3\,dx$
$$\int 2\sin(3x - 5)\,dx = 2 \cdot \frac{1}{3}\int \sin(3x - 5)\,3\,dx = \frac{2}{3}\int \sin u\,du$$
$$= -\frac{2}{3}\cos u + c = -\frac{2}{3}\cos(3x - 5) + c$$

(d) Let $u = mx + n$, then $du = mx$
$$\int e^{mx+n}\,dx = \frac{1}{m}\int e^{mx+n}\,m\,dx = \frac{1}{m}\int e^u\,du = \frac{1}{m}e^u + c = \frac{1}{m}e^{mx+n} + c$$

(e) $F(x) = \int x\sqrt{x}\,dx = \int x^{\frac{3}{2}}dx = \dfrac{x^{\frac{5}{2}}}{\left(\frac{5}{2}\right)} + c = \frac{2}{5}x^{\frac{5}{2}} + c$

$F(1) = \frac{2}{5}1^{\frac{5}{2}} + c = \frac{2}{5} + c = 2 \Rightarrow c = \frac{8}{5}$

Therefore $F(x) = \frac{2}{5}x^{\frac{5}{2}} + \frac{8}{5}$

Examples 16.4 and 16.5 make it clear that Table 16.3 is limited in scope, because we cannot use the integrals directly to evaluate composite integrals. We therefore need to revise some of the derivative formulae.

| | Derivative formula | Integration formula |
|---|---|---|
| 1 | $\frac{d}{dx}(u(x)) = u'(x) \Rightarrow du = u'(x)dx$ | $\int du = u + c$ |
| 2 | $\frac{d}{dx}\left(\frac{u^{n+1}}{(n+1)}\right) = u^n u'(x),\, n \ne -1 \Rightarrow d\left(\frac{u^{n+1}}{(n+1)}\right) = u^n u'(x)\,dx$ | $\int u^n\,du = \frac{u^{n+1}}{n+1} + c,\, n \ne -1$ |
| 3 | $\frac{d}{dx}(\sin(u)) = \cos(u)u'(x) \Rightarrow d(\sin(u)) = \cos(u)u'(x)\,dx$ | $\int \cos u\,du = \sin u + c$ |
| 4 | $\frac{d}{dx}(\cos(u)) = -\sin(u)u'(x) \Rightarrow d(\cos(u)) = -\sin(u)u'(x)dx$ | $\int \sin u\,du = -\cos u + c$ |
| 5 | $\frac{d}{dt}(\tan u) = \frac{1}{\cos^2 u}u'(t) \Rightarrow d(\tan u) = \frac{1}{\cos^2 u}u'(t)\,dt$ | $\int \frac{1}{\cos^2 u}\,du = \tan u + c$ |
| 6 | $\frac{d}{dx}(e^u) = e^u u'(x) \Rightarrow d(e^u) = e^u u'(x)\,dx$ | $\int e^u\,du = e^u + c$ |
| 7 | $\frac{d}{dx}(\ln|u|) = \frac{1}{u}u'(x) \Rightarrow d(\ln|u|) = \frac{1}{u}u'(x)\,dx$ | $\int \frac{1}{u}\,du = \ln|u| + c$ |

Table 16.4  Integration formulae

## Example 16.6

Evaluate:

(a) $\int \sqrt{6x + 11}\,dx$

(b) $\int x^2(5x^3 + 2)^8\,dx$

(c) $\int \frac{x^3 - 2}{\sqrt[5]{x^4 - 8x + 13}}\,dx$

(d) $\int x\sin^4(3x^2)\cos(3x^2)\,dx$

## Solution

(a) Let $u = 6x + 11$ then calculate $du$:

$$u = 6x + 11 \Rightarrow du = 6\,dx$$

Since $du$ contains the factor 6, the integral is still not in the proper form $\int f(u)\,du$. However, here we can use one of two approaches, introduce the factor 6, as we have done before, i.e.,

$$\int \sqrt{6x + 11}\,dx = \frac{1}{6}\int \sqrt{6x + 11}\;6\,dx$$

$$= \frac{1}{6}\int \sqrt{u}\,du = \frac{1}{6}\int u^{\frac{1}{2}}\,du$$

$$= \frac{1}{6}\frac{u^{\frac{3}{2}}}{\frac{3}{2}} + c = \frac{2}{18}u^{\frac{3}{2}} + c = \frac{1}{9}(6x + 11)^{\frac{3}{2}} + c$$

or since $u = 6x + 11 \Rightarrow du = 6\,dx \Rightarrow dx = \dfrac{du}{6}$, then

$$\int \sqrt{6x + 11}\,dx = \int \sqrt{u}\,\frac{du}{6} = \frac{1}{6}\int u^{\frac{1}{2}}\,du$$

and we follow the same steps as before.

(b) Let $u = 5x^3 + 2$, then $du = 15x^2\,dx$. This means that we need to introduce the factor 15 into the integrand

$$\int (5x^3 + 2)^8 x^2\,dx = \frac{1}{15}\int (5x^3 + 2)^8\;15x^2\,dx$$

$$= \frac{1}{15}\int u^8\,du = \frac{1}{15}\frac{u^9}{9} + c = \frac{1}{135}(5x^3 + 2)^9 + c$$

(c) Let $u = x^4 - 8x + 13 \Rightarrow du = (4x^3 - 8)dx = 4(x^3 - 2)dx$

$$\int \frac{x^3 - 2}{\sqrt[5]{x^4 - 8x + 13}}\,dx = \frac{1}{4}\int \frac{4(x^3 - 2)dx}{\sqrt[5]{x^4 - 8x + 13}} = \frac{1}{4}\int \frac{du}{u^{\frac{1}{5}}}$$

$$= \frac{1}{4}\int u^{-\frac{1}{5}}\,du = \frac{1}{4}\frac{u^{\frac{4}{5}}}{\frac{4}{5}} + c = \frac{5}{16}(x^4 - 8x + 13)^{\frac{4}{5}} + c$$

(d) We let $u = \sin(3x^2) \Rightarrow du = \cos(3x^2)6x\,dx$ using the chain rule.

$$\int \sin^4(3x^2)\cos(3x^2)\,x\,dx = \frac{1}{6}\int \sin^4(3x^2)\cos(3x^2)\;6x\,dx$$

$$= \frac{1}{6}\int u^4\,du = \frac{1}{6}\frac{u^5}{5} + c = \frac{1}{30}\sin^5(3x^2) + c$$

## Some applications to economics

Three functions of importance to an economist are:

- **the cost function**: $C(x) = $ total cost of producing $x$ units of a product during some time period
- **the revenue function**: $R(x) = $ total revenue from selling $x$ units of the product during the time period

- **the profit function**: $P(x)$ = total profit obtained by selling $x$ units of the product during the time period

If all units produced are sold, then

$$P(x) = R(x) - C(x)$$

profit = revenue − cost

The total cost $C(x)$ of producing $x$ units is made up of two parts: **overhead** or **fixed** costs, $a$, and **variable** cost or **manufacturing** cost, $M(x)$

$$C(x) = a + M(x)$$

If a firm can sell all the items it produces for $p$ units of money (€, £, or $, …) each, its total revenue $R(x)$ will be

$$R(x) = px$$

Also, economists call $P'(x)$, $R'(x)$, and $C'(x)$ the **marginal profit**, **marginal revenue**, and **marginal cost**; and they interpret these quantities as the additional profit, revenue, or cost that result from producing and selling an additional unit of the product when production/sales levels are at $x$ units. You usually invest to make a profit. Thus, $R(x) - C(x) \geqslant 0$. An investment is valuable until $R'(x) = C'(x)$ since beyond that point, the marginal cost will be more than the revenue.

---

## Example 16.7

A company produces fans with a maximum capacity of 500 units per day. Their marginal cost per day is given by the following model

$$MC(x) = 0.0012x^2 - 0.018x + 25$$

where $x$ is the number of fans produced.

(a) Given that fixed costs are €339, find the cost function, $C(x)$.

(b) Find the average cost per unit produced and graph $AC(x)$ and $MC(x)$ on the same set of axes.

(c) Find the minimum average cost, and the number of units produced at that level.

(d) What is marginal cost per unit when the number of units is as given in (c)?

(e) The company plans to produce 400 units per day. What should the selling price of each unit be to have a profit of €20 per unit?

---

### Solution

(a) $MC$ is the rate of change of cost, thus, cost itself is an antiderivative of $MC$.

$$C(x) = \int(0.0012x^2 - 0.018x + 25)dx = 0.0004x^3 - 0.09x^2 + 25x + c$$

Since fixed costs are €339, then $C(0) = 339$ and thus, $c = 339$

Therefore, the cost function is

$$C(x) = 0.0004x^3 - 0.09x^2 + 25x + 339$$

---

> *MC* is the cost for every new unit produced. Total cost will include the cost of producing $x$ units plus fixed costs which are costs involved in running the operation, regardless of the number of units produced.
> The average cost,
> $$AC(x) = \frac{C(x)}{x}$$

(b) $AC(x) = \dfrac{C(x)}{x}$

$$= \dfrac{0.0004x^3 - 0.09x^2 + 25x + 339}{x}$$

$$= 0.0004x^2 - 0.09x + 25 + \dfrac{339}{x}$$

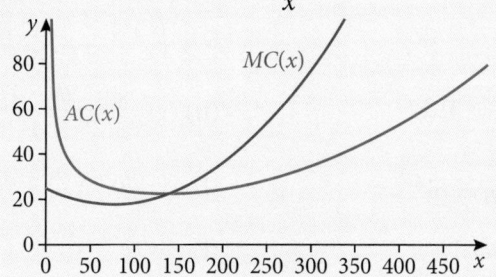

(c) To find the minimum average cost, we find the derivative of the $AC$ function and equate it to zero.

$$AC'(x) = 0.0008x - 0.09 - \dfrac{339}{x^2} = 0$$

$$\Rightarrow x \approx 136 \Rightarrow AC(136) \approx 23$$

(d) $MC(136) = 0.0012(136)^2 - 0.018(136) + 25 \approx 23$

(e) At 400 units, the average cost is: $AC(400) \approx 54$
Thus, the company should charge €74 per fan.

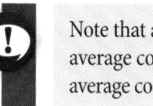

Note that at minimum average cost,
average cost = marginal cost!

## Exercise 16.1

1. Find the most general antiderivative of each function.

(a) $f(x) = x + 2$

(b) $f(t) = 3t^2 - 2t + 1$

(c) $g(x) = \dfrac{1}{3} - \dfrac{2}{7}x^3$

(d) $f(t) = (t - 1)(2t + 3)$

(e) $g(u) = u^{\frac{2}{5}} - 4u^3$

(f) $f(x) = 2\sqrt{x} - \dfrac{3}{2\sqrt{x}}$

(g) $h(\theta) = 3\sin\theta + 4\cos\theta$

(h) $f(t) = 3t^2 - 2\sin t$

(i) $f(x) = \sqrt{x}(2x - 5)$

(j) $g(\theta) = 3\cos\theta - \dfrac{2}{\cos^2\theta}$

(k) $h(t) = e^{3t-1}$

(l) $f(t) = \dfrac{2}{t}$

(m) $h(u) = \dfrac{t}{3t^2 + 5}$

(n) $h(\theta) = e^{\sin\theta}\cos\theta$

(o) $f(x) = (3 + 2x)^2$

2. For each of these derivatives, find $f$.

(a) $f''(x) = 4x - 15x^2$

(b) $f''(x) = 1 + 3x^2 - 4x^3,\ f'(0) = 2,\ f(1) = 2$

(c) $f''(t) = 8t - \sin t$

(d) $f'(x) = 12x^3 - 8x + 7, f(0) = 3$

(e) $f'(\theta) = 2\cos\theta - \sin(2\theta)$

**3.** Evaluate:

**(a)** $\int x(3x^2 + 7)^5 \, dx$

**(b)** $\int \dfrac{x}{(3x^2 + 5)^4} \, dx$

**(c)** $\int 2x^2 \sqrt[4]{5x^3 + 2} \, dx$

**(d)** $\int \dfrac{(3 + 2\sqrt{x})^5}{\sqrt{x}} \, dx$

**(e)** $\int t^2 \sqrt{2t^3 - 7} \, dt$

**(f)** $\int \left(2 + \dfrac{3}{x}\right)^5 \left(\dfrac{1}{x^2}\right) dx$

**(g)** $\int \sin(7x - 3) \, dx$

**(h)** $\int \dfrac{\sin(2\theta - 1)}{\cos(2\theta - 1) + 3} \, d\theta$

**(i)** $\int \dfrac{1}{\cos^2(5\theta - 2)} \, d\theta$

**(j)** $\int \cos(\pi x + 3) \, dx$

**(k)** $\int x e^{x^2 + 1} \, dx$

**(l)** $\int \sqrt{t}\, e^{2t\sqrt{t}} \, dt$

**(m)** $\int \dfrac{2}{\theta} (\ln\theta)^2 \, d\theta$

**(n)** $\int \dfrac{dz}{z \ln 2z}$

**4.** Evaluate:

**(a)** $\int t\sqrt{3 - 5t^2} \, dt$

**(b)** $\int \dfrac{\theta^2}{\cos^2 \theta^3} \, d\theta$

**(c)** $\int \dfrac{\sin \sqrt{t}}{2\sqrt{t}} \, dt$

**(d)** $\int \dfrac{\tan^5 2t}{\cos^2 2t} \, dt$

**(e)** $\int \dfrac{dx}{\sqrt{x}(\sqrt{x} + 2)}$

**(f)** $\int \dfrac{x + 3}{x^2 + 6x + 7} \, dx$

**(g)** $\int \dfrac{k^3 x^3}{\sqrt{a^2 - a^4 x^4}} \, dx$

**(h)** $\int 3x\sqrt{x - 1} \, dx$

**(i)** $\int \sqrt{1 + \cos\theta}\, \sin\theta \, d\theta$

**(j)** $\int t^2 \sqrt{1 - t} \, dt$

**(k)** $\int \dfrac{r^2 - 1}{\sqrt{2r - 1}} \, dr$

**(l)** $\int \dfrac{e^{x^2} - e^{-x^2}}{e^{x^2} + e^{-x^2}} x \, dx$

**(m)** $\int \dfrac{t^2 + 2}{\sqrt{t - 5}} \, dt$

**5.** A company produces artificial leather wallets with maximum capacity of 800 units per day. Their marginal cost per day is given by the following model:

$$MC(x) = 0.0006x^2 - 0.02x - 10$$

where $x$ is the number of wallets produced.

**(a)** Given that fixed costs are €2500, find the cost function, $C(x)$.

**(b)** Find the average cost per unit produced and graph $AC(x)$ and $MC(x)$ on the same set of axes.

**(c)** Find the minimum average cost, and the number of units produced at that level.

**(d)** What is marginal cost per unit if the number of units produced is as in **(c)**?

**(e)** The company plans on producing 400 units per day. What should be the selling price of each unit in order to have a profit of €2 per unit?

6. A chemical plant is adding pollution to a lake at the rate of $40\sqrt{t^3}$ tons per year, where $t$ is the number of years since the plant has been in operation.

   (a) Find a model that describes the amount of pollution in the lake after $t$ years of the plant's operations. You are given that the only source of pollution to this lake comes from the plant.

   (b) Use your model to find how much pollution entered the lake in the first three years of the plant's operation.

   (c) Life in the lake will cease when 400 tons of pollution have entered the lake. How long will this life in the lake survive if the plant does not stop polluting it?

7. An influenza epidemic hits a large city and spreads at the rate of $12e^{0.2t}$ new cases per day, where $t$ is the number of days since the epidemic began. The epidemic started with 4 cases.

   (a) Find a formula for the total number of cases in the first $t$ days of the epidemic.

   (b) Find the number of cases during the first 30 days.

8. World consumption of tin is $0.22e^{0.01t}$ million tons per year, where $t$ is the number of years since 2013. Find a formula for the total tin consumption within $t$ years of 2013 and estimate when the known world reserves of 156 million tons will be exhausted. (Assume that consumption continues to follow the model and that no new resources are discovered.)

9. World consumption of silver is $12e^{0.015t}$ thousand tons per year where $t$ is measured in years and $t = 0$ corresponds to 2017.

   Find a formula for the total amount of silver that will be consumed within $t$ years from 2017 and estimate when the known world reserves of 533 thousand metric tons will be exhausted. (Assume that consumption continues to follow the model and that no new resources are discovered.)

10. The marginal average cost of producing $x$ Swiss sports watches is given by
    $$\overline{C}'(x) = -\frac{1000}{x^2}, \text{ and } \overline{C}(100) = 25$$
    where $\overline{C}(x)$ is the average cost in Swiss Francs (SF).

    (a) Find the average cost function and the cost function.

    (b) What are the fixed costs?

11. In 2000, the percentage of renewable energy consumption out of primary energy consumption in Germany was 2.42%. Since 2000, the renewables consumption percentage has been growing at a rate given by
    $$f'(t) = 0.42 + 0.1062t - 0.00747t^2$$
    where $t$ is years after 2000. Find $f(t)$ and estimate the percentage of renewables in 2020.

12. Using production and geological data, an oil company estimates that oil will be pumped from a field at the rate given by

$$r(t) = \frac{100}{t + 1} + 5 \quad 0 < t \leqslant 20$$

where $r(t)$ is the rate of production (in 1000 barrels per year) and $t$ is the number of years after production begins. How many barrels of oil will the field produce:

(a) in the first $t$ years of its useable life

(b) in its entire useable life?

13. The weekly marginal cost of producing $x$ pairs of tennis shoes is given by

$$C'(x) = 12 + \frac{500}{x + 1} \text{ where } C(x) \text{ is cost in euros.}$$

(a) Given that the fixed costs are €2000 per week, find

(i) the cost function

(ii) the average cost per pair when 1000 pairs are produced each week.

(b) The weekly marginal revenue from the sale of $x$ tennis shoes is given by $R'(x) = 40 - 0.02x + \frac{200}{x + 1}$ where $R(x)$ is revenue in euros. Find the revenue function.

(c) Find the profit function, give the level of production that maximises profit, and the amount of profit.

## 16.2 Area and the definite integral

The function $f(x)$ is continuous and non-negative on an interval $[a, b]$. How do we find the area between the graph of $f(x)$ and the interval $[a, b]$ on the $x$-axis? (Figure 16.2)

We divide the base interval $[a, b]$ into $n$ equal subintervals, and over each subinterval construct a rectangle that extends from the $x$-axis to any point on the curve $y = f(x)$ that is above the subinterval; the particular point does not matter – it can be above the centre, above one endpoint, or any other point in the subinterval. In Figure 16.3 it is above the centre.

For each $n$, the total area of the rectangles can be viewed as an approximation to the exact area in question. As $n$ increases, these approximations will get better and better and will eventually approach the exact area as a limit.

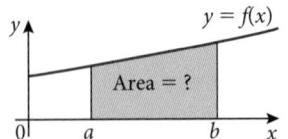

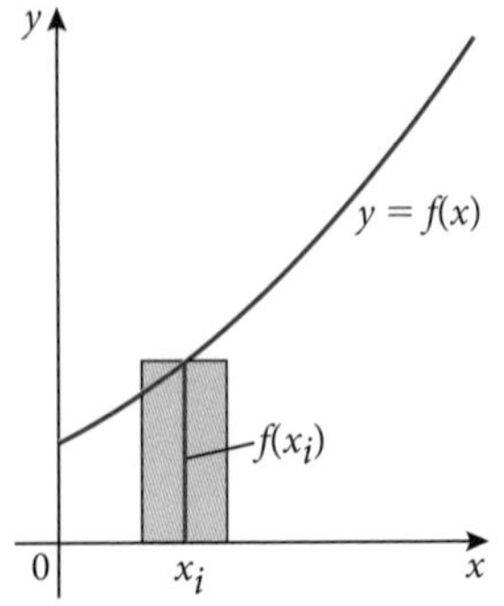

**Figure 16.2** Area under $f(x)$

**Figure 16.3** Area of a rectangular strip

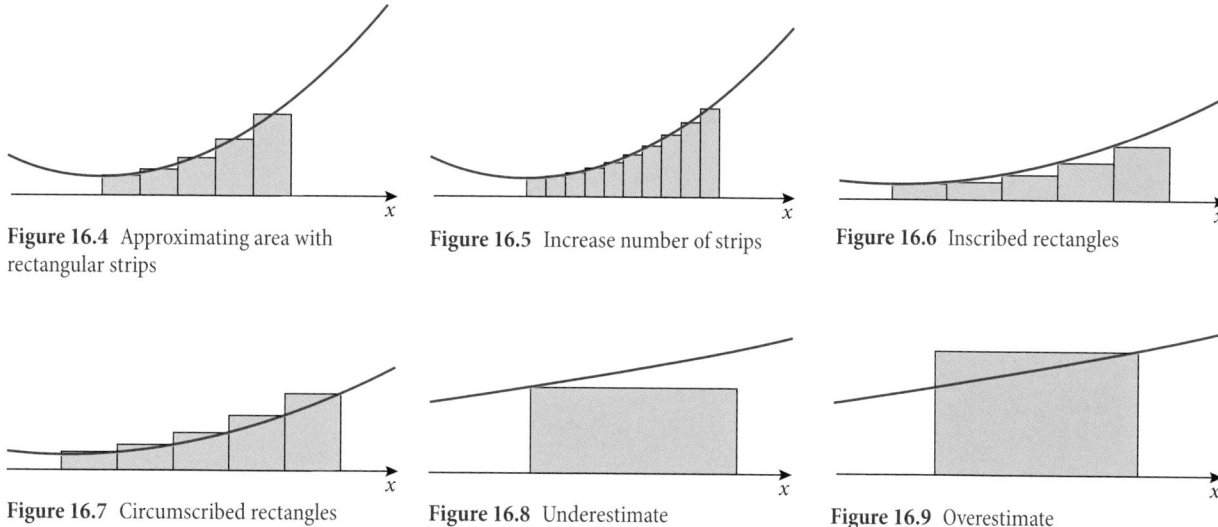

**Figure 16.4** Approximating area with rectangular strips

**Figure 16.5** Increase number of strips

**Figure 16.6** Inscribed rectangles

**Figure 16.7** Circumscribed rectangles

**Figure 16.8** Underestimate

**Figure 16.9** Overestimate

A traditional approach to this is to study how the choice of where to put the rectangular strip does not affect the approximation as the number of intervals increases. You can construct inscribed rectangles which, at the start give you an underestimate of the area (Figures 16.6 and 16.8). On the other hand you can construct circumscribed rectangles that, at the start overestimate the area (Figures 16.7 and 16.9).

As the number of intervals increases, the difference between the over-estimates and the under-estimates will approach 0.

Figures 16.6 and 16.7 show $n$ inscribed and subscribed rectangles and Figure 16.10 shows the difference between the over- and under-estimates.

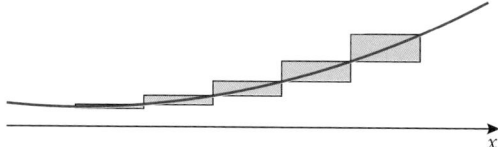

**Figure 16.10** Difference between over- and under-estimates

Figure 16.10 demonstrates that as the number $n$ increases, the difference between the estimates will approach 0. Since the way we set up our rectangles in the first instance is by choosing a point inside the interval, the areas of the rectangles will lie between the over- and under-estimates, and hence, as the difference between the extremes approaches zero, the rectangles we construct will give the area of the region required.

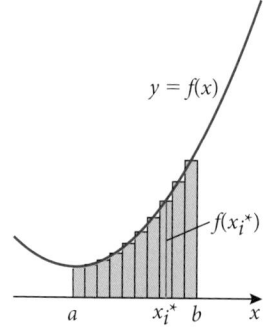

**Figure 16.11** Area of one rectangle of width $=\Delta x$

If we consider the width of each interval to be $\Delta x$, then the area of any rectangle is given as

$$A_i = f(x_i^*)\Delta x$$

The total area of the rectangles so constructed is

$$A_n = \sum_{i=0}^{n} f(x_i^*)\Delta x$$

where $x_i^*$ is an arbitrary point within any subinterval $[x_{i-1}, x_i]$, $x_0 = a$, and $x_n = b$

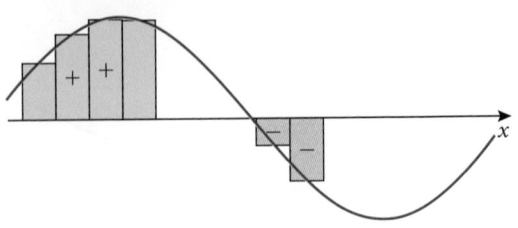

**Figure 16.12** Signed areas

In the case of a function $f(x)$ that has both positive and negative values on $[a, b]$, it is necessary to consider the signs of the areas in the following sense.

On each subinterval, we have a rectangle with width $\Delta x$ and height $f(x^*)$. If $f(x^*) > 0$, then this rectangle is above the $x$-axis; if $f(x^*) < 0$, then this rectangle is below the $x$-axis. We will consider the sum defined above as the sum of the signed areas of these rectangles. That means the total area on the interval is the sum of the areas above the $x$-axis minus the sum of the areas of the rectangles below the $x$-axis.

We now are ready to look at a loose definition of the definite integral:

$f(x)$ is a continuous function defined for $a \leqslant x \leqslant b$, and we divide the interval $[a, b]$ into $n$ subintervals of equal width $\Delta x = \left(\frac{b - a}{n}\right)$. We let $x_0 = a$, and $x_n = b$ and we choose $x_1^*, x_2^*, \ldots, x_n^*$ in these subintervals, so that $x_i^*$ lies in the *i*th subinterval $[x_{i-1}, x_i]$. Then the definite integral of $f(x)$ from $a$ to $b$ is

$$\int_a^b f(x)\,dx = \lim_{n\to\infty} \sum_{i=1}^n f(x_i^*)\,\Delta x$$

In this notation $\int_a^b f(x)\,dx$, in addition to the known integrand and differential, $a$ and $b$ are called the limits of integration: $a$ is the lower limit and $b$ is the upper limit.

Because we have assumed that $f(x)$ is continuous, it can be proved that the limit definition above always exists and gives the same value no matter how we choose the points $x_i^*$. If we take these points at the centre, at two thirds the distance from the lower endpoint or at the upper endpoint, the value is the same. This why we will state the **definition of the integral** from now on as

$$\int_a^b f(x)\,dx = \lim_{n\to\infty} \sum_{i=1}^n f(x_i)\,\Delta x$$

Calling the area under the function an integral is no coincidence.

### Example 16.8

Find the area $A(x)$ between the graph of the function $f(x) = 3$ and the interval $[-1, x]$, and find the derivative $A'(x)$ of this area function.

**Solution**

The area in question is

$$A(x) = 3(x - (-1)) = 3x + 3$$

$$A'(x) = 3 = f(x)$$

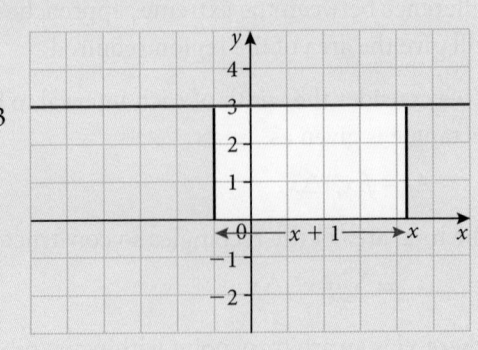

## Example 16.9

Find the area $A(x)$ between the graph of the function $f(x) = 3x + 2$ and the interval $[-\frac{2}{3}, x]$, and find the derivative $A'(x)$ of this area function.

### Solution

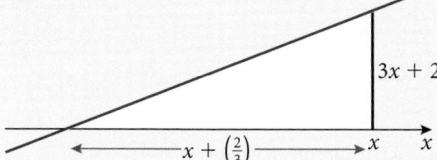

The area in question is

$$A(x) = \frac{1}{2}\left(x + \frac{2}{3}\right)(3x + 2) = \frac{1}{6}(3x + 2)^2$$

since this is the area of a triangle. Hence

$$A'(x) = \frac{1}{6} \times 2(3x + 2) \times 3 = 3x + 2 = f(x)$$

## Example 16.10

Find the area $A(x)$ between the graph of the function $f(x) = x + 2$ and the interval $[-1, x]$, and find the derivative $A'(x)$ of this area function.

### Solution

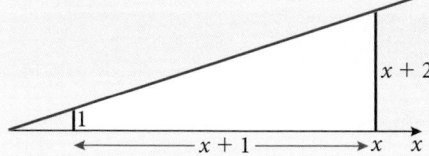

This is a trapezium, so the area is

$$A(x) = \frac{1}{2}(1 + (x + 2))(x + 1) = \frac{1}{2}(x^2 + 4x + 3) \text{ and}$$

$$A'(x) = \frac{1}{2} \times (2x + 4) = x + 2 = f(x)$$

Note that in Examples 16.8 to 16.10, $A'(x) = f(x)$

That is, the derivative of the area function $A(x)$ is the function whose graph forms the upper boundary of the region. It can be shown that this is true, not only for linear functions but for all continuous functions. Thus, to find the area function $A(x)$, we can look instead for a particular function whose derivative is $f(x)$. This is simply the antiderivative of $f(x)$.

So, intuitively as we have seen above, we define the area function as

$A(x) = \int_{a}^{x} f(t)\, \mathrm{d}t$, that is $A'(x) = f(x)$

This is the trigger to the **fundamental theorem of calculus** .

We will now look at some of the properties of the definite integral.

## Basic properties of the definite integral

- $\int_a^b f(x)\,dx = -\int_b^a f(x)\,dx$

  When we defined the definite integral $\int_a^b f(x)\,dx$, we implicitly assumed $[a < b]$. When we reverse $a$ and $b$, $\Delta x$ changes from $\left(\dfrac{b-a}{n}\right)$ to $\left(\dfrac{a-b}{n}\right)$. Therefore the result above follows.

- $\int_a^a f(x)\,dx = 0$

  When $a = b$, then $\Delta x = 0$, and so, the result above follows.

- $\int_a^b c\,dx = c(b-a)$

- $\int_a^b [f(x) \pm g(x)]\,dx = \int_a^b f(x)\,dx \pm \int_a^b g(x)\,dx$

- $\int_a^b c f(x)\,dx = c\int_a^b f(x)\,dx$, where $c$ is any constant.

- $\int_a^b f(x)\,dx = \int_a^c f(x)\,dx + \int_c^b f(x)\,dx$

This property can be demonstrated as follows. The area from $a$ to $b$ is the sum of the two areas, that is $A(x) = A_1 + A_2$ (Figure 16.13). Additionally, even if $c > b$ the relationship holds because the area from $c$ to $b$ in this case will be negative.

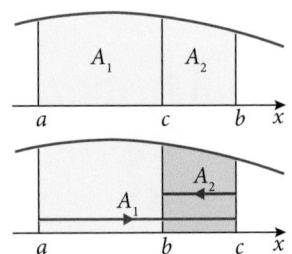

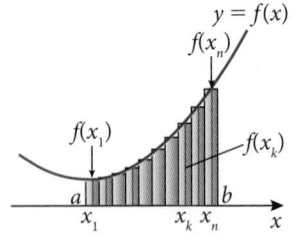

**Figure 16.13** $A(x) = A_1 + A_2$

## Average value of a fuction

From statistics, the average value of a variable is

$$\overline{X} = \frac{\sum_{i=1}^{n} X_i}{n}$$

We can also think of the average value of a function in the same manner. Consider a continuous function $f(x)$ defined over a closed interval $[a, b]$. We partition this interval into $n$ subintervals of equal length in a fashion similar to the previous discussion. Each interval has a length $\Delta x = \dfrac{b-a}{n}$

Since $\Delta x = \dfrac{b-a}{n} \Rightarrow \dfrac{1}{n} = \dfrac{\Delta x}{b-a}$, the average value of $f(x)$ is given by

$$\text{av}(f) = \frac{\sum_{k=1}^{n} f(x_k)}{n} = \frac{1}{n}\sum_{k=1}^{n} f(x_k) = \frac{1}{b-a}\sum_{k=1}^{n} f(x_k)\,\Delta x$$

which leads to the fact on the left about the average value of a function $f(x)$ over an interval $[a, b]$.

**Figure 16.14** Approximate total area under $f(x)$

The **average (mean value)** of an integrable function $f(x)$ over an interval $[a, b]$ is given by

$$\text{av}(f) = \frac{1}{b-a}\int_a^b f(x)\,dx$$

## Max-min inequality

If $f_{max}$ and $f_{min}$ represent the maximum and minimum values of a non-negative continuous differentiable function $f(x)$ over an interval $[a, b]$, then the area under the curve lies between the area of the rectangle with base $[a, b]$ and $f_{min}$ as height and the rectangle with $f_{max}$ as height.

That is $(b - a)f_{min} \leq \int_a^b f(x)\, dx \leq (b - a)f_{max}$

with the assumption that $b > a$, this in turn is equivalent to

$$f_{min} \leq \frac{1}{b - a}\int_a^b f(x)\, dx \leq f_{max}$$

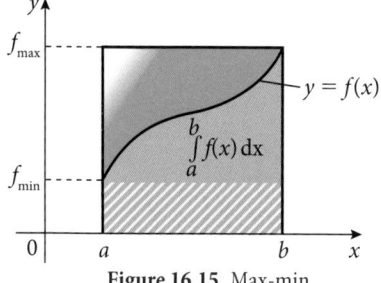

**Figure 16.15** Max-min inequality

Now using the intermediate value theorem, we can ascertain that there is at least one point

$$c \in [a, b] \text{ where } f(c) = \frac{1}{b - a}\int_a^b f(x)\, dx$$

The value $f(c)$ in this theorem is in fact the average value of the function. This allows us to interpret the average value as the height of a rectangle whose base is $[a, b]$ and that has the same area under the function $f(x)$.

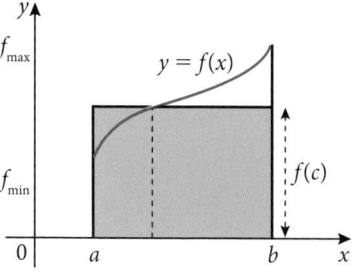

**Figure 16.16** Average value

## The first fundamental theorem of integral calculus

Our understanding of the definite integral as the area under the curve of $f(x)$ helps us establish the basis for the fundamental theorem of integral calculus.

In the definition of the definite integral, let's make the upper limit, a variable, say $x$. Then we will call the area between $a$ and $x$, $A(x)$, i.e.,

$$A(x) = \int_a^x f(t)\, dt$$

From Examples 16.8−16.10, we can claim that

$$\frac{d}{dx}(A(x)) = A'(x) = \frac{d}{dx}\left(\int_a^x f(t)\, dt\right) = f(x)$$

This, very powerful statement is called the **first fundamental theorem of integral calculus**. In essence what it says is that the processes of integration and derivations are inverses of one another.

The discussion leading to the fundamental theorem is for information only. It is not examinable.

It is important to remember that $\int_a^x f(t)\, dt$ is a function of $x$.

### Example 16.11

Find each derivative.

(a) $\dfrac{d}{dx}\displaystyle\int_{-e}^x \frac{1}{\cos^2 t}\, dt$  (b) $\dfrac{d}{dx}\displaystyle\int_0^x \frac{dt}{1 + t^4}$  (c) $\dfrac{d}{dx}\displaystyle\int_x^\pi \frac{1}{1 + t^4}\, dt$

### Solution

(a) This is a direct application of the fundamental theorem:

$$\frac{d}{dx}\int_{-e}^x \frac{1}{\cos^2 t}\, dt = \frac{1}{\cos^2 x}$$

683

(b) This is also straightforward:

$$\frac{d}{dx}\int_0^x \frac{dt}{1+t^4} = \frac{1}{1+x^4}$$

(c) We need to rewrite the expression before we perform the calculation:

$$\frac{d}{dx}\int_x^\pi \frac{1}{1+t^4}\,dt = \frac{d}{dx}\int_\pi^x -\frac{1}{1+t^4}\,dt = -\frac{d}{dx}\int_x^\pi \frac{1}{1+t^4}\,dt = \frac{-1}{1+x^4}$$

## The second fundamental theorem of integral calculus

Recall that $A(x) = \int_a^x f(t)\,dt$

If $F(x)$ is any antiderivative of $f(x)$, then applying what we learned earlier

$F(x) = A(x) + c$ where $c$ is an arbitrary constant.

Now, $\qquad F(b) = A(b) + c = \int_a^b f(t)\,dt + c$

and $\qquad F(a) = A(a) + c = \int_a^a f(t)\,dt + c = 0 + c \Rightarrow F(a) = c$

hence, $\qquad F(b) - F(a) = \int_a^b f(t)\,dt + c - c$

$$= \int_a^b f(t)\,dt$$

The second fundamental theorem of calculus states:

$\int_a^b f(t)\,dt = F(b) - F(a)$

The theorem is also known as the **evaluation theorem**. Also, since we know that $F'(x)$ is the rate of change in $F(x)$ with respect to $x$ and that $F(b) - F(a)$ is the change in $y$ when $x$ changes from $a$ to $b$, we can reformulate the theorem in words to read:

The integral of a rate of change is the **total change**: $\int_a^b F'(x)\,dx = F(b) - F(a)$

Here are some examples where this applies:

- If $V'(t)$ is the rate at which a liquid flows into or out of a container at time $t$, then $\int_{t_1}^{t_2} V'(t)\,dt = V(t_2) - V(t_1)$ is the change in the amount of liquid in the container between time $t_1$ and $t_2$

- If the rate of growth of a population is $n'(t)$, then $\int_{t_1}^{t_2} n'(t)\,dt = n(t_2) - n(t_1)$ is the increase (or decrease) in population during the time period from $t_1$ to $t_2$

**Notation**
If we know that $F(x)$ is an antiderivative of $f(x)$, then we will write
$\int_a^b f(t)\,dt = F(x)\Big|_a^b$
$= F(b) - F(a)$

This theorem has many other applications in calculus and several other fields. It is a very powerful tool to deal with problems involving area, volume, and work. In this book we will apply it to finding areas between functions and volumes of revolution as well as displacement problems.

## Example 16.12

Evaluate

(a) $\displaystyle\int_{-1}^{3} x^5\,dx$    (b) $\displaystyle\int_{0}^{4}\sqrt{x}\,dx$    (c) $\displaystyle\int_{\pi}^{2\pi}\cos\theta\,d\theta$    (d) $\displaystyle\int_{1}^{2}\frac{4+u^2}{u^3}\,du$

### Solution

(a) $\displaystyle\int_{-1}^{3} x^5\,dx = \frac{x^6}{6}\Big|_{-1}^{3} = \frac{3^6}{6} - \frac{1}{6} = \frac{364}{3}$

(b) $\displaystyle\int_{0}^{4}\sqrt{x}\,dx = \frac{2}{3}x^{\frac{3}{2}}\Big|_{0}^{4} = \frac{2}{3}4^{\frac{3}{2}} - 0 = \frac{16}{3}$

(c) $\displaystyle\int_{\pi}^{2\pi}\cos\theta\,d\theta = \sin\theta\Big|_{\pi}^{2\pi} = 0 - 0 = 0$

(d) $\displaystyle\int_{1}^{2}\frac{4+u^2}{u^3}\,du = \int_{1}^{2}\left(\frac{4}{u^3}+\frac{1}{u}\right)du = \left[4\cdot\frac{u^{-2}}{-2}+\ln|u|\right]_{1}^{2} = [-2u^{-2}+\ln u]_{1}^{2}$

$$= (-2\cdot2^{-2}+\ln2) - (-2\cdot1+\ln1) = \frac{3}{2}+\ln2$$

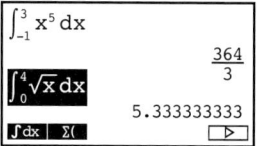

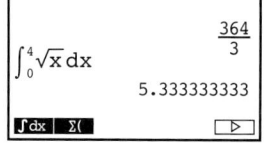

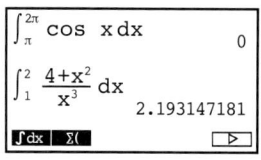

**Figure 16.17** GDC screens for the solutions to Example 16.12

## Using substitution with the definite integral

In section 16.1 we discussed the use of substitution to evaluate integrals in cases that are not easily recognised. We established that

$$\int f(u(x))\cdot u'(x)\,dx = \int f(u)\,du = F(u(x)) + c = F(x) + c$$

When evaluating definite integrals by substitution, two methods are available.

Evaluate the indefinite integral first, revert to the original variable, then use the fundamental theorem. For example, to evaluate

$$\int_{0}^{\frac{\pi}{3}}\frac{\tan^5 x}{\cos^2 x}\,dx$$

we find the indefinite integral

$$\int\frac{\tan^5 x}{\cos^2 x}\,dx = \int u^5\,du = \frac{1}{6}u^6 = \frac{1}{6}\tan^6 x,$$

then we use the fundamental theorem, i.e.,

$$\int_{0}^{\frac{\pi}{3}}\frac{\tan^5 x}{\cos^2 x}\,dx = \frac{1}{6}\tan^6 x\Big|_{0}^{\frac{\pi}{3}} = \frac{1}{6}(\sqrt{3})6 = \frac{27}{6} = \frac{9}{2}$$

Or we can use the following substitution rule for definite integrals

$$\int_{a}^{b} f(u(x))\,u'(x)\,dx = \int_{u(a)}^{u(b)} f(u)\,du$$

Thus, to evaluate

$$\int_{0}^{\frac{\pi}{3}}\frac{\tan^5 x}{\cos^2 x}\,dx,$$

Let $u = \tan x \Rightarrow u\left(\frac{\pi}{3}\right) = \sqrt{3}$, $u(0) = 0$, and so

$$\int_{0}^{\frac{\pi}{3}}\frac{\tan^5 x}{\cos^2 x}\,dx = \int_{0}^{\sqrt{3}} u^5\,du = \frac{1}{6}u^6\Big|_{0}^{\sqrt{3}} = \frac{9}{2}$$

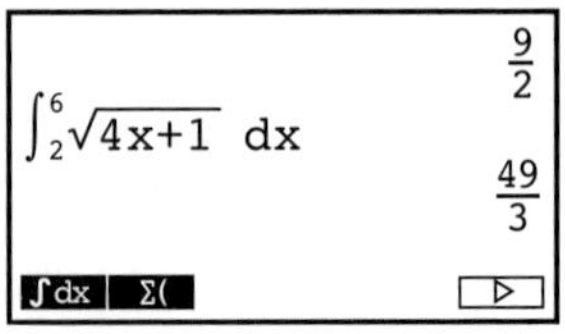

### Example 16.13

Evaluate $\int_2^6 \sqrt{4x+1}\, dx$

$\int_2^6 \sqrt{4x+1}\ dx$     $\frac{9}{2}$

$\frac{49}{3}$

∫dx   Σ(

**Figure 16.18** GDC screen for the solution to Example 16.13

### Solution

Let $u = 4x + 1$, then $du = 4\,dx$. The limits of integration are $u(2) = 9$, and $u(6) = 25$, therefore

$$\int_2^6 \sqrt{4x+1}\, dx = \frac{1}{4}\int_9^{25} \sqrt{u}\, du = \frac{1}{4}\left(\frac{2}{3}u^{\frac{3}{2}}\right)\Big]_9^{25}$$

$$= \frac{1}{6}(125 - 27) = \frac{49}{3}$$

Using this method, we do not return to the original variable of integration. We simply evaluate the new integral between the appropriate values of $u$.

## Numerical integration: the trapezoidal rule

The discussion in Section 16.2 started with finding the area under a curve using inscribed rectangles which, at the start give you an underestimate of the area (Figure 16.6), or circumscribed rectangles that, at the start overestimate the area (Figure 16.7). Then we developed the second fundamental theorem of calculus, which we claimed, will enable us to evaluate the definite integral of continuous functions over specified intervals.

There are, however, functions whose antiderivative are nonelementary. An **elementary function** is one that can be expressed in terms of polynomial, trigonometric, exponential, or logarithmic functions.

For example, the function $f(x) = e^{-x^2}$ has no elementary antiderivative. Consequently, we cannot use the fundamental theorem of calculus to evaluate an integral such as

$$\int_0^1 e^{-x^2}\, dx$$

An approximation of the area can be provided by dividing the region into trapezoids with equal heights of $\Delta x$ and bases the values of the function $y_i$ at each point $x_i$ as shown in Figure 16.19:

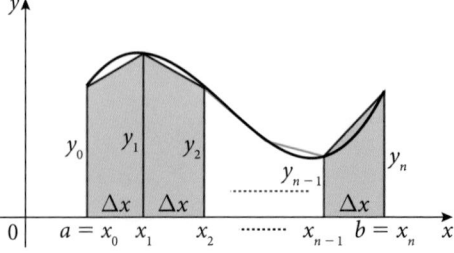

**Figure 16.19** Dividing the region into trapezoids

The area of every trapezoid is of the form:

$$\frac{1}{2} \cdot \text{height} \cdot (\text{base } 1 + \text{base } 2) = \frac{1}{2} \Delta x (y_i + y_{i+1})$$

Thus, the total area is

$$\text{Area} = \frac{\Delta x}{2}(y_0 + y_1) + \frac{\Delta x}{2}(y_1 + y_2) + \frac{\Delta x}{2}(y_2 + y_3) + \cdots + \frac{\Delta x}{2}(y_{n-1} + y_n)$$

$$= \frac{\Delta x}{2}(y_0 + y_1 + y_1 + y_2 + y_2 + y_3 + \cdots + y_{n-1} + y_n)$$

$$= \frac{\Delta x}{2}(y_0 + 2(y_1 + y_2 + y_3 + \cdots + y_{n-1}) + y_n)$$

This last formula gives us the general form for the method of trapezoidal approximation of a definite integral.

To approximate $\int_a^b f(x)\, dx$ by using $n$ trapezoids:

1. Calculate $\Delta x = \dfrac{b-a}{n}$
2. Find the numbers $x_0, x_1, x_2, \ldots, x_n$ starting with $x_0 = a$ and successively adding $\Delta x$, ending with $x_n = b$.
3. Evaluate $f(x)$ at each $x_i$ found in step 2.
4. The approximation for the integral is

$$\int_a^b f(x)\, dx \approx T_n = \frac{\Delta x}{2}(y_0 + 2(y_1 + y_2 + y_3 + \cdots + y_{n-1}) + y_n)$$

$$= \frac{b-a}{2n}(y_0 + 2(y_1 + y_2 + y_3 + \cdots + y_{n-1}) + y_n)$$

## Example 16.14

Calculate the approximation to the integral $\int_0^1 e^{-x^2}\, dx$ by using five intervals.

### Solution

First find $\Delta x = \dfrac{1-0}{5} = 0.2$, then the endpoints of intervals:

$$x_0 = 0,\ x_1 = 0.2,\ x_2 = 0.4,\ x_3 = 0.6,\ x_4 = 0.8,\ x_5 = 1$$

Then we evaluate the function at each point:

$$y_0 = e^{-0^2},\ y_1 = e^{-0.2^2},\ \ldots,\ y_5 = e^{-1^2}$$

and substitute the values into the formula above.

To check your answer, evaluate the definite integral, shown here in the GDC screen.

The error is 0.0024558, which is 0.33%

Increasing the number of intervals will improve the accuracy.

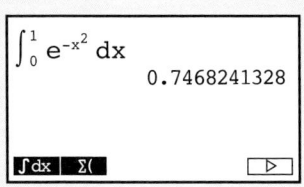

Reminder: You need to learn how to use your GDC!

## Example 16.15

Instruments onboard a ship travelling on a path heading due north showed the following readings for the velocity in $km\,h^{-1}$ over an interval of 10 hours.

| $T$ | 0 | 1 | 2 | 3 | 4 | 5 | 6 | 7 | 8 | 9 | 10 |
|----|----|----|----|----|----|----|----|----|----|----|----|
| $V$ | 12 | 14 | 17 | 21 | 22 | 21 | 15 | 11 | 11 | 14 | 17 |

Give an estimate of the distance travelled.

### Solution

In Chapter 13, we mentioned that velocity is the derivative of displacement. Thus, displacement is an antiderivative of velocity. So, we want to calculate

$$s(t) = \int_0^{10} v(t)\,dt$$

Using the trapezoidal approximation with $n = 10$ and $\Delta x = \dfrac{10 - 0}{10} = 1$ we obtain

$$s(t) = \int_0^{10} v(t)\,dt \approx T_{10} = \frac{\Delta x}{2}(y_0 + 2(y_1 + y_2 + y_3 + \cdots + y_{n-1}) + y_n)$$

$$= \frac{1}{2}(12 + 2(14 + 17 + 21 + 22 + 21 + 15 + 11 + 11 + 14) + 17)$$

$$= 160.5\,km$$

## Continuous money flow

Suppose an investment brings in an income of €8000 per year. How much income can be obtained in 5 years? Obviously, the answer is €40,000. This is an income flow.

However, not every income flow is that simple. For example, you buy a share in a salt mine projected to give you income at a rate of $€1000e^{-0.1t}$ per year where $t$ is measured in years. How much income can be obtained in 5 years? This is an income from a continuous income flow. The $€1000e^{-0.1t}$ is a rate of change, and thus the total over 5 years will be $\int_0^5 1000e^{-0.1t}\,dt$

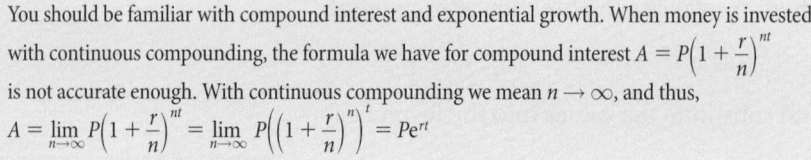

You should be familiar with compound interest and exponential growth. When money is invested with continuous compounding, the formula we have for compound interest $A = P\left(1 + \dfrac{r}{n}\right)^{nt}$ is not accurate enough. With continuous compounding we mean $n \to \infty$, and thus,

$$A = \lim_{n\to\infty} P\left(1 + \frac{r}{n}\right)^{nt} = \lim_{n\to\infty} P\left(\left(1 + \frac{r}{n}\right)^n\right)^t = Pe^{rt}$$

When we say money is being compounded continuously, we are not suggesting that the financial institution is actually continuously placing money in our account. Rather, we calculate what is in our account as if this were in fact happening. Then we can think of the account as a **continuous flow of money**. Think about it as an electricity company continuously totalling the amount of electricity that they are selling you, etc.

We then assume that we have a continuous flow of money and that the positive function $f(t)$ represents the rate of change of this flow. If we let $A(t)$ be the total amount of income obtained from this flow, then $A'(t) = f(t)$. Also, since the amount of money from this flow is zero when $t = 0$, then $A(0) = 0$.

Thus $A(t) = \int_0^t f(x)\,dx$

## Present value and future value of a continuous income flow

Let $f(t)$ be the continuous rate of change of total income on an interval $[0, T]$. If we are able to invest an amount of money now at a continuous interest rate $r$, over the same interval $[0, T]$, and generate the same income flow then this amount is called the **present value $P(T)$** of this income flow is given by

$$P(T) = \int_0^T f(t)e^{-rt}\,dt$$

That is, the capital value of an asset that produces a continuous flow of income is the sum of the present value of all future earnings from the asset. The present value is also called the capital value.

The **future value** of a continuous flow is given by

$$FV(T) = \int_0^T f(t)e^{r(T-t)}\,dt = e^{rT}\int_0^T f(t)e^{-rt}\,dt$$

That is, the future value of a continuous income stream is the total value of all money produced by the continuous income stream at the end of $T$ years including interest earned.

You need to observe the differences in terms:

**Total income:** $\int_a^b f(t)dt$

over an interval $[a, b]$,

or $\int_0^T f(t)dt$, over the life of the income flow.

**Future value:**

$$FV(T) = \int_0^T f(t)e^{r(T-t)}\,dt$$

$$= e^{rT}\int_0^T f(t)e^{-rt}\,dt$$

## Example 16.16

The rate of change of income, in thousands of dollars per year from a silver mine is projected to be $f(t) = 100e^{-0.04t}$, where $t$ is in years.

(a) Find the total amount of money expected from this mine over its lifetime of 10 years.

(b) What is the future value of this income flow given that the current interest rate is 8%?

(c) What is the amount of interest received?

(d) The mine is being offered for sale now at a price of $600,000. Would you buy it? Explain your answer.

### Solution

(a) We need the total flow of money from the mine:
$$A = \int_0^{10} 100e^{-0.04t}\,dt = 2500(1 - e^{-0.4}) \approx \$824,000$$

(b) $FV(T) = e^{rT}\int_0^T f(t)e^{-rt}\,dt = e^{0.08\times10}\int_0^{10} 100e^{-0.04t} \times e^{-0.08t}\,dt$

$$= e^{0.8}\int_0^{10} 100e^{-0.12t}\,dt$$

$$\approx \$1,296,000$$

You can also make the comparison by finding the future value of $600,000 with continuous compounding, i.e., $600e^{0.08 \times 10} \approx \$1,335,000$ which is more than the future value of the income from the mine. Had the interest rate been 6% instead, the present value would have been $632,000 which means that the income flow has more value than the $600,000 required for its purchase.

(c) Interest $= \$1,296,000 - \$824,000 = \$472,000$

(d) In order to compare the value of the flow of income to the investment of $600,000, we calculate the present value of the income flow:

$$P(T) = \int_0^T f(t)e^{-rt}\,dt = \int_0^{10} 100e^{-0.04t}e^{-0.08t}\,dt = \frac{2500}{3}(1 - e^{-0.12}) \approx \$582,000$$

This is not a good investment since you can invest the $600,000 and generate more income at current interest rates.

## Exercise 16.2

**1.** Evaluate each integral.

(a) $\displaystyle\int_{-2}^{1} (3x^2 - 4x^3)\,dx$

(b) $\displaystyle\int_{2}^{7} 8\,dx$

(c) $\displaystyle\int_{1}^{5} \frac{2}{t^3}\,dt$

(d) $\displaystyle\int_{2}^{2} (\cos t - \tan t)\,dt$

(e) $\displaystyle\int_{1}^{7} \frac{2x^2 - 3x + 5}{\sqrt{x}}\,dx$

(f) $\displaystyle\int_{0}^{\pi} \cos\theta\,d\theta$

(g) $\displaystyle\int_{0}^{\pi} \sin\theta\,d\theta$

(h) $\displaystyle\int_{3}^{1} (5x^4 + 3x^2)\,dx$

(i) $\displaystyle\int_{1}^{3} \frac{u^5 + 2}{u^2}\,du$

(j) $\displaystyle\int_{1}^{e} \frac{2\,dx}{x}$

(k) $\displaystyle\int_{1}^{3} \frac{2x}{x^2 + 2}\,dx$

(l) $\displaystyle\int_{1}^{3} (2 - \sqrt{x})^2\,dx$

(m) $\displaystyle\int_{0}^{\frac{\pi}{4}} \frac{3}{\cos^2\theta}\,d\theta$

(n) $\displaystyle\int_{0}^{1} (8x^7 + \sqrt{\pi})\,dx$

(o) $\displaystyle\int_{0}^{2} |3x|\,dx$

(p) $\displaystyle\int_{-2}^{0} |3x|\,dx$

(q) $\displaystyle\int_{-2}^{2} |3x|\,dx$

(r) $\displaystyle\int_{0}^{\frac{\pi}{2}} \sin 2x\,dx$

(s) $\displaystyle\int_{1}^{9} \frac{1}{\sqrt{x}}\,dx$

(t) $\displaystyle\int_{-2}^{2} (e^x - e^{-x})\,dx$

**2.** Evaluate:

(a) $\displaystyle\int_{0}^{4} \frac{x^3\,dx}{\sqrt{x^2 + 1}}$

(b) $\displaystyle\int_{1}^{\sqrt{e}} \frac{\sin(\pi \ln x)}{x}\,dx$

(c) $\displaystyle\int_{e}^{e^2} \frac{dt}{t \ln t}$

(d) $\displaystyle\int_{-1}^{2} 3x\sqrt{9 - x^2}\,dx$

(e) $\displaystyle\int_{-\frac{\pi}{3}}^{\frac{2\pi}{3}} \frac{\sin x}{\sqrt{3 + \cos x}}\,dx$

(f) $\displaystyle\int_{e}^{e^2} \frac{\ln x}{x}\,dx$

(g) $\displaystyle\int_{-\ln 2}^{\ln 2} \frac{e^{2x}}{e^{2x} + 9}\,dx$

(h) $\displaystyle\int_{0}^{\frac{\pi}{4}} \frac{\sqrt{\tan x}}{\cos^2 x}\,dx$

(i) $\displaystyle\int_{0}^{\sqrt{\pi}} 7x \cos x^2\,dx$

(j) $\displaystyle\int_{\pi^2}^{4\pi^2} \frac{\sin\sqrt{x}}{\sqrt{x}}\,dx$

**(k)** $\displaystyle\int_0^{\frac{\pi}{6}}(1-\sin 3t)\cos 3t\,dt$      **(l)** $\displaystyle\int_0^{\frac{\pi}{4}}e^{\sin 2\theta}\cos 2\theta\,d\theta$

**(m)** $\displaystyle\int_0^{\frac{\pi}{8}}(3+e^{\tan 2t})\frac{1}{\cos^2 2t}\,dt$      **(n)** $\displaystyle\int_0^{\sqrt{\ln\pi}}4t\,e^{t^2}\sin(e^{t^2})\,dt$

**3.** Find the average value of the given function over the given interval.

   **(a)** $x^4,[1,2]$                           **(b)** $\cos x,\left[0,\dfrac{\pi}{2}\right]$

   **(c)** $\dfrac{1}{\cos^2 x},\left[\dfrac{\pi}{6},\dfrac{\pi}{4}\right]$             **(d)** $e^{-2x},[0,4]$

**4.** Find the indicated derivative.

   **(a)** $\dfrac{d}{dx}\displaystyle\int_2^x\dfrac{\sin t}{t}\,dt$            **(b)** $\dfrac{d}{dt}\displaystyle\int_t^3\dfrac{\sin x}{x}\,dx$

   **(c)** $\dfrac{d}{dt}\displaystyle\int_{-\pi}^t\dfrac{\cos y}{1+y^2}\,dy$

**5. (a)** Find $\displaystyle\int_0^k\dfrac{dx}{3x+2}$, giving your answer in terms of $k$.

   **(b)** Given that $\displaystyle\int_0^k\dfrac{dx}{3x+2}=1$, calculate the value of $k$.

**6.** Given that $p,q\in\mathbb{N}$, show that:

$$\int_0^1 x^p(1-x)^q\,dx=\int_0^1 x^q(1-x)^p\,dx$$

   Do not attempt to evaluate the integrals.

**7.** Given that $k\in\mathbb{N}$, evaluate each integral.

   **(a)** $\displaystyle\int x(1-x)^k\,dx$           **(b)** $\displaystyle\int_0^1 x(1-x)^k\,dx$

**8.** Let $F(x)=\displaystyle\int_3^x\sqrt{5t^2+2}\,dt$. Find:

   **(a)** $F(3)$          **(b)** $F'(3)$               **(c)** $F''(3)$

**9.** Show that the function $f(x)=\displaystyle\int_x^{3x}\dfrac{dt}{t}$ is constant over the set of positive real numbers.

**10.** Although it is clear that human intelligence cannot be measured by a single number, IQ tests are still widely used. The distribution of IQs of a certain country follows a normal distribution with a mean IQ score of 100 and a standard deviation of 15. The normal model used to calculate the proportion of citizens with IQs between $a$ and $b$ is given by the integral:

$$\int_{\left(\frac{a-100}{15}\right)}^{\left(\frac{b-100}{15}\right)}\dfrac{e^{-\frac{x^2}{2}}}{\sqrt{2\pi}}\,dx$$

   Using $n\geqslant 4$, estimate the proportion of citizens who have IQs between:

   **(a)** 115 and 145              **(b)** 100 and 130

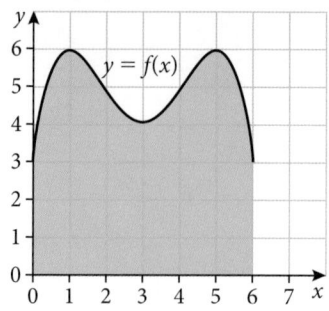

**Figure 16.21** Diagram for question 12

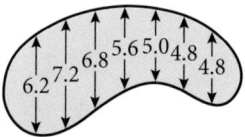

**Figure 16.22** Diagram for question 13

11. The cable of a suspension bridge hangs in a parabolic curve.

 The length of the cable is given by $\int_{-400}^{400}\sqrt{1 + \left(\frac{x}{1000}\right)^2}\, dx$

 Approximate the length of this cable. Use successively higher values for $n$ until the answers agree to the nearest whole number.

12. Estimate the area under the graph of function $f(x)$ in Figure 16.21.

13. Figure 16.22 shows a sketch of a small pond. The width was measured at 2-metre intervals. Estimate the surface area of this pond.

14. During training sessions for skiers in the Alps region, a radar gun is used to record the speed of a skier during the first 5 seconds of a downhill race. Estimate the distance this skier covered during those 5 seconds.

| $t$ (s) | $v$(m s$^{-1}$) | $t$ (s) | $v$(m s$^{-1}$) |
|---|---|---|---|
| 0 | 0 | 3.0 | 10.51 |
| 0.5 | 4.67 | 3.5 | 10.67 |
| 1.0 | 7.34 | 4.0 | 10.76 |
| 1.5 | 8.86 | 4.5 | 10.81 |
| 2.0 | 9.73 | 5.0 | 10.81 |
| 2.5 | 10.22 | | |

15. An oil well produces income at a constant rate $r(t) =$ US$240,000 per year for a lifetime $T = 10$ years. Consider the continuous interest rate $i = 0.06$

 **(a)** Write down the formula for the capital value of the asset for its lifetime $T$.

 **(b)** Use your formula to find the capital value of this specific oil well.

16. An oil well with a rate of change of income of $100e^{-0.1t}$, in thousands of dollars, and a useful life of 5 years, is being offered for sale at $310,000. Would you buy it when the current continuous compound interest rate is 10%?

17. In degrees Celsius, the average temperature for each of the months of the year in New Delhi, India is shown in the table.

| Jan | Feb | Mar | Apr | May | Jun |
|---|---|---|---|---|---|
| 15 | 20 | 25 | 31 | 32.5 | 32 |

| Jul | Aug | Sep | Oct | Nov | Dec |
|---|---|---|---|---|---|
| 30 | 30 | 28 | 23 | 18 | 15 |

 Find an approximate value for the mean temperature during the year in the city. Discuss your model for estimation.

**18.** A large corporation based in London wishes to set up a special endowment fund at a nearby university into which money will flow continuously at a rate of £20,000 per year for the coming 10 years. In order to achieve this, the corporation manages to invest the initial fund with continuous interest rate of 8.5% per year. What should be the size of the initial fund?

**19.** A vending machine is installed at a large school with the expectation that the income flow from this machine will be at the rate $f(t) = 5000e^{0.04t}$ where $t$ is time in years since the installation of the machine and money is in pounds sterling.

    **(a)** Find the total income produced by the machine during the first 5 years of operation.

    **(b)** Find the future value of this income stream when it is invested at 12%, compounded continuously for 5 years, and find the total interest earned.

## 16.3 Areas

We have seen how the area between a curve defined by $y = f(x)$ and the $x$-axis can be computed by the integral $\int_a^b f(x)\, dx$ on an interval $[a, b]$ where $f(x) \geqslant 0$.

In this section, we shall use integration to find the area of more general regions between curves.

### Areas between curves of functions of the form $y = f(x)$ and the $x$-axis

If the function $y = f(x)$ is always above the $x$-axis, finding the area is a straightforward computation of the integral $\int_a^b f(x)\, dx$

## Example 16.17

Find the area under the curve $f(x) = x^3 - x + 1$ and the $x$-axis over the interval $[-1, 2]$.

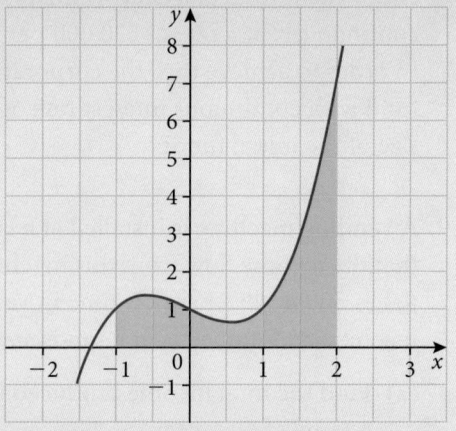

### Solution

This area is:

$$\int_{-1}^{2} (x^3 - x + 1)\, dx = \left[ \frac{x^4}{4} - \frac{x^2}{2} + x \right]_{-1}^{2}$$

$$= (4 - 2 + 2) - \left( \frac{1}{4} - \frac{1}{2} - 1 \right) = 5\frac{1}{4}$$

Using a GDC, this is done for example by simply choosing the MATH menu, then the $\int dx$ menu item and then typing in the function with the integration limits.

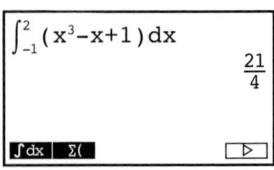

**Figure 16.23** GDC screen for the solution to Example 16.17

In some cases, you will have to adjust how to work. This is the case when the graph intersects the $x$-axis. Since we are interested in the area bounded by the curve and the interval $[a, b]$ on the $x$-axis, we do not want the two areas to cancel each other. This is why we have to split the process into different subintervals where we take the absolute values of the areas found and add them.

## Example 16.18

Find the area under the curve $f(x) = x^3 - x - 1$ and the $x$-axis over the interval $[-1, 2]$.

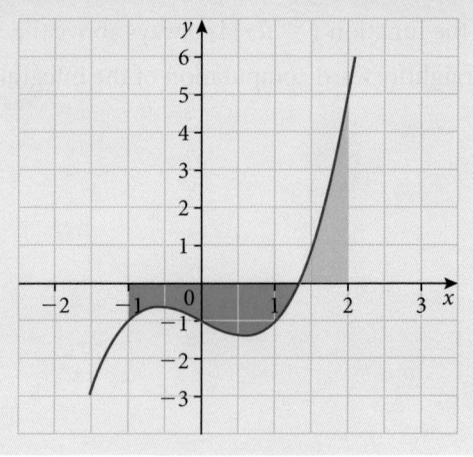

## Solution

As you see from the diagram, part of the graph is below the $x$-axis, and its area will be negative. If you try to integrate this function without paying attention to the intersection with the $x$-axis, here is what you get:

$$\int_{-1}^{2} (x^3 - x - 1)\, dx = \left[\frac{x^4}{4} - \frac{x^2}{2} - x\right]_{-1}^{2}$$

$$= (4 - 2 - 2) - \left(\frac{1}{4} - \frac{1}{2} + 1\right) = -\frac{3}{4}$$

This integration has to be split before we start. However, this is a function where you cannot find the intersection point. So, we either use a GDC to find the intersection or we just take the absolute values of the different parts of the region. This is done by integrating the absolute value of the function:

$$\text{Area} = \int_{a}^{b} \left|f(x)\right|\, dx$$

hence,

$$\text{Area} = \int_{-1}^{2} \left|(x^3 - x - 1)\right|\, dx$$

This is not easy to find, given the difficulty with the $x$-intercept. It is best to use a GDC.

$$\int_{-1}^{2} |x^3 - x - 1|\, dx$$
$$3.614515769$$

**Figure 16.24** GDC screen for the solution to Example 16.18

## Example 16.19

Find the area enclosed by the graph of the function $f(x) = x^3 - 4x^2 + x + 6$ and the $x$-axis.

## Solution

This function intersects the $x$-axis at three points where $x = -1, 2,$ and $3$. To find the area, we split it into two and then add the absolute values:

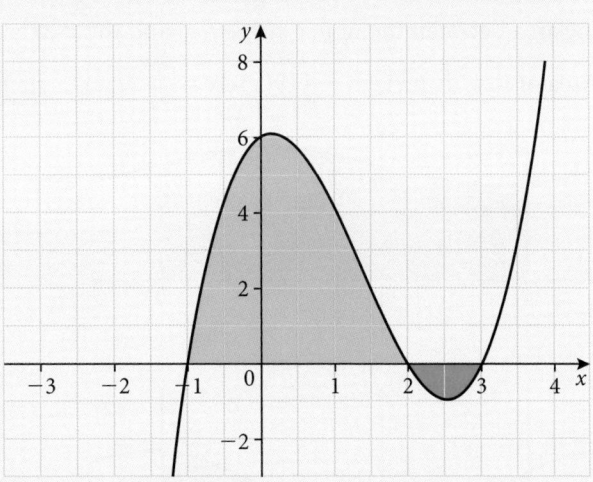

$$\text{Area} = \int_{-1}^{3} \left|f(x)\right|\, dx = \int_{-1}^{2} f(x)\, dx + \int_{2}^{3} (-f(x))\, dx$$

$$= \int_{-1}^{2} x^3 - 4x^2 + x + 6\, dx + \int_{2}^{3} -x^3 + 4x^2 - x - 6\, dx$$

$$= \left[\frac{x^4}{4} - \frac{4x^3}{3} + \frac{x^2}{2} + 6x\right]_{-1}^{2} + \left[-\frac{x^4}{4} + \frac{4x^3}{3} - \frac{x^2}{2} - 6x\right]_{2}^{3}$$

$$= \frac{45}{4} + \frac{7}{12} = \frac{71}{6}$$

695

Here are two possible GDC outputs for the solution to Example 16.19.

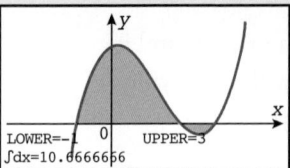 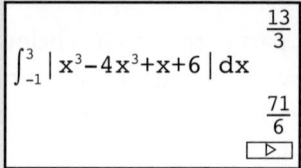 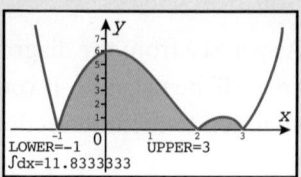

Notice how the first screen did not give the correct answer. This is because the function did not specify the absolute value, and the area below the $x$-axis was subtracted from the area above it.

## Area between curves

In some practical problems, you may have to compute the area between two curves. Suppose $f(x)$ and $g(x)$ are functions such that $f(x) \geqslant g(x)$ on the interval $[a, b]$ (Figure 16.25). Note that we do not insist that both functions are non-negative.

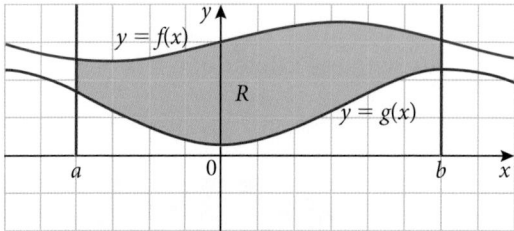

**Figure 16.25** Area between two curves

To find the area of the region $R$ between the curves from $x = a$ to $x = b$, we subtract the area between the lower curve $g(x)$ and the $x$-axis from the area between the upper curve $f(x)$ and the $x$-axis; that is

$$\text{Area of } R = \int_a^b f(x)\, dx - \int_a^b g(x)\, dx = \int_a^b [f(x) - g(x)]\, dx$$

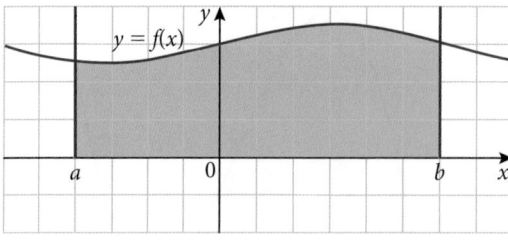

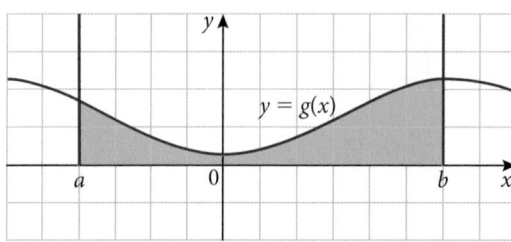

**Figure 16.26** Areas under functions $f$ and $g$

This fact applies to all functions, not only positive functions.
These facts are used to define the area between curves.

If $f(x)$ and $g(x)$ are functions such that $f(x) \geqslant g(x)$ on the interval $[a, b]$, then the area between the two curves is given by

$$A = \int_a^b [f(x) - g(x)]\, dx$$

## Example 16.20

Find the area of the region between the curves $y = x^3$ and $y = x^2 - x$ on the interval $[0, 1]$.

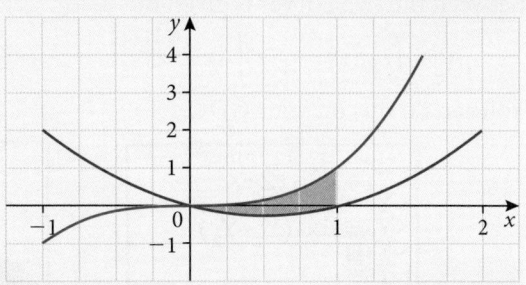

### Solution

$y = x^3$ appears to be higher than $y = x^2 - x$ with one intersection at $x = 0$
Thus, the required area is

$$A = \int_0^1 [x^3 - (x^2 - x)]\,dx = \left[\frac{x^4}{4} - \frac{x^3}{3} + \frac{x^2}{2}\right]_0^1 = \frac{5}{12}$$

In some cases you must be very careful of how you calculate the area. This is the case where the two functions intersect at more than one point.

## Example 16.21

Find the area of the region bounded by the curves
$y = x^3 + 2x^2$ and
$y = x^2 + 2x$

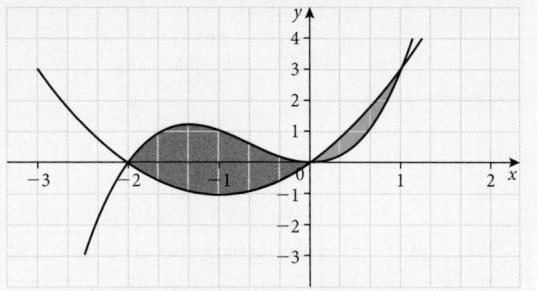

### Solution

The two curves intersect when

$x^3 + 2x^2 = x^2 + 2x \Rightarrow x^3 + x^2 - 2x = 0 \Rightarrow x(x + 2)(x - 1) = 0$
This is when $x = -2, 0$, or $1$

The area is equal to

$$A = \int_{-2}^0 [x^3 + 2x^2 - (x^2 + 2x)]\,dx + \int_0^1 [x^2 + 2x - x^3 + 2x^2]\,dx$$

$$= \int_{-2}^0 [x^3 + x^2 - 2x]\,dx + \int_0^1 [-x^2 + 2x - x^3]\,dx$$

$$= \left[\frac{x^4}{4} + \frac{x^3}{3} - x^2\right]_{-2}^0 + \left[-\frac{x^4}{4} - \frac{x^3}{3} + x^2\right]_0^1$$

$$= 0 - \left[\frac{16}{4} - \frac{8}{3} - 4\right] + \left[-\frac{1}{4} - \frac{1}{3} + 1\right] - 0 = \frac{37}{12}$$

This discussion leads us to stating the general expression you should use in evaluating areas between curves.

If $f(x)$ and $g(x)$ are functions continuous on the interval $[a, b]$, then the area between the two curves is given by

$$A = \int_a^b \left| f(x) - g(x) \right| dx$$

You can do this on your GDC.

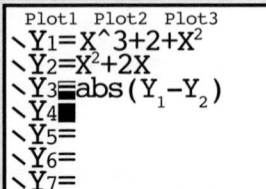 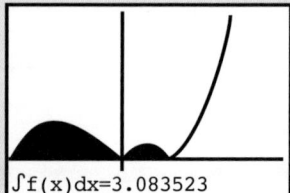

## Areas along the $y$-axis (optional)

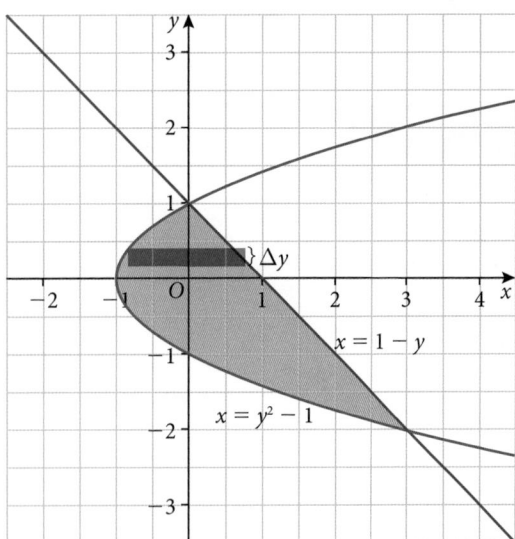

**Figure 16.27** Areas along the $y$-axis

To find the area enclosed by $y = 1 - x$ and $y^2 = x + 1$, it is best to treat the region between them by regarding $x$ as a function of $y$ (Figure 16.27).

The area of the shaded region can be calculated using the integral:

$$A(y) = \int_{-2}^1 \left| (1 - y) - (y^2 - 1) \right| dy$$

$$= \int_{-2}^1 \left| 2 - y - y^2 \right| dy = \left[ 2y - \frac{y^2}{2} - \frac{y^3}{3} \right]_{-2}^1 = \frac{9}{2}$$

If we use $y$ as a function of $x$, then the calculation involves calculating the area by dividing the interval into two: $[-1, 0]$ and $[0, 3]$.

In the first part, the area is enclosed between $y = \sqrt{x + 1}$ and $y = -\sqrt{x + 1}$, and the area in the second part is enclosed by $y = 1 - x$ and $y = -\sqrt{x + 1}$

$$2\int_{-1}^0 \sqrt{x + 1}\ dx + \int_0^3 \left( (1 - x) - (-\sqrt{x + 1}) \right) dx$$

## Some applications of area in economics

### Demand and supply: consumers' and producers' surpluses

Consumers' surplus measures the benefit that consumers derive from an economy in which competition keeps prices relatively low.

The price-demand curve $d(x)$ for a product gives the price at which $x$ units will be sold. Producers' surplus measures the total benefit that producers derive from being able to sell at the market price.

Obviously, as the price of an item rises, so will the quantity that producers are willing to supply at that price. The relationship between the price of an item and the quantity that producers are willing to supply at that price is called the price-supply curve.

Where the supply and demand curves intersect is the market equilibrium.

From Figure 16.28:

consumers' surplus $= \int_0^A (d(x) - B) \, dx$, and

producers' surplus $= \int_0^A (B - s(x)) \, dx$

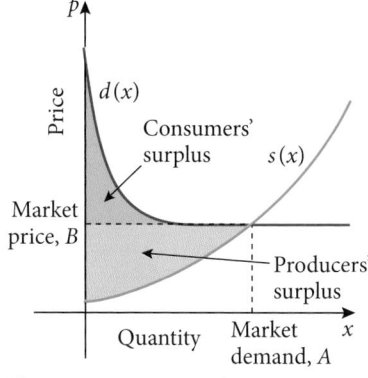

**Figure 16.28** Price-supply curve

The **demand** function can be viewed from two perspectives.
- The demand is usually defined as a schedule of quantities that buyers are willing and able to purchase at a schedule of prices in a given time interval. That is, the quantity, $Q$, is a function of price: $Q = f(P)$ This is known in economics as **ceteris paribus**, given that all other factors are kept constant.
- Demand can also be perceived as the maximum price buyers are willing and able to pay for each unit of output, **ceteris paribus**. That is, price, $P$, is a function of quantity: $P = f(Q)$. This is what we will call price-demand curve.

Note that these two functions are inverses of each other and hence they give you the same information. With this in mind, you can imagine getting one graph from the other can simply be achieved by flipping the horizontal and vertical axes, and this is exactly what economists do.

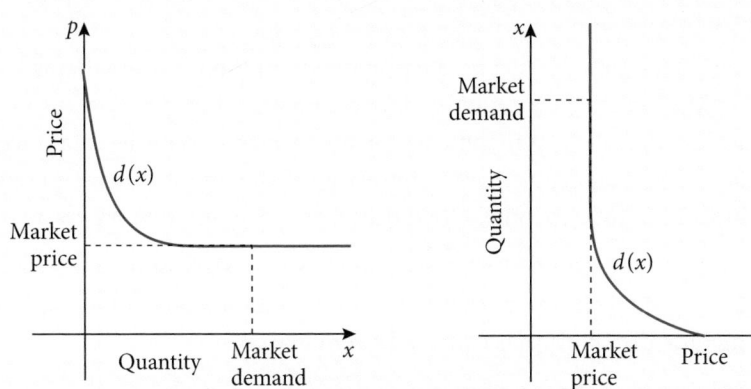

Occasionally, rate of change of demand is called 'marginal price at demand' and rate of change of supply is called 'marginal price at supply'.

## Example 16.22

The market research department of a supermarket chain has determined that the marginal price $d'(x)$ at $x$ thousand tubes per month for a certain brand of toothpaste is given by $d'(x) = -0.015e^{-0.01x}$

(a) Find the price-demand equation if we know that the monthly demand is 50 thousand tubes when the price of the tube is €4.35

(b) Find the monthly demand when the price of a tube is set at €3.89

(c) The marginal price $s'(x)$ at a supply level of $x$ thousand tubes per month for this brand is given by $s'(x) = 0.001e^{0.01x}$

Find the price-supply equation given that suppliers are willing to supply 100 thousand tubes per month for a price of €3.65 each.

(d) Under these conditions, what would the market equilibrium be?

### Solution

(a) $d(x) = \int -0.015e^{-0.01x}\,dx = 1.5e^{-0.01x} + c$, with $d(50) = €4.35$

$1.5e^{-0.01 \times 50} + c = 4.35 \Rightarrow c = €3.44$, so, $d(x) = 1.5e^{-0.01x} + 3.44$

(b) To find the demand when the price is €3.89, we solve $d(x) = €3.89$ for $x$:

$d(x) = 1.5e^{-0.01x} + 3.44 = 3.89 \Rightarrow 1.5e^{-0.01x} = 0.45 \Rightarrow x \approx 120\,000$ tubes.

(c) $s(x) = \int 0.001e^{0.01x}\,dx = 0.1e^{0.01x} + c$ with $s(100) = €3.98$

$s(100) = 0.1e^{0.01 \times 100} + c = €3.98 \Rightarrow c = 3.38$, so $s(x) = 0.1e^{0.01x} + 3.38$

(d) Market equilibrium is achieved when demand = supply

$1.5e^{-0.01x} + 3.44 = 0.1e^{0.01x} + 3.38 \Rightarrow x \approx 143\,000$ tubes per month at a price of $d(143) = 1.5e^{-0.01(143)} + 3.44 \approx €3.80$

### Lorentz curves

For a given population, let $f(x)$ be the proportion of total income that is received by the lowest-paid $x\%$ of income recipients. Thus, $f(0.2) = 0.15$ means that the lowest-paid 20% of income recipients receive 15% of the total income, or if $f(0.5) = 0.4$, this means that the lowest-paid 50% of income recipients receive 40% of the total income. Obviously, $0 \leqslant x \leqslant 1$ and $0 \leqslant f(x) \leqslant 1$.

$f(0) = 0$ and $f(1) = 1$ and $f(x) \leqslant x$, since the lowest $100x\%$ of the income recipients cannot receive more than 100% of the total income. The graph of such an income distribution is called a **Lorentz curve**.

Suppose a Lorentz curve for a certain country is given by $f(x) = x^3$
Since $f(0.5) = 0.125$, the lowest-paid 50% receive 12.5% of the total income.

A perfect equality of income distribution is represented by the curve $f(x) = x$. In this case the lowest-paid 5% would receive 5% of total income, and lowest-paid 10% would receive 10%, and so on. Deviation from this perfect equality represents an inequality of income distribution (Figure 16.29).

The **coefficient of inequality** is the ratio of the area between the graph of $f(x)$ and $y = x$ to the area under $y = x$ on $[0, 1]$.

This means

$$L = \frac{\int_0^1 (x - f(x))\,dx}{\int_0^1 x\,dx} = \frac{\int_0^1 (x - f(x))\,dx}{\frac{1}{2}} = 2\int_0^1 (x - f(x))\,dx$$

This is also known as the **Gini index**.

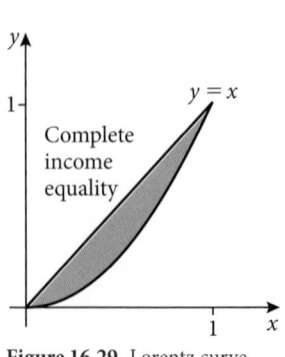

**Figure 16.29** Lorentz curve

Note that with perfect equality, $f(x) = x$, and thus $L = 0$ and in complete inequality $L = 1$, which means: the higher the index, the more inequality there is.

## Some applications in probability

You have seen in Chapter 15 that when a random variable $X$ is discrete, you assign a positive probability to each value that $X$ can take and get the probability distribution for $X$. The sum of all the probabilities associated with the different values of $X$ is 1.

We represent the probabilities corresponding to the different values of the random variable $X$ with a probability histogram (relative frequency histogram), where the area of each bar corresponds to the probability of the specific value it represents.

Consider now a continuous random variable $X$, such as height, mass, and length of life of a particular product – a TV set for example. Because it is continuous, the possible values of $X$ are over an interval. Moreover, there are an infinite number of possible values of $X$. Hence, we cannot find a probability distribution function for $X$ by listing all the possible values of $X$ along with their probabilities. If we try to assign probabilities to each of these uncountable values, the probabilities will no longer sum to 1. Therefore, you must use a different approach to generate the probability distribution for such random variables. As you have seen so far, the approach is to use integrals to calculate areas corresponding to different probabilities instead. The function describing the probability distribution in continuous cases is called the **probability density function.**

### Example 16.23

The lifetime $Y$, in 1000 hours, of LED light bulbs produced by a certain company has a probability density function defined by

$$f(y) = \begin{cases} \dfrac{3}{500} y (10 - y), & 0 \leqslant y \leqslant 10 \\ 0, & \text{otherwise} \end{cases}$$

(a) Find the mean life of a light bulb.

(b) Find the median life of a light bulb.

(c) Find the standard deviation of the life of a light bulb.

(d) Find the probability that a light bulb will last more than 8000 hours.

### Solution

(a) Recall that the mean, $E(X) = \sum x p(x)$, and thus in the continuous case
$$E(X) = \int x f(x) \, dx$$
$$E(Y) = \int y f(y) \, dy = \int_0^{10} \frac{3}{500} y^2 (10 - y) \, dy = 5 \text{ i.e., 5000 hours}$$

(b) The median is the number that has 50% of the data below it. Thus
$$\int_0^m \frac{3}{500} y (10 - y) \, dy = 0.5 \Rightarrow \frac{1}{500} (m^2(15 - m)) = 0.5 \Rightarrow m = 5$$
i.e., 5000 hours.

There are 3 solutions to the equation. Only one of them is in the domain [0, 10]

(c) We first find the variance. The standard deviation will be the square root of the answer.
$$\text{Var}(Y) = \int (y - E(Y))^2 f(y) \, dy = \int_0^{10} y^2 f(y) \, dy - (E(Y))^2 = 30 - 25 = 5$$
$$\Rightarrow \text{Standard deviation} = \sqrt{5} \approx 2.236, \text{ i.e., the standard deviation}$$
is 2236 hours.

(d) $P(y > 8000) = \int_8^{10} \frac{3}{500} y (10 - y) \, dy = 0.104$

### Exercise 16.3

1. Find the area of the region bounded by the given curves. Sketch the region and then compute the required area.

   (a) $y = x + 1, y = 7 - x^2$

   (b) $y = \cos x, y = x - \frac{\pi}{2}, x = -\pi$

   (c) $y = 2x, y = x^2 - 2$

   (d) $y = x^3, y = x^2 - 2, x = 1$

   (e) $y = x^6, y = x^2$

   (f) $y = 5x - x^2, y = x^2$

   (g) $y = 2x - x^3, y = x - x^2$

   (h) $y = \sin x, y = 2 - \sin x$ (one period)

   (i) $y = \frac{x}{2}, y = \sqrt{x}, x = 9$

   (j) $y = \frac{x^4}{10}, y = 3x - x^3$

   (k) $y = \frac{1}{x}, y = \frac{1}{x^3}, x = 8$

   (l) $y = 2 \sin x, y = \sqrt{3} \tan x, -\frac{\pi}{4} \leqslant x \leqslant \frac{\pi}{4}$

   (m) $y = x - 1$ and $y^2 = 2x + 6$

   (n) $x = 2y^2$ and $x = 4 + y^2$

   (o) $4x + y^2 = 12$ and $y = x$

   (p) $x - y = 7$ and $x = 2y^2 - y + 3$

   (q) $x = y^2$ and $x = 2y^2 - y - 2$

   (r) $y = x^3 + 2x^2, y = x^3 - 2x, x = -3,$ and $x = 2$

   (s) $y = x^3 + 1$ and $y = (x + 1)^2$

   (t) $y = x^3 + x$ and $y = 3x^2 - x$

   (u) $y = 3 - \sqrt{x}$ and $y = \frac{2\sqrt{x} + 1}{2\sqrt{x}}$

2. Find the area of the shaded region of the graph in Figure 16.30.

3. Find the area of the region enclosed by $y = e^x$, $x = 0$, and the tangent to $y = e^x$ at $x = 1$

4. Find the area of the region enclosed by $x = 3y^2$ and $x = 12y - y^2 - 5$

5. Find the area of the region enclosed by $y = (x - 2)^2$ and $y = x(x - 4)^2$

6. Find a value for $m > 0$ such that the area under the graph of $y = e^{2x}$ over the interval $[0, m]$ is 3 square units.

7. Find the area of the region bounded by $y = x^3 - 4x^2 + 3x$ and the $x$-axis.

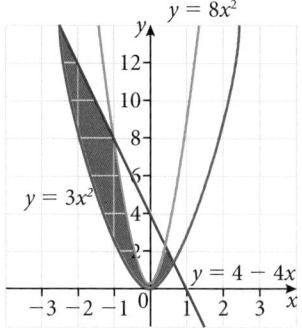

**Figure 16.30** Graph for question 2

8. A 175 m long dam, with a cross section as shown in the diagram, is built to collect water from a river.

   Estimate the amount of concrete needed to build this dam.

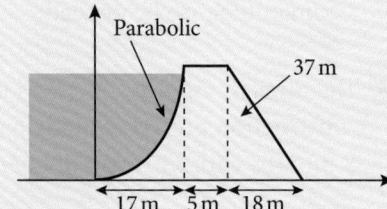

9. The demand and supply functions for a certain product are given. Find the market demand, consumers' surplus, as well as producers' surplus:

   (a) $d(x) = 360 - 0.03x^2$   $s(x) = 0.006x^2$

   (b) $d(x) = 300e^{-0.01x}$   $s(x) = 100 - 100e^{-0.02x}$

   (c) $d(x) = 400e^{-0.01x}$   $s(x) = 0.01x^{2.1}$

10. The table shows the income distribution of a European country in 2014.

| Lowest-paid $x\%$ of income recipients | 0 | 20 | 40 | 60 | 80 | 100 |
|---|---|---|---|---|---|---|
| Income share in total income | 0 | 0.087 | 0.226 | 0.405 | 0.633 | 1 |

   Calculate the Gini index for this country.

11. The lifetime of batteries used in remote control units for one type of TV set has a probability density function defined by

$$f(x) = \begin{cases} \dfrac{15}{76}(x^4 - 2x^2 + 2), & 0 \leqslant x \leqslant 1 \\ -\dfrac{15}{8056}(15x - 121), & 1 < x \leqslant 8\dfrac{1}{15} \\ 0, & \text{otherwise} \end{cases}$$

   where $x$ is measured in tens of hours.

   (a) Find the mean life of such batteries.

   (b) What is the probability that a battery will last at least 20 hours?

12. The weekly amount of oil pumped out of an oil well, in hundreds of barrels, has probability density function $f(y)$ defined by

$$f(y) = \begin{cases} \frac{1}{8}y^2, & 0 \leqslant y < 2 \\ \frac{y}{8}(4-y), & 2 \leqslant y \leqslant 4 \\ 0, & \text{otherwise} \end{cases}$$

(a) Sketch the graph of the probability density function.

(b) Find the mean production per week of this well.

(c) When the production falls below 10% of the weekly production, some maintenance will have to be done in terms of replacing the pumps with more specialised ones. At what level of production will maintenance be needed?

13. The time, in months, in excess of one year to complete a building construction project is modelled by a continuous random variable $Y$ months with a probability density function

$$f(y) = \begin{cases} ky^2(5-y), & 0 \leqslant y \leqslant 5 \\ 0, & \text{otherwise} \end{cases}$$

(a) Show that $k = \frac{12}{625}$

(b) Find the mean, median, and mode of this distribution. Give your answers to 1 decimal place.

(c) What proportion of the projects is completed in less than three months of excess time?

(d) Find the standard deviation of the excess time.

(e) What proportion of the projects are finished within 1 standard deviation of the mean excess time?

14. (a) The marginal price for a monthly demand of $x$ bottles (in thousands) of a brand of shampoo is given by $d'(x) = \dfrac{-6000}{(3x+50)^2}$

Find the price-demand equation if the monthly demand is 150 000 bottles when the price of the bottle is £8.

(b) What is the monthly demand when the price is £6.50?

(c) The marginal price for monthly supply of $x$ bottles (in thousands) of this shampoo is given by $s'(x) = \dfrac{300}{(3x+25)^2}$

Find the price-supply equation if the monthly supply is 75 000 bottles when the price of the bottle is £5.

(d) Under these conditions, what would the market equilibrium be?

**15.** The life expectancy (in years) of a microwave oven is a continuous random variable with probability density function:

$$f(t) = \begin{cases} \dfrac{2}{(t+2)^2}, & t \geq 0 \\ 0, & \text{otherwise} \end{cases}$$

Find the probability that a randomly chosen oven lasts:

**(a)** at most 6 years

**(b)** from 6 to 12 years

**(c)** at most $x$ years, and then find the value of that probability as $x \to \infty$

# 16.4 Volumes with integrals

The underlying principle for finding the area of a plane region is to divide the region into thin strips, approximate the area of each strip by the area of a rectangle, and then add the approximations and take the limit of the sum to produce an integral for the area. The same strategy can be used to find the volume of a solid.

The idea is to divide the solid that stretches over an interval $[a, b]$ into thin slices, approximate the volume of each slice, add the approximations and take the limit of the sum to produce an integral of the volume.

We start by taking cross-sections perpendicular to the $x$-axis (Figure 16.31) as shown. Each slice will be approximated by a cylindrical solid whose volume will be equal to the product of its base times its height (Figure 16.32).

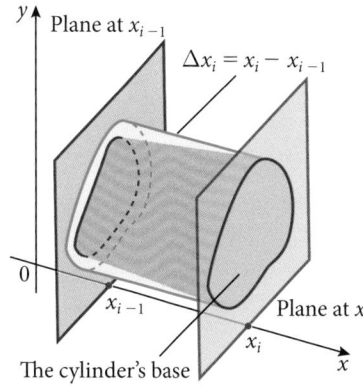

Figure 16.31 Area of a cross-section

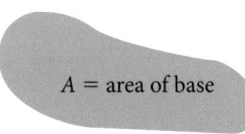

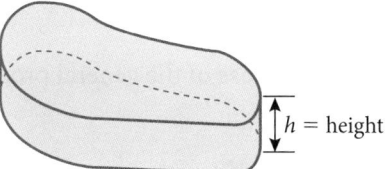

Figure 16.32 Volume of a cylindrical strip

If we call the volume of the slice $v_i$ and the area of its base $A(x)$, then

$$v_i = A(x_i) \cdot h = A(x_i) \cdot \Delta x_i$$

Using this approximation, the volume of the whole solid can be found by

$$V \approx \sum_{i=1}^{n} A(x_i) \Delta x_i$$

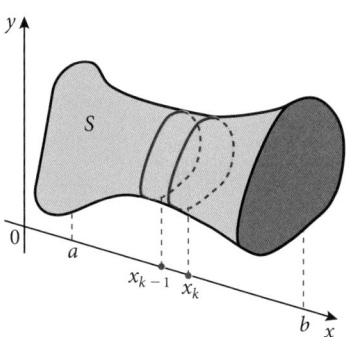

Figure 16.33 Subintervals of $[a, b]$

Figure 16.34 A cylindrical strip

Taking the limit as $n$ increases and the widths of the subintervals approach zero yields the definite integral

$$V = \lim_{n\to\infty} \sum_{i=1}^{n} A(x_i)\,\Delta x_i = \int_a^b A(x)\,dx$$

If we place the solid along the $y$-axis and take the cross-sections perpendicular to that axis, we will arrive at a similar expression for the volume of the solid:

$$V = \lim_{n\to\infty} \sum_{i=1}^{n} A(y_i)\,\Delta y_i = \int_a^b A(y)\,dy$$

If the region bounded by a closed interval $[a, b]$ on the $x$-axis and a function $f(x)$ is rotated about the $x$-axis, the volume of the resulting **solid of revolution** is given by

$$V = \int_a^b \pi(f(x))^2\,dx$$

If the region bounded by a closed interval $[c, d]$ on the $y$-axis and a function $g(y)$ is rotated about the $y$-axis, the volume of the resulting solid of revolution is given by

$$V = \int_c^d \pi(g(y))^2\,dy$$

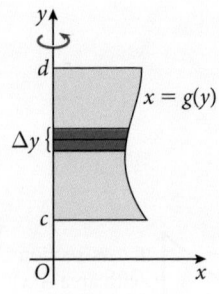

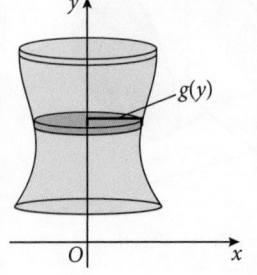

### Example 16.24

Consider the solid formed when the graph of the parabola $y = \sqrt{2x}$ over $[0, 4]$ is rotated around the $x$-axis through an angle of $2\pi$ radians as shown.

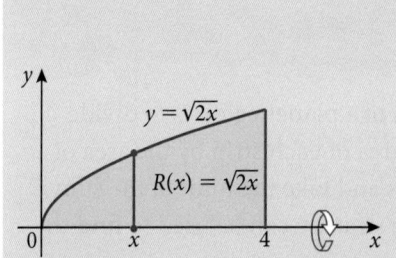

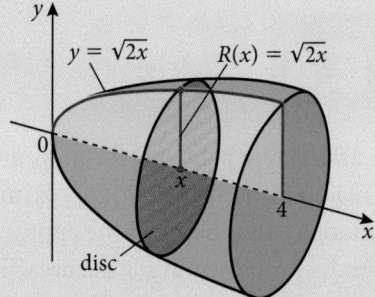

**Solution**

The cross-section is a circular disc whose radius is $y = \sqrt{2x}$

Therefore $A(x) = \pi R^2 = \pi(\sqrt{2x})^2 = 2\pi x$

The volume is then

$$V = \int_0^4 A(x)\,dx = \int_0^4 2\pi x\,dx = 2\pi \frac{x^2}{2}\Big]_0^4 = 16\pi \text{ cubic units}$$

Example 16.25 is a special case of the general process for finding volumes of solids of revolution.

### Example 16.25

Find the volume of a sphere with radius $R = a$

**Solution**

If we place the sphere with its centre at the origin, then the equation of the circle is

$$x^2 + y^2 = a^2 \Rightarrow y = \pm\sqrt{a^2 - x^2}$$

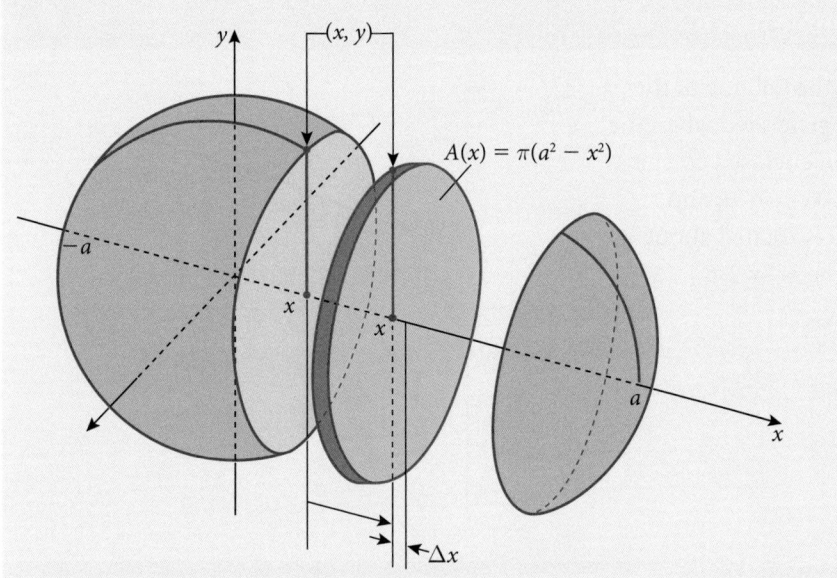

The cross-section of the sphere, perpendicular to the $x$-axis, is a circular disc with radius $y$. So the area is:

$$A(x) = \pi R^2 = \pi y^2 = \pi(\sqrt{a^2 - x^2})^2 = \pi(a^2 - x^2)$$

The volume of the sphere is:

$$V = \int_{-a}^{a} \pi(a^2 - x^2)\, dx = \pi\left[a^2 x - \frac{x^3}{3}\right]_{-a}^{a} = \pi\left(a^3 - \frac{a^3}{3}\right) - \pi\left(-a^3 + \frac{a^3}{3}\right)$$

$$= \pi\left(2a^3 - 2\frac{a^3}{3}\right) = \frac{4\pi a^3}{3}$$

If we want to rotate the right-hand region of the circle around the $y$-axis, then the cross-section of the sphere, perpendicular to the $y$-axis, is a circular disc with radius $x$. Solving the equation for $x$ instead:

$x^2 + y^2 = a^2 \Rightarrow x = \pm\sqrt{a^2 - y^2}$, and hence the area is

$A(y) = \pi R^2 = \pi x^2 = \pi(\sqrt{a^2 - y^2})^2 = \pi(a^2 - y^2)$, and the volume of the sphere is

$$V = \int_{-a}^{a} \pi(a^2 - y^2)\, dy = \pi\left[a^2 y - \frac{y^3}{3}\right]_{-a}^{a} = \pi\left(a^3 - \frac{a^3}{3}\right) - \pi\left(-a^3 + \frac{a^3}{3}\right)$$

$$= \pi\left(2a^3 - 2\frac{a^3}{3}\right) = \frac{4\pi a^3}{3}$$

The same result is given in Example 16.26

## Example 16.26

Find the volume of the
solid generated when the
region enclosed by
$y = \sqrt{3x}$, $x = 3$, and
$y = 0$ is rotated about
the $x$-axis by $2\pi$.

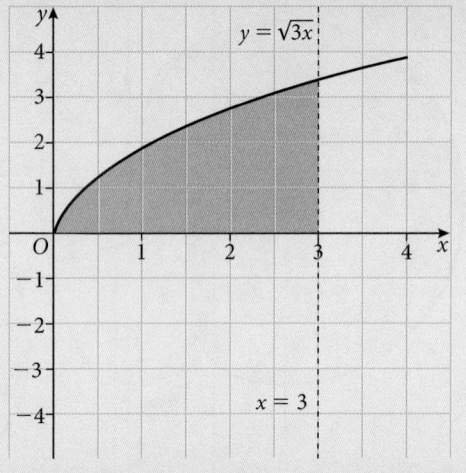

### Solution

$$V = \int_0^3 \pi(f(x))^2\,dx = \pi\int_0^3(\sqrt{3x})^2\,dx = 3\pi\left[\frac{x^2}{2}\right]_0^3 = \frac{27\pi}{2}$$

## Example 16.27

Find the volume of the solid generated when the region enclosed by
$y = \sqrt{3x}$, $y = 3$, and $x = 0$ is rotated about the $y$-axis by $2\pi$.

### Solution

Here, we first find $x$ as a function of $y$.

$y = \sqrt{3x} \Rightarrow x = \frac{y^2}{3}$, the interval on the $y$-axis is $[0, 3]$

So, the volume required is

$$V = \int_0^3 \pi\left(\frac{y^2}{3}\right)^2\,dy = \frac{\pi}{9}\int_0^3 y^4\,dy = \frac{\pi}{9}\left[\frac{y^5}{5}\right]_0^3 = \frac{27\pi}{5}$$

## Washers

Consider the region $R$ between two curves, $y = f(x)$ and $y = g(x)$ from $x = a$
to $x = b$ where $f(x) > g(x)$. Rotating $R$ through $2\pi$ about the $x$-axis generates a
solid of revolution $S$. How can we find the volume of $S$?

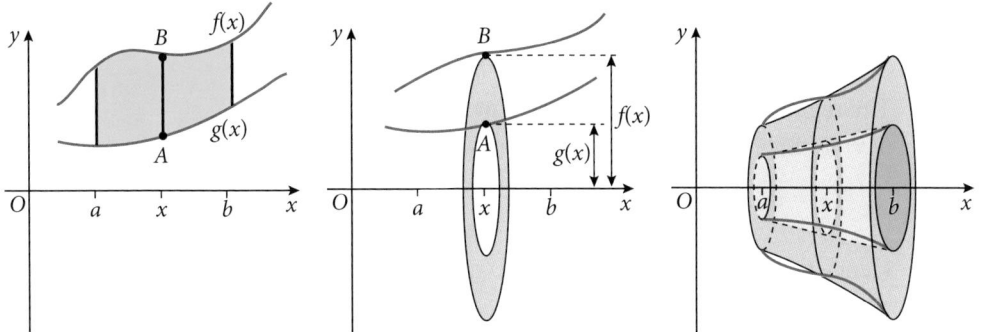

**Figure 16.35** Generating washers

Consider an arbitrary point $x$ in the interval $[a, b]$. The segment $AB$ represents the difference $f(x) - g(x)$. When we rotate this slice through $2\pi$, the cross-section perpendicular to the $x$-axis is going to look like a washer whose area is

$$A = \pi(R^2 - r^2) = \pi\left((f(x))^2 - (g(x))^2\right)$$

So, the volume of $S$ is

$$V = \int_a^b A(x)\,dx = \pi\int_a^b \left((f(x))^2 - (g(x))^2\right)dx$$

If you are rotating about the $y$-axis, a similar formula applies.

$$V = \pi\int_c^d \left((p(y))^2 - (q(y))^2\right)dy$$

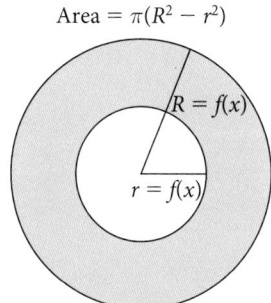

**Figure 16.36** Area of a typical washer

Area $= \pi(R^2 - r^2)$

$R = f(x)$

$r = f(x)$

To understand the washer more, think of it in the following manner:

Let $P$ be the solid generated by rotating the curve $y = f(x)$ and $Q$ be the solid generated by rotating the curve $y = g(x)$. Then $S$ can be found by removing the solid of revolution generated by $y = f(x)$ from the solid of revolution generated by $y = g(x)$.

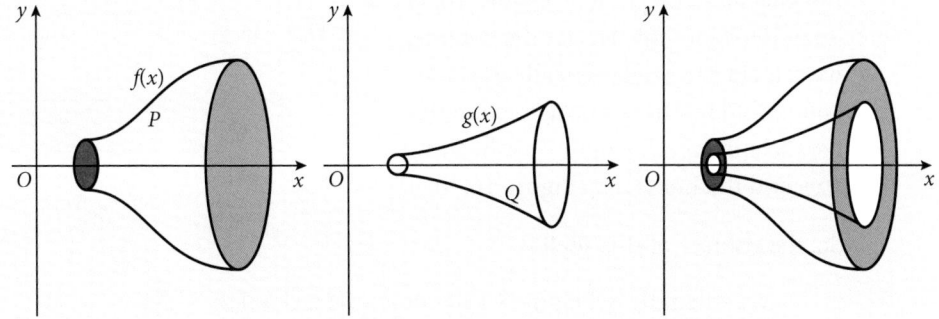

Volume of $S$ = Volume of $P$ − Volume of $Q$ which justifies the formula:

$$V = \pi\int_a^b (f(x))^2\,dx - \pi\int_a^b (g(x))^2\,dx = \pi\int_a^b \left((f(x))^2 - (g(x))^2\right)dx$$

## Example 16.28

The region in the first quadrant between $f(x) = 6 - x^2$ and $h(x) = \dfrac{8}{x^2}$ is rotated about the $x$-axis. Find the volume of the generated solid.

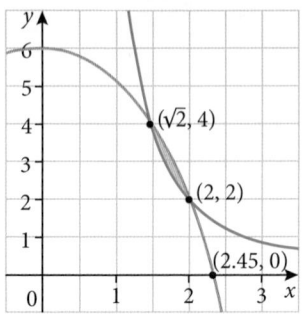

**Figure 16.37** Diagram for Example 16.28

The points shown on the graph: $(\sqrt{2}, 4)$, $(2, 2)$, $(2.45, 0)$

### Solution

The region to rotate (shaded) is shown in Figure 16.37. $f(x)$ is larger than $h(x)$ in this interval and the two curves intersect at

$$\frac{8}{x^2} = 6 - x^2 \Rightarrow x = \sqrt{2}, x = 2$$

Hence the volume of the solid of revolution is

$$V = \pi \int_{\sqrt{2}}^{2} \left( (6 - x^2)^2 - \left( \frac{8}{x^2} \right)^2 \right) dx$$

$$= \pi \int_{\sqrt{2}}^{2} \left( x^4 - 12x^2 + 36 - \frac{64}{x^4} \right) dx$$

$$= \pi \left[ \frac{x^5}{5} - 4x^3 + 36x + \frac{64}{3x^3} \right]_{\sqrt{2}}^{2}$$

$$= \frac{736 - 512\sqrt{2}}{15} \pi$$

### An alternative method: Volumes by cylindrical shells

Consider the region $R$ under the curve of $f(x)$. Rotate $R$ about the $y$-axis by $2\pi$. We divide $R$ into vertical strips of width $\Delta x$ each (Figure 16.38). When we rotate a strip through $2\pi$ around the $y$-axis, we generate a cylindrical shell of $\Delta x$ thickness and height $f(x)$ (Figure 16.39). To understand how we get the volume, we can cut the shell vertically as shown and unfold it. The resulting rectangular parallelepiped has length $2\pi x$, height $f(x)$, and thickness $\Delta x$ (Figure 16.40).

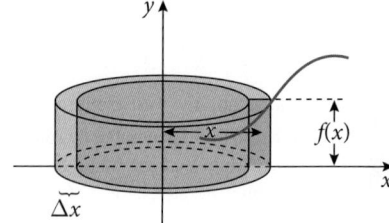

**Figure 16.38** Divide $R$ into vertical strips of width $\Delta x$ each

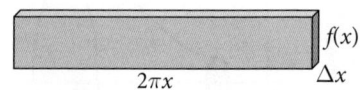

**Figure 16.40** The resulting rectangular parallelepiped

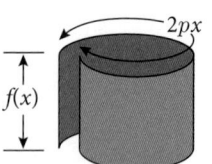

**Figure 16.39** Cylindrical shell of $\Delta x$ thickness and height $f(x)$

So, the volume of this shell is

$$\Delta v_i = \text{length} \times \text{height} \times \text{thickness} = (2\pi x) \times f(x) \times \Delta x$$

The volume of the whole solid is the sum of the volumes of these shells as the number of shells increases, and consequently

$$V = \lim_{n \to \infty} \sum_{i=1}^{n} \Delta v_i = \lim_{\Delta x \to 0} \int (2\pi x) \times f(x) \times \Delta x = 2\pi \int_a^b x f(x) \, dx$$

In many problems, involving rotation about the $y$-axis, this would be more accessible than the disc-washer method.

## Example 16.29

Find the volume of the solid generated when the region under $f(x) = \dfrac{2}{1 + x^2}$, $x = 0$, and $x = 3$, is rotated through $2\pi$ about the $y$-axis.

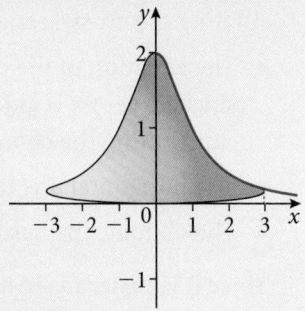

### Solution

Using the shell method, we have

$$V = 2\pi \int_0^3 x \times \frac{2}{1 + x^2}\, dx$$

$$= 2\pi \int_0^3 \frac{2x}{1 + x^2}\, dx = 2\pi \int_0^{10} \frac{du}{u}$$

$$= 2\pi [\ln u]_1^{10} = 2\pi \ln 10$$

## Example 16.30

A dam is to be constructed with a cross-section as shown in the diagram. Engineers designed it in a way so that the whole body of the dam will be generated by taking the cross-section and rotating it by an angle of 150° around a vertical axis as shown.

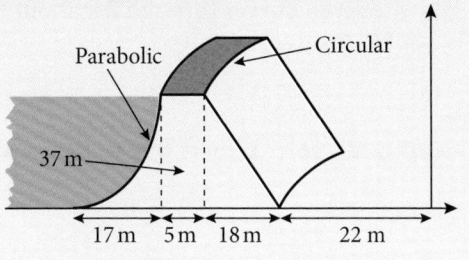

(a) Estimate the volume of concrete required to build this dam.

(b) A mixing truck carries 8.5 m³ of concrete. Approximately how many truck loads would be needed to build the dam?

### Solution

(a) Place the $y$-axis along the vertical axis shown. The cross-section can be divided into three parts:

   (i)   a parabola with vertex at $(-62, 0)$ and passing through $(-45, 37)$

   (ii)  a horizontal line between $(-45, 37)$ and $(-40, 37)$

   (iii) a line through $(-40, 37)$ and $(-22, 0)$.

The dam will constitute $\frac{150}{360} = \frac{5}{12}$ of the solid of revolution created by rotating the cross-section region around the $y$-axis.

- The equation of the parabola will be of the form $y = a(x - h)^2$, and since its vertex is at $(-62, 0)$, then it is of the form $y = a(x + 62)^2$, by substituting the coordinates $(-45, 37)$ we find that $a = 0.1280$
- The horizontal line has the equation $y = 37$
- The oblique line has the equation $y = -\frac{37}{18}(x + 22)$

Since it is easier to set up, we will use the shell method with volume formula $V = 2\pi \int_a^b x f(x)\, dx$, thus, the volume needed is

$$V = \left| \frac{5}{12} \left( 2\pi \int_{-62}^{-45} 0.1280x\,(x + 62)^2\, dx \right. \right.$$

$$\left. \left. + 2\pi \int_{-45}^{-40} 37x\, dx + 2\pi \int_{-40}^{-22} -\frac{37}{18}x\,(x + 22)dx \right) \right|$$

$$\approx 77\,253 \text{ m}^3$$

(b) $\frac{77253}{8.5} \approx 9089$ truckloads will be needed.

## Exercise 16.4

1. Find the volume of the solid obtained by rotating the region bounded by the given curves through $2\pi$ about the $x$-axis. Sketch the region, the solid and a typical disc.

(a) $y = 3 - \frac{x}{3}, y = 0, x = 2, x = 3$      (b) $y = 2 - x^2, y = 0$

(c) $y = \sqrt{16 - x^2}, y = 0, x = 1, x = 3$      (d) $y = \frac{3}{x}, y = 0, x = 1, x = 3$

(e) $y = 3 - x, y = 0, x = 0$

(f) $y = \sqrt{\sin x}, y = 0, 0 \leqslant x \leqslant \pi$

(g) $y = \sqrt{\cos x}, y = 0, -\frac{\pi}{2} \leqslant x \leqslant \frac{\pi}{3}$      (h) $y = 4 - x^2, y = 0$

(i) $y = x^3 + 2x + 1, y = 0, x = 1$      (j) $y = -4x - x^2, y = x^2$

(k) $y = 1 - x^2, y = x^3 + 1$      (l) $y = \sqrt{36 - x^2}, y = 4$

(m) $x = \sqrt{y}, y = 2x$

(n) $y = \sin x, y = \cos x, x = \frac{\pi}{4}, x = \frac{\pi}{2}$

(o) $y = 2x^2 + 4, y = x, x = 1, x = 3$

(p) $y = \sqrt{x^4 + 1}, y = 0, x = 1, x = 3$

(q) $y = 16 - x, y = 3x + 12, x = -1$

(r) $y = \frac{1}{x}, y = \frac{5}{2} - x$

2. Find the volume resulting from a rotation of the region shown in the diagram through $2\pi$ about:

   (a) the $x$-axis

   (b) the $y$-axis.

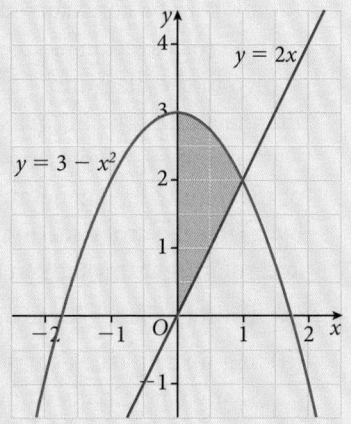

3. Find the volume of the solid obtained by rotating the region bounded by the given curves about the $y$-axis through $2\pi$. Sketch the region, the solid and a typical disc/shell.

   (a) $y = x^2$, $y = 0$, $x = 1$, and $x = 3$

   (b) $y = x$, $y = \sqrt{9 - x^2}$, $x = 0$

   (c) $y = x^3 - 4x^2 + 4x$, $y = 0$

   (d) $y = \sqrt{3x}$, $x = 5$, $x = 11$, $y = 0$

   (e) $y = x^2$, $y = \dfrac{2}{1 + x^2}$

   (f) $y = \sqrt{x^2 + 2}$, $x = 3$, $y = 0$, $x = 0$

   (g) $y = \dfrac{7x}{\sqrt{x^3 + 7}}$, $x = 3$, $y = 0$

   (h) $y = \sin x$, $y = \cos x$, $x = \dfrac{\pi}{4}$, $x = \dfrac{\pi}{2}$

   (i) $y = 2x^2 + 4$, $y = x$, $x = 1$, $x = 3$

   (j) $y = \sin(x^2)$, $y = 0$, $x = 0$, $x = \sqrt{\pi}$

   (k) $y = 5 - x^3$, $y = 5 - 4x$

4. The region shown in the diagram is rotated through $2\pi$ about the $y$-axis to form a solid. Estimate the volume of this solid.

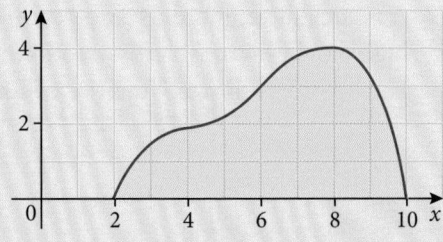

It is practical to use cylindrical shells.

5. A dam is to be constructed with a cross-section as shown in the diagram. Engineers designed it in a way so that the whole body of the dam will be generated by taking the cross-section

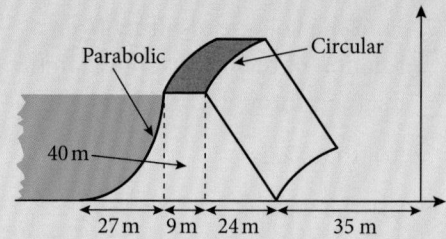

and rotating it through an angle of 180° around a vertical axis as shown.

   (a) Estimate the volume of concrete required to build this dam.

   (b) If a mixing truck carries 8.5 m³ of concrete, approximately how many truck loads would be needed to build the dam?

6. Find the volume of a ring which is created by boring a hole of radius 3 cm through the centre of a solid sphere of radius 5 cm.

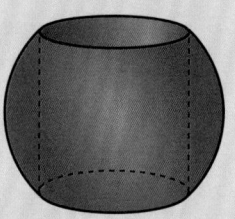

7. The paraboloid generated by rotating the region between the parabola $x^2 = 2py$ and the $y$-axis through $2\pi$ around the $y$-axis is shown in Figure 16.41. Show that the volume of the paraboloid is one-half that of the circumscribed cylinder.

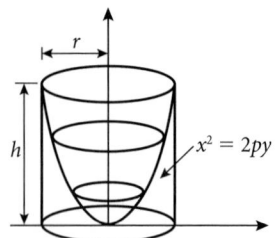

Figure 16.41 Diagram for question 7

8. A contractor wants to bid for the job of levelling an 18-metre hill. It will cost €9.80 per m³ to remove the material from the hill. The table, based on surveying data, shows areas of horizontal cross-sections of the hill at 3-metre intervals. Estimate how much this job should cost.

| Height (m) | 0 | 3 | 6 | 9 | 12 | 15 | 18 |
|---|---|---|---|---|---|---|---|
| Area (m²) | 141 | 82 | 35 | 25 | 14 | 5 | 0 |

9. A vertical hole of radius 3 and centred on the axis of a wooden 25-cm high paraboloid with cross-section $y = 25 - x^2$ is bored through the paraboloid as shown. Find the volume of the wood that remains.

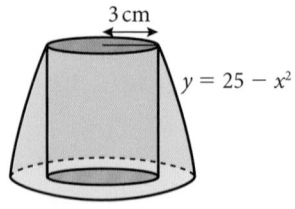

Figure 16.42 Diagram for question 9

## 16.5 Modelling linear motion

So far, our mathematical models have considered the motion of an object only along a straight line. For example, projectile motion (e.g. a ball being thrown) is often modelled by a position function that simply gives the height (displacement) of the object. In that way, we are modelling the motion as if it was restricted to a vertical line.

In this section we will again analyse the motion of an object as if its motion takes place along a straight line in space. This can only make sense if the mass (and thus, size) of the object is not considered. Hence, the object is modelled by a particle whose mass is considered to be zero. This study of motion, without reference either to the forces that cause it or to the mass of the object, is known as **kinematics**.

## Displacement and total distance travelled

Recall from Chapter 14 that given time $t$, displacement $s$, velocity $v$ and acceleration $a$, we have:

$$v = \frac{ds}{dt} \text{ and } a = \frac{dv}{dt}, \text{ also } a = \frac{d}{dt}\left(\frac{ds}{dt}\right) = \frac{d^2s}{dt^2}$$

It is important to understand the difference between displacement and distance travelled. Consider a couple of simple examples of an object moving along the $x$-axis.

Assume that the object does not change direction during the interval $0 \leqslant t \leqslant 5$. If the position of the object at $t = 0$ is $x = 2$ and at $t = 5$ its position is $x = -3$, then its displacement, or change in position, is $-5$ because the object changed its position by 5 units in the negative direction. This can be calculated by final position $-$ initial position $= -3 - 2 = -5$. However, the distance travelled would be the absolute value of displacement, calculated by

$$\left|\text{final position} - \text{initial position}\right| = \left|-3 - 2\right| = 5$$

Assume that the initial and final positions of another object are the same as in the first example – that is, at $t = 0$ its position is $x = 2$, and at $t = 5$ its position is $x = -3$. However, the object changed direction in that it first travelled to the left (negative velocity) from $x = 2$ to $x = -5$ during the interval $0 \leqslant t \leqslant 3$, and then travelled to the right (positive velocity) from $x = -5$ to $x = -3$. The object's displacement is $-5$; the same as in the first example because its net change in position is just the difference between final and initial positions. However, it's clear that the object has travelled further than in the first example. But, we cannot calculate it the same way as we did in the first example. We will have to make a separate calculation for each interval where the direction changed. Hence, total distance travelled $= \left|-5 - 2\right| + \left|-3 - (-5)\right| = 7 + 2 = 9$

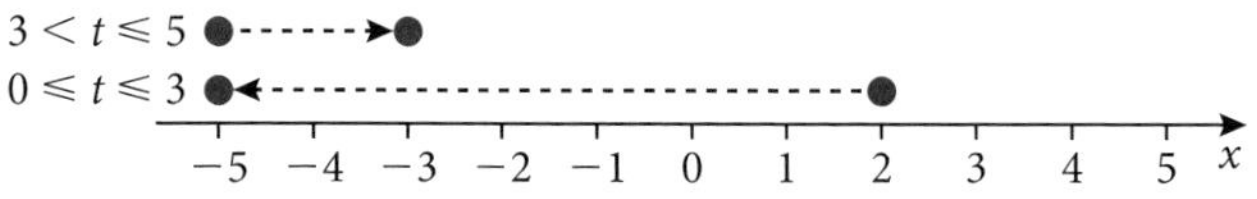

$3 < t \leqslant 5$ ●------→●
$0 \leqslant t \leqslant 3$ ●◄------------------------●

$$\begin{array}{ccccccccccc} & & & & & & & & & & x \\ -5 & -4 & -3 & -2 & -1 & 0 & 1 & 2 & 3 & 4 & 5 \end{array}$$

**Figure 16.43** Travelled distances

The **velocity** $v = \frac{ds}{dt}$ of a particle is a measure of how fast it is moving and of its direction of motion relative to a fixed point.

The **speed** $|v|$ of a particle is a measure of how fast it is moving that does not indicate direction. Thus, speed is the magnitude of the velocity and is always positive.

The **acceleration** $a = \frac{dv}{dt}$ of a particle is a measure of how fast its velocity is changing.

### Example 16.31

The displacement $s$ of a particle on the $x$-axis, relative to the origin, is given by the position function $s(t) = -t^2 + 6t$ where $s$ is in centimetres and $t$ is in seconds.

(a) Find a function for the velocity of the particle $v(t)$ in terms of $t$. Graph the functions $s(t)$ and $v(t)$ on separate axes.

(b) Find the position of the particle at the following times:

$$t = 0, 1, 3 \text{ and } 6 \text{ s}$$

(c) Find the displacement of the particle for the following intervals:

$$0 \leqslant t \leqslant 1,\ 1 \leqslant t \leqslant 3,\ 3 \leqslant t \leqslant 6 \text{ and } 0 \leqslant t \leqslant 6$$

(d) Find the total distance travelled by the particle for the following intervals:

$$0 \leqslant t \leqslant 1,\ 1 \leqslant t \leqslant 3,\ 3 \leqslant t \leqslant 6 \text{ and } 0 \leqslant t \leqslant 6$$

### Solution

(a) $v(t) = \dfrac{d}{dt}(-t^2 + 6t) = -2t + 6$

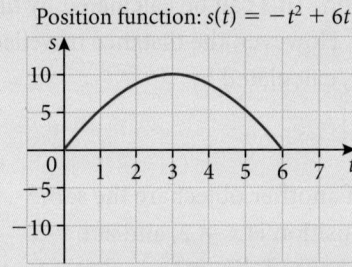

Position function: $s(t) = -t^2 + 6t$

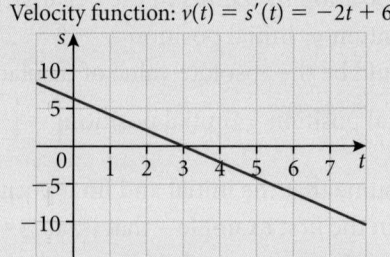

Velocity function: $v(t) = s'(t) = -2t + 6$

(b) The position of the particle at:

- $t = 0$ is $s(0) = 0$ cm
- $t = 1$ is $s(1) = 5$ cm
- $t = 3$ is $s(3) = 9$ cm
- $t = 6$ is $s(6) = 0$ cm

(c) The displacement of the particle for the interval:

- $0 \leqslant t \leqslant 1$ is $\Delta\text{position} = s(1) - s(0) = 5$ cm
- $1 \leqslant t \leqslant 3$ is $\Delta\text{position} = s(3) - s(1) = 4$ cm
- $3 \leqslant t \leqslant 6$ is $\Delta\text{position} = s(6) - s(3) = -9$ cm
- $0 \leqslant t \leqslant 6$ is $\Delta\text{position} = s(6) - s(0) = 0$ cm

This last result makes sense considering the particle moved to the right 9 cm then at $t = 3$ changed direction and moved to the left 9 cm, ending where it started – thus, no change in net position.

(d) The total distance travelled by the particle for the interval:

- $0 \leqslant t \leqslant 1$ is $|s(1) - s(0)| = |5 - 0| = 5$
- $1 \leqslant t \leqslant 3$ is $|s(3) - s(1)| = |9 - 5| = 4$
- $3 \leqslant t \leqslant 6$ is $|s(6) - s(3)| = |0 - 9| = |-9| = 9$
- motion of the object changed direction (velocity$= 0$) at $t = 3$
- $0 \leqslant t \leqslant 6$ is
  $$|s(3) - s(0)| + |s(6) - s(3)| = |9 - 0| + |0 - 9| = 9 + 9 = 18$$

Since differentiation of the position function gives the velocity function (i.e. $v = \dfrac{ds}{dt}$), we expect that the inverse of differentiation, integration, will lead us in the reverse direction – that is, from velocity to position. When velocity is constant, we can find the displacement with the formula:

displacement = velocity × change in time

If we drove a car at a constant velocity of $50 \, \text{km h}^{-1}$ for 3 hours then our displacement (same as distance travelled in this case) is $150 \, \text{km}$. If a particle travelled to the left on the $x$-axis at a constant rate of $-4$ units $\text{s}^{-1}$ for 5 seconds then the displacement of the particle is $-20$ units.

The velocity–time graph in Figure 16.44 depicts an object's motion with a constant velocity of $5 \, \text{cm s}^{-1}$ for $0 \leqslant t \leqslant 3$

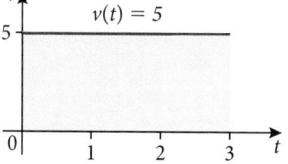

Clearly, the object's displacement is $5 \, \text{cm s}^{-1} \times 3 \, \text{s} = 15$ cm for this interval.

The rectangular area ($3 \times 5 = 15$) under the velocity curve is equal to the displacement of the object.

**Figure 16.44** Velocity–time graph

Consider the area under the graph of $v(t)$ from $t = 0$ to $t = 3$ in Example 16.31. The area is the area of a triangle $= \dfrac{1}{2} \times 3 \times 6 = 9$

The area under the velocity curve for a certain interval is equal to the displacement for that interval. We can argue that just as the total area can be found by summing the areas of narrow rectangular strips, the displacement can be found by summing small displacements ($v \cdot \Delta t$). Consider:
displacement = velocity × change in time $\Rightarrow s = v \cdot \Delta t \Rightarrow s = v \cdot dt$

We already know that when $f(x) \geqslant 0$ the definite integral $\int_a^b f(x) \, dx$ gives the area between $y = f(x)$ and the $x$-axis from $x = a$ to $x = b$. And if $f(x) \leqslant 0$ then $\int_a^b f(x) \, dx$ gives a number that is the opposite of the area between $y = f(x)$ and the $x$-axis from $a$ to $b$.

Given that $v(t)$ is the velocity function for a particle moving along a line, then:
$\int_a^b v(t) \, dt$ gives the displacement from $t = a$ to $t = b$
$\int_a^b |v(t)| \, dt$ gives the total distance travelled from $t = a$ to $t = b$

Applying integration to find the displacement and distance travelled for the two intervals $3 \leqslant t \leqslant 6$ and $0 \leqslant t \leqslant 6$ in Example 16.31:

For $3 \leqslant t \leqslant 6$:

Displacement $= \int_3^6 (-2t + 6) \, dt = \left| -t^2 + 6t \right|_3^6 = 0 - 9 = -9$

Distance travelled $= \int_3^6 |(-2t + 6)| \, dt = \left| -t^2 + 6t \right|_3^6 = |0 - 9| = 9$

For $0 \leqslant t \leqslant 6$:

Displacement $\int_0^6 (-2t + 6)\,dt = -t^2 + 6t\big|_0^6 = 0$

Distance travelled $= \int_0^3 \big|(-2t + 6)\big|\,dt + \int_3^6 \big|(-2t + 6)\big|\,dt$ (particle changed direction at $t = 3$)

$$= \left|-t^2 + 6t\right|_0^3 + \left|-t^2 + 6t\right|_3^6 = 9 + 9 = 18$$

## Example 16.32

The function $v(t) = \sin(\pi t)$ gives the velocity in m s$^{-1}$ of a particle moving along the $x$-axis.

(a) Determine when the particle is moving to the right, to the left, and when it is stationary. At any time it stops, determine if it changes direction at that time.

(b) Find the displacement of the particle for the time interval $0 \leqslant t \leqslant 3$

(c) Find the total distance travelled by the particle for the time interval $0 \leqslant t \leqslant 3$

### Solution

(a) $v(t) = \sin(\pi t) = 0$

$\Rightarrow \sin(k \cdot \pi) = 0$ for $k \in \mathbb{Z} \Rightarrow \pi t = k\pi \Rightarrow t = k, k \in \mathbb{Z}$

for $0 \leqslant t \leqslant 3$, $t = 0, 1, 2, 3$

Therefore, the particle is stationary at $t = 0, 1, 2, 3$

Since $t = 0$ and $t = 3$ are endpoints of the interval, the particle can only change direction at $t = 1$ or $t = 2$

$v\left(\dfrac{1}{2}\right) = \sin\left(\pi \cdot \dfrac{1}{2}\right) = 1$; $v\left(\dfrac{3}{2}\right) = \sin\left(\pi \cdot \dfrac{3}{2}\right) = -1 \Rightarrow$ direction changes at $t = 1$

$v\left(\dfrac{3}{2}\right) = \sin\left(\pi \cdot \dfrac{3}{2}\right) = -1$; $v\left(\dfrac{5}{2}\right) = \sin\left(\pi \cdot \dfrac{5}{2}\right) = 1 \Rightarrow$ direction changes again at $t = 2$

Note that displacement $= A_1 - A_2 + A_3$, and distance travelled $= A_1 + A_2 + A_3$

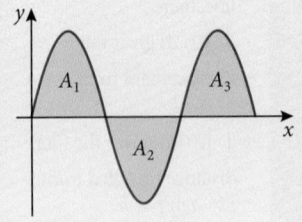

(b) displacement $= \displaystyle\int_0^3 \sin(\pi t)\,dt = -\dfrac{1}{\pi}\cos(\pi t)\Big|_0^3$

$= -\dfrac{1}{\pi}\cos(3\pi) - \left(-\dfrac{1}{\pi}\cos(0)\right) = \dfrac{2}{\pi} \approx 0.637\text{ m}$

(c) total distance travelled $= \displaystyle\int_0^1 \big|\sin(\pi t)\big|\,dt + \int_1^2 \big|\sin(\pi t)\big|\,dt + \int_2^3 \big|\sin(\pi t)\big|\,dt$

$= \left|\dfrac{2}{\pi}\right| + \left|-\dfrac{2}{\pi}\right| + \left|\dfrac{2}{\pi}\right| = \dfrac{6}{\pi} \approx 1.91\text{ m}$

## Position and velocity from acceleration

If we can obtain position from velocity by applying integration then we can also obtain velocity from acceleration by integrating.

## Example 16.33

The motion of a parachutist is modelled as linear motion by considering that the parachutist is a particle moving along a line whose positive direction is vertically downwards. The parachute is opened at $t = 0$ at which time the parachutist's position is $s = 0$

According to the model, the acceleration function for the parachutist's motion for $t > 0$ is given by: $a(t) = -54e^{-1.5t}$

(a) At the moment the parachute opens, the parachutist has a velocity of $42 \text{ m s}^{-1}$. Find the velocity function of the parachutist for $t > 0$

What does the model say about the parachutist's velocity as $t \to \infty$?

(b) Find the position function of the parachutist for $t > 0$

## Solution

(a) $v(t) = \int a(t)\, dt = \int (-54e^{-1.5t})\, dt$

$\quad = -54\left(\dfrac{1}{-1.5}\right)e^{-1.5t} + c$

$\quad = 36e^{-1.5t} + c$

Since $v = 42$ when $t = 0$, then $42 = 36e^0 + c \Rightarrow 42 = 36 + c \Rightarrow c = 6$

Therefore, after the parachute opens ($t > 0$) the velocity function is $v(t) = 36e^{-1.5t} + 6$

Since $\lim\limits_{t\to\infty} e^{-1.5t} = \lim\limits_{t\to\infty} \dfrac{1}{e^{1.5t}} = 0$, then $\lim\limits_{t\to\infty} v(t) = 6 \text{ m s}^{-1}$

(b) $s(t) = \int v(t)\, dt = \int (36e^{-1.5t} + 6)\, dt$

$\quad = 36\left(\dfrac{1}{-1.5}\right)e^{-1.5t} + 6t + c$

$\quad = -24e^{-1.5t} + 6t + c$

Since $s = 0$ when $t = 0$, then $0 = -24e^0 + 6(0) + c \Rightarrow 0 = -24 + c$

$\Rightarrow c = 24$

Therefore, after the parachute opens ($t > 0$) the position function is $s(t) = -24e^{-1.5t} + 6t + 24$

> The limit of the velocity as $t \to \infty$, for a falling object, is called the **terminal velocity** of the object. While the limit $t \to \infty$ is never attained as the parachutist eventually lands on the ground, the velocity gets close to the terminal velocity very quickly. For example, after just 8 seconds, the velocity is $v(8) = 36e^{-1.5(8)} + 6$
> $\approx 6.0002 \text{ m s}^{-1}$

## Uniformly accelerated motion

Motion under the effect of gravity in the vicinity of the Earth (or other planets) is an important case of rectilinear motion, called **uniformly accelerated motion**.

If a particle moves with constant acceleration along the $s$-axis, and if we know the initial speed and position of the particle, then it is possible to have specific formulae for the position and speed at any time $t$.

Assume acceleration is constant, i.e., $a(t) = a$, $v(0) = v_0$ and $s(0) = s_0$

$v(t) = \int a(t)\, dt = at + c$, however, we know that $v(0) = v_0$, then

$v(0) = v_0 = a \times 0 + c \Rightarrow c = v_0$, hence $v(t) = at + v_0$

$s(t) = \int v(t)\, dt = \int (at + v_0)\, dt = \frac{1}{2}at^2 + v_0 t + c$, and, as above substituting

$s(0) = s_0$ into the equation, we have

$s(t) = \frac{1}{2}at^2 + v_0 t + s_0$

When this is applied to free-fall model ($s$-axis vertical), then

$v(t) = -gt + v_0$, and

$s(t) = -\frac{1}{2}gt^2 + v_0 t + s_0$, where $g = 9.8 \text{ m s}^{-2}$

### Example 16.34

A tennis player hits a serve from a height of 244 cm at an initial speed of 54 m s$^{-1}$ and at an angle of 5° below the horizontal. The serve is in if the tennis ball clears a 91.4 cm high net that is 1190 cm away and hits the ground in front of the service line 1829 cm away. Determine whether the serve is in.

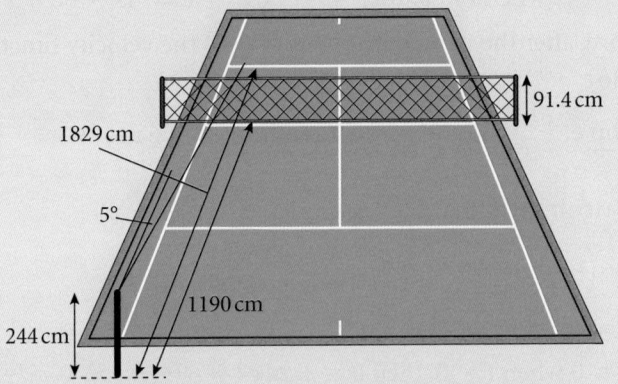

### Solution

We have two components for the motion of the ball: a vertical one and a horizontal one.

The vertical component has accaleration of $-9.8 \text{ m s}^{-2}$ due to gravity,

i.e., $y''(t) = -9.8$

The vertical component of the velocity is

$v(0) = 54\sin(-5°) = -4.68 \text{ m s}^{-1}$

This means $y'(t) = \int -9.8\, dt = -9.8t + c \Rightarrow -4.68 = y'(0) = c$

$\Rightarrow y'(t) = -9.8t - 4.68$

Since the initial height is 2.44 m,

$$y(t) = \int y'(t)\, dt = \int (-4.9t - 4.68)\, dt = -4.9t^2 - 4.68t + c$$

$$\Rightarrow y(t) = -4.9t^2 - 4.68t + 2.44$$

The horizontal component is determined by $x''(0) = 0$ with initial velocity

$$x'(0) = 54 \cos(-5°) \approx 53.4 \text{ m s}^{-1}$$

So $x(t) = \int 53.4\, dt = 53.4t$

To clear the net, the height of the ball, $y$ must be at least 0.914 m when $x = 11.90$ m

That means $x(t) = 53.4t = 11.90 \Rightarrow t = 0.22285$

At this time, the height is

$$y(0.22285) = -4.9(0.22285)^2 - 4.68(0.22285) + 2.44 \approx 1.154$$

Thus, the ball is high enough to clear the net.

The second condition is that we need $x \leq 18.29$ when the ball hits the ground ($y = 0$).

$$y(t) = -4.9t^2 - 4.68t + 2.44 = 0 \Rightarrow t \approx 0.375$$

However, $x(0.375) = 53.4(0.375) \approx 20.03$ which is behind the line at 18.29

The ball is out.

## Exercise 16.5

1. The velocity of a particle along a rectilinear path is given by the equation $v(t)$ in m s$^{-1}$. Find both the net distance and the total distance it travels between the times $t = a$ and $t = b$

   (a) $v(t) = t^2 - 11t + 24$, $a = 0$, $b = 10$

   (b) $v(t) = t - \dfrac{1}{t^2}$, $a = 0.1$, $b = 1$

   (c) $v(t) = \sin 2t$, $a = 0$, $b = \dfrac{\pi}{2}$

   (d) $v(t) = \sin t + \cos t$, $a = 0$, $b = \pi$

   (e) $v(t) = t^3 - 8t^2 + 15t$, $a = 0$, $b = 6$

   (f) $v(t) = \sin\left(\dfrac{\pi t}{2}\right) + \cos\left(\dfrac{\pi t}{2}\right)$, $a = 0$, $b = 1$

2. The acceleration of a particle along a rectilinear path is given by the equation $a(t)$ in m s$^{-2}$ and the initial velocity $v_0$ m s$^{-1}$ is also given. Find the velocity of the particle as a function of $t$ and both the net distance and the total distance it travels between the times $t = a$ and $t = b$

(a) $a(t) = 3, v_0 = 0, a = 0, b = 2$

(b) $a(t) = 2t - 4, v_0 = 3, a = 0, b = 3$

(c) $a(t) = \sin t, v_0 = 0, a = 0, b = \dfrac{3\pi}{2}$

(d) $a(t) = \dfrac{-1}{\sqrt{t+1}}, v_0 = 2, a = 0, b = 4$

(e) $a(t) = 6t - \dfrac{1}{(t+1)^3}, v_0 = 2, a = 0, b = 2$

3. The velocity and initial position of an object moving along a coordinate line are given. Find the position of the object at time $t$.

(a) $v = 9.8t + 5, s(0) = 10$  (b) $v = 32t - 2, s(0.5) = 4$

(c) $v = \sin \pi t, s(0) = 0$  (d) $v = \dfrac{1}{t+2}, t > -2, s(-1) = \dfrac{1}{2}$

4. The acceleration is given as well as the initial velocity and initial position of an object moving on a coordinate line. Find the position of the object at time $t$.

(a) $a = e^t, v(0) = 20, s(0) = 5$  (b) $a = 9.8, v(0) = -3, s(0) = 0$

(c) $a = -4\sin 2t, v(0) = 2, s(0) = -3$

(d) $a = \dfrac{9}{\pi^2}\cos\dfrac{3t}{\pi}, v(0) = 0, s(0) = -1$

5. An object moves with a speed of $v(t)$ m s$^{-1}$ along the $s$-axis. Find the displacement and the distance travelled by the object during the given time interval.

(a) $v(t) = 2t - 4; 0 \leqslant t \leqslant 6$  (b) $v(t) = |t - 3|; 0 \leqslant t \leqslant 5$

(c) $v(t) = t^3 - 3t^2 + 2t; 0 \leqslant t \leqslant 3$  (d) $v(t) = \sqrt{2} - 2; 0 \leqslant t \leqslant 3$

6. An object moves with an acceleration $a(t)$ m s$^{-2}$ along the $s$-axis. Find the displacement and the distance travelled by the object during the given time interval.

(a) $a(t) = t - 2, v_0 = 0, 1 \leqslant t \leqslant 5$

(b) $a(t) = \dfrac{1}{\sqrt{5t+1}}, v_0 = 2, 0 \leqslant t \leqslant 3$

(c) $a(t) = -2, v_0 = 3, 1 \leqslant t \leqslant 4$

7. The velocity of an object moving along the $s$-axis is $v = 9.8t - 3$ Find the object's displacement between $t = 1$ and $t = 3$ given that:

(a) $s(0) = 5$.  (b) $s(0) = -2$  (c) $s(0) = s_0$

8. The displacement $s$ metres of a moving object from a fixed point $O$ at time $t$ seconds is given by $s(t) = 50t - 10t^2 + 1000$

   (a) Find the velocity of the object in m s$^{-1}$.

   (b) Find its maximum displacement from $O$.

9. A particle moves along a line so that its speed $v$ at time $t$ is given by

$$v(t) = \begin{cases} 5t, & 0 \leqslant t < 1 \\ 6\sqrt{t} - \frac{1}{t}, & t \geqslant 1 \end{cases}$$

   where $t$ is in seconds and $v$ is in cm s$^{-1}$. Estimate the time(s) at which the particle is 4 cm from its starting position.

10. A projectile is fired vertically upwards with an initial velocity of 49 m s$^{-1}$ from a platform 150 m high.

    (a) How long will it take the projectile to reach its maximum height?

    (b) What is the maximum height?

    (c) How long will it take the projectile to pass its starting point on the way down?

    (d) What is the velocity when it passes its starting point on the way down?

    (e) How long will it take the projectile to hit the ground?

    (f) What will its speed be at impact?

11. The graph of the acceleration $a(t)$ of a car measured in m s$^{-2}$ is shown in Figure 16.45. Estimate the increase in the velocity of the car during the 6-second time interval.

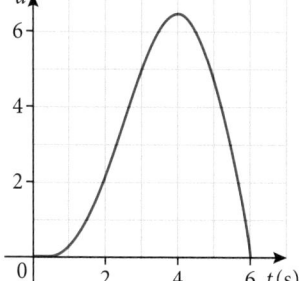

**Figure 16.45** Diagram for question 11

12. Figure 16.46 shows a pendulum with length $L$ that makes a maximum angle $\theta_0$ with the vertical. Using Newton's second law, it can be shown that the period $T$ (the time for one complete swing) is given by

$$T = 4\sqrt{\frac{L}{g}} \int_0^{\pi/2} \frac{dt}{\sqrt{1 - k^2 \sin^2 t}}$$

   where $k = \sin\left(\frac{1}{2}\theta_0\right)$ and $g$ is the acceleration due to gravity.

   If $L = 1$ m and $\theta_0 = 40°$, estimate the period of this pendulum to the nearest second.

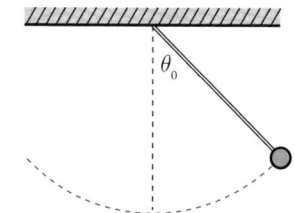

**Figure 16.46** Diagram for question 12

13. A car moving initially at 44 m s$^{-1}$ decelerates at the constant rate of 4 m s$^{-2}$. Determine how far the car travels before coming to a stop.

14. A car that has stopped at a toll booth leaves the booth with a constant acceleration of 4 m s$^{-2}$. At that time, there is a truck 1.5 km ahead moving with a constant speed of 50 km h$^{-1}$. How long will take the car to catch up with the truck, and how far will they be from the toll booth at that time?

15. An athlete in a 100-metre race starts out with an acceleration of $4 \text{ m s}^{-2}$ and maintains that acceleration for 2 seconds. His acceleration then drops to zero and he continues the race at a fixed speed. What is his time for the race?

16. The tennis player in Example 16.34 hits the serve from a height of 244 cm. Determine whether the serve is in.

    (a) At an initial speed of $54 \text{ m s}^{-1}$ but at an angle of 6° below the horizontal.

    (b) At an initial speed of $52 \text{ m s}^{-1}$ but at an angle of 5° below the horizontal.

## Chapter 16 practice questions

1. The graph represents the function $y = e^x$ and the tangent to the graph passing through the origin.

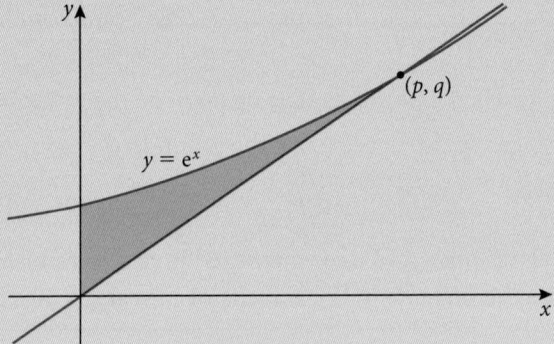

Find:
   (a) the values of $p$ and $q$    (b) the area of the shaded region.

2. The diagram shows part of the graph of $y = 2e^x$ and $y = e^{2-x}$

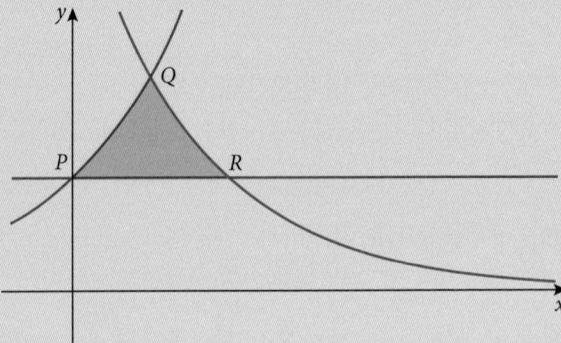

   (a) Find the coordinates of the points $P$, $Q$ and $R$.

   The shaded region is rotated through 360° about the $x$-axis.

   (b) Write down an integral which represents the volume of the solid obtained.

3. The diagram shows part of the graph of $y = \dfrac{1}{x^2}$

The area of the shaded region is $\dfrac{5}{3}$ square units.

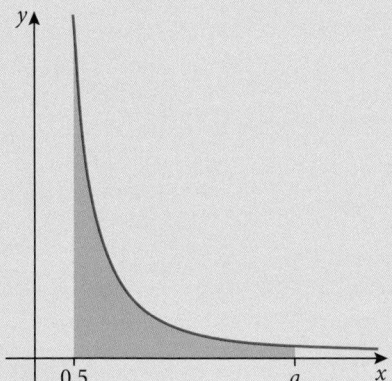

Find the exact value of $a$.

4. (a) Find the equation of the tangent line to the curve $y = \ln x$ at the point $(e, 1)$, and verify that the origin is on this line.

(b) Show that $(x \ln x - x)' = \ln x$

(c) The diagram shows the region enclosed by the curve $y = \ln x$, the tangent line in part (a), and the line $y = 0$.

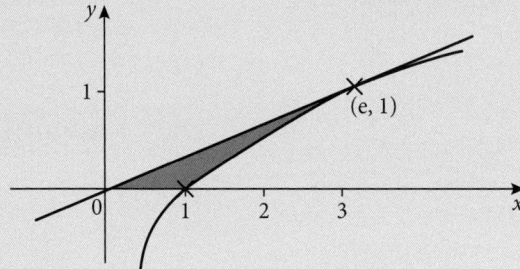

Use the result of part (b) to show that the area of this region is $\dfrac{1}{2}e - 1$

5. The main runway at Concordville airport is 2 km long. An aeroplane, landing at Concordville, touches down at point $T$, and immediately starts to slow down. The point $A$ is at the southern end of the runway. A marker is located at point $P$ on the runway.

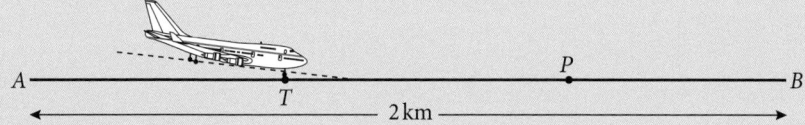

As the aeroplane slows down, its distance, $s$, from $A$, is given by

$s = c + 100t - 4t^2$

where $t$ is the time in seconds after touchdown, and $c$ metres is the distance of $T$ from $A$.

(a) The aeroplane touches down 800 m from $A$, (i.e. $c = 800$).

    (i) Find the distance travelled by the aeroplane in the first 5 s after touchdown

    (ii) Write down an expression for the velocity of the aeroplane at time $t$ s after touchdown, and hence find the velocity after 5 s.

    The aeroplane passes the marker at $P$ with a velocity of $36 \, m \, s^{-1}$. Find:

    (iii) how many seconds after touchdown it passes the marker

    (iv) the distance from $P$ to $A$.

(b) Show that if the aeroplane touches down before reaching the point $P$, it can stop before reaching the northern end, $B$, of the runway.

6. (a) Sketch the graph of $y = \pi \sin x - x$, $-3 \leqslant x \leqslant 3$, on millimetre square paper, using a scale of 2 cm per unit on each axis.

    Label and number both axes and indicate clearly the approximate positions of the $x$-intercepts and the local maximum and minimum points.

(b) Find the solution of the equation $\pi \sin x - x = 0$, $x > 0$

(c) Find the indefinite integral $\int (\pi \sin x - x) \, dx$ and hence, or otherwise, calculate the area of the region enclosed by the graph, the $x$-axis and the line $x = 1$

7. Using production and geological data, an oil company estimates that oil will be pumped from a field at the rate given by

$$r(t) = \frac{120t}{t^2 + 1} + 3, \quad 0 < t \leqslant 20$$

where $r(t)$ is the rate of production (in 1000 barrels per year) and $t$ is the number of years after production begins.

(a) When is the rate of production greatest?

(b) How many barrels of oil will the field produce in the first $t$ years of its useable life?

(c) It is estimated that after producing 250 000 barrels, the field will have covered the cost of the entire operation of drilling and set up. How long will that take?

(d) How many barrels of oil will the field produce in its entire useable life?

8. Note: radians are used in this question.

(a) (i) Sketch the graph of $y = x^2 \cos x$, for $0 \leqslant x \leqslant 2$ making clear the approximate positions of the positive intercept, the maximum point and the endpoints.

    (ii) Write down the approximate coordinates of the positive $x$-intercept, the maximum point and the endpoints.

**(b)** Find the exact value of the positive $x$-intercept for $0 \leqslant x \leqslant 2$

Let $R$ be the region in the first quadrant enclosed by the graph and the $x$-axis.

**(c)** **(i)** Shade $R$ on your diagram.

　　**(ii)** Write down an integral which represents the area of $R$.

**(d)** Evaluate the integral in part **(c)(ii)**, either by using a graphic display calculator, or by using the following information.

$$\frac{d}{dx}(x^2 \sin x + 2x \cos x - 2 \sin x) = x^2 \cos x$$

**9.** Note: radians are used in this question.

The function $f$ is given by
$f(x) = (\sin x)^2 \cos x$

The diagram shows part of the graph of $y = f(x)$

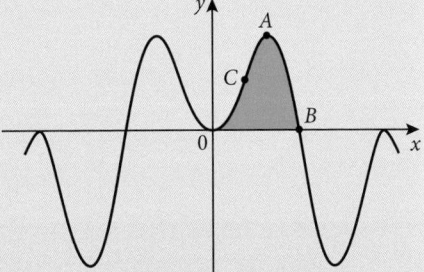

The point $A$ is a maximum point, the point $B$ lies on the $x$-axis, and the point $C$ is a point of inflection.

**(a)** Give the period of $f$.

**(b)** From consideration of the graph of $y = f(x)$, find to an accuracy of 1 significant figure the range of $f$.

**(c)** **(i)** Find $f'(x)$.

　　**(ii)** Hence show that at the point $A$, $\cos x = \sqrt{\dfrac{1}{3}}$

　　**(iii)** Find the exact maximum value.

**(d)** Find the exact value of the $x$-coordinate at the point $B$.

**(e)** **(i)** Find $\int f(x)\,dx$

　　**(ii)** Find the area of the shaded region in the diagram.

**(f)** Given that $f''(x) = 9(\cos x)^3 - 7 \cos x$, find the $x$-coordinate at the point $C$.

**10.** An automobile company is ready to introduce a new line of hybrid cars through a Europe wide sales campaign. After test marketing the line in a carefully selected group of cities, the marketing research department estimates that sales (in millions of euros) will increase at the monthly rate of $S'(t) = 10 - 10e^{-0.1t}$, $0 \leqslant t \leqslant 24$, $t$ months after the campaign has started.

**(a)** What will the total sales $S(t)$ be $t$ months after the beginning of the campaign if we assume that sales are zero at the beginning of the campaign?

**(b)** What are the estimated total sales for the first 12 months of the campaign?

**(c)** When will the estimated total sales reach €100 million?

**Table 16.5** Table for question 11

| $t$ | Increase in maintenance |
|---|---|
| 1 | 4000 |
| 2 | 4500 |
| 3 | 5000 |
| 4 | 6000 |
| 5 | 7000 |
| 6 | 8000 |
| 7 | 8000 |
| 8 | 10 000 |
| 9 | 10 500 |
| 10 | 12 000 |

11. Maintenance costs for an apartment building generally increase as the building gets older. Maintenance costs (in euros per year) for the last ten years for a particular apartment building are given in Table 16.5, where $t$ is the number of years after the building has been finished.

    (a) Set up a model to represent the rate of increase of maintenance costs as a function of years since its construction. Justify why you chose the specific model.

    (b) Use your model to give the total cost of maintenance from the end of the second year until the tenth year.

12. In the table below, $x$ is the cumulative percentage of families and $y$ is the cumulative percentage of total family income received for the USA in 2014.

| $x$ | 0.00 | 0.20 | 0.40 | 0.60 | 0.80 | 1.00 |
|---|---|---|---|---|---|---|
| $y$ | 0.000 | 0.009 | 0.059 | 0.158 | 0.344 | 1.000 |

    (a) Find an equation for a Lorentz curve for the data.

    (b) Use the equation found in (a) to approximate the Gini index for the USA in 2014.

13. A drone flying at an altitude of 78 m needs to drop supplies to a specific location on the ground. The drone is flying horizontally at a speed of 30.5 m s$^{-1}$. How far away from the target should the drone release the supplies in order to reach the target location?

14. A particle moves along a straight line. When it is a distance $s$ from a fixed point, where $s > 1$, the velocity $v$ is given by $v = \dfrac{3s + 2}{2s - 1}$
    Find the acceleration when $s = 2$.

15. An amusement company maintains records for each video game it installs in an arcade. Let $C(t)$ and $R(t)$ represent the total accumulated costs and revenues (in thousands of euros) $t$ years after a certain game has been installed. It is estimated that the marginal cost and the marginal revenue are given by $C'(t) = 2t, R'(t) = 5te^{-0.1t^2}$

    (a) Find the useful life of this game.

    (b) Find the area between the graphs of $C'(t)$ and $R'(t)$ over the useful life of the game and interpret the results.

**Table 16.6** Table for question 16

| X | Present | New |
|---|---|---|
| 0.20 | 0.0247 | 0.104 |
| 0.40 | 0.1215 | 0.256 |
| 0.60 | 0.3089 | 0.456 |
| 0.80 | 0.5986 | 0.704 |
| 1 | 1 | 1 |

16. The government of a small country is planning changes in the tax structure to provide a more equitable distribution of income. Table 16.6 gives the present distribution of income and the new one.

    (a) Find a Lorentz curve for each distribution and justify your choice of function.

    (b) Find the Gini index for each curve and determine whether the proposed changes provide a more equitable income distribution.

**17.** The acceleration, $a(t)$ m s$^{-2}$, during the first 80 s of motion of a train is given by $a(t) = -\frac{1}{20}t + 2$ where $t$ is the time in seconds.

If the train starts from rest at $t = 0$, find the distance travelled in the first minute of its motion.

**18.** In the diagram, $PTQ$ is an arc of the parabola $y = a^2 - x^2$, where $a$ is a positive constant, and $PQRS$ is a rectangle. The area of the rectangle $PQRS$ is equal to the area between the arc $PTQ$ of the parabola and the $x$-axis.

Find, in terms of $a$, the dimensions of the rectangle.

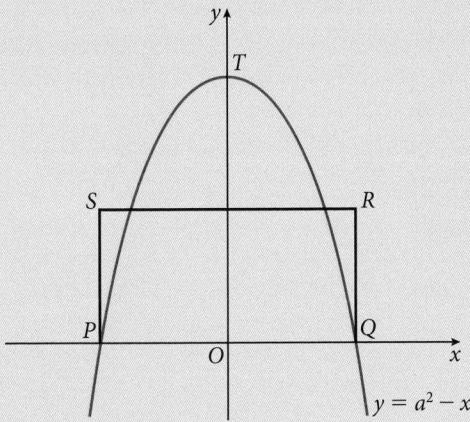

**19.** Consider the function $f_k(x) = \begin{cases} x \ln x - kx, & x > 0 \\ 0, & x = 0 \end{cases}$, where $k \in \mathbb{N}$

    **(a)** Find the derivative of $f_k(x)$, $x > 0$

    **(b)** Find the interval over which $f(x)$ is increasing.

The graph of the function $f_k(x)$ is shown.

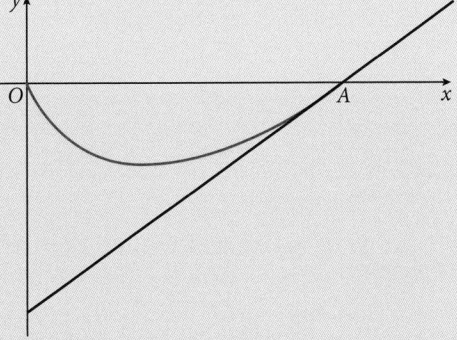

    **(c) (i)** Show that the stationary point of $f_k(x)$ is at $x = e^{k-1}$

        **(ii)** One $x$-intercept is at $(0, 0)$. Find the coordinates of the other $x$-intercept.

    **(d)** Find the area enclosed by the curve and the $x$-axis.

    **(e)** Find the equation of the tangent to the curve at $A$.

    **(f)** Show that the area of the triangular region created by the tangent and the coordinate axes is twice the area enclosed by the curve and the $x$-axis.

    **(g)** Show that the $x$-intercepts of $f_k(x)$ for consecutive values of $k$ form a geometric sequence.

20. World consumption of lead is $5.3e^{0.01t}$ million metric tons per year where $t$ is measured in years and $t = 0$ corresponds to 2016.

   Find a formula for the total amount of lead that will be consumed within $t$ years from 2016 and estimate when the known world reserves of 130 million metric tons will be exhausted. (Assume that consumption continues to follow the model and that no new resources are discovered.)

21. A uranium mine produces income at a constant rate $r(t) = 560\,000\sqrt{t}$ US\$ per year for a lifetime $T = 20$ years. Consider the interest rate $i = 0.05$ and is continuous. Find the capital value of this mine.

22. (a) The price-demand function and price supply function for a certain product is given by $D(x) = e^{-x^2}$ and $S(x) = e^{x^2} - 1$ where quantity $x$ is in 100 units and price is in tens of euros.

       Sketch both curves and shade the areas corresponding to consumer's surplus and producer's surplus.

    (b) Find the market equilibrium.

    (c) Find the values of consumer's surplus and producer's surplus.

23. Table 16.7 shows the income distribution of a certain country in 2014. $x$ represents the lowest-paid portion of all income recipients, and $y$ represents their income share in total income. Thus 0.2 represents the 20% lowest-paid income recipients and they get 12.47% of total income.

    (a) Draw a scatter plot of this data.

    (b) Suggest a model that describes the relationship between $x$ and $y$. Justify your model.

    (c) Use your model to draw a Lorentz curve for this country. Comment on its accuracy.

    (d) Calculate the coefficient of inequality for this country.

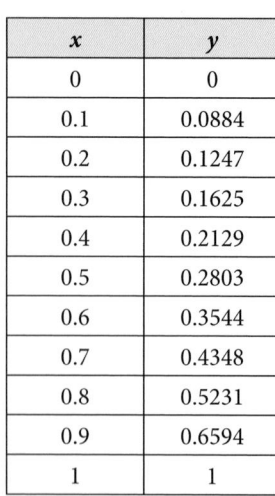

| x | y |
|---|---|
| 0 | 0 |
| 0.1 | 0.0884 |
| 0.2 | 0.1247 |
| 0.3 | 0.1625 |
| 0.4 | 0.2129 |
| 0.5 | 0.2803 |
| 0.6 | 0.3544 |
| 0.7 | 0.4348 |
| 0.8 | 0.5231 |
| 0.9 | 0.6594 |
| 1 | 1 |

**Table 16.7** Table for question 23

24. A particle is moving along a straight line so that $t$ seconds after passing through a fixed point $O$ on the line, its velocity $v(t)\,\mathrm{m\,s^{-1}}$ is given by

    $$v(t) = t \sin\left(\frac{\pi}{3}t\right)$$

    (a) Find the values of $t$ for which $v(t) = 0$, given that $0 \leqslant t \leqslant 6$

    (b) (i) Write down a mathematical expression for the total distance travelled by the particle in the first six seconds after passing through $O$.

       (ii) Find this distance.

25. A particle is projected along a straight-line path. After $t$ seconds, its velocity $v$ in metres per second is given by $v = \dfrac{1}{2 + t^2}$

    (a) Find the distance travelled in the first second.

    (b) Find an expression for the acceleration at time $t$.

26. An investor is presented with a choice of two investments: an established clothing store and a new computer store. Each choice requires the same initial investment and each produces a continuous income stream of 4%, compounded continuously. The rate of flow of income of the clothing store is $f(t) = 120\,000$, and the rate of flow of income from the computer store is expected to be $g(t) = 100\,000e^{0.05t}$.

Compare the future values of these investments to determine which is the better choice:

(a) over the next 5 years            (b) over the next 10 years.

27. A particle moves in a straight line with velocity, in metres per second, at time $t$ seconds, given by $v(t) = 6t^2 - 6t, t \geqslant 0$

Calculate the total distance travelled by the particle in the first two seconds of motion.

28. A particle moves in a straight line. Its velocity $v\,\text{m s}^{-1}$ after $t$ seconds is given by $v = e^{-\sqrt{t}} \sin t$

Find the total distance travelled in the time interval $[0, 2\pi]$

29. Consider the function $f(x) = \dfrac{1}{x^2 + 5x + 4}$

(a) Sketch the graph of the function, indicating the equations of the asymptotes, intercepts and extreme values.

(b) Find $\displaystyle\int_0^1 f(x)\,dx$

(c) Sketch the graph of $f(|x|)$ and hence determine the area of the region between this graph, the $x$-axis and the lines $x = -1$, and $x = 1$

30. A manufacturer of electric kettles guarantees their products for 1 year. If a kettle fails within one year, it gets replaced with a new one at no extra charge. The time to failure of the kettles after it is sold is given by the probability density function

$$f(t) = \begin{cases} 0.01\,e^{-0.1t}, & t \geqslant 0 \\ 0, & \text{otherwise} \end{cases}$$

where $t$ is in months. Find the probability that a randomly chosen kettle will fail:

(a) during the warranty period

(b) during the second year after purchase.

The manufacturer does not want to replace more than 6% of the kettles.

(c) For how long should they guarantee the kettles?

31. Figure 16.47 shows a tract of land with measurement in metres. A surveyor has measured its widths $w_i$ at 50-metre intervals with the results listed in the table. Estimate the area of this tract.

| X | 0 | 50 | 100 | 150 | 200 | 250 | 300 | 350 | 400 | 450 | 500 | 550 | 600 |
|---|---|----|-----|-----|-----|-----|-----|-----|-----|-----|-----|-----|-----|
| W | 0 | 165 | 192 | 146 | 63 | 42 | 84 | 155 | 224 | 270 | 267 | 215 | 0 |

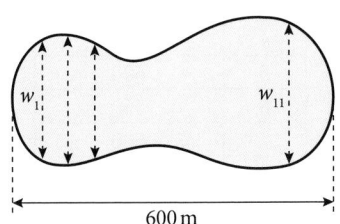

**Figure 16.47** Diagram for question 31

| $t$ | Consumption |
|---|---|
| 1 | 61.23 |
| 2 | 59.40 |
| 3 | 57.81 |
| 4 | 57.59 |
| 5 | 58.86 |
| 6 | 59.25 |
| 7 | 60.99 |
| 8 | 62.29 |
| 9 | 64.27 |
| 10 | 65.60 |
| 11 | 66.74 |
| 12 | 66.86 |
| 13 | 67.88 |
| 14 | 67.62 |
| 15 | 69.21 |
| 16 | 70.38 |
| 17 | 71.91 |
| 18 | 74.05 |
| 19 | 74.45 |
| 20 | 76.00 |
| 21 | 76.80 |
| 22 | 77.68 |
| 23 | 78.57 |
| 24 | 80.31 |
| 25 | 83.16 |
| 26 | 84.47 |
| 27 | 85.63 |
| 28 | 87.10 |
| 29 | 86.52 |
| 30 | 85.59 |
| 31 | 88.53 |
| 32 | 89.56 |
| 33 | 90.51 |
| 34 | 92.09 |
| 35 | 92.99 |
| 36 | 94.84 |

**Table 16.8** Data for question 32

| $t$ | Depreciation (£) |
|---|---|
| 1 | 910 |
| 2 | 800 |
| 3 | 700 |
| 4 | 660 |
| 5 | 600 |
| 6 | 550 |
| 7 | 500 |
| 8 | 450 |
| 9 | 400 |
| 10 | 370 |

**Table 16.9** Table for question 35

32. Table 16.8 shows world oil consumption in millions of barrels per year. $t$ is the time after 1979.

   (a) Looking at a scatter plot of the data, it appears that a cubic model may be a good fit. Give an estimate of the model and justify its appropriateness.

   (b) Use your model to estimate the total consumption of oil between 2000 and 2010.

33. Jim is planning a project for his retirement in 8 years. He will need £50,000 in his account in 8 years and has an arrangement with a financial institution that he deposits money into his account at a constant rate throughout the 8-year period and they will give him 6% interest continuously compounded.

   (a) At what rate (pounds per year) must he deposit each year in order to meet his goal?

   (b) If he decides, instead, to make one lump sum now, what amount should he deposit now?

34. The rate at which natural gas is extracted often increases at first until the easily accessible part is exhausted. Then the rate of extraction tends to decline.
   In a new field, gas is being extracted at a rate of $\dfrac{4t}{t^2 + 1}$ millions of cubic metres per year, where $t$ is in years since the field was opened. If this field has an estimated 16 million cubic metres of gas, how long will it take to exhaust this field at the given rate of extraction?

35. Table 16.9 shows the amounts of annual depreciation for the first 10 years of a piece of machinery purchased for £20,000.

   (a) Draw a scatter plot of the depreciation and suggest a model that best approximates it. Justify your choice.

   Note that depreciation is considered as a negative number, and its absolute value decreases with time.

   (b) Use your model to find a model that approximates the value of this machinery after $t$ years of service.

   (c) To keep the machinery working under good conditions, it needs to be maintained. The machinery must be replaced when its value reaches a value equal to its maintenance cost. The maintenance is annually increasing at the rate $200e^{0.2t}$ at the end of each year. During its first year ($t = 0$) it will need £1000 in initial maintenance. When should this piece of machinery be replaced?

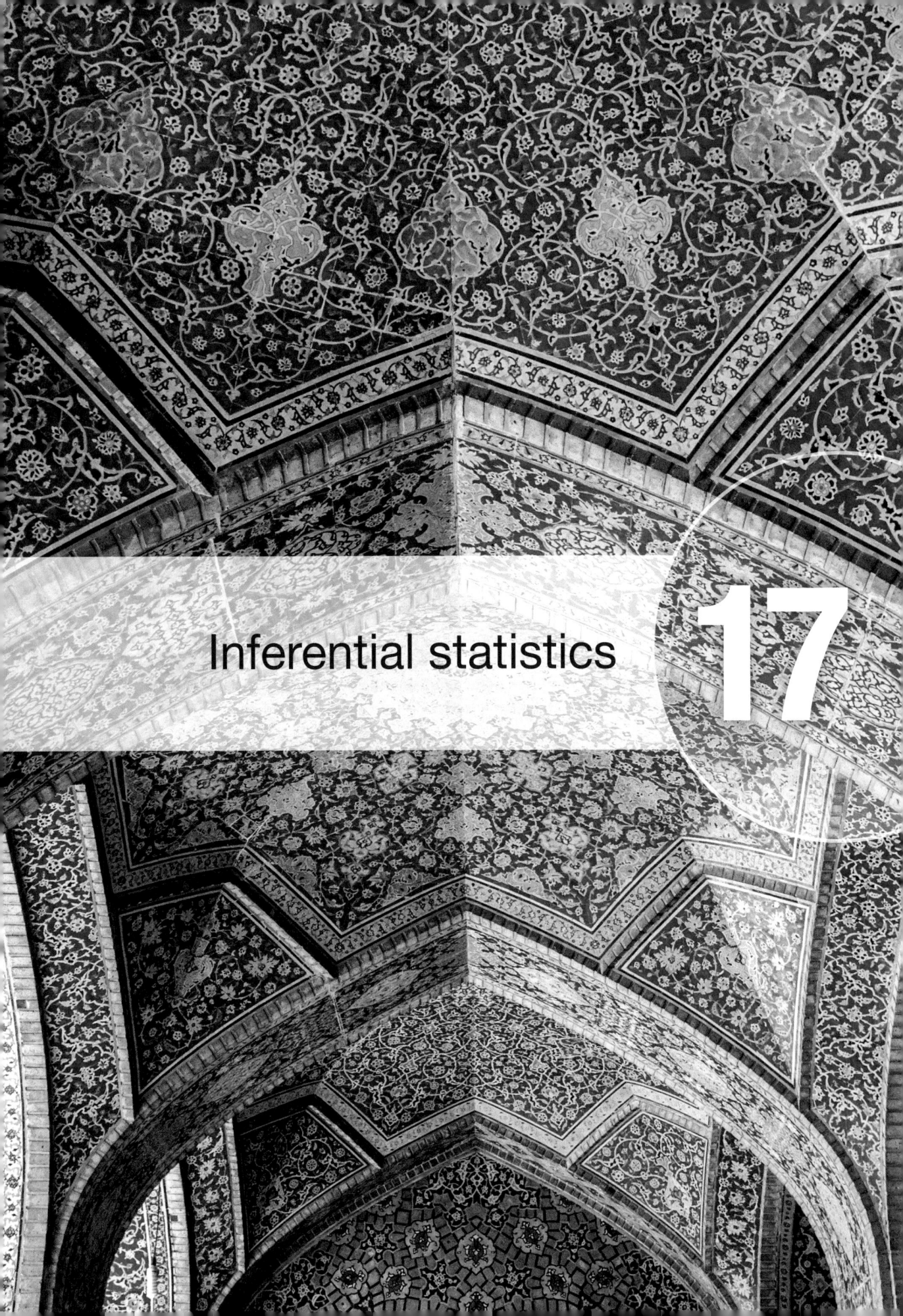

Inferential statistics

17

In this chapter, you will see a number of topics all related to **repeated sampling**, particularly in ensuring that statistical measures obtained through sampling represent the population parameters accurately.

## 17.1 Statistical inference, reliability, and validity

In **descriptive statistics**, data are analysed in terms of their distribution, cluster, and spread, and the results of the analyses are described visually through graphs, charts, and diagrams. In contrast, **inferential statistics** attempts to make a statement about an unknown or perhaps unknowable population by taking samples and generalising the findings from a sample to the entire population. It is merely a probabilistic model, based on the premise that a carefully selected random sample represents the population.

### Arriving at a generalisation

Statistical surveys are conducted to make a claim about a population parameter by analysing a sample from within the population. Surveys regarding upcoming elections, community issues, food and lifestyle preferences, and the economy are all conducted by selecting representative samples from which a probabilistic estimate of the population mean, proportion, or variance is constructed. The next time you read a newspaper article that claims it is based on an opinion poll, you are reading about an **inference** made from a survey sample. Where students and professionals differ is mainly in the generation of a representative sample. Professional organisations have the resources to enable them to select a random sample that students do not.

As noted earlier in Chapter 10, **reliability** is a measure of the consistency in data collection. If data and results are reproducible, the random errors in collection are deemed minimal, and the results are considered to be reliable.

Whether the collection of data addresses the survey objective is another matter entirely, and is a measure of the **validity** of the survey and conclusions obtained. Obviously, without reliability there is no validity; however, reliability does not necessarily validate the conclusions reached. For example, if your bathroom scales give you your mass consistently, you have reliability. However, if you use the results to measure your shoe size, you don't have validity.

There is no simple empirical measure of reliability or validity. However, as validity requires reliability, every study must start by ensuring that results are reproducible. There are two common strategies that address reliability to minimise errors in survey design.

- **Retesting:** to determine whether the initial results can be replicated, a survey or study can be conducted a second time to determine if the results are consistent.

- **Parallel testing:** a second version of the same survey or study is conducted using questions that test the same issues or concerns. You're probably familiar with this notion. Your teachers often design parallel quizzes and tests for different classes.

After a retest or parallel test, the results obtained can be assessed for **correlation** to quantify the degree to which the two sets of results may or may not be consistent. Correlation will be discussed in Chapter 19.

## Example 17.1

To determine the proportion of cars that are imports, a group of students conduct a survey at the shopping centre and record their findings. Before making a statement about their findings, they decide to ensure the reliability of their data. Give some of the ways they can do this.

### Solution

In order to ensure reliability, they should

- take a larger sample
- retest by varying the location and/or time of the study, thereby taking multiple samples.

## Exercise 17.1

1. Determine whether the following are descriptive statistics or inferential statistics.
   (a) One hundred students are randomly selected and the mean distance from home to school is calculated.
   (b) Thirty students are randomly selected and the mean distance from home to school is used to estimate the average distance for all students.
   (c) The average number of hours spent gaming by all students at your school is found by taking a random sample of 12 students.
   (d) Voter support for six candidates is found and displayed as a bar chart.

2. A company that sells sugarless chewing gum includes in its advertisements that three out of four dentists recommend sugarless gum. How might the company ensure that the statement is reliable?

3. A university nursing program requires top marks in Mathematics before students can apply successfully for admission. Comment on the validity of this requirement.

4. Public opinion regarding a new aquatic centre produced some unexpected results.

   (a) How might retesting be helpful in establishing the reliability of the results?

   (b) How might parallel testing be useful in doing the same?

5. Safety-conscious car buyers seem to prefer white vehicles over other colours. To test this hypothesis, you decide that you will check the colours of cars parked at the mall. How could reliability be addressed to ensure your findings can be justified?

6. Smart traffic lights are programmed to ensure that in the absence of traffic volume, the lights remain or switch to green in the predominant direction of travel. The predominant direction is determined by noting how frequently or infrequently vehicles trigger the sensors set into the pavement at intersections. If the number of cars at a particular intersection is recorded at 05:00 in the morning, what could you do to ensure the reliability of the findings?

7. Apparently, the eighth grade boys at your school are of the opinion that using men's cologne enhances their appeal and that the more they use, the better. Comment on the validity of this hypothesis.

## 17.2 Unbiased estimators

To make inferences about population parameters based on a single sample, we need to use **unbiased estimators**. Statisticians have found in practice that the sample mean $\bar{x}$ is as good an estimator of $\mu$ as one can find, but the population parameter $\sigma^2$ is better estimated by

$$s_{n-1}^2 = \frac{n}{n-1}s_n^2 = \frac{\sum_{i=1}^{k} f_i(x_i - \bar{x})^2}{n-1}$$

or its computational equivalent $s_{n-1}^2 = \frac{\sum_{i=1}^{k} f_i x_i^2}{n-1} - \frac{n}{n-1}\bar{x}^2$

The standard deviation would be $s_{n-1}$, but may be noted otherwise as $s$, $s_x$, $\sigma_{n-1}$, particularly on scientific calculators and in GDC output.

Fortunately, these unbiased estimators for the mean and standard deviation are readily available on your GDC. In the screenshot, the unbiased estimators of the mean and standard deviation are given as $\bar{x} = 4$ and $s_x \approx 2.16$

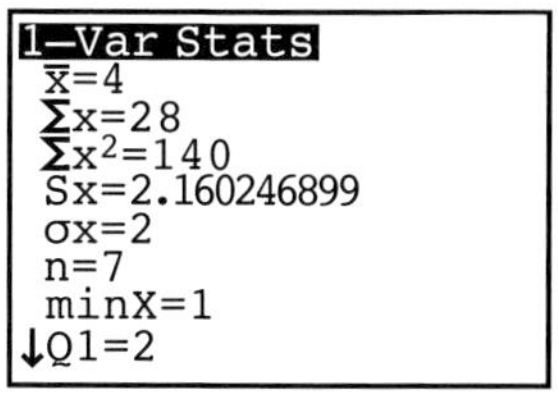

```
1-Var Stats
x̄=4
Σx=28
Σx²=140
Sx=2.160246899
σx=2
n=7
minX=1
↓Q1=2
```

**Figure 17.1** GDC output for mean and standard deviation

## Example 17.2

A sample of 30 boys' heights is taken from a population of 12th grade boys and is assumed to be normal. The measurements obtained in centimetres are:

{173, 181, 164, 183, 186, 178, 190, 178, 175, 180, 168, 169, 177, 166, 171, 184, 170, 169, 191, 178, 171, 165, 178, 182, 183, 170, 168, 173, 182, 180}.

Determine the unbiased estimates of the population mean and variance.

### Solution

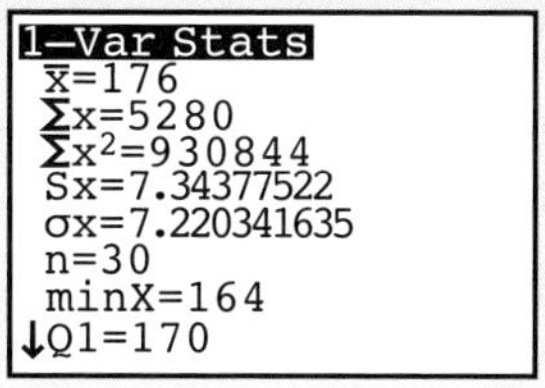

Using a GDC, we find

$\bar{x} = 176$ and $s_x \approx 7.34$, so

$s_{n-1}^2 \approx 53.9$

```
1-Var Stats
x̄=176
Σx=5280
Σx²=930844
Sx=7.34377522
σx=7.220341635
n=30
minX=164
↓Q1=170
```

As noted earlier in this chapter, you must be aware of the variability of sampling results. At best, these measures are **point estimates**. We will take a look at **interval estimates** that will account for quantifiable **margins of error** in section 17.4.

## Exercise 17.2

1. Your GDC produced two values, 4.3 and 4.15, each as the standard deviation. Which one is the unbiased estimate?

2. A sample yields the data {11, 12, 13, 14, 15, 16, 17, 18, 19}. Find the unbiased estimates of the population mean $\mu$ and variance $\sigma^2$, assuming the data are normally distributed.

3. A large number of students take part in a mathematics competition, and the scores out of 60 marks are considered to be normally distributed. Find the unbiased estimates of the population mean $\mu$ and variance $\sigma^2$ from the random sample of 30 scores shown in the table.

| Score | 35 | 38 | 40 | 43 | 47 | 51 | 52 | 54 | 55 |
|-------|----|----|----|----|----|----|----|----|----|
| Frequency | 1 | 4 | 5 | 3 | 7 | 4 | 2 | 2 | 2 |

4. A company that bottles soda plans to fill its 500 ml bottles with 505 ml of soda to make sure that the bottles contain at least 500 ml of soda. Their records indicate that the variance is $\sigma^2 = 2.6$ ml². A sample of 12 bottles contained the following amounts: {496.4, 500.2, 501.3, 507.3, 498.0, 500.3, 504.7, 508.2, 505.3, 491.2, 490.5, 507.4}

Find the unbiased estimates of the population mean $\mu$ and variance $\sigma^2$.

5. Using your GDC, you find that $\sigma = 6.1$ in a data set of size $n = 15$. What is the unbiased estimator $s_{n-1}^2$ of the variance?

6. You recorded the values of the two standard deviations that your GDC gave to 3 significant figures as $s_x \approx 5.31$ and $\sigma_x \approx 5.22$, but forgot to note the sample size. What would it have been?

7. Determine the size of the sample at which point the unbiased estimate of the population variance is within 1% of the true population variance.

## 17.3 Distribution of sample means

Consider a simple uniform distribution such as the distribution of outcomes in the repeated rolls of a fair dice. To obtain experimental data, roll a dice 6 times. The distribution of the outcomes may or may not look uniform in this first experiment; however, in theory, if you keep rolling the dice another 6 times, then another 6 times, and so on, the cumulative distribution will begin to look uniform.

Now, instead of merely keeping a tally, after rolling the dice 6 times, find the mean of the outcomes. It may or may not be the expected value of 3.5. Roll the dice another 6 times and record the mean again. If you generate many more sets of rolls and record their means, you will eventually have a **distribution of sample means**. What shape will this distribution take? The distribution of the means is clustered around the expected value. This is an interpretation of the **central limit theorem**. If you start with any other distribution, including ones for which there were no descriptions available prior to sampling, the results would have been the same: the distribution of sample means would be approximately normal, provided a large number of samples are taken.

When you sample from a large population, with values represented by the random variable $X$, the sample means have a distribution given by the random variable $\overline{X} \sim N\left(\mu, \dfrac{\sigma^2}{n}\right)$

Consider the random variable $X$ which has a distribution with $E(X) = \mu$ and $\text{Var}(X) = \sigma^2$

If we take $n$ independent observations $X_1, X_2, \ldots, X_n$ from $X$, then
$$E(X_1) = E(X_2) = \cdots = E(X_n) = \mu$$
$$\text{Var}(X_1) = \text{Var}(X_1) = \cdots = \text{Var}(X_1) = \sigma^2$$

Since $\overline{X}$ is the sum of the sample means, $\overline{X} = \dfrac{\sum_i X_i}{n} = \dfrac{1}{n}\sum_i X_i$

Hence
$$E(\overline{X}) = E\left(\frac{1}{n}\sum_i X_i\right) = \frac{1}{n}E\left(\sum_i X_i\right) = \frac{1}{n}\sum_i E(X_i) = \frac{1}{n}\cdot n\mu = \mu$$

and
$$\text{Var}(\overline{X}) = \text{Var}\left(\frac{1}{n}\sum_i X_i\right) = \frac{1}{n^2}\text{Var}\left(\sum_i X_i\right) = \frac{1}{n^2}\sum_i \text{Var}(X_i) = \frac{1}{n^2}\cdot n\sigma^2 = \frac{\sigma^2}{n}$$
$$\Rightarrow \sigma_{\overline{X}} = \frac{\sigma}{\sqrt{n}}$$

## Example 17.3

Samples of size 20 are taken ten times and their means recorded as
$$\overline{X} = \{5.6, 5.9, 5.1, 4.9, 5.4, 6.1, 5.7, 5.2, 4.8, 5.3\}$$
Determine the estimated population mean and variance from these means.

### Solution

Use your GDC, but keep in mind that $n = 10$, not 20.

You should find that $\mu = \overline{x} = 5.4$

and since $\sigma_{\overline{X}}^2 = \dfrac{\sigma^2}{n} \Rightarrow 0.402^2 = \dfrac{\sigma^2}{10}$

$$\Rightarrow \sigma^2 \approx 1.62$$

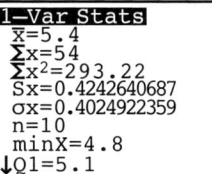

```
1-Var Stats
x̄=5.4
Σx=54
Σx²=293.22
Sx=0.4242640687
σx=0.4024922359
n=10
minX=4.8
↓Q1=5.1
```

## Exercise 17.3

1. Use your GDC to generate a random sample of size 20 from the uniformly distributed random variable $X = \{0, 1, 2, 3, 4, 5, 6, 7, 8, 9\}$ Find the mean of the sample and represent it by the random variable $\overline{X}$. Repeat this 29 more times so that you have 30 measures for $\overline{X}$. Estimate the population mean and variance from $\overline{X}$.

2. You participated in a health study in which you measured your mass at 07:30 each morning, then found the average (mean) body mass after two weeks. Twenty other randomly chosen girls do the same and the results in kilograms are given here:

   {51.2, 55.4, 65.3, 49.4, 54.6, 53.1, 50.5, 70.6, 75.1, 58.3,

   60.1, 47.7, 55.8, 49.9, 55.6, 57.3, 63.3, 61.1, 50.7, 73.6, 47.5}

   Find the unbiased estimates of the population mean and variance. Assume that body masses are normally distributed.

3. Reaction times to push a button in a video game are measured. Several measurements are taken for each of 30 players, but only their mean value is recorded using the random variable $\overline{X}$. Given that the mean values for 30 players has a variance of $\sigma_{\overline{X}}^2 = 0.03 \, \text{s}$, what is the value of the population variance $\sigma^2$?

4. A random sample of 50 waiting times for taxis is recorded and the mean calculated. This is repeated for ten days. Given that the standard deviation is 6.2 minutes, find an estimate for the population variance.

5. Prior to a national competition, 22 martial arts competitors, all in the same weight class, were asked to keep a record of their carbohydrate intake for the 7 days before their matches. Their **average** carbohydrate intake in grams were recorded. The means, represented by $\overline{X}$, are given in Table 17.1. Estimate the population mean and variance from $\overline{X}$.

| Average carbohydrate intake (g) | |
| --- | --- |
| 9 | 0 |
| 8 | 5 0 5 0 1 2 |
| 7 | 0 2 6 3 6 3 4 |
| 6 | 8 5 8 5 3 7 6 3 |

Table 17.1  Key: 8|5 means 85 g

The central limit theorem explains how repeated sampling from a known distribution results in an array of possible means, with a tendency to cluster, but with a variability that may not have been seen in the original distribution.

Consider a very large binomial population in which precisely 70% are marked. In a random sample of size 20, how many of the 20 items would be marked? This can be simulated by generating a random sample with your GDC by setting $n = 20$ and $p = 0.7$. The expected number of marked items is $E(X) = 14$ so it is likely that a value fairly close to 14 is obtained. If this is repeated another 99 times the results may look like Table 17.2.

The most likely number of marked items are clustered around the expected value and the least likely outcomes are the farthest from it. If you exclude the least likely 10% of outcomes, you can create a **90% box plot**. You may have to settle for a sum as close to 90% as possible, given the randomness of your sample. In the data set in Table 17.2, the 90% box plot is found by excluding the frequencies for 8, 9, 10 and 11 (1, 0, 2 and 5) on the left and the frequency for 18 (3) on the right. This gives a total of total of 11 or 11%.

This illustration looks much like a traditional box plot, except that 90% of the most likely outcomes are intended to lie within the box (Figure 17.2).

| Number marked in 20 | Frequency |
| --- | --- |
| 8 | 1 |
| 9 | 0 |
| 10 | 2 |
| 11 | 5 |
| 12 | 12 |
| 13 | 19 |
| 14 | 21 |
| 15 | 18 |
| 16 | 11 |
| 17 | 8 |
| 18 | 3 |

**Table 17.2** Result of simulating 100 trials

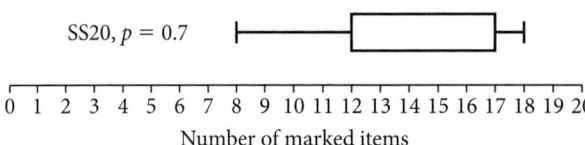

**Figure 17.2** 90% box plot of simulating 100 trials with $p = 0.7$

What this 90% box plot says is that even though the known probability of marked items is 70%, sampling observations show that results ranging from 12 out of 20 (60%) to 17 out of 20 (85%) occurred 90% of the time. This box plot shows the **probability interval** for sample size 20 (SS20) and $p = 0.7$

The 90% box plot generated by using your GDC for sample size 20 with $p = 0.8$ may look quite similar to Figure 17.3.

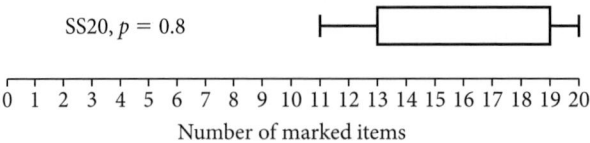

**Figure 17.3** 90% box plot for sample size 20 with $p = 0.8$

On the other hand, a 90% box plot for $n = 20$ and $p = 0.6$, would have shifted the box plot the other way and should look more like Figure 17.4.

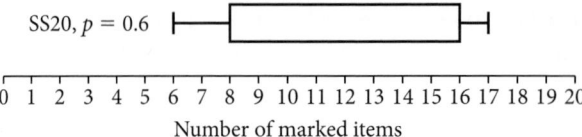

**Figure 17.4** 90% box plot for $n = 20$ and $p = 0.6$

The population proportions may be known to be precisely 60%, 70%, and 80%, but sampling produces a range of possible outcomes. Your results will likely be similar to those in Figures 17.2–17.4.

The set of box plots in increments of 2% calculated by using a binomial distribution is presented in Figure 17.5.

We look at probability intervals in order to make connections to inferences. Instead of looking at the 90% box plot chart **horizontally**, we will now see what we can conclude by looking at the chart **vertically**.

In a survey of an unknown population, if there are 14 marked items out of 20, the point estimate of the population value would be $p = 0.7$. However, with a sample as small as 20, reliability may be an issue. Knowing that there is a range of possible distributions from which a result of 14 out of 20 could have come 90% of the time, we can simply look vertically along 14 out of 20 to find them.

You'll see (Figure 17.6) that 14 out of 20 could have come from populations where the known percentage of marked items ranged from $0.5 \leqslant p \leqslant 0.86$. The point estimate of $p = 0.7$ will be called $\hat{p}$ and is around the middle of this range which is called the **confidence interval**. We are 90% confident that a survey result of 14 marked items out of 20 comes from a population in which 50% to 86% are marked.

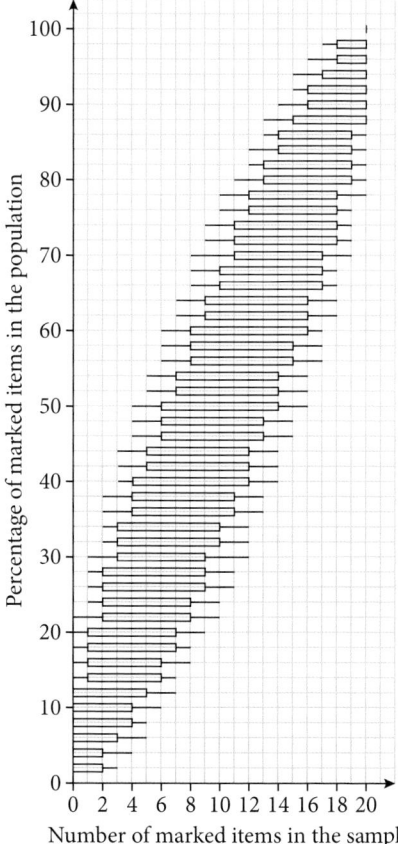

Figure 17.5  90% box plots for sample size 20

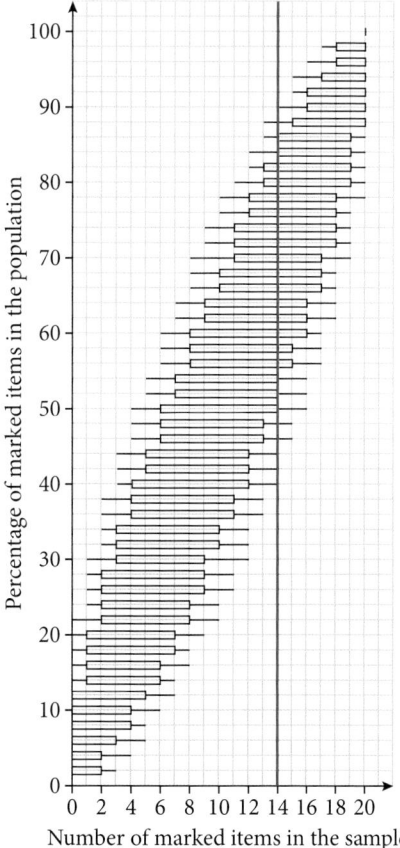

Figure 17.6  90% box plots for sample size 20

## Example 17.4

A random sample of twenty students shows that 10 students wear corrective lenses. Using the box plot in Figure 17.5, find the 90% confidence interval for the true percentage of corrective lens wearers.

## Solution

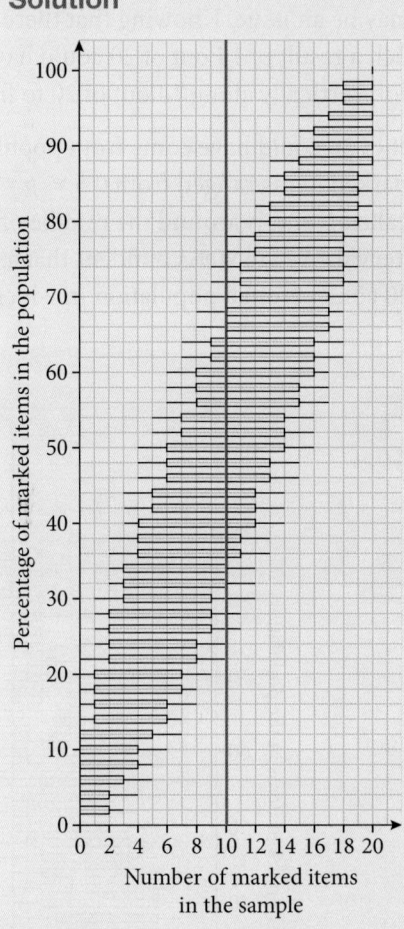

The lowest box plot that 10 marked out of 20 touches is 0.32, and the highest is 0.68. The point estimate is $\hat{p} = 0.5$, but the 90% confidence interval is $0.32 \leqslant p \leqslant 0.68$

The results should give reason to ponder the size of the interval. As 68% is clearly twice 32%, our findings are imprecise and less than definitive! The problem is with the limited sample size of 20. Our experience with the central limit theorem would suggest that repeated sampling or larger sample sizes should help. Compare this confidence interval to one that would be found if the sample size were doubled to 40, keeping $\hat{p} = 0.5$ (Figure 17.7). This would mean 20 marked items out of 40. Draw a vertical line at 20.

**Figure 17.7** 90% box plots for sample size 40

The confidence interval here is $0.38 \leqslant p \leqslant 0.62$

If the sample size is increased further to 100, the 90% confidence interval should be narrower still.

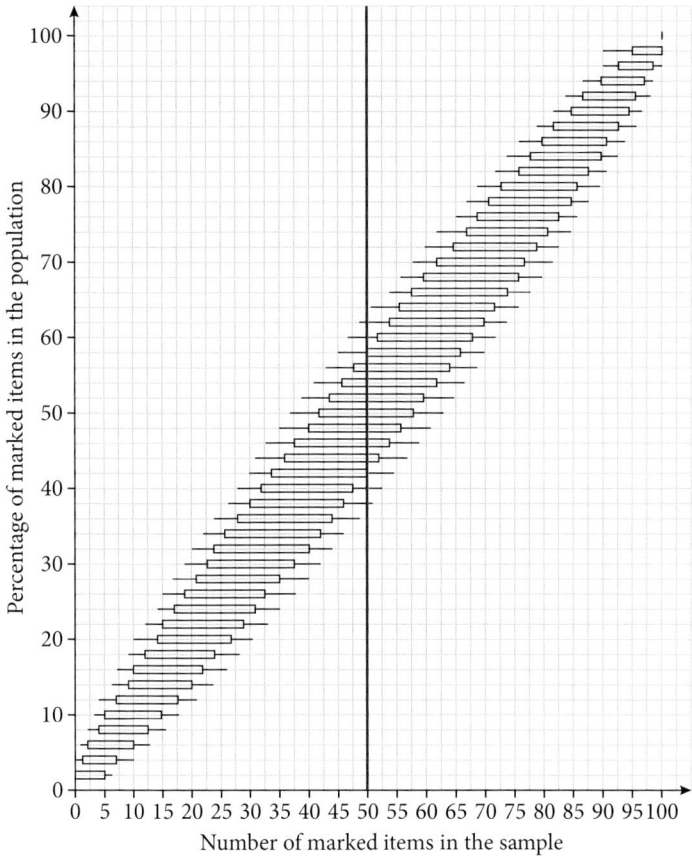

**Figure 17.8** 90% box plots for sample size 100

The 90% confidence interval with $\hat{p} = 0.5$ is now $0.42 \leqslant p \leqslant 0.58$ and is clearly much narrower. The difference between $\hat{p} = 0.5$ and the ends of the interval is $\pm 0.08$. This is called the **margin of error**. As you have seen, the margin of error is usually reduced by increasing the sample size. Of course, if instead of 90% box plots, you used a 40% box plot, thereby making the boxes shorter, the vertical line would intersect significantly fewer boxes. The problem is that you would only be 40% confident of your smaller interval, and that isn't much of an improvement!

## Confidence intervals made to any specification

You can find a confidence interval for any level of confidence. The size of the interval depends on the sample size.

In Figure 17.9, the screenshots show a sample size of $n = 100$, $\hat{p} = 0.5$, and a 90% confidence level which confirm what you've seen with the box plots.

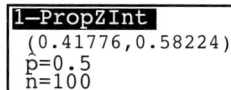

**Figure 17.9** The GDC screens confirm what you've seen with the box plots

```
1-PropZInt
x=50
n=100
C-Level:0.4
Calculate
```

```
1-PropZInt
(0.47378,0.52622)
p̂=0.5
n=100
```

**Figure 17.10** Confidence interval of 40% with $n = 100$ and $\hat{p} = 0.5$

```
1-PropZInt
x=66
n=110
C-Level:0.99
Calculate
```

```
1-PropZInt
(0.47968,0.72032)
p̂=0.6
n=110
```

**Figure 17.11** Solution to Example 17.5

If you had, in fact, wanted a confidence interval of 40% with $n = 100$ and $\hat{p} = 0.5$, your GDC would have provided the results shown in Figure 17.10.

The interval is noticeably smaller, but correct only 40% of the time, making it a poor predictor. A popular confidence level for many statistical inferences is 95%.

## Example 17.5

Find the 99% confidence interval for $p$, given that $\hat{p} = 0.6$ in a sample of size $n = 110$

### Solution

Use a GDC to find the confidence interval.

The 99% confidence interval is $0.48 \leqslant p \leqslant 0.72$

## Margin of error

Reducing the size of the confidence interval simply means reducing the margin of error. As you have seen, this is accomplished by adjusting the sample size. Again, since the central limit theorem allows us to consider $\hat{p}$ to be normally distributed, we can consider $\hat{p}$ to be the mean with a standard deviation. Instead of standard deviation this measure is called the **standard error of the proportion,** designated either as $\sigma_{\hat{p}}$ or $SE_p$

$$SE_p = \sqrt{\frac{\hat{p}(1 - \hat{p})}{n}}$$

## Example 17.6

A mayoral candidate commissions a polling firm to determine her level of voter support. A random sample of size $n = 300$ reveals the candidate's support to be at $\hat{p} = 0.35$. What is the standard error of the proportion in a sample of this size?

For the mayoral candidate, the margin of error to assess the margin by which she is leading or trailing her political opponents is just as important as the proportion of voter support.

### Solution

$\hat{p} = 0.35$, so $SE_p = \sqrt{\frac{0.35(1 - 0.35)}{300}} \approx 0.0275$

Once you have calculate $\sigma_{\hat{p}}$, all you need to find is the number of units of $\sigma_{\hat{p}}$ that would give you the margin of error to match the confidence level.

If you wish to find the margin of error for a 90% confidence interval, use the invNorm feature of your GDC to find that 90% of a standard normal distribution lie within $-1.645 < z < 1.645$, so the margin of error is

```
invNorm(.9,0,1,CENTER)
{-1.644853626 1.644853626}
```

**Figure 17.12** Solution to Example 17.6

$$\pm 1.645 \times \sqrt{\frac{\hat{p}(1 - \hat{p})}{n}}$$

## Example 17.7

(a) Find the margin of error at a 90% level of confidence, given $\hat{p} = 0.35$ and $n = 300$

(b) Just by looking at the formula for the margin of error, what sample size would be required if a margin of error half as large is desired?

### Solution

(a) $\pm 1.645 \times \sqrt{\dfrac{0.35(1 - 0.35)}{300}} \approx \pm 0.0453$

The margin of error is approximately $\pm 4.53\%$

(The support is approximately $30.5\% \leqslant p \leqslant 39.5\%$)

(b) The sample size $n$ will have to increase four-fold for the denominator to double; hence, 1200 people would have to be surveyed.

As noted earlier, a popular confidence level in many surveys is 95%. Since a 95% confidence interval is wider than one at 90%, if the mayoral candidate above wanted a smaller margin of error, but at a higher level of confidence, the survey sample size would have to be larger still.

To find the margin of error and confidence interval at a 95% level, use the **invNorm** feature of your GDC to find that 95% of a standard normal distribution lies within $-1.96 < z < 1.96$, so the margin of error is

$\pm 1.96 \times \sqrt{\dfrac{\hat{p}(1 - \hat{p})}{n}}$ and the 95% confidence interval is

$\hat{p} - 1.96 \times \sqrt{\dfrac{\hat{p}(1 - \hat{p})}{n}} < p < \hat{p} + 1.96 \times \sqrt{\dfrac{\hat{p}(1 - \hat{p})}{n}}$

```
invNorm(0.95,0,1,CENTER)
{-1.959963986  1.959963986}
```

**Figure 17.13** Using the invNorm function

A 90% confidence interval, derived from the 90% probability interval chart, excludes the least likely 10% of the outcomes even though such outcomes were observed. This 10% value is assigned the variable $\alpha$ and is called the **level of significance**. For a 90% confidence interval, $\alpha = 10\%$; at 95%, $\alpha = 5\%$. We will revisit $\alpha$ in Chapter 18.

## Confidence interval for means

We can now address surveys which produce a sample mean, where $\bar{x}$ is the point estimate, and the sample size determines the confidence interval for the population mean $\mu$ in the same manner that $\hat{p}$ and the standard error defined the confidence interval for the true population proportion, $p$.

As you saw with $\hat{p}$, there is a margin of error for $\bar{x}$ that depends on the standard deviation of the sample means, called the **standard error of the mean,** $\sigma_{\bar{x}}$ or $SEM = \dfrac{\sigma}{\sqrt{n}}$. A 90% confidence interval for $\mu$ would be similarly established as above, $\bar{x} - 1.645 \times \dfrac{\sigma}{\sqrt{n}} < \mu < \bar{x} + 1.645 \times \dfrac{\sigma}{\sqrt{n}}$

The discussion about confidence intervals for proportions is meant for clarification of the concept only. You will be examined on confidence intervals for the mean which is discussed next.

The only problem in using the *SEM* is that $\sigma$ (the population standard deviation) is in the formula and must be known. This is rarely the case when sampling, except in testing a hypothesis in which population parameters have been established from past data. For now, just keep in mind that in finding the confidence interval for the true population mean $\mu$, its standard deviation (or variance) must be known. In Chapter 18, we will analyse the more typical situation where the standard deviation is not known, and your understanding of confidence intervals will be important in the transition to testing conclusions.

### Example 17.8

A quality-control inspector is given the task of determining the amount of soft drink that is contained in a company's 1-litre bottles. Old inspection records provided some parameters: $\mu = 1010$ ml and $\sigma = 20$ ml. A random sample of 100 bottles obtained produced the statistic $\bar{x} = 1005$ ml. Determine the 95% confidence interval for $\mu$.

### Solution

Using your GDC, you should find the results shown in the screenshots.

The confidence interval is:

$$1005 - 1.96 \times \frac{20}{\sqrt{100}} < \mu < 1005 + 1.96 \times \frac{20}{\sqrt{100}} \Rightarrow 1001.1 < \mu < 1008.9$$

```
ZInterval
 Inpt:Data Stats
 σ=20
 x̄=1005
 n=100
 C-Level:0.95
 Calculate
```

```
ZInterval
 (1001.1,1008.9)
 x̄=30000
 n=30
```

**Figure 17.14** GDC screens for the solution to Example 17.8

### Exercise 17.4

1. Use the invNorm feature of your GDC to find the confidence interval within which the given percentage of the standard normal distribution, centered about the mean would lie.

   (a) 86%          (b) 95%          (c) 99%

2. In a random sample of 80, the number of defective items is found to be 8.

   (a) Use your GDC to find:
      (i) the sample proportion $\hat{p}$
      (ii) the 90% confidence interval for the population proportion $p$
      (iii) the margin of error at a 90% level of confidence

   (b) Additional sampling resulted in exactly 40 defective items in a total sample of 400, the same proportion as before. Compare the 90% confidence interval for this sample size to the one in part (a) (ii) to see if other statistics are also unchanged.

3. A civic survey finds that 65% of 400 public transit users surveyed would pay more if a more frequent schedule of buses and trains were available. Determine:

   (a) the 95% confidence interval for the population proportion $p$
   (b) the standard error and the margin of error at this level of confidence.

4. A software company wishes to commission a survey to determine how well known their company brand has become.
   (a) At the 95% level of confidence, determine the largest possible margin of error in a sample of size $n = 300$.
   (b) With the digit '8' being associated with 'fortune', the new Asian owners would be happier if the level of confidence were 88%. Find the margin of error with the same sample size.

5. Marine biologists in 2011 found that their best estimate for the western North Atlantic adult female loggerhead turtle population was $\mu = 38\,334$ and $\sigma = 2287$. Shortly thereafter, if a further 30 estimates had resulted in a mean of $\bar{x} = 30\,000$, determine the 95% confidence interval for the true population of these adult female sea turtles.

6. Inspectors are hired to determine if the production modifications specified by management resulted in any significant improvement to production time. Old records show that the company had been manufacturing industrial pumps for a few years with $\mu = 47$ minutes and $\sigma = 3.7$ minutes. The production times of 30 pumps yields $\bar{x} = 45$ minutes. Find the 95% confidence interval for $\mu$ and explain why the changes were or were not effective.

7. An amateur golf player's score for a round over the past two years have averaged 120 with a standard deviation of $\sigma = 5.1$. This year, in 10 rounds, his average was $\bar{x} = 115$. By finding a 95% confidence interval, determine if there was notable improvement.

## Chapter 17 practice questions

1. Indicate whether these sampling activities use descriptive or inferential statistics.
   (a) Fifty students are randomly selected and asked about the amount of spending money they are given in a week. The mean and variance are then found.
   (b) Thirty students are randomly selected and the mean of their spending money is used to estimate the average amount for all students.
   (c) The research and development team of a car manufacturer deliberately crashes 20 new vehicles all manufactured under the same specifications to determine the average damage done to them at various speeds. The average cost of repair damage is included in their final report and projected to all vehicles of this specification in sales presentations.
   (d) The annual fair sells hotdogs, each associated with a political party. Consumers can purchase the liberal hotdog, the conservative hotdog, or socialist hotdog. The results (with tongue-in-cheek), are published as percentage support for the three parties.

2. The school cafeteria is offering a new dessert that they claim is the most delicious, low-fat, low-carbohydrate, high-protein dessert ever put on the menu. How would you design a study to validate their claim?

3. A classroom sample resulted in the finding that 8 out of 20 students skipped breakfast. Comment on the reliability and validity of the inference that 40% of the students are from families that do not eat together.

4. The school administration did not expect the response from a student sample on the issue of relaxing the dress code for school-sponsored activities.

   (a) How might re-testing be helpful in establishing the reliability of the results?

   (b) How might parallel testing be useful in doing the same?

5. The boys have noticed that there are more girls in the gym recently. The gym supervisor keeps a record of students when they sign in, so the number of boys and girls can easily be tallied for any time period in the school year. How would you select a random sample to determine the gender divide? What would you do to ensure the reliability of your findings?

6. Some students were conducting a test of two popular sodas in the hallway as part of their statistics assignment. While blindfolded, each subject was given soda A then soda B and asked to indicate a preference. How would you enhance the reliability of your findings in this test?

7. The 2018 Canadian national men's hockey team was comprised of 25 professionals whose ages are {19, 25, 22, 24, 28, 28, 27, 25, 29, 21, 26, 20, 20, 23, 25, 35, 28, 31, 21, 23, 25, 21, 27, 25, 21}.

   Find the unbiased estimate of the population mean $\mu$ and variance $\sigma^2$ of Canadian professional hockey players, assuming the data are normally distributed.

8. With an approaching hurricane, storm monitoring stations located along the Atlantic coast of North America measure the wind speed and pass them on to the National Hurricane Center in the US. The following data were recorded of the maximum sustained winds in kilometres per hour for one hurricane, two days after an advisory was issued.

   {140, 148, 130, 144, 150, 142, 153, 143, 145, 150}.

   Determine the unbiased estimate of the true mean and variance.

9. Carrots are sold in one kilogram bags, but the actual mass will vary, naturally. The growers maintain that they check to ensure that their bags provide consistency to the customer. A sample of 10 bags contained the following amounts in kilograms: {0.966, 1.202, 1.121, 1.130, 1.115, 1.097, 0.896, 1.240, 1.104, 1.089}. Find the unbiased estimate of the true population mean and standard deviation.

10. Using your GDC, you find that $\sigma = 4.3$ in a data set of size $n = 30$. What is the value of the unbiased estimator $s_{n-1}$?

11. The values of the two standard deviations that your GDC gave to 3 significant figures with its notations are $s_x \approx 2.66$ and $\sigma_x \approx 2.61$. What is the sample size?

12. Rainfall is typically measured at prominent locations, such as the airport or city hall by multiple devices and the average reading found. However, when a municipality is large with varied terrain, the same process is also undertaken in different parts of the city and each average rainfall can be represented by the random variable $\overline{X}$. If such measurements are collected in 20 locations with an average of $\overline{x} = 5.2\,\text{mm}$ and $\sigma_{\overline{X}}^2 = 0.4\,\text{mm}^2$ on one day, estimate the population mean and variance from $\overline{X}$.

13. Laboratory rats are given a maze to complete, with the lab assistant recording the time until completion. Several attempts are made by each rat and the mean of the recorded time is noted as a representative value by the random variable $\overline{X}$. Twenty rats are tested this way and their mean times recorded. The mean of the sample means is $\overline{x} = 38.5\,\text{s}$ and the sample standard deviation is $\sigma_{\overline{X}} = 5.3\,\text{s}$. Determine the population mean and variance.

14. Strawberry pickers working in a field are fast, but careful not to damage the ripe fruit or plants. As their collection bins are weighed several times a day, an average mass of collected berries per weighing can be found. If 20 pickers were able to bring back the following amounts in kilograms of strawberries per bin, determine the true mean and variance of the weight of a bin of berries:
    {10.4, 12.5, 11.3, 12.2, 12.8, 10.7, 13.0, 11.6, 12.1, 11.5, 12.3, 12.5, 11.4, 10.5, 12.3, 13.4, 11.5, 10.5, 12.3, 12.3}

15. Use your GDC to find the confidence interval within which the given percentage of the standard normal distribution, centered about the mean would lie.
    (a) 80%              (b) 90%              (c) 97.5%

16. In a random sample of 100 pennies, a coin collector finds that the number of pennies minted in 2017 is 18. Use your GDC to find:
    (a) the sample proportion $\hat{p}$
    (b) the 90% confidence interval for the population proportion $p$ of all pennies minted in 2017 currently in circulation
    (c) the margin of error at a 95% level of confidence.

17. An opinion poll finds that 58% of 200 public library users surveyed would favour the proposal that the library add more resources in languages other than English. Determine:
    (a) the 95% confidence interval for the population proportion $p$.
    (b) the standard error and the margin of error at this level of confidence

18. A company designing Canadian down jackets wishes to commission a survey to determine how well their company brand has become known in Europe.

    (a) At the 95% level of confidence, determine the margin of error in a sample of size $n = 100$

    (b) Find the sample size necessary to reduce the margin of error by half.

19. The artistry of the glassblower belies the incredibly harsh conditions under which he works. The glass is formed in a crucible from raw ingredients at a temperature around 1300°C, whereas the blowing is done at a 'cooler temperature' around 1000°C. The temperature within the first furnace is kept constant, but inevitably, there is some variability.

    In order to strive for consistency, records were kept which showed the mean temperature was 1320°C with a standard deviation of 60°C. The last 20 measurements this year showed a mean temperature of $\bar{x} = 1240$°C, perhaps indicating a need for repairs. Construct a 95% confidence interval for the true mean temperature to determine whether the intended 1300°C temperature is being achieved.

# Statistical tests and analyses

# 18

By the end of this chapter, you should be familiar with...

- the *t*-test and degrees of freedom
- testing a hypothesis
- type I and type II errors
- the *t*-test to compare two means
- the chi-squared test of goodness of fit
- the chi-squared test of independence.

With our understanding of probability distributions and statistical inference, we will now look at the analysis and testing of sample data. Statistical tests are conducted within a strict framework, called hypothesis testing, that requires each analysis to consider sample size, known and unknown population parameters, and alternative hypotheses. All of these tests can be done with the help of your GDC.

## 18.1 The Student *t*-test

The *t*-distribution was developed by William Gosset who analysed results of small sample batches at the Guinness Brewery, and published his papers under the pseudonym of Student. The *t*-distribution is the analysis of choice, particularly when small samples are taken from a normal distribution while the population variance is unknown. The *t*-distribution is standardised with a mean of $t = 0$ and is very similar to the standard normal $z$-distribution, except that the sample size makes a difference.

When samples are taken to establish a true population parameter, such as the mean or proportion about which there is a speculated value, the population variance will not be known. Hence, it will be lacking in one measurement called a **degree of freedom**. If the sample size is $n$, this means that there will be $n - 1$ degrees of freedom, noted as $v$.

In statistical analysis, much like in mathematical proof, one counter-example is sufficient to reject a claim or generalisation; however, one result is not adequate to substantiate it. The best we can do is either reject a claim or fail to reject a claim.

As you saw in Chapter 17, where a margin of error provided a feasible set of outcomes defined as a confidence interval, there was a set of outcomes that were rejected as being too extreme. At a level of significance of $\alpha = 0.1$, the confidence interval left behind 10% of the extreme, least likely outcomes represented by $\alpha$. In Figures 18.1–18.3, $\alpha$ identifies the size of the **rejection region**. If rejecting a claim requires dismissing values at one extreme only, a **one-tailed test** is used and if it requires dismissing values at both extremes, a **two-tailed test**, where $\alpha$ will need to be split into two parts, each of size $\frac{\alpha}{2}$ is used. A confidence interval is an example of a two-tailed test.

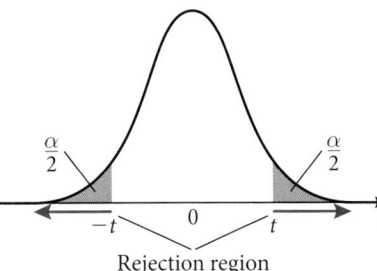

**Figure 18.1** Two-tailed test

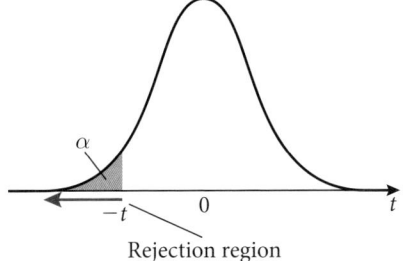

**Figure 18.2** Lower (or left)-tailed test

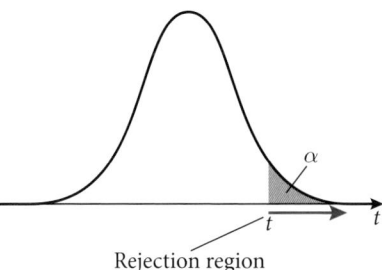

**Figure 18.3** Upper (or right)-tailed test

## Confidence interval for means revisited

In Chapter 17, you used your GDC to find that 95% of the values in a standard normal distribution lie within $-1.96 < z < 1.96$ (Figure 18.4)

Similarly, your GDC can provide you with the $t$-values within which lie 95% of the distribution for any value of $v$. For example Figure 18.5 shows the output with a sample size of $n = 10$, the population variance unknown and $v = 9$

The output states that the rejection region starts at $t \approx 1.83$, but that the single region encompasses the entire 5%. The GDC screen shows the input in the form invT$(1 - \alpha, v)$ to represent $t_{v, \alpha}$

If you need to establish rejection regions at extreme values found at both ends of the spectrum, use $\dfrac{\alpha}{2} = 0.025$ to obtain the result in Figure 18.6.

Here, it shows $t_{9, 0.025} \approx 2.26$, indicating that the 2.5% rejection region in the right-tail starts at $t \approx 2.26$

```
invNorm(0.95,0,1,CENTER)
{-1.959963986 1.959963986}
```

**Figure 18.4** 95% confidence interval

```
invT(.95,9)
                1.833112923
```

**Figure 18.5** $t_{9, 0.05}$

```
invT(.975,9)
                2.262157158
```

**Figure 18.6** $t_{9, 0.025}$

### Example 18.1

To test a manufacturer's claim that the mean breaking point of their canvas shopping bags is 50 kg, a sample of ten such bags was tested, yielding a mean of $\bar{x} = 47.5$ and a standard deviation of $s_{n-1} = 5$

Determine the 95% confidence interval for the true population mean $\mu$ to determine whether the manufacturer's claim can be justified.

### Solution

The true population mean $\mu$ is not known. The speculated population mean, which we will call $\mu_0$, is given, but the population variance $\sigma^2$ is not, so there are $v = 10 - 1 = 9$ degrees of freedom. Using the values given above, $v = 9, \dfrac{\alpha}{2} = 0.25$, and $t_{9, 0.025} \approx 2.26$, the 95% confidence interval correct to 3 significant figures is:

$$\bar{x} - 2.26 \times \frac{s_{n-1}}{\sqrt{n}} < \mu < \bar{x} + 2.26 \times \frac{s_{n-1}}{\sqrt{n}}$$

$$47.5 - 2.26 \times \frac{5}{\sqrt{10}} < \mu < 47.5 + 2.26 \times \frac{5}{\sqrt{10}}$$

$$43.9 < \mu < 51.1$$

The calculation is built into the GDC which can be used to provide the confidence interval shown in the screenshots.

```
TInterval
 Inpt:Data Stats
 x̄=47.5
 Sx=5
 n=10
 C-Level:0.95
 Calculate
```

```
TInterval
 (43.923,51.077)
 x̄=47.5
 Sx=5
 n=10
```

Since $\mu_0$ is within this interval for $\mu$, we have collaborating evidence and do not reject the claim that $\mu = 50$

## Using the $t$-test with a single sample

By now, you should be familiar with using a confidence interval to determine whether sampling data produce an interval within which a claim might lie to support the claim, or to reject the claim if the interval does not. The testing can be even simpler. If the probability of a sample outcome is less than $\alpha$, then it is unlikely and the claim can be rejected. Such a result is known as being **statistically significant**. This reasoning is particularly useful if the rejection region lies to one side only, based on claims that involve an inequality, such as $\mu > a$ or $\mu < a$

### Example 18.2

The manufacturer of a wireless game controller claims that the mean life of the battery it uses is at least 1000 hours. Testing 20 game controllers, the mean was found to be $\bar{x} = 970$ hours and the standard deviation was $s_{n-1} = 100$ hours. Determine whether or not the sample statistics support their claim at the $\alpha = 0.05$ level of significance.

### Solution

The manufacturer's claim is that $\mu \geqslant 1000$; hence, statistical evidence showing a mean greater than 1000 merely provides collaborating evidence, but nothing conclusive. There is only one way to reject the claim: produce statistics that show the mean is significantly less than the one in the claim. How much less? Is 970 sufficiently less than 1000? Start by considering the least likely 5% of sample means of size 20 if the true mean is in fact equal to $\mu_0$.

Using a GDC, we can find that $t_{19,\,0.05} \approx 1.73$ so if the true mean is 1000, the rejection region would be found at a $t$-score that is approximately 1.73 below the mean or $t \approx -1.73$. This value is

$$\mu_0 - 1.73 \times \frac{s_{n-1}}{\sqrt{n}} = 1000 - 1.73 \times \frac{100}{\sqrt{20}} \approx 961$$

The results show that the claim, $\mu \geqslant 1000$, can be collaborated by a sample mean of 961 or greater. Since 970 falls above it, there is no reason to reject the claim.

```
invT(0.95,19)
           1.729132792
```

**Figure 18.7** GDC screen for the solution to Example 18.2

Note that the sample mean could have been as low as 961 to provide collaborating evidence that the true mean is $\mu \geqslant 1000$. This may not be intuitive at all. A little reflection on the nature of a probability model with $n = 20$ and $s_{n-1} = 100$ should help.

In Example 18.2, we did not reject the claim that $\mu \geqslant 1000$ since the sample mean of $\bar{x} = 970$ fell outside the one-tailed rejection region with $\alpha = 0.05$

The sample mean seemed to be a contradiction of the claim. Use a GDC to run a one-tailed $t$-test. Make sure that the rejection region is on the negative side, highlighted on the third to last line of the screenshot, to see how likely or unlikely the result is.

The screenshot shows the rejection region to be $\mu < 1000$. The **p-value**, the probability of an outcome as extreme as $\bar{x} = 970$, when the true mean is 1000, is shown in the middle of the screen and at the bottom of the next screen. As you can see, its $p$-value at approximately 0.0978 is above the level of significance, $\alpha = 0.05$, thereby giving no reason to reject the claim.

Later, in Section 18.4, you will use the $t$-test again to compare two different samples as well as to analyse two samples involving the same respondents, taken at different times by measuring the differences.

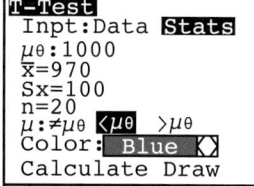

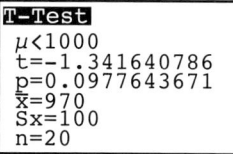

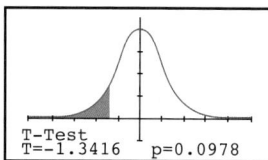

**Figure 18.8** The input, output, and graph of the solution to Example 18.2

## Exercise 18.1

1. Find the number of degrees of freedom $v$ in each instance.

   (a) A sample of size $n = 20$ is taken from a normally distributed population whose mean is $\mu = 92$, but the variance is unknown.

   (b) Twelve randomly selected individuals are monitored to see how long it takes before for medication that dilates the pupils wears off to verify a study that claims $\mu = 6.1$ hours.

   (c) A supplier of orchids selects 30 plants randomly to determine whether or not there is an average of at least 8 blooms on each plant as advertised. (The Guinness world record is 106!)

2. Determine the critical $t$-values given the sample size $n$ and the level of significance $\alpha$.

   (a) $n = 13$ and $\alpha = 0.05$, right-tailed

   (b) $n = 19$ and $\alpha = 0.01$, left-tailed

   (c) $n = 25$ and $\alpha = 0.05$, two-tailed

   (d) $n = 13$ and $\alpha = 0.05$, two-tailed

3. Determine whether these results obtained are statistically significant.

   (a) Right-tailed test, $\alpha = 0.05$ and $p$-value $= 0.01$

   (b) Left-tailed test, $\alpha = 0.05$ and $p$-value $= 0.001$

   (c) Two-tailed test, $\alpha = 0.05$ and $p$-value $= 0.03$

   (d) Two-tailed test, $\alpha = 0.01$ and $p$-value $= 0.006$

   (e) Right-tailed test, $t$-statistic $> t$ critical value

   (f) Left-tailed test, $t$-statistic $> t$ critical value

4. Find the 95% confidence interval for the true mean, given
   $n = 18$, $\bar{x} = 120$, and $s_{n-1} = 15$

5. Find the 99% confidence interval for the true mean, given
   $n = 18$, $\bar{x} = 120$, and $s_{n-1} = 15$

6. From a normally distributed population, a sample of size $n = 26$ yields
   $\bar{x} = 540$, and $s_{n-1} = 80$. At a level of significance of $\alpha = 0.05$, do the
   statistics support or refute the claim that the true population mean is
   $\mu = 500$?

7. A cordless screwdriver is advertised as having a charger that will fully
   charge the tool in an average time of 4.5 hours. A random sample of 10
   such chargers are tested and found to have a mean charging time of
   $\bar{x} = 4.8$, and $s_{n-1} = 1.6$ hours. Test the claim, at the level of significance
   of $\alpha = 0.05$

## 18.2 Hypothesis testing of means

If a claim is made that contains an equality, it is called the **null hypothesis**
with null implying no difference, hence an equality and designated by $H_0$.
Its opposite is called the **alternative hypothesis**, and is noted as $H_1$. The claim
or its opposite, whichever contains the equality is $H_0$, and in your analysis,
your task is to determine whether to reject $H_0$ or to not reject it. Table 18.1
shows some possible claims and how hypothesis testing is carried out.

| Claim | Opposite | Hypotheses | Rejection region(s) indicated by $H_1$. Consequence of rejection |
|---|---|---|---|
| $\mu > 20$ | $\mu \leqslant 20$ | $H_0: \mu \leqslant 20$ <br> $H_1: \mu > 20$ | Right-tailed, the size of the rejection region is $\alpha$. If $H_0$ is rejected, we have evidence to support the claim. |
| $\mu \geqslant 20$ | $\mu < 20$ | $H_0: \mu \geqslant 20$ <br> $H_1: \mu < 20$ | Left-tailed, the size of the rejection region is $\alpha$. If $H_0$ is rejected, we have evidence to reject the claim. |
| $\mu < 20$ | $\mu \geqslant 20$ | $H_0: \mu \geqslant 20$ <br> $H_1: \mu < 20$ | Left-tailed, the size of the rejection region is $\alpha$. If $H_0$ is rejected, we have evidence to support the claim. |
| $\mu = 20$ | $\mu \neq 20$ | $H_0: \mu = 20$ <br> $H_1: \mu \neq 20$ | Two-tailed, the size of each rejection region is $\frac{\alpha}{2}$. If $H_0$ is rejected, we have evidence to reject the claim. |

**Table 18.1** How hypothesis testing is carried out

It is also possible to test a claim at one tail only. For example, if a claim is that
$\mu = 350$, but you wish to test only the alternative that $\mu > 350$, then the entire
rejection region of size $\alpha$ will lie in the right tail.

## Example 18.3

A security firm claims that their personnel are dispatched to investigate incidents more than 30 times per week. In a random sample of 12 weeks of data, the mean number of such investigations is $\bar{x} = 31$ with a standard deviation of $s_{n-1} = 5.6$. State the null and alternative hypotheses and test the data at the $\alpha = 0.05$ level of significance.

### Solution

Since the claim is $\mu > 30$, its opposite is $\mu \leqslant 30$. As the opposite contains the equality, it is the null hypothesis $H_0$. The claim is assigned to the alternative $H_1$.

$H_0: \mu \leqslant 30$

$H_1: \mu > 30$

The rejection region is always given by $H_1$, hence it is right-tailed.

Enter the relevant information into a GDC as shown in Figure 18.9.

Using either the calculation or graphic screen, note that the $p$-value is approximately 0.274 and larger than $\alpha$; hence, we do not reject the null hypothesis $H_0$, and reject $H_1$, the alternative. Since $H_1$ is the claim, it is rejected. Note how the observed sample mean appeared to collaborate the claim, yet statistically, with the standard deviation being as large as it was, it is rejected.

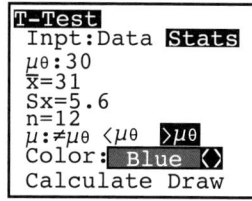

Figure 18.9 GDC screen for solution to Example 18.3

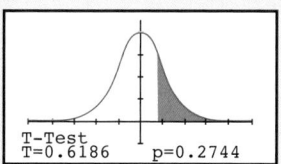

As a quick comparison, consider Example 17.3 with $s_{n-1} = 1.5$ The results would have been quite different with the $p$-value less than $\alpha$, prompting us to reject the null hypothesis.

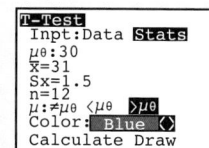

## Exercise 18.2

1. State the null and alternative hypotheses for each given claim about the true value of the population mean.

   (a) $\mu \geqslant 128$    (b) $\mu < 100$    (c) $\mu > 42$    (d) $\mu \leqslant 13$

2. The organisers of a mathematics competition taken by hundreds of students claim that the mean score on this year's competition was at least 48 out of 75. In a random sample of 16 students who participated in the competition, $\bar{x} = 43.1$, and $s_{n-1} = 6.2$. State $H_0$ and $H_1$ then test the organisers' claim at a level of significance of $\alpha = 0.05$

3. The manufacturers of digital calipers claim that the over-measurement error is at most 0.1 mm. Assume that the errors are normally distributed. A random sample of 12 calipers shows over-measurement errors with $\bar{x} = 0.09$ mm, and $s_{n-1} = 0.08$ mm. State $H_0$ and $H_1$ then test the manufacturers' claim at a level of significance of $\alpha = 0.05$

4. The distributor of frozen Alaskan salmon, sold by mass, claims that there are 7 to 9 pieces of salmon in a package. A random sample of 8 packages shows there were $\bar{x} = 9.5$ pieces of fish with $s_{n-1} = 3.1$.
Use a confidence interval with a level of significance $\alpha = 0.05$ to determine if there were more pieces than claimed. Hence, determine whether or not the pieces were smaller than expected.

5. There is a route up Grouse Mountain in Vancouver called the Grouse Grind that is used by about 100 000 people a year. The route is 2.9 km long with a vertical climb of 853 m. According to a website, the average time to walk up it is 1.5 to 2 hours. A random sample of 20 walking times shows the mean walking time in minutes is $\bar{x} = 140$, with $s_{n-1} = 20$. State the null and alternative hypotheses and test the claim that 2 hours is a reasonable upper limit.

6. A sport fishing company guarantees at least 23 kg of salmon and halibut are caught each day. A random sample of 10 catches indicate $\bar{x} = 21$ kg, with $s_{n-1} = 3.5$ kg. Test the company's claim at the level of significance of $\alpha = 0.05$

## 18.3 Type I and type II errors

Using a probabilistic model to determine what is not unlikely 95% of the time discounts the least likely $\alpha = 5\%$ of outcomes still possible due to chance alone. The probability of rejecting an observation that is actually true, given by $\alpha$ is called the **type I error**. The opposite error, in not rejecting an outcome that is not true is called a **type II error** and is given the designation $\beta$. Considering how the two errors are defined, it will be a logical deduction to conclude that as one decreases, the other increases.

Generally, the level of significance is set before statistical analysis, so $\alpha$ is known. The calculation of $\beta$ requires the comparison of two different means, not just the mutually exclusive null and alternative hypotheses, and critical values referenced in Section 18.2.

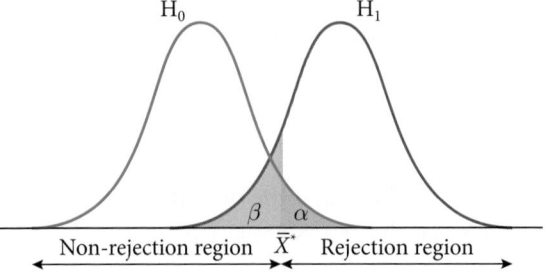

**Figure 18.10** The relationship between type I and type II errors visualised

## Example 18.4

According to the visitors' bureau, the mean number of visitors to a particular city per day is $\mu = 20\,000$. A random survey on 20 days produced a sample mean of $\bar{x} = 21\,000$ with a standard deviation of $s_{n-1} = 1800$

(a) State the null and alternative hypotheses.

(b) Test the hypotheses at the $\alpha = 0.05$ significance level.

(c) Indicate the size of the type I error.

(d) If the transportation commission claims that the true mean is actually $\mu = 21\,400$, state the hypotheses to test this new claim against the original one from the visitors' bureau.

(e) Find the size of the type II error.

## Solution

(a) $H_0 : \mu = 20\,000$

$H_1 : \mu \neq 20\,000$

(b) Using a GDC, we can find that the $p$-value $\approx 0.0105$

Since there are two possible rejection regions that combine to form the 5% of the least likely outcomes, $\alpha$ should be split into two equal parts of 0.025, but $p$-value $< \dfrac{\alpha}{2}$, and the null hypothesis should be rejected.

(c) Since the level of significance is $\alpha = 0.05$, the type I error is also 0.05.

(d) Let the originally proposed mean be $\mu_0 = 20\,000$ and the new (alternative) mean be $\mu_1 = 21\,400$, leading to the hypotheses

$H_0 : \mu_0 = 20\,000$

$H_1 : \mu_1 = 21\,400$

(e) To utilise a GDC efficiently, but more importantly, not to lose precision in calculation, store values.

Let $C$: critical $t$-value at $\alpha = 0.05$, $n = 20$

$E$: standard error of the mean (SEM), $SEM = \dfrac{1800}{\sqrt{20}} \approx 402$

$V$: critical value $\bar{x}^*$ for $\mu_0$

$T$: $t$-value for $\mu_1 = 21\,400$

$\beta \approx 0.0482$

```
invT(.95,19)→C
              1.729132792
1800/√20→E
              402.4922359
C*E+20000→V
              20695.96252
(V-21400)/E→T
              -1.749195173
```

```
tcdf(-1E99,T,19)
              0.0481968505
```

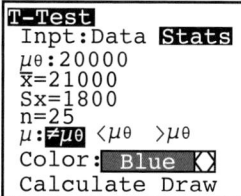

```
T-Test
Inpt:Data Stats
μ0:20000
x̄=21000
Sx=1800
n=25
μ:≠μ0 <μ0 >μ0
Color: Blue  ◇
Calculate Draw
```

```
T-Test
μ≠20000
t=2.777777778
p=0.0104543751
x̄=21000
Sx=1800
n=25
```

**Figure 18.11** Using a GDC to find the $p$-value $\approx 0.0105$

In a normal distribution, the measure of variability commonly used is the **standard deviation**. In a $t$-distribution, the equivalent measure is called the **standard error**. If the variability is centred about the mean, it is called the **standard error of the mean** or *SEM*. If it is centered about a proportion, it is called the standard error of the proportion or $SE_p$.

In Example 18.4, the population variance was not given or perhaps not known. This is not unusual. However, when the true population mean and variance are both known, the calculations of the type I and II errors require the use of the normal distribution and $z$-scores. Since degrees of freedom $v$ are not required, the calculation of the type II error requires fewer steps.

### Example 18.5

Let's modify the earlier example regarding the number of visitors to the city. Suppose both parameters are known, with the population mean $\mu_0 = 20\,000$ and standard deviation $\sigma = 1600$, and a sample of size $n = 20$ yields a sample mean $\bar{x} = 21\,000$

An alternative mean, $\mu_1 = 21\,400$, is to be tested against the original mean at the 0.05 significance level. Determine the size of the type II error.

### Solution

Since the population standard deviation $\alpha$ is known, it will be used regardless of whether or not the sample standard deviation is known. The values needed, $C$ and $E$, are found in the same way as in the earlier $t$-test; however, we will use those values with the normal distribution and $z$-test.

We do not need to find the critical $z$-value since $v$ is not required. The sample standard deviation is stored again on the GDC as $E$. The rest is done with the available GDC features $\beta \approx 0.0117$.

```
1600/√20→E
                357.7708764
invNorm(0.95,20000,E,LEFT▶
                20588.48072
V
                20588.48072
normalcdf(-1E99,V,21400,E)
                0.0116564841
```

**Figure 18.12** GDC output for Example 18.5

```
20000-21400→D
                -1400
invNorm(0.95,0,1,LEFT)    ▶
                1.644853626
Ans+D/E→Z
                -2.268265335
normalcdf(-1E99,Z,0,1)
                0.0116564841
```

**Figure 18.13** GDC output for the alternative solution

**Alternative solution**

Instead of calculating the critical value $\bar{x}^*$ which is given as $V$ in the process above, we can bypass it and determine the number of units of $SEM$, which we called $E$, that there are in the difference between $\mu_0$ and $\mu_1$

Let $D = \mu_0 - \mu_1$ and let $Z$ be the $z$-score for $\mu_1 = 21\,400$

Our GDC calculations are now:
- $D = 20\,000 - 21\,400 = -1400$
- For $\alpha = 0.05$, find the $z$-score $\approx 1.645$
- Add the number of $SEM$ units represented by $D$.
- Find the area under the standard normal curve, $\beta \approx 0.0117$

## The power of the test

Since $\beta$ is the probability that we fail to reject a false null hypothesis, its complement, $1 - \beta$, is the probability that we correctly reject a false null hypothesis and is called the **power of the test**. As you can see in the overlapping distribution graph shown in Figure 18.10, as you decrease $\alpha$, you increase $\beta$, thereby decreasing the power of the test.

1. According to the board of trade, historical weather data for Vancouver indicate that the average precipitation in November, measured in millimetres, is $\mu_0 = 158.2$, whereas the government weather office claims that the average is $\mu_1 = 167.2$

   A random selection of 16 years of data for past Novembers indicate a mean of $\bar{x} = 165.5$ with $s_{n-1} = 50.4$

   Assume that the average precipitation is normally distributed and compare the two claims at a level of significance of $\alpha = 0.05$

   (a) State the null and alternative hypotheses.

   (b) Determine the size of the type II error in the analysis.

   (c) What is the power of the test?

2. The players' union claims that the mean annual salary of hockey players is US$2.4 million, whereas the owners' association claims that it is US$3.0 million. A random sample of 40 players' salaries shows $\bar{x} = $ US$ 2.3 million with $s_{n-1} = $ US$ 0.325 million. Assuming players' salaries to be normally distributed, compare the two claims at a level of significance of $\alpha = 0.05$

   (a) State the null and alternative hypotheses.

   (b) Calculate the type II error.

3. Bottled water is filled by a machine calibrated to dispense water with a mean of $\mu = 500$ ml and a standard deviation of $\sigma = 5$ ml, and the amount dispensed is assumed to be normally distributed. Given that the true mean is 502.4, determine, at a level of significance of $\alpha = 0.05$

   (a) the type I error         (b) the type II error.

4. A Canadian newspaper reported that a certain breakfast cereal is deceptively high in sodium, with a mean of $\mu_0 = 230$ mg of sodium per cup of cereal. A random sample of 20 different boxes of the same product, taken a cup at a time, shows $s_{n-1} = 20.2$ mg of sodium. Given that the actual mean is $\mu_1 = 241.6$ mg of sodium, what is the probability of a type II error for a hypothesis test at a level of significance of $\alpha = 0.05$?

## 18.4   *t*-test of two means

The *t*-test is ideally suited to comparisons of two sample means. There are two applications that we will discuss in this section: **paired** and **unpaired** samples. Paired samples involve data taken twice, perhaps as a way of testing reliability, but more often as a way to test the results taken before and after a study. Unpaired samples are taken from two different groups, both assumed to be normally distributed. Their sample sizes need not be equal; for instance, comparing the mean yearly salary of 20 men to the yearly salary of 30 women.

## Paired $t$-tests

In a paired $t$-test, the differences $d$ between the first and second sample results are important. Consequently, the null and alternative hypotheses involve $d$. Generally, in many before-and-after analyses, the claim would be that $d > 0$, referring to increased performance, or $d < 0$, referring to a reduction in time or cost.

### Example 18.6

A personal trainer guarantees an improvement in performance compared to other training methods for deadlifts with a claim that the client will be able to lift at least 20 kg more after three weeks of training. State the null and alternative hypotheses and test this claim at a level of significance of $\alpha = 0.05$ using this data set in kilograms.

| Before | 105 | 155 | 200 | 110 | 95 | 115 | 185 | 145 | 110 | 100 | 120 | 170 | 110 |
|--------|-----|-----|-----|-----|----|-----|-----|-----|-----|-----|-----|-----|-----|
| After | 120 | 180 | 210 | 135 | 110 | 135 | 210 | 170 | 100 | 110 | 140 | 160 | 130 |

**Solution**

$H_0 : d \geqslant 20$

$H_1 : d < 20$

Note that the claim is $H_0$

Determine the differences between each individual's before-and-after measurements.

| Before | 105 | 155 | 200 | 110 | 95 | 115 | 185 | 145 | 110 | 100 | 120 | 170 | 110 |
|--------|-----|-----|-----|-----|----|-----|-----|-----|-----|-----|-----|-----|-----|
| After | 120 | 180 | 210 | 130 | 110 | 135 | 200 | 170 | 100 | 110 | 140 | 160 | 130 |
| $d$ | 15 | 25 | 10 | 20 | 15 | 20 | 15 | 25 | $-10$ | 10 | 20 | $-10$ | 20 |

Enter the differences into a list on a GDC, then find the $p$-value at the $\alpha = 0.05$ level of significance.

Since the $p$-value $\approx 0.0308$ and is less than $\alpha$, we reject the null hypothesis and the claim.

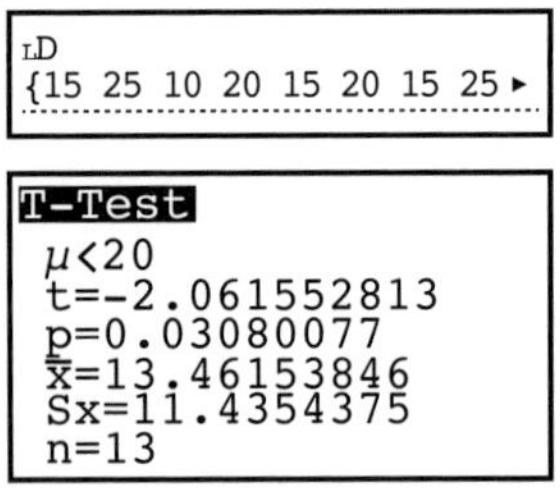

L D
{15 25 10 20 15 20 15 25 ▶

```
T-Test
μ<20
t=-2.061552813
p=0.03080077
x̄=13.46153846
Sx=11.4354375
n=13
```

**Figure 18.14** GDC screens for the solution to Example 18.6

## Unpaired $t$-tests (2-sample $t$-test)

When two different samples of related data need to be analysed, an unpaired $t$-test should be used. The number of data in the two samples need not be the same. If the sample standard deviations are comparable, the samples can be **pooled**; otherwise, they will be analysed as **unpooled** data. Generally, if the larger standard deviation of one sample is less than twice the other, they can be pooled.

A GDC has both pooled and unpooled sample options. In IB examinations, the variances will be given to be the same; hence, we will only look at the pooled, unpaired $t$-test. Given two samples of sizes $n_1$ and $n_2$, a pooled sample will have a degree of freedom $v = n_1 + n_2 - 2$

If correct statistical analysis is to be conducted, before using a 2-sample $t$-test, you will need to carry out a test of variances, called the **$F$-test**, to determine whether the samples can be pooled or not. Although a GDC can perform an $F$-test, the IB syllabus does not require it.

## Example 18.7

Two brands of tyres are tested for stopping distance to make a comparison, with a full set of 4 tyres used with the test vehicle in each test.

The statistics obtained are as follows:

$n_1 = 16, \bar{x}_1 = 37.3 \text{ m}, s_1 = 6.1 \text{ m}$

$n_2 = 10, \bar{x}_2 = 40.1 \text{ m}, s_2 = 5.7 \text{ m}$

Test at a level of significance of $\alpha = 0.05$ whether the two brands have equivalent stopping distances.

The *F*-test would show that comparing the two variances for equivalence would have produced a *p*-value $\approx 0.864$; hence, the variances can be considered equal.

### Solution

The hypotheses are:

$H_0 : \mu_1 = \mu_2$

$H_1 : \mu_1 \neq \mu_2$

Use a GDC to perform a 2-sample *t*-test. Since the variances can be considered to be equal, the samples can be pooled as seen in the first two screens on the right.

The results show a *p*-value $\approx 0.255$ which, being geater than $\alpha$, means we do not reject $H_0$

Note that the second screen shows the pooled degrees of freedom, $v = n_1 + n_2 - 2 = 24$

If we had mistakenly selected the unpooled option, as shown in the third and fourth screens, the degrees of freedom would have been a non-integer value, $v \approx 20.2656$

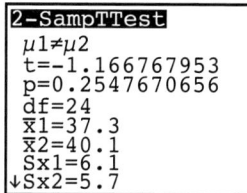

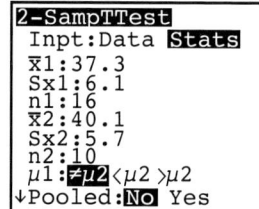

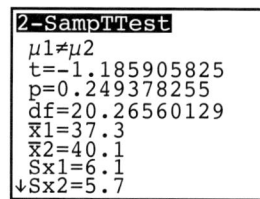

**Figure 18.15** GDC screens for the solution to Example 18.7

## Exercise 18.4

1. Performance enhancement training, conducted with 9 randomly selected sprinters in the 100 m sprint in the city championship, found that their times, in seconds, before and after training were as shown below. Assume that the times for all sprinters in the championship are normally distributed.

| Before | 12.1 | 13.4 | 13.8 | 14.0 | 13.6 | 12.8 | 13.4 | 14.2 | 13.9 |
|--------|------|------|------|------|------|------|------|------|------|
| After  | 11.6 | 12.8 | 13.8 | 13.7 | 13.8 | 12.9 | 12.9 | 13.5 | 13.6 |

The trainers' claim is that their program reduced the sprinters' times.

(a) State the null and alternative hypotheses.

(b) Test the claim that the training improved their times at $\alpha = 0.05$

2. A new mass-reduction program claims that clients will lose at least 10 kg by following their exercise routine and using their prescribed nutrition products. A random sample of 11 clients' masses are shown, before and after following the program.

| Before | 90.2 | 106.1 | 85.6 | 103.4 | 91.4 | 85.1 | 87.3 | 100.5 | 78.2 | 81.1 | 85.5 |
|--------|------|-------|------|-------|------|------|------|-------|------|------|------|
| After  | 82.2 | 95.6  | 80.1 | 92.8  | 89.3 | 80.1 | 78.2 | 102.1 | 70.0 | 71.8 | 77.7 |

(a) State the null and alternative hypotheses.

(b) Test the company's claim at a level of significance of $\alpha = 0.05$.

3. A tutoring service claims to improve achievement in mathematics in a series of weekly sessions over two months. On two forms of a standardised test, 10 students were tested before and after the two months of tutoring. Their scores are given in the table.

| Before | 62 | 58 | 75 | 53 | 81 | 46 | 51 | 91 | 63 | 45 |
|--------|----|----|----|----|----|----|----|----|----|----|
| After  | 63 | 64 | 72 | 57 | 85 | 51 | 55 | 87 | 63 | 49 |

(a) State the null and alternative hypotheses.

(b) Test the claim at a level of significance of $\alpha = 0.05$

4. The IB English teacher tried a speed-reading program at school to improve students' speed and comprehension. The 11 test subjects' composite scores are given before and after the program.

| Before | 94  | 102 | 143 | 120 | 128 | 98  | 111 | 131 | 150 | 134 | 106 |
|--------|-----|-----|-----|-----|-----|-----|-----|-----|-----|-----|-----|
| After  | 100 | 113 | 145 | 125 | 130 | 107 | 111 | 133 | 145 | 130 | 115 |

(a) The claim is that the program improves the readers' speed and comprehension. State the null and alternative hypotheses.

(b) Test the claim at the $\alpha = 0.05$ level of significance.

5. An IQ test is given to 12 randomly selected boys and 16 randomly selected girls, with the following results:

boys : $n_1 = 12, \bar{x}_1 = 102.5, s_1 = 13.2$

girls : $n_2 = 16, \bar{x}_2 = 107.0, s_2 = 16.5$

At the level of significance of $\alpha = 0.05$, test the claim that the mean score for girls is higher than for boys. Assume that the variances are equal at this level of significance.

6. An entrance exam designed by the Mathematics faculty at a local university produces the following results from two schools

school 1 : $n_1 = 26, \bar{x}_1 = 86.4, s_1 = 11.2$

school 2 : $n_2 = 23, \bar{x}_2 = 95.2, s_2 = 16.3$

Test the claim that the scores are equivalent at the level of significance of $\alpha = 0.05$. The $F$-test produces a $p$-value $\approx 0.0721$, hence, the variances can be considered equal.

# 18.5 | Chi-squared test of the goodness of fit (GOF)

When there are only two outcomes in an experiment, it is a binomial experiment. When there are many, it is called a **multinomial experiment**. The **chi-squared ($\chi^2$) test of the goodness of fit** is designed to determine whether or not observed outcomes of a multinomial experiment fit an expected model.

The calculation of the $\chi^2$ statistic can be done entirely on a GDC, but an explanation may be useful. Let's take a look at the fairness of a dice. The fit or model that is expected of a fair dice is that the outcomes from 1 to 6 are equally likely. Hence, in 60 rolls of the dice, the expected frequencies are 10 for each face. If you roll the dice 60 times, you would actually be quite surprised if each face showed up exactly 10 times. On the other hand, a high frequency of any one of the outcomes, or a low frequency of others may give you reason to suspect a biased dice.

Table 18.2 shows the observed frequencies in 60 rolls. However, instead of actually rolling a dice 60 times, assume that some frequencies may not be unlikely. For example, if you assume the frequencies of five faces as 11, 8, 9, 13, and 12, the sixth value is the difference between 60 and the cumulative total. You would have had five free choices, but the sixth would have to produce a cumulative total of 60. In terms of a $\chi^2$ analysis, the **degrees of freedom** is one less than the number of outcomes. In this example, $v = 5$

Now, in order to determine how different the observed frequencies are from the expected frequencies, find the difference by subtracting the expected value from the observed value.

| Outcome | 1 | 2 | 3 | 4 | 5 | 6 | Total frequency |
|---|---|---|---|---|---|---|---|
| Observed | 11 | 8 | 9 | 13 | 12 | 7 | 60 |
| Expected | 10 | 10 | 10 | 10 | 10 | 10 | 60 |
| Difference | 1 | $-2$ | $-1$ | 3 | 2 | $-3$ | |

**Table 18.2** Subtracting the expected value from the observed value

Now square the differences. The sum is shown in Table 18.3.

| Difference | 1 | $-2$ | $-1$ | 3 | 2 | $-3$ | 0 |
|---|---|---|---|---|---|---|---|
| Squared | 1 | 4 | 1 | 9 | 4 | 9 | 28 |

**Table 18.3** Square the differences

Next, we normalise the differences by dividing each by the expected value.

| Squared | 1 | 4 | 1 | 9 | 4 | 9 | 28 |
|---|---|---|---|---|---|---|---|
| Normalised | $\frac{1}{10}$ | $\frac{4}{10}$ | $\frac{1}{10}$ | $\frac{9}{10}$ | $\frac{4}{10}$ | $\frac{9}{10}$ | $\frac{28}{10}$ |

**Table 18.4** Normalise the differences

The opposite faces of a hexagonal dice add up to 7; the pits for the dots on each face should have removed an equal amount of the material from each face so that the faces are balanced. If, for example, you suspect a dice is biased due to the unexpected large frequency of one outcome, its opposite face is also likely to have a low frequency of occurrences.

We can do this calculation using the formula.

$$\chi^2 = \sum_{k=1}^{n} \frac{(O_k - E_k)^2}{E_k}$$ where $O_k$ is an observed value, and $E_k$ is the corresponding expected value. Note that in calculating the $\chi^2$ statistic, there was no reference to the population parameters $\mu$ or $\sigma$. For this reason, the $\chi^2$ statistic is called a **nonparametric statistic**.

The sum of these values is the $\chi^2$ statistic. In this case it is 2.8. Is this value small or is this large? What exactly do we do with this value? We compare this statistic against the critical $\chi^2$ value for $v = 5$ to determine if there indeed is a goodness of fit (GOF). You can use a GDC to provide you a $p$-value to compare against $\alpha$ for a decision.

> Note that the $\chi^2$ statistic is always a positive value. If the observed frequencies are close to the expected frequencies, the $\chi^2$ value is small; otherwise, as the differences grow larger, so will the $\chi^2$ value. Hence, it is always a right-tailed test.

With a GDC, first enter the observed and expected values into two lists. $L_1$ is the default for the observed values and $L_2$ is the default for the expected values.

Run the $\chi^2$ GOF-test to find the $\chi^2$ statistic of 2.8 as well as the $p$-value.

With a $p$-value $\approx 0.731$, the $\chi^2$ statistic is much greater than our typical significance level; hence, there is no difference in terms of fit with the expected model.

**The standard procedure in analysing multinomial outcomes is:**
- start with a claim
- establish the null and alternative hypotheses
- select an appropriate level of significance
- test the GOF by using your GDC to enter the data and/or statistics to find the $p$-value
- reject or do not reject the null hypothesis
- reject or do not reject the claim

```
L1      L2          2
11      10
8       10
9       10
13      10
12      10
7       10
L2(7)=
```

```
χ²GOF-Test
Observed:L1
Expected:L2
df:5
Color: Blue
Calculate Draw
```

```
χ²GOF-Test
χ²=2.8
p=0.7307864867
df=5
CNTRB={0.1 0.4 0.1 0.9 0...
```

**Figure 18.16** $\chi^2$ test

## Example 18.8

A spinner is designed with three outcomes $A$, $B$, and $C$, encompassing $\frac{1}{2}$, $\frac{1}{3}$, and $\frac{1}{6}$ of the circle into which it is divided. The needle is spun 120 times with the outcomes shown in the table. Test at a level of significance of $\alpha = 0.05$ whether the data fits the intended spinner model.

| Outcome | $A$ | $B$ | $C$ | Total frequency |
|---|---|---|---|---|
| Observed | 72 | 33 | 15 | 120 |

## Solution

The claim: the data support the intended probability model.

The hypotheses:

$H_0$: There is no (statistically significant) difference between the model and the data.

$H_1$: There is a difference that cannot be attributed to chance alone.

Calculate the expected values, then enter the observed and expected values into a GDC.

| Outcome | A | B | C | Total frequency |
|---|---|---|---|---|
| Observed | 72 | 33 | 15 | 120 |
| Expected | 60 | 40 | 20 | 120 |

Since the $p$-value $\approx 0.0874$ hence greater than $\alpha$, we do not reject $H_0$ and do not reject the claim.

The calculation of a statistic and its $p$-value can be done entirely by a GDC. The statistic and its $p$-value are valid, but only if every expected value is at least 5. If not, two (or more) outcomes will have to be combined. The observed values are not under any restriction.

## Example 18.9

The strength of chili peppers is given in terms of Scoville heat units (SHU). To plan their inventory, a new chili shop intends to offer six peppers for tasting on opening day and a purchase limited to a single item. They expect the following sales to 100 customers:

| Pepper (SHU) | Jalapeno (10 000) | Habanero (350 000) | White Habanero (500 000) | Dorset Naga (900 000) | Ghost (1 000 000) | Scorpion (2 000 000) |
|---|---|---|---|---|---|---|
| E | 30 | 40 | 19 | 6 | 4 | 1 |

Test their model at a significance level of $\alpha = 0.05$ against the observed values below, kept as a tally of the first 100 customers' purchases.

| Pepper (SHU) | Jalapeno (10 000) | Habanero (350 000) | White Habanero (500 000) | Dorset Naga (900 000) | Ghost (1 000 000) | Scorpion (2 000 000) |
|---|---|---|---|---|---|---|
| O | 28 | 44 | 24 | 4 | 0 | 0 |

## Solution

Note that two of the expected values are less than 5, but can be combined before we proceed. Unless we can generate a much larger sample so that the expected values are all larger than 5, we'll have to reclassify the last two, perhaps calling it Ghost/Scorpion.

| Pepper | Jalapeno | Habanero | White Habanero | Dorset Naga | Ghost/Scorpion |
|---|---|---|---|---|---|
| O | 28 | 44 | 24 | 4 | 0 |
| E | 30 | 40 | 19 | 6 | 5 |

The claim: The data support the distribution model.

The hypotheses:

$H_0$: There is no (statistically significant) difference between the model and the data.

$H_1$: There is a difference that cannot be attributed to chance alone.

As before, entering the observed values into one list and the expected values into the other, now with $v = 4$, run the $\chi^2$ test.

With the $p$-value $\approx 0.111$ hence greater than $\alpha$, we do not reject $H_0$ and do not reject the claim.

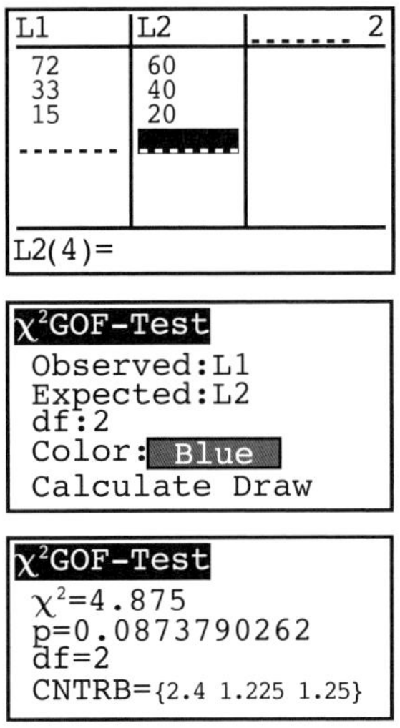

**Figure 18.17** GDC screens for the solution to Example 18.8

The calculation of a $\chi^2$ statistic and its $p$-value can be done entirely by a GDC. The $\chi^2$ statistic and its $p$-value are valid, but only if every expected value is at least 5. If not, two (or more) outcomes will have to be combined. The observed values are not under any restriction.

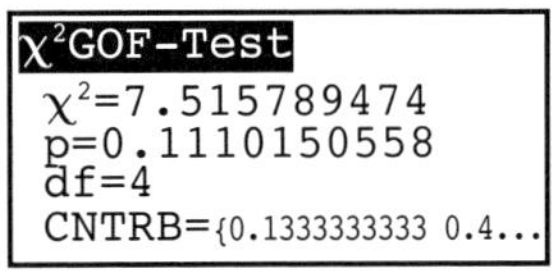

**Figure 18.18** GDC screen for the solution to Example 8.9

The grouping of outcomes need not be dictated by low expected values. It can certainly be a conscious decision on your part to analyse pairings or groupings as you deem appropriate.

1. A regular dice is rolled 60 times and the results are given in the table. At the level of significance of $\alpha = 0.05$, determine if the dice is fair.

| Outcome | 1 | 2 | 3 | 4 | 5 | 6 | Total frequency |
|---|---|---|---|---|---|---|---|
| Observed | 7 | 10 | 12 | 14 | 8 | 9 | 60 |

2. An upright spinner with 4 equal regions, $A$, $B$, $C$, and $D$, is suspect due to the unbalanced weight of the spinner needle. The observed frequencies of the outcomes in 40 spins are shown in the table.

| Outcome | $A$ | $B$ | $C$ | $D$ | Total frequency |
|---|---|---|---|---|---|
| Observed | 13 | 7 | 9 | 11 | 40 |

   Determine at a level of significance of $\alpha = 0.05$ whether or not the four spinner outcomes can be considered to be equally likely.

3. A biased dice is produced to sell as a novelty item with expected outcomes, given as percentages, shown in the table.

| Outcome | 1 | 2 | 3 | 4 | 5 | 6 | Total probability |
|---|---|---|---|---|---|---|---|
| Expected % | 20 | 10 | 10 | 10 | 10 | 40 | 100 |

   In 120 rolls of the dice, the observed outcomes are as follows.

| Outcome | 1 | 2 | 3 | 4 | 5 | 6 | Total frequency |
|---|---|---|---|---|---|---|---|
| Observed | 27 | 15 | 16 | 18 | 13 | 31 | 120 |

   At the level of significance of $\alpha = 0.05$, test whether the data support the design.

4. A college instructor uses a predetermined grade distribution in all of her classes, regardless of the students who comprise the classes.

| Grade | A | B+ | B | B− | C | D | E | Total |
|---|---|---|---|---|---|---|---|---|
| Planned % | 10 | 10 | 10 | 20 | 20 | 20 | 10 | 100 |

   The actual distribution of marks for a class of 50 students is given below.

| Grade | A | B+ | B | B− | C | D | E | Total |
|---|---|---|---|---|---|---|---|---|
| Actual (observed) | 3 | 4 | 5 | 9 | 10 | 12 | 7 | 50 |

   Calculate the expected values then determine whether the data fit the intended model at a level of significance of $\alpha = 0.05$

5. An experiment consists of flipping a coin 5 times and recording the number of heads. When the experiment is repeated 120 times, the following number of heads is noted in the 160 trials.

| Heads | 0 | 1 | 2 | 3 | 4 | 5 | Total frequency |
|---|---|---|---|---|---|---|---|
| Observed | 2 | 18 | 41 | 62 | 28 | 9 | 160 |

   Assuming the coin is fair and using the binomial distribution to find the expected numbers of outcomes, determine at a level of significance of $\alpha = 0.05$ whether or not the observations follow the theoretical binomial distribution.

**6.** The director of human resources is inclined to believe that the absenteeism from work depends on the day of the week.

(a) Given the number of absences within the company over the past four months (refer to Table 18.5), determine if she is correct at the level of significance of $\alpha = 0.05$

(b) The director is still convinced that either side of weekends, the absences are significantly higher. Test the claim that Friday and Monday absences combined are statistically higher than during the rest of the week at the level of significance of $\alpha = 0.05$

| Day | Absences |
|---|---|
| Monday | 24 |
| Tuesday | 14 |
| Wednesday | 12 |
| Thursday | 14 |
| Friday | 26 |
| Total absences | 90 |

Table 18.5 Data for question 6

# 18.6 Chi-squared test of independence

When multinomial outcomes from different groups need to be compared, the test to use is the **chi-squared test of independence**. The null hypothesis would claim that there is no difference between the distributions of outcomes of the groups being compared. Unlike the $\chi^2$ GOF test, there will be two or more sets of observed values, each matched to the expected outcomes, assuming all groups could be combined as one.

## Degrees of freedom

For example, suppose 30 girls and 45 boys were surveyed to find their favourite TV crime drama from one of four which we'll call $A$, $B$, $C$, and $D$. By comparing the preferences of the girls and of the boys against all respondents combined, it can be determined whether the preferences of the girls and boys are statistically equivalent. Start by noting the preferences of all respondents combined and listing them in a format called a **contingency table** or a **two-way table** to determine the degrees of freedom.

| Preferred TV show | A | B | C | D | Total |
|---|---|---|---|---|---|
| Girls | | | | | 30 |
| Boys | | | | | 45 |
| Total | 30 | 17 | 13 | | 75 |

Table 18.6 Two-way table for TV show preferences

As you have seen with the $\chi^2$ GOF test, if we were randomly entering values while being mindful of the total number of respondents, once the first three numbers such as 30, 17, and 13 are placed in the bottom row with 75 as the total, the last value in the row is pre-determined. Consequently, having four TV shows produces 3 degrees of freedom.

| Preferred TV show | A | B | C | D | Total |
|---|---|---|---|---|---|
| Girls | | | | | 30 |
| Boys | | | | | 45 |
| Total | 30 | 17 | 13 | 15 | 75 |

**Table 18.7** Updated two-way table for TV show preferences

Similarly, as soon as the first three values for the girls are placed in their row, not only is their entry for *D* already pre-determined, all of the values for the boys are determined since the column totals are known, producing no additional degrees of freedom.

| Preferred TV show | A | B | C | D | Total |
|---|---|---|---|---|---|
| Girls | 12 | 4 | 8 | 6 | 30 |
| Boys | 18 | 13 | 5 | 9 | 45 |
| Total | 30 | 17 | 13 | 15 | 75 |

**Table 18.8** Complete two-way table for TV show preferences

Just as in the $\chi^2$ GOF test, you've had the freedom to manipulate the values, but the restrictions based on the totals meant just the values 12, 4, and 8 noted in the row for girls make $v = 3$ for the entire contingency table.

If there were three groups being compared, perhaps girls, boys, and teachers, to see if their preferences were comparable, the first three values for the boys would also have to be entered into this table before the additional row of entries for teachers would be pre-determined. In general, the number of degrees of freedom, given $m$ groups and $n$ outcomes is $v = (m - 1)(n - 1)$

## Calculating the chi-squared statistic

We will now use the algorithm to calculate a $\chi^2$ statistic, using the values above as actual observed data. First of all, we need to state a claim and the hypotheses before we continue by finding the expected values.

### Example 18.10

Thirty girls and forty-five boys were surveyed to determine if there is any difference in their preference amongst four TV crime dramas. At a level of significance of $\alpha = 0.05$, determine whether the choice of crime drama is independent of the gender, given the observed data in Table 18.8.

### Solution

The null hypothesis is always the one that states that there is no difference. Therefore,

$H_0$: There is no statistically significant difference in the girls' and boys' preferences

$H_1$: There is a difference that cannot be attributed to chance alone.

Since Table 18.8 lists only the observed values, we need to calculate the expected values if the preferences expressed by the two separate groups were the same as the preferences expressed by the groups as a whole. This implies that the ratios expected from the girls and from the boys should match the ratios for the entire group.

$$A: \frac{30}{75}, B: \frac{17}{75}, C: \frac{13}{75}, \text{ and } D: \frac{15}{75}$$

For the 30 girls surveyed, if their preferences were to match those for the group as a whole, we would expect to find

$$A: \frac{30}{75} \times 30 = 12, B: \frac{17}{75} \times 30 = 6.8, C: \frac{13}{75} \times 30 = 5.2, \text{ and } D: \frac{15}{75} \times 30 = 6$$

For the 45 boys surveyed, we would expect

$$A: \frac{30}{75} \times 45 = 18, B: \frac{17}{75} \times 45 = 10.2, C: \frac{13}{75} \times 45 = 7.8, \text{ and } D: \frac{15}{75} \times 45 = 9$$

These results can be entered into two rows of expected values, one for the girls and the second for the boys.

Observed values:

| 12 | 4 | 8 | 6 |
|----|----|----|----|
| 18 | 13 | 5 | 9 |

Expected values:

| 12 | 6.8 | 5.2 | 6 |
|----|-----|-----|---|
| 18 | 10.2 | 7.8 | 9 |

In order to calculate the $\chi^2$ statistic, we can use the formula

$$\chi^2 = \sum_{k=1}^{n} \frac{(O_k - E_k)^2}{E_k}$$

However, after calculating the $\chi^2$ statistic, we still need to compare it against the $\chi^2$ critical value for $v = 3$. A GDC can complete the entire analysis just by entering the matrix representing the observed values only.

Enter the observed values into a $2 \times 4$ matrix $A$. A GDC will calculate and enter the expected values into matrix $B$ before calculating the $\chi^2$ statistic.

$\chi^2 \approx 4.43$ and its $p$ − value $\approx 0.218$. Since $p$-value $> \alpha$, we do not reject the null hypothesis. The choice of crime drama is independent of gender.

Incidentally, if we wish to verify the calculations of the expected values, we can look into matrix $B$.

```
[A]
                    [12  4  8  6]
                    [18 13  5  9]
```

```
χ²GOF-Test
  Observed:[A]
  Expected:[B]
  Color: Blue
  Calculate Draw
```

```
χ²GOF-Test
  χ²=4.43438914
  p=0.2182178223
  df=3
```

```
[A]
                    [12  4  8  6]
                    [18 13  5  9]

[B]
                    [12  6.8  5.2  6]
                    [18 10.2  7.8  9]
```

**Figure 18.19** GDC screens for the solution to Example 18.10

If you have a $2 \times 2$ contingency table, there is only 1 degree of freedom. Unfortunately, this situation makes the $\chi^2$ statistic unreliable. **Yates' continuity correction**, which reduces the positive difference between each pair of observed and expected values by 0.5 before squaring, then dividing by the expected value, has been considered for quite some time, although some statisticians feel that it over-compensates. Of course, since a GDC will not perform this calculation, it would have to be done by hand. An alternative approach would be to use a test of two proportions, provided the sample size is large. Like the 2-sample $t$-test, this can be done quickly with a GDC.

Note that each expected value must be at least 5, as with the $\chi^2$ GOF test; otherwise, the outcomes will need to be combined.

1. In a random survey of 9th and 10th grade students, their favourite music genre of four choices was tallied. The table below indicates their responses.

| Genre | Pop | Classical | Dance | Metal | Total |
|---|---|---|---|---|---|
| 9th grade | 7 | 8 | 6 | 11 | 32 |
| 10th grade | 9 | 11 | 15 | 8 | 43 |

The claim is made that the choice of music is independent of the student's grade.

(a) State the null and alternative hypothesis, indicating which of the two reflects the claim.

(b) Indicate the degrees of freedom, $v$, in this analysis.

(c) At the level of significance of $\alpha = 0.01$, test the claim.

2. The mathematics marks of last year's HL students were compared to the predicted marks for this year's candidates, with the results shown below.

| IB mark | 7 | 6 | 5 | 4 | 3 | 2 | Total frequency |
|---|---|---|---|---|---|---|---|
| Last year | 8 | 14 | 13 | 9 | 8 | 1 | 53 |
| This year | 11 | 12 | 8 | 5 | 4 | 0 | 40 |

To test the claim that the marks are independent of the groups:

(a) state the null and alternative hypotheses and identify the claim

(b) identify the number of degrees of freedom, $v$

(c) find the $\chi^2$ statistic

(d) determine at the level of significance of $\alpha = 0.05$ whether or not the achievement level distributions and the groups of students are independent.

3. The difference in voting preferences between three areas was found through random sampling by students in the Geography class and is summarised below.

| Area | Liberal | Conservative | Socialist | Total |
|---|---|---|---|---|
| Westside | 47 | 30 | 10 | 87 |
| Eastside | 25 | 26 | 34 | 85 |
| City centre | 30 | 25 | 18 | 73 |

The claim is that the voting preferences are not independent of the area.

(a) State the null and alternative hypotheses.

(b) Indicate the number of degrees of freedom, $v$.

(c) At the level of significance of $\alpha = 0.05$, test the claim.

4. The number of social media postings per week by students and business owners is presented below.

| Postings | 0–4 | 5–9 | 10–14 | 15–19 | 20–24 | 25–30 | 31+ | Total |
|---|---|---|---|---|---|---|---|---|
| Students | 2 | 16 | 20 | 23 | 40 | 35 | 12 | 148 |
| Business owners | 5 | 12 | 16 | 18 | 16 | 8 | 4 | 79 |

The claim by the business owners is that the distributions are independent of the sample groups.

(a) State the null and alternative hypotheses. Identify the claim.

(b) Indicate the number of degrees of freedom, $v$.

(c) Test the claim at the level of significance of $\alpha = 0.05$

5. Before repainting the interior walls, the school administration sought some colour preferences from the teachers and students. The Maths HL class surveyed random samples of the groups, and their findings are given in the table below.

| Colour | blue | green | beige | pink | yellow | Total frequency |
|---|---|---|---|---|---|---|
| Teachers | 9 | 8 | 11 | 6 | 6 | 40 |
| 12th grade | 12 | 7 | 5 | 6 | 9 | 39 |
| 11th grade | 11 | 10 | 8 | 7 | 9 | 45 |
| 10th grade | 10 | 9 | 5 | 8 | 9 | 41 |

Test the claim that the colour preferences are independent of the school grade or group at the level of significance of $\alpha = 0.05$

(a) State the null and alternative hypotheses. Identify the claim.

(b) Indicate the number of degrees of freedom, $v$.

(c) Test the claim at the level of significance of $\alpha = 0.05$

6. The school counsellor would like to test whether mark distribution in IB subjects are independent in the following subjects: Mathematics HL, Biology HL, Chemistry HL, and Physics HL. Data from the previous graduating class are given in the table.

| IB mark awarded | 7 | 6 | 5 | 4 | 3 | 2 | 1 | Total |
|---|---|---|---|---|---|---|---|---|
| Mathematics | 4 | 6 | 3 | 5 | 3 | 2 | 2 | 25 |
| Biology | 2 | 4 | 7 | 9 | 4 | 2 | 1 | 29 |
| Chemistry | 2 | 7 | 10 | 7 | 3 | 1 | 1 | 31 |
| Physics | 6 | 4 | 9 | 10 | 3 | 3 | 0 | 35 |

Test her claim at the level of significance of $\alpha = 0.05$

7. At the top of the league standings in La Liga and Bundesliga in 2017–18 were Barcelona and Bayern Munich respectively. Over the past 10 years, they have amassed impressive records. Their games played (GP), won (W), drawn (D), and lost (L) over ten seasons from 2008–2009 to 2017–2018 are shown in the table. Test the claim that the outcomes and teams are independent at the level of significance of $\alpha = 0.05$

| Football, 10 seasons | W | D | L | GP |
|---|---|---|---|---|
| Barcelona | 290 | 58 | 32 | 380 |
| Bayern Munich | 245 | 54 | 41 | 340 |

8. (Optional) Use the 2-sample proportion test on your GDC to test whether or not the percentage of wins of the two teams in question 7 are statistically comparable at the level of significance of $\alpha = 0.05$

## Chapter 18 practice questions

1. Determine the number of degrees of freedom $v$ in each situation.
   (a) A random sample of size $n = 18$ is taken from a population assumed to be normal, to test a claim that the average hours of sleep IB students get per week night is $\mu = 5.5$. The variance is unknown.
   (b) A preference survey is taken where respondents are offered six choices, each of which is initially considered to be as attractive an option as the rest.
   (c) The enrolments in three faculties are compared for five universities with similar programs.

2. Determine the critical $t$-values given the sample size $n$ and the level of significance $\alpha$.
   (a) $n = 25$ and $\alpha = 0.05$, right-tailed
   (b) $n = 25$ and $\alpha = 0.05$, left-tailed
   (c) $n = 20$ and $\alpha = 0.05$, two-tailed
   (d) $n = 18$ and $\alpha = 0.01$, two-tailed

3. Determine whether the following results are statistically significant.
   (a) Right-tailed test, $\alpha = 0.05$ and $p$-value$= 0.08$
   (b) Left-tailed test, $\alpha = 0.05$ and $p$-value$= 0.01$
   (c) Two-tailed test, $\alpha = 0.05$ and $p$-value$= 0.01$
   (d) Right-tailed test, $t$-statistic, $t$ critical value

4. Find the 95% confidence interval for the true mean, given $n = 30, \bar{x} = 64$, and $s_{n-1} = 12$

5. A sample of size $n = 23$ from a normally distributed population shows $\bar{x} = 100$ and $s_{n-1} = 12$. At a level of significance of $\alpha = 0.05$, do the statistics support the claim that the true population mean is $\mu > 110$?

6. Along with map directions, a popular phone app indicates that customers at the new grocery superstore spend an average of 30 minutes in the store. A random sample of 20 customers' times shows a mean of $\bar{x} = 20$, and $s_{n-1} = 16$ minutes. At the level of significance of $\alpha = 0.05$, test the statistic given by the app.

7. State the null and alternative hypotheses for each given claim about the true value of the population mean.

   (a) $\mu = 62.4$     (b) $\mu \geq 13$     (c) $\mu > 72.3$     (d) $\mu < 102$

8. The Center for Disease Control (CDC) in the US claims that the average BMI for adult American men is 28.6, while rating a BMI in the range [18.5, 24.9] as normal.
   In a random sample of 20 men, $\bar{x} = 25.3$, and $s_{n-1} = 7.1$

   Assume that men's BMI are normally distributed. State $H_0$ and $H_1$, then test the CDC's claim at a level of significance of $\alpha = 0.05$

9. A product liability lawsuit awarded a consumer a substantial sum of money for third-degree burns from coffee which was served at a temperature between 82°C and 88°C, purchased at a drive-through restaurant. The restaurant chain claims that their coffee is now given to the consumer at an average temperature of 79°C. A random sample of 10 cups of coffee, purchased and immediately measured for temperature, showed $\bar{x} = 80.1$°C, and $s_{n-1} = 1.0$°C

   Assume the coffee temperatures are normally distributed. State $H_0$ and $H_1$ then test restaurant's claim at a level of significance of $\alpha = 0.05$

10. The fitness room at the community centre is experiencing a surge of members wishing to use the treadmills, particularly in inclement weather. In their submission to management, the fitness room staff claim that the average wait time for an available treadmill at peak times is in excess of 20 minutes. During those peak times, a random sample was taken 8 times that shows $\bar{x} = 22$ minutes, with $s_{n-1} = 5$ minutes. Assume that wait times during peak hours are normally distributed. State $H_0$ and $H_1$, and test the claim of the fitness room staff at the level of significance of $\alpha = 0.05$

11. One type of fuse used in cars is rated at 2A, but a random sample of 23 fuses indicates $\bar{x} = 2.2$A, with $s_{n-1} = 0.15$A. State $H_0$ and $H_1$, then at the level of significance of $\alpha = 0.05$, test whether or not these fuses are manufactured to specification. Assume that the rating of the fuses is normally distributed in manufacturing.

12. A highly-regarded cooking magazine tested oven temperatures that while set at the same temperature, the internal temperature varied by as much as 32°C. If 15 randomly selected ovens set to 204°C shows $\bar{x} = 200$°C, with $s_{n-1} = 10$°C, test the magazine's claim that the ovens are not reaching the indicated temperature.

13. A shoe retailer notes that their regular customers replace their work dress shoes every 1.5 years on average, with quite a large variability. The footwear industry report claims $\mu_0 = 1.2$ years, but a consumer report has it at $\mu_1 = 1.6$ years. No measure of variability is given in either report. A random sample of 20 work dress shoe wearers shows $\bar{x} = 1.0$ year, with $s_{n-1} = 1.8$ years. Assuming that the life of work dress shoes are normally distributed, compare the two claims at the level of significance of $\alpha = 0.05$

    (a) State the null and alternative hypotheses.

    (b) Determine the size of the type I and type II errors in the analysis.

    (c) What is the power of the test?

14. An investment newspaper reported that the vacancy rate for office space in Vancouver averaged 0.045 in Q1 2018, while a commercial real estate company had the average at 0.054. Compare the two means at the level of significance of $\alpha = 0.05$, given that an independent random sample of 12 measurements in Q1 found $s_{n-1} = 0.003$

    (a) State the null and alternative hypotheses.

    (b) Determine the size of the type I and type II errors in the analysis.

    (c) What is the power of the test?

15. An experiment is conducted with racing car drivers to improve their reaction times. Measurements are taken without vehicles, but with the use of a virtual reality simulator capable of measuring reaction times by recording eye movements. Initially, reaction times are taken at 10 points in the simulation, then after training, re-taken to measure any differences in total reaction times. The results, in seconds, are given below for eight drivers. The trainers' claim is that reaction times are reduced. Assume that the distribution of total reaction times is normally distributed.

| Before training | 2.1 | 1.6 | 1.3 | 1.2 | 2.3 | 1.6 | 1.2 | 1.5 |
|---|---|---|---|---|---|---|---|---|
| After training | 1.9 | 1.4 | 1.2 | 1.8 | 2.2 | 1.7 | 1.3 | 1.1 |

    (a) State the null and alternative hypotheses.

    (b) Test the claim that the training reduced their times at $\alpha = 0.05$

16. An experimental drug is tested with individuals with high blood pressure. The readings are taken before the drug is administered, then two weeks afterwards. The results given below are of the systolic blood pressure, given in mm Hg.

| Before | 142 | 150 | 140 | 145 | 152 | 140 | 146 | 148 |
|---|---|---|---|---|---|---|---|---|
| After drug use | 131 | 140 | 144 | 147 | 142 | 130 | 150 | 142 |

    Assume that the systolic blood pressure readings are normally distributed amongst individuals with high blood pressure, and test the claim that the drug lowers blood pressure by more than 6 mm Hg.

    (a) State the null and alternative hypotheses.

    (b) Test the claim at the level of significance of $\alpha = 0.05$

17. The typing speed, in words per minute (wpm), is tested in a quiet office space, then again in the same office space with music in the background. The results for 10 typists are indicated below, showing their average wpm found over a 5-minute interval.

| Without music | 42.0 | 35.0 | 53.4 | 49.2 | 31.2 | 26.4 | 50.2 | 35.0 | 40.2 | 45.2 |
|---|---|---|---|---|---|---|---|---|---|---|
| With music | 41.4 | 39.2 | 55.2 | 51.0 | 33.2 | 26.0 | 48.0 | 27.2 | 38.4 | 47.0 |

The office manager claims that having music in the background improves the typing speed.

(a) State the null and alternative hypotheses.

(b) Test the claim at the level of significance of $\alpha = 0.05$ that the background music improved their typing speed.

18. Two laptop brands were tested to determine which laptop had the battery that lasted longer, starting at a full charge and under the same usage conditions. Eight randomly selected identical laptops from brand A were tested against six from brand D, with the claim that they were statistically equivalent. The results were as follows.

laptop A: $n_1 = 8$, $\bar{x}_1 = 16$ hours, $s_1 = 3$ hours

laptop D: $n_2 = 6$, $\bar{x}_2 = 12$ hours, $s_2 = 6$ hours

Assume the laptop battery life is normally distributed within both models, and test the claim at the level of significance of $\alpha = 0.05$. The variances can be considered equal, since the $F$-test produces a $p$-value $\approx 0.0984$

19. Random samples of the monthly salaries of men and women in a company were obtained to determine if there is statistical significance in the difference between them. The claim by the union is that the men's salaries are notably higher than those of women. The sample data yield the following statistics

men: $n_1 = 20$, $\bar{x}_1 = $ US\$6000, $s_1 = $ US\$600

women: $n_2 = 13$, $\bar{x}_2 = $ US\$5000, $s_2 = $ US\$800

Assume the salaries of all men and women are both normally distributed, and test the union's claim. The $F$-test produces a $p$-value $\approx 0.254$; hence, the variances can be considered equal.

20. From a standard deck of cards, four cards are chosen with replacement, and the number of clubs are recorded. Repeating the process until 100 sets of 4 cards are chosen, recording the number of clubs out of four results in this distribution.

| Clubs | 0 | 1 | 2 | 3 | 4 | Total cards |
|---|---|---|---|---|---|---|
| Frequency | 30 | 45 | 20 | 4 | 1 | 100 |

Use your GDC to find the binomial probabilities of each of the five expected frequencies, then find the $p$-value of the $\chi^2$ statistic to determine whether or not the observed outcomes match the expected binomial distribution at the level of significance of $\alpha = 0.05$

**21.** A college instructor gave out the following marks to her 80 students last semester.

| Mark | A | B | C | D | F | Total |
|---|---|---|---|---|---|---|
| Students | 12 | 17 | 20 | 20 | 11 | 80 |

Determine at the level of significance of $\alpha = 0.05$ whether or not the distribution was uniform.

**22.** There is a radio station that plays an eclectic and totally unexpected mix of music. On a television news interview, the director disclosed that the reggae, dance, classical, jazz, and country selections are actually programmed in the ratio $2:8:2:3:1$. In the course of one afternoon, the station played the following selections.

| Type | Reggae | Dance | Classical | Jazz | Country | Total played |
|---|---|---|---|---|---|---|
| Played | 7 | 45 | 9 | 15 | 4 | 80 |

At the level of significance of $\alpha = 0.05$, determine if the afternoon selections were in keeping with the station's intended ratio.

There are several ways to answer question 23.

**23.** One hundred flips of a coin resulted in 59 heads and 41 tails. Test whether the coin is fair at the level of significance of $\alpha = 0.05$

**24.** Students in 9th grade were surveyed to determine if gender makes a difference to the amount of TV watched on school nights.

| Hours of TV | Under 10 hours | Between 10 and 20 | More than 20 | Total number of students |
|---|---|---|---|---|
| Boys | 8 | 31 | 28 | 67 |
| Girls | 11 | 15 | 13 | 39 |

At the level of significance of $\alpha = 0.05$, determine if gender and the hours spent watching TV are independent.

**25.** A school is considering the elimination of school uniforms. A random survey of students in the middle years (MYP), their parents, students in the upper classes (DP), and their parents revealed the following opinions.

| Uniform | Yes | No | Total surveyed |
|---|---|---|---|
| MYP students | 26 | 22 | 48 |
| MYP parents | 28 | 14 | 42 |
| DP students | 16 | 25 | 41 |
| DP parents | 18 | 19 | 37 |

**(a)** Determine whether or not the preference for uniforms and the groups surveyed are independent at the level of significance of $\alpha = 0.05$

**(b)** Look closely at the responses and suggest a possible re-examination of the analysis.

**26.** A polling firm prepared a survey to determine the level of support for tax reforms that will be debated, and conducted the survey in key regions with different socio-economic backgrounds. The survey results were as follows.

| Tax reform support | For | Against | Undecided | Total surveyed |
|---|---|---|---|---|
| Metro (core) | 71 | 100 | 50 | 221 |
| Metro (outlying) | 37 | 32 | 12 | 81 |
| Interior | 34 | 19 | 14 | 67 |
| North Coast | 15 | 11 | 8 | 34 |

(a) Calculate the $\chi^2$ statistic for the data in the contingency table, and at the level of significance of $\alpha = 0.05$, determine whether or not opinions the tax reform and voting regions are independent.

(b) Recalculate the $\chi^2$ statistic by comparing the Metro (core) response to the others combined.

**27.** To help teachers during cold and flu season, the school district health department tried an experiment with three cough remedies to test their effectiveness. The teachers were randomly selected to try one of three medications identified only as brands A, B, and C to avoid any bias from prior use. The results are shown in the table.

| Relief | None | Some | Complete | Teachers per brand |
|---|---|---|---|---|
| Brand A | 13 | 25 | 6 | 44 |
| Brand B | 12 | 24 | 9 | 45 |
| Brand C | 7 | 22 | 11 | 40 |

At the level of significance of $\alpha = 0.05$, do the following.

(a) Test whether or not the effectiveness of the three medications are independent of the brands used.

(b) The results for Brand C look suspiciously different from the other two. Test its results against the other two combined.

**28.** Eleven students who had under-performed in a philosophy practice examination were given extra tuition before their final examination. The differences between their final examination marks and their practice examination marks were:

$$10, -1, 6, 7, -5, -5, 2, -3, 8, 9, -2$$

Assume that these differences form a random sample from a normal distribution with mean $\mu$ and variance $\sigma^2$.

(a) Determine unbiased estimates of $\mu$ and $\sigma^2$

(b) (i) State suitable hypotheses to test the claim that extra tuition improves examination marks.

(ii) Calculate the $p$-value of the sample.

(iii) Determine whether or not the above claim is supported at the 5% significance level.

**29.** A manufacturer of stopwatches employs a large number of people to time the winner of a 100-metre sprint. It is believed that if the true time of the winner is $\mu$ seconds, the times recorded are normally distributed with mean $\mu$ seconds and standard deviation 0.03 seconds.

(a) Calculate a 99% confidence interval for $\mu$. Give your answer correct to 3 decimal places.

(b) Interpret the result found in (a).

(c) Find the confidence level of the interval that corresponds to halving the width of the 99% confidence interval. Give your answer as a percentage to the nearest whole number.

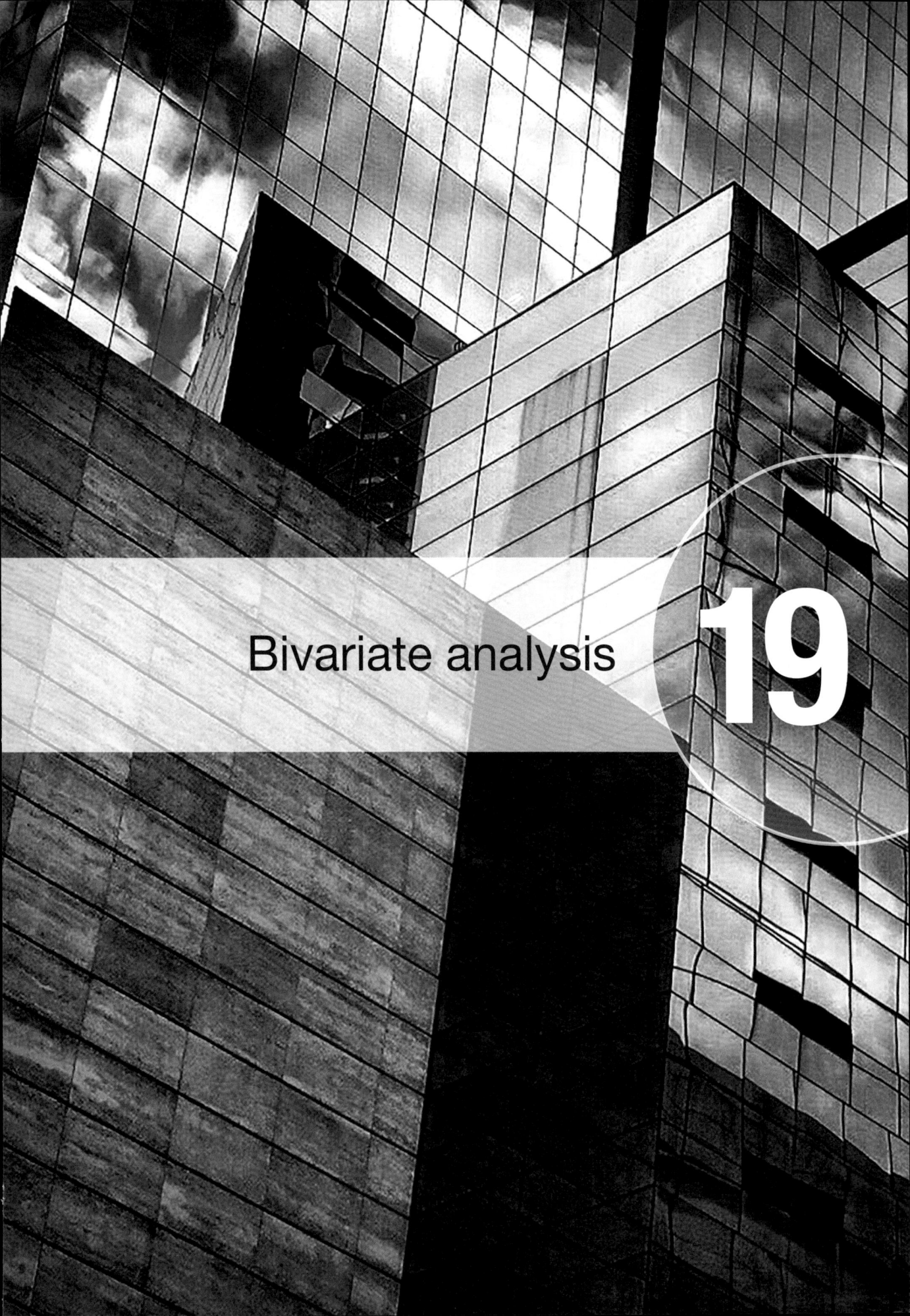

Bivariate analysis

19

## Learning objectives

By the end of this chapter, you should be familiar with...

- the key features of a scatter diagram and estimating a best-fit line by eye
- the difference between causation and correlation
- calculating (using a GDC) and interpreting Pearson's product-moment correlation coefficient for linear associations ($r$);
- calculating (using a GDC) and interpreting Spearman's rank-order correlation coefficient
- the limitations of and differences between Pearson's $r$ and Spearman's $r_s$
- and the least-square regression model and a piecewise linear model
- interpreting $SS_{res}$ as measure of the fit of a model
- the coefficient of determination $R^2$ and its relationship to Pearson's $r$
- non-linear regression using a GDC
- non-linear regression by linearising data to find exponential, logarithmic, or power models
- log-log and log-lin graphs.

We are often interested in whether one variable produces a change in another, and, if so, by how much. For example:

- Do students who study more than others get better test scores?
- Do heavier cars have worse fuel economy?
- Do plants that are fertilised more grow bigger or faster?

All of these questions involve investigating the relationship between two variables: they are **bivariate** statistics. We will start by examining data graphically, and then examine several quantitative methods to measure the relationship between two variables.

## 19.1  Scatter diagrams

Suppose we are interested in whether students who spend more time studying for a test get better scores on that test. We collect data on a class of ten Mathematics students and obtain these results.

| Student | Tim | Joon | Jim | Kyle | Steve | Niki | Henry | Anton | Cindy | Lukas |
|---|---|---|---|---|---|---|---|---|---|---|
| **Hours ($x$)** | 4.0 | 4.5 | 6.0 | 3.5 | 3.0 | 5.0 | 5.5 | 6.5 | 7.0 | 6.5 |
| **Score** | 65 | 80 | 83 | 61 | 55 | 79 | 85 | 79 | 92 | 95 |

**Table 19.1**  Mathematics test scores for ten students

For this analysis:

- hours is the **explanatory** or **independent variable**
- score is the **response** or **dependent variable**.
- the students are the **subjects** of this study/experiment

Deciding which variable is the explanatory/independent variable is largely up to us. Does the number of hours spent studying explain their test score, or does the test score explain the number of hours spent studying? The second statement seems pretty silly. In experiments, the explanatory variable is often the variable that is directly manipulated, and then we observe a response in the response variable. If both variables are observed (neither is directly manipulated or controlled), then it is mostly about the question we want to ask or the model we want to create: does one variable explain the other? Do we want a model for *a* in terms of *b*, or *b* in terms of *a*?

To begin examining the relationship between hours spent studying and test scores, start by drawing a scatter diagram, as in Figure 19.1.

It appears that, in general, as the number of hours spent studying increases, so does the score. Therefore, we say that the two variables are **associated**.

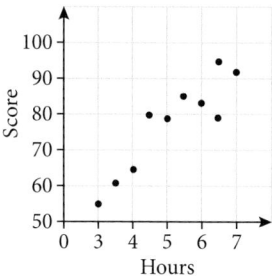

**Figure 19.1** Scatter diagram

Two variables measured on the same subjects are associated if specific values of one variable tend to occur in connection with particular values of the other variable.

When we examine a scatter diagram, we should look for form, direction, strength, and unusual features such as outliers or clusters.

## Form

Here we look at the shape of the graph. This is often the first step in determining a likely model. In general, start by deciding if the form is linear or non-linear.

**(a)** **(b)**

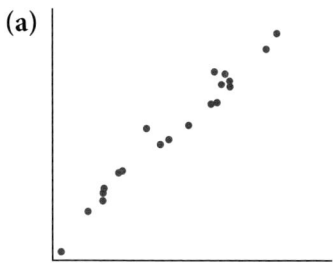

**Figure 19.2** Linear forms

**(a)** **(b)** **(c)**

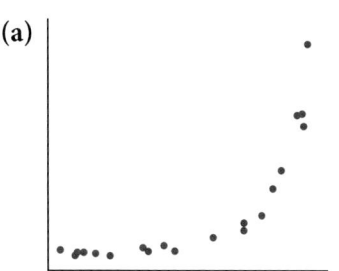
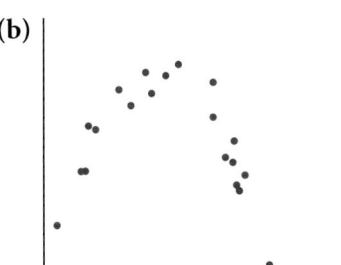
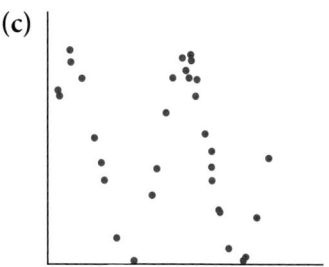

**Figure 19.3** Non-linear forms

## Direction

Direction tells us how the response variable changes in relation to the explanatory variable. If the response variable increases along with the explanatory variable, then the association is positive. If the response variable decreases while the explanatory variable increases, then the association is negative. Graphically, this is analogous to the gradient.

Figure 19.2(a) shows a positive association and Figure 19.2(b) shows a negative association.

If the response variable does not change as the explanatory variable changes, or if the response variable changes in ways that appear random, then there is zero or no association (Figure 19.4).

## Strength

Strength refers to how tightly the data appears to fit the form.

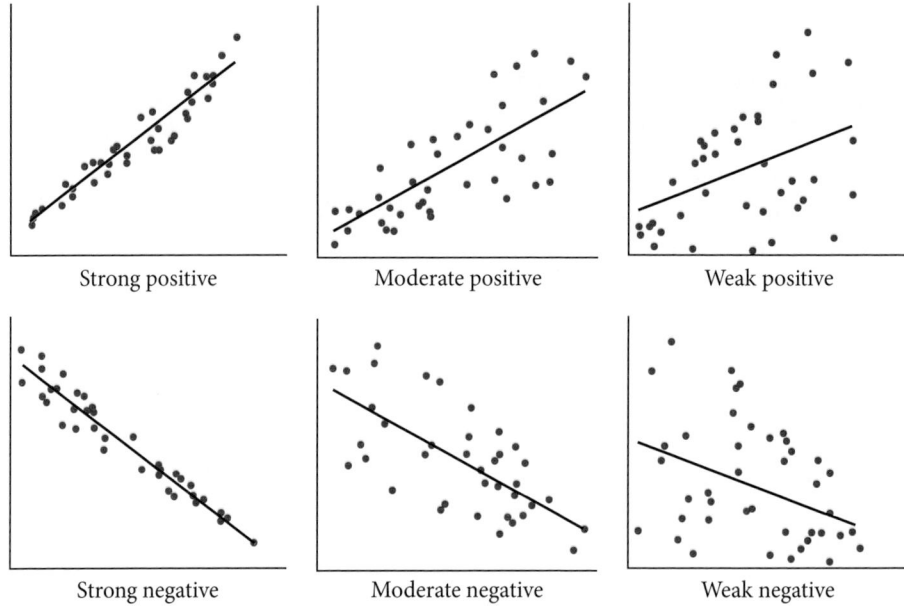

**Figure 19.5** Strong, moderate, and weak associations

In Figure 19.5, the best-fitting linear model for the data set is also shown. As you can see, weak associations are widely scattered around the best-fit line, while strong associations show data tightly clustered around the best-fit line. Notice that both positive and negative associations can be strong (and both can be weak).

## Unusual features

We want to make note of any unusual features we see. Outliers are data points that do not fit the pattern or form of the other data points, as shown in Figure 19.6.

You will see many sources refer to correlation instead of association when describing direction and strength. We will use correlation primarily in the next section when discussing **correlation coefficients**, which are quantitative measures of specific types of association.

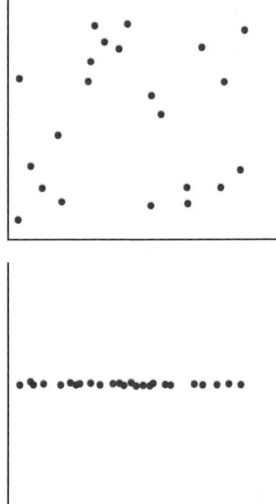

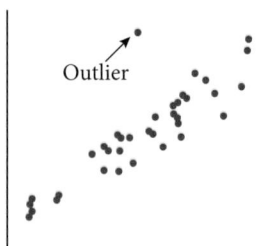

**Figure 19.4** Two examples of zero association

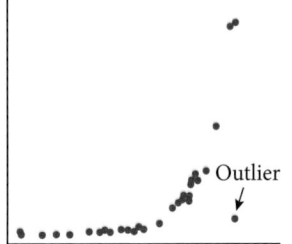

**Figure 19.6** Outliers are a departure from the predominant pattern

We may also see clusters, gaps, or other patterns in the data that should be investigated, as shown in Figure 19.7.

In general, unusual features in your data are always worth investigating: What caused the unusual feature? Is it a problem with the way the data was generated or collected? Was it due to an error in entering the data? Is there something in the phenomena being investigated that is creating the feature?

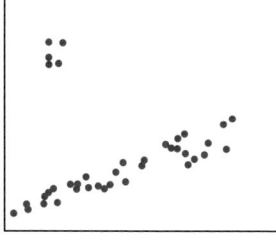

**(a)** A cluster of outliers

**(b)** Two different directions

**(c)** This cluster fits the overall direction and form, but why the gap?

**Figure 19.7** Unusual features in scatter diagrams

## Example 19.1

Describe each scatter diagram.

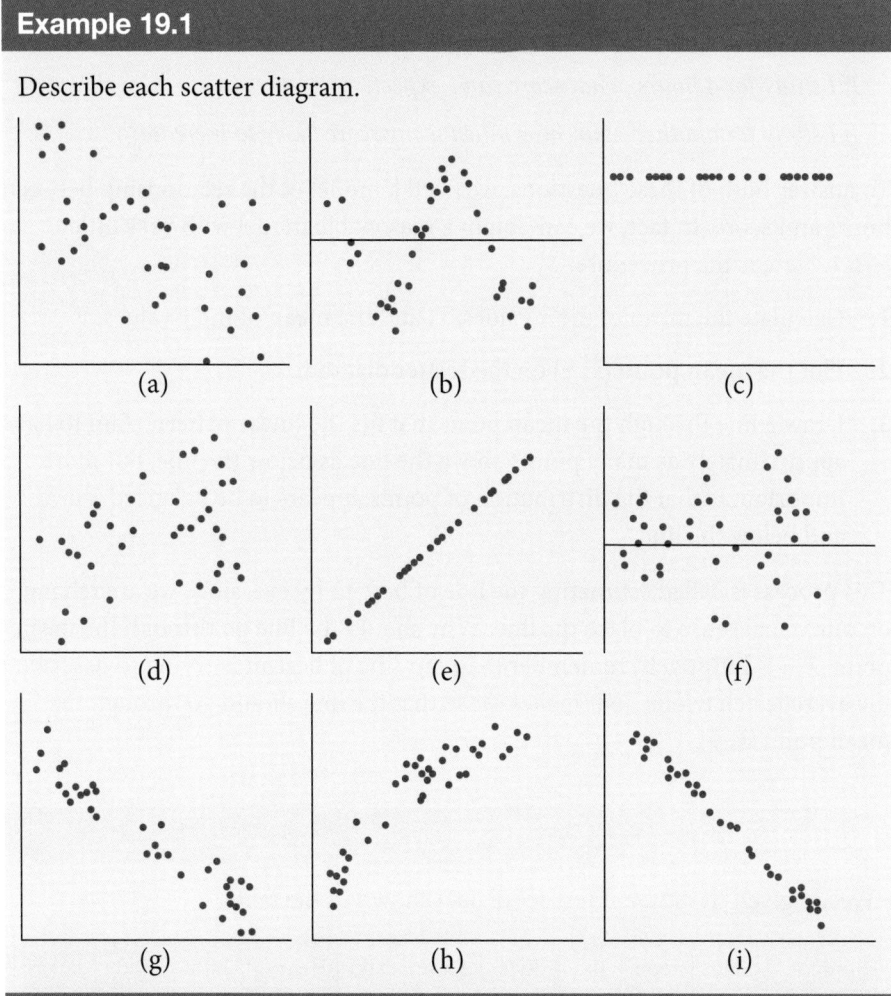

### Solution

(a) Moderate negative linear association
(b) Sinusoidal form
(c) No correlation
(d) Moderate positive linear association
(e) Perfect positive linear association
(f) No correlation
(g) Strong negative linear association
(h) Strong positive non-linear association
(i) Very strong negative linear association

## Estimating the line of best-fit

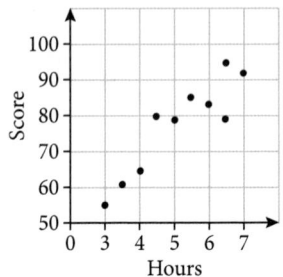

**Figure 19.8** Scatter diagram for the data in Table 19.1

Later in this chapter we will learn methods for finding the best-fitting linear model. Until then, it is worthwhile to think about what we are trying to achieve and see what we can do already. Consider the class of Mathematics students again; the scatter diagram is in Figure 19.8.

The association appears approximately linear, positive, and strong. Students in this class might ask questions such as:

*If I study for 4 hours, what score can I expect?*

*If I study for another hour, how much is my score likely to increase?*

To answer both of these questions, we need a model of the relationship between hours and score. In fact, we can obtain a reasonable model with very little effort. Here is the procedure:

1. Calculate the mean of the $x$ values, $\bar{x}$, and the mean of the $y$ values, $\bar{y}$.

2. Plot the **mean point** $(\bar{x}, \bar{y})$ on the scatter diagram.

3. Draw a line through the mean point that fits the linear pattern. Aim to have approximately as many points above the line as below the line, but more important is that the distribution of points appears to be balanced above and below the line.

This process is called estimating the line of best-fit by eye, since we are relying on our visual sense to place the line. Why should the line go through the mean point $(\bar{x}, \bar{y})$? Intuitively, remember than our line of best-fit is trying to describe the average behaviour, so it makes sense that the line should go through the mean point $(\bar{x}, \bar{y})$.

### Example 19.2

We are given the student test score data shown in the table.

| Student | Tim | Joon | Jim | Kyle | Steve | Niki | Henry | Anton | Cindy | Lukas |
|---|---|---|---|---|---|---|---|---|---|---|
| Hours ($x$) | 4 | 4.5 | 6 | 3.5 | 3 | 5 | 5.5 | 6.5 | 7 | 6.5 |
| Score ($y$) | 65 | 80 | 83 | 61 | 55 | 79 | 85 | 79 | 92 | 95 |

(a) Draw a scatter diagram and draw the line of best-fit by eye.

(b) Use your best-fit line to predict the score of a student who studies 4 hours.

(c) Use your best-fit line to find the equation of a linear model.

(d) Interpret your linear model to estimate how much a student's score will improve for an additional hour of study.

## Solution

(a) Plot the scatter diagram. Then calculate the coordinates of the mean point:

$$\left.\begin{array}{l} \bar{x} = \dfrac{1}{n}\sum x_i = \dfrac{1}{10}(51.5) = 5.15 \\[2mm] \bar{y} = \dfrac{1}{n}\sum y_i = \dfrac{1}{10}(774) = 77.4 \end{array}\right\} \Rightarrow (\bar{x}, \bar{y})$$

$$= (5.15, 77.4)$$

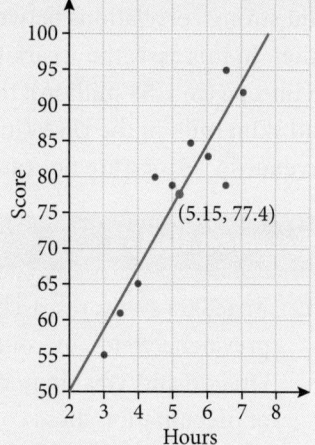

Then, plot the mean point on the scatter diagram, and chose a line of best fit by eye. One possible line is shown.

Notice that although the line does not have the same number of data points above and below the line, it appears to be balanced.

(b) To predict the score of a student who studies 4 hours, we simply read off the graph. The best-fit line appears to pass through the point $(4, 67)$, so we predict that a student who studies 4 hours would likely earn, on average, a score of 67.

(c) To find the equation of our best-fit line, first find the gradient of the best-fit line. Pick two points on the best-fit line near the left and right sides of the data. On the left side, our best-fit line passes through $(2, 50)$, so that is an easy point to choose. On the right side, our best-fit line appears to pass through approximately $(7.2, 95)$. Therefore, the gradient is

$$\frac{y_1 - y_2}{x_1 - x_2} = \frac{50 - 95}{2 - 7.2} \approx 8.7$$

Using the point $(2, 50)$, we can find the linear model:

$$y - 50 = 8.7(x - 2) \Rightarrow y = 8.7x + 32.6$$

(d) Recall that in this context, the gradient of the linear model tells us the change in score for each additional hour of study. Therefore, we predict that a student's score will increase by 8.7 for each additional hour of study.

> **!** Do not use existing data points to make predictions unless they happen to fall exactly on the best-fit line. Instead, use the best-fit line. For example, Tim earned a score of 65 with 4 hours of study. However, since Tim is not exactly on the best-fit line, he does not represent the average behaviour. Instead we predict that a student who studies 4 hours would receive, on average, a score of 67.
>
> Similarly, do not use data points to estimate the slope of the best fit line, unless they happen to fall exactly on the best-fit line.

## Association and correlation are not causation

It is important to remember than no matter how strong an association appears to be, changes in the explanatory variable do not necessarily cause changes in the response variable. It could be that the two variables are entirely unrelated. For example, consider the scatter diagram shown in Figure 19.9 (the axes labels have been deliberately omitted).

It is perfectly correct to say that there is a strong, positive, linear correlation. However, do changes in the $x$ variable cause changes in the $y$ variable? No matter how strong the correlation appears to be, the data alone cannot prove that changes in $x$ cause changes in $y$. To establish a causal relationship, we would need to design an experiment to deliberately manipulate the explanatory

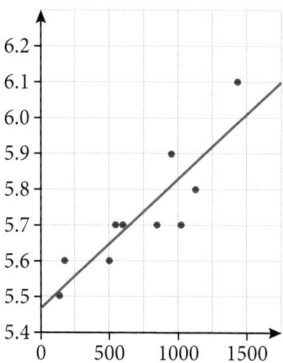

**Figure 19.9** The data alone cannot prove that changes in $x$ cause changes in $y$

787

variable and look for changes in the response variable. In fact, it's not hard to find strong correlations between variables that have nothing to do with each other. In this case, the *x* variable is revenue from commercial space launches in Europe (in US$ million); the *y* variable is per capita consumption of 1% and skim milk in the USA (in US gallons). Clearly, increases in space launch revenue in Europe are not causing people to drink more milk in the USA!

### Exercise 19.1

1. In a 2008 study, researchers from Cornell University found that autism prevalence rates, by county, for school-aged children in California, Oregon and Washington in 2005 were positively related to the amount of precipitation these counties received.

   (a) Identify the explanatory and response variables from this statement.

   (b) Does this mean that rainfall causes autism? Explain.

2. Describe each scatter diagram.

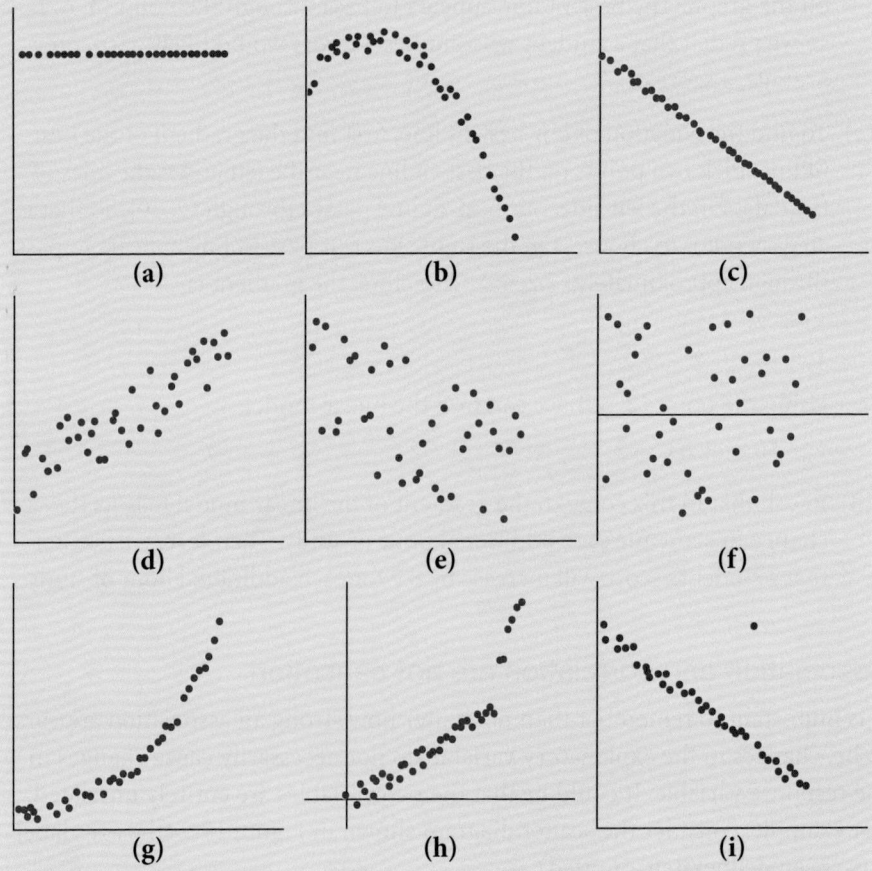

(a)  (b)  (c)

(d)  (e)  (f)

(g)  (h)  (i)

3. For each set of data, create a best-fit line by eye for *y* on *x*, and find the equation of your line.

   (a)

   | x | 10 | 21 | 29 | 40 | 50 | 60 |
   |---|----|----|----|----|----|----|
   | y | 163 | 130 | 109 | 67 | 74 | 26 |

**(b)**

| x | 1.0 | 1.2 | 1.2 | 2.2 | 2.8 | 3.9 | 5.0 | 6.0 | 7.0 | 8.0 | 9.0 | 10.0 |
|---|---|---|---|---|---|---|---|---|---|---|---|---|
| y | 24.1 | 16.6 | 6.5 | 15.4 | 27.2 | 31.8 | 41.2 | 22.2 | 44.9 | 31.1 | 42.4 | 61.9 |

**(c)**

| x | 10 | 20 | 30 | 40 | 50 | 60 |
|---|---|---|---|---|---|---|
| y | 0 | 27 | 9 | 60 | 85 | 95 |

4.  Lena is interested in whether the number of pages in a book determines how long it takes her to read it. She keeps track of the number of hours it takes her to read a book, and the number of pages in the book. Here is the data for the last six books she read:

| Pages | 150 | 330 | 600 | 450 | 200 | 250 |
|---|---|---|---|---|---|---|
| Hours | 5 | 13 | 19 | 14 | 9 | 9 |

   **(a)** Identify the explanatory and response variables in this context.

   **(b)** Generate a scatter diagram and describe the relationship between the number of pages and the time required to read a book.

   **(c)** Create a best-fit line by eye and find the equation of your line for the number of hours $t$ in terms of the number of pages $n$.

   **(d)** Interpret the gradient and $t$-intercept in context.

   **(e)** Predict how long it would take Lena to read a 500-page book.

5.  William is doing a study in his class concerning the lean mass and metabolic rate of a random sample of 12 of his school mates. Metabolic rate, the rate at which the body consumes energy, is important in studies concerning weight loss, exercise and dieting. The table gives the data William collected.

   **(a)** Identify the explanatory variable. Draw a scatter diagram of the data.

   **(b)** Find the mean mass and the mean rate.

| Student | Mass (kg) | Rate |
|---|---|---|
| 1 | 36.1 | 995 |
| 2 | 54.6 | 1425 |
| 3 | 48.5 | 1396 |
| 4 | 42.0 | 1418 |
| 5 | 50.6 | 1502 |
| 6 | 42.0 | 1256 |
| 7 | 40.3 | 1189 |
| 8 | 33.1 | 913 |
| 9 | 42.4 | 1124 |
| 10 | 34.5 | 1052 |
| 11 | 51.1 | 1347 |

   **(c)** Draw a line of best-fit on your diagram.

   **(d)** Describe the strength of the relationship and interpret the gradient of the line.

   **(e)** Liz has a mass of 40 kg. What metabolic rate should she expect to see?

   **(f)** Kevin has a mass of 70 kg. Can you use the model to predict his metabolic rate?

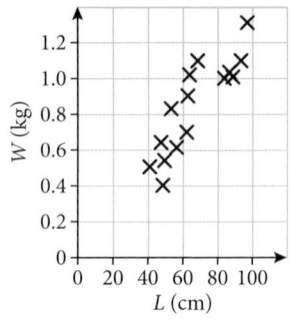

**Figure 19.10** Diagram for question 6

6. A sample of 15 fish was weighed. The weight $W$ was plotted against length $L$ as shown in Figure 19.10.

   Exactly two of the following statements about the plot could be correct. Identify the two correct statements.

   (a) The value of $r$, the correlation coefficient, is approximately 0.871

   (b) There is an exact linear relation between $W$ and $L$.

   (c) The line of regression of $W$ on $L$ has equation $W = 0.012L + 0.008$

   (d) There is negative correlation between the length and weight.

   (e) The value of $r$, the correlation coefficient, is approximately 0.998

   (f) The line of regression of $W$ on $L$ has equation $W = 63.5L + 16.5$

7. The manager of a factory wonders how production costs increase with production levels. She records the number of items produced and the production costs for 12 months. Here is her data.

| Number of items (thousands) | 18 | 36 | 45 | 22 | 69 | 72 | 13 | 33 | 59 | 79 | 10 | 53 |
|---|---|---|---|---|---|---|---|---|---|---|---|---|
| Production cost (thousands GBP) | 37 | 54 | 63 | 42 | 84 | 91 | 33 | 49 | 79 | 98 | 32 | 71 |

   (a) Identify the explanatory and response variables in this context.

   (b) Draw a scatter diagram for production cost versus number of items.

   (c) Describe the association between production cost and the number of items produced.

   (d) Calculate the coordinates of the mean point.

   (e) Draw a line of best-fit by eye on your diagram.

   (f) Use your line of best-fit to predict the production cost when 70 000 items are produced.

   (g) Find the gradient of your line of best-fit and interpret it in context.

   (h) Estimate the $y$-intercept of your line of best-fit and interpret it in context. Give a reason why we should be cautious about this value.

   (i) The manager would like to predict the production cost when 100 000 items are produced. Give a reason why she should be cautious about this prediction.

8. When cars are driven, their tires heat up. The table below shows tire temperature for a specific tire model when driven consistently at the given speeds.

| Speed (km h⁻¹) | 20 | 30 | 40 | 50 | 60 | 70 | 80 | 90 |
|---|---|---|---|---|---|---|---|---|
| Temperature (°C) | 45 | 52 | 64 | 66 | 91 | 86 | 98 | 105 |

   (a) Identify the explanatory and response variables.

   (b) Generate a scatter diagram for speed versus temperature and describe the association you see.

**(c)** Calculate the coordinates of the mean point.

**(d)** Draw a line of best-fit by eye on your diagram.

**(e)** Use your best-fit line to predict the production cost when driving $60 \, \text{km h}^{-1}$.

**(f)** Find the gradient of your best-fit line and interpret it in context.

**(g)** Estimate the $y$-intercept of your best-fit line and interpret it in context. Give a reason why we should be cautious about this value.

**(h)** Give a reason why we should not use this best-fit line fit to predict the tire temperature when driving $150 \, \text{km h}^{-1}$.

# 19.2 Measures of correlation

So far, we have used our visual sense to describe the association we see in a scatter diagram. Can we quantify that association? In this section, we will look at two methods for doing so: Pearson's product-moment correlation coefficient ($r$) and Spearman's rank correlation coefficient ($r_s$).

## Pearson's product-moment correlation coefficient ($r$)

Consider the two scatter diagrams in Figure 19.11.

Which diagram shows a stronger correlation, diagram A or diagram B?

In fact, both diagrams use the exact same data; only the scales are different.

For this reason, we should try to quantify the strength and direction of an association. This is exactly what **Pearson's product-moment correlation coefficient** aims to do. By convention, we refer to this statistic simply as **$r$**.

In this course, we will rely on technology to find the value of $r$. Our goal is to be able to interpret and use $r$. Consult your GDC manual to learn how to find $r$.

Diagram A

Diagram B

**Figure 19.11** Which diagram shows a stronger correlation?

**Facts about $r$**
- The value of $r$ is always in the interval $-1 \leqslant r \leqslant +1$
- The sign of $r$ tells us the direction of the correlation: positive or negative or zero.
- The size of $r$ tells us the strength of the linear correlation, as shown.

| Perfect negative correlation | No correlation | Perfect positive correlation |
|---|---|---|
| $r = -1$ | $r = 0$ | $r = +1$ |

$$-1 \qquad\qquad\qquad 0 \qquad\qquad\qquad 1$$

| Strong negative correlation | Weak negative correlation | Weak positive correlation | Strong positive correlation |
|---|---|---|---|

- If $r = +1$, there is a perfect positive linear correlation; all the points fall on a line with positive slope.
- If $r = 0$, there is zero correlation.
- If $r = -1$, there is a perfect negative linear correlation; all the points fall on a line with negative slope.
- $r$ has no units, and is not a percentage.

791

Remember: Pearson's correlation coefficient $r$ applies only to linear relationships; that is, it measures the strength of a linear correlation (if one exists).

There are no strict rules about what intervals of $r$ indicate strong, moderate or weak correlations. Some authors claim $|r| > 0.5$ indicates a strong correlation, while others want $|r| > 0.87$. In reality, the interpretation of $r$ depends on context: even a very low value of $r$ may indicate a useful correlation in some contexts. Since in this course we are using only one explanatory variable, it is usually the case that $r$ is not close to 1: in most (interesting) real-world situations there is more than one explanatory variable at work.

Here are some examples of $r$:

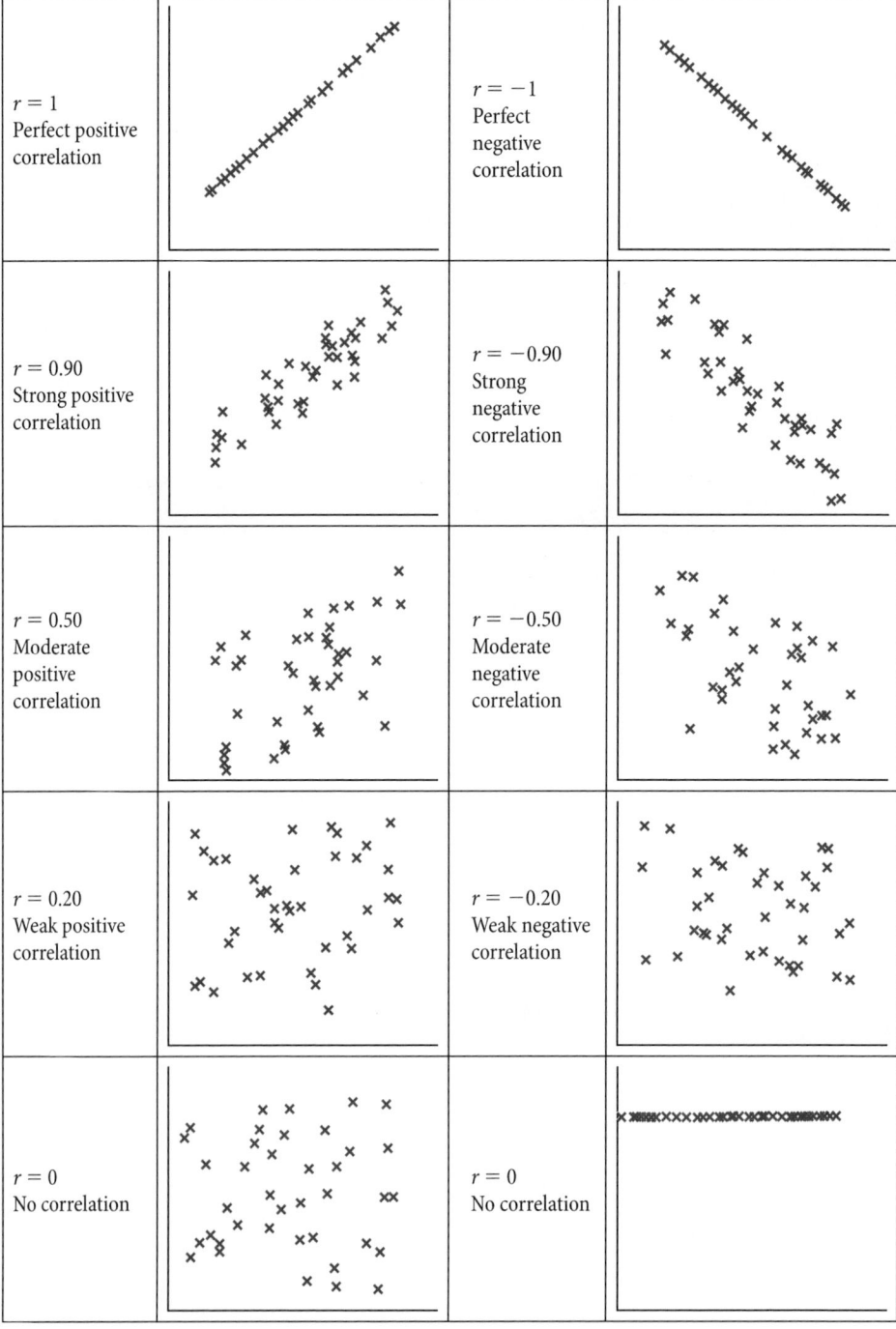

**Figure 19.12** Examples of $r$

There is a famous set of scatter diagrams referred to as Anscombe's quartet, devised in 1973 by statistician Francis Anscombe. All four data sets share the same *r* value, but look very different, as shown in Figure 19.13.

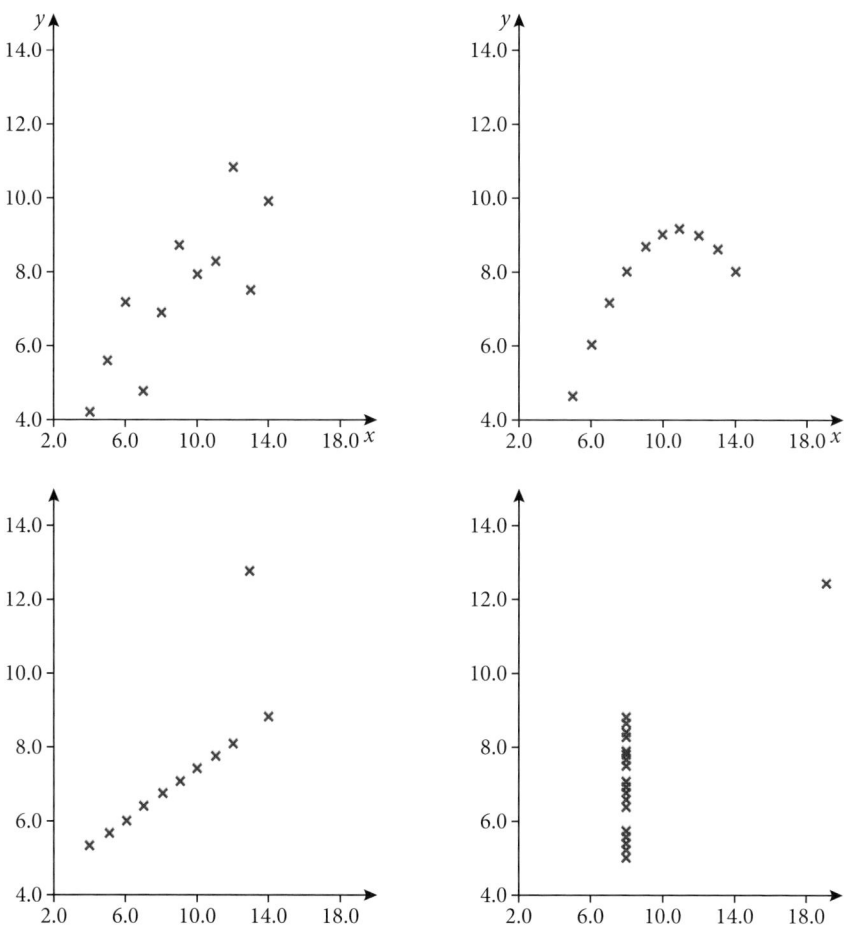

Figure 19.13  Four different scatter diagrams, all with $r = 0.816$

The lesson here is to always draw a scatter diagram; never rely solely on *r*.

It is always important to draw a scatter diagram. Do not rely on *r* to find an association, since there several ways that *r* can show that there is a correlation when there is not, or that there is no linear correlation when a non-linear association exists. Here are some examples:

**Strong positive correlation, $r = 0.75$**

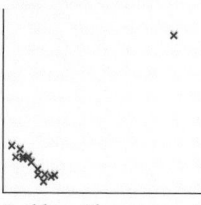

Problem: There seems to be a negative association for most of the data, but a dramatic outlier fools *r* into indicating a positive linear correlation.

**No correlation, $r = -0.02$**

Problem: A non-linear pattern does not have a linear correlation, but there is clearly some sort of association between *x* and *y*.

## Example 19.3

In an experiment, a metal bar is heated and its length is measured. The table shows measurements taken during the experiment.

| Temperature (°C) | 40 | 45 | 50 | 55 | 60 | 65 | 70 | 75 | 80 |
|---|---|---|---|---|---|---|---|---|---|
| Length (mm) | 20 | 20.12 | 20.20 | 20.21 | 20.25 | 20.25 | 20.34 | 20.47 | 20.61 |

(a)  Create a scatter diagram for this data.

(b)  Describe the association, calculating and interpreting *r* if appropriate.

**Solution**

(a)

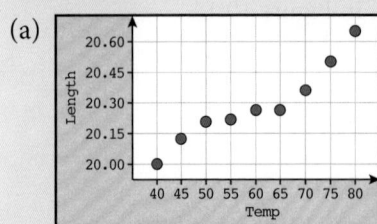

(b) The association appears to be strong and positive. Since the association appears to be approximately linear, it is appropriate to proceed with calculating $r$. Our GDC gives $r = 0.955$, which we interpret as a strong, positive linear correlation.

Sometimes the data we obtain has unusual features such as an outlier. Depending on the context, we may decide to analyse the data without the outlier.

**Example 19.4**

A student is interested in whether the mass of a vehicle affects the fuel consumption. The tables contain data on 15 small cars.

| Mass (kg) | 1120 | 1170 | 1180 | 1180 | 1220 | 1250 | 1400 | 1400 |
|---|---|---|---|---|---|---|---|---|
| Fuel use (litres/100 km) | 20.3 | 21.8 | 23.4 | 21.0 | 24.4 | 22.6 | 11.3 | 26.5 |

| Mass (kg) | 1460 | 1480 | 1510 | 1560 | 1740 | 1840 | 1960 |
|---|---|---|---|---|---|---|---|
| Fuel use (litres/100 km) | 30.5 | 27.7 | 29.0 | 32.1 | 33.8 | 35.8 | 38.1 |

(a) Create a scatter diagram for this data.

(b) Describe the association, calculating and interpreting $r$ if appropriate.

**Solution**

(a) There is a clear outlier at (1400, 11.3) This vehicle is much more efficient than the others, for its weight. Why is that? In fact, this data point represents a fuel-electric hybrid car. So, in this context, it is appropriate to

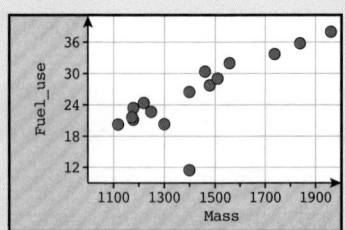

remove that data point from our observation and focus only on the conventional fuel-powered vehicles. Here is the new scatter diagram:

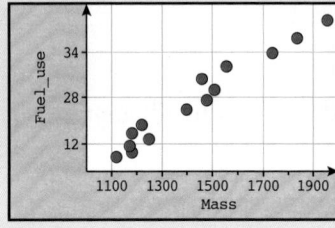

(b) The scatter diagram suggests a strong, positive, and approximately linear association, so we will proceed to calculate $r$. With the outlier removed, a GDC shows that $r = 0.980$, confirming a very strong, positive correlation.

Remember that $r$ only reports the strength of a linear correlation. For non-linear correlation, $r$ is not useful.

### Example 19.5

A student in a Physics class collects data on the height, in metres, of an object at specific times, in seconds, after it is dropped from a height of 35 m. The data is given in the table.

| Time (s) | 0.00 | 0.25 | 0.50 | 0.75 | 1.00 | 1.25 | 1.50 | 1.75 | 2.00 | 2.25 | 2.50 |
|---|---|---|---|---|---|---|---|---|---|---|---|
| Height (m) | 34.83 | 34.62 | 34.11 | 32.58 | 30.21 | 27.25 | 23.85 | 20.27 | 15.52 | 10.1 | 4.38 |

(a) Create a scatter diagram for this data.

(b) Describe the association, calculating and interpreting $r$ if appropriate.

### Solution

(a)
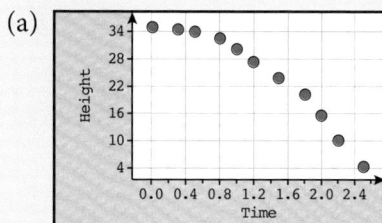

(b) The scatter diagram looks to have a strong, negative, and non-linear association. We know that falling objects can be modelled with quadratic functions and the scatter diagram suggests that a quadratic model could be appropriate. Since the data appears to have a non-linear association, it is not useful to calculate $r$.

## Spearman's rank correlation coefficient ($r_s$)

Smartphones often lose their resale value quickly after they are purchased. Data on the change in value for one particular model are given in Table 19.2.

| Months since purchase | 0 | 2 | 4 | 6 | 8 | 10 | 12 | 14 | 16 | 18 | 20 | 22 | 24 |
|---|---|---|---|---|---|---|---|---|---|---|---|---|---|
| Change in value (%) | 0 | −29 | −47 | −58 | −69 | −72 | −71 | −77 | −80 | −82 | −83 | −85 | −87 |

Table 19.2 Change in value for a particular model of smartphone

How strong is the association between the age of the smartphone and depreciation? Draw a scatter diagram to check for form before calculating Pearson's $r$. The scatter diagram is shown in Figure 19.14.

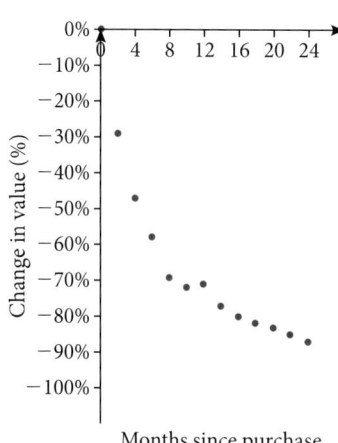

Figure 19.14 Association between smartphone age and depreciation

795

| Month rank | Change in value rank |
|------------|----------------------|
| 13 | 1 |
| 12 | 2 |
| 11 | 3 |
| 10 | 4 |
| 9 | 5 |
| 8 | 7 |
| 7 | 6 |
| 6 | 8 |
| 5 | 9 |
| 4 | 10 |
| 3 | 11 |
| 2 | 12 |
| 1 | 13 |

**Table 19.3** Rank ordering of smartphone data from Table 19.2

The association appears to be strong, negative, and non-linear. Since Pearson's $r$ measures the strength of a linear correlation, it will likely underestimate the strength of this association.

So how can we quantify the strength and direction of the association? Suppose we want to measure whether value consistently decreases as the age of the phone increases.

Consider this: if it is true that the value decreases consistently as time progresses, then we could rank the loss in value data and the ranks should be in strictly descending order. We will do the same with the month data to generate the rank ordering shown in Table 19.3.

We observe that the rank-order is almost perfect. It turns out we can now quantify the association using **Spearman's rank correlation coefficient**, $r_s$. Using a GDC, we find that $r_s = -0.99$. The interpretation for Spearman's $r_s$ is the same as Pearson's $r$, as shown below.

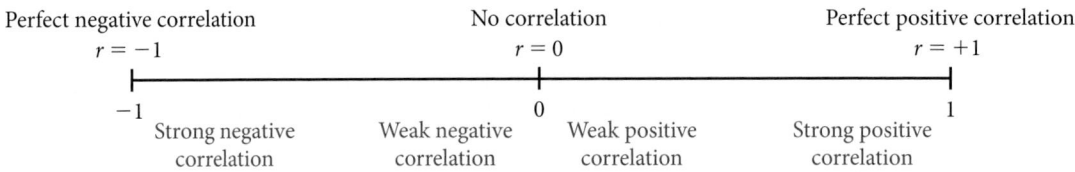

Perfect negative correlation $r = -1$ | No correlation $r = 0$ | Perfect positive correlation $r = +1$

$-1$ ..... $0$ ..... $1$

Strong negative correlation | Weak negative correlation | Weak positive correlation | Strong positive correlation

Most GDCs do not include an option for calculating Spearman's $r_s$. Don't worry! The calculation for $r_s$ is exactly the same as Pearson's $r$. Therefore, just use your GDC's tool for finding $r$ after you enter the rank order data. Make sure that you interpret the resulting $r$ value as Spearman's $r_s$, not Pearson's $r$.

**Facts about $r_s$**
- The value of $r_s$ is always in the interval $-1 \leqslant r \leqslant +1$
- The sign of $r_s$ tells us the direction of the rank-order correlation: positive or negative or zero.
- The size of $r_s$ tells us the strength of the rank-order correlation.
- If $r_s = +1$, there is a perfect positive rank-order correlation.
- If $r = 0$, there is zero correlation.
- If $r = -1$, there is a perfect negative rank-order correlation.
- $r$ has no units, and is not a percentage.
- $r$ is a measure of monotonicity.

Therefore, we conclude that there is a nearly perfect negative rank-order correlation between the age of this smartphone model and its loss in value.

## Differences between Pearson's $r$ and Spearman's $r_s$

Even though the calculations for Pearson's $r$ and Spearman's $r_s$ are identical, there are two differences to keep in mind. Both of these have to do with the fact that, for Spearman's $r_s$, we convert the data set into ranks before calculating the correlation.

### Linear and monotonic functions

As you have seen, both Pearson's $r$ and Spearman's $r_s$ attempt to measure the direction and strength of a correlation. However, while Pearson's $r$ measures the strength of a linear correlation, Spearman's is measuring the strength of a rank-order correlation. Spearman's $r_s$ measures whether the response variable consistently increases or decreases as the explanatory variable increases. This is called **monotonicity**. Monotonic functions only increase or decrease, as illustrated in Figure 19.15. Non-monotonic functions are shown in Figure 19.16.

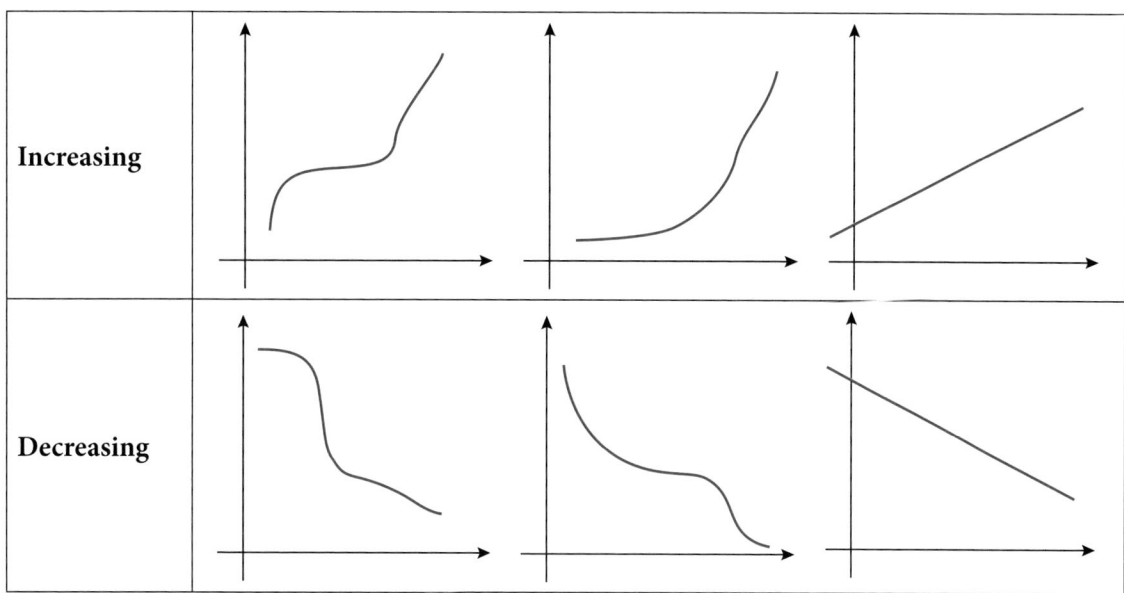

**Figure 19.15** Monotonic functions

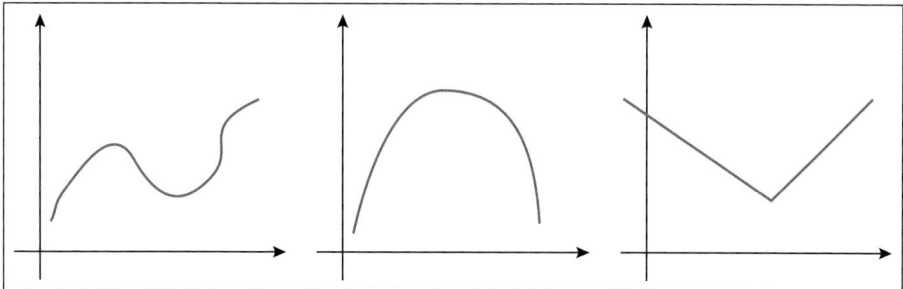

**Figure 19.16** Non-monotonic functions

## Effect of outliers

Recall that Pearson's $r$ is sensitive to outliers. As you have seen previously, the direction of Pearson's $r$ can change completely when there are dramatic outliers. However, for Spearman's $r_s$, since we convert the data set into ranks before we calculate $r_s$, the effect of outliers is much reduced.

For example, consider the data shown in the scatter diagram in Figure 19.17.

The value of Pearson's $r$ is 0.77, indicating a strong positive correlation. From the scatter diagram, we can see that is not the case. If we rank the data and calculate Spearman's $r_s$, we obtain $r_s = -0.48$. Spearman's $r_s$ is a much better indicator of the direction and strength that we see in the scatter diagram. (Of course, with a dramatic outlier like this, we should investigate the cause of that outlier and then decide what to do: Remove the outlier? Collect more or different data?)

When ranking data to calculate Spearman's $r_s$, we are not allowed to have tied ranks. But what we do if there are tied ranks? We simply average the rank position for the tied data values.

**Figure 19.17** A data set with extreme outlier(s) can 'fool' Pearson's $r$, but Spearman's $r_s$ is more resistant to outliers

### Example 19.6

The table shows how much time a group of students spent studying for a mathematics test and their scores.

| Student | Tim | Joon | Jim | Kyle | Steve | Niki | Henry | Anton | Cindy | Lukas | Julie |
|---|---|---|---|---|---|---|---|---|---|---|---|
| Hours (x) | 4.0 | 4.5 | 6.0 | 3.5 | 3.5 | 5.0 | 5.5 | 6.5 | 7.0 | 6.5 | 6.5 |
| Score (y) | 65 | 80 | 83 | 61 | 55 | 79 | 85 | 79 | 92 | 95 | 80 |

To calculate the rank order for this data set, note that there are two students, Kyle and Steve, who spent 3.5 hours studying and three students, Anton, Lukas and Julie, who spent 6.5 hours studying.

| Student | Tim | Joon | Jim | Kyle | Steve | Niki | Henry | Anton | Cindy | Lukas | Julie |
|---|---|---|---|---|---|---|---|---|---|---|---|
| Hours Rank | 9 | 8 | 5 | 10 | 10 | 7 | 6 | 2 | 1 | 2 | 2 |
| Score Rank | 9 | 5 | 4 | 10 | 11 | 7 | 3 | 7 | 2 | 1 | 5 |

Kyle and Steve are tied for rank 10, while Anton, Lukas, and Julie are tied for rank 2. Therefore, Anton, Lukas, and Julie are effectively occupying ranks 2, 3, and 4; the next rank (Joon) is rank 5. Therefore, we assign Anton, Lukas, and Julie a rank equal to the mean rank they are occupying: $\text{Rank} = \dfrac{2 + 3 + 4}{3} = 3$

Likewise, Kyle and Steve are taking up ranks 10 and 11, so we will assign the mean rank they are occupying as well: $\text{Rank} = \dfrac{10 + 11}{2} = 10.5$

Our new rank order is shown in the table.

| Student | Tim | Joon | Jim | Kyle | Steve | Niki | Henry | Anton | Cindy | Lukas | Julie |
|---|---|---|---|---|---|---|---|---|---|---|---|
| Hours Rank | 9 | 8 | 5 | 10.5 | 10.5 | 7 | 6 | 3 | 1 | 3 | 3 |
| Score Rank | 9 | 5 | 4 | 10 | 11 | 7 | 3 | 7 | 2 | 1 | 5 |

We can then calculate Spearman's $r_s$ as usual.

### Exercise 19.2

1. Explain the differences between Spearman's $r_s$ and Pearson's $r$.

2. Assign each scatter diagram to one of the following $r$ values: $r = 0.90$, $r = 0.74$, $r = 0.51$, $r = -0.03$, $r = -0.40$, $r = -0.58$, $r = -0.95$, $r = -0.99$

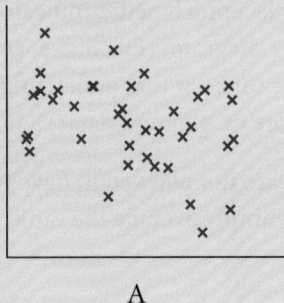

A

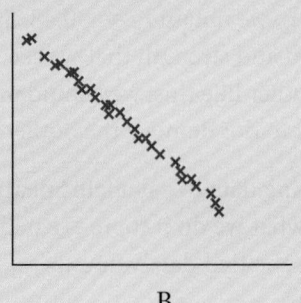

B

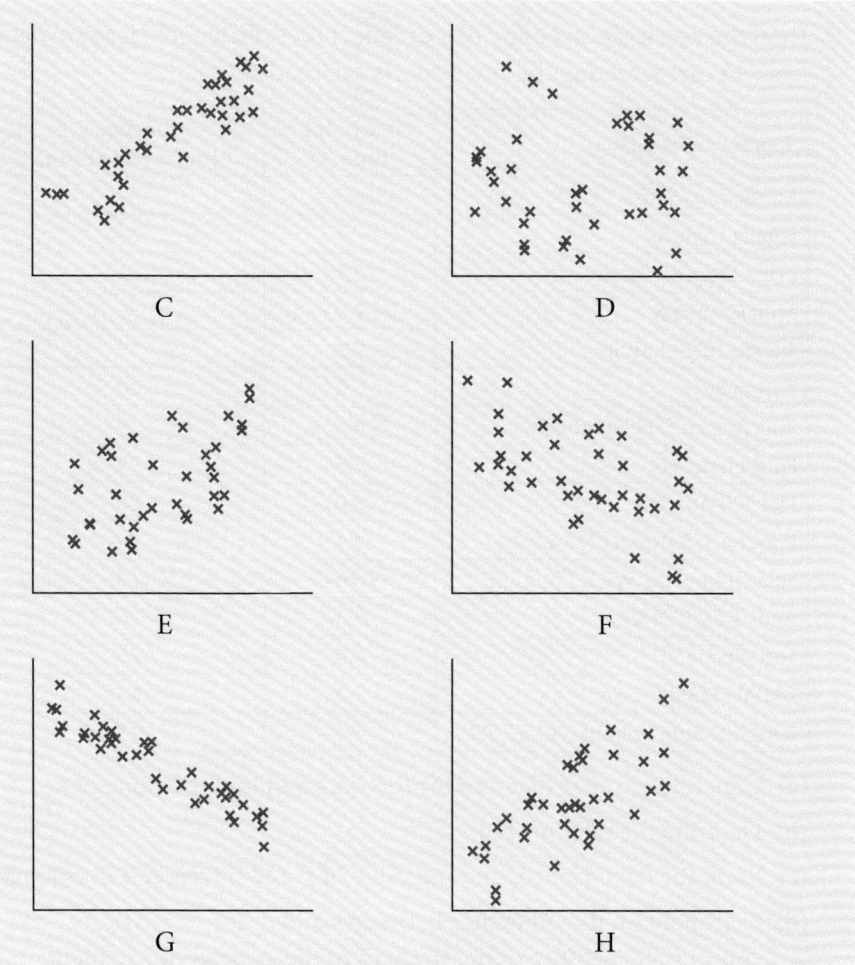

C

D

E

F

G

H

3. Six metal plates are immersed in an acid bath. After some time, they are removed and the mass is measured to determine the mass that was lost. The table gives data from this experiment.

| Time (hours) | 150 | 200 | 200 | 300 | 450 | 500 |
|---|---|---|---|---|---|---|
| Mass lost (%) | 0.761 | 1.44 | 1.15 | 1.65 | 2.56 | 2.44 |

(a) Generate a scatter diagram and calculate the value of $r$.

(b) Describe the association between mass lost and hours.

4. The following table shows the cost in AUD of seven paperback books chosen at random, together with the number of pages in each book.

| Book | 1 | 2 | 3 | 4 | 5 | 6 | 7 |
|---|---|---|---|---|---|---|---|
| Number of pages, $x$ | 50 | 120 | 200 | 330 | 400 | 450 | 630 |
| Cost, $y$ (AUD) | 6.00 | 5.40 | 7.20 | 4.60 | 7.60 | 5.80 | 5.20 |

(a) Plot these pairs of values on a scatter diagram.

(b) Write down the linear correlation coefficient, $r$, for the data.

(c) Stephen wishes to buy a paperback book that has 350 pages in it. He plans to draw a line of best-fit to determine the price. State whether or not this is an appropriate method in this case and justify your answer.

**5.** Does the acceleration ability of a car affect fuel efficiency? Here are data from 15 cars targeting the small sports car/roadster market.

| Sports cars/roadsters | 0–60 mph time (s) | Efficiency (miles per gallon) |
|---|---|---|
| Mazda MX-5 Miata Club | 6.7 | 34 |
| Honda Civic Si | 7.3 | 34 |
| Fiat 124 Spider Lusso | 7.1 | 31 |
| Mini Cooper S | 7.0 | 30 |
| Subaru BRZ Premium | 7.2 | 30 |
| Toyota 86 | 7.2 | 30 |
| Volkswagen GTI Autobahn | 6.6 | 29 |
| Ford Fiesta ST | 7.3 | 29 |
| Fiat 500 Abarth | 8.0 | 28 |
| Porsche 718 Boxster (base) | 4.4 | 26 |
| Subaru Impreza WRX Premium | 6.0 | 26 |
| Audi TT 2.0T (AT) | 6.3 | 26 |
| Ford Focus ST | 6.6 | 26 |
| BMW M235i | 5.2 | 25 |
| Ford Mustang Premium (2.3T, AT) | 6.4 | 25 |

**(a)** Generate a scatter diagram for this data and find the value of $r$.

**(b)** Describe the relationship.

**(c)** Rank the data. Hence, find the value of $r_s$ and interpret it in context.

**(d)** Comment on the difference between $r$ and $r_s$. Is $r$ or $r_s$ a more appropriate measure of the strength of the association? Explain.

**6.** Here is the same data from question **5**, but in litres per 100 km.

| Sports cars/roadsters | 0–60 mph time (s) | Efficiency (litres per 100 km) |
|---|---|---|
| Mazda MX-5 Miata Club | 6.7 | 6.92 |
| Honda Civic Si | 7.3 | 6.92 |
| Fiat 124 Spider Lusso | 7.1 | 7.59 |
| Mini Cooper S | 7.0 | 7.84 |
| Subaru BRZ Premium | 7.2 | 7.84 |
| Toyota 86 | 7.2 | 7.84 |
| Volkswagen GTI Autobahn | 6.6 | 8.11 |
| Ford Fiesta ST | 7.3 | 8.11 |
| Fiat 500 Abarth | 8.0 | 8.4 |
| Porsche 718 Boxster (base) | 4.4 | 9.05 |
| Subaru Impreza WRX Premium | 6.0 | 9.05 |
| Audi TT 2.0T (AT) | 6.3 | 9.05 |
| Ford Focus ST | 6.6 | 9.05 |
| BMW M235i | 5.2 | 9.41 |
| Ford Mustang Premium (2.3T, AT) | 6.4 | 9.41 |

(a) Generate a scatter diagram and calculate the value of $r$.

(b) Describe the association and compare to question 5.

(c) Convert this data into ranked data.

(d) Find and interpret the value of $r_s$, then compare to question 5.

7. Fast food is often considered unhealthy because of high contents of fat and sodium. Are items that are high in fat also high in sodium? Here are some data on some popular menu items from fast-food restaurants.

| Fat (g) | 21 | 29 | 34 | 34 | 38 | 41 | 44 |
|---|---|---|---|---|---|---|---|
| Sodium (mg) | 900 | 1510 | 1320 | 830 | 1200 | 960 | 1210 |

(a) Generate a scatter diagram and calculate the value of $r$.

(b) Describe the association between fat and sodium for these menu items.

8. Fast food is often considered unhealthy because of high contents of fat and calories. Are items that are high in fat also high in calories? Here are some data on some popular menu items from fast-food restaurants.

| Fat (g) | 21 | 29 | 34 | 34 | 38 | 41 | 44 |
|---|---|---|---|---|---|---|---|
| Calories (kcal) | 410 | 580 | 570 | 560 | 660 | 680 | 670 |

(a) Generate a scatter diagram and calculate the value of $r$.

(b) Describe the association between fat and calories for these menu items.

9. Ian is doing a science experiment where he controls three explanatory variables ($a$, $b$, and $c$) and one response variable ($S$). After collecting the data and calculating Pearson's $r$, he obtains the following statistics.

| Variable | $a$ | $b$ | $c$ |
|---|---|---|---|
| $r$ | 0.86 | 0.32 | 0.55 |

(a) Can Ian conclude that $a$ is the strongest explanatory variable?

(b) Do changes in $a$ cause changes in $S$?

10. An anemometer measures wind speed by the rotation a set of small cups called vanes. Anemometres may be calibrated by using a wind tunnel and measuring the rotational speed of the vanes. One such calibration generated the following data.

| Wind speed (m s$^{-1}$) | 1.0 | 1.2 | 1.4 | 1.6 | 1.8 | 2.0 |
|---|---|---|---|---|---|---|
| Rotations per minute (RPM) | 29 | 42 | 65 | 91 | 110 | 121 |

(a) Generate a scatter diagram and calculate the value of $r$.

(b) Describe the association between wind speed and RPM.

(c) Give a reason why a linear model may not be the best for this situation.

**11.** Many universities calculate a measure of academic performance called a grade point average (GPA). Do students who work more hours in off-campus jobs have worse GPAs? Data from ten randomly sampled students at a certain university are given below.

| GPA | 3.14 | 2.75 | 3.68 | 3.22 | 2.45 | 2.8 | 3.00 | 2.23 | 3.14 | 2.9 |
|-----|------|------|------|------|------|-----|------|------|------|-----|
| Time (hours) | 25 | 30 | 11 | 18 | 22 | 40 | 15 | 29 | 10 | 0 |

(a) Generate a scatter diagram for this data.

(b) Calculate the value of $r_s$ and interpret it in context.

(c) Calculate Pearson's $r$. Compare to $r_s$ – give a reason why $r$ and $r_s$ are not significantly different.

**12.** The specific weight of water varies with temperature according to the data given:

| Temperature (°C) | 0 | 0.1 | 1 | 4 | 10 | 15 | 20 |
|------------------|---|-----|---|---|----|----|----|
| Specific weight $(\text{k N m}^{-3})$ | 9.8050 | 9.8052 | 9.8057 | 9.8064 | 9.8040 | 9.7980 | 9.7890 |

(a) Generate a scatter diagram for this data.

(b) Calculate the value of $r_s$ and interpret it in context.

(c) Give a reason why $r_s$ is not the best measure of association for this data.

# 19.3 Linear regression

So far we have found linear best-fit models by eye, using the mean point. Is there a more rigorous method? Yes, there is. In this course we will use GDCs to find the linear models and we will focus on correct use and interpretation. The process of fitting a model to data is called **regression**.

In the same way that we would like to measure the strength of correlation, we would like to measure the predictive power or fit of any model we create. We will do this using the **coefficient of determination ($R^2$)**.

## Least-squares regression line

### Example 19.7

In Zambia, a staple of the diet is ground maize, known as mealie meal. In one market, bags of the brand *Mother's Meal*, are priced as shown in the table. Prices are given in Zambian Kwacha (ZMW).

| Mass (kg) | 5 | 10 | 25 | 50 |
|-----------|---|----|----|----|
| *Mother's Meal* price (ZMW) | 20 | 34 | 79 | 140 |

(a) Explain why it is appropriate to use a linear model for this data.

(b) Find the best-fit linear model for this data to predict the price when given the mass. Use $P$ to represent the price in ZMW and $m$ to represent the mass in kg.

(c) Interpret the gradient and $P$-intercept of your model in context.

(d) Use your model to predict the price of a 50 kg bag. Does your model over-predict or under-predict the actual price? Give a reason why this may be the case.

(e) Write down a reasonable domain for your model and give a reason for your choice.

## Solution

(a) To decide whether a linear model is appropriate, start by drawing a scatter diagram.

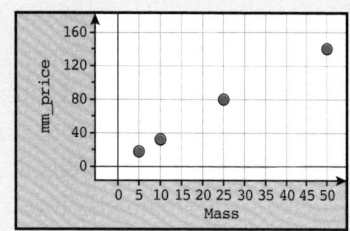

We see that the data appears to have a strong, positive, linear association. To 3 significant figures, Pearson's $r = 0.999$. A linear model is appropriate.

(b) Use a GDC to find the least-squares regression line. To 3 significant figures, the best-fitting linear model is $P = 2.67m + 8.14$

(c) The gradient of 2.67 suggests that, on average, the price for each additional kg of mealie meal is 2.67 ZMW. The $P$-intercept of 8.14 suggests that an empty bag of mealie meal costs 8.14 ZMW. This is a fixed cost representing distribution, packaging, and other costs.

(d) According to our model, the price of a 50 kg bag of mealie meal would be $P = 2.67(50) + 8.14 = 142$ ZMW (3 s.f.). This is a very slight over-prediction, which suggests that the 50 kg bag is a relatively good price compared to the other bags. This price is probably set to encourage consumers to purchase in greater volume.

(e) We can be confident predicting prices for bags with mass $5 \leqslant m \leqslant 50$ since any prediction in that interval would be an interpolation. Therefore, the domain for our model should be $5 \leqslant m \leqslant 50$

A prediction based on a mass outside that interval would be an extrapolation and is therefore unreliable.

There are two main methods for finding a best-fit linear model. In this course, we will use the least-squares regression line. This is the default in most GDCs, but you may also see a median-median line – we will not use that method in this course.

You may also see two different options for the least-squares regression line:
$y = mx + b$ and
$y = a + bx$
Mathematically, they are identical, but be careful to interpret them correctly: in the first, $m$ represents the gradient and $b$ represents the $y$-intercept. In the second, $a$ represents the gradient and $b$ represents the $y$-intercept.

Why is it called least-squares regression? As you will see in the next section, this method of regression minimises the sum of the squared residuals, $SS_{res}$.

## What can go wrong?

When using linear regression models, or any other model, we need to be careful to avoid a few common pitfalls.

The least-squares regression line for $y$ on $x$ will be identical to the regression line for $x$ on $y$ if and only if $|r| = 1$. In other words, the difference between the two models is small when the linear correlation is strong, so predictions of $x$ from $y$ are more reliable. For moderate and weak correlations, however, the difference between the models can be significant, therefore predictions of $x$ from $y$ are unreliable.

**Don't try to predict $x$ from $y$ using the model (especially if the correlation is not strong).** Once we have a linear model, it's tempting to use the model in reverse. For example, in Example 19.7, we could try to predict the mass of a bag priced at 50 ZMW. What's the problem? Algebraically, we can certainly solve the equation to express $x$ in terms of $y$. It is very tempting to do so! However, because of the process used to find the least-squares regression line, if we swap $x$ and $y$ and re-do the regression on our GDC, we will get a model that is not algebraically equivalent to the previous model. Therefore, your least-squares regression line should only be used in one direction: to predict a value of the response variable ($y$) given a value of the explanatory variable ($x$). If you are asked to predict $x$ from $y$ using a model for $y$ in terms of $x$, make sure you proceed with extreme caution!

See Chapter 9 for more details about the dangers of extrapolation.

**Don't extrapolate.** As we've emphasised before in this textbook, extrapolating assumes that the pattern you've observed will continue. You may have good reason to believe that is the case, but is it always risky, even when the linear correlation is very strong.

**The model is only as good as the strength of the correlation.** Just because you can make a model, doesn't mean you should. You need to examine the scatter diagram and make sure a linear model is appropriate. Once you decide that the form is approximately linear, you may proceed with calculating the least-squares regression line. However, remember that the predictive power of your least-squares regression line is only as good as the strength of the correlation. If the linear correlation is weak to begin with, then you can't expect your model to make accurate predictions. We will talk more about this in the next section, when we will quantify the predictive power of a model.

### Piecewise linear models

For more information about piecewise models, refer to Chapter 9.

Sometimes you have data that very clearly has two linear trends visible. In this case, we can construct a piecewise linear model by partitioning the data set into two or more subsets.

| Distance from shore (km) | Depth (m) |
|---|---|
| 0.2 | 2.0 |
| 0.4 | 1.9 |
| 0.6 | 2.3 |
| 0.8 | 2.3 |
| 1.0 | 2.2 |
| 1.2 | 2.8 |
| 1.4 | 3.3 |
| 1.6 | 4.0 |
| 1.8 | 4.6 |
| 2.0 | 5.1 |

**Table 19.4** Data for Example 19.8

### Example 19.8

A research vessel is mapping the floor of the sea in a certain area. To do this, the boat travels directly out from the shore and measures the depth of the water continuously. The data in the table shows a sample of the data collected.

(a) Explain why a piecewise linear model is appropriate to model depth as a function of distance from shore.

(b) Generate a piecewise linear model, using $d$ for distance from shore and $h$ for depth.

(c) Interpret the gradients in your piecewise model.

(d) Use your model to predict the depth 500 m from the shore and 1.5 km from the shore.

## Solution

(a) To determine if a piecewise linear model is appropriate, examine the data using a scatter diagram.

It appears that there are two linear forms; a piecewise linear model appears appropriate. Furthermore, it appears that the breakpoint between the linear models should be at a distance of 1.0 km.

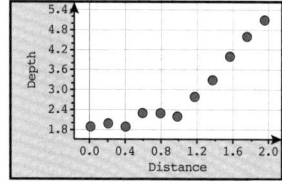

**Figure 19.18** Scatter diagram for solution to Example 19.8(a)

(b) To generate the two pieces, we partition the data into two parts, before and after the breakpoint we have chosen visually. Note that the data point (1.0, 2.2) appears to fit in both parts, so we will include it in both least-squares regression line calculations.

| Distance from shore (km) | 0.2 | 0.4 | 0.6 | 0.8 | 1.0 | 1.2 | 1.4 | 1.6 | 1.8 | 2.0 |
|---|---|---|---|---|---|---|---|---|---|---|
| Depth (m) | 2.0 | 1.9 | 2.3 | 2.3 | 2.2 | 2.8 | 3.3 | 4.0 | 4.6 | 5.1 |

Then, enter the data as two separate data sets into a GDC and calculate the least-squares regression line for each. The GDC shows that the best-fit model for the first part of the data is $d = 0.400h + 1.90$, and the best-fit model for the second part of the data is $d = 2.94h - 0.747$

Now, we combine these two into a piecewise model:

$$d = \begin{cases} 0.400h + 1.90, \, 0 \leqslant h \leqslant 1.0 \\ 2.94h - 0.747, \, 1.0 < h \leqslant 2.0 \end{cases}$$

Graphically, our model appears to fit the data quite well.

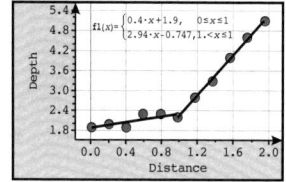

**Figure 19.19** GDC screen for solution to Example 19.8(b)

(c) The two pieces of the model suggest that the depth of the sea is increasing at a rate of $0.4 \, \text{m km}^{-1}$ for the first km, then the rate increases to $2.94 \, \text{m km}^{-1}$.

(d) To use the model to make predictions, proceed as with any piecewise model.

For 500 m from the shore: $h = 0.5 \Rightarrow d = 0.4(0.5) + 1.9 = 2.1 \, \text{m deep}$

For 1.5 km from the shore: $h = 1.5 \Rightarrow d = 2.94(1.5) - 0.747 = 3.66 \, \text{m deep}$

## Coefficient of determination

So far we have examined two statistics that quantify the strength of a correlation. We may also want to measure the predictive power of a particular model. How can we do that? Remember that the purpose of a model is to predict the values of the response variable, given values of the explanatory variable. Another way to think about that is to consider the variation in the response variable: do changes (variation) in the explanatory variable, using a given model, predict changes (variation) in the response variable?

How can we quantify the variation in the response variable and how well does a given model predict that variation? We will look at the data about test scores and the number of hours the student studied again.

| Student | Hours ($x$) | Score ($y$) |
|---------|-------------|-------------|
| Tim     | 4.0         | 65          |
| Joon    | 4.5         | 80          |
| Jim     | 6.0         | 83          |
| Kyle    | 3.5         | 61          |
| Steve   | 3.0         | 55          |
| Niki    | 5.0         | 79          |
| Henry   | 5.5         | 85          |
| Anton   | 6.5         | 79          |
| Cindy   | 7.0         | 92          |
| Lukas   | 6.5         | 95          |

**Table 19.5** Data on studying and test scores

Is the number of hours the student studies the only determining factor in the test score? Probably not. There are almost certainly other factors such as their course load, the free time they have to study, their previous math courses, etc. The question, then, is how much of the variation in the students' scores does the number of hours studying account for? Is it 100%? 90%? 50%? This is the question we will answer.

First, look at the variation in the response variable, score.

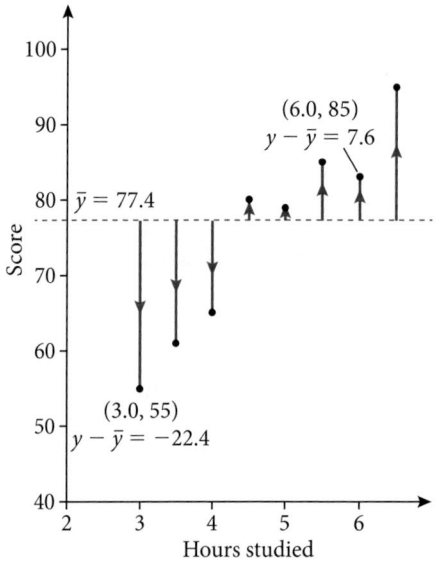

**Figure 19.20** Variation from mean ($\bar{y}$) in the response variable

In Figure 19.20, the variation in the response variable is shown as deviations from the mean response, $\bar{y}$. In this context, those are deviations from the mean score of 77.4. To illustrate, two of these deviations are labelled in Figure 19.20: the data point (3.0, 55) has a deviation from the mean of $-22.4$ and the data point (6.0, 85) has a deviation of $+7.6$.

Now, we are interested in measuring total variation, so we should sum the deviations – but, since these are variations from the mean, their sum is always zero. Therefore, we will do the same thing we do when we are finding standard deviation: we will first square all the deviations, and then add them up. Since this is the total sum of the squared deviations, we will call it $SS_{tot}$. For this data set, $SS_{tot} = \sum(y - \bar{y})^2 = 1550$ (3 s.f.)

Next, how much of the total variation does our least-squares regression line successfully predict? We can use a GDC to find that the equation of the least-squares regression line is $g = 8.69t + 32.7$

If this model predicted the response variable perfectly, i.e. no other factors affected the students' scores, then the data points would fall exactly on the regression line. Of course, the data points don't fall exactly on the regression line, which means some other factors are involved.

When we add the least-squares regression line to the scatter diagram, we see that it divides each deviation into two parts: the successfully predicted variation

Recall that variance is calculated as the mean sum of the squared deviations,

$$\sigma^2 = \frac{1}{n}\sum(y - \bar{y})^2.$$

So, another way to find the total variation $SS_{tot}$ is to multiply the variance by $n$:

$$SS_{tot} = \sigma^2 \times n$$
$$= \left(\frac{1}{n}\sum(y - \bar{y})^2\right) \times n$$
$$= \sum(y - \bar{y})^2$$

and the remaining, unpredicted variation. These are shown in Figure 19.21, with the predicted variation in green and the unpredicted variation in red.

Sometimes the green (successfully predicted) and red (unpredicted) portions overlap: this happens when the regression line is not between $\bar{y}$ and the observed data point. The unpredicted variations are the differences between the observed $y$ values and the predicted $y$ values. These differences (observed − predicted) are called the **residuals** (Figure 19.22).

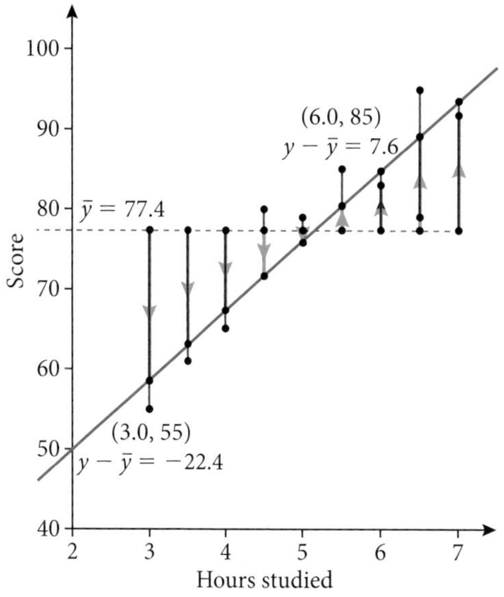

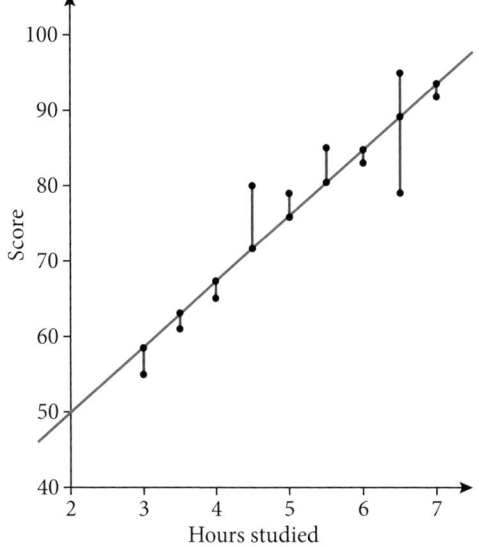

**Figure 19.21** The successfully predicted variation is shown in green, and the unpredicted, remaining variation is shown in red

**Figure 19.22** The residuals are the observed response values minus the predicted response values

Clearly, the best model will minimise the residuals. Now, look closely at the residuals, shown in Table 19.6.

| Student | Tim | Joon | Jim | Kyle | Steve | Niki | Henry | Anton | Cindy | Lukas |
|---|---|---|---|---|---|---|---|---|---|---|
| **Hours ($x$)** | 4.0 | 4.5 | 6.0 | 3.5 | 3.0 | 5.0 | 5.5 | 6.5 | 7.0 | 6.5 |
| **Score ($y$)** | 65 | 80 | 83 | 61 | 55 | 79 | 85 | 79 | 92 | 95 |
| **Predicted score** | 67.4 | 71.8 | 84.8 | 63.1 | 58.7 | 76.1 | 80.4 | 89.1 | 93.5 | 89.1 |
| **Residual (observed − predicted)** | −2.41 | 8.25 | −1.78 | −2.07 | −3.72 | 2.90 | 4.56 | −10.1 | −1.47 | 5.87 |

**Table 19.6** Predicted scores and residual for each student

To measure the fit of the model, we could perhaps add up all the residuals. The problem with this is that any line that passes through the mean point will have a sum of residuals equal to zero. So, as with the squared deviations, we square all the residuals and then add them up. This value is called the sum of the squared residuals, or $SS_{res}$. Here, $SS_{res} = \sum(y_{observed} - y_{predicted})^2 = 264$ (3 s.f.). It turns out that $SS_{res}$ is a good measure of the predictive power of our model. If the model was perfect, $SS_{res}$ would be zero. The bigger that $SS_{res}$ is, the worse our model is at predicting and fitting the response variable.

However, when $SS_{res}$ is not zero, what should we compare it to? Does the number $SS_{res} = 264$ represent a good or poor model?

Since $SS_{res}$ is the portion of total variation that was not predicted, it is a fraction of $SS_{tot}$. Therefore, to measure the predictive power of our model, we can calculate a ratio:

$$\frac{SS_{res}}{SS_{tot}} = \frac{264}{1550} = 0.170$$

This ratio represents the portion of the variation in the response variable that our model did not predict. But it is better to look at what the model did achieve. By convention we look at the portion of variation the model predicted successfully, $R^2$:

$$R^2 = 1 - 0.170 = 0.830$$

$R^2$ is the **coefficient of determination**. And, since this is a ratio of the successfully predicted variation, we can interpret it as a percentage: 83% of the variation in students' scores is successfully predicted by the least-squares regression line $g = 8.69t + 32.7$ using hours of study as the explanatory variable.

Why do we label it $R^2$? Well, if you use your GDC to find Pearson's $r$ for this data, you will find: $r = 0.911$ and $r^2 = 0.830 = R^2$

You also may have noticed that your GDC tells you the value of $R^2$ whenever it performs a linear regression (though it may call it $r^2$). In fact, we will rely on a GDC to find $R^2$, or we will use the property $R^2 = r^2$

The method of regression your GDC uses is called **least-squares** because it minimises $SS_{res}$, the sum of the squared residuals.

Why do we capitalise $R^2$ here? For single-variable linear regressions, $R^2 = r^2$, but for multi-variable regressions there are other ways of calculating $R^2$, so it is not always equal to $r^2$. In this course, it is always true that $R^2 = r^2$.

The **coefficient of determination**, $R^2$, is given by

$$R^2 = 1 - \frac{SS_{res}}{SS_{tot}}$$

- For linear regressions, Pearson's $r^2 = R^2$.
- $R^2$ is the portion of the variation in the response variable that is successfully predicted by a given model.
- Since $R^2$ is a ratio, it can be interpreted as a percentage.
- $0 \leqslant R^2 \leqslant 1$; values near 0 indicate that most of the variation in the response variable is due to other factors, while values near 1 indicate that most of the variation in the response variable is predicted by the explanatory variable.
- The sum of the squared residuals, $SS_{res}$, can also be used on its own to measure the fit of a model.
- While $R^2$ is a good measure of the predictive power of a model, it is not the only criteria you should use for choosing a model. Model choice should be based on contextual understanding of the phenomena and usefulness of the model as well.

Look up how to find $R^2$ directly from data using your GDC. Most GDCs calculate it automatically whenever a linear regression is done, but you may need to turn on a calculator setting such as 'diagnostics' to see it.

Note that although we have explored the derivation of $R^2$ as a ratio of the successfully predicted variation, we will rarely calculate it directly. We will use a GDC to do it, or we will square $r$.

### Example 19.9

The scatter diagram represents a random sample of IB students who went through four years of university. The data contains each student's overall score on the IB exams they took and their grade point average (GPA) in university (on a scale of 1 to 4).

For this data, we have $r = 0.758$ and a least-squares regression line of
GPA $= 0.151$(IB score) $- 1.51$

(a) Describe the scatter diagram.

(b) Interpret the gradient of the least-squares regression line.

(c) Calculate and interpret the coefficient of determination, $R^2$

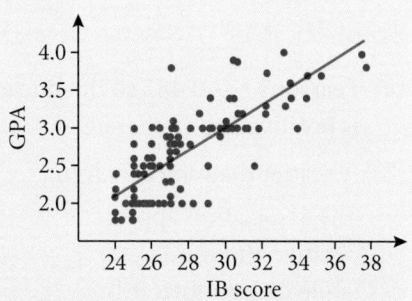

### Solution

(a) The scatter diagram shows a positive, moderately strong, approximately linear association. There is one possible outlier at approximately (27, 3.75) but it does not appear extreme enough to affect our analysis.

(b) The gradient 0.151 indicates that for each additional point in a student's overall IB score, their predicted university GPA increases by 0.151 points.

(c) $R^2 = r^2 = 0.758^2 = 0.575$. The least-squares regression line successfully predicts 57.5% of the variation in students' university GPAs from the variation in IB scores.

Note that if 57.5% of the variation in students' university GPA is successfully predicted by changes in IB scores, then 42.5% of the variation is not accounted for. In other words, there are factors other than just IB scores that predict university GPAs. That is not too surprising, but the power of $R^2$ lies in quantifying exactly how strong the predictive power of an explanatory variable and corresponding model is.

### Example 19.10

The table represents the volume in mm³ and mass in grams of a certain fruit studied by a biologist.

| Volume (mm³) | 223 | 236 | 242 | 226 | 223 | 221 | 233 | 222 | 222 | 218 | 232 | 223 |
|---|---|---|---|---|---|---|---|---|---|---|---|---|
| Mass (g) | 165 | 171 | 173 | 170 | 168 | 172 | 168 | 167 | 162 | 166 | 164 | 164 |

(a) Calculate $r$ and interpret its value.

(b) Explain why a linear regression model is appropriate for this data.

(c) Calculate the least-squares regression line for mass in terms of volume and interpret it.

(d) Find the coefficient of determination, $R^2$, and interpret it.

(e) Predict the mass of a fruit with a volume of 230 g.

(f) A student uses the regression model from (c) to predict the volume of a 170 g fruit. Give a reason why this may not be reliable.

# 19 Bivariate analysis

## Solution

(a) Pearson's $r = 0.487$ so the linear correlation between these two variables is positive and moderate.

(b) A scatter diagram is helpful. The association appears positive and moderate; while it's not obviously linear there are no other strong patterns. A linear model seems appropriate so we will proceed with caution.

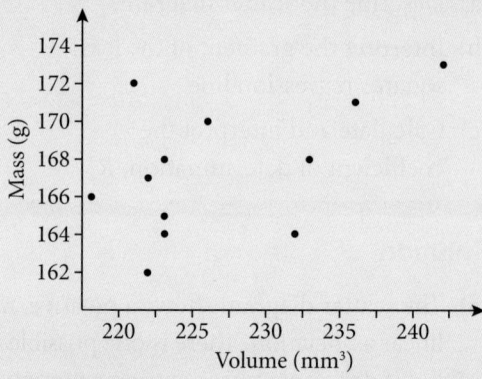

(c) Using a GDC, we obtain a least-squares regression line of $y = 0.233x + 115$

On average, the mass of these fruits increases by 0.233 g for each additional mm³.

(d) Using a GDC, $R^2 = 23.7$

This suggests that 23.7% of the variation in the mass of the fruits is successfully predicted by the least-squares regression line using volume. There are clearly other factors causing variation in the mass of fruits.

(e) Using our least-squares regression line, $y = 0.233(230) + 115 = 168$ g

(f) The regression model in (c) is for $y$ on $x$, and should only be used to predict $y$ values from $x$ values, i.e. we should only use the model to predict mass, given volume. This is especially true since the coefficient of determination, $R^2$, is not close to 1.

## Exercise 19.3

1. Given a linear model with $r = 0.6$, write down the value of $R^2$.

2. Given a linear model with $r = -0.8$, write down the value of $R^2$.

3. Given a linear model with $R^2 = +0.81$, write down the two possible values of $r$.

4. Given a linear model with $R^2 = +0.49$ and a negative correlation, write down the value of $r$.

5. In a project, a Mathematics student develops two different models for one set of data. Model A has $SS_{res} = 321$; model B has $SS_{res} = 120$. Which model appears to fit the data better?

6. State whether each statement is true or false.

   (a) A model with $SS_{res}$ close to zero fits the data very well.

   (b) If two different models have the same $SS_{res}$ for a single data set, they are equally appropriate for the data.

7. Six metal plates are immersed in an acid bath. After some time, they are removed and the mass is measured to determine the mass that was lost. The table gives data from this experiment.

| Time (hours) | 150 | 200 | 200 | 300 | 450 | 500 |
|---|---|---|---|---|---|---|
| Mass lost (%) | 0.761 | 1.44 | 1.15 | 1.65 | 2.56 | 2.44 |

   (a) Find the equation of the least-squares regression line for mass lost, $L$, in terms of hours immersed, $h$. Interpret your model in context.

   (b) Find the percentage of mass that was lost for each additional 100 hours a metal plate was in this acid bath.

   (c) Find the value of $R^2$ for your least-squares regression line and interpret it in context.

   (d) Use your least-squares regression line to predict the mass loss of a metal plate immersed for 400 hours.

   (e) Is it appropriate to use your model to predict the number of hours a metal plate with 1.3% mass loss was immersed? Explain.

8. Does the acceleration ability of a car affect fuel efficiency? Here are data from 15 cars.

| Sports cars/roadsters | 0-60 mph time (s) | Efficiency (miles per gallon) |
|---|---|---|
| Mazda MX-5 Miata Club | 6.7 | 34 |
| Honda Civic Si | 7.3 | 34 |
| Fiat 124 Spider Lusso | 7.1 | 31 |
| Mini Cooper S | 7.0 | 30 |
| Subaru BRZ Premium | 7.2 | 30 |
| Toyota 86 | 7.2 | 30 |
| Volkswagen GTI Autobahn | 6.6 | 29 |
| Ford Fiesta ST | 7.3 | 29 |
| Fiat 500 Abarth | 8.0 | 28 |
| Porsche 718 Boxster (base) | 4.4 | 26 |
| Subaru Impreza WRX Premium | 6.0 | 26 |
| Audi TT 2.0T (AT) | 6.3 | 26 |
| Ford Focus ST | 6.6 | 26 |
| BMW M235i | 5.2 | 25 |
| Ford Mustang Premium (2.3T, AT) | 6.4 | 25 |

   (a) Find the equation of the least-squares regression line for efficiency, $E$, in terms of acceleration time, $t$. Interpret your model in context.

   (b) Find the value of $R^2$ for your least-squares regression line and interpret it in context.

   (c) Use your least-squares regression line to predict the efficiency of a car with a 0–60 mph time of 5.7 s. Give a reason why this prediction may not be reliable.

   (d) Is it appropriate to use your model to predict the efficiency of a car with a 0–60 time of 9 s? Explain.

9. Do bigger aeroplanes use more fuel? The tables contain data on the number of passenger seats and fuel consumption of aeroplanes currently in widespread use.

| Seats | 405 | 296 | 288 | 258 | 240 | 230 | 193 | 188 | 148 |
|---|---|---|---|---|---|---|---|---|---|
| Fuel consumption (litres per minute) | 224 | 140 | 138 | 94 | 100 | 95 | 87 | 62 | 79 |

| Seats | 142 | 131 | 122 | 115 | 112 | 103 | 102 | 78 |
|---|---|---|---|---|---|---|---|---|
| Fuel consumption (litres per minute) | 56 | 46 | 54 | 70 | 51 | 40 | 51 | 48 |

(a) Generate a scatter diagram and describe the association between the number of seats and the fuel consumption.

(b) Find the equation of the least-squares regression line for fuel consumption, $F$, in terms of the number of seats, $n$. Interpret your model in context.

(c) Find the value of $R^2$ for your least-squares regression line and interpret it in context.

(d) Use your model to predict the fuel consumption of an aeroplane with 350 seats.

(e) Give a reason why a linear model may not be the most appropriate model for this data.

10. A computer-based workout program ranks its workouts according to the training stress score. Another training website calculates a separate relative effort value for each workout that is uploaded to the site. Are training stress score and relative effort related? Table 19.7 shows a sample of one cyclist's last 12 workouts.

(a) Draw a scatter diagram and describe the association between training stress score and relative effort.

(b) Find the equation of the least-squares regression line for relative effort, $E$, in terms of the training stress score, $S$. Interpret your model in context.

(c) Find the value of $R^2$ for your least-squares regression line and interpret it in context.

(d) Use your model to predict the relative effort for a workout with a training stress score of 60.

(e) Write down a valid domain for your model.

(f) If you removed the possible outlier at (48, 37), how would your answers to (d) and (e) change?

(g) Remove the outlier at (48, 37) and recalculate the least-squares regression line and $R^2$. Did the values change significantly?

| Training stress score | Relative effort |
|---|---|
| 82 | 59 |
| 96 | 70 |
| 48 | 37 |
| 79 | 60 |
| 79 | 58 |
| 79 | 51 |
| 88 | 67 |
| 112 | 79 |
| 82 | 48 |
| 74 | 44 |
| 74 | 46 |
| 96 | 84 |

**Table 19.7** Data for question 10

11. It is well known that many people who eat more don't gain weight as quickly as others do. Could non-exercise activity (fidgeting, maintaining posture, etc.) explain the difference? In a 1999 study, researchers deliberately overfed 16 non-obese volunteers and measured the change in their calorie expenditure through non-exercise activity (NEA) as well as their weight-gain after 8 weeks. The data collected are shown below.

| Change in NEA (kcal) | −94 | −57 | −29 | 135 | 143 | 151 | 245 | 355 |
|---|---|---|---|---|---|---|---|---|
| Change in mass (kg) | 4.2 | 3.0 | 3.7 | 2.7 | 3.2 | 3.6 | 2.4 | 1.3 |

| Change in NEA (kcal) | 392 | 473 | 486 | 535 | 571 | 580 | 620 | 690 |
|---|---|---|---|---|---|---|---|---|
| Change in mass (kg) | 3.8 | 1.7 | 1.6 | 2.2 | 1.0 | 0.4 | 2.3 | 1.1 |

(a) Generate a scatter diagram and describe the association between change in non-exercise activity and change in mass.

(b) Find the equation of the least-squares regression line for change in mass, $M$, in terms of the change in non-exercise activity, $N$. Interpret your model in context.

(c) Find the value of $R^2$ for your least-squares regression line and interpret it in context.

(d) Use your model to predict the change in mass for an individual with a change in non-exercise activity of 200 kcal.

(e) Explain what a negative value for change in non-exercise activity means in this context.

(f) Write down a valid domain for your model.

(g) Making a prediction for change in mass for a change in non-exercise activity of 1100 kcal is an extrapolation. Give another reason why it is particularly nonsensical in this context.

12. A company believes that its workers make more mistakes when they work faster. To test this theory, they collect data on the production rate and the number of mistakes at that rate. This data is given in Table 19.8.

(a) Identify the explanatory and response variables in this context.

(b) Generate a scatter diagram and describe the association between mistakes and production rate.

(c) Find the equation of the least-squares regression line for mistakes, $M$, in terms of the production rate, $P$. Interpret your model in context.

(d) Find the value of $R^2$ for your least-squares regression line and interpret it in context.

(e) Would you advise this company to lower production rates to reduce mistakes?

(f) Remove the outlier from the data and repeat parts (b)–(e).

(g) Is it acceptable to remove the outlier in this case? Explain.

| Mistakes | Production rate (units per hour) |
|---|---|
| 20 | 400 |
| 30 | 450 |
| 10 | 350 |
| 20 | 375 |
| 30 | 400 |
| 25 | 400 |
| 30 | 450 |
| 20 | 300 |
| 10 | 300 |
| 40 | 300 |

Table 19.8 Data for question 12

13. To track the number of active users for a website, it is common to count the number of users who have logged in at least once in a given month. This count is called monthly active users. The number of monthly active users for one social media service is shown in the scatter diagram.

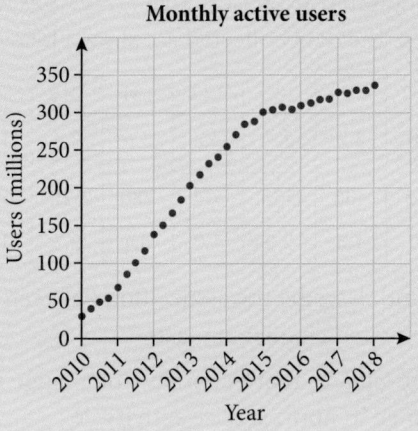

**Monthly active users**

(a) Describe the association shown in the scatter diagram.

(b) A piecewise linear model may be appropriate. Give the domain for each piece of a piecewise linear model, using years since 2010 as the explanatory variable.

14. A dairy farmer would like to investigate whether temperature has an effect on the butterfat content of the milk his dairy cows produce. Over the course of a year, he records the butterfat content of the milk they produce and the air temperature. His data are shown in the tables.

| Temperature (°C) | 3.0 | 3.0 | 4.0 | 5.0 | 7.0 | 8.0 | 13.0 | 13.0 | 13.5 | 13.5 |
|---|---|---|---|---|---|---|---|---|---|---|
| Butterfat (%) | 4.87 | 5.09 | 4.97 | 4.52 | 4.83 | 4.85 | 4.77 | 4.48 | 4.23 | 4.85 |

| Temperature (°C) | 14.0 | 14.5 | 15.0 | 15.5 | 16.0 | 16.5 | 18.0 | 18.0 | 18.5 | 18.5 |
|---|---|---|---|---|---|---|---|---|---|---|
| Butterfat (%) | 4.51 | 4.74 | 4.45 | 4.7 | 4.65 | 4.45 | 4.63 | 4.65 | 4.65 | 4.59 |

(a) Identify the explanatory and response variables in this context.

(b) Generate a scatter diagram and describe the association between temperature and butterfat content.

(c) Find the equation of the least-squares regression line for butterfat, $F$, in terms of the temperature, $T$. Interpret your model in context.

(d) Find the value of $R^2$ for your least-squares regression line and interpret it in context.

(e) If the farmer wants to increase the butterfat content of the cows' milk, would investing in a climate-controlled barn for the cows be a good choice?

# 19.4   Non-linear regression and models

So far in this chapter we have looked at data that appears to have an approximately linear association. What do we do when the data is clearly non-linear? There are two approaches.

1. Make an educated guess at a likely model and trust our calculator to work it out. Your GDC should be able to perform linear, quadratic, cubic, exponential, power, and sinusoidal regressions.

2. Re-express the data using logarithms until the scatter diagram is approximately linear, use a linear regression, then use properties of logarithms and exponents to rewrite the linear model as a non-linear model. This approach is what your calculator does to find exponential, logarithmic, and power models.

We will now see how these two approaches work.

Fishing line is sold according to its breaking strength. Since stronger fishing line is thicker, less line can be put on a spool. Table 19.9 shows similarly-priced spools available at an online fishing shop. What is the relationship between the strength of the fishing line and the amount on the spool?

 We cannot just let a GDC do the work all the time. Sophisticated analysis is more straightforward when dealing with a linear regression and linear model. The mathematics required to linearise and un-linearise a data set is straightforward, so it allows us to use advanced linear analysis techniques even on non-linear phenomena.

Refer to Chapter 9 for more details on linear, quadratic, cubic, exponential, and sinusoidal/trigonometric models.

| Strength (kg) | Length (m) |
|---|---|
| 4.5 | 1370 |
| 6.8 | 820 |
| 11.3 | 545 |
| 13.6 | 400 |
| 9.1 | 595 |
| 5.4 | 1010 |
| 9.1 | 615 |
| 5.4 | 1330 |
| 9.1 | 770 |
| 45.4 | 135 |
| 36.3 | 275 |
| 45.4 | 275 |
| 29.5 | 275 |
| 22.7 | 455 |
| 22.7 | 230 |

**Table 19.9** Fishing line data

## Getting started

First, generate a scatter diagram on a GDC (Figure 19.23).

There appears to be a negative, strong, non-linear association. There are no major outliers or other unusual features. From the shape of the data, we suspect an exponential or power model might work well. In this context, since we are dealing with length and volume (thicker line is stronger), a power model is a likely candidate.

 In the real world, we choose an appropriate model not just by the shape of the data but also by the nature of the phenomenon we are modelling. See Chapter 9 for further discussion.

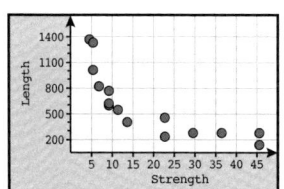

**Figure 19.23** Scatter diagram for the fishing line data

## Using a GDC

On a GDC, we can try both an exponential and a power model to examine the predictive power of each. An exponential regression result is shown in Figure 19.24.

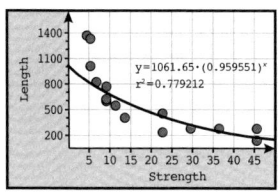

**Figure 19.24** Exponential regression on a GDC

The GDC finds that the best-fitting exponential model is $L = 1060(0.960)^S$ where $L$ is the predicted length and $S$ is the strength. The coefficient of determination $R^2 = 0.779$. This model is quite strong. However, we should not choose a model based solely on the value of $R^2$. We must also consider the overall fit of the model, the context of the data, and the usefulness of the model. While this model works well for many of the data points, it seems to be severely under-predicting the lengths for low strength values.

Try a power model, as shown in Figure 19.25.

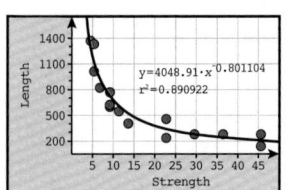

**Figure 19.25** Power model

A GDC shows that the best-fitting power model is $L = 4050(S)^{-0.801}$, with a coefficient of determination of $R^2 = 0.891$. Not only has the predictive power of the model increased compared to the exponential model, but, visually, the power model appears to follow the trend of points much more closely. The power model appears to be a better choice for this data.

How can we interpret the rate of change in this model? Well, if $S$ increases by 1 percent:

$$L = 4050(1.01S)^{-0.801}$$
$$= 4050(S)^{-0.801}(1.01)^{-0.801}$$
$$= 4050(S)^{-0.801}(0.992)$$

This will have the effect of multiplying the original expression by 0.992. Multiplying by 0.992 is the same as decreasing by $1 - 0.992 = 0.008 - (0.8\%)$. We can therefore say that a 1 percent increase in the breaking strength implies a 0.8% decrease in line length.

Next, we will see how we can use logarithms to linearise data and find exponential, logarithmic, and power models.

## Linearising data using logarithms

How does the GDC find the exponential and power models? The process is:

1. Linearise the data using logarithms

2. Fit an intermediate linear model to the re-expressed data

3. Use properties of logarithms and exponents to find the non-linear model.

**Step 1:** Linearise the data using logarithms

First, we re-express the data by applying a logarithm to all the data points. You should not do this by hand for each data point – instead, look up how to apply a function to a list on your GDC as shown in Figure 19.26.

Next, we look at three different scatter diagrams to check which re-expression did the best job of linearising the data.

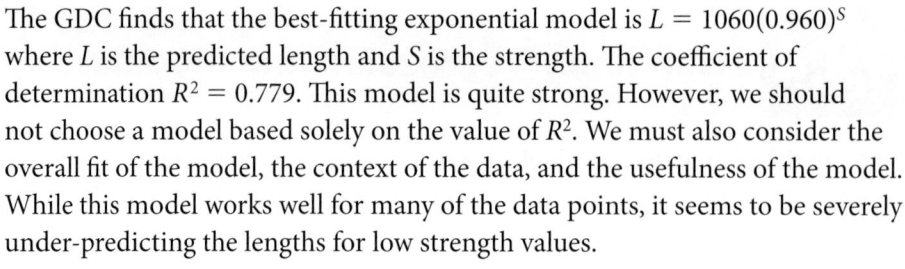

**Figure 19.26** Linearising data using logarithms

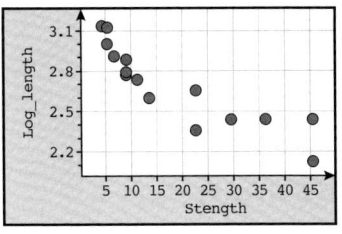 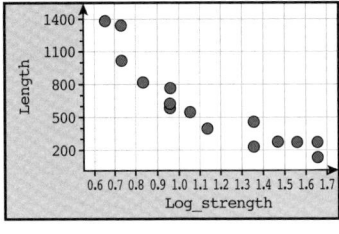 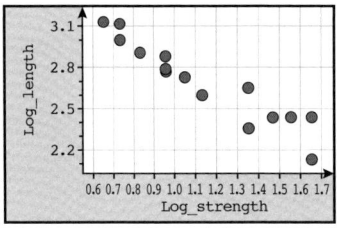

(a) log(Length) vs. Strength          (b) Length vs. log(Strength)          (c) log(Length) vs. log(Strength)

**Figure 19.27** Compare each re-expression to see which is most linear, but also consider the context

From the GDC images we can see that graph (c), the log-log re-expression, did the best job at linearising the data, so we will use that re-expression.

**Step 2:** Fit an intermediate linear model to the re-expressed data

Now, we ask the GDC to do a linear regression to find the least-squares regression line for the log-log data (Figure 19.28).

Be careful. The equation that the GDC generated is not $L = -0.801(S) + 3.61$

Since this equation is fit to the re-expressed data, the GDC has actually found the model

$$\log(L) = -0.801 \times \log(S) + 3.61$$

This is nice, but we have a problem: we are not interested in predicting the logarithm of $L$. We want to predict $L$.

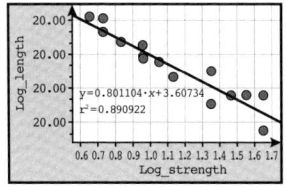

**Figure 19.28** Asking the GDC to do a linear regression

**Step 3:** Use properties of logarithms and exponents to find the non-linear model

$$\log(L) = -0.801 \times \log(S) + 3.61$$

$$\Rightarrow \quad 10^{\log(L)} = 10^{-0.801 \times \log(S) + 3.61}$$

since $10^{\log(L)} = L$:

$$\Rightarrow \quad L = 10^{\log(S) \times -0.801 + 3.61}$$

since $x^{a+b} = x^a \times x^b$ and $x^{ab} = (x^a)^b$

$$\Rightarrow \quad L = (10^{\log(S)})^{-0.801} \times 10^{3.61}$$

since $10^{\log(S)} = S$:

$$\Rightarrow \quad L = S^{-0.801} \times 10^{3.61}$$

$$\Rightarrow \quad L = 4050 \times S^{-0.801}$$

Note that the model we obtained is exactly what the GDC had generated for a power model. Also, the $R^2$ value of the linear log-log model, 0.891, is identical to the $R^2$ value that the GDC automatically generated for the power model. This makes sense: since $R^2$ is a ratio of the successfully-predicted variation in the response variable, it is not changed by re-expressing the data for a given model.

Depending on which of the three scatter diagrams you choose, you will obtain either an exponential model, logarithmic model, or power model. For each of these the interpretation of the rate of change is different – and, most importantly, be careful to not interpret any of them as a linear rate of change! This is summarised in Table 19.10.

| Linearisation method | Linear model | Non-linear model | Rate of change interpretation |
|---|---|---|---|
| $\log(y)$ vs $x$ | $\log(y) = mx + b$ | Exponential $y = 10^b(10^m)^x$ | 1 unit increase in $x$ $\Downarrow$ $y$ increases by a factor of $10^m$ |
| $y$ vs $\log(x)$ | $y = m\log(x) + b$ | Logarithmic $y = m\log(x) + b$ | $x$ increases by a factor of 10 $\Downarrow$ $y$ increases by $m$ units |
| $\log(y)$ vs. $\log(x)$ | $\log(y) = m\log(x) + b$ | Power $y = 10^b x^m$ | 1 percent increase in $x$ $\Downarrow$ $y$ increases by $1.01^m$ percent (approximately $m$ percent) |

**Table 19.10** Logarithmic linearisation methods and associated non-linear models

## Example 19.11

To determine the age of a tree, it is possible to count the growth rings of the tree. However, this involves either cutting down the tree, which kills the tree, or extracting a core sample, which can also injure the tree. To avoid injuring trees, a forester examined 20 trees of a certain species that had been cut down. She determined the age of each tree, and then measured the diameter of the tree trunk. A sample of the data collected are given in the tables.

| Diameter (cm) | 4.6 | 5.6 | 11.2 | 14.0 | 16.8 | 19.6 | 19.6 | 25.1 | 25.7 | 27.4 |
|---|---|---|---|---|---|---|---|---|---|---|
| Age (years) | 5 | 8 | 8 | 14 | 8 | 10 | 13 | 16 | 18 | 12 |

| Diameter (cm) | 30.7 | 32.5 | 33.5 | 36.3 | 36.3 | 39.1 | 39.1 | 39.1 | 41.9 | 44.7 |
|---|---|---|---|---|---|---|---|---|---|---|
| Age (years) | 20 | 22 | 28 | 25 | 34 | 30 | 35 | 38 | 42 | 33 |

(a) Give a reason why a linear model is not appropriate for the relationship between diameter and age for this tree species.

(b) Use linearisation to develop an appropriate model to predict the age $A$ of this tree species, given its diameter $D$. Find and interpret the coefficient of determination.

(c) Interpret the parameters of your model in context.

(d) Use your model to predict the age of a tree of this species with a trunk diameter of 23 cm.

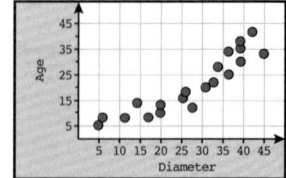

**Figure 19.29** GDC screen for the solution to Example 19.11(a)

| | A diame... | B age | C log_di... | D log_age | E |
|---|---|---|---|---|---|
| ▬ | | | =log(diam | =log(age) | |
| 1 | 4.6 | 5 | 0.662758 | 0.69897 | |
| 2 | 5.6 | 8 | 0.748188 | 0.90309 | |
| 3 | 11.2 | 8 | 1.04922 | 0.90309 | |
| 4 | 14 | 14 | 1.14613 | 1.14613 | |
| 5 | 16.8 | 8 | 1.22531 | 0.90309 | |
| 6 | 19.6 | 10 | 1.29226 | 1 | |
| D1 | =0.69897000433602 | | | ◄ ► | |

**Figure 19.30** GDC screen for the solution to Example 19.11(b)

## Solution

(a) To explore this relationship, generate a scatter diagram.

There appears to be a strong, positive relationship. There is a clear curve to the data, so a linear model is not appropriate.

(b) First, re-express the data using logarithms.

Then generate the three re-expressed scatter diagrams.

Of these three, the first, log(age) vs diameter, appears to have the most linear form. Use a GDC to find the intermediate linear model.

Our GDC reports that log(age) = 0.0211(diameter) + 0.677 is the best-fitting linear model, with a coefficient of determination of $R^2 = 0.898$, which implies that this model successfully predicts 89.8% of the variation in log(age) from the variation in diameter. To find the non-linear model to predict age, use laws of exponents and logarithms:

$$\log(A) = 0.0211(D) + 0.677$$
$$\Rightarrow \quad A = 10^{0.0211(D)+0.677}$$
$$= 10^{0.0211(D)}10^{0.677}$$
$$= 1.05^D \times 4.75$$
$$= 4.75(1.05)^D$$

(c) The parameter 4.75 is the $y$-intercept; it indicates that a tree with $D = 0$ will have an age of 4.75 years. In this context, that doesn't make sense. The parameter of 1.05 is the exponential growth rate; it indicates that for each additional cm in diameter, age increases by a factor of 1.05. In other words, the age grows by about 5% for each additional cm of diameter.

(d) We predict that a trunk diameter of 23 cm will have an age of
$A = 4.75(1.05)^{23} = 14.5$ years

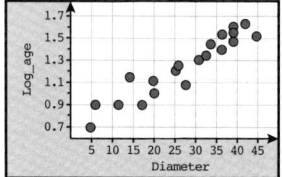
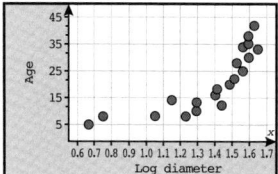
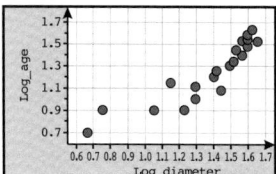
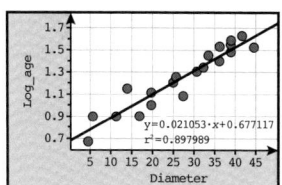

**Figure 19.31** GDC screens for the solution to Example 19.11

## Interpreting graphs with logarithmic axes

Because logarithmic transformations are so useful for linearising data, you may also see graphs drawn with logarithmic axes. This is especially common when the data appears to follow an exponential or logarithmic trend.

For example, Moore's law suggests that the number of transistors in integrated circuits doubles approximately every two years. This sort of exponential growth quickly runs off normal linear axes, and/or makes the initial growth very difficult to see. Here is a scatter diagram of integrated circuits from 1971 until 2017:

Integrated circuits are the main processing unit in many of our computing devices such as desktops, laptops, smartphones, and GPS units.

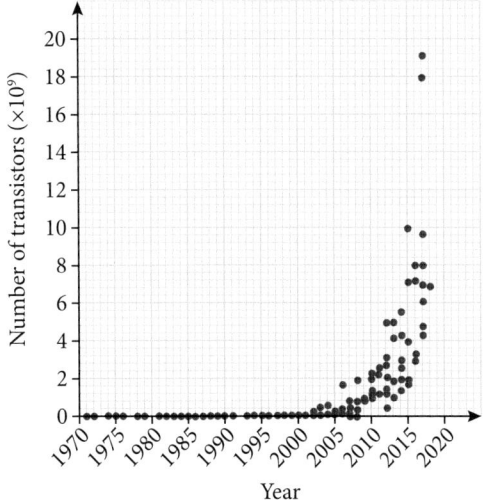

**Figure 19.32** A graph with linear axes depicting data that increases exponentially

The problem with Figure 19.32 is that it is impossible to read any of the early values from the graph: every data point before the year 2000 appears to have zero transistors. We know that we could linearise this data with a logarithmic transformation, but we will do that in a slightly different way by scaling the vertical axis (number of transistors) logarithmically. We will leave the horizontal axis (year) with a linear scale, since it is not growing exponentially.

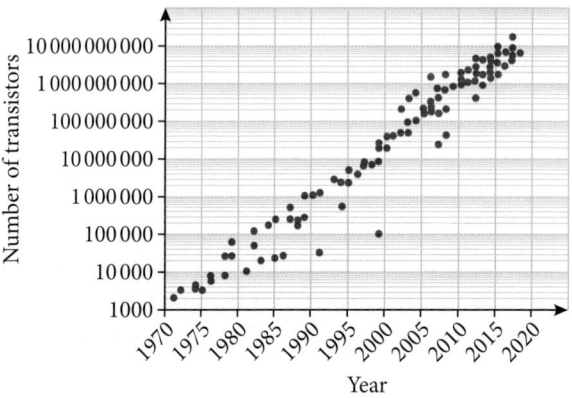

**Figure 19.33** A log-lin graph depicting data that increases exponentially

After we re-scale the vertical axis to have a logarithmic scale, we see that the pattern is linearised and we can now read the data values from the vertical axis. To read logarithmic graphs, calculate the values of gridlines as equal divisions between the nearby labelled gridlines. For example, if we zoom in on the bottom left corner of Figure 19.33, we can see Figure 19.34.

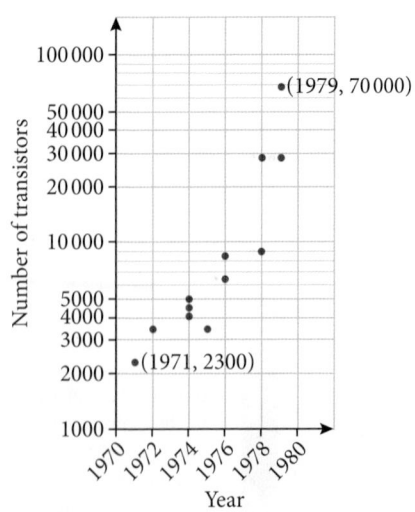

**Figure 19.34** Zooming in on the bottom left corner of Figure 19.33

In Figure 19.34, on the vertical axis, between 1000 and 10 000, each grid line represents an additional 1000 transistors. Between 10 000 and 100 000, each gridline represents an additional 10 000 transistors. Therefore, we can see that in the 1970s, early integrated circuits had around 2300 transistors, and by the end of the decade, integrated circuits had around 70 000 transistors.

Graphs such as those in Figure 19.33 and Figure 19.34 are called **log-lin** graphs, since the $y$-axis is logarithmic and the $x$-axis is linear. We can also have **log-log** and **lin-log** graphs, in the same way that we might chose to transform the $x$ or $y$ or both variables when we linearise data.

For example, consider the fishing line data from earlier in this section. We applied logarithms to both variables to obtain the best linearisation, so it is appropriate to plot that data on a log-log graph (Figure 19.35).

Note that the horizontal axis (strength) is now in factors of 10, while the vertical axis (length) is in factors of 2. The graph works exactly the same; just interpolate the values from the nearby axis labels. For example, we can read point $A$ as having a breaking strength of about 7 kg with spool length of about 800 m. Point $B$ has a breaking strength of about 45 kg and spool length of about 270 m.

As with logarithmic transformations, be very careful interpreting slope on a graph with logarithmic axes: even though the data appears linear, the rate of change is not. Use the guidance for logarithmic transformations given earlier in Example 19.10 to interpret the rate of change.

Non-linear regression is useful for many types of models. Sometimes we even know what type of model to expect, and we can use a GDC to find the model and then interpolate unknown data.

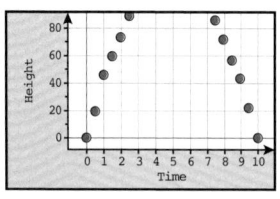

Figure 19.35 A log-log graph of fishing line length per spool and breaking strength

## Example 19.12

Some students are collecting data on a model rocket launch. They use video analysis software to record the height and time of the rocket at various points. Unfortunately, their camera is aimed wrong and the rocket goes above the viewing frame of the camera. Here are the data they collected.

| Time (s) | 0 | 0.5 | 1.0 | 1.5 | 2.0 | 2.5 | 7.5 | 8.0 | 8.5 | 9.0 | 9.5 | 10.0 |
|---|---|---|---|---|---|---|---|---|---|---|---|---|
| Height (m) | 0 | 19.5 | 46.3 | 60.1 | 74.0 | 89.5 | 86.4 | 72.4 | 57.2 | 43.6 | 22.1 | 0 |

(a) Describe the association between height and time for this data.

(b) The falling-object model is given as $s(t) = \frac{1}{2}at^2 + v_0t + s_0$ where $s(t)$ is the position at time $t$, $a$ is total acceleration due to gravity, $v_0$ is the initial velocity, and $s_0$ is the initial position. Use your GDC to find the best-fit values of $a$, $v_0$, and $s_0$.

(c) Interpret the values of $v_0$ and $s_0$ in context.

(d) On Earth, at sea level, acceleration due to gravity is the constant $g = 9.81 \text{ m s}^{-2}$. Compare to the value of $a$ and interpret the difference.

(e) Interpret the coefficient of determination for the model.

(f) Use your model to predict the maximum height of the model rocket and the time at which this occurred.

## Solution

(a) Since we are modelling the position of a falling object, we presume that the model should be quadratic. We can see some evidence of that here, although with the missing data we cannot be sure. In the scatter diagram itself, we see a strong positive association in the first part and a strong negative association in the second part. There is some evidence of a non-linear relationship (but maybe only because we are looking for it!).

(b) We are told to fit a quadratic model. Therefore, use the built-in function on the GDC to obtain the graph in the diagram.

Our GDC suggests that the model should be $s(t) = -4.65t^2 + 46.4t + 0.687$ with a coefficient of determination of $R^2 = 0.996$. Therefore we have $\frac{1}{2}a = -4.65 \Rightarrow a = -9.3$, $v_0 = 46.4$, and $s_0 = 0.687$

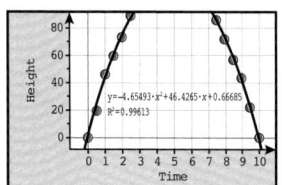

Figure 19.36 GDC screens for the solution to Example 19.12

(c) The initial velocity, when the rocket is launched, is $v_0 = 46.4 \text{ m s}^{-1}$. The initial position is $s_0 = 0.687$ m; this could be the height of the launch pad. However, our data indicates that the height of the rocket at time $t = 0$ is zero, so there must be error present in the data collection or in the model.

(d) The acceleration due to gravity of the model rocket is $a = -9.3 \text{ m s}^{-2}$. This is less that the gravitational constant of $g = 9.81 \text{ m s}^{-2}$. The gravitational constant only describes the rate of acceleration if no other forces are present. In this case, we can assume that air resistance is acting against the model rocket and causing its net acceleration to be less than $g$.

(e) The coefficient of determination of $R^2 = 0.996$ indicates that the quadratic model accounts for 99.6% of the variation in the height, using time as the explanatory variable.

(f) The maximum height can be found at the vertex of the parabola, at

$$t = -\frac{46.4}{2(-4.65)} = 4.97 \text{ s}$$

The height at that time is

$$s(4.97) = -4.65(4.97)^2 + 46.4(4.97) + 0.687 = 116 \text{ m}$$

> Recall that the $x$-coordinate of the vertex of a parabola with equation $y = ax^2 + bx + c$ is given by $x = -\dfrac{b}{2a}$

## Exercise 19.4

1. Identify the most likely logarithmic linearisation and the corresponding model for each scatter diagram, or state that logarithmic linearisation should not be used and give a reason.

(a)
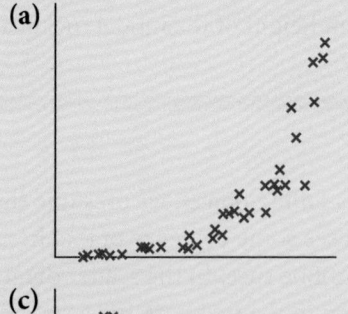

(b)
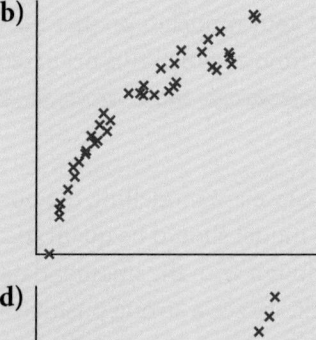

(c)
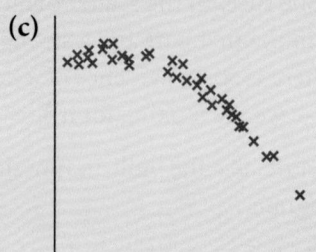

(d)
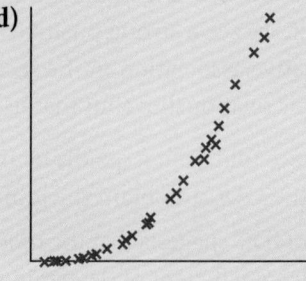

2. The specific weight of water varies with temperature according to the data given in Table 19.11.

(a) Generate a scatter diagram for this data and describe the association between temperature and specific weight.

(b) Give a reason why linearisation using logarithms should not be used for this data.

(c) Use your GDC to find an appropriate model for this data, with $w$ as specific weight in terms of temperature $T$.

(d) Use your model to predict the specific weight of water at 12 °C to 4 significant figures.

(e) Use your model to predict the maximum specific weight of water and the temperature at which this occurs.

| Temperature (°C) | Specific weight (k N m⁻³) |
|---|---|
| 0 | 9.8050 |
| 0.1 | 9.8052 |
| 1 | 9.8057 |
| 4 | 9.8064 |
| 10 | 9.8040 |
| 15 | 9.7980 |
| 20 | 9.7890 |

Table 19.11 Data for question 2

3. Do faster aeroplanes use more fuel? The tables contain data on the average speed and fuel consumption of different aeroplanes on mid-range journeys.

| Speed (km h⁻¹) | 836 | 802 | 779 | 741 | 762 | 770 | 765 | 723 | 687 |
|---|---|---|---|---|---|---|---|---|---|
| Fuel consumption (litres per minute) | 224 | 140 | 138 | 94 | 100 | 95 | 87 | 62 | 79 |

| Speed (km h⁻¹) | 670 | 665 | 609 | 679 | 625 | 580 | 607 | 605 |
|---|---|---|---|---|---|---|---|---|
| Fuel consumption (litres per minute) | 56 | 46 | 54 | 70 | 51 | 40 | 51 | 48 |

(a) Generate a scatter diagram and describe the association between speed and fuel consumption.

(b) Linearise the data using logarithms to find the best-fit model for this data, using $v$ for speed and $F$ for fuel consumption.

(c) Interpret your model and $R^2$ in context.

(d) Use your model to predict the fuel consumption of a journey with an average speed of 700 km h⁻¹.

4. Some researchers drove a compact car at various speeds around a racing track to see if fuel consumption relates to speed. Their data are given in Table 19.12.

(a) Generate a scatter diagram for this data.

(b) Give a reason why linearisation using logarithms should not be used for this data.

(c) Use your GDC to find an appropriate model for this data, with fuel consumption $C$ in terms of speed $v$.

(d) Use your model to predict the fuel consumption of this car at 95 km h⁻¹.

| Speed (km h⁻¹) | Fuel consumption (litres per 100 km) |
|---|---|
| 50 | 9.6 |
| 60 | 8.9 |
| 70 | 8.3 |
| 80 | 8.0 |
| 90 | 8.1 |
| 100 | 8.7 |
| 110 | 9.8 |
| 120 | 10.4 |

Table 19.12 Data for question 4

(e) Give a reason why it is not reasonable to use this model to predict the fuel consumption at $140\,\mathrm{km\,h^{-1}}$.

(f) Use your model to predict the best fuel efficiency (minimum fuel consumption) and the speed at which this occurs. Comment on the reasonableness of this prediction.

5.  Medical researchers are interested in how long it takes the body to remove a certain drug from the bloodstream. In one experiement, a patient is given a drug and the patient's blood is tested at regular intervals to determine the concentration in millimoles per litre $(\mathrm{mmol\,L^{-1}})$. The patient is tested just before being given the drug, then twice in 30 minutes, then every six hours. The data for this experiment are given in the table.

| Time since administration (h) | 0 | 0.25 | 0.5 | 6 | 12 | 18 | 24 | 30 | 36 | 42 | 48 |
|---|---|---|---|---|---|---|---|---|---|---|---|
| Concentration $(\mathrm{mmol\,L^{-1}})$ | 0 | 3.8 | 8.0 | 4.9 | 3.2 | 1.7 | 1.3 | 0.6 | 0.3 | 0.2 | 0.1 |

(a) Generate a scatter diagram for the data. Describe the feature of the data that makes linearisation difficult.

(b) Hence, consider an appropriate subset of the data. Given the form of the scatter diagram, what model is likely to be appropriate? Which linearisation via logarithms should be used?

(c) Hence, perform an appropriate linearisation via logarithms, based on your answer to part **(b)**, and find an appropriate least-squares model for concentration $C$ in terms of time $t$.

(d) Write down a suitable domain for your model.

(e) Interpret your model in context.

(f) Find and interpret the value of $R^2$.

(g) Use your model to predict the blood concentration of this drug 4 hours after it is given.

(h) The half-life of a drug is the length of time that it takes the body to remove half of the drug from the bloodstream. Use your model to find the half-life of this drug.

6.  The orbital period for a planet is related to the planet's distance from the Sun. Use the scatter diagram to answer the questions.

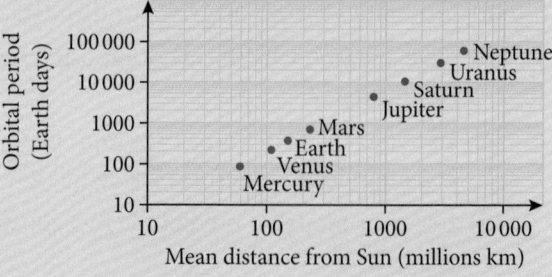

(a) Estimate the average distance from the Sun and orbital period for Jupiter.

**(b)** Estimate the average distance from the Sun and orbital period for Mercury.

**(c)** Write down what type of non-linear model appears to be appropriate for this data and give a reason for your answer.

7. The data from question **6** is often expressed in different units: Astronomical units (1 AU is equal to the mean distance from the Earth to the Sun) and the number of Earth years as given in the tables.

| Planet | Mercury | Venus | Earth | Mars |
|---|---|---|---|---|
| Mean distance from Sun (AU) | 0.390 | 0.723 | 1 | 1.52 |
| Orbital period (Earth years) | 0.241 | 0.616 | 1 | 1.88 |

| Planet | Jupiter | Saturn | Uranus | Neptune |
|---|---|---|---|---|
| Mean distance from Sun (AU) | 5.20 | 9.58 | 19.2 | 30.1 |
| Orbital period (Earth years) | 11.9 | 29.6 | 84.1 | 165 |

**(a)** Generate a scatter diagram and describe the association between average distance from the Sun and orbital period.

**(b)** Linearise the data using logarithms to find the best-fit model, using $P$ for orbital period and $D$ for average distance from the Sun.

**(c)** Interpret your model and $R^2$ in context.

**(d)** There is a large gap between Mars and Jupiter. The largest object in this region is the dwarf planet Ceres, located 2.77 AU from the Sun. Use your model to predict the orbital period of Ceres.

**(e)** The actual orbital period of Ceres is 4.60 Earth years. Calculate the percentage error of your prediction.

8. The tables give the temperature in a growing chamber at various times since midnight.

| Hours since midnight | 0 | 2 | 4 | 6 | 8 | 10 | 12 |
|---|---|---|---|---|---|---|---|
| Temperature (°C) | 23.3 | 23.7 | 24.2 | 24.7 | 24.9 | 25.0 | 24.7 |

| Hours since midnight | 14 | 16 | 18 | 20 | 22 | 24 |
|---|---|---|---|---|---|---|
| Temperature (°C) | 24.3 | 23.8 | 23.3 | 23.0 | 23.1 | 23.3 |

**(a)** Generate a scatter diagram for the data and describe it.

**(b)** Give a reason why logarithms should not be used to linearise these data.

**(c)** Hence, use your GDC to find an appropriate model for temperature $T$ in terms of hours since midnight $h$.

**(d)** Based on your model, write down the average temperature in the growing chamber.

**(e)** Use your model to predict the temperature at 17:00.

**9.** The torque produced by an engine is dependent on the rotational speed (RPM; revolutions min$^{-1}$). The tables give data for a Chevy 350 engine.

| RPM | 1500 | 2000 | 2500 | 3000 | 3500 |
|---|---|---|---|---|---|
| Torque (Nm) | 380 | 461 | 475 | 508 | 529 |

| RPM | 4000 | 4500 | 5000 | 5500 | 6000 |
|---|---|---|---|---|---|
| Torque (Nm) | 515 | 488 | 475 | 420 | 339 |

(a) Generate a scatter diagram for the data and describe it.

(b) Give a reason why logarithms should not be used to linearise this data.

(c) Hence, use your GDC to find an appropriate model for torque $N$ in terms of RPM.

(d) Based on your model, predict the rotational speed that produces maximum torque.

## Chapter 19 practice questions

**1.** What is wrong with each statement?

(a) The value of Pearson's correlation coefficient for the time it takes an athlete to run 5 km and the time for her to cycle 30 km is $r = 1.21$

(b) The fuel economy of passenger cars decreases linearly with the mass of the car with Pearson's $r = 0.78$

(c) Among mammals, those with greater average body mass have longer life expectancy; the rank correlation coefficient is $r_s = -0.85$

(d) For a set of $(x, y)$ data, the least-squares regression line is $y = 25 - 3.52x$ with $r = 0.64$

**2.** For each of the following situations, would you expect Spearman's rank correlation coefficient $r_s$ to be close to $+1$, 0, or $-1$?

(a) The number of hours an IB diploma student studies and the amount they spend on lunch each week

(b) The tax on fuel and the number of litres sold

(c) The age of children and their mass

(d) The speed of a motor vehicle and the distance required to brake to a stop

(e) The amount spent on snack food and the average number of years a person spends in education in various countries

(f) The outside temperature and money spent on heating

**3.** The intensity of radiation, $I$, from a certain source is measured at regular intervals. The measurements are given in Table 19.13.

(a) Show that the relationship between $\ln(I)$ and $t$ is approximately linear.

(b) Hence find the least-squares regression line for $\ln(I)$ in terms of $t$.

(c) Find the value of $R^2$ and interpret it in context.

(d) Hence, find the non-linear model $I = I_0 e^{-kt}$ where $I_0$ and $k$ are constants to be determined.

| Time, $t$ (min) | $\ln(I)$ |
|---|---|
| 0.2 | 0.508 |
| 0.4 | 0.212 |
| 0.6 | −0.051 |
| 0.8 | −0.387 |
| 1.0 | −0.444 |

**Table 19.13** Data for question 3

4. The time it takes for a runner to complete a marathon can be predicted from the time it takes to run 5 km (Table 19.14).
   (a) Generate a scatter diagram, find the value of $r$, and describe the relationship between 5 km time and marathon time.
   (b) Find the least-squares regression line for marathon time ($M$) based on 5 km time ($T$). Interpret the gradient and intercept in context.
   (c) Use your least-squares regression model to predict the marathon time for a runner with a 5 km time of 20 minutes.
   (d) The distance of a marathon is 42.195 km. If runners ran at the same pace for marathons as they do for 5 km, the gradient of the least-squares model for marathon time based on 5 km time would be $k$.
      (i) Find the value of $k$.
      (ii) Find the percentage difference between $k$ and the gradient from part (b).
      (iii) Explain the meaning of the percentage difference in context.

| 5 km time (min) | Marathon time (min) |
|---|---|
| 15 | 144 |
| 17 | 163 |
| 19 | 182 |
| 21 | 201 |
| 23 | 220 |
| 25 | 238 |
| 27 | 256 |
| 29 | 274 |

Table 19.14  Data for question 4

5. The width and length of several leaves from the same tree were measured. The results are given in Table 19.15.
   (a) Generate a scatter diagram, find the value of $r$, and describe the relationship between width and length for these leaves.
   (b) Find the least-squares regression line for length ($L$) based on width ($W$). Interpret the gradient and intercept in context.
   (c) Write down a suitable domain for your model.
   (d) Use your least-squares regression model to predict the length of a leaf with a width of 47 mm.
   (e) Give a reason why we should not use the model to predict the length of a leaf that is 60 mm wide.

| Width (mm) | Length (mm) |
|---|---|
| 44 | 102 |
| 43 | 99 |
| 45 | 103 |
| 48 | 111 |
| 52 | 119 |
| 43 | 100 |
| 46 | 106 |
| 50 | 114 |

Table 19.15  Data for question 5

6. In a 2017 study, researchers used data from US national heath surveys from 1997 to 2009, involving 333 247 participants, and found that 'light and moderate' consumption of alcohol correlated with fewer deaths from cardiovascular disease. Does this mean that drinking alcohol is beneficial?

7. Is the flying speed of animals related to their overall body length? A sample of data for some flying species are given in the table below.

| Animal | Length (cm) | Speed (cm s$^{-1}$) |
|---|---|---|
| Fruit fly | 0.2 | 190 |
| Horse fly | 1.3 | 660 |
| Hummingbird | 8.1 | 1120 |
| Dragonfly | 8.5 | 1000 |
| Bat | 11 | 690 |
| Common swift (bird) | 17 | 2550 |
| Flying fish | 34 | 1560 |
| Pintail duck | 56 | 2280 |
| Swan | 120 | 1880 |
| Pelican | 160 | 2280 |

(a) Generate a scatter diagram for these data and describe it.

(b) Are there any outliers in the data? Assuming the data is correct, is it appropriate to remove the outliers?

(c) Calculate the value of Pearson's $r$ and interpret it in context.

(d) Rank the data, then calculate the value of Spearman's $r_s$ and interpret it in context.

(e) Is one measure of association, $r$ or $r_s$, more appropriate for this data? Give a reason for your answer.

(f) Linearise the data using logarithms, and find the most appropriate model for flying speed $v$ in terms of body length $L$.

(g) Find and interpret the value of $R^2$ for your model.

(h) An albatross has a body length of 130 cm. Predict its flying speed.

8. The table below lists the gold medal winners from the 2004 Olympics in Men's weightlifting.

| Winner (country) | Weight class (kg) | Mass lifted (kg) |
|---|---|---|
| Halil Mutlu (Turkey) | 56 | 295 |
| Zhiyong Shi (China) | 62 | 325 |
| Guozheng Zhang (China) | 69 | 347.5 |
| Taner Sagir (Turkey) | 77 | 375 |
| George Asanidze (Georgia) | 85 | 382.5 |
| Milen Dobrev (Bulgaria) | 94 | 407.5 |
| Dmitry Berestov (Russia) | 105 | 425 |

(a) Generate a scatter diagram for these data and describe it.

(b) Find the most appropriate non-linear model using linearisation with logarithms to model mass lifted $m$ in terms of weight class $w$.

(c) Find and interpret the value of $R^2$ for your model.

(d) Based on your model:

    (i) Which athlete performed better than we would expect? By how much?

    (ii) Which athlete performed worse than we would expect? By how much?

9. The atmospheric concentration of carbon dioxide ($CO_2$) in parts per million (ppm) in the month of June at the Mauna Loa observatory in Hawaii, USA, is given in the table.

| Year | 1960 | 1970 | 1980 | 1990 | 2000 | 2010 | 2017 | 2018 |
|---|---|---|---|---|---|---|---|---|
| $CO_2$ concentration (ppm) | 320 | 328 | 341 | 356 | 372 | 392 | 409 | 411 |

(a) Generate a scatter diagram for these data and describe it.

(b) Calculate the value of Pearson's $r$ and interpret it in context.

(c) When constructing models, we often modify dates to be 'years since' a certain date.

    (i) Modify the date value to be years since 1960.

    (ii) Recalculate the value of Pearson's $r$ and interpret it in context. Has it changed?

    (iii) Find a least-squares regression line for $CO_2$ concentration $C$ based on $y$, years since 1960. Interpret the slope and intercept in context.

    (iv) Write down an appropriate domain for your model in context.

    (v) Use your model to predict $CO_2$ concentration in the year 2005.

    (vi) Give a reason why this model should not be used to predict the $CO_2$ concentration in the year 2050.

10. In 1881, the Russian chemist D. Mendeleev investigated the solubility of sodium nitrate ($NaNO_3$) at different water temperatures. His data are shown in Table 19.16.

(a) Generate a scatter diagram for these data and describe it.

(b) Calculate the value of Pearson's $r$ and interpret it in context.

(c) Find a least-squares regression line to predict solubility $S$ based on temperature $T$ and interpret it in context.

(d) Use your least-squares regression line to predict the solubility of $NaNO_3$ at 25°C.

(e) Give a reason why your least-squares regression line should not be used to predict solubility of $NaNO_3$ at 95°C.

(f) A certain solution has a $NaNO_3$ concentration of 100 g per 100 ml. Predict the temperature of this solution, or give a reason why you cannot.

| Temp (°C) | Solubility (g per 100 ml) |
|---|---|
| 0.0 | 66.7 |
| 4.0 | 71.0 |
| 10 | 76.3 |
| 15 | 80.6 |
| 21 | 85.7 |
| 29 | 92.9 |
| 36 | 99.4 |
| 51 | 114 |
| 68 | 125 |

Table 19.16 Data for question 10

11. The mean monthly temperatures for Vienna, Austria, are given in the tables below.

| Month | Jan | Feb | Mar | Apr | May | Jun |
|---|---|---|---|---|---|---|
| Mean temperature (°C) | −0.6 | 1.1 | 5.0 | 10.0 | 14.8 | 18.1 |

| Month | Jul | Aug | Sep | Oct | Nov | Dec |
|---|---|---|---|---|---|---|
| Mean temperature (°C) | 19.9 | 19.5 | 15.2 | 10 | 4.8 | 1.0 |

(a) Generate a scatter diagram for mean temperature versus months since January, such that January is month 0.

(b) Give a reason why a sinusoidal model is appropriate for this data.

(c) Use a GDC to find a model for this data in the form $T = a\sin(bM - c) + d$, where $T$ is the mean temperature for month $M$, January is month 0, and $a, b, c, d$ are constants to be determined.

| Time (s) | Velocity (cm s$^{-1}$) |
|---|---|
| 0 | 0 |
| 0.1 | −91.3 |
| 0.2 | −189 |
| 0.3 | −316 |
| 0.4 | −425 |
| 0.5 | −500 |
| 0.6 | −571 |
| 0.7 | −740 |
| 0.8 | −836 |
| 0.9 | −940 |
| 1.0 | −1020 |

**Table 19.17** Data for question 12

| Year | Wind power (GW) |
|---|---|
| 1980 | 0.01 |
| 1985 | 1.02 |
| 1988 | 1.58 |
| 1990 | 1.93 |
| 1992 | 2.51 |
| 1995 | 4.82 |
| 1997 | 7.64 |
| 1999 | 13.8 |
| 2001 | 23.9 |
| 2002 | 31.1 |
| 2003 | 39.4 |
| 2004 | 47.6 |
| 2005 | 69.1 |
| 2006 | 74.2 |
| 2007 | 93.9 |
| 2008 | 121 |
| 2009 | 158 |
| 2010 | 198 |
| 2011 | 238 |
| 2012 | 282 |
| 2013 | 319 |
| 2014 | 370 |
| 2015 | 432 |
| 2016 | 488 |
| 2017 | 540 |

**Table 19.18** Data for question 13

(d) The air temperature in a cave does not vary and is usually equal to the mean temperature of the region. Write down the expected air temperature for a cave near Vienna.

(e) Find the period of the model.

(f) Hence give a reason why the period of the GDC model is not ideal.

12. In Physics class, a student dropped an object and measured its velocity every 0.1 s for 1 s. Her data is shown in Table 19.17.

(a) Generate a scatter diagram for the data and describe it.

(b) Calculate the value of Pearson's $r$ and interpret it in context.

(c) Find a least-squares regression line to predict velocity $v(t)$ based on time $t$.

(d) Interpret the slope and intercept of your least-squares regression line, with appropriate units.

(e) Use your least-squares regression line to predict the velocity of this object at 0.15 s.

(f) The rate of change in velocity (acceleration) due to gravity should be 981 cm sec$^{-2}$ on Earth. Does this experiment agree with that theoretical value? Explain your answer.

(g) Use your model for velocity to create a model for the position $s(t)$ at time $t$, assuming $s(0) = 0$

(h) Hence, predict the distance travelled by the object during the time $0 \leqslant t \leqslant 1$

13. The global wind power production in GW (gigawatts) by year is given in Table 19.18.

(a) Generate and describe a scatter diagram for these data.

(b) Give a reason why log(wind power) vs year is the most appropriate logarithmic linearisation.

(c) Hence, generate a new scatter diagram showing log(wind power) versus year.

(d) Hence, identify an outlier and interpret it in context.

(e) Find an appropriate non-linear model for the global production of wind power $W$ in terms of $t$, years since 1980, using only the data from 1985 onwards.

(f) Interpret the growth rate of your model.

(g) Find and interpret the value of $R^2$ for your model.

(h) The number of years it takes for global wind power production to double is $n$. Use your model to find the value of $n$.

**14.** The tables give data on the monthly video game sales (in thousands) for a particular video game released in June. Each data point represents the sales for that month, where month 1 is July.

| Months since release (1=July) | 1 | 2 | 4 | 6 | 7 |
|---|---|---|---|---|---|
| Monthly video game sales (thousands) | 15.0 | 18.4 | 22.3 | 28.6 | 23.2 |

| Months since release (1=July) | 8 | 10 | 12 | 14 | 16 |
|---|---|---|---|---|---|
| Monthly video game sales (thousands) | 25.4 | 26.9 | 28.4 | 27.7 | 29.3 |

**(a)** Generate a scatter diagram.

    **(i)** Identify a possible outlier and give a possible reason for this outlier in this context.

    **(ii)** Give a reason why plotting $y$ vs $\log(x)$ is likely to be the best linearisation method.

**(b)** Hence, find an appropriate model for monthly video game sales $S$ in terms of months since release $m$. Give a reason why this model makes sense in this context.

**(c)** Predict the number of video game sales in November (month 5).

**(d)** Total video game sales for the first year of release can be estimated with $\int_a^b S \, dm$

    **(i)** Write down the values of $a$ and $b$.

    **(ii)** Hence, use your model to estimate the total video game sales for the first year.

**15.** The average number of minutes of daylight per day for Bangalore, India are given in the tables.

| Month | Jan | Feb | Mar | Apr | May | Jun |
|---|---|---|---|---|---|---|
| Minutes of daylight per day | 686 | 701 | 722 | 744 | 762 | 771 |

| Month | Jul | Aug | Sep | Oct | Nov | Dec |
|---|---|---|---|---|---|---|
| Minutes of daylight per day | 767 | 752 | 731 | 708 | 690 | 681 |

**(a)** Generate a scatter diagram for these data with January as month 0.

**(b)** Give a reason why a sinusoidal model is appropriate for these data.

**(c)** Use a GDC to find a model for these data in the form $T = a\sin(bM - c) + d$, where $T$ is the mean temperature for month $M$, January is month 0, and $a, b, c, d$ are constants to be determined.

**(d)** Write down the annual average number of minutes of daylight per day.

**(e)** Find the period for your model and comment on this value.

**(f)** The latitude of Bangalore is about 13° north of the equator. The latitude of Ndola, Zambia is about 13° south of the equator. Write down a model for the number of minutes of daylight in Ndola in the form $N = a_1\sin(bM - c) + d$ where $a_1$ is a constant to be determined and $b, c, d$ are equal to the model for daylight minutes in Bangalore.

| RPM | kW |
|------|------|
| 1500 | 52.2 |
| 2000 | 93.2 |
| 2500 | 116 |
| 3000 | 164 |
| 3500 | 194 |
| 4000 | 209 |
| 4500 | 235 |
| 5000 | 254 |
| 5500 | 257 |
| 6000 | 224 |
| 6500 | 185 |

**Table 19.19** Data for question 16

16. The amount of power in kW that an engine can produce varies according to the rotational speed in revolutions per minute (RPM). Table 19.19 gives data for a Chevy 350 engine.
    (a) Generate a scatter diagram for the data and describe it.
    (b) Give a reason why logarithms should not be used to linearise this data.
    (c) Hence, use your GDC to find an appropriate model for power $P$ in terms of RPM.
    (d) Based on your model, predict the rotational speed that produces maximum power.
    (e) Give a reason why your model might not predict maximum power accurately.

17. The mean monthly temperature in Lusaka, Zambia is given in the tables.

| Month | Jan | Feb | Mar | Apr | May | Jun |
|-------|-----|-----|-----|-----|-----|-----|
| Mean temperature (°C) | 21.4 | 21.4 | 21.0 | 20.1 | 18.0 | 16.2 |

| Month | Jul | Aug | Sep | Oct | Nov | Dec |
|-------|-----|-----|-----|-----|-----|-----|
| Mean temperature (°C) | 16.0 | 18.3 | 21.8 | 24.4 | 23.2 | 21.7 |

    (a) Generate a scatter diagram for mean temperature versus month since January, such that January is month 0.
    (b) Give a reason why a sinusoidal model is not appropriate for these data.
    (c) Give a reason why neither cubic nor quadratic models are appropriate for these data.

18. The mean monthly precipitation (rainfall) in Lusaka, Zambia is given in the tables.

| Month | Jan | Feb | Mar | Apr | May | Jun |
|-------|-----|-----|-----|-----|-----|-----|
| Mean precipitation (mm) | 215 | 183 | 87 | 28 | 4 | 0 |

| Month | Jul | Aug | Sep | Oct | Nov | Dec |
|-------|-----|-----|-----|-----|-----|-----|
| Mean precipitation (mm) | 0 | 0 | 2 | 17 | 91 | 204 |

    (a) Generate a scatter diagram for mean precipitation versus months since January, such that January is month 0.
    (b) Use a GDC to find a model in the form $P = am^2 + bm + c$, where $P$ is the mean precipitation for month $m$, January is month 0, and $a$, $b$, $c$ are constants to be determined.
    (c) Find and interpret the value of $R^2$ for your model.
    (d) Use your model to predict the minimum precipitation.
    (e) Hence, give a reason why this model is not ideal.

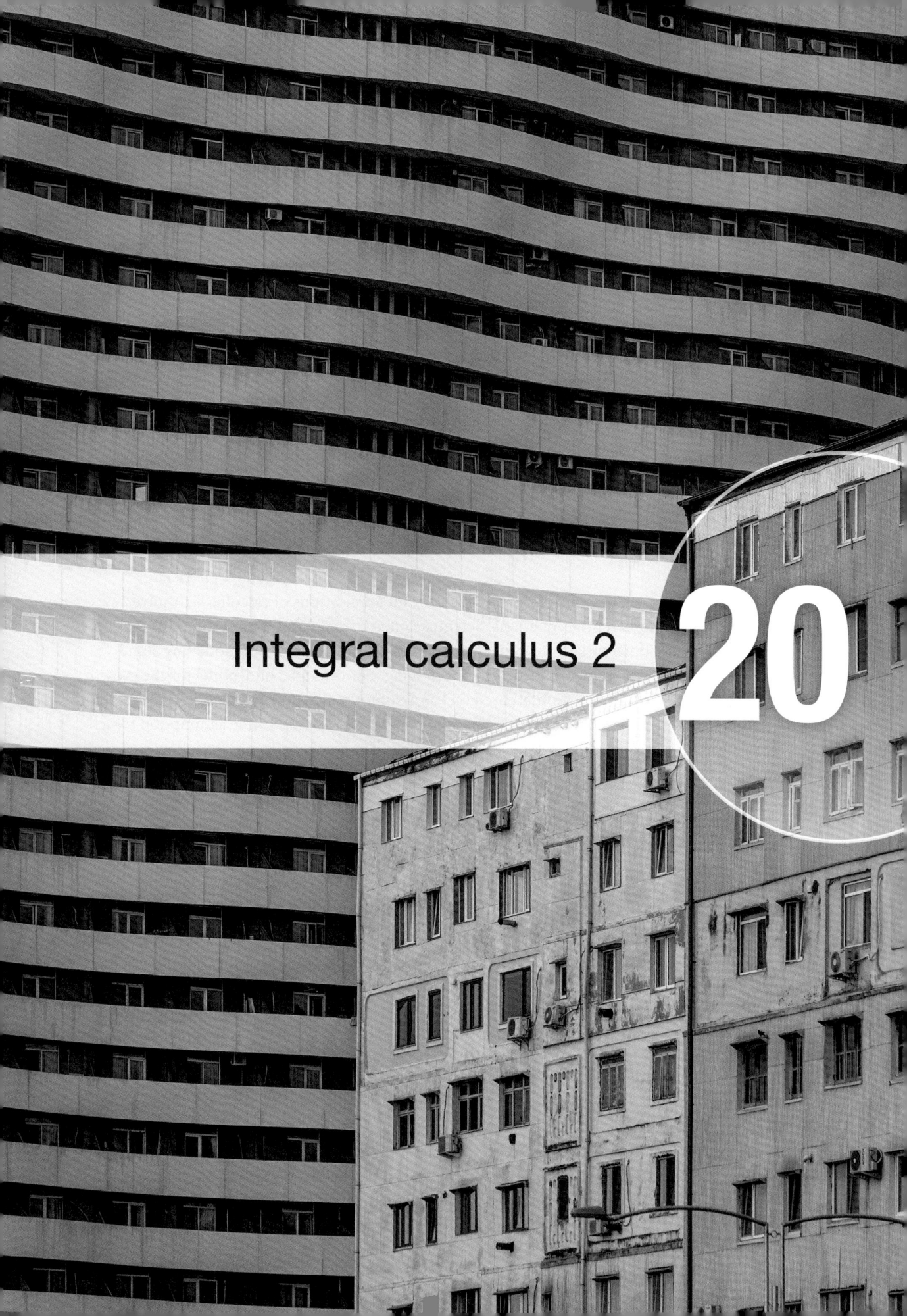

Integral calculus 2

20

## Learning objectives

By the end of this chapter, you should be familiar with...

- working with first order differential equations:
  - solving differential equations by separating the variables
  - setting up differential equations from context
- slope fields
- finding numerical solutions of differential equations using Euler's method
- finding numerical solutions for a coupled system $\dfrac{dx}{dt} = f_1(x, y, t)$ and $\dfrac{dy}{dt} = f_2(x, y, t)$
- using a phase portrait to solve coupled differential equations of the form $\dfrac{dx}{dt} = ax + by$ and $\dfrac{dy}{dt} = cx + dy$
- solving $\dfrac{d^2x}{dt^2} = f(x, \dfrac{dx}{dt}, t)$ by Euler's method
- solutions to $\dfrac{d^2x}{dt^2} + a\dfrac{dx}{dt} + b = 0$ by use of the phase portrait method.

Possibly one of the most significant applications of calculus is to solve differential equations. Scientists often use calculus to find solutions for differential equations that have arisen in the process of modelling some phenomenon that they are studying. For example, one model for the growth of a population is based on the assumption that the population grows at a rate proportional to the size of the population. It is a sensible assumption for a population of bacteria or animals under ideal conditions, to follow an equation such as $\dfrac{dP}{dt} = kP$, where $t$ stands for time and $P$ for the number of individuals, bacteria, or animals at any time $t$.

Although it is often impossible to find an explicit formula for the solution of a differential equation, we will see that graphical and numerical approaches can provide the information required.

In this chapter you will study equations involving an unknown function and its derivative(s). These are called differential equations. Differential equations come in a wide variety of forms, and many different procedures – analytical, graphical and numerical – exist for finding their solutions.

## 20.1 Differential equations

In general, a **differential equation** is an equation that contains an unknown function and one or more of its derivatives. The **order** of a differential equation is the order of the highest derivative that occurs in the equation. A first order differential equation is an equation that involves an unknown function and its first derivative.

Here are some examples (where $y$ is an unknown function of $x$ or $t$):

$$x\frac{dy}{dx} + y\frac{dy}{dx} - y = 0 \quad \text{first order differential equation} \quad F\left(x, y, \frac{dy}{dx}\right) = 0$$

$$\frac{dy}{dx} + \frac{y^2 - y}{x^2} = 0 \qquad \text{first order differential equation} \quad F\left(x, y, \frac{dy}{dx}\right) = 0$$

$$\frac{d^2y}{dx^2} + 3\frac{dy}{dx} - 5y = 0 \quad \text{second order differential equation} \quad F\left(x, y, \frac{dy}{dx}, \frac{d^2y}{dx^2}\right) = 0$$

$$\frac{dy}{dt} + y\sin t - e^{\cos t} = 0 \text{ first order differential equation} \quad F\left(t, y, \frac{dy}{dt}\right) = 0$$

In a first order differential equation, the first derivative, $\frac{dy}{dx}$, of the unknown function can be isolated on one side of the equation. Hence, a simpler general form for first order differential equations is $\frac{dy}{dx} = F(x, y)$ where $\frac{dy}{dx}$ is expressed as a function in terms of $x$ and $y$. Note that the first order differential equations can all be re-written in this form. For example,

$$x\frac{dy}{dx} + y\frac{dy}{dx} - y = 0 \Rightarrow \frac{dy}{dx} = \frac{y}{x + y}$$

## Solution of a differential equation

A function $f$ is a **solution** of a differential equation if the equation is satisfied for all values of $x$ in some interval when $y = f(x)$ and its derivatives are substituted into the equation. Thus, $y = f(x) = x - \frac{1}{x}$ is a solution of $x\frac{dy}{dx} + y = 2x$ over the set of non-zero real numbers.

Since $\frac{dy}{dx} = 1 + \frac{1}{x^2} \Rightarrow x\left(1 + \frac{1}{x^2}\right) + x - \frac{1}{x} = x + \frac{1}{x} + x - \frac{1}{x} = 2x$

and thus $x\frac{dy}{dx} + y = 2x$ for all values over this interval.

In algebra we usually seek the unknown variable values that satisfy an equation such as $3x^2 - 2x - 5 = 0$. By contrast, in solving a differential equation, we are looking for the unknown functions $y = y(x)$ for which an identity such as $y'(x) = 3x^2 y(x)$ holds on some interval of real numbers. If possible, we ideally want to find all the solutions of the differential equation.

### Example 20.1

Verify that $y(x) = Ce^{x^3}$ is a solution to the differential equation $\frac{dy}{dx} = 3x^2 y$

## Solution

Since $C$ is a constant in $y(x) = Ce^{x^3}$, then

$$\frac{dy}{dx} = C(3x^2 e^{x^3}) = 3x^2(Ce^{x^3}) = 3x^2 y$$

Consequently, every function $y(x)$ of the form $y(x) = Ce^{x^3}$ satisfies, and thus is a solution of, the differential equation $\frac{dy}{dx} = 3x^2 y$ for all real $x$.

In fact, $y(x) = Ce^{x^3}$ defines an infinite family of different solutions to this differential equation, one for each value of the arbitrary constant $C$.

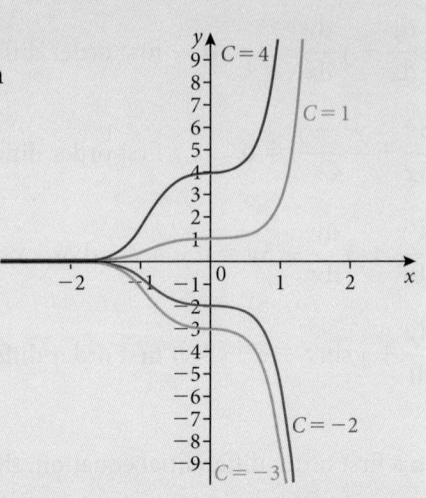

A differential equation may use symbols for the independent and dependent variables other than $x$ and $y$. Also note that we are using $F$ to represent a two-variable function that when set equal to $\frac{dy}{dx}$ is the differential equation, and $f$ to represent the unknown function whose slope at the point $(x, y)$ is $\frac{dy}{dx}$

## Example 20.2

Verify that $y(x) = -\dfrac{1}{2x^4 + 3}$ is a solution to the differential equation

$$\frac{dy}{dx} = 8x^3 y^2 \text{ over the interval } ]-\infty, \infty[$$

## Solution

Note that the denominator in $y(x)$ is never zero and that $y(x)$ is differentiable everywhere. Furthermore, for all real numbers $x$

$$\frac{d}{dx} y(x) = \frac{d}{dx}\left(-\frac{1}{2x^4 + 3}\right) = \frac{8x^3}{(2x^4 + 3)^2} = 8x^3\left(-\frac{1}{2x^4 + 3}\right)^2 = 8x^3 y^2$$

Thus, $y(x) = -\dfrac{1}{2x^4 + 3}$ is a solution of the given differential equation.

## Example 20.3

The solution of a differential equation is the (initially unknown) function $y = f(x)$ whose derivative is $\dfrac{dy}{dx}$. Find a solution to the differential equation

$$\frac{dy}{dx} = \frac{1}{x + 1}, x \neq -1$$

## Solution

Every solution of this equation is an anti-derivative of $\dfrac{1}{x + 1}$

$$y = \int \frac{1}{x + 1} dx = \ln|x + 1| + C, x \neq -1$$

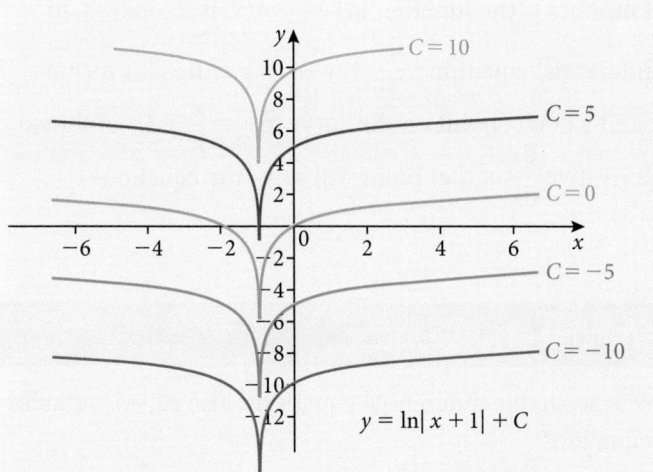

$$y = \ln|x + 1| + C$$

So, the solution of the differential equation $\dfrac{dy}{dx} = \dfrac{1}{x+1}$ is the explicitly defined function $y = \ln|x + 1| + C$, where $C$ is an arbitrary constant. This is called a general solution because it is not a single function, but an infinite family of functions dependent on the constant $C$.

When applying differential equations, we are usually not as interested in finding a family of solutions (the general solution) as we are in finding a solution that satisfies some additional requirement. In many physical problems we need to find the **particular** solution that satisfies a condition of the form $y(x_0) = y_0$. This is called an **initial condition**, and the problem of finding a solution of the differential equation that satisfies the initial condition is called an **initial-value problem (IVP)**. For instance, in Example 20.3, if we are given the initial condition that $y = 5$ when $x = 0$, then we can solve for $C$, giving $C = 5$ and the particular solution of $y = \ln|x + 1| + 5$

> In general, we wish to find the explicit solution of a differential equation written in the form $y = f(x)$ where $f$ is a known function. However, it is sometimes not possible to solve for $y$. In such a case we must settle for an implicit solution written in the form $g(y) = f(x)$ where $g$ and $f$ are known functions and $g(y) \neq y$

For example, the general solution to the equation $x\dfrac{dy}{dx} + y\dfrac{dy}{dx} - y = 0$ is $\ln y = \dfrac{x}{y} + C$

It is an equation relating $x$ and $y$ and implies a function exists that defines $y$ as a function of $x$.

To verify that this is a solution, we differentiate: using implicit differentiation and the product rule:

$$\frac{d}{dx}(\ln y) = \frac{d}{dx}\left(\frac{x}{y} + C\right) \Rightarrow \frac{1}{y}\frac{dy}{dx} = \frac{d}{dx}(xy^{-1}) + \frac{d}{dx}(C)$$

$$\Rightarrow \frac{1}{y}\frac{dy}{dx} = y^{-1} + x\left(-y^{-2}\frac{dy}{dx}\right) + 0 \Rightarrow \frac{1}{y}\frac{dy}{dx} = \frac{1}{y} - \frac{x}{y^2}\frac{dy}{dx}$$

$$\Rightarrow y^2\left(\frac{1}{y}\frac{dy}{dx}\right) = y^2\left(\frac{1}{y} - \frac{x}{y^2}\frac{dy}{dx}\right) \Rightarrow x\frac{dy}{dx} + y\frac{dy}{dx} - y = 0$$

Therefore, for any real number $C$ the function $\ln y = \dfrac{x}{y} + C$ is a solution, in implicit form, to the differential equation $x\dfrac{dy}{dx} + y\dfrac{dy}{dx} - y = 0$. This means that the coordinates $x$ and $y$ of any point on the curve $\ln y = \dfrac{x}{y} + C$ combined with the value of the derivative $\dfrac{dy}{dx}$ at that point will solve the equation $x\dfrac{dy}{dx} + y\dfrac{dy}{dx} - y = 0$

## Separable differential equations

In this section, we look at separable differential equations, also called variables-separable differential equations.

The first order differential equation:

$$\frac{dy}{dx} = F(x, y)$$

is called variables separable when the function $F(x, y)$ can be factored into a product or quotient of two functions such as

$$\frac{dy}{dx} = p(x)q(y) \text{ or } \frac{dy}{dx} = \frac{g(x)}{h(y)}$$

In such cases, the variables $x$ and $y$ can be separated by writing

$$\frac{dy}{q(y)} = p(x)\,dx \text{ or } h(y)\,dy = g(x)\,dx$$

and then simply integrating both sides with respect to $x$. That is

$$\int \frac{dy}{q(y)} = \int p(x)\,dx + c \text{ or } \int h(y)\,dy = \int g(x)\,dx + c$$

> You need to remember that $h(y)$ is a continuous function of $y$ alone and $g(x)$ is a continuous function of $x$ alone. This also applies to $q(y)$ and $p(x)$.

Here are some examples of differential equations that are separable:

| Original differential equation | Rewritten with variables separated |
|---|---|
| $(x^2 + 4)y' = 3xy$ | $\dfrac{dy}{y} = \dfrac{3x}{x^2 + 4}\,dx$ |
| $\dfrac{3xe^y y'}{1 + e^{2y}} = 5$ | $\dfrac{3e^y}{1 + e^{2y}}\,dy = \dfrac{5}{x}\,dx$ |
| $\dfrac{dy}{dx} = xy + 4$ | Not separable |
| $3x^2 + y\dfrac{dy}{dx} = 7$ | $y\,dy = (7 - 3x^2)\,dx$ |
| $y^2\dfrac{dy}{dx} + x^2 = xy^2$ | Not separable |

**Table 20.1** Examples of differential equations that are separable or not separable

> It is not always obvious if a differential equation is separable. Some algebraic manipulation is needed to confirm that the differential equation can, in fact, be written in the form $\dfrac{dy}{dx} = p(x)q(y)$ or $\dfrac{dy}{dx} = \dfrac{g(x)}{h(y)}$. For example, $\dfrac{dy}{dx} = \dfrac{3}{xy} - \dfrac{x^2}{y}$ is separable because it can be written as $\dfrac{dy}{dx} = \dfrac{1}{y}\left(\dfrac{3}{x} - x^2\right)$; and $\dfrac{\tan x}{y}\dfrac{dy}{dx} = \dfrac{2}{\ln y}$ is also separable because it can be written as $\dfrac{dy}{dx} = \dfrac{2y}{\ln y}\cot x$. However, the equations $\dfrac{dy}{dx} = x^2 + y^2$ and $\dfrac{dy}{dx} = 1 + xy$ are not separable.

## Example 20.4

Find the general solution of the differential equation $y' - 9x^2y^2 = 5y^2$

### Solution

We first factorise the equation to separate the variables.

$$\frac{dy}{dx} = 5y^2 + 9x^2y^2 \Rightarrow \frac{dy}{dx} = y^2(5 + 9x^2)$$

$$\Rightarrow \frac{dy}{y^2} = (5 + 9x^2)\,dx$$

$$\Rightarrow -\frac{1}{y} = 5x + 3x^3 + c$$

$$\Rightarrow y = \frac{-1}{5x + 3x^3 + c}$$

This is a general solution of the differential equation. In this case we are able to express this function in explicit form.

## Example 20.5

Solve $\dfrac{dy}{dx} = \dfrac{3x^2y}{1 + 4y^2}$

### Solution

First separate the variables

$$\frac{1 + 4y^2}{y}\,dy = 3x^2\,dx$$

then integrate both sides

$$\int \frac{1 + 4y^2}{y}\,dy = \int 3x^2\,dx \Rightarrow \int\left(\frac{1}{y} + 4y\right)dy = \int 3x^2\,dx$$

$$\ln|y| + 2y^3 = x^3 + c$$

The solution is exact, but implicit, as it cannot be written in an explicit form $y = f(x)$. Some of the solution curves for a few values of $c$ are shown here:

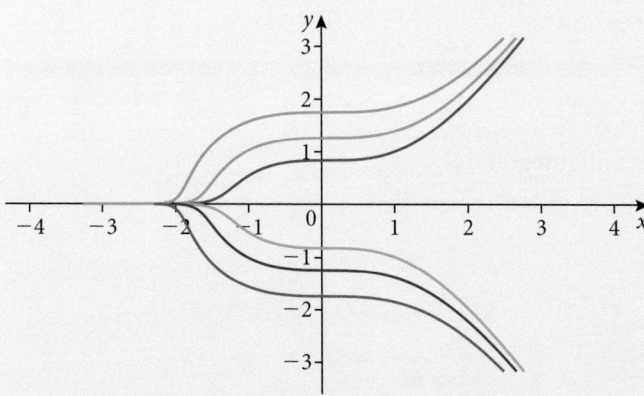

**To solve equations by separating the variables:**

- Write the differential equation in the standard form $\dfrac{dy}{dx} = f(x, y)$
- Can you separate the variables? That is, is $\dfrac{dy}{dx} = g(x)\,h(y)$ or $\dfrac{dy}{dx} = \dfrac{p(x)}{q(y)}$
- If so, separate the variables, to get $\dfrac{dy}{h(y)} = g(x)\,dx$ or $q(y)\,dy = p(x)\,dx$
- Integrate both parts, to get $\displaystyle\int \dfrac{dy}{h(y)} = \int g(x)\,dx + c$ or $\displaystyle\int q(y)\,dy = \int p(x)\,dx + c$
- Evaluate the integrals if you can and don't forget the arbitrary constant. Even though we have two integrals, one on the left and one on the right, you can combine both arbitrary constants into one.
- If possible, resolve the resulting equation with respect to $y$, to obtain your equation in explicit form $y = f(x)$.

## Example 20.6

Find the general solution of the differential equation
$$x^2 y \frac{dy}{dx} = x + 1,\ x > 0,\ y > 0$$

### Solution

The equation is separable because you can rearrange the equation as:

$$\frac{dy}{dx} = \frac{1}{y}\left(\frac{x+1}{x^2}\right) \text{ which is in the form } \frac{dy}{dx} = p(x)\,q(y)$$

Now separate the variables and integrate:

$$y\,dy = \frac{x+1}{x^2}\,dx \Rightarrow \int y\,dy = \int \frac{x+1}{x^2}\,dx$$

$$\Rightarrow \frac{1}{2}y^2 = \ln x - \frac{1}{x} + c$$

This is the general solution of the differential equation in implicit form. In this case, we can also find the explicit form

$$y = \sqrt{2\ln x - \frac{2}{x} + c}$$

## Example 20.7

Solve the initial value problem $\dfrac{dy}{dx} = \dfrac{y}{x+1}$, $y(1) = 4$

### Solution

Separate the variables and integrate:

$$\frac{dy}{dx} = \frac{y}{x+1}$$

$$\Rightarrow \frac{dy}{y} = \frac{dx}{x+1}$$

$$\Rightarrow \int \frac{dy}{y} = \int \frac{dx}{x+1}$$

$$\ln|y| = \ln|x + 1| + c$$
$$\Rightarrow |y| = e^{\ln|x+1|+c} = e^{\ln|x+1|}e^c = |x + 1|e^c$$

Now, since $c$ is an arbitrary constant, we can replace $e^c$ with a constant $C$, and our solution becomes:

$$|y| = C|x + 1|$$

Using the initial condition:

$$4 = C|1 + 1| \Rightarrow C = 2$$

and the particular solution is $|y| = 2|x + 1|$

that is $y = \pm 2(x + 1)$

## Example 20.8

Solve the differential equation $x\,dx + e^{(x+y)}\cos y\,dy = 0$

### Solution

The equation can be separated with a few steps:

$$x\,dx = e^y \cos y\,dy \Rightarrow e^y \cos y\,dy = -xe^{-x}\,dx$$
$$\Rightarrow \int e^y \cos y\,dy = -\int xe^{-x}\,dx$$

Using integration by parts and simplifying we get the implicit general solution:

$$\Rightarrow \frac{1}{2}e^y(\sin y + \cos y) = e^{-x}(x + 1) + c$$

## Differential equations as mathematical models

The following examples illustrate typical cases where scientific principles are translated into differential equations.

## Example 20.9

Find the general solution of the basic population growth model

$$\frac{dP}{dt} = kP$$

Population growth rate in cases where the birth and death rates are not variable is proportional to the size of the population. That is $\frac{dP}{dt} = kP$ where $k$ is a constant.

### Solution

Separate the variables:

$$\frac{dP}{P} = k\,dt$$

Integrate both sides to get

$$\int \frac{dP}{P} = \int k\,dt$$

$$\Rightarrow \ln|P| = kt + c$$

where $c$ is an arbitrary constant. This last equation can be simplified to give an explicit expression for $P$:

$$\ln|P| = kt + c$$

$$\Rightarrow |P| = e^{kt+c} = e^{kt}e^c = Ae^{kt}$$

where $e^c = A$, thus

$$P = Ae^{kt} \text{ or } P = -Ae^{kt}$$

This is the general solution. All solutions to this problem will be in this form.

If the constant $k$ is positive, the model describes population growth; if it is negative, it describes population decay.

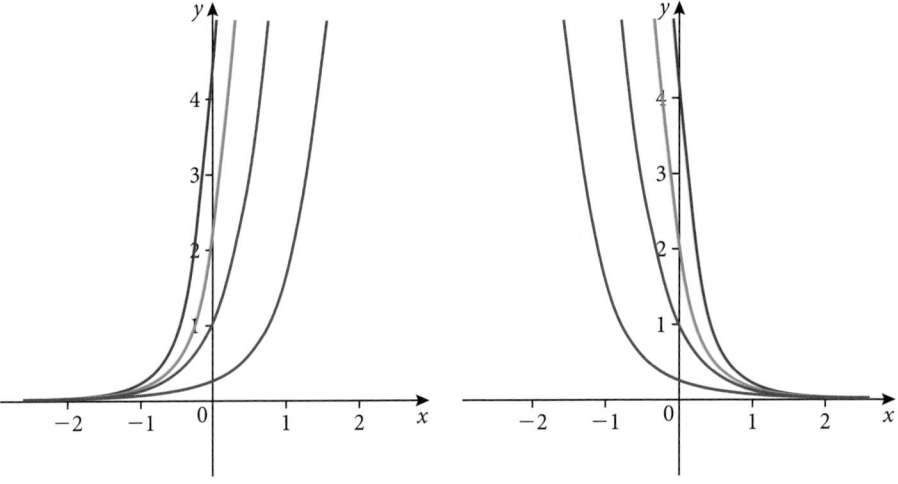

**Figure 20.1** Population growth

**Figure 20.2** Population decay

The first graph corresponds to positive values of $k$ (Figure 20.1) and the second to negative values of $k$ (Figure 20.2).

If the population growth model in Example 20.9 had the additional initial value that at $t_0$ the population is $P_0$, then this particular population satisfies

$$P = Ae^{kt}$$

and hence

$$P_0 = Ae^{kt_0} \Rightarrow A = \frac{P_0}{e^{kt_0}} = P_0 e^{-kt_0}$$

and the solution to the initial value problem is

$$P = Ae^{kt} = P_0 e^{-kt_0}e^{kt} = P_0 e^{k(t-t_0)}$$

There is a very important special case when $t_0 = 0$, the solution becomes

$$P = P_0 e^{k(t-t_0)} = P_0 e^{kt}$$

This is the usual growth model which starts at time $t = 0$ with initial population $P_0$.

---

## Example 20.10

A cold object is placed in a warmer medium that is kept at a constant temperature $S$. The rate of change of the temperature $T(t)$ with respect to time is proportional to the difference between the surrounding medium and the object and hence it satisfies

$$\frac{dt}{dt} = k(S - T) \text{ and } T(0) = T_0$$

where $k > 0$ and $T_0 < S$, i.e., the initial temperature is less than the temperature of the surrounding medium. Find the solution to the initial value problem.

Newton's law of cooling states that the rate of change of the temperature $T$ of an object is proportional to the difference between $T$ and the temperature of the surrounding medium $S$. That is $\frac{dT}{dt} = k(T - S)$ where $k$ is a constant and $S$ is usually considered constant.

### Solution

The variables can be separated, so:

$$\frac{dT}{dt} = k(S - T) \Leftrightarrow \frac{dT}{S - T} = k \, dt$$

Integrate and find the general solution:

$$\int \frac{dt}{S - T} = \int k \, dt \Rightarrow -\ln\left|S - T\right| = kt + c \Rightarrow \ln\left|S - T\right| = -kt - c$$

where $c$ is an arbitrary constant. Now since we know that the temperature $T$ is less than the surrounding temperature, then

$$\ln\left|S - T\right| = \ln(S - T)$$

The general solution then is

$$\ln(S - T) = -kt - c$$
$$\Rightarrow S - T = e^{-kt-c}$$
$$\Rightarrow T = S - e^{-kt-c}$$

The initial condition implies

$$T = S - e^{-kt-c}$$
$$\Rightarrow T_0 = S - e^{0-c}$$
$$\Rightarrow e^{-c} = S - T_0$$
$$\Rightarrow c = -\ln(S - T_0)$$

Therefore, substituting this value in the general solution

$$\ln(S - T) = -kt - c = -kt + \ln(S - T_0)$$
$$\Rightarrow \ln(S - T) - \ln(S - T_0) = -kt$$
$$\Rightarrow \ln\left(\frac{S - T}{S - T_0}\right) = -kt$$
$$\Rightarrow \frac{S - T}{S - T_0} = e^{-kt}$$
$$\Rightarrow S - T = (S - T_0)e^{-kt}$$
$$\Rightarrow T = S - (S - T_0)e^{-kt}$$

This is an example of limited growth. This is because the maximum value that $T$ can achieve is $S$. For example, if a can of soda is taken out of a fridge and left in a room with constant temperature of 21°C, the temperature of the soda will increase to reach the room temperature. In fact, since $k > 0$ and $S$ is a constant, then

$$T = S - (S - T_0)e^{-kt} \Rightarrow \frac{dT}{dt} = (S - T_0)e^{-kt}$$

Also, since $T_0 < S$, then

$$\frac{dT}{dt} = (S - T_0)e^{-kt} > 0$$

The temperature will always increase. However, as time passes,

$$\lim_{t \to \infty} e^{-kt} = 0 \Rightarrow \lim_{t \to \infty} T = \lim_{t \to \infty}(S - (S - T_0)e^{-kt}) = S$$

The diagram shows how the temperature climbs to 21°C but does not exceed it.

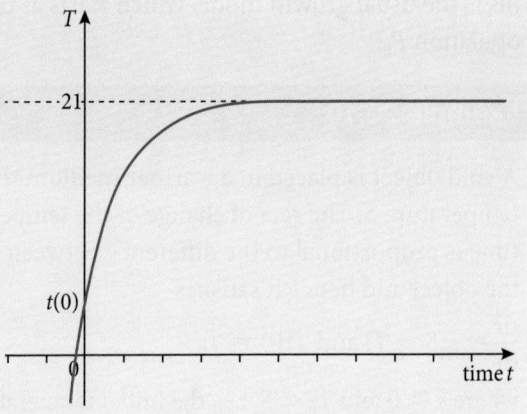

## Example 20.11

The rate of decay of a substance $y$ at any time $t$ is directly proportional to the amount of $y$ and also directly proportional to the amount of another substance $x$. The constant of proportionality is $-\dfrac{1}{2}$ and the value of $x$ at any time $t$ is given by $x = \dfrac{4}{(1 + t)^2}$

(a) Given the initial conditions that $y = 10$ when $t = 0$, find $y$ as an explicit function of $t$.

(b) Determine the amount of the substance remaining as $t$ becomes very large.

## Solution

(a) The rate of decay of substance $y$ is proportional to the product $xy$, and with the constant of proportionality having a value of $-\dfrac{1}{2}$ and $x = \dfrac{4}{(1 + t)^2}$ which gives

$$\frac{dy}{dt} = -\frac{1}{2}\left(\frac{4}{(1 + t)^2}\right)y$$

Separating the variables gives:

$$\frac{dy}{y} = \frac{-2}{(1 + t)^2}dt \Rightarrow \int \frac{dy}{y} = -2\int \frac{1}{(1 + t)^2}dt$$

$$\Rightarrow \ln y = \frac{2}{1 + t} + c$$

$$e^{\ln y} = e^{\frac{2}{1+t}+c} \Rightarrow y = e^{\frac{2}{1+t}}e^c = Ae^{\frac{2}{1+t}}$$

Now, $y = 10$ when $t = 0$, so:

$$10 = Ae^{\frac{2}{1+0}} \Rightarrow A = 10e^{-2}$$

Therefore the explicit form is $y = Ae^{\frac{2}{1+t}} = 10e^{-2}e^{\frac{2}{1+t}} = 10e^{\frac{-2t}{1+t}}$

(b) $\displaystyle\lim_{t \to \infty}\frac{-2t}{1 + t} = -2 \Rightarrow y = 10e^{-2} \approx 1.36$

## Logistic differential equations

Example 20.9 looked at the exponential growth model. In that model, we used the fact that the rate of change of a variable $y$ ($P$ in the example) is proportional to the value of $y$. We observed that the differential equation is of the form $\dfrac{dy}{dt} = ky$ and found that the general solution was $y = Ae^{kt}$

Exponential growth is unlimited and does not provide an accurate model for the growth of a population over a long time period. To obtain a more realistic model we need to take account of the fact that as the population increases, quite a few factors will begin to affect the growth rate. For example, there will be increased competition for the limited resources that are available, increases in disease, and overcrowding of the limited available space, all of which would serve to slow the growth rate. In order to model this situation mathematically, we amend the differential equation leading to the modest exponential growth law we introduced by addition of a term that slows the growth down as the population increases. If we consider a closed environment (neglecting factors such as immigration and emigration), then the rate of change of population can be modelled by the differential equation

$$\frac{dy}{dt} = (b - d)y$$

where $b$ and $d$ denote the birth rate and death rate.

In a more general situation, the increased competition as the population grows will result in a corresponding increase in the death rate. One way to take this into account is to assume that the death rate is directly proportional to the present population, and that the birth rate remains constant. The resulting initial-value problem leading to the population growth can now be written as

$$\frac{dy}{dt} = (b_0 - d_0 y)y$$

where $b_0$ and $d_0$ are positive constants representing specific birth and death rates. It is useful to write the differential equation in the equivalent form

$$\frac{dy}{dt} = k\left(1 - \frac{y}{L}\right)y \quad (1)$$

where $k = b_0$, and $L = \dfrac{b_0}{d_0}$

Equation (1) is called the **logistic equation**, and the corresponding population model is called the **logistic model**. The differential equation (1) is separable and can be easily solved.

Before doing that, however, we give a qualitative analysis of the differential equation. The constant $L$ is called the **carrying capacity** of the population.

We see from Equation (1) that if $y < L$, then $\dfrac{dy}{dt} > 0$ and the population increases, whereas if $y > L$, then $\dfrac{dy}{dt} < 0$ and the population decreases.

We can therefore interpret $L$ as representing the maximum population that the environment can sustain.

The graph of the function $y$ is called the **logistic curve**.

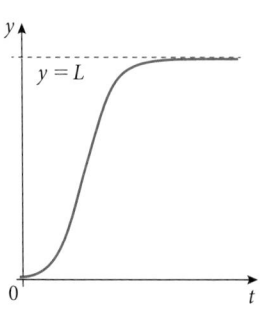

**Figure 20.3** Logistic curve

### Example 20.12

Solve the logistic differential equation $\dfrac{dy}{dt} = ky\left(1 - \dfrac{y}{L}\right)$

#### Solution

First separate the variables:

$$\frac{dy}{dt} = ky\left(1 - \frac{y}{L}\right) \Rightarrow \frac{dy}{y\left(1 - \dfrac{y}{L}\right)} = k\,dt$$

Then integrate both sides using partial fractions:

$$\int \frac{dy}{y\left(1 - \dfrac{y}{L}\right)} = \int k\,dt \Rightarrow \int\left(\frac{1}{y} + \frac{1}{L-y}\right)dy = \int k\,dt$$

$$\Rightarrow \ln|y| - \ln|L - y| = kt + C$$

$$\Rightarrow \ln\left|\frac{L-y}{y}\right| = -kt - C \Rightarrow \frac{L-y}{y} = \pm e^{-C}e^{-kt} = be^{-kt}$$

$$\Rightarrow y = \frac{L}{1 + be^{-kt}}$$

For population growth models, we also have an alternative form for the logistic equation. We replace $y$ by $P(t)$ and impose the initial condition that $P(0) = P_0$. We can conclude that solutions of the logistic population model may be put in the general form

$$\frac{dP}{dt} = rP(L - P),\ P(0) = P_0 \Rightarrow P(t) = \frac{LP_0}{P_0 + (L - P_0)e^{-rLt}}$$

This formula enables us to make an important observation:
In the long term, as $t \to \infty$, the population does indeed tend to the carrying capacity $L$.

$$\lim_{t\to\infty} P(t) = \lim_{t\to\infty} \frac{LP_0}{P_0 + (L - P_0)e^{-rLt}} = \frac{LP_0}{P_0} = L$$

### Example 20.13

An official wildlife commission releases 20 deer into a protected forest. After 5 years, the deer population is 208. The commission believes that the environment cannot support more than 400 deer.

(a)  Write a model for the deer population in terms of the number of years, $t$.

(b)  Use the model to estimate the deer population after 10 years.

(c)  Discuss the long-term trend of the deer population.

#### Solution

The growth of the deer population $P$ is

$$\frac{dP}{dt} = rP(400 - P),\quad 20 \leqslant P \leqslant 400$$

where $t$ is the number of years.

(a) We know that $L = 400$. Thus, the solution of the equation is of the form

$$P = \frac{400 \times 20}{20 + (400 - 20)\,e^{-400rt}}$$

Because $P = 208$ when $t = 5$, we can solve for $r$

$$208 = \frac{400 \times 20}{20 + 380e^{-5 \times 400r}} \Rightarrow r \approx 0.001512$$

Therefore, the model for this deer population is

$$P(t) = \frac{400 \times 20}{20 + (400 - 20)\,e^{-400 \times 0.001512 \times t}} = \frac{400}{1 + 19e^{-0.6049t}}$$

(b) To estimate the deer population after 10 years, substitute 10 for $t$ in the

model $p = \dfrac{400}{1 + 19e^{-0.6049 \times 10}} \approx 383$

(c) As $t$ increases without bound, the denominator in the model will approach 1 and so:

$$\lim_{t \to \infty} p = \lim_{t \to \infty} \frac{400}{1 + 19e^{-0.6049t}} = 400$$

## Differential equations reducible to variables separable

Some differential equations that are not separable in $x$ and $y$ can be made separable by a change of variables. This is true for differential equations of the form $y' = f(x, y)$, where $f$ is a **homogeneous function**. The function given by $f(x, y)$ is **homogeneous of degree $n$** if $f(tx, ty) = t^n f(x, y)$, where $n$ is an integer.

For example

- $f(x, y) = 2x^2y - 5x^3 + 7xy^2$ is homogeneous of degree 3 because

$$f(tx, ty) = 2(tx)^2(ty) - 5(tx)^3 + 7(tx)(ty)^2$$
$$= t^3(2x^2y - 5x^3 + 7xy^2) = t^3 f(x, y)$$

- $f(x, y) = 2x^2y - 5x^2y^2 + 7xy$ is not homogeneous because
$f(tx, ty) = t^2(2tx^2y - 5t^2x^2y^2 + 7xy) \neq t^n(2x^2y - 5x^2y^2 + 7xy)$

- $f(x, y) = \dfrac{4x}{7y}$ is a homogeneous function of degree 0 because

$$f(tx, ty) = \frac{4tx}{7ty} = t^0 \frac{x}{y}$$

A differential equation can be reduced to separable if it is of the form
$M(x, y)\,dx + N(x, y)\,dy = 0$
where $M$ and $N$ are homogeneous functions of the same degree.

Equivalently, when a first order differential equation can be written as $y' = f(x, y)$, then it is reducible if the function $f$ has the property $f(tx, ty) = f(x, y)$ for all $t \neq 0$. This condition is equivalent to saying that $f$ can be expressed as a function of $\dfrac{y}{x}$. For if $f$ is reducible, then

$$f(x, y) = f\left(x, x\left(\frac{y}{x}\right)\right) = f\left(1, \frac{y}{x}\right) = g\left(\frac{y}{x}\right), \quad (x \neq 0)$$

For example

- $(3x^2 - 2xy)\,dx + 2y^2 = 0$ is reducible.

- $(2x^2 + 3)\,dx + 3y^2\,dy = 0$ is not reducible.

A differential equation can be transformed into a variables separable differential equation by substituting $y = vx$ (or $x = uy$), where $v$ is a differentiable function of $x$. Using this substitution, $dy = v\,dx + x\,dv$ and with a few steps of algebraic manipulation, it is possible to express the resulting equation as a separable equation in $x$ and $v$, as seen in Examples 20.14 and 20.15.

### Example 20.14

Solve the differential equation $xyy' = 3x^2 + y^2$

#### Solution

This equation is reducible since:
$$y' = f(x, y) = \frac{3x^2 + y^2}{xy} \Rightarrow f(tx, ty) = \frac{3(tx)^2 + (ty)^2}{(tx)(ty)} = \frac{3x^2 + y^2}{xy} = y'$$

Several approaches are possible, one of which is to substitute $y = vx$ and $dy = v\,dx + x\,dv$ directly into the given equation:

$$y' = \frac{dy}{dx} = \frac{3x^2 + y^2}{xy} \Rightarrow \frac{v\,dx + x\,dv}{dx} = \frac{3x^2 + v^2x^2}{x(vx)} = \frac{3 + v^2}{v}$$

$$\Rightarrow v^2\,dx + xv\,dv = 3\,dx + v^2\,dx \Rightarrow v\,dv = \frac{3}{x}\,dx \Rightarrow \int v\,dv = \int \frac{3}{x}\,dx + C$$

$$\Rightarrow \frac{v^2}{2} = 3\ln|x| + C \Rightarrow v^2 = 6\ln|x| + 2C$$

By substituting the value of $v$ back and simplifying

$$y^2 = 6x^2\big(6\ln|x| + 2C\big)$$

### Example 20.15

Find the general solution of $(x^2 - y^2)\,dx + 3xy\,dy = 0$

#### Solution

Since $(x^2 - y^2)$ and $3xy$ are reducible, we can use the method suggested above. Substitute $y = vx$ and $dy = v\,dx + x\,dv$ into the given equation

$$(x^2 - y^2)\,dx + 3xy\,dy = 0 \Rightarrow (x^2 - x^2v^2)\,dx + 3x^2v(v\,dx + x\,dv) = 0$$

$$\Rightarrow x^2(1 + 2v^2)\,dx + 3x^3v\,dv = 0$$

Divide by $x^2$, simplify and integrate

$$(1 + 2v^2)\,dx + 3xv\,dv = 0 \Rightarrow \int \frac{dx}{x} = -3\int \frac{v\,dv}{1 + 2v^2}$$

$$\ln|x| = -\frac{3}{4}\ln(1 + 2v^2) + C_1 \Rightarrow 4\ln|x| = -3\ln(1 + 2v^2) + 4C_1$$

Now let $4C_1 = \ln C$ and simplify

$$\ln x^4 = \ln(C(1 + 2v^2)^{-3}) \Rightarrow x^4 = C(1 (+ 2)v\,2)^{-3}$$

Substitute $v = \dfrac{y}{x}$ into this equation and simplify

$$x^4 = C\left(1 + 2\frac{y^2}{x^2}\right)^{-3} \Rightarrow x^4\left(1 + 2\frac{y^2}{x^2}\right)^3 = C \Rightarrow (x^2 + 2y^2)^3 = Cx^2$$

## Exercise 20.1

1. (a) Show that every member of the family of functions $f(x) = \frac{1}{x}(\ln x + k)$ is a solution to the differential equation $x^2 \frac{dy}{dx} + xy - 1 = 0$

   (b) Write the equation in the form $\frac{dy}{dx} = F(x, y)$, and illustrate part (a) by graphing two members of the family on the same axes.

   (c) Find a solution that satisfies the initial conditions $y(1) = 2$ and $y(2) = 1$

2. Solve each differential equation.

   (a) $y'e^{y-x} = 1$

   (b) $\frac{dy}{dx} = y^2 x + x$

   (c) $e^{x-y} dy = x\,dx$

   (d) $y' = xy^2 - x - y^2 + 1$

   (e) $(xy \ln x)y' = (y + 1)^2$

   (f) $\frac{dy}{dx} = \frac{1 + 2y^2}{y \sin^2 x}$

   (g) $(1 + \tan y)y' = x^2 + 1$

   (h) $\frac{dy}{dt} = \frac{te^t}{y\sqrt{y^2 + 1}}$

   (i) $y \sec \theta\, dy = e^y \sin^2 \theta$

   (j) $\frac{dy}{dx} = e^x(1 + y^2)$

3. Solve each initial value problem.

   (a) $x^{-3} dy = 4y\,dx,\ y(0) = 3$

   (b) $\frac{dy}{dx} = xy,\ y(0) = 1$

   (c) $y' - xy^2 = 0,\ y(1) = 2$

   (d) $y' - y^2 = 0,\ y(2) = 1$

   (e) $\frac{dy}{dx} - e^y = 0,\ y(0) = 1$

   (f) $\frac{dy}{dx} = y^{-2}x + y^{-2},\ y(0) = 1$

   (g) $x\,dy - y^2\,dx = -dy,\ y(0) = 1$

   (h) $y^2\,dy - x\,dx = dx - dy,\ y(0) = 3$

   (i) $yy' = xy^2 + x,\ y(0) = 0$

   (j) $y' = \frac{xy - y}{y + 1},\ y(2) = 1$

   (k) $\frac{dy}{dx} = x\sqrt{\frac{1 - y^2}{1 - x^2}},\ y(0) = 0$

   (l) $y'(1 + e^x) = e^{x-y},\ y(1) = 0$

   (m) $(y + 1)\,dy = (x^2 y - y)\,dx,\ y(3) = 1$

   (n) $\cos y\,dx + (1 + e^{-x})\sin y\,dy = 0,\ y(0) = \frac{\pi}{4}$

   (o) $xy' - y = 2x^2 y,\ y(1) = 1$

   (p) $xy\,dx + e^{-x^2}(y^2 - 1)\,dy = 0,\ y(0) = 1$

   (q) $x \cos x = (2y + e^{3y})y',\ y(0) = 0$

   (r) $\frac{dy}{dx} = e^x - 2x,\ y(0) = 3$

4. Solve each equation using an appropriate substitution.

   (a) $2x\frac{dy}{dx} = x + y$

   (b) $(x + y)\,dy = (x - y)\,dx$

   (c) $(x^2 - y^2)y' = xy$

849

**(d)** $x\,dy - \left(2xe^{-\frac{y}{x}} + y\right)dx = 0, \quad y(1) = 0$

**(e)** $\left(x\sec\frac{y}{x} + y\right)dx = x\,dy, \quad y(1) = 0$ **(f)** $\left(x^2 + y^2\right)dx = \left(xy - x^2\right)dy$

**(g)** $y\,dx = \left(x + \sqrt{xy}\right)dy$

5. The temperature $T$ in °C of a kettle in a room satisfies the differential equation $\dfrac{dT}{dt} = m(T - 21)$, where $t$ is in minutes and $m$ is a constant.

   **(a)** Solve the differential equation showing that $T = Ce^{mt} + 21$, where $C$ is an arbitrary constant.

   **(b)** Given that $T(0) = 99$, and $T(15) = 69$, find:

   **(i)** the value of $m$ and $C$

   **(ii)** $t$ when $T = 39$

6. Studies indicate that there may be as few as 3200 tigers (Panthera tigris) left in the wild. The WWF arranged for the release of 25 tigers into a wildlife reserve. After 2 years there were 39 tigers in the reserve. The reserve has a capacity of 200 tigers.

   **(a)** Write a logistic equation that models the population of tigers in the reserve.

   **(b)** Find the population after 5 years.

   **(c)** When will the population reach 100?

7. The logistic equation $P(t) = \dfrac{2100}{1 + 29e^{-0.75t}}$ models the growth of a certain population.

   **(a)** Find the initial population.

   **(b)** Find when the population will reach 50% of its carrying capacity.

   **(c)** Write a logistic differential equation that has the solution $P(t)$.

8. The number of bacteria in a certain culture grows at a rate that is proportional to the number of bacteria present.

   The number increases from 500 to 2000 in two hours. Determine:

   **(a)** the number present after 12 hours

   **(b)** the doubling time.

9. In the year 2000, the population of a city near Antwerp in Belgium was 20 000. In 2010 it had increased to 50 870, and in 2015 it was 78 680.

   **(a)** Set up a logistic model to predict the population after $t$ years.

   **(b)** What is the maximum capacity that this city can have?

   **(c)** Predict the city's population in 2030.

10. The number of bacteria in a culture grows at a rate proportional to the number present. Initially there were 10 bacteria in the culture. Given that the doubling time of the culture is 3 hours, find the number of bacteria that are present after 24 hours.

11. A certain cell culture has a doubling time of 4 hours. Initially there were 2000 cells present. Assuming an exponential growth law, determine the time it takes for the culture to contain 1 000 000 cells.

12. The population in a small village in 2005 was 500. After 5 years this had grown to 800 and after 10 years to 1000. Using the logistic population model, determine the population in 2020.

13. Consider the following differential equation:
$$\frac{dy}{dt} = -by + d,$$ with $b$ and $d$ real positive constants.
    (a) Solve the differential equation.
    (b) Sketch the solution for several initial conditions.
    (c) Describe the behaviour of the solutions under each of these conditions:
        (i) $b$ increases but $d$ remains the same
        (ii) $d$ increases but $b$ remains the same
        (iii) both $b$ and $d$ increase but the ratio $d : b$ remains the same.

14. The heating system in a building fails during the time when the interior temperature reaches 21°C. The external temperature is $-12$°C.
    (a) Use Newton's law of cooling $\frac{dT}{dt} = -k(T - S)$ to find the interior temperature, $T$, at time $t$. $S$ is the surrounding's temperature, and $k$ is a constant that depends on several factors including insulation material. Consider $k$ to be 0.15 for this environment.
    (b) How long will it take the interior temperature to fall to 0°C?

## 20.2 More applications of differential equations

### Electric circuits

An electromotive force (usually supplied by a battery or generator) produces a potential difference of $E(t)$ volts (V) and a current of $I(t)$ amperes (A) at time $t$. The circuit also contains a resistor with a resistance of $R$ ohms (W) and an inductor with an inductance of $L$ henries (H).

Ohm's law gives the potential difference across the resistor as $RI$. The potential difference across the inductor is $L\left(\frac{dI}{dt}\right)$

One of Kirchhoff's laws says that the sum of the potential differences is equal to the supplied voltage $E(t)$.

Thus, we have $L\frac{dI}{dt} + RI = E(t)$

which is a first order linear differential equation. The solution gives the current $I$ at time $t$.

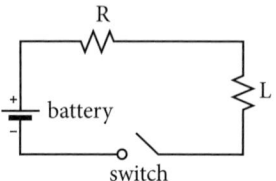

R

L

battery

switch

**Figure 20.4** Circuit diagram for Example 20.16

Multiply both sides of the model by $e^{3t}$ and simplify noting that the left-hand side is of the form $d(uv)$.

## Example 20.16

Suppose that in the simple circuit shown, the resistance is 12 W and the inductance is 4 H. The power supply provides a constant potential difference of 60 V and the switch is closed when $t = 0$ so the current starts with $I(0) = 0$. Find:

(a) $I(t)$

(b) the current after one second

(c) the limiting value of the current.

## Solution

(a) The situation is modelled by the differential equation $L\dfrac{\mathrm{d}I}{\mathrm{d}t} + RI = E(t)$

In this case $L = 4$, $R = 12$, and $E(t) = 60$

$$L\frac{\mathrm{d}I}{\mathrm{d}t} + RI = E(t) \Rightarrow 4\frac{\mathrm{d}I}{\mathrm{d}t} + 12I = 60 \Rightarrow \frac{\mathrm{d}I}{\mathrm{d}t} + 3I = 15, \quad I(0) = 0$$

Multiply by $e^{3t}$ and to solve the differential equation:

$$e^{3t}\frac{\mathrm{d}I}{\mathrm{d}t} + 3e^{3t}I = 15e^{3t} \Rightarrow \frac{\mathrm{d}}{\mathrm{d}t}(e^{3t}I) = 15e^{3t} \Rightarrow \int \mathrm{d}(e^{3t}I) = \int 15e^{3t}\mathrm{d}t$$

$$\Rightarrow e^{3t}I = 5e^{3t} + C \Rightarrow I = \frac{1}{e^{3t}}(5e^{3t} + C)$$

With the initial condition of $I(0) = 0$, we have $0 = 5 + C$, so $C = -5$

Therefore, $I(t) = \dfrac{1}{e^{3t}}(5e^{3t} - 5) = 5(1 - e^{-3t})$

(b) After 1 second the current is $I(1) = 5(1 - e^{-3}) \approx 4.75$ A

(c) The limiting value of the current is given by $\displaystyle\lim_{t\to\infty} I(t) = \lim_{t\to\infty} 5(1 - e^{-3t})$

$$= 5$$

## Example 20.17

Suppose that in the circuit in Example 20.16, the resistance and inductance remain constant. A generator produces a variable potential difference of $E(t) = 120 \sin 60t$ V and the switch is closed when $t = 0$ so the current starts with $I(0) = 0$. Find $I(t)$.

## Solution

Again, multiply both sides of the model by $e^{3t}$ and simplify noting that the left-hand side is of the form $d(uv)$.

The differential equation is now:

$$L\frac{\mathrm{d}I}{\mathrm{d}t} + RI = E(t) \Rightarrow 4\frac{\mathrm{d}I}{\mathrm{d}t} + 12I = 120\sin 60t \Rightarrow \frac{\mathrm{d}I}{\mathrm{d}t} + 3I = 30\sin 60t, \; I(0) = 0$$

$$e^{3t}\frac{\mathrm{d}I}{\mathrm{d}t} + 3e^{3t}I = 30e^{3t}\sin 60t \Rightarrow \frac{\mathrm{d}}{\mathrm{d}t}(e^{3t}I) = 30e^{3t}\sin 60t$$

$$\Rightarrow \int \mathrm{d}(e^{3t}I) = \int 30e^{3t}\sin 60t \, \mathrm{d}t$$

$$\Rightarrow I = \frac{1}{e^{3t}}\left(e^{3t}\left(\frac{10\sin 60t}{401} - \frac{200\cos 60t}{401}\right) + C\right)$$

$$= \frac{10}{401}(\sin 60t - 2\cos 60t) + Ce^{-3t}$$

With the initial condition of $I(0) = 0$, we have $0 = \dfrac{10}{401}(-2) + C \Rightarrow C = \dfrac{20}{401}$

Therefore, $I(t) = \dfrac{10}{401}(\sin 60t - 2\cos 60t) + \dfrac{20}{401}e^{-3t}$

$$= \frac{10}{401}(\sin 60t - 2\cos 60t + 2e^{-3t})$$

## Mixture problems

Consider a tank containing a solution of water and salt. As shown in Figure 20.5, there is an inflow of the solution at a rate of $r_{in}$ litres per second with a concentration $c_{in}$ of grams per litre. The amount of salt flowing in is therefore $r_{in} \times c_{in}$ grams.

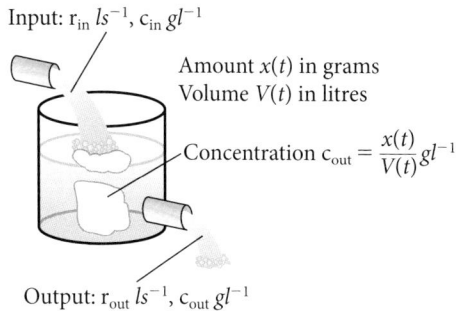

Figure 20.5 Tank containing a solution of water and salt

The amount of solution flowing out of the tank has a rate of $r_{out}$ litres per second with a concentration $c_{out}$ of grams per litre. The amount of salt leaving is therefore $r_{out} \times c_{out}$ grams. We want to find the amount $x(t)$ of salt in the tank at time $t$. We are given that the amount is $x(0) = x_0$ and the volume is $v(0) = v_0$ when $t = 0$.

The amount of salt in the tank at any time $t$ is therefore
$$\{\text{salt in}\} - \{\text{salt out}\} = r_{in} \times c_{in} - r_{out} \times c_{out}$$

That is to say the change in amount of salt is given by

$$\frac{\mathrm{d}x}{\mathrm{d}t} = r_{in} \times c_{in} - r_{out} \times c_{out}$$

with $r_{in}$, $r_{out}$ and $c_{in}$ being constants and $c_{out} = \dfrac{x(t)}{v(t)}$

Thus the amount of salt in the tank is the solution to the differential equation

$$\frac{\mathrm{d}x}{\mathrm{d}t} = r_{in} \times c_{in} - r_{out} \times \frac{x(t)}{v(t)} = r_{in} \times c_{in} - \frac{r_{out}}{v(t)}x(t)$$

where $v(t) = v_0 + (r_{in} - r_{out})t$

The equation here need not be memorised. It is the process we went through to set up the equation that you should strive to understand, because it is a useful tool for obtaining different differential equations representing real situations.

853

## Example 20.18

A tank contains 50 litres of a solution composed of 90% water and 10% hydrochloric acid. A second solution containing 50% water and 50% hydrochloric acid is being poured into the tank at a rate of 4 litres per minute. At the same time, the tank is being drained at a rate of 5 litres per minute.

(a) How much acid will there be in the tank after $t$ minutes?

(b) How much acid will there be in the tank after 10 minutes?

### Solution

Multiply both sides of the model by $\dfrac{1}{(50 - t)^5}$ and simplify, noting that the left-hand side is of the form $\mathrm{d}(uv)$.

Let $y$ be the number of litres of acid in the tank at any time $t$. Since acid is 10% of the contents, the number of litres in the tank is $y = 5$ when $t = 0$. Since the tank loses 1 litre of solution per minute, the number of litres of solution in the tank at any time $t$ is $50 - t$, and the tank loses 5 litres of solution per minute, it must lose $\dfrac{5}{50 - t}y$ litres of acid every minute.

Furthermore, the tank is also gaining 2 litres of acid per minute, the rate of change of acid in the tank is

$$\frac{\mathrm{d}y}{\mathrm{d}t} = 2 - \frac{5}{50 - t}y \Rightarrow \frac{\mathrm{d}y}{\mathrm{d}t} + \frac{5}{50 - t}y = 2$$

(a) Multiply both sides of the differential equation by $\dfrac{1}{(50 - t)^5}$

$$\frac{1}{(50 - t)^5}\left(\frac{\mathrm{d}y}{\mathrm{d}t} + \frac{5}{50 - t}y\right) = 2\left(\frac{1}{(50 - t)^5}\right)$$

$$\Rightarrow \left(\frac{1}{(50 - t)^5}\mathrm{d}y + \frac{5}{(50 - t)^6}y\,\mathrm{d}t\right) = 2\left(\frac{1}{(50 - t)^5}\right)\mathrm{d}t$$

$$\Rightarrow \int \mathrm{d}\left(\frac{1}{(50 - t)^5}y\right) = \int 2\left(\frac{1}{(50 - t)^5}\right)\mathrm{d}t$$

$$\Rightarrow \frac{1}{(50 - t)^5}y = \frac{1}{2(50 - t)^4} + c$$

So, the general solution is $y = \dfrac{50 - t}{2} + C(50 - t)^5$

Because $y = 5$ when $t = 0$, we have $C = -\dfrac{20}{50^5}$

Thus, there will be $y = \dfrac{50 - t}{2} - 20\left(\dfrac{50 - t}{50}\right)^5$

(b) When $t = 10$, the amount of acid is

$$y = \frac{50 - 10}{2} - 20\left(\frac{50 - 10}{50}\right)^5 \approx 13.45 \text{ litres}$$

## Population growth

Example 20.19

In 1940, the population of a country was 100 million and was then growing at the rate of 1 million per year.

A study in 1950 showed that the population growth rate was still 1 million per year and that the population was 110 million.

Predict this country's population for the year 2020 assuming that conditions did not change.

### Solution

The model has the same form discussed in Section 20.1

$$\frac{dP}{dt} = rP(L - P), P(0) = P_0$$

$$\Rightarrow P(t) = \frac{LP_0}{P_0 + (L - P_0)e^{-rLt}}$$

For 1940, $P = 100$ and $\frac{dP}{dt} = 1$

Substituting these values in the differential equation:

$$\frac{dP}{dt} = rP(L - P) \Rightarrow 1 = k \times 100(L - 100) \Rightarrow r = \frac{1}{100(L - 100)}$$

For 1950, $P = 110$ and $\frac{dP}{dt} = 1$

Substituting these values in the differential equation:

$$\frac{dP}{dt} = rP(L - P) \Rightarrow 1 = r \times 110(L - 110) \Rightarrow r = \frac{1}{110(L - 110)}$$

Solve the system of two equations: $L = 210$ million, and $r = 0.000091$

Considering 1940 to be the initial year, $P_0 = 100$, the model is then

$$P(t) = \frac{210 \times 100}{100 + 100e^{-0.000091 \times 210t}} = \frac{210}{1 + e^{-0.0191t}}$$

In 2020: $P(t) = \dfrac{200}{1 + e^{-0.0191 \times 60}} \approx 151.8$ million

## Falling objects

Another application is Newton's second law. We will consider an example which helps us solve problems concerning falling objects.

When an object falls through the air, there is force due to air resistance, known as drag, which is assumed to be proportional to velocity.

**Drag coefficients** are used in the design of cars and aircraft. For example, the drag coefficient of a Boeing 747 is 0.031 and that for Airbus 380 is 0.0265

Example 20.20

An object of mass $m = 5\,\text{kg}$ is allowed to fall near the Earth's surface.
The drag force has a magnitude $kv$, where $k$ is the drag coefficient.
The drag coefficient, $k$ of this object is 0.5. Assume $g = 9.8\,\text{m s}^{-2}$

(a) Develop a differential equation to calculate the velocity of the object at any time $t$.

(b) If the object is released from a height of 400 m, find how long it will take it to reach the ground and at what speed.

**Solution**

In fact, $v(t) = 98$ is a solution to the differential equation

$$v = 98 \Rightarrow \frac{\mathrm{d}v}{\mathrm{d}t} = 0$$
$$= 9.8 - 0.1 \times 98$$

$v(t) = 98$ is called an **equilibrium solution**. The approach in Example 20.20 can be applied to the general differential equation

$$m\frac{\mathrm{d}v}{\mathrm{d}t} = mg - kv$$

by dividing by $m$ and solving the resulting equation and simplifying, we get

$$v(t) = \frac{mg}{k} + Ce^{-\frac{kt}{m}}$$

which gives $\lim\limits_{t\to\infty} v(t)$

$$= \lim_{t\to\infty}\left(\frac{mg}{k} + Ce^{-\frac{kt}{m}}\right)$$
$$= \frac{mg}{k} + 0 \text{ leading}$$

to the equilibrium

solution $v(t) = \dfrac{mg}{k}$

Alternatively, without solving the differential equation we can argue along the same lines in the example, i.e.,

$$\frac{\mathrm{d}v}{\mathrm{d}t} = 0 \Rightarrow m\frac{\mathrm{d}v}{\mathrm{d}t}$$
$$= mg - kv = 0$$
$$\Rightarrow v = \frac{mg}{k}$$

(a) The net force on the object is the difference between its weight, i.e., due to gravity, $mg$ and drag force, $kv$.

$$F = m\frac{\mathrm{d}v}{\mathrm{d}t} = mg - kv$$

Substitute the given values

$$5\frac{\mathrm{d}v}{\mathrm{d}t} = 5 \times 9.8 - 0.5v \Rightarrow \frac{\mathrm{d}v}{\mathrm{d}t} = 9.8 - 0.1v$$

Solve by integrating

$$\frac{\mathrm{d}v}{\mathrm{d}t} = 9.8 - 0.1v \Rightarrow \int\frac{\mathrm{d}v}{9.8 - 0.1v} = \int \mathrm{d}t \Rightarrow -10\ln|9.8 - 0.1v| = t + c$$

With a few steps of algebraic manipulation, the solution can be reduced to be of the form

$v = 98 + Ce^{\frac{-t}{10}}$, where $C$ is an arbitrary constant.

To find $C$, substitute the initial condition that $v = 0$ when the object is released, i.e., $t = 0$.

$0 = 98 + Ce^{\frac{-0}{10}} \Rightarrow C = -98$, and so, the solution is $v = 98 - 98e^{\frac{-t}{10}}$

Graphs of the general solution are shown in the diagram, with the particular solution shown by the blue curve. Note how all solutions approach the equilibrium solution of $v = 98$

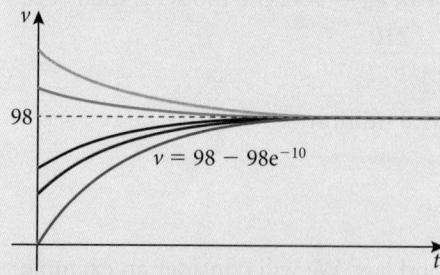

(b) We need to find the time it takes the object to fall 400 m. Distance, $s$, is related to velocity by $\dfrac{\mathrm{d}s}{\mathrm{d}t} = v$, or $\dfrac{\mathrm{d}s}{\mathrm{d}t} = 98 - 98e^{\frac{-t}{10}}$

Integrating both sides:

$s(t) = 98t + 980e^{\frac{-t}{10}} + c$

Since the object starts to fall when $t = 0$, then we know that $s = 0$

Thus $c = -980$ and the distance fallen at any time $t$ is

$s(t) = 98t + 980e^{\frac{-t}{10}} - 980$

The object hits the ground when it has fallen 400 m. To find the time $t$, solve the equation:

$$400 = 98t + 980e^{\frac{-t}{10}} - 980$$

The value can be approximated using a GDC, with the result $t \approx 10.64$ s

The speed is then $v \approx 98 - 98e^{\frac{-10.64}{10}} = 64.2 \text{ m s}^{-1}$

## Exercise 20.2

1. All radioactive substances decay at a rate proportional to the amount of the substance that exists at any time. The half-life of radium is 1620 years. How much (accurate to 3 significant figures) of a 10-gram specimen of the radioactive radium will remain after 25 years?

2. The rate, in degrees Celsius per minute at which the temperature of a cup of tea decreases is given by $-r(T - 20)$ where $r$ is a constant. When the tea is poured into the cup its temperature is 70°C, and after 10 minutes it goes down to 50°C.

   (a) Find an equation for the temperature in terms of time $t$.

   (b) When does the tea reach room temperature of 21°C?

3. The brakes of a car are applied when it is travelling at 120 km h$^{-1}$ and provide a deceleration of 10 m s$^{-2}$. How far does the car travel before coming a stop?

4. The skid marks made by a car involved in an accident indicated that brakes were fully applied for a distance of 75 m before it came to a stop. The car is known to have a constant deceleration of 20 m s$^{-2}$ when the brakes are applied fully. Under these conditions, how fast, in km h$^{-1}$ was the car travelling when the brakes were first applied?

5. According to data listed in the United Nations Department of Economic and Social Affairs, the world's total population reached 7.6 billion people in 2017 and was then increasing at the rate of about 225 000 people each day. Assume that natural population growth continues at this rate.

   (a) What is the annual growth rate in billion persons per year?

   (b) What will the world population be at the beginning of the 22nd century?

   (c) Some demographers believe that the maximum population for which the planet can provide adequate resources is 60 billion. When will the population reach 60 billion?

6. A specimen of charcoal found at a monument site turns out to contain 65% as much $^{14}$C as a sample of present-day charcoal of equal mass. The rate of decay of $^{14}$C is known to be 0.0001216 grams per year. What is the age of the sample?

$^{14}$C is a common way of writing Carbon 14, a radioactive isotope of carbon.

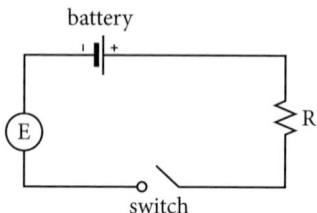

battery

E

R

switch

**Figure 20.6** Circuit diagram for question 7

Multiply both sides of the differential equation (1) by $e^{4t}$ and note that the left side of the equation is of the form $d(uv)$ and integrate.

7. The diagram shows a circuit containing an electromotive force, a capacitor with a capacitance of $C$ farads (F), and a resistor with a resistance of $R$ ohms ($\Omega$). The potential difference across the capacitor is $\frac{Q}{C}$, where $Q$ is the charge (in coulombs). In this case Kirchhoff's law gives $RI + \frac{Q}{C} = E(t)$

Given that $I = \frac{dQ}{dt}$ we have

$$R\frac{dQ}{dt} + \frac{1}{C}Q = E(t) \quad (1)$$

Suppose the resistance is $5\,\Omega$, the capacitance is 0.05 F, a battery gives a constant potential difference of 60 V, and the initial charge is $Q(0) = 0\,C$. Find the charge and the current at time $t$.

8. The growth of bacteria in a culture is proportional to the number present. In a laboratory experiment, 10 000 bacteria were present at the start. After a few minutes, $t_1$ minutes, there were 20 000, and 10 minutes after that there were 100 000.

(a) In terms of $t$, find the number of bacteria in the culture at any time $t$ minutes.

(b) How many bacteria were there after 20 minutes?

(c) Work out the time $t_1$.

9. A chef in a restaurant places a 2-kilogram roast, initially at 10°C, in a 200°C oven. After 75 minutes it is found that the internal temperature of the roast is 52°C. When will the roast reach 70°C?

10. In 1990, the population of the German town of Tegel was 25 000 people. In the 2000 census, the population was 30 000 people. If the population continues growing at the same rate, what will its population be in 2030?

11. Veterinary physicians use sodium pentobarbital to anaesthetise dogs when treating them. A dog is anaesthetised when its bloodstream contains at least 45 mg of sodium pentobarbital per kilogram of the dog's body mass. The material is eliminated exponentially from the dog's bloodstream, with a half-life of 5 hours. What single dose should be administered in order to anaesthetise a 50-kilogram dog for one hour?

12. A certain piece of doubtful information about cadmium in the drinking water began to spread one day in a city with population of 100 000. Within 7 days, 10 000 people have heard the rumour. The rate of increase of the number who have heard the rumour is proportional to the number who have not yet heard it. How long will it be until half of the city has heard the rumour?

**13.** Experiments show that the intensity of light at a depth of $x$ metres below the surface of a lake, $I(x)$, varies proportionally with the intensity of light. The surface intensity of light, measured in lux, at a certain lake is 108 lux. At 25 m below the surface it is measured to be 10.8 lux.

    **(a)** Find the equation giving you the relationship between intensity, $I(x)$ and depth $x$.

    **(b)** A phytoplankton needs 8.1 lux to produce more energy than it uses. What is the maximum depth this organism can live?

**14.** For large falling objects, it is assumed that drag is proportional to the square of the velocity of the object.

    **(a)** Set up a differential equation to calculate the velocity of an object at any time $t$ under this assumption.

    **(b)** Determine the limiting velocity.

    **(c)** If $m = 10\,\text{kg}$, find the drag coefficient so that the limiting speed is $49\,\text{m s}^{-1}$.

**15.** The diagram shows an electric circuit containing a capacitor, resistor, and battery. The charge $Q(t)$ on the capacitor satisfies one of Kirchhoff's laws $RI + \dfrac{Q}{C} = V$ where $R$ is the resistance, $I$ is the intensity, $C$ is the capacitance, and $V$ is the constant potential difference supplied by the battery. It is known that $I = \dfrac{dQ}{dt}$

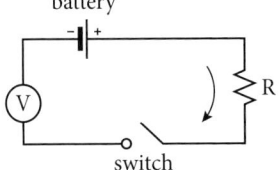

    **(a)** If $Q(0) = 0$, find $Q(t)$ at any time $t$, and sketch the graph of $Q$ versus $t$.

    **(b)** Find the limiting value $Q_L$ that $Q(t)$ approaches after a long time.

    **(c)** At time $t_1$, $Q(t)$ reaches a value very close to $Q_L$. The switch is opened, the battery is removed, the circuit closed again and the switch is closed. Find $Q(t)$ for $t > t_1$, sketch its graph and interpret what happens.

**Figure 20.7** Circuit diagram for question 15

Multiply both sides of the equation by $e^{(t/RC)}$ and note that the left side of the equation is of the form $d(uv)$ and then integrate.

**16.** At time $t = 0$, a tank contains $q_0$ kg of salt dissolved in 300 litres of water. Water containing 0.25 kg of salt per litre is being poured into the tank at the rate of $s$ litres per minute. The mixed solution is being drained out of the tank at the same rate.

    **(a)** Set up the initial value problem that describes this flow process.

    **(b)** Find the amount of salt $q(t)$ in the tank at any time $t$.

    **(c)** Find the equilibrium/limiting amount $q_L$ of salt present in the long run.

You need to multiply both sides of the differential equation with $e^{\frac{st}{300}}$, and then integrate.

# 20 Integral calculus 2

## 20.3 Numerical solutions: slope fields and Euler's method

### Slope fields

Often the primary objective when solving a first order differential equation is to find an explicit solution. However, many differential equations used in mathematical models cannot be solved by means of an analytical method. For such equations, we must resort to graphical and/or numerical methods. They can be carried out by hand or by technology and provide us with rough qualitative information about the graph of a solution to a differential equation.

A first order differential equation in the form $\dfrac{dy}{dx} = F(x, y)$ specifies the slope of the **solution curve** $y = f(x)$ at each point in the $xy$-plane where $F$ is defined. We can use this fact to draw a short line segment whose slope is $F(x, y)$ at any point $(x, y)$ in the plane. A plot of these line segments showing the slope (or direction) of the solution curve is a called a **slope field** (or direction field) for the first order differential equation. As a rule, the segments are drawn at representative points, evenly spaced in both directions. Figure 20.8 shows a slope field for the equation

$$\frac{dy}{dx} = x - y \quad (1)$$

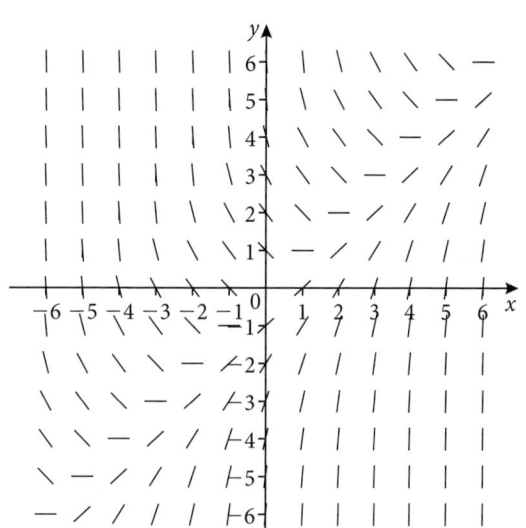

**Figure 20.8** Slope field for $\dfrac{dy}{dx} = x - y$

If a particular solution curve passes through a given point $(x_0, y_0)$, then its gradient (slope) at that point is $F(x_0, y_0)$. Thus, the direction (slope or gradient) field gives an indication of the behaviour of the family of solutions of a differential equation. Figure 20.9 shows the direction field for the differential equation **(1)** with a few members of the family sketched in red. One real particular solution 'through $(0, 0)$' is drawn in blue. At each point in the plane, the slope of the short line segment is $x - y$. For example, at $(0, 0)$ the segment is horizontal, at $(1, 3)$, the line has a slope $-2$. Usually, we do not draw all these segments by hand; instead, we use software.

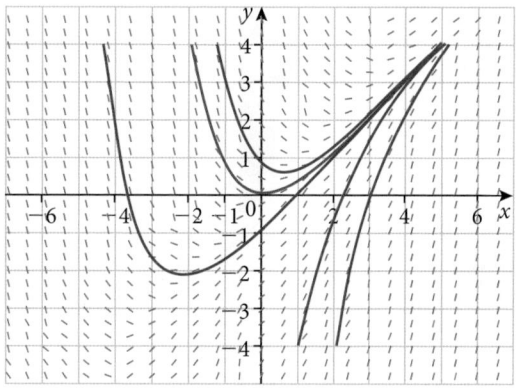

**Figure 20.9** Slope field with a few solutions

Solutions to a differential equation can be sketched by drawing in curves that are at each point tangent to the line segment at that point. Thus, a family of solution curves can be produced. To use a slope field to sketch a particular solution all we need to know is one point (an initial condition) that the solution curve passes through.

## Example 20.21

Draw a solution curve to the differential equation $y' = x + e^{-y}$ that passes through the point $(-4, 2)$.

## Solution

Using the slope field for this equation (created by computer), we draw the curve tangent to the line at $(-4, 2)$, shown in blue on the diagram, and continue using the neighbouring tangent lines as guides.

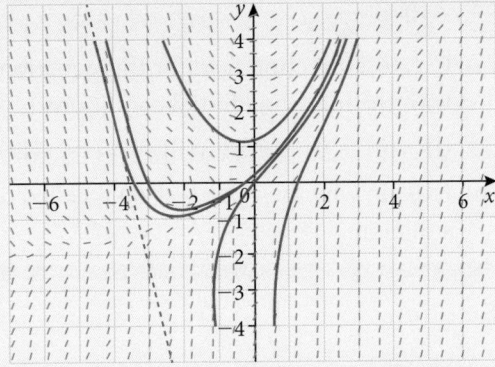

Note that a small change in the initial condition causes quite different behaviour.

## Example 20.22

A tennis ball is thrown straight downwards from a helicopter hovering at an altitude of 900 m. With what speed will hit the ground? Consider the air resistance proportionality factor to be $0.1$ kg s$^{-1}$, and the acceleration due to gravity to be $g = 9.8$ m s$^{-2}$

## Solution

If we consider the $y$-axis to be directed downwards, then the ball's velocity $v = \dfrac{dy}{dt}$ and the gravitational acceleration $g$ are both positive. The acceleration due to air resistance is negative and it is proportional to the velocity itself. Thus, the total acceleration is of the form $\dfrac{dv}{dt} = 9.8 - 0.1v$

We can estimate the speed from any altitude by looking at the slope field of the above equation.

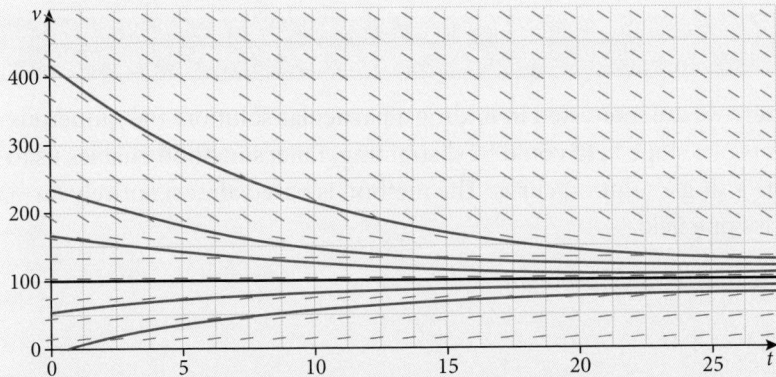

Note how all these solution graphs appear to approach the horizontal line $v = 100$ as an asymptote. This implies that, however you throw it, the tennis ball should approach the limiting velocity of 100 m s$^{-1}$.

## Example 20.23

The acceleration $\dfrac{dv}{dt}$, in metres per second squared, at time $t$ seconds of a 75 kg skydiver falling from an aircraft is given by the equation

$$\frac{dv}{dt} = 10 - \frac{v^2}{360}$$

(a) Create a direction field for this equation and sketch the solution curves with the following initial conditions.

    (i) $v(0) = 0$     (ii) $v(0) = 35$     (iii) $v(0) = 90$

(b) Explain why the value $v = 60$ is called the terminal velocity for this situation.

## Solution

(a) Sketches as shown.

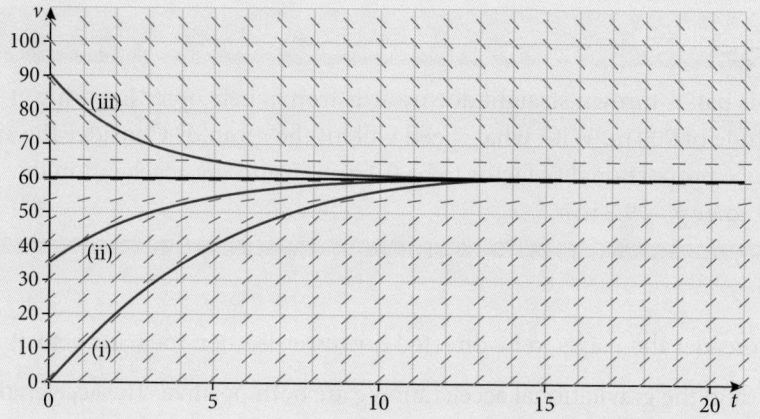

(b) From the slope field it appears that all solutions have a limiting value of 60 as $t$ gets very large. Due to increasing air resistance the skydiver reaches a maximum velocity, or terminal velocity, of 60 m s$^{-1}$.

## Euler's method

In cases where we are interested in finding a particular solution, the numerous line segments of a slope field could be distracting. Euler's method enables us to approximate a single solution curve. The method is based almost entirely on the idea of a slope field.

Consider the initial-value problem:

$$\frac{dy}{dx} = x + y, y(2) = 0$$

Figure 20.10 shows the slope field for the differential equation $\dfrac{dy}{dx} = x + y$

An approximation to the particular solution can be sketched by drawing a smooth curve through the point $(2, 0)$ that follows the slopes in the slope field.

Let $y(x)$ represent the solution curve. To approximate a value of $y$ for a specific value of $x$, for example $y$ when $x = 3$, we could make an educated guess from the sketch of $y$ made with the aid of the slope field. But if we want a more accurate approximation then we need to use a more refined method. The simplest numerical method is called **Euler's method** which uses the basic idea behind the construction of slope fields to find numerical approximations to solutions of differential equations. Let's illustrate the method with the initial-value problem that we have just been considering.

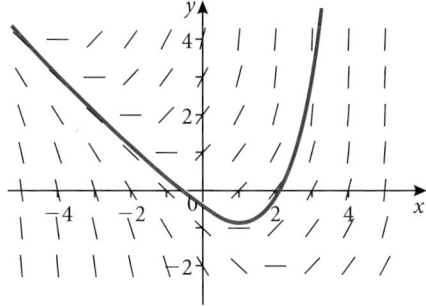

**Figure 20.10** Slope field with one solution

We know from the differential equation that the slope of the solution curve is 2 at the point $(2, 0)$ because

$$\frac{dy}{dx} = x + y = 2 + 0 = 2$$

Hence, the line tangent to the solution curve at $(2, 0)$ has the equation: $y - 0 = 2(x - 2) \Rightarrow y = 2x - 4$. We can use this tangent line as a rough approximation to the solution curve (see Figure 20.11). This approximation clearly becomes less accurate as we move away from the point of tangency $(2, 0)$.

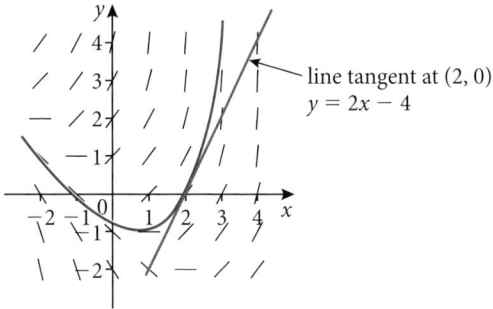

**Figure 20.11** Tangent line approximation at $(2, 0)$

Euler's method improves this approximation by moving a short horizontal distance (the **step size** $h$) along this tangent line and then changing direction according to the slope field. In this way we build an approximation to the curve by attaching little line segments together, each having the slope of the solution curve at its starting point.

In general, after being presented with an initial value problem $\frac{dy}{dx} = F(x, y)$, $y(x_0) = y_0$, we choose a step size $h$. Starting at the point $(x_0, y_0)$, for the interval $x_0 \leqslant x \leqslant x_0 + h$, we approximate the solution curve with the tangent line, i.e. the line with slope $F(x_0, y_0)$. This takes us as far as the point $(x_1, y_1)$, whose coordinates are calculated as follows:

$$x_1 = x_0 + h, \; y_2 = y_1 + hF(x_1, y_1)$$

Repeating this process we get an approximation to the solution curve consisting of line segments joining the points $(x_0, y_0)$, $(x_1, y_1)$, $(x_2, y_2)$, ....

Each computed value $y_n$ is an estimate of the corresponding true solution $y$ at $x = x_n$. The accuracy of the estimates depends on the choice of the step size $h$ and the overall number of steps (iterations). Decreasing the step size while increasing the number of steps leads to increasingly more accurate estimates for solution values.

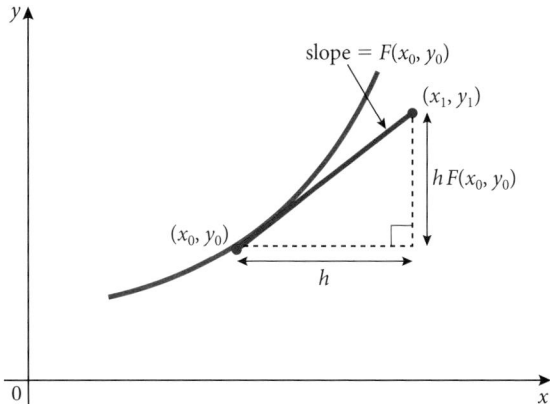

**Figure 20.12** One iteration of Euler's method

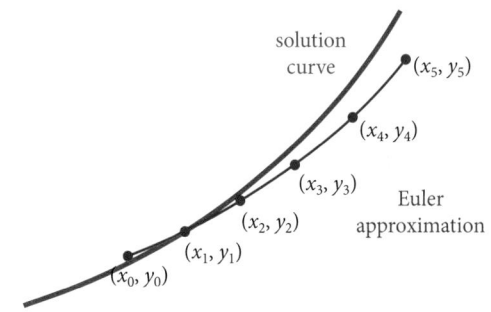

**Figure 20.13** Euler's approximation

## Euler's numerical method

Let's now apply Euler's method to solve the equation at the start of this section.

### Example 20.24

For the differential equation $\dfrac{dy}{dx} = x + y$ such that $y(2) = 0$, use Euler's method with a step value of 0.2 to find an approximate value of $y$ when $x = 3$, giving your answer to 2 decimal places.

**Solution**

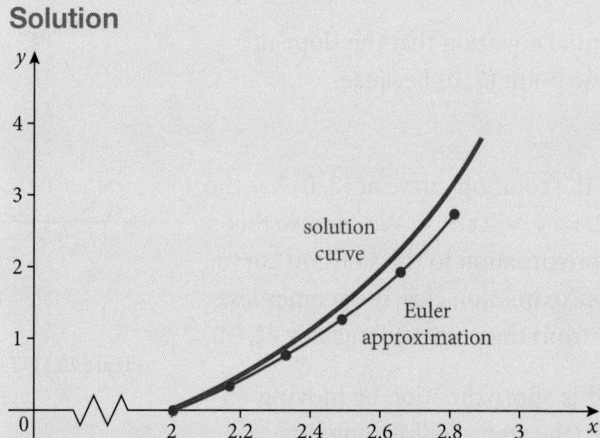

We use Euler's method to build an approximation to the true solution curve starting at $x = 2$ and finishing at $x = 3$ by piecing together five short segments. We are given that $h = 0.2$, $x_0 = 2$, $y_0 = 0$ and $F(x, y) = x + y$. Using the appropriate formulae for $x_n$ and $y_n$ and iterating five times, we have

$x_1 = x_0 + h = 2 + 0.2 = 2.2$ $\quad y_1 = y_0 + hF(x_0, y_0) = 0 + 0.2(2 + 0) = 0.4$

$x_2 = x_1 + h = 2.2 + 0.2 = 2.4$ $\quad y_2 = y_1 + hF(x_1, y_1) = 0.4 + 0.2(2.2 + 0.4) = 0.92$

$x_3 = x_2 + h = 2.4 + 0.2 = 2.6$ $\quad y_3 = y_2 + hF(x_2, y_2) = 0.92 + 0.2(2.4 + 0.92) = 1.584$

$x_4 = x_3 + h = 2.6 + 0.2 = 2.8$ $\quad y_4 = y_3 + hF(x_3, y_3) = 1.584 + 0.2(2.6 + 1.584) = 2.4208$

$x_5 = x_4 + h = 2.8 + 0.2 = 3$ $\quad y_5 = y_4 + hF(x_4, y_4) = 2.4208 + 0.2(2.8 + 2.4208) = 3.46496$

This process leads to a value of $y = 3.46$ (2 d.p.) when $x = 3$.

Because we will perform most of the calculations for each iteration on our GDC, it is often sufficient to simply display relevant results for each iteration in a table.

| $N$ | $x_n$ | $y_n$ | $h\,F(x_n, y_n)$ | $x_{n+1}$ | $y_{n+1}$ |
|---|---|---|---|---|---|
| 0 | 2 | 0 | 0.4 | 2.2 | 0.4 |
| 1 | 2.2 | 0.4 | 0.52 | 2.4 | 0.92 |
| 2 | 2.4 | 0.92 | 0.664 | 2.6 | 1.584 |
| 3 | 2.6 | 1.584 | 0.8368 | 2.8 | 2.4208 |
| 4 | 2.8 | 2.4208 | 1.04416 | 3.0 | 3.46496 |

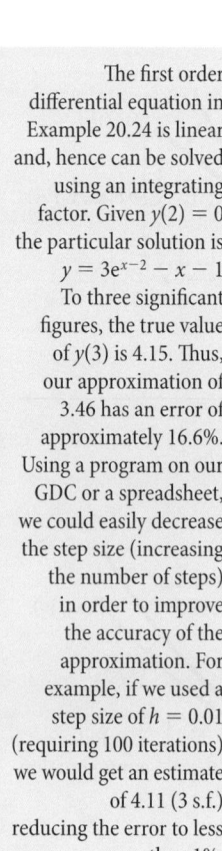

For the differential equation $\dfrac{dy}{dx} = F(x, y)$ with the initial condition $y(x_0) = y_0$, the recursive formulae for generating the coordinates of the unknown $(n + 1)$st point $(x_{n+1}, y_{n+1})$ from the known $n$th point $(x_n, y_n)$ on the approximate solution curve (Euler approximation) are: $x_{n+1} = x_n + h$, $y_{n+1} = y_n + h\,F(x_n, y_n)$, for $n = 1, 2, 3, \dots N$, where $h$, the step size, is a constant, and $N$ is the total number of steps (iterations).

The first order differential equation in Example 20.24 is linear and, hence can be solved using an integrating factor. Given $y(2) = 0$ the particular solution is $y = 3e^{x-2} - x - 1$ To three significant figures, the true value of $y(3)$ is 4.15. Thus, our approximation of 3.46 has an error of approximately 16.6%. Using a program on our GDC or a spreadsheet, we could easily decrease the step size (increasing the number of steps) in order to improve the accuracy of the approximation. For example, if we used a step size of $h = 0.01$ (requiring 100 iterations) we would get an estimate of 4.11 (3 s.f.) reducing the error to less than 1%.

A numerical method like Euler's is especially useful when applied to a differential equation that can't be solved by any known analytical methods.

## Example 20.25

Given that $\dfrac{dy}{dx} = \dfrac{x+1}{xy+2}$ and $y = 1$ when $x = 0$, use Euler's method with step size $h = 0.25$ to approximate the value of $y$ when $x = 1$. Give the approximation to 3 significant figures.

## Solution

We have that $x_0 = 0$, $y_0 = 1$, $h = 0.25$ and $F(x, y) = \dfrac{x+1}{xy+2}$

Thus the recursive formula for $y_n$ is

$$y_{n+1} = y_n + hF(x, y) = y_n + (0.25)\frac{x_n + 1}{x_n y_n + 2} \Rightarrow y_{n+1} = y_n + \frac{x_n + 1}{4x_n y_n + 8}$$

$n = 0$: $x_1 = x_0 + h = 0 + 0.25 = 0.25$

$$y_1 = y_0 + \frac{x_0 + 1}{4x_0 y_0 + 8} = 1 + \frac{0 + 1}{4(0)(1) + 8} = \frac{9}{8} = 1.125$$

$n = 1$: $x_2 = x_1 + h = 0.25 + 0.25 = 0.5$

$$y_2 = y_1 + \frac{x_1 + 1}{4x_1 y_1 + 8} = 1.125 + \frac{0.25 + 1}{4(0.25)(1.125) + 8} \approx 1.261986$$

$n = 2$: $x_3 = x_2 + h = 0.5 + 0.25 = 0.75$

$$y_3 = y_2 + \frac{x_2 + 1}{4x_2 y_2 + 8} = 1.261986 + \frac{0.5 + 1}{4(0.5)(1.261986) + 8} \approx 1.404518$$

$n = 3$: $x_4 = x_3 + h = 0.75 + 0.25 = 1$

$$y_4 = y_3 + \frac{x_3 + 1}{4x_3 y_3 + 8} = 1.404518 + \frac{0.75 + 1}{4(0.75)(1.404518) + 8} \approx 1.547801$$

Therefore, the approximate value of $y$ when $x = 1$ is $y \approx 1.55$

## Exercise 20.3

In all the exercises for this section a GDC, CAS, or computer is highly recommended.

1. Draw a direction field for each differential equation and determine the behaviour of the solution curves.

   (a) $\dfrac{dy}{dx} = 3 - 2y$      (b) $\dfrac{dy}{dx} = -1 - 2y$

   (c) $y' = 5 + 2y$      (d) $\dfrac{dy}{dx} = y(3 - y)$

2. Use Euler's method with step size 0.1 to estimate $y(0.5)$, where $y(x)$ is the solution of the initial-value problem $y' = y + xy$, $y(0) = 1$

3. **(a)** Use Euler's method and a GDC or computer to calculate $y(1)$, where $y(x)$ is the solution of the initial value problem

$$\frac{dy}{dx} = 6x^2 - 3x^2y, \quad y(0) = 3$$

Use these step sizes:

   **(i)** $h = 0.1$

   **(ii)** $h = 0.01$

   **(iii)** $h = 0.001$

   **(b)** Verify that $y = e^{-x^3} + 2$ is the exact solution of the differential equation.

   **(c)** Find the errors in using Euler's method to compute $y(1)$ with the step sizes in part **(a)**. What happens to the error when the step size is divided by 10?

4. Use Euler's method with step size 0.5 and four iterations to compute the approximate $y$-values of the solution of the initial-value problem

$$y' = 3x - 2y + 1, \quad y(1) = 2$$

5. Use Euler's method with step size 0.2 to estimate $y(1)$, where $y(x)$ is the solution of the initial-value problem $\dfrac{dy}{dx} = x + y^2, \quad y(0) = 0$

6. Use Euler's method with step size 0.1 to estimate $y(0.5)$, where $y(x)$ is the solution of the initial-value problem $\dfrac{dy}{dx} = x^2 + y^2, \quad y(0) = 1$

7. Use Euler's method to find approximate values for the solution of the given initial value problem at $t = 0.5$, 1.5, and 3, with:

   **(i)** $h = 0.1$      **(ii)** $h = 0.05$      **(iii)** $h = 0.01$

   **(a)** $\dfrac{dy}{dt} = y(3 - ty), \quad y(0) = 0.5$

   **(b)** $y' = 5 - 3\sqrt{y}, \quad y(0) = 2$

   **(c)** $\dfrac{dy}{dt} = \dfrac{4 - ty}{1 + y^2}, \quad y(0) = -2$

8. Copy and complete the table using an exact solution of the differential equation and two approximations obtained using Euler's method to approximate the particular solution of the equation.

   Use $h = 0.2$ and $h = 0.1$ and give your answer to 4 decimal places.

$$\frac{dy}{dx} - y = \cos x, \quad y(0) = 0$$

| $x$ | | | | | | |
|---|---|---|---|---|---|---|
| $y(x)$ [exact] | | | | | | |
| $y(x)$ [$h = 0.2$] | | | | | | |
| $y(x)$ [$h = 0.1$] | | | | | | |

9. The diagram in Figure 20.14 shows the slope field for a certain explicit first order differential equation.

   (a) Use this slope field to sketch the solution curve satisfying $y(0) = 1$

   (b) Given that $y(0) = 1$, use your answer to part (a) to estimate the value of $y(3)$

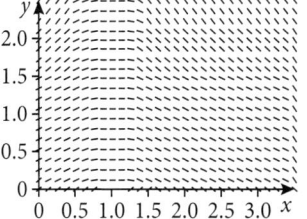

Figure 20.14 Diagram for question 9

10. Consider the differential equation:

$$\frac{dy}{dx} = \begin{cases} -1, & x < 1 \\ 0, & 1 \leqslant x \leqslant 2 \\ 1, & x > 2 \end{cases}$$

   (a) Sketch the slope field for this equation. Your sketch should include the range $0 \leqslant x \leqslant 4$ and $0 \leqslant y \leqslant 3$

   (b) Given that $y(0) = 1.5$, find the value of $y(4)$

11. Match each differential equation with the corresponding slope field.

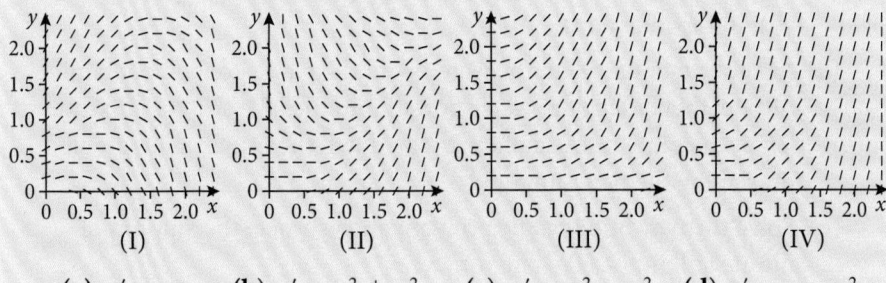

(I)      (II)      (III)      (IV)

   (a) $y' = xy$    (b) $y' = x^2 + y^2$    (c) $y' = x^2 - y^2$    (d) $y' = y - x^2$

12. Match each slope field with its differential equation.

(a) 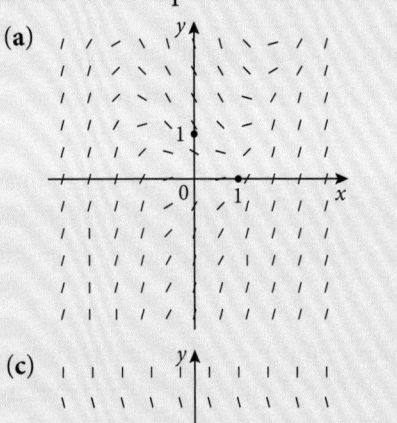     (b)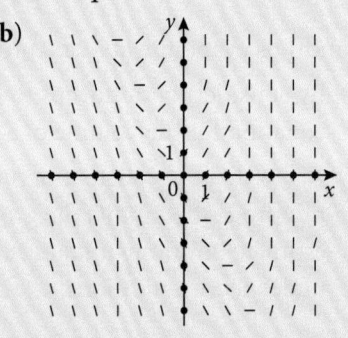

(c) 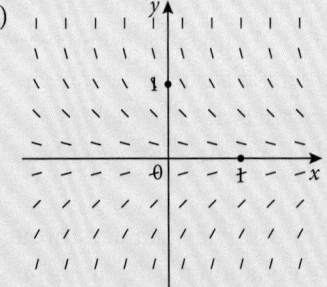     (d)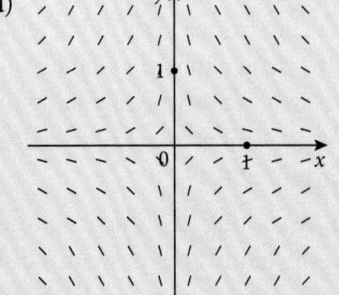

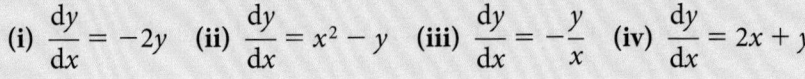

(i) $\dfrac{dy}{dx} = -2y$    (ii) $\dfrac{dy}{dx} = x^2 - y$    (iii) $\dfrac{dy}{dx} = -\dfrac{y}{x}$    (iv) $\dfrac{dy}{dx} = 2x + y$

13. An artificial lake next to a large chemical plant contains 1 000 000 litres of water and is initially free of pollutants. Water containing 0.01 g per litre of nitrate pollutant flows into the lake from the plant at a rate of 300 litres per hour. The mixture flows out of the lake at the same rate, so that the amount of water in the lake remains constant. Assume that the chemical mixes quickly with the water and is uniformly distributed in the lake.

    (a) Write a differential equation for the amount of nitrate in the lake at any time.

    (b) Without solving the differential equation, what amount of the nitrate will be in the lake after a very long time? Does this limiting amount depend on the amount that was present initially?

    (c) Solve the differential equation and estimate the amount of pollutant in the lake:

        (i) after 1 year

        (ii) in the long term.

    (d) At the end of one year, the water from the plant stopped containing nitrate; thereafter pure water flowed into the lake and the mixture left at the same rate as before. Write down the initial value problem describing the new situation.

    (e) Solve the new initial value problem and estimate the amount of nitrate still in the lake after one year.

    (f) How long will it take to get the level of nitrate to the acceptable level of 10 g?

14. Raindrops evaporate at a rate proportional to their surface area.

    (a) By considering raindrops to be spherical, write a differential equation for the volume of a raindrop as a function of time, $t$.

    (b) Given that the constant of proportionality is 0.1, and the size of a raindrop when it hits the ground is $0.034 \text{ cm}^3$, use a slope field to estimate how long it takes such a raindrop to completely evaporate under these conditions.

15. Ampicillin is being administered intravenously to a hospital patient suffering from a serious bacterial infection. Fluid containing $5 \text{ mg cm}^{-3}$ of the drug enters the patient's bloodstream at a rate of $100 \text{ cm}^3 \text{ h}^{-1}$. The drug is absorbed by body tissues or otherwise leaves the bloodstream at a rate proportional to the amount present, with a rate constant of 0.4

    (a) Assuming that the drug is always uniformly distributed throughout the bloodstream, write a differential equation for the amount of drug that is present in the bloodstream at any time.

    (b) How much of the drug is present in the bloodstream in the long term?

# 20.4 Coupled differential equations

Many real-life situations are governed by a system of differential equations. Consider the population problems that we looked at in sections 20.1 and 20.2. In these problems we looked only at a population of one species which lives alone in an environment. When we considered the population exponential growth or the logistic model, we considered the growth of the species' population to be a constant proportion of the size of the population. In this section, we will study models that consider two species interacting in the same habitat. One population is the prey, with ample food supply, and the second species, the predators, feed on the prey.

We start with an example using rabbits as prey with population size $R(t)$, and foxes with population size $F(t)$ at any time $t$.

With ample food supply and in the absence of predators, the rabbit population is an exponential model

$$\frac{dR}{dt} = kR, \text{ where } k \text{ is a positive constant.}$$

In the absence of food supply, namely the prey, we assume that the fox population will decline at a constant rate.

$$\frac{dF}{dt} = -rF, \text{ where } r \text{ is a positive constant.}$$

To be more realistic, with both species present, we assume that the principal cause of death among rabbits is being eaten by foxes, and the survival rate of foxes depends on their available food supply, rabbits. We also assume that the two species encounter each other at a rate proportional to both populations and is therefore, proportional to their product $RF$. (The more there are of one population, the more encounters there are likely to be. Recall the probability of the intersection of two events from Chapter 11: $P(A \cap B) = P(A)P(B)$).

In other words, we need to know something about one population to find the other population. So, to find the population of either the prey or the predator we need to solve a system of at least two differential equations:

$$\frac{dR}{dt} = kR - aRF, \quad \frac{dF}{dt} = -rF + bRF$$

where $a$, $b$, $k$, and $r$ are positive constants. Note that the term $-aRF$ decreases the growth rate of the rabbits while the term $bRF$ increases the growth rate of the foxes.

A solution of the system of equations is a pair of functions $R(t)$ and $F(t)$ that describe two populations of prey and predator as functions of time. Because the system is coupled ($R$ and $F$ occur in both equations), we have to solve them simultaneously. Unfortunately, it is usually not possible to find explicit formulae for $R$ and $F$ as functions of $t$. We can, however, use graphical methods to analyse the equations.

Before you start this section, check your knowledge of eigenvalues and eigenvectors from Chapter 7.

These equations are known as the predator-prey equations, or the Lotka-Volterra equations.

Assume that populations of rabbits and foxes are defined by the Lotka-Volterra equations with $a = 0.002$, $b = 0.00004$, $k = 0.07$, and $r = 0.03$. Time $t$ is measured in months.

(a) Find the constant solutions (called the **equilibrium solutions**) and interpret the result.

(b) Use the system of differential equations to find an expression for the rate of change of foxes with respect to rabbits, i.e. $\dfrac{dF}{dR}$

(c) Draw a direction field for the resulting differential equation in the $RF$-plane. Then use the direction field to sketch a few solution curves. Describe what happens in one of the solution curves.

(d) Assume that at some point in time there are 1000 rabbits and 30 foxes. Draw the corresponding solution curve and use it to describe the changes in both population levels.

**Solution**

(a) With the values given, the Lotka-Volterra equations are

$$\frac{dR}{dt} = 0.07R - 0.002RF. \quad \frac{dF}{dt} = -0.03F + 0.00004RF$$

$R$ and $F$ are constant if their derivatives are zero, that is

$$\frac{dR}{dt} = 0.07R - 0.002RF = 0 \Rightarrow R = 0 \text{ or } F = 35$$

$$\frac{dF}{dt} = -0.03F + 0.00004RF = 0 \Rightarrow F = 0 \text{ or } R = 750$$

One obvious solution is when $R = 0$ and $F = 0$, which is trivial since when there are no rabbits nor foxes, the populations will not increase or decrease!

The other constant solution is with 35 foxes and 750 rabbits. This means that 750 rabbits are enough to maintain a population of 35 foxes. There aren't too many foxes that will result in fewer rabbits and not too few foxes that will result in more rabbits.

(b) We can use the chain rule to eliminate the parameter $t$:

$$\frac{dF}{dt} = \frac{dF}{dR} \cdot \frac{dR}{dt} \Rightarrow \frac{dF}{dR} = \frac{\dfrac{dF}{dt}}{\dfrac{dR}{dt}} = \frac{-0.03F + 0.00004RF}{0.07R - 0.002RF}$$

(c) Drawing a direction field by hand is extremely time consuming. It is best to use software to do this.

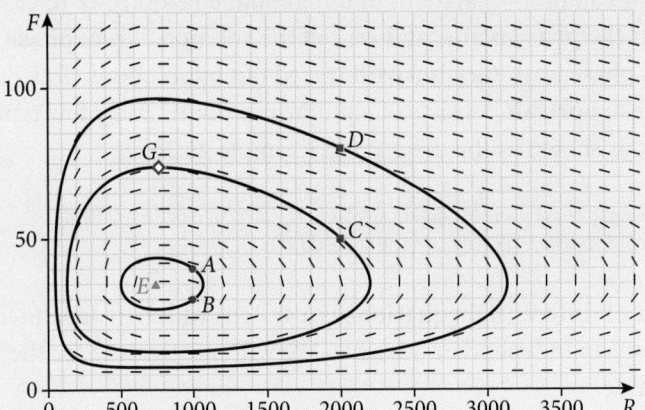

If we move along a solution curve, the second in the diagram for example, at point $C$, the rabbit population is 2000 and the fox population 50. At this point $\dfrac{dR}{dt} = -60$ and $\dfrac{dF}{dt} = 2.5$, which means that the rabbit population is decreasing at an average of 60 rabbits per month while the fox population is still increasing at an average rate of 2.5 foxes per month. At point $G$, the rabbit population is still on the decrease and is at 775 rabbits now, while the fox population appears to have reached the maximum of about 74 foxes. After that both populations decrease, and the cycle goes on until the rabbit population start to increase followed by the fox population.

(d) Starting with 1000 rabbits and 30 foxes, corresponds to the smallest solution curve and point $B$. At $B$, $t = 0$, and the rabbit population is increasing because its gradient is 10, which is positive. Letting $t$ increase, we move counter clockwise on the curve. The rest of the argument is similar to part (c).

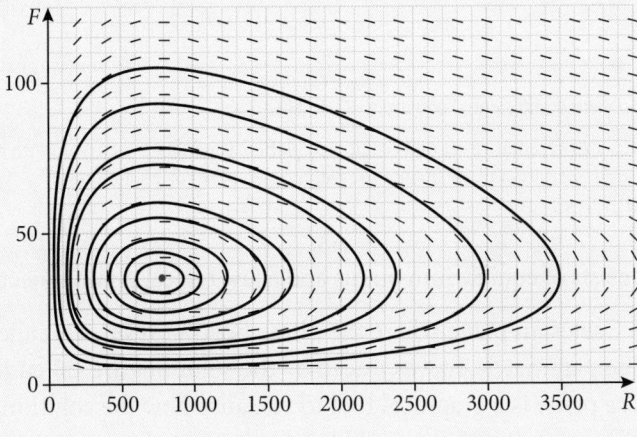

Note the point $E$ in the centre of the diagram from Example 20.26, corresponding to 775 rabbits and 35 foxes. This is the equilibrium point.

When we represent solutions of systems of differential equations we refer to the $R$-$F$ plane as the **phase plane**, and we call the solution curves **phase trajectories**. So, a phase trajectory is a path traced out by solutions $(R, F)$ as time passes. A **phase portrait** consists of equilibrium points and a few typical phase trajectories, as shown in the diagram in Example 20.26.

> It is preferable for the phase plane to contain arrows in the direction of the gradients. Unfortunately, some software packages do and some don't.

A **coupled system of differential equations** is a system with two dependent variables coupled by an independent variable. We will mostly use $t$ for the independent variable.

Coupled systems of equations are of the form

$$\frac{dx}{dt} = f(x, y, t)$$

$$\frac{dy}{dt} = g(x, y, t)$$

where $f$ and $g$ are functions of $x$, $y$ and $t$. Moreover, in this course we will mostly focus on linear systems such as

$$\frac{dx}{dt} = ax + by$$

$$\frac{dy}{dt} = cx + dy$$

where $x$ and $y$ are functions of $t$.

A useful and compact notation is to write $X = \begin{pmatrix} x(t) \\ y(t) \end{pmatrix}$, and $M = \begin{pmatrix} a & b \\ c & d \end{pmatrix}$ thus the linear equations can be written as

$$\frac{dx}{dt} = X' = MX$$

Given a coupled system of differential equations $\dfrac{dx}{dt} = MX$,

a **phase portrait** is a representative set of its solutions, plotted as parametric curves (with $t$ as the parameter) on the Cartesian plane tracing the path of each particular solution $(x, y) = (x(t), y(t))$, $-\infty < t < \infty$

A phase portrait consists of equilibrium points and a few typical phase trajectories.

In this context, the Cartesian plane where the phase portrait resides is called the **phase plane**. The parametric curves traced by the solutions are called their **trajectories**. A phase plane is a graphical tool to visualise how the solutions of a given system of differential equations would behave in the long term.

It is quite labour-intensive, but it is possible to sketch the phase plane by hand without first having to solve the system of equations that it represents. Just like a direction field, a phase plane can be a tool to predict the behaviours of a system's solutions. To do so, we draw a grid on the x-y plane. Then, at each grid point $\begin{pmatrix} x \\ y \end{pmatrix} = \begin{pmatrix} x_0 \\ y_0 \end{pmatrix}$, we can calculate the solution trajectory's instantaneous direction of motion at that point by using the given system of equations to compute the tangent/velocity vector, $X'$. Namely, substitute $\begin{pmatrix} x \\ y \end{pmatrix} = \begin{pmatrix} x_0 \\ y_0 \end{pmatrix}$ to compute $\dfrac{dx}{dt} = MX$

For example, the phase plane and portrait of the system $\dfrac{dx}{dt} = \begin{pmatrix} 5 & -3 \\ -6 & 2 \end{pmatrix}\begin{pmatrix} x \\ y \end{pmatrix}$ are given below.

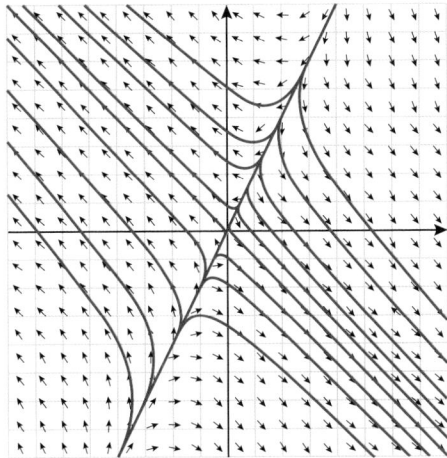

**Figure 20.15** Phase plane and portrait of the system $\dfrac{dx}{dt} = \begin{pmatrix} 5 & -3 \\ -6 & 2 \end{pmatrix}\begin{pmatrix} x \\ y \end{pmatrix}$

## Analytical solution of coupled systems

We start with the system written in matrix form:

$X' = MX$ where, in general, $M$ is a $n \times n$ matrix and $X$ is a vector whose components are the unknown functions in the system.

If we consider the case when $n = 1$, then the system will be reduced to a familiar type of differential equation

$X' = mX$ which is a linear differential equation that has the solution in the form $X(t) = Ce^{kt}$

When we consider the matrix form above, we can think of the solution to be parallel to this where the constant scalar $C$ is replaced by a vector $v$, i.e., $X(t) = ve^{kt}$

To see if this guess is a solution to the equation, then

$X(t) = ve^{kt} \Rightarrow X'(t) = kve^{kt}$ (remember that $v$ is constant)

Now, substitute this into the matrix equation:

$X' = MX \Rightarrow kve^{kt} = Mve^{kt}$

$\qquad\qquad \Rightarrow (Mv - kv)e^{kt} = 0$

$\qquad\qquad \Rightarrow (M - kI)ve^{kt} = 0$

If $X_1(t) = C_1 v_1 e^{kt}$, and $X_2(t) = C_2 v_2 e^{kt}$ are solutions to the system (called homogeneous) $X' = MX$, then $X(t) = C_1 v_1 e^{kt} + C_2 v_2 e^{kt}$ is also a solution to the system.

Since $e^{kt} \neq 0$, then in order for $X(t) = \boldsymbol{v}e^{kt}$ to be a solution, we are left with

$$(\boldsymbol{M} - k\boldsymbol{I})\boldsymbol{v} = 0$$

That is, $k$ and $\boldsymbol{v}$ are the eigenvalue and eigenvector of the matrix $\boldsymbol{M}$.

### Example 20.27

Solve the system of coupled differential equations

$$\begin{cases} \dfrac{dx}{dt} = 5x - 3y \\ \dfrac{dy}{dt} = -6x + 2y \end{cases}$$

### Solution

The system can be written in matrix form

$$\begin{pmatrix} \dfrac{dx}{dt} \\ \dfrac{dy}{dt} \end{pmatrix} = \begin{pmatrix} 5 & -3 \\ -6 & 2 \end{pmatrix} \begin{pmatrix} x \\ y \end{pmatrix}$$

The eigenvalues and their corresponding eigenvectors are

$$k_1 = -1 \rightarrow \boldsymbol{v}_1 = \begin{pmatrix} 1 \\ 2 \end{pmatrix}; \quad k_2 = 8 \rightarrow \boldsymbol{v}_2 = \begin{pmatrix} 1 \\ -1 \end{pmatrix}$$

$$\begin{pmatrix} x \\ y \end{pmatrix} = C_1 e^{k_1 t} \boldsymbol{v}_1 + C_2 e^{k_2 t} \boldsymbol{v}_2 = C_1 e^{-t} \begin{pmatrix} 1 \\ 2 \end{pmatrix} + C_2 e^{8t} \begin{pmatrix} 1 \\ -1 \end{pmatrix}$$

$$\Rightarrow \begin{cases} x = C_1 e^{-t} + C_2 e^{8t} \\ y = 2C_1 e^{-t} - C_2 e^{8t} \end{cases}$$

The diagram shows a phase portrait containing some of the trajectories and the equilibrium solution. Compare it with Figure 20.15.

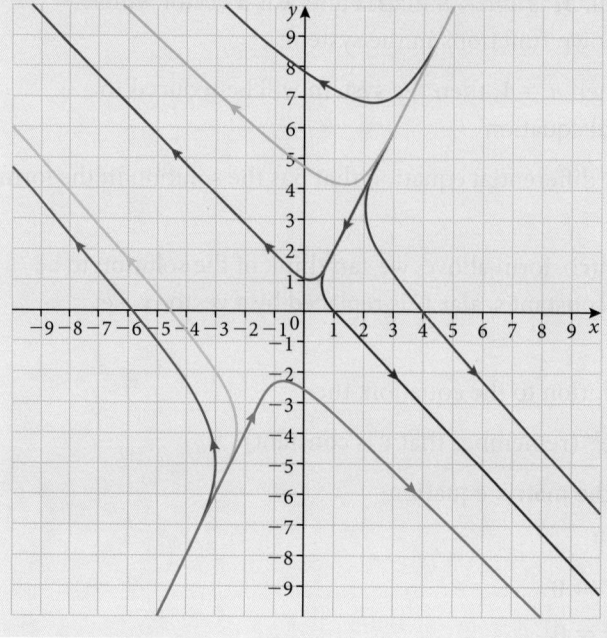

**Important:**
The eigenvectors in a solution give the direction of curves, while the eigenvalues give the sense of that direction: positive or negative. That is if an eigenvalue is negative and the corresponding eigenvector is $\boldsymbol{v} = \begin{pmatrix} a \\ b \end{pmatrix}$ then the curves will be parallel to $\begin{pmatrix} a \\ b \end{pmatrix}$ and going towards the origin as $t$ increases. If, on the other hand, the eigenvalue is positive, then the curves will still be parallel to $\begin{pmatrix} a \\ b \end{pmatrix}$ but, they will be heading away from the origin as $t$ increases. When both eigenvalues are in the solution we have a combination of these behaviours. For large negative $t$ the solution is dominated by the portion that has the negative eigenvalue since in these cases the exponent will be large and positive. Solutions for large positive $t$ are dominated by the portion with the positive eigenvalue.

Note that one eigenvalue is positive while the other is negative. The system's solution as apparent from the plane and the portrait, has all trajectories heading towards the origin and then turning away. This makes the system unstable. The equilibrium point is a **saddle point**.

## Example 20.28

Solve the system of couple differential equations

$$\begin{cases} \dfrac{dx}{dt} = x + 3y \\ \dfrac{dy}{dt} = x - y \end{cases}$$

### Solution

The system can be written in matrix form

$$\begin{pmatrix} \dfrac{dx}{dt} \\ \dfrac{dy}{dt} \end{pmatrix} = \begin{pmatrix} 1 & 3 \\ 1 & -1 \end{pmatrix}\begin{pmatrix} x \\ y \end{pmatrix}$$

The eigenvalues and their corresponding eigenvectors are

$$\lambda_1 = 2 \rightarrow v_1 = \begin{pmatrix} 3 \\ 1 \end{pmatrix}; \quad \lambda_2 = -2 \rightarrow v_2 = \begin{pmatrix} -1 \\ 1 \end{pmatrix}$$

$$\begin{pmatrix} x \\ y \end{pmatrix} = C_1 e^{\lambda_1 t} v_1 + C_2 e^{\lambda_2 t} v_2 = C_1 e^{2t}\begin{pmatrix} 3 \\ 1 \end{pmatrix} + C_2 e^{-2t}\begin{pmatrix} -1 \\ 1 \end{pmatrix}$$

$$\Rightarrow \begin{cases} x = 3C_1 e^{2t} - C_2 e^{-2t} \\ y = C_1 e^{2t} + C_2 e^{-2t} \end{cases}$$

Again, one eigenvalue is **positive** while the other is **negative**. The system's solution trajectories are heading towards the origin and then turning away. So, the system is **unstable** and the equilibrium point is a **saddle point**.

The diagram shows a phase portrait containing some of the trajectories and equilibrium solution.

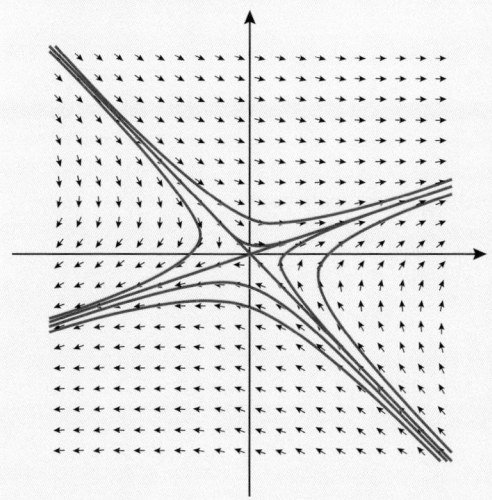

### Example 20.29

Solve the system of coupled differential equations

$$X' = \begin{pmatrix} 1 & -4 \\ 2 & -5 \end{pmatrix}\begin{pmatrix} x \\ y \end{pmatrix}$$

### Solution

The eigenvalues and their corresponding eigenvectors are

$$\lambda_1 = -1 \rightarrow v_1 = \begin{pmatrix} 2 \\ 1 \end{pmatrix}, \quad \lambda_2 = -3 \rightarrow v_2 = \begin{pmatrix} 1 \\ 1 \end{pmatrix}$$

thus, the solution is

$$\begin{pmatrix} x \\ y \end{pmatrix} = C_1 e^{\lambda_1 t} v_1 + C_2 e^{\lambda_2 t} v_2 = C_1 e^{-t}\begin{pmatrix} 2 \\ 1 \end{pmatrix} + C_2 e^{-3t}\begin{pmatrix} 1 \\ 1 \end{pmatrix}$$

$$\Rightarrow \begin{cases} x = 2C_1 e^{-t} - C_2 e^{-3t} \\ y = C_1 e^{-t} + C_2 e^{-3t} \end{cases}$$

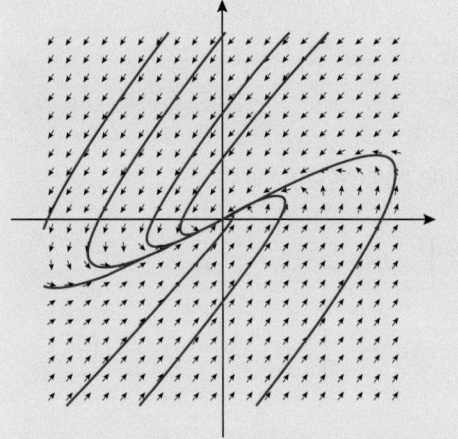

The diagram shows the phase plane and the phase portrait. Note that the eigenvalues are both negative. The system's solution trajectories are heading towards the origin. This makes the system stable.

### Example 20.30

Solve the system of coupled differential equations

$$X' = \begin{pmatrix} 1 & -3 \\ 0 & 2 \end{pmatrix}\begin{pmatrix} x \\ y \end{pmatrix}$$

### Solution

The eigenvalues and their corresponding eigenvectors are

$$\lambda_1 = 1 \rightarrow v_1 = \begin{pmatrix} 1 \\ 0 \end{pmatrix}, \quad \lambda_2 = 2 \rightarrow v_2 = \begin{pmatrix} -3 \\ 1 \end{pmatrix}$$

thus, the solution is

$$\begin{pmatrix} x \\ y \end{pmatrix} = C_1 e^{\lambda_1 t} v_1 + C_2 e^{\lambda_2 t} v_2 = C_1 e^{t}\begin{pmatrix} 1 \\ 0 \end{pmatrix} + C_2 e^{2t}\begin{pmatrix} -3 \\ 1 \end{pmatrix}$$

$$\Rightarrow \begin{cases} x = C_1 e^{t} - 3C_2 e^{2t} \\ y = C_2 e^{2t} \end{cases}$$

Note that the eigenvalues are both positive. The system's solution trajectories are heading away from the origin. This makes the system unstable.

The diagram shows the phase plane and the phase portrait.

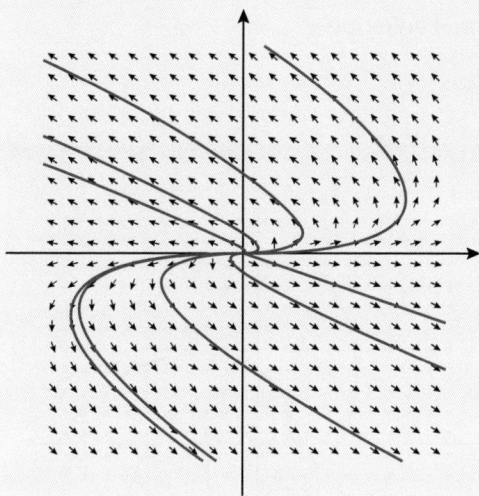

Example 20.31

Solve the system of coupled differential equations

$$X' = \begin{pmatrix} 1 & 2 \\ -9 & -5 \end{pmatrix} \begin{pmatrix} x \\ y \end{pmatrix}$$

**Solution**

The eigenvalues are

$$\lambda_1 = -2 + 3i, \quad \lambda_2 = -2 - 3i$$

Finding the eigenvectors for this case is beyond this syllabus. We will only sketch the phase plane and portrait and observe the behaviour of the system.

The diagram shows the phase plane and the phase portrait.

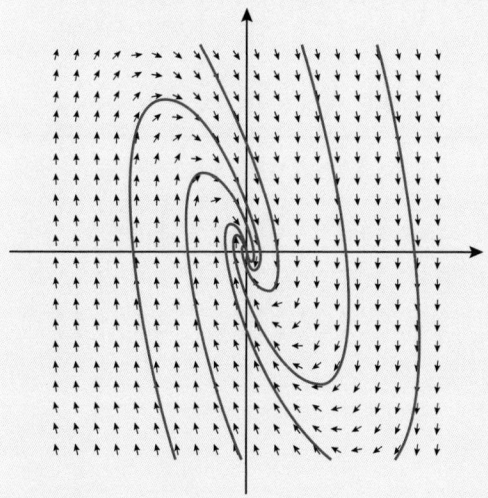

When the eigenvalues are complex and the real part is negative, the trajectories approach the origin in a spiral manner. The origin is an asymptotically stable focus, also known as a spiral sink.

**Example 20.32**

Solve the system of coupled differential equations

$$X' = \begin{pmatrix} 2 & 3 \\ -3 & 2 \end{pmatrix}\begin{pmatrix} x \\ y \end{pmatrix}$$

**Solution**

The eigenvalues are

$$\lambda_1 = 2 + 3i, \quad \lambda_2 = 2 - 3i$$

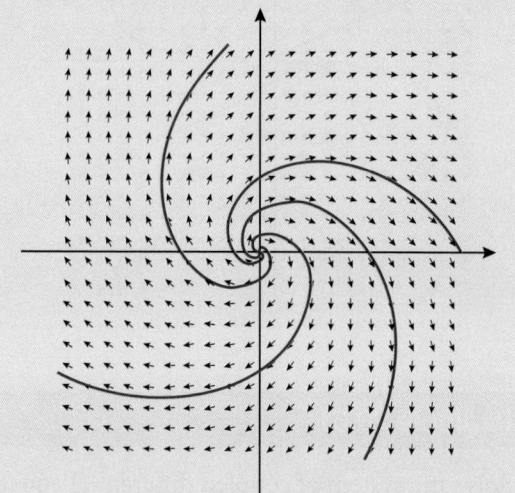

> When the eigenvalues are complex and the real part is positive, the trajectories move away from the origin in a spiral manner. The origin is an unstable focus, also known as a spiral source.

**Example 20.33**

Solve the system of coupled differential equations

$$X' = \begin{pmatrix} 2 & 2 \\ -4 & -2 \end{pmatrix}\begin{pmatrix} x \\ y \end{pmatrix}$$

**Solution**

The eigenvalues are

$$\lambda_1 = 2i, \quad \lambda_2 = -2i$$

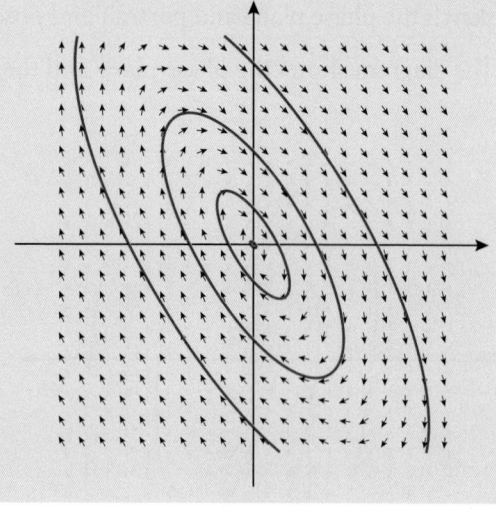

> When the eigenvalues are purely imaginary, the origin is a stable equilibrium point (but not asymptotically stable). It is also called a centre.

## Numerical solution of coupled systems

Euler's method can be used to find approximate solutions to coupled systems in the same manner as in linear equations in section 20.3.

### Example 20.34

Consider the initial value problem

$x' = 3x - 2y$

$y' = 5x - 4y$

with initial conditions $x(0) = 3$ and $y(0) = 6$

Apply Euler's method (3 steps) to estimate the solution to this system using $h = 0.1$

### Solution

Euler's iterative formulae are

$$\begin{cases} x' = f(t, x, y) \\ y' = g(t, x, y) \end{cases} \Rightarrow \begin{cases} x_{n+1} = x_n + hf(t_n, x_n, y_n) \\ y_{n+1} = y_n + hg(t_n, x_n, y_n) \end{cases}$$

Substitute the initial values for this system into the equations

$$\begin{cases} x_{n+1} = x_n + h(3x_n - 2y_n) \\ y_{n+1} = y_n + h(5x_n - 4y_n) \end{cases}$$

**First step**

$$\begin{cases} x_1 = x_0 + h(3x_0 - 2y_0) = 3 + 0.1(3 \cdot 3 - 2 \cdot 6) = 2.7 \\ y_1 = y_0 + h(5x_0 - 4y_0) = 6 + 0.1(5 \cdot 3 - 4 \cdot 6) = 5.1 \end{cases}$$

**Second step**

$$\begin{cases} x_2 = x_1 + h(3x_1 - 2y_1) = 2.7 + 0.1(3 \cdot (2.7) - 2 \cdot (5.1)) = 2.49 \\ y_2 = y_1 + h(5x_1 - 4y_1) = 5.1 + 0.1(5 \cdot (2.7) - 4 \cdot (5.1)) = 4.41 \end{cases}$$

**Third step**

$$\begin{cases} x_3 = x_2 + h(3x_2 - 2y_2) = 2.49 + 0.1(3 \cdot (2.49) - 2 \cdot (4.41)) = 2.355 \\ y_3 = y_2 + h(5x_2 - 4y_2) = 4.41 + 0.1(5 \cdot (2.49) - 4 \cdot (4.41)) = 3.891 \end{cases}$$

Note that if you solve the system the solution is

$$\begin{cases} x(t) = 2e^{-2t} + e^t \\ y(t) = 5e^{-2t} + e^t \end{cases}$$

When you evaluate the exact solution at $t = 0.3$ (starting with $t = 0$ and going forwards with 3 steps with $h = 0.1$, we arrive at $t = 0.3$) $x(0.3) \approx 2.447$ and $y(0.3) \approx 4.094$

## Solution of second-order differential equations using coupled systems

In general, we can express a second-order differential equation $y'' = f(x, y, y')$ as a system of two first order equations using an appropriate substitution such as $y' = u$, which in turn means that $y'' = u'$, and the equation becomes

$y' = u$

$u' = f(x, y, u)$

Additionally, you can use the substitution above to solve the second order differential equation using Euler's method by running the method parallel on both equations:

$$y_{n+1} = y_n + hu_n$$
$$u_{n+1} = u_n + hf(x_n, y_n, u_n)$$

For example,

$$y'' + xy' + y = 0, y(0) = 1, y'(0) = 2$$

We use the substitution $y' = u$, which transforms our equation into

$$y' = u \Rightarrow y'' = u'$$

Substitute this value into the original equation

$$y'' + xy' + y = 0 \Rightarrow y'' = -xy' - y \Rightarrow u' = -xu - y$$

Thus, we have a coupled system

$$y' = u$$
$$u' = -xu - y$$

Using Euler's method, we have

$$y_{n+1} = y_n + hy'_n = y_n + hu_n$$
$$u_{n+1} = u_n + hu'_n = u_n + h(-x_nu_n - y_n)$$

Now, if we choose $h = 0.1$, $y_0 = 1$, $u_0 = 2$, we find

$$y_1 = y_0 + 0.1u_0 = 1 + 0.1 \times 2 = 1.2$$
$$u_1 = u_0 + 0.1(-x_0u_0 - y_0) = 2 + 0.1(-0 \times 2 - 1) = 1.9$$
$$y_2 = y_1 + 0.1u_1 = 1.2 + 0.1 \times 1.9 = 1.39$$
$$u_2 = u_1 + 0.1(-x_1u_1 - y_1) = 1.9 + 0.1(-0.1 \times 1.9 - 1.2) = 1.761$$
$$\Rightarrow y(0.2) \approx 1.39, y'(0.2) \approx 1.761$$

### Example 20.35

(a) Convert this differential equation into a system then solve the system and use your solution to find the solution to the original differential equation.

$$2y'' + 5y' - 3y = 0, \quad y(0) = -4, y'(0) = 5$$

(b) Use two steps of Euler's method to estimate $y(0.2)$ and compare your answer to the exact solution you found in part (a). Use $h = 0.1$

### Solution

(a) We use the substitution $y' = u$, which implies that $y'' = u'$, thus the original differential equation becomes $2u' + 5u - 3y = 0$
and $u' = \frac{3}{2}y - \frac{5}{2}u$
Now we have the system

$$y' = u$$
$$u' = \frac{3}{2}y - \frac{5}{2}u$$

This can be written in matrix form as

$$X' = \begin{pmatrix} y' \\ u' \end{pmatrix} = \begin{pmatrix} 0 & 1 \\ \dfrac{3}{2} & -\dfrac{5}{2} \end{pmatrix} \begin{pmatrix} y \\ u \end{pmatrix}$$

To find the eigenvalues we need to set $\det(M - kI) = \begin{vmatrix} -k & 1 \\ \dfrac{3}{2} & -k - \dfrac{5}{2} \end{vmatrix} = 0$

Thus, the eigenvalues are $k_1 = -3,\ k_2 = \dfrac{1}{2}$

The respective eigenvectors are $v_1 = \begin{pmatrix} 1 \\ -3 \end{pmatrix},\ v_2 = \begin{pmatrix} 2 \\ 1 \end{pmatrix}$

The general solution is then $X = \begin{pmatrix} y \\ u \end{pmatrix} = C_1 e^{-3t} \begin{pmatrix} 1 \\ -3 \end{pmatrix} + C_2 e^{\frac{t}{2}} \begin{pmatrix} 2 \\ 1 \end{pmatrix}$

with initial conditions

$$\begin{pmatrix} y \\ u \end{pmatrix} = \begin{pmatrix} y \\ y' \end{pmatrix} = \begin{pmatrix} -4 \\ 5 \end{pmatrix} = C_1 \begin{pmatrix} 1 \\ -3 \end{pmatrix} + C_2 \begin{pmatrix} 2 \\ 1 \end{pmatrix} \Rightarrow C_1 = -2, C_2 = -1$$

Since we need $y$ to solve the initial differential equation, then the first row of the solution is the value we need $y = -2e^{-3t} - 2e^{\frac{t}{2}}$

(b) Using Euler's method, we have

$$y_{n+1} = y_n + hy'_n = y_n + hu_n$$
$$u_{n+1} = u_n + hu'_n = u_n + h\left( \frac{3}{2}y_n - \frac{5}{2}u_n \right)$$

Thus,

$$y_1 = y_0 + hu_0 = -4 + 0.1 \times 5 = -3.5$$
$$u_1 = u_0 + hu'_0 = 5 + 0.1\left( \frac{3}{2} \times (-4) - \frac{5}{2} \times 5 \right) = 3.35$$
$$y_2 = y_1 + hu_1 = -3.5 + 0.1 \times 3.35 = -3.165$$
$$u_2 = u_1 + hu'_1 = 3.35 + 0.1\left( \frac{3}{2} \times (-3.5) - \frac{5}{2} \times 3.35 \right) = 1.9875$$

Exact value: $y(0.2) = -2e^{-3(0.2)} - 2e^{\frac{0.2}{2}} \approx -3.31$

Note that as a check in this case, the second row must be the derivative of the first row.

## Exercise 20.4

A GDC, CAS, or computer are highly recommended for these questions.

1. Find the general solution to each system. Sketch a phase portrait with at least three trajectories. Describe the general behaviour of solutions to each system.

(a) $\begin{pmatrix} \dfrac{dx}{dt} \\ \dfrac{dy}{dt} \end{pmatrix} = \begin{pmatrix} 2 & 3 \\ 2 & 1 \end{pmatrix} \begin{pmatrix} x \\ y \end{pmatrix}$

(b) $\begin{cases} \dfrac{dx}{dt} = -10x + 6y \\ \dfrac{dy}{dt} = 15x - 19y \end{cases}$

(c) $\begin{cases} \dfrac{dx}{dt} = x + 3y \\ \dfrac{dy}{dt} = 5x + 3y \end{cases}$

(d) $\begin{pmatrix} \dfrac{dx}{dt} \\ \dfrac{dy}{dt} \end{pmatrix} = \begin{pmatrix} 1 & 2 \\ 4 & 3 \end{pmatrix} \begin{pmatrix} x \\ y \end{pmatrix}$

**(e)** $\begin{pmatrix} \dfrac{dx}{dt} \\ \dfrac{dy}{dt} \end{pmatrix} = \begin{pmatrix} -4 & 2 \\ -\dfrac{5}{2} & 2 \end{pmatrix} \begin{pmatrix} x \\ y \end{pmatrix}$

**(f)** $\begin{pmatrix} \dfrac{dx}{dt} \\ \dfrac{dy}{dt} \end{pmatrix} = \begin{pmatrix} 10 & -5 \\ 8 & -12 \end{pmatrix} \begin{pmatrix} x \\ y \end{pmatrix}$

**(g)** $\begin{cases} \dfrac{dx}{dt} = 7x - 3y \\ \dfrac{dy}{dt} = 6x - 4y \end{cases}$

**(h)** $\begin{cases} \dfrac{dx}{dt} = -3x + y\sqrt{2} \\ \dfrac{dy}{dt} = x\sqrt{2} - 2y \end{cases}$

**(i)** $X' = \begin{pmatrix} -4 & -3 \\ 6 & 5 \end{pmatrix} X$

2. Sketch a phase portrait for each system, including at least three trajectories. Analyse future paths according to your phase portrait.

**(a)** $\begin{cases} \dfrac{dx}{dt} = 6x - y \\ \dfrac{dy}{dt} = 5x + 4y \end{cases}$

**(b)** $\begin{cases} \dfrac{dx}{dt} = 2x + 8y \\ \dfrac{dy}{dt} = -x - 2y \end{cases}$

**(c)** $\begin{cases} \dfrac{dx}{dt} = -y \\ \dfrac{dy}{dt} = 1.01x - 0.2y \end{cases}$

**(d)** $\begin{pmatrix} \dfrac{dx}{dt} \\ \dfrac{dy}{dt} \end{pmatrix} = \begin{pmatrix} -\dfrac{1}{2} & 1 \\ -1 & -\dfrac{1}{2} \end{pmatrix} \begin{pmatrix} x \\ y \end{pmatrix}$

**(e)** $\begin{pmatrix} \dfrac{dx}{dt} \\ \dfrac{dy}{dt} \end{pmatrix} = \begin{pmatrix} 3 & -2 \\ 4 & -1 \end{pmatrix} \begin{pmatrix} x \\ y \end{pmatrix}$

**(f)** $\begin{pmatrix} \dfrac{dx}{dt} \\ \dfrac{dy}{dt} \end{pmatrix} = \begin{pmatrix} 2 & -5 \\ 1 & -2 \end{pmatrix} \begin{pmatrix} x \\ y \end{pmatrix}$

3. Find the general solution to each system. Sketch a phase portrait with at least three trajectories. Describe the general behaviour of solutions to each system. Then solve each initial value problem and use Euler's method to estimate the solution using $h = 0.1$ and two iterations. Compare your estimate to the exact value.

**(a)** $\begin{cases} \dfrac{dx}{dt} = y \\ \dfrac{dy}{dt} = 6x - y \end{cases}$ $\qquad x(0) = 3, y(0) = 1$

**(b)** $\begin{cases} \dfrac{dx}{dt} = -y \\ \dfrac{dy}{dt} = 10x - 7y \end{cases}$ $\qquad x(0) = 2, y(0) = -7$

**(c)** $\begin{pmatrix} \dfrac{dx}{dt} \\ \dfrac{dy}{dt} \end{pmatrix} = \begin{pmatrix} 1 & 2 \\ 3 & 2 \end{pmatrix} \begin{pmatrix} x \\ y \end{pmatrix}$ $\qquad X(0) = \begin{pmatrix} 0 \\ -4 \end{pmatrix}$

**(d)** $\begin{pmatrix} \dfrac{dx}{dt} \\ \dfrac{dy}{dt} \end{pmatrix} = \begin{pmatrix} -5 & 1 \\ 4 & -2 \end{pmatrix}\begin{pmatrix} x \\ y \end{pmatrix}$    $X(0) = \begin{pmatrix} 1 \\ 2 \end{pmatrix}$

**(e)** $X' = \begin{pmatrix} 2 & 4 \\ 4 & 2 \end{pmatrix}X$        $X(0) = \begin{pmatrix} 5 \\ -1 \end{pmatrix}$

4.  Convert each differential equation into a system of coupled equations, solve the system and use this solution to get the solution to the original differential equation. Then use two iterations of Euler's method to estimate a solution to each equation.

    **(a)** $u'' + 4u' + 3u = 0, \quad u(0) = u'(0) = 2$

    **(b)** $y'' - y' - 2y = 0, \quad\quad y(0) = 3, y'(0) = 8$

    **(c)** $2y'' - 5y' + 2y = 0, \quad y(0) = -2, y'(0) = -\dfrac{7}{4}$

    **(d)** $u'' + 3u' + 2u = 0, \quad u(0) = -1, u'(0) = -5$

    **(e)** $y'' + 6y' + 5y = 0, \quad\quad y(0) = 3, y'(0) = -1$

5.  Consider two tanks connected as shown in the diagram. Tank 1 contains $x(t)$ grams (g) of salt in 100 litres ($l$) of water and tank 2 contains $y(t)$ g of salt in $200\,l$ of water. The saline solution in each tank is kept uniform by stirring, and saline solution is pumped from each tank to the other at the rates indicated in the figure. In addition, fresh water flows into tank 1 at $20\,l\,\text{min}^{-1}$, and the saline solution in tank 2 flows out at $20\,l\,\text{min}^{-1}$, so the total volume of saline in the two tanks remains constant. The amounts of salt in each tank at the start of the process are $x(0) = 50\,\text{g}$ and $y(0) = 100$

    **(a)** What are the salt concentrations in the two tanks at any time $t$?

    **(b)** Show that, when we compute the rates of change of the amount of salt in the two tanks, we get the system of differential equations that $x(t)$ and $y(t)$ must satisfy:

    $$\frac{dx}{dt} = \frac{-3}{10}x + \frac{1}{20}y$$
    $$\frac{dy}{dt} = \frac{3}{10}x - \frac{3}{20}y$$

    **(c)** Hence, find the amounts $x(t)$ and $y(t)$ of salt in the two tanks at time $t$.

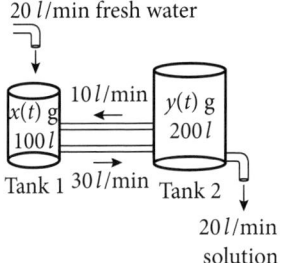

20 $l$/min fresh water

$10\,l$/min

$x(t)$ g
$100\,l$

$y(t)$ g
$200\,l$

Tank 1   $30\,l$/min   Tank 2

20 $l$/min
solution

**Figure 20.16** Diagram for question 5

A CAS is highly recommended for the solution.

---

## Chapter 20 practice questions

1.  Solve the differential equation $\dfrac{dy}{dx} = \dfrac{xy}{\sqrt{1 + x^2}}$

    Given that $y = 1$ when $x = 0$ express $y$ as an explicit function of $x$.

2.  Find the particular solution to the differential equation
    $\dfrac{dy}{dx} = \sin x \cos^2 y$ given that $y = \dfrac{\pi}{4}$ when $x = \dfrac{\pi}{2}$

3. The solution curve to the differential equation $x\dfrac{dy}{dx} = y(3 - y)$ passes through the point $(2, 2)$. Find $y$ as an explicit function of $x$.

4. Show that the general solution to the differential equation
$x\dfrac{dy}{dx} = y\ln x$ is $y = Cx\ln\sqrt{x}$

5. The half-life of radium-226 is 1620 years. The activity of a sample is 800 Bq. What is the activity (accurate to 3 significant figures) after 25 years?

6. Solve each separable differential equation.

   (a) $\dfrac{dy}{dx} = \dfrac{2x}{y}$         (b) $\dfrac{dy}{dx} = \dfrac{y^2}{x^2}$

   (c) $x^2\dfrac{dy}{dx} = y^2 - y$      (d) $x\dfrac{dy}{dx} = \tan y$

   (e) $\dfrac{dy}{dx} = xy$           (f) $\sqrt{x^2 + 1}\dfrac{dy}{dx} = \dfrac{x}{y}$

   (g) $\dfrac{dy}{dx} = \dfrac{y^2 - 1}{e^x}$      (h) $\ln y\dfrac{dy}{dx} = 1$

7. Using the method of separation of variables, show that an implicit solution for the differential equation
$\dfrac{dy}{dx} = \dfrac{xy + y}{xy + x}$ is $ye^y = Axe^x$, where $A$ is an arbitrary constant.

8. Find the general solution, in explicit form, to the differential equation
$y\dfrac{dy}{dx} = \cos x$
Comment on the possible values of the constant $C$.

9. The equation for the rate of change of the population (in thousands), $p$, of a certain species is given by $\dfrac{dp}{dt} = 5p - 2p^2$

   (a) The initial population is 4000, that is, $p(0) = 4$
   What is the limiting value of the population, that is, $\lim_{t\to\infty} p(t)$?

   (b) When $p(0) = 0.5$, what is $\lim_{t\to\infty} p(t)$?

   (c) Comment on the long-term behaviour of the species' population growth.

10. Solve the initial-value problem $(1 + x^2)\dfrac{dy}{dx} = 1 + y^2$, $y(2) = 3$
Write the solution in explicit form, expressing $y$ in terms of $x$.

11. (a) Use the method of partial fractions to express $\dfrac{1}{x^2 - x - 2}$ as the sum of two fractions.

   (b) Consider the differential equation
   $\dfrac{dy}{dx} = \dfrac{y^2}{x^2 - x - 2}$, $x > 2$ such that $y = 1$ when $x = 5$
   Show that the solution is $2e^{\frac{3-3y}{y}} = \dfrac{x + 1}{x - 2}$

12. Consider the differential equation $\dfrac{dy}{dx} = \dfrac{x + 2y}{3y - 2x}$, for $x > 0$

   (a) Use the substitution $y = vx$ to show that $v + x\dfrac{dv}{dx} = \dfrac{1 + 2v}{3v - 2}$

   (b) Hence find the solution of the differential equation, given that $y = 0$ when $x = 1$

13. Use the substitution $y = vx$ to show that the general solution to the differential equation $y^2 - x^2 + xy\dfrac{dy}{dx} = 0$ is $2x^2y^2 - x^4 = C$, where $C$ is a constant.

14. Consider the differential equation $\dfrac{dy}{dx} = \dfrac{y^2 + y}{x}$

   (a) Find the general solution.

   (b) Given that $y = 1$ when $x = 1$, find a particular.

   (c) Use Euler's method with step size $h = 0.2$ to approximate the solution at $x = 1.2, 1.4, 1.6,$ and $1.8$.

   (d) Compute the percentage error for each of the approximate solutions found in part (c) compared to the solution for same value of $x$ found using the explicit solution found in part (b).

15. Given that $\dfrac{dy}{dx} = xy^2$ and $y = 1$ at $x = 0$, use Euler's method with five steps to approximate the value of $y$ at $x = 1$

16. Use Euler's method with step size $h = 0.1$ to approximate the value of $y$ when $x = 1$ for the differential equation $\dfrac{dy}{dx} = e^{xy}$, given that the solution curve passes through the point $(0, 1)$.

17. Use the substitution $y = vx$ to find the general solution to the differential equation $\dfrac{dy}{dx} = \dfrac{x^2 + 3y^2}{2xy}$

18. Given that $\dfrac{dy}{dx} = x\sqrt{y}$, and $y = 4$ when $x = 1$, use Euler's method with step size $h = 0.1$ to approximate the solution at $x = 1.1, 1.2, 1.3, 1.4$ and $1.5$

19. Consider the initial-value problem $\dfrac{dy}{dx} = x - y$, $y(0) = 0$

   (a) Show that the solution is $y = e^{-x} + x - 1$

   (b) Use Euler's method with five steps to find an approximate value of $y$ when $x = 1$

   (c) Use Euler's method with 10 steps to find another approximation for $y(1)$

   (d) Compare the approximate values for $y(1)$ found in parts (b) and (c) to the actual value using the solution $y = e^{-x} + x - 1$

20. The rate, in degrees Celsius per minute, at which the temperature of a cup of tea decreases is given by $-k(\alpha - 20)$ where $\alpha$ is the temperature in degrees Celsius and $k$ is a constant. When $t = 0$ minutes, $\alpha = 70°\,C$; and when $t = 10$ minutes, $\alpha = 50°C$. Find an equation for the temperature in terms of time $t$.

21. The radioactive isotope $^{14}C$ is known to decay slowly with a decay constant $k \approx 0.0001216$ when time is measured in years.

   (a) Show that the half-life of this isotope is approximately 5700 years.

   (b) Carbon-14 in an ancient skull has only one-quarter as much $^{14}C$ as a present-day bone. Estimate the age of the ancient skull.

22. Carbon-14 taken from an alleged artefact of the time of the Roman empire found near Rome contained $4.3 \times 10^{10}$ atoms of $^{14}C$ per gram. Carbon extracted from a present-day specimen of the same substance contained $5.1 \times 10^{10}$ atoms of $^{14}C$ per gram. Estimate the approximate age of the relic. What is your opinion as to its authenticity?

23. Consider the differential equation
   $$\frac{dy}{dx} = x^2 + y^2, \text{ where } y = 2 \text{ when } x = 0$$

   (a) Use Euler's method with step size 0.25 to find an approximate value of $y$ when $x = 3$

   (b) Write down, giving a reason, whether your approximate value for $y$ is greater or less than the actual value of $y$.

24. A tank containing 240 litres of pure water is used in a lab experiment. A mixture containing a concentration of $s$ grams per litre (g/$l$) of salt enters the tank at a rate of $4\,l$ per minute, and the stirred mixture leaves the tank at the same rate.

   (a) Write down the initial value problem describing the process.

   (b) Find an expression, in terms of $s$, for the amount of salt in the tank at any time $t$.

   (c) Find the equilibrium amount of salt in the tank.
   That is, the amount as $t \to \infty$

25. The population of flies in a farming area increases at a rate proportional to the current population, and in the absence of other factors, doubles each week. This summer, there are 200 000 flies in the area initially, and predators (bats, spiders, etc.) eat 140 000 flies per week. Set up an initial value problem and solve it to determine the population of flies in this area at any time (in weeks).

Multiply both sides of the equation by $e^{\frac{t}{30}}$

Multiply both sides of the differential equation by $e^{-(\ln 2)t}$

# Internal assessment

Internal assessment (IA) is an important component of the Applications and Interpretation HL course and contributes 20% to your final grade. It is a significant part of the overall assessment for the course and should be taken seriously. It should also be pointed out that your work in completing the IA component differs in important ways from the written exams (external assessment) for the course.

- Unlike written examinations, you do *not* perform IA work under strict time constraints.

- You have some freedom to decide which mathematical topic you wish to explore.

- Your IA work involves writing about mathematics, not just using mathematical procedures.

- Regular discussion with, and feedback from, your teacher will be essential.

- You should endeavour to explore a topic in which you have a genuine personal interest.

- You will be rewarded for evidence of creativity, curiosity, and independent thinking.

## Mathematical exploration

To satisfy the IA component, you are required to complete a piece of written work on a mathematical topic that you choose in consultation with your teacher. This piece of written work is formally referred to as the mathematical exploration. It will be referred to simply as the 'exploration' throughout this chapter. Your primary objective is to *explore* a mathematical topic in which you are *genuinely interested* and that is at an *appropriate level* for the course. A fundamental aspect of your exploration must be the *use of mathematics* in a manner that clearly demonstrates your knowledge and understanding of the relevant mathematics. Your teacher may provide you with a list of ideas (or 'stimuli') to help you in the process of finding a suitable topic.

It is your responsibility to determine whether or not you are sufficiently interested in a particular topic – and it is your teacher's responsibility to determine if an exploration of the topic can be conducted at a mathematical level that is suitable for the course. Your teacher will help you determine if an exploration of a certain topic can potentially address the five assessment criteria satisfactorily. Your exploration should be approximately 12 to 20 pages long with double line spacing.

See the list of 200 ideas included in the eBook. You may find a suitable topic in the list, or the list may help you find or develop your own ideas for a mathematical topic to explore.

# Internal assessment

## Internal assessment criteria

Your exploration will be assessed by your teacher according to the following five criteria.

### A Presentation

This criterion assesses the organisation and coherence of the exploration. A well-organised exploration has an introduction, a rationale (which includes a brief explanation of why the topic was chosen), a description of the aim of the exploration, and a conclusion.

### B Mathematical communication

This criterion assesses to what extent you are able to:
- use appropriate mathematical language (notation, symbols, terminology)
- clearly define key terms, variables, and parameters
- use multiple forms of mathematical representation, such as formulae, diagrams, tables, charts, graphs, and models
- apply a deductive approach in general, and present any proofs in a logical manner.

### C Personal engagement

This criterion assesses the extent to which you engage with the exploration and present it in such a way that clearly shows *your own personal approach*. Personal engagement may be recognised in several different ways. These may include – but are not limited to – thinking independently and/or creatively, addressing personal interest, presenting mathematical ideas in your own words and diagrams, developing your own ideas and testing them, and creating your own examples to illustrate important results.

### D Reflection

This criterion assesses how well you *review, analyse,* and *evaluate* the exploration. Although reflection may be seen in the conclusion to the exploration, you should also give evidence of reflective thought throughout the exploration. Reflection can be demonstrated by consideration of limitations and/or extensions, commenting on what you've learned, or comparing different mathematical methods and approaches.

### E Use of mathematics

This criterion assesses to what extent and how well you use mathematics in your exploration. The mathematical working in your exploration needs to be *sufficiently sophisticated* and *rigorous*. The chosen topic should involve mathematics in the Applications and Interpretation HL syllabus or at a similar level. Sophistication and rigour can include understanding and use of challenging mathematical concepts, looking at a problem from different perspectives, mathematical arguments expressed clearly in a logical manner, or seeing underlying structures to link different areas of mathematics.

Your exploration will earn a score out of a total of 20 possible marks. The five criteria do not contribute equally to the overall score for your exploration. For example, criterion E (Use of mathematics) is 30% of the overall score, whereas crtieria C (Personal engagement) and D (Reflection) contribute 15% each.

It is very important that you familiarise yourself with the assessment criteria for the Applications and Interpretation HL exploration and refer to them while you are writing your exploration. The achievement levels for each criteria and associated descriptors are as follows:

| A | Presentation |
|---|---|
| 0 | The exploration does not reach the standard described by the descriptors below. |
| 1 | The exploration has some coherence or some organisation. |
| 2 | The exploration has some coherence and shows some organisation. |
| 3 | The exploration is coherent and well organised. |
| 4 | The exploration is coherent, well organised, and concise. |

| B | Mathematical communication |
|---|---|
| 0 | The exploration does not reach the standard described by the descriptors below. |
| 1 | The exploration contains some relevant mathematical communication that is partially appropriate. |
| 2 | The exploration contains some relevant appropriate mathematical communication. |
| 3 | The mathematical communication is relevant, appropriate, and is mostly consistent. |
| 4 | The mathematical communication is relevant, appropriate, and consistent throughout. |

| C | Personal engagement |
|---|---|
| 0 | The exploration does not reach the standard described by the descriptors below. |
| 1 | There is evidence of some personal engagement. |
| 2 | There is evidence of significant personal engagement. |
| 3 | There is evidence of outstanding personal engagement. |

| D | Reflection |
|---|---|
| 0 | The exploration does not reach the standard described by the descriptors below. |
| 1 | There is evidence of limited reflection. |
| 2 | There is evidence of meaningful reflection. |
| 3 | There is substantial evidence of critical reflection. |

| E | Use of mathematics |
|---|---|
| 0 | The exploration does not reach the standard described by the descriptors below. |
| 1 | Some relevant mathematics is used. Limited understanding is demonstrated. |
| 2 | Some relevant mathematics is used. The mathematics explored is partially correct. Some knowledge and understanding are demonstrated. |
| 3 | Relevant mathematics commensurate with the level of the course is used. The mathematics explored is correct. Some knowledge and understanding are demonstrated. |

| 4 | Relevant mathematics commensurate with the level of the course is used. The mathematics explored is correct. Good knowledge and understanding are demonstrated. |
|---|---|
| 5 | Relevant mathematics commensurate with the level of the course is used. The mathematics explored is correct and demonstrates sophistication or rigour. Thorough knowledge and understanding are demonstrated. |
| 6 | Relevant mathematics commensurate with the level of the course is used. The mathematics explored is precise and demonstrates sophistication and rigour. Thorough knowledge and understanding are demonstrated. |

## Guidance

Conducting an in-depth individual exploration into the mathematics of a particular topic can be an interesting and very rewarding experience. It is important to take all stages of your work on the exploration seriously – not only because it is worth 20% of your final grade for the course but also because of the opportunity to pursue your own personal interests without the pressure of examination conditions. The exploration should *not* be approached as simply an extended homework assignment. The task of writing the exploration will require you to analyse, think, write, edit, and use mathematics in a readable and focused manner. Hopefully, it will also be enjoyable, thought-provoking, and satisfying, and it should give you the opportunity to gain a deeper appreciation for the beauty, power, and usefulness of mathematics.

Although it is required that your exploration is completely your own work, you should consult with your teacher on a regular basis. You are allowed to work collaboratively with fellow students, but this should be limited to the following: selecting a topic, finding resources, understanding relevant mathematical knowledge and skills, and receiving peer feedback on your writing. While you are encouraged to *talk* through your ideas with others, it is not appropriate for you to *work* with others on your exploration. Your teacher should provide support and advice during the planning and writing stages of your exploration. Both you and your teacher will need to verify the authenticity of your exploration.

Any text, diagrams, images, mathematical working, or ideas that are not your own must be cited where they appear in your exploration. Otherwise, all of the work connected with your exploration must be your own. Your exploration must provide the reader with the exact sources of quotations, ideas, and points of view with a complete and accurate bibliography. There are a number of acceptable bibliographic styles. Whichever style you choose, it must include all relevant source information and be applied consistently. Group work is not allowed. Also, if you are writing an extended essay for mathematics, you are not allowed to submit the same or similar piece of work for the exploration – and you should not write about the same mathematical topic for both.

In organising a successful exploration, consider the following suggestions:

1. Select a topic in which you are *genuinely interested*. Include a brief explanation in the early part of your exploration about why you chose your topic – including why you find it interesting.

Your teacher will provide oral and/or written advice on a draft of your exploration pertaining to how it can be improved. Your teacher will also write thorough and descriptive comments on the final version of your exploration to assist IB moderators in confirming the criteria scores they've awarded.

Warning: Failure to properly cite any text, diagrams, images, mathematical working, or ideas that are not your own may result in your exploration being reviewed for malpractice, which could have serious consequences.

If you are uncertain about the formatting and style of citations and a bibliography (not the same thing), then you should consult with teacher(s) at your school who have expertise in this area – such as an English teacher or librarian. A bibliography is required but it does not replace the need for appropriate citations (inline or footnotes) at the pertinent location in the exploration.

2. Consult with your teacher to confirm that the topic is at the *appropriate level of mathematics* – namely, that it is at the same or similar level of the mathematics in the HL syllabus.

3. Find as much *information* about the topic as possible. Although information found on websites can be very helpful, try to also find information in books, journals, textbooks, and other printed material.

4. Although there is no requirement that you present your exploration to your classmates, it should be written so that they can follow it without trouble. Your exploration needs to be *logically organised* and use appropriate mathematical terminology and notation.

5. The most important aspects of your exploration should be about *mathematical communication and using mathematics*. Although other aspects of your topic – for example, historical, personal, cultural – can be discussed, be careful to keep focus on the mathematical features.

6. Two of the assessment criteria – Personal engagement and Reflection – are about *what you think about the topic* you are exploring. Don't hesitate to pose your own relevant and insightful questions as part of your exploration – and then to address these questions using mathematics at a suitably sophisticated level along with sufficient written commentary.

7. Although your teacher will expect and require you to work independently, you are allowed to *consult with your teacher* – and your teacher is allowed to give you advice and feedback to a certain extent while you are working on your exploration. It is especially important to check with your teacher that any *mathematics in your exploration is correct*. Your teacher will not give mathematical answers or corrections but can indicate where any errors have been made or where improvement is needed.

## Mathematical exploration - HL student checklist

- Is your exploration written entirely by yourself? Have you avoided simply replicating work and ideas from sources you found?  ☐ Yes  ☐ No

- Have you strived to apply your personal interest, develop your own ideas, and use critical thinking skills during your exploration?  ☐ Yes  ☐ No

- Did you refer to the five assessment criteria while writing your exploration?  ☐ Yes  ☐ No

- Does your exploration focus on good mathematical communication – and does it read like an article from a mathematical journal?  ☐ Yes  ☐ No

- Does your exploration have a clearly identified introduction and conclusion?  ☐ Yes  ☐ No

- Have you provided appropriate citation for any ideas, mathematical working, images, graphs, etc. that are not your own at the point they appear in your exploration?  ☐ Yes  ☐ No

- Not including the bibliography, is your exploration 12 to 20 pages?  ☐ Yes  ☐ No

- Are graphs, tables, and diagrams sufficiently described and labelled?  ☐ Yes  ☐ No

- To the best of your knowledge, have you used mathematics that is at the same level, or similar, to that studied in Applications and Interpretation HL?  ☐ Yes  ☐ No

- Have you attempted to discuss mathematical ideas, and use mathematics, with a sufficient level of sophistication and rigour?  ☐ Yes  ☐ No

- Are formulae, graphs, tables, and diagrams in the main body of text? (Preferably no full-page graphs, and no separate appendices.)  ☐ Yes  ☐ No

- Have you used technology – such as a GDC, spreadsheet, mathematics software, drawing and word-processing software – to enhance mathematical communication?  ☐ Yes  ☐ No

- Have you used appropriate mathematical language (notation, symbols, terminology) and defined key terms?  ☐ Yes  ☐ No

- Is the mathematics in your exploration performed precisely and accurately?  ☐ Yes  ☐ No

- Has calculator/computer notation and terminology been used? ($y = x^2$, not $y = x\wedge2$; $\pi$, not <pi>; $\approx$, not 'approximately equal to'; $|x|$, not abs($x$); etc)  ☐ Yes  ☐ No

- Have you included reflective and explanatory comments about the topic being explored throughout your exploration?  ☐ Yes  ☐ No

## Finding, developing, and choosing a topic for your exploration

It is fair to say that the *most important stage* of completing your exploration is determining the mathematical topic you are going to investigate, write about, and apply. Your exploration is much more likely to be successful – and gratifying – if it focuses on a mathematical topic in which you have a genuine interest, is at a suitable level for the Applications and Interpretation HL course, and for which you are confident that you can discuss and use the relevant mathematics in a manner that demonstrates thorough knowledge and understanding. There is no single approach for determining an exploration topic that is guaranteed to be successful for all students. Your teacher will provide helpful advice and support. Your teacher may supply you with a short list of some broad stimuli to start the process of finding a much narrower topic. Many teachers have found that starting with a sufficiently narrow topic is often more successful than starting with a very broad topic that requires a significant effort to reduce to the extent that it can be explored in less than 20 pages (double spaced).

 Avoid choosing a topic that is too broad and/or too complicated.

In the eBook for this textbook you will find a list of 200 mathematical topics. Some of the topics in the list are broad but many are already quite narrow in scope. It is possible that some of these 200 topics could be the focus of an exploration, while others will require you to investigate further to develop a narrower focus to explore. Do not restrict yourself to the topics in the list. This list is only the tip of the iceberg with regard to potential topics for your exploration. Reading through this list may stimulate you to think of some other topic(s) that you may find interesting to explore. Many of the items in the list may be unfamiliar to you. A quick search on the internet should give you a better idea what each is about and help you determine if you're interested enough to investigate further – and to see if it might be a suitable topic for your exploration.

# Theory of knowledge

At the start of his wonderful book *Nature's Numbers,* the mathematician Ian Stewart writes:

> *'We live in a universe of patterns. Every night the stars move in circles across the sky. The seasons cycle at yearly intervals. No two snowflakes are ever exactly the same, but they all have sixfold symmetry. Tigers and zebras are covered in patterns of stripes, leopards and hyenas are covered in patterns of spots. Intricate trains of waves march across the oceans; very similar trains of sand dunes march across the desert. Coloured arcs of light adorn the sky in the form of rainbows, and a bright circular halo sometimes surrounds the moon on winter nights. Spherical drops of water fall from clouds.'*

We could add to Stewart's list. Wallpaper is patterned (there are surprisingly only 17 different distinct groups of possible patterns); buildings often exhibit mirror symmetry and their structure is carefully proportioned; the digital traces on memory sticks or hard drives are patterned in a way that makes them suitable for storing data; mechanical devices such as clocks and engines depend on symmetry and patterning for their smooth movement; the day is divided into equal parts that are represented using angles or digits; music possesses horizontal and vertical symmetries – and human behaviour is patterned.

It is no accident that the world is full of patterns. Symmetry in a building is not only easy on the eye but it ensures that the design is simple. Pattern is a labour-saving strategy. The same plan can be used for each window, or the plan for one side of a building can be used in reverse for the other side. These informational shortcuts can be found both in the man-made world and in nature. The same blueprint for generating twig patterns can be used for bigger branches, or one plan can be used for all the petals in a flower. It is a sort of design efficiency. The wealth of patterns in the world is a series of cost-effective solutions to problems – and that is why these patterns are worth studying.

Mathematics is one way in which human beings formally study patterns. While the natural sciences study patterns by going out into the world, collecting examples and analysing them, mathematics studies patterns in the abstract. Mathematics in its purest form is not fieldwork or experiment. Its raw materials are abstract structures specified by symbols, and mathematicians arrive at conclusions through their manipulation. In this sense, mathematics is a little 'other-wordly' – a characteristic that makes it interesting from a ToK perspective. It means that in some sense, mathematics is more like an art than a science. There is in this suggestion more than a hint of a deep reliance on creativity and imagination. A comparison with the arts and the sciences is instructive and reveals the truly special place that mathematics occupies in human knowledge.

In this chapter, we will investigate mathematics using the basic structure of the knowledge framework: Perspectives, methods and tools, and the link to the individual.

Under 'Perspectives', we will look at the orientation of mathematics within the academy. There are a number of key questions to be answered here:

- What is mathematics about?
- How should we think of mathematics: as a human construction or something in the world?
- Why is mathematics useful?

Under 'Methods and Tools' we will discuss exactly what mathematicians do – how they arrive at mathematical knowledge and what counts as facts and truth in mathematics. This is where we unpack the key conceptual building blocks of mathematical thought.

The final section deals with mathematics and the individual. What is the link between mathematics and supposedly subjective phenomena such as beauty? How reliable are our mathematical intuitions? Is mathematics a personal journey or is it something that we collaborate on?

On the way, we will have fun with infinite numbers, self-similar patterns and security codes. While it might be removed from the physical world, the world of mathematics is just as fascinating, if not more so. Enjoy!

What role does mathematics play in your life?

# Perspectives

## Mathematics and number

As a first definition, let's say that mathematics is the formal study of patterns. In this section we will see how far this basic idea takes us.

Imagine a simple pattern in the world – a set of similar objects, for example, a field of animals. Let's say that the animals are of the same kind – they are cows. To recognise that a group of different things all belong to the same kind is already remarkable. It means ignoring all the things that mark out individual animals and focusing only on what they have in common. Grouping a set of things together by common characteristics is a powerful technique in the sciences. If such a classification is effective, it might yield understanding, generalisations and predictions. We call groups that have these properties **natural kinds** – it is something that might be expected to happen in biology. But mathematics goes one step further. Suppose that we make a mark 'I' on a clay tablet for every cow in the field. We end up with a mark 'IIIIIII'. What we have done now is to abstract away everything about the animals in the field: the fact that they are animals, that they are cows, that they are eating grass. What is left is their number.

So, the simplest pattern that we can deal with abstractly is number. In a somewhat magical way, the inscriptions of the tablet **represent** the cows in the field. They are a convenient stand-in for the real world. If we want to find out what happens when we remove 'III' cows from the field. We can either move them physically or we simply separate the 'cow' symbols: 'IIIIII    III'. Manipulating the symbols is clearly easier to perform. Mathematics manipulates representations rather than the real world because it is easier.

We do not know if something like this story is accurate at the beginning of the long history of mathematics. But we do know that imprints on a Sumerian clay tablet led eventually to the astounding sophistication of the proof of Fermat's last theorem and to modern algebra, analysis, and geometry. Mathematics has been shaped by the job it is expected to perform and through countless quirks of culture. Improvised methods designed to deliver a temporary solution to an unforeseen problem become permanent. If they work well, they get passed on and take on a life of their own. Less good solutions eventually fall into disuse in a sort of Darwinian selection of competing ideas. We could call histories like this **cultural evolution**.

But has the counting of cows in a field really got anything to do with modern mathematics? Let's examine the example more closely. We add an 'I' on the tablet for each cow in the field, subject to two strict rules: no cow should be 'counted' more than once and all the cows in the field are counted. Although these rules are quite natural to us, they are mathematically sophisticated. Mathematically, we are establishing a mapping between the marks on the tablet and the cows in the field that is a **one-to-one correspondence**. This means a mapping links a mark to a unique cow (injective) and that all cows in the field are linked (surjective). While these early users of mathematics might not have understood it quite in these terms, they nonetheless needed to use these properties when counting. But there is something else at work here. The compound symbol 'IIIIIII' stands for the whole field of cows. It is a property of the whole set. It expresses the size of the set or its **cardinality**. The counting of cows in a field has a lot to do with the deep nature of mathematics itself.

Indeed, there are three more ideas illustrated by this simple example. The first is the power of numbers to create ordering: I  II  III  IIII is such an ordering. This is called the **ordinal** property of number. Second, it illustrates the special place of sets and mappings in mathematics. We focused on the set of cows and the set of marks on the tablet. Third, we counted the first set by establishing a one-to-one correspondence with the second. This is a technique that works with any sets, including those that have infinitely many members. Mathematics is truly about sets and the mappings between them.

By representing the real world by marks bearing a special relation to their targets, human beings initiated perhaps the most extraordinary technical advance in their history: the invention of symbolic representation. Manipulating symbols is easier than manipulating objects in the world. Moreover, symbols allow this information to be communicated over distance and time. But the most powerful feature of symbols is that they can be used to represent states of affairs that are not physically present. Symbols can represent past worlds, possible worlds, and desired future worlds. Symbols allow us to tell stories, write histories, and make plans. Symbols that do not actually correspond with the world are called **counterfactuals**. They describe 'what if' situations. What if the Allies had lost World War II? What if we add sulfuric acid to copper? What if we wake up one morning to discover that we have been transformed into a giant insect? What if parallel lines could actually meet? What if there was a solution to the equation $x^2 = -1$? The power of symbolic representation is that it allows us to build abstract worlds – virtual realities where the 'what if' conditions are true.

There is a sense in which the world of mathematics is one such virtual universe, containing all manner of exciting and weird things. Mathematicians discuss 11-dimensional hypercubes, infinite sets of numbers, infinite numbers, surfaces that turn you from being right-handed to left-handed as you traverse them, spaces where the angles of a triangle add up to more than 180 degrees, spaces where parallel lines diverge, systems where the order of the operation matters (where $A * B$ is not the same as $B * A$), vectors in infinite-dimensional space, series that go on forever, and geometric figures that are self-similar called fractals (where you can take a small piece of the original figure then enlarge it and it looks identical – truly identical – to the original). And all this started with the making of a simple mark on a clay tablet.

Mathematics uses symbols to describe these amazing structures in the basic language of sets and the mappings between them. Because symbols are abstract and not limited to representing things in the world, mathematicians can use their imaginations to create a virtual reality following its own rule system unhindered by what the world is really like, a **counterfactual world**. In this world, mathematicians can explore the patterns they encounter.

Yet mathematics is remarkably useful in this world. From building bridges to controlling strategy in football, mathematics lies at the heart of the modern world. If mathematics really is so other-worldy, how come it has so much to say about this one?

This is an important question that motivates much of what follows.

If symbolic representation is the most significant technical advance in history, what would you put in second place?

## Purpose: mathematics for its own sake

ToK uses the map metaphor; knowledge is taken to be like a map that is used for a particular purpose, such as solving a particular problem or answering a question. The map is a simplified picture of the world and its simplicity is its strength. It ensures that we get the job done with the least cognitive cost. If this is right, then it is natural to ask about the purpose of this particular map. What problems does it solve or what questions does it answer? There seem to be two categories: those questions that occur strictly within the virtual reality of mathematics itself (mathematics for its own sake) and those that occur in the world outside (mathematics as a tool). These categories broadly correspond to two subdivisions of mathematics that are often two different departments within a university: pure mathematics and applied mathematics.

Let's start with pure mathematics. A typical example of a problem in this category is how to solve a particular type of equation.

An example of a problem in pure mathematics might be how to solve the equation

$$(1) \quad x^3 - 2x^2 - x + 2 = 0$$

The task is to find a value for $x$ that satisfies the equation. In books like this, there are many such equations and, in this context, they often have simple integer solutions. An initial strategy might be to try a value for $x$ to see if it fits. If we try $x = 0$, then equation (1) gives us:

$$0^3 - 2 \cdot 0^2 - 0 + 2 = 0, \text{ i.e. } 2 = 0$$

which is clearly not true. So, we can say that $x = 0$ is not a solution to the equation.

But if we try $x = 1$, then equation (1) gives us:

$$1^3 - 2 \cdot 1^2 - 1 + 2 = 0$$

In other words, $1 - 2 - 1 + 2 = 0$ is true. So, $x = 1$ is a solution to the equation.

The trick now, as you know, is to factor out $(x - 1)$ from equation (1) to give:

$$(2) \quad (x - 1)(x^2 - x - 2) = 0$$

We can now try to find values of $x$ that make the second bracket in (2) equal to 0. This can be done either by trying out hopeful values of $x$ (2 seems to be a good bet, for example) or using the quadratic formula. We end up with $x = 2$ or $x = -1$

The equation therefore has three solutions: $x = 1$ or $x = -1$ or $x = 2$

The history of these problems illustrates the great attraction of pure mathematics. Certainly, these problems were of interest from the 7th century in what is now the Middle East – the home of algebra. The great 11th century Persian mathematician and poet Omar Khayyam wrote a treatise about similar so-called cubic equations and realised they could have more than one solution. By the 16th century, cubic equations were of public interest. In Italy, contests were held to showcase the ability of mathematicians to solve cubic equations, often with a great deal of money at stake. One such contest took place in 1635 between Antonio Fior and Niccolò Tartaglia. Fior was a student of Scipione del Ferro, who had found a method for solving equations of the type $x^3 + ax = b$, which is known as the 'unknowns and cubes problem' (where $a$ and $b$ are given numbers).

Del Ferro kept his method secret until just before his death when he passed the method on to his student. Fior began to boast that he knew how to solve cubics. Tartaglia also announced that he had been able to solve a number of cubic equations successfully. Fior immediately challenged Tartaglia to a contest. Each was to give the other a set of 30 problems and put up a sum of money. The person who had solved the most after 30 days would take all the money.

Tartaglia had produced a method to solve a different type of cubic $x^3 + ax^2 = b$. Fior was confident that his ability to solve cubic equations would defeat Tartaglia and submitted 30 problems of the 'unknowns and cubes' type, but Tartaglia submitted a variety of different problems. Although Tartaglia could not initially solve the 'unknowns and cubes' type of equation, he worked hard and discovered a method to solve this type of problem. He then managed to solve all of Fior's problems in less than two hours. In the meantime, Fior had made little progress with Tartaglia's problems and it was obvious who was the winner. Tartaglia did not take Fior's money though; the honour of winning was enough.

Tartaglia represents the essence of the pure mathematician: someone who is intrigued by puzzles and has a deep desire to solve them. It is the problem itself that is the motivation, not possible real-world applications.

> What other knowledge is worth pursuing for its own sake?

A modern example is the solution of Fermat's conjecture by Andrew Wiles. The French mathematician Pierre de Fermat wrote the conjecture in 1627 as a short observation in his copy of *The Arithmetics of Diophantus*.

The conjecture is that the equation

$$A^n + B^n = C^n$$

where $A$, $B$, $C$ are positive integers and $n > 2$ has no solution. Despite a large number of attempts to prove it, the conjecture remained unproved for 358 years until Wiles published his successful proof in 1995. The proof is way beyond the scope of this book, but there have been a number of interesting books and TV programmes made about it, including Simon Singh *Fermat's Last Theorem* (1997) and the BBC Horizon programme *Fermat's Last Theorem* (1996). As mathematician Roger Penrose remarked, '*QED: how to solve the greatest mathematical puzzle of your age. Lock self in room. Emerge 7 years later*'.

## Purpose: mathematical models

Unlike pure mathematics, which is about the solution of exclusively mathematical puzzles, applied mathematics is about solving real-world problems. The mathematics it produces can be just as interesting from an insider's viewpoint as the problems of pure mathematics (and often the two are inseparable), but a piece of applied mathematics is judged by whether it can be usefully applied in the world.

Here is an example of applied mathematics at work. This is a problem that could have been posed in this book or, indeed (and this is the point), in a physics course.

*A stone is dropped down a 30 m well. How long will it take the stone to reach the bottom of the well, neglecting the effect of air resistance?*

The typical way to solve this type of problem is to use what we call a **mathematical model**. The essence of mathematical modelling is to produce a description of the problem where the main physical features become variables in an equation which is then solved and translated back into the real world.

To model the situation above:

We know that the acceleration due to gravity is 9.8 m s$^{-2}$, and we also know that the distance travelled $s$ is given by the equation:

$$s = \frac{1}{2}at^2, \text{ where } a = \text{acceleration and } t = \text{time}$$

So we substitute the known values into the equation and get:

$$30 = \frac{1}{2}(9.8)t^2$$

Rearranging the equation gives us:

$$\frac{60}{9.8} = t^2, \text{ so } t = \sqrt{\frac{60}{9.8}} = 2.47 \text{ seconds (3 s.f.)}$$

There are a number of points to make about the process here that are typical of mathematical models.

(1) The model neglects factors that are known to operate in the real-world situation. There are two big assumptions made: that the stone will not experience air resistance, which will act as a significant drag force, and that the acceleration due to gravity is constant.

(2) The model appeals to a law of nature. In this case, the law of acceleration due to gravity.

(3) The model uses values for constants that are established empirically. In this case, the acceleration due to gravity at the Earth's surface.

We know that neither of the assumptions in (1) is true. The effect of air resistance can be highly significant. We know that if you have the misfortune to fall from an airplane above 100 m or so, the height does not matter – the speed of impact with the ground will be the same, around 150 km h$^{-1}$, because of the effect of air resistance (of course, it matters how you fall). The changing strength of gravitational force is a less important factor for normal wells.

But if we are dealing with a well that is 4000 km deep, then this factor would be significant. The point is that the model is actually fictional (it even breaks a major law of physics). It could never be true in the sense of exactly corresponding to reality. However, it is a sort of idealisation that we accept because the model provides an approximation to the behaviour of the stone (although not such a good one for deeper wells) and more importantly it gives us understanding of the system. If we were to make the modelling assumptions more realistic, the mathematics in the model would become too complicated to solve easily. Points (2) and (3) show us that the actual content of the model depends on something outside mathematics – namely some well-established results in physics. The mathematics is only a tool, albeit an important one. A model is a mathematical map – a simplified picture of reality that is useful.

Another beautiful example is the Lotka–Volterra model of prey–predator population dynamics in biology. This model was proposed by Alfred Lotka in 1925 and independently by Vito Volterra in 1926.

The model assumes a closed environment where there are only two species, prey and predator, and no other factors. The rate of growth of prey is assumed to be a constant proportion $A$ of the population. The rate at which predators eat prey is $B$, which is assumed to be a constant proportion of the product of predators and prey. The death rate of predators, $C$, is assumed to be a constant proportion of the population, and there is a rate of generation of new predators, $D$, dependent on the product of prey and predators.

These modelling assumptions give rise to a pair of coupled differential equations:

(1) $\dfrac{dx}{dt} = Ax - Bxy$

(2) $\dfrac{dy}{dt} = -Cy + Dxy$

A modern computer package gives the following evolution of prey and predators over time:

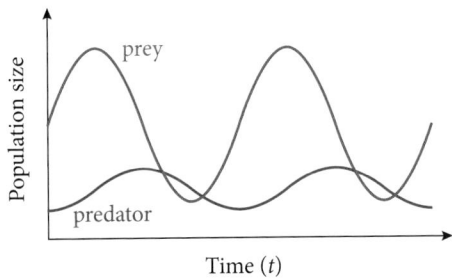

Figure 1 Evolution of prey and predator populations over time

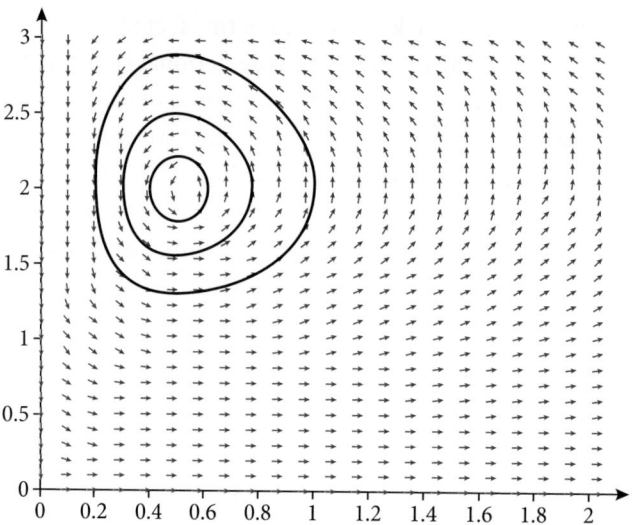

**Figure 2** A phase space diagram. Number of prey (in units of 1000) on the *x*-axis, number of predators on the *y*-axis

It is interesting to look at a phase space diagram that represents each point $(x, y)$ as a combination of numbers of prey and predators. Here the evolution of the system over time appears as a closed loop around the stationary point $\left(\dfrac{C}{D}, \dfrac{A}{B}\right)$, which is an 'attractor' of the dynamical system. (You could try to prove that this is a stationary point – it is not hard.) The position of an orbit around the attractor depends on the initial numbers of prey and predator. Notice that starting the model with too great a population of prey could end up with an extinction of predators (Figure 2) because the very high prey numbers leads to overpopulation of predators for whom there is not enough prey left to eat. The system itself is a nice example of circular causality.

As with the previous example, the modelling assumptions ensure that the mathematics of the model remains tractable, but the cost is that the model is not realistic. It is assumed that the prey do not die from natural causes or that the predators do not come into existence except through the provision of food. There is no competition between either prey or predators. Nonetheless, the model provides some important and powerful insights about the nature of population dynamics. As the model becomes more sophisticated and more factors are taken into consideration, not only does the mathematics become rapidly more difficult, but we lose sight of clear trends in the model (such as orbits around stationary points in phase space). We gain accuracy but lose understanding. This is a characteristic of both models and maps. A map that is as detailed as the territory it depicts is no use to anyone. It is precisely the simplification (literally what makes it false) that makes it useful. Virginia Woolf said about art, '*Art is not a copy of the world; one of the damn things is enough*', and the same could be said about models.

The distinction between pure and applied mathematics becomes blurred in the hands of someone like the great Carl Friedrich Gauss (1777–1855). He was perhaps happiest in the realm of number theory which he called the 'queen of mathematics', and the idea that queens stay in their rarified towers and do not dirty their hands in the ways of the world was perhaps not so far from his thinking. He found great satisfaction in working with patterns and sequences of numbers. It is the same Gauss who, as a young man, enabled astronomers to rediscover the minor planet Ceres after they had lost it in the glare of the sun, by calculating its orbit from the scant data that had been collected on its initial discovery in 1801 and then predicting where in the sky it would be found more than a year later. This feat immediately brought Gauss to the attention of the scientific community. His skills as a number theorist presented him with the opportunity of solving a very real scientific problem.

Who would have guessed that recent work in prime number theory would give rise to a system of encoding data that is used by banks all over the world? The system is called 'dual key cryptography'. The key to the code is a very large number that is the product of two primes. The bank holds one of the primes and the client's computer the other. The key can be made public because in order for it to work it has to be split up into its component prime factors. This task is virtually impossible for large numbers. For example, present computer programs would take longer than the 13.8 billion years since the big bang to find the two prime factors of the number:

25 195 908 475 657 893 494 027 183 240 048 398 571 429 282 126 204 032
027 777 137 836 043 662 020 707 595 556 264 018 525 880 784 406 918 290
641 249 515 082 189 298 559 149 176 184 502 808 489 120 072 844 992 687
392 807 287 776 735 971 418 347 270 261 896 375 014 971 824 691 165 077
613 379 859 095 700 097 330 459 748 808 428 401 797 429 100 642 458 691
817 195 118 746 121 515 172 654 632 282 216 869 987 549 182 422 433 637
259 085 141 865 462 043 576 798 423 387 184 774 447 920 739 934 236 584
823 824 281 198 163 815 010 674 810 451 660 377 306 056 201 619 676 256
133 844 143 603 833 904 414 952 634 432 190 114 657 544 454 178 424 020
924 616 515 723 350 778 707 749 817 125 772 467 962 926 386 356 373 289
912 154 831 438 167 899 885 040 445 364 023 527 381 951 378 636 564 391
212 010 397 122 822 120 720 357

But this number is indeed of the form of the product of two large primes. If you know one of them, it takes an ordinary computer a fraction of a second to do the division and find the other.

Just as pure research in the natural sciences produced results that could also be used for technological or engineering applications, so in mathematics, problems motivated purely from within the most abstract recesses of the subject (pure mathematics) give rise to very useful techniques for solving problems with strong applications in the world outside of mathematics. Mathematicians often practise their art as art for its own sake. They are motivated by the internal beauty and elegance of their subject. Nevertheless, it often happens that pure mathematics created for no other purpose than solving internal mathematical problems turns out to have some extraordinary and very practical applications.

Can you think of an example of a model that does not represent the world well but is nonetheless useful?

What other examples are there of pure research that end up having immense practical benefit?

## Constructivist view of mathematics

Having thought a little about what the purpose of mathematics could be, let's move on to the question of whether it is best thought of as an invention or as something out there in the world.

Broadly speaking, the **constructivist** views mathematics as a human invention. The vision we had of mathematics as a vast virtual reality limited only by the imagination and the rules that are installed there is a constructivist view. However, we are then bound to ask why mathematics has so many useful applications in the real world. Why is mathematics important when it comes to building bridges, doing science and medicine, economics and even playing basketball? Chess is also a game invented by humans, but it does not have very much use in the outside world. Constructivism cannot account for the success of mathematics in the outside world.

On this view, mathematics is what might be called a **social fact**. A social fact is true by virtue of the role that it plays in our social lives. Social facts do have real causal power in the world. That a particular piece of paper is money is a social fact that does make things happen. That piece of paper acquires its status ultimately from a whole set of social agreements. In the end, social facts are produced by **language acts** – performances that change the social world. A language act would be a registry officer saying 'I pronounce you married'. The use of language in a **performative** manner creates social facts. Social facts are no less real or definite than those about the natural world. The statement 'John is married' is definitely either true or false. One is reminded of the story about the little boy who, when asked by his grandmother what day it will be tomorrow, replies, 'Let's wait and see'. Social facts do not require us to wait and see. They rely on social agreements, not on empirical evidence.

The mathematician Reuben Hersh argues for a type of constructivism that he calls **Humanism**. For Hersh, numbers and other mathematical objects are social facts. Hersh defends this view on the Edge website:

*'[Mathematics] … is neither physical nor mental, it's social. It's part of culture, it's part of history, it's like law, like religion, like money, like all those very real things, which are real only as part of collective human consciousness. Being part of society and culture, it's both internal and external: internal to society and culture as a whole, external to the individual, who has to learn it from books and in school. That's what math is.'*

Hersh called his theory of mathematics humanism because it's saying that mathematics is something human. *'There's no math without people. Many people think that ellipses and numbers and so on are there whether or not any people know about them; I think that's a confusion.'*

Hersh points out that we do use numbers to describe physical reality and that this seems to contradict the idea that numbers are a social construction.

It is important to note here that we use numbers in two distinct ways: as nouns and as adjectives. When we say nine apples, nine is an adjective.

If it's an objective fact that there are nine apples on the table, that's just as objective as the fact that the apples are red, or that they're ripe or anything else about them; that's a fact. The problem occurs when we make a subconscious switch to 'nine' as an abstract noun in the sort of problems we deal with in Mathematics class. Hersh thinks that this is not really the same nine. They are connected, but the number nine is an abstract object as part of a number system. It is a result of our mathematics game – our deduction from axioms. It is a human creation.

Hersh sees a political and pedagogical dimension to his thinking about mathematics. He thinks that a humanistic vision of mathematics chimes in with more progressive politics. How can politics enter mathematics? As soon as we think of mathematics as a social construction then the exact arrangements by which this comes about – the institutions that build and maintain it – become important. These arrangements are political. Particularly interesting for us here is how a different view of mathematics can bring about changes in teaching and learning.

*'Humanism sees mathematics as part of human culture and human history. It's hard to come to rigorous conclusions about this kind of thing, but I feel it's almost obvious that Platonism and Formalism are anti-educational, and interfere with understanding, and Humanism at least doesn't hurt and could be beneficial. Formalism is connected with rote, the traditional method, which is still common in many parts of the world. Here's an algorithm; practise it for a while; now here's another one. That's certainly what makes a lot of people hate mathematics (…) There are various kinds of Platonists. Some are good teachers, some are bad. But the Platonist idea, that, as my friend Phil Davis puts it, Pi is in the sky, helps to make mathematics intimidating and remote. It can be an excuse for a pupil's failure to learn, or for a teacher's saying "some people just don't get it". The humanistic philosophy brings mathematics down to earth, makes it accessible psychologically, and increases the likelihood that someone can learn it, because it's just one of the things that people do.'*

There is a possibility that the arguments explored in this section might cast light on an aspect of mathematics learning that has seemed puzzling – why it is that mathematical ability is seen to be closely correlated with a certain type of intelligence. There is a widespread view that mathematics polarises society into two distinct groups: those who can do it and those who cannot. Those who cannot do it often feel the stigma of failure and that there is an exclusive club whose membership they have been denied. Those who can do it often find themselves labelled as 'nerds' or as people who are, in some sense, socially deficient. Is Hersh correct in attributing this to a formalistic or Platonic view? Is he right to suggest that if mathematics is just a meaningless set of formal exercises, then it will not be valued by society? If we deny that mathematics is out there to be discovered, it takes the stigma away from the particular individual who does not make the discovery. It is interesting to speculate on how consequences in the classroom flow from a humanist view of mathematics.

What are the strengths and weaknesses of mathematical humanism?

## Platonic view of mathematics

One way to explain why mathematics applies so well to things like bridges and planets is simply to take mathematics as being out there in the world, independent of human beings. As with other things in the natural world, it is our task to discover it (literally to 'lift the cover'). This is called the **Platonic** view because the philosopher Plato (427–347 BC) took the view that mathematical objects belonged to the real world, underlying the world of appearances in which we lived. Mathematical objects such as perfect circles and numbers existed in this real world; circles on Earth were mere inferior shadows. Many mathematicians have at least some sympathy with this view. They talk about mathematical objects as though they had an existence independent of us and that we are accountable to mathematical truths in the same way as we are accountable to physical facts about the universe. They feel that there really is a mathematical world out there and that they are trying to discover truths about it, much like natural science discovers truth about the physical world.

This view is itself not entirely without problems. In ToK we might want to ask: 'If mathematics is out there in the world, where is it?' We do not see circles, triangles, $\sqrt{2}\,\pi$, i, e, and other mathematical objects obviously floating around in the world. We have to do a great deal of work to find them through inference and abstraction.

While this might be true, there is some evidence that mathematics is hidden not too far below the surface of our reality. Take prime numbers as an example. The Platonist might want to try to find them somewhere in nature. One place where she might start is in Tennessee. In the summer of 2016, the forests were alive with a cicada that exploits a property of prime numbers for its own survival. These cicadas have a curious life cycle. They stay in the ground for 13 years. Then they emerge and enjoy a relatively brief period courting and mating before laying eggs in the ground and dying. There is another species of cicada that has the same cycle and no fewer than 12 types that have a cycle of 17 years. There are, to add to the puzzle, none that have cycles of 12, 14, 15, 16 or 18 years. The clue is that 13 and 17 are prime numbers. There is a predator wasp that has evolved to have a similar life cycle. But if a predator had a life cycle of 6 years, the prey and the predator would only meet every $6 \times 17 = 112$ years. Whereas, if the cicada had a life cycle of 12 years, the prey and predator would meet every cicada cycle. Nature has discovered prime numbers through the cicada life cycles by evolutionary trial and error.

The relationship of nature to geometry was explored by the Scottish biologist D'Arcy Wentworth Thompson in his magnificent book of 1917, *On Growth and Form*. He explored the formation of shells and the wings of dragonflies, and examined the skeletons of dinosaurs through the eyes of a civil engineer constructing bridges and wondered about the formation of bee cells and the arrangement of sunflower seeds.

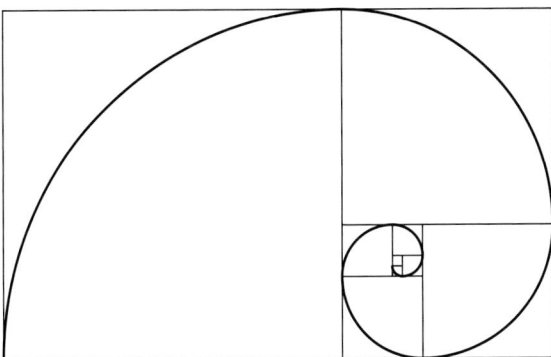

**Figure 3** Spirals in nature

Many spirals in nature are formed, like the one in Figure 3, from the sequence:

1, 1, 2, 3, 5, 8, 13, 21, 34, …

This is called the Fibonacci sequence after the Italian mathematician Leonardo Pisano Bigolio (1170–1250), known as Fibonacci. The Fibonacci sequence is related to the golden number $\varphi$. The interested reader is referred to the many excellent sources on the internet.

Do you think that the mathematics teaching you have experienced reflects a Platonist or constructivist view of mathematics?

# The methods and tools of mathematics

## The language and concepts of mathematics

Knowledge in mathematics is like a map representing some aspect of the world. Like other areas of knowledge, it possesses a specialised vocabulary naming important concepts to build this map. Unlike some areas, this vocabulary is very precisely defined. This makes sense. If the world of mathematics is populated by some rather esoteric objects that are literally like nothing on Earth, then it is very important that these objects are precisely specified.

The other chapters in this book are all about establishing and using this very special vocabulary and becoming fluent in the methods that connect mathematical concepts into meaningful mathematical sentences. We will not spend too much time on these matters here, but there are a few aspects to highlight.

## Notation

Since mathematical objects are abstract and we cannot point to them, we have to represent them with symbols. But the symbol and the idea are different things – there is a danger that we confuse them. Take representations of fractions. The symbols $\frac{1}{3}$, $\frac{2}{6}$, $\frac{3}{9}$, 0.3333… all represent the same number despite appearing to be quite different. (Perhaps the infinite number of ways of representing fractions is one of the reasons why some students have so much difficulty with them.) Some symbols such as $\frac{1}{0}$, $\sin^{-1}(1.2)$ or $\log(-2)$ have no meaning at all. More worrying is that an expression such as 'the smallest real number larger than 1' doesn't actually mean anything either. This is because there is no smallest real number larger than 1. (Think carefully about this.)

In a similar vein, the fact that there are different conventions for writing mathematics does not mean that the mathematics is different. Some conventions represent the number $\frac{3}{10}$ by the decimal 0.3, others by 0,3.

Either way, the mathematics is the same and these do not really count as different mathematical cultures. Carl Friedrich Gauss, one of the greatest mathematicians of all time, said '*non notations, sed notions*' – not notations but notions.

## Algebra

A staple method used in mathematics is the substitution of letters for numbers. In fact, mathematicians use letters for many sorts of mathematical objects, not just numbers. The reason is that they want to make generalised statements. By using a letter, they do not have to commit to making a statement about a specific number, but instead can make one about all numbers of a particular kind at once. This is a very powerful tool.

This is illustrated by a worked example. Imagine we want to prove that if we add an odd number to another odd number we get an even number. We hope to show that this is true for any choice of odd numbers. We could proceed by trying out different pairs of odd numbers and checking that the result is even:

$1 + 3 = 4$ even

$5 + 7 = 12$ even

$13 + 9 = 22$ even

$131 + 257 = 388$ even

You can see that this method will not serve as a proof because we would have to check every possible pair of odd numbers and, since this set is infinite, we would never finish. What we need is to define a general odd number without committing to a particular one. For example, we can define 'odd' by being 'one more than an even number'.

If $k$ is an even number, then we can write $k = 2j$ for some whole number $j$.

If $m$ is an odd number, then we can write $m = 2j + 1$ for some whole number $j$.

All we have to do now is to add two of these general odd numbers together.

So, we want to take two odd numbers, let's say $m = 2j + 1$ and $n = 2i + 1$ where $j$ and $i$ are whole numbers. There is a subtlety here because we use different letters $j$ and $i$ for the whole numbers in the expressions above because we want to allow $m$ and $n$ the possibility of being different odd numbers.

If we used the same letter, say $j$, in the expressions for $m$ and $n$ then we would be making our odd numbers equal and we would only have proved that if we add together two equal odd numbers, then the result is even.

Now we have to use some symbolic rules.

$$m + n = (2j + 1) + (2i + 1)$$

We can remove the brackets and rearrange to give:

$$m + n = 2j + 2i + 2$$

Finally, we can use the fact that 2 is a common factor of all terms in the expression to place it outside a bracket.

$$m + n = 2(j + i + 1)$$

But $j$, $i$ and 1 are all whole numbers so $j + i + 1$ is also a whole number. Technically, this comes from the fact that the whole numbers are **closed under the operation of addition** because they form an important structure called a **group**. Let's call this whole number $p$.

So, we have that $m + n = 2p$. But this is precisely the definition of an even number that we started with. An even number is 2 times a whole number. This proves that any two odd numbers added together gives an even number.

The big chain of reasoning above is called a proof. It is immensely powerful because it covers an infinite number of situations. There is an infinite number of possible pairs of odd numbers to which the result applies. This is the power and beauty of using letters for numbers — a practice that was developed in Baghdad and Damascus about 1000 years ago. In one sense, mathematicians have a god-like ability when it comes to dealing with infinite sets.

## Proof

Proof is the central concept in mathematics because it guarantees mathematical truth. When something is proved, we can say that it is true.

This type of truth is independent of place and time. In contrast to the science of the day, the mathematical truths of Pythagoras are just as true today as they were then – indeed his famous theorem is still taught today as can be seen in this book. But the science of the time has long been rejected. There were four chemical elements in the 4th century BC, and Aristotle thought that the heart was the organ for thinking. Actually, we do not have to go far back in time to find textbooks in the natural sciences that contain statements that we would dispute today. The truths of the natural sciences are always subject to revision, but mathematical truths are eternal.

But there is something even more striking about mathematical truths — that is, mathematical statements that have been proved. A statement such as 'odd + odd = even' has such power that we can say that it is certain. This is not just a matter of confidence – we are not talking about psychological certainty here.

It is certain because it cannot be otherwise. The negation of a mathematical truth (like 'odd + odd = odd') is to utter a self-contradiction or absurdity. Let's reflect on the power of this statement. This means that there is no possible world in which 'odd + odd = odd' (given the standard meanings of these terms). A story that makes this statement is describing a world that is self-contradictory — that is, an absurd and unintelligible world. Such a story is just not credible. But this means that mathematics is really radically different from other areas of knowledge, including the natural sciences. It is not a contradiction to say that the moon is composed of green cheese. There could be universes where this is true, but it just happens not to be in ours. Mathematics deals in what we call **necessary truths**, while the sciences deal mainly in **contingent truth**.

This is something that students of ToK should think about carefully. What is it about mathematical truth that makes it immune to revision and provides the basis for certainty and makes the negation of a mathematical truth a contradiction?

Recall that the constructivist sees mathematics as a big abstract game played by human beings according to invented rules. The hero of *The Glass Bead Game*, a novel by the German writer Hermann Hesse, must learn music, mathematics, and cultural history to play the game. On this view, mathematics is just like the glass bead game. There are parallels we can draw between a game like chess and mathematical proof. First, chess is played on a special board with pieces that can move in a particular way. The pieces must be set up on the board in a particular fashion before the game can begin. The same is true of mathematical proof. It starts with a collection of statements in mathematical language called **axioms**. They themselves cannot be proved. They are simply taken as self-evidently true and form the starting point for mathematical reasoning.

Once the game is set up, we can start playing. A move in chess means transforming the position of the pieces on the board by applying one of the game's rules that govern movement. Typically in chess, a move involves the movement of only one piece. (Can you think of an exception?) If the state of the pieces before the move was legitimate and the move was made according to the rules of the game, then the state of the pieces after the move is also legitimate. The same is true of a mathematical proof. One applies the rules (these are rules of algebra typically) to a line in the proof to get the next line. The whole proof is a chain of such moves.

Finally, the chess game ends. Either one of the players has achieved checkmate, or a stalemate (a draw) has been agreed. Similarly, a mathematical proof has an end. This is a point where the proof arrives at the required result at the end of the chain of reasoning. This result is called a **theorem**.

Once a proof of a mathematical statement is produced, we have a logical duty to believe the result, however unlikely. This is illustrated with a famous example.

Many people do not believe that $1 = 0.99999999\ldots$
(The three dots indicate that the 9's continue indefinitely).

The proof is straightforward.

Let $\qquad\qquad\qquad\qquad x = 0.9999999\ldots$

Then $\qquad\qquad\qquad\quad\; 10x = 9.9999999\ldots$

Subtract both equations $\quad 10x - x = 9.99999999\ldots - 0.9999999\ldots$

This implies $\qquad\qquad\qquad 9x = 9$

Giving $x = 1$ as required.

$0.999999\ldots$ really does look very different to 1 but if the proof works then we are forced to believe that they are the same.

Are you happy with every stage of this proof?

## Sets

A set is a collection of elements that can themselves be sets. They can be combined in various ways to produce new sets. The concepts of a set and membership of a set are **primitive**. This means that they cannot be explained in terms of more simple ideas. These seem to be rather modest beginnings on which to build the complexities of modern mathematics. Nevertheless, in the 20th century there were a number of projects that were designed to do just that: reduce the whole of mathematics to set theory. The most important work here was by Quine, von Neumann and Zermelo, and Bertrand Russell and Alfred North Whitehead in the three volumes of their *Principia Mathematica* of 1910–1913. Starting out with the notion of the empty set and the idea that no set can be a member of itself, we can construct the whole number system.

## Mappings between sets

Once we have established sets in our mathematical universe, we want to do something useful with them. One of the most important ideas in the whole of mathematics is that of a mapping. A mapping is a rule that associates every member of a set with a member of a second set. This is what we were doing when we started this chapter by counting cows. We set up a one-to-one correspondence between a set of numbers and a set of cows.

## Infinite sets

Consider the function $f(x) = 2x$ defined over the natural numbers.

Clearly it sets up a one-to-one correspondence between the set of natural numbers and the set of even numbers (check this yourself). So, this means that there are as many even numbers as there are natural numbers.

This is rather strange because we would think intuitively that there were more natural numbers than even numbers – they are after all the result of taking away an infinite number of odd numbers from the original set. But we are saying that the set that is left over has as many members as the original set. This strangeness is characteristic of infinite sets (indeed it can be used to define what we mean by infinite). Infinite sets can be put in a one-to-one correspondence with a proper subset of themselves.

But the story doesn't stop here. Using sets and mappings we can show that there are many different types of infinity. The set of natural numbers contains the smallest type of infinity, usually denoted by $\aleph_0$, which we call 'aleph nought'. In the 19th century, the German mathematician Georg Cantor showed by an ingenious argument that the number of numbers between 0 and 1 is a bigger type of infinity than aleph nought.

It turns out that there is an infinity of different types of infinity — a whole hierarchy of infinities, in fact — and this probably does not surprise you anymore, there are more infinities than finite cardinal numbers.

The methods and concepts of mathematics, therefore, are quite unlike anything to be found in the sciences, although they do seem to bear a strong resemblance to the arts in terms of the setting of the rules of the game and the use of the imagination. This is something we will explore in the next section.

What is it about the difference between the methods of the natural sciences and mathematics that accounts for the radical difference in types of knowledge produced?

## Mathematics and the knower

English poet John Keats said, *"Beauty is truth, truth beauty – that is all / Ye know on earth, and all ye need to know."*

In this section we will see how mathematics impinges on our personal thinking about the world. One of the more surprising aspects of mathematics is the two-way link to the arts and beauty.

### Beauty by the numbers

There is a long-held view that we find certain things beautiful because of their special proportions or some other intrinsic mathematical feature. This is the thinking that has inspired architects since the times of ancient Egypt and generations of painters, sculptors, musicians, and writers. Mathematics seems to endow beauty with a certain eternal objectivity. Things are beautiful because of the mathematical relationships between their parts. Moreover, this is a very public beauty because it can be dissected and discussed.

Let's take the example of the builders of the Parthenon. They were deeply interested in symmetry and proportion. In particular, they were interested in how to divide a line so that the proportion of the shorter part to the longer part is the same as that of the longer part to the whole. You can check that you get the quadratic equation $x^2 + x - 1 = 0$. One solution to this equation is the golden ratio $x = \dfrac{-1 + \sqrt{5}}{2} = 0.61803398875\ldots = \varphi$.

This proportion features significantly in the design of the Parthenon and many other buildings of the period. Since it is also related to the Fibonacci sequence, you will find $\varphi$ turning up anywhere where there are spirals. It is used quite self-consciously in painting (Piet Mondrian, for example) and in music (particularly the music of Debussy). There are those who go as far as saying that it is present in the proportions of the perfect human figure and that we have a predisposition towards this ratio.

See if you can spot the connection between the golden ratio and the Fibonacci sequence. Hint: write down a difference equation for generating the sequence.

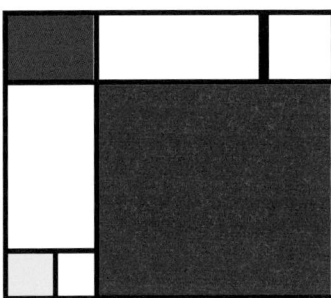

**Figure 4** *Composition with Red, Blue and Yellow* (1926) Piet Mondrian. The proportions of some of the rectangles in this painting is $\varphi$

## Beauty in numbers

Keats also put it the other way around: the beautiful is the true. Could we allow ourselves to be guided to truth in mathematics because of the beauty of the equations? This is a position taken by surprisingly many mathematicians. They look for beauty and elegance as an indicator of truth. Many mathematical physicists were guided in the 20th century by considerations of beauty and elegance.

Einstein suggested that the most incomprehensible thing about the universe was that it was comprehensible. From a ToK point of view, the most incomprehensible thing about the universe is that it is comprehensible in the language of mathematics. Galileo wrote, *'Philosophy is written in this grand book, the universe… It is written in the language of mathematics, and its characters are triangles, circles and other geometric figures…'*

Perhaps what is more puzzling is not just that we can describe the universe in mathematical terms, but that the mathematics we need to do this is mostly simple, elegant, and even beautiful.

To illustrate this, let's look at some of the famous equations of physics. Most people will be familiar with Einstein's field equations and Maxwell's equations.

EINSTEIN'S FIELD EQUATION
for general relativity

$$R_{\mu v} - Rg_{\mu v} = \frac{8\pi G}{c^4} T_{\mu v}$$

**Figure 5** Einstein's field equation

1. $\nabla \cdot \mathbf{D} = \rho_V$

2. $\nabla \cdot \mathbf{B} = 0$

3. $\nabla \times \mathbf{E} = -\dfrac{\partial \mathbf{B}}{\partial t}$

4. $\nabla \times \mathbf{H} = \dfrac{\partial \mathbf{D}}{\partial t} + \mathbf{J}$

**Figure 6** Maxwell's equations

It is perplexing that the whole crazy complex universe can be described by such simple, elegant, and even beautiful equations. It seems that our mathematics fits the universe rather well. It is difficult to believe that mathematics is just a mind game that we humans have invented.

But the argument from simplicity and beauty goes further. Symmetry in the underlying algebra led mathematical physicists to propose the existence of new fundamental particles, which were subsequently discovered. In some cases, beauty and elegance of the mathematical description have even been used as evidence of truth. The physicist Paul Dirac said, *'It seems that if one is working from the point of view of getting beauty in one's equations, and if one has really a sound insight, one is on a sure line of progress'*.

Dirac's own equation for the electron must rate as one of the most profoundly beautiful of all. Its beauty lies in the extraordinary neatness of the underlying mathematics – it all seems to fit so perfectly together:

$$(i\partial\!\!\!/ - m)\psi = 0$$

**Figure 7** Dirac's equation of the electron

The physicist and mathematician Palle Jorgensen wrote:

> *'[Dirac] ... liked to use his equation for the electron as an example stressing that he was led to it by paying attention to the beauty of the math, more than to the physics experiments.'*

It was because of the structure of the mathematics in particular that there were two symmetrical parts to the equation — one representing a negatively charged particle (the electron) and the other a similar particle but with a positive charge — that scientists were led to the discovery of the positron. It seems fair to say that the mathematics did really come first here.

We will leave the last word on this subject to Dirac himself, writing in Scientific American in 1963:

> 'I think there is a moral to this story, namely that it is more important to have beauty in one's equations than to have them fit experiment.'

By any standards this is an extraordinary statement for a mathematical physicist to make.

## Mathematics and personal intuitions

Sometimes our intuition can let us down badly when it comes to making judgments of probability. Here is an example to illustrate how we might have to correct our intuition by careful mathematical reasoning.

Consider the following case. There is a rare genetic disease among the population. Very few people have the disease. As a precaution, a test has been developed to detect whether particular individuals have the disease. Although the test is quite good, it is not perfect — it is only 99% accurate. Person $X$ takes the test and it shows positive. The question for your mathematical intuition is: 'What is the probability that $X$ actually has the disease?' (You should recognise this as being a problem of conditional probability.)

Think about this for a moment before we continue.

Many of the students (and teachers) we have worked with in the past give the same answer: the probability that $X$ actually has the disease given a positive test result is about 99%. Did you say the same? If you did, then your mathematical intuition let you down – very badly.

Let's put some numbers into the problem to illustrate this. For the sake of simplicity, assume that the country in which the test takes place has a population of 10 million. We are told that the disease is very rare. Assume that only 100 people in the whole country have the disease. We are told that the test is 99% accurate so, of the 100 cases of the disease the test would show positive in 99 cases and negative in one. So far so good.

Now consider the 9 999 900 people who don't have the disease. In 99% of these cases the test does its job and records a negative result. In 1% of the cases however it gets it wrong and produces a positive result. 1% of 9 999 900 is 99 999. So, of the whole population tested there would be a total of $99\,999 + 99 = 100\,098$ positive results. But of these only 99 have the disease. Therefore, the probability of having the disease given a positive result is $\frac{99}{100\,098} = 0.0989\%$ or about 1 in 1000. This is quite a big difference from the 990 in 1000 that we expected intuitively. That is out by a whopping 99 000%.

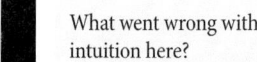 What went wrong with intuition here?

## Mathematics and personal qualities

There are undoubtedly special qualities well-suited to doing mathematics. There are a host of great mathematicians from Archimedes, Euclid, Hypatia, through to Andrew Wiles, Grigori Perelman, and Maryam Mirzakhani, who contributed significantly to the area. Maryam was the first woman to receive the Fields medal (the equivalent of the Nobel prize in mathematics). Although mathematics is collaborative in the sense that mathematicians build on the work of others and take on the challenges that the area itself has recognised as being important, it is nevertheless largely a solitary pursuit. It requires great depth of thought, imaginative leaps, careful and sometimes laborious computations, innovative ways of solving very hard problems, and, most of all, great persistence. Mathematicians need to develop their intuition and their nose for a profitable strategy. They are guided by emotion and by hunches — they are a far cry from the stereotype of the coldly logical thinker who is closer to computer than human.

## Conclusion

We have seen that mathematics is really one of the crowning achievements of human civilisation. Its ancient art has been responsible for some of the most extraordinary intellectual journeys taken by humankind, and its methods have allowed the building of great cities, and the production of great art, and it has been the language of great science.

From a ToK perspective, mathematics, with its absolute and unchanging notion of necessary truth, makes a good contrast to the natural sciences with their reliance on observation of the external world, experimental method, and provisional nature of its results.

Two countering arguments should be set against this view of mathematics. The idea that the axioms of mathematics (the rules of the game) are arbitrary both deprives mathematics of its status as something independent of human beings, and makes it vulnerable to the charge that its results cannot ever be entirely relevant to the world outside mathematics.

Platonists would certainly argue that mathematics is out there in the universe, with or without human beings. They would argue that it is built into the structure of the cosmos – a fact that explains why the laws of the natural sciences lend themselves so readily to mathematical expression.

Both views produce challenging questions in ToK. The constructivist is a victim of the success of mathematics in fields such as the natural sciences. She has to account for why mathematics is so supremely good at describing the outside world to which, according to this view, it should ultimately be blind. The Platonist, on the other hand, finds it hard to identify mathematical structures embedded in the world or has a hard time explaining why they are there.

We have seen how mathematics is closely integrated into artistic thinking; perhaps because both are abstract areas of knowledge indirectly linked to the world and not held to account through experiment and observation, but instead, open to thought experiment and leaps of imagination. Mathematics can challenge our intuitions and can push our cognitive resources as individual knowers. Infinity is not something that the human mind can fathom in its entirety. Instead, mathematics gives us the tools to deal with it in precisely this unfathomed state. We can be challenged by results that seem counter to our intuition, but ultimately, the nature of mathematical proof is that it forces us to accept them nonetheless. In turn, individuals can, through their insight and personal perspectives, make ground-breaking contributions that change the direction of mathematics forever. The history of mathematics is a history of great thinkers building on the work of previous generations to do ever more powerful things using ever more sophisticated tools.

The Greek thinkers of the 4th century BC thought that mathematics lay at the core of human knowledge. They thought that mathematics was one of the few areas in which humans could apprehend the eternal forms only accessible to pure unembodied intellect. They thought that in mathematics they could glimpse the very framework on which the world and its myriad processes rested. Maybe they were right.

# Answers

## Chapter 1

### Exercise 1.1
1. (a) 26    (b) 1    (c) 1201    (d) 83
2. (a) 27.05    (b) 800.01    (c) 3.14    (d) 0.00
3. (Answers will vary)
   - (a) height
   - (b) swimming competition times
   - (c) distance between cities
   - (d) photo file size
   - (e) conversion of 0.5 inches to cm
   - (f) grocery purchase
   - (g) cold day in Singapore
   - (h) current in amps
4. (a) 0.78%    (b) 63.93%    (c) 0.02%    (d) 0.56%
   (a) 0.01%    (b) 0.00%    (c) 0.05%    (d) 100%
5. (a) 3    (b) 4    (c) 2    (d) 5
   (e) 4    (f) 5    (g) 1    (h) 2
6. (a) 5630    (b) 3100    (c) 4 760 000
   (d) 3.14    (e) 0.000 207    (f) 100
   (g) 0.020 1    (h) 0.020 0

### Exercise 1.2
1. $3^6$    2. $3^2$    3. $3^4$    4. $3^{-8}$
5. $3^8$    6. $3^3$    7. $3^{10}$    8. $3^{16}$
9. $3^0$    10. $3^{\frac{3}{2}}$    11. $3^4$    12. $3^{\frac{7}{3}}$
13. $3^{\frac{5}{6}}$    14. $3^1$    15. $3^1$    16. $3^0$
17. $3^{\frac{9}{2}}$    18. $3^{\frac{11}{6}}$    19. $3^{\frac{7}{6}}$    20. $3^3$
21. $3^{\frac{11}{2}}$    22. $3^{\frac{4}{3}}$    23. $3^0$    24. $3^{\frac{-4}{3}}$

### Exercise 1.3
1. (a) $1.203 \times 10^3$   (b) $7 \times 10^9$   (c) $3.01 \times 10^{-4}$
   (d) $2.001 \times 10^1$   (e) $2 \times 10^3$   (f) $7.0 \times 10^{-4}$
   (g) $1.203 \times 10^1$   (h) $1.0006 \times 10^4$   (i) $1.0001 \times 10^1$
   (j) $1 \times 10^{100}$
2. (a) $2 \times 10^6$     (b) $4 \times 10^{-2}$
   (c) $2 \times 10^{12}$    (d) $2 \times 10^1$
3. (a) $1.07 \times 10^9$    (b) $2.15 \times 10^9$
   (c) $2.31 \times 10^1$    (d) $2.25 \times 10^1$

### Exercise 1.4
1. (a) $3 = \log 1000$    (b) $3 = \log_4 64$
   (c) $\frac{3}{2} = \log_{100} 1000$    (d) $\frac{1}{2} = \log_9 3$
   (e) $\frac{1}{2} = \log_8 2\sqrt{2}$    (f) $0 = \log 1$
   (g) $0 = \ln 1$    (h) $-2 = \log_6 \frac{1}{36}$
   (i) $-2 = \log_{\sqrt{2}} \frac{1}{2}$    (j) $-\frac{1}{2} = \log_3 \frac{1}{\sqrt{3}}$
   (k) $-3 = \log_{\frac{1}{2}} 8$    (l) $-\frac{1}{2} = \log_8 \frac{\sqrt{2}}{4}$
   (m) not possible    (n) $-1 = \log_{0.01} 100$
   (o) $3 = \log_{\frac{\sqrt{2}}{2}} \frac{\sqrt{2}}{4}$

2. (a) $x = \log_2 y$    (b) $x = \log y$
   (c) $x = \ln y$    (d) $x = \frac{1}{3}\log_2 y$
   (e) $x = \log_2 \frac{y}{3}$    (f) $x = \log_2(5 - y)$
   (g) $x = \frac{1}{2}\log_3 y$    (h) $x = 2\log_3 y$
   (i) $x = \frac{1}{2}\ln y$    (j) $x = \log_2(y) + 3$
   (k) $x = 2\ln y$    (l) $x = \frac{1}{2}\ln 2y$
3. Yes, $31^{st}$ square
4. (a) $R = 6.2$    (b) $R = 5.5$

### Exercise 1.5
1. (a) 4    (b) $\frac{1}{4}$    (c) 8
   (d) $\frac{1}{2}$    (e) not possible    (f) $\frac{3}{2}$
   (g) 3    (h) $\frac{2}{3}$    (i) 2
   (j) $-1$    (k) $-2$    (l) $\frac{1}{2}$
2. (a) 3   (b) $\frac{1}{2}$   (c) 3   (d) $\frac{2}{3}$   (e) $\frac{3}{2}$
   (f) $\frac{1}{4}$   (g) $\frac{1}{6}$   (h) $\frac{3}{4}$   (i) 1   (j) $\frac{1}{2}$
   (k) 1   (l) $\frac{5}{2}$
3. (a) $x = \frac{1}{9}$    (b) $x = 35$
   (c) $x = \frac{1}{\sqrt{3}}$    (d) $x = 0, x = -2$

### Chapter 1 Practice questions
1. (a) 4   (b) 2   (c) 1   (d) 5
   (e) 5   (f) 1   (g) 3
2. (a) 58 300    (b) 6110    (c) 124 000
   (d) 1.62    (e) 0.00305    (f) 400
3. (a) $2^8$   (b) $2^6$   (c) $2^6$   (d) $2^{-9}$   (e) $2^6$
   (f) $2^6$   (g) $2^{10}$   (h) $2^{24}$   (i) $2^6$   (j) $2^{\frac{3}{2}}$
   (k) $2^4$   (l) $2^3$   (m) $2^{\frac{5}{6}}$   (n) $2^{\frac{5}{4}}$   (o) $2^{-1}$
   (p) $2^{\frac{3}{2}}$   (q) $2^{\frac{9}{2}}$   (r) $2^{\frac{11}{6}}$   (r) $2^3$   (t) $2^1$
   (u) $2^{\frac{1}{2}}$   (v) $2^{-\frac{5}{3}}$   (w) $2^0$   (x) $2^{-1}$
4. (a) $5.227 \times 10^4$    (b) $1.31401 \times 10^1$
   (c) $6.04 \times 10^{-5}$    (d) $9 \times 10^{-3}$
   (e) $9.0 \times 10^{-3}$    (f) $3.2001 \times 10^1$
   (g) $5.00003 \times 10^5$    (h) $1.0000 \times 10^2$
   (i) $1 \times 10^{-6}$ m
5. (a) $1.00 \times 10^{18}$    (b) $1.52 \times 10^1$
   (c) $1.00 \times 10^{-9}$ s    (d) $1.62 \times 10^0$
6. (a) $5 = \log_3 243$    (b) $8 = \log_2 256$
   (c) $\frac{1}{2} = \log_{100} 10$    (d) $\frac{1}{6} = \log_{64} 2$
   (e) $\frac{5}{2} = \log_3 9\sqrt{3}$    (f) $-3 = \log 0.001$
   (g) $0 = \ln 1$    (h) $-3 = \log_5\left(\frac{1}{125}\right)$
   (i) $-2 = \log_{3\sqrt{3}}\left(\frac{1}{27}\right)$    (j) $-\frac{1}{2} = \log_8\left(\frac{1}{2\sqrt{2}}\right)$
   (k) $-3 = \log_{\frac{1}{4}} 64$    (l) $-\frac{1}{2} = \log_{27}\left(\frac{\sqrt{3}}{9}\right)$

**(m)** not possible   **(n)** $-2 = \log_{0.1} 100$

**(o)** $3 = \log_{\frac{\sqrt{3}}{3}}\left(\dfrac{\sqrt{3}}{9}\right)$   **(p)** $-3 = \log_{\frac{1}{\sqrt{2}}} 2\sqrt{2}$

**7. (a)** $x = \log_5 y$   **(b)** $x = \log y$

**(c)** $x = \ln y$   **(d)** $x = \dfrac{1}{2}\log_2 y$

**(e)** $x = \log_3 y - 1$   **(f)** $x = \log_3(y - 7)$

**(g)** $x = -\dfrac{1}{2}\log_2 y$   **(h)** $x = 3\log_2 y$

**(i)** $x = 2\ln y$   **(j)** $x = \log_5 y - 3$

**(k)** $x = \ln y + 1$   **(l)** $x = -\dfrac{1}{2}\ln y$

**8. (a)** 5   **(b)** $\dfrac{1}{5}$   **(c)** $-4$

**(d)** $\dfrac{3}{2}$   **(e)** not possible   **(f)** $\dfrac{3}{4}$

**(g)** 3   **(h)** 3   **(i)** $-2$

**(j)** $-\dfrac{1}{2}$

**9. (a)** 1   **(b)** $\dfrac{1}{2}$   **(c)** 3   **(d)** $\dfrac{2}{3}$   **(e)** 6

**(f)** $\dfrac{1}{4}$   **(g)** 5   **(h)** $\dfrac{2}{3}$   **(i)** 2   **(j)** $-\dfrac{3}{2}$

**10. (a)** $\dfrac{1}{125}$   **(b)** 2   **(c)** $x = 4, x = -2$

**11. (a)** $m = 3, n = 4$   **(b)** $x = \dfrac{15}{2}$

**12.** $a = 5$
**13.** $y = x^2, y = x^{-2}$

## Chapter 2

### Exercise 2.1

**1. (i)** (a) I, (b) J, (c) H, (d) K, (e) L, (f) B, (g) C, (h) G, (i) F
**(ii)** (d) is not a function, it fails the vertical line test.
**2. (a)** 4
**(b) (i)** 5   **(ii)** 7
**(c)** $y \geqslant 0$
**3.** Area $= 4x^2 + 60x$

**4. (a)** $T(x) = \dfrac{\sqrt{x^2 + 1.44}}{15} + \dfrac{4 - x}{25}$

**(b)** 0.226 hours
**(c)** e.g. There is no fence between the field and the road, so it is possible to drive onto the road at any point; no time is lost driving onto the road; no obstacles in the field.
**5. (a)** 0.0346   **(b)** 6 mm
**6. (a)** $a(t) = 500 - 2t, 0 \leqslant t \leqslant 250$
**(b)** 225 minutes
**7. (a)** 93.5 cm$^3$   **(b)** 2.04 cm

**8. (a)** $t(x) = \dfrac{3\sqrt{x^2 + 160\,000}}{1100} + \dfrac{\sqrt{x^2 - 960x + 270\,400}}{200}$

**(b)** $0 \leqslant x \leqslant 480$
**(c)** 2.62 minutes

### Exercise 2.2

**1. (a)** $3x - y - 4 = 0$
**(b)** $2x - y - 7 = 0$
**(c)** $3x + y + 4 = 0$

**2. (a)** $F = \dfrac{9}{5}C + 32$   **(b)** 98.6 °F

**(c)** 5505 °C   **(d)** $F = \dfrac{9}{5}K - 459.4$

**(e) (i)** $-459.4$ °F   **(ii)** $-273$ °C
**3. (a)** $T = \dfrac{5}{8}P$   **(b)** 832 PHP   **(c)** 800 THB
**4. (a)** $C_A = 400 + 50n$, where $n$ = number of months
**(b)** $C_B = 150 + 80n$
**(c)** Model A is cheaper by $110
**5. (a)** $C = 0.8d + 4$
**(b)** $13.60
**(c)** a charge of $1 for every km
**(d)** a fixed charge of $2

**6. (a)** $-\dfrac{1}{3}$   **(b)** $-5$

**7. (a)** $-\dfrac{5}{2}$   **(b)** $-9$

**8. (a)** 75   **(b)** 156
**9. (a)** at $x = 5$ both parts of $f(x) = 0$
**(b)**

**(c)** $-10.4$   **(d)** 20   **(e)** 1.55, 5.21, 8.79
**10. (a)**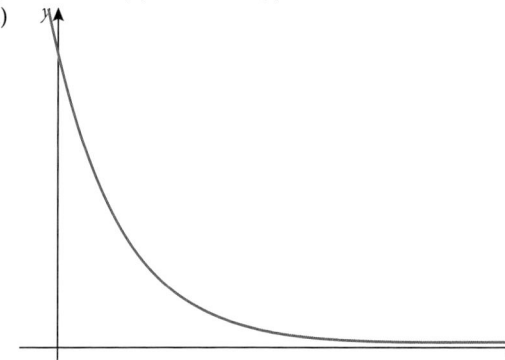

**(b)** 30 Bq   **(c)** 0.631 days
**(d)** $m = 0$   **(e)** decreases towards zero
**11. (a)**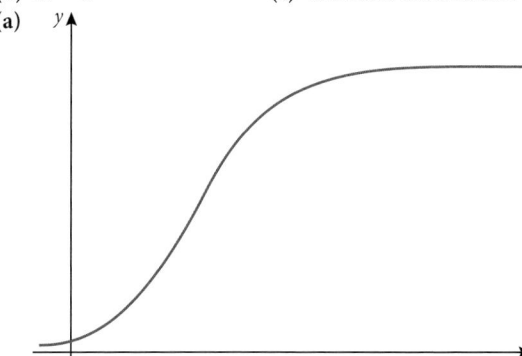

**(b)** 5   **(c)** 1.61 years
**(d)** $P = 100$   **(e)** increases towards 100

# Answers

**12. (a)**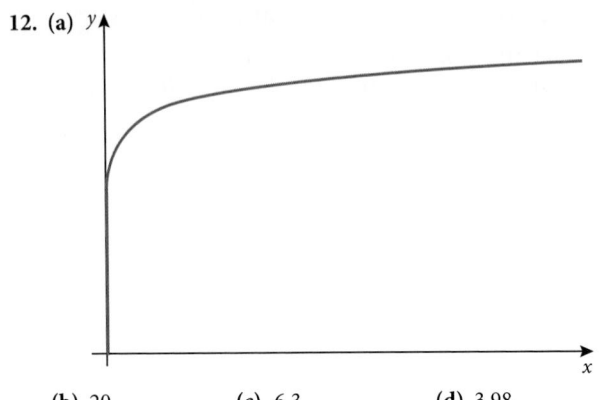

**(b)** 20    **(c)** 6.3    **(d)** 3.98

## Exercise 2.3

**1. (a) (i)** 1    **(ii)** $\dfrac{1}{7}$

**(b) (i)** $\dfrac{2}{x-3}$    **(ii)** $\dfrac{1}{2x-3}$

**2. (a) (i)** 1    **(ii)** $-7$    **(iii)** 7
   **(iv)** $-47$    **(v)** $-1$    **(vi)** $-79$
**(b) (i)** $1-2x^2$    **(ii)** $12x-4x^2-7$
   **(iii)** $4x-9$    **(iv)** $4x^2-x^4-2$

**3. (a)** $12x+7, x \in \mathbb{R}$    **(b)** $4x^2+1, x \in \mathbb{R}$

**(c)** $x+2, x \geqslant -1$    **(d)** $\dfrac{2}{x+3}, x \neq -3$

**(e)** $x, x \in \mathbb{R}$    **(f)** $x^2+1, x \in \mathbb{R}$

**(g)** $\dfrac{2}{4x^2-1}, x \neq 0, x \neq \pm\dfrac{1}{2}$    **(h)** $\dfrac{2x-8}{3x-7}, x \neq 4, x \neq \dfrac{7}{3}$

**4.** $-1.19, 0.543, 1.46, 3.19$

**5. (a)** 5    **(b)** 4    **(c)** $-1$

**6. (a)** $x+6$    **(b)** $x+9$    **(c)** $x+12$
   **(d)** $x+3n$    **(e)** $-130$

**7. (a)** $4x$    **(b)** $8x$    **(c)** $16x$
   **(d)** $2^n x$    **(e)** 18

**8. (a)** $9x-8$    **(b)** $27x-26$    **(c)** $81x-80$
   **(d)** $3^n x - 3^n + 1$    **(e)** 10

**9. (a)** 48 metres of fabric produced every 130 minutes

**(b)** $S(t) = \dfrac{3456}{13}t + 10\sqrt{\dfrac{288}{13}t}$

**(d)** $D(t) = \dfrac{1728}{13}t + 10\left(\sqrt{\dfrac{432}{13}t} - \sqrt{\dfrac{288}{13}t}\right)$

**(e)** \$1 165 396

## Exercise 2.4

**1. (a)** 2    **(b)** 6    **(c)** $-1$    **(d)** b
**2.** 4
**3.** 6
**4. (a)** $\dfrac{x+3}{2}, x \in \mathbb{R}$

**(b)** $4x-7, x \in \mathbb{R}$
**(c)** $x^2, x \geqslant 0$

**(d)** $\dfrac{1}{x}-2, x \neq 0$

**(e)** $\sqrt{4-x}, x \leqslant 4$

**(f)** $x^2+5, x \geqslant 0$

**(g)** $\dfrac{x-b}{a}, x \in \mathbb{R}$

**(h)** $-1-\sqrt{1+x}, x \geqslant -1$

**(i)** $-\sqrt{\dfrac{x+1}{1-x}}, -1 \leqslant x \leqslant 1$

**5. (a)** $\sqrt[3]{x-1}$

**(b)**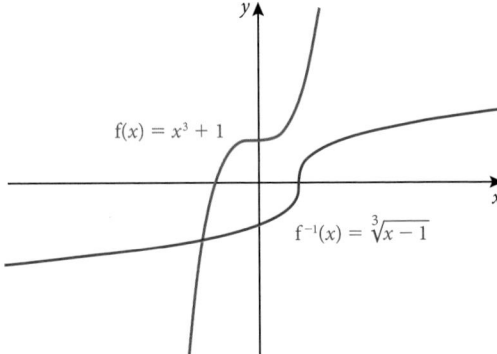

**(c)** $y = x$

**6. (a)**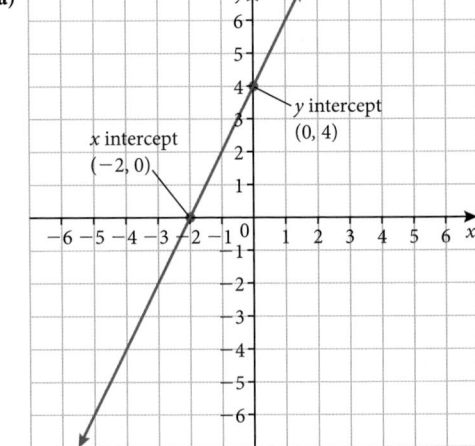

**(b)**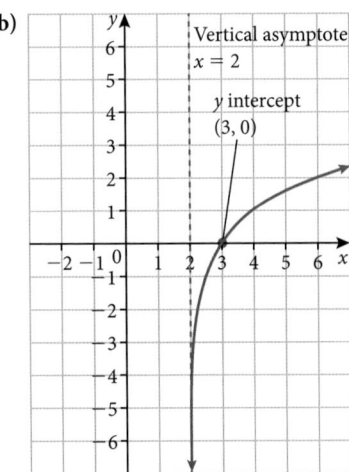

**(c)**

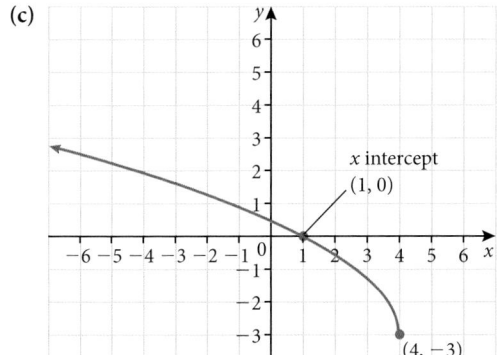

7. **(a)** $-1$ **(b)** $-2$ **(c)** $3$ **(d)** $-1$
8. **(a)** $4x^2 + 4x + 4, y \geqslant 3$ **(b)** $2 < x < 5$
9. **(a)** $-2$ **(b)** $3^{\frac{x+4}{5}} - 2, y > -2$
10. **(a)**

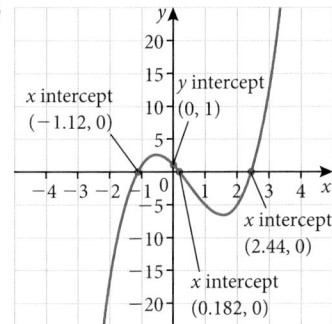

**(b)** it is not one-to-one
**(c)** $-0.541$
**(d)**

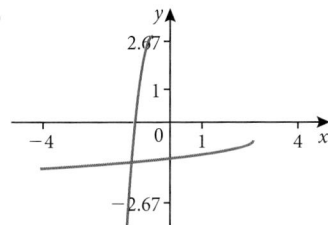

11. **(a)** $\dfrac{0.15}{1.15} = 13\%$

    **(b)** $C(x) = \dfrac{100x}{1000 + x}$

    **(c)** $C^{-1}(x) = \dfrac{1000x}{100 - x}$

    **(d)** 667 ml

12. **(a)** See Worked Solutions

    **(b)** $S^{-1}(x) = \dfrac{60x}{80 - x}$

    **(c)** $46.7 \text{ km h}^{-1}$

13. **(a)** See Worked Solutions

    **(b)** $S^{-1}(t) = \dfrac{20t - 300}{22 - t}$

    **(c)** 50 minutes

14. **(a)** 336 km

    **(b)** $d^{-1}(h) = \sqrt{h^2 + r^2} - r, h > 0$

    **(c)** 3.14 km

1. **(a)**

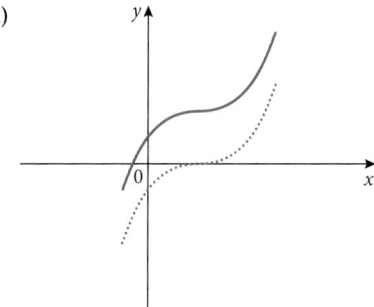

**(b)**

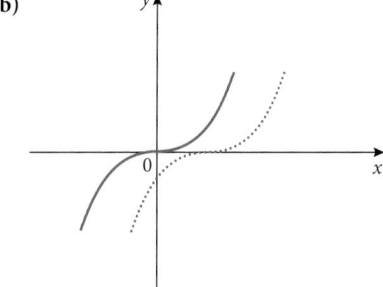

**(c)**

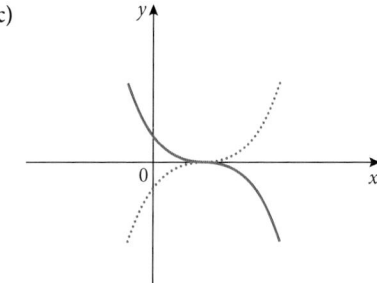

**(d)**

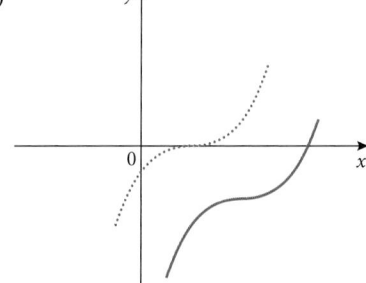

**(e)**

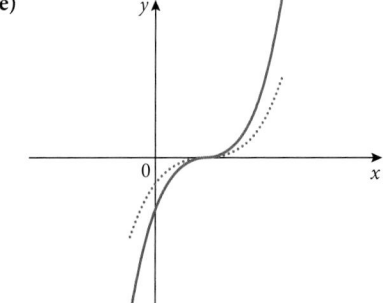

# Answers

**(f)**

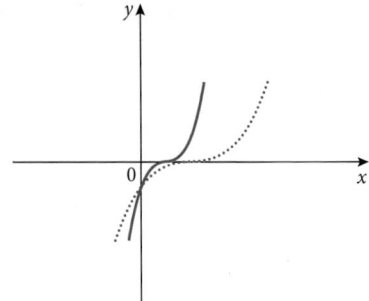

**(g)**

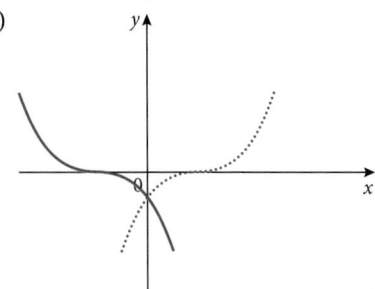

**(h)**

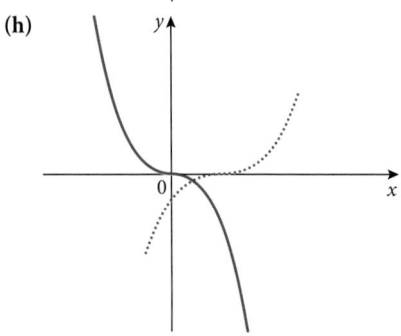

**2. (a)** $g(x) = 12x^2 - 8x$

**(b)** $g(x) = \dfrac{3}{4}x^2 - x$

**(c)** $g(x) = 3x^2 + 4x + 4$

**(d)** $g(x) = 3x^2 - 14x + 15$

**(e)** $g(x) = -3x^2 + 2x$

**(f)** $g(x) = 3x^2 + 2x$

**3.** $y = -x^3 + 6x^2 - 8x$

**4.** $y = -x^3 + 4x + 2$

**5. (a)** A reflection in the $x$-axis, a stretch scale factor 2 with $x$-axis invariant, a translation $\begin{pmatrix} -4 \\ 0 \end{pmatrix}$. (any order is acceptable)

**(b)** A translation $\begin{pmatrix} 3 \\ -2 \end{pmatrix}$.

**(c)** A reflection in the $y$-axis, followed by a stretch scale factor $\dfrac{1}{2}$ with $y$-axis invariant, followed by a translation $\begin{pmatrix} 2 \\ 0 \end{pmatrix}$.

**(d)** A stretch scale factor 2 with $x$-axis invariant, followed by a translation $\begin{pmatrix} 0 \\ 4 \end{pmatrix}$.

**(e)** A reflection in the $x$-axis, followed by a stretch scale factor 3 with $x$-axis invariant, followed by a translation $\begin{pmatrix} 0 \\ 6 \end{pmatrix}$.

**6. (a)** 2.22 m      **(b)** 3 m

**(c)** $h(d) = 1 + 2d - \dfrac{1}{2}d^2$

**7. (a)** 4

**(b)**

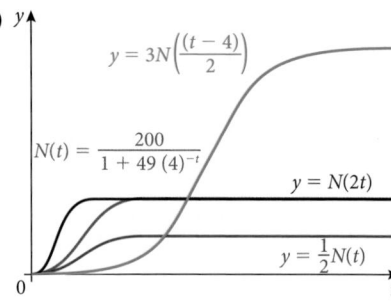

**(c)** $y = 200$

**(d)** See graph

**(e) (i)** $N(0) = 2, y = 100$

**(ii)** $N(0) = 4, y = 200$

**(iii)** $N(0) = 0.76, y = 600$

**(f)** $a = 600, b = 784, c = 2$

**8. (a) (i)** \$100      **(ii)** \$1400

**(b)** $T_1(x) = \begin{cases} 240 & x \leqslant 6000 \\ \dfrac{3(x - 5000)^2}{100\,000} + 240 & 6000 < x \leqslant 10\,000 \\ \dfrac{3x}{25} - 480 & 10\,000 < x \end{cases}$

$T_2(x) = \begin{cases} 0 & x \leqslant 5000 \\ \dfrac{(x - 5000)^2}{40\,000} & 5000 < x \leqslant 9000 \\ \dfrac{x}{10} - 600 & 9000 < x \end{cases}$

$T_3(x) = \begin{cases} 0 & x \leqslant 2500 \\ \dfrac{(x - 2500)^2}{10\,000} & 2500 < x \leqslant 4500 \\ \dfrac{x}{5} - 500 & 4500 < x \end{cases}$

**(c) (i)** $T_1(8000) = 360, T_2(8000) = 100, T_3(8000) = 1100$

**(ii)** $T_1(20\,000) = 1920, T_2(20\,000) = 1400,$
$T_3(20\,000) = 3500$

**9. (a)** 85 °C

**(b)**

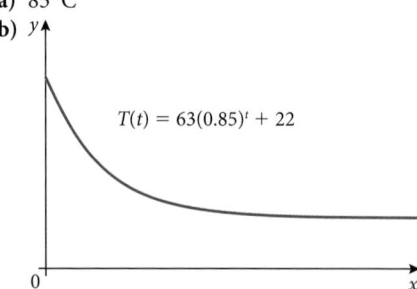

$T(t) = 63(0.85)^t + 22$

**(c)** 31.0 °C

**(d)** 22 °C

**(e)** $T_2(t) = 63(0.922)^t + 22$

**(g)** $T_3(t) = 63(0.614)^t + 22$

**(h)** 4.23 minutes

## Chapter 2 Practice questions

**1. (a)** $M = \dfrac{19}{4}E$

**(b)** 807.50 MYR

**(c)** 736.84 EUR

**2.** $\dfrac{x-5}{3x-2}$

**3.** **(a)** See Worked Solutions

    **(b)** $\dfrac{108-9x}{2x-36}$          **(c)** 9

**4.** **(a)** 6.61 m

    **(b)** 6.5 m

    **(c)** $h(d) = 2 + 6d - 2d^2$

**5.** **(a)**

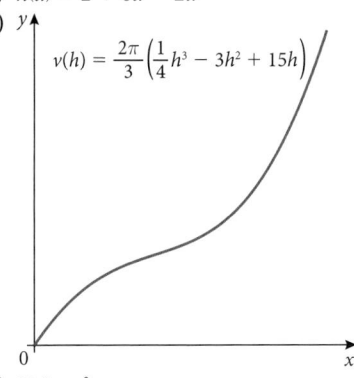

$$v(h) = \frac{2\pi}{3}\left(\frac{1}{4}h^3 - 3h^2 + 15h\right)$$

    **(b)** 209 cm³

    **(c)** **(i)** 7.56 cm   **(ii)** 5.5 cm

    **(d)** $C(h) = \dfrac{5183}{v(h)}$

    **(e)** 25.9%

**6.** **(a)** **(i)** $-4 \le y \le -2$

       **(ii)** $-5 \le y \le -1$

       **(iii)** $\dfrac{3}{2} \le x \le \dfrac{11}{2}$

    **(b)** **(i)** it is not one-to-one

       **(ii)** $-1 \le x \le 3$

       **(iii)** $y$-intercept at $(0, 1)$. Endpoints at $(1, 3)$ and $(-1, -1)$.

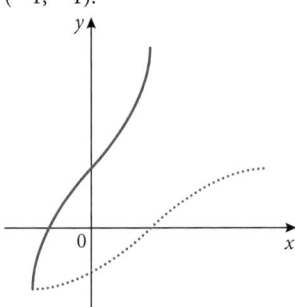

    **(c)** **(i)** $h^{-1}(x) = \dfrac{5 + xd}{2 - x}$

       **(ii)** $-2$

       **(iii)** $k(x) = \dfrac{x + 5}{2}$

**7.** **(a)** **(i)** 0     **(ii)** 6     **(iii)** 0

    **(b)** It is many-to-one, so the inverse would fail the vertical line test.

    **(c)** $-1, 5$

**8.** **(a)**

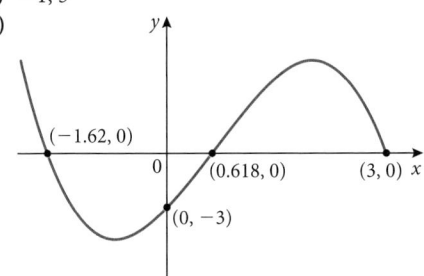

**(b)** It is many-to-one, so the inverse would fail the vertical line test.

**(c)** $-0.667$          **(d)** 1.21

**9.** **(a)** $a > 0$ and $a \ne 1$   **(b)** $y = x^3, y = \dfrac{1}{x^3}$

**10.** **(a)**

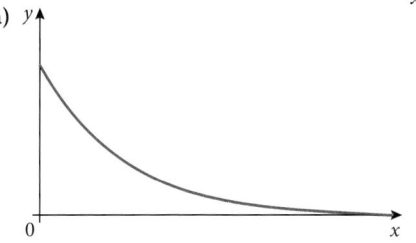

    **(b)** 80 Bq          **(c)** 6.38 days

    **(d)** $y = 0$           **(e)** It approaches zero.

    **(f)** 5.40 days      **(g)** 13.4 days

**11.** **(a)** $\lim\limits_{x \to 2^-} f(x) = 1 - 2(2) = -3$ and

       $\lim\limits_{x \to 2^+} f(x) = \dfrac{3}{4}(2 - 2)^2 - 3 = -3$. So $f$ is continuous.

    **(b)** $g(x) = \begin{cases} 2x - 3, & x \ge 0 \\ \dfrac{3}{4}x^2 - 3, & x < 0 \end{cases}$

**12.** **(a)** 5            **(b)** \$14.60

**13.** **(a)**

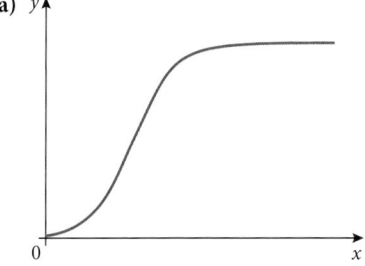

    **(b)** 2           **(c)** 2.66 years

    **(d)** $y = 130$      **(e)** It approaches 130

**14.** **(a)**

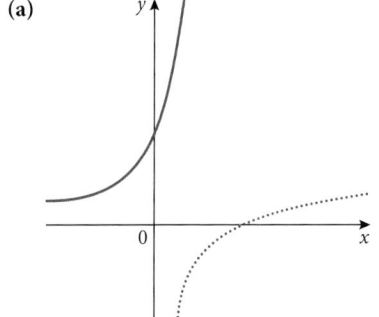

    **(b)** $y > 1$         **(c)** $a = \dfrac{1}{3}, b = -\dfrac{1}{3}$

**15.** **(a)** $y \ge \dfrac{1}{25}$ and $y \ge 0$

    **(b)** $\dfrac{9x^2 - 24x + 166}{3750}$

    **(c)** **(i)** $\sqrt{\dfrac{75x - 3}{2}}$

       **(ii)** $x \ge \dfrac{1}{25}$ and $y \ge 0$

**16.** **(a)** $C(x) = 6\sqrt{x^2 + 2500} + 2\sqrt{x^2 - 240x + 16\,000}$

    **(b)** \$538 million

**17.** $\ln\left(\dfrac{e^2}{x - 3}\right)$

**18.** $f(x) = -\dfrac{1}{2}x^2 + \dfrac{7}{2}x - 4$

# Answers

## Chapter 3

### Exercise 3.1

**1.** **(a)** $a_n = 2n - 3$, $a_1 = -1$, $a_2 = 1$, $a_3 = 3$, $a_4 = 5$, $a_5 = 7$

**(b)** $b_n = 2^n - 3$, $b_1 = -1$, $b_2 = 1$, $b_3 = 5$, $b_4 = 13$, $b_5 = 29$

**(c)** $c_n = 3(2)^{-n}$, $c_1 = \frac{3}{2}$, $c_2 = \frac{3}{4}$, $c_3 = \frac{3}{8}$, $c_4 = \frac{3}{16}$, $c_5 = \frac{3}{32}$

**(d)** $\begin{cases} d_1 = 5 \\ d_n = d_{n-1} + 3, \quad \text{for } n > 1, \end{cases}$
$d_1 = 5$, $d_2 = 8$, $d_3 = 11$, $d_4 = 14$, $d_5 = 17$

**(e)** $e_n = (-1)^n 2^n + 3$, $e_1 = 1$, $e_2 = 7$, $e_3 = -5$, $e_4 = 19$, $e_5 = -29$

**(f)** $\begin{cases} f_1 = 5 \\ f_n = f_{n-1} + 2n, \quad \text{for } n > 1, \end{cases}$
$f_1 = 3$, $f_2 = 7$, $f_3 = 13$, $f_4 = 21$, $f_5 = 31$

**2.** **(a)** $a_n = 2 - 5n$, $a_1 = -3$, $a_2 = -8$, $a_3 = -13$, $a_4 = -18$, $a_5 = -23$, $a_{50} = -248$

**(b)** $b_n = 2 \times 3^{n-1}$, $b_1 = 2$, $b_2 = 6$, $b_3 = 18$, $b_4 = 54$, $b_5 = 162$, $b_{50} = 2 \times 3^{49} = 4.79 \times 10^{23}$

**(c)** $u_n = (-1)^{n-1} \frac{2n}{n^2 + 2}$, $u_1 = \frac{2}{3}$, $u_2 = -\frac{2}{3}$, $u_3 = \frac{6}{11}$, $u_4 = -\frac{4}{9}$, $u_5 = \frac{10}{27}$, $u_{50} = -\frac{50}{1251}$

**(d)** $a_n = n^{n-1}$, $a_1 = 1$, $a_2 = 2$, $a_3 = 9$, $a_4 = 64$, $a_5 = 625$, $a_{50} = 50^{49} = 1.78 \times 10^{83}$

**(e)** $a_n = 2a_{n-1} + 5$, $a_1 = 3$, $a_2 = 11$, $a_3 = 27$, $a_4 = 59$, $a_5 = 123$, $a_{50} = 4.50 \times 10^{15}$

**(f)** $u_{n+1} = \frac{3}{2u_n + 1}$, $u_1 = 0$, $u_2 = 3$, $u_3 = \frac{3}{7}$, $u_4 = \frac{21}{13}$, $u_5 = \frac{39}{55}$, $u_{50} = 1.0000000039208$

**(g)** $b_n = 3b_{n-1}$, $b_1 = 2$, $b_2 = 6$, $b_3 = 18$, $b_4 = 54$, $b_5 = 162$, $b_{50} = 2 \times 3^{49} = 4.79 \times 10^{23}$

**(h)** $a_n = a_{n-1} + 2$, $a_1 = -1$, $a_2 = 1$, $a_3 = 3$, $a_4 = 5$, $a_5 = 7$, $a_{50} = 97$

**3.** **(a)** $u_n = \frac{1}{4} u_{n-1}$ and $u_1 = \frac{1}{3}$

**(b)** $u_n = \frac{4a^2}{3} u_{n-1}$ and $u_1 = \frac{a}{2}$

**(c)** $u_n = u_{n-1} + a + k$ and $u_1 = a - 5k$

**4.** **(a)** $u_n = n^2 + 3$     **(b)** $u_n = 3n - 1$

**(c)** $u_n = \frac{2n - 1}{n^2}$     **(d)** $u_n = \frac{2n - 1}{n + 3}$

**5.** **(a)**

| $n$ | $F_n$ | $a_n$ |
|---|---|---|
| 1 | 1 | 1 |
| 2 | 1 | 2 |
| 3 | 2 | 1.5 |
| 4 | 3 | 1.666666667 |
| 5 | 5 | 1.6 |
| 6 | 8 | 1.625 |
| 7 | 13 | 1.61538461538462 |
| 8 | 21 | 1.61904761904762 |
| 9 | 34 | 1.61764705882353 |
| 10 | 55 | 1.61818181818182 |
| 11 | 89 | 1.61797752808989 |
| 12 | 144 | 1.61805555555556 |
| 13 | 233 | 1.61802575107296 |
| 14 | 377 | 1.61803713527851 |
| 15 | 610 | 1.61803278688525 |
| 16 | 987 | 1.61803444782168 |
| 17 | 1597 | 1.61803381340013 |
| 18 | 2584 | 1.61803405572755 |
| 19 | 4181 | 1.61803396316671 |
| 20 | 6765 | 1.61803399852180 |
| 21 | 10946 | 1.61803398501736 |
| 22 | 17711 | 1.61803399017560 |
| 23 | 28657 | 1.61803398820532 |
| 24 | 46368 | 1.61803398895790 |
| 25 | 75025 | 1.61803398867044 |
| 26 | 121393 | 1.61803398878024 |
| 27 | 196418 | 1.61803398873830 |
| 28 | 317811 | 1.61803398875432 |
| 29 | 514229 | 1.61803398874820 |
| 30 | 832040 | 1.61803398875054 |

**(b)** $\lim_{n \to \infty} a_n = 1.61803$ (correct to 5 decimal places)

**6.** **(a)**

| $n$ | $G_n$ | $b_n$ |
|---|---|---|
| 1 | 1 | 1 |
| 2 | 1 | 3 |
| 3 | 3 | 2.33333333333333 |
| 4 | 7 | 2.42857142857143 |
| 5 | 17 | 2.41176470588235 |
| 6 | 41 | 2.41463414634146 |
| 7 | 99 | 2.41414141414141 |
| 8 | 239 | 2.41422594142259 |
| 9 | 577 | 2.41421143847487 |
| 10 | 1393 | 2.41421392677674 |
| 11 | 3363 | 2.41421349985132 |
| 12 | 8119 | 2.41421357310014 |
| 13 | 19601 | 2.41421356053263 |
| 14 | 47321 | 2.41421356268887 |
| 15 | 114243 | 2.41421356231892 |
| 16 | 275807 | 2.41421356238239 |
| 17 | 665857 | 2.41421356237150 |
| 18 | 1607521 | 2.41421356237337 |
| 19 | 3880899 | 2.41421356237305 |
| 20 | 9369319 | 2.41421356237310 |
| 21 | 22619537 | 2.41421356237309 |
| 22 | 54608393 | 2.41421356237310 |
| 23 | 131836323 | 2.41421356237309 |
| 24 | 318281039 | 2.41421356237309 |
| 25 | 768398401 | 2.41421356237309 |
| 26 | 1855077841 | 2.41421356237309 |
| 27 | 4478554083 | 2.41421356237309 |
| 28 | 10812186007 | 2.41421356237309 |
| 29 | 26102926097 | 2.41421356237309 |
| 30 | 63018038201 | 2.41421356237309 |

**(b)** $\lim_{n \to \infty} b_n = 2.41421$ (correct to 5 decimal places)

### Exercise 3.2

**1.** **(a)** Yes. $d = 2$, $a_{50} = 97$

**(b)** Yes. $d = 1$, $b_{50} = 52$

**(c)** Yes. $d = 2$, $c_{50} = 97$

**(d)** No, difference between terms is not constant.

**(e)** Yes. $d = -7$, $u_{50} = -341$

**2.** **(a)** **(i)** 26    **(ii)** $u_n = 4n - 6$

**(iii)** $u_n = u_{n-1} + 4$ and $u_1 = -2$

**(b)** **(i)** 9.23    **(ii)** $u_n = 10.19 - 0.12n$

**(iii)** $u_n = u_{n-1} - 0.12$ and $u_1 = 10.07$

**(c) (i)** 79    **(ii)** $u_n = 103 - 3n$
    **(iii)** $u_n = u_{n-1} - 3$ and $u_1 = 100$
**(d) (i)** $-\dfrac{27}{4}$   **(ii)** $u_n = \dfrac{13}{4} - \dfrac{5}{4}n$
    **(iii)** $u_n = u_{n-1} - \dfrac{5}{4}$ and $u_1 = 2$

**3.** $a_n = 4n - 14$
**4.** $a_n = \dfrac{11}{3}n - 51$
**5. (a)** 88    **(b)** 36    **(c)** 16    **(d)** 11
**6.** $u_n = 4n + 27$
**7.** Yes, the 3271$^{th}$ term
**8. (a)** $d = -7$, so $u_{110} = 9689 - 7(109) = 8926$.
   **(b)** Yes, 1 is the 1385$^{th}$ term.
**9.** No, 995 is not a term in this sequence.
**10. (a)** $d = 4.33$     **(b)** $u_n = 4.33n + 0.07$
   **(c) (i)** 30.4   **(ii)** 0.330%
**11. (a)** $d = -27$     **(b)** $u_n = 307 - 27n$
   **(c) (i)** 307 g
    **(ii)** An overestimate. The remaining mass includes the mass of the tub, so the amount of ice cream will likely be less than 307 g
   **(d) (i)** 12
    **(ii)** An underestimate. When there is only a small amount of ice cream left, it will be difficult to fill the spoon. So the final few spoonfuls will likely be smaller, meaning more spoonfuls will be required.
**12. (a)** $580      **(b)** 94 years
**13. (a)** $9900     **(b)** $1650
**14.** 3.5%
**15.** 9.30 years

## Exercise 3.3

**1. (a)** Geometric   **(b)** $r = 3^a$   **(c)** $u_{10} = 3^{9a+1}$
**2. (a)** Arithmetic   **(b)** $d = 3$   **(c)** $a_{10} = 27$
**3. (a)** Geometric        **(b)** $r = 2$
   **(c)** $b_{10} = 2^{12} = 4096$
**4. (a)** Neither
**5. (a)** Geometric   **(b)** $r = 3$   **(c)** 78732
**6. (a)** Geometric   **(b)** $r = 2.5$   **(c)** 7629.39
**7. (a)** Geometric   **(b)** $r = -2.5$   **(c)** $-7629.39$
**8. (a)** Arithmetic   **(b)** $d = 0.75$   **(c)** 8.75
**9. (a)** Geometric   **(b)** $r = -\dfrac{2}{3}$   **(c)** $-\dfrac{1024}{2187}$
**10. (a)** Arithmetic   **(b)** $d = 3$   **(c)** 79
**11. (a)** Geometric   **(b)** $r = -3$   **(c)** 19683
**12. (a)** Geometric   **(b)** $r = 2$   **(c)** 51.2
**13. (a)** Neither
**14. (a)** Neither
**15. (a)** Arithmetic   **(b)** $d = 1.3$   **(c)** 14.1
**16. (a)** 32          **(b)** $u_n = 5n - 8$
   **(c)** $u_n = u_{n-1} + 5$ and $u_1 = -3$
**17. (a)** $-9$         **(b)** $u_n = 23 - 4n$
   **(c)** $u_n = u_{n-1} - 4$ and $u_1 = 19$
**18. (a)** 69          **(b)** $u_n = 11n - 19$
   **(c)** $u_n = u_{n-1} + 11$ and $u_1 = -8$
**19. (a)** 9.35
   **(b)** $u_n = 10.15 - 0.1n$
   **(c)** $u_n = u_{n-1} - 0.1$ and $u_1 = 10.05$
**20. (a)** 93          **(b)** $u_n = 101 - n$
   **(c)** $u_n = u_{n-1} - 1$ and $u_1 = 100$

**21. (a)** $-8.5$        **(b)** $u_n = 3.5 - 1.5n$
   **(c)** $u_n = u_{n-1} - 1.5$ and $u_1 = 2$
**22. (a)** 384        **(b)** $u_n = 3(2)^{n-1}$
   **(c)** $u_n = 2u_{n-1}$ and $u_1 = 3$
**23. (a)** 8748       **(b)** $u_n = 4(3)^{n-1}$
   **(c)** $u_n = 3u_{n-1}$ and $u_1 = 4$
**24. (a)** $-5$        **(b)** $u_n = 5(-1)^{n-1}$
   **(c)** $u_n = -u_{n-1}$ and $u_1 = 5$
**25. (a)** $-384$      **(b)** $u_n = 3(-2)^{n-1}$
   **(c)** $u_n = -2u_{n-1}$ and $u_1 = 3$
**26. (a)** $-\dfrac{4}{9}$       **(b)** $u_n = 972\left(-\dfrac{1}{3}\right)^{n-1}$
   **(c)** $u_n = -\dfrac{1}{3}u_{n-1}$ and $u_1 = 972$
**27. (a)** $\dfrac{2187}{64}$      **(b)** $u_n = -2\left(-\dfrac{3}{2}\right)^{n-1}$
   **(c)** $u_n = -\dfrac{3}{2}u_{n-1}$ and $u_1 = -2$
**28. (a)** $\dfrac{390625}{117649}$
   **(b)** $u_n = 35\left(\dfrac{5}{7}\right)^{n-1}$
   **(c)** $u_n = \dfrac{5}{7}u_{n-1}$, $u_1 = 35$
**29. (a)** $-\dfrac{3}{64}$      **(b)** $u_n = -6\left(\dfrac{1}{2}\right)^{n-1}$
   **(c)** $u_n = \dfrac{1}{2}u_{n-1}$ and $u_1 = -6$
**30. (a)** 1216       **(b)** $u_n = 9.5(2)^{n-1}$
   **(c)** $u_n = 2u_{n-1}$ and $u_1 = 9.5$
**31. (a)** $\dfrac{893871739}{12800000} = 69.8$    **(b)** $u_n = 100\left(\dfrac{19}{20}\right)^{n-1}$
   **(c)** $u_n = \dfrac{19}{20}u_{n-1}$ and $u_1 = 100$
**32. (a)** $\dfrac{2187}{1048576}$      **(b)** $u_n = 2\left(\dfrac{3}{8}\right)^{n-1}$
   **(c)** $u_n = \dfrac{3}{8}u_{n-1}$ and $u_1 = 2$
**33.** 234375
**34.** 63
**35.** $\dfrac{2187}{16}$
**36.** $\pm\dfrac{125}{36} = \pm 3.47$
**37.** 11
**38.** $2940.97
**39.** 20.8%
**40. (a)** 197.1 million     **(b)** 59.2%
**41. (a)** $3809.01      **(b)** 12
**42.** $945.23
**43. (a)** $5057.31      **(b)** 178
**44. (a)** 1.389 billion    **(b)** 1.407 billion
   **(c)** 5.86 years, so towards the end of 2021
**45.** 2.3%
**46.** 1.46%

## Exercise 3.4

**1.** 11280      **2.** 940      **3.** 6578
**4.** 42625     **5.** $\dfrac{11718744}{78125}$    **6.** $-\dfrac{105469}{1024}$
**7.** 0.7       **8.** $\dfrac{10}{7}$      **9.** 0.588
**10.** $ 37008.79    **11.** $\dfrac{n}{2}(3n + 7)$

**12.** 17

**13.** 30

**14. (a)** 125250        **(b)** 83501

**15.** $u_1 = 1$ and $d = 5$

**16. (a)** 2890

    **(b)** $\dfrac{52147336773594269}{179528798935698303} = 0.290$

    **(c)** $-\dfrac{479796225641557869184786090396628981226 17}{2324012507657082572151126964510410118 5600}$
    $= -2.06$

**17. (a)** 1.95 m        **(b)** 152 m

**18. (a)** $\dfrac{819}{128} = 6.3984375 \text{ cm}^2$    **(b)** $\dfrac{32}{5} = 6.4 \text{ cm}^2$

**19. (a)** $\dfrac{1023}{8} = 127.875 \text{ cm}^2$    **(b)** $128 \text{ cm}^2$

## Exercise 3.5

**1.** $ 4877.36

**2. (a)** $60 is invested at the beginning of every year for 30 years at 1.5% interest, compounded annually. Solving for the future value gives $2286.11.

    **(b)** A regular payment is invested at the beginning of every year for 15 years at 3% interest, compounded annually. The future value is $1000. Solving for the annual payment gives $52.20.

    **(c)** $100 is invested at the beginning of every year in an account earning 1.2% interest, compounded annually. The future value is $2500. Solving for the number of years gives 21.76.

    **(d)** $50 is invested at the beginning of every month for 30 months at 4% interest, compounded semi-annually. Solving for the future value gives $1579.38.

**3.** 17.3

**4.** 8.26%

**5. (a)** $10524.81

    **(b)** $10383.56. In both cases the annual investment is equal. However, the future value is greater in part **(a)** because the investment earns interest for longer.

**6. (a)** $24 643.63      **(b)** $35 204.98 more than in part **(a)**

    **(c)** $56.81

**7. (a)** 10 years and 7.3 months      **(b)** $267 372

    **(c)** It now takes about 9 years and 7.66 months, with a total payment of $266 018. If the family can afford the extra monthly payment, then they can pay back the loan more quickly and with a smaller total payment.

## Chapter 3 Practice questions

**1. (a)** $\dfrac{a + 2d}{a + 6d} = \dfrac{a}{a + 2d}$ which simplifies to $2d(2d - a) = 0$

    so $d = \dfrac{a}{2}$, since $d \neq 0$.

    **(b)** 32

**2. (a)** 2835      **(b)** $\displaystyle\sum_{n=1}^{27} 7n + 7$      **(c)** 335

**3. (a)** Substitute $a + ar = 10$ into $a + ar + ar^2 + ar^3 = 30$ twice and simplify to get $r^2 = 2$.

    **(b) (i)** $\dfrac{10}{1 + \sqrt{2}}$      **(ii)** 310

**4. (a)** $u_1 = 13$ and $d = 7$      **(b)** 40

**5.** $\dfrac{2}{3}$

**6. (a)** $1.5n + 0.1 - 3(1.2)^{n-1}$      **(b)** $3 \leqslant n \leqslant 9$

    **(c)** 1.642

**7.** 0.0286 m

**8. (a)** 3.26 m      **(b)** 28      **(c)** 156 m

**9. (a)** $\dfrac{1}{3}$      **(b)** 7

**10. (a)** $u_n = \dfrac{7 - a}{7}\left(\dfrac{a}{7}\right)^{n-1}$

    **(b)** $u_1 = \dfrac{7 - a}{7}$ and $r = \dfrac{a}{7}$

    **(c) (i)** $0 < a < 7$      **(ii)** $\dfrac{7}{7 - a}$

**11.** $n = 18$ and $d = \dfrac{6}{17}$

**12. (a) (i)** $S_1 = P\left(1 + \dfrac{I}{100}\right) - R$

    $S_2 = \left(P\left(1 + \dfrac{I}{100}\right) - R\right)\left(1 + \dfrac{I}{100}\right) - R$

    **(ii)** $S_n = P\left(1 + \dfrac{I}{100}\right)_n$

    $- R\left(1 + \left(1 + \dfrac{I}{100}\right) + \dots + \left(1 + \dfrac{I}{100}\right)_{n-1}\right)$

    $= P\left(1 + \dfrac{I}{100}\right)_n - \dfrac{100R}{I\left(\left(1 + \dfrac{I}{100}\right)_{n-1}\right)}$

    **(b) (i)** $111.22      **(ii)** $3651.95

**13. (a)** $298468

    **(b)** $P \times 1.02^{19} + P \times 1.02^{18} + \dots + P$

    **(c)** $12284

    **(d) (i)** $S_n = Q \times 1.028^n - 5000(1 + 1.028 + \dots + 1.028^{n-1})$

    $Q = \dfrac{5000}{1.028} + \dfrac{5000}{1.028^2} + \dots + \dfrac{5000}{1.028^n}$

    **(ii)** $178572

# Chapter 4

## Exercise 4.1

**1. (a)** $VW = 10$    **(b)** $KL = 4\sqrt{5}$    **(c)** $TG = 17$

**2. (a)** $(1, 1)$    **(b)** $(2, 3)$    **(c)** $(-2, -0.5)$

**3. (a)** $4, -\dfrac{1}{4}$    **(b)** $-2, \dfrac{1}{2}$    **(c)** $-\dfrac{2}{3}, \dfrac{3}{2}$

**4. (a)** $(4, 11)$    **(b)** $(3, 9)$    **(c)** $(7, -2)$

    **(d)** $\left(\dfrac{1}{3}, 3\right)$    **(e)** $(30, 23)$    **(f)** $(2, -3)$

**5.** $y = 0.5x - 5$

**6.** $y = -\dfrac{2}{3}x - 5$

**7.** $y = -\dfrac{1}{2}x + 4$

**8.** $y = \dfrac{3}{2}x - 1$

**9.** $y = 2x + 7$

**10.** $y = -2x + 7$

**11.** The treasure is buried at $(5, 2)$

    Regardless of where the point $X$ has moved, the treasure will still lie somewhere along the line $(GC)$. A trench of indeterminate length may have to be dug, however.

## Exercise 4.2

**1. (a)** $\dfrac{24}{25}$    **(b)** $\dfrac{24}{25}$    **(c)** $\dfrac{7}{24}$

    **(d)** $74°$    **(e)** $16°$    **(f)** $16°$

2. **(a)** 53.1°  **(b)** 61.9°  **(c)** 45°

3. With trigonometry: $\theta = \arctan\dfrac{1.62}{2} \approx 39.0°$. Since the distance from the tip of the shadow to the base of the streetlight is 5 m, $x \approx 5 \cdot \tan 39.0° \approx 4.05$ m

   Without trigonometry: Since there are two similar triangles, $\dfrac{1.62}{2} = \dfrac{x}{5} \Rightarrow x = 4.05$ m

4. 49 m

5. Height $= 2 \cdot \cos(36°) \approx 1.618$ m. This is $\Phi$ (the Golden Ratio)

6. **(a)** **(i)** $h_1 = 50 \tan 39° \approx 40$ m
   **(ii)** $h_2 = 50 \tan 50° \approx 60$ m
   **(iii)** 19 m
   **(b)** The part of the diagram containing 11° is not a right-angled triangle

7. 27.7 m

8. 151 m

9. The ratio of the known opposite side to the hypotenuse gives $\sin\theta = \dfrac{18}{30} = \dfrac{3}{5}$, making this a familiar 3-4-5 right-angled triangle. The distance along the ground is equal to $4 \times 6 = 24$ m

10. The original distance is $d_1 = \dfrac{50}{\tan 4°}$ and the second distance: $d_2 = \dfrac{50}{\tan 12°}$; hence, the distance travelled in 5 minutes is $d_1 - d_2 = \dfrac{50}{\tan 4°} - \dfrac{50}{\tan 12°} \approx 479.8$ m

    Its speed would then be approximately 5.7576 km h$^{-1}$ $\approx 3.11$ knots

## Exercise 4.3

1. **(a)** **(ii)** Find side $a$ with the cosine rule, the angle $B$ with sine rule, then angle $C$ by subtraction
   Area $= \dfrac{1}{2}b(c \sin A)$
   **(b)** **(ii)** Find angle $B$ by subtraction, then the missing sides with the sine rule
   Area $= \dfrac{1}{2}b(c \sin A)$
   **(c)** **(ii)** Find the angle $C$ by subtraction, then the two missing sides with the sine rule
   Area $= \dfrac{1}{2}\left(\dfrac{c^2 \sin A \sin B}{\sin C}\right)$
   **(d)** **(ii)** Find angle $A$ with the cosine rule, angle $B$ with the sine rule, then angle $C$ by subtraction
   Once angle $A$ is found, Area $= \dfrac{1}{2}b(c \sin A)$

2. **(a)** $\widehat{C} = 78°$, $a = 5.11$, $b = 9.72$
   **(b)** $\widehat{C} = 72°$, $b = c = 12.9$
   **(c)** $c = 8.72$, $\widehat{A} = 37°$, $\widehat{B} = 83°$
   **(d)** $\widehat{A} = 41°$, $\widehat{B} = 83°$, $\widehat{C} = 56°$

3. **(a)** 24.3 cm$^2$   **(b)** 49.2 cm$^2$
   **(c)** 26.0 cm$^2$   **(d)** 89.3 cm$^2$

4. **(a)** The height of $\triangle PQR$ is 7 cm. Since $p = 10$ cm and $7 < p < 14$, there are two triangles possible.
   $\widehat{R} = 44.4°$ or 135.6°(its supplement)
   **(b)** If $\widehat{R} = 44.4°$, then $\widehat{Q} = 105.6°$ and $q = 19.3$ cm
   If $\widehat{R} = 135.6°$, then $\widehat{Q} = 14.4°$ and $q = 4.98$ cm

5. If $\widehat{R} = 44.4°$, then the area of $\triangle PQR$ is 67.4 cm$^2$
   If $\widehat{R} = 135.6°$, then the area of $\triangle PQR$ is 17.4 cm$^2$

## Exercise 4.4

1. **(a)** 800 cm$^3$
   **(b)** The volume would be the same.

2. Approximately 47.5 cm$^2$

3. **(a)** 262 cm$^3$
   **(b)** 32.7 cm$^3$ (or 11.1%)

4. Since the radius of the hemisphere is limited to the radius of the cylindrical blank, it can be no greater than 2.5 cm; the height of the inverted cone is therefore no greater than 7.5 cm.
   Total volume $= \dfrac{4}{3}\pi \cdot (2.5)^3 + \dfrac{1}{3}\pi(2.5)^2(7.5) \approx 114.54$ cm$^3$, which is 38.9%

5. $\dfrac{(6 - \pi)}{6} \approx 47.6\%$

6. **(a)** Flying along the edge of the grid from $O$ to $P$ is 50 m. Directly back along the diagonal from $P$ to $O$ is 30 m. The total is 80 m.
   **(b)** Three-quarters of the total distance is 60 m, which is 10 m on the way back along the diagonal and $\dfrac{2}{3}$ along the diagonal from $O$ along $OP$. Hence, $\dfrac{2}{3}(10, 20, 20) = \left(\dfrac{20}{3}, \dfrac{40}{3}, \dfrac{40}{3}\right)$

## Chapter 4 Practice questions

1. **(a)** 13   **(b)** 10   **(c)** 6

2. **(a)** $(1, 2)$   **(b)** $(-4, 3)$   **(c)** $\left(\dfrac{9}{2}, -2\right)$

3. **(a)** parallel   **(b)** intersecting   **(c)** coincident

4. **(a)** $(8, 11)$   **(b)** $(-10, -16)$   **(c)** $(-1, 2)$

5. $y = \dfrac{-2}{3}x - 2$

6. $y = -\dfrac{3}{2}x + \dfrac{7}{2}$

7. $y = 2x + 7$

8. $y = \dfrac{1}{2}x - 6$

9. $y = -2x + 10$

10. **(a)** $\dfrac{9}{41}$   **(b)** $\dfrac{40}{41}$   **(c)** $\dfrac{9}{40}$
    **(d)** 77°   **(e)** 25°

11. 49 m

12. 20.4 cm

13. 38.0 m

14. 0.823 m

15. 10 m

16. 26 m

17. **(a)** $\widehat{A} = 105°$, $b = 5.18$ cm, $c = 7.32$ cm
    **(b)** $\widehat{C} = 30°$, $c = 4.26$ cm, $b = 5.47$ cm
    **(c)** $b = 10.9$ cm, $\widehat{A} = 76°$, $\widehat{C} = 68°$
    **(d)** $\widehat{A} = 117°$, $\widehat{B} = 26°$, $\widehat{C} = 37°$

18. **(a)** 18.3 cm$^2$   **(b)** 10.9 cm$^2$
    **(c)** 63.5 cm$^2$   **(d)** 21.3 cm$^2$

19. Either $r = 14 \sin 36°$ precisely (approximately 8.23 cm) or $r > 14$ cm

20. **(a)** 150°   **(b)** $3\pi$ cm$^2$

21. **(a)** 41.8°, 138°   **(b)** $BC = 13.1$ cm

22. $\dfrac{2}{\sqrt{7}}$

# Answers

## Chapter 5

### Exercise 5.1

**1. (a)** arc length $= \dfrac{120}{360}(2\pi)(6) = 4\pi$ cm.

**(b)** arc length $= \dfrac{70}{360}(2\pi)(12) = \dfrac{14}{3}\pi$ cm

**2. (a)** area of sector $= \dfrac{30}{360}\pi(10)^2 = \dfrac{25}{3}\pi$ cm$^2$,

arc length $= \dfrac{30}{360}(2\pi)(10) = \dfrac{5}{3}\pi$ cm

**(b)** area of sector $= \dfrac{45}{360}\pi(8)^2 = 8\pi$ m$^2$,

arc length $= \dfrac{45}{360}(2\pi)(8) = 2\pi$ m

**(c)** area of sector $= \dfrac{52}{360}\pi(180)^2 = 4680\pi$ mm$^2$,

arc length $= \dfrac{52}{360}(2\pi)(180) = 52\pi$ cm

**(d)** area of sector $= \dfrac{n}{360}\pi(15)^2 = \dfrac{5\pi n}{8}$ cm$^2$,

arc length $= \dfrac{n}{360}(2\pi)(15) = \dfrac{\pi n}{12}$ cm

**3.** $12 = \dfrac{\theta}{360}(2\pi)(8) \Rightarrow 270 = \pi\theta \Rightarrow \theta = \dfrac{270}{\pi} \approx 85.9°$

**4. (a)** $1.5 \times 360 = 540°\,\text{s}^{-1}$

**(b)** Bicycle speed is equal to the speed of a point along the circumference of the wheel.

Speed $= 1.5 \times 2\pi(35) \approx 330\,\text{cm s}^{-1}$,

$330\,\dfrac{\text{cm}}{\text{s}} \times \dfrac{60\,\text{s}}{1\,\text{min}} \times \dfrac{60\,\text{min}}{1\,\text{h}} \times \dfrac{1\,\text{m}}{100\,\text{cm}} \times \dfrac{1\,\text{km}}{1000\,\text{m}}$

$= 11.9\,\text{km h}^{-1}$

**5.** $\dfrac{16}{5}\pi = \dfrac{\theta}{360}\pi(4)^2 \Rightarrow \dfrac{1}{5} = \dfrac{\theta}{360} \Rightarrow \theta = 72°$

**6.** $A = l \Rightarrow \dfrac{\theta}{360}\pi r^2 = \dfrac{\theta}{360}(2\pi)r \Rightarrow r^2 = r \Rightarrow r = 0$ or $r = 1$

**7.** Maria $= \dfrac{112}{360}(2\pi)(230) \approx 450$ m

Norbert $= \dfrac{66}{360}(2\pi)(500) \approx 576$ m

$\therefore$ Norbert walks $576 - 450 = 126$ m farther.

**8.** $\dfrac{1}{24}\pi(400)^2 \approx 20\,944 \approx 21\,000\,\text{m}^2\,\text{h}^{-1}$

**9. (a)** The watered area is 11.8 m$^2$, so the water requirement is $2.5\,\text{mm} \times \dfrac{1\,\text{m}}{1000\,\text{mm}} \times 11.8 = 0.0295\,\text{m}^3$. In cm$^3$,

$0.0295\,\text{m}^3 \times \dfrac{100^3\,\text{cm}^3}{1\,\text{m}^3} = 29\,500\,\text{cm}^3$

**(b)** $\dfrac{29\,500}{800} \approx 36.9$ min

**(c)** New area $= \dfrac{100}{360}\pi(3)^2 \approx 7.85$ m$^2$

Water requirement $= 7.85 \times \dfrac{2.5}{1000} \approx 19\,600\,\text{cm}^3$

Required time $= \dfrac{19\,600}{800} \approx 24.5$ min

**10. (a)** 1.85 km  **(b)** 10 000 km

**11.** $\dfrac{329\pi}{4} = 258$ m$^2$

### Exercise 5.2

**1.** $\dfrac{\pi}{3}$  **2.** $\dfrac{5\pi}{6}$  **3.** $-\dfrac{3\pi}{2}$  **4.** $\dfrac{\pi}{5}$

**5.** $\dfrac{3\pi}{4}$  **6.** $\dfrac{5\pi}{18}$  **7.** $-\dfrac{\pi}{4}$  **8.** $\dfrac{20\pi}{9}$

**9.** $-\dfrac{8\pi}{3}$  **10.** 135°  **11.** $-630°$  **12.** 115°

**13.** 210°  **14.** $-143°$  **15.** 300°  **16.** 15°

**17.** 90.0°  **18.** 480°  **19.** 390°, $-330°$

**20.** $\dfrac{7\pi}{2}, -\dfrac{\pi}{2}$  **21.** 535°, $-185°$

**22.** $\dfrac{11\pi}{6}, -\dfrac{13\pi}{6}$  **23.** $\dfrac{11\pi}{3}, -\dfrac{\pi}{3}$

**24.** $3.25 + 2\pi = 9.53$, $3.25 - 2\pi = -3.03$

**25.** $-180°$, 180°

**26.** $\dfrac{\pi}{3}, \dfrac{7\pi}{3}$

**27.** $\pi, -\pi$

**28.** $s = 6\left(\dfrac{2\pi}{3}\right) = 4\pi$

**29.** $s = 1(12) = 12$

**30.** $12 = 8\theta \Rightarrow \theta = \dfrac{2}{3}$ or $\theta = 38.2°$

**31.** $15 = r\left(\dfrac{2\pi}{3}\right) \Rightarrow r = \dfrac{45}{2\pi}$

**32.** $A = \dfrac{1}{2}r^2\theta = \dfrac{1}{2}(4^2)(1.5) = 12$

**33.** $A = \dfrac{1}{2}r^2\theta = \dfrac{1}{2}(10^2)\left(\dfrac{5\pi}{6}\right) = \dfrac{125\pi}{3}$

**34.** $60 = 20\alpha \Rightarrow \alpha = 3 \approx 172°$

**35.** $s = r\theta = 2(16) = 32$

**36.** $24 = \dfrac{60}{360}\pi r^2 \Rightarrow r = \sqrt{\dfrac{144}{\pi}} = \dfrac{12}{\sqrt{\pi}}$ cm $\approx 6.77$ cm

**37. (a)** area $= \dfrac{200\pi}{3} = 209$ mm$^2$, arc length $= \dfrac{20\pi}{3} = 20.9$ mm

**(b)** area $= \dfrac{1225\pi}{3} = 1280$ m$^2$, arc length $= \dfrac{35\pi}{3} = 36.7$ m

**(c)** area $= 9$ km$^2$, arc length $= 6$ km

**(d)** area $= 22.5$ cm$^2$, arc length $= 15$ cm

**38. (a)** $\theta = 2\pi$  **(b)** $\theta = 15\pi$  **(c)** $\theta = 2$  **(d)** $\theta = 20$

**(e)** $\theta = 3\pi$  **(f)** $\theta = \dfrac{\pi}{6}$  **(g)** $\theta = 3$  **(h)** $\theta = 5$

**39. (a)** $l = 3\sqrt{5}$ cm  **(b)** $r = l = 3\sqrt{5}$ cm

**(c)** $2\pi \times 3 = 6\pi$ cm  **(d)** $\theta = \dfrac{2\pi\sqrt{5}}{5}$

### Exercise 5.3

**1.** *After finding the solutions to* **(a)** *using a 30-60-90° right-angled triangle in the unit circle, the solutions to* **(b)**-**(d)** *can be found using symmetry.*

**(a)** $\sin 30° = \dfrac{1}{2}$, $\cos 30° = \dfrac{\sqrt{3}}{2}$, $\tan 30° = \dfrac{\sqrt{3}}{3}$

**(b)** $\sin 150° = \dfrac{1}{2}$, $\cos 150° = -\dfrac{\sqrt{3}}{2}$, $\tan 150° = -\dfrac{\sqrt{3}}{3}$

**(c)** $\sin 210° = -\dfrac{1}{2}$, $\cos 210° = -\dfrac{\sqrt{3}}{2}$, $\tan 210° = \dfrac{\sqrt{3}}{3}$

**(d)** $\sin 330° = -\dfrac{1}{2}$, $\cos 330° = \dfrac{\sqrt{3}}{2}$, $\tan 330° = -\dfrac{\sqrt{3}}{3}$

**2.** *After finding the solutions to* **(a)** *using a 30-60-90° right-angled triangle in the unit circle, the solutions to* **(b)**-**(d)** *can be found using symmetry.*

**(a)** $\sin 60° = \dfrac{\sqrt{3}}{2}$, $\cos 60° = \dfrac{1}{2}$, $\tan 60° = \sqrt{3}$

**(b)** $\sin 120° = \dfrac{\sqrt{3}}{2}$, $\cos 120° = -\dfrac{1}{2}$, $\tan 120° = -\sqrt{3}$

**(c)** $\sin 240° = -\dfrac{\sqrt{3}}{2}$, $\cos 240° = -\dfrac{1}{2}$, $\tan 240° = \sqrt{3}$

**(d)** $\sin 300° = -\dfrac{\sqrt{3}}{2}$, $\cos 300° = \dfrac{1}{2}$, $\tan 300° = -\sqrt{3}$

**3. (a)** $\sin 0° = 0$, $\cos 0° = 1$, $\tan 0° = 0$
   **(b)** $\sin 90° = 1$, $\cos 90° = 0$, $\tan 90°$ is undefined
   **(c)** $\sin 180° = 0$, $\cos 180° = -1$, $\tan 180° = 0$
   **(d)** $\sin 270° = -1$, $\cos 270° = 0$, $\tan 270°$ is undefined

**4. (a)** Start by drawing a right-angled triangle in the coordinate plane with the hypotenuse as the radius of a circle and $\theta$ in standard position:

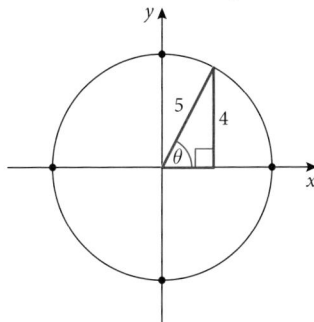

The length of the missing side must be 3 (Pythagorean theorem or by recognising the Pythagorean triple $3:4:5$).
Therefore, $\cos\theta = \dfrac{3}{5}$ and $\tan\theta = \dfrac{4}{3}$.

**5.** *Same method as #4:*

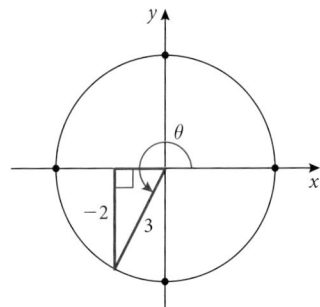

The missing side is $x^2 + 2^2 = 3^2 \Rightarrow x = \sqrt{9-4}$
$= \sqrt{5} \Rightarrow -\sqrt{5}$.
Therefore, $\cos\theta = -\dfrac{\sqrt{5}}{3}$ and $\tan\theta = \dfrac{-2}{-\sqrt{5}} = \dfrac{2\sqrt{5}}{5}$.

**6.** *Same method as #4.*
If $\cos\theta = \dfrac{3}{4}$, the angle is in quadrants I or IV. Since $\pi < \theta < 2\pi$ implies the angle is in quadrants III or IV, we conclude the angle is in quadrant IV.

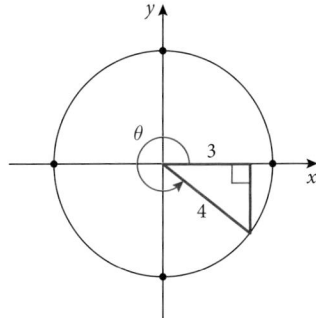

The missing side is $y^2 + 3^2 = 4^2 \Rightarrow x = \sqrt{16-9}$
$= \sqrt{7} \Rightarrow -\sqrt{7}$.
Therefore, $\sin\theta = -\dfrac{\sqrt{7}}{4}$ and $\tan\theta = -\dfrac{\sqrt{7}}{3}$.

**7. (a)** I     **(b)** $\left(\cos\left(\dfrac{\pi}{6}\right), \sin\left(\dfrac{\pi}{6}\right)\right) = \left(\dfrac{\sqrt{3}}{2}, \dfrac{1}{2}\right)$

**8. (a)** IV     **(b)** $\left(\dfrac{1}{2}, -\dfrac{\sqrt{3}}{2}\right)$

**9. (a)** IV     **(b)** $\left(\dfrac{\sqrt{2}}{2}, -\dfrac{\sqrt{2}}{2}\right)$

**10. (a)** on the $y$-axis   **(b)** $(0, 1)$

**11. (a)** II     **(b)** $(-0.416, 0.909)$

**12. (a)** IV     **(b)** $\left(\dfrac{\sqrt{2}}{2}, -\dfrac{\sqrt{2}}{2}\right)$

**13. (a)** IV     **(b)** $(0.540, -0.841)$

**14. (a)** III     **(b)** $\left(-\dfrac{\sqrt{2}}{2}, -\dfrac{\sqrt{2}}{2}\right)$

**15. (a)** III     **(b)** $(-0.929, -0.369)$

**16.** $\sin\dfrac{\pi}{3} = \dfrac{\sqrt{3}}{2}$, $\cos\dfrac{\pi}{3} = \dfrac{1}{2}$, $\tan\dfrac{\pi}{3} = \sqrt{3}$

**17.** $\sin\dfrac{5\pi}{6} = \dfrac{1}{2}$, $\cos\dfrac{5\pi}{6} = -\dfrac{\sqrt{3}}{2}$, $\tan\dfrac{5\pi}{6} = -\dfrac{\sqrt{3}}{3}$

**18.** $\sin\left(-\dfrac{3\pi}{4}\right) = -\dfrac{\sqrt{2}}{2}$, $\cos\left(-\dfrac{3\pi}{4}\right) = -\dfrac{\sqrt{2}}{2}$, $\tan\left(-\dfrac{3\pi}{4}\right) = 1$

**19.** $\sin\dfrac{\pi}{2} = 1$, $\cos\dfrac{\pi}{2} = 0$, $\tan\dfrac{\pi}{2}$ is undefined

**20.** $\sin\left(-\dfrac{4\pi}{3}\right) = \dfrac{\sqrt{3}}{2}$, $\cos\left(-\dfrac{4\pi}{3}\right) = -\dfrac{1}{2}$, $\tan\left(-\dfrac{4\pi}{3}\right) = -\sqrt{3}$

**21.** $\sin 3\pi = 0$, $\cos 3\pi = -1$, $\tan 3\pi = 0$

**22.** $\sin\dfrac{3\pi}{2} = -1$, $\cos\dfrac{3\pi}{2} = 0$, $\tan\dfrac{3\pi}{2}$ is undefined

**23.** $\sin\left(-\dfrac{7\pi}{6}\right) = \dfrac{1}{2}$, $\cos\left(-\dfrac{7\pi}{6}\right) = -\dfrac{\sqrt{3}}{2}$, $\tan\left(-\dfrac{7\pi}{6}\right) = -\dfrac{\sqrt{3}}{3}$

**24.** $\sin 1.25\pi = -\dfrac{\sqrt{2}}{2}$, $\cos 1.25\pi = -\dfrac{\sqrt{2}}{2}$, $\tan 1.25\pi = 1$

**25.** $\dfrac{13\pi}{6} - 2\pi = \dfrac{\pi}{6}$, $\sin\dfrac{\pi}{6} = \dfrac{1}{2}$, $\cos\dfrac{\pi}{6} = \dfrac{\sqrt{3}}{2}$

**26.** $\dfrac{10\pi}{3} - 2\pi = \dfrac{4\pi}{3}$, $\sin\dfrac{4\pi}{3} = -\dfrac{\sqrt{3}}{2}$, $\cos\dfrac{4\pi}{3}, = -\dfrac{1}{2}$

**27.** $\dfrac{15\pi}{4} - 2\pi = \dfrac{7\pi}{4}$, $\sin\dfrac{7\pi}{4} = -\dfrac{\sqrt{2}}{2}$, $\cos\dfrac{7\pi}{4} = \dfrac{\sqrt{2}}{2}$

**28.** $\dfrac{17\pi}{6} - 2\pi = \dfrac{5\pi}{6}$, $\sin\dfrac{5\pi}{6} = \dfrac{1}{2}$, $\cos\dfrac{5\pi}{6} = -\dfrac{\sqrt{3}}{2}$

## Exercise 5.4

**1.** GDC screen shown below.

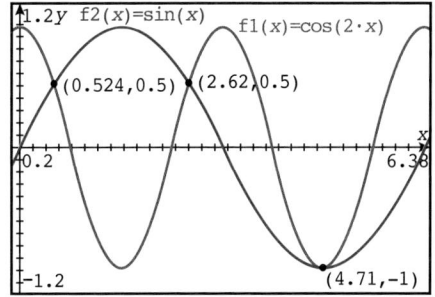

Therefore $x = \{0.524, 2.62, 4.71\}$

# Answers

**2.** GDC screen shown below.

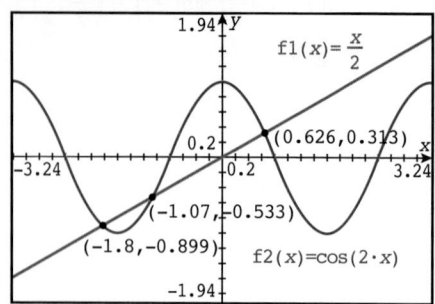

Therefore $x = \{-1.80, -1.07, 0.626\}$

**3.** $x = \{22.0, 38.0\}$ (to 3 significant figures)

**4.** $x = \{1.23, 1.91\}$

**5.** By the Pythagorean identity, $\cos^2 x + \sin^2 x = 1$, hence $\cos^2 x + \sin^2 x = 2$ has no solution since $1 \neq 2$.

**6. (a)** $x = 2.80, 4.34$

   **(b)** $x = 1.57, 4.71$

   **(c)** $x = 0.427, 2.71$

   **(d)** $x = 0.524, 2.62$

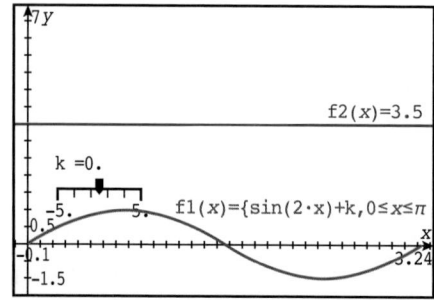

**7.** We first graph the equation with $k = 0$:

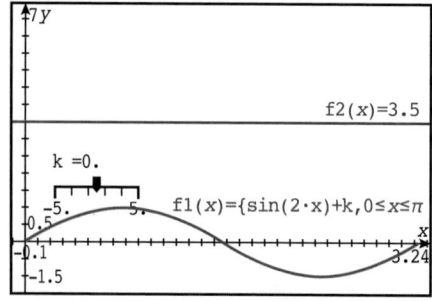

**(a)** Knowing that $k$ is producing a vertical translation, we can see that the equation will have one solution when the maximum or minimum of $\sin(2x) + k$ is at 3.5. Since the maximum is 1 when $k = 0$, one solution will be when the maximum is translated up $k = 3.5 - 1 = 2.5$ units. Verify this graphically:

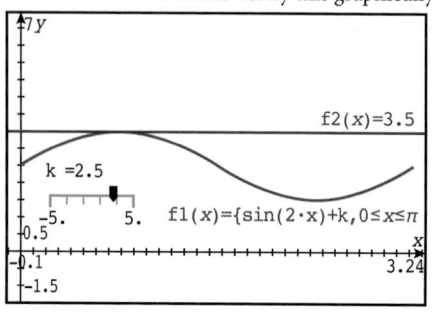

**(b)** The next solution is when the minimum of $-1$ is translated to 3.5, so $k = 3.5 - (-1) = 4.5$ Verify graphically:

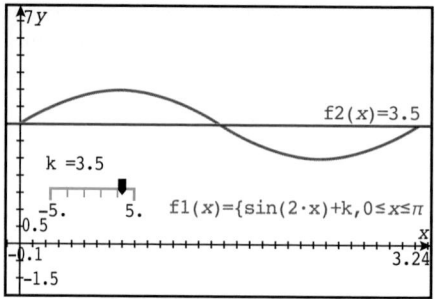

Therefore, the equation has one solution when $k \in \{2.5, 4.5\}$.

Using the reasoning above, the equation will have two solutions when $2.5 < k < 4.5, k \neq 3.5$.

**(c)** Using the reasoning above, the equation will have three solutions when $k = 3.5$. Verify graphically:

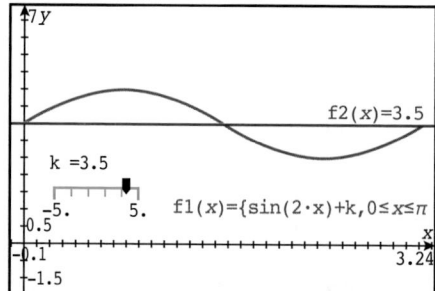

**(d)** Using the reasoning above, the equation will have no solutions when $k < 2.5$ or $k > 4.5$.

**8.** We first graph the equation with $k = 0$ and find maxima and minima:

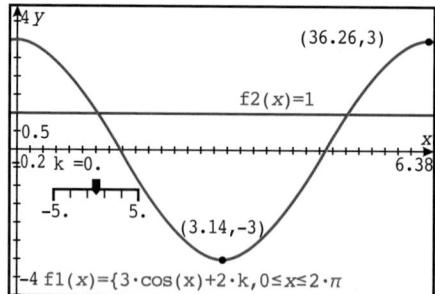

**(a)** Knowing that $2k$ is producing a vertical translation, we see that the equation will have two solutions when the maximum of $3\cos(x) + 2k$ is at 1. Since the maximum is 3 when $2k = 0$, one solution will be when the maximum is translated $2k = 1 - 3 \Rightarrow k = -1$.

Verify this graphically:

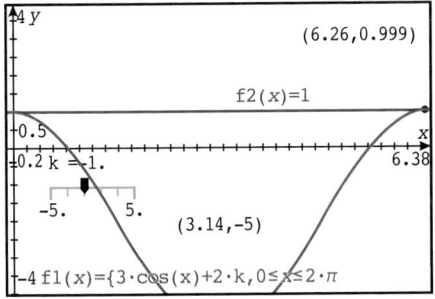

As $k$ increases, the function will translate up, and will continue to have two solutions until the minimum at $-3$ is reached. Hence, $2k = 1 - (-3) \Rightarrow k = 2$.

Verify this graphically:

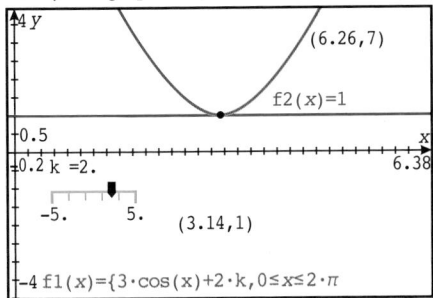

Therefore, the equation has two solutions when $-1 \leqslant k < 2$.

**(b)** Using the reasoning above, the equation will have one solution when $k = 2$.

**(c)** Using the reasoning above, the equation will have no solutions when $k < -1$ or $k > 2$.

## Exercise 5.5

**1. (a)** $F$        **(b)** $y = x + 8$

**(c)** The cells for sites $F$, $B$, $C$, $G$, and $H$ will be changed.

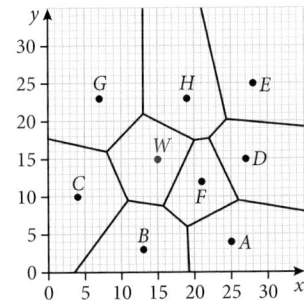

**(d)**

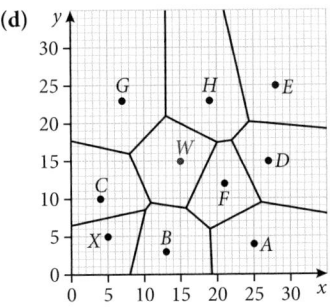

No, the new cell will have at least two rays as edges.

**2.** The process is shown below.

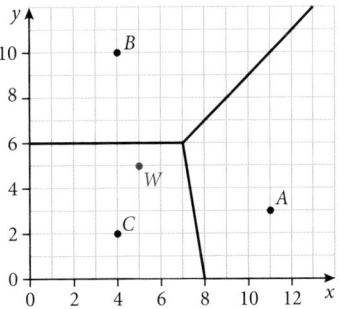

Plot the new site $W$.

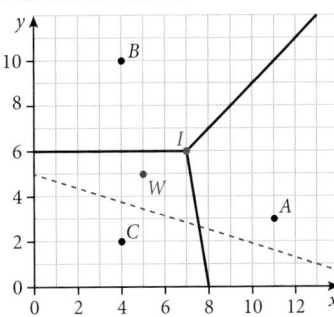

Draw the perpendicular bisector with containing cell's site ($C$).

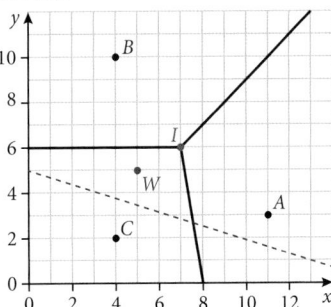

Add ray or line segment between $W$ and $C$.

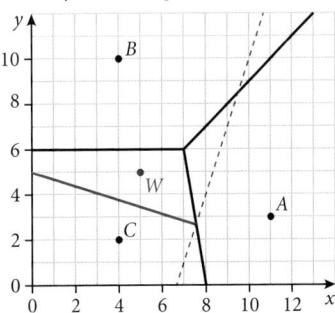

Find perpendicular biector between $W$ and next adjacent cell site ($A$).

# Answers

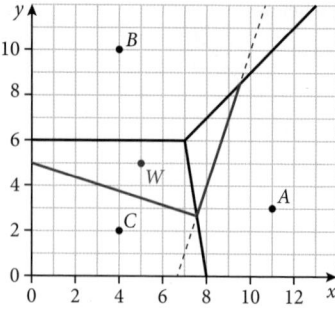

Add ray or line segment between $W$ and $A$.

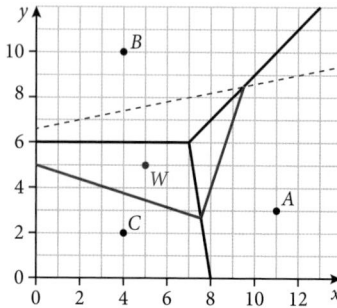

Find perpendicular bisector between $W$ and next adjacent cell site ($B$)

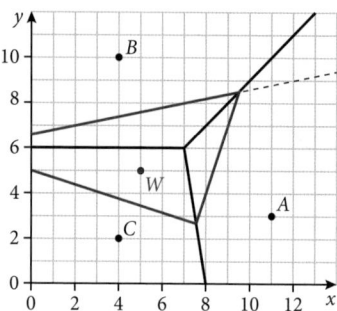

Add ray or line segment between $W$ and $B$.

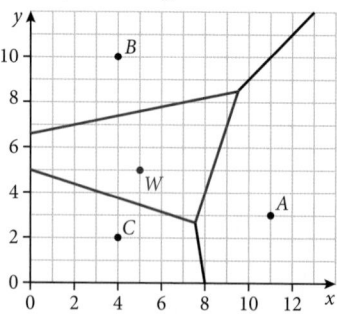

Remove edges inside $W$'s new cell.

3. **(a)** We can find $Q$ by reflecting $P$ across the edges shared with the adjacent, empty cell, which gives $Q(4, 1)$.

**(b)** $A = \frac{1}{2}(4)(3) = 6 \text{ km}^2$

**(c)** The point $X(9, 2)$ is in the cell of site $R$. Therefore, we find the distance to point $R(5, 4)$ using the distance formula: $d = \sqrt{(9-5)^2 + (2-4)^2} = \sqrt{4^2 + (-2)^2}$

$= \sqrt{20} \approx 4.47 \text{ km}$

4. **(a)** $A = \frac{1}{2}(9)(6) = 27 \text{ km}^2$

**(b)** Triangle $BLN$ is a special right-angled triangle with sides in the ratio $1 : \sqrt{3} : 2$, therefore $BLN = 60°$.

**(c)** The region can be seen as a triangle (shown in green) and the sector of a circle (shown in orange).

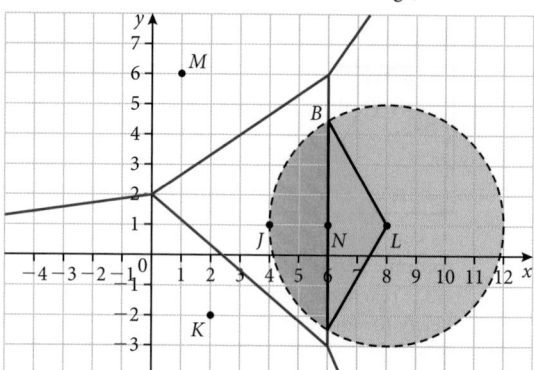

Area = Area of triangle + Area of sector

$$= \frac{1}{2}(2 \times 2\sqrt{3})(2) + \frac{360 - 120}{360}\pi(4^2)$$

$$= 4\sqrt{3} + \frac{32}{3}\pi$$

$$\approx 40.4 \text{ km}^2$$

5. **(a)** The partial diagrams are shown below.

**(i)**

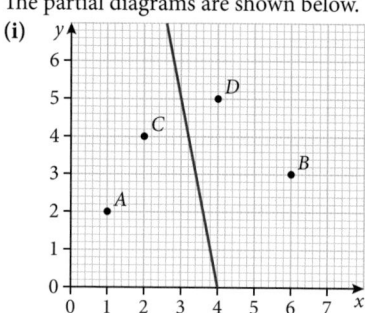

**(ii)**

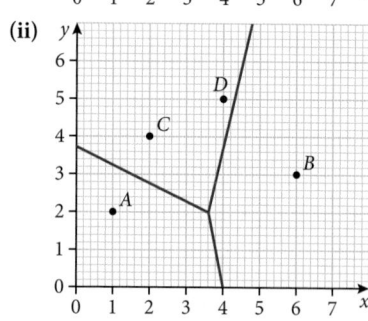

**(iii)**

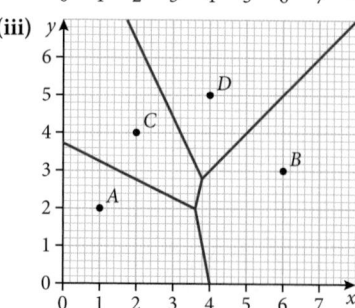

**(b)** Point $M(3.5, 2.5)$ is in the cell for site $C$. Therefore, it is likely to have soil type **silt**.

(c) $N(5, 4)$ is on the boundary between the cells for sites $B$ and $D$. Since is it equidistant to sites $B$ and $D$, it is not possible to determine the likely soil type.

(d) Cells for sites $A$ and $B$ would be divided. Cell $B$ would be divided because $E(6, 0)$ is in the cell $B$. Cell $A$ would be divided because the perpendicular bisector of segment $EB$ intersects the boundary of cell $A$. There are no more adjacent cells, so no other cells would be divided.

(e) No, the answer to (b) does not change because the cell for site $C$, which contains $M$, does not change.

(f) The study area likely to be loam is contained within cell $E$. After we add site $E$ to the Voronoi diagram, we have the following:

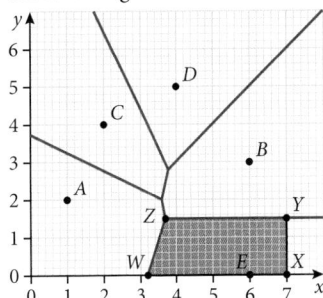

(The purple shaded are is the cell for site $E$.)
Cell $E$ forms a trapezoid. To find the area we must determine the height and length of the two bases. The height $= 1.5$ (from the graph). The base lengths require more work: we must find the equation of the boundary between sites $A$ and $E$ to find exact coordinates of the points $Z$ and $W$.

Midpoint of $AE = \left(\dfrac{6+1}{2}, \dfrac{0+2}{2}\right) = (3.5, 1)$

Slope of $AE = \dfrac{2-0}{1-6} = -\dfrac{2}{5} \Rightarrow$ slope of $ZW = \dfrac{5}{2}$

$\therefore$ equation of $ZW$: $y - 1 = \dfrac{5}{2}(x - 3.5) \Rightarrow y = \dfrac{5}{2}x - 7.75$

$\therefore$ when $y = 0 \Rightarrow 0 = \dfrac{5}{2}x - 7.75 \Rightarrow x = 3.1 \Rightarrow W(3.1, 0)$

$\therefore$ when $y = 1.5 \Rightarrow 1.5 = \dfrac{5}{2}x - 7.75 \Rightarrow x = 3.7$

$\Rightarrow Z(3.7, 1.5)$

Therefore the bottom base is $7 - 3.1 = 3.9$, top base is $7 - 3.7 = 3.3$.

Area of cell $E = \dfrac{1}{2}(b_1 + b_2)(h) = \dfrac{1}{2}(3.9 + 3.3)(1.5)$

$= 5.4 \text{ m}^2$

6. The centre of the LEC must be found by checking distances to the nearest site of likely candidates. Since all possible centres are vertices adjacent to site E, we can find the distance of each vertex to site E:

$V(10, 16)$: $d = \sqrt{(10 - 11)^2 + (16 - 8)^2} = \sqrt{1 + 64} = \sqrt{65}$

$W(15, 15)$: $d = \sqrt{(15 - 11)^2 + (15 - 8)^2} = \sqrt{16 + 9} = \sqrt{65}$

$X(16, 14)$: $d = \sqrt{(16 - 11)^2 + (14 - 8)^2} = \sqrt{25 + 36} = \sqrt{61}$

$Y(16, 8)$: $d = \sqrt{(16 - 11)^2 + (8 - 8)^2} = \sqrt{25 + 0} = \sqrt{25}$

$Z(6, 8)$: $d = \sqrt{(6 - 11)^2 + (8 - 8)^2} = \sqrt{25 + 0} = \sqrt{25}$

Since $V(10, 16)$ and $W(15, 15)$ are both $\sqrt{65}$ units from their nearest site, there are two largest empty circles, centred at each point, both with radius $\sqrt{65} \approx 8.06$.

7. (a) After adding site $H$, the Voronoi diagram should look like this:

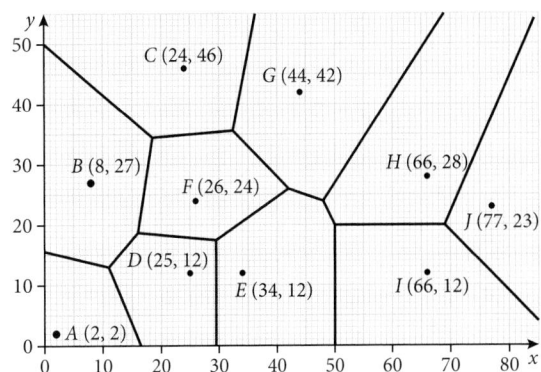

(b) The region bounded by points $E$, $G$, $H$, and $I$ contains three vertices, labelled as $K$, $L$, and $M$ on the diagram below:

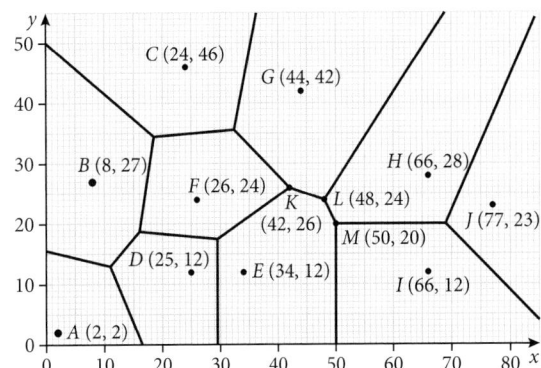

Since all three are adjacent to vertex $E$, we will find the distance of each to vertex $E$. For $K$,
$d_K = \sqrt{(42 - 34)^2 + (26 - 12)^2} = \sqrt{260}$; for $L$,
$d_L = \sqrt{(48 - 34)^2 + (24 - 12)^2} = \sqrt{340}$; for $M$,
$d_M = \sqrt{(50 - 34)^2 + (20 - 12)^2} = \sqrt{320}$. Therefore, the best location for a new bank branch is at $L(48, 24)$.

## Chapter 5 Practice questions

1. (a) $x = 2.79$      (b) $x = 3.32, 5.41$

2. $x = 1.57, 4.71, 7.85$

3. (a) $\dfrac{10\pi}{3}$      (b) $6\pi$

4. (a) $6 \text{ cm}$      (b) $9 \text{ cm}$

5. (a) $r = 13.5 \text{ cm}$      (b) $27 + 3\pi = 36.4 \text{ cm}$
   (c) $20.25\pi = 63.6 \text{ cm}$

6. $r = 4, \theta = \dfrac{\pi}{6}$

7. $r = 15, \theta = 1.6$

8. (a) $12.5$ hours      (b) $3 \text{ m}$
   (c) approximately $2 \times 3.25 = 6.5$ hours

9. (a)

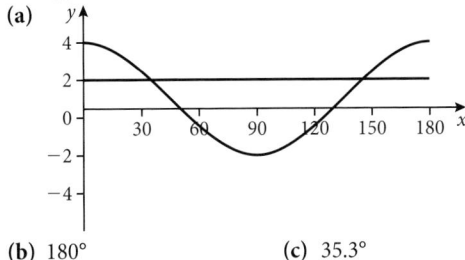

(b) $180°$      (c) $35.3°$

933

# Answers

**10. (a)**

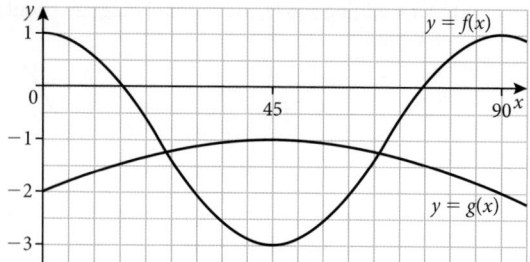

**(b)** 2

**(c)** any value $0° \leqslant x < 24.3°$ or $65.7° < x \leqslant 90°$

**(d) (i)** $a = 24.3°$      **(ii)** $b = 65.7°$

**11. (a)**

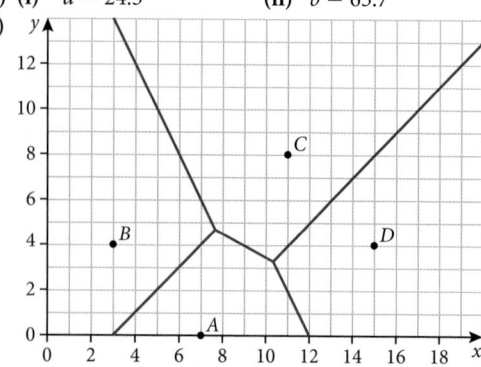

**(b)**

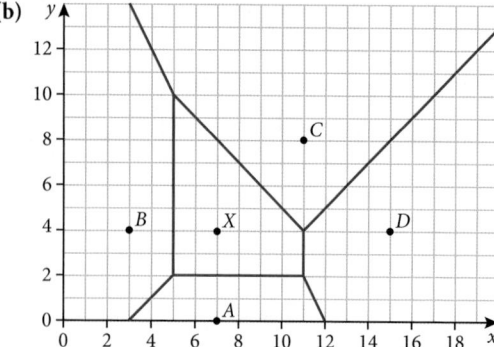

**(c)** $y = 2$      **(d)** $y = -x + 15$

**(e)**

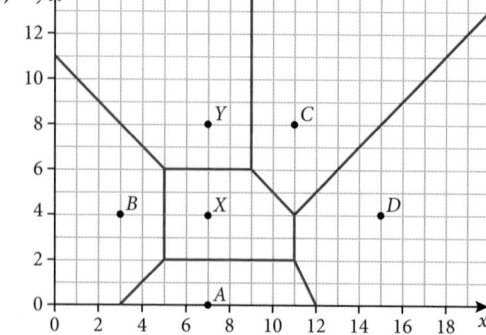

**(f)** $(11, 2)$

**(g)**

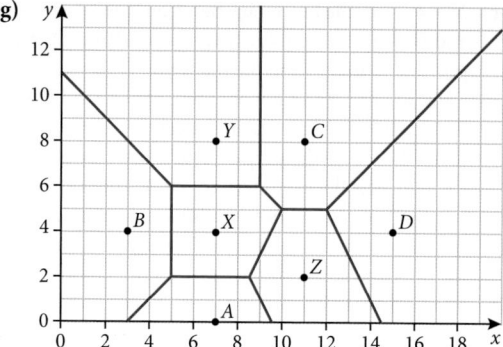

**(h)** $2x + y = 29$

**12. (a)**

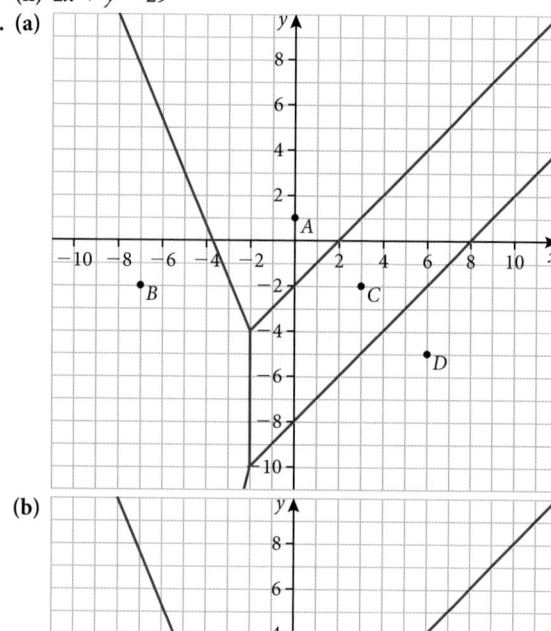

**(b)**

**(c)** $x = 1.5$

934

**(d)**

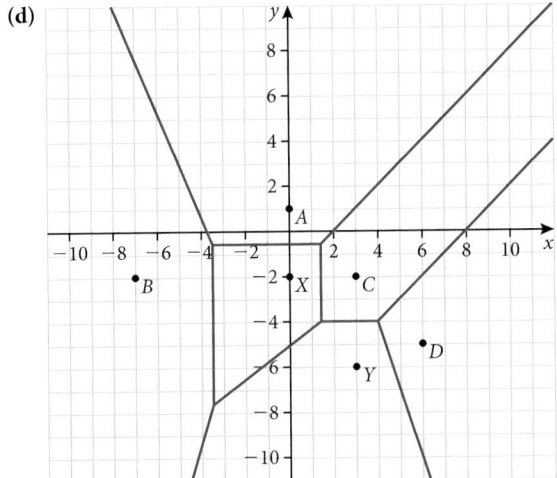

**(e)** $6x - 8y = 41$

**13. (a)** the largest empty circle (LEC) must be centred on a vertex

**(b)** All vertices in the diagram are adjacent to cell $K$ and $(15, 8)$ is the vertex farthest from site $K$, therefore it must be the centre of the largest empty circle.

**14. (a)** The LEC must be at $(6, 6)$. The LEC must be on a vertex and $(6, 6)$ is the vertex farthest from any site.

**(b)**

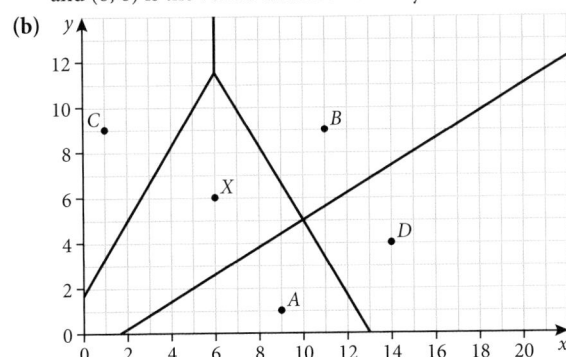

**15.** Since $A$, $B$, $C$, and $D$ are collinear, the perpendicular bisectors between each pair of points must be parallel. Hence, the edges in the Voronoi diagram are parallel which implies there are no vertices.

**16. (a)** The Voronoi diagram has 3 edges that meet in a single vertex as shown.

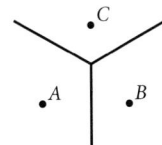

**(b)** The Voronoi diagram has 4 edges that meet in a single vertex as shown.

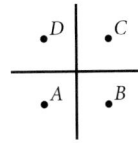

**(c)** The Voronoi diagram for a regular $n$-gon will have $n$ edges that meet at a single vertex in the centre of the $n$-gon. The angle between each pair of adjacent edges will be $\dfrac{360°}{n}$.

# Chapter 6

## Exercise 6.1

**1. (a)** $6i$     **(b)** $2\sqrt{3}i$     **(c)** $3\sqrt{7}i$     **(d)** $-12$

**2. (a)** $4 + 3i$     **(b)** $-3 - 2i$     **(c)** $-18 + 3\sqrt{2}i$
    **(d)** $4\sqrt{2} - 2\sqrt{2}i$     **(e)** $2i$     **(f)** $12 + 2\sqrt{3}i$
    **(g)** $-7$     **(h)** $-2 + 2i$

**3. (a)** $-i$     **(b)** $-1$     **(c)** $i$     **(d)** $1$

**4. (a)** $x = -1 \pm 3i$     **(b)** $x = 2 \pm \sqrt{3}i$

    **(c)** $x = -1 \pm \dfrac{\sqrt{6}}{2}i$     **(d)** $x = 1 \pm 3i$

    **(e)** $x = -3 \pm i$     **(f)** $x = -\dfrac{3}{2} \pm \dfrac{\sqrt{3}}{2}i$

**5. (a)** $1, -1, i, -i$

    **(b)** $1, -\dfrac{1}{2} + \dfrac{\sqrt{3}}{2}i, \dfrac{1}{2} - \dfrac{\sqrt{3}}{2}i$

## Exercise 6.2

**1. (a)** $-1 - 2i$     **(b)** $2 - i$     **(c)** $\sqrt{2} - i$

**2. (a)** $17$     **(b)** $-13$     **(c)** $25$

**3. (a)** $25$     **(b)** $169$     **(c)** $36$

**4. (a)** $2 + i$     **(b)** $-\dfrac{3}{5} - \dfrac{4}{5}i$     **(c)** $-\dfrac{14}{13} + \dfrac{8}{13}i$

**5. (a)** $y = x^2 - 2x + 2$     **(b)** $y = x^2 + 14x + 50$
    **(c)** $y = x^2 + 4\sqrt{3}x + 15$

**6. (a)** $y = x^2 - 4x + 1$     **(b)** $y = x^2 - x - 1$

    **(c)** $y = x^2 + 2x + 5$     **(d)** $y = x^2 - 3x + \dfrac{17}{2}$

    **(e)** $y = x^2 - x + \dfrac{3}{2}$     **(f)** $y = x^2 + 4\sqrt{3}x + 15$

**7. (a)** $y = x^2 - (8 + i)x + (17 + i)$
    **(b)** $y = x^2 - 5 - 12i$     **(c)** $y = x^2 - 7 - 6\sqrt{2}i$

**8.** $\dfrac{17}{13} - \dfrac{19}{13}i$

**9.** $x = -\dfrac{1}{2}, y = -2$ or $x = 1, y = 1$

**10. (a)** $-8$
    **(b)** $z^{6n} = (z^3)^{2n} = (-8)^{2n} = (-1)^{2n} \cdot (8)^{2n} = 8^{2n}$
    **(c)** $2^{48}$

**11. (a)** $-4i$
    **(b)** $z^{4k} = (z^2)^{2n} = (-4i)^{2n} = (16i^2)^n = (-16)^k$
    **(c)** $2^{46}i$

**12.** $2$

**13.** $\dfrac{9 - \sqrt{2}}{3} + \dfrac{2}{3}i$

**14.** $x = \dfrac{-2}{65}, y = \dfrac{29}{65}$

**15. (a)** $\dfrac{1}{2} + \dfrac{1}{2}i$     **(b)** $2 - i$ or $-2 + i$

## Exercise 6.3

**1.**

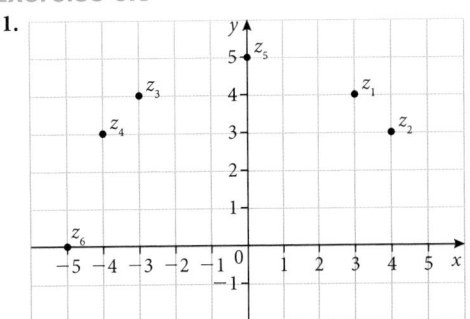

    **(a)** All have $|z| = 5$     **(b)** $-4 - 3i$
    **(c)** Yes     **(d)** $z = \pm\sqrt{7} + 3\sqrt{2}i$

# Answers

**2. (a)** $\sqrt{2}$    **(b)** 2    **(c)** 2    **(d)** 2
   **(e)** 13    **(f)** 13    **(g)** 29    **(h)** $6\sqrt{3}$

**3. (a)**

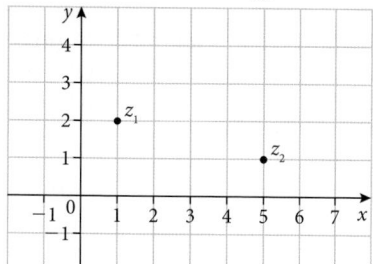

**(b)** $|z_1| = \sqrt{5}$
   $|z_2| = \sqrt{26}$

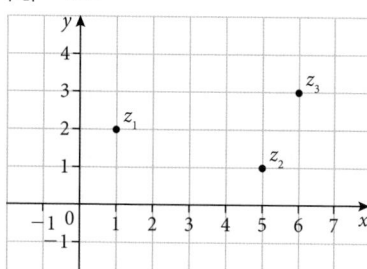

**(c) and (d)**

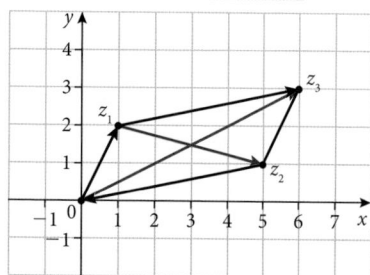

**(e)** $Oz_3$ is the sum, $(z_1 + z_2)$
    $z_2 z_1$ is the difference, $(z_1 - z_2)$

**4.** Consider the diagram in Question 3d. By the Triangle Inequality, the sum of the lengths of any two sides exceeds the length of the third.

**5.** $22 - 2\sqrt{3} \approx 18.5$

**6. (a)** circle with centre $(0, 0)$ and radius 3
   **(b)** $y$-axis
   **(c)** line $x=4$
   **(d)** circle with centre $(3, 0)$ and radius 2
   **(e)** a line segment between $(1, 0)$ and $(3, 0)$

## Exercise 6.4

**1. (a)** $z = 3\sqrt{2}\operatorname{cis}\dfrac{\pi}{4}$    **(b)** $z = 3\sqrt{2}\operatorname{cis}\dfrac{3\pi}{4}$

   **(c)** $z = 3\sqrt{2}\operatorname{cis}\dfrac{7\pi}{4}$    **(d)** $z = 3\sqrt{2}\operatorname{cis}\dfrac{5\pi}{4}$

   **(e)** $z = 10\operatorname{cis}\dfrac{\pi}{3}$    **(f)** $z = 10\operatorname{cis}\dfrac{\pi}{6}$

   **(g)** $z = 10\operatorname{cis}\dfrac{2\pi}{3}$    **(h)** $z = 10\operatorname{cis}\dfrac{7\pi}{6}$

**2. (a)** $15i$    **(b)** $-4 - 4\sqrt{3}\,i$

**3. (a)** $\dfrac{3}{2} + \dfrac{3\sqrt{3}}{2}i$    **(b)** $-2 - 2\sqrt{3}\,i$
   **(c)** $2\sqrt{2} - 2\sqrt{2}\,i$    **(d)** $4\sqrt{3} - 4i$

**4. (a)** $4 - 4i$    **(b)** $-4$
   **(c)** $-8 - 8\sqrt{3}\,i$    **(d)** $-64$

**5. (a)** $2, -1 \pm \sqrt{3}\,i$    **(b)** $\pm\left(\dfrac{\sqrt{6}}{2} + \dfrac{\sqrt{2}}{2}i\right)$
   **(c)** $\sqrt{2} \pm \sqrt{2}\,i, -\sqrt{2} \pm \sqrt{2}\,i$

**6. (a)** $6\sqrt{2}\,e^{\frac{3\pi}{4}i}$    **(b)** $4\sqrt{3}\,e^{\frac{\pi}{3}i}$    **(c)** $2e^{\frac{5\pi}{3}i}$

**7. (a)** $6e^{\frac{\pi}{2}i}$    **(b)** $12\,e^{\frac{5\pi}{4}i}$

**8. (a)** $-2\sqrt{2 - \sqrt{2}} + 2\sqrt{24 + \sqrt{2}}\,i$
   **(b)** $-4\sqrt{2} - 4\sqrt{2}\,i$

**9. (a)** $2$    **(b)** $\pm\left(\dfrac{\sqrt{2}}{2} + \dfrac{\sqrt{2}}{2}i\right)$
   **(c)** No
     $i = e^{\frac{\pi}{2}i}, -i = e^{\frac{3\pi}{2}i}$ and by de Moivre's law,
$$\frac{\dfrac{\pi}{2} + 2k\pi}{2} \neq \frac{\dfrac{3\pi}{2} + 2k\pi}{2}$$

**10.** $z = -1 + 0i \Rightarrow \begin{cases} z = -1 \\ z = e^{\pi i} \end{cases} \Rightarrow -1 = e^{\pi i} \Rightarrow e^{\pi i} + 1 = 0$

## Exercise 6.5

**1. (a)** $Z = 12 + 9j$    **(b)** $Z = 15\,e^{0.644i}$
**2.** $Z = 4 - 3j$     **3.** $|V| = 15$V
**4.** $|V| = 10$V     **5.** $|V| = 60$V
**6.** $|I| = 20$A     **7.** $Z \approx 6.36 - 1.28j\ \Omega$
**8.** $6.25\ \Omega$

## Chapter 6 Practice questions

**1. (a)** $7i$   **(b)** $3\sqrt{2}\,i$   **(c)** $-3$   **(d)** $-18$
**2. (a)** $4i$   **(b)** $25 + 5i$   **(c)** $5 + 5i$
   **(d)** $-3\sqrt{2} + 3\sqrt{2}\,i$   **(e)** $2\sqrt{3} - 2\sqrt{3}\,i$   **(f)** $-i$
   **(g)** $-1$   **(h)** $-3 - 3i$
**3. (a)** $-1$   **(b)** $i$   **(c)** $1$   **(d)** $-i$
**4. (a)** $x = -2 \pm 2i$   **(b)** $x = 3\pm i$   **(c)** $x = 4 \pm 3i$
   **(d)** $x = 2 \pm 2i$   **(e)** $x = 5 \pm 2i$   **(f)** $x = -4 \pm 4i$
**5. (a)** $1 + 2i$   **(b)** $2 + 3i$   **(c)** $3\sqrt{5} - 2i$
**6. (a)** $52$    **(b)** $-53$    **(c)** $28$
**7. (a)** $225$    **(b)** $100$    **(c)** $27$
**8. (a)** $3-i$    **(b)** $2\sqrt{3} + 4i$    **(c)** $\dfrac{-1}{3}$
**9. (a)** $y = x^2 - 2x + 5$    **(b)** $y = x^2 - 8x + 25$
   **(c)** $y = x^2 + 6\sqrt{2}\,x + 20$
**10. (a)** $y = x^2 + 2x + 2$    **(b)** $y = x^2 - 4x + 6$
   **(c)** $y = x^2 - (7 + i)x + 14 + 5i$
**11.** For example,

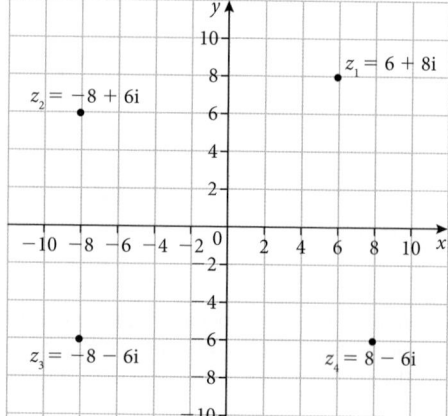

**12.** $\pm 2\sqrt{23}$

**13.** The blue diagonal represents the sum; the red, the difference.

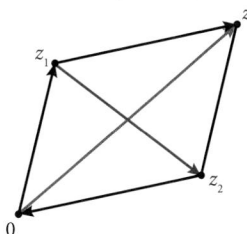

**14. (a)** $2\sqrt{2}\,\text{cis}\dfrac{7\pi}{4}$  **(b)** $2\sqrt{2}\,\text{cis}\dfrac{5\pi}{4}$

**(c)** $2\sqrt{2}\,\text{cis}\dfrac{\pi}{4}$  **(d)** $2\sqrt{2}\,\text{cis}\dfrac{3\pi}{4}$

**(e)** $4\text{cis}\dfrac{\pi}{3}$  **(f)** $4\text{cis}\dfrac{5\pi}{3}$

**(g)** $4\text{cis}\dfrac{2\pi}{3}$  **(h)** $4\text{cis}\dfrac{4\pi}{3}$

**15. (a)** $-6\sqrt{2}-6\sqrt{2}\,i$  **(b)** $16i$

**16. (a)** $3i$  **(b)** $\dfrac{-5\sqrt{2}}{2}+\dfrac{5\sqrt{2}}{2}i$

**(c)** $\dfrac{\sqrt{3}}{4}+\dfrac{1}{4}i$  **(d)** $2\sqrt{2}-2\sqrt{2}i$

**17. (a)** $-4$  **(b)** $-4$
**(c)** $8$  **(d)** $-128+128\sqrt{3}\,i$

**18. (a)** $-2,\,1\pm\sqrt{3}\,i$  **(b)** $4,\,-2\pm2\sqrt{3}\,i$

**(c)** $\pm\left(\dfrac{\sqrt{6}}{2}+\dfrac{3\sqrt{2}}{2}i\right)$

**19. (a)** $\sqrt{2}\,e^{\frac{-7\pi}{4}i}$  **(b)** $2\sqrt{2}\,e^{\frac{-5\pi}{4}i}$  **(c)** $2\sqrt{3}\,e^{\frac{2\pi}{3}i}$

**20. (a)** $5-5\sqrt{3}\,i$  **(b)** $-4\sqrt{3}+4i$

**21. (a)** $\dfrac{-5\sqrt{2}}{2}-\dfrac{5\sqrt{2}}{2}i$  **(b)** $2\sqrt{2}-2\sqrt{2}i$

**22. (a)** $Z=12-5j$  **(b)** $Z=13\,e^{-0.394°i}$

**23.** $Z=4+4j$  **24.** $|V|=13\text{V}$
**25.** $|V|=10\text{V}$  **26.** $|V|=10\text{V}$
**27.** $|I|=5\text{A}$  **28.** $Z\approx5.15+0.355j\,\Omega$
**29. (a)** $13+13i$  **(b)** $13\sqrt{2}\,e^{\frac{-\pi}{4}i}$

**30. (a)** $\left\{\dfrac{3}{2}\text{cis}\dfrac{\pi}{3},\,-\dfrac{3}{2}\text{cis}\pi,\,\dfrac{3}{2}\text{cis}\dfrac{5\pi}{3}\right\}$

**(c)** Area $=\dfrac{1}{2}ab\sin C=\dfrac{1}{2}\times\dfrac{3\sqrt{3}}{2}\times\dfrac{3\sqrt{3}}{2}\times\dfrac{\sqrt{3}}{2}=\dfrac{27\sqrt{3}}{16}$

**31. (a)** $w^7=\left(\text{cis}\dfrac{2\pi}{7}\right)^7=\text{cis}\,2\pi=1$

**(b) (i)** $(w-1)(1+w+w^2+w^3+w^4+w^5+w^6)$
$=w^7-1$
**(ii)** $w^7-1=0$ and $w-1\neq0$
$\Rightarrow 1+w+w^2+w^3+w^4+w^5+w^6=0$

**(c)**

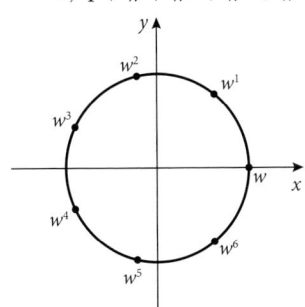

Roots are $w^k=\text{cis}\dfrac{2k\pi}{7}$

**(d) (i)** $\alpha^*=(w+w^2+w^4)^*=w^*+(w^2)^*+(w^4)^*$
since $w^*=w^6,\,(w^2)^*=w^5,\,(w^4)^*=w^3,$
$\Rightarrow\alpha^*=w^6+w^5+w^3$
**(ii)** $b=1,c=2$

**(e)** $\dfrac{\sqrt{7}}{2}$

**32.** $a=-4,b=13$

## Chapter 7

### Exercise 7.1

**1. (a) (i)** $\begin{pmatrix}x-1 & x-3\\ y+3 & y+1\end{pmatrix}$

**(ii)** $\begin{pmatrix}-x-7 & 3x+3\\ 3y-7 & 11-y\end{pmatrix}$

**(iii)** Not possible

**(b)** $x=-3,\,y=5$
**(c)** $x=3,\,y=-3$

**(d)** $AB=\begin{pmatrix}2x-2 & xy-2x+6\\ xy-x+y+11 & -3\end{pmatrix};$

$BA=\begin{pmatrix}-2x-3y+1 & x^2+x-9\\ y^2-3y-6 & 4x+3y-6\end{pmatrix}$

**(e)** Not possible

**2. (a)** $x=2,\,y=-10$
**(b)** $p=2,\,q=-4$
**(c)** $a=-3,\,b=8,\,c=1,\,d=-11$

**3. (a)** $\begin{pmatrix}0&1&0&0&1&2&0\\1&0&1&1&1&1&0\\0&1&0&2&0&0&2\\0&1&2&0&1&0&0\\1&1&0&1&0&1&0\\2&1&0&0&1&0&0\\0&0&2&0&0&0&0\end{pmatrix}$

**(b)** $\begin{pmatrix}6&3&1&2&3&2&0\\3&5&2&3&3&3&2\\1&2&9&1&3&1&0\\2&3&1&6&1&2&4\\3&3&3&1&4&3&0\\2&3&1&2&3&6&0\\0&2&0&4&0&0&4\end{pmatrix}$ This represents the number of routes via one other city

**4. (a)** $A+C=\begin{pmatrix}x+1 & 10 & y+1\\ 0 & -x-3 & y+3\\ 2x+y+7 & x-3y & -x+2y-1\end{pmatrix}$

**(b)** $AB=\begin{pmatrix}17m+2 & -6\\ 4-9m & 9\\ 7m-2 & -17\end{pmatrix}$

**(c)** Not possible  **(d)** $x=3,\,y=1$
**(e)** Not possible  **(f)** $m=3$

**5.** $a=-3,\,b=3,\,c=2$
**6.** $x=4,\,y=-3$
**7.** $m=2,\,n=3$
**8.** Shop A: €18.77

**9. (a)** $\begin{pmatrix}2 & 4\\ -2 & 12\end{pmatrix}$  **(b)** Associative

**(c)** $\begin{pmatrix}-22 & 16\\ 60 & -7\end{pmatrix}$  **(d)** Associative

**10.** $AB=\$88\,142$ total profit
**11.** $r=7,\,s=-22$

# Answers

**12. (a)** $\begin{pmatrix} 1 & 2 \\ 0 & 1 \end{pmatrix}$; $\begin{pmatrix} 1 & 3 \\ 0 & 1 \end{pmatrix}$; $\begin{pmatrix} 1 & 4 \\ 0 & 1 \end{pmatrix}$; $\begin{pmatrix} 1 & n \\ 0 & 1 \end{pmatrix}$

**(b)** $\begin{pmatrix} 9 & 18 \\ 0 & 9 \end{pmatrix}$; $\begin{pmatrix} 27 & 81 \\ 0 & 27 \end{pmatrix}$; $\begin{pmatrix} 81 & 324 \\ 0 & 81 \end{pmatrix}$; $\begin{pmatrix} 3^n & n \times 3^n \\ 0 & 3^n \end{pmatrix}$

**13.** $x = \dfrac{11}{3}, y = \dfrac{8}{3}$

**14.** $x = 1, y = -4$

**15.** 5

**16.** $x = 5, y = 1$

## Exercise 7.2

**1. (a)** $\begin{pmatrix} -9 & -7 \\ 4 & 3 \end{pmatrix}$

**(b)** $M = \begin{pmatrix} -9 & -7 \\ 4 & 3 \end{pmatrix}\begin{pmatrix} 2 & 1 \\ 3 & 5 \end{pmatrix}$

**(c)** $\begin{pmatrix} -39 & -44 \\ 17 & 19 \end{pmatrix}$

**(d) (i)** $N = \begin{pmatrix} 2 & 1 \\ 3 & 5 \end{pmatrix}\begin{pmatrix} -9 & -7 \\ 4 & 3 \end{pmatrix}$

**(ii)** $N = \begin{pmatrix} -14 & -11 \\ -7 & -6 \end{pmatrix}$

**(e)** If $AB = C$ then $B = A^{-1}C$, while if $BA = C$, then $B = CA^{-1}$. Also, $A^{-1}C \neq CA^{-1}$.

**2.** $\begin{pmatrix} 1 & -\dfrac{3}{5} \\ 0 & 1 \end{pmatrix}$

**3. (a)** $|A| = -5 \neq 0$

**(b)** $\begin{pmatrix} \dfrac{9}{5} & \dfrac{11}{5} & -\dfrac{8}{5} \\ \dfrac{6}{5} & \dfrac{9}{5} & -\dfrac{7}{5} \\ 1 & 1 & -1 \end{pmatrix}$

**(c)** $x = \dfrac{1}{2}, y = -1, z = \dfrac{1}{5}$

**4. (a)** $\begin{pmatrix} \dfrac{\sqrt{3}}{2} & \dfrac{1}{2} \\ -\dfrac{1}{2} & \dfrac{\sqrt{3}}{2} \end{pmatrix}$

**(b)** $\begin{pmatrix} \dfrac{3}{a} + 1 & -1 \\ -a - 2 & a \end{pmatrix}$

**5.** $x = 2$ or $x = 3$

**6.** $n = 0.5$

**7. (a)** $X = \begin{pmatrix} \dfrac{1}{2} & 0 \\ \dfrac{3}{4} & -\dfrac{7}{6} \end{pmatrix}$

**(b)** $Y = \begin{pmatrix} 1 & \dfrac{13}{12} \\ -1 & -\dfrac{5}{3} \end{pmatrix}$

**(c)** $X \neq Y$ — Not commutative

**8. (a)** $PQ = \begin{pmatrix} 5 & 0 & 3 \\ 33 & -11 & -1 \\ 2 & 1 & 2 \end{pmatrix}$

$QP = \begin{pmatrix} 4 & -5 & -8 \\ 8 & 0 & -4 \\ -5 & -10 & -8 \end{pmatrix}$

**(b) (i)** $P^{-1} = \begin{pmatrix} 1 & 0 & -1 \\ -\dfrac{7}{5} & \dfrac{1}{5} & \dfrac{11}{5} \\ 1 & 0 & -2 \end{pmatrix}$

**(ii)** $Q^{-1} = \begin{pmatrix} 0 & \dfrac{1}{4} & 0 \\ -\dfrac{1}{3} & \dfrac{1}{3} & -\dfrac{1}{3} \\ \dfrac{2}{3} & -\dfrac{5}{12} & -\dfrac{1}{3} \end{pmatrix}$

**(iii)** $P^{-1}Q^{-1} = \begin{pmatrix} -\dfrac{2}{3} & \dfrac{2}{3} & \dfrac{1}{3} \\ \dfrac{7}{5} & -\dfrac{6}{5} & -\dfrac{4}{5} \\ -\dfrac{4}{3} & \dfrac{13}{12} & \dfrac{2}{3} \end{pmatrix}$

**(iv)** $Q^{-1}P^{-1} = \begin{pmatrix} -\dfrac{7}{20} & \dfrac{1}{20} & \dfrac{11}{20} \\ -\dfrac{17}{15} & \dfrac{1}{15} & \dfrac{26}{15} \\ \dfrac{11}{12} & -\dfrac{1}{12} & -\dfrac{11}{12} \end{pmatrix}$

**(v)** $(PQ)^{-1} = \begin{pmatrix} -\dfrac{7}{20} & \dfrac{1}{20} & \dfrac{11}{20} \\ -\dfrac{17}{5} & \dfrac{1}{15} & \dfrac{26}{15} \\ \dfrac{11}{12} & -\dfrac{1}{12} & -\dfrac{11}{12} \end{pmatrix}$

**(vi)** $(QP)^{-1} = \begin{pmatrix} -\dfrac{2}{3} & \dfrac{2}{3} & \dfrac{1}{3} \\ \dfrac{7}{5} & -\dfrac{6}{5} & -\dfrac{4}{5} \\ -\dfrac{4}{3} & \dfrac{13}{12} & \dfrac{2}{3} \end{pmatrix}$

**(c)** $(PQ)^{-1} = Q^{-1}P^{-1}$ and $(QP)^{-1} = P^{-1}Q^{-1}$

**9. (a)** $\begin{pmatrix} -7 \\ 3 \\ -2 \end{pmatrix}$ **(b)** $x = -7, y = 3, z = -2$

**10.** $x = -1$

**11. (a)** $x = 1, y = 2$ **(b)** $x = 0, y = 1$
**(c)** $x = -3, y = -29; x = 0, y = 1$

**12. (a)** $17x - 8y + 37 = 0$ **(b)** $y + 2 = 0$
**(c)** $x + 5 = 0$

**13. (a)** 165
**(b)** 80
**(c)** 136

**14. (a)** $x = \dfrac{89}{8}$ or $x = \dfrac{129}{8}$
**(b)** $x = -4$ or $x = -2$ or $x = -3 \pm \sqrt{21}$

**15. (a)** $-3$
**(b)** 3

**16. (a)** $-25$
**(b)** $x^2 - 7x - 25$, constant $= \det(A)$
**(c)** $-(a + d)$
**(d)** $f(A) = 0$
**(e)** $ad - bc$; $x^2 - (a + d)x + (ad - bc)$, constant $= \det(A)$; $f(A) = 0$

**17. (a)** $-22$
**(b)** $x^3 - x^2 - 22x + 22$, constant $= -\det(A)$
**(c)** opposite of the sum of the main diagonal
**(d)** $f(A) = 0$
**(e)** Students' working

**18. (a)** GAUSS IS A GREAT MATHEMATICIAN
**(b)** Answers vary

## Exercise 7.3

1. $m = 2$ or $m = 3$
2. (a) $a = 7, b = 2$
   (b) $(-1, 2, -1)$
3. $m = 2$
4. (a) $(-1, 3, 2)$
   (b) $(5, 8, -2)$
   (c) $\left(\dfrac{13}{16} + 5t, \dfrac{11}{16} + 19t, 16t\right)$
   (d) $(-7, 3, -2)$
   (e) $(-1 - t, 2 - 3t, t)$
   (f) inconsistent
   (g) $(-2, 4, 3)$
   (h) $(4, -2, 1)$
5. (a) $k \neq \dfrac{-1 \pm \sqrt{33}}{4}$
   (b) $k = 1$
   (c) $\begin{pmatrix} 1 & 0 & 0 & -2 & -3 & 1 \\ 0 & 1 & 0 & 3 & 3 & -1 \\ 0 & 0 & 1 & -2 & -4 & 1 \end{pmatrix}$
6. (a) $k \neq \dfrac{71 \pm i\sqrt{251}}{42}$      (b) $k = 2$
   (c) $\begin{pmatrix} 1 & 0 & 0 & \frac{3}{5} & \frac{1}{5} & 2 \\ 0 & 1 & 0 & \frac{2}{5} & \frac{4}{5} & -3 \\ 0 & 0 & 1 & \frac{3}{5} & \frac{6}{5} & -5 \end{pmatrix}$
7. (a) $\begin{pmatrix} 1 & 0 & 0 & \frac{1}{2} & -1 & -\frac{1}{2} \\ 0 & 1 & 0 & \frac{1}{2} & -\frac{2}{3} & -\frac{5}{6} \\ 0 & 0 & 1 & 0 & \frac{2}{3} & \frac{1}{3} \end{pmatrix}$;
   (b) $\begin{pmatrix} 1 & 0 & 0 & 2 & -\frac{16}{13} & -\frac{19}{13} \\ 0 & 1 & 0 & 1 & -\frac{11}{13} & -\frac{9}{13} \\ 0 & 0 & 1 & -1 & \frac{12}{13} & \frac{11}{13} \end{pmatrix}$; $B$ is the inverse of $A$
8. (a) $f(x) = 4x^2 - 6x - 5$
   (b) $f(x) = \dfrac{1}{2}(7 - m)x^2 - \dfrac{1}{2}(17 - m)x + m$
9. $m = 2,\ \begin{pmatrix} -7t - \frac{3}{5} \\ -t - \frac{19}{5} \\ 5t \end{pmatrix}$
10. $m = -1,\ \begin{pmatrix} 7t - \frac{9}{5} \\ 3t - \frac{11}{5} \\ 5t \end{pmatrix}$
11. (a) $3$      (b) $\begin{pmatrix} 3 & -4 & -6 \\ 0 & -2 & -3 \\ 0 & 0 & -\frac{1}{2} \end{pmatrix}$
    (c) $3$      (d) $-1672$
    (e) $\begin{pmatrix} 2 & 1 & -3 & 5 \\ 0 & 1 & 2 & -16 \\ 0 & 0 & 36 & -184 \\ 0 & 0 & -0 & -\frac{209}{9} \end{pmatrix}$

## Exercise 7.4

1. (a) $\lambda^2 - 3\lambda + 2; \lambda = 1$ or
   $\lambda = 2; \begin{pmatrix} t \\ 2t \end{pmatrix}, \begin{pmatrix} t \\ t \end{pmatrix}; \begin{pmatrix} 1 & 1 \\ 2 & 1 \end{pmatrix}\begin{pmatrix} 1 & 0 \\ 0 & 2 \end{pmatrix}\begin{pmatrix} -1 & 1 \\ 2 & -1 \end{pmatrix}$
   (b) $\lambda^2 - 9; \lambda = \pm 3; \begin{pmatrix} 2t \\ t \end{pmatrix}, \begin{pmatrix} -t \\ t \end{pmatrix};$
   $\begin{pmatrix} 2 & -1 \\ 1 & 1 \end{pmatrix}\begin{pmatrix} 3 & 0 \\ 0 & -3 \end{pmatrix}\dfrac{1}{3}\begin{pmatrix} 1 & 1 \\ -1 & 2 \end{pmatrix}$
   (c) $\lambda^2 - 5\lambda + 6; \lambda = 2$ or $\lambda = 3; \begin{pmatrix} t \\ t \end{pmatrix}, \begin{pmatrix} 0 \\ t \end{pmatrix};$
   $\begin{pmatrix} 1 & 0 \\ 1 & 1 \end{pmatrix}\begin{pmatrix} 2 & 0 \\ 0 & 3 \end{pmatrix}\begin{pmatrix} 1 & 0 \\ -1 & 1 \end{pmatrix}$
   (d) $\lambda^2 - 2\lambda - 3; \lambda = -1$ or $\lambda = 3; \begin{pmatrix} t \\ t \end{pmatrix}, \begin{pmatrix} -t \\ 3t \end{pmatrix};$
   $\begin{pmatrix} 1 & -1 \\ 1 & 3 \end{pmatrix}\begin{pmatrix} -1 & 0 \\ 0 & 3 \end{pmatrix}\dfrac{1}{4}\begin{pmatrix} 3 & 1 \\ -1 & 1 \end{pmatrix}$
   (e) $\lambda^2 - 2\lambda - 3; \lambda = -1$ or $\lambda = 3; \begin{pmatrix} t \\ -t \end{pmatrix}, \begin{pmatrix} t \\ t \end{pmatrix};$
   $\begin{pmatrix} 1 & 1 \\ -1 & 1 \end{pmatrix}\begin{pmatrix} -1 & 0 \\ 0 & 3 \end{pmatrix}\dfrac{1}{2}\begin{pmatrix} 1 & -1 \\ 1 & 1 \end{pmatrix}$
   (f) $\lambda^2 - 3\lambda + 2; \lambda = 1$ or $\lambda = 2; \begin{pmatrix} t \\ 0 \end{pmatrix}, \begin{pmatrix} 0 \\ t \end{pmatrix};$
   $\begin{pmatrix} 1 & 0 \\ 0 & 1 \end{pmatrix}\begin{pmatrix} 1 & 0 \\ 0 & 2 \end{pmatrix}\begin{pmatrix} 1 & 0 \\ 0 & 1 \end{pmatrix}$
   (g) $\lambda^2 - 6\lambda + 5; \lambda = 1$ or $\lambda = 5; \begin{pmatrix} 3t \\ -t \end{pmatrix}, \begin{pmatrix} t \\ t \end{pmatrix};$
   $\begin{pmatrix} 3 & 1 \\ -1 & 1 \end{pmatrix}\begin{pmatrix} 1 & 0 \\ 0 & 5 \end{pmatrix}\dfrac{1}{4}\begin{pmatrix} 1 & -1 \\ 1 & 3 \end{pmatrix}$
   (h) $\lambda^2 - 9\lambda + 20; \lambda = 4$ or $\lambda = 5; \begin{pmatrix} t \\ 3t \end{pmatrix}, \begin{pmatrix} t \\ 2t \end{pmatrix};$
   $\begin{pmatrix} 1 & 1 \\ 3 & 2 \end{pmatrix}\begin{pmatrix} 4 & 0 \\ 0 & 5 \end{pmatrix}\begin{pmatrix} -2 & 1 \\ 3 & -1 \end{pmatrix}$
   (i) $\lambda^2 - 11\lambda + 10; \lambda = 1$ or $\lambda = 10; \begin{pmatrix} -3t \\ t \end{pmatrix}, \begin{pmatrix} 3t \\ 2t \end{pmatrix};$
   $\begin{pmatrix} -3 & 3 \\ 1 & 2 \end{pmatrix}\begin{pmatrix} 1 & 0 \\ 0 & 10 \end{pmatrix}\dfrac{1}{9}\begin{pmatrix} -2 & 3 \\ 1 & 3 \end{pmatrix}$
   (j) $\lambda^2 - \lambda - 1; \lambda = \dfrac{1 + \sqrt{5}}{2}$ or $\lambda = \dfrac{1 - \sqrt{5}}{2};$
   $\begin{pmatrix} \frac{1 + \sqrt{5}}{2} \\ 1 \end{pmatrix}, \begin{pmatrix} \frac{1 - \sqrt{5}}{2} \\ 1 \end{pmatrix};$
   $\begin{pmatrix} \frac{1 + \sqrt{5}}{2} & \frac{1 - \sqrt{5}}{2} \\ 1 & 1 \end{pmatrix}\begin{pmatrix} \frac{1 + \sqrt{5}}{2} & 0 \\ 0 & \frac{1 - \sqrt{5}}{2} \end{pmatrix}\dfrac{1}{\sqrt{5}}\begin{pmatrix} 1 & \frac{\sqrt{5} - 1}{2} \\ -1 & \frac{1 + \sqrt{5}}{2} \end{pmatrix}$
   (k) $\lambda^2 - (a + b)\lambda + ab; \lambda = a$ or $\lambda = b; \begin{pmatrix} t \\ 0 \end{pmatrix}, \begin{pmatrix} 0 \\ t \end{pmatrix};$
   $\begin{pmatrix} 1 & 0 \\ 0 & 1 \end{pmatrix}\begin{pmatrix} a & 0 \\ 0 & b \end{pmatrix}\begin{pmatrix} 1 & 0 \\ 0 & 1 \end{pmatrix}$
   (l) $\lambda^2 - 6\lambda + 5; \lambda = 5$ or $\lambda = 1; \begin{pmatrix} t \\ t \end{pmatrix}, \begin{pmatrix} -t \\ t \end{pmatrix};$
   $\begin{pmatrix} 1 & -1 \\ 1 & 1 \end{pmatrix}\begin{pmatrix} 5 & 0 \\ 0 & 1 \end{pmatrix}\dfrac{1}{2}\begin{pmatrix} 1 & 1 \\ -1 & 1 \end{pmatrix}$
   (m) $\lambda^2 - a^2 b^2; \lambda = ab$ or $\lambda = -ab; \begin{pmatrix} a \\ b \end{pmatrix}, \begin{pmatrix} -a \\ b \end{pmatrix};$
   $\begin{pmatrix} a & -a \\ b & b \end{pmatrix}\begin{pmatrix} ab & 0 \\ 0 & -ab \end{pmatrix}\dfrac{1}{2ab}\begin{pmatrix} b & a \\ -b & a \end{pmatrix}$
   (n) $\lambda^2 - 3\lambda - 10; \lambda = 5$ or $\lambda = -2; \begin{pmatrix} 3t \\ 4t \end{pmatrix}, \begin{pmatrix} t \\ -t \end{pmatrix};$
   $\begin{pmatrix} 3 & 1 \\ 4 & -1 \end{pmatrix}\begin{pmatrix} 5 & 0 \\ 0 & -2 \end{pmatrix}\dfrac{1}{7}\begin{pmatrix} 1 & 1 \\ 4 & -3 \end{pmatrix}$

# Answers

**(o)** $\lambda^2 - 13\lambda - 5$; $\lambda = \dfrac{13 + 3\sqrt{21}}{2}$ or $\lambda = \dfrac{13 - 3\sqrt{21}}{2}$;

$\begin{pmatrix} \dfrac{3\sqrt{21} - 7}{10} \\ 1 \end{pmatrix}, \begin{pmatrix} \dfrac{-7 - 3\sqrt{21}}{10} \\ 1 \end{pmatrix}$;

$\begin{pmatrix} \dfrac{3\sqrt{21} - 7}{10} & \dfrac{-7 - 3\sqrt{21}}{10} \\ 1 & 1 \end{pmatrix}$

$\begin{pmatrix} \dfrac{13 + 3\sqrt{21}}{2} & 0 \\ 0 & \dfrac{13 - 3\sqrt{21}}{2} \end{pmatrix} \dfrac{5}{3\sqrt{21}} \begin{pmatrix} 1 & \dfrac{7 + 3\sqrt{21}}{10} \\ -1 & \dfrac{3\sqrt{21} - 7}{10} \end{pmatrix}$

**(p)** $\lambda^2 - 81$; $\lambda = 9$ or $\lambda = -9$; $\begin{pmatrix} 4t \\ 3t \end{pmatrix}, \begin{pmatrix} t \\ 3t \end{pmatrix}$;

$\begin{pmatrix} 4 & 1 \\ 3 & 3 \end{pmatrix}\begin{pmatrix} 9 & 0 \\ 0 & -9 \end{pmatrix}\dfrac{1}{9}\begin{pmatrix} 3 & -1 \\ -3 & 4 \end{pmatrix}$

**(q)** $\lambda^2 - 2\lambda - 3$; $\lambda = -1$ or $\lambda = 3$; $\begin{pmatrix} t \\ t \end{pmatrix}, \begin{pmatrix} t \\ 2t \end{pmatrix}$;

$\begin{pmatrix} 1 & 1 \\ 1 & 2 \end{pmatrix}\begin{pmatrix} -1 & 0 \\ 0 & 3 \end{pmatrix}\begin{pmatrix} 2 & -1 \\ -1 & 1 \end{pmatrix}$

**(r)** $\lambda^2 + 3\lambda + 2$; $\lambda = -2$ or $\lambda = -1$; $\begin{pmatrix} -3t \\ 5t \end{pmatrix}, \begin{pmatrix} -2t \\ 3t \end{pmatrix}$;

$\begin{pmatrix} -3 & -2 \\ 5 & 3 \end{pmatrix}\begin{pmatrix} -2 & 0 \\ 0 & -1 \end{pmatrix}\begin{pmatrix} 3 & 2 \\ -5 & -3 \end{pmatrix}$

**2. (a)** Given that state 1 has occurred, the chance that it happens again is 30%.

**(b)** $TX_0 = \begin{pmatrix} 0.48 \\ 0.52 \end{pmatrix}$

**3.** $\dfrac{10}{13}$

**4. (a)** $T = \begin{pmatrix} 0.80 & 0.05 & 0.10 \\ 0.05 & 0.90 & 0.15 \\ 0.15 & 0.05 & 0.75 \end{pmatrix}$

**(b)** $X_1 = TX_0 = \begin{pmatrix} 0.80 & 0.05 & 0.10 \\ 0.05 & 0.90 & 0.15 \\ 0.15 & 0.05 & 0.75 \end{pmatrix}\begin{pmatrix} 0.4 \\ 0.3 \\ 0.3 \end{pmatrix} = \begin{pmatrix} 0.365 \\ 0.335 \\ 0.300 \end{pmatrix}$

**(c)** $X_2 = TX_1 = T^2X_0 = \begin{pmatrix} 0.3388 \\ 0.3648 \\ 0.2965 \end{pmatrix}$

**5. (a)** vote is evenly split    **(b)** Liberal with 55%
    A   0.3790             A   0.2395

**6. (a)** B   0.4195        **(b)** B   0.4971
    C   0.2015             C   0.2634

**7. (a)** $X_1 = TX_0 = \begin{pmatrix} 0.8 & 0.3 \\ 0.2 & 0.7 \end{pmatrix}\begin{pmatrix} 0 \\ 1 \end{pmatrix} = \begin{pmatrix} 0.3 \\ 0.7 \end{pmatrix}$;

$X_2 = TX_1 = \begin{pmatrix} 0.8 & 0.3 \\ 0.2 & 0.7 \end{pmatrix}\begin{pmatrix} 0.3 \\ 0.7 \end{pmatrix} = \begin{pmatrix} 0.45 \\ 0.55 \end{pmatrix}$

$X_3 = TX_2 = \begin{pmatrix} 0.8 & 0.3 \\ 0.2 & 0.7 \end{pmatrix}\begin{pmatrix} 0.45 \\ 0.55 \end{pmatrix} = \begin{pmatrix} 0.525 \\ 0.475 \end{pmatrix}$

**(b)** $X_{11} = T^{11}X_0 = \begin{pmatrix} 0.8 & 0.3 \\ 0.2 & 0.7 \end{pmatrix}^{11}\begin{pmatrix} 0 \\ 1 \end{pmatrix} = \begin{pmatrix} 0.600 \\ 0.400 \end{pmatrix}$

long term will stabilise around 60% donation and 40% no donations.

**8.** $X_n = T^nX_0 \approx \begin{pmatrix} 0.56 \\ 0.23 \\ 0.21 \end{pmatrix}$ 56% Geneva, 23% Zurich, 21% Basel

**9. (a)**

| | Now | 1 year | 2 years | 3 years | 4 years | 5 years |
|---|---|---|---|---|---|---|
| City | 200 000 | 19 0750 | 182 240 | 174 411 | 167 208 | 160 581 |
| Suburbs | 25 000 | 34 250 | 42 760 | 50 589 | 57 792 | 64 419 |

**(b)** City population will approach 84 375 and the suburbs population will approach 140 625.

## Exercise 7.5

**1. (a)** $y$-axis reflection. $(0, 0), (-3, 0), (-3, 1)$

**(b)** dilation in both directions of magnitude 2. $(0, 0), (6, 0), (6, 2)$

**(c)** $x$-axis reflection. $(0, 0), (3, 0), (3, -1)$

**(d)** reflection in $y = x$. $(0, 0), (0, 3), (1, 3)$

**(e)** dilation in $y$-direction of magnitude 3. $(0, 0), (3, 0), (3, 3)$

**(f)** Composition: reflection in $y = x$ and reflection in $y$-axis since

$\begin{pmatrix} -1 & 0 \\ 0 & 1 \end{pmatrix}\begin{pmatrix} 0 & 1 \\ 1 & 0 \end{pmatrix} = \begin{pmatrix} 0 & -1 \\ 1 & 0 \end{pmatrix}$ (or rotation 90 degrees anticlockwise about $(0, 0)$). $(0, 0), (0, 3), (-1, 3)$

**2.** Since $\begin{pmatrix} 0 & 2 \\ 2 & 0 \end{pmatrix} = \begin{pmatrix} 0 & 1 \\ 1 & 0 \end{pmatrix}\begin{pmatrix} 2 & 0 \\ 0 & 2 \end{pmatrix}$, then it is a dilation followed by reflection in $y = x$

**3.** A cannot represent a rotation since there is no angle with $\sin\theta = \dfrac{1}{2}$ and $\sin\theta = -\dfrac{1}{2}$ at the same time!

**4. (a)** The image is $3x + 2y = 18$

**(b)** image is $2x + 3y = 18$

**5.** $\begin{pmatrix} -1 & 0 \\ 0 & 1 \end{pmatrix}\begin{pmatrix} 1 & 0 \\ 0 & -1 \end{pmatrix} = \begin{pmatrix} -1 & 0 \\ 0 & -1 \end{pmatrix}$

**6. (a)** $x^2 + (y - 2)^2 = 36$; a circle with centre at $(0, 1)$ and radius 3 is transformed into a circle with centre at $(0, 2)$ and radius 6.

**(b)** $3x - 4y = 12$; A line with slope $-1.5$ and $y$-intercept 3 is transformed into a line with slope 0.75 and $y$-intercept $-3$.

**7. (a)** product of matrices $= \begin{pmatrix} 0 & -1 \\ 1 & 0 \end{pmatrix} = \begin{pmatrix} \cos 90 & -\sin 90 \\ \sin 90 & \cos 90 \end{pmatrix}$

**(b)** product of matrices $= \begin{pmatrix} 0 & 1 \\ -1 & 0 \end{pmatrix} =$

$\begin{pmatrix} \cos(-90) & -\sin(-90) \\ \sin(-90) & \cos(-90) \end{pmatrix}$

**8. (a)** $\begin{pmatrix} 4 \\ 9 \end{pmatrix}, \begin{pmatrix} 6 \\ 9 \end{pmatrix}, \begin{pmatrix} 6 \\ 12 \end{pmatrix}, \begin{pmatrix} 4 \\ 12 \end{pmatrix}$

**(b)** $\begin{pmatrix} 2 \\ 3 \end{pmatrix}, \begin{pmatrix} 4 \\ 3 \end{pmatrix}, \begin{pmatrix} 4 \\ 6 \end{pmatrix}, \begin{pmatrix} 2 \\ 6 \end{pmatrix}$

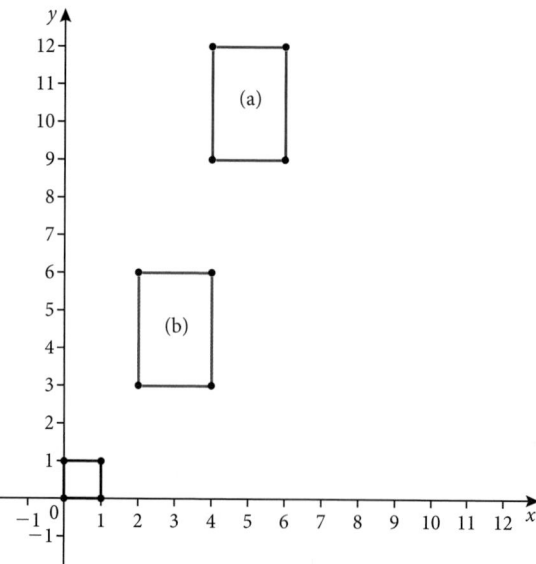

**9.** $y = 6$. Answers will differ.

For example $\begin{pmatrix} 2 & 1 & 0 \\ 0 & 3 & 6 \end{pmatrix} \mapsto \begin{pmatrix} 2 & 1 & 0 \\ 6 & 6 & 6 \end{pmatrix}$

**10. (a)** Eigenvectors: $\begin{pmatrix} k_1 \\ 0 \end{pmatrix}$ and $\begin{pmatrix} 0 \\ k_2 \end{pmatrix}$. Images of vectors along the $x$- and $y$-axes are multiples of the original vectors.

**(b)** Eigenvectors: $\begin{pmatrix} k \\ k \end{pmatrix}$ and $\begin{pmatrix} k \\ -k \end{pmatrix}$. Images of vectors along the $y = x$ and $y = -x$ are multiples of the original vectors.

**(c)** Eigenvectors: $\begin{pmatrix} -k \\ k \end{pmatrix}$ and $\begin{pmatrix} k \\ k \end{pmatrix}$. Images of vectors along the $y = x$ and $y = -x$ are multiples of the original vectors.

**(d)** Eigenvectors: $\begin{pmatrix} -2k \\ 3k \end{pmatrix}$ and $\begin{pmatrix} k \\ 2k \end{pmatrix}$. Images of vectors along $y = -\dfrac{3}{2}x$ and $y = 2x$ are multiples of the original vectors.

**11. (a)** $\begin{pmatrix} \cos(\alpha + \beta) & -\sin(\alpha + \beta) \\ \sin(\alpha + \beta) & \cos(\alpha + \beta) \end{pmatrix}$.

**(b)** $M(\beta)M(\alpha) =$

$\begin{pmatrix} \cos\alpha \cos\beta - \sin\alpha \sin\beta & -\sin\alpha \cos\beta - \cos\alpha \sin\beta \\ \cos\alpha \sin\beta + \sin\alpha \cos\beta & \cos\alpha \cos\beta - \sin\alpha \sin\beta \end{pmatrix}$

**(c)** By comparison we recognise the formulas for $\cos(\alpha + \beta)$ and $\sin(\alpha + \beta)$.

**12. (a)** $\left(\dfrac{4}{3}\right)^n$

**(b)** $\lim\limits_{n \to \infty} \left(\dfrac{4}{3}\right)^n = \infty$

**(c)** $3 \cdot 4^n$

**(d)** $\left(\dfrac{1}{3}\right)^n$

**(e)** $3 \cdot 4^n \cdot \left(\dfrac{1}{3}\right)^n = 3\left(\dfrac{4}{3}\right)^n$; $\quad \lim\limits_{n \to \infty} 3\left(\dfrac{4}{3}\right)^n = \infty$

**(f)** $3 \cdot 4^{n-1}$

**(g)** $a_0 \cdot \left(\dfrac{1}{9}\right)^n$, $a_0 = $ area of original triangle $= \dfrac{\sqrt{3}}{4}$

**(h)** $a_0 + a_0\left(\sum\limits_1^n 3 \cdot 4^{i-1} \cdot \left(\dfrac{1}{9}\right)^i\right) = a_0\left(1 + \sum\limits_1^n \dfrac{3}{4} \cdot \left(\dfrac{4}{9}\right)^i\right)$

$\Rightarrow \lim\limits_{n \to \infty} = a_0\left(1 + \dfrac{3}{4} \cdot \dfrac{\dfrac{4}{9}}{1 - \dfrac{4}{9}}\right) = \dfrac{8}{5} \cdot \dfrac{\sqrt{3}}{4} = \dfrac{2\sqrt{3}}{5}$

## Chapter 7 Practice questions

**1.** $x = 1$ or $x = -7$

**2. (a)** $M^2 = \begin{pmatrix} a^2 + 4 & 2a - 2 \\ 2a - 2 & 5 \end{pmatrix}$

**(b)** $a = -1$

**(c)** $M^{-1} = \begin{pmatrix} \dfrac{1}{3} & \dfrac{2}{3} \\ \dfrac{2}{3} & \dfrac{1}{3} \end{pmatrix}$, $x = 1, y = -1$

**3.** $B = \begin{pmatrix} 1 & 3 \\ 4 & 12 \end{pmatrix}$

**4.** $a = \dfrac{28}{33}$ $b = \dfrac{59}{33}$ $c = \dfrac{20}{33}$ $d = \dfrac{28}{33}$

**5. (a)** $A^{-1} = \begin{pmatrix} \dfrac{1}{19} & \dfrac{2}{19} \\ -\dfrac{7}{19} & \dfrac{5}{19} \end{pmatrix}$

**(b) (i)** $X = (C - B)A^{-1}$

**(ii)** $X = \begin{pmatrix} 2 & -3 \\ -4 & 1 \end{pmatrix}$

**6. (a)** $A + B = \begin{pmatrix} a + 1 & b + 2 \\ c + d & c + 1 \end{pmatrix}$

**(b)** $AB = \begin{pmatrix} a + bd & 2a + bc \\ c + d & 3c \end{pmatrix}$

**7. (a)** $A^{-1} = \begin{pmatrix} \dfrac{1}{10} & \dfrac{2}{5} & \dfrac{1}{10} \\ -\dfrac{7}{10} & \dfrac{1}{5} & \dfrac{3}{10} \\ -\dfrac{6}{5} & \dfrac{1}{5} & \dfrac{4}{5} \end{pmatrix}$

**(b)** $x = \dfrac{6}{5}, y = \dfrac{3}{5}, z = \dfrac{8}{5}$

**8. (a)** $\begin{pmatrix} -3 & 2 \\ 1 & \dfrac{14 - a}{3} \end{pmatrix}$

**(b)** $\begin{pmatrix} -14 & 4a - 4 \\ -2 & 7a - 2 \end{pmatrix}$

**(c)** $\begin{pmatrix} \dfrac{a}{5a + 2} & -\dfrac{2}{5a + 2} \\ \dfrac{1}{5a + 2} & \dfrac{5}{5a + 2} \end{pmatrix}$

**9. (a)** $a = 7, b = 2$ **(b)** $x = -1, y = 2, z = -1$

**10. (a)** $B = A^{-1}C$

**(b)** $DA = \begin{pmatrix} 1 & 0 & 0 \\ 0 & 1 & 0 \\ 0 & 0 & 1 \end{pmatrix}$, $B = \begin{pmatrix} 1 \\ -1 \\ 2 \end{pmatrix}$

**(c)** $(1, -1, 2)$

**11. (a)** $\text{Det} = 0$

**(b)** $\lambda = 5$

**(c)** $(2 - 3t, 1 + t, t)$

**12.** $\lambda = -3$ or $\lambda = 4$; $\begin{pmatrix} 2t \\ -5t \end{pmatrix}, \begin{pmatrix} t \\ t \end{pmatrix}$

**13. (a)** $0.772, 0.129, 0.099$

**(b)** $0.756, 0.139, 0.106$

**14. (a)** $P = \begin{pmatrix} \dfrac{1}{3} & 0 \\ 1 & 1 \end{pmatrix} \Rightarrow D = P^{-1}AP = \begin{pmatrix} 1 & 0 \\ 0 & -1 \end{pmatrix}$

**(b)** $A^8 = PD^8P^{-1} = \begin{pmatrix} 1 & 0 \\ 0 & 1 \end{pmatrix}$

**15.** 54% in region 1, 17% in region 2, and 29% in region 3

**16. (a)** 45% Channel 1, 55% Channel 2

**(b)**

|  | 1 | 2 | 3 | 4 | 5 |
|---|---|---|---|---|---|
| Channel 1 | 0.45 | 0.415 | 0.391 | 0.373 | 0.361 |
| Channel 2 | 0.55 | 0.585 | 0.609 | 0.627 | 0.639 |

**(c)** Channel 1 $\approx \dfrac{1}{3}$

Channel 2 $\approx \dfrac{2}{3}$

**17. (a)** 81, 243

**(b)** $3^n$

**(c)** $a, \dfrac{3}{4}a, \dfrac{9}{16}a, \dfrac{27}{64}a, \cdots, \left(\dfrac{3}{4}\right)^n a$, area tends to zero.

**(d)** $\begin{pmatrix} \dfrac{1}{2} & 0 \\ 0 & \dfrac{1}{2} \end{pmatrix}\begin{pmatrix} x \\ y \end{pmatrix}$; $\begin{pmatrix} \dfrac{1}{2} & 0 \\ 0 & \dfrac{1}{2} \end{pmatrix}\begin{pmatrix} x \\ y \end{pmatrix} + \begin{pmatrix} \dfrac{1}{2} \\ 0 \end{pmatrix}$; $\begin{pmatrix} \dfrac{1}{2} & 0 \\ 0 & \dfrac{1}{2} \end{pmatrix}\begin{pmatrix} x \\ y \end{pmatrix} + \begin{pmatrix} \dfrac{1}{4} \\ \dfrac{\sqrt{3}}{4} \end{pmatrix}$

# Answers

**18.**

| $N=0$ | $N=1$ | $N=2$ | $N=3$ | $N=4$ | $N=5$ |
|---|---|---|---|---|---|
| 0.12 | 0.23 | 0.29 | 0.31 | 0.32 | 0.33 |
| 0.32 | 0.34 | 0.34 | 0.34 | 0.33 | 0.33 |
| 0.56 | 0.42 | 0.37 | 0.35 | 0.34 | 0.34 |

## Chapter 8

### Exercise 8.1

**1. (a)** $v = \begin{pmatrix} 10 \\ -4 \end{pmatrix}$ **(b)** $v = \begin{pmatrix} -15 \\ 6 \end{pmatrix}$ **(c)** $v = \begin{pmatrix} 15 \\ -6 \end{pmatrix}$

**2. (a)** $w = \begin{pmatrix} 4 \\ 1 \\ -5 \end{pmatrix}$ **(b)** $w = \begin{pmatrix} 10 \\ -2 \\ 1 \end{pmatrix}$ **(c)** $w = \begin{pmatrix} 10 \\ 4 \\ -17 \end{pmatrix}$

**3. (a)** $\begin{pmatrix} \frac{12}{13} \\ -\frac{5}{13} \end{pmatrix}$ **(b)** $\begin{pmatrix} \frac{-1}{\sqrt{10}} \\ \frac{3}{\sqrt{10}} \end{pmatrix}$ **(c)** $\begin{pmatrix} \frac{1}{3} \\ \frac{2}{3} \\ \frac{2}{3} \end{pmatrix}$

**4.** $(6, 8)$

**5.** $x = 3, y = 5$

**6.** $r = 6, s = 2$

**7.** $\begin{pmatrix} 4 \\ 7 \end{pmatrix} = \frac{5}{2}\begin{pmatrix} 2 \\ 3 \end{pmatrix} - \frac{1}{2}\begin{pmatrix} 2 \\ 1 \end{pmatrix}$

**8.** $\begin{pmatrix} 5 \\ -5 \end{pmatrix} = r\begin{pmatrix} 1 \\ -1 \end{pmatrix} + (r-5)\begin{pmatrix} -1 \\ 1 \end{pmatrix}$

**9.** $\begin{pmatrix} -11 \\ 0 \end{pmatrix} = 2\begin{pmatrix} 2 \\ 5 \end{pmatrix} - 5\begin{pmatrix} 3 \\ 2 \end{pmatrix}$

**10. (a)** $\begin{pmatrix} 8 \\ -6 \end{pmatrix}$ **(b)** $\begin{pmatrix} -\frac{50}{13} \\ -\frac{120}{13} \end{pmatrix}$ **(c)** $\begin{pmatrix} \frac{20}{3} \\ -\frac{10}{3} \\ -\frac{20}{3} \end{pmatrix}$

### Exercise 8.2

**1. (a)** $\dfrac{x-1}{4} = \dfrac{y+3}{3}$ **(b)** $\dfrac{x-6}{4} = y+1$

**(c)** $\dfrac{x+9}{2} = \dfrac{y-5}{-3}$ **(d)** $\dfrac{x-7}{5} = \dfrac{y-1}{-2}$

**2. (a)** $x-6 = \dfrac{y-7}{-2} = \dfrac{z}{2}$ **(b)** $\dfrac{x+3}{-3} = \dfrac{y+2}{3} = \dfrac{z-9}{6}$

**3. (a)** $r = \begin{pmatrix} 1 \\ 0 \\ 2 \end{pmatrix} + k\begin{pmatrix} 3 \\ -4 \\ 5 \end{pmatrix}$ **(b)** $r = \begin{pmatrix} -2 \\ 3 \\ 0 \end{pmatrix} + k\begin{pmatrix} 1 \\ -1 \\ 2 \end{pmatrix}$

**4. (a)** $(6, 11)$ **(b)** $(6, 0)$ **(c)** $(6, -2, 2)$ **(d)** $(6, -1, 5)$

**5. (a)** $r = \begin{pmatrix} 3 \\ -2 \\ 1 \end{pmatrix} + k\begin{pmatrix} 2 \\ -2 \\ 4 \end{pmatrix}$ **(b)** $r = \begin{pmatrix} 0 \\ 3 \\ -1 \end{pmatrix} + k\begin{pmatrix} 3 \\ -2 \\ 1 \end{pmatrix}$

**(c)** $r = \begin{pmatrix} -3 \\ 0 \\ -2 \end{pmatrix} + k\begin{pmatrix} \frac{1}{2} \\ \frac{5}{2} \\ -2 \end{pmatrix}$ **(d)** $r = \begin{pmatrix} a \\ b \\ c \end{pmatrix} + k\begin{pmatrix} m \\ n \\ p \end{pmatrix}$

**6. (a)** $r = \begin{pmatrix} 3 \\ -1 \\ -5 \end{pmatrix} + k\begin{pmatrix} 2 \\ -2 \\ 4 \end{pmatrix}$ **(b)** $r = \begin{pmatrix} 3 \\ -1 \\ -5 \end{pmatrix} + k\begin{pmatrix} 3 \\ 2 \\ -4 \end{pmatrix}$

**(c)** $r = \begin{pmatrix} 3 \\ -1 \\ -5 \end{pmatrix} + k\begin{pmatrix} 3 \\ -1 \\ 5 \end{pmatrix}$ **(d)** $r = \begin{pmatrix} 3 \\ -1 \\ -5 \end{pmatrix} + k\begin{pmatrix} 7 \\ 1 \\ -2 \end{pmatrix}$

**7.** $a = \begin{pmatrix} 2 \\ 4 \\ 6 \end{pmatrix}$, $b = \begin{pmatrix} 4 \\ -1 \\ 7 \end{pmatrix}$, and $c = \begin{pmatrix} -2 \\ -3 \\ 1 \end{pmatrix}$

**8. (a)** $r = \begin{pmatrix} -3 \\ 7 \end{pmatrix} + k\begin{pmatrix} 2 \\ 4 \end{pmatrix}$ **(b)** $r = \begin{pmatrix} 2 \\ -5 \end{pmatrix} + k\begin{pmatrix} -4 \\ 6 \end{pmatrix}$

**(c)** $r = \begin{pmatrix} 8 \\ -2 \\ 1 \end{pmatrix} + k\begin{pmatrix} -11 \\ 12 \\ 6 \end{pmatrix}$ **(d)** $r = \begin{pmatrix} 0 \\ -6 \\ -3 \end{pmatrix} + k\begin{pmatrix} 7 \\ 5 \\ 3 \end{pmatrix}$

**(e)** $r = \begin{pmatrix} 3 \\ -4 \\ 5 \end{pmatrix} + k\begin{pmatrix} 2 \\ 2 \\ 0 \end{pmatrix}$ **(f)** $r = \begin{pmatrix} -7 \\ -4 \\ 2 \end{pmatrix} + k\begin{pmatrix} 9 \\ 0 \\ 10 \end{pmatrix}$

**9. (a)** $\dfrac{x+3}{2} = \dfrac{y-7}{4}$ **(b)** $\dfrac{x-2}{-4} = \dfrac{y+5}{6}$

**(c)** $\dfrac{x-8}{-11} = \dfrac{y+2}{12} = \dfrac{z-1}{6}$ **(d)** $\dfrac{x}{7} = \dfrac{y+6}{5} = \dfrac{z+3}{3}$

**(e)** $\dfrac{x-3}{2} = \dfrac{y+4}{2}, z = 5$ **(f)** $\dfrac{x+7}{9} = \dfrac{z-2}{10}, y = -4$

**10.** $(2, 3)$

**11.** $(1, -1, 2)$

### Exercise 8.3

**1. (a)** $\sqrt{5.44} \approx 2.33\,\text{m}$ **(b)** $2\sqrt{5.44} \approx 4.66\,\text{m}$

**(c)** $10\sqrt{5.44} \approx 23.3\,\text{m}$ **(d)** $t\sqrt{5.44} \approx 2.33t\,\text{m}$

**2. (a)** $\sqrt{120.64} \approx 11.0\,\text{m}$

**(b)** $\sqrt{108.16} = 10.4\,\text{m}$

**3.** The minimum distance is 0 at $t = 6.875$ minutes.

**4. (a)** $56\,\text{m}$

**(b)** $30\,\text{m}$

**(c)** $\sqrt{3396} \approx 58.3\,\text{m}$

**5. (a)** $d = \begin{pmatrix} 14 \\ 16 \\ 46 \end{pmatrix} + t\begin{pmatrix} 2 \\ 1 \\ -2 \end{pmatrix}$ **(b)** $48.1\,\text{m}$

**6. (a)** $(1900, 3400)$ **(b)** $t = 30$ seconds

**(c)** Yes. **(d)** $80.9\,\text{m}$

**(e)** $t = 31.2$ seconds

### Exercise 8.4

**1. (a)** $3$ **(b)** $0$ **(c)** $0$ **(d)** $0$

**2. (a)** $46.8$ **(b)** $67.9$ **(c)** $33.1$ **(d)** $127$

**3.** $\vec{F} \cdot \vec{AB} = \begin{pmatrix} 30 \\ 150 \end{pmatrix} \cdot \begin{pmatrix} 15 \\ 40 \end{pmatrix} = 6450J$

**4. (a)** $\begin{pmatrix} 1 \\ 0 \\ 0 \end{pmatrix}$ $(-1, 0, 0)$ **(b)** $\begin{pmatrix} 0 \\ 1 \\ 0 \end{pmatrix}$ $(0, -1, 0)$

**(c)** $\begin{pmatrix} 14 \\ -20 \\ -19 \end{pmatrix}$ $(-14, 20, 19)$ **(d)** $\begin{pmatrix} 18 \\ 48 \\ -10 \end{pmatrix}$ $(-18, -48, 10)$

### Exercise 8.5

**1. (a)** $90°$ **(b)** $85°$ **(c)** $109°$ **(d)** $32°$

**2. (a)** neither

**(b)** orthogonal

**(c)** orthogonal

**3. (a)** $26.6°, 63.4°, 90°$

**(b)** $41.4°, 74.4°, 64.1°$

**(c)** $41.6°, 116.6°, 21.8°$

**4.** $\begin{pmatrix} 5k \\ -3k \end{pmatrix}$

**5.** No

**6.** $\pm\sqrt{6}$

**7.** $\dfrac{48 \pm 25\sqrt{3}}{39}$

**8.** Since the sides are the vectors $a$ and $b$ with $|a| = |b|$, the diagonals are the vectors $a + b$ and $a - b$

$(a + b)(a - b) = |a^2| - |b^2| = 0$

**9. (a)** $(x - 1)(x - 3) + (y - 2)(y - 4) = 0$

**(b)** $(x - 3)(x + 1) + (y - 4)(y + 7) = 0$

## Chapter 8 Practice questions

1. (a) $v = \begin{pmatrix} -8 \\ 2 \end{pmatrix}$    (b) $v = \begin{pmatrix} 4 \\ -1 \end{pmatrix}$    (c) $v = \begin{pmatrix} 12 \\ -3 \end{pmatrix}$

2. (a) $w = \begin{pmatrix} -8 \\ 10 \\ -10 \end{pmatrix}$    (b) $w = \begin{pmatrix} 8 \\ -10 \\ 10 \end{pmatrix}$    (c) $w = \begin{pmatrix} 34 \\ -26 \\ 41 \end{pmatrix}$

3. (a) $\dfrac{x+2}{5} = \dfrac{y-2}{3}$    (b) $\dfrac{x-4}{3} = \dfrac{y+1}{-2}$

   (c) $\dfrac{x}{2} = \dfrac{y-4}{3}$    (d) $\dfrac{x-11}{3} = \dfrac{y-7}{-4}$

4. (a) $\dfrac{x-3}{2} = \dfrac{y}{-3} = \dfrac{z+2}{-6}$    (b) $\dfrac{x+4}{5} = \dfrac{z}{-2}, y = 4$

5. (a) $r = \begin{pmatrix} -1 \\ 3 \\ -2 \end{pmatrix} + k\begin{pmatrix} 2 \\ -3 \\ 6 \end{pmatrix}$    (b) $r = \begin{pmatrix} -9 \\ 3 \\ -3 \end{pmatrix} + k\begin{pmatrix} 2 \\ 0 \\ -2 \end{pmatrix}$

6. (a) $\sqrt{5} \approx 2.24$ m    (b) $5.03$ m
   (c) $22.4$ m    (d) $2.24t$ m

7. (a) $\sqrt{19^2 + 2^2} \approx 19.1$ m    (b) $\sqrt{5t^2 - 40t + 400}$ m

8. (a) (i) $\sqrt{22^2 + 1^2} \approx 22.0$ m
     (ii) $\sqrt{10t^2 - 150t + 625}$
   (b) No (minimum $\approx 7.91$ m)

9. (a) $-8$    (b) $0$    (c) $0$    (d) $-16$

10. (a) $\dfrac{77}{2}$    (b) $-28$

   (c) $\dfrac{81\sqrt{2}}{2} \approx 57.3$    (d) $71.8$

11. (a) $\begin{pmatrix} 0 \\ 0 \\ -1 \end{pmatrix}$   (b) $\begin{pmatrix} 0 \\ 0 \\ 1 \end{pmatrix}$   (c) $\begin{pmatrix} 2 \\ 2 \\ -3 \end{pmatrix}$   (d) $\begin{pmatrix} 14 \\ -34 \\ 1 \end{pmatrix}$

12. (a) $90°$   (b) $109°$   (c) $116°$   (d) $76°$

13. (a) $6 \text{ ms}^{-1}$    (b) $8$ m

   (c) Let $a = \begin{pmatrix} -4 \\ 2 \\ 4 \end{pmatrix}$ and $b = \begin{pmatrix} 4 \\ -6 \\ 7 \end{pmatrix}$. Since $a \cdot b = 0$,

     they are perpendicular.

   (d) 3 seconds later

14. (a) (i) $\overrightarrow{OA} = \begin{pmatrix} -2 \\ 4 \\ 3 \end{pmatrix}$ and $\overrightarrow{OB} = \begin{pmatrix} -1 \\ 3 \\ 1 \end{pmatrix}$ so $\overrightarrow{AB} = \begin{pmatrix} 1 \\ -1 \\ -2 \end{pmatrix}$

     (ii) $|\overrightarrow{AB}| = \sqrt{6}$

   (b) $r = \begin{pmatrix} -2 \\ 4 \\ 3 \end{pmatrix} + t\begin{pmatrix} 1 \\ -1 \\ -2 \end{pmatrix}$

   (c) $\overrightarrow{AC} = \begin{pmatrix} 2 \\ y-4 \\ -4 \end{pmatrix} = k \cdot \begin{pmatrix} 1 \\ -1 \\ -2 \end{pmatrix} \Rightarrow k = 2 \Rightarrow y = 2$

   (d) (i) $0$    (ii) $90°$

   (e) $\sqrt{\dfrac{30}{2}}$

15. $m = \dfrac{1}{\sqrt{2}}$ and $n = -\dfrac{1}{\sqrt{2}}$, $m = -\dfrac{1}{\sqrt{2}}$ and $n = \dfrac{1}{\sqrt{2}}$

16. (a) $\overrightarrow{AB} = \begin{pmatrix} 9 \\ 6 \\ -3 \end{pmatrix}$

   (b) Let $\overrightarrow{OC} = \begin{pmatrix} x \\ y \\ z \end{pmatrix} \Rightarrow \overrightarrow{AC} = \begin{pmatrix} x+3 \\ y+2 \\ z-2 \end{pmatrix}$

     $\overrightarrow{CB} = \begin{pmatrix} 6-x \\ 4-y \\ -1-z \end{pmatrix} \Rightarrow \begin{pmatrix} x+3 \\ y+2 \\ z-2 \end{pmatrix} = \begin{pmatrix} 12-2x \\ 8-2y \\ -2-2z \end{pmatrix}$

     $\Rightarrow \begin{cases} x+3 = 12-2x \\ y+2 = 8-2y \\ z-2 = -2-2z \end{cases} \Rightarrow \overrightarrow{OC} = \begin{pmatrix} 3 \\ 2 \\ 0 \end{pmatrix}$

(c) $15.5°$

(d) (i) $\sin\theta = \dfrac{DE}{CD}$ and $\overrightarrow{OC} + \overrightarrow{CD} = \overrightarrow{OD}$

     $\Rightarrow |\overrightarrow{DE}| = |k \cdot \overrightarrow{OC} - \overrightarrow{OC}|\sin\theta \Rightarrow |\overrightarrow{DE}|$

     $= (k-1) \cdot |\overrightarrow{OC}|\sin\theta$

   (ii) $|\overrightarrow{DE}| < 3 \Rightarrow (k-1)\sqrt{13}\sin 15.50... < 3$

     $\Rightarrow k-1 \approx 3.11... \Rightarrow 1 < k < 4.11$

# Chapter 9

## Exercise 9.1

1. (a) $d = 3000 - 900t$
   (b) $(0, 3000)$; 3000 km represents the distance remaining at time $t = 0$
   (c) $t = 3.33$; 3 hours and 20 minutes remaining until the plane reaches its destination.
   (d) Domain: $0 \leqslant t \leqslant \dfrac{10}{3}$; Range: $0 \leqslant d \leqslant 3000$

2. (a) $d = -800t + 5000$
   (b) The plane is travelling at 800 km h$^{-1}$
   (c) $(0, 5000)$; 5000 km represents the distance remaining at time $t = 0$
   (d) $t = 6.25$; 6 hours and 15 minutes remaining until the plane reaches its destination.
   (e) Domain: $0 \leqslant t \leqslant 6.25$; Range: $0 \leqslant d \leqslant 5000$

3. (a) EU = USA + 33
   (b) 45
   (c) 11
   (d) The EU size increases by 1 for every USA size increase by 1
   (e) $6 \leqslant$ USA $\leqslant 16$, $39 \leqslant$ EU $\leqslant 49$

4. (a) $t = \dfrac{7}{5}n + 14$
   (b) 1410 minutes
   (c) The gradient of $\dfrac{7}{5}$ implies that it takes 7 minutes to read 5 pages $\Rightarrow$ 1.4 pages min$^{-1}$; the $t$-intercept of 14 suggests that it takes an additional 14 minutes to start/finish a book
   (d) $n > 0$; $t > 14$

5. (a) $F = \dfrac{9}{5}C + 32$
   (b) Every degree Celsius is equal to 1.8 degrees Fahrenheit.
   (c) $0°C$ is equal to $32°F$
   (d) $0°F$ is equal to $-17.8°C$
   (e) $50°F$
   (f) $-40$
   (g) $C > -273$; $F > -459.4$

6. (a) $C = 8.5n + 350$
   (b) $n > 0$; $C > 350$
   (c) (i) 1200 ZAR      (ii) 2050 ZAR
     (iii) 3750 ZAR
   (d) (i) 12 ZAR cup$^{-1}$    (ii) 10.25 ZAR cup$^{-1}$
     (iii) 9.375 ZAR cup$^{-1}$
   (e) The average cost per cup decreases when more cups are ordered.
   (f) $D = 8n + 550$
   (g) $n > 0$; $D > 550$
   (h) It costs 8 ZAR for each additional cup ordered.
   (i) 5350 ZAR

**(j)** $x = 400$

**(k)** $C = \begin{cases} 8.5n + 350, & x < 500 \\ 8.5n, & x \geq 500 \end{cases}$

**(l)** $a = 400$, $b = 499$, $k = 1100$

**7. (a)** $C = \begin{cases} 2.6, & m \leq 234.8 \\ 2.6 + 0.2\left(\dfrac{m - 234.8}{117.4}\right), & 234.8 < m \leq 9656.1 \\ 18.65 + 0.2\left(\dfrac{m - 9656.1}{86.9}\right), & 9656.1 < m \end{cases}$

**(b)** 2.60 GBP, 10.72 GBP, 30.95 GBP

**(c)** $D = \begin{cases} 2.6, & t \leq 50.4 \\ 2.6 + 0.2\left(\dfrac{t - 50.4}{25.2}\right), & 50.4 < t \end{cases}$

**(d)** 2.60 GBP, 4.58 GBP, 9.34 GBP

**(e) (i)** 11.72 GBP **(ii)** 9.01 GBP

**8. (a)** $h(t) = -4.9t^2 + 5t + 60$

**(b) (i)** 61.3 m **(ii)** 4.05 s **(iii)** $0 \leq t < 2.03$

**9. (a)** 106 m **(b)** $h = 44.9$ m

**(c)** $t = 4.91$ s **(d)** $t = 5.39$ s

**(e)** Jane's model appears to fit the data better and Kevin's red curve should have $y$-intercept 100.

$f1(x) := -0.25 \cdot x^3 - 2.32 \cdot x^2 + 1.93 \cdot x + 106$

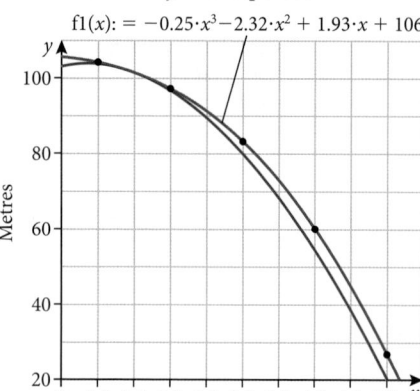

**10. (a)** 11 km **(b)** 26.6 km

**(c)** The distance ship A travels $t$ hours after noon is $15(t - 1)$; the distance ship B travels $t$ hours after noon is $11t$. The distance between them is therefore
$$s(t) = \sqrt{(15(t-1))^2 + (11t)^2}$$
$$= \sqrt{225(t^2 - 2t + 1) + 121t^2}$$
$$= \sqrt{346t^2 - 450t + 225}$$

**(d)**

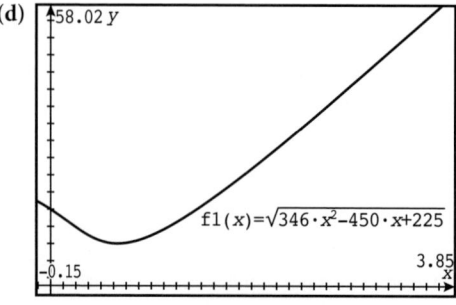

**(e)** The GDC tells us the minimum distance between the boats is more than 8 km, so the captain cannot see ship B between noon and 16:00:

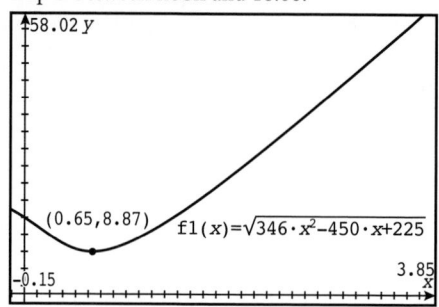

**11. (a)** Profit = Income − Cost
$$P(x) = 150x - 0.6x^2 - (2600 + 0.4x^2)$$
$$= 150x - 0.6x^2 - 2600 - 0.4x^2 = -x^2 + 150x - 2600$$

**(b)** maximum profit when
$$x = -\frac{-150}{2 \times -1} \text{ or } x = -\frac{20 + 130}{2} = 75 \text{ machines}$$

**(c)** $I(75) = 150(75) - 0.6(75)^2 = \$7875$
Selling price per machine $= \dfrac{7875}{75} = \$105$

**(d)** $P(x) = 0$ or $(x - 20)(130 - x) = 0 \Rightarrow x = 20, x = 130$
$\Rightarrow$ Smallest number must be 21.

**12. (a)** 300 m

**(b) (i)** 221.6 m **(ii)** $t_0 = 4$
**(iii)** $k = 249.6$
**(iv)** 295.1 m, 280.4 m, 193.6 m
**(v)** 21.4 s **(vi)** 35.7 s
**(vii)** $0 \leq t \leq 35.7$, $0 \leq h(t) \leq 300$

**(c) (i)** $x = 6.39$ s
**(ii)** $g(t) = \begin{cases} -4.9t^2 + 300, & t < 6.39 \\ -7t + 145, & t \geq 6.39 \end{cases}$
**(iii)** 20.7 s

**(d) (i)** 5.82 s **(ii)** $d = 134$ m
**(iii)** $f(t) = \begin{cases} -4.9t^2 + 300, & t < 5.82 \\ -7t + 175, & t \geq 5.82 \end{cases}$
**(iv)** 25.0 s

## Exercise 9.2

**1. (a)** $v = 56(0.840)^t$ **(b)** 16.5 m s$^{-1}$
**(c)** 13.9 s **(d)** 23.1 s

**2. (a)** 59 bacteria **(b)** 8.77 hours
**(c)** $k = \dfrac{\ln 2.5}{4} = 0.229$ **(d)** $n > 4.05 \Rightarrow n = 5$

**3. (a)** $A(0) = 10$ **(b)** $A(50) = 6.16$
**(c)** $t = 333$ minutes later $\Rightarrow$ 18:33

**4. (a)** $L = 16(1.08)^t$ **(b)** 20.2 cm
**(c)** 5.80 years

**5. (a)** $k = -0.0462$ **(b)** 45.6 units
**(c)** 88.2 units **(d)** 23.8 years

**6. (a)** The initial quantity of glyphosate is 500 units.
**(b)** $t$ represents the number of 45-day half-lives.
**(c)** $A(1) = 250$, there are 250 units of glyphosate left after 45 days.

**7. (a)** the temperature in the oven/the maximum possible temperature of the cake mix
**(b)** $a = 132$
**(c)** $t = 34.8$ minutes

**8. (a)** $a = 810$, $b = 1.917$    **(b)** 44 fish

**(c)** August    **(d)** $p = 40$

**9. (a)** $a$ is the location of the horizontal asymptote. The temperature of the water cannot fall below room temperature, 20°C, so there must be a horizontal asymptote at $T = 20$, hence $a = 20$.

**(b)** $b = 80$

$$85 = 20 + 80k^{-1} \Rightarrow \frac{65}{80} = k^{-1} \Rightarrow k = \frac{80}{65} = 1.23$$

**(c)** $T = 48.3$°C

**(d)** $m = 8.06 \Rightarrow$ 8 minutes and 4 seconds

**10. (a)** 10 °C    **(b)** 124 °C

**(c)** $k = 5.64$ minutes

**11. (a)** 1.5

**(b)** $C = 2.5$

**(c)** 3 hours and 19 minutes

**12. (a)** $p + q = 47$, $4p + q = 53$

**(b)** $p = 2$, $q = 45$

**(c)** $C = 109$

## Exercise 9.3

**1. (a)** 6.25 hours (6 hours 15 minutes)

**(b)** 0.9 m    **(c)** $p = 0.45$

**(d)** $q = \dfrac{4\pi}{25} = 0.503$    **(e)** $r = 1.05$

**(f)** 23:00    **(g)** 469 minutes

**2. (a)** $p = 2.2$    **(b)** $q = \dfrac{\pi}{7}$    **(c)** 7.01 m

**3. (a)** 2 m    **(b)** $k = 1.48$    **(c)** 5.2 minutes

**4. (a) (i)** 100 m   **(ii)** 50 m

**(b) (i)** $a = -50$, $b = \dfrac{\pi}{10}$, $c = 50$

    **(ii)** $d = -5$

**5. (a)** $a = 4$    **(b)** $b = \dfrac{\pi}{15}$    **(c)** 10 seconds

**6. (a)** 101 m

**(b)** 35.5 m

**(c)** $a = 50$, $b = \dfrac{\pi}{15}$, $c = 75$, $d = 51$

**(d)** $t = 12.8$ minutes

**7. (a)** 2    **(b)** 15

**(c)** $-3$    **(d)** $16 \leqslant t \leqslant 20$

**8. (a)** 35 cm    **(b)** 5 cm

**(c)** 15 cm    **(d)** $A = 15$, $C = 20$

**(e)** 4 seconds    **(f)** $b = 90$

**(g)** $t = 0.595$    **(h)** 15

**9. (a)** 4 m    **(b)** 11 m

**(c)** 12 hours    **(d)** 20:00

**10. (a)** $a = 19$, $b = 2$    **(b)** $c = 30$

**(c)** $3.48 \leqslant x \leqslant 8.52$

## Exercise 9.4

**1. (a)** $P(t) = \dfrac{97.2}{1 + 91.2\,e^{-1.17t}}$

**(b)** $L = 97.2\,\text{km}\,\text{h}^{-1}$

**(c)** 7.76 seconds

**2. (a)** $L = 500$    **(b)** $C = 499$    **(c)** $k = 0.621$

**(d)**

**(e)** 13.5 days

**3. (a)** $C = 18.7$, $L = 19\,680$    **(b)** 19 680 rats

**(c)** 14.6 years

## Exercise 9.5

**1. (a)** always    **(b)** sometimes

**(c)** sometimes    **(d)** never

**2. (a)** never    **(b)** never

**(c)** never    **(d)** never

**3. (a)** $y = 210$    **(b)** $x = 16$

**4. (a)** $y = 160$    **(b)** $x = 10$

**5. (a)** $y = 1024$    **(b)** $x = 4$

**6. (a)** $y = 2.5$    **(b)** $x = 100$

**7. (a)** $y = 0.625$    **(b)** $x = 10$

**8. (a)** $y = 2$    **(b)** $x = 3$

**9. (a)** $S = 0.75d$    **(b)** 5.25 m    **(c)** 13.3 m

**10. (a)** 9.8    **(b)** 39.2    **(c)** 5.7 seconds

**11. (a)** 1.62    **(b)** 40.5 m    **(c)** 11.1 s

**12. (a)** $4.00 \times 10^{14}$    **(b)** 7560 m s$^{-1}$    **(c)** $6.25 \times 10^{6}$ m

**13. (a)** 7.664    **(b)** 3920 cm$^3$    **(c)** 2.35 m

**14. (a)** 0.613    **(b)** 1060 W    **(c)** 14.8 m s$^{-1}$

## Exercise 9.6

**1. (a)** logistic

**(b)** inverse variation

**(c)** linear/direct linear variation

**(d)** logistic

**(e)** quadratic/direct quadratic variation

**(f)** trigonometric

**(g)** cubic

**(h)** logistic

**(i)** exponential (growth)

**(j)** trigonometric

**(k)** exponential (decay)

**(l)** inverse variation

**(m)** linear (probably NOT direct linear variation because there will be a fixed cost as well)

**2. (a)** $2 \leqslant t \leqslant 6$

**(b)** $11 \leqslant x \leqslant 100$

**(c)** $1.3 \leqslant n \leqslant 8.5$

**3. (a)** A constant rate of change; data in a linear pattern

**(b)** Data that appears parabolic

**(c)** Data that is cyclical/periodic

**(d)** Values that increases quickly, with bound, or decreases to approach a fixed value. The rate of change is a constant factor.

**(e)** Values that increase quickly then slow as they approach a maximum

# Answers

(f) As one value increases, the other decreases; data appears hyperbolic.

## Chapter 9 Practice questions

**1. (a)** $T = 493$
**(b)** $n = 6.12 \Rightarrow$ in the year 2007
**(c)** $P = 39636$
**(d)** At the end of 7 years, $P = 46807 < 25600 \times 2$, therefore not doubled.
**(e)** $\dfrac{25600}{280} = \dfrac{640}{7} = 91.4$
**(f)** $n = 9.31 \Rightarrow$ after 10 years

**2. (a)**

| $x$ | 0 | 10 | 20 | 30 | 40 | 50 | 60 | 70 | 80 | 90 |
|---|---|---|---|---|---|---|---|---|---|---|
| $P$ | $-30$ | 15 | **50** | 75 | 90 | **95** | 90 | 75 | 50 | **15** |

**(b)**

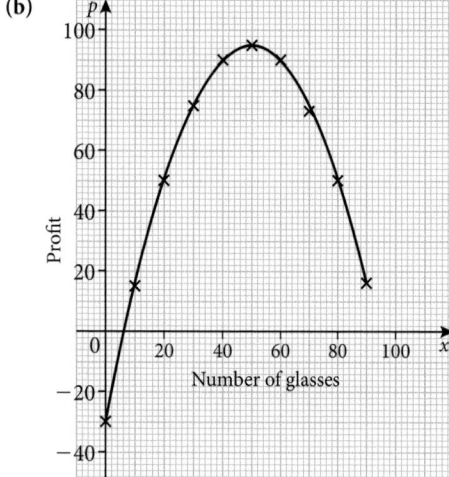

**(c) (i)** maximum profit = 95 swiss francs
**(ii)** 50 glasses
**(iii)** 67 or 33 (from graph)
**(iv)** 30 swiss francs
**(v)** Fiona's share = $\dfrac{3}{6}$;
Profit from 40 glasses = 90 swiss francs;
Fiona's profit = $\dfrac{1}{2} \times 90 = 45$ swiss francs

**3. (a)** $Q(0) = 25$
**(b)** $Q(20) = 0.003(20)^2 - 0.625(20) + 25 = 13.7$
**(c)** Average rate of loss = $\dfrac{Q(10) - Q(20)}{20 - 10} = \dfrac{19.05 - 13.7}{10}$
$= 0.535$ units per minute
**(d)** $0.003t^2 - 0.625t + 25 = 0 \Rightarrow$ Energy runs out after 54.0 minutes

**4. (a)** 400 USD    **(b)** $t = 8.64 \Rightarrow$ 9 months
**(c)** $8500(0.95)^2 - (400 \times 2 + 2000) = 4871.25$ USD

**5. (a)** $3600\,\text{m}^2$    **(b)** $180\,\text{m}$
**(c) (i)** 100    **(ii)** $10000\,\text{m}^2$
**(d)** $m = 3600, n = 10000$

**6. (a)** $100\,°\text{C}$    **(b)** $95.3\,°\text{C}$    **(c)** $h = 8820\,\text{m}$
**7. (a)** $T(0) = 20 + 70 \times 2.72^{-0.4\times0} = 90$
**(b)** $21.3\,°\text{C}$    **(c)** 1.66 minutes
**(d)** $20\,°\text{C}$; that is the horizontal asymptote of the model/the room temperature.

**(e)** It indicates by how much the temperature increases per minute.
**(f)** $m = 3.8$; this is the time when the temperature of the soup and the temperature of the water are equal.
**(g)** $3.8 < m \leqslant 6$
**8. (a)** 25000 USD    **(b)** 19601.32 USD    **(c)** 8.55 years
**9. (a)** 1800 bacteria    **(b)** 145800 bacteria    **(c)** 33.5 hours
**10. (a)** 45 euros    **(b)** 8 or 25 people
**(c) (i)** $n = 14$    **(ii)** 28.29 euros
**11. (a)** $135\,\text{m}$
**(b)** period = $\dfrac{60 \text{ minutes}}{2.4 \text{ rotations}} = 25$ minutes
**(c)** $b = \dfrac{2\pi}{25}$    **(d)** $a = -61$
**(e)**

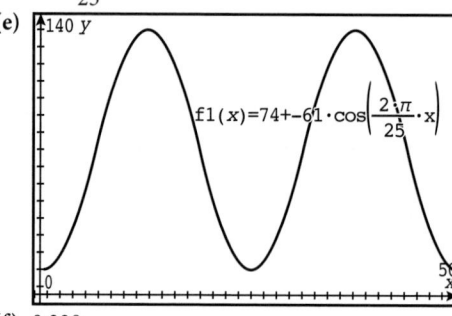

**(f)** 0.330

# Chapter 10

## Exercise 10.1

**1. (a)** No experimental units. Total number of students. Students in one school. Qualitative.
**(b)** No units. All 10th grade students. One school or class. Quantitative
**(c)** Units of length (m, inches etc). All new born children. Children at one hospital or on one day. Quantitative.
**(d)** No units. All children under 14. One school or one geographical location. Qualitative.
**(e)** Units of time (hours, mins etc). All commuters. Workers in one city or a one company. Quantitative.
**(f)** No units. Country's population. Subset of registered voters (age group, location etc). Qualitative.
**(g)** No units. All students at international schools. One year group or from one school. Qualitative.
**2. (a) (i)** 1176 students in grades 10–12.
**(ii)** Random sampling, appropriate stratified random sampling, quota sampling or convenience sampling of first 30 students at a location.
**(b) (i)** All bolts produced.
**(ii)** Random sampling from shift or container or testing each nth bolt from shift or container.
**3. (a)** Qualitative    **(b)** Quantitative, discrete.
**(c)** Quantitative, continuous **(d)** Quantitative, continuous
**(e)** Quantitative, continuous **(f)** Quantitative, discrete
**(g)** Quantitative, continuous **(h)** Qualitative
**(i)** Qualitative
**4. (a)** Numerical, continuous    **(b)** Numerical, discrete
**(c)** Numerical, continuous    **(d)** Numerical, continuous
**(e)** Numerical, continuous    **(f)** Numerical, continuous
**(g)** Categorical
**5. (a)** Inferential.    **(b)** Descriptive

**6. (a)** Inferential    **(b)** Descriptive    **(c)** Descriptive
   **(d)** Inferential    **(e)** Inferential
**7. (a)** Non-random      **(b)** Judgement sample
   **(c)** Non-sampling errors: there will be bias in the sample as no randomisation was used to obtain it. Not all members of the population have the same chance of being selected, this will limit generalisations of the results.
   **(d)** Random      **(e)** Simple random
   **(f)** Sampling errors: differences between the sample and the population.
**8.** Stratified sampling
**9.** Any appropriate stratified sampling plan.
**10.** Randomly selected quota sampling. Equal quotas of men and women but the individuals are selected randomly.

## Exercise 10.2

**1. (a)** $x = 168$, $y = 200$      **(b)** 168
   **(c)** 18
**2. (a)**

| Height $h$ (m) | Frequency |
|---|---|
| $h < 1.25$ | 3 |
| $1.25 \leqslant h < 1.50$ | 3 |
| $1.50 \leqslant h < 1.75$ | 12 |
| $1.75 \leqslant h \leqslant 2.00$ | 12 |

or other appropriate intervals.

**(b)**

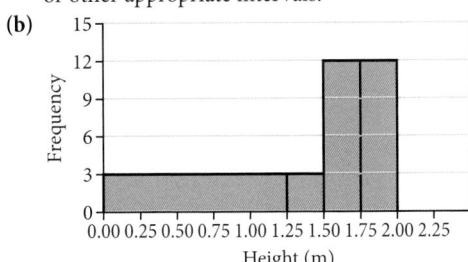

Or appropriate graph based on students' intervals

**(c)**

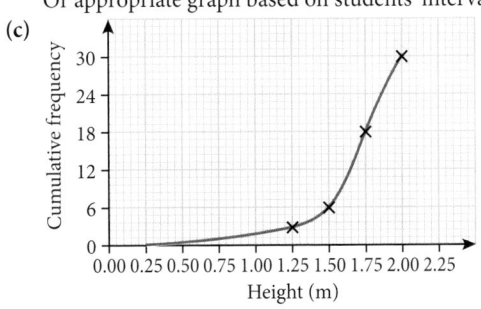

Or appropriate graph based on students' intervals
   **(d)** The majority of students are over 1.50 m tall equally split between 1.50-1.75 and 1.75-2.00 m.
**3. (a)** $p = 11$, $q = 46$, $r = 14$
   **(b)**

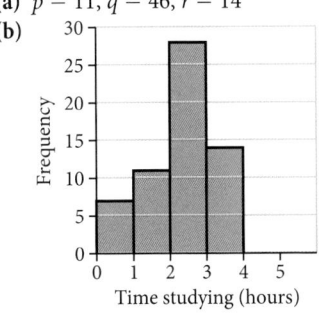

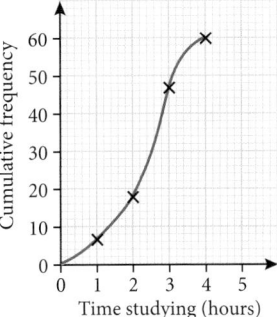

   **(c)** Data is left skewed with the majority of students studying more than 2 hours a night. Very few students studied less than 1 hour a night.
**4. (a)** 55                  **(b)** 7.27%
   **(c)** 8, 14.5%      **(d)** 44 students
**5. (a)**

| Mass of students $M$ (kg) | Frequency |
|---|---|
| $15 \leqslant M < 30$ | 13 |
| $30 \leqslant M < 45$ | 17 |
| $45 \leqslant M < 60$ | 7 |

   **(b)** 37             **(c)** $30 \leqslant M < 45$
   **(d)** 27             **(e)** 27 kg
**6. (a)** 40.4 cm to 54 cm      **(b)** 30
   **(c)** 11             **(d)** 2%
   **(e)** In the second sample fewer pike were shorter than 42 cm (4), more pike were longer than 47 cm (70) and a larger percentage are big (40.4%). There were also fewer pike in the sample.
**7. (a)**

| Absences (days) | Frequency | Cumulative Frequency |
|---|---|---|
| $0 \leqslant x \leqslant 5$ | 30 | 30 |
| $6 \leqslant x \leqslant 11$ | 62 | 92 |
| $12 \leqslant x \leqslant 17$ | 61 | 153 |
| $18 \leqslant x \leqslant 23$ | 30 | 183 |
| $19 \leqslant x \leqslant 29$ | 17 | 200 |

   **(b)**

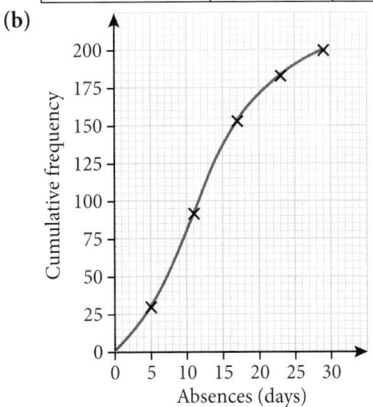

# Answers

**8.**

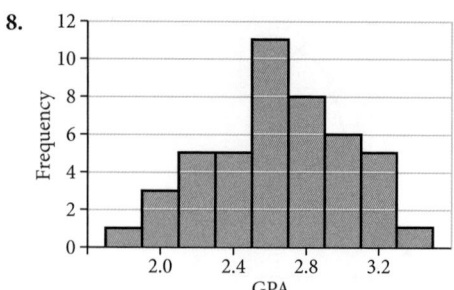

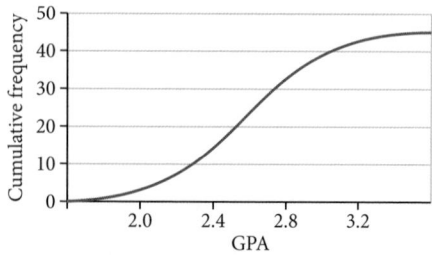

The majority of the results are grouped with a mode of 2.6 with a slight positive skew.

**9.**

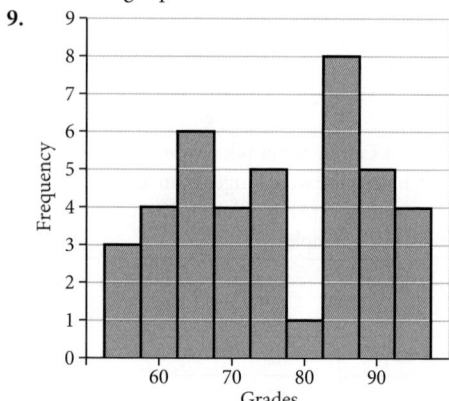

The grades appear to be divided into two groups, one with mode around 65 and the other around 85. No outliers are detected.

**10. (a)**

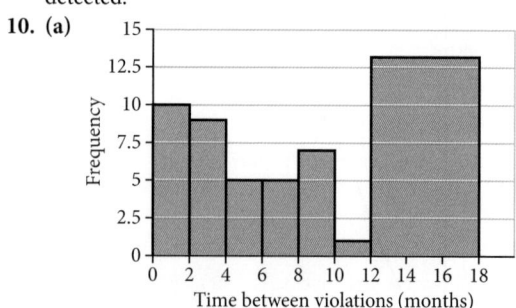

**(b)** Depends on how the intervals are selected.

**(c)** 72%

**11. (a)**

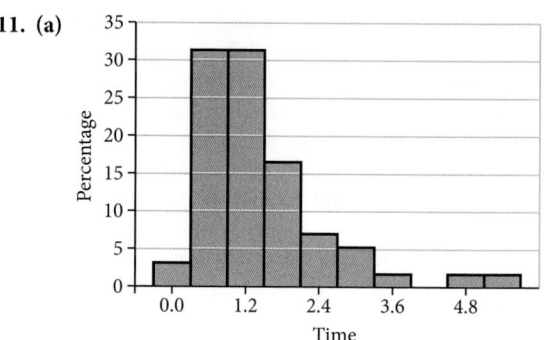

**(b)**

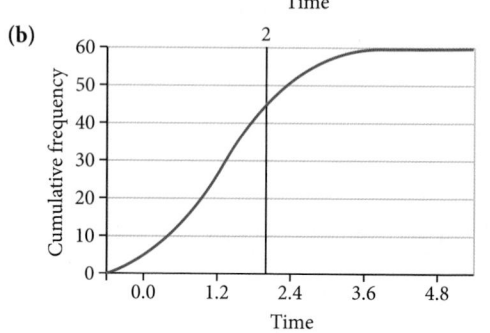

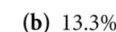

**(c)** Drawing a vertical line at 2 minutes, it intersects the cumulative frequency graph at approximately 45 out of 60. About 15 customers have to wait more than 2 minutes.

**12. (a)** $30 \text{ km h}^{-1}$  **(b)** 13.3%

**(c)** $k = 45$  **(d)** $48 \text{ km h}^{-1}$

**13. (a)**

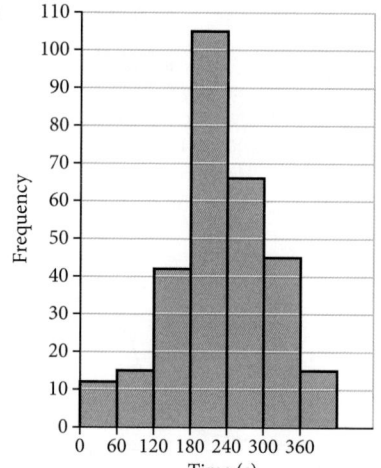

**(b)**

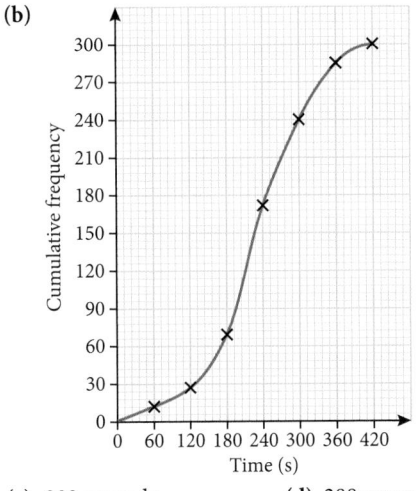

**(c)** 288 seconds      **(d)** 288 seconds

**14. (a)** Skewed to the right, there is a mode at about 7 days stay, and a few extremes that stayed more than 20 days. A good proportion stayed for about 3 days.

**(b)**

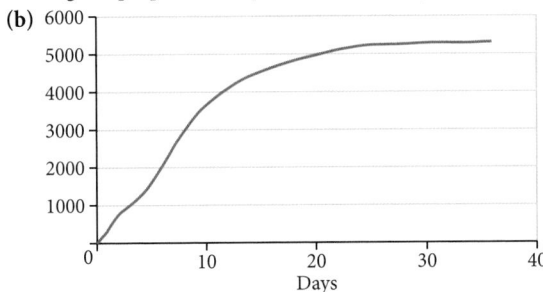

Approximately 35% of the patients.

**15. (a)**

| Speed, $s$ (km h$^{-1}$) | Frequency | Cumulative Frequency |
|---|---|---|
| $60 \leqslant s < 75$ | 70 | 70 |
| $75 \leqslant s < 90$ | 110 | 180 |
| $90 \leqslant s < 105$ | 150 | 330 |
| $105 \leqslant s < 120$ | 70 | 400 |
| $120 \leqslant s < 135$ | 40 | 440 |
| $135 \leqslant s$ | 10 | 450 |

**(b)**

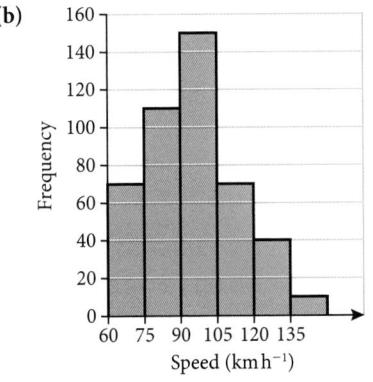

**(c)**

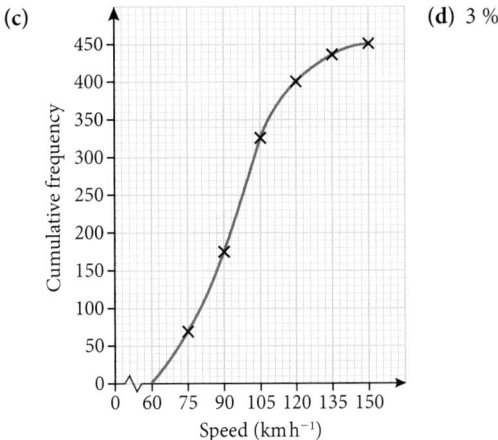

**(d)** 3 %

Exercise 10.3

**1. (a)** $n = 4$      **(b)** $\mu = 9{,}080$, $\sigma = 2{,}040$
    **(c)** 24%      **(d)** 7,000

**2. (a)** mean = 11, median = 3, mode = 1
    **(b)** 2314      **(c)** range = 50, $\sigma^2 = 257$, $\sigma = 16.0$
    **(d)**

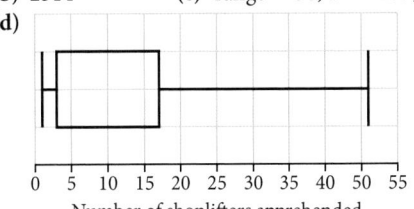

Lower limit is equal to $Q_1$.

**(e)** Outliers = median $\pm$ 1.5 IQR
    IQR = $17 - 1 = 16$
    Outliers $< 3 - 16$ and $> 3 + 16$
    $51 > 3 + 16$, therefore 51 is an outlier.

**3. (a)** $\mu = 55.8$, median = 60, mode = 10, 30, 60 and 70 (all occur twice)
    **(b)** range = 130, $\sigma = 36.6$
    **(c)**

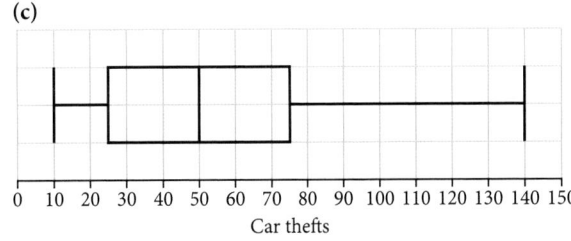

**(d)** Outliers = median $\pm$ 1.5 IQR
    IQR = $75 - 25 = 50$
    Outliers $< 60 - 50$ and $> 60 + 50$
    $140 > 60 + 50$, therefore 140 is an outlier

**4. (a)** $\mu = 71.5$, $\sigma = 7.04$
    **(b)**

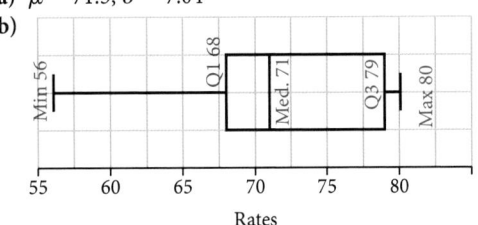

**(c)** IQR = 11. There are no outliers.

5. (a) $\mu = 162.6$, $\sigma = 23.11$

(b)

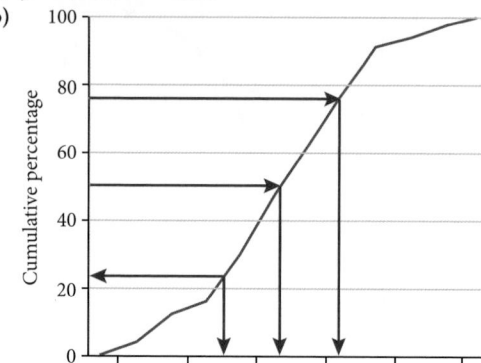

$Q_1 \approx 150$, median $\approx 165$, $Q_3 \approx 182$

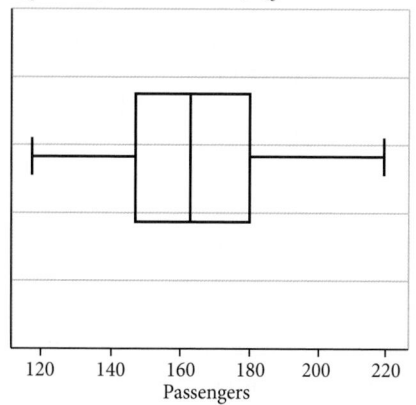

(c) Your GDC shows: $Q_1 = 146.75$, $Q_3 = 179.25$, IQR $= 32.5$ No outliers.

6. (a)

| Daily commuting time (minutes) | Frequency | Cumulative Frequency |
|---|---|---|
| $0 \leqslant t < 10$ | 8 | 8 |
| $10 \leqslant t < 20$ | 18 | 26 |
| $20 \leqslant t < 30$ | 12 | 38 |
| $30 \leqslant t < 40$ | 8 | 46 |
| $40 \leqslant t < 50$ | 4 | 50 |

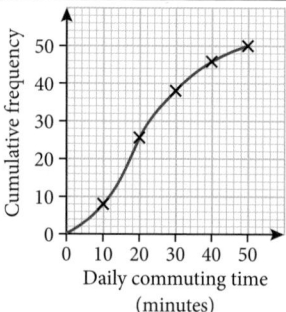

(b) $Q_1 \approx 14$, median $\approx 20$, $Q_3 \approx 30$, IQR $\approx 16$

(c) $\mu = 21.4$, $\sigma = 11.6$

7. (a) $\mu = €108.4$

(b) $\sigma = €35.6$

(c) $\mu = €84.4$, $\sigma = €35.6$

8. (a) $\mu = 106{,}500$, median $= 76{,}000$, $\sigma = 93{,}200$, IQR $= 37{,}000$

(b) Yes 382,000 is an outlier

$\mu = 75{,}900$, median $= 74{,}000$, $\sigma = 16{,}600$, IQR $= 21{,}500$

Standard deviation, as it uses the square of the difference from the mean so outliers make up a greater contribution.

(c) The second set, they are more representative of the data set.

9. (a) $x = 5$     (b) median $= 5$, $\sigma = 1.41$

(c) 15     (d) 4.73

10. (a) $\mu = 31.4$, $\sigma = 9.35$, $Q_1 = 25.5$, median $= 28.5$, $Q_3 = 36$, IQR $= 10.5$

(b) More than 50% of customers spend less than the mean time on coffee and dessert, this means that a few customers are the main cause of losses.

11. (a) minimum $= 20$, $Q_1 = 34$, median $= 41$, $Q_3 = 48.5$ and maximum $= 64$. IQR $= 14.5$

(b)

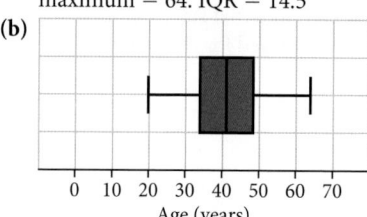

The data is very nearly symmetrical about the median.

(c)

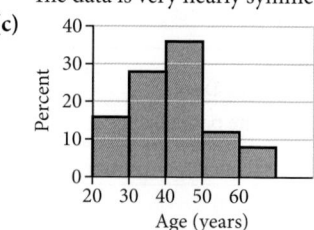

(d)

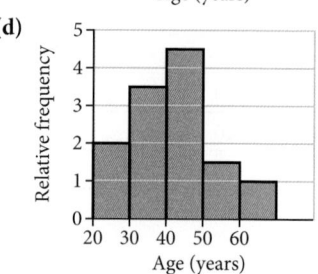

The most frequent interval is 40–50 and the majority of the jockeys are under 50.

## Chapter 10 Practice questions

1. (a) 30°C     (b) 30°C, 24°C and 12°C

(c) 22.5°C

2. (a) (i) $\mu = 13.7$     (ii) $\sigma = 2.52$

(b) 13.1

3. (a) $a = 4$     (b) $b = 2$

4. (a) $x = 30$

(b) (i) $\mu = 2.6$    (ii) 3    (iii) $\sigma = 1.06$

(c) (i) $\mu = 4.6$    (ii) 5    (iii) $\sigma = 1.06$

5. (a) 1.75     (b) 2

6. (a) 61    (b) 14    (c) 20    (d) 53 kg

**7. (a) (i)** 13 seconds

**(ii)** $Q_1 = 10$ seconds, $Q_3 = 16$ seconds

**(iii)** 6 seconds

**(b)** 1400      **(c)** 14 seconds

**(d) (i)** 500      **(ii)** 150

**(e) (i)** 13.25 seconds      **(ii)** 4.41 seconds

**(f)** No, required time is 8.84 seconds.

**8. (a)** $170 \leqslant h < 180$      **(b)** $\mu = 171$ cm, $\sigma = 11.1$

**(c)** 171 cm      **(d)** 13.3 cm

**(e)** 28%

**9. (a)**

| Examination score $x$ (%) | $0 \leqslant x < 20$ | $20 \leqslant x < 40$ | $40 \leqslant x < 60$ |
|---|---|---|---|
| Frequency | 14 | 26 | 56 |

| Examination score $x$ (%) | $60 \leqslant x < 80$ | $80 \leqslant x < 100$ |
|---|---|---|
| Frequency | 18 | 6 |

**(b)** 50      **(c)** 46

# Chapter 11

## Exercise 11.1

**1.** $\dfrac{1}{2}$

**2. (a) (i)** $\dfrac{5}{18}$      **(ii)** $\dfrac{1}{6}$

**(b)** 139

**3. (a)** $\dfrac{1}{5}$      **(b)** $\dfrac{2}{5}$      **(c)** $\dfrac{1}{2}$

**4. (a) (i)** $\dfrac{1}{6}$      **(ii)** $\dfrac{1}{6}$      **(iii)** $\dfrac{1}{3}$

      **(iv)** $\dfrac{1}{9}$      **(v)** $\dfrac{35}{36}$

**(b)** 33

**5. (a)** $\dfrac{3}{8}$      **(b)** $\dfrac{3}{8}$

**6.** 0.73

**7. (a)** $\dfrac{1}{4}$      **(b)** $\dfrac{3}{13}$      **(c)** $\dfrac{51}{52}$

**8. (a)** $\dfrac{4}{5}$      **(b)** $\dfrac{11}{30}$      **(c)** 1

**9. (a) (i)** $\dfrac{1221}{4318}$      **(ii)** $\dfrac{480}{2159}$

**(b) (i)** 16.8 million      **(ii)** 8.25 million

**10. (a) (i)** $\dfrac{1}{20}$      **(ii)** $\dfrac{47}{60}$

**(b)** 12

**11. (a)** Drug A is $\dfrac{25}{29} = 0.862$ , drug B is $\dfrac{62}{69} = 0.899$

**(b)** Drug B

**(c)** 172

**12. (a)** $\dfrac{519}{1046} = 0.496$      **(b)** $\dfrac{611}{7322} = 0.0834$

**(c)** $\dfrac{318}{3661} = 0.0869$

**13. (a)** $\dfrac{14675}{62893} = 0.233$      **(b)** $\dfrac{261634}{440251} = 0.594$

**14. (a)**

| Region | Relative Frequency |
|---|---|
| Bangkok | 0.692 |
| Central | 0.530 |
| North | 0.414 |
| Northeast | 0.360 |
| South | 0.463 |
| Whole Kingdom | 0.475 |

**(b)** Possible answers include: better infrastructure, easier access to internet, more people who can afford it.

**(c)** 29.8 million

**15. (a)** $79      **(b)** $126.40

## Exercise 11.2

**1.** 0.35

**2. (a)** 0.5      **(b)** 0.2      **(c)** 0.8

**(d)** 0.1      **(e)** 0.7

**3.** $P(A \cap B) = \dfrac{1}{9} \neq 0$, so not mutually exclusive.

$P(A)P(B) = \dfrac{2}{27} \neq \dfrac{1}{9}$, so not independent.

**4.** $\dfrac{29}{35}$

**5.** 0.9

**6. (a)** 0.92

**(b) (i)** 0.0064   **(ii)** 0.1536   **(iii)** 0.1472

**(c)** 0.487

**7. (a)** 10000      **(b)** 0.9      **(c)** 0.3439

**8.** $\dfrac{3}{8}$

**9. (a)** $\dfrac{11}{36}$      **(b)** $\dfrac{11}{12}$      **(c)** $\dfrac{1}{3}$

**10. (a) (i)** $\dfrac{7}{15}$   **(ii)** $\dfrac{11}{75}$   **(iii)** $\dfrac{4}{15}$   **(iv)** $\dfrac{46}{75}$

**(b)** $\dfrac{11}{20}$

**(c)** Not independent, because the ratio of male:female in each grade is not equal.

**11. (a) (i)** 0.56      **(ii)** 0.15

**(b)** 0.268

**(c)** Not independent, because the proportion using glasses is different in each case.

**12. (a)** 0.17      **(b)** Not independent

**13. (a)** 0.65      **(b)** 0.35      **(c)** 0.52

**14. (a)** $\dfrac{14}{25}$      **(b)** $\dfrac{1}{10}$

**15. (a)** $\dfrac{1}{216}$      **(b)** $\dfrac{91}{216}$      **(c)** $\dfrac{25}{72}$

**16. (a)** 0.21      **(b)** 0.441      **(c)** 0.657

**17. (a)** $\dfrac{23}{144}$      **(b)** $\dfrac{11}{144}$      **(c)** $\dfrac{5}{48}$

**(d) (i)** $\dfrac{5}{72}$   **(ii)** $\dfrac{115}{144}$   **(iii)** $\dfrac{67}{72}$   **(iv)** $\dfrac{29}{144}$

**18. (a)** $\dfrac{4805}{46656}$      **(b)** $\dfrac{148955}{1679616}$      **(c)** $\dfrac{36}{67}$

**19. (a) (i)** 0.2601   **(ii)** 0.7599

**(b)** 5.958 million

# Answers

**20. (a)**

| Driving Centre | Pass Rate |
|---|---|
| Singapore Safety | 0.529 |
| Bukit Batok | 0.599 |
| Comfort | 0.535 |

Bukit Batok driving centre has the highest pass rate.

**(b) (i)** 0.280 **(ii)** 0.779

**(c) (i)** 0.321 **(ii)** 0.493

**(d)** 10147

**21. (a) (i)** $\frac{1}{4}$ **(ii)** $\frac{1}{2}$

**(b)** 15

**(c) (i)** 0 **(ii)** $\frac{1}{2}$

**(d) (i)** $\frac{3}{16}$ **(ii)** $\frac{3}{16}$ **(iii)** $\frac{9}{16}$

**(e)** 2

**(f) (i)** $\frac{41}{80}$ **(ii)** $\frac{21}{80}$

## Exercise 11.3

**1. (a)** $\frac{23}{131}$ **(b)** $\frac{2}{3}$ **(c)** $\frac{23}{70}$

**(d)** $\frac{75}{262}$ **(e)** $\frac{21}{61}$

**2. (a)** $\frac{9}{32}$ **(b)** $\frac{1}{23}$ **(c)** $\frac{34}{63}$ **(d)** 92

**3. (a)** 30.5 million **(b)** $\frac{29}{100}$ **(c)** 0.27

**(d)** $\frac{70}{143}$ **(e)** $\frac{64}{157}$

**4. (a)** $\frac{1}{110}$ **(b)** $\frac{8}{25}$

**(c)** Since the answer to part **(b)** is much greater than part **(a)**, the experiment indicates that group pressure does have an impact on a person to conform.

**5.** $\frac{1}{3}$

**6. (a)** $\frac{32}{157}$ **(b)** $\frac{32}{125}$ **(c)** $\frac{31}{38}$

**7. (a)** $\frac{5}{18}$ **(b)** $\frac{39}{43}$ **(c)** $\frac{689}{3065}$

**8. (a) (i)** 0.18 **(ii)** 0.9

**(b)** 1613

**(c)** 0.208

**9. (a) (i)** $\frac{2}{5}$ **(ii)** $\frac{16}{25}$ **(iii)** $\frac{5}{8}$

**(b) (i)** $\frac{7}{19}$ **(ii)** $\frac{62}{95}$ **(iii)** $\frac{19}{31}$

**10. (a)** 0.21 **(b)** 0.286

**11. (a)** 0.565 **(b)** 0.0956

**12. (a) (i)** 0.0133 **(ii)** 0.987

**(iii)** $3.42 \times 10^{-5}$ **(iv)** 0.749

**(b)** The probability of a false negative is very low. This is a good thing, because it means that only a small number of infected people are not detected by the test.

**(c)** The probability of a false positive is very high. Almost 75% of people testing positive are actually healthy. This could cause unnecessary anxiety for these people.

**13. (a) (i)** 0.0872 **(ii)** 0.9128

**(iii)** 0.00153 **(iv)** 0.213

**(b)** About 1 or 2 drivers out of 1000 will have a false negative result, meaning that they are drink driving, but get away undetected.

**(c)** About 213 drivers will test positive, even though their BAC is less than 0.5. Clearly this is not acceptable, as the police will be punishing innocent drivers. The police will need to consider using a second test, to more accurately identify those drivers who have a BAC greater than 0.5.

**14. (a)** 0.88 **(b)** 0.74

**15.** 0.0333

**16. (a)** 0.64 **(b)** 0.234

**17. (a)** 0.5 **(b)** 0.35 **(c)** 0.75

## Chapter 11 Practice questions

**1. (a)** 0.2 **(b)** 0.76 **(c)** 0.8

**2. (a)** 0.6

**(b)** $P(A \cap B) \neq 0$

**(c)** $P(B|A) = P(B)$, so they are independent.

**(d)** 0.36

**3. (a)** $\frac{3}{32}$ **(b)** $\frac{3}{4}$ **(c)** $\frac{5}{32}$

**4. (a)** $\frac{1}{11}$ **(b)** $\frac{12}{121}$

**5.** 0.8

**6. (a)** $\frac{11}{36}$ **(b)** $\frac{1}{11}$

**7. (a)**

| | Males | Females | Totals |
|---|---|---|---|
| Unemployed | 20 | 40 | 60 |
| Employed | 90 | 50 | 140 |
| Totals | 110 | 90 | 200 |

**(b) (i)** $\frac{1}{5}$ **(ii)** $\frac{9}{14}$

**8. (a)** 0.00002 **(b)** 0.99998 **(c)** 0.167

**9. (a) (i)** $\frac{121}{125}$ **(ii)** $\frac{21}{25}$

**(b)** $\frac{8}{41}$

**10. (a) (i)** $\frac{7}{44}$ **(ii)** $\frac{11}{18}$

**(b) (i)** $\frac{5525}{27434}$ **(ii)** $\frac{21909}{27434}$

**11. (a) (i)** $\frac{63}{185}$ **(ii)** $\frac{2}{185}$ **(iii)** $\frac{178}{185}$ **(iv)** $\frac{3}{10}$

**(b)** Quality of component from machines I and II appears to be the same, but lower quality from machine III.

**12. (a)** 29.7%

**(b) (i)** 0.0437 **(ii)** 0.433 **(iii)** 0.144

**(c)** 0.256

**(d)** 22.5 million

**13. (a)** $\frac{3}{190}$

**(b) (i)** 0.277 **(ii)** 0.00788

**(iii)** 0.277 **(iv)** 0.000898

**(c)** 0.0728

**14. (a)** $860 **(b)** $1290 **(c)** 0.0645

**15. (a) (i)** 0.0106 **(ii)** 0.944

**(iii)** 0.989 **(iv)** $3.03 \times 10^{-6}$

**(b) (i)** 0.000694 **(ii)** 0.144

**16. (a)** $1.5 \times 10^{-5}$ **(b)** 0.00797 **(c)** 0.376

**17. (a)** 0.9996 **(b)** 0.0004 **(c)** 0.0004

**18. (a) (i)** 0.273     **(ii)** 0.245
    **(b) (i)** 0.0233     **(ii)** 0.0827
    **(c) (i)** $1.05 \times 10^{-7}$     **(ii)** 0.893

**19.** 0.000104

**20. (a)** $\frac{2}{3}$     **(b)** $\frac{2}{9}$     **(c)** $\frac{3}{4}$

**21. (a) (i)** $\frac{15}{22}$   **(ii)** $\frac{259}{484}$   **(iii)** $\frac{22}{29}$
    **(b) (i)** $\frac{15}{21}$   **(ii)** $\frac{6}{11}$   **(iii)** $\frac{3}{4}$

**22. (a)**

| Test centre | Pass rate |
|---|---|
| Aberdeen | 0.537 |
| Bangor | 0.469 |
| Cheltenham | 0.554 |
| Doncaster | 0.460 |

    Cheltenham has the highest overall pass rate
    **(b)** $\frac{3938}{10477}$
    **(c) (i)** 0.313   **(ii)** $\frac{361}{929}$
    **(d) (i)** 0.254   **(ii)** 0.501
    **(e)** 10184

**23. (a) (i)** $\frac{1}{2}$   **(ii)** $\frac{1}{2}$
    **(b)** 5
    **(c) (i)** $\frac{3}{4}$   **(ii)** $\frac{1}{2}$
    **(d) (i)** $\frac{1}{4}$   **(ii)** $\frac{3}{8}$   **(iii)** $\frac{1}{8}$
    **(e)** 6
    **(f) (i)** $\frac{4}{15}$   **(ii)** $\frac{9}{40}$   **(iii)** $\frac{27}{82}$

**24. (a)** 99.8%     **(b)** 0.0247

## Chapter 12

### Exercise 12.1

**1. (a)** 4; 11; {5, 6, 5, 6}     **(b)** 4; 6; {3, 3, 3, 3}
    **(c)** 5; 5; {2, 1, 3, 2, 2}

**2. (a)** No;     **(b)** Yes, $K_5$

**3.** $n - 1$

**4.** $\frac{n(n - 1)}{2}$

**5. (a)** 7, 12     **(b)** 30, 221     **(c)** $m + n, mn$

**6.** 8, 16

**7. (a)** 8
    **(b)** Yes: $r = 2, |v| = 14$, or : $r = 4, |v| = 7$
    **(c)** $\frac{p}{2}$
    **(d)** see exercises 3 & 4.

**8.** If graph with $n$ vertices is connected then the possible degrees will be any number from 1 to $n - 1$. Since we have $n$ vertices, by the pigeon hole principle at least 2 vertices must occupy the same number!

**9.** A subgraph must contain elements from both components. You cannot have a subgraph from one side only because each vertex is connected to a vertex from the other component.

**10.** use different colours for the vertices. No two adjacent vertices can have the same colour: (a) and (c).

**11.** 12

**12. (a)** No, $|E|$ is not even
    **(b)** Yes,                 **(c)** Yes

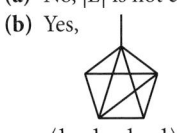
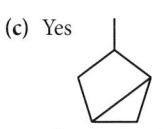

**13. (a)** $\begin{pmatrix} 1 & 1 & 1 & 1 \\ 1 & 1 & 1 & 2 \\ 1 & 1 & 0 & 3 \\ 1 & 2 & 3 & 0 \end{pmatrix}$    **(b)** $\begin{pmatrix} 0 & 1 & 1 & 1 \\ 1 & 0 & 1 & 1 \\ 1 & 1 & 0 & 1 \\ 1 & 1 & 1 & 0 \end{pmatrix}$

    **(c)** $\begin{pmatrix} 1 & 0 & 0 & 0 & 0 \\ 0 & 0 & 1 & 1 & 0 \\ 0 & 1 & 0 & 1 & 0 \\ 0 & 1 & 1 & 0 & 1 \\ 0 & 0 & 0 & 1 & 0 \end{pmatrix}$

**14. (a)**     **(b)**     **(c)**
    **(d)**     **(e)**     **(f)**

**15.** adjacency matrices are similar with degrees: 2, 2, 2, 3, 3, 3

**16.** $A$ and $C$ are the same with degree sequence: 1, 1, 2, 3, 3.
    $B$ and $D$ are the same with degree sequence: 2, 2, 2, 3, 3.

### Exercise 12.2

**1.** Vertices have even degrees.
    **(a)** 123174263456751     **(b)** 1234543251

**2. (a)** 1234214241     **(b)** 12345241
    **(c)** vertices 2 and 5 have degree 5 each.

**3. (a)** When $n$ is odd
    **(b)** When $m$ and $n$ are both even.

**4.** 1(a) Hamiltonian: 12345671; 1(b) Hamiltonian: 123451
    2(a) Hamiltonian: 12341; 2(b) Hamiltonian path: 12345;
    2(c) neither.

**5. (a)** (10, 9, 6, 5, 9, 8, 5, 4, 8, 7, 4, 2, 5, 3, 2,1, 3, 6, 10)
    **(b)** (10, 9, 8, 7, 4, 5, 2, 1, 3, 6, 10)
    **(c)** An Eulerian circuit is always possible ($n \geqslant 3$) because the degree of every vertex is even. Hamiltonian cycle is also possible using the same plan as above: Visit all vertices except one side, and then go back along that side.

**6.** Length 1 = 0; Length 2 = 2; Length 3 = 3, and length 4 = 10.

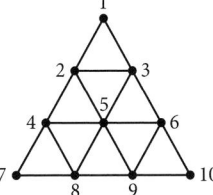

**7. (a)** 51 between vertices not on the main diagonal, 52 for vertices on the diagonal.
    **(b)** 205 between vertices not on the main diagonal, 204 for vertices on the diagonal.
    **(c)** 819 between vertices not on the main diagonal, 820 for vertices on the diagonal.

# Answers

**8. (a)** 48 among vertices of the 3-part, and 36 among the 4-part.
**(b)** 144 from vertices of 3-part to vertices of 4-part.
**(c)** 576 among vertices of the 3-part, and 432 among the 4-part.
**(d)** 1728 from vertices of 3-part to vertices of 4-part.

**9. (a)** No cycle – If you start at the left, you will need to visit $c$ and $d$ twice.
Path: $abcdef$.
**(b)** Cycle: $abcdea$
**(c)** No Cycle since $f$ has degree 1.
Path: $eabcdf$.
**(d)** Neither cycle nor path: 3 vertices with degree 1
**(e)** No Cycle, because in any of them $a$ or $d$ would have to be visited twice.
Path: $eacdb$.
**(f)** Cycle: $ahgfedcbia$.

## Exercise 12.3

**1. (a)** Planar. Redraw.

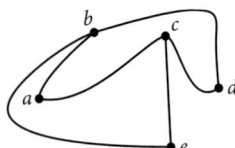

**(b)** Planar. Redraw.

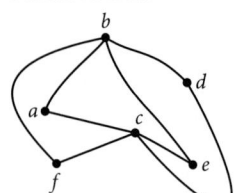

**(c)** Planar. Redraw.

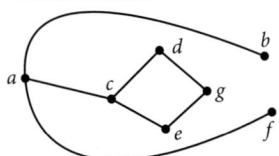

**(d)** Not planar. $bf$ and $ce$ must cross. So must $ae$ and $bd$.
**2.** 15
**3. (a)** 15     **(b)** 18
**4. (a)** 7     **(b)** 9
**5.** 6
**6. (a)** not planar     **(b)** planar
**7. (a)** label them as 1: inscholastics, 2: ismtf, 3: Pearson, 4: wazir-garry

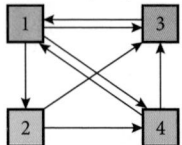

**(b)**
$$\begin{pmatrix} 0 & 0 & 1 & \frac{1}{2} \\ \frac{1}{3} & 0 & 0 & 0 \\ \frac{1}{3} & \frac{1}{2} & 0 & \frac{1}{2} \\ \frac{1}{3} & \frac{1}{2} & 0 & 0 \end{pmatrix}$$

**(c)**
$$\begin{pmatrix} 0.38 \\ 0.12 \\ 0.29 \\ 0.19 \end{pmatrix}$$

## Exercise 12.4

**1. (a)** 5, 7, 10, 11, 13, 14, 16, 17
**(b)** 3, 1, 9
**(c)** 3: 4, 5, 11, 12, 13, 14; 7: No descendants; 15: 16, 17
**(d)** 4: 4, 5, 11, 12; 7: No siblings; 9: No siblings.
**2.** $|u| = 18, |v| = 36, |f| = 35$
**3.** 31
**4.** $\binom{n}{2}$
**5. (a)** These are the only 2 non-isomorphic trees.

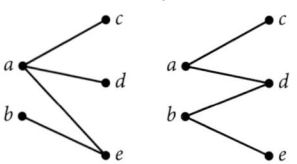

**(b)** $\dfrac{n+1}{2}$

**6. (a)**

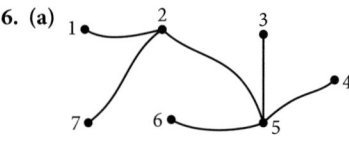

**(b)**

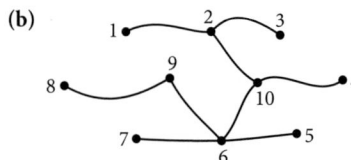

**(c)**

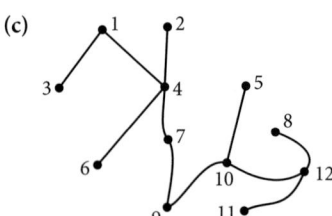

**7. (a)** 12, 23, 34, 45, 56, 67
**(b)** 12, 23, 34, 45, 56, 67, 78, 89, 9 10
**(c)** 12, 24, 45, 58, 8(12), (12)(11), (11)9, 9(10), 47, 76, 63
**8. (a)** 13, 34, 45, 58, 89, 46, 67, 7 (10), 12
**(b)** 17, 78, 89, 9(10), (10)(11), (11)6, 65, 54, (10)(14), 9(13), 83, 32
**(c)** 12, 23, 34, 46, 65, 5(10), (10)9, 98, 87, (10)(11), (11)(12), (12)(13), (13)(14), (14)(15), (10)(16), (16)(17), (17)(19), (19)(20), (20)(18)

**(d)**

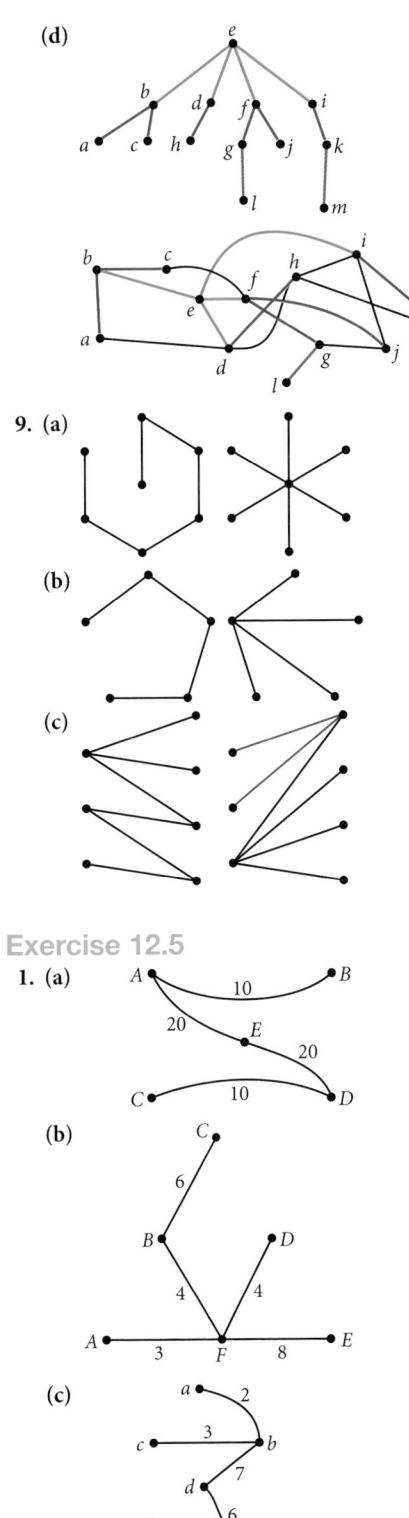

**9. (a)**

**(b)**

**(c)**

**1. (a)**

**(b)**

**(c)**

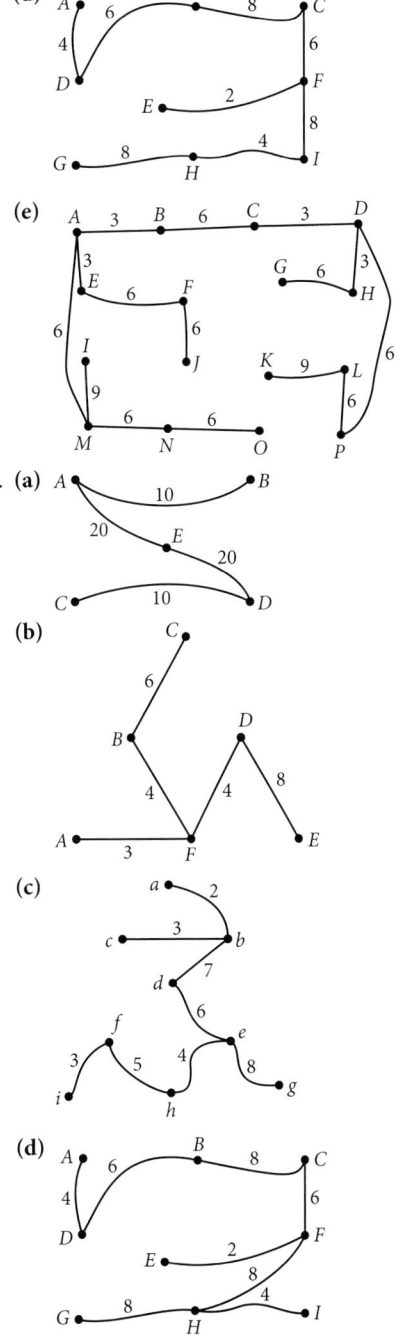

**(d)**

**(e)**

**2. (a)**

**(b)**

**(c)**

**(d)**

**(e)** A few shapes are possible, one of which is similar to 5.

**3.** 1 and 6 have the same final tree. However, when building the tree with Kruskal's, *AB* and *CD* were added first. In Using Prim's, *AB* was followed by *AE*, *ED*, and then *CD*. 2 and 7, there is no apparent difference. The different shapes are due to random choices.

3 and 8 have the same final tree too. In Kruskal's, the order of addition to the tree is the following: *ab*, *bc*, *fi*, *he*, *fh*, *ed*, *bd*, and *eg*. In Prim's: *ab*, *bc*, *bd*, *ed*, *he*, *fh*, , *fi* and *eg*.

4 and 9 may have the same tree too. However, Kruskals, the order of edge addition is: *ef*, *ad*, *hi*, *cf*, *db*, *bc*, *fi*, and *gh*. In Prim"s: *ef*, *fc*, *fh*, *ih*, *cb*, *bd*, *da*, and *gh*.

5 and 10 may have the same tree too. However, Kruskals, the order of edge addition is: *AB, AE, CD, DH, BC, ...*. Prim's order is: *AB, AE, BC, CD, DH, ...*

**4.**

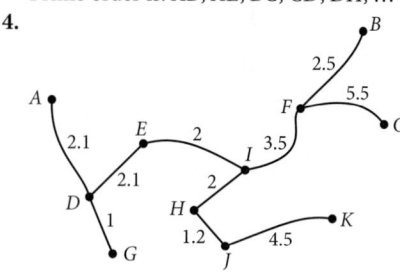

## Exercise 12.6

1. *abedf*, 70
2. *ACDEGH*, 48
3. 32, *acfimpsu*.
4. *abed*
5. *ACDF, BCDEGH*
6. *ADBCA*, 85.
7. *EDCABE* or *DEBACD*: 400
8. Vienna-Frankfurt-Prague-Moscow-Milan-Vienna: €1070.
9. New York-Paris-London-Vienna-Milan-New York: €1215.
10. *DACBED*: 550
11. *age*: 19
12. *abdfhl*: 21; *acehl*: 13.
13. Without visiting any city twice: *ESYFITAPGE*: 926.
    Visiting *Y* twice: *EGYSYFITAPE*: 871.
14. *abcdhghcgbfgfea*: 8300.
15. *abcdecjfefibjfgihgha*: 9200.

## Chapter 12 Practice questions

1. *BAGF*: 16
2. **(a)**

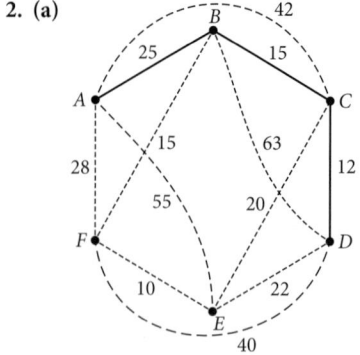

   **(b)** *ABCD*: $52
3. *ACEDFGHIBA*: 8.6 km
4. **(a)**

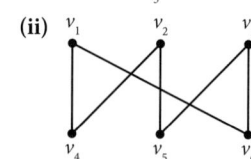

   **(b)** Sample: *ACBDCAEFGA* with 130 000 which she can afford.

5. He will have 20 minutes break.
6. **(a)** $C_n$ has *n* edges. Additionally, $W_n$ has another *n* edges connecting the extra vertex with the *n* vertices.
   **(b) (i)** using adjacency matrices, odd powers will give $2^{n-1}$ for adjacent vertices
   **(ii)** even powers will give same to non-adjacent.
7. Sample: Kruskal's: *BC, AB, AE, CF, GH, AD, DH, EI*; Prim's: *BC, AB, AE, CF, AD, DH, GH, EI*; 26

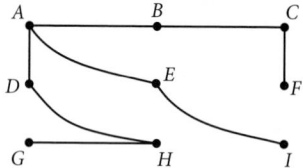

8. *PT, SU, RU, PQ, TR* for a total distance of 719.
9. 1043 cents (10.43 Dollars or Euros)
10. **(a)**

Adam / Flor / Bernard / Eva / Cecile / Donatella

   **(b)** Through Adam.
   **(c)** Bernard, without him, Eva is isolated.
11. **(a) (i)**

$K_5$

   **(ii)** $v_1$ $v_2$ $v_3$ $v_4$ $v_5$ $v_6$

   **(b)** Each vertex has degree 2, so a Hamiltonian cycle is possible.
12. No. More than 2 vertices have an odd degree.
13. Tree: *h, e, d, a, i, g*, weight = 31
14. **(a)**

|   | A | B | C | D | E | F |
|---|---|---|---|---|---|---|
| A | 0 | 1 | 2 | 2 | 2 | 1 |
| B | 1 | 0 | 1 | 2 | 3 | 2 |
| C | 2 | 1 | 0 | 1 | 2 | 1 |
| D | 2 | 2 | 1 | 0 | 2 | 1 |
| E | 2 | 3 | 2 | 2 | 0 | 1 |
| F | 1 | 2 | I | 1 | 1 | 0 |

   **(b) (i)** City *F* is the most accessible since its index is 1.5. City *E* is the least accessible since its index is 10.
   **(ii)** Cities *C* and *F* are the most accessible since their index is 1.5 each.
   City *E* is still the least accessible since its index is 10.

**15. (a)**

$$
\begin{array}{c|cccccccc}
U & 1 & 2 & 3 & 4 & 5 & 6 & 7 & 8 \\
\hline
1 & 0 & 1 & 0 & 1 & 0 & 0 & 0 & 1 \\
2 & 1 & 0 & 1 & 0 & 0 & 0 & 1 & 0 \\
3 & 0 & 1 & 0 & 1 & 0 & 1 & 0 & 0 \\
4 & 1 & 0 & 1 & 0 & 1 & 0 & 0 & 0 \\
5 & 0 & 0 & 0 & 1 & 0 & 1 & 0 & 1 \\
6 & 0 & 0 & 1 & 0 & 1 & 0 & 1 & 0 \\
7 & 0 & 1 & 0 & 0 & 0 & 1 & 0 & 1 \\
8 & 1 & 0 & 0 & 0 & 1 & 0 & 1 & 0 \\
\end{array}
$$

**(b)**

$$
\begin{array}{c|cccccccc}
 & A & E & B & F & C & G & D & H \\
\hline
A & 0 & 1 & 0 & 1 & 0 & 0 & 0 & 1 \\
E & 1 & 0 & 1 & 0 & 0 & 0 & 1 & 0 \\
B & 0 & 1 & 0 & 1 & 0 & 1 & 0 & 0 \\
F & 1 & 0 & 1 & 0 & 1 & 0 & 0 & 0 \\
C & 0 & 0 & 0 & 1 & 0 & 1 & 0 & 1 \\
G & 0 & 0 & 1 & 0 & 1 & 0 & 1 & 0 \\
D & 0 & 1 & 0 & 0 & 0 & 1 & 0 & 1 \\
H & 1 & 0 & 0 & 0 & 1 & 0 & 1 & 0 \\
\end{array}
$$

**(c)** $V$ is planar as $U$ is planar.

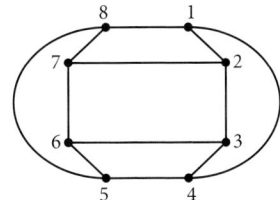

**16. (a)**

| Vertices added to the Tree | Edge added | Weight |
|---|---|---|
| 3 | ∅ | 0 |
| 5 | 3, 5 | 10 |
| 6 | 3, 6 | 20 |
| 7 | 5, 7 | 30 |
| 10 | 6, 10 | 30 |
| 1 | 3, 1 | 40 |
| 2 | 1, 2 | 30 |
| 11 | 2, 11 | 30 |
| 9 | 1, 9 | 40 |
| 4 | 6, 4 | 40 |
| 8 | 7, 8 | 40 |
| | | 310 |

**(b)** Any of two paths: 1 - 3 - 4 - 5 - 6 - 8- 10 -11 or 1 - 3 - 4 - 5 - 6 - 9 - 11 with weight 80.

**17. (a)** No, not all vertices are even.

**(b)**

$$
\begin{array}{c|cccccc}
 & A & B & C & D & E & U \\
\hline
A & 0 & 1 & 0 & 1 & 0 & 0 \\
B & 1 & 0 & 1 & 1 & 0 & 1 \\
C & 0 & 1 & 0 & 0 & 1 & 1 \\
D & 1 & 1 & 0 & 0 & 1 & 1 \\
E & 0 & 0 & 1 & 1 & 0 & 1 \\
F & 0 & 1 & 1 & 1 & 1 & 0 \\
\end{array}
$$

There is 1 walk of length 2.

**(c)** Tree: { AD, CE, CU, BU, AB} with weight of 28.

**18. (a)** Kruskal's algorithm:

T: = empty graph
for I : =1 to $n - 1$
begin
    e : = any edge in G with smallest weight that does not form a simple circuit when added to T
    T : = T with e added
end {T is a minimum spanning tree of G.}

**(b)** AJ, GM, MF, DE, JH, LK, FE, GL, KC, CB, AI, HG. Total weight 39

**(c)**

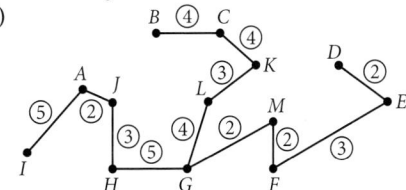

**19. (a)** BAEBCEFCDF

**(b)** All vertices have even degree.

**20.**

Weight = 17. (other trees are also possible)

**21. (a)**

**(b)** Yes. The graph has exactly two vertices (B and C) with odd degree. It means that there is a path (starting at B or C) that will go once and only once through every door.

**(c)** Yes. $O \rightarrow D \rightarrow A \rightarrow C \rightarrow B \rightarrow O$ is a Hamiltonian cycle. It means that there is a path (starting anywhere) that will go once and only once through every room before returning to its starting point.

**22. (a)**

**(b)** Degree of every vertex is even.

**(c)** AEBACDBDCEDA

# Answers

**(d)**

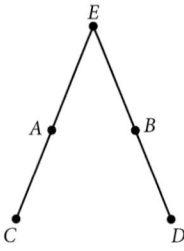

**23. (a)** *MQ, QL, MP, PN, NR.*
   **(b)** 11
**24. (a)** $G_2$ does not have since 4 vertices have odd degrees.
   **(b)** *BABCECFEFBDEDADC.*
**25. (a)** $e = 9 \not\leq 2v - 4$.
   **(b)** delete *AD*.
   **(c)** Subgraphs would have 8 edges which satisfies the condition $\leq 2v - 4$
**26. (a)** 24.
   **(b) (i)** *BDEC*
       **(ii)** 33
   **(c)** *DBAEC* is a min. spanning tree of 26 weight. Upper bound $= 26 \times 2 = 52$
   **(d)** A minimum tour is 34. 33 cannot be achieved.
**27. (a)** Example

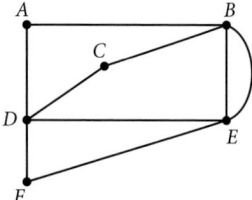

   **(b)** All vertices are of even order. *BEDABCDFEB* (not unique).
   **(c)** *ABCDEF*
**28. (a) (i)** Eulerian circuit: $V_1, V_2, V_3, V_4, V_2, V_6, V_5, V_4, V_6, V_1$.
       **(ii)** Hamiltonian cycle: $V_1, V_2, V_3, V_4, V_5, V_6, V_1$.
   **(b)** There is no Eulerian circuit since $V_2$ and $V_6$ are now odd degree. There is a Hamiltonian cycle still, same as above.
   **(c) (i)** An Eulerian trail: $V_2, V_3, V_4, V_2, V_6, V_5, V_4, V_6, V_1$.
       **(ii)** A Hamiltonian path: $V_2, V_3, V_4, V_5, V_6, V_1$.
**29.** 35
**30. (a)** One upper bound is the length of any cycle, eg *ABCDEA* gives 73. Other methods also apply.
   **(b) (i)** *AB, AD, BC* in that order.
       **(ii)** Weight 33. Lower bound = 60
**31. (a)** Not planar. $e = 15 \not\leq 3v - 6 = 12$.
   **(b)** *BD, DF, FA, FE, EC*, in that order. Weight = 12
**32. (a)**

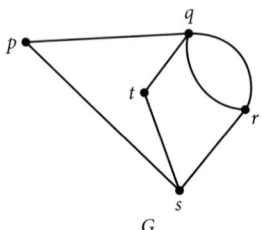 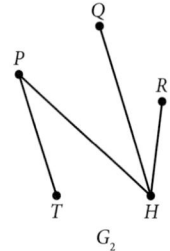

   **(b) (i)** $G_1$ is not simple, $G_2$ is simple.
       **(ii)** Both are connected.

**(iii)** Both are bipartite: $G_1$: components are {p, r, t} and {q, s},
   $G_2$: components are {P, R, Q} and {T, S}
**(iv)** $G_1$ is not a tree, it has a cycle. $G_2$ is a tree.
**(v)** $G_1$ contains an Eulerian trail *rqpsrqts*. $G_2$ does not have since 4 certices have odd degrees.
**33. (a)** *FD, FC, CB, BA, CE.*
   **(b)** 76
**34. (a) (i)** *D, E*
       **(ii)** *EBD*
       **(iii)** Example: *ABEFGCBDBEGDFCA.*
       **(iv)** 36
   **(b)** Example: *ABEFDGCA.*
**35. (a)** *HP, KQ, QF, FE, PB, ER, PQ, BC, CD*; 31.
   **(b)** 48
   **(c)** $\dfrac{n(n-1)!}{2}$ is the maximum number of cycles to examine. 3 is the number of possible cycles for a graph with 3 vertices.
   **(d)** # < 1814400.
**36. (a) (i)** Odd degree vertices.
       **(ii)** Bipartite: components are {B, D} and {A, C, E}.
       **(iii)** $G_A = \begin{pmatrix} 0 & 2 & 0 & 1 & 0 \\ 2 & 0 & 1 & 0 & 2 \\ 0 & 1 & 0 & 1 & 0 \\ 1 & 0 & 1 & 0 & 1 \\ 0 & 2 & 0 & 1 & 0 \end{pmatrix}$, 36.
   **(b)** *PUQTRS*, 18.
**37. (a)** *AD, DB, BC, CF, FE, EH, HI, HG, HG*
   **(b)**

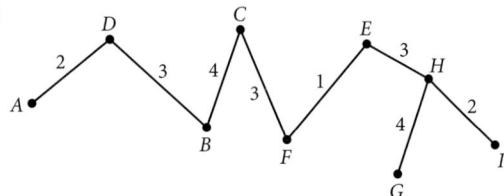

   Weight = 22
**38. (a)** Every edge creates 2 degrees, with $e$ edges there are $2e$ degrees.
   **(c) (i)** $(n, d) = (1, 6), (2, 5), (3, 4), (5, 2)$ or $(6, 1)$
       **(ii)**

(1, 6)          (2, 5)

(3, 4)          (5, 2)

(6, 1)

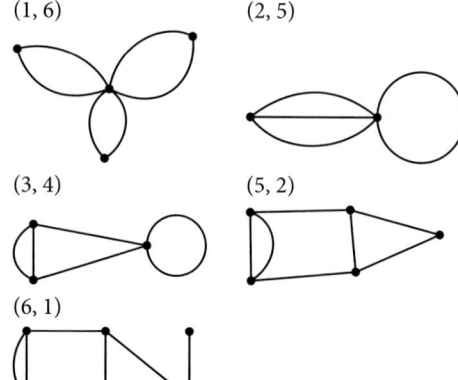

**39 (a)** 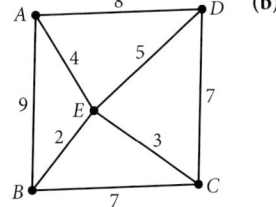 **(b)** 1

**(c)** $ABCDEA$, weight 32; $ABCEDA$, weight 32; $ABECDA$, weight 29; $AEBCDA$, weight, 28, which is the one with the least weight.

**40 (a) (i)**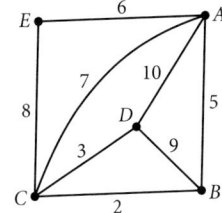

**(ii)** Possible, since 2 vertices have odd degree.

**(iii)** A possible walk is $ACBDABCDCEA$. Length = 55.

**(b)** Every edge connects two vertices.
The degree of a vertex is equal to the number of edges connected to this vertex.
If we add the degree for every vertex, we add the number of edges twice, so the sum of the degrees of all the vertices must be even.
If there is only one vertex with an odd degree the sum of the degrees of all the vertices is odd. This is a contradiction

**41 (a) (i)** See Worked Solutions

**(ii)** Number of paths from $v_i$ to $v_j$ with a maximum length of 3.

**(b)** If there is a bipartition $(A, B)$ of the graph, where $A, B$ are independent sets of vertices and the vertices $\{x, y, z\}$ form a triangle. By cardinality either $A$ or $B$ must contain at least two vertices from the triangle. However, any pair of vertices in the triangle have an edge between them, contradicting the fact that $A$ is independent. Thus a bipartite graph cannot contain a triangle.

**(c)** A bipartite graph is divided into two pieces, of size $p$ and $q$, where $p + q = n$. Then the maximum number of edges is $pq$. Using calculus this product is maximal when $p = q$, in which case it is equal to $\dfrac{n^2}{4}$

**42 (a)** Not bipartite.

**(b) (i)**
$$\begin{pmatrix} 0 & 1 & 0 & 0 & 0 & 1 \\ 1 & 0 & 1 & 1 & 1 & 0 \\ 0 & 1 & 0 & 1 & 0 & 0 \\ 0 & 1 & 1 & 0 & 1 & 1 \\ 0 & 1 & 0 & 1 & 0 & 1 \\ 1 & 0 & 0 & 1 & 1 & 0 \end{pmatrix}$$

**(ii)** 13

**(c)**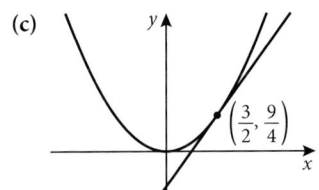

# Chapter 13

## Exercise 13.1

**1.** Achilles can get arbitrarily close to the tortoise. Select any distance between the tortoise and Achilles and after some number of intervals Achilles must be that close. Therefore, the distance between the tortoise and Achilles approaches zero.

**2. (a)** 4    **(b)** $-5$    **(c)** 6
   **(d)** $+\infty$    **(e)** 3    **(f)** 0.354

**3. (a) (i)** $31.0 \, \text{cm s}^{-1}$
     **(ii)** $92.5 \, \text{cm s}^{-1}$
     **(iii)** $153 \, \text{cm s}^{-1}$

**(b)** $92.5 \, \text{cm s}^{-1}$; use average velocity for $t = 2$ to $t = 4$, or find the average of the average velocity for $t = 2$ to $t = 3$ and for $t = 3$ to $t = 4$

**4. (a) (i)** $3.75 \, \text{km h}^{-1}$
     **(ii)** $12.5 \pm 2.5 \, \text{km h}^{-1}$ (depending on line drawn by hand)
     **(iii)** no, gradient is never zero

**(b) (i)** $8 \, \text{km h}^{-1}$
     **(ii)** $16 \pm 6 \, \text{km h}^{-1}$ (depending on line drawn by hand)
     **(iii)** no, gradient is never zero

**(c) (i)** $9 \, \text{km h}^{-1}$
     **(ii)** $0 \, \text{km h}^{-1}$
     **(iii)** bicyclist is not moving from $t = 1.5$ to $t = 2.5$; during this interval the gradient of the tangent is horizontal – hence, rate of change of distance (velocity) is zero

**5. (a)** Using $(0, 100)$ and $(7, 23)$, $-11°\text{C min}^{-1}$

**(b)** $-19 \pm 4°\text{C min}^{-1}$ (depending on line drawn by hand)

**(c)** $-4.2 \pm 0.8 \, °\text{C min}^{-1}$ (depending on line drawn by hand)

**6. (a)** The slope is very negative, then zero, then very positive. (Answers will vary depending on line drawn by hand.)
at $x = -1.5$, about $-1.1 \pm 0.25 \, \text{mm}^{-1}$
at $x = -1$, about $-0.6 \pm 0.15 \, \text{mm}^{-1}$
at $x = 0$, $0 \, \text{mm}^{-1}$
at $x = 1$, about $-0.6 \pm 0.15 \, \text{mm}^{-1}$
at $x = 1.5$, about $1.1 \pm 0.25 \, \text{mm}^{-1}$

**(b)** The curve is symmetric over the $y$-axis so the slope at $x$ is the negative of the slope at $-x$.
Vertical/undefined.

**7.** At $A$, $m = 0.37$; at $B$, $m = 1$; at $C$, $m = 2.7$. It appears that at $A$, $m = \dfrac{1}{e}$ and at $C$, $m = e$. This suggests that the value of the derivative at any point is equal to the value of the function at that point.

# Answers

## Exercise 13.2

**1. (a)** 2.86 s

**(b)** 140 cm s$^{-1}$

**(c)**

| $T$ | distance travelled (cm) | velocity (cm s$^{-1}$) |
|---|---|---|
| 0 | 0 | 40 |
| 1 | 75 | 110 |
| 2 | 220 | 180 |
| 3 | 435 | 250 |

**(d)** The distance computed for $t = 3$ is longer than the length of the ramp

**(e)** $d'(2.86) = 240$ cm s$^{-1}$

**(f)** 40 cm s$^{-1}$

**(g)** 1.89 s

**(h)** $d'(1.89) = 172$ cm s$^{-1}$

**(i)** 1.29 s

**2. (a)** $100\pi = 314$ cm$^2$    **(b)** cm$^2$ cm$^{-1}$

**(c)** $20\pi = 63$ cm$^2$ cm$^{-1}$    **(d)** $40\pi = 126$ cm$^2$ cm$^{-1}$

**3. (a)** 17.7 km

**(b)** $d(5) = \frac{2}{3}(5)^3 - \frac{35}{6}(5)^2 + \frac{35}{2}(5) = 25$ km

**(c)** 5 km h$^{-1}$

**(d)** km h$^{-1}$

**(e)** at $t = 0$, speed is 17.5 km h$^{-1}$ at $t = 5$, speed is approximately 9.17 km h$^{-1}$

**(f)** maximum speed of 17.5 km h$^{-1}$ at $t = 0$

**(g)** minimum speed of 0.486 km h$^{-1}$ at $t \approx 2.92$

**4. (a)** m s$^{-1}$

**(b)** 2 m s$^{-1}$

**(c)** 1.73 m s$^{-1}$,    1.22 m s$^{-1}$,    1.0 m s$^{-1}$

**(d)** 3.46 m    4.90 m,    6 m

**5. (a)** Between A and B

**(b) (i)** A, B, and F    **(ii)** D and E    **(iii)** C

**(c)** Pair B and D, and pair E and F

**6.**

| Function | Derivative diagram |
|---|---|
| $f_1$ | d |
| $f_2$ | e |
| $f_3$ | b |
| $f_4$ | a |

**7. (a)** increasing when $1 < x < 5$; decreasing when $0 < x < 1$ and $5 < x < 6$

**(b)** increasing when $0 < x < 1$ and $3 < x < 5$; decreasing when $1 < x < 3$ and $5 < x < 6$

**8. (a)** decreasing when $-3 < x < -2$ and $1 < x < 3$; decreasing when $0 < x < 1$ and $5 < x < 6$

**(b)**

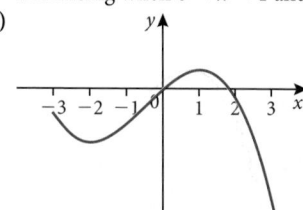

**9. (a)** If $T'(r) > 0$, then the torque is increasing as RPM increases.

**(b)** Graph:

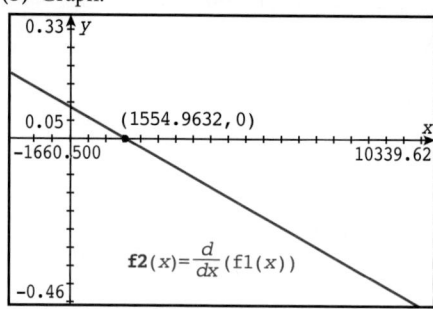

$$f2(x) = \frac{d}{dx}(f1(x))$$

**(c)** Since $T'(r) > 0$ for $0 \leqslant r < 1555$, the torque is increasing on that interval. Since $T'(r) < 0$ for $r > 1555$, the torque is decreasing on that interval. The maximum torque is 505.8 Nm at 1555 RPM. $T'(r) = 0$ at this point.

**10. (a)**

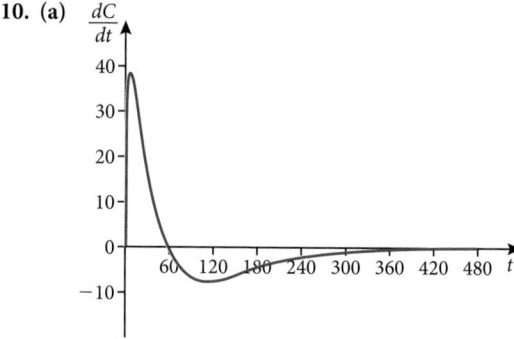

**(b)** Increasing on $0 < t < 57.0$, decreasing on $t > 57.0$

**(c)** 110 min, $-7.51$ nl min$^{-1}$

**(d)** 57.0 min

**11. (a)** thousands of people per year

**(b)**

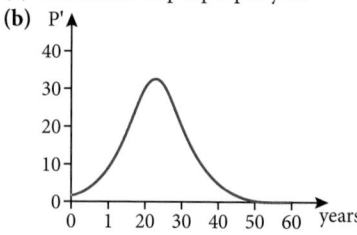

**(c)** The derivative is always positive

**(d)** There is a rapid increase ('boom') in population, then it levels off

**(e)** 8830 people per year

**(f)** 1995 to 2010

**(g)** in the year 2002, 32 500 people per year

## Exercise 13.3

**1. (a) (i)** $y' = 6x - 4$    **(ii)** $-4$

**(b) (i)** $y' = -2x - 6$    **(ii)** 0

**(c) (i)** $y' = -\frac{6}{x^4}$    **(ii)** $-6$

**(d) (i)** $y' = 5x^4 - 3x^2 - 1$    **(ii)** 1

**(e) (i)** $y' = 2x - 4$    **(ii)** 0

**(f) (i)** $y' = 2 - \frac{1}{x^2} + \frac{9}{x^4}$    **(ii)** 10

**(g) (i)** $y' = 1 - \frac{2}{x^3}$    **(ii)** 3

**2. (a)** $(0, 0)$  
**(b)** $(2, 8)$ and $(-2, -8)$  
**(c)** $\left(\dfrac{5}{2}, -\dfrac{21}{4}\right)$  
**(d)** $(1, -2)$

**3.** $a = -5, \ b = 2$

**4.** The power rule gives us $y' = 1 \times mx^{1-1} + 0 = m$; therefore, the gradient of $y = mx + b$ is always $m$.

**5. (a)** $g'(x) = \dfrac{1}{3}x^{-\frac{2}{3}} = \dfrac{1}{3\sqrt[3]{x^2}}$

**(b) (i)** $\dfrac{1}{12}$  **(ii)** $y = \dfrac{1}{12}x + \dfrac{4}{3}$

**(c) (i)** $g'(0)$ is undefined.  
**(ii)** The gradient is vertical at the origin so the tangent is $x = 0$.

**6.** $y = \dfrac{1}{2}x - \dfrac{3}{2}$

**7. (a)** $-13.0\,°\text{C min}^{-1}$  **(b)** $-\dfrac{273}{32} = 8.53\,°\text{C min}^{-1}$

**8. (a)** $\dfrac{14}{3} = 4.67\,°\text{C h}^{-1}$  **(b)** $C'(t) = 3\sqrt{t}$

**(c)** $t = \dfrac{196}{81} = 2.42\,\text{s}$

**9. (a)** cases per year  
**(b)** $\dfrac{dC}{dt} = -666t^2 + 14520t - 12\,700$  
**(c)** At $t = 7, \dfrac{dC}{dt} = 56\,300$ cases per year. At the beginning of 1990, the cumulative number of cases was increasing at a rate of 56 300 cases per year.  
**(d)** In 1990, there were 5 8700 new cases reported. This represents the average rate of change during 1990, while the result in **(c)** is the instantaneous rate of change at the beginning of 1990.  
**(e)** For $0 \leqslant t < 0.913, \dfrac{dC}{dt} < 0$. Therefore, during most of 1983 the cumulative number of cases decreased. For $t > 0.913, \dfrac{dC}{dt} > 0$, so from late 1983 until 1998 the cumulative number of cases increased.

**10. (a)** $\dfrac{dP}{dt} = 34t - 3t^2$  
**(b)** For $0 \leqslant t < 11.3, \dfrac{dP}{dt} > 0 \Rightarrow$ population is increasing. For $11.3 < t \leqslant 20, \dfrac{dP}{dt} < 0 \Rightarrow$ population is decreasing.  
**(c)** 40 bacteria $\text{min}^{-1}$  
**(d)** 15 min  
**(e)** 96.3 bacteria $\text{min}^{-1}$ at 5.67 min  
**11. (a)** $h'(t) = -9.8t + 16$, this function gives the velocity of the tennis ball in m s$^{-1}$  
**(b)** 6.20 m s$^{-1}$

## Exercise 13.4

**1. (a)** $y' = 12(3x - 8)^3$  
**(b)** $y' = -\dfrac{1}{2\sqrt{1-x}}$  
**(c)** $y' = -\dfrac{9}{x^4}$  
**(d)** $y' = x - \dfrac{1}{x^2}$  
**(e)** $y' = -\dfrac{4x}{(x^2 + 4)^3}$  
**(f)** $y' = \dfrac{1}{(x + 1)^2}$  
**(g)** $y' = \dfrac{-1}{2\sqrt{(x+2)^3}}$ $\left[\text{or } \dfrac{-1}{(2x+4)\sqrt{x+2}}\right]$  
**(h)** $y' = 12x(2x^2 - 1)^2$

**(i)** $y' = \dfrac{-3x + 2}{2\sqrt{(1-x)}}$  
**(j)** $y' = \dfrac{-6x + 5}{(3x^2 - 5x + 7)^2}$  **(k)** $y' = \dfrac{2}{\sqrt[3]{(2x+5)^2}}$  
**(l)** $y' = 2(2x - 1)^2(7x^4 - 2x^3 + 3)$  
**(m)** $y' = \dfrac{3x}{\sqrt{3x^2 - 2}}$  **(n)** $y' = \dfrac{x^2 + 4x}{(x+2)^2}$  **(o)** $y' = \dfrac{-2}{(x-1)^2}$

**2. (a)** $\dfrac{dy}{dx} = -\dfrac{4}{x^5}$  **(b)** $\dfrac{d^2y}{dx^2} = \dfrac{20}{x^6}$

**3. (a)** $y = 4x^3 - 3x - 1; \dfrac{dy}{dx} = 12x^2 - 3$  
**(b)** $\dfrac{dy}{dx} = (2x + 1)^2 + 4(x - 1)(2x + 1) = 12x^2 - 3$

**4. (a)** $f'(x) = \dfrac{(x + 1)^2(2x - 3) - 2(x + 1)(x^2 - 3x + 4)}{(x + 1)^4}$

$= \dfrac{(x + 1)(2x - 3) - 2(x^2 - 3x + 4)}{(x + 1)^3}$

$= \dfrac{5x - 11}{(x + 1)^3}$

**(b)** $f''(x) = \dfrac{5(x + 1)^3 - 3(x + 2)^2(5x - 11)}{(x + 1)^6}$

$= \dfrac{5(x + 1) - 3(5x - 11)}{(x + 1)^4}$

$= \dfrac{-10x + 38}{(x + 1)^4}$

**5.** $f'(x) = \dfrac{2a}{(x + a)^2}; f''(x) = \dfrac{-4a}{(x + a)^3}$

**6.** See Worked Solutions

**7.** $\dfrac{d^3y}{dx^3} = 24x > 0$ for $0 < x < 1$.

**8. (a)** $h = \sqrt{b^2 + (b + 1)^2} = \sqrt{2b^2 + 2b + 1}$  
**(b)** $\dfrac{dh}{db} = \left(\dfrac{1}{2}\right)(2b^2 + 2b + 1)^{-\frac{1}{2}}(4b + 2)$

$= \dfrac{2b + 1}{\sqrt{2b^2 + 2b + 1}}$

## Exercise 13.5

**1. (a)** $y' = x^2e^x + 2xe^x$  
**(b)** $y' = -4\sin(4x)$  
**(c)** $y' = -2e^{1-2x}$  
**(d)** $y' = \dfrac{1 + \sin(x) - x\cos(x)}{(1 + \sin x)^2}$  
**(e)** $y' = \dfrac{e^x(x - 1)}{x^2}$  
**(f)** $y' = 2\sin(2x)\cos(2x) = \sin(4x)$  
**(g)** $y' = \dfrac{-xe^x + e^x - 1}{(e^x - 1)^2}$  
**(h)** $y' = \cos x$  
**(i)** $y' = \dfrac{\sin(x)}{2\sqrt{3 - \cos(x)}}$

**2. (a)** $y = \dfrac{1}{6e}x + \ln 2$  **(b)** $y = \dfrac{3}{2}x - \dfrac{3}{2} + \ln 2$  
**(c)** $y = x - \dfrac{\ln 2}{2} - \dfrac{\pi}{4}$  **(d)** $y = \dfrac{1}{2}x$  
**(e)** $y = x - e$  **(f)** $y = -\dfrac{1}{e}x + 2$

# Answers

**3. (a)** $f'(x) = e^x - 3x^2$;  $f''(x) = e^x - 6x$
  **(b)** $x \approx 3.73$ or $x \approx 0.910$ or $x \approx -0.459$
  **(c)** decreasing on $(-\infty, -0.459)$ and $(0.910, 3.73)$
     increasing on $(-0.459, 0.910)$ and $(3.73, \infty)$
**4. (a)** The two functions intersect for all $x$ such that $\cos x = 1$,
     i.e. $x = k \cdot 2\pi, k \in \mathbb{Z}$ The derivatives for the two
     functions are $y' = -e^{-x}$ and $y' = -e^{-x}(\cos x + \sin x)$.
     The derivatives are equal whenever $x = k \cdot 2\pi, k \in \mathbb{Z}$.
     Therefore the functions are tangent at all of the
     intersection points.
  **(b)**

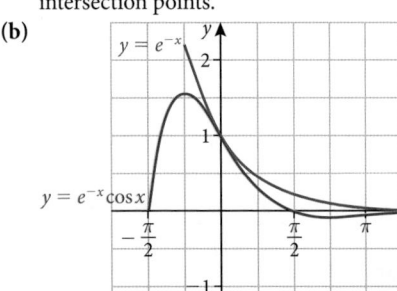

**5.** $8\,\mathrm{m\,s^{-1}}$
**6.** $h'(x) = \dfrac{-x^2 + 2x + 3}{e^x}$

**7. (a)** $\dfrac{d^2y}{dx^2} = e^{-x}$    **(b)** $\dfrac{d^2y}{dx^2} = \dfrac{1}{x^2}$

  **(c)** $\dfrac{d^2y}{dx^2} = -2\sin^2(x) + 2\cos^2(x) = 2\cos(2x)$

  **(d)** $\dfrac{d^2y}{dx^2} = 2\cos(x^2) - 4x^2\sin(x^2)$

**8. (a)** $\sin\theta = \dfrac{a}{a+3}$
     $\Rightarrow a\sin(\theta) + 3\sin(\theta) = a$
     $\Rightarrow 3\sin\theta = a - a\sin\theta$
     $\Rightarrow 3\sin\theta = a(1 - \sin\theta)$
     $\Rightarrow a = \dfrac{3\sin\theta}{1 - \sin\theta}$

  **(b)** $\dfrac{da}{d\theta} = \dfrac{3\cos\theta}{(1 - \sin\theta)^2}$

## Chapter 13 Practice questions
**1. (a)** $t = 3, t = 5$        **(b)** $v(t) = 3t^2 - 14t + 7$
  **(c)** $t \approx 0.569, t \approx 4.10$    **(d)** $a(t) = 6t - 14$
  **(e) (i)** $-14\,\mathrm{cm\,s^{-2}}$    **(ii)** $19\,\mathrm{cm\,s^{-2}}$
**2. (a)** gradient $= 3$        **(b)** $y = 3x - \dfrac{9}{4}$
  **(c)**

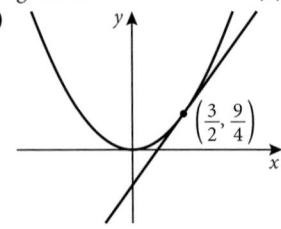

  **(d)** $Q\left(\dfrac{3}{4}, 0\right),\ R\left(0, -\dfrac{9}{4}\right)$
  **(e)** Midpoint of $[PR] = \left(\dfrac{\frac{3}{2} + 0}{2}, \dfrac{\frac{9}{4} - \frac{9}{4}}{2}\right) = \left(\dfrac{3}{4}, 0\right)$
     $\Rightarrow$ point $Q$
  **(f)** $y = 2ax - a^2$

**(g)** $T\left(\dfrac{a}{2}, 0\right),\ U(0, -a^2)$
**(h)** Midpoint of $[SU] = \left(\dfrac{a+0}{2}, \dfrac{a^2 - a^2}{2}\right) = \left(\dfrac{a}{2}, 0\right)$
     $\Rightarrow$ point $T$
**3.** $a = 1$
**4. (a) (i)** $x = 0$        **(ii)** $y = 3$
  **(b)** $\dfrac{dy}{dx} = \dfrac{2}{x^2}$
  **(c)** increasing for all $x$, except $x = 0$
**5.** $b = 2,\ c = 3$
**6. (a)** $10\,\mathrm{m\,s^{-1}}$  **(b)** $10\,\mathrm{sec}$    **(c)** $50$ metres
**7. (a)** $-\dfrac{5}{3}$    **(b)** $\dfrac{1}{4}$        **(c)** $3$        **(d)** $-\infty$
**8. (a)** $f'(x) = \dfrac{3x - 4}{2\sqrt{x}}$

  **(b)** $f'(x) = -\dfrac{1}{x^2} + \dfrac{1}{2}$

  **(c)** $f'(x) = -\dfrac{91}{3x^{14}}$

**9. (a)** $(2, 20)$    **(b)** $(4, 16)$
**10. (a) (i)** $v(0) = 0\,\mathrm{m\,s^{-1}}$    **(ii)** $v(10) \approx 51.3\,\mathrm{m\,s^{-1}}$
  **(b) (i)** $a(t) = 9.9e^{-0.15t}$    **(ii)** $a(0) = 9.9\,\mathrm{m\,s^{-2}}$
  **(c) (i)** $66\,\mathrm{m\,s^{-1}}$    **(ii)** $0\,\mathrm{m\,s^{-2}}$
  **(iii)** as an object falls it approaches terminal velocity
**11. (a) (i)** $g'(x) = -\dfrac{3}{e^{3x}}$
  **(ii)** $e^{3x} > 0$ for all $x$, hence $-\dfrac{3}{e^{3x}} < 0$
     for all $x$ — therefore, $f(x)$ is decreasing for all $x$
  **(b) (i)** $e + 2$        **(ii)** $-3e$
**12. (a)** $\dfrac{dy}{dx} = -\dfrac{4}{(2x + 3)^3}$
  **(b)** $\dfrac{dy}{dx} = 5\cos(5x)e^{\sin(5x)}$
  **(c)** $\dfrac{dy}{dx} = 4x\tan(x^2)(\tan^2(x^2) + 1)$
  **(d)** $\dfrac{dy}{dx} = \dfrac{-xe^x + e^x - 1}{(e^x - 1)^2}$
  **(e)** $\dfrac{dy}{dx} = e^x(2\cos(2x) + \sin(2x))$
  **(f)** $\dfrac{dy}{dx} = 2x\ln x + 2x\ln 3 + x - \dfrac{1}{x}$
**13. (a)** E        **(b)** A        **(c)** C
**14.** $\dfrac{dy}{dx} = 2x\ln x + x$
**15.** $\sin x = \dfrac{1}{2},\ \sin x = -1$
**16. (a)** $f'(x) = \dfrac{2}{2x - 1}$
  **(b)** $x = \dfrac{1 + \sqrt{17}}{4} = 1.28$
**17. (a)** $\dfrac{dy}{dx} = \sec^2 x - 8\cos x$
  **(b)** $\cos x = \dfrac{1}{2}$
**18.** $\dfrac{dy}{dx} = \cos(kx)(k) - k\cos(kx) - kx(-\sin(kx)(k))$
     $= k^2 x\sin(kx)$

**19.** $\dfrac{d^2y}{dx^2} = \dfrac{-4}{(2x-1)^2}$

**20 (a)** $k = \dfrac{\ln 2}{20}$

   **(b)** 510 bacteria per minute

**21. (a)** $g'(x) = \dfrac{2(x^2+6) - 2x(2x)}{(x^2+6)^2} = \dfrac{2x^2 + 12 - 4x^2}{(x^2+6)^4}$

          $= \dfrac{12 - 2x^2}{(x^2+6)^2}$

   **(b)** $b = \sqrt{6}$

**22. (a)**

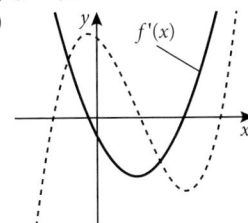

   **(b)**

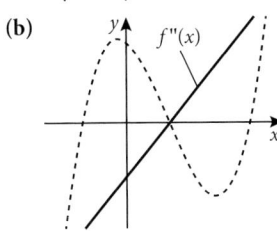

**23. (a)** $c'(x) = 6x^2 - 24x + 30$

   **(b)** $c'(1) = 12$, the marginal cost when producing 100 baseball caps is 12 000 THB.

   **(c)** Since $c'(x)$ is positive (it is a quadratic opening up, with no $x$-intercepts) for all $x > 0$, $c(x)$ is positive for all $x > 0$:

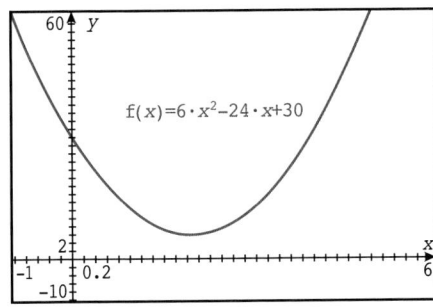

   **(d)** $p(x) = -2x^3 + 12x^2 - 15x$

   **(e)** $p'(x) = -6x^2 + 24x - 15$

   **(f)** $p(x)$ increases for $0.775 < x < 3.22$, decreasing elsewhere

   **(g)** Profit of 9350 THB for producing 322 baseball caps.

**24. (a)** $-9000\,\text{l}\,\text{min}^{-1}$      **(b)** $-6000\,\text{l}\,\text{min}^{-1}$

   **(c)** Since the rate at which the water is draining is a negative quantity, it is increasing (slowing down).

**25. (a)** 200 students

**(b)**

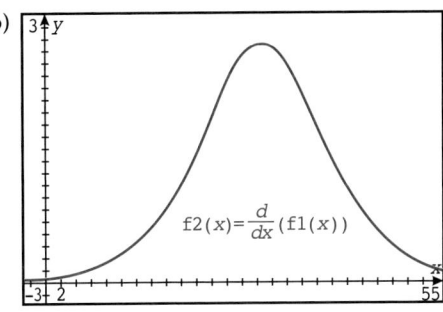

   **(c)** 4.94 students day$^{-1}$

   **(d)** 1.57 students day$^{-1}$

   **(e)** 9.12 students day$^{-1}$ on day 29

   **(f)** day 49

**26. (a)** $M'(t) = -\dfrac{4500}{t^2}$

   **(b)** Since $t^2$ is always positive, $M'(t)$ is always negative

       $\Rightarrow M(t)$ is a decreasing function.

   **(c)** After 2.12 years      **(d)** At $t = 1$

   **(e)** After 9.49 years, $1220

**27. (a)** $P'(n) = -6n^2 + 12n + 1$. $P'(n)$ is a quadratic function opening downward with zeros at $-0.0801$ and $2.08$, hence $P'(n) > 0$ on $0 \leqslant n < 2.08$, so $P(n)$ is increasing on that interval.

   **(b)** 2080 t-shirts, $10 000 profit

**28. (a)** $E'(t) = -0.0021\,t^2 + 0.0556t - 0.0843$

   **(b)** $a = 1972$, $b = 1995$

**29. (a)** CD sales increased sharply and then decreased.

   **(b)** During 1993

   **(c)** At the very end of 1998

   **(d)** During 2007, CD sales decreased by 158 million.

   **(e)** graph **A**

**30 (a)** 4.42 cm

   **(b)** $v(t) = -0.463b \sin\!\left(\dfrac{\pi}{30}bt\right)$;

       $a(t) = -0.0485\,b^2 \cos\!\left(\dfrac{\pi}{30}bt\right)$

   **(c)** Maximum velocity is 255 cm s$^{-1}$; maximum acceleration is 14700 cm s$^{-1}$

   **(d)** $k = m^2$

   **(e)** Since the maximum acceleration force increases by a factor of $m^2$ when RPM increases by a factor of $m$, the forces from acceleration grow quickly, Therefore RPM must not grow too large, which is why engines have a maximum RPM known as the 'redline'.

**31. (a) (i)** when $d = 110.8$ the rate of change is the greatest $\Rightarrow$ day 111

     **(ii)** late April, in the middle of spring

     **(iii)** on average, the next day is 0.2 °C warmer.

   **(b) (i)** when $d = 293.3$ the rate of change is the smallest $\Rightarrow$ day 294

     **(ii)** late October, in the middle of fall

     **(iii)** on average, the next day is 0.2°C colder

   **(c) (i)** When $d = 19.5$ or $d = 202$, $T'(d) = 0$

     **(ii)** This is late January and late July, the middle of winter and middle of summer, respectively.

**32.** $\dfrac{dV}{dx} = \pi x(40 - x)\,\text{cm}^3$ per cm of depth

**33. (a)** $0.889 < n < 11.3 \Rightarrow$ from late in the first year until early in the 12$^{\text{th}}$ year.

**(b)** $\dfrac{dC}{dn} = 1300(0.95)^n(\ln 0.95) - 700(0.75)^n(\ln 0.75)$

$= -66.7(0.95)^n - 201(0.75)^n$

**(c)** Increasing for $0 \leqslant n < 4.68$, decreasing for $n > 4.68$

**(d)** $0 \leqslant n < 2.1$ from the start until early in the third year.

**(e)** $n = 11.97 \Rightarrow$ at the very end of the 12th year

**(f) (i)** $n = 63.5 \Rightarrow$ in the middle of the 64th year

   **(ii)** Since $\dfrac{dC}{dn} < 0$ for all $n > 4.68$, the pollutant concentration will continue to decrease.

**34. (a)** $M = -0.35\cos\left(\dfrac{2\pi}{5.4}t\right) + 4$

**(b)** $\dfrac{dM}{dt} = 0.130\pi\sin\left(\dfrac{2\pi}{5.4}t\right)$

**(c)** Increasing for $0 < t < 2.7$, $5.4 < t \leqslant 8.1$, decreasing for $2.7 < t < 5.4$

**35. (a)** $\dfrac{dC}{dt} = \sin\left(\dfrac{\pi}{12}t\right) + 3$

**(b)** from 2 to 10 hours after sunset

**(c)** 4000 chirps per hour, 6 hours after sunset.

# Chapter 14

## Exercise 14.1

**1. (a)** $\dfrac{dy}{dx} = 2x - 2 = 0 \Rightarrow x = 1$,

$y = 1^2 - 2(1) - 6 = -7 \therefore (1, -7)$

**(b)** $y' = 8x + 12 = 0 \Rightarrow x = -1.5$,

$y = 4(-1.5)^2 + 12(-1.5) + 17 = 8 \therefore (-1.5, 8)$

**(c)** $\dfrac{dy}{dx} = -2x + 6 = 0 \Rightarrow x = 3$,

$y = -(3)^2 + 6(3) - 7 = 2 \therefore (3, 2)$

**2. (a) (i)** $\dfrac{dy}{dx} = 6x^2 + 6x - 72 = 0 \Rightarrow (-4, 213)$,

$(3, -130)$ are stationary points

   **(ii)** Sign diagram:

Therefore $(-4, 213)$ is local maximum, $(3, -130)$ is a local minimum.

   **(iii)**

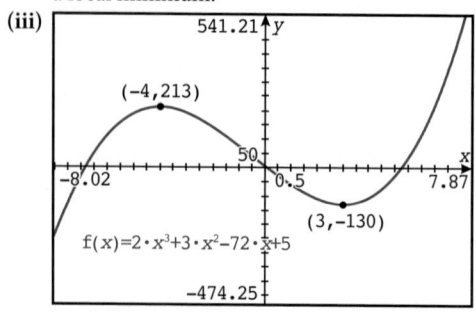

**(b) (i)** $y' = \dfrac{1}{2}x^2 \Rightarrow (0, -5)$ is stationary

   **(ii)** Sign diagram:

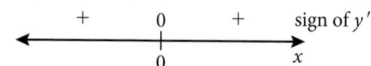

Therefore $(0, -5)$ is neither a minimum nor a maximum.

   **(iii)**

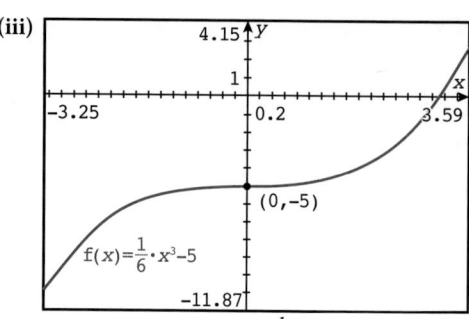

**(c) (i)** $y = -x^3 + 6x^2 - 9x \Rightarrow \dfrac{dy}{dx} = -3x^2 + 12x - 9$

$\Rightarrow (1, -4), (3, 0)$

   **(ii)** Sign diagram:

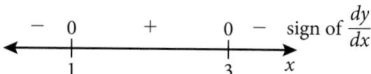

Therefore $(1, -4)$ is a local minimum and $(3, 0)$ is a local maximum.

   **(iii)**

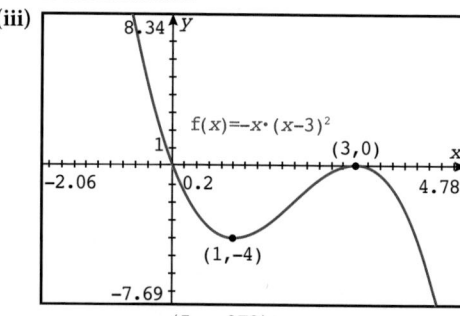

**(d) (i)** $(-1, 4), (0, 6), \left(\dfrac{5}{2}, -\dfrac{279}{16}\right)$

   **(ii)** $(-1, 4)$ is a local minimum since $f'(x)$ changes from negative to positive at $x = -1$; $(0, 6)$ is a local maximum since $f'(x)$ changes from postive to negative at $x = 0$; $\left(\dfrac{5}{2}, -\dfrac{279}{16}\right)$ is a local minimum since $f'(x)$ changes from negative to positive at $x = \dfrac{5}{2}$.

   **(iii)**

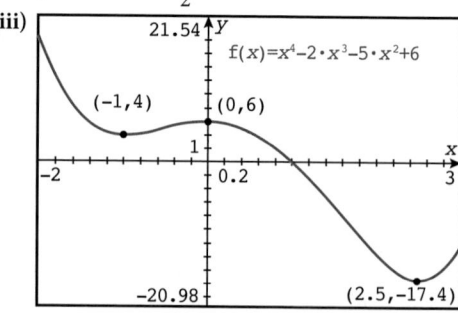

**(e) (i)** $y = x - x^{\frac{1}{2}} \Rightarrow \dfrac{dy}{dx} = 1 - \dfrac{1}{2}x^{-\frac{1}{2}} = 1 - \dfrac{1}{2\sqrt{x}} =$

$0 \Rightarrow x = \dfrac{1}{4}, \therefore$ stationary point at $\left(\dfrac{1}{4}, -\dfrac{1}{4}\right)$

   **(ii)** $\dfrac{dy}{dx}$ changes sign from negative to positive at

$x = \dfrac{1}{4}$ so $\left(\dfrac{1}{4}, -\dfrac{1}{4}\right)$ is a local minimum.

**(iii)**

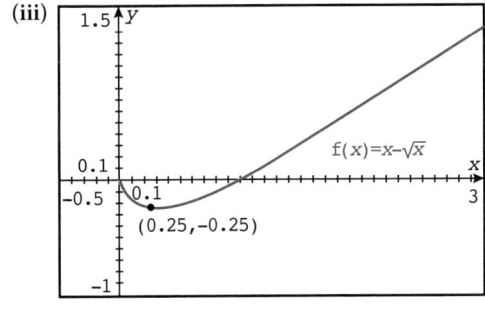

**3. (a)** Local max $(-2, 16)$; local min $(2, -16)$
**(b)** Local max $(0, 0)$, local minima $(-2, -4)$ and $(2, -4)$
**(c)** Local min $(2, 4)$, local max $(-2, -4)$
**(d)** Local max $(1, 2)$, local min $(-1, -2)$
**(e)** Local max $(0, 5)$, local minima $(-1, 0)$ and $(2, -27)$

**4. (a) (i)**

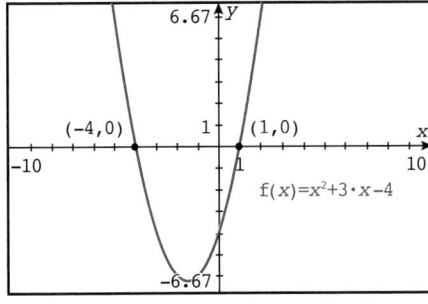

**(ii)** Local maximum at $x = -4$ since $f'(x)$ changes from positive to negative; local minimum at $x = 1$ since $f'(x)$ changes from negative to positive.

**(b) (i)**

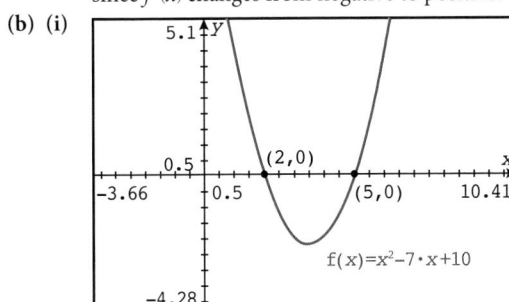

**(ii)** Local maximum at $x = 2$ since $f'(x)$ changes from positive to negative; local minimum at $x = 5$ since $f'(x)$ changes from negative to positive.

**(c) (i)**

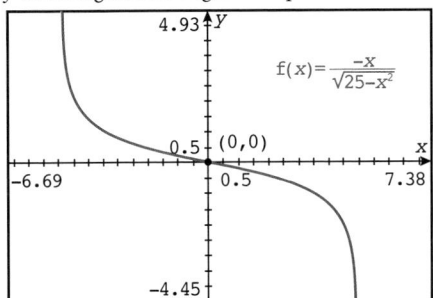

**(ii)** Local maximum at $x = 0$ since $f'(x)$ changes from positive to negative.

**5. (a) (i)**

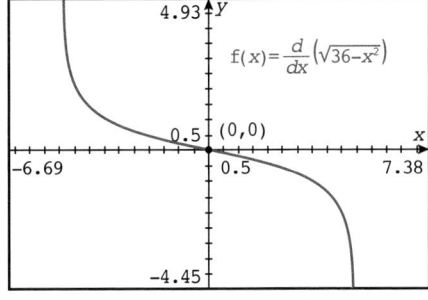

**(ii)** Local maximum at $x = 0$ since $f'(x)$ changes from positive to negative; maximum is at $(0, 6)$.

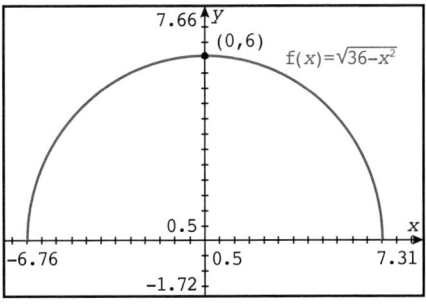

**(b) (i)**

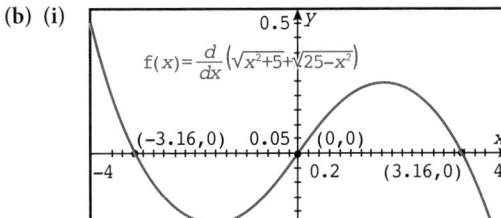

**(ii)** Local maximum at $x = -3.16$ since $f'(x)$ changes from positive to negative; local minimum at $x = 0$ since $f'(x)$ changes from negative to positive; local maximum at $x = 3.16$ since $f'(x)$ changes from positive to negative. Graph of function:

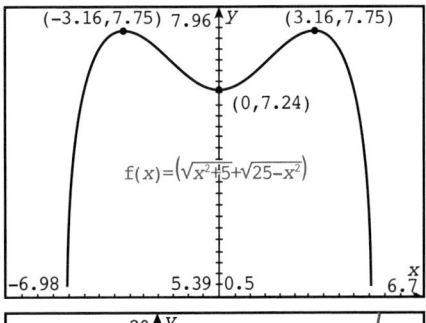

**(c) (i)**

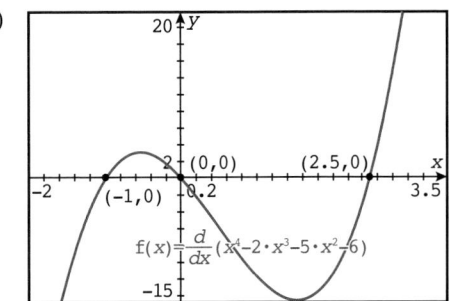

**(ii)** Local minimum at $x = -1$ since $f'(x)$ changes from negative to positive; local maximum at $x = 0$ since $f'(x)$ changes from positive to negative; local minimum at $x = 2.5$ since $f'(x)$ changes from negative to positive.

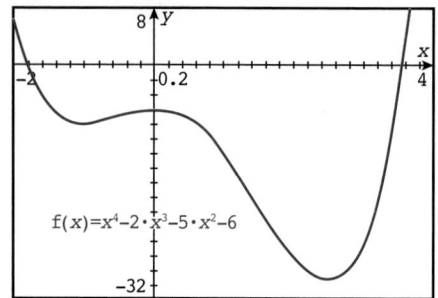

$f(x) = x^4 - 2 \cdot x^3 - 5 \cdot x^2 - 6$

**6. (a) (i)** Increasing on $1 < x < 5$; decreasing elsewhere.
**(ii)** Local min at $x = 1$, local max at $x = 5$.
**(b) (i)** Increasing on $0 < x < 1$ or $3 < x < 5$, decreasing on $1 < x < 3$ or $x > 5$
**(ii)** Local maxima at $x = 1$ and $x = 5$; local min at $x = 3$

**7. (a)** $v(t) = 3t^2 - 8t + 1$; $a(t) = 6t - 8$
**(b)** Max displacement is $-6.88$ m; $t = \dfrac{4 + \sqrt{13}}{3} = 2.54$ s
**(c)** Min velocity $= -4.33$ m s$^{-1}$; $t = 1.33$ s$^{-1}$
**(d)** Starting from $-6$ m, the object moves toward the origin, passing the origin at $t = 0$ s, before reaching its maximum positive displacement at $0.131$ s, then moving in the negative direction reaching minimum displacement of $-6.88$ m at $2.54$ seconds, where it changes directions again.

**8. (a)** Max: $(-2, 18)$, $(2, 18)$; Min: $(0, 10)$
**(b)** Min: $(2, 12)$
**(c)** Min: $(-3, 18)$, $(3, 18)$
**(d)** Max: $(0, 0)$; Min: $(-1, -4)$, $(1, -4)$
**(e)** Max: $(0, 1)$, $(2.51, 1)$; Min: $(1.77, -1)$
**(f)** Min: $\left(-1, -\dfrac{1}{e}\right)$
**(g)** Max: $(2.03, 1.82)$; Min: $(0, 0)$, $(4.91, -4.81)$

**9. (a)** local maximum    **(b)** neither
**(c)** local minimum    **(d)** local minimum
**(e)** local maximum    **(f)** cannot be determined

**10. (a)** $v(0) = 27$ m s$^{-1}$; $a(0) = -66$ m s$^{-1}$
**(b)** $v(3) = 45$ m s$^{-1}$; $a(3) = 78$ m s$^{-1}$
**(c)** $0.5$ s or $2.25$ s. These are times when the displacement is at a relative maximum or minimum.
**(d)** $t = \dfrac{11}{8} = 1.375$ s; where acceleration is zero

**11. (a)** $x = 5.77$ tonnes; $D = 34.6 = 34600$ dollars
**(b)** second derivative $= \dfrac{200}{x^3}$ is positive at $x = 5.77 \Rightarrow$ local minimum.

**12. (a)** $(10 - t)$ m s$^{-1}$ **(b)** $10$ s    **(c)** $50$ m
**13. (a)** $v = 14 - 9.8t$
**(b)** Max height is $10$ m at $1.43$ s
**(c)** Velocity at max height is zero.
**14. (a)** Vertical velocity is given by
$$h'(t) = -5\pi \cos\left(\frac{\pi}{10}(t + 5)\right),\ \text{cosine is maximized}$$

when $\dfrac{\pi}{10}(t + 5) = 2\pi \Rightarrow t = 15$, cosine is minimized
when $\dfrac{\pi}{10}(t + 5) = \pi \Rightarrow t = 5$.

**(b)** Vertical acceleration is given by
$$h''(t) = \frac{\pi^2}{2}\sin\left(\frac{\pi}{10}(t + 5)\right),\ \text{sine is maximized when}$$
$$\frac{\pi}{10}(t + 5) = \frac{\pi}{2}\ \text{or}\ \frac{\pi}{10}(t + 5) = \frac{5\pi}{2} \Rightarrow t = 0, 20,$$
sine is minimized when $\dfrac{\pi}{10}(t + 5) = \dfrac{3\pi}{2} \Rightarrow t = 10$

**15.** 467 spinners; 9650 RMB profit.
**16. (a)** 122 wheels, 16600 EUR    **(b)** 170 wheels, 112 EUR
**17.** 16.4 minutes
**18.** 47.5 seconds
**19. (a)** $C'(t) = -0.26\left(\dfrac{2t^2 - 3}{(2t^2 + 3)^2}\right)$

**(b)** At time $t = \sqrt{\dfrac{3}{2}} = 1.22$, the bloodstream concentration is $0.0531$ mg cm$^{-3}$. The first derivative is positive to the left and negative to the right of $t = \sqrt{\dfrac{3}{2}}$

**(c)** The bloodstream concentration is increasing the fastest at time $t = 0$ (rate of change: $\dfrac{13}{150} = 0.0867$ mg cm$^{-3}$ h$^{-1}$) and decreasing the fastest at time $t = \dfrac{3\sqrt{2}}{2} = 2.12$ hours (rate of change: $-\dfrac{13}{1200} = 0.0108$ mg cm$^{-3}$ h$^{-1}$)

## Exercise 14.2

**1. (a) (i)** $y = -4x$    **(ii)** $y = \dfrac{1}{4}x$
**(b) (i)** $y = 10$    **(ii)** $x = -3$
**(c) (i)** $y = -4x - 8$    **(ii)** $y = \dfrac{1}{4}x + \dfrac{19}{4}$
**(d) (i)** $y = \dfrac{4}{27}$    **(ii)** $x = -\dfrac{2}{3}$
**(e) (i)** $y = -x + 1$    **(ii)** $y = x + 1$
**(f) (i)** $y = -2x + 4$    **(ii)** $y = \dfrac{1}{2}x + \dfrac{11}{4}$
**(g) (i)** $y = \sqrt{2}(x) + \dfrac{\sqrt{2}(4 - \pi)}{8} = 1.41x + 0.152$
   **(ii)** $y = -\dfrac{\sqrt{2}}{2}x + \dfrac{\pi + 8}{8\sqrt{2}} = -0.707x + 0.985$
**(h) (i)** $y = e^2(x - 1)$    **(ii)** $y = -\dfrac{1}{e^2}x + \dfrac{2 + e^4}{e^2}$

**2. (a)** $x = 0$    **(b)** $x = \pm 2$    **(c)** $x = 2$
**3.** $x = 0, \pi, 2\pi$
**4.** $y = 2x,\ y = -x + 1,\ y = 2x - 4$
**5.** $y = -2x$
**6.** $a = -2,\ b = -4$
**7.** $a = -3,\ b = 2$
**8.** $(3, 6)$
**9.** $y = 7x - 16,\ y = 7x + 16$
**10.** $y = \dfrac{1}{2}x - \dfrac{7}{2}; \left(-\dfrac{1}{2}, -\dfrac{15}{4}\right)$
**11. (a)** At $x = 1, y' = -3$
**(b)** Tangent: $y = -3x + 3$; Normal: $y = \dfrac{1}{3}x - \dfrac{1}{3}$
**12. (a)** $y = 2x + \dfrac{5}{2}$    **(b)** $\left(\dfrac{2}{3}, \dfrac{41}{27}\right)$
**13.** Tangent: $y = -\dfrac{3}{4}x + 1$; Normal: $y = \dfrac{4}{3}x - \dfrac{22}{3}$

**14. (a)**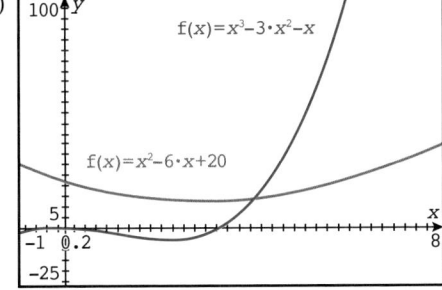

**(b)** $x = 1$

**(c)** Tangent to $y = x^2 - 6x + 20$ is $y = -4x + 19$
Tangent to $y = x^3 - 3x^2 - x$ is $y = -4x + 1$

**15. (a)** $(-2, 4); (2, -20)$      **(b)** $y = -2x, y = -10x$

**16.** $y = 11x - 25, y = -x - 1$

**17.** $y = -0.731x + 1$

**18.** $y = -\dfrac{\sqrt{3}}{4}x + \dfrac{6 + \pi\sqrt{3}}{24}$

## Exercise 14.3

**1. (a)** Maximum volume is 66.1 in².
**(b)** Box should be 5.33 by 7.83 by 1.59 inches.

**2. (a)** 10.9 m      **(b)** 37.2 m

**3.** The lifeguard should run 283 m and swim the remaining distance; the total time will be 99.2 s

**4.** $\sqrt{2}$ by $\dfrac{\sqrt{2}}{2}$

**5.** $13\dfrac{1}{3}$ cm by $6\dfrac{2}{3}$ cm

**6.** $\dfrac{\sqrt{5}}{2}$

**7.** $x = 5\sqrt{2\pi} = 12.5$ cm

**8.** $x = 2.64$ km

**9.** $h = R\sqrt{2}, r = \dfrac{R\sqrt{2}}{2}$

**10. (a)** $h = \dfrac{27 - r^2}{r}; V = \pi r(27 - r^2)$
**(b)** $r = 3$ cm

**11.** $a = -2, b = 8, c = 10$

**12. (a)** $\dfrac{4}{25}x^2$
**(b)** $C(x) = \dfrac{4}{25}x^2 + 85x + 2150$
**(c)** $C'(x) = \dfrac{8}{25}x + 85$, this is positive for all $x > 0$
hence production costs increase for all $x > 0$.
**(d)** $\overline{C}(x) = \dfrac{0.16x^2 + 85x + 2150}{x}$
**(e)** The minimum average manufacturing cost per device is $122; 116 devices should be produced to minimize average cost per device.

**13.** $x = 11.5$ cm, max volume = 403 cm³

**14.** 6 nautical miles

## Exercise 14.4

**1. (a)** $-18.1$ cm min⁻¹    **(b)** $-6.79$ cm min⁻¹
**2. (a)** 0.298 cm s⁻¹    **(b)** 0.439 cm s⁻¹
**3. (a)** $2\pi$ cm hr⁻¹    **(b)** $8\pi$ cm hr⁻¹
**4.** $\dfrac{d\theta}{dt} = \dfrac{3}{34} = .0882$ radian min⁻¹

**5.** 4.8 m s⁻¹
**6.** $\dfrac{8}{3} = 2.67$ m s⁻¹
**7.** 69.5 km hr⁻¹
**8.** $\dfrac{dy}{dt} = \dfrac{12}{\sqrt{10}} = 3.79$
**9.** 0.01 m s⁻¹
**10.** 30 mm³ s⁻¹
**11.** 45 km hr⁻¹
**12.** $\dfrac{8\sqrt{3}}{3} = 4.62$ cm s⁻¹
**13.** 1.5 units s⁻¹
**14.** 222 m s⁻¹ = 800 km hr⁻¹
**15. (a)** 115 degrees s⁻¹
     **(b)** 57 degrees s⁻¹
**16.** $-499$ km hr⁻¹

## Chapter 14 Practice questions

**1. (a)** The slant height is equal to the radius of the semicircle. Using $A_{\text{semicircle}} = \dfrac{1}{2}\pi r^2$ we get
$$39.27 = \dfrac{1}{2}\pi l^2$$
$$25 = l^2$$
$$5 = l$$

**(b) (i)** The circumference of the base of the cone is equal to the arc length of the semicircle. Since $r = l = 5$, we have $C = \dfrac{1}{2}(2\pi r) = \pi(5) = 5\pi = 15.7$ m.

**(ii)** For the distance $C$ to be formed into a circle, it must satisfy $C = 2\pi r \Rightarrow 5\pi = 2\pi r \Rightarrow r = \dfrac{5}{2}$.

**(iii)** $r^2 + h^2 = l^2 \Rightarrow \left(\dfrac{5}{2}\right)^2 + h^2 = 5^2 \Rightarrow$
$$h = \sqrt{5^2 - \left(\dfrac{5}{2}\right)^2} = \dfrac{5\sqrt{3}}{2} = 4.33$$

**(c)** $h + 2r = 9.33 \Rightarrow h = 9.33 - 2r$

**(d)** Using the general model for the volume of a cone, $V = \dfrac{1}{3}Bh$, we have $V = \dfrac{1}{3}\pi r^2 h \Rightarrow$
$$V = \dfrac{1}{3}\pi r^2(9.33 - 2r) = 3.11\pi r^2 - \dfrac{2}{3}\pi r^3$$

**(e)** $\dfrac{dV}{dr} = 3.11\pi(2)r - \dfrac{2}{3}\pi(3)r^2 = 6.22\pi r - 2\pi r^2$

**(f) (i)** We need to find where $\dfrac{dV}{dr} = 0$ hence
$6.22\pi r - 2\pi r^2 = 0 \Rightarrow (\pi r)(6.22 - 2r) = 0 \Rightarrow r = 0$ or $r = 3.11$. We discard $r = 0$.
Sign analysis shows that $\dfrac{dV}{dr}$ is positive to the left of $r = 3.11$ and negative to the right, so the volume is maximized at $r = 3.11$ m.

**(ii)** $V = 3.11\pi(3.11)^2 - \dfrac{2}{3}\pi(3.11)^3 = 31.5$ m³

**2. (a)** $f'(x) = 5x^4$      **(b)** 80
**(c)** $y - 32 = -\dfrac{1}{80}(x - 2) \Rightarrow 80y - 2560 = -x + 2 \Rightarrow$
$x + 80y - 2562 = 0$

**3. (a)** $\dfrac{dy}{dx} = 3x^2 + 2kx$    **(b)** $k = -\dfrac{9}{2}$
**(c)** $y = (3)^3 - \dfrac{9}{2}(3)^2 = -\dfrac{27}{2}$

**4. (a)** $(0, 7)$     **(b)** $(-2, 3)$     **(c)** $]-2, 0[$

# Answers

**(d)** $f'(x) = -3x^2 - 6x \Rightarrow f'(-3) = -9$
$f(-3) = 7$
$\therefore y - 7 = -9(x + 3)$ or $y = -9x - 20$

**(e)** See graph.

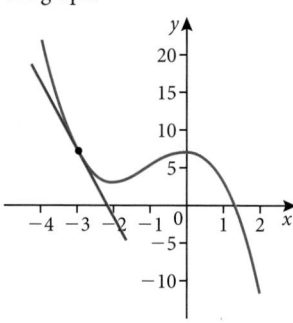

**5. (a)** $f'(x) = 9x^2 - 7 - \dfrac{4}{x^2}$     **(b)** $(1.08, 9.92)$

**(c)** $(-1.08, 10.1)$

**6. (a)** $T = 2\pi r + 4r + 4l$

**(b)** $V = Bh = \dfrac{1}{2}\pi r^2 l \Rightarrow 0.75 = \dfrac{1}{2}\pi r^2 l \Rightarrow 1.5 = \pi r^2 l$

**(c)** From (b), $1.5 = \pi r^2 l \Rightarrow l = \dfrac{1.5}{\pi r^2}$. From (a),

$T = 2\pi r + 4r + 4l \Rightarrow T = 2\pi r + 4r + 4\left(\dfrac{1.5}{\pi r^2}\right)$

$\Rightarrow T = 2\pi r + 4r + \dfrac{6}{\pi r^2}$

$\Rightarrow T = (2\pi + 4)r + \dfrac{6}{\pi r^2}$

**(d)** $\dfrac{dT}{dr} = 2\pi + 4 - \dfrac{12}{\pi r^3}$

**(e)** We solve $\dfrac{dT}{dr} = 2\pi + 4 - \dfrac{12}{\pi r^3} = 0$

$\Rightarrow r = \sqrt[3]{\dfrac{12}{\pi(2\pi + 4)}} = 0.719\,\text{m}$

**(f)** $l = \dfrac{1.5}{\pi r^2} = \dfrac{1.5}{\pi(0.719)^2} = 0.924\,\text{m}$

**(g)** $T = (2\pi + 4)(0.719) + \dfrac{6}{\pi(0.719)^2} = 11.1\,\text{m}$

**7. (a)** $V = lwh \Rightarrow V = 20lw$

**(b)** $V = 20lw \Rightarrow 3000 = 20lw \Rightarrow l = \dfrac{150}{w}$

**(c)** $S = 2(20) + 4w + 2l = 40 + 4w + 2\left(\dfrac{150}{w}\right)$
$= 40 + 4w + \dfrac{300}{w}$

**(d)**

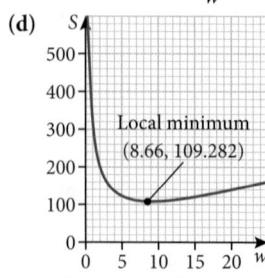

**(e)** $\dfrac{dS}{dW} = 4 - \dfrac{300}{w^2}$

**(f)** $\dfrac{dS}{dW} = 4 - \dfrac{300}{w^2} = 0 \Rightarrow w = \sqrt{75} = 8.66\,\text{cm}$

**(g)** $l = \dfrac{150}{w} = \dfrac{150}{\sqrt{75}} = 17.3\,\text{cm}$

**(h)** $S = 40 + 4w + \dfrac{300}{w} = 40 + 4(\sqrt{75}) + \dfrac{300}{\sqrt{75}}$
$= 40 + 40\sqrt{3} = 110\,\text{cm}$

**8. (a)** $16x^4 - 27x$

**(b)** $f'(x) = 64x^3 - 27$

**(c)** $x = \dfrac{3}{4}$

**9. (a)** $f(-2) = \dfrac{3}{4}(-2)^4 - (-2)^3 - 9(-2)^2 + 20 = 4$

**(b)** $f'(x) = 3x^3 - 3x^2 - 18x$

**(c)** $f'(3) = 3(3)^3 - 3(3)^2 - 18(3) = 0$ therefore $x = 3$ is a stationary point.
Now, we must show that this stationary point is in fact a local minimum. We can do this by using the first derivative test or sketching the graph on our GDC.
First derivative test:
$f'(2) < 0, f'(4) > 0$ therefore there is a local minimum at $x = 3$.

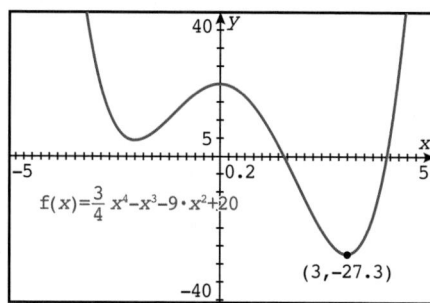

**(d)**

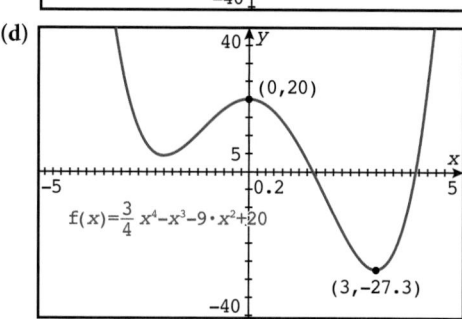

**(e)** $(0, 20)$

**(f)** $f'(2) = 3(2)^3 - 3(2)^2 - 18(2) = -24$

**(g) (i)** $\dfrac{1}{24}$

     **(ii)** $y + 12 = \dfrac{1}{24}(x - 2) \Rightarrow 24y + 288 = x - 2$
$\Rightarrow -x + 24y + 290 = 0 \Rightarrow x - 24y - 290 = 0$
$\therefore b = -24, c = -290$

**10. (a)** $f''(x) = 3ax^2 - 3$      **(b)** $f'(0) = -3$

**(c)** Local maximum implies $f'(x) = 0 \Rightarrow f'(-2)$
$= 3a(-2)^2 - 3 = 0 \Rightarrow a = \dfrac{1}{4}$.

**11. (a)** $f(1) = \dfrac{14}{(1)} + (1) - 6 = 9$

**(b)** $f'(x) = -\dfrac{14}{x^2} + 1$

**(c)** $f'(x) = -\dfrac{14}{x^2} + 1 = 0 \Rightarrow x^2 = 14 \Rightarrow x = \sqrt{14} = 3.7$
We see from the given graph that the stationary point at $x = 3.7$ must be a local minimum.

**(d)** From the graph, minimum occurs at

$x = \sqrt{14} \Rightarrow f(\sqrt{14}) = \dfrac{14}{\sqrt{14}} + \sqrt{14} - 6 = 1.48331.$

Therefore range is $1.48 \leqslant f(x) \leqslant 9$.

**(e)** $f(7) = \dfrac{14}{(7)} + (7) - 6 = 3$

**(f)** $m = \dfrac{9 - 3}{1 - 7} = -1$

**(g)** $M = \left(\dfrac{1 + 7}{2}, \dfrac{9 + 3}{2}\right) = (4, 6)$

**(h)** $f'(4) = -\dfrac{14}{4^2} + 1 = \dfrac{1}{8}$

**(i)** Point on function is $(4, f(4))$, $f(4) = \dfrac{14}{(4)} + (4) - 6$

$= 1.5 \Rightarrow (4, 1.5)$

Therefore equation of $L$ is $y - 1.5 = \dfrac{1}{8}(x - 4)$

$\Rightarrow y = \dfrac{1}{8}x + 1.$

**12. (a)** $y = -\dfrac{75^2}{10} + \dfrac{27}{2} \times 75 = 450$ therefore $A$ is on the track.

**(b)** $\dfrac{dy}{dx} = -\dfrac{2x}{10} + \dfrac{27}{2} = -0.2x + 13.5$

**(c)** Stationary point(s) occur at $\dfrac{dy}{dx} = -0.2x + 13.5 = 0$

$\Rightarrow x = 67.5$. Since $A$ has $x$-coordinate 75, it cannot be the farthest point north.

**(d) (i)** $M = \left(\dfrac{0 + 100}{2}, \dfrac{0 + 350}{2}\right) = (50, 175)$

**(ii)** $m = \dfrac{350 - 0}{100 - 0} = 3.5$

**(e)** $y - 150 = 3.5(x - 0) \Rightarrow 3.5x - y = -150$

**(f)** Use your GDCs Intersection feature to find the first point of intersection is at $(18.4, 214)$.

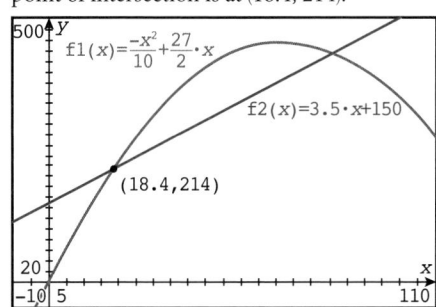

**13. (a)** $f'(x) = 15x^2 - 8x + 1$

**(b)** $f'(x) = 15x^2 - 8x + 1 = 0 \Rightarrow x = \dfrac{1}{5}, \dfrac{1}{3}.$

Use GDC graph or first derivative test to find which stationary point is a local minimum and which is a local maximum. GDC graph:

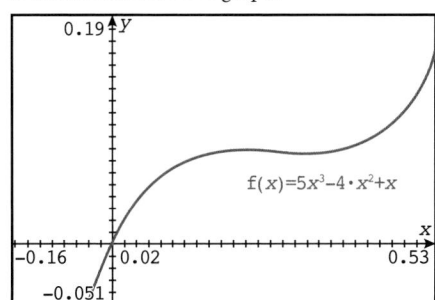

Or, via first derivative test, using sign diagram:

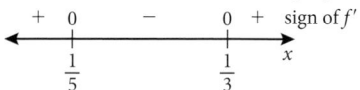

Therefore,

**(i)** local maximum is at $x = \dfrac{1}{5}$

**(ii)** local minimum is at $x = \dfrac{1}{3}$

**14. (a)** Vertical asymptote occurs when denominator equals zero; therefore, $x^2 = 0 \Rightarrow x = 0$.

**(b)** $g'(x) = b - \dfrac{2}{x^3}$

**(c)** $3 = b - \dfrac{2}{(1)^3} \Rightarrow b = 5$

**(d)** Point of tangency is $g(1) = 3 \Rightarrow (1, 3)$, gradient is 3 (given), equation of $T$ is $y - 3 = 3(x - 1) \Rightarrow y = 3x$.

**(e)** $x$-intercept is at $(-0.439, 0)$. Graph:

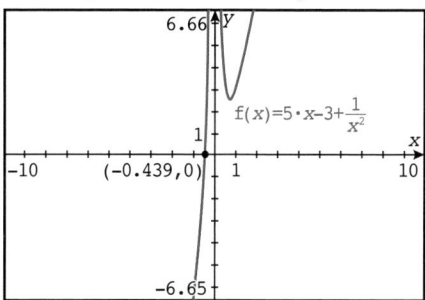

**(f) (i)** and **(ii)** shown on graph below:

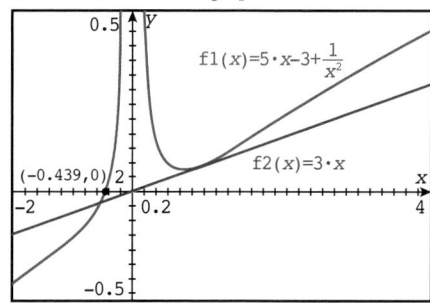

**(g)** $(0.737, 2.53)$

**(h)** $0.737 < x < 5$

**15. (a)** $\dfrac{dy}{dx} = 4x - 5$

**(b)** $L : 6x + 2y = -1 \Rightarrow y = -3x - \dfrac{1}{2} \Rightarrow$ Gradient of $L$ is $m = -3$.

$\therefore \dfrac{dy}{dx} = 4x - 5 = -3 \Rightarrow x = \dfrac{1}{2}.$

**16. (a)**

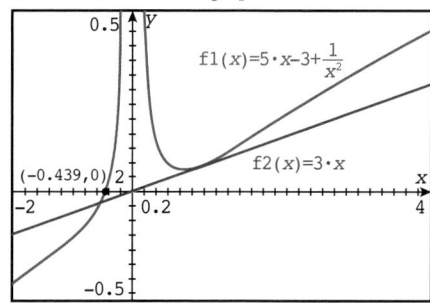

# Answers

**(b)** $f(-1) = -\frac{1}{3}(-1)^3 + \frac{5}{3}(-1)^2 - (-1) - 3 = 0$

**(c)** $(0, -3)$

**(d)** $f'(x) = -x^2 + \frac{10}{3}x - 1$

**(e)** $f'(-1) = -(-1)^2 + \frac{10}{3}(-1) - 1 = -\frac{16}{3}$

**(f)** $f'(-1)$ gives the gradient of the tangent to the curve at the point where $x = -1$.

**(g)** $y - 0 = -\frac{16}{3}(x + 1) \Rightarrow y = -\frac{16}{3}x - \frac{16}{3}$

**(h)**

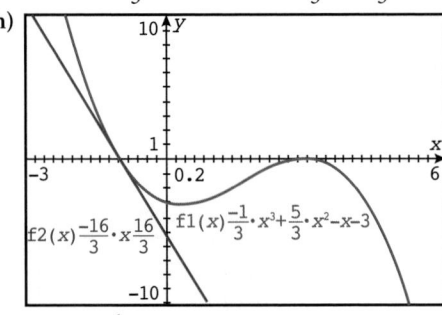

**(i)** **(i)** $a = \frac{1}{3}$  **(ii)** $b = 3$

**(j)** $f(x)$ is increasing on the interval $a < x < b$.

**17. (a)** $4(2x) + 4(x) + 4(y) = 48 \Rightarrow 3x + y = 12$
$\Rightarrow y = 12 - 3x$

**(b)** $V = lwh = (x)(2x)(y) = 2x^2 y = 2x^2(12 - 3x)$
$= 24x^2 - 6x^3$

**(c)** $\frac{dV}{dx} = 48x - 18x^2$

**(d)** $\frac{dV}{dx} = 48x - 18x^2 = 0 \Rightarrow x = 0, \frac{8}{3}$. $x = 0$ is nonsensical, so maximum volume occurs at $x = \frac{8}{3} = 2.67$ m.

**(e)** $V = 24\left(\frac{8}{3}\right)^2 - 6\left(\frac{8}{3}\right)^3 = 56.9$ m³.

**(f)** length $= 2x = 5.33$ m, height $= y = 12 - 3\left(\frac{8}{3}\right) = 4$ m

**(g)** SA $= 2 \times \frac{16}{3} \times 4 + 2 \times \frac{8}{3} \times 4 + 2 \times \frac{16}{3} \times \frac{8}{3} = 92.4$ m².
Therefore $\frac{92.4}{15 \times 4} = 1.54 \Rightarrow 2$ tins are required.

**18. (a)** $y = 0.5x - 1$

**(b)** Equation is of the form $y = mx$ since it passes through the origin. Angle with $x$ axis is $\frac{34.92°}{2} = 17.46°$.
Gradient $m = \tan(17.46°) = 0.315$; therefore equation is $y = 0.315x$.

**(c)** $(5.41, 1.70)$

**(d)** The throw is not valid, since the point $Q$ is 5.65 m from the centre of the throwing circle, which implies that the discus will land outside the sector.

**(e)** **(i)** $R(0.795, -0.585)$  **(ii)** $y = 0.340x - 0.855$

**19.** $2.36$ m s⁻¹

**20.** $-1.23$ m s⁻¹

**21.** $720$ kPa s⁻¹

**22. (a)** $\sqrt{2^2 + 6^2} = \sqrt{40} = 2\sqrt{10}$ km

**(b)** $r = \sqrt{(6 - w)^2 + 4}$

**(c)** $T = \frac{\sqrt{(6 - w)^2 + 4}}{3} + \frac{w}{5}$

**(d)** $w = 4.5$ km, $r = 2.5$ km

**(e)** $1.73$ hours

**23. (a)** 3.94 min, 43.2 nanograms ml⁻¹ min⁻¹

**(b)** 53.3 min

**(c)** 88.8 min, $-4.36$ nanograms ml⁻¹ min⁻¹

**(d)** From 0.594 min to 12.3 min

**24. (a)** thousands of new subscribers per year

**(b)** When $t = 11.8 \Rightarrow$ late in 2011 at a rate of 7370 new subscribers per year.

**(c)** For $5.51 < t < 21.3 \Rightarrow$ mid-2005 to early 2021

**(d)** $S'(t)$ is positive for all $t \in \mathbb{R}$.

**25. (a)** Radius $= 32.6$ cm, Height $= 65.1$ cm

**(b)** 217 litres

## Chapter 15

### Exercise 15.1

**1. (a)** continuous  **(b)** discrete  **(c)** continuous
**(d)** discrete  **(e)** discrete  **(f)** continuous

**2.** (Answers will vary.) The grains of sand in a bucket.

**3.** $E(X) = 3.5$, $Var(X) \approx 2.92$

**4.** $E(X) = 5$, $Var(X) \approx 9.67$

**5.** $E(X) \approx 2.67$, $Var(X) \approx 2.22$

**6.** $E(X) = 6.5$, $Var(X) \approx 11.9$

**7.** $b = 0.2 - \frac{2}{3}a$

**8.** $E(X) = 4$, $Var(X) = 4$

**9.** $E(X) = 3$, $Var(X) = 2$

**10.** $E(X) = 2$, $Var(X) = 1$

**11. (a)** $E(X) = \frac{14}{9}$, $Var(X) = \frac{13}{162}$  **(b)** 0.583

**12. (a)** Because $\int_0^3 \frac{1}{9}x^2 \, dx = 1$

**(b)** **(i)** $E(X) = \frac{9}{4}$, $Var(X) = \frac{27}{80}$  **(ii)** $\frac{7}{27}$

**13.** $\frac{3}{26}$

**14. (a)** 0.564

**(b)** **(i)** 0.564  **(ii)** 0.564

**(c)**

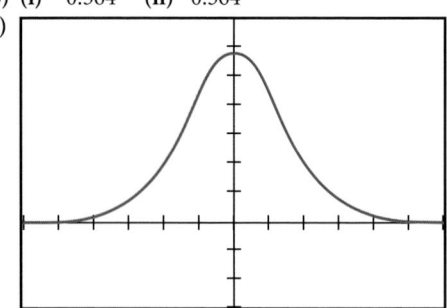

It looks like a normal curve.

### Exercise 15.2

**1. (a)** $E(X) = 10$, $Var(X) = 5$  **(b)** 0.176

**(c)** $\left\{ \begin{array}{l} 9.54 \times 10^{-7}, 1.91 \times 10^{-5}, 1.81 \times 10^{-4}, 0.00109, \\ 0.00462, 0.0148, 0.0370, 0.0739, 0.120, 0.160, 0.176, \\ 0.160, 0.120, 0.0739, 0.0370, 0.0148, 0.00462, \\ 0.00109, 1.81 \times 10^{-4}, 1.91 \times 10^{-5}, 9.54 \times 10^{-7} \end{array} \right\}$

**2. (a)** $E(X) = 25$, $Var(X) = 12.5$  **(b)** 0.112

**3. (a)** $E(X) = 8$, $Var(X) = 1.6$  **(b)** 0.302

**(c)** $1 - P(X = 0) \approx 1.00$

4. (a) (i) $E(X) = 10$, $\text{Var}(X) \approx 8.33$
   (ii) 0.137
   (iii) $1.77 \times 10^{-5}$
   (iv) $2.05 \times 10^{-47}$
(b)

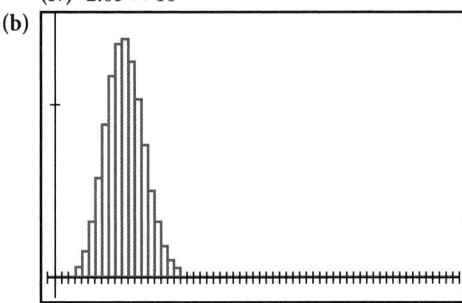

5. (a) $E(X) = 24$, $\text{Var}(X) = 4.8$
   (b) 0.179
6. (a) $E(X) = 1.8$, $\text{Var}(X) = 1.26$
   (b) 0.302
   (c) $1 - [P(X = 6)] \approx 0.999$
7. (a) 0.008 No
   (b) $E(X) = 18$, $\text{Var}(X) = 12.6$
   (c) The standard deviation with the sample 10 times larger is $\sqrt{10}$ times larger.
8. (a) $E(X) = 2$, $\text{Var}(X) = 1.8$
   (b) $3.56 \times 10^{-4}$
   (c) $1 - [P(X = 0) + P(X = 1) + \dots + P(X = 7)] \approx 4.16 \times 10^{-4}$
9.

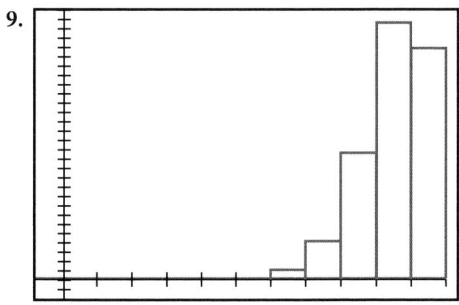

10. $1 - [P(X = 0) + P(X = 1)] \approx 0.181$
11. (a) $1 - [P(X = 0) + P(X = 1) + \dots + P(X = 7)] \approx 0.678$
    (b) 0.107

## Exercise 15.3

1. (a) 0.0758    (b) 0.0126    (c) 0.607
2. 0.161
3. 0.184
4. $4.95 \times 10^{-5}$
5. 0.202
6. 0.125
7. (a) 0.190    (b) 0.0164    (c) 0.984
8. 0.184
9. $1 - [P(X = 0) + P(X = 1)] \approx 0.0902$
10. 0.0572

## Exercise 15.4

1. According to the empirical rule, we find that approximately 68% of those young men have a height between 173 cm and 189 cm, 95% of them between 165 cm and 197, and 99.7% between 157 cm and 205 cm.

2. 0.560
3. 0.737
4. $1 - P(X \leqslant 180) \approx 0.0534$
5. (a) 42.5 km
   (b) 43.5 km
6. By the empirical rule, approximately 68% of the data falls within one standard deviation of the mean.
7. 169 cm
8. 0.645
9. $P(\text{not fail within 3 years}) = 1 - P(\text{it will fail within 3 years}) \approx 0.252$
10. Let the unknown variance be $x^2$, where $x$ if the unknown standard deviation. Since the area under the normal curve [100, 140] is 0.19, using the Solver, we have normalcdf(100, 140, 100, $x$) = 0.19 $\Rightarrow$ $x \approx 80.7 \Rightarrow \sigma^2 \approx 6500$

## Exercise 15.5

1. (a) Since $E(X) = 5$, $\text{Var}(X) = 5$
      $\Rightarrow$ $E(X + 5) = 10$, $\text{Var}(X + 5) = 5$
   (b) $E(3X) = 15$, $\text{Var}(3X) = 45$
2. (a) Since $E(X) = 100$, $\text{Var}(X) = 16$,
      $\Rightarrow$ $E(X + 5) = 105$, $\text{Var}(X + 5) = 16$
   (b) $E(3X) = 300$, $\text{Var}(3X) = 144$
3. (a) $X \sim N(100, 13^2)$
   (b) $Y \sim N(2 \cdot 67 - 33, 4 \cdot 144 + 25) \Rightarrow Y \sim N(101, 601)$
4. Consider $Y = 2X_2 - X_4 \Rightarrow Y \sim N(4.6 - 4.4, 0.16 + 0.04)$
   $\Rightarrow YN(0.2, 0.2)$
   $P(Y > 0) = 1 - P(Y < 0) = 1 - \text{normalcdf}(-\text{E99}, 0, 0.2, 0.2) \approx 0.841$
5. $X \sim N(2 \cdot 13 + 12, 4 \cdot 0.4^2 + 0.3^2) \Rightarrow X \sim N(38, 0.73)$, so $P(X \leqslant 40) \approx 0.997$
6. Let $Y = X_1 + X_2 \sim Po(7)$, $P(Y = 5) = \dfrac{e^{-7} \cdot 7^5}{5!} \approx 0.128$
7. (a) $X_1 \sim Po(40)$, $X_2 \sim Po(20)$ where $X = X_1 + X_2$
   (b) (i) $E(X_1) = 5$
      (ii) $E(X_2) = 2.5$
      (iii) $E(X) = 7.5$
      (iv) $P(X = 10) = \dfrac{e^{-7.5} \cdot 7.5^{10}}{10!} \approx 0.0858$
8. (a) 0.184    (b) 0.271
   (c) 0.168 Not equivalent to the sum.    (d) 0.815

## Exercise 15.6

1. (a) (i) [0.68    0.18    0.14]
      (ii) [0.672    0.182    0.146]
         [0.6688    0.1838    0.1474]
      (iii) [0.66667    0.185183    0.148147]
   (b) [0.68    0.18    0.14]
      [0.666667    0.185185    0.148148]
2. (a) (i) [0.3    0.25    0.3    0.15]
      (ii) [0.3    0.25    0.3    0.15]
   (b) [0.315    0.21    0.3    0.175]
      [0.295    0.228    0.303    0.174]
      Fluctuations with decreases in B and D, increases in A and C.
      (iii) [0.30225    0.22433    0.30037    0.173006] Yes.
         [0.302281    0.224335    0.30038    0.173004]

# Answers

**3. (a) (i)** $[0.88 \quad 0.025 \quad 0.095]$
**(ii)** $[0.88 \quad 0.025 \quad 0.095]$
$[0.888 \quad 0.0215 \quad 0.0905]$
$[0.8888 \quad 0.02025 \quad 0.09095]$
**(b)** $[0.888889 \quad 0.020202 \quad 0.090909]$

**4. (a)** The transition matrix $T$ is deterministic. The initial state makes little difference.
**(b)** $[0.289474 \quad 0.368421 \quad 0.342105]$- brand B
**(c)** Since each state is multiplied by $T$ to generate the next state, the previous one is found by multiplying by $T^{-1}$.
$[0.25 \quad 0.5 \quad 0.25]$
**(d)** $x_n \cdot T^{100} = [0.3 \quad 0.35 \quad 0.35] \cdot \begin{bmatrix} 0.5 & 0.3 & 0.2 \\ 0.3 & 0.3 & 0.4 \\ 0.1 & 0.5 & 0.4 \end{bmatrix}^{100}$
$= [0.289474 \quad 0.368421 \quad 0.342105]$

**5. (a)** $\begin{bmatrix} 0.8 & 0.05 & 0.15 \\ 0.05 & 0.9 & 0.05 \\ 0.1 & 0.15 & 0.75 \end{bmatrix}$
**(b)** $[0.365 \quad 0.335 \quad 0.3]$
**(c)** $[0.33875 \quad 0.36475 \quad 0.2965]$

**6. (a)** $[0.5 \quad 0.5]$ – tied for vote
**(b)** $[0.45 \quad 0.55]$ – liberals

**7. (a)** $[0.379 \quad 0.4195 \quad 0.2015]$
**(b)** $[0.239497 \quad 0.49706 \quad 0.263444]$

## Chapter 15 Practice questions

**1.** $E(X) = 4.5$
**2. (a)** 0.26 **(b)** 0.37 **(c)** 0.77
**(d)** $E(X) = 16.29$ **(e)** $\mathrm{Var}(X) = 8.126$
**3.** Since the probabilities add up to 1, $a = 0.1$, making the probabilities $\{0.1, 0.15, 0.2, 0.25, 0.3\}$. $E(X) = 3.5$, $\mathrm{Var}(X) \approx 1.75$
**4.** $E(X) = 9.5\%$, $\mathrm{Var}(X) \approx 69.8\%$
**5. (a)** $k = \dfrac{1}{10}$ **(b)** $\dfrac{37}{60}$ **(c)** $\dfrac{19}{30}$
**(d)** $E(X) = 16$, standard deviation $= 7$
**6.** Since $\displaystyle\sum_{2}^{3} x^3\, dx = \dfrac{65}{4}$, $k = \dfrac{4}{65}$,
**7.** Since $\displaystyle\sum_{0}^{2} x^2(2-x)\, dx = \dfrac{4}{3}$, $k = \dfrac{3}{4}$
**8. (a)** Since $\displaystyle\sum_{0}^{5}(-x^2 + 2x + 15)\, dx = \dfrac{175}{3}$, $k = \dfrac{3}{175}$
**(b)** mean $= \dfrac{55}{28}$, mode $= 1$
**(c)** Solving for $m$: $\displaystyle\sum_{0}^{m}\dfrac{3}{175}(-x^2 + 2x + 15)\, dx = 0.5 \Rightarrow$
$m \approx 1.857$
**9.** $E(X) = 1.2$, $\mathrm{Var}(X) = 0.4$

**10. (a) (i)** $P(X = 24) \approx 0.105$
**(ii)** $P(X = 22) \approx 0.0925$, $P(X = 26) \approx 0.0902$
**(iii)** $P(X = 20) \approx 0.0616$, $P(X = 28) \approx 0.0595$
**(b)** As the probability is the highest at the expected value and those adjacent gradually decrease, the distribution will likely fit a normal curve.

**11.** $P(X \geqslant 10) = 1 - P(X \leqslant 9) \approx 0.0139$
**12. (a)** 0.817 **(b)** 1 **(c)** 0.0162
**13. (a)** 0.0338 **(b)** 0.0245 **(c)** 0.783
**14. (a)** 3 **(b)** 0.101 **(c)** 0.000215
**15. (a)** 0.0025 **(b)** 0.9826 **(c)** 0.9999
**16. (a)** 0.1396 **(b)** 0.1912 **(c)** 0.9576
**17. (a)** 0.8187 **(b)** 0.5488
**18. (a)** 0.9877 **(b)** 0.999998 **(c)** 0.0000244
**19. (a)** 0.0334 **(b)** 646 **(c)** 98th
**20.** $P(X \geqslant 7) = 1 - P(X < 7) \approx 0.212$ (Note: In a continuous distribution, for any specific value $a$, $P(X = a) = 0$, so the cumulative normal distribution on your GDC can extend to include $X = 7$ without concern.)
**21. (a)** 0.655 **(b)** 0.00820 **(c)** 82
**22. (a)** 0.227 **(b)** 0.55% **(c)** 29.7 **(d)** 229
**23. (a)** Not likely, $p=0.13\%$ **(b)** 15.9%
**(c)** 68.3% **(d)** 5396 **(e)** 43785
**24.** 0.0447
**25. (a) (i)** $[0.41 \quad 0.265 \quad 0.325]$
**(ii)** $[0.41 \quad 0.265 \quad 0.325]$
$[0.417 \quad 0.259 \quad 0.324]$
$[0.4186 \quad 0.2583 \quad 0.3231]$
**(iii)** $[0.419345 \quad 0.258067 \quad 0.322588]$
**(b)** $[0.419355 \quad 0.258065 \quad 0.322581]$
**26.** Steady state: $[0.667 \quad 0.333]$
**(a)** $[0.5 \quad 0.5] \cdot \begin{bmatrix} 0.75 & 0.25 \\ 0.5 & 0.5 \end{bmatrix}^{50} \approx [0.667 \quad 0.333]$
**(b)** $[0 \quad 1] \cdot \begin{bmatrix} 0.75 & 0.25 \\ 0.5 & 0.5 \end{bmatrix}^{50} \approx [0.667 \quad 0.333]$
**(c)** $[1 \quad 0] \cdot \begin{bmatrix} 0.75 & 0.25 \\ 0.5 & 0.5 \end{bmatrix}^{50} \approx [0.667 \quad 0.333]$
**27.** $[0.333 \quad 0.667]$
**28. (a) (i)** 0.191 **(ii)** 0.0880
**(b)** 0.0364
**(c) (i)** 19.1 **(ii)** 0.0869
**29. (a)** 0.819
$z = 1.95996$
**(b)** $\dfrac{53.92 - 50}{\sigma} = 1.95996 \Rightarrow \sigma \approx 2.00004$
**(c)** $t = 48.7$
**(d) (i)** 0.360 **(ii)** 0.924

**30. (a)** 0.513

$\dfrac{P(\text{more than 10 on Monday}) \cdot P(\text{more than 40 over the week})}{P(\text{more than 10 on Monday})}$

**(b)** $= \dfrac{P(X > 40) + P(X = 11) \times P(Y > 29) + P(X = 12) \times P(Y > 28) + \ldots + P(X = 40) \times P(Y > 0)}{P(X > 40)}$

$= \dfrac{P(X > 40) + \displaystyle\sum_{r=11}^{40} P(X = r) \cdot P(Y > 40 - r)}{P(X > 40)}$

**31.** 0.0879

# Chapter 16

## Exercise 16.1

**1.** (a) $\dfrac{x^2}{2} + 2x + c$

(b) $t^3 - t^2 + t + c$

(c) $\dfrac{x}{3} - \dfrac{x^4}{14} + c$

(d) $\dfrac{2t^3}{3} + \dfrac{t^2}{2} - 3t + c$

(e) $\dfrac{5u^{\frac{7}{5}}}{7} - u^4 + c$

(f) $\dfrac{4x\sqrt{x}}{3} - 3\sqrt{x} + c$

(g) $-3\cos\theta + 4\sin\theta + c$

(h) $t^3 + 2\cos t + c$

(i) $\dfrac{4x^2\sqrt{x}}{5} - \dfrac{10x\sqrt{x}}{3} + c$

(j) $3\sin\theta - 2\tan\theta + c$

(k) $\dfrac{1}{3}e^{3t-1} + c$

(l) $2\ln|t| + c$

(m) $\dfrac{1}{6}\ln(3t^2 + 5) + c$

(n) $e^{\sin\theta} + c$

(o) $\dfrac{(2x + 3)^3}{6} + c$

**2.** (a) $-\dfrac{5x^4}{4} + \dfrac{2x^3}{3} + cx + k$

(b) $-\dfrac{x^5}{5} + \dfrac{x^4}{4} + \dfrac{x^2}{2} + 2x - \dfrac{11}{20}$

(c) $\dfrac{4t^3}{3} + \sin t + ct + k$

(d) $3x^4 - 4x^2 + 7x + 3$

(e) $2\sin\theta + \dfrac{1}{2}\cos 2\theta + c$

**3.** (a) $\dfrac{(3x^2 + 7)^6}{36} + c$

(b) $-\dfrac{1}{18(3x^2 + 5)^3} + c$

(c) $\dfrac{8\sqrt[4]{(5x^3 + 2)^5}}{75} + c$

(d) $\dfrac{(2\sqrt{x} + 3)^6}{6} + c$

(e) $\dfrac{\sqrt{(2t^3 - 7)^3}}{9} + c$

(f) $-\dfrac{(2x + 3)^6}{18x^6} + c$

(g) $-\dfrac{\cos(7x - 3)}{7} + c$

(h) $-\dfrac{1}{2}\ln(\cos(2\theta - 1) + 3) + c$

(i) $\dfrac{1}{5}\tan(5\theta - 2) + c$

(j) $\dfrac{1}{\pi}\sin(\pi x + 3) + c$

(k) $\dfrac{1}{2}e^{x^2+1} + c$

(l) $\dfrac{1}{3}e^{2t\sqrt{t}} + c$

(m) $\dfrac{2}{3}(\ln\theta)^3 + c$

(n) $\ln|\ln 2z| + c$

**4.** (a) $-\dfrac{1}{15}\sqrt{(3 - 5t^2)^3} + c$

(b) $\dfrac{1}{3}\tan\theta^3 + c$

(c) $-\cos\sqrt{t} + c$

(d) $\dfrac{1}{12}\tan^6 2t + c$

(e) $2\ln(\sqrt{x} + 2) + c$

(f) $\dfrac{1}{2}\ln(x^2 + 6x + 7) + c$

(g) $-\dfrac{k^3}{2a^4}\sqrt{a^2 - a^4x^4} + c = -\dfrac{k^3}{2|a|^3}\sqrt{1 - a^2x^4} + c$

(h) $\dfrac{2}{5}(3x^2 - x - 2)\sqrt{x - 1} + c$

(i) $-\dfrac{2}{3}\sqrt{(1 + \cos\theta)^3} + c$

(j) $\dfrac{2}{105}(15t^3 - 3t^2 - 4t - 8)\sqrt{1 - t} + c$

(k) $\dfrac{1}{15}(3r^2 + 2r - 13)\sqrt{2r - 1} + c$

(l) $\dfrac{1}{2}\ln(e^{x^2} + e^{-x^2}) + c$

(m) $\dfrac{2}{15}(3t^2 + 20t + 230)\sqrt{t - 5} + c$

**5.** (a) $C(x) = 0.0002x^3 - 0.01x^2 - 10x + 2500$

(b) 

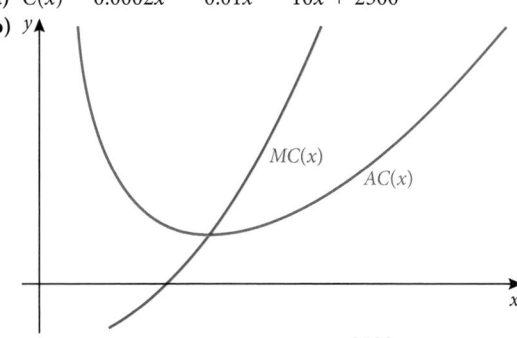

$AC(x) = 0.0002x^2 - 0.01x - 10 + \dfrac{2500}{x}$

(c) $x \approx 193.\ AC(193) \approx 8.5$

(d) $MC(193) \approx 8.5$

(e) $AC(400) \approx 24.25$, They should sell it for €26.25.

**6.** (a) $P(t) = 16t^{\frac{5}{2}}$    (b) 249 tons    (c) 3.6 years

**7.** (a) $f(t) = 60e^{0.2t} - 56$    (b) $f(30) \approx 24150$

**8.** $C(t) = \int 0.22e^{0.01t}\,dt = 22e^{0.01t} + c;\ C(0) = 0 \Rightarrow c = -22$

$22e^{0.01t} - 22 = 156 \Rightarrow t \approx 209$ years, i.e., in year 2222.

**9.** $C(t) = \int 12e^{0.015t}\,dt = 800e^{0.015t} + c;\ C(0) = 0 \Rightarrow c = -800$

$800e^{0.015t} - 800 = 533 \Rightarrow t \approx 34$ years, i.e., in year 2051.

**10.** (a) $\overline{C}(x) = \int -\dfrac{1000}{x^2}\,dx = \dfrac{1000}{x} + c$, with $\overline{C}(100) = 25$,

$c = 15$

$C(x) = x \times \overline{C}(x) = 15x + 1000$

(b) $C(0) = 1000$ SF

**11.** $f(t) = 2.42 + 0.42t + 0.0531t^2 - 0.00249t^3$

$f(2020) = 12.18\%$

**12.** (a) Quantity produced $= \int\left(\dfrac{100}{t + 1} + 5\right)dt$

$= 100\ln(t + 1) + 5t + c$, with initial production at 0,

$c = 0$

(b) approximately 404 thousand barrels

**13.** (a) (i) $C(x) = 12x + 500\ln(x + 1) + 2000$

(ii) $C(1000) = 17.45$

(b) $R(x) = 40x - 0.01x^2 + 200\ln(x + 1)$

(c) $P(x) = 28x - 0.01x^2 - 300\ln(x + 1) - 2000$

1389 pairs and 15428 Euros profit.

## Exercise 16.2

**1.** (a) 24

(b) 40

(c) $\dfrac{24}{25}$

(d) 0

(e) $\dfrac{176\sqrt{7} - 44}{5}$

(f) 0

(g) 2

(h) $-268$

(i) $\dfrac{64}{3}$

(j) 2

(k) $\ln\left(\dfrac{11}{3}\right)$

(l) $\dfrac{44}{3} - 8\sqrt{3}$

(m) 3

(n) $\sqrt{\pi} + 1$

(o) 6

(p) 6

(q) 12

(r) 1

(s) 4

(t) 0

**2.** (a) $\dfrac{14\sqrt{17} + 2}{3}$

(b) $\dfrac{1}{\pi}$

(c) $\ln(2)$

(d) $16\sqrt{2} - 5\sqrt{5}$

(e) $\sqrt{14} - \sqrt{10}$

(f) $\dfrac{3}{2}$

(g) $-\dfrac{1}{2}\ln\left(\dfrac{37}{52}\right)$

(h) $\dfrac{2}{3}$

(i) 0

(j) $-4$

(k) $\dfrac{1}{6}$

(l) $\dfrac{e - 1}{2}$

(m) $1 + \dfrac{e}{2}$

(n) $2\cos(1) + 2$

# Answers

3. (a) $\dfrac{31}{5}$    (b) $\dfrac{2}{\pi}$    (c) $\dfrac{12-4\sqrt{3}}{\pi}$    (d) $\dfrac{e^8-1}{8e^8}$

4. (a) $\dfrac{\sin x}{x}$    (b) $-\dfrac{\sin x}{x}$    (c) $\dfrac{\cos t}{1+t^2}$

5. (a) $\dfrac{1}{3}\ln\left(\dfrac{3k+2}{2}\right)$    (b) $k=\dfrac{2(e^3-1)}{3}$

6. Substitute $u=1-x$

7. (a) $-(1-x)^{k+1}\left(\dfrac{1}{k+1}+\dfrac{x-1}{k+2}\right)$

  (b) $\dfrac{1}{(k+1)(k+2)}$

8. (a) 0    (b) $\sqrt{47}$    (c) $\dfrac{15\sqrt{47}}{47}$

9. $f(x)=\ln x$

10. (a) $\dfrac{1}{\sqrt{2\pi}}\displaystyle\int_1^3 e^{-\frac{x^2}{2}}\,dx\approx0.16$    (b) $\dfrac{1}{\sqrt{2\pi}}\displaystyle\int_0^2 e^{-\frac{x^2}{2}}\,dx\approx0.477$

11. 821

12. 29

13. $\Delta x=2m\Rightarrow$
$\text{area}\approx\dfrac{2}{2}(0+2(6.2+7.2+6.8+5.6+5+4.8+4.8)+0)$
$=80.8\,\text{m}^2$

14. $\Delta x=\dfrac{1}{2}\Rightarrow$
$\text{distance}\approx\dfrac{1}{2\times2}(0+2(4.67+7.34+8.86+9.73$
$+10.22+10.51+10.67+10.76+10.81)+10.81)$
$=44.49\,\text{m}$

15. (a) $\left(\begin{array}{c}\text{Capital}\\\text{value}\end{array}\right)=\displaystyle\int_0^T r(t)e^{-it}\,dt$

  (b) $\left(\begin{array}{c}\text{Well's}\\\text{capital value}\end{array}\right)=\displaystyle\int_0^{10}240\,000e^{-0.06t}\,dt\approx\$1804753$

16. $\left(\begin{array}{c}\text{Well's}\\\text{capital value}\end{array}\right)=P(T)=\displaystyle\int_0^5 100e^{-0.1t}e^{-0.1t}\,dt$
$=500(1-e^{-1})\approx316$, Yes since this is
more than the price to be paid.

17. Remember what the mean value of a function is and considering $b=12$ and $a=1$, then the mean temperature is modelled by
$\dfrac{1}{11}\displaystyle\int_1^{12}f(x)dx$. We use Trap. method to estimate
$\displaystyle\int_1^{12}f(x)dx$ with $\Delta x=1$
$\displaystyle\int_1^{12}f(x)dx\approx\dfrac{1}{2}(15+2(20+25+31+32.5+32+30$
$+30+28+23+18)+15)$
$=284.5$
Thus, mean temp $=\dfrac{284.5}{11}=25.9°C$. This is the average temperature. There is also the maximum and minimum that are usually reported. Also, some may consider the average to be $\dfrac{284.5}{12}=23.7$ since the 284.5 is the total for 12 months and we average by dividing by 12.

18. The initial gift is the *present value* of the income stream of £20 000 per year. Thus, the amount of the gift should be
Present value $=\displaystyle\int_0^{10}20\,000e^{-0.085t}\,dt\approx£134\,726$

19. (a) Vending machine produces a total
income $=\displaystyle\int_0^5 5000e^{0.04t}\,dt\approx£27\,675$

  (b) The future value of the income stream at 12% compounded continuously is
$FV=e^{0.12\times5}\displaystyle\int_0^5 5000e^{0.04t}e^{-0.12t}\,dt\approx£37\,545$
Total interest $=£37545-£27675=£9870$

## Exercise 16.3

1. (a) $\dfrac{125}{6}$    (b) $\dfrac{9\pi^2}{8}+1$    (c) $4\sqrt{3}$

  (d) $\dfrac{10}{3}$    (e) $\dfrac{8}{21}$    (f) $\dfrac{125}{24}$

  (g) $\dfrac{13}{12}$    (h) $4\pi$    (i) $\dfrac{43}{12}$

  (j) approx. 361.95 (4 points of intersection!)

  (k) $3\ln2-\dfrac{63}{128}$

  (l) $\left(\text{between }-\dfrac{\pi}{6}\text{ and }\dfrac{\pi}{6}\right)\sqrt{3}\ln\left(\dfrac{3}{4}\right)-2\sqrt{3}+4$

  (m) 18    (n) $\dfrac{32}{3}$    (o) $\dfrac{64}{3}$

  (p) 9    (q) $\dfrac{9}{2}$    (r) 19

  (s) $\dfrac{37}{12}$    (t) $\dfrac{1}{2}$    (u) $\dfrac{2\sqrt{2}}{3}$

2. $\dfrac{269}{54}$

3. $\dfrac{e}{2}-1$

4. $\dfrac{16}{3}$

5. 25.36

6. $m=0.973$

7. $\dfrac{37}{12}$

8. Parabola passes through $(17,37)$: $y=0.128x^2\Rightarrow$ area
$=\displaystyle\int_0^{17}0.1280x^2\,dx=209.62\,\text{m}^2$.
Rectangle area $=5\times37$, Triangle area $=\dfrac{1}{2}\times18\times37$
Total area of cross section $=727.62$
Thus volume needed is $727.62\times175=127\,000\,\text{m}^3$

9. (a) $x=100$,
$\left(\begin{array}{c}\text{Consumers'}\\\text{surplus}\end{array}\right)=\displaystyle\int_0^{100}(360-0.03x^2-60)dx=20000$
$\left(\begin{array}{c}\text{Producers'}\\\text{surplus}\end{array}\right)=\displaystyle\int_0^{100}(60-0.006\,x^2)dx=4000$

  (b) $x=119.48$
$\left(\begin{array}{c}\text{Consumers'}\\\text{surplus}\end{array}\right)=\displaystyle\int_0^{119.48}(300e^{-0.01x}-90.83)dx$
$\approx10\,065$
$\left(\begin{array}{c}\text{Producers'}\\\text{surplus}\end{array}\right)=\displaystyle\int_0^{119.48}(90.83-100+100e^{-0.02x})dx$
$\approx3446$

  (c) $x=97.63$
$\left(\begin{array}{c}\text{Consumers'}\\\text{surplus}\end{array}\right)=\displaystyle\int_0^{97.63}(400e^{-0.01x}-150.69)dx$
$\approx10\,220$

10. $L=2\displaystyle\int_0^1(x-f(x))dx=2\displaystyle\int_0^1 xdx-2\displaystyle\int_0^1 f(x)dx$
$=1-2\displaystyle\int_0^1 f(x)dx$
It is enough to estimate $\displaystyle\int_0^1 f(x)dx$, multiply by 2 and then subtract from 1.

We will use Trap. method

$$\int_0^1 f(x)dx \approx \frac{0.2}{2}(0 + 2(0.087 + 0.226 + 0.405 + 0.633) + 1)$$

$$= 0.3702$$

$L = 1 - 2 \times 0.3702 = 0.2596$

**11. (a)** 24.7 hours          **(b)** 0.514

**12. (a)**

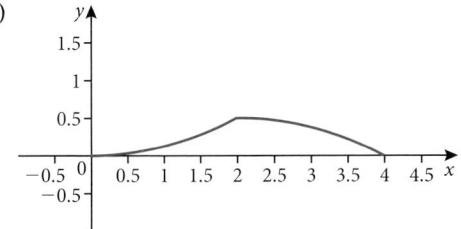

**(b)** $\frac{7}{3}$             **(c)** 134 barrels

**13. (a)** $\int_0^5 ky^2(5 - y)dy = 1$     **(b)** 3, 3.1, 3.3

**(c)** 0.475     **(d)** 1     **(e)** 0.64

**14. (a)** $d(x) = \dfrac{2000}{3x + 50} + 4;$     **(b)** 250 thousand bottles

**(c)** $s(x) = \dfrac{-100}{3x + 25} + 5.4$

**(d)** 484 thousand bottles at £5.33 each.

**15. (a)** 0.75     **(b)** 0.11

**(c)** $\int_0^x \dfrac{2}{(t + 2)^2}dt = \dfrac{x}{x + 2}$ , $\lim\limits_{x \to \infty} \dfrac{x}{x + 2} = 1$

## Exercise 16.4

**1. (a)** $\dfrac{127\pi}{27}$    **(b)** $\dfrac{64\sqrt{2}\,\pi}{15}$    **(c)** $\dfrac{70\pi}{3}$

**(d)** $6\pi$    **(e)** $9\pi$    **(f)** $2\pi$

**(g)** $\left(\dfrac{\sqrt{3}}{2} + 1\right)\pi$    **(h)** $\dfrac{512\pi}{15}$    **(i)** $\approx 5.937\pi$

**(j)** $\dfrac{32\pi}{3}$    **(k)** $\dfrac{23\pi}{210}$    **(l)** $\dfrac{160\pi\sqrt{5}}{3}$

**(m)** $\dfrac{64}{15}\pi$    **(n)** $\dfrac{\pi}{2}$    **(o)** $\dfrac{1778}{5}\pi$

**(p)** $\dfrac{252}{5}\pi$    **(q)** $\dfrac{656\pi}{3}$    **(r)** $\dfrac{9}{8}\pi$

**2. (a)** $\dfrac{88}{15}\pi$          **(b)** $\dfrac{7}{6}\pi$

**3. (a)** $40\pi$          **(b)** $\pi(18 - 9\sqrt{2})$

**(c)** $\dfrac{32}{15}\pi$          **(d)** $\dfrac{4}{5}\pi(121\sqrt{33} - 25\sqrt{15})$

**(e)** $2\pi\left(\ln 2 - \dfrac{1}{4}\right)$       **(f)** $2\pi\left(\dfrac{11}{3}\sqrt{11} - \dfrac{2}{3}\sqrt{2}\right)$

**(g)** $\dfrac{28}{3}\pi(\sqrt{34} - \sqrt{7})$     **(h)** $\pi\left(\dfrac{1}{2}\sqrt{2}\,\pi - \pi + 2\right)$

**(i)** $\dfrac{284}{3}\pi$     **(j)** $2\pi$     **(k)** $\dfrac{256}{15}\pi$

**4.** $V = 2\pi\int_2^{10} xf(x)dx$; we use Trap. Method to estimate the integral

$$\Delta x = \frac{10 - 2}{8} = 1;$$

$$\int_2^{10} xf(x)dx \approx \frac{1}{2}(2 \cdot f(2) + 2(3 \cdot f(3) + 4 \cdot f(4) + \cdots$$

$$+ \ 9 \cdot f(9)) + 10 \cdot f(10))$$

$$= \frac{1}{2}(2 \cdot 0 + 2(3 \cdot 1.5 + 4 \cdot 2 + \cdots \ + 9 \cdot 3) + 10 \cdot 0) \approx 127$$

volume $\approx 2\pi \times 127 \approx 798$ cubic units.

**5. (a)** $V = \left|\dfrac{1}{2}\left(2\pi\int_{-95}^{-68} 0.0549x(x + 95)^2 dx + 2\pi\int_{-68}^{-59} 40xdx\right.\right.$

$$\left.\left. + \ 2\pi\int_{-59}^{-35} -\frac{5}{3}x(x + 35)dx\right)\right|$$

$$\approx 233\,310\,\text{m}^3$$

**(b)** $\dfrac{233310}{8.5} \approx 27449$

**6.** $V = \pi\int_{-4}^4 (25 - y^2 - 9)dy = \dfrac{256\pi}{3} \approx 268.08$ cm

**7.** $V = \pi\int_0^h x^2\,dy = \pi\int_0^h 2py\,dy = \left(\pi py^2\right]_0^h = \pi ph^2$

R is the radius of the cylinder and is on the parabola

$\Rightarrow r^2 = 2ph$ and volume of cylinder $= \pi r^2 h = \pi \cdot 2ph \cdot h$

$$= 2\pi ph^2$$

**8.** Trap. Method: vol $= \dfrac{3}{2}(141 + 2(82 + 35 + 25 + 14 + 5) + 0)$

$$\approx 694.5\,\text{m}^3$$

Cost $= 694.5 \times 9.8 = €6806.1$

**9.** Use cross sections: $V = \pi\int_0^{16} (25 - y - 9)dy$

$$= 128\pi \approx 402\,\text{cm}^3$$

Use cylindrical shells: $V = 2\pi\int_3^5 x(25 - x^2)dx = 128\pi$

## Exercise 16.5

**1. (a)** $\dfrac{70}{3}$ m, 65 m     **(b)** 8.5 m to the left, 8.5 m.

**(c)** 1 m, 1 m     **(d)** 2 m, $2\sqrt{2}$ m.

**(e)** 18 m, 28.67 m     **(f)** $\dfrac{4}{\pi}$ m , $\dfrac{4}{\pi}$ m

**2. (a)** 3t, 6 m, 6 m     **(b)** $t^2 - 4t + 3$, 0, 2.67 m

**(c)** $1 - \cos t, \left(\dfrac{3\pi}{2} + 1\right)$m, $\left(\dfrac{3\pi}{2} + 1\right)$m

**(d)** $4 - 2\sqrt{t + 1}$ , 2.43 m, 2.91 m

**(e)** $3t^2 + \dfrac{1}{2(1 + t)^2} + \dfrac{3}{2}$, 11.3 m, 11.3 m

**3. (a)** $4.9t^2 + 50t + 10$     **(b)** $16t^2 - 2t + 1$

**(c)** $\dfrac{1}{\pi} - \dfrac{\cos\pi t}{\pi}$     **(d)** $\ln(t + 2) + \dfrac{1}{2}$

**4. (a)** $e^t + 19t + 4$     **(b)** $4.9t^2 - 3t$

**(c)** $\sin(2t) - 3$     **(d)** $-\cos\left(\dfrac{3t}{\pi}\right)$

**5. (a)** 12; 20     **(b)** $\dfrac{13}{2}; \dfrac{13}{2}$

**(c)** $\dfrac{9}{4} ; \dfrac{11}{4}$     **(d)** $2\sqrt{3} - 6; 6 - 2\sqrt{3}$

**6. (a)** $-\dfrac{10}{3}; \dfrac{17}{3}$    **(b)** $\dfrac{204}{25}$    **(c)** $-6; \dfrac{13}{2}$

**7. (a)** $\dfrac{166}{5}$    **(b)** $\dfrac{166}{5}$    **(c)** $\dfrac{166}{5}$

**8. (a)** $50 - 20t$    **(b)** 1062.5

**9.** 1.0041 s

**10. (a)** 5 s     **(b)** 272.5 m     **(c)** 10 s

**(d)** $-49\,\text{m s}^{-1}$     **(e)** 12.46 s     **(f)** $-73.08\,\text{m s}^{-1}$

**11.** By the Net Change Theorem, the increase in velocity is equal to $\int_0^6 a(t)dt$

We use Trap. Method with $n = 6$ and $\Delta t = \dfrac{6 - 0}{6} = 1$

$$\int_0^6 a(t)dt \approx \frac{1}{2}(0 + 2(0.5 + 2.1 + 5 + 6.5 + 4.8) + 0) \approx 19\,\text{ms}^{-1}$$

# Answers

**12.** With $\theta_0 = 40°$ and $f(t) = \dfrac{1}{\sqrt{1 - (k \sin t)^2}}$, we use $n = 10$,

$\Delta t = \dfrac{\dfrac{\pi}{2} - 0}{10} = \dfrac{\pi}{20}$ and obtain

$T_n = \dfrac{\Delta t}{2}\left(1 + 2[f(1) + \cdots + f(9)] + \dfrac{1}{\sqrt{1 - (0.342 \sin 40)^2}}\right)$

$\approx 1.57$

We multiply by $4\sqrt{\dfrac{1}{9.8}}$ to find the period $T$ of the pendulum to be approx. 2 seconds.

**13.** 242 m

**14.** Car: acc = 4, $v = 4t$, $s = 2t^2$. Truck: $s = 1500 + 13.89t$
$2t^2 = 1500 + 13.89t$; $t \approx 31$ s, $s \approx 1922$ m

**15.** $a(t) = \begin{cases} 4 & t \leqslant 2 \\ 0 & t > 2 \end{cases} \Rightarrow v(t) = \begin{cases} 4t & t \leqslant 2 \\ 8 & t > 2 \end{cases}$

$s(t) = \begin{cases} 2t^2 & t \leqslant 2 \\ 8t - 8 & t > 2 \end{cases} \Rightarrow 8t - 8 = 100 \Rightarrow t = 13.5$ s

**16. (a)** Ball is in.
**(b)** Ball clears the net but is out.

## Chapter 16 Practice questions

**1. (a)** $p = 1$, $q = e$
**(b)** $\dfrac{e}{2} - 1$ square units

**2. (a)** $(0, 2)$, $\left(1 - \dfrac{\ln 2}{2}, e\sqrt{2}\right)$, $(2 - \ln 2, 2)$

**(b)** $V = \pi \displaystyle\int_0^{1 - \frac{\ln 2}{2}} ((2e^x)^2 - 2^2)dx +$

$\pi \displaystyle\int_{1 - \frac{\ln 2}{2}}^{2 - \ln 2} ((e^{2-x})^2 - 2^2)dx$

**3.** $a = 3$

**4. (a)** $y = \dfrac{x}{e}$     **(b)** $\ln x + 1 - 1$
**(c)** $\dfrac{1}{2} \cdot e \cdot 1 - \displaystyle\int_1^e \ln x\, dx$

**5. (a) (i)** 400 m     **(ii)** $v = 100 - 8t$, 60 m s$^{-1}$
**(iii)** 8 s     **(iv)** 1344 m
**(b)** Distance needed 625 m

**6. (a)** See Worked Solutions
**(b)** 2.31
**(c)** $-\pi \cos x - \dfrac{x^2}{2} + c$; 0.944

**7. (a)** $r'(t) = \dfrac{120(1 - t^2)}{(t^2 + 1)^2} = 0 \Rightarrow t = 1$, graphically it is seen that at $t = 1$, production is greatest.
**(b)** $Q(t) = \int r(t)dt = 60 \ln(t^2 + 1) + 3t + c$ and with initial cons. $c = 0$
**(c)** $60 \ln(t^2 + 1) + 3t = 250 \Rightarrow t \approx 6.7$ years.
**(d)** approx... 420 thousand barrels

**8. (a) (i)** See Worked Solutions
**(ii)** $(1.57, 0)$; $(1.1, 0.55)$; $(0, 0)$, $(2, -1.66)$
**(b)** $x = \dfrac{\pi}{2}$
**(c) (i)** See Worked Solutions
**(ii)** $\displaystyle\int_0^{\frac{\pi}{2}} x^2 \cos x\, dx$
**(d)** $\dfrac{\pi^2}{4} - 2 \approx 0.4674$

**9. (a)** $2\pi$
**(b)** range: $\{y | -0.4 < y < 0.4\}$

**(c) (i)** $-3 \sin^3 x + 2 \sin x$
**(ii)** $f' = 0 \Rightarrow \sin x = 0$ or $\sin x = \dfrac{2}{3}$

$\sin x = \dfrac{2}{3} \Rightarrow \cos x = \sqrt{\dfrac{1}{3}}$

**(iii)** $\dfrac{2\sqrt{3}}{9}$

**(d)** $\dfrac{\pi}{2}$

**(e) (i)** $\dfrac{1}{3} \sin^3 x + c$     **(ii)** $\dfrac{1}{3}$

**(f)** $\arccos \dfrac{\sqrt{7}}{3} \approx 0.491$

**10. (a)** $S(t) = 10t + 100e^{-0.1t} - 100$, $\quad 0 \leqslant t \leqslant 24$
**(b)** €50 million
**(c)** 18 to 19 months.

**11. (a)** Using scatter plot, second degree polynomial appears appropriate. The equation by running regression is
$M'(t) = \displaystyle\int_{10}^2 (4009 + 423.2t + 51.2t^2)dt \approx 69316$
An appropriate model since $r^2 = 95\%$. (A cubic model is also possible)
**(b)** $M(t) = \displaystyle\int_2^{10} (4009 + 423.2t + 51.2t^2)dt \approx 69316$

**12. (a)** A cubic model seems to fit best with $r^2 = 99.3\%$.
$y = -0.01214 + 0.6177x - 2.235x^2 + 2.615x^3$
This is an estimate and may have some discrepancy, since it does not pass through $(0, 0)$, nor $(1, 1)$
**(b)** Gini index $= 2\displaystyle\int_0^1 (x - (-0.01214 + 0.6177x - 2.235x^2$
$+ 2.615x^3))dx \approx 0.589$

**13.** $\approx 122$ metres.

**14.** $a = -\dfrac{56}{27}$

**15. (a)** $C'(t) = R'(t) \Rightarrow 2t = 5te^{-0.1t^2} \Rightarrow t \approx 3$ years
**(b)** Area $= \displaystyle\int_0^3 (R'(t) - C'(t))dt = \int_0^3 (5te^{-0.1t^2} - 2t)dt$
$\approx 5.836$
The total profit over the useful life of the game is approximately €5836.

**16. (a)** Drawing scatter plots indicate that the present model is a power function and the new one is a quadratic model. Running regression for both, we get the following, present: $y = x^{2.3}$,
New: $y = 0.4x + 0.6x^2$ both with $r^2 = 1$.
**(b)** Gini index for the present system: 0.394
Gini index for the new system: 0.2
Interpretation: Yes, income will be more equally distributed after the changes.

**17.** 1800 m

**18.** $2a$ by $\dfrac{2}{3}a^2$

**19. (a)** $\ln x + 1 - k$     **(b)** $x > \dfrac{1}{e}$
**(c) (i)** $\ln x = k - 1$   $x = e^{k-1}$
**(ii)** $(e^k, 0)$
**(d)** $\dfrac{e^{2k}}{4}$     **(e)** $y = x - e^k$
**(f)** Verify     **(g)** Common ratio $= e$

**20.** $C(t) = \int 5.3e^{0.01t} dt = 530e^{0.01t} + c$; $C(0) = 0 \Rightarrow c = -530$
$530e^{0.01t} - 530 = 130 \Rightarrow t \approx 22$ years, i.e., in year 2038.

**21.** $\left(\dfrac{\text{Mine's}}{\text{capital value}}\right) = \displaystyle\int_0^{20} 560\,000\sqrt{t}\, e^{-0.05t} dt \approx \$18\,980\,552$

**22. (a)**

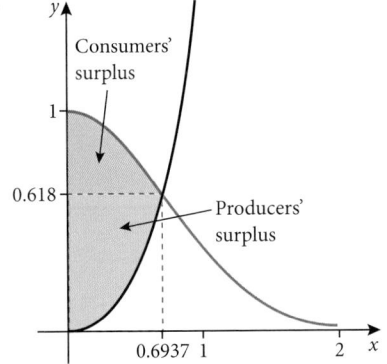

Consumers' surplus

0.618

Producers' surplus

0.6937  1        2    $x$

**(b)** (0.6937, 0.618), i.e, 694 units at 6.90 Euros.

**(c)** Consumer's: $\int_0^{0.6937} (e^{-x^2} - 0.618)dx = 0.168$

Producer's: $\int_0^{0.6937} (0.618 - (e^2 - 1)) = 0.299$

**23. (a)**

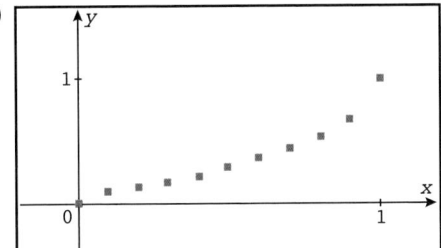

**(b)** A cubic model is a good fit since $r^2 = 0.991$. the model is
$y = 1.858x^3 - 1.931x^2 + 1.040x - 0.005$

**(c)**

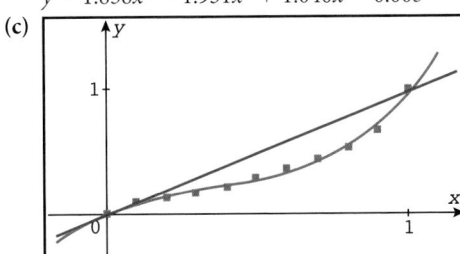

We have to accept slight discrepancies since this is not an exact function.

**(d)** $L = 2\int_0^1 (x - f(x))dx \approx 0.3184$ as shown in GDC output.

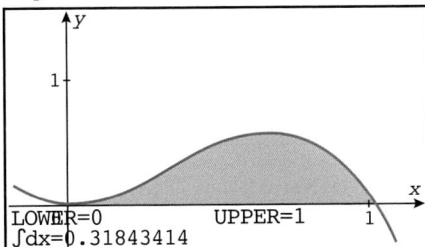

LOWER=0        UPPER=1
∫dx=0.31843414

**24. (a)** $t = 0$, 3, or 6

**(b) (i)** $\int_0^6 \left| t \sin\left(\frac{\pi}{3}\right)t dt \right|$

**(ii)** 11.5 m

**25. (a)** 0.435       **(b)** $\dfrac{-2t}{(2 + t^2)^2}$

**26. (a)** Clothing: $FV = e^{0.04 \times 5} \int_0^5 120\,000 e^{-0.04t}\, dt \approx 664\,208$

Computer: $FV = e^{0.04 \times 5} \int_0^5 100\,000 e^{0.05t} e^{-0.04t}\, dt$
$\approx 626\,227$

The clothing store is the better investment over 5 years.

**(b)** Clothing: $FV = e^{0.04 \times 10} \int_0^{10} 120\,000 e^{-0.04t}\, dt \approx 1475474$

Computer: $FV = e^{0.04 \times 10} \int_0^{10} 100\,000 e^{0.05t} e^{-0.04t}\, dt$
$\approx 1\,568\,966$

The computer store is the better investment over 10 years.

**27.** 6 m

**28.** 0.852

**29. (a)**

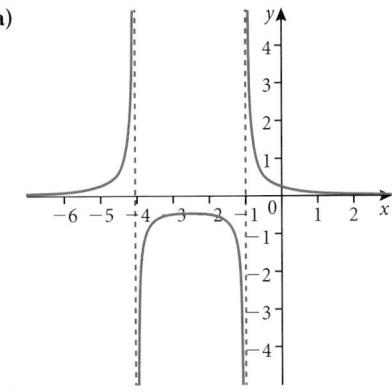

**(b)** 0.157

**(c)** area = 0.313

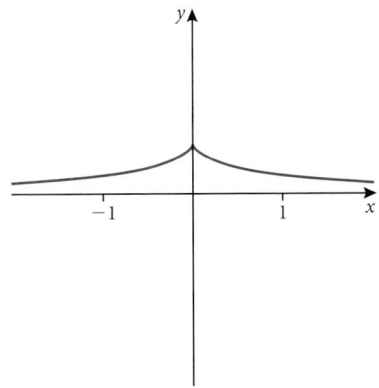

**30. (a)** $\int_0^{12} 0.01 e^{-0.1t}\, dt \approx 0.07$

**(b)** $\int_{12}^{24} 0.01 e^{-0.1t}\, dt \approx 0.021$

**(c)** $\int_0^n 0.01 e^{-0.1t}\, dt \leqslant 0.06 \Rightarrow n = 9$

**31.** $\Delta x = 50$, $n = 12$
$T_n = \dfrac{50}{2}(0 + 2(165 + 192 + \cdots + 215) + 0) \approx 91\,150\, m^2$

**32. (a)** A possible model using GDC:
$y = -0.00083t^3 + 0.0531t^2 + 0.1434t + 58.2689$
$r^2 = 0.99$, which means that 99% of changes in consumption can be explained by this model.

**(b)** $\int_{21}^{31} (-0.000\,83t^3 + 0.0531t^2 + 0.1434t + 58.2689)dt$
$\approx 832.0782$ million barrels

**33. (a)** This is a continuous income flow.

$$FV = e^{0.06 \times 8} \int_0^8 S e^{-0.06t}\, dt = 50000 \Rightarrow S \approx £4870 \text{ per year}$$

**(b)** This is a present value calculation.

$$P(T) = \int_0^8 4870 e^{-0.06t}\, dt \approx 30942 \text{ or equivalently, since}$$

we know the future value:

$50\,000 = P e^{0.06 \times 8} \Rightarrow P \approx 30\,939.17$ the discrepancy is due to our rounding up the answer in (a).

**34.** $\int_0^T \dfrac{4t}{t^2 + 1}\, dt = 16 \Rightarrow T$ is between 54 and 55 years.

**35. (a)**

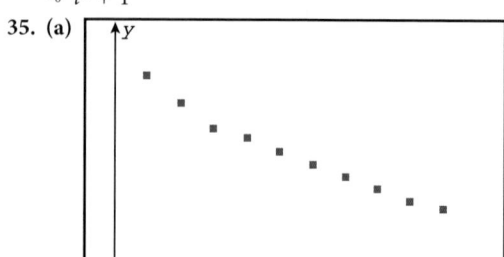

This appears to be an exponential decay. If we run Exponential regression we get: Depreciation $\approx -976.3$ $e^{-0.0998t}$ with $r^2 = 0.996$ indicating a very good fit. (an attempt to fit a cubic model will also have a high $r^2$. However, when it comes to estimating the value of the machine, we observe that its value first decreases but it then increases, which does not fit a realistic situation. A quadratic model is also a good fit, but depreciation starts to increase in absolute value.)

**(b)** Value$(t) \approx \int -976.3 e^{-0.0998t}\, dt = 9782.6 e^{-0.0998t} + C.$
With initial value of 20 000, the model will be:
Value$(t) \approx 9782.6 e^{-0.0998t} + 10217$

**(c)** Maintenance $= \int 200 e^{0.2t}\, dt = 1000 e^{0.2t} + c,$
with initial condition of £1000, $c = 0$
Solve $9782.6 e^{-0.0998t} + 10217 = 1000 e^{0.2t}$ for $t$.
$t = 12.8$ years

## Chapter 17

### Exercise 17.1

**1. (a)** descriptive  **(b)** inferential
**(c)** descriptive  **(d)** descriptive
**2.** Repeated sampling would ensure consistency and reliability.
**3.** The validity of the requirement is difficult to establish. Even if there is a correlation, there would be little causation as the relevance of the mathematical understanding in the nursing program is not evident.
**4. (a)** Retesting would increase the total sample size and may produce results closer to what might have been expected.
**(b)** Parallel testing may address any issues about the way the survey questions were phrased.
**5.** Retesting, at different times of the day and days of the week, will provide a larger sample and better randomisation.
**6.** Although the effectiveness of the prioritisation of traffic direction may be easily measured early in the morning or late at night when there is little traffic, the efficiency of the smart traffic lights should also be retested during peak hours.
**7.** Simply ask the girls whether more is better.

### Exercise 17.2

**1.** Since the unbiased standard deviation is found after division by $n - 1$ instead of , it is the larger one, 4.3.
**2.** $x' = 15$ and $s_{n-1}{}^2 = 7.5$
**3.** $x' = 45.7$ and $s_{n-1}{}^2 \approx 36.2$
**4.** $x' = 500.9$ and $s_{n-1}{}^2 \approx 36.6$
**5.** $s_{n-1}{}^2 \approx 39.9$
**6.**

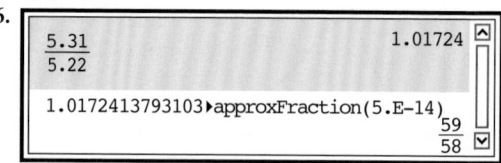

$n = 59$

**7.**

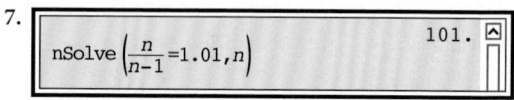

So, when $n \geq 102$, $s_{n-1}{}^2$ will be within 1% of $\sigma^2$.

### Exercise 17.3

**1.** (Answers will vary.) Remember to set a seed for your random number generator on your GDC since two GDCs starting with the same seed will generate the same pseudo-random sequence. Use a command like **RandSeed $n$** (where is a uniquely chosen value).
**2.** $\mu = 57.4$ and $\sigma_{X'}{}^2 = \dfrac{\sigma^2}{n} \Rightarrow 8.01^2 = \dfrac{\sigma^2}{21} \Rightarrow \sigma^2 \approx 1350$
**3.** 0.9
**4.** $6.2^2 = \dfrac{\sigma^2}{10} \Rightarrow \sigma^2 \approx 384$
**5.** $\mu = 737$ and $76.9^2 = \dfrac{\sigma^2}{22} \Rightarrow \sigma^2 \approx 130040$

### Exercise 17.4

**1. (a)** $(-1.48, 1.48)$  **(b)** $(-1.96, 1.96)$  **(c)** $(-2.58, 2.58)$
**2. (a) (i)** $p''' = 0.1$
  **(ii)** $(0.0448, 0.155)$
  **(iii)** 0.0552
**(b)** $(0.0753, 0.125)$ The larger sample reduced the margin of error to 0.025
**3. (a)** $(0.603, 0.697)$
**(b)** $SE_p = \sqrt{\dfrac{0.65(1 - 0.65)}{400}} \approx 0.0238$ and the margin of error $\approx 0.047$
**4. (a)** The maximum product of $p'''(1 - p''')$ is found when $p''' = 0.5$; consequently, the maximum margin of error at a 95% level of confidence with a sample of size $n = 300$
is $1.96 \times \sqrt{\dfrac{0.5 \times 0.5}{300}} \approx 0.0566$
**(b)** Using the **invNorm** feature of your GDC, you will find that 88% of a standard normal distribution will lie within $-1.55 < z < 1.55$.
The margin of error is $1.55 \times \sqrt{\dfrac{0.5 \times 0.5}{300}} \approx 0.0447$
**5.**

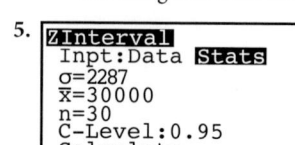

(29182, 30818)

**6.** (43.7, 46.3) The previous mean exceeds the confidence interval, so the changes were effective.

**7.** The 95% confidence interval is (111.8, 118.2). Since his previous average is above the interval, there is notable improvement.

## Chapter 17 Practice questions

**1. (a)** descriptive      **(b)** inferential
    **(c)** inferentia      **(d)** descriptive

**2.** In the absence of laboratory analysis, compare the ingredients of the new dessert to those of the others to determine if their claim is valid in terms of fat, carbohydrate, and protein content. As for 'most delicious', a random sample of consumers would be useful.

**3.** The sample result does translate to 40%; however, to establish reliability, further sampling would be useful. In terms of validity, that "students skipped breakfast" does not translate to "students are from families that do not eat together".

**4. (a)** (Answers will vary.) Re-testing allows other students to be selected, perhaps on a different day, time, or location, and would provide a larger overall sample.
    **(b)** (Answers will vary.) Parallel testing would allow for slight modifications to the wording of the survey which may better define the question posed.

**5.** (Answers will vary.) With the help of the supervisor, a random number generator can be used to select specific dates, times, and a tally of the number of boys and girls using the gym. Retesting on different dates and times would enhance the reliability of the results.

**6.** (Answers will vary.) To eliminate bias, each subject should indicate if there is a preference prior to taking the test, since some students may be able to recognize the taste of a familiar brand.

**7.** $x' \approx 24.8$ and $s_{n-1}{}^2 \approx 14.8$

**8.** $x' \approx 144.5$ and $s_{n-1}{}^2 \approx 42.7$

**9.** $x' = 1.096$ and $s_{n-1}{}^2 \approx 0.0101$

**10.** $s_{n-1} \approx 4.37$

**11.**

$$\boxed{\text{nSolve}\left((2.66)^2=(2.61)^2 \cdot \frac{n}{n-1}, n\right) \qquad 26.8524}$$

Rounded to the nearest integer, $n = 27$

**12.** $\mu = 5.2$ and $\sigma_{X'}{}^2 = \dfrac{\sigma^2}{n} \Rightarrow 0.4^2 = \dfrac{\sigma^2}{20} \Rightarrow \sigma^2 \approx 3.2$

**13.** $\mu = 38.5$ and $\sigma_{X'}{}^2 = \dfrac{\sigma^2}{n} \Rightarrow 5.3^2 = \dfrac{\sigma^2}{20} \Rightarrow \sigma^2 \approx 562$

**14.** $\mu \approx 11.9$ and $\sigma_{X'}{}^2 = \dfrac{\sigma^2}{n} \Rightarrow 0.847^2 = \dfrac{\sigma^2}{20} \Rightarrow \sigma^2 \approx 14.3$

**15. (a)** $(-1.28, 1.28)$   **(b)** $(-1.64, 1.64)$   **(c)** $(-2.24, 2.24)$

**16. (a)** $p''' = 0.18$      **(b)** $(0.117, 0.243)$
    **(c)** If you find the 95% confidence interval $(0.1047, 0.2553)$ first, you can easily find the margin of error $= 0.0753$

**17. (a)** $(0.512, 0.648)$

    **(b)** $SE_p = \sqrt{\dfrac{p'''(1 - p''')}{n}} \approx 0.0349$
    margin of error $\approx 0.0684$

**18. (a)** The maximum margin of error at a 95% level of confidence with a sample of size
$$n = 100 \text{ is } 1.96 \times \sqrt{\frac{0.5 \times 0.5}{100}} \approx 0.098$$
    **(b)** 400

**19.** The 95% confidence interval is approximately (1214, 1266), so the 1300°C temperature is not achieved.

## Chapter 18
### Exercise 18.1

**1. (a)** 19      **(b)** 4      **(c)** 14

**2. (a)** 1.78      **(b)** $-2.55$
    **(c)** $\pm 2.06$      **(d)** $\pm 2.18$

**3. (a)** Yes      **(b)** Yes      **(c)** No
    **(d)** No      **(e)** Yes      **(f)** No

**4.** (112.5, 127.5)

**5.** (109.8, 130.3)

**6.** The value $\approx 0.0173$ which is less than 0.05. The statistics refute the claim.

**7.** The value $\approx 0.568$ which is greater than 0.05. The statistics support the claim.

### Exercise 18.2

**1. (a)** $H_0: \mu \geqslant 128$ (the claim)
       $H_1: \mu < 128$
    **(b)** $H_0: \mu \geqslant 100$
       $H_1: \mu < 100$ (the claim)
    **(c)** $H_0: \mu < 42$
       $H_1: \mu > 42$ (the claim)
    **(d)** $H_0: \mu < 13$ (the claim)
       $H_1: \mu > 13$

**2.** $H_0: \mu \geqslant 48$ (the claim)   Since the $p$-value $\approx 0.00323$, we
   $H_1: \mu < 48$           reject $H_0$ and the claim at the level of significance of $\alpha = 0.05$.

**3.** $H_0: \mu < 0.1$ (the claim)   Since the $p$-value $\approx 0.663$, we do
   $H_1: \mu > 0.1$          not reject $H_0$ and the claim at the level of significance of $\alpha = 0.05$.

**4.** Confidence interval: (6.91, 12.1). The advertised maximum number of pieces is entirely within this interval; hence, the pieces are not smaller than expected.
Alternatively, if the pieces were small, there would be more than 9 pieces of fish, and the hypotheses would have been
$H_0: \mu < 9$ (the claim)
$H_1: \mu > 9$.
The $p$-value $\approx 0.331$, so we would do not reject $H_0$ and the claim at the level of significance of $\alpha = 0.05$.

**5.** $H_0: \mu < 120$ (the claim)   The $p$-value $\approx 0.000131$, so we
   $H_1: \mu > 120$         reject the at the level of significance of $\alpha = 0.05$.

**6.** $H_0: \mu \geqslant 23$ (the claim)   Since the $p$-value $\approx 0.0521$, we do
   $H_1: \mu < 23$          not reject $H_0$ and claim at the level of significance of $\alpha = 0.05$.

### Exercise 18.3

**1. (a)** $H_0: \mu_0 = 158.2$
       $H_1: \mu_1 = 167.2$
    **(b)** $\beta \approx 0.842$
    **(c)** $1 - \beta \approx 0.158$

**2. (a)** $H_0: \mu_0 = 2.4\,M$
       $H_1: \mu_1 = 3.0\,M$
    **(b)** $\beta \approx 1.31 \times 10^{-12}$

**3. (a)** 0.05      **(b)** $\beta \approx 0.878$

**4.** $\beta \approx 0.206$

# Answers

## Exercise 18.4

**1. (a)** $H_0: d < 0$ (claim: 'before' minus 'after' is positive).
$H_1: d > 0$

**(b)** The $p$-value $\approx 0.0137$. Since the $p$-value is less than $\alpha$, reject $H_0$ and do not reject the claim at the level of significance of $\alpha = 0.05$.

**2. (a)** $H_0: d \geq 10$ (claim)
$H_1: d < 10$

**(b)** The $p$-value $\approx 0.00871$. Since the $p$-value is less than $\alpha$, reject $H_0$ and reject the claim at the level of significance of $\alpha = 0.05$.

**3. (a)** $H_0: d \geq 0$ (claim: 'before' – 'after' should be negative)
$H_1: d < 0$

**(b)** The $p$-value $\approx 0.0431$. Since the $p$-value is less than $\alpha$, reject $H_0$ and do not reject the claim at the level of significance of $\alpha = 0.05$.

**4. (a)** $H_0: d \geq 0$ (claim: 'before' – 'after' should be negative)
$H_1: d < 0$

**(b)** The $p$-value $\approx 0.0292$. Since the $p$-value is less than $\alpha$, reject $H_0$ and do not reject the claim at the level of significance of $\alpha = 0.05$.

**5.** $H_0: \mu_{girls} < \mu_{boys}$ (claim)
$H_1: \mu_{girls} > \mu_{boys}$
The $p$-value $\approx 0.222$. Since the $p$-value is greater than $\alpha$, we do not reject $H_0$ but reject the claim at the level of significance of $\alpha = 0.05$.

**6.** $H_0: \mu_1 = \mu_2$ (claim)
$H_1: \mu_1 \neq \mu_2$
The $p$-value $\approx 0.0310$. Since the $p$-value is less than $\alpha$, reject $H_0$ and reject the claim at the level of significance of $\alpha = 0.05$.

## Exercise 18.5

**1.**

| Outcome | 1 | 2 | 3 | 4 | 5 | 6 | Total frequency |
|---|---|---|---|---|---|---|---|
| Observed | 7 | 10 | 12 | 14 | 8 | 9 | 60 |
| Expected | 10 | 10 | 10 | 10 | 10 | 10 | 60 |

$H_0$: The observed outcomes fit the intended model.
$H_1$: The observed outcomes do not fit the intended model.
The $p$-value $\approx 0.639$. Since the $p$-value is greater than $\alpha$, do not reject $H_0$ at the level of significance of $\alpha = 0.05$. (The data support the notion that the die is fair.)

**2.**

| Outcome | A | B | C | D | Total frequency |
|---|---|---|---|---|---|
| Observed | 13 | 7 | 9 | 11 | 40 |
| Expected | 10 | 10 | 10 | 10 | 40 |

$H_0$: The observed outcomes fit the intended model.
$H_1$: The observed outcomes do not fit the intended model.
The $p$-value $\approx 0.572$. Since the $p$-value is greater than $\alpha$, do not reject $H_0$ at the level of significance of $\alpha = 0.05$. (The data support the intended probability distribution.)

**3.**

| Outcome | 1 | 2 | 3 | 4 | 5 | 6 | Total frequency |
|---|---|---|---|---|---|---|---|
| Observed | 27 | 15 | 16 | 18 | 13 | 31 | 120 |
| Expected | 24 | 12 | 12 | 12 | 12 | 48 | 120 |

$H_0$: The observed outcomes fit the intended model.
$H_1$: The observed outcomes do not fit the intended model.

The $p$-value $\approx 0.0413$. Since the $p$-value is less than $\alpha$, reject $H_0$ at the level of significance of $\alpha = 0.05$. (The data do not support the intended probability model.)

**4.**

| Grade | A | B+ | B | B− | C | D | E | Total |
|---|---|---|---|---|---|---|---|---|
| Observed | 3 | 4 | 5 | 9 | 10 | 12 | 7 | 50 |
| Expected | 5 | 5 | 5 | 10 | 10 | 10 | 5 | 50 |

$H_0$: The observed outcomes fit the intended model.
$H_1$: The observed outcomes do not fit the intended model.
The $p$-value $\approx 0.890$. Since the $p$-value is greater than $\alpha$, do not reject $H_0$ at the level of significance of $\alpha = 0.05$.

**5.** The expected values are found using the binomial probability feature on your GDC:
$160 \times \text{binomPdf}(5, 0.5) = \{5, 25, 50, 50, 25, 5\}$

| Heads | 0 | 1 | 2 | 3 | 4 | 5 | Total frequency |
|---|---|---|---|---|---|---|---|
| Observed | 2 | 18 | 41 | 62 | 28 | 9 | 160 |
| Expected | 5 | 25 | 50 | 50 | 25 | 5 | 160 |

$H_0$: The observed outcomes fit the intended model.
$H_1$: The observed outcomes do not fit the intended model.
The $p$-value $\approx 0.0373$. Since the $p$-value is less than $\alpha$, reject $H_0$ at the level of significance of $\alpha = 0.05$. (The observations do not fit the binomial model of a fair die tossed 5 times.)

**6. (a)**

| Day | Mon | Tue | Wed | Thu | Fri | Total absences |
|---|---|---|---|---|---|---|
| Observed | 24 | 14 | 12 | 14 | 26 | 90 |
| Expected | 18 | 18 | 18 | 18 | 18 | 90 |

$H_0$: The observed outcomes fit the intended model.
$H_1$: The observed outcomes do not fit the intended model.
The $p$-value $\approx 0.0533$. Since the $p$-value is greater than $\alpha$, do not reject $H_0$ at the level of significance of $\alpha = 0.05$. (There is insufficient evidence that absenteeism is dependent on the day of the week.)

**(b)**

| Day | Mon/Fri | Tue | Wed | Thu | Total absences |
|---|---|---|---|---|---|
| Observed | 24 | 14 | 12 | 14 | 90 |
| Expected | 18 | 18 | 18 | 18 | 90 |

$H_0$: The observed outcomes fit the intended model.
$H_1$: The observed outcomes do not fit the intended model.
The $p$-value $\approx 0.0265$. Since the $p$-value is less than $\alpha$, reject $H_0$ at the level of significance of $\alpha = 0.05$. (Friday and Monday absences combined are statistically higher than during the rest of the week.)

## Exercise 18.6

**1. (a)** $H_0$: There is no difference in the preferences of the 9th and 10th graders. (claim)
$H_1$: There is a difference that cannot be attributed to chance alone.

**(b)** 3

**(c)** The $p$-value $\approx 0.319$. Since the $p$-value is greater than $\alpha$, we do not reject $H_0$ at the level of significance of $\alpha = 0.05$.

**2. (a)** $H_0$: There is no difference in the marks between last year's and this year's class. (claim)
$H_1$: There is a difference that cannot be attributed to chance alone.
**(b)** 5
**(c)** $\chi^2 \approx 3.55$
**(d)** The p-value $\approx 0.616$. Since the p-value is greater than $\alpha$, do not reject $H_0$ at the level of significance of $\alpha = 0.05$. The achievement level distributions and the groups of students are independent.

**3. (a)** $H_0$: There is no difference between the voting preferences between areas.
$H_1$: There is a difference that cannot be attributed to chance alone. (claim)
**(b)** 4
**(c)** p-value $\approx 0.00043$. Since the p-value is less than $\alpha$, reject $H_0$ at the level of significance of $\alpha = 0.05$. The voting preferences are not independent of the ridings.

**4. (a)** $H_0$: There is no difference in the posting distributions between the groups. (claim)
$H_1$: There is a difference that cannot be attributed to chance alone.
**(b)** The expected values, shown below to 1 decimal place, contain one value that is less than 5.
$$\begin{bmatrix} 4.6 & 18.3 & 23.5 & 26.7 & 36.5 & 28.0 & 10.4 \\ 2.4 & 9.7 & 12.5 & 14.3 & 19.5 & 15.0 & 5.6 \end{bmatrix}$$
The values in the first and second columns will be combined. The observed matrix will become
$$\begin{bmatrix} 18 & 20 & 23 & 40 & 35 & 12 \\ 17 & 16 & 18 & 16 & 8 & 4 \end{bmatrix}$$ and the expected values to 1 decimal place will become
$$\begin{bmatrix} 22.8 & 23.5 & 26.7 & 36.5 & 28.0 & 10.4 \\ 12.2 & 12.5 & 14.3 & 19.5 & 15.0 & 5.6 \end{bmatrix}.$$
So $v = 5$.
**(c)** The p-value $\approx 0.0285$. Since the p-value is less than $\alpha$, reject $H_0$ at the level of significance of $\alpha = 0.05$. The distributions of postings by the groups are not independent.

**5. (a)** $H_0$: There is no difference in the colour preferences between the groups. (claim)
$H_1$: There is a difference that cannot be attributed to chance alone.
**(b)** $v = 12$
**(c)** The p-value $\approx 0.9$. Since the p-value is greater than $\alpha$, do not reject $H_0$ at the level of significance of $\alpha = 0.05$. The colour preferences are independent of the groups.

**6.** The expected values include a few whose values are less than 5, notably in columns 1, 6, and 7.

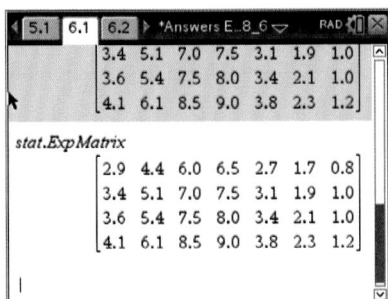

The first two columns as well as the last three columns will be combined so that the $\chi^2$ onditions are met.
The new matrix of observed values is
$$\begin{bmatrix} 10 & 3 & 5 & 7 \\ 6 & 7 & 9 & 7 \\ 9 & 10 & 7 & 5 \\ 10 & 9 & 10 & 6 \end{bmatrix}$$
and the expected values will become
$$\begin{bmatrix} 7.3 & 6.0 & 6.5 & 5.2 \\ 8.5 & 7.0 & 7.5 & 6.0 \\ 9.0 & 7.5 & 8.0 & 6.5 \\ 10.2 & 8.5 & 9.0 & 7.3 \end{bmatrix}.$$
The value $\approx 0.7$ and $v = 9$. Since the value is greater than $\alpha$, do not reject $H_0$ at the level of significance of $\alpha = 0.05$. The mark distributions are independent of the groups.

**7.** The p-value $\approx 0.2$ and $v = 2$. Since the p-value is greater than $\alpha$, do not reject $H_0$. The league outcomes are independent of the teams.

**8.**

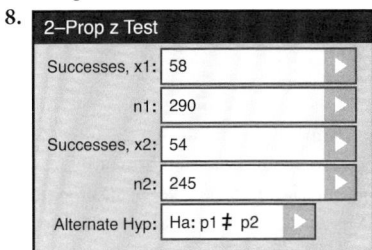

The p-value $\approx 0.6$, we do not reject the null hypothesis. There is no difference to the proportion of wins.

## Chapter 18 Practice questions

**1. (a)** 17 **(b)** 5 **(c)** 8
**2. (a)** 1.71 **(b)** $-1.71$
**(c)** $\pm 2.10$ **(d)** $\pm 2.90$
**3. (a)** No **(b)** Yes
**(c)** Yes **(d)** No
**4.** (59.5, 68.5)
**5.** Since the 95% confidence interval (94.8, 105) does not contain any value equal to or greater than 110, we reject the claim.
**6.** The 95% confidence interval (12.5, 27.5) does not contain the average of 30 minutes. The statistics do not support the average given.
**7. (a)** $H_0: \mu = 62.4$ (the claim)
$H_1: \mu \neq 62.4$
**(b)** $H_0: \mu \geq 13$ (the claim)
$H_1: \mu < 13$
**(c)** $H_0: \mu < 72.3$
$H_1: \mu > 72.3$ (the claim)
**(d)** $H_0: \mu \geq 102$
$H_1: \mu < 102$ (the claim)
**8.** $H_0: \mu = 28.6$ (the claim)
$H_1: \mu \neq 28.6$
The p-value $\approx 0.0514$, so we do not reject the null hypothesis and claim at the level of significance of $\alpha = 0.05$. The average American BMI is 28.6.
**9.** $H_0: \mu = 79° C$ (the claim)
$H_1: \mu \neq 79° C$
The p-value $\approx 0.0000952$, so we reject the null hypothesis and the claim at the level of significance of $\alpha = 0.05$. The coffee is not being served at $79° C$

# Answers

10. $H_0: \mu < 20$ minutes

    $H_1: \mu > 20$ minutes (the claim)

    The $p$-value $\approx 0.148$, so we do not reject the null hypothesis, but we reject the claim at the level of significance of $\alpha = 0.05$. The mean wait times are not in excess of 20 minutes.

11. $H_0: \mu = 2A$ (the claim)

    $H_1: \mu \neq 2A$

    The $p$-value $\approx 0.00000196$, so we reject the null hypothesis and claim at the level of significance of $\alpha = 0.05$. The fuses are not manufactured to specification.

    ```
    invT(0.05,14)
                        -1.761310111
    ```

    ```
    T-Test
      μ<204
      t=-1.549193338
      p=0.07182
      x̄=200
      Sx=10
      n=15
    ```

12. $H_0: \mu = 204°\,C$

    $H_1: \mu < 204°\,C$ (the claim)

    The $p$-value $\approx 0.0718$, so we do not reject the null hypothesis, but reject the claim at the level of significance of $\alpha = 0.05$. The ovens are reaching the temperature set.

13. (a) $H_0: \mu_0 = 1.2$ years

    $H_1: \mu_1 = 1.6$ years

    (b) $\alpha = 0.05$ and $\beta \approx 0.764$

    (c) $1 - \beta \approx 0.236$

14. (a) $H_0: \mu_0 = 0.045$

    $H_1: \mu_1 = 0.054$

    (b) $\alpha = 0.05$ and $\beta \approx 0.000002$

    (c) $1 - \beta \approx 1.0$

15. (a) $H_0: d < 0$

    $H_1: d > 0$ (claim: 'before' minus 'after' is positive)

    (b) The $p$-value $\approx 0.589$. Since the value is greater than $\alpha$, do not reject $H_0$ and reject the claim at the level of significance of $\alpha = 0.05$. The training did not reduce the reaction times.

16. (a) $H_0: d < 6$

    $H_1: d > 6$ (claim)

    'lowers by more than 6' means 'before' minus 'after' is more than 6

    (b) The $p$-value $\approx 0.708$. Since the $p$-value is greater than $\alpha$, do not reject $H_0$ and reject the claim. The drug does not lower blood pressure by more than 6mm Hg.

17. (a) $H_0: d \geqslant 0$

    $H_1: d < 0$ (claim)

    'improves their typing speed' means 'without' minus 'with' is negative

    (b) The $p$-value $\approx 0.544$. Since the $p$-value is greater than $\alpha$, do not reject $H_0$ and reject the claim at the level of significance of $\alpha = 0.05$. There is no improvement to the typing speed.

18. The $p$-value $\approx 0.126$. Since the $p$-value is greater than $\alpha$, there is no difference to the mean battery lives of the laptop computers.

19. The claim is that the mean salary of men is greater than that of women, $\mu_1 > \mu_2$.

    $H_0: \mu_1 < \mu_2$

    $H_1: \mu_1 > \mu_2$ (claim)

Since the $p$-value $\approx 1.00$, we do not reject the null hypothesis, and reject the claim. The statistics do not support the claim that men's salaries are significantly higher than women's at the level of significance of $\alpha = 0.05$.

20. To one decimal, the expected values are found on your GDC as

    ```
    100·binomPdf(4,0.25)
        {31.6,42.2,21.1,4.7,0.4}
    ```

    Since the last two values are each less than 5, they will be combined.

    ```
    100-100·binomCdf(4,0.25,0,2)   5.1
    ```

    Caution: Do not re-enter these expected values to one decimal by hand, but simply store the original expected values into a list and combine the last two entries as you will for the observed values; alternatively, enter the first three in the form $100 \cdot \text{binompdf}(4, 0.25, x)$ with $x \in \{0, 1, 2\}$, then the last cell as $100 -$ sum of those three values.

    The $p$-value $\approx 0.954$ with $v = 3$, so there is no difference between the observed and expected values at the level of significance of $\alpha = 0.05$.

21. If the distribution were uniform, the expected values would each be equal to 16.

    The $p$-value $\approx 0.328$ with $v = 4$, so there is no difference between the observed and expected values, and the distribution can be considered to be uniform at the level of significance of $\alpha = 0.05$.

22. The intended ratio $2:8:2:3:1$ means the expected values are $\{10, 40, 10, 15, 5\}$.

    The $p$-value $\approx 0.768$ with $v = 4$, so there is no difference between the observed and expected values, and the distribution can be considered to match the intended ratio at the level of significance of $\alpha = 0.05$.

23. (The method of solution will vary, since confidence intervals, test of proportions, or a $\chi^2$ test with Yates' Continuity Correction are all possibilities.) Using a test of proportions,

    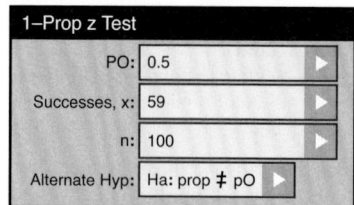

    yields a $p$-value $\approx 0.0719$, so we do not reject the hypothesis that there is no difference between this outcome and what is expected at the level of significance of $\alpha = 0.05$. The coin is fair.

24. First, establish the hypotheses.

    $H_0$: There is no difference between genders in the amount of TV watched. (claim)

    $H_1$: There is a difference that cannot be attributed to chance alone.

    The $p$-value $\approx 0.109$, so we do not reject the null hypothesis at the level of significance of $\alpha = 0.05$. Gender and hours of TV watched are independent.

**25. (a)** $H_0$: There is no difference between the groups in their option of uniforms. (claim)

$H_1$: There is a difference that cannot be attributed to chance alone.

The $p$-value $\approx 0.0844$, so we do not reject the null hypothesis at the level of significance of $\alpha = 0.05$. Whether or not the school uniform should be eliminated is independent of the groups.

**(b)** There is a notable difference in the response from MYP parents compared to the rest. Isolating that group from the others may provide a good statistical analysis of the difference. [Note that isolating one group from the rest combined will result in a $2 \times 2$ contingency table which requires the use of Yates' Continuity Correction. A test of two proportions may be an alternative.]

**26. (a)** The hypotheses:

$H_1$: here is no difference in opinion regarding tax reform between voting regions. (claim)

$H_1$: There is a difference that cannot be attributed to chance alone.

The $\chi^2$ statistics is approximately 12.3 and the $p$-value $\approx 0.0547$, so we do not reject the null hypothesis at the level of significance of $\alpha = 0.05$. Opinions on tax reform are independent of the voting regions.

**(b)** When the Metro (core) region is compared to the other regions combined, the observed matrix becomes

$\begin{bmatrix} 71 & 100 & 50 \\ 86 & 62 & 34 \end{bmatrix}$. The $\chi^2$ statistic is approximately 9.71 and the value $\approx 0.00779$, so we can reject the null hypothesis that there is no difference between the Metro (core) region's opinions compared to the others combined at the level of significance of $\alpha = 0.05$.

**27. (a)** The hypotheses are

$H_1$: There is no difference between the brands in their effectiveness. (claim)

$H_1$: There is a difference that cannot be attributed to chance alone.

The $p$-value $\approx 0.498$, so we do not reject the null hypothesis at the level of significance of $\alpha = 0.05$. There is no difference between brands and their effectiveness.

**(b)** When the other two are combined, the observed matrix becomes $\begin{bmatrix} 25 & 49 & 15 \\ 7 & 22 & 11 \end{bmatrix}$

The $p$-value $\approx 0.247$ so we do not reject the null hypothesis that the distributions are same.

**28. (a)** unbiased estimate of $\mu$ is 2.36

unbiased estimate of $\sigma^2$ is 33.65

**(b) (i)** $H_0: \mu = 0$

$H_1: \mu > 0$

**(ii)** $p$-value $= 0.103$

**(iii)** $1.03 > 0.05$. There is insufficient evidence at the 5% level to support the claim that extra tuition improves examination marks.

**29. (a)** $(9.761, 9.825)$

**(b)** If this process is carried out a large number of times, approximately 99% of the intervals will contain $\mu$.

**(c)** 80%

# Chapter 19

## Exercise 19.1

**1 (a)** Precipitation is the explanatory variable; autism prevalence rate is the response variable

**(b)** No; no matter how strong the association is, it does not prove that precipitation causes autism, only that they are associated.

**2 (a)** No association.

**(b)** Strong nonlinear association, possibly quadratic.

**(c)** Nearly perfect negative linear association.

**(d)** Strong positive linear association.

**(e)** Moderate negative linear association.

**(f)** No association.

**(g)** Strong positive nonlinear association, possibly exponential.

**(h)** Mostly strong positive linear association, but a cluster of outliers is a departure from the major pattern.

**(i)** Very strong negative linear association with one outlier.

**3 (a)** The best fit line is approximately $y = -2.6x + 185$

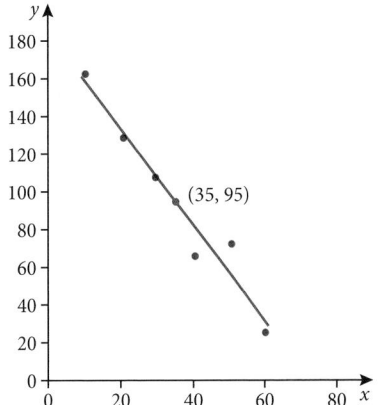

**(b)** The best fit line is approximately $y = 4x + 12$

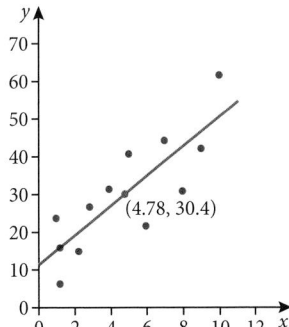

# Answers

**(c)** The best fit line is approximately $y = 2x - 24$

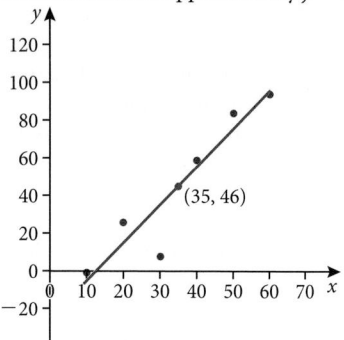

**4 (a)** Number of pages is the explanatory variable, hours to finish is the response variable.

**(b)** Scatter diagram is below. The association is strong, positive, and approximately linear.

**(c)** The best fit line is approximately $t = 0.03n + 2.22$

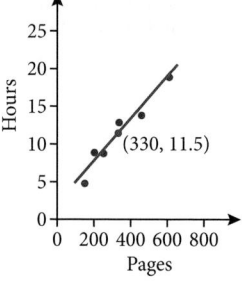

**(d)** The gradient of 0.03 indicates that each additional page adds about 0.03 hours to the reading time, or an additional 100 pages adds 3 hours. The $t$–intercept of 2.22 indicates that a book with zero pages will take 2.22 hours to read: this indicates the extra time it takes to get the book and find the last page and any other task that is not related to the number of pages.

**(e)** About 16.3 hours.

**5 (a)** The explanatory variable is mass.

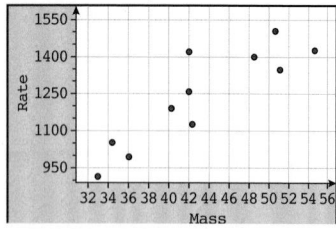

**(b)** $M' = 43.2$, $R' = 1240$

**(c)** A possible best–fit line is shown below.

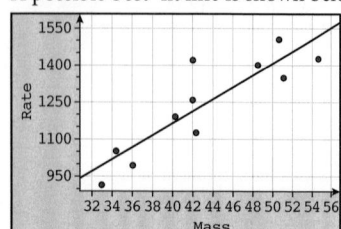

**(d)** There appears to be a strong, positive, linear correlation. The gradient of the line is about 24, which indicates that metaboloic rate increases by about 24 units for each additional kg in mass.

**(e)** About 1160.

**(f)** Kevin's mass is beyond the range of the observed data; we cannot predict his metabolic rate as it would be an extrapolation.

**6 (a)** and **(c)** are correct.

**7 (a)** Number of items produced is the explanatory variable; production cost is the response variable.

**(b)** A scatter diagram is shown below.

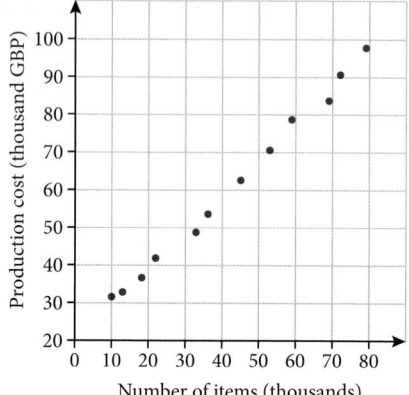

**(c)** Strong, positive, approximately linear, no outliers visible.

**(d)** (42.4, 61.1)

**(e)** As shown below.

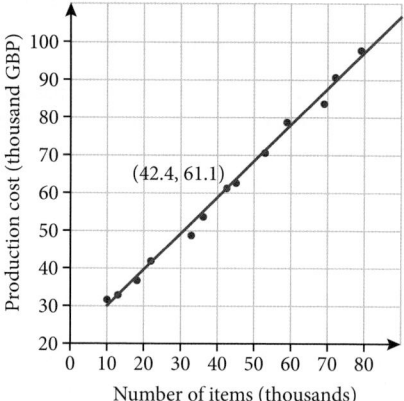

**(f)** 88 thousand GBP

**(g)** 1.0 (approximate); best fit is 0.966. This suggests that that the production cost increases by about 1000 GBP for each additional 1000 items produced.

**(h)** 20 (approximate); this could be interpreted as the fixed costs to keep the factory operating. This is an extrapolation.

**(i)** Predicting production cost for 100 thousand items would be an extrapolation.

**8 (a)** Speed is the explanatory variable; temperature is the response variable.

**(b)** The association is strong, positive, and approximately linear.

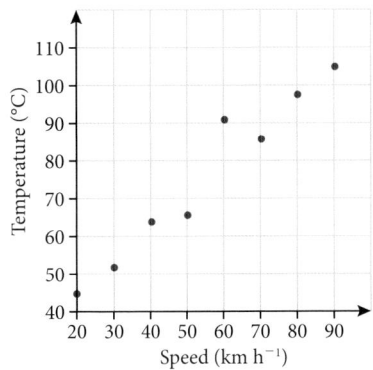

**(c)** (55, 75.9)  **(d)** Shown below.

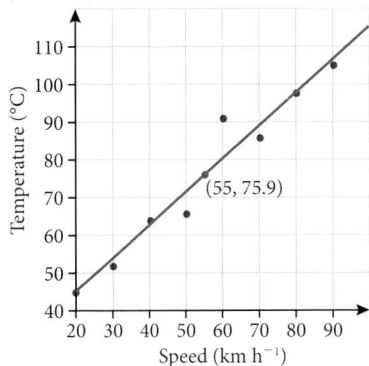

**(e)** About 80 km h$^{-1}$

**(f)** About 0.9 (best fit is 0.882). This indicates that tire temperatures increase by about 0.9°C for each additional km h$^{-1}$.

**(g)** about 27 (best fit is 27.4). This suggests that the initial tire temperature before being driven was about 27°C. This is an extrapolation.

**(h)** A prediction from a speed of 150 km h$^{-1}$ would be an extrapolation.

## Exercise 19.2

**1** Spearman's $r_s$ measures the strength of a **monotonic** (continually increasing or decreasing) relationship. It is not sensitive to outliers. Pearson's $r$ measures the strength of a **linear** relationship, and is sensitive to outliers.

**2** A: $r = -0.40$, B: $r = -0.99$, C: $r = 0.90$, D: $r = -0.03$, E: $r = 0.51$, F: $r = -0.58$, G: $r = -0.95$, H: $r = 0.74$

**3 (a)**

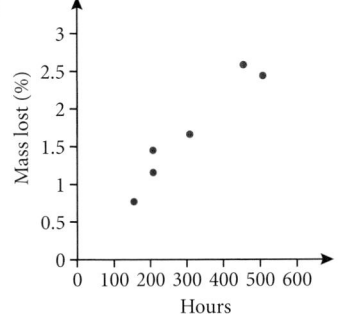

$r = 0.967$

**(b)** There is a strong positive correlation between hours in the acid bath and mass of metal lost.

**4 (a)**

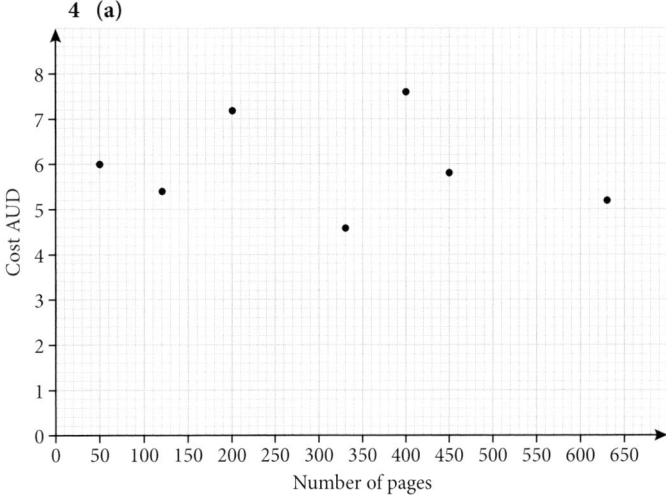

**(b)** $r = -0.141$

**(c)** No, the linear correlation is too weak for reliable predictions, as seen on the scatter plot and shown by the value of $r$.

**5 (a)** $r = 0.561$

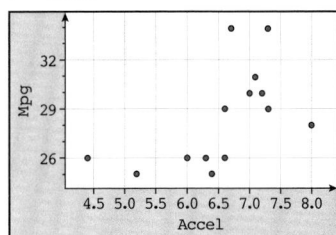

**(b)** There is a moderate positive correlation between acceleration and fuel efficiency; as cars get slower the efficiency appears to increase. It appears to be only somewhat linear, which is supported by the value of $r$.

**(c)** The ranked data is shown below.
$r_s = 0.670$; in general there is a moderate positive rank correlation between acceleration and fuel efficiency in MPG; as acceleration times increase MPG increases as well. Cars that accelerate slower are more efficient in general.

**(d)** Person's $r$ measure the strength of a linear correlation. Since this data appears to be somewhat nonlinear, Spearman's $r_s$ is a more appropriate measure since it measures the strength of the monotonic association.

**6 (a)** $r = -0.598$

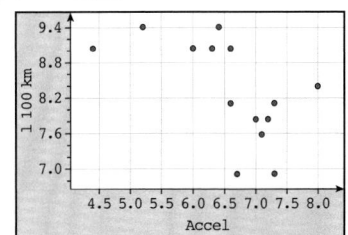

**(b)** There is a moderate negative linear correlation. As cars get slower, the L per 100km required decreases. The linear correlation is slightly stronger than in the

# Answers

previous exercise. By converting to L per 100km, we had to find the reciprocal of the units which may have made the association slightly more linear.

**(c)** The ranked data is shown below.

**(d)** $r_s = -0.670$; in general there is a moderate negative rank correlation between acceleration and fuel efficiency in L per 100 km; as acceleration times increase, L per 100 km decreases. Cars that accelerate slower are more efficient in general. The value of $r_s$ has the opposite sign from the previous exercises as the direction of correlation has reversed.

**7 (a)** $r = 0.0852$

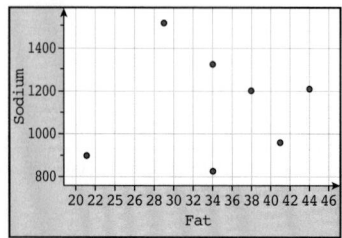

**(b)** There is almost no association between fat and sodium for these menu items.

**8 (a)** $r = 0.940$

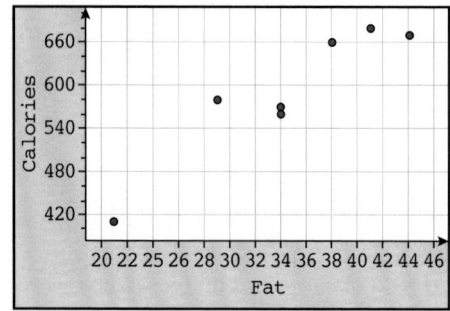

**(b)** There is a very strong positive linear correlation between fat and calorie content for these menu items.

**9 (a)** While *a* has the strongest linear correlation, we should not make any conclusions without looking at a scatter diagram. It could be that the design of the experiment causes outliers in variable *a* which are causing the relatively strong linear correlation.

**(b)** Without knowing more about the nature and design of the experiment, we cannot conclude that changes in *a* **cause** changes in *S*. However, if Ian deliberately manipulated the values of *a* and measured *S*, we may have strong evidence for causality (we would also need to suspect an underlying relationship between the two variables).

**10. (a)** $r = 0.993$

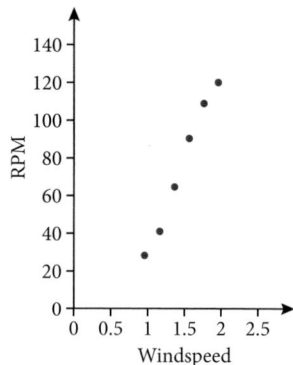

**(b)** There is a strong positive linear correlation between wind speed and RPM.

**(c)** The scatter diagram appears to have a slight curve to it and it is reasonable to suspect that the force applied by wind may be nonlinear. A logistic or power model may be more appropriate.

**11. (a)** The association appears moderate, negative, and approximately linear.

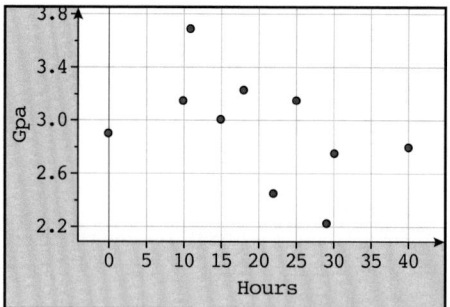

**(b)** $r_s = -0.533$; there is a moderate negative rank correlation between GPA and hours worked.

**(c)** $r = -0.461$; since the form of the association is approximately linear with no strong outliers, $r$ is relatively close to $r_s$.

**12. (a)** The association is generally negative, but is nonlinear with a possible quadratic form.

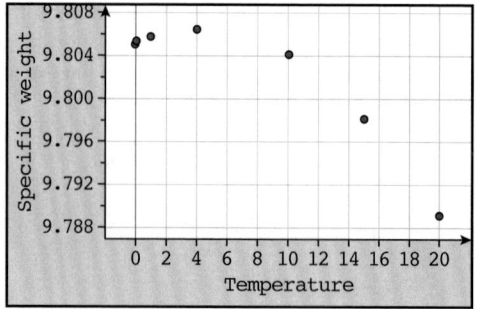

**(b)** $r_s = -0.643$; there is a moderate negative rank correlation between temperature and the specific weight of water.

**(c)** The form is non−monotonic.

### Exercise 19.3

**1** $R^2 = 0.36$

**2** $R^2 = 0.64$

**3** $r = \pm 0.9$

**4** $r = -0.7$

**5** Model B.

**6 (a)** True

**(b)** False: although the models both fit the data equally, we would need to examine the data visually and in context to determine with model is more appropriate.

**7 (a)** $L = 0.00475h + 0.242$. On average, each additional hour increases mass loss by 0.00475%. The $L$ intercept of 0.242 is not meaningful in this context (it is an extrapolation, and it suggests a mass loss of 0.242% for 0 hours).

**(b)** 0.475%

**(c)** $R^2 = 0.935$. The LSRL predicts 93.5% of the variation in mass loss using time in bath as the explanantory variable.

**(d)** 2.14%

**(e)** Although the value of $R^2$ is relatively close to 1, the LSRL we generated should only be used to predict mass loss from hours in the acid bath. Using the model 'in reverse' lowers our confidence in the prediction significantly.

**8 (a)** $E = 1.85t + 16.4$. On average, for each additional second in 0–60 mph time, fuel efficiency increases my 1.85 miles $\text{gal}^{-1}$. The $E$–intercept of 16.4 is not meaningful in this context,; it is an extrapolation and a car would have to have instantaneous acceleration to 60 mph!

**(b)** $R^2 = 0.315$. The LSRL predicts 31.5% of the variation in efficiency using 0–60 mph time as the explanatory variable.

**(c)** 26.9 miles $\text{gal}^{-1}$. Since the LSRL predicts only 31.5% of the variation in efficiency, there are major other factors affecting efficiency.

**(d)** 9 seconds is beyond the range of the observed value of the explanatory variable; any prediction would be an extrapolation.

**9 (a)**

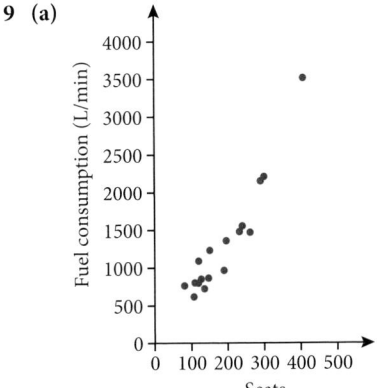

The association between the number of seats and fuel consumption is strong, positive, and approximately linear. There is one possible outlier but it appears to fit the general trend.

**(b)** $F = 7.87n - 126$. The fuel consumption increases by 7.87 L $\text{min}^{-1}$ for each additional seat. The $F$–intercept of −126 is not meaningful in this context.

**(c)** $R^2 = 0.898$. The LSRL predicts 89.8% of the variation in fuel consumption using the number of seat as the explanatory variable.

**(d)** 2630 L $\text{min}^{-1}$

**(e)** The data appears to have a slight nonlinear form in the scatter diagram. It would make sense that fuel consumption would increase in a faster–than–linear rate as the limits of current technology are reached, so a nonlinear model may not be the most appropriate.

**10. (a)** The scatter diagram shows a strong, positive, linear association. There is a possible outlier at (48, 37).

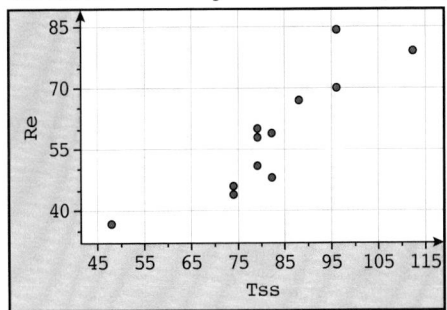

**(b)** $E = 0.815(S) - 8.62$. For each additional unit increase in Training Stress Score, Relative Effort increases by 0.815 units. The $E$–intercept of −8.62 is not meaningful in this context.

**(c)** $R^2 = 0.771$. The LSRL predicts 77.1% of the variation in Relative Effort using the Training Stress Score as the explanatory variable.

**(d)** 40.3

**(e)** $48 \leqslant S \leqslant 112$

**(f)** Since the low outlier is the minimum value in the domain, if we remove the outlier we would need to adjust the domain to $74 \leqslant S \leqslant 112$ and a prediction based on a Relative Effort of 60 would then become an extrapolation.

**(g)** Without (48,37), the LSRL is $E = 0.996(S) - 24.7$ with $R^2 = 0.756$. The LSRL has changed significantly but the value of $R^2$ is only slightly different.

**11. (a)** The scatter diagram shows a strong negative approximately linear association between change in NEA and change in mass.

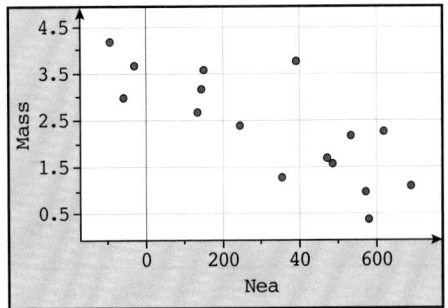

**(b)** $M = -0.00344N + 3.51$. On average, for each additional 100 calories in change of nonexercise activity, change in mass decreases by 0.344 kg. The $M$–intercept suggests that when nonexercise activity does not change, mass will increase by 3.51 kg.

# Answers

(c) $R^2 = 0.606$. The LSRL predicts 60.6% of the variation in change in mass using change in nonexercise activity as the explanatory variable.

(d) 4.20 kg

(e) A negative change in NEA suggests that the individual did less nonexercise activity after being overfed.

(f) $-94 \leqslant N \leqslant 690$

(g) The change in mass for $N = 1100$ would be negative, suggesting that the person was able to lose weight by eating more!

12. (a) Production Rate is the explanatory variable; Defects is the response variable.

(b) There appears to be a weak positive association between production rate and defects. However, an outlier at (300, 40) may be affecting the strength of the association.

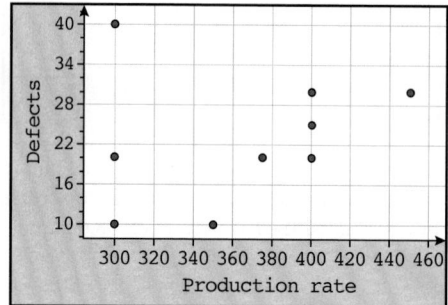

(c) $D = 0.0479R + 5.67$; The number of defects increases by about 5 for each 100 unit increase in production rate. The $D$-intercept of 5.67 is not meaningful in this context.

(d) $R^2 = 0.0872$; the LSRL predicts 8.72% of the variation in the number of defects using production rate as the explanatory variable.

(e) The low value of $R^2$ suggests that there are more significant factors determining the variation in the number of defects. It is better to try to find other factors instead.

(f) There appears to be a moderate positive approximately linear association between production rate and defects.

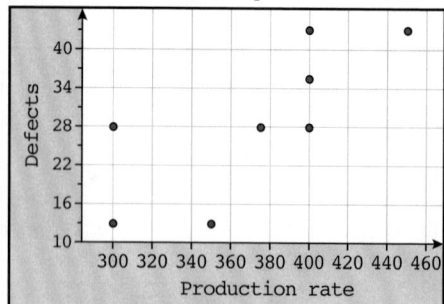

$D = 0.113R - 21.3$; The number of defects increases by about 11 for each 100 unit increase in production rate. The $D$-intercept of $-21.3$ is not meaningful in this context.

$R^2 = 0.630$; the LSRL predicts 63.0% of the variation in the number of defects using production rate as the explanatory variable.

The production rate appears to be a significant predictor of the number of defects. Lower production rates should be attempted and further data can then be recorded.

(g) It is OK to remove the outlier to investigate how influential it is. However, we must be careful to not put too much confidence in our predictions and recommendations using the revised data. Instead, we must investigate the outlier and see what caused it: was it a particularly unskilled worker? Mis–entered data? A power failure during production or other failure? Something else? If there are no unusual causes for the outlier, we should not remove it in our final analysis.

13. (a) There is a strong positive association between year and millions of active monthly users. The form appears nonlinear or piecewise linear.

(b) Suitable domains would be $0 \leqslant y < 5$ and $5 \leqslant y \leqslant 8$.

14. (a) Temperature is the explanatory variable, butterfat is the response variable.

(b) The association appears to be negative and moderate and approximately linear.

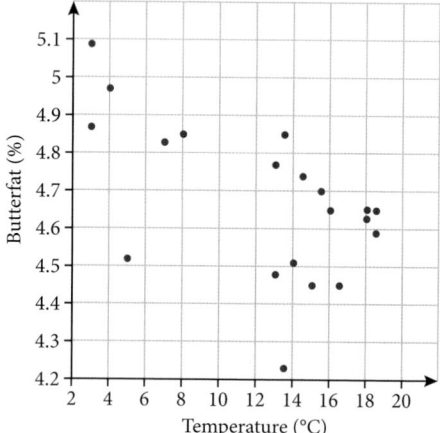

There is a gap in the data; why are there no data values for temperatures in the interval $8 < T < 13$? We should proceed with caution.

(c) The LSRL is $F = -0.0216T + 4.94$. The butterfat content decreases by 0.02% for each increase of 1°C. The $F$-intercept of 4.94 indicates that at a temperature of 0°C, we predict that the butterfat content would be 4.94% − beware, this is an extrapolation.

(d) $R^2 = 0.318$; 31.8% of the variation in butterfat content is predicted by the LSRL using temperature as the explanatory variable.

(e) Correlation is not causation: while temperature and butterfat may have a moderate correlation, we cannot claim that lowering the temperature will cause increased butterfat content. Further research is needed; an investment in a climate–controlled barn may be an expensive experiment.

## Exercise 19.4

1. (a) $\log(y)$ vs $x$; exponential model
   (b) $y$ vs $\log(x)$; logarithmic model
   (c) logarithmic linearization should not be used since data appears to have a turning point
   (d) $\log(y)$ vs $\log(x)$; power model

2. **(a)** A scatter diagram with an appropriate model is shown below.

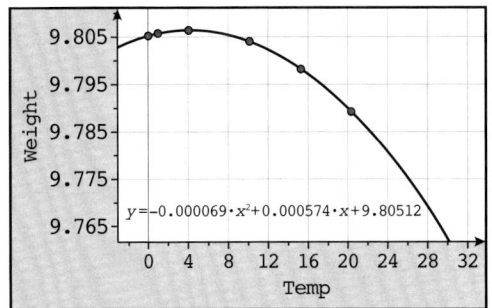

**(b)** The data does not appear to fit an exponential, logarithmic or power model; the data has a turning point, so linearization with logarithms cannot be used.

**(c)** The data appears to have a quadratic form, so we will use quadratic regression. The quadratic model is $w = -0.000069\,T^2 + 0.000574T + 9.80512$.

**(d)** 9.802

**(e)** According to our model, the maximum specific weight is 9.81 at 4.15°C.

3. **(a)** The association is strong, positive, and nonlinear. There are no dramatic outliers.

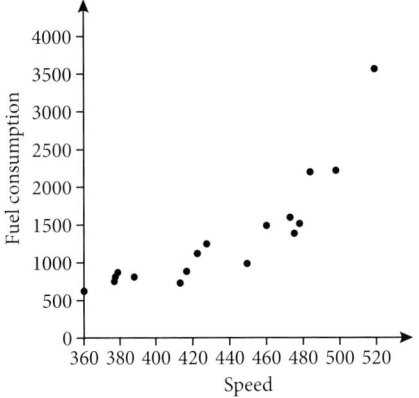

**(b)** log(fuel) vs speed produces the best linearization, with $\log(F) = 0.00241v + 0.192 \Rightarrow F = 1.55(1.00556)^v$.

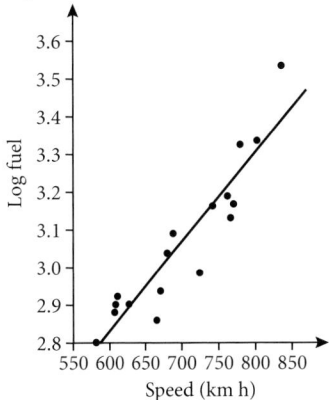

**(c)** The model suggests that fuel consumption increases by 0.556% for each additional $\text{km h}^{-1}$ increase in speed. The initial value ($F$–intercept) of 24.5 is not meaningful in this context; it is an extrapolation and it does not

make sense to have fuel consumption at a speed of 0. $R^2 = .862$; 86.2% of the variation in fuel consumption is predicted by the exponential model with speed as the explanatory variable.

**(d)** $75.3\,\text{L min}^{-1}$

4. **(a)** A scatter diagram with an appropriate model is shown below.

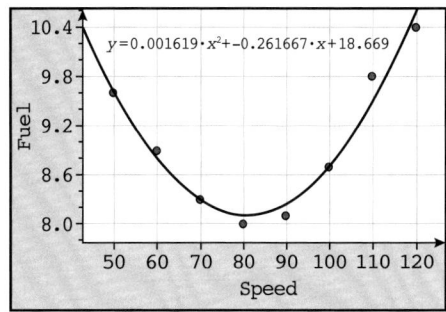

**(b)** The data appears to have a quadratic form, so we will use quadratic regression.

**(c)** The quadratic model is $C = 0.001619v^2 - 0.261667v + 18.669$.

**(d)** 8.42 L per 100 km

**(e)** $140\,\text{km h}^{-1}$ is outside the range of observed values, hence the prediction would be an extrapolation.

**(f)** According to our model, maximum fuel efficiency (minimum fuel consumption) of 8.1 L per 100 km occurs at $80.8\,\text{km h}^{-1}$. This seems reasonable, although it should be noted that the observed data includes a value of 8.0 L per 100 km, which is less than the theoretical minimum fuel consumption.

5. **(a)** The data appears to have a strong negative nonlinear form, except for the first two data points. This is probably because this is the time immediately after the drug was given, so blood concentration of the drug is increasing during this time.

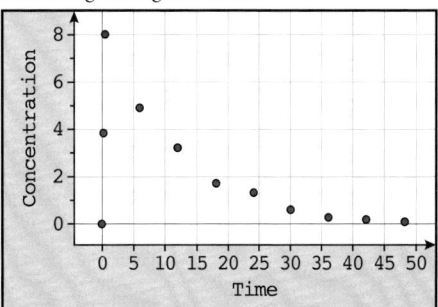

The data has a clear turning point at 0.5 hours; it is increasing up until that time and decreasing after. An exponential model appears appropriate.

**(b)** We will use the data with $t \geqslant 0.5$. An exponential model appears appropriate, so we will use a $\log(y)$ vs $x$ linearization.

# Answers

**(c)** The linearized scatter diagram is shown below.

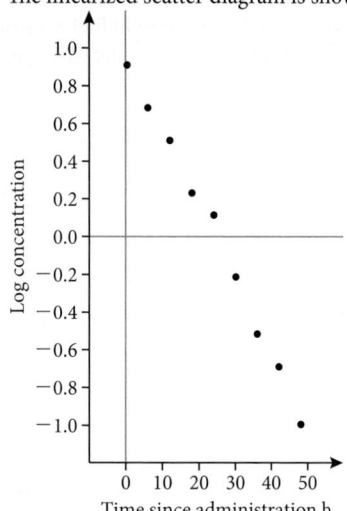

Time since administration h

The linear model is
$$\log(C) = -0.0399t + 0.960 \Rightarrow C = 9.12(0.912)^t.$$

**(d)** $t \geqslant 0.5$

**(e)** For each additional hour, the blood concentration of this drug decreases by 8.8%.

**(f)** $R^2 = 0.994$; 99.4% of the variation in the blood concentration is predicted by the exponential model using time as the explanatory variable.

**(g)** 6.31 mmol $L^{-1}$

**(h)** 7.52 hours

**6. (a)** Average distance from sun $\approx$ 800 million km, length of year $\approx$ 4000 days

**(b)** Average distance from sun $\approx$ 60 million km, length of year $\approx$ 90 days

**(c)** A power model appears to be appropriate since the data appears linear on a log–log graph.

**7. (a)** There is a strong, positive nonlinear association between average distance from the Sun and the orbital period.

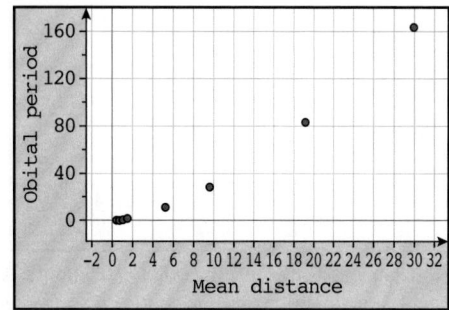

Mean distance

**(b)** The best linearization is given by the log–log scatter diagram:

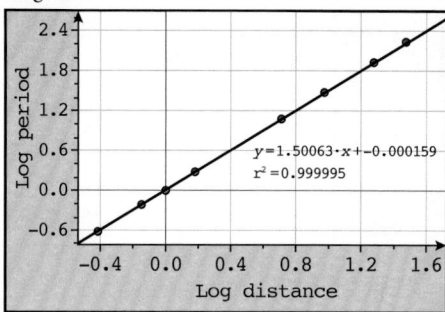

Log distance

The model is $P = D^{1.5}$.

**(c)** The orbital period of equal to the cube of the square root $\left(\dfrac{3}{2}\text{ power}\right)$ of the distance from the sun.

With $R^2 = 1.00$ (3 s.f.), this model predict 100% of the variation in orbital period using distance from the sun as the explanatory variable.

**(d)** 4.61 years

**(e)** 0.217%

**8. (a)** There appears to be a strong sinusoidal association between temperature and time.

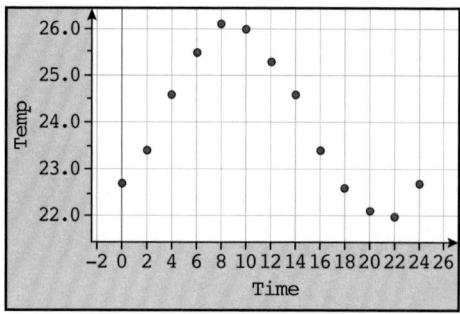

Time

**(b)** The pattern appears sinusoidal/there are turning points.

**(c)** $T = 2.03 \sin(0.263x - 0.782) + 24.0$

**(d)** 24°C

**(e)** 22.9°C

**9. (a)** There appears to be a strong nonlinear relationship between RPM and torque.

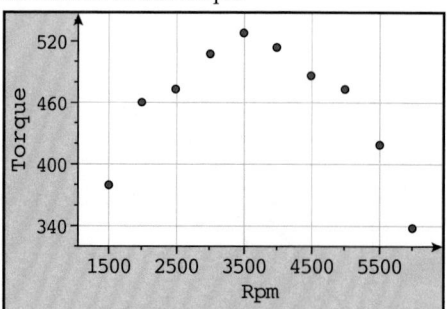

Rpm

**(b)** There is a turning point in the data.

**(c)** $N = -0.0000309\,r^2 + 0.223r + 122$

**(d)** Maximum torque is 523 Nm at 3610 RPM.

## Chapter 19 Practice questions

**1. (a)** The value of $r$ in the interval $-1 \leqslant r \leqslant 1$,
**(b)** If the association is negative, the value of $r$ must be negative.
**(c)** If life expectancy increases as body mass increases, then $r_s$ must be positive.
**(d)** The gradient in the LSRL is negative, but $r$ is positive.

**2. (a)** 0      **(b)** $-1$      **(c)** $+1$
**(d)** $+1$      **(e)** 0      **(f)** $-1$

**3. (a)** The scatter diagram shows an strong, negative, approximately linear asscoiation, with $r = -0.984$.

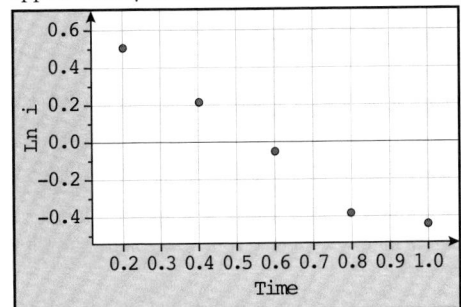

**(b)** $\ln(I) = -1.25t + 0.719$
**(c)** $R^2 = 0.968$; 96.8% of the variation in $\ln(I)$ is predicted by the LSRL using time as the explanatory variable.
**(d)** $I = 2.05\,e^{-1.25t}$; $I_0 = 2.05, k = -1.25$

**4. (a)** The scatter diagram shows a nearly perfect, positive linear correlation with $r = 1.00$.

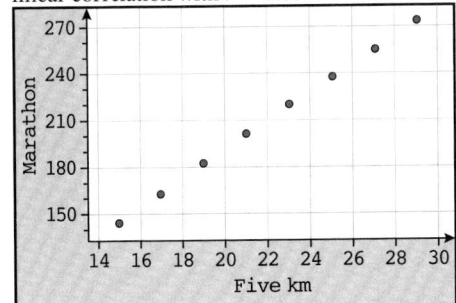

**(b)** $M = 9.30T + 5.20$. For each additional minute in 5 km time, we estimate an additional 9.3 minutes in marathon time. The $M$ intercept of 5.20 is not meaningful in this context.
**(c)** 191 minutes
**(d) (i)** The slope gradient be $\dfrac{42.195}{5} = 8.44\,\text{min}\,\text{min}^{-1}$.
    **(ii)** The model gradient from part (b) is 10% more than this theoretical gradient.
    **(iii)** We see that, on average, runners' pace is about 10% slower on a marathon than on a 5 km run.

**5. (a)** There is a very strong, positive, linear correlation between leaf width and length, with $r = 0.997$.

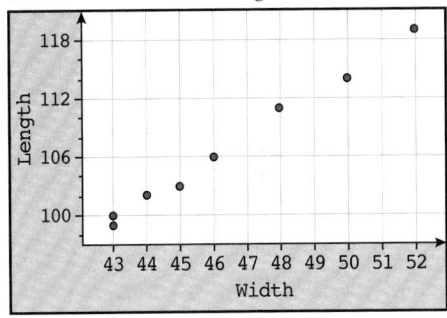

**(b)** $L = 2.15W + 6.85$. For each additional mm in width, we estimate an additional 2.15–mm in length. The $L$ intercept of 6.85 is not meaningful in this context.
**(c)** $43 \leqslant W \leqslant 52$
**(d)** 108 mm
**(e)** 60 mm is outside the domain for our model/it would be an extrapolation.

**6.** No, the study was observational and so did not control alcohol intake and then observe responses. There may be other factors that are linked with alcohol intake that reduce the risk of cardiovascular disease.

**7. (a)** There is a positive association. It appears somewhat nonlinear. There appears to be an outlier at (17, 2550) – the Common swift.

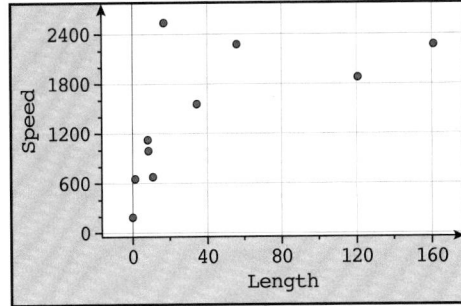

**(b)** There is one outlier (Common swift). Assuming the data is correct, we cannot justify removing this data point simply because it doesn't 'fit.'
**(c)** $r = 0.620$. There is a moderate positive linear correlation between body length and flying speed.
**(d)** $r_s = 0.820$. There is a strong positive rank correlation between body length and flying speed.
**(e)** The data appears to have a slight nonlinear curve, so Spearman's $r_s$ is more appropriate. However, the data is approximately linear so tis is also acceptable to measure the strength of the linear association with Pearson's $r$.
**(f)** A log–log linearization appears to produce the most linear pattern; the model is $v = 364\,L^{0.354}$
**(g)** $R^2 = 0.818$; About 81.8% of the variation in flying speed is accounted for by the power model with body length as the explanatory variable.
**(h)** $2590\,\text{cm}\,\text{s}^{-1}$

# Answers

**8. (a)** There appears to be a strong, positive association between weight class and weight lifted. The association is linear or slightly nonlinear.

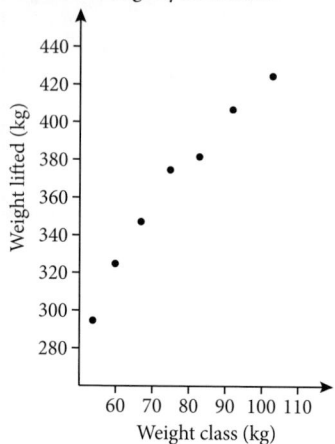

**(b)** $m = 464 \log_{(w)} - 510$

**(c)** $R^2 = 0.986$; 98.6 of the variation in weight lifted is predicted by the model using weight class as the explanatory variable.

**(d) (i)** Taner Sagir performed better than the model would predict; he lifted 375 kg but the model predicts he would lift 366 kg for a difference of +9 kg.

**(ii)** Halil Mutlu performed worse than the model would predict; he lifted 295 kg but the model predicts he would lift 302 kg for difference of −7 kg.

**9. (a)** There appears to be a strong, positive association, approximately linear but with some evidence of an nonlinear association, with no outliers.

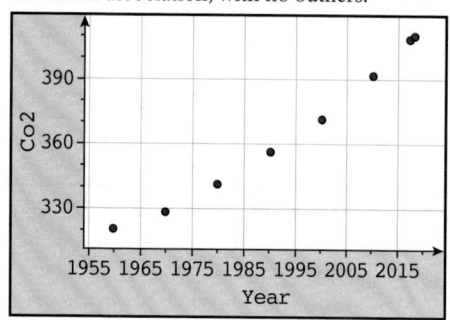

**(b)** $r = 0.991$, there is a strong positive linear correlation between year and $CO_2$ concentration.

**(c) (i)**

| Year | 0 | 10 | 20 | 30 | 40 | 50 | 57 | 58 |
|---|---|---|---|---|---|---|---|---|
| $CO_2$ concentration (ppm) | 320 | 328 | 341 | 356 | 372 | 392 | 409 | 411 |

**(ii)** $r = 0.991$, there is a strong positive linear correlation between years since 1960 and $CO_2$ concentration. The correlation coefficient has not changed by subtracting 1960 from each date value.

**(iii)** $C = 1.62y + 313$. The slope of 1.62 ppm year$^{-1}$ indicates that $CO_2$ concentration is increasing by 1.62 ppm for each additional year. The intercept of 313 indicates the ppreedicted $CO_2$ concentration in 1960.

**(iv)** years from 1960 to 2018

**(v)** 385 ppm

**(vi)** The year 2050 is beyond the range of the observed data; it would be an extrapolation.

**10. (a)** There appears to be a nearly perfect positive linear association between temperature and solubility of $NaNO_3$.

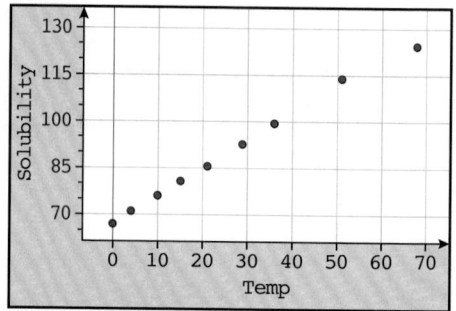

**(b)** $r = 0.999$, There is a nearly perfect positive linear association between temperature and solubility of $NaNO_3$.

**(c)** $S = 0.872T + 67.5$. The gradient of 0.872 g per 100ml per°C indicates that for each additional°C, we expect an additional 0.872 g per 100 ml can be dissolved. The $S$−intercept of 67.5 indicates that at 0°C, we expect solubility of 67.5 g per 100 ml.

**(d)** At $T = 25$°C, we expect solubility of $S = 89.3$ g per 100 ml.

**(e)** 95°C is beyond the range of observed data; it would be an extrapolation.

**(f)** We should not use this model to predict temperature from solubility. (Also we do not know if the solution given has the maximum amount of $NaNO_3$ dissolved for that temperature.)

**11. (a)**

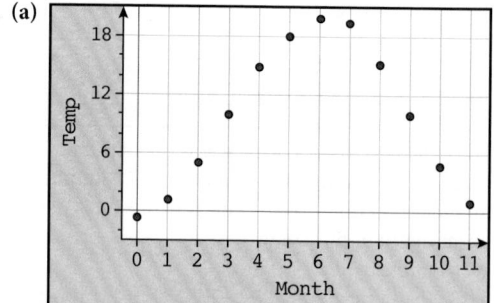

**(b)** We know the temperature will be periodic by year and it appears to have a sinusoidal form.

**(c)** $T = 10.3 \sin(0.515M - 1.53) + 9.72$

**(d)** 9.72°C

**(e)** 12.2 months

**(f)** We know that the temperature model should have a period of 12 months.

**12. (a)** The appears to be a very strong negative linear correlation between time and velocity.

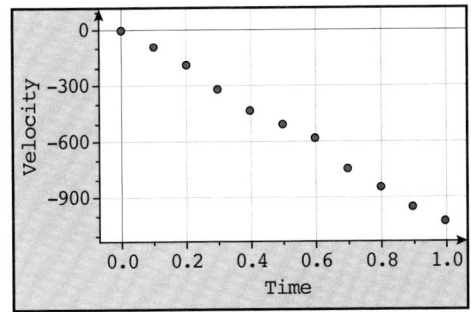

**(b)** $r = -0.998$. There is an almost perfect negative linear correlation between time and velocity.

**(c)** $v(t) = -1040t + 7.87$

**(d)** The gradient of $-1040 \text{ cm sec}^{-2}$ indicates that for each additional second, velocity changes by $-1040 \text{ cm sec}^{-1}$. The intercept of 7.87 suggests that the object was moving upward at $7.87 \text{ cm sec}^{-1}$ at the start of the experiment.

**(e)** $-148 \text{ cm sec}^{-1}$

**(f)** The rate of change of velocity predicted by this experiment is $-1040 \text{ cm sec}^{-2}$. This is grater that the possible acceleration due to gravity; most likely there are errors in measurement.

**(g)** $s(t) = \int v(t) \, dt = -520t^2 + 7.87t$

**(h)** 512 cm

**13. (a)** There is a strong, positive, nonlinear associate between year and global wind energy production.

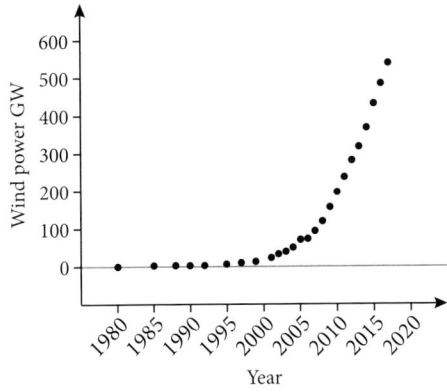

**(b)** The data has a clear exponential pattern.

**(c)**

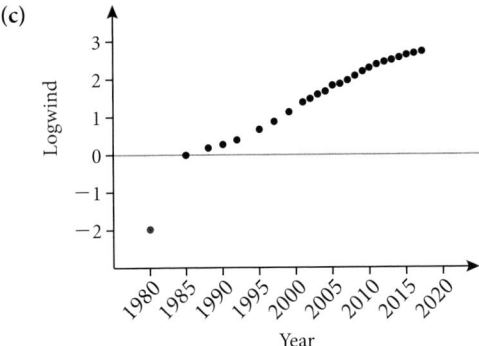

**(d)** Wind power production in 1980 is an outlier; wind energy increased significantly more from 1980 to 1985 than any other comparable period.

**(e)** $W = 0.159(1.26)^t$

**(f)** Global wind energy production has increased an average of by 26% per year since 1985.

**(g)** $R^2 = 0.959$; 95.9% of the variation in global wind power production is predicted by the model using year as the explanatory variable.

**(h)** $n = 3$ years

**14. (a)**

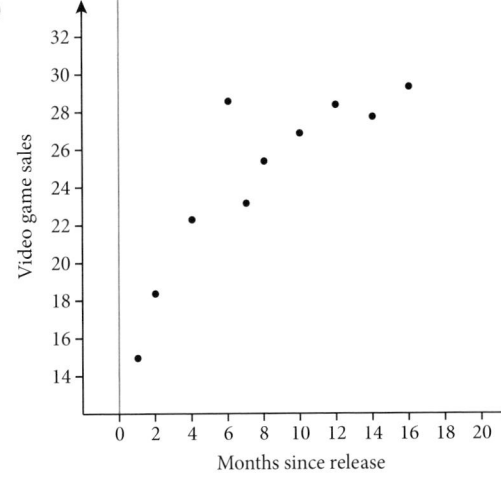

**(i)** At month 6 (December), there appears to be an outlier. Since it is greater than the surrounding months, it may be increased sales during the holidays.

**(ii)** There appears to be a logarithmic pattern to the data.

**(b)** $S = 11.7 \log(m) + 15.4$. Logarithmic growth decreases over time, which makes sense in this context: we expect sales to increase rapidly and then the rate of growth should slow.

**(c)** 73.7 thousand sales

**(d) (i)** $a = 0, b = 12$ **(ii)** 275 thousand sales

**15. (a)**

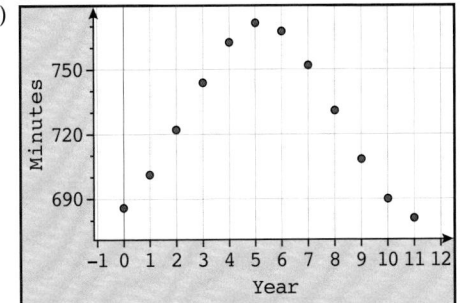

**(b)** We know that the length of the day is cyclical.

**(c)** $L = 44.8 \sin(0.519M - 1.12) + 726$

**(d)** 726 minutes

**(e)** 12.1 months. This value is reasonable as we expect the period of to be near 12 months.

**(f)** $N = -44.8 \sin(0.519M - 1.12) + 726$

**16. (a)** There appears to be a strong association between power and RPM. A turning point suggests the association is nonlinear.

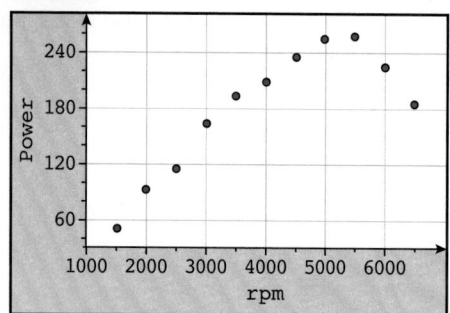

**(b)** There is a turning point in the pattern.
**(c)** $P = -0.0000157\,s^2 + 0.159s - 165$
**(d)** Maximum power is 237 kW at 5060 RPM.
**(e)** The model appears to under-predict maximum power since 257 kW was observed at 5500 RPM.

**17. (a)**

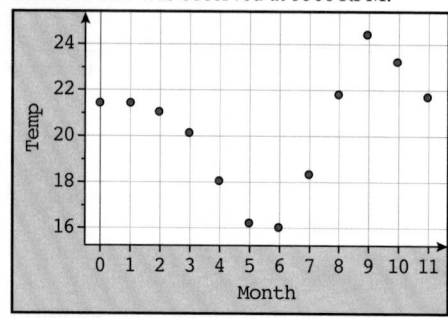

**(b)** The pattern is not sinusoidal because there appear to be unequal local maxima.
**(c)** There are 3 turning points; a cubic model requires at most 2 turning points; a quadratic model requires at most 1 turning point.

**18. (a)**

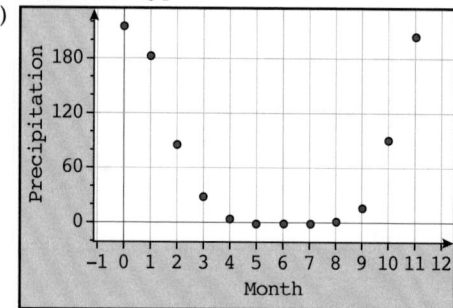

**(b)** $P = 7.34m^2 - 86.2m + 234$
**(c)** $R^2 = 0.944$; the model accounts for 94.4% of the variation in precipitation using month as the explanatory variable.
**(d)** $-19.2$ mm
**(e)** It is not possible to have negative precipitation.

# Chapter 20

## Exercise 20.1

**1. (a)** $f(x) = \dfrac{1}{x}(\ln x + k) \Rightarrow \dfrac{dy}{dx} = \dfrac{1 - k - \ln x}{x^2}$

$\Rightarrow x^2 \times \dfrac{1 - k - \ln x}{x^2} + x\left(\dfrac{1}{x}(\ln x + k)\right) - 1$

$= 1 - k - \ln x + \ln x + k - 1 = 0$

**(b)** $\dfrac{dy}{dx} = \dfrac{1 - xy}{x^2}$

**(c)** $f(x) = \dfrac{1}{x}(\ln x + 2),\ f(x) = \dfrac{1}{x}(\ln x - 2 - \ln 2)$

**2. (a)** $y = \ln(e^x - C)$     **(b)** $y^2 = Ce^{x^2} - 1$

**(c)** $y = x + \ln\dfrac{1}{x + Ce^x + 1}$    **(d)** $\dfrac{y - 1}{y + 1} = e^{(x-1)^2} + c$

**(e)** $\ln(y + 1) + y\ln(y + 1) + 1 = (y + 1)(\ln(\ln x) + c)$

**(f)** $y^2 = Ce^{-4\cot x} - \dfrac{1}{2}$

**(g)** $y + \ln|\sec y| = \dfrac{1}{3}x^3 + x + c$

**(h)** $\sqrt{(y^2 + 1)^3} = 3e^t(t - 1) + c$

**(i)** $e^{-y}(y + 1) = -\dfrac{1}{3}\sin^3\theta + c$

**(j)** $y = \tan(e^x + C)$

**3. (a)** $y = 3e^{x^4}$       **(b)** $y = e^{\frac{1}{2}x^2}$

**(c)** $y = \dfrac{2}{2 - x^2}$      **(d)** $y = \dfrac{1}{3 - x}$

**(e)** $y = \ln\left(\dfrac{e}{1 - ex}\right)$    **(f)** $y^3 = \dfrac{3(x + 1)^2}{2} - \dfrac{1}{2}$

**(g)** $y = \dfrac{1}{\ln|x + 1| + 1}$

**(h)** $2y^3 + 6y = 3x^2 + 6x + 72$

**(i)** $y^2 = e^{x^2} - 1$

**(j)** $y + \ln|y| = \dfrac{1}{2}x^2 - x + 1$

**(k)** $\arcsin y = 1 - \sqrt{1 - x^2}$

**(l)** $y = \ln\left(\ln\dfrac{e(e^x + 1)}{1 + e}\right)$

**(m)** $y + \ln y = \dfrac{x^3}{3} - x - 5$

**(n)** $y = \cos^{-1}\left(\dfrac{e^x + 1}{2}\right)$

**(o)** $|y| = |x|\,e^{x^2 - 1}$

**(p)** $2\ln|y| - y^2 = e^{x^2} - 1$

**(q)** $e^{3y} + 3y^2 = 3(-\sin x + x\sin x) + 1$

**(r)** $y = e^x - x^2 + 2$

**4. (a)** $|x| = C(x - y)^2$    **(b)** $|y^2 + 2xy - x^2| = C$

**(c)** $y = Ce^{-\frac{x^2}{2y^2}}$      **(d)** $e^{\frac{y}{x}} = 1 + \ln x^2$

**(e)** $x = e^{\sin\left(\frac{y}{x}\right)}$

**(f)** $\ln\left|\dfrac{(x + y)^2}{cx}\right| = \dfrac{y}{x}$ or $(x + y)^2 = cxe^{y/x}$

**(g)** $4x = y(\ln|y| - c)^2$

**5. (a)** See Worked Solutions
**(b) (i)** $C = 78$

    **(ii)** $m = \dfrac{1}{15}\ln\dfrac{8}{13}$

**6. (a)** $p = \dfrac{200}{1 + 7e^{-0.2640t}}$   **(b)** 70 tigers   **(c)** 7.37 yr.

**7. (a)** 70               **(b)** 4.49 yr

**(c)** $\dfrac{dP}{dt} = 0.75P\left(1 - \dfrac{P}{2100}\right)$

**8. (a)** $\dfrac{dP}{dt} = kP \Rightarrow P(t) = 500e^{t\ln 2} \Rightarrow P(12) = 2\,048\,000$

**(b)** $1000 = 500e^{t\ln 2} \Rightarrow t = 1$ hour.

**9. (a)** $P(t) = \dfrac{10\,007.4}{20 + 480.37e^{-0.1t}}$

**(b)** 500.37        **(c)** 227.87

**10.** 2560

**11.** $t \approx 35.86$ hrs.

**12.** 1091

**13. (a)** $y = ce^{-bt} + (d/b)$

**(b)**

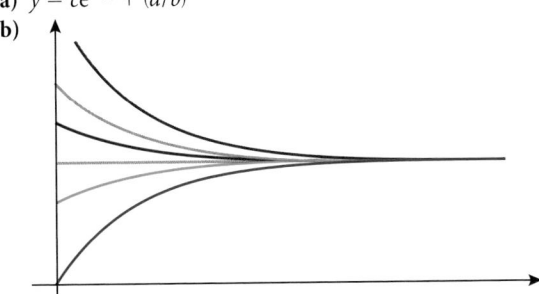

**(c) (i)** equilibrium is lower and is approached more rapidly

**(ii)** equilibrium is higher

**(iii)** equilibrium stays the same but is approached more rapidly

**14. (a)** $\dfrac{dT}{dt} = -k(T - S) \Rightarrow T = S + (T_0 - S)e^{-kt}$

$= -12 + 33e^{-0.15t}$

**(b)** $-12 + 33e^{-0.15t} = 0 \Rightarrow t = \dfrac{\ln(12/33)}{-0.15} \approx 6.7$ hours.

## Exercise 20.2

**1.** 9.89 grams.

**2. (a)** $T = 20 + 50\,e^{-\frac{t}{10}\ln\frac{5}{3}}$. **(b)** 76.6 minutes

**3.** $a = -10\,\text{m s}^2$ and $\nu_0 = 120\,\text{km h}^{-1} \approx 33.33\,\text{m s}^{-1}$
$\Rightarrow$ distance $\approx 55.56$ m.

**4.** $197\,\text{km h}^{-1}$

**5. (a)** $\dfrac{dP}{dt} = kP \Rightarrow$ when $t = 0$, $k = \dfrac{\left(\dfrac{dP}{dt}\right)}{P} = \dfrac{0.08213}{7.6}$
$= 0.01081$

**(b)** $P(t) = 7.6\,e^{0.01081t} = 7.6\,e^{0.01081(83)} = 18.64$ billion

**(c)** $60 = 7.6e^{0.01081t} \Rightarrow t = \dfrac{\ln 7.895}{0.01081}$
$\approx 191$ years, in the year 2208.

**6.** $\dfrac{dN}{dt} = -kN \Rightarrow N = N_0\,e^{-kt}, t = \dfrac{-\ln 0.65}{0.0001216} \approx 3543$ years.

**7.** $Q(t) = 3(1 - e^{-4t})$, $\quad I(t) = \dfrac{dQ}{dt} = 12\,e^{-4t}$

**8. (a)** $N(t) = 10\,000\,e^{\frac{\ln 5}{10}t}$.

**(b)** 250 000 **(c)** $\dfrac{10\ln 2}{\ln 5}$

**9.** $\dfrac{dT}{dt} = k(200 - T) \Rightarrow T = 200 - Ce^{-kt}$, with initial conditions: $T = 200 - 190e^{-0.00333t}$; time needed is approximately 114 minutes.

**10.** $P(t) = 25e^{0.01823t} \Rightarrow P(40) = 25e^{0.01823\times40} \approx 51.840$, about 52 000 people.

**11.** Let A be the amount per kg.
$\dfrac{dA}{dt} = -kA \Rightarrow A(t) = A_0e^{-kt}, k = 0.13863,$
$A(0) = 50(45) = 2250$ mg
$A(1) = 2250 = A_0e^{-0.13863\times1} \Rightarrow A_0 = 2585$ mg to anaesthetise the dog properly.

**12.** $\dfrac{dN}{dt} = k(100 - N), N(0) = 0 \Rightarrow \ln(100 - N) = -kt + c.$
With initial conditions $N(t) = 100(1 - e^{-0.01505t})$
$\Rightarrow t \approx 46.05$ days

**13. (a)** $\dfrac{dI}{dx} = -kI \Rightarrow I(x) = 108e^{-0.092x}$

**(b)** Approximately at 28.16 m

**14. (a)** $m\dfrac{dv}{dt} = mg - kv^2$

**(b)** $\dfrac{dv}{dt} = 0 \Rightarrow m\dfrac{dv}{dt} = mg - kv^2 = 0 \Rightarrow v = \sqrt{\dfrac{mg}{k}}$

**(c)** $K = 2/49$

**15. (a)** $Q(t) = CV(1 - e^{-t/RC})$

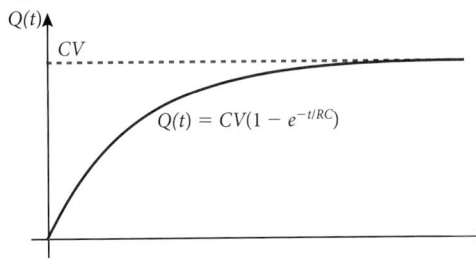

**(b)** $Q_L(t) = CV$

**(c)** When the battery is removed, $R\dfrac{dQ}{dt} + \dfrac{Q}{C} = 0$ which can be solved to give $Q(t) = Ke^{-t/RC}$, and initial conditions yield $K = CVe^{t_1/RC}$, substitute this into the result and we have $Q(t) = CVe^{-(t-t_1)/RC}$

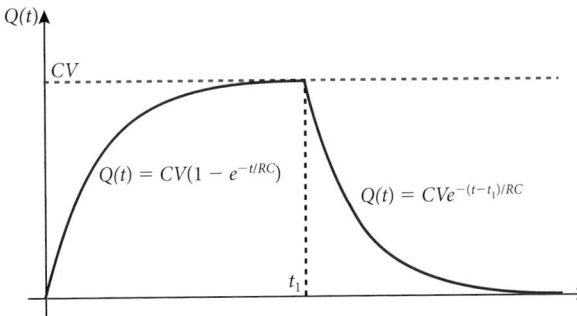

When there is no battery, the capacity will go down till it is completely exhausted.

**16. (a)** $\dfrac{dq}{dt} = 0.25s - \dfrac{sq}{300}$

**(b)** $q(t) = 75 + ce^{-st/300}$, with initial conditions $q(t) = 75 + (q_0 - 75)e^{-st/300}$

**(c)** 75

## Exercise 20.3

**1. (a)**

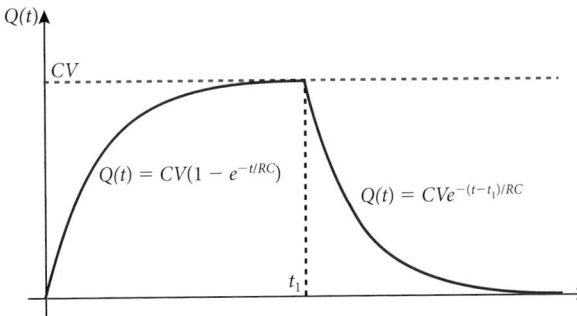

Solutions tend to 3/2 as $x \to \infty$

# Answers

**(b)**

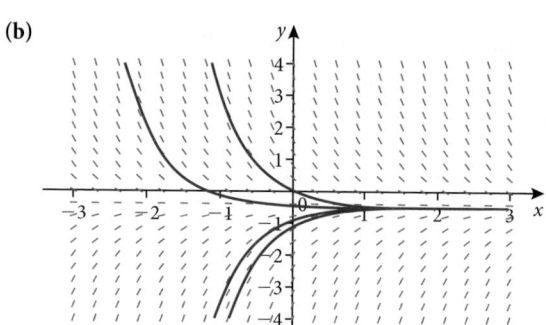

Solutions tend to $-1/2$ as $x \to \infty$

**(c)**

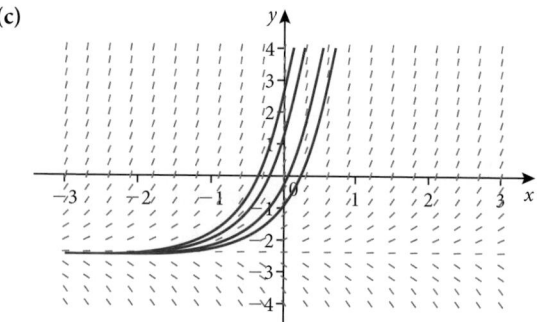

Solutions tend to diverge from $-5/2$ as $x \to \infty$

**(d)**

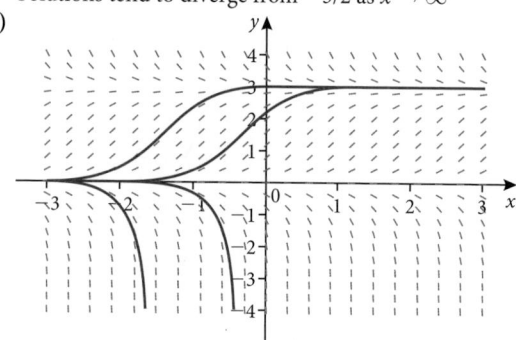

Solutions tend to 3 if initial values are positive and diverge from zero if initial values are negative.

**2.** 1.7616

**3. (a) (i)** 2.3928    **(ii)** 2.3701    **(iii)** 2.3681
   **(b)** $-3x^2e^{-x^3} = 6x^2 - 3x^2(e^{-x^3} + 2)$
   **(c) (i)** $-0.0249$      **(ii)** $-0.0022$
     **(iii)** $2.3681 - 0.0002$; error is divided approx. by 10

**4.** $y_1 = 2$, $y_2 = 2.75$, $y_3 = 3.5$, $y_4 = 4.25$

**5.** 0.4150

**6.** 1.8371

**7. (a)** $t = 0.5$
   **(i)** 1.70308    **(ii)** 1.79548    **(iii)** 1.84579
   $t = 1.5$
   **(i)** 2.44030    **(ii)** 2.43292    **(iii)** 2.42905
   $t = 3$
   **(i)** 1.11925    **(ii)** 1.12191    **(iii)** 1.12328

   **(b)** $t = 0.5$
   **(i)** 2.30800    **(ii)** 2.30167    **(iii)** 2.29686
   $t = 1.5$
   **(i)** 2.60023    **(ii)** 2.59352    **(iii)** 2.5883
   $t = 3$
   **(i)** 2.73521    **(ii)** 2.73209    **(iii)** 2.72959

   **(c)** $t = 0.5$
   **(i)** $-1.48849$   **(ii)** $-1.46909$   **(iii)** $-1.45865$
   $t = 1.5$
   **(i)** 1.04687    **(ii)** 1.05351    **(iii)** 1.05715
   $t = 3$
   **(i)** 1.51971    **(ii)** 1.50549    **(iii)** 1.49879

**8.** Exact equation: $y = \dfrac{1}{2}(\sin x - \cos x + e^x)$

| $x$ | 0 | 0.2 | 0.4 | 0.6 | 0.8 | 1.0 |
|---|---|---|---|---|---|---|
| $y(x)$ [exact] | 0.0000 | 0.2200 | 0.4801 | 0.7807 | 1.1231 | 1.5097 |
| $y(x)$ [$h = 0.2$] | 0.0000 | 0.2000 | 0.4360 | 0.7074 | 1.0140 | 1.3561 |
| $y(x)$ [$h = 0.1$] | 0.0000 | 0.2095 | 0.4568 | 0.7418 | 1.0649 | 1.4273 |

**9. (a)**

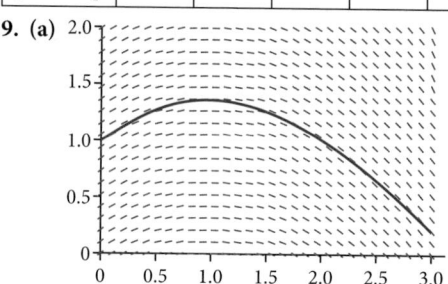

   **(b)** $y(3)$ is somewhere between 0 and 0.5.

**10. (a)**

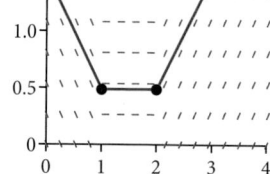

   **(b)** $Y(4) = 2.5$

**11. (a)** III     **(b)** IV     **(c)** II     **(d)** I

**12. (a)** II     **(b)** IV     **(c)** I     **(d)** III

**13. (a)** $\dfrac{dq}{dt} = 300(0.01 - 10^{-6}q)$; in grams and in hours, $q(0) = 0$.
   **(b)** $q \to 10^4$, no
   **(c)** $q(t) = 10^4(1 - e^{-0.0003t})$.
     **(i)** After one year the amount of pollutant will be approximately 9278 g
     **(ii)** $\lim_{t \to \infty} q(10^4(1 - e^{-0.0003t})) = 10^4$
   **(d)** $\dfrac{dq}{dt} = 300(-10^{-6}q)$;   $q(0) = 9278$
   **(e)** $q(t) = 9278(e^{-0.0003t})$, after one year there will be approximately 670 g
   **(f)** 2.6 years

**14. (a)** $\dfrac{dV}{dt} = -k\,V^{2/3}$;   where $k = c\sqrt[3]{36\pi}$ and is a constant of proportionality that depends on several factors like temperatutre and friction.

**(b)**

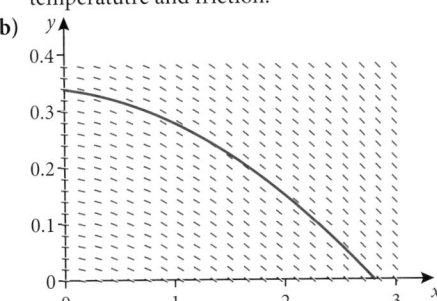

Approx. 2.8 minutes.

**15. (a)** $\dfrac{dq}{dt} = 500 - 0.4q$; $q$ in mg, $t$ in h.

**(b)** We use a slope field

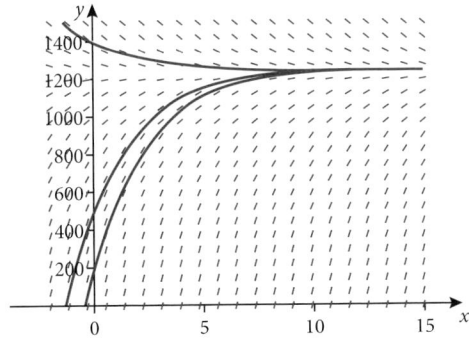

The amount will stabilise at 1250 mg.

## Exercise 20.4

**1. (a)** Eigenvalues: $4, -1$, eigenvectors $\begin{pmatrix}3\\2\end{pmatrix}, \begin{pmatrix}1\\-1\end{pmatrix}$

General solution: $\begin{cases} x = 3C_1e^{4t} + C_2e^{-t} \\ y = 2C_1e^{4t} - C_2e^{-t} \end{cases}$

Trajectories approach equilibrium and then move away. Saddle point. Unstable.

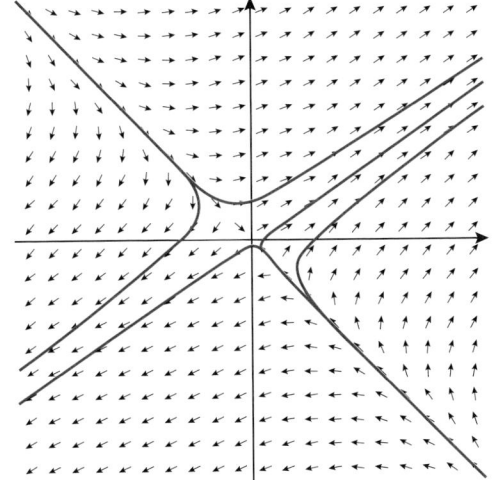

**(b)** Eigenvalues: $-4, -25$, eigenvectors $\begin{pmatrix}1\\1\end{pmatrix}, \begin{pmatrix}2\\-5\end{pmatrix}$

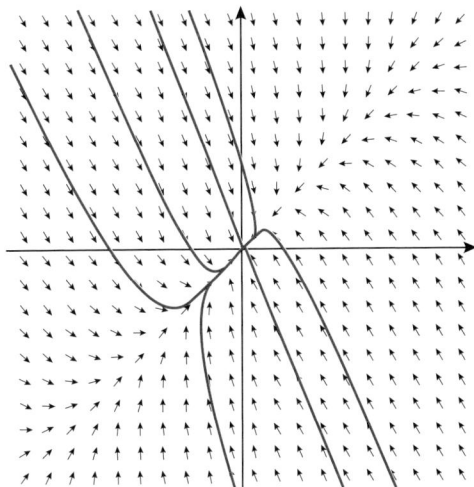

General solution: $\begin{cases} x = C_1e^{-4t} + 2C_2e^{-25t} \\ y = C_1e^{-4t} - 5C_2e^{-25t} \end{cases}$

Stable system. Trajectories move towards the equilibrium point which is called stable node in this case.

**(c)** Eigenvalues: $-2, 6$, eigenvectors $\begin{pmatrix}1\\-1\end{pmatrix}, \begin{pmatrix}3\\5\end{pmatrix}$

General solution: $\begin{cases} x = C_1e^{-2t} + 3C_2e^{6t} \\ y = -C_1e^{-2t} + 5C_2e^{6t} \end{cases}$

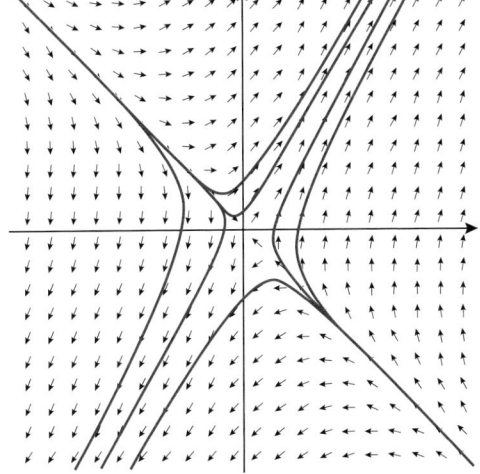

Trajectories approach equilibrium and then move away. Saddle point. Unstable.

**(d)** Eigenvalues: 5, $-1$, eigenvectors $\begin{pmatrix} 1 \\ 2 \end{pmatrix}, \begin{pmatrix} -1 \\ 1 \end{pmatrix}$

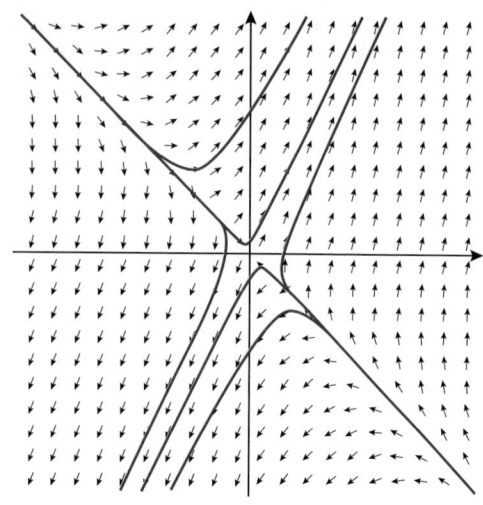

General solution: $\begin{cases} x = C_1 e^{5t} - C_2 e^{-t} \\ y = 2C_1 e^{5t} + C_2 e^{-t} \end{cases}$

Trajectories approach equilibrium and then move away. Saddle point. Unstable.

**(e)** Eigenvalues: $-3$, 1, eigenvectors $\begin{pmatrix} 2 \\ 1 \end{pmatrix}, \begin{pmatrix} 2 \\ 5 \end{pmatrix}$

General solution: $\begin{cases} x = 2C_1 e^{-3t} + 2C_2 e^{t} \\ y = C_1 e^{-3t} + 5C_2 e^{t} \end{cases}$

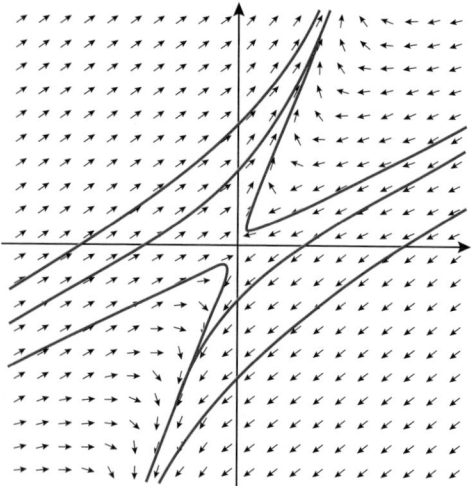

Trajectories approach equilibrium and then move away. Saddle point. Unstable.

**(f)** Eigenvalues: 8, $-10$, eigenvectors $\begin{pmatrix} 5 \\ 2 \end{pmatrix}, \begin{pmatrix} 1 \\ 4 \end{pmatrix}$

General solution: $\begin{cases} x = 5C_1 e^{8t} + C_2 e^{-10t} \\ y = 2C_1 e^{8t} + 4C_2 e^{-10t} \end{cases}$

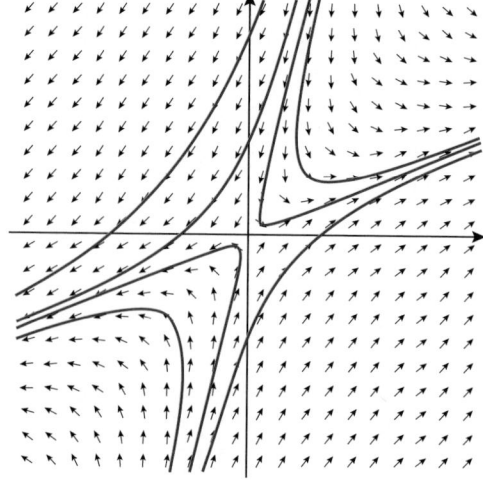

Trajectories approach equilibrium and then move away. Saddle point. Unstable.

**(g)** Eigenvalues: $-2$, 5, eigenvectors $\begin{pmatrix} 1 \\ 3 \end{pmatrix}, \begin{pmatrix} 1 \\ \frac{2}{3} \end{pmatrix}$

General solution: $\begin{cases} x = C_1 e^{-2t} + C_2 e^{5t} \\ y = 3C_1 e^{-2t} + \frac{2}{3} C_2 e^{5t} \end{cases}$

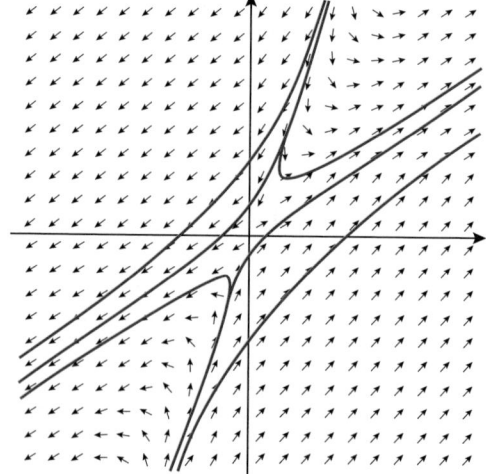

Trajectories approach equilibrium and then move away. Saddle point. Unstable.

**(h)** Eigenvalues: $-1, -4$, eigenvectors $\begin{pmatrix} 1 \\ \sqrt{2} \end{pmatrix}, \begin{pmatrix} -\sqrt{2} \\ 1 \end{pmatrix}$

General solution: $\begin{cases} x = C_1 e^{-t} - \sqrt{2}\, C_2 e^{-4t} \\ y = \sqrt{2}\, C_1 e^{-t} + C_2 e^{-4t} \end{cases}$

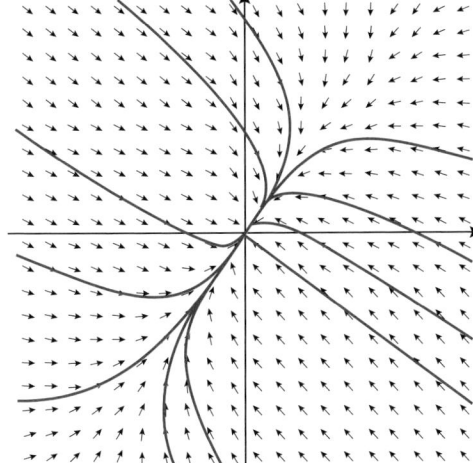

With both eigenvalues negative, the curves move towards the equilibrium. Stable system with the origin as a **node**.

**(i)** Eigenvalues: $-1, 2$, eigenvectors $\begin{pmatrix} -1 \\ 1 \end{pmatrix}, \begin{pmatrix} -1 \\ 2 \end{pmatrix}$

General solution: $\begin{cases} x = -C_1 e^{-t} - C_2 e^{2t} \\ y = C_1 e^{-t} + 2C_2 e^{2t} \end{cases}$

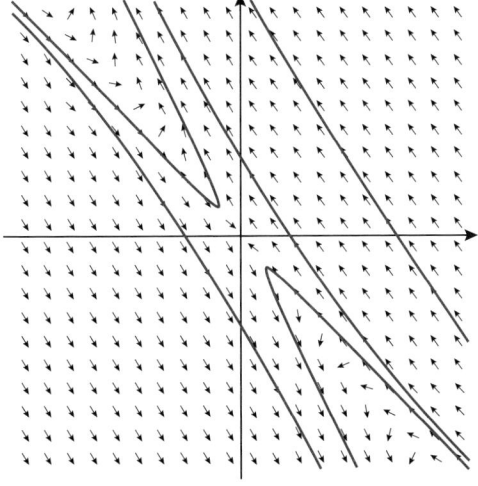

Trajectories approach equilibrium and then move away. Saddle point. Unstable.

**2. (a)** Eigenvalues: $5 \pm 2i$.

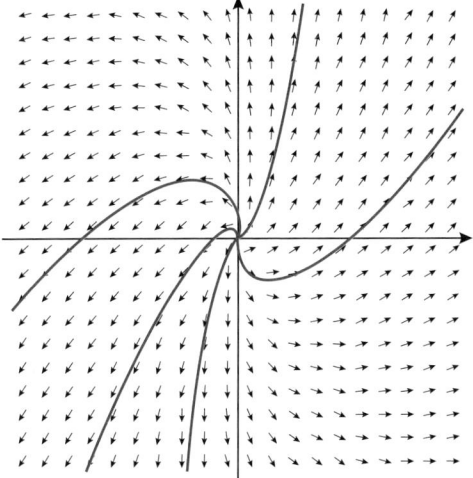

Eigenvalues are complex and the real part is positive, the trajectories move away from the origin in a spiral manner, the origin is an **unstable focus**, also known as a **spiral source**.

**(b)** Eigenvalues: $\pm 2i$.

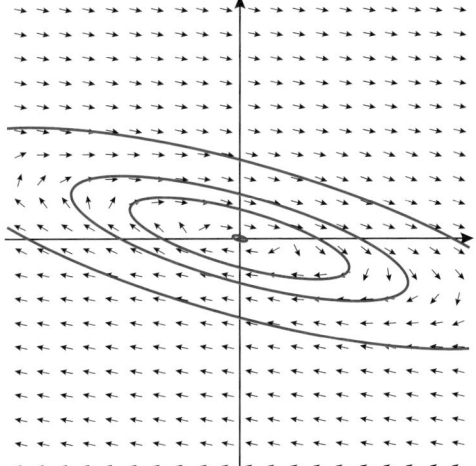

Eigenvalues are purely imaginary; the origin is a **stable equilibrium point** (but not asymptotically stable). It is also called a **center**.

**(c)** Eigenvalues: $-0.1 \pm i$.

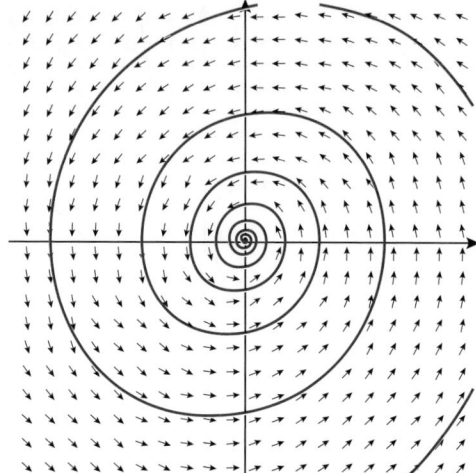

Eigenvalues are complex and the real part is negative, the trajectories approach the origin in a spiral manner. The origin is an ***asymptotically stable focus***, also known as a ***spiral sink***.

**(d)** Eigenvalues: $-\dfrac{1}{2} \pm i$

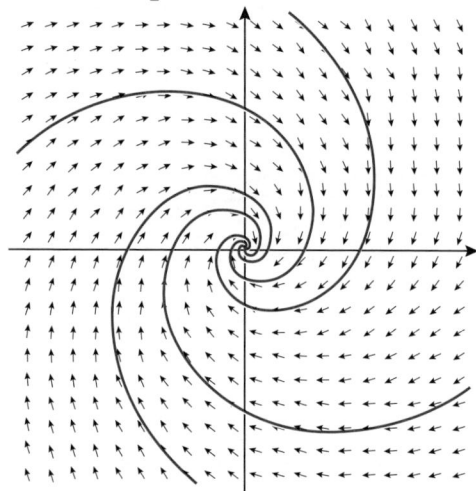

Eigenvalues are complex and the real part is negative, the trajectories approach the origin in a spiral manner. The origin is an ***asymptotically stable focus***, also known as a ***spiral sink***.

**(e)** Eigenvalues: $1 \pm i$

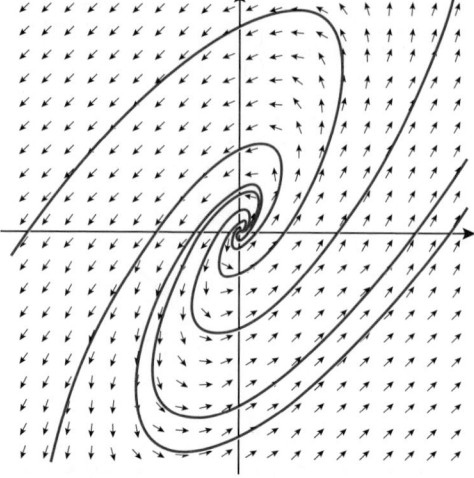

Eigenvalues are complex and the real part is positive, the trajectories move away from the origin in a spiral manner, the origin is an ***unstable focus***, also known as a ***spiral source***.

**(f)** Eigenvalues: $\pm i$

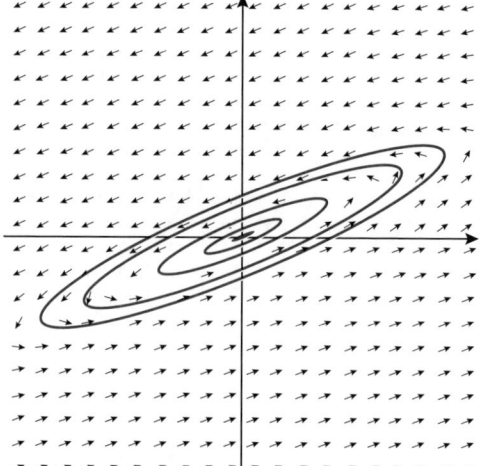

Eigenvalues are purely imaginary; the origin is a ***stable equilibrium point*** (but not asymptotically stable). It is also called a ***center***.

**3. (a)** Eigenvalues: $-3, 2$, eigenvectors $\begin{pmatrix} 1 \\ -3 \end{pmatrix}, \begin{pmatrix} 1 \\ 2 \end{pmatrix}$

General solution: $\begin{cases} x = C_1 e^{-3t} + C_2 e^{2t} \\ y = -3C_1 e^{-3t} + 2C_2 e^{2t} \end{cases}$

Particular solution: $\begin{cases} x = e^{-3t} + 2e^{2t} \\ y = -3e^{-3t} + 4e^{2t} \end{cases}$

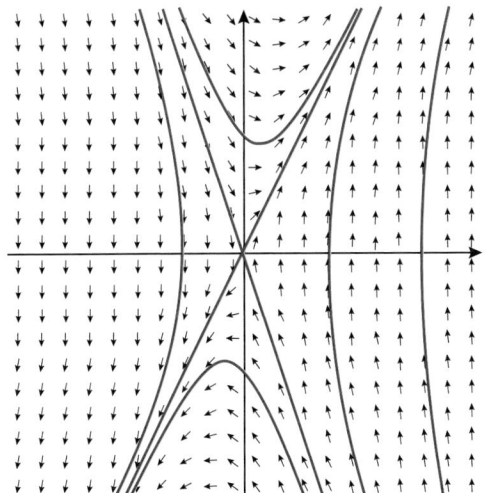

Trajectories approach equilibrium and then move away.
Saddle point. Unstable.

Euler: $\begin{cases} x_{n+1} = x_n + h(y_n) \\ y_{n+1} = y_n + h(6x_n - y_n) \end{cases} \Rightarrow \begin{cases} x_2 = 3.37 \\ y_2 = 4.29 \end{cases}$

Exact: $\begin{cases} x = e^{-3(0.2)} + 2e^{2(0.2)} \approx 3.53 \\ y = -3e^{-3(0.2)} + 4e^{2(0.2)} \approx 4.32 \end{cases}$

**(b)** Eigenvalues: $-2, -5$, eigenvectors $\begin{pmatrix} 1 \\ 2 \end{pmatrix}, \begin{pmatrix} 1 \\ 5 \end{pmatrix}$

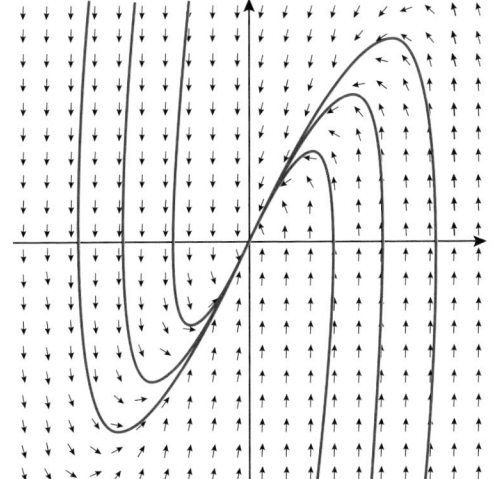

General solution: $\begin{cases} x = C_1 e^{-2t} + C_2 e^{-5t} \\ y = 2C_1 e^{-2t} + 5C_2 e^{-5t} \end{cases}$

Particular solution: $\begin{cases} x = \dfrac{17}{3} e^{-2t} - \dfrac{11}{3} e^{-5t} \\ y = \dfrac{34}{3} e^{-2t} - \dfrac{55}{3} e^{-5t} \end{cases}$

With both eigenvalues negative, the curves move
towards the equilibrium. Stable system.

Euler: $\begin{cases} x_{n+1} = x_n + h(-y_n) \\ y_{n+1} = y_n + h(10x_n - 7y_n) \end{cases} \Rightarrow \begin{cases} x_2 = 1.29 \\ y_2 = 1.27 \end{cases}$

Exact: $\begin{cases} x = \dfrac{17}{3} e^{-2(0.2)} - \dfrac{11}{3} e^{-5(0.2)} \approx 2.45 \\ y = \dfrac{34}{3} e^{-2(0.2)} - \dfrac{55}{3} e^{-5(0.2)} \approx 0.85 \end{cases}$

**(c)** Eigenvalues: $-1, 4$, eigenvectors $\begin{pmatrix} -1 \\ 1 \end{pmatrix}, \begin{pmatrix} 2 \\ 3 \end{pmatrix}$

General solution: $\begin{cases} x = -C_1 e^{-t} + 2C_2 e^{4t} \\ y = C_1 e^{-t} + 3C_2 e^{4t} \end{cases}$

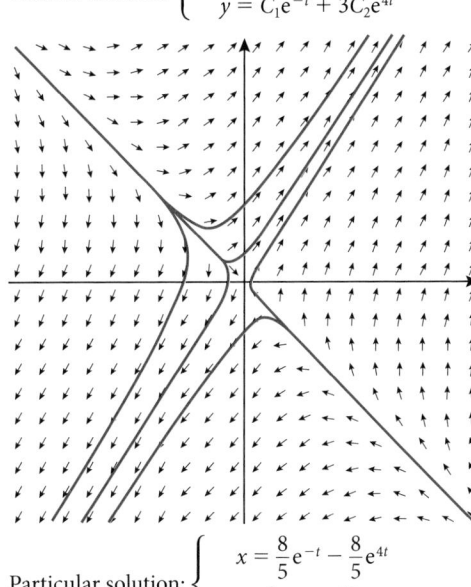

Particular solution: $\begin{cases} x = \dfrac{8}{5} e^{-t} - \dfrac{8}{5} e^{4t} \\ y = -\dfrac{8}{5} e^{-t} - \dfrac{12}{5} e^{4t} \end{cases}$

Trajectories approach equilibrium and then move away.
Saddle point. Unstable.

Euler: $\begin{cases} x_{n+1} = x_n + h(x_n + 2y_n) \\ y_{n+1} = y_n + h(3x_n + 2y_n) \end{cases} \Rightarrow \begin{cases} x_2 = -1.84 \\ y_2 = -6 \end{cases}$

Exact: $\begin{cases} x = \dfrac{8}{5} e^{-(0.2)} - \dfrac{8}{5} e^{4(0.2)} \approx -0.42 \\ y = -\dfrac{8}{5} e^{-(0.2)} - \dfrac{12}{5} e^{4(0.2)} \approx -3.91 \end{cases}$

**(d)** Eigenvalues: $1, -6$, eigenvectors $\begin{pmatrix} 1 \\ 4 \end{pmatrix}, \begin{pmatrix} -1 \\ 1 \end{pmatrix}$

General solution: $\begin{cases} x = C_1 e^{-t} - C_2 e^{-6t} \\ y = 4C_1 e^{-t} + C_2 e^{-6t} \end{cases}$

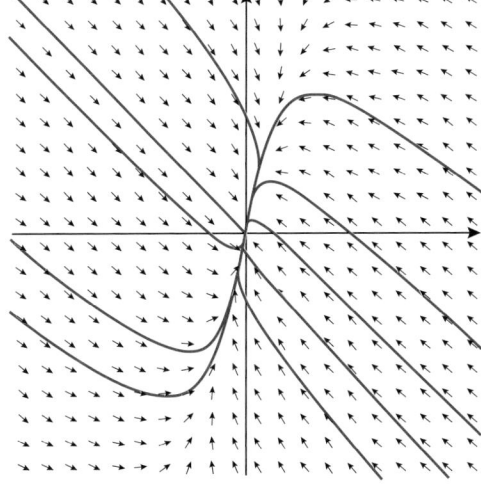

Particular solution: $\begin{cases} x = \dfrac{3}{5} e^{-t} + \dfrac{2}{5} e^{-6t} \\ y = \dfrac{12}{5} e^{-t} - \dfrac{2}{5} e^{-6t} \end{cases}$

With both eigenvalues negative, the curves move towards the equilibrium. Stable system with the origin as a **node**.

Euler: $\begin{cases} x_{n+1} = x_n + h(-5x_n + y_n) \\ y_{n+1} = y_n + h(4x_n - 2y_n) \end{cases} \Rightarrow \begin{cases} x_2 = 0.55 \\ y_2 = 1.88 \end{cases}$

Exact: $\begin{cases} x = \dfrac{3}{5}e^{-(0.2)} + \dfrac{2}{5}e^{-6(0.2)} \approx 0.61 \\ y = \dfrac{12}{5}e^{-(0.2)} - \dfrac{2}{5}e^{-6(0.2)} \approx 1.84 \end{cases}$

(e) Eigenvalues: $-2, -$, eigenvectors $\begin{pmatrix} -1 \\ 1 \end{pmatrix}, \begin{pmatrix} 1 \\ 1 \end{pmatrix}$

General solution: $\begin{cases} x = -C_1 e^{-2t} + C_2 e^{6t} \\ y = C_1 e^{-2t} + C_2 e^{6t} \end{cases}$

Particular solution: $\begin{cases} x = 3e^{-2t} + 2e^{6t} \\ y = -3e^{-2t} + 2e^{6t} \end{cases}$

Euler: $\begin{cases} x_{n+1} = x_n + h(2x_n + 4y_n) \\ y_{n+1} = y_n + h(4x_n + 2y_n) \end{cases} \Rightarrow \begin{cases} x_2 = 6.4 \\ y_2 = 3.2 \end{cases}$

Exact: $\begin{cases} x = 3e^{-2(0.2)} + 2e^{6(0.2)} \approx 8.65 \\ y = -3e^{-2(0.2)} + 2e^{6(0.2)} \approx 4.63 \end{cases}$

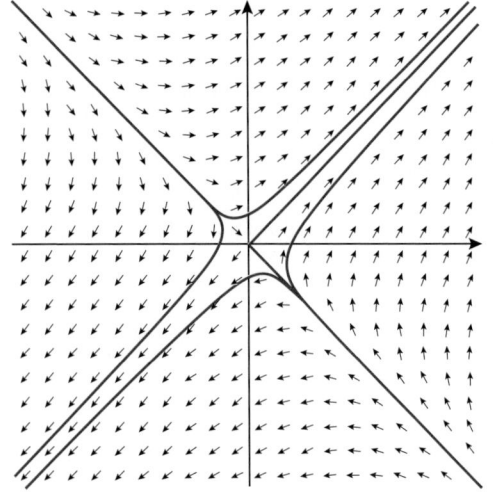

4. (a) $u'' = x; x'' = -4x - 3u; u = 4e^{-t} - 2e^{-3t}; (0.2) = 2.26$

(b) $y'' = u; u'' = 2y + u; y = \dfrac{11}{3}e^{2t} - \dfrac{2}{3}e^{-t}; (0.2) = 5.5$

(c) $y'' = u; u'' = -y + \dfrac{5}{2}u; y = -\dfrac{3}{2}e^{\frac{t}{2}} - \dfrac{1}{2}e^{2t};$ $(0.2) = -2.374$

(d) $u'' = v; v'' = -2u - 3v;$ $u = -7e^{-t} + 6e^{-2t} (0.2) = -1.83$

(e) $y'' = u; u'' = -5y - 6u; y = \dfrac{7}{2}e^{-t} - \dfrac{1}{2}e^{-5t}; (0.2) = 2.71$

## Chapter 20 Practice questions

1. $y = \dfrac{1}{e}e^{\sqrt{1+x^2}}$

2. $\tan y = 1 - \cos x$

3. $y = \dfrac{3x}{x+1}$

4. $x\dfrac{dy}{dx} = y\ln x \Rightarrow \dfrac{dy}{y} = \dfrac{\ln x}{x}dx \Rightarrow \ln|y| = \dfrac{1}{2}(\ln x)^2 + c$ $\Rightarrow y = Ce^{\frac{1}{2}(\ln x)^2} = Cx\ln\sqrt{x}$

5. 791 Bq

6. (a) $2x^2 - y^2 = C$

(b) $y = \dfrac{x}{1 - Cx}$

(c) $\ln(y - 1) - \ln y + C_1 = -\dfrac{1}{x}$ or $\dfrac{y}{y - 1} = Ce^{-\frac{1}{x}}$

(d) $x = C\sin y$ or $y = \arcsin(cx)$

(e) $y = Ce^{\frac{x^2}{2}}$

(f) $y^2 = 2\sqrt{x^2 + 1} + C$

(g) $\ln\sqrt{\dfrac{y - 1}{y + 1}} = -e^{-x} + C$

(h) $x = y\ln y - y + C$

7. $\int\dfrac{y + 1}{y}dy = \int\dfrac{x + 1}{x}dx \Rightarrow \int\left(1 + \dfrac{1}{y}\right)dy = \int\left(1 + \dfrac{1}{x}\right)dx$

$\Rightarrow y + \ln|y| = x + \ln|x| + c$

$\Rightarrow e^{y + \ln|y|} = e^{x + \ln|x| + c} \Rightarrow ye^y = Axe^x$

8. $y = \pm\sqrt{2\sin x + C}$
The constant cannot be completely arbitrary because $2\sin x + C \geqslant 0$
If $C < -2$ then $2\sin x + C$ will always be negative, regardless of the value of $x$. If $C > 2$, $2\sin x + C$ will always be positive. If $-2 < C < 2$, then whether $2\sin x + C$ is positive will depend on the value of $x$.

9. (a) $\dfrac{5}{2}$

(b) $\dfrac{5}{2}$

(c) Regardless of the initial value of the population, as time increases, the population stabilises at 2500.

10. $y = \dfrac{7x + 1}{7 - x}$

11. (a) $\dfrac{1}{3(x - 2)} - \dfrac{1}{3(x + 1)}$

(b) $\dfrac{dy}{dx} = \dfrac{y^2}{x^2 - x - 2} \Rightarrow \dfrac{dy}{y^2} = \dfrac{dx}{x^2 - x - 2}$

$= \dfrac{dx}{3(x - 2)} - \dfrac{dx}{3(x + 1)}$

$-\dfrac{1}{y} = \dfrac{1}{3}\ln\left(\dfrac{x - 2}{x + 1}\right) + c$

12. (a) Substitute $y = vx$ and $\dfrac{dy}{dx} = v + x\dfrac{dv}{dx}$, simplify and separate variables.

(b) $x^2 + 4xy - 3y^2 - 1 = 0$

13. See Worked Solutions

14. (a) $\left|\dfrac{y}{y + 1}\right| = C|x|$

(b) $\left|\dfrac{y}{y + 1}\right| = \dfrac{1}{2}|x|$

(c)

| $x_n$ | $y_n$ |
|-------|-------|
| 1.2 | 1.400 |
| 1.4 | 1.960 |
| 1.6 | 2.789 |
| 1.8 | 4.110 |

**(d)**

| $x_n$ | Approx. $y_n$ | exact $y_n$ | % error |
|-------|---------------|-------------|---------|
| 1.2 | 1.400 | 1.5 | 6.6667 |
| 1.4 | 1.960 | 2.333 | 16 |
| 1.6 | 2.789 | 4 | 30.3 |
| 1.8 | 4.110 | 9 | 54.3 |

**15.** $y = 1.5405$ at $= 1$.

**16.** $y = 5.9584$ at $= 1$.

**17.** $y^2 = Cx^3 - x^2$

**18.**

| | |
|-----|---------|
| 1.1 | 4.2 |
| 1.2 | 4.42543 |
| 1.3 | 4.67787 |
| 1.4 | 4.95904 |
| 1.5 | 5.27081 |

**19. (a)** substitute $y = e^{-x} + x - 1$ and $\dfrac{dy}{dx} = -e^{-x} + 1$ into equation.

**(b)** $y(1) \approx 0.32768$

**(c)** $y(1) \approx 0.348678\,4401$

**(d)** Actual value is 0.367 879 4412. Using more steps/smaller step size gives better approximation.

**20.** $\alpha = 20 + 50e^{-\frac{t}{10}\ln\frac{5}{3}}$

**21. (a)** $\dfrac{dN}{dt} = -kN \Rightarrow N(t) = N_0 e^{-kt}$, so

$$\frac{N_0}{2} = N_0 e^{-kt} \Rightarrow t = \frac{\ln 2}{k} = \frac{\ln 2}{0.0001216} \approx 5700$$

**(b)** $\dfrac{N_0}{4} = N_0 e^{-kt} \Rightarrow t = \dfrac{\ln 4}{0.0001216} \approx 11400$

**22.** Similar to Q21: $4.3 \times 10^{10} = 5.1 \times 10^{10} e^{-0.0001216t}$

$\Rightarrow t = \dfrac{\ln(5.1/4.3)}{0.0001216} \approx 1403$ years. Given that the Roman empire existed about 2000 years ago, this appears not to be genuine from the era.

**23. (a)** $y \approx 5.32$

**(b)** Less than actual value; $\dfrac{dy}{dx} > 0$ so solution curve is curving upward; short segments from Euler's method to approximate solution curve will be below the actual solution curve.

**24. (a)** The rate of salt going in $= s\,\text{g/L} \times 4\,\text{L/min} = 4s\,\text{g/min}$. The rate of salt going out is $(4/120)Q\,\text{g/min}$.

$$\frac{dQ}{dt} = 4s - \frac{4}{120}Q; \quad Q(0) = 0$$

**(b)** $Q(t) = 120s\left(1 - e^{-\frac{t}{30}}\right)$

**(c)** $120\,s$

**25** $\dfrac{dP}{dt} = (\ln 2)P - 140\,000 \Rightarrow P \approx 201977.31 - 1977.31e^{t\ln 2}$

$$= 201977.31 - 1977.31$$
$$\times 2^t; \quad 0 \leqslant t \leqslant 6.67 \text{ weeks}$$

# Index